This book is to be returned on or before
the last date stamped below.

INTERNATIONAL SOCIETY
FOR ROCK MECHANICS

SOCIÉTÉ INTERNATIONALE
DE MÉCANIQUE DES ROCHES

INTERNATIONALE GESELLSCHAFT
FÜR FELSMECHANIK

International Congress on Rock Mechanics

Congrès international de mécanique des roches

Internationaler Kongress über Felsmechanik

PROCEEDINGS / COMPTES-RENDUS / BERICHTE

VOLUME / TOME / BAND 1

MELBOURNE / 1983

Proceedings
Comptes-rendus
Berichte

**Fifth Congress of the International Society
for Rock Mechanics**

**Cinquième congrès de la Société Internationale
de Mécanique des Roches**

**Fünfter Kongress der Internationalen Gesellschaft
für Felsmechanik**

MELBOURNE (AUSTRALIA) 1983

Rock Mechanics for Resource Development, Mining and Civil
Engineering

La mécanique des roches en rapport avec l'exploitation de
ressources naturelles, l'industrie minière et le génie civil

Felsmechanik für den Aufschluss von Bodenschätzen,
Bergbau und Tiefbau

VOLUME 1 **Theme A** **Site Exploration and Evaluation**
Theme B **Surface and Near-surface Excavation**

TOME 1 **Thème A** **Exploration et evaluation in situ**
Thème B **Excavation et surface et à faible profondeur**

BAND 1 **Thema A** **Untersuchung und Beurteilung des Betriebspunktes**
Thema B **Übertägige und oberflächennahe Felsbauwerke**

A. A. BALKEMA / ROTTERDAM / 1983

and   Australian Geomechanics Society

et   Société Australienne de Géomécanique

und   Australische Gesellschaft für Geomechanik

For the complete set of three volumes, ISBN 90 6191 236 9

For volume 1, ISBN 90 6191 237 7
For volume 2, ISBN 90 6191 238 5
For volume 3, ISBN 90 6191 239 3

Published by A.A. Balkema, P.O. Box 1675, Rotterdam, Netherlands
Distributed in U.S.A. and Canada by M.B.S., 99 Main Street, Salem, NH 03079

Printed in Australia by Brown Prior Anderson Pty Ltd
5 Evans Street Burwood Victoria Australia 3125

THEMES VOLUME 1

Theme A Site Exploration and Evaluation

1 – Geophysical testing and exploration
2 – In situ and laboratory testing
3 – Classification, prediction, observation and monitoring
4 – Hydro-geology

Theme B Surface and Near-surface Excavations

1 – Stability of rock slopes
2 – Foundations on and in rock, including dam foundations
3 – Near-surface construction especially in cities

THEMES TOME 1

Thème A Exploration et evaluation in situ

1 – Exploration et essais géophysiques
2 – Essais in situ et en laboratoire
3 – Classification, prédiction, observation et contrôle
4 – Géologie hydrique

Thème B Excavation en surface et à faible profondeur

1 – Stabilité des pentes rocheuses
2 – Fondations dans et sur les roches, y compris fondations de barrage
3 – Construction à faible profondeur surtout dans les villes

THEMEN BAND 1

Thema A Untersuchung und Beurteilung des Betriebspunktes

1 – Geophysikalische Prüfung und Untersuchung
2 – Prüfung an Ort und Stelle und im Labor
3 – Klassifizierung, Voraussage, Beobachtung und Überwachung
4 – Hydrogeologie

Thema B Übertägige und oberflächennahe Felsbauwerke

1 – Standsicherheit von Felsabhängen
2 – Fundamente auf und im Fels, einschliesslich Stauwerkfundamente
3 – Oberflächennahe Konstruktionen, insbesondere in Städten

Themes and contents of the volumes
Thèmes et contenu des tomes
Themen und Inhalt der Bände

Foreword

The 5th International Congress of the International Society for Rock Mechanics was held in Melbourne, Australia in April 1983. The overall theme for the Congress was:

"Rock Mechanics for Resource Development,
Mining and Civil Engineering"

This theme was chosen having regard to new developments and applications of rock mechanics throughout the world, and particular interests in rock engineering in Australia.

Each national group of the Society in the fifty-five countries participating in ISRM was responsible for the selection and review of the papers from their country. The papers have been reproduced in these volumes from the manuscripts as received.

The first two volumes of the proceedings contain the papers received in each of the five main themes of the Congress.

The third volume of the proceedings is to contain the General Reports, together with reports of discussions presented by Congress participants, and reports of the Opening and Closing Ceremonies. These General Reports, prepared by internationally recognised specialists, represent an international survey of the state of the art in rock mechanics at the time of the Congress.

The International Society for Rock Mechanics is represented in Australia by the Australian Geomechanics Society, which is jointly sponsored by the mining and engineering communities in Australia represented by the learned bodies, The Australasian Institute of Mining and Metallurgy, and The Institution of Engineers, Australia.

The Australian Geomechanics Society also represents in Australia the International Society for Soil Mechanics and Foundation Engineering, and the International Association of Engineering Geologists. In Australia we have found it professionally valuable for all those having an interest in geomechanics — mining engineers, civil engineers and geologists — to share joint learned activities in geomechanics on a national and state level.

The Australian Geomechanics Society was honoured to be invited by our international colleagues in rock mechanics to host the 5th International Congress in Australia in 1983. The arrangements for the Congress were in the hands of an Organising Committee of the Australian Geomechanics Society, which was supported very strongly indeed by the resources of The Australasian Institute of Mining and Metallurgy, who kindly accepted the invitation to provide the secretariat for the Congress.

Professor H. G. Poulos,
Chairman,
Australian Geomechanics
Society

Professor L. A. Endersbee, AO
Congress Chairman

April 1983

Avant-propos

Le 5ᵉ Congrès International de la Société de Mécanique des Roches s'est tenu à Melbourne, Australie, en avril 1983, et a eu pour thème général:

"Rôle de la Mécanique des Roches dans l'Exploitation des Ressources Naturelles, l'Industrie Minière et le Génie Civil"

Ce thème a été choisi pour sa pertinence aux innovations et aux applications récentes en mécanique des roches à travers le monde et en rapport avec les intérêts spécifiques en mécanique des roches en Australie.

Chaque groupe national de la Société dans les cinquante-cinq pays membres de la SIMR s'est chargé de selectionner et revoir les communications soumises par leurs groupements nationaux. Ces communications sont reproduites dans les volumes telles qu'elles ont été reçues.

Les deux premiers volumes des comptes rendus regroupent les communications ayant trait à chacun des thèmes majeurs du Congrès.

Le troisième volume doit comporter les Rapports Généraux, ensemble avec des rapports de discussions des participants au Congrès ainsi que des rapports des Cérémonies d'Ouverture et de Fermeture du Congrès. Ces Rapports Généraux, préparés par des spécialistes mondialement connus, représentent un exposé international du stade d'avancement de la mécanique des roches au moment du Congrès.

La Société Internationale de Mécanique des Roches est représentée en Australie par la Société Australienne de Géomécanique qui est sous l'égide conjointe des communautés minière et d'ingénierie qui sont elles-mêmes représentées par l'Institut Australasien des Mines et de la Métallurgie et l'Institut d'Ingénieurs d'Australie.

La Société de Géomécanique Australienne représente également la Société Internationale de Mécanique des Sols et du Génie des Fondations de même que l'Association Internationale des Géologues-Ingénieurs. En Australie, nous avons trouvé qu'au niveau national, et au niveau des Etats, la mise en commun des activités de recherche en géomécanique par les ingénieurs des mines, les ingénieurs du génie civil et les géologues peut être d'une aide professionnelle précieuse.

La Société Australienne de Géomécanique est sensible à l'honneur d'avoir été invitée par nos collègues internationaux d'être la Société hôte du 5ᵉ Congrés International en Australie, en 1983.

L'organisation du Congrès a été le fait d'un Comité Organisateur fourni par la Société Australienne de Géomécanique. Ce Comité a été pleinement soutenu par l'Institut Australasien des Mines et de la Métallurgie qui a eu l'amabilité d'accepter l'invitation d'assurer les travaux de Secrétariat du Congrès.

Le Professeur H. G. Poulos
Président
Société Australienne de Géomécanique

Le Professeur L. A. Endersbee, AO
Président du Congrès

Avril 1983

Vorwort

Der 5. Internationale Kongress der Internationalen Gesellschaft für Felsmechanik fand in Melbourne, Australien, im April 1983 statt. Das Generalthema des Kongresses lautete:

"Felsmechanik für den Aufschluss von Bodenschätzen,
Bergbau und Tiefbau"

Dieses Thema wurde nicht nur mit Bezug auf neue Entwicklungen und Anwendungen auf dem Gebiet der Felsmechanik auf der ganzen Welt, sondern auch insbesondere mit Bezug auf das Felsingenieurwesen in Australien gewählt.

Jede Nationalgruppe der Gesellschaft in den fünfundfünfzig Mitgliedsländern der IGFM war für die Auswahl und Rezension der Beiträge aus dem eigenen Lande verantwortlich, Beiträge, welche in diesen Bänden in der Originalfassung abgedruckt worden sind.

Die ersten zwei Bände des Tätigkeitsberichtes enthalten die Beiträge zu jedem der fünf Hauptthemen des Kongresses.

Der dritte Band des Tätigkeitsberichtes wird die Allgemein-Berichte sowie Einzelheiten der Diskussionen der Kongressteilnehmer als auch Berichte über die Eröffnungs- und Abschluss-Sitzungen enthalten. Diese von international anerkannten Fachleuten verfassten Allgemein-Berichte bieten einen Überblick über den Entwicklungsstand auf der ganzen Welt auf dem Gebiet der Felsmechanik zur Zeit des Kongresses.

Die Internationale Gesellschaft für Felsmechanik wird in Australien von der Australischen Gesellschaft für Geomechanik vertreten, welche gemeinsam von dem australischen Bergbau- und Ingenieurwesen durch die Fachvereinigungen "Australisches Institut für Bergbau und Metallurgie" und "Gesellschaft Australischer Ingenieure" unterstützt wird.

Die Australische Gesellschaft für Geomechanik ist gleichzeitig die australische Mitgliedsvereinigung der Internationalen Gesellschaft für Bodenmechanik und für das Fundament-Ingenieurwesen, als auch der Internationalen Vereinigung für Ingenieurgeologie. In Australien hat es sich für sämtliche an der Geomechanik interessierten Fachleute, wie z.B. Bergwerks- und Bauingenieure und Geologen, als zweckmässig erwiesen, ihre beruflichen Interessen in der Geomechanik gemeinschaftlich auf nationaler und auch auf regionaler Grundlage zu verfolgen.

Der Australischen Gesellschaft für Geomechanik wurde die Ehre zuteil, von ihren internationalen Kollegen auf dem Gebiet der Felsmechanik eingeladen zu werden, den 5. Internationalen Kongress im Jahre 1983 in Australien abzuhalten. Die Kongress-Vorbereitungen wurden von einem Organisationskomitee der Australischen Gesellschaft für Geomechanik getroffen. Dieses genas die volle Unterstützung des "Australischen Instituts für Bergbau und Metallurgie", welches freundlicherweise auch das Sekretariat für den Kongress zur Verfügung stellte.

Professor H. G. Poulos
Vorsitzender
Australische Gesellschaft für Geomechanik

Professor L. A. Endersbee, AO
Vorsitzender des Kongresses

April 1983

Organization of the fifth ISRM Congress

CONGRESS ORGANISING COMMITTEE

W. E. Bamford (Chairman of Committee)
Professor L. A. Endersbee, AO (Congress Chairman)
Dr. J. R. Barrett
Dr. A. G. Bennet
W. J. Cuming
Professor I. B. Donald
J. R. Enever
Dr. R. S. Evans

Dr. I. W. Johnston
Dr. M. Kurzeme
Professor H. G. Poulos
W. M. G. Regan
E. D. J. Stewart
W. E. Vance (Executive Secretary)
Mrs Judy Webber (Secretary to the Committee)

ADVISORY COMMITTEE

President of the ISRM: W. Wittke
Secretary General of the ISRM: A. Silverio

Vice-President for
Africa:	A.Chaoui	Europe:	S. Uriel Romero
Asia:	M. Yoshida	North America:	T. C. Atchison
Australasia:	W. E. Bamford	South America:	O. Moretto

GENERAL REPORTERS

Theme A
Professor David H. Stapledon, South Australian Institute of Technology, Adelaide, Australia
Dr. Peter Rissler of Ruhrtalsperrenverein, Essen, West Germany

Theme B
Professor Richard E. Goodman, University of California, Berkeley, U.S.A.
Professor Klaus W. John, Ruhruniversität, Bochum, West Germany

Theme C
Professor Charles Fairhurst and Dr. Barry H. G. Brady, University of Minnesota, Minnesota, U.S.A.
Professor Yoshio Hiramatsu, Kyoto University, Kyoto, Japan

Theme D
Dr. Per Anders Persson, Director, Research and Development, Nitro Nobel AB, Sweden
Roger Holmberg, President, Swedish Detonic Research Foundation Stockholm, Sweden

Theme E
Professor François G. Cornet, University of Pierre and Marie Curie, Paris, France

Organisation du cinquième Congrès de la SIMR

COMITE D'ORGANISATION DU CONGRES

W. E. Bamford (Président du Comité)
Professeur L. A. Endersbee, AO (Président du Congrès)
Dr. J. R. Barrett
Dr. A. G. Bennet
W. J. Cuming
Professeur I. B. Donald
J. R. Enever
Dr. R. S. Evans

Dr. I. W. Johnston
Dr. M. Kurzeme
Professeur H. G. Poulos
W. M. G. Regan
E. D. J. Stewart
W. E. Vance (Secrétaire exécutif)
Mme Judy Webber (Secrétaire du comité)

COMITE CONSULTATIF

Président de la SIMR: W. Wittke
Secrétaire Général de la SIMR: A. Silverio

Vice-Président pour
l'Afrique:	A.Chaoui	l'Europe:	S. Uriel Romero
l'Asie:	M. Yoshida	l'Amérique du Nord:	T. C. Atchison
l'Australasie:	W. E. Bamford	l'Amérique du Sud:	O. Moretto

RAPPORTEURS GÉNÉRAUX

Thème A

Professeur David H. Stapledon, Institut de Technologie de l'Australie du Sud, Adelaïde, Australie
Dr. Peter Rissler, Ruhrtalsperrenverein, Essen, République Fédérale d'Allemagne

Thème B

Professeur Richard E. Goodman, Université de Californie, Berkeley, E.U.
Professeur Klaus W. John, Université de la Ruhr, Bochum, République Fédérale d'Allemagne

Thème C

Professeur Charles Fairhurst et Dr. Barry H. G. Brady, Université de Minnesota, E.U.
Professeur Yoshio Hiramatsu, Université de Kyoto, Japan

Thème D

Dr. Pers Anders Persson, Directeur, Recherche et Développement, Nitro Nobel AB, Suède
Roger Holmberg, Président, Fondation Suèdoise pour la Recherche Détonique, Stockholm, Suède

Thème E

Professeur François G. Cornet, Université de Pierre et Marie Curie, Paris, France

Organisation des fünften Kongresses der IGFM

ORGANISATIONSKOMITEE DES KONGRESSES

W. E. Bamford (Vorsitzender des Komitees)

Professor L. A. Endersbee, AO (Vorsitzender des Kongresses)

Dr. J. R. Barrett

Dr. A. G. Bennet

W. J. Cuming

Professor I. B. Donald

J. R. Enever

Dr. R. S. Evans

Dr. I. W. Johnston

Dr. M. Kurzeme

Professor H. G. Poulos

W. M. G. Regan

E. D. J. Stewart

W. E. Vance (Geschäftsführer)

Frau Judy Webber (Schriftführerin des Komitees)

BERATENDES KOMITEE

Der Präsident der IGFM: W. Wittke

Der Generalsekretär der IGFM: A. Silverio

Vizepräsident für

Afrika:	A.Chaoui	Europa:	S. Uriel Romero
Asien:	M. Yoshida	Nordamerika:	T. C. Atchison
Australien und Ozeanien: W. E. Bamford		Südamerika:	O. Moretto

HAUPTREFERENTEN

Thema A

Professor David H. Stapledon, South Australian Institute of Technology, Adelaide, Australien

Dr. Peter Rissler des Ruhrtalsperrenvereins, Essen, Westdeutschland

Thema B

Professor Richard E. Goodman, University of California, Berkeley, U.S.A.

Professor Klaus W. John, Ruhruniversität, Bochum, Westdeutschland

Thema C

Professor Charles Fairhurst und Dr. Barry H. G. Brady, University of Minnesota, Minnesota, U.S.A.

Professor Yoshio Hiramatsu, Kyoto Universität, Kyoto, Japan

Thema D

Dr. Per Anders Persson, Direktor, Forschung und Entwicklung, Nitro Nobel AB, Schweden

Roger Holmberg, Präsident der Swedish Detonic Research Foundation, Stockholm, Schweden

Thema E

Professor François G. Cornet, Université de Pierre et Marie Curie, Paris, Frankreich

Contents / Contenu / Inhalt
Volume / Tome / Band 1

A Theme / Thème / Thema

Site Exploration and Evaluation
Exploration et evaluation in situ
Untersuchung und Beurteilung des Betriebspunktes

1 Geophysical Testing and Exploration
Exploration et essais géophysiques
Geophysikalische Prüfung und Untersuchung

2 In Situ and Laboratory Testing
Essais in situ et en laboratoire
Prüfung an Ort und Stelle und im Labor

3 Classification, prediction, observation and monitoring
Classification, prédiction, observation et contrôle
Klassifizierung, Voraussage, Beobachtung und Überwachung

B   Theme   Surface and Near-surface Excavations
Thème   Excavation en Surface et à Faible Profondeur
Thema   Übertägige und oberflächennahe Felsbauwerke

1   Stability of rock slopes
Stabilité des pentes rocheuses
Standsicherheit von Felsabhängen

2   Foundations on and in rock, including dam foundations
Fondations dan et sur les roches, y compris fondations de barrage
Fundamente auf und im Fels, einschliesslich Stauwerkfundamente

3 Near-surface construction especially in cities
Construction à faible profondeur surtout dan les villes.
Oberflächennache Konstruktionen, insbesondere in Städten

Index to Authors
Index des Auteurs
Inhaltsverzeichnis nach Schriftstellern

Volume / Tome / Band 1

CABRIL DAM — CONTROL OF THE GROUTING EFFECTIVENESS BY GEOPHYSICAL SEISMIC TESTS

Barrage du Cabril — Contrôle de l'efficacité des injections par des essais géophysiques séismiques

Staudamm Cabril — Kontrolle der Injektionswirksamkeit durch geophysikalische seismische Versuche

L. Fialho Rodrigues
Research Officer, LNEC (Laboratório Nacional de Engenharia Civil), Lisboa, Portugal

Ricardo Oliveira
Head Site Exploration Division
LNEC (Laboratório Nacional de Engenharia Civil), Lisboa, Portugal

A. Correia de Sousa
Principal Engineer, EDP (Electricidade de Portugal), Porto, Portugal

SYNOPSIS

In order to evaluate the effectiveness of the grout treatment for reinforcement of the Cabril dam foundation, the longitudinal wave velocity has been measured at eight zones of the rock mass, before and after grouting work, by means of crosshole and uphole seismic tests. An increase in the average longitudinal wave velocity from 2 to 20% was measured and a good correspondence between this increase in the average velocity and the decrease in permeability, as well as in grout take, was confirmed.

RESUME

Afin d'évaluer l'efficacité de la méthode de jointoiement pour le renforcement de la fondation du barrage de Cabril on a mesuré, avant et après jointoiement, la vitesse longitudinale d'onde en huit zones de cisaillement dans la masse rocheuse au moyen d'essais séismiques dans des trous de sonde horizontaux et verticaux. Une augmentation de la vitesse moyenne de l'onde longitudinale allant de 2 à 20% a été enregistrée et on a pu établir une bonne concordance entre cette augmentation de la vitesse et une diminution de la perméabilité de la prise du jointoiement.

ZUSAMMENFASSUNG

Um die Wirksamkeit der Vermörtelungsbehandlung für die Verstärkung des Cabril Stauwehrfundamentes zu bewerten, wurde die Fortpflanzungsgeschwindigkeit der longitudinalen Welle in acht Zonen der Gesteinsmasse vor und nach der Vermörtelung mittels seismischer Kreuzloch- und Vertikallochprüfungen gemessen. Eine Erhöhung in der mittleren Fortpflanzungsgeschwindigkeit der longitudinalen Welle von 2 bis 20% wurde festgestellt, und eine gute Übereinstimmung zwischen dieser Erhöhung der mittleren Geschwindigkeit und der Erniedrigung der Durchlässigkeit sowie der Vermörtelungswirksamkeit wurde bestätigt.

1. INTRODUCTION

The Cabril dam, built 30 years ago, is situated on the Zêzere river, in the center of Portugal and is founded on a granitic rock mass, which is affected by some local faults and shear zones.

An important complementary grouting treatment, enviseaging the reinforcement of the dam foundation, was carried out in the general programme of reparing the dam.

For the execution of this treatment the rock mass foundation was divided into ten blocks and the following programme was carried out for each block:
- grouting of two retaining curtains in the up stream and downstream limits of the blocks,
- grouting for the general consolidation of the rock mass foundation,
- grouting of a deep impermeabilization curtain.

Besides permeability, dilatometer and other tests, performed to control the effectiveness of grouting, the longitudinal wave velocity was measured in eight of the ten blocks, before and after the treatment work.

This paper deals with the description of the geophysical testing procedures and the presentation and discussion of these results.

2. TESTING PROCEDURES

Very few examples of application of seismic methods for the control of grouting effectiveness in the treatment of rock masses are shown in the literature. Most of them however deal with refraction seismic methods applied from the ground surface (KNILL et al 1972).

In the case of Cabril dam the crosshole and up hole seismic methods were applied, the measurement of the longitudinal wave velocities being

to be performed in each of the eight blocks be
fore the beginning of grouting, after the end of
the consolidation treatment of the dam found
ation and finally after the execution of the
deep impermeabilization curtain.

However, the tight schedule for the planning and
performance of the treatment, didn't allow the
execution of the geophysical tests in three of
the eight blocks at the end of the consolidation
stage.

Crosshole and downhole seismic methods were used
to measure the longitudinal wave velocity. In
Fig. 1 a general layout of the consolidation
blocks with the locations of the boreholes and
geophones for seismic tests and the boreholes
for permeability tests can be seen.

whole time of the programme.
- The borehole and geophone positions were care
fully surveyed.
- The reception of the seismic waves inside the
boreholes was achieved by the use of hydrophones
instead of geophones in order to minimize any
error arising from the different coupling con
ditions, of the geophones to the borehole walls,
from test to test.
- To prevent any eventual damage of the bore
holes, instantaneous electrical caps, instead
of explosives, were used as a source of elastic
waves. Moreover all the detonations were per
formed after filling the boreholes with water,
in order to maintain similar shot conditions and
a further hydrophone was located at a fixed dis
tance from the shot point to control the zero

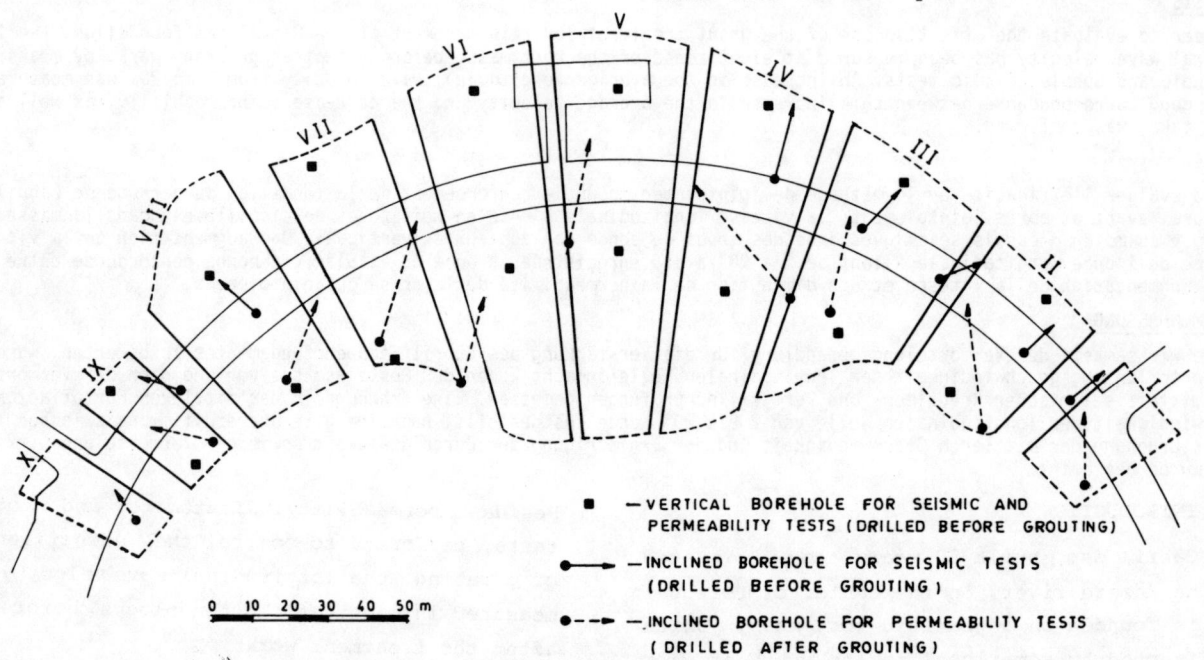

0 10 20 30 40 50 m

■ —VERTICAL BOREHOLE FOR SEISMIC AND
PERMEABILITY TESTS (DRILLED BEFORE GROUTING)

●→ —INCLINED BOREHOLE FOR SEISMIC TESTS
(DRILLED BEFORE GROUTING)

●--→ —INCLINED BOREHOLE FOR PERMEABILITY TESTS
(DRILLED AFTER GROUTING)

Fig. 1 - <u>Location of treatment blocks, boreholes and geophones for geophysical and permeability tests</u>

In order to maintain similar conditions for the
performance of the geophysical tests at the dif
ferent stages of the study the following pro
cedure was adopted in all the experiments:
- The boreholes and the respective test depths
were the same before the treatment and after
the stages of grouting.
- The geophones were fixed at the same position
in the general drainage gallery, during the

time.
- Recording equipment allowing a precision of
about 0.1 ms in the picking up of the arrival
time of elastic waves on the records was used,
therefore for an average longitudinal wave ve
locity of about 5000 m/s, a precision of about
1% in calculations was achieved.

3. TEST RESULTS AND DISCUSSION

The results of the seismic tests, given   in terms of longitudinal wave velocity before and after the treatment (including grouting of the retaining curtains, general consolidation grouting and deep impermeabilization curtain)   are presented in table I.

TABLE I - Results of seismic geophysical tests

Consolidation block	Longitudinal wave velocity, C (m/s)		Number of tests
	Before grouting	After grouting	
II	3650 - 4520 x̄ = 4120 ρ = 419	4600 - 5200 x̄ = 4938 ρ = 220	20
III	4100 - 5100 x̄ = 4637 ρ = 374	4800 - 5450 x̄ = 5173 ρ = 172	10
IV	3840 - 5470 x̄ = 4773 ρ = 256	3950 - 5600 x̄ = 5103 ρ = 302	90
V	4890 - 5500 x̄ = 5330 ρ = 203	5010 - 5650 x̄ = 5410 ρ = 213	30
VI	4810 - 5420 x̄ = 5270 ρ = 183	4890 - 5600 x̄ = 5440 ρ = 288	44
VII	4240 - 5370 x = 4450 ρ = 445	4780 - 5800 x̄ = 5153 ρ = 468	6
VIII	3440 - 5450 x̄ = 4369 ρ = 535	3760 - 5650 x̄ = 4780 ρ = 503	50
IX	3220 - 5110 x̄ = 4256 ρ = 517	3800 - 5430 x̄ = 4618 ρ = 515	43

3220 - 5110 - extreme values
x̄ - mean value
ρ - standard deviation

These results show that the granitic rock mass, of an average longitudinal wave velocity ranging from 4232 to 5500 m/s, had a high mechanical quality (see for example SJØGREN. 1972) before the present treatment work, partly due to the grouting carried out during the construction of the dam, 30 years ago.

The present treatment of the dam foundation caused a general improvement of the rock mass quality, with an increase in both the mean and extreme values of the longitudinal wave velocity

ty in all the blocks tested.

In table II the increase in the average longitudinal wave velocity and the decrease in permeability after the grouting works, as well as the respective grout take are presented.

TABLE II - Longitudinal wave velocity and permeability before and after grouting and grout take

Block	Mean longitudinal wave velocity (m/s)		Velocity increase (%)	Mean permeability (lugeons)		Permeability decrease (%)	Mean grout take (kg/m)
	Before grouting	After grouting		Before grouting	After grouting		
II	4120	4938	20	5.4	0.6	89	23.1
III	4637	5173	12	-	-	-	37.3
IV	4773	5103	7	3.2	1.2	63	16.4
V	5330	5410	2	0.0	0.0	-	2.1
VI	5270	5440	3	0.1	0	100	8.9
VII	4450	5153	16	27.1*	0.4*	98	14.4
VIII	4369	4780	9	1.6	1.0	38	27.8
IX	4256	4618	9	6.5	1.0	85	44.7

* Permeability test performed outside of the zone interested by the geophysical survey

These results show that the grouting work increased the mean longitudinal wave velocity from about 2 to 20% and decreased the mean permeability from about 38 to 100% indicating the improvement of the mechanical properties of the   rock mass and the corresponding decrease of permeability introduced by grouting.

In general the results exhibit a good agreement between the increase in longitudinal wave velocity and the corresponding decrease in permeability. Furthermore to the general increase in longitudinal wave velocity, corresponds a larger grout take (ranging from 2.1 to 44.7 kg/m).

The lowest values of these parameters have been determined in block V, located at the river bed zone and the highest values in the blocks situated at higher elevation.

In seismic tests performed in block V, no increase in longitudinal wave velocity was detect ed after grouting, as can be seen in fig. 3.

BLOCK V

DETONATIONS IN BOREHOLE SJ5
GEOPHONES AT DRAINAGE GALLERY

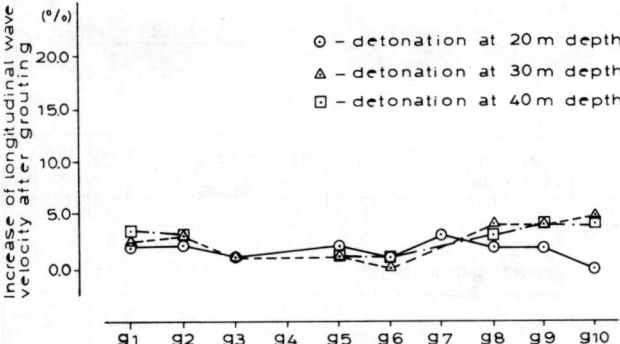

Fig. 3 - <u>Results of seismic tests in</u>
<u>block V</u>

In this block a zero permeability was also mea
sured, before and after the treatment, in the
borehole located downstream of the axis of the
dam. These measurements indicate very good me
chanical and hydraulic properties of the rock
mass in this zone of the foundation.

After the treatment a new borehole for the eva
luation of the final permeability of the rock
mass has been drilled in each block (Fig. 1).

The results of these tests are presented in
table III.

TABLE III - <u>Permeability values obtained</u>
<u>in control boreholes drilled</u>
<u>after treatment</u>

Block	Permeability (lugeons)
II	1.1
III	1.3
IV	0.3
V	0.0
VII	0.4

TABLE III (Cont.)

Block	Permeability (lugeons)
VIII	1.0
IX	0.3

These results are of the same order of magni
tude with those obtained in the boreholes lo
cated downstream and upstream of the
dam axis, showing the low permeability of the
rock mass achieved after the treatment, with
a zero value being again obtained in block V.

4. CONCLUSIONS

Crosshole and uphole seismic methods with very
accurate and well controled testing procedures
can be used to assess the effectiveness of
grouting in granitic rock masses even when they
show a relatively high mechanical quality be
fore the treatment.

The increase in longitudinal wave velocity
seems an adequate parameter to evaluate the
improvement on the mechanical properties introduced by grouting and indirectly the reduc
tion in permeability achieved.

The advantage of such a method, apart from be
ing very fast, lies on allowing the possibility
of studying large volumes of rock mass to be
submited to grout treatment.

5. REFERENCES

KNILL, J.L. and PRICE, D.G. (1972) - Seismic
evaluation of rock masses. 24th International
Geological Congress, Montreal.

SJØGREN, B.; ØFSTHUS, A. and SANDBERG, J.
(1979) - Seismic classification of rock mass
qualities. Geophysical Prospecting. Nº 27,
pp. 409-442.

IN-SITU DYNAMIC DEFORMATION PROPERTIES OF SOFT ROCK MASSES

Les Propriétés de la déformation dynamique in-situ du corps de la roche douce

Das dynamische Verformungsverhalten von Halbfestgesteinen

Koji Ishikawa and Keiji Miyajima
Chuo Kaihatsu Corporation, Osaka and Tokyo, Japan

Satoshi Hibino and Yoshikazu Fujiwara
Central Research Institute of Electric Power Industry, Abiko, Japan

Mamoru Yamagata
Honshu Shikoku Bridge Authority, Tokyo, Japan

SYNOPSIS
This paper summarizes the results of in-situ dynamic loading tests in order to estimate the deformation behaviour of foundation rocks during earthquakes. From the results their dynamic elasto-plastic and cisco-elastic properties were classfied.

RESUME
Ce rapport présente les resultats d'essai de charge dynamique in-situ pour estimer les effets de déformation des roches de fondations au cours de tremblements de terre. Une classification de leurs propriétés dynamiques élasto-plastiques et visco-élastiques à été élaborée.

ZUSAMMENFASSUNG
Die Ergebnisse aus den dynamischen Belastungsversuchen des felsigen Bauuntergrundes, die als Studie zum Verformungsverhalten des Bodens während des Erdbebens durchgeführt worden sind, werden hier kurz dargestellt. Durch diese Versuche wurden die dynamischen elasto-plastischen und visko-elastischen Eigenschaften des felsigen Bodens geklärt.

1. CHARACTERISTICS OF GROUND AT TEST SITE

Tests were performed inside a test adit of cross section of about 2.5 x 2.5m.

The geological section and physical properties of ground near the test site are shown in Figs. 1(a) and (b). The geology of Site A consists of weathered granite. The density γ_t of the rock mass is roughly 2.10 tf/m³ and is seen to be isotropic and homogeneous, but the dynamic modulus of elasticity Edf obtained by velocity logging increases cumulatively in the direction of depth and it is thought to be a heterogeneous multiple-layered ground. The geology of Site B comprises alternations of Neogene sandstone and mudstone. The vicinity of the ground surface cosists of mudstone of little deformability, whereas the underlying sandstone is of rock character of high density and large deformability. This is thought to be a multiple-layered ground interbedded by a weak layer in the direction of depth.

2. METHOD OF TESTING

As shown in Fig. 2, the testing apparatus can produce any sinusoidal waveform of f = 0.5 Hz

loads were applied in the vertical direction directly to the ground by circular rigid loading plates of diameters 60 to 80cm. Ground surface displacements, subsurface strains, etc. were measured under various static load, dynamic load and frequency conditions.

Fig. 3 shows loading patterns in dynamic loading tests. The dynamic loading tests consisted of repetitively applying loads of Δp = 1.0 -10.0 kgf/cm² at any frequency of sinusoidal waveform in a range of f = 0.1 -5.0 Hz for 1 to 2 minutes.

The dynamic elastic modulus Ed was obtained by the equations below from the solutions of Bousinesq and Love in case of rigid body loading considering the ground to be semi-infinite, isotropic and homogeneous. This was defined as secant elastic modulus at the apex of the hysteresis loop.

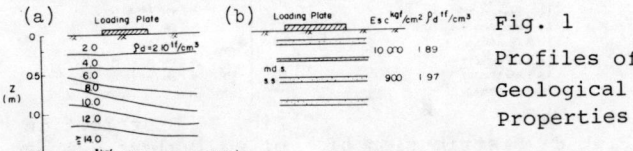

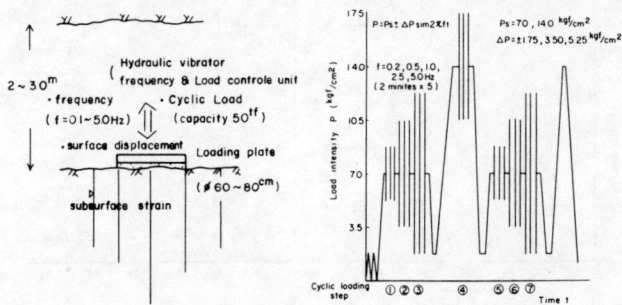

Fig. 1
Profiles of
Geological
Properties

in the vertical direction and cyclic loads up to dynamic loading internsity of Δp = 10 kgf/cm² made to fit conditions during earthquake by servovalve and vibrator. Cyclic

Fig. 2 In-situ cyclic loading testing system

Fig. 3 An example of loading pattern

From displacement on loading plate,

$$Ed = \frac{(1-\nu d)^2 \Delta P}{2a \cdot \delta d} \qquad (2.1)$$

From subsurface strain:

$$Ed = \frac{(1+\nu d) \cdot \Delta P}{\pi \cdot a^2 \cdot \varepsilon d}(-\frac{1}{2})[(1-2\nu d)\bar{J_0}^1 + \xi \bar{J_0}^2] \qquad (2.2)$$

Where, νd is dynamic Poisson's ratio, Δp is dynamic load, a is loading plate radius and δd is dynamic displacement.

Next, the equivalent damping ratio heq is based on damping energy and elastic strain energy expressed by the area of the hysteresis loop and was obtained as defined by Eq. (2.3).

$$heq = \frac{1}{2\pi} \cdot \frac{W_Y}{W_E} \qquad (2.3)$$

Where, W_Y: damping energy of hysteresis loop

W_E: elastic strain energy

3. TEST RESULTS

3.1 Relationship with Dynamic Loading Intensity

Fig. 4 shows the relationship between values of dynamic loading intensity Δp and the dynamic moduli of elasticity Ed obtained at the same time. Fig. 5 shows the relationship between Δp and equivalent damping ratio heq. According to these figures, dynamic moduli of elasticity Ed for both the granite and the alternations decrease linearly on semi-logarithmic graph when Δp is increased. The gradient indicating

the trend of decrease is greater for the alternating layers. The equivalent damping ratio heq is smaller on the whole for the alternating layers. As regads heq, it shows a trend of more decrease the larger the value of Δp regardless of frequency, in both of the cases. However, in the case of granite, a distinct trend of increase is shown in the results of Steps 5 -7 subjected to load hysteresis.

3.2 Relationship with Frequency

Figs. 6 and 7 respectively show the relationships of the various frequencies f with Ed and heq. According to these figures, the value of Ed tends to be incresed the higher that frequency becomes for both granite and the alternations. Particularly, the trend of increase in Ed is conspicuous in the range of f = 0.1 to 1.0 Hz. However, at $f \geq 1.0$ Hz, this trend is not prominent. Further, the influence due to the extent of Δp is not prominent.

The value of heq, in the case of granite, shows a trend of more increase the larger the values of f at the various dynamic load levels. On the other hand, in case of the alternating layers, a trend of decrease is indicated. The different trends of the two are prominent especially in the range of f = 0.1 -1.0 Hz.

3.3 Distributions and Variations of Ed and heq in Direction of Depth

At granitic ground, besides dynamic displacements on the loading plate, subsurface dynamic strains were also measured. Based on these, Ed and heq were determined using Eq. (2.2) and Eq. (2.3).

Fig. 8 shows the variations of Ed in the direction of depth by loading hysteresis and by dynamic load at specific frequency obtained from loading plate displacement and subsurface strain.

According to the figure, there is a distinct trend for Ed to gradually increase cumulatively in the direction of depth similarly to the dynamic modulus of elasticity Edf obtained in

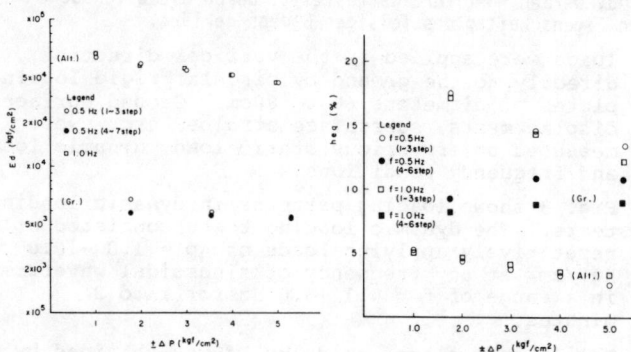

Fig. 4 Relationship between dynamic secant elastic modulus and dynamic loading intensity

Fig. 5 Relationship between equivalent damping ratio and dynamic loading intensity

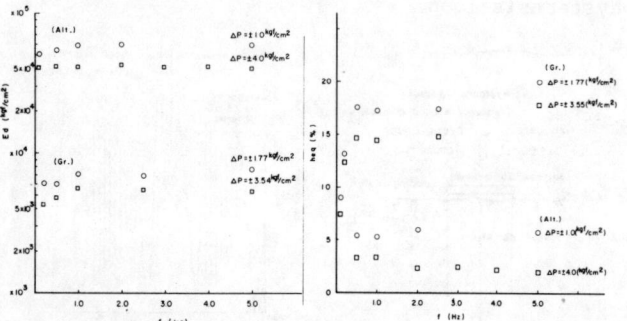

Fig. 6 Relationship between dynamic secant elastic modulus and frequency of cyclic loading

Fig. 7 Relationship between equivalent damping ratio and frequency of cyclic loading

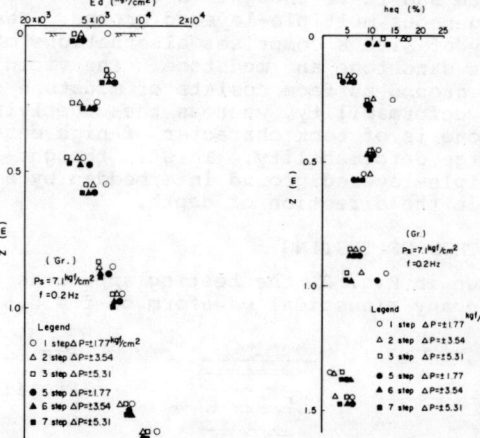

Fig. 8 Distribution of dynamic secant elastic modulus in the foundation beneath the loading plate by each step (dynamic loading intensity)

Fig. 9 Distribution of equivalent damping ratio in the foundation beneath the loading plate by each step (dynamic loading intensity)

velocity logging. However, Ed obtained by loading plate displacement coincides approximately with Ed obtained at the subsurface close to the ground surface (about z = 2.0a). The correspondence of the Ed values of these two is seen in the variation in dynamic load and in the loading hysteresis, and the Ed value obtained on the loading plate and its variations were surmised to be susceptible mainly to the influences of the physical properties and behaviors of the ground in the vicinity of the surface layer.

The variations in Ed value according to dynamic load level are correspondent for the ground surface and the subsurface, and are also correspondent when subjected to loading hysteresis.

As for computation of Ed at a certain point in the ground, it was performed by inverse calculations based on measured strains using the solution considering the ground to be isotropic and homogeneous. However, these Ed values are also in correspondence with Ed value obtained from ground surface displacement similarly assuming isotropic and homogeneous natures, while they are correspondent again also with Edf distribution in the direction of depth obtained from velocity logging (strain level x 10^{-6} - 10^{-7}, frequency about 100 Hz), and it is thought to be close to the true value.

Fig. 9 shows the variations of heq in the direction of depth for various conditions similarly to the previously-mentioned Ed values.

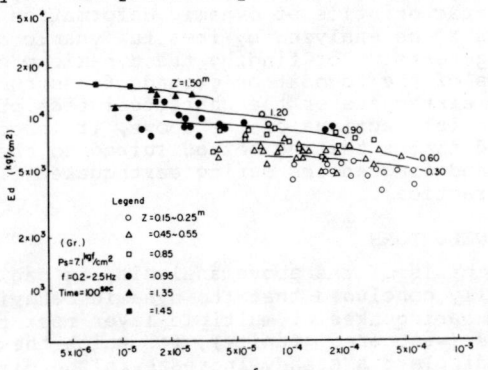

Fig. 10 Relationship between vertical strain component and dynamic secant elastic modulus (calculated)

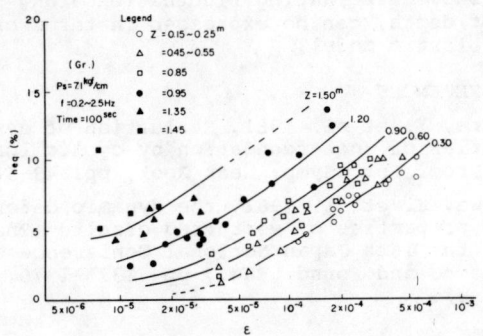

Fig. 11 Relationship between vertical strain component and equivalent damping ratio (calculated)

Such variations in Ed and heq values are thought to be due to predominance of plastic properties particularly because of the influence of loosening near the surface layer caused by stress relief accompanying excavation and the influence of confining effect due to loading hysteresis.

3.4 Strain Dependency

In case of weathered granite the elastic and damping characteristics of the materials indicate strain dependencies.

Figs. 10 and 11 show the relationships between measured values of subsurface strains and the values of Ed and heq obtained by calculations expressed with depth z (= correspondent to confining pressure) as the parameter. According to this, Ed value, when depth z is established, is determined readily by strain level, the stress-strain relationship is non-linear and a plastic propety is predominant, and materials characteristics are shown to possess strain dependency. Similarly, the heq-ε relationship is also readily determined if depth z is established, and materials characteristics show strain dependency where a plastic property is predominant.

3.5 Velocity Dependency

In the case of the alternating layer ground the elastic and damping characteristics of the material will show velocity dependencies.

Fig. 12 shows the relationship between loading rate $\overline{\Delta p}$ (= $\Delta p \cdot w$) (or displacement ratio $\overline{\delta}$ (= $\delta \cdot w$)) and the value of Ed obtained by calculations with dynamic load Δp and frequency f as parameters. According to this, Ed value is readily determined by velocity if Δp and f are definite. In this case, if dynamic load is made constant while frequency is increased, Ed shows a trend of increase accompanying increase in rate of loading, and if a visco-elastic model is set up, behavior which can be explained by the Maxwell model or the three-elements model is seen. In effect, it is indicated that materials characteristics show a velocity dependency.

Fig. 13 shows the relationship between the average rate of loading $\overline{\Delta p}$ (= $\Delta p \cdot w$) and heq obtained by calculations. According to this, heq is readily determined by $\overline{\Delta p}$.

4. CONSIDERATIONS

4.1 Case of Weathered Rock

Strongly-weathered granite is uniformly deteriorated to an extent that natural joints have become unnoticeable, and it appers at first glance as an isotropic and homogeneous ground.

However, according to these test results, loosening has occurred in the vicinity of the surface layer affected by stress relief accompanying excavation, and as a result the weathered granite rock mass can be considered as a heterogeneous multiple-layered ground in which elastic characteristics gradually increase in the direction of depth. The deformation properties of ground of a semi-infinite state in the case of such a ground structure may be considered as main representing ground characteristics to a depth of about z = 2.0a. The influence of this loosening near the surface layer becomes evident as cyclic effect on elasticity characteristics and damping characteristics.

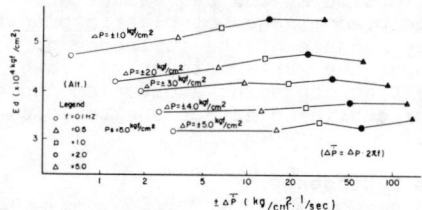

Fig. 12 Relationship between dynamic loading velocity and dynamic secant elastic modulus (calculated)

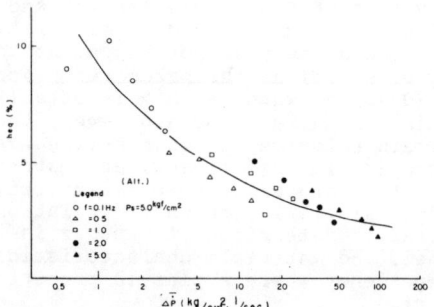

Fig. 13   Relationship between dynamic loading velocity and equivalent damping ratio (calculated)

The elastic characteristics become more prominent influenced by confining pressure in the direction of depth, while on the other hand, damping characteristics are weakened.

In the case of weathered granite, vibration tests by sample is greatly affected by the influece of stress relief at the time of sampling, and the testing mehod is considered as being unsuitable. The dynamic deformation properties of ground obtained by in-situ cyclic loading tests may be thought to be influenced by stress relief at the beginning, but it may be considered that a figure close to actual has been obtained due to cyclic effects. According to this, the dynamic deformation propeties of ground materials do not indicate viscous damping while indicating a behavior which is not that of a linear elastic body. In general, the properties may be looked upon as those of a non-linear body in which the properties of a plastic material are predominant.

As described above, the dynamic deformation properties of the weathered granite rock mass during earthquake are such that the rock mass is thought to be one where elasto-plastic properties are relatively predominant.

4.2 Case of Alternating Layer Ground

The dynamic deformation properties of alternating layer ground with heterogeneous weak layers underlying, are those of a multiple-layered ground having the nature of being susceptible to the influences of dynamic loading level and velocity effect.

The damping characteristics are those of a visco-elastic ground, and depending on the test conditions, the underground loss damping of the vibration system is inconspicuous and the viscous damping characteristics as a ground material indicate a predominant relationship.

Consequently, a frequency dependency is seen in the equivalent damping ratio heq. A trend of heq being reduced more the higher the dynamic load level is seen. This phenomenon is thought to be due to velocity effect. In this case, a trend of Ed increasing and heq decreasing the higher the dynamic loading rate and the deformation rate - in other words, a trend of elastic properties being predominant rather than viscous properties - is seen. It is conceivable from this that either of the above properties is predominantly indicated depending on the lengths of retardation time E/η and loading time T in case of adopting the Maxwell model or the three-elements model from among visco-elastic models.

In the case of a multiple-layered ground with a weak layer intercalated toward the bottom, the dynamic deformation properties are prominently affected by the properties of the hard ground near the surface layer while the dynamic loading level is low, but when the dynamic loading level becomes high, or when frequency becomes high, it is considered that the influence of the properties of the underlying weak layer and the time lag in strain behavior at the vicinity of the boudary between weak and strong layers disappear so that the phenomenon is that of deformation occurring in an intergrated form.

The results of analyses performed on the dynamic deformation properties of two different types of multiple-layered ground at soft rock masses have been described. In case of such multiple-layered ground, it is considered necessary for the characteristics of dynamic deformation behaviors to be analyzed by in-situ dynamic cyclic loading tests. For finding the dynamic properties of the foundation ground of a structure during earthquake or the characteristics of dynamic interactions of the above, it is considered that a testing method suited to the loads and frequencies during earthquake will be more practical.

5. CONCLUSIONS

On the basis of the above analysis, it can be generally concluded that the dynamic behavior during earthquakes of multiple layer rock formations (weathered granite), for which the E value displays a steady increase in the direction of depth, can be simulated by means of the elasto-plastic model; while that of the composite layer type rock mass (interbedded layers of sandstone and mudstone), where the E value shows alternating fluctuation along the axis of depth, can be expressed in terms of the visco-elastic model.

6. REFERENCES

Fujiwara, Y. et al. 1981, Evaluation of dynamic properties of rock foundation by cyclic loading test, Proc. Int. Symp. Weak Rock, pp. 49-54.

Ishikawa, K. et al. 1981, the dynamic deformation properties of weathered granite (2nd report), the 16th Japan National Conference on Soil Mech. and Found. Eng., pp. 1373-1376, (in Japanese).

GEOPHYSICAL EXPLORATION FOR A PROJECT IN THE HIMALAYAS

Exploration géophysique pour un projet dans l'Himalaya
Geophysikalische Untersuchungen für ein Projekt im Himalaya

A. K. Dhawan
Deputy Director, Central Soil and Materials Research Station, New Delhi – 110016, India

J. M. Kate
Asst. Professor, Indian Institute of Technology, New Delhi – 110016, India

A. B. Joshi
Director, Central Water Commission, New Delhi – 110022, India

SYNOPSIS

The paper presents the results of sub-surface investigations carried out for a hydroelectric power project in the Himalayas. The field studies comprised a seismic refraction survey, in-situ rock testing and bore hole drilling at critical locations to understand the state of the ground. The interpretation of the results has been carried out to locate various geological contacts and irregularities etc. Various sub-surface horizons have been identified on the basis of seismic wave velocity contrast and are compared with a log of bore holes. In addition an attempt has been made to obtain correlation of seismic wave velocities with in-situ properties.

RESUME

On a présenté les résultats obtenus lors des études souterraines effectuées en vue d'un aménagement hydroélectrique dans l'Himalaya. Les essais sur le terrain ont porté entre autre sur les levés de réfraction sismique, essais des rochers in-situ et forages de puits dans des endroits critiques, en vue de connaître le terrain. L'interprétation des données a permis de déterminer les divers contacts et irrégularités géologiques. On a identifié divers horizons souterrains par contraste de la vitesse de l'onde sismique et on les a étudiés par rapport aux strates révélées par le sondage. On a également essayé d'établir une correlation entre les vitesses des ondes sismiques et les propriétés des rochers in-situ.

ZUSAMMENFASSUNG

Der Aufsatz beschreibt die Ergebnisse der unterirdischen Untersuchungen für ein hydroelektrisches Kraftwerkprojekt im Himalaya. Die Felduntersuchungen beziehen sich auf seismische Refraktionsaufnahme, Gesteinsprüfungen in-situ und Bohrungen an kritischen Stellen, um die Erkenntnisse der unterirdischen Lagen zu gewinnen. Die Auswertung der Ergebnisse wurde durchgeführt, um verschiedene geologische Kontakte, Verwerfungen usw. zu bestimmen. Verschiedene unterirdische Horizonte wurden durch unterschiedliche seismische Wellengeschwindigkeiten identifiziert und mit den Bohrlochmessungen verglichen. Weiterhin wurde ein Versuch unternommen, um ein Verhältnis zwischen seismischer Wellengeschwindigkeit und den Eigenschaften der Gesteine in-situ zu bestimmen.

1. INTRODUCTION

It is necessary to investigate vast area while planning for hydroelectric power projects. The recent advancement in technology provides Geophysical methods for exploring large areas within a short duration.

The hydroelectric power project under study envisages utilization of a major river for the generation of power in the north-western part of India.

The project layout consists of a 125 m high concrete arch gravity dam, 6.5 km long and 9.5 m diameter concrete lined power tunnel on the right bank of the river, a 25 m dia and 103 m high surge shaft and an under-ground power house; housing three generating units of 180 MW each. The tailrace tunnel after crossing under the main river, discharges 1.75 km downstream into the main river. The tailrace system comprises of 9.5 m dia & 1.77 km long tunnel, a 0.36 km long cut and cover section & a 95 m long open channel.

2. GEOLOGY

2.1 Regional geology

The project area is occupied by variety of rock types comprising granite-gneiss, granite, phyllitic quartzitic slate, limestone, carbonaceous phyllite, quartzite & volcanics ranging in age from precambrian to upper tertiary. The entire sequence is highly folded & faulted. Several major regional thrusts are exposed in the project area. These are the Jutogh thrust (phyllite/granite gneiss contact) in the reservoir about 1.5km upstream of the proposed dam axis & normal to the river course and the shali thrust (phyllite/volcanic contact) which cuts across the power tunnel. The main boundary fault has been located on the south-west of the project. The project is located 45 km from the epicentre of earthquakes of magnitude 6 and 60 km from the earthquake of magnitude 8.

2.2 Geological features at the dam site

The bedrock at the dam site consists of variable bands of grey to dark grey quartzitic and slaty phyllites and their foliation trends in N20°W-S20°E to N10°E-S10°W and dips in the order of 45°-65° in upstream direction. The phyllites were subjected to excessive tectonic activity which eventually caused tight folding, shearing and jointing. No large faults have been detected below the river bed.

3. GEOPHYSICAL INVESTIGATIONS

The seismic refraction method has been used throughout the investigation, both on land and over water. The areas surveyed included river bottom, dam abutments, intake and diversion area, parts of power tunnel, power house and tailrace.

In this paper, the results of seismic refraction survey carried out at the dam site has only been presented. The seismic surveys were initially conducted to define any geological anomalies so that the boreholes could be better located. The borehole results were in turn used to calibrate and interpret seismic refraction test data. The seismic survey included determination of both 'P' & 'S' wave velocities (v_P & v_S respectively).

A 12 channel seismorgraph with amplification and recording system was used for seismic refraction survey. Five shot points were used for each profile giving a double control (direct & reverse shots) of the refracting horizons. Hawkins (1961) method has been adopted for detailed interpretation and the results were cross checked by critical distance method. The standard hydrophone technique was tried to get seismic profile of the river cross section. However, this method didn't succeed due to very swift (10 m/s) current in the river. An alternative technique of implanting geophones on both banks and locating shot points at the surface of the river has been used to get the time-distance graph. This method although gave the same amount of information but was too cumbersome and time consuming.

4. ROCK TESTING

4.1 In-situ tests

Thirteen NX holes totalling a depth of around 1100 m were drilled at the dam site. Some of the holes were drilled inclined from both the river banks, as it was not possible to drill into river bed. Goodman jack tests were performed in the boreholes for obtaining deformability characteristics of rock at various depths. These tests were performed both parallel as well as normal to the dam axis.

The in-situ tests such as Plate Jack tests, Flat Jack tests & Shear tests were conducted in a drift located along the dam axis on the left bank. It may be mentioned here that extensive seismic survey has been carried out in this drift with the specific aim of establishing correlations, if any, between seismic wave velocities & in-situ properties of rock.

4.2 Laboratory Tests

Representative rock specimens collected from the drift were tested in the laboratory to obtain density (ρ) modulus of elasticity (E) & poisson's ratio (ν). The tests were conducted both on dry & saturated rock specimens.

5. RESULTS, DISCUSSIONS & CORRELATIONS

5.1 Seismic survey & ground truth

The 'P' wave velocities (v_P) of bedrock at various locations were computed and the values are shown in Table 1.

A typical section of the seismic profile taken across the river downstream of the dam axis along with the borehole log is shown in Fig. 1. The drill holes near the seismic lines show the bedrock level varying from 625-640 metres. At most of the locations, the depth of bedrock as given by seismic survey & the drill hole data are in agreement with each other within ± 10%.

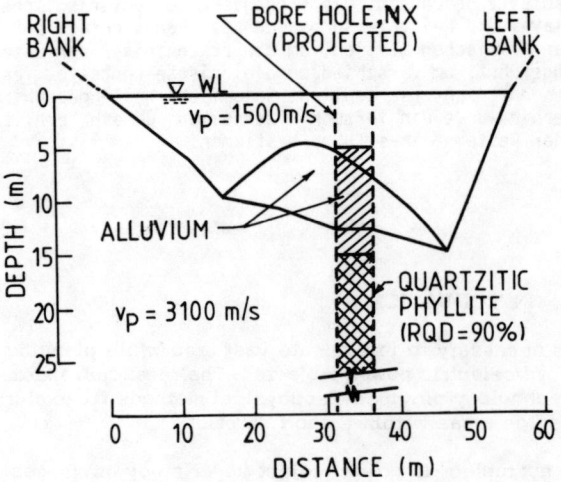

Fig. 1 Typical subsurface profile along with the log of nearest borehole.

TABLE 1. Summary of seismic survey results

Location of Seismic Line	Depth to bedrock (m)	Bedrock velocity $(10^3 m/s)$
U/S of dam axis across the river	1.6 - 8.0	1.9 - 4.5
D/S of dam axis across the river	0.0 - 11.9	2.3 - 3.5
D/S of dam axis across the river	0.0 - 14.6	3.1
At the dam axis partly on land and partly across the river	5.0 - 15.0	3.6 - 7.2
Toe of the dam across the river	2.0 - 6.5	3.0 - 4.0
Left bank d/s of dam axis	9.9 - 27.7	2.8 - 4.5
Left bank d/s of dam axis	10.8 - 30.3	2.5 - 3.4
Left bank d/s of dam axis	10.1 - 23.6	1.8 - 4.4
Left bank d/s of dam axis	12.1 - 19.7	3.0 - 4.5
Right bank d/s of dam axis	2.0 - 6.5	5.1
Right bank d/s of dam axis	1.5 - 6.5	4.5
Left bank normal to dam axis	11.0 - 16.5	4.2
Across the dam axis at river level	4.5 - 10.0	3.1 - 5.0
In the drift on the left bank along the dam axis.	1.0 - 4.0	2.85

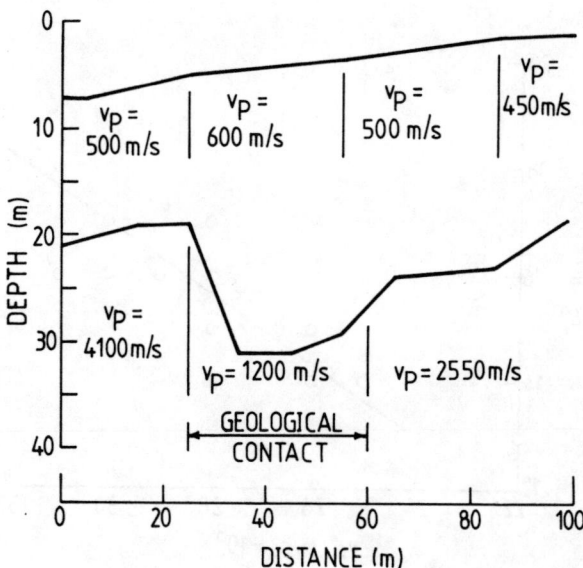

Fig.2   Subsurface profile exhibiting geological Contact

Table 2.   Classification of geological formations.

Formation	Range of 'P' wave velocity (m/s)
River Boulders	1300 - 1900
Unconsolidated soil	400 - 625
Overburden/weathered Phyllites	925 - 1200
Unweathered Phyllites	2520 - 4500

In addition, the contact between various geological formations were located by laying continuous seismic profiles across the anticipated contact after studying seismic velocity contrast. Fig.2 illustrates one such case for locating gelogical contact on the basis of seismic wave velocities contrast. It shows a sand-witched zone of low velocity between two comparatively higher velocity zones of different formations indicating a gelogical contact.

Based upon the seismic survey results & geological interpretation of the area, the classifications of various gelogical formations has been obtained as given in Table 2.

5.2   Correlations

The relationship between 'P' & 'S' wave velocities is shown in Fig. 3. On an average, the ratio between v_P

and v_S for the formations studied here is of the order of 1.79. The value of poisson's ratio (ν) calculated from this v_P/v_S ratio is 0.273. The average value of ν obtained from laboratory tests is 0.23 giving a ratio of v(field)/v(lab.) of 1.187.

The values of Dynamic Modulus of Elasticity (Edy) were calculated from v_P data by equation given by Robertshaw & Brown (1953). The Static Modulii of Elasticity (E) & Deformation (D) were determined by Goodman jack tests.

The variation between v_P & E is shown in Fig. 4. The relationship between E_{dy} computed from these

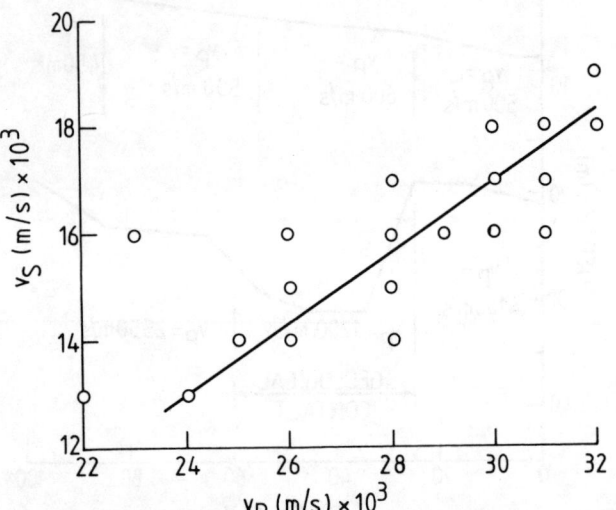

Fig. 3 Variation between 'P' wave and 'S'
wave velocities

values of v_P & E can be expressed by the following
equation.

$$E_{dy} = 0.4167 + 4.94 \times E \quad \ldots \ldots \quad (1)$$

The correlation coefficient in this case works out to
be 0.586.

Similarly, the plot between v_P & D is shown in Fig. 5
and the relationship between E_{dy} & D is expressed by:

$$E_{dy} = 1.402 + 3.566 \times D \quad \ldots \ldots \ldots \quad (2)$$

Which gives a correlation coefficient of only 0.2417
indicating a poor correlation.

The summary of results of in-situ tests carried out in
drift on the left bank of the dam is presented in Table
3. The subsurface profile interpreted from the seismic
refraction survey in the drift is shown in Fig. 6.

Table 3 Results of in-situ tests

Test	No. of tests,	D (10^5kN / m^2)		E (10^5kN / m^2)	
		Min.	Max.	Min.	Max.
Plate Jack	4	6.15	44.5	18.1	65.2
Flat Jack	3	1.5	4.4	2.1	6.7

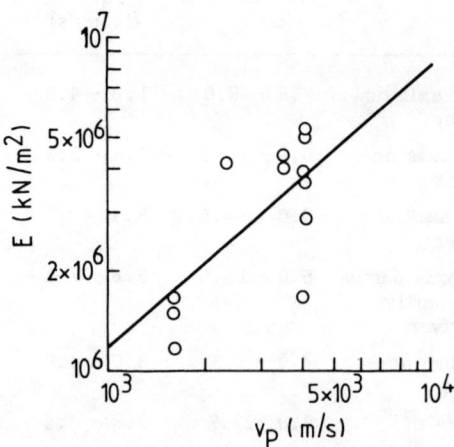

Fig. 4 Modulus of Elasticity versus Velocity
of 'P' waves.

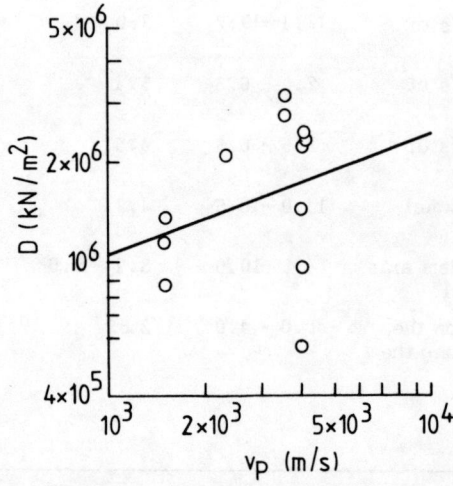

Fig. 5 Modulus of Deformation Vs. Velocity of
'P' waves.

The values of D & E found out from plate jack tests
are comparable with the first layer seismic velocity
of 1400 m/s (which gives E = 26.1×10^5kN/m^2). The
pressure bulbs due to plate jack tests have not penetrated into the bedrock. It is interesting to note that
the values of D & E obtained by Flat jack test are
nowhere comparable with those from seismic wave
velocity indicating that Flat jack tests should be
used with caution for obtaining D & E.

Fig. 7 Shows the variation between RQD and v_P which
gives a correlation coefficient of 0.55.

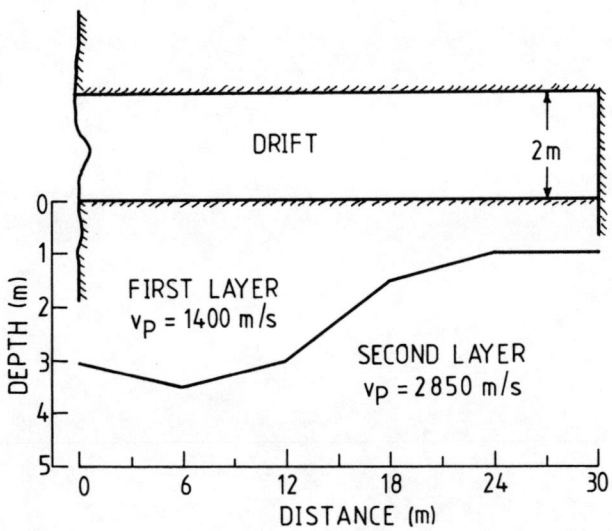

Fig. 6   Subsurface profile in the drift interpreted
by seismic refraction survey.

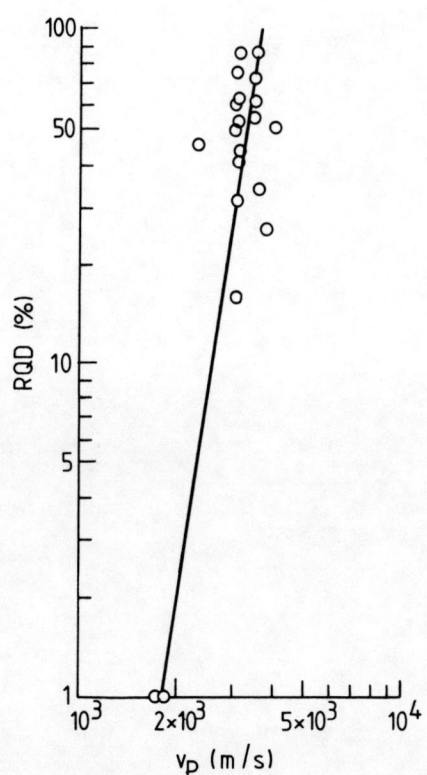

Fig. 7   Relation between RQD & v_p

6.   CONCLUSIONS

The in-situ rock tests being very costly and time consuming cannot be performed at many places and their correlation with seismic refraction data is of great help to designers for effective design of the structures eliminating many assumptions usually made.

7.   ACKNOWLEDGEMENTS

The authors express their sincere thanks to M/s GPR International, Canada for their guidance in carrying out the above work and to Messers R.Sundramurthy, V.U.Koundanaya, V.K.Malhotra, M.C.Meena, A.B. Thammaih, S.S.Singh, A.Chakravarty and Kishan Singh who as the members of the geophysical investigation team carried out the field work. The authors are also grateful to M/s National Hydroelectric Power Corporation, N.Delhi (India) who financed these investigations. The authors also express their thanks to Director, Central Soil & Materia's Research Station, New Delhi for his permission to present this paper.

8.   REFERENCES

Aalok, B.K.,et al.(1981). Progress Report No. 1 on the geological investigations of a hydro-electric power project in north-western India : Unpublished.

Dhawan,A.K. (1982). Correlation between seismic wave velocities and certain in-situ properties of subsurface strata - a field study: M. Tech. Thesis, Indian Institute of Technology, Delhi.

Hawkins, L.V.(1961). Reciprocal method of routine shallow seismic refraction investigations: Geophysics (26), 6, 806-819.

National Hydroelectric Power Corporation(India),(1982). Feasibility Report for a hydroelectric project in north western India (volume 1 to 5): (Unpublished).

Robertshaw, J., Brown, P.D.(1953). The in-situ measurement of young's modulus for rock by a dynamic method: Geotechnique, (3), 7, 283.

ACOUSTIC AND MECHANICAL PROPERTIES OF JOINTED ROCK
Le comportement séismique des roches solides fissurés
Akustische und mechanische Eigenschaften des Gebirges

Chikaosa Tanimoto and Kazuhiko Ikeda
Dr.Eng., Lecturer, Department of Civil Engineering, Kyoto University, Sakyo, Kyoto 606,
Japan
Dr.Eng., Chairman, Chishitsu-Keisoku Co., Tokyo, Japan

SYNOPSIS

The authors studied the relationships between wave velocities, fracture frequency, aperture, moisture content, contact pressure at joints and so on, underground at 104 tunneling sites, on several sites at the ground surface, and in laboratory tests. By considering 'fracture frequency' as the main parameter, the authors have established the practical relationship between seismic behavior and mechanical properties of rock mass. In general the velocity ratio k, defined by the ratio of P-wave velocity in-situ or through a cracked model to that of intact rock, and fracture frequency n, defined by the number of joints per meter, can be expressed by an equation $n = 5.0/k^2 - 4.0$. Also an aperture wider than 1 mm and a decrease of contact pressure to less than 3 MPa markedly influence the seismic behavior of the rock mass.

RESUME

Les auteurs ont étudié les rapports entre la vitesse des ondes, la fréquence des fissures, l'ouverture, l'humidité, la pression de contact aux joints, etc. à 104 sites souterrains de tunnels, sur plusieurs sites à la surface du sol, et par essais de laboratoire. En considérant la fréquence des fissures comme paramètre principal, les auteurs, ont établi le rapport de vitesse k, qui se définit par le rapport de vitesse des ondes-P "in-situ", ou à travers le modèle fissuré, à celui des roches intactes; la fréqunce des fissures n, qui se définit par la quantité des joints par mètre, peut s'exprimer par l'équation $n = 5.0/k^2 - 4.0$. En outre, si l'ouverture est plus large que 1 mm et si la pression de contact se réduit à moins de 3 MPa, le comportement séismique des masses de roches sera influencé d'une manière notable.

ZUSAMMENFASSUNG

Die Autoren haben den Zusammenhang zwischen Wellengeschwindigkeit, Klüftigkeit, Kluftöffnung, Wassergehalt, Normaldruck an Klüften etc., untertage an 104 Tunnelprojekten und mehreren Stellen an der Erdoberfläche, sowie an Versuchen im Laboratorium untersucht. Bei Betrachtung der Klüftigkeit als Hauptparameter wurde von den Autoren ein einfaches Verhalten zwischen akustischen und mechanischen Eigenschaften des Gebirges ermittelt. Das Geschwindigkeitsverhältnis k ist als Verhältnis der in-situ P-Welle für Klüftigkeit n definiert. Die Klüftigkeit, ausgedrückt als Anzahl von Klüften pro Meter, kann mit der folgenden Formel ermittelt werden: $n = 5.0/k^2 - 4.0$. Die durchgeführten Versuche zeigen Kluftöffnungen größer als 1.0 mm und Normaldruckänderungen kleiner als 3 MPa. Sie haben einen großen Einfluß auf das akustische Verhalten des Gebirges.

1. INTRODUCTION

Seismic surveys have been extensively developed in Japan and have become conventional in both investigations at the planning stage and field measurement/monitoring at the construction stage. Though the propagation velocity of a seismic wave is an overall index of the dynamic behavior of a rock mass, its relationship to the mechanical properties of rock including discontinuities has not yet been clarified.

The following parameters were chosen to describe the behavior of P-waves in jointed rocks: velocity (V_p), fracture frequency (n), aperture (b), amplitude (A), joint contact pressure (p_c), moisture content of the joint fill (w) and velocity ratio (k), defined by the ratio of P-wave velocity in-situ or through a cracked sample to that of intact rock.(Ikeda, 1979; Hata,Tanimoto et al,1980) They are linked to each other as shown in Fig. 1. Fracture frequency can be defined by number of joints per unit length (1/L), by total length of joints in unit area ($L/L^2=1/L$), or by total area of discontinuous planes per unit volume ($L^2/L^3 = 1/L$). In practice, the relationship between V_p and fracture frequency, n , can be determined with relative ease. Seismic surveys at the ground surface, in a tunnel, or in a borehole are fairly easy and convenient to perform and they supply global information about the rock mass properties. In comparison, it is difficult to carry out direct tests of the deformability and strength of the rock mass at typical construction sites.

Therefore, the objective of this study is to estimate mechanical properties of rock masses through seismic surveys without performing expensive and time-consuming field tests, which tend to be impractical especially in tunneling.

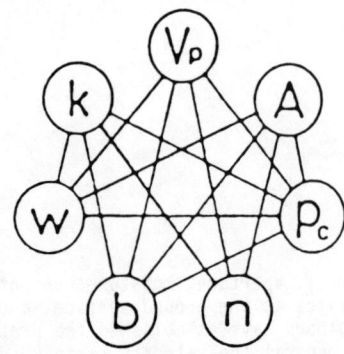

Fig. 2   Experimental Set-up

2.   EXPERIMENTAL METHOD

Cylindrical specimens of intact rock obtained from sites where seismic surveys were carried out and plaster cylinders of various lengths were used for laboratory tests. The samples were constructed by stacking individual cylinders of the rock or plaster to produce a "jointed" sample with joints normal to the axial direction. Filter paper was used to produce the desired joint aperture and moisture content. A typical experimental set-up is shown in Fig. 2. The required fracture frequency was obtained by changing the length of the individual cylinder comprising the test specimens and the axial load was used to control the contact pressure in the joints.

Seismic (or acoustic) events were generated by a piezoelectric transducer glued directly to the end of the sample. An identical transducer was used to pick up the signal at the opposite end of the specimen. As a result the induced waves propagated in the direction of the axial stress and normal to the joints. The frequency of the generating transducer was constant at 400 Hz.

The rock types used in the study were: rhyolite, sandstone, granite, grano-diorite, and tuff. The plaster was carefully mixed always with the same amount of water, at the same temperature, and was cured for the same length of time. The diameter of all specimens was 50 mm and their total, stacked, length was 100 mm. Individual cylinders with lengths of 100, 50, 33.3, 25, and 20 mm were used to produce joint frequencies of 0, 10, 20, 30, and 40 joints per meter. The

cylinders had a smooth finish with a tolerance of 0.01 mm at both ends. Apertures of the joints were measured by dial gauges with electric read-out to an accuracy of 0.002 mm.

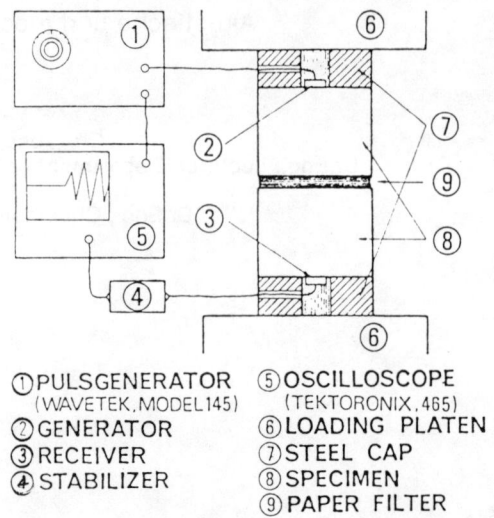

① PULSGENERATOR   ⑤ OSCILLOSCOPE
  (WAVETEK, MODEL 145)   (TEKTORONIX, 465)
② GENERATOR   ⑥ LOADING PLATEN
③ RECEIVER   ⑦ STEEL CAP
④ STABILIZER   ⑧ SPECIMEN
  ⑨ PAPER FILTER

Fig. 1   Parameters in Seismic Survey

3.   RELATIONSHIPS BETWEEN RESPECTIVE PARAMETERS

3.1   Fracture Frequency, n, Velocity Ratio, k, and Amplitude, A

In Fig. 3, the P-wave velocity and its peak amplitude are denoted by V_p and A, and, V_p and A_o for jointed and intact specimens respectively. These results show that fracture frequency does not significantly influence the P-wave velocity and the velocity ratio, k, is constant with n. Also, an aperture of less than 0.01 mm does not have an influence on the wave propagation even at contact pressure as low as 1 - 2 MPa. The amplitude, however, is very sensitive to the fracture frequency, and the amplitude ratio, A/A_o (which is 1.0 for n=0) remarkably decreases to 0.85 for n=10, 0.5 for n=20, and 0.45 for n=30. It means that amplitude should be measured in order to estimate a state of joints (including a magnitude of aperture) as well as velocity, which is taken solely into account in a conventional seismic survey.

3.2   Aperture, b, P-wave Velocity, Vp, and Contact Pressure, p_c

In practice, many joints may have aperture wider than 0.01 mm. When an aperture is wide enough, for example 1 mm, rock may make contact at only a few points in the joint and waves are forced to detour (longer path makes apparent reduction of propagating velocity, but actual velocity is not changed). In this situation an increase in fracture frequency correlates well with reduction of amplitude of received wave (attenuation of the wave). Thus, contact area is an important parameter. Also, in cases of highly fractured zones joints may contain infilling material such as clay which behaves in quite different a manner than the intact rock.

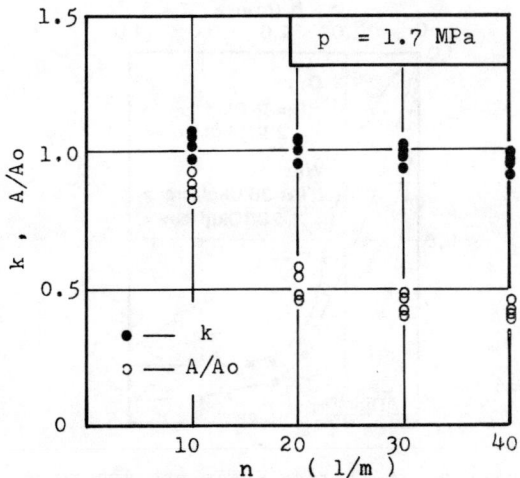

Fig. 3 Fracture Frequency (n), Velocity Ratio (k) and Amplitude Ratio (A/Ao) (Ao: A for intact rock.)

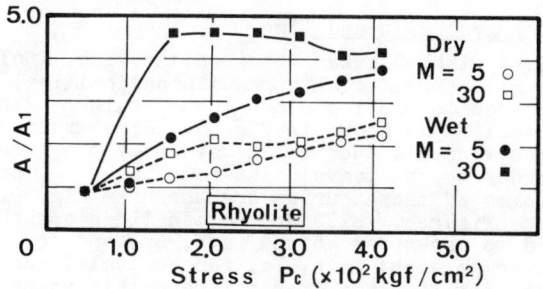

Fig. 5 Contact Pressure (p) and Amplitude Ratio (A/A) (A : A for p_=5 MPa)

In order to simulate this situation, we inserted sheets of filter-paper between stacked specimens which produced apertures in the range of 0.1 - 3.4 mm. Fig. 4 and 5 show the relationships between aperture, P-wave velocity, amplitude and contact pressure, where A_1 is a peak amplitude

under p_c = 5 MPa. When joint containing water is subject to compression, water content in the joint decreases as contact pressure becomes higher. Water contents for p_c = 5, 20, and 40 MPa were 48-55, 26-34, and 19-25 % respectively. In general, apparent velocity varies in wider range at a low stress level, and increases as contact pressure does. A water saturated joint gives higher velocity than a dry joint.

3.3 Velocity Ratio, k, Amplitude Ratio, A/Ao, and Aperture, b

Fig. 6 and 7 show the relationships between k and b, and between A/Ao and b. Since we found that Vp and A do not show any difference in the case of b<0.04 mm, it can be assumed that an aperture less than 0.04 mm is similar to intact rock in seismic behavior. According to the results such as Fig. 6 and 7, velocity ratio is inversely proportional to aperture in the range of 5 MPa < p_c < 25 MPa and remains constant in the range of p_c > 25 MPa. The lower the contact pressure, the more is the velocity ratio influenced by the magnitude of aperture. Also, an amplitude of propagating wave is remarkably sensitive to aperture, and the amplitude is observed to be less than half of that in intact rock in case of b > 0.5 mm.

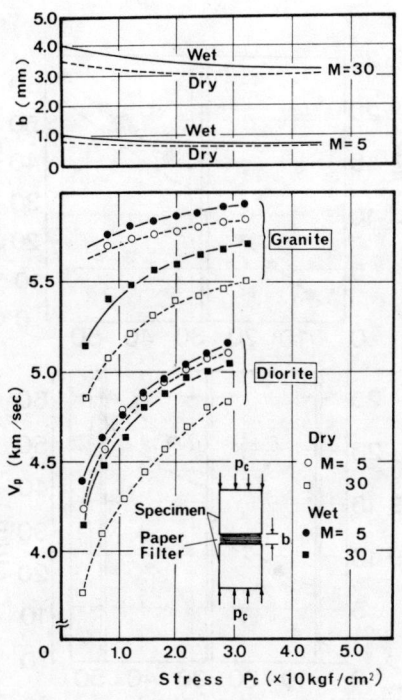

Fig. 4 P-wave Velocity (Vp), Aperture (b), and Joint Contact Pressure (p)

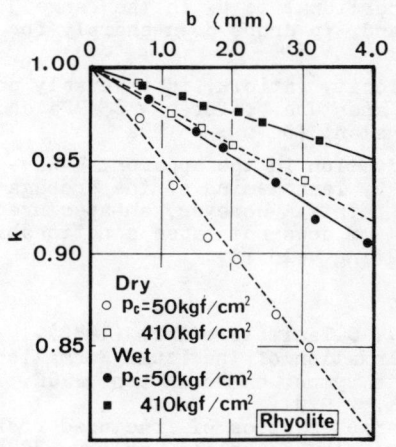

Fig. 6 Velocity Ratio and Aperture

3.4 Travel Path and Time

Several plaster cubes 50 mm on the side, including slits (artificial known discontinuities), were used to provide other than straight path propagation as shown in Fig. 8. Nine transducers were attached to the sides of the cubes in order to get travel time-path curves. Some examples of these curves are shown in Fig. 9. In the figures, solid and dotted lines correspond to cases for actual path, Lr, and for assumed straight path, La, respectively. Both cases show the same gradient, and this means that the apparent reduction of P-wave velocity comes from assuming discontinuous rock mass is a uniform body and only the wave path varies depending on characteristics of the joints. Therefore, when Vp and n can be observed in field, it becomes possible to estimate aperture, and/or change of contact pressure (stress redistribution in or thickness of relaxed/decomposed zone around excavated surface), and then not only joint (distribution) but also state of joint alteration can be evaluated.

4. CONCLUSIONS

Following conclusions were obtained from the laboratory tests:

1) The correlation between P-wave velocity,Vp, and fracture frequency,n, indicates that an aperture norrower than 0.04 mm does not influence the value of velocity even if fracture frequency varies, and fractures may not be detected under high pressure, e.g. it is difficult to find new extensions of fractures caused by high stress-concentration around an underground opening using a seismic survey. Since in most cases apertures are considered wider than 0.04 mm, a value of the velocity ratio suggests joint frequency and its change will provide some quantitative information on aperture and stress in rock mass if the joint frequency is known.

2) The attenuation of amplitude is very sensitive to contact area in a joint, and it is a measure of the magnitude of aperture or degree of weathering.

3) There is a clear correlation between the P-wave velocity,Vp, and joint contact pressure,p_c; Vp is proportional to p_c in the range 3 MPa < p_c < 20 MPa and, Vp drops down sharply for p_c < 3 MPa.

4) The velocity ratio,k, is inversely proportional to aperture,b, for p_c < 25 MPa and remains almost constant for p > 25 MPa.

5) The reduction in the apparent P-wave velocity is caused by lengthening of the propagation path caused by joints. However, an aperture narrower than 3 - 4 mm does not cause a noticeable change in propagating velocity.

References

Hata, S., C.Tanimoto et al. (1980).
    Relaxation of the Izumi Formation: Proc. 13th Sympo. on Rock Mech.,JSCE, 106-110.
Ikeda, K. (1979).
    Characteristics of fractured rock and strength, J.Applied Geology, 20-4,158-170.

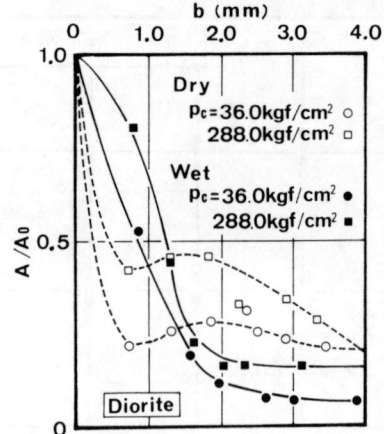

Fig. 7 Amplitude Ratio and Aperture

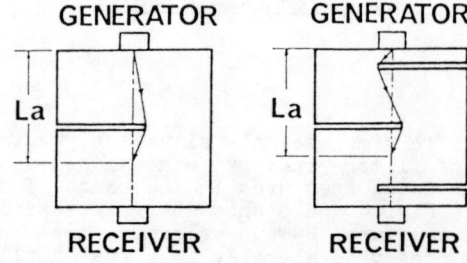

Fig. 8 Travel Path and Sample with Slits

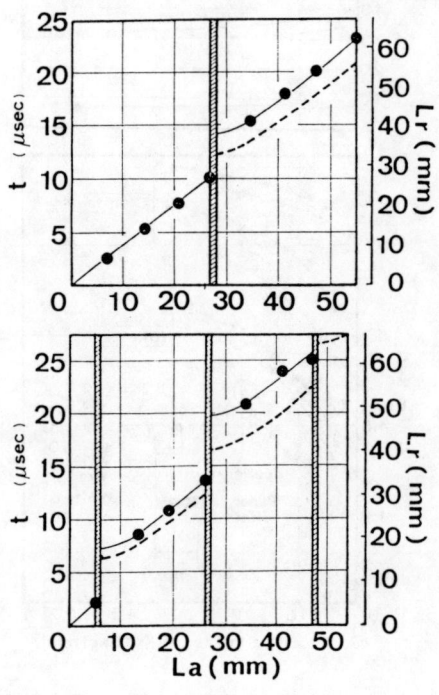

Fig. 9 Observed Travel Path and Time

GEOPHYSICAL STUDIES OF ROCK MASSES

Etudes géophysiques des massifs rocheux

Geophysikalische Untersuchungen von Fels

A. I. Savich
Dr.Sc.(Phys.-Math.), Chief of Geophysical Department

A. D. Mikhailov
Dipl.Eng., Chief Expert of Geophysical Department

V. I. Koptev
Cand.Sc.(Tech.), Chief Expert of Geophysical Department

M. M. Iljin
Cand.Sc. (Phys.-Math.), Senior Geophysicist of Geophysical Department
"Hydroproject" Institute, Moscow, USSR

SYNOPSIS

The paper presents the basic results and methods of geophysical studies of rock masses in the USSR. Three main objectives of studies are considered: studies of the structure, properties and the state of natural rock masses, studies of change of their properties as a result of mining and construction work and studies of the dynamics of deformation processes going on in the foundation rock of the structures in operation.

RESUME

Le rapport traite de principaux résultats et de la méthodologie d'études géophysiques des massifs rocheux en URSS, effec-tuées dans les trois directions principales: étude de la constitution, des propriétés et de l'état naturel des massifs rocheux, étude de modifications de ces propriétés au cours des travaux d'excavation et de construction, étude de la dyna-mique des déformations dans les fondations d'ouvrages au cours de l'exploitation.

ZUSAMMENFASSUNG

Im Beitrag werden Hauptergebnisse der geophysikalischen Untersuchungen sowie Felsuntersuchungsmethodik in der UdSSR be-handelt. Es sind drei Hauptrichtungen der Untersuchungen zu verzeichnen: Studien von Felsbeschaffenheit im natürlichen Zustand, Erfassung von Beschaffenheitsänderungen im Felsbau, Studien der Dynamik der Verformungsvorgänge in Gründungen von Bauwerken beim Betrieb.

1. INTRODUCTION

The designing, construction and operation of modern structures founded on rocks or cut in rocks foreordains the urgent problem of comprehensive studies of rock masses. Judging by the problems arising hereat and the methods of tackling the problems one may distinguish the following objectives:
- studies of the structure, properties and state of natural rock masses;
- assessment of changes in properties and state of rocks in the process of mining and construction work and of various kinds of treatment for improvement of rock masses in the zones of their interaction with structures;
- studies of deformation processes in rock masses and prediction of the resultant phenomena during the service life of the engineering structures.

The paper deals with the principles, methods and outcomes of the solution of some typical problems of the above studies, using geophysical and, mainly, seismoacoustic methods. The report is based on the **generalized** data of geophysical survey conducted in the areas of large hydroelectric projects in the USSR, considering the data of analogous studies obtained in other fields of the engineering geology, mining and exploration geophysics.

2. STUDIES OF THE STURCTURE, PROPERTIES AND STATE OF NATURAL MASSES

The typical problems of this trend, as well as the main geophysical methods, used to solve them, are summed up in Table I. As a rule, to solve these problems the methods of engineering geophysics are used in close relation with traditional methods of engineering survey: engineering-geological, hydrological and geotechnical survey methods. As the engineering geophysics developed and improved, its role in solution of the problems in question is steadily increasing, while the volumes of laborious and expensive traditional surveys are going down.

Physical prerequisites for solution of the problems listed in Table I by the geophysical methods are the theoretical and experimental relations between the parameters of different geophysical fields (elastic waves fields, electric and magnetic fields and other physical fields in rocks) and characteristics of

Table I. System of geophysical methods used in engineering-geological studies of rock masses.

Purpose of investigation	Methods and types of investigation									
	seismic methods	ultrasonic and seismic methods	electrical profiling and vertical electrical sounding	methods of induced polarization and self-potential method	magnetic searching and micro magnetic survey	standard electrical logging	resistivimetry and double-solution method	nuclear methods	tele-and-photo logging	seismometric methods
1. Investigation of rock mass structure.										
1.1 Lithological zonation	+	+	+	+	+	+	-	+	+	-
1.2 Singling out of fault zones	+	+	+	+	+	+	+	+	+	+
1.3 Estimation of thickness and nature of weathering zone	+	+	+	-	-	+	-	+	+	-
2. Determination of physical-and-mechanical properties and state of the rocks.										
2.1 Studying of jointing intensity	+	+	+	-	+	+	-	-	+	-
2.2 Assessment of elastic and deformability indices	+	+	+	-	-	+	-	-	-	-
2.3 Determination of strength parameters	+	+	-	-	-	+	-	-	-	-
2.4 Studying of permeability properties	-	+	-	-	-	+	+	+	-	-
2.5 Assessment of water saturation, water content and porosity	-	+	-	+	-	+	+	+	-	-
2.6 Assessment of rock preservation "in situ"	+	+	+	-	-	+	-	-	-	+
2.7 Assessment of anisotropy and heterogeneity	+	+	+	-	-	-	+	-	-	-
2.8 Studying of state of stress	+	+	-	-	-	-	-	-	-	-

the media under study (Goryainov, Lyakhovitski, 1979; Nikitin, 1981; Ogilvi, 1962; Savich et al., 1969). The solutions sought are mainly based on the absolute quantities of measured geophysical parameters (time of wave run, propagation velocities and indices of elastic waves attenuation, electric resistance values, magnetic susceptibility etc.) and also on the regularities of their distribution in space. To obtain the data rather complicated schemes of measurements are applied, the observations conducted on the suface of rock masses and in the interior points of the medium being closely combined (Goryainov, Lyakhovitski, 1979; Savich et al., 1969)For interpreting the obtained data the machine processing methods based on the theoretical solutions developed in the areal and research geophysics and also in engineering geophysics, are finding ever increasing use (Lyakhovitski, Napadenski, 1976; Pysetski et al., 1978).

2.1 Rock masses mapping.

To do this the most promising systems are the areal and volumetric observation systems when several geophysical methods characterizing the medium by a combination of various geophysical parameters are used at a time. This ensures not only a greater degree of reliability in distinguishing the rocks with different physical properties in the rock mass, but also increases the trustworthiness of the engineering and geological interpretation of the geophysical survey data.

As an example to the above statement, on Fig.1 is shown a geological-geophysical section obtained as a result of a complex of areal (over the mass surface) seismic prospecting and electrometric and magnetometer surveys conducted. The drilling that followed the surveys has proved that the errors in defining the main engineering-geological boundaries are not in excess of 5% at practically full agreement of the quantitative and qualitative characteristics of properties of rocks in the distinguished structural elements of the mass. These

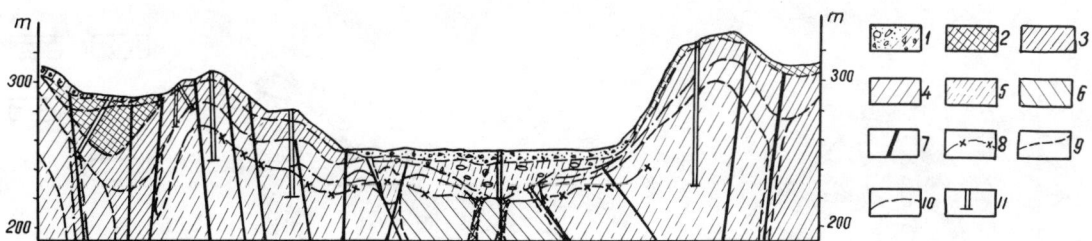

Fig.1 Geological-geophysical section at the dam site

    1 – talus-and-alluvial deposits; 2-6 – quartzites and quartzitic schists with interlayers of micaceous sandstones and clay-micaceous schists having respective values of U_p : 1.4-2.0; 2.0-2.8; 2.8-3.5; 3.5-4.7; 4.7-5.3 km/sec; 7 – zones of tectonic dislocations; 8 – footing of intensive weathering zone; 9 – lithologic boundaries; 10 – boundaries of rocks having various degrees of preservation; 11 – boreholes.

sections evidently contain a considerable amount of engineering-geological information and may serve a basis for elaboration of the respective engineering designs.

A more detailed differentiation of the rock mass and, especially, of its inner parts can be obtained using the volumetric observation systems and in particular, the multipoint sounding. The numerical methods of solution of primal and inverse problems of sounding (Yefimova, 1973; Karus et al., 1980), widely used in the USSR, make it possible to single out zones with anomalous values of geophysical parameters in the rock pillars between the mining excavations (Fig.2). The combination of these data

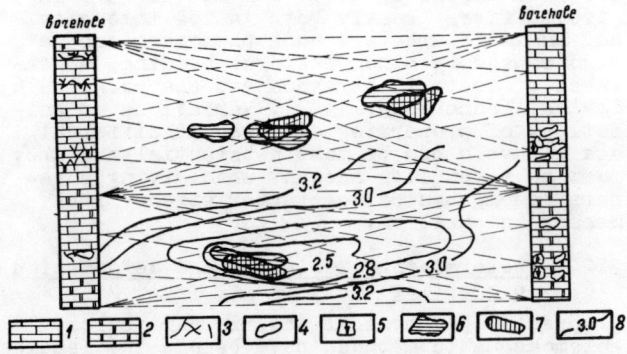

Fig.2 Results of acoustic sounding of carbonate rock between boreholes (acc. to Karus, 1980)

    1 – limestone; 2 – dolomite; 3 – fissures; 4 – inclusions; 5 – drop zone of drilling tools; 6 – zone of abnormal great intake values; 7 – zone of low velocities; 8 – isolines of velocities, km/s.

and the results of the profile geophysical measurements makes it possible to build free

volumetric engineering-geological models (see Fig.3). Such models practically give a comprehensive characteristic of the mass structure and predetermine the authenticity of the results of subsequent studies of properties and state of rocks in situ.

2.2 Studies of properties and state of rocks

The last years are marked by certain achievements in the use of geophysical methods for studying the physical and mechanical properties of rocks and state of rock masses. It was promoted by the development of the required equipment and methods of measuring of the same geophysical parameters of the medium at different scale levels: from samples of some cm³ in volume to large geostructural blocks of hundred meters and first kms in size. The mentioned possibility to use the engineering geophysics allowed one to realize the following principle of the technique of studying the rocks "in situ" conducting of measurements at different scale levels and an integrated interpretation of the data of different-scale measurements (Savich et al., 1969). Such studies resulted in more comprehensive analysis of the factors which govern the variability of physical-and-mechanical properties of rock masses and in the development of the methods of quantitative estimation of properties with the help of geophysical methods. Some of them are considered below.

2.2.1 Estimation of rock mass heterogeneity

Parallel with traditional statistical methods, the different-scale geophysical measurements allow using a radically new approach to the estimate of heterogeneity of rock masses, namely: to estimate the nature and degree of heterogeneity of the medium by the shape and slope of scale curves*) for the geophysical parameters to be measured.

An alternative of this method, where the elastic wave velocities are used as initial parameters, has been in application for a number of years in the USSR for estimation of heterogeneity of rock foundations of hydroelectric structures (Savich et al., 1969; Lykoshin, 1971). In accordance with the data obtained, rock masses which are composed of different types of rocks and located in various natural conditions, feature particularly specific laws of variation of mean values of velocities of elastic waves $\bar{V}$ depending on the scale of single

---

*)i.e. curves which represent the variation of mean (or most probable) values of the parameter in question against the given scale of measurements W_i .

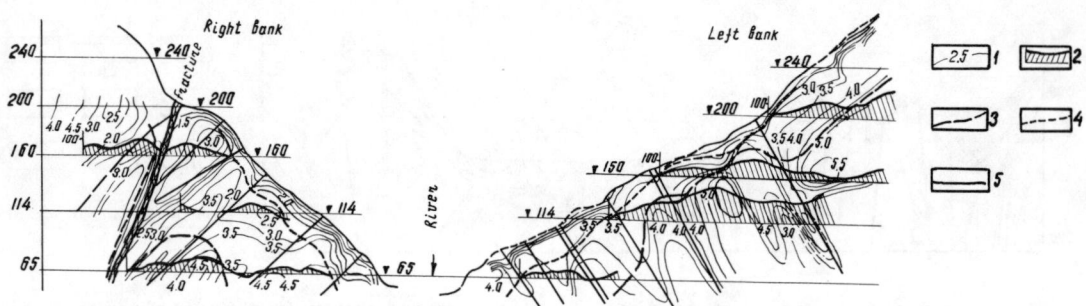

Fig.3 <u>Character of spatial distribution of</u>
<u>longitudinal waves velocities v_p and</u>
<u>stresses σ_y in rock mass at the Inguri</u>
<u>arch dam foundation (section along the</u>
<u>line of drainage galleries)</u>

    1 - isoline of velocity v_p km/sec; 2 -
σ_y curves in mining openings; 3 - isoline of $\sigma_y = 200 \cdot 10^5 N/m^2$; 4 - isoline
$\sigma_y = 100 \cdot 10^5 N/m^2$; 5 - large tectonic
fissure

measurements W_i , which is determined by the
approximate expression (1).
$$W \approx \pi \ell (\alpha \lambda)^2 \qquad (1)$$
where λ - length of elastic waves; $\lambda = v/f$,
  f - frequency of elastic vibrations; $\alpha \approx 0.25$;
  ℓ - base (interval) of velocity measuring.

A series of such   relationships is given in
Fig.4. It was found out that the slopes of

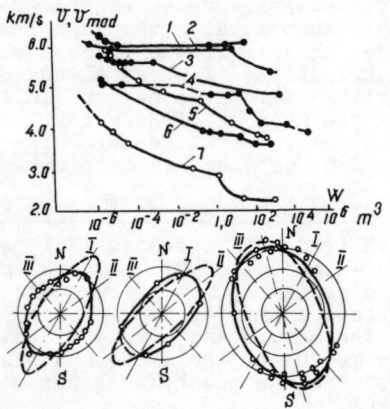

Fig. 4 <u>Alteration of average values (a) and</u>
<u>anisotropy (b) of elastic waves veloci-</u>
<u>ties depending on the scale of investi-</u>
<u>gations</u>

    1 - diabases; 2 - granites; 3 - porphy-
ric basalts; 4 - aleurolites; 5 -
gneisses, crystalline schists; 6 - limestones; 7 - limestones; Volumes 1-3
respectively 1; 10; $10^4 m^3$.

curves $\bar{v} = f(W_i)$ indicate the degree of heterogeneity of rock masses in question, while the
breakpoints show the prevailing degree of heterogeneities of different levels and orders
(Savich et al., 1969). Judging from the data
presented in Fig.4 different rock masses have
their own clearly defined heterogeneity charac-

teristics. The knowledge of these characteristics is necessary both for the choice of optimal technique of studying the properties of the
rock mass, (in order to choose the measurement
bases, frequencies of elastic vibrations etc),
and for the estimation of the scale effect
as applied to the moduli of elasticity and deformation, strength indices etc.

2.2.2 <u>Studies of anisotropy</u>

It is well known that anisotropy of rock masses is governed by various natural factors and
each factor has a tangible influence on the
media properties only at a certain scale level
(Lykoshin et al., 1970). For practical purposes it is important to have an estimate of
anisotropy of the mass at a level which corresponds to the scale of the structure under
designing or its separate elements. The problem can be solved by means of azimuthal measurements of geophysical parameters provided
there is a strong correlation between the technique and scale of studies. In Fig. 4 are
given examples of such measurements for one
and the same rock mass at three scale levels:
the anisotropy of the carbonate rock mass in
question, caused by the effect of diversely
oriented tectonic fissures of various order,
for rock blocks of about 1 m and 20-25 m in
size, differs greatly both in its intensity
and in its character. Such phenomena can be
also caused by other factors: bedding, appearance of tectonic stresses etc (Lavrova, 1976;
Savich, Yashchenko, 1979). Variations of anisotropy of properties of rocks at different
scale levels can be most substantial and they
must be taken into account when giving a general geomechanical characteristic of rock
masses in question.

2.2.3 <u>Determining of elastic and deformation</u>
<u>properties</u>

At present geophysical and, in particular,
seismoacoustic methods have become the leading
methods of quantitative studies of elastic and
deformation properties of rock masses (Goryainov, Lyakhovitski, 1979; Nikitin, 1981; Savich,
Yashchenko, 1979). The methods used in the
USSR to solve the problems are the following:
- numerous measurements of the geophysical parameters in different parts of the given mass;
- schematic zonation of the rock mass on the
basis of geophysical and engineering-geological studies;
- determination of reasons which govern the
variation of measured geophysical parameters
and singling out of the zones within which

the effect of one of factors is predominating;
- establishment of correlation relations between the geophysical parameters and measured static modulus of elasticity E_e or modulus of deformation D;
- conversion of geophysical parameters to moduli E_e and D, statistical processing of the data and estimation of the generalized parameters of deformability (Savich, Yashchenko, 1979).

The seismoacoustic methods are used most successfully. The generalization of the obtained results made it possible to elicit clear defined regularities in relations between the elastic waves velocities $v_p(v_s)$ and their usage for estimating the values of dynamic modulus of elasticity E_{dyn} and the sought values E_e and D. The interrelations between the given parameters are governed mainly by the type of rocks under study and by the conditions of estimation of the static moduli E_e and D_i : duration of application of static load δ, maximum value $\delta = \delta_{max}$, number of loading cycles (Savich, Yashchenko, 1979). Under fixed conditions of definition of static parameters the relations between moduli E_{dyn}, D and E_e are very authentically given by expressions (2) – (3) (Savich, 1979).

$$\lg E_e = A_E e^{-\alpha_E \delta_{max}} \lg E_{dyn} - B_E e^{-\beta_E \delta_{max}} \qquad (2)$$

$$\lg D = A_D e^{-\alpha_D \delta_{max}} \lg E_{dyn} - B_D e^{-\beta_D \delta_{max}} \qquad (3)$$

Values A, B, α and β for the rocks of the definite geological complexes are rather close. The statistical processing of the obtained data has indicated that rocks can be divided into four groups, each group having constant values of the above coefficients. The rock groups are the following: sedimentary carbonate rocks, sedimentary clastic and dust-like rocks, igneous and metamorphic rocks *). The values of the A, B, α and β coefficients for the mentioned groups according to Savich, (1979) are given in Table II. Using these data it is

Table II Summary table of coefficients for
equations 2 and 3 for various types
of rocks**)

Sl. Nos.	Rocks	static moduli E, D	A	α	B	β
1.	I.Carbonate rocks	D_Σ	0.970	0.0119	6.153	0.0110
2.		E_e	0.344	0.0176	2.425	0.0149
3.	II. Igneous rocks	D_Σ	0.944	0.0188	5.730	0.0148
4.		E_e	0.462	0.0260	2.775	0.0200
5.	III.Crystalline schists	D_Σ	0.728	0.0090	5.004	0.0076
6.	gneiss	E_e	0.216	0.0160	1.614	0.0100
7.	IV.Sandstone aleurolite tuff-	D_Σ	0.460	0.0140	3.000	0.0100
8.	breccia	E_e	0.180	0.0185	1.230	0.0100

*)This gradation coincides with division of rocks by their properties, adopted in engineering geology (Sergeiev, 1978).
**)The coefficients shown correspond to values D, E_e, E_{dyn} having the dimensions in 10^5 N/m^2.

very easy to obtain the relation between the values D , E_e and D for any type of rock. For instance, for carbonate rocks, the relation between the modulus E_{dyn} and summary modulus of deformation D_Σ , corresponding to the value $\delta_{max} \approx 8$ MPa*), is described by the following expression*)

$$\lg D_\Sigma = 0,970\, e^{-0,119 \cdot 8,0}\, \lg E_{dyn} - 6,153\, e^{-0,110 \cdot 8,0} = (4)$$

$$= 1,515\, \lg E_{dyn} - 3,427$$

The use of such correlating equations allows one to determine the sought values of the static parameters of deformability without conducting of labour-consuming and expensive geotechnical tests. With a probability 0.9 the error when calculating the moduli D and E is below 30 per cent, which is good enough for solution of most of practical problems.

2.2.4 Determining of permeability properties
To give a quantitative characteristic of the permeability properties of rocks different electric radioactive and other borehole investigations are successfully used (Dakhnov, 1975; Melkovitski et al., 1982; Ogilvie, 1962).

The defining of the direction and velocity of the seepage flow is based both on the data of the inter-hole studies and the data of observations carried out in one borehole. The inter-hole studies are carried on with the help of thermometric, calorimetric and, most often, electrometric and radioisotope methods. The last ones have been known for a long time, however, the modern measuring devices and the methods of pulse excitation of the borehole allow obtaining far more reliable data. When conducting studies in one borehole the tracer method (Feronski, 1977) and electrometric and flowmetering methods (Grinbaum, 1975) are used.

The main factor which complicates the interpretation of the geophysical data is a non-uniform and complex nature of the seepage flow conditioned by the rocks jointing. That is why, it is necessary to define the permeability properties for particular quasi-homogeneous rock mass blocks, singled out during the engineering-geological zonation.

2.2.5 Estimate of a degree of preservation
of rocks in the mass
In the practice of hydroelectric resources development in the USSR the qualitative classification of rock masses by their preservation has found wide application. The term "preservation" covers the degree of variability of the indices of the main physical and mechanical properties of the rocks, such as parameters of jointing, strength and deformability under the action of natural and technogeneous factors combination. Objective characteristic of the preservation degree can be obtained from the data of geophysical investigation and particularly on the basis of measuring the longitudinal wave velocities v_p. As the absolute v_p values are considerably dependent on the lithology of the rock masses under investigation, that is why the parameter $q = v_p / v_{p\,max}$ is used

*) The dimensions of D and E_{dyn} are 0.1 MPa.

as the preservation degree characteristic; here $v_{p\,max}$ is the velocity of longitudinal waves in non-weathered monolithic rock varieties. This parameter serves as the objective characteristic of the degree of preservation if conditions of the v_p and $v_{p\,max}$ determination are standardized. For this reason, on assessing the parameter "q", the value of v_p is determined usually as the mean value of the velocity from the seismic data, characterizing the generalized indices of the properties of the rock blocks with linear sizes n(10 to 100)m, the $v_{p\,max}$ being defined as the mean maximum value of the velocity for the given variety of the rocks by the results of ultrasonic logging. Further the parameter "q" is used for singling out zones and rock blocks considerably differing in the degree of their preservation.

To classify the rocks by the categories of their preservation, the generalized empirical relations between the velocities of longitudinal waves v_p and deformation modulus D (Savich Yashchenko, 1979), compressive strength $R_{compr.}$ in rock mass (Shaumyan, 1972), joint voids volume ratio η_{jv} and other indices are employed. Boundaries of the rocks of differing categories are found from condition that indices of the strength and deformability properties for two neighbouring categories differ from 2 to 2.5 times (see Fig.5). Table III shows the relative indices of the properties of the rocks of different categories. The set forth approach to the assessment of the rock mass preservation has been employed at the projects in the USSR and abroad.

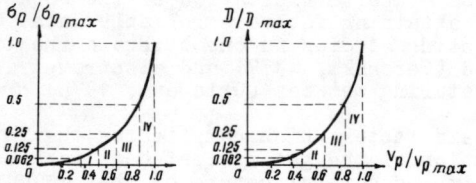

Fig.5 To estimation of limiting values of v_p for rocks of various degrees of preservation

a – "strength" criterion (the graph is taken from the paper by Shaumyan, 1972); b – deformation modulus criterion (acc. to the data taken from the paper by Savich et al., 1969); I-IV – degrees of preservation.

2.2.6 Study of rock jointing

Geophysical methods in engineering survey are used both when studying the general nature of rock mass jointing and when making the quantitative assessment of their jointing. Usually, the data of seismoacoustic methods and results of some types of logging are used for these purposes.

Relation between velocities of elastic waves and value and nature of joint void ratio for the rocks (Lykoshin et al., 1971; Salganik, 1973) serves as the physical basis for handling the mentioned problems. In the first approximation this relation can be described by

Table III. Relative indices of the properties of the rocks of different categories of preservation

Category of preservation	Ratio D/D_{max} %%	Ratio $R_c/R_{c\,max}$ %%	Ratio $q = v_p/v_{p\,max}$ %%	q^2 %%	Characteristics of rock condition
IV$_2$	100-72	100-72	100-97	100-95	Superior
IV$_1$	72-52	72-46	97-87	95-75	Good
III	52-25	46-19	87-65	75-44	Normal
II	25-11	19-8	65-50	44-25	Poor
I	11-4	8-24	50-35	25-12.5	Very poor
O$_2$	4-1.5	2.4-0.5	35-25	12.5-6.2	Semi rocks
O$_1$	1.5	0.5	25	6.2	Rock debris, alluvium, gruss

the equations similar in form to that of "mean time", if in this case the parameter "a_f" which means a certain effective velocity in void and joint filler is happily chosen:

$$\frac{1}{v} = \frac{\eta_{jv}}{a_f} + \frac{1 - \eta_{jv}}{v_{max}} \qquad (5)$$

Here v is the velocity of elastic waves in a jointed medium. v_{max} is the velocity of waves in monolithic rocks. All things being equal, the "a_f" value depends on the frequency of elastic vibrations used for determination of η_{jv} values, thus the problem of η_{jv} assessment is closely related to the problem of determining the parameter "a_f" for the various conditions of measuring velocities of the elastic waves. As a rule, the value of parameter "a_f" by seismic data is 1.3 to 2.0 times lower than those obtained by the ultrasonic investigation.

Criterion of correctness of assessing the joint void ratio is the direct correlation of the values η_{jv}, determined by geophysical methods, with the results of geological documentation. Example of such correlation is shown in Fig.6, where the values η_{jv} calculated by the data of ultrasonic logging are compared with the values of areal joint void ratio K, defined when examining the exploration pits & by visual inspection of the mining openings. Proceeding from these data, the value η_{jv} determined by geophysical investigation is 1.7 to 2 times more than the value K determined by geological tests. The latter finds its explanation in the fact that, firstly, the geophysical methods yield the value of joint voids volume, while the geologic methods allow, as a rule, for determination of the relative value of joints area; besides, the ultrasonic measurements take account of not only visible joints but also of microjoints which cannot be identified in boreholes visually. The diagram 2 shows the comparison of joints void volume data obtained

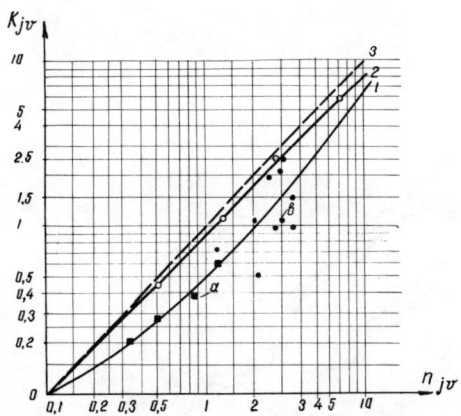

Fig.6 Correlation between joint voids acc. to
the data of geological survey K_{jv} (ratio
of joint voids) and acc. to seismic data
n_{jv}.

1 - ultrasonic logging, a - Ust-Ilim
hydropower station, b - Inguri hydropower station; 2 - seismic survey (Ust-
Ilim hydropower station); 3 - the line
of equal values of $n_{jv} = K_{jv}$.

in seismic surveys and by visual documentation
of boreholes. These data are evedencing that,
provided the initial parameters are selected
correctly, the ultrasonic methods and the logging give the values of the joints void volume
fully agreing with data of geologic documentation.

2.2.7 Estimation of rock mass state of stress

First applications of geophysical methods in
studying rock mass state of stress were made
in the USSR in the Institute of Geophysics,
Academy of Sciences,USSR, in the 60-s under
Riznichenko guidance (Riznichenko, 1956, 1967).
By the present time,enough experience of the
similar investigation has been gained and the
range of typical problems has been delineated.
The main problems are as follows:
1 - determination of dimensions and nature of
the relief zone and the stress concentration
around the underground excavations, river valley slopes, construction pits, etc.;
2 - estimation of tensor and orientation of
principal axis of stresses;
3 - determination of variability of behaviour
of stress fields of variable levels and orders
with time.

When handling the first problem, the nature of
variation of geophysical parameters within the
area of underground workings and slopes is
studied. As usual, elastic wave velocities or
electrical resistivities are used.

By variability of $v_p(v_s)$ and β_K values the
qualitative information on the structure of the
stress field around underground workings (Fig.7)
and in the slopes can be gained and zones of
stress concentration and relief can be singled
out (Lavrov, 1981; Riznichenko et al., 1956,
1967; Savich, Koptev, 1981).

A quantitative solution of the problem is based
on the relation between the elastic wave velo-

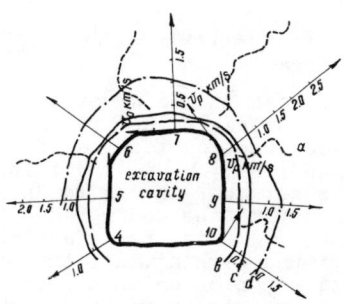

Fig.7 Alteration of the longitudinal waves velocities in the rock mass around excavation

1, 2, 3, ... 10 - shot holes;
a - graph of v_p alteration along the shot
hole; b - boundaries of a zone of intensively relieved rocks; c - boundaries of
a zone of weakened rocks; d - external
boundary of zone of increased stresses.

cities and pressure applied to the rock. The
relation is usually used as the "calibrating"
one for determination of the stresses by elastic wave velocities. The methods of stress
evaluation based on the similar relations are
termed "correlating" ones.

Relatively widespread is the method by which
the calibrating curves are obtained by mass
measurements on rock samples and then   used
for assessment of the stress values in any
point of the rock mass. Nevertheless, the calibrating curves obtained on the samples do
not represent the process of the natural rock
mass deformation development, either by the
scale,   or by the loading pattern. That is why
there is a trend to change sample testing for
field loading plate testing. In sufficiently
homogenous rocks utilization of this procedure
yields in satisfactory results (Miachkin, 1978;
Turchaninov, Panin, 1978).

The procedure of plotting the system of calibrating curves of the form $\Delta v/v_0 = f(v_0, 6)$,
where v_0 is the velocity under no-load condition, with heterogeneity of rocks and scale
effects taken into account, has been recently
developed in the "Hydroproject" Institute
(Moscow, USSR) on the basis of empirical and
experimental studies of velocity variation during the rock deformation process. The procedure and the results obtained are presented in
detail (Balavadze, 1981; Savich, Koptev, 1981).

Alongside with correlating methods, a method,
termed the "seismoacoustic version of stress
relief method" is widely used. The method is
based on the following main provisions: each
underground working is considered as an original field test of the rock mass relief. On determining the radial deformations ε_r developing in the case and with the E_{dyn} and D values
known one can assess the stresses existing in
the rock mass. On the basis of Khachikian's solution for the deformation of the walls of the
cylindrical underground workings the equation
was obtained which relates the indices of elastic and deformability properties of the rocks

around the underground workings to the radial
stress σ_r value:

$$\sigma_r = \frac{D_m \varepsilon_r h_0}{(1+\mu_m)(\tau_{ef}+h_0)} \left[1+(1+\mu_m)\frac{D_m}{D_0}\ln\left(1+\frac{h_0}{\tau_{ef}}\right)\right]^{-1} \quad (6)$$

where D_m and D_0 are the deformation modulus of
the rocks undisturbed in the process of driving
and the mean deformation modulus of the rocks
in the weakened zone; h_0 is the thickness of
the weakened zone; μ_m is the Poisson ratio
for the rocks beyond the underground working
influence zone; τ_{ef} is the effective radius
of the underground working. All the parameters
in this equation can be defined from the seismoacoustic survey data (Savich, Koptev, 1981).
Fig. 3 contains the data of determining the
horizontal component of stress σ_r (parallel
to the river valley slope) in the area of the
Inguri Dam using the above method (Balavadze
et al., 1981)

3. STUDIES OF CHANGES IN PROPERTIES AND STATE OF ROCKS DURING CONSTRUCTION AND MINING WORK

With the help of geophysical methods, and mainly with the help of seismoacoustic methods many
problems which arise in the process of construction are being successfully solved.The
following problems are the most widely encountered:
- study of rock properties and conditions in
the vicinity of mine excavations (tunnels,
shafts. underground machine rooms, etc.);
- estimation of overburden stripping work upon
the rock mass properties and study of the surface zone of the construction pits;
- control of quality and estimation of efficiency of grout injection and, in particular, of
consolidation grouting;
- estimation of variability of the physical
and mechanical properties of the rock mass
which is brought by mining and construction
work.

Let us have a look upon some examples which
confirm the efficiency of this trend   in
engineering geophysics.

3.1 Study of the surface overburden stripping zone in construction pits

Such study is carried out on the basis of detailed seismic researches at the network of
the surface profiles and observations effected
in construction excavations (Savich et al.,
1974). As an example, in Fig.8   are shown
seismic-geological sections at geophysical
profiles which have been used for drawing of
maps of isolines of depths of rocks having various elastic properties, as well as the maps
of rock properties below the weakening zone.
As a result of such investigations at the Inguri Dam site it was found that after execution of stripping work and removal of the
weathered layer in the foundation pit everywhere was formed the "secondary" weakening
zone, 2 to 15 m deep. This zone can be divided
into some subzones, and the rocks of those subzones significantly differ in their properties.
The structure and thickness of the secondary
weakening zone is being governed by the rock
properties in the unchanged part of the rock
mass, as well as by   gypsometric position of
the area under investigation and by   the

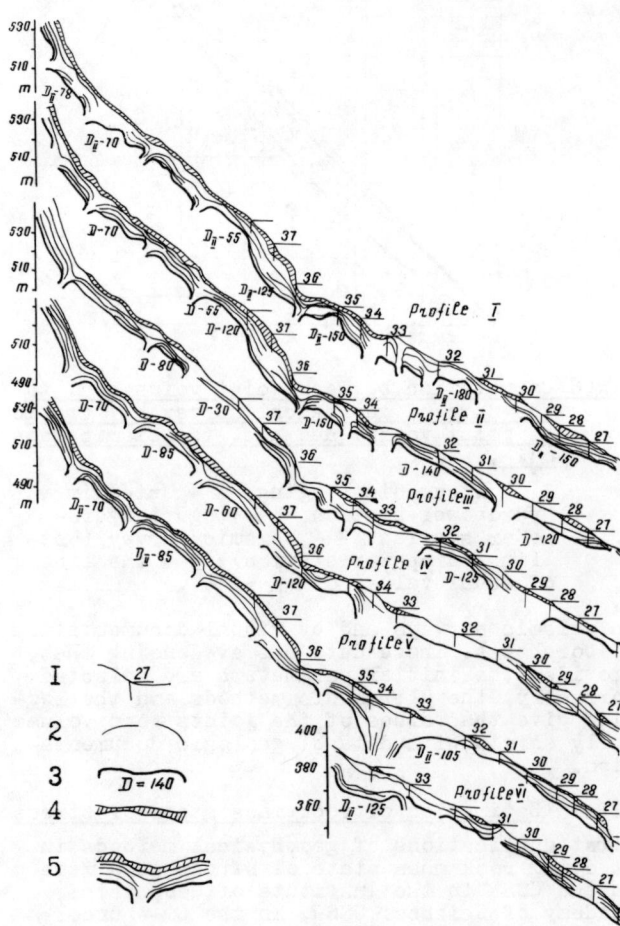

Fig.8 Seismic-geological sections of the foundation pit of the Inguri arch dam

1 - earth surface, section boundary and
its No.; 2 - isolines of longitudinal
waves velocities; 3 - boundary of rocks
unchanged by the relief and values of
deformation moduli $D \cdot 10^5 \text{N/m}^2$ and the
velocities of longitudinal waves
v_p km/s ; 4 - zone of rocks broken by
explosions, with velocities of longitudinal waves not less than 1,0 km/s .

steepness of the slope.

Similar work executed in the foundation pit of
the Ust-Ilim hydropower station have allowed
to find with great precision and details the
thickness of the zone of rock to be removed
(Mikhailov, et al., 1977). The control drilling and the recording documentation on the
excavation have confirmed the results of the
geophysical researches (Fig.9).

The geophysical work carried on in the construction pits of the structures enabled to
give the detailed properties of rocks in the
top part of the section, which allows to spe-
cify the quantity of rock to be removed, to
localize the most weak areas of the rock mass,

and finally, to make the rock foundation preparation work cheaper.

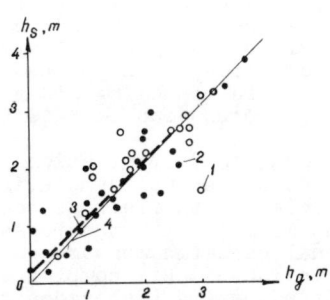

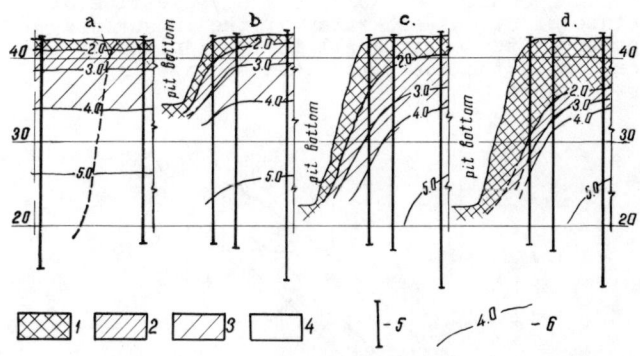

Fig.9 <u>Comparison of data from seismic (h_S)
and geological (h_g) methods on the
depths of stripping in the pit of the
Ust-Ilim hydropower station</u>

1 - boreholes-pits (d=915 mm); 2 - core
drilling boreholes; 3 - empiric dependence between h_S and h_g ; 4 - line of
equal values of h_S and h_g .

3.2 <u>Study of slope relief zone</u>

At the slopes of construction pits, at the slopes of deep navigation locks excavations and
of water passage structures the said geophysical methods are being successfully used for
controlling the dynamics of relief process, as
well as for estimation of the slopes stability.
The researches which have been carried out in
the foundation pit of the Dnieper navigation
lock can serve an example of such work (Andreiev, Lavrov, 1976), excavated in granite. Under
the combined influence of explosions and stress
relief in the pit slopes, zones of technogeneous
weakening were formed, characterized with various velocities of longitudinal waves υ_p and
with indices of relative preservation q .
(Fig.10) With the help of repeated observations
the dynamics of the technogeneous relief in
time has been studied. It turned out that during three years after completion of excavation
the velocities of longitudinal waves in the
rocks of the relieved slope gradually decreased.
Total reduction of velocities in the relieved
rock was equal to 200 to 300 per cent, the deformation modulus - by 4 to 6 times, and the
ratio of jointing voids by 10 to 20 times.

3.3 <u>Grouting quality control</u>

The geophysical methods are being used as well
also for quality control of consolidation and
anti-seepage grouting of rock foundations. When
doing so, as a rule, the consolidation grouting
is controlled with seismic-acoustic methods,
while the anti-seepage grouting is controlled
with electrical logging (Savich et al., 1979).

Efficiency of grouting is estimated from the
relative alteration of velocities of the longitudinal waves or of the specific electrical resistances. Usually, the following problems are
to be solved: to find the possible efficiency
of grouting in the given type of rocks and at
the specified technology of grouting; to set
up the grouting quality degrees; to locate the
zones and areas of poor quality grouting; gene-

Fig.10 <u>Generalized seismologic-geological cross-
sections across the construction pit
slopes for the cases of the pit still
unexcavated (a), for the depth of the
pit 7 m(b) and 22 m(c,d). Time interval
between cases of c and d is one year
(acc. to the paper by Andreiev and Lavrova, 1976</u>

1 - intensively relieved rock mass,
η-0.8; 2 - strongly relieved rock mass,
η = 0.8-0.5; 3 - relieved rock mass,
η = 0.5-0.3; 4 - poorly relieved rock
mass, η =0.3-0.1; 5 - borehole;
6 - isoline of velocity, km/s .

ral and differented estimation of grouting
efficiency.

The first out of the above mentioned problems
is solved in the process of grouting at the experimental grounds, and as a result of this
curves are drawn interrelating the quality of
grouting and the values of corresponding changes of geophysical parameters. The graphs of
relative change of the longitudinal waves velocities which are used for grouting quality estimation at the foundation of the Inguri arch
dam (Fig.11) can serve an example of this. Estimation of grouting quality is effected on the
basis of repeated measurements of velocities
of the longitudinal waves, prior and after
grouting, with the help of seismic sounding of
the rock between the boreholes.

Control over consolidation grouting quality is
also effected behind the lining of pressure
tunnels with the help of the geophysical methods. Work in the Inguri power station power
tunnel (Lavrov, 1981) can serve an example of
such researches where the control was effected
with radial boreholes which have been drilled
prior and after grouting.

4. INVESTIGATION OF ALTERATION OF ELASTIC WAVES
VELOCITIES IN ROCKS AND DEVELOPMENT OF DE-
FORMATION PROCESS IN TIME

This problem has come into being in connection
with construction of big engineering structures and, mainly, high dams with deep and vast
storage reservoirs which brought with themselves
active influence of man upon properties and
conditions of environment. The cases of large
landslides, downfalls, significant deformations
of the foundations and powerful local earthquakes which have been provoked by technoge-

neous factors predetermined the arrangement of special complex of work to study dynamics of state of stress-and-strain of the subsurface parts of the earth shell in the regions of location of the most important engineering structures.

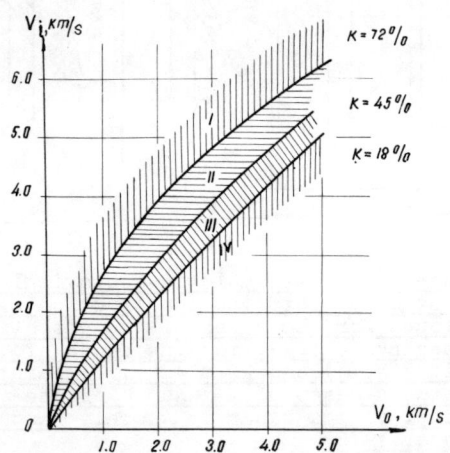

Fig.11 Nomograph for estimation of grouting quality according to the velocity of longitudinal waves measured prior to grouting (V_0) and after grouting (V_1)

I - "excellent" grouting zone; II - "good" grouting zone; III - "satisfactory" grouting zone; IV - "unsatisfactory" grouting zone.

A component part of this complex is longterm geophysical observations which allow on the basis of study of the alteration of the parameters of various geophysical fields with time to get some certain information on deformation processes which take place at different parts of the rock mass under study. The procedure of longterm geophysical investigations employed in the USSR provides for their use in combination with the traditional (geotechnical and geodetic) field studies. These studies are being effected in their full scope only in some individual (reference) parts of the rock mass and serve a basis for correct and reliable quantitative interpretation of geophysical data. The task of geophysical methods is to get information on the deformation processes which develop in time at various scales and in different (at the surface and inside) parts of the rock mass.

The most complete complex of alike longterm observations is being made in the USSR at present in the region of the Inguri arch dam, 271.5 m high, and deep Inguri storage reservoir which is located in Western Georgia in the zone of 8-points seismicity (Balavadze et al., 1981).

This complex of work consists of the following:
1 - detailed seismic, electromagnetic and ultrasonic investigations at the different points of the arch dam foundation in the range of depths from 0 to 200-250 m, where the processes which take place in the structural blocks, 1.0-10.0 m³ to 1000-10,000 m³ in volume are studied;
2 - seismic studies at the submerged profile in

the head part of the reservoir where the processes which take place in geostructural blocks having line sizes n·100 m, at the depths down to 1.0 km, are studied;
3 - regional geophysical (electrometric, gravimetric and magnetic-metering) as well as the instrumental seismological observations which render information on the processes which take place in the blocks having linear sizes of n·km at the depths down to 5-10 km.

The above mentioned observations are combined with high accurate inclinometering and deformometric studies, with field geotechnical observations as well as with the repeated detailed and regional geodetic measurements. The detailed description of this complex can be found in the published papers (Balavadze et al., 1981; Savich et al., 1979). The observations over changes of parameters of the geophysical fields in the region of the Inguri power station are being made during 6-8 years, and thanks to this work the most general regularities of deformation processes development were revealed.

The general trends of alteration of the media properties during construction of the dam prior to filling the reservoir are distinctly shown in the curves of the variation of the longitudinal waves velocities U_p against electric resistances ρ (Fig.12) These data confirm that placing concrete in the dam body and creation of a surplus surcharge on the foundation have brought with themselves a significant growth of U_p and ρ values, and at the background of this growth there have taken place some relative variations of those parameters, connected with alterations of season temperatures, water content in the rock mass and with some occasional factors.

Judging by the results of the different-scale investigations the mentioned deformation processes in the larger blocks occur with a greater intensity. It means that the compaction of the rock mass which brings with itself the growth of U_p and ρ values happens, mainly, thanks to closure of large fissures. It is notable that, depending of the depth, the variations of U_p and ρ values attenuate, and deeper than 80 m those variations are practically not found. But, as far as those depths coincide with the boundary of significant settlement of rock, the above mentioned boundary can be interpreted as the boundary of zone which is an active deformed zone under the structure.

Filling of the reservoir has brought with itself a sharp change in the nature of deformation processes and the corresponding alterations in parameters of the various geophysical fields. In the area of the arch dam it has caused a reduction of the elastic waves (longitudinal and transverse) velocities as well as a reduction of electrical resistances (Fig.13). The most intensive reduction of the U_p and ρ values took place in the periods of water level rise in the reservoir, and it was a little bit weaker during the periods when the water level was being kept constant. The analysis has shown that this phenomenon has been caused by entering of water inside the rock mass and by development of pressure inside the fissures. The last phenomenon, in particular, is con-

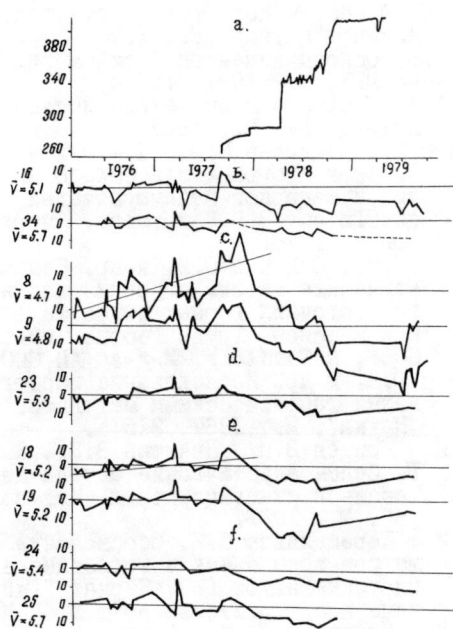

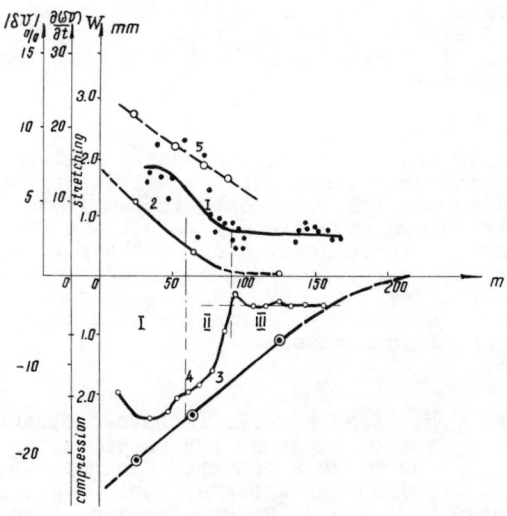

Fig.12 Variations of relative velocities of
elastic waves in seismic range of frequencies v_s at the intermidiate zone of
the middle part of the left-bank slope
in different directions in respect to
the Inguri river valley

a - reservoir filling graph; b - variation of velocities along OX axis;
c - variation of velocities along OY
axis; d - variation of velocities along
OZ axis; c - variation of velocities
parallel to the slope; f - alteration
of velocities normally to the slope;
16, 8,25 - Nos of observation points;
$\bar{v}$ = 5120 - average values of zero
cycle velocities.

Fig.13 Alteration of variations intensity for
velocities of seismic waves and rocks
deformation in the middle part of the
left-bank abutment of the arch dam depending on the depth from the earth
surface in the rock mass

1 - I - /δv/ = f(h)/δv/ = $\frac{v_i - v_0}{v_0}$, where v_0 is
velocity v_p in the cycles which took
place prior to filling the water sto-
rage reservoir; 2,3 - $\partial(\delta v)/\partial t = f$(h),
where $\partial(\delta v)/\partial t$ is a temporary gradient of value δv in (%) (year)$^{-1}$;
data prior to (2) and (3) in the period
of reservoir filling; 4,5 - graphs of
deformations in the rock mass under investigation prior to (4) and in the pe-
rod of (5) filling of reservoir (acc.
to data provided by Kouznetsov); h -
depth from the earth surface along the
normal to the slope; I-III - zones of
various intensity of v_p variations.

firmed by the shift in time of the reduction
of the value v_p in different parts of the rock
mass, this shift being governed by water filtration velocities inside the rocks under investigation.

The observed reduction of the velocities v_p
and of electric resistances is accompanied with
a sharp reduction of the foundation settlement;
this phenomenon is interrelated with the process of a relative de-consolidation of rock
mass which is caused by the hydrostatic pressure of water which has entered the rock mass.
The boundary of the zone, where this process
takes place under the dam, is at a depth of
some 250 m, while at the head part of the reservoir it is at the depth of some 500 m.

The influence of the water storage reservoir
upon the deformation processes has manifested
itself distinctly in the peculiarities of alteration of the local seismicity of the region
(Balavadze et al., 1981). The variations of the
regional seismicity which accompanied the filling of the Inguri power station water storage
reservoir, have reached their top one and a
half year after beginning of water level rise,

and one year since the relative stabilization
of the water level at 130 m. However, the
first (weak enough) stage of seismic activity
began one month since the said filling of the
reservoir. It has manifested itself in a series of seismic shocks with the magnitude
M = 0.5 to 1.7 under the head part of the reservoir. Later on the foci of earthquakes moved
to south-west along tectonic structures which
separate the water-bearing limestones of the
lower body from the watertight strata of Jurassic deposits. This process was finalized with
a powerful outbreak of seismic activity. Within
the limited area at a distance of 10 to 30 km
south-west from the dam, in a period of time
since December of 1979 to March of 1980 there
have happened more than 100 earthquakes with a
magnitude M~1.7-4.3, and the most powerful of
them have caused local seismic effects of 6-7
points . It was found that one of the main factors which determines the said seismic activity within the region of the Inguri hydropower
station is the inside pressure in joints. The
above mentioned migration of the earthquake
foci is a good illustration for spreading of
water along the fissures to the periphery of
the water reservoir.

Thus, widening of the zone of influence of the reservoir manifests itself in the variations of different geophysical fields, and it is characterized at various scale levels with some peculiarities.

CONCLUSION

The paper deals with some standard engineering-geological problems which are solved in the USSR with the help of geophysical methods. The examples shown here clearly illustrate the efficiency of those geophysical methods in solving various problems of geology and rock mechanics, as well as advantages of further development of the engineering geophysics for studying of rock masses.

REFERENCES

Андреев В.Н., Лаврова Л.Д. Геолого-геофизическое определение зон выветривания и разгрузки в гранитах Днепрогэс-П. Труды Гидропроекта, вып.50, М., 1976.

Балавадзе Б.К./ред./ Геолого-геофизические исследования в районе Ингурской ГЭС. Труды I-го координационного совещания, май 1979 г. Изд-во "Мецниереба", Тбилиси, 1981.

Горяинов Н.Н., Ляховицкий Ф.М. Сейсмические методы в инженерной геологии. Изд-во "Недра", М., 1979.

Гринбаум И.И. Геофизические методы определения фильтрационных свойств горных пород. "Недра", М., 1975, 271 с.

Дахнов В.Н. Геофизические методы определения комплекторских свойств и нефтегазоносности горных пород. "Недра", М., 1975.

Ефимова Е.А. Решение прямой задачи сейсмического просвечивания численными методами. Вестник МГУ, вып.IУ, № 5, 1973.

Карус Е.В. /ред/ Методические указания по проведению межскважинного прозвучивания и интерпретации его результатов при решении инженерно-геологических задач. М., ВНИИЯГ, 1980, 58 с.

Лаврова Л.Д. Изучение анизотропии скального массива сейсмоакустическими методами. Изв.АН СССР, сер "Физика Земли", № II, М., 1976.

Лавров В.Е. Применение сейсмоакустических методов для исследований в гидротехнических тоннелях в процессе строительства. Труды Гидропроекта, вып.78, М., 1981.

Лыкошин А.Г. /ред/ Труды Гидропроекта, сб.21, Инженерная геофизика, М., 1971.

Ляховицкий Ф.М. Нападенский Г.Б. Опыт автоматической обработки данных малоглубинной сейсморазведки МПВ. В кн.: "Разведочная геофизика", М., "Недра", 1976, вып. 70.

Мельковицкий И.М., Ряполова В.А., Хордикайнен М.А. Методика геофизических исследований при поисках и разведке месторождений пресных вод. "Недра", М., 1982, 102-145 с.

Михайлов А.Д., Ященко З.Г., Даниленко С.Д. Изменения под нагрузкой состояния скального массива в основании плотин Братской и Усть-Илимской ГЭС по данным геофизических исследований. Сб. "Отражение современных полей напряжений и свойств пород в состоянии скальных массивов", Апатиты, 1977.

Мячкин В.И. Процессы подготовки землетрясений. Изд-во "Наука", М., 1978.

Никитин В.Н. Основы инженерной сейсмики. Изд-во МГУ, М., 1981, 176 с.

Огильви А.А. Геофизические методы исследований. Изд-во МГУ, М., 1962.

Писецкий В.Б., Бондарев В.Н., Крылатков С.М. Система программ для обработки данных инженерной сейсморазведки. Труды СГИ им. В.В.Вахрушева, Свердловск, 1978.

Ризниченко Ю.В., Силаева С.И. и др. Сейсмоакустические методы изучения напряженного состояния горных пород на образцах и в массиве. Труды Геофиз. ин-та АН СССР, № 34 (161), Изд-во АН СССР, 1956.

Ризниченко Ю.В. и др. Исследование горного давления геофизическими методами. Изд-во "Наука", М., 1967, 215 с.

Савич А.И., Коптев В.И., Никитин В.Н., Ященко З.Г. Сейсмоакустические методы изучения массивов скальных пород. Изд-во "Недра", М., 1969.

Савич А.И., Кереселидзе С.Б. Обоснование параметров зоны съема в котловане арочной плотины Ингури ГЭС. Журнал "Гидротехническое строительство", № 6, М., 1974.

Савич А.И., Ященко З.Г. Исследование упругих и деформационных свойств горных пород сейсмоакустическими методами. Изд-во "Недра", М., 1979.

Савич А.И. Коптев В.И. Изучение напряженного состояния массивов скальных пород сейсмоакустическими методами в связи со строительством подземных гидротехнических сооружений. Сб. Труды Гидропроекта, вып. 78, 1981, 47-65 с.

Савич А.И., Коптев В.И., Михайлов А.Д. Применение геофизических методов для изучения свойств и состояния массивов горных пород. "Сб. научных трудов Гидропроекта", вып. 76, М., 1981.

Савич А.И., Ященко З.Г., Горбунов А.А. Опыт оценки качества укрепительной цементации скальных пород сейсмографическими методами на Ингури ГЭС. Гидротехническое строительство, 1977, № 9.

Салганик Р.Л. Механика тел с большим числом трещин. Изв. АН СССР, МТТ, 1973, № 4.

Сергеев Е.М. Инженерная геология. Изд-во МГУ, М., 1978.

Турчанинов И.А., Панин В.И. Инженерные геофизические методы определения и контроля напряженно-деформированного состояния массивов пород. Изд-во "Наука", Л., 1975, 112 с. с ил.

Феронский В.Н. Радиоизотопные методы исследования в инженерной геологии и гидрогеологии. Атомиздат, 1977, 130-291 с.

Шаумян Л.В. Физико-механические свойства массивов скальных горных пород. Изд-во "Наука", М., 1972.

Lykoshin A.G. et al. Studies of Properties and Conditions of Rock Massifs by Seismic-Acoustic Methods. R. on the Second Congress of the ISRM, Beograd, 1970.

Savich A.I., Koptev V.I., Iljin M.M, Zamakhaev A.M. Lipskaya A.E. Geophysical methods for study of deformation processes in foundations of large hydraulic structures and storage reservoirs. Bull. of the Int. Ass. of Engineering Geology, №20, 1979, pp. 58-61

GEODYNAMICS — A VALUABLE TOOL FOR ROCK QUALITY TESTING

La géodynamique — Une méthode précieuse pour l'évaluation des propriétés
de roches

Geodynamik — Eine wertvolle Methode zur Bestimmung der Felsgüte

T. L. By

Research Engineer, Norwegian Institute of Rock Blasting Techniques, Oslo, Norway

SYNOPSIS

Rock quality evaluation by means of geodynamic methods has already for many years been used for pre-investigation purposes. The sound velocity has been the only quality parameter. It is possible by means of appropriate sensors for borehole installation, to give a fair in-situ rock quality estimation. The presupposition is that one controls an analysing method that reflects the rock mass conditions. In such a method the sound velocity is only one of several parameters. A mathematical model for stress wave propagation in a medium built up from elastic shells or layers is developed. For verification of the model's validity in rocks, a field experimental program is accomplished.

RESUME

L'évaluation des propriétés de roches par la méthode géodynamique se pratique depuis de nombreuses années déjà, mais le seul paramètre qualitatif en a été la vitesse du son. On peut effectuer une assez bonne in situ évaluation des propriétés de roches en se servant de sondes adéquates — mais cela suppose qu'on sache la bonne méthode d'analyse des données. Dans une telle méthode la vitesse du son n'est qu'un paramètre parmi plusieurs. L'auteur a développé un modèle mathématique de la diffusion des ondes de contrainte dans un milieu artificiel composé de couches ou des strates élastiques. Afin de vérifier la précision du modèle on a effectué un programme d'essais sur le terrain.

ZUSAMMENFASSUNG

Vorbestimmung der Stabilität/Klüftigkeit einer Fels-Partie mit Hilfe geodynamischer Voruntersuchungsmethoden ist seit Jahren in Gebrauch. Bei den bisher verwendeten Methoden ist die Schallgeschwindigkeit der einzige nutzbare Parameter. Mit Hilfe geeigneter Sonden, installiert in Felsbohrlöchern, ist es möglich, aus den aufgezeichneten Daten eine reelle Qualitätsbestimmung der untersuchten Felspartie zu geben. Dies unter der Voraussetzung, daß die Ergebnisse der Untersuchungsmethode die Felsbedingungen widerspiegelt. In einer solchen Analysemethode sind Daten wie Maximal-Amplitude, Steigegeschwindigkeit der Bodenwellen, Pulsdauer und Frequenz wichtige Parameter. Ein mathematisches Modell für die Ausbreitung der Spannungswellen in einem elastischen, geschichteten Medium ist in dieser Arbeit aufgestellt worden. Die Theorie ist durch praktische Versuchssprengungen im Fels mit Aufzeichnung der Meßdaten und Messungen geprüft worden.

1. INTRODUCTORY REMARKS

Rock quality evaluation by means of geodynamic methods has for several years been used as a valuable contribution in geological preinvestigations. A more or less reliable picture of the signal velocity in the rock has been registrated. From this velocity one has made conclusions concerning nature of rock, ore occurence, rock joints, joint spacing, faults and material properties.

From one single parameter - the signal velocity - one has obtained series of information about the rock mass. But can we consider the signal velocity to be a characteristic parameter for the rock mass?

Stress wave registrations in rocks gives more information than the travel time alone. Attenuation characteristics, rising time, pulse duration, pulse form and frequency response can be correlated with mechanical properties and fracture frequency, fig. 1.

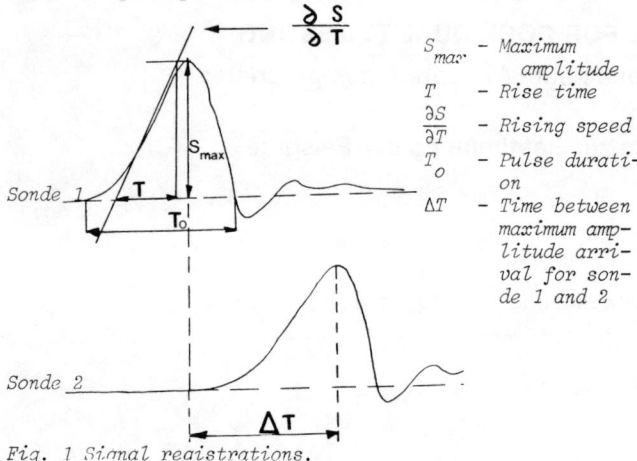

S_{max} — Maximum amplitude
T — Rise time
$\frac{\partial S}{\partial T}$ — Rising speed
T_o — Pulse duration
ΔT — Time between maximum amplitude arrival for sonde 1 and 2

Fig. 1 Signal registrations.

The information collected we can call "the response function" of the rock mass.

Development of a more effective borehole seismic will mean a great progress in rock quality evaluation. The method gives us more reliable information about rock parameters and properties than other methods now in use.

2. MATERIAL MODEL

It is no commonplace matter to find the connection between the registrated signals and rock quality. To give a reliable statement about the rock mass, and even yield a quality evaluation, one has to assume that *the correct* analysis method is ruled. Consequently one has to develop a material model which reflects the in-situ conditions satisfactory.

Previous, Prof. Leif N. Persen at the University of Trondheim, The Norwegian Technical Highschool, has described stress wave propagation in rocks from the assumption that a viscoelastic model can be used, Persen, 1974. This is proved to be a suitable model for describing wave propagation in homogenous rocks in the area near a detonating charge, that is in the area where blasting vibrations can cause damages on structures.

This article shal deal with another material model. The intention is eventually to be able to characterize the rock mass according to a normal cracking index, which can be used for practical purposes. A model for such an index will be introduced and explained.

Rocks in laboratory scale shows elastic properties at stress loads. Therefor one choosed to build up a model as an elastic, layered medium, fig. 2.

The theory discussing propagation of spherical waves in this medium, leads to attenuation curves, represented as arched lines in double logarithmic scale, fig. 3. The exact numerical solution for a rectangular input pulse, spherical waves, travelling through a medium built up from

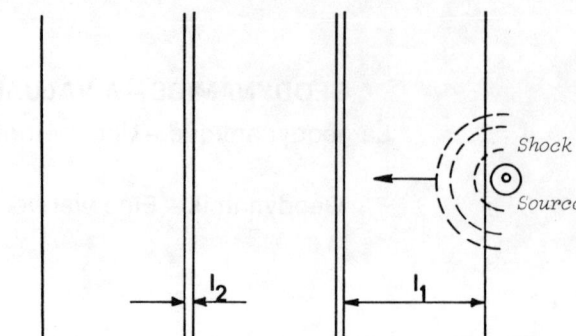

Fig. 2 Propagation of spherical waves in layered medium.

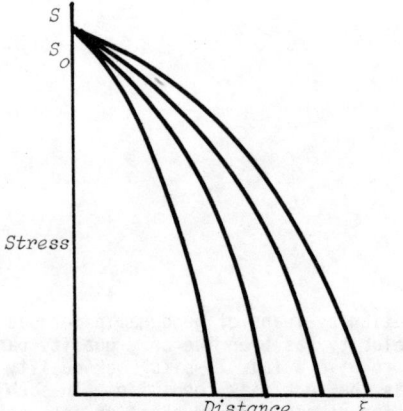

Fig. 3 Theoretical attenuation curves double logarithmic scale.

elastic plates or shells is expressed in Appendix I. Within one percent accuracy the exact solution can be represented in the following analytic form:

$$S = S_o + \lambda \cdot \ln\xi \qquad (1)$$

S = Maximal stress amplitude
S_o = Constant, gives the calculated reduced stress at the shock source, depending on l_1 and l_2, fig. 2
λ = Attenuation factor, depending on the relation between l_1 and l_2, fig. 2
ξ = Distance from the source.

The quantities are all non-dimensional.

For verification of the theoretical results, a field experimental program was planed and carried out in a quarry north of Oslo, the capitol of Norway. These experiments and examination and evaluation of the results will be given a nearer scrutiny.

3. FIELD EXPERIMENTS.

The experiments took place in a layered limestone with a slope of 75°. A standard experimental layout is shown in fig. 4. Pickups with strain gauges and accelerometers were located in boreholes at depths of 6 - 10 meters, fig. 5. In other boreholes, all gathered on the same line as the pickup holes, small charges of high explosives were detonated. Registrations of the

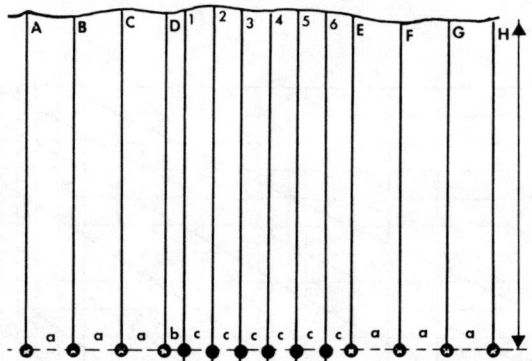

Fig. 4 *Standard experimental layout.* *A - H are*
blasting holes. *1 - 6 pickup holes.*

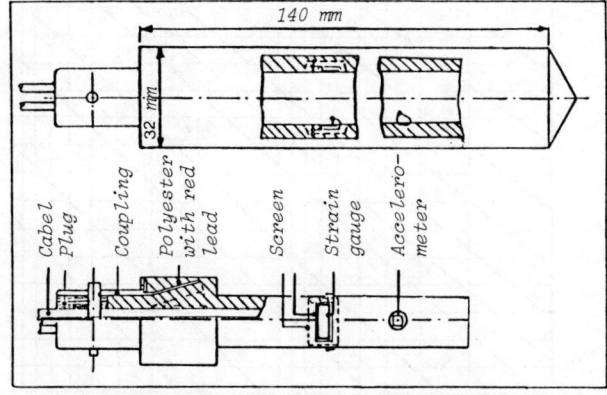

Fig. 5 *Sketch of pickup with strain gauge and*
accelerometer.

shock wave as it passed the different pickups
were stored on a 14 channel taperecorder, 0 -40
kHz frequency range.

Three experimental lines were located perpendicular to the bedding and one line along the bed-
ing. The limestone had a characteristic cubic
fracturing, fig. 6.

Each charge consisted of 0,5 kg Pentrite Wax.
This explosive has a very high detonation velocity and gives a concise in put pulse.

As an interesting novelty, a new type of hydraulic rock sonde was tested during the experiments,
fig. 7. This pick up was equiped with accelerometer. Fastening in the hole was done with hydraulic clamps. Correct orientation was attained by means of entering rods.

4. RESULTS

Signals from strain gauges and accelerometers
were stored in a 14 channel Ampex taperecorder
with linear frequency response up to 40 kHz.

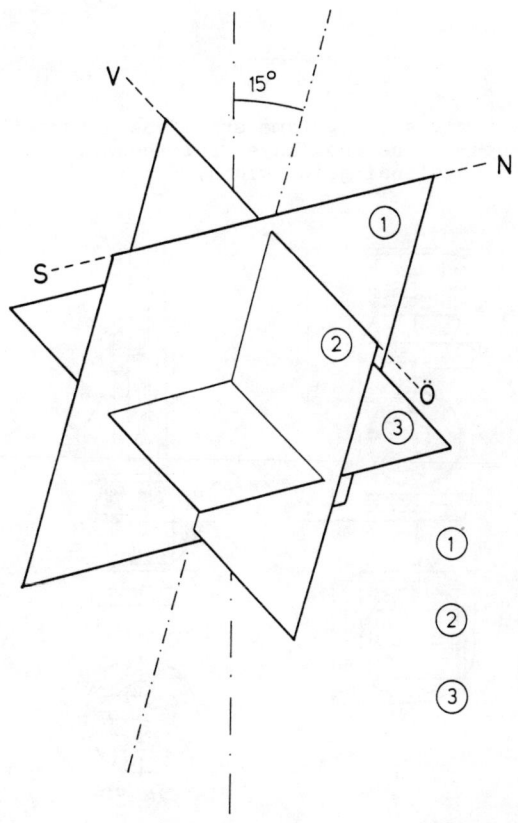

Fig. 6 *Sketch of principle, fracturing planes.*
Plane 1, 2 and 3 are perpendicular on each other.

The stored signals were analysed with regard to
maximum amplitudes, pulse durations, rising
times and signal velocitys. So far frequency
analysis are not accomplished.

The most important data from the experiments,
are the signal attenuation and the signal velocity. Fig. 8 shows data from one of the experimental lines in double logarithmic scale.
Logarithmic curve fitting on the data gives the
attenuation curve. This is the attenuation of
the maximum stress amplitudes as a function of
the distances from the shock source.

Let us for a moment return to fig. 3, the theoretical attenuation curves for stress pulse propagation in the stratified, elastic medium.
Depending on the combination of layer thickness
and joint width, one can obtain one curve that
gives the best agreement with the experimental
results. The actual combination of layer thickness, l_1 and joint width, l_2 can be expressed as
the normal cracking index:

$$q_n = \frac{l_1}{l_2}$$

The figures S_0 and and λ in eq. (1) is calculated
for the actual attenuation curve and then expressed as the characteristic damping parameter:

$$\alpha = \frac{S_o}{\lambda}$$

Fig. 9 gives the parameter "α" as a function of the non-dimensional layer thickness, ξ_b and the non-dimensional joint width, δ.

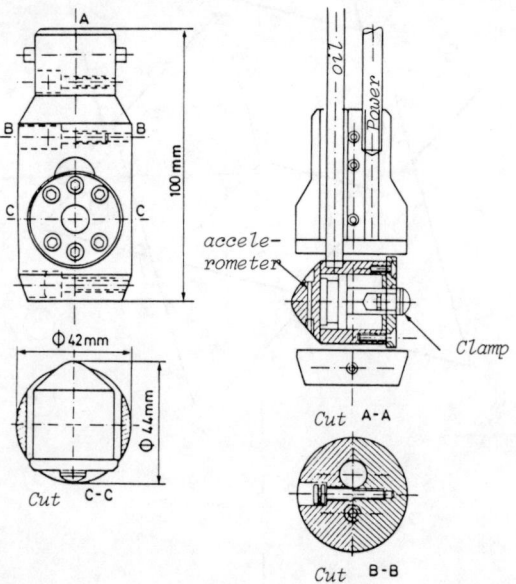

Fig.7 *Rock pickup for hydraulic fastening in boreholes.*

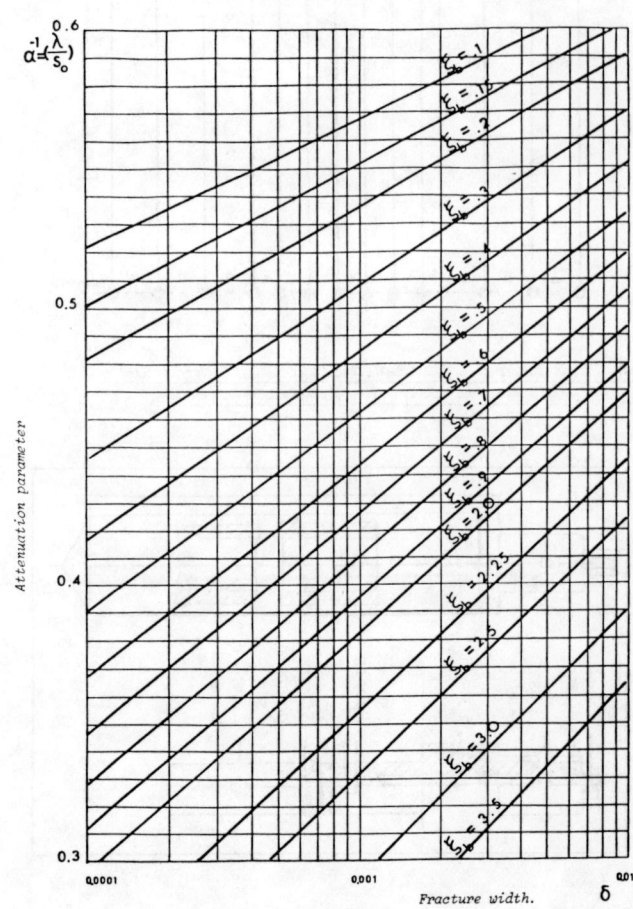

Fig. 9 *The attenuation depends on the layer thickness, ξ_b, and fracture width, δ.*

Assertion:

One can by means of the described data scrutiny, characterize the rock mass after a normal cracking index. The examined rock area gives a signal attenuation as if it had the calculated normal fracturing.

Simultaneously with the shock wave registration in-situ in boreholes of length 6 - 8 meters, surface vibrations were registrated be means of accelerometers and seismometers, as shown in fig. 10. Data from this experiment were analysed with regard to a comparison between in-situ and surface vibrations. The experiments sustains the idea that surface registrations of stress waves can be a reliable method for rock quality evaluation. Data from the surface fits well with data from in-hole registrations.

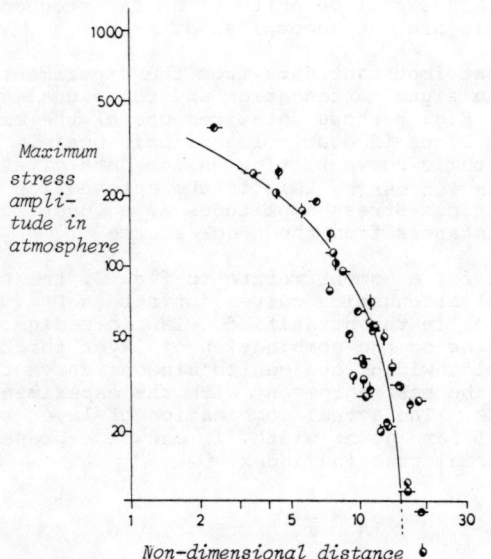

Fig. 8 *Data from field experiments in layered limestone. Logarithmic curve fitting gives the attenuation curve in the figure.*

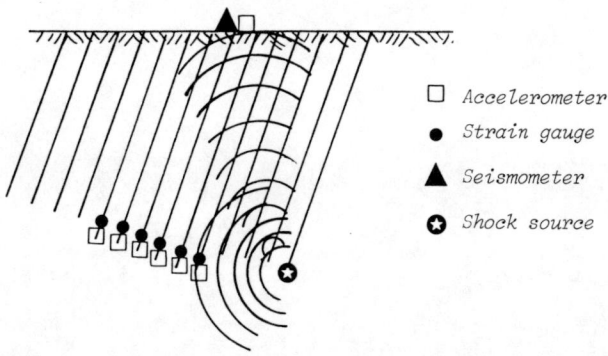

- □ *Accelerometer*
- ● *Strain gauge*
- ▲ *Seismometer*
- ✪ *Shock source*

Fig. 10 Comparisson between surface and in-situ registrations.

5. Conclusions.

Geodynamics for preinvestigations and rock quality evaluation and control, for instance pillar stability, is an interesting subject in expansion. The Norwegian Institute of Rock Blasting Techniques will support a research program on the theme. The purpose is to be able to give sufficent pre-information for underground projects and to develop a system for continious control of rock quality and stability.

Reference

Persen, L. N., (1974). Rock Dynamics and Geophysical Exploration.
276 pp. Amsterdam: Elsevier.

APPENDIX I.

Exact solution rectangular input pulse, spherical waves, elastic slab.

$$
S_r = \begin{cases}
\dfrac{\xi^{-2}\xi_o^{-1}-\xi^{-3}}{\xi_o^{-1}-1} + 2\sum_{n=1}^{\infty}\dfrac{1}{\zeta_{pol\,n}} \cdot \dfrac{a\,\sin(\zeta_{pol\,n}(\xi-\xi_o)) - b\,\cos(\zeta_{pol\,n}(\xi-\xi_o))}{c\,\sin(\zeta_{pol\,n}(1-\xi_o)) + d\,\cos(\zeta_{pol\,n}(1-\xi_o))} \cdot \\
\qquad\qquad\qquad \cos(\zeta_{pol\,n}\tau) \qquad \text{for } \tau < \tau_o \\[2ex]
2\sum_{n=1}^{\infty}\dfrac{1}{\zeta_{pol\,n}} \cdot \dfrac{a\,\sin(\zeta_{pol\,n}(\xi-\xi_o)) - b\,\cos(\zeta_{pol\,n}(\xi-\xi_o))}{c\,\sin(\zeta_{pol\,n}(1-\xi_o)) + d\,\cos(\zeta_{pol\,n}(1-\xi_o))} \cdot \\[1ex]
\qquad\qquad \left[\cos(\zeta_{pol\,n}\tau) - \cos(\zeta_{pol\,n}(\tau-\tau_o))\right] \quad \text{for } \tau > \tau_o
\end{cases}
$$

a, b, c, d are depending on $\zeta_{pol\,n}$, ξ and ξ_o

S_r = Non dimensional stress

ξ = Non dimensional distance

ξ_o = Non dimensional slab thickness

τ = Non dimensional time

τ_o = Non dimensional pulse duration

$\zeta_{pol\,n}$ = Pole number "n", from the solution of the Laplace transform of the stress pulse.

SIMULTANEOUS INFLUENCE OF SEVERAL PARAMETERS OVER ROCK AND ROCK MASS PROPERTIES

Influence simultanée de plusieurs paramètres sur les propriétés des roches et des massifs rocheux

Simultaner Einfluss verschiedener Parameter auf die Eigenschaften der Felsen und der Gebirge

F. Peres-Rodrigues

Principal Research Officer, Head Rock Foundations Division, Laboratório Nacional de Engenharia Civil, Lisbon, Portugal

SYNOPSIS

This paper is intended as a contribution to the knowledge of the joint influence of several parameters such as anisotropy , heterogeneity, scale effect, time and temperature, over the properties of rocks and rock masses. This influence will be dealt with following a statistical approach. General expressions will be presented to reproduce such an influence, which as a rule are sets of families of anisotropy surfaces. Lastly an instance of application to the study of the deformability is presented, taking into account the simultaneous influence of anisotropy and scale effect.

RESUME

Cette communication a le dessein de contribuer à la connaissance de l'influence conjointe de plusieurs paramètres, tels que l'anisotropie, l'hétérogénéité, l'effet d'échelle, le temps et la température, sur les propriétés des roches et des massifs rocheux. Cette influence sera traitée en termes statistiques et on présentera des expressions génériques qui la traduisent: en règle générale, ce seront des ensembles de familles de surfaces d'anisotropie. On termine par une application à l'étude de la déformabilité, prenant en considération l'influence simultanée de l'anisotropie et de l'effet d'échelle.

ZUSAMMENFASSUNG

Es wird versucht, einen Beitrag über die Kentnisse der zusammenwirkenden Einflüsse verschiedener Parameter, bzw. Anisotropie, Heterogenität, Größeneffekte, Zeit und Temperatur, auf die Eigenschaften von Gestein und Gebirge zu geben. Dieser Einfluß wird statistisch behandelt. Es werden allgemeine Ausdrücke erwähnt, die diese Einflüsse beschreiben. Diese sind, im allgemeinen, Gruppen von anisotropischen Flächenscharen. Anschließend wird ein Beispiel über die Anwendung im Rahmen der Untersuchung der Verformbarkeit gegeben, wobei der simultane Einfluß der Anisotropie und des Maßstabseffektes berücksichtigt worden ist.

1. GENERAL CONSIDERATIONS

Generally speaking, the anisotropy LEKHNITSKII (1963), and the heterogeneity are the main causes leading rocks and rock masses to depart from elastic, homogeneous and isotropic behaviour. Thus the characteristic parameters of the majority of physical properties that are worth determining in rocks and rock masses depend on the direction of determination and on the volume of material considered. On account of these facts, Rock Mechanics may be envisaged as a statistical mechanics, where the properties under study can be characterized by a mean or a most probable value and by a deviation, which can be the standard deviation if the values determined can be considered to belong to a normal distribution. Thus Statistics has an ever increasing application in Rock Mechanics. The influence of the volume of material tested, i.e. the scale effect, over the statistical parameters that define each property, has to be investigated so that such influence may be taken into account. It is accepted that this influence may obey different laws depending on the property

under study. It is practically taken for granted that there are fairly satisfactory correlations between the parameters that define a given property, obtained with samples whose size may vary from that of a rock crystal DOUGLAS et al. (1969) and PERES-RODRIGUES et al. (1970), to the very rock mass containing it. The statistical approach to Rock Mechanics on one hand adds to the difficulty of its study, on the other hand leads to values closer to reality. Safety in the structure -rock mass joint behaviour ought to be assessed in statistical terms through the optimization of the cost-hazard relations. Such a relation should consider the most economical life span of the structure taking into account social, human and technical factors.

In Rock Mechanics decisions to be taken more and more involve parameters of different nature, of which those exerting a stronger influence are not always the technical and economic ones; regrettably decisions rooted on political convenience sometimes step over technical and economic consi derations.

A promising approach whose results are still far from giving satisfaction consists in the search, now more systematized, of general statistical laws that can represent the behaviour of rocks and rock masses, taking into account the aniso - tropy and heterogeity as well as the time factor. A number of researchers are engaged in studies of this kind, and in a fairly near future their res tricted contributions are expected to concur to the elaboration of theories more or less general, that would define one or several properties of rock and rock masses. All these contributions should take a statistical form and permit the de termination of values with a given probability of occurrence, complying with a certain statisti cal distribution. Among the statistical distribu tions that are most usual in Rock Mechanics, men - tion should be made of the normal, the binomial, the exponential, the logarithmic, the Poisson's and the extreme distributions.

The fact that powerful highly sophisticated compu ters are now available has made it possible to develop and work out statistical theories with high degree of complexity. Nevertheless the methods and test techniques used at present for the characterization of rocks and rock masses are a drawback for the progress of Rock Mechanics on account of the small precision and sensitivity with which results are obtained. After an im - pressive development of the computer operating techniques, we have reached a stage in which one might expect a similar development of the experi mental and measuring techniques for rock and rock mass testing. In this field some progress has already been achieved which, however, is not enough to cope with our present needs ROCHA et al. (1970).

Among the properties that should be studied in rocks and rock masses, two types can be singled out: one which is essentially related to mass and volume, the other which is essentially concerned with plans and surfaces. As examples of the for - mer type, let us mention the modulus of elastici ty LOUREIRO-PINTO (1966), LOUREIRO-PINTO (1970), PERES-RODRIGUES (1966) and PERES-RODRIGUES (1970), permeability, unit weight, volume coefficient of expansion, etc., whereas the shear strength of the material or of the corresponding joints, the tensile ultimate strength, etc., fit into the latter type.

Properties related to mass or volume keep their mean value regardless of the volume of the test piece, provided the number of test pieces is enough for the sampling to be considered representative. Deviations in each set of test pieces with the same volume decrease as the volume augments in each test piece, and they tend to zero as the volume of the test pieces tends to the volume of the rock mass. In such a case, therefore, a single test piece would be tested and the deviation would be null. This fact has already been confirmed as regards deformability and the modulus of elasti - city PERES-RODRIGUES (1974).

As concerns the properties of plane type, the evo lution of their mean values and deviations with the size of the surface tested is less known. Nevertheless there are some experiments that show that the mean values and the deviation tend res - pectively to increasing and to decreasing as the size of the tested surface augments.

With reference to studies taking into account the time factor in the variation of rock and rock mass properties, only a few are known in which that influence was accelerated through chemical and physical processes. As such studies must extend for a period longer than the human life span, they are obviously difficult to perform;on the other hand Rock Mechanics is not such old a science that we could have the results of those studies if they had been undertaken.

There are numerous studies on the influence of the time factor in slow tests of the creep or stress-relaxation types, since the duration of these is considerably smaller than that of the above tests. These studies have been carried out on various rocks ranging from sound to consider - ably altered. The three types of creep (retarded, viscous and accelerated) have been recognized and ascertained, and often failure has been reach ed. The degree of moisture is a parameter that has also been considered in these studies,mainly in altered or very altered rocks. As regards the influence of anisotropy and heterogeneity in tests of this kind, little or nothing has been investigated and the same can be said about the influence of the size of the test pieces.

The strength of rocks and rock masses evaluated through shear tests has been given sustained attention in studies where the effect of aniso - tropy has been considered as well as that of the size of test pieces, i.e, that of the area tested. As for the time factor, it practically has not been taken into account and nothing is known about the creep behaviour, that is, about the time evolu tion of normal and shear strains due to the appli cation of steady normal and shear stresses. As far as we know this study, which seems of much interest, has not yet been attempted so as to provide information about possible creep stages and occurrence of failure after a given time. It seems that the consideration of the time factor in the long run in order to take into account the alterability of materials is a problem still far from solution within the study of shear strength.

Other parameter that hardly has been studied is temperature PERES-RODRIGUES et al. (1982), which until recently could be considered unimportant; however with the advent of nuclear installations for peaceful purposes and of very deep boreholes, its influence over the properties of rocks and of rock masses raises increasing interest.

Some of the properties whose determination is of major interest in rocks and rock masses have been presented, as well as the parameters that to some extent may exert more influence on their varia - tion: anisotropy, heterogeneity, scale effect and time factor.

The formulation of a theory that would take into account the simultaneous influence of these para meters over the value of each property of rocks and rock masses is far from having being enunciat ed or so much as delineated. Its difficulty and complexity have so far hindered attempts to do that, the more so as test techniques and methods that might supply results with the required pre cision are sometimes lacking.

Notwithstanding all these reasons it seems already possible in terms still vague to enunciate prin ciples and formulate some laws which will provide a start for the study and discussion of the pro - blems.

2. FORMULATION OF PROBABLE LAWS

Let us consider a given property of the rock or rock mass, for which a value, v, is to be determined with a given probability of occurrence. Let the set of the values obtained obey a statistical distribution whatsoever.

As said, that value, v, will be given in statistical terms by:

$$v = v_m + \Delta v \tag{1}$$

where v_m and Δv are characteristic values of the property under study obtained through a significant number of tests that would permit to obtain the type of the most probable statistical distribution; in the case of a normal distribution, for instance, v_m corresponds to the mean value and Δv to the positive or negative deviation with a desired probability of occurrence.

The characteristic values v_m and Δv are given by expressions relating the unknowns P, α, d, t, T,... one another by means of coefficients:

$$\begin{cases} v_m = f_1(P,\alpha, d, t, T,\ldots,A_1,\ldots,A_i,\ldots,A_m) \\ \Delta v = f_2(P,\alpha, d, t, T,\ldots,B_1,\ldots,B_j,\ldots,B_p) \end{cases} \tag{2}$$

where:

P - any point of the rock or rock mass that determines the influence of the heterogeneity;

α - any direction passing through point P that determines the influence of the anisotropy;

d - any dimension, either of volume or of surface, of the test piece, that determines the influence of the scale effects;

t - any time that determines the influence of the alterability in quick tests and the influence of creep and stress-relexation in slow tests;

T - any temperature that influences the property under study;

A_i - any coefficient representing the m coefficients that relate the unknowns one another;

B_j - any coefficient representing the p coefficients that relate the unknowns one another.

By substituting (2) for (1), one can eliminate the characteristic values and express the value of v direct, as a function of the unknown and of a number of coefficients, as a rule equal to m +p since:

$$v = f(P,\alpha, d, t, T,\ldots,C_1,\ldots,C_k,\ldots,C_{m+p}) \tag{3}$$

The determination of the most probable values of the coefficients, contained in expressions (2) and (3), should be carried out on basis of values obtained with a number of tests higher than the number of coefficients existing in each expression, so that a statistical method can be applied, for instance the least-square method.

The study of rock and rock mass properties is still far from progressing as outlined, but it is expected to evolve following these lines.

It is possible to establish intermediate stages in which first the influence of one or more unknowns would be taken into account, the remaining being considered constant or invariable.

Thus if in expressions (2) and (3) only α is considered to be unknown and the remaining are constant, we shall have

$$\begin{cases} v_m = f_1(\alpha, A_1,\ldots,A_i,\ldots,A_m) \\ \Delta v = f_2(\alpha, B_1,\ldots,B_j,\ldots,B_p) \end{cases} \tag{4}$$

$$v = f(\alpha, C_1,\ldots,C_k,\ldots,C_{m+p}) \tag{5}$$

with P, d, t, T,..., = const.

For a given point P of the rock or rock mass, expression (5) will represent the surface of anisotropy of the property being studied with a given probability of occurrence, since for each direction α passing through point P it will give the corresponding value of the property.

Expressions (4) in statistical terms will represent the surfaces of anisotropy of the characteristic values of the property under study; if the probability of occurrence is considered a parameter, thus to each point P a family of surfaces of anisotropy will correspond.

Now in expressions (2) and (3) assume P and α to be unknowns and the remaining to be constant. Then:

$$\begin{cases} v_m = f_1(P,\alpha, A_1,\ldots,A_i,\ldots,A_m) \\ \Delta v = f_2(P,\alpha, B_1,\ldots,B_j,\ldots,B_p) \end{cases} \tag{6}$$

$$v = f(P,\alpha, C_1,\ldots,C_k,\ldots,C_{m+p}) \tag{7}$$

with d, t, T,..., = const.

With reference to the rock or rock mass expression (7) will represent a field of surfaces of anisotropy of the property being studied, with a given probability of occurrence, the influence of the heterogeneity being considered here through the unknown P.

Expression (6) in statistical terms will represent a field of families of surfaces of anisotropy, in which the probability of occurrence is a parameter.

The simultaneous study of a larger number of unknowns makes the visualization of this study more and more complex, calling even for the consideration of hyperspaces and hypersurfaces, i.e. the consideration of geometries in general heterometric with dimensions larger than 3.

The influence of the scale effect, expressed by the unknown d, should be approached together with that of anisotropy. In this case the expressions (2) and (3) take the specific form:

$$\begin{cases} v_m = f_1(\alpha, d, A_1,\ldots,A_i,\ldots,A_m) \\ \Delta v = f_2(\alpha, d, B_1,\ldots,B_j,\ldots,B_p) \end{cases} \tag{8}$$

$$v = f(\alpha, d, C_1,\ldots,C_k,\ldots,C_{m+p}) \tag{9}$$

with P, t, T,... = const.

With reference to a given point P of the rock or rock mass, expression (9) will represent a family of surfaces of anisotropy, taking the scale effect d as a parameter.

Expressions (8) in statistical terms will represent a field of families of surface of anisotro

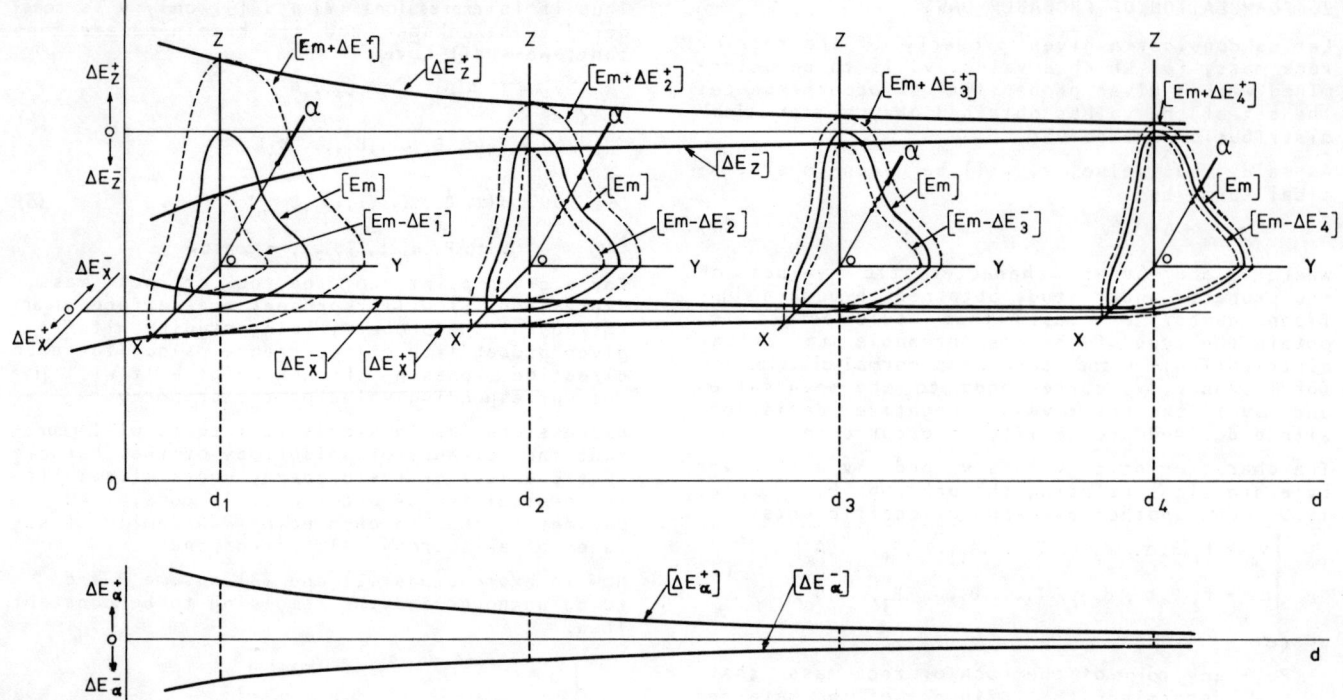

py with reference to a given point P, taking the probability of occurrence as a parameter.

3. APPLICATION TO A SPECIFIC CASE

This joint influence was already studied with reference to the property of deformability, by assuming that the moduli of elasticity, E, complied with a law of the closed quartic type. Thus if we assume that the modulus of elasticity, E, is given in space by Cartesian co-ordinates by:

$$E^2 = X^2 + Y^2 + Z^2 \qquad (10)$$

the anisotropy being considered following a closed quartic type law, we shall be able to write expression (9) in parametric terms:

$$\frac{X^2}{\left[A(d)\right]^2} + \frac{Y^2}{\left[B(d)\right]^2} + \frac{Z^2}{\left[C(d)\right]^2} + \frac{X^2 Z^2}{\left[E(d)\right]^4} + \frac{Y^2 Z^2}{\left[F(d)\right]^4} = 1 \qquad (11)$$

and so expressions (8):

$$\frac{X_m^2}{A_1^2} + \frac{Y_m^2}{B_1^2} + \frac{Z_m^2}{C_1^2} + \frac{X_m^2 Z_m^2}{E_1^4} + \frac{Y_m^2 Z_m^2}{F_1^4} = 1$$

$$\Delta E^+ = E_m \left[\frac{1 - e^{\frac{1}{A_2 d^2 + B_2 d + C_2}}}{e^{\frac{1}{A_2 d^2 + B_2 d + C_2}}} \right] \qquad (12)$$

$$\Delta E^- = E_m \left[\frac{1 - e^{\frac{1}{A_2 d^2 + B_2 d + C_2}}}{2 - e^{\frac{1}{A_2 d^2 + B_2 d + C_2}}} \right]$$

where:

$$E = E_m + \Delta E^+ \quad \text{or} \quad E = E_m - \Delta E^- \qquad (13)$$

depending on the values of E being higher or lower than the characteristic value E_m.

Expressions given in (11), (12) and (13) resulted from experimental studies carried out on three types of rock, in which the scale effect was represented by the volume of the test pieces; the ratio between the minimum and the maximum volumes of the test pieces was 1:4096. For each volume a significant number of test pieces were available which highlighted some important conclusions, namely:

a) The set of unit strains, ε, measured on test pieces with the same volume, very satisfactorily fit a normal distribution, reason why it made sense to determine a mean value ε_m given by the arithmetic mean, and a standard deviation, $\Delta \varepsilon_p$;

b) The mean values, ε_m, were independent of the volume of the test piece;

c) The deviations, $\Delta \varepsilon$, with a given probability of occurrence, very satisfactorily varied with the volume of the test piece following an exponential distribution, and presented a tendency towards zero when the volume of the test piece tended towards the overall volume of the rock or rock mass;

d) Since the moduli of elasticity, E, are proportional to the inverses of unit strains, ε, the mean value of the set of the moduli of elasticity, E_m, obtained from the set of unit strains, ε, will be given by the harmonic mean of E, i.e.

$$\frac{1}{E_m} = \frac{\sum\limits_{i=1}^{n} \frac{1}{E_i}}{n} \qquad (14)$$

n being the number of E values in the set;

e) The mean values, E_m, still are independent of the volume of the test piece, as in b).;

f) Deviations ΔE, with a given probability of occurrence, will be different as the values of E are higher or lower than E_m, and will be given by the expressions indicated in (12); this is due to the fact that the set of the values of E belongs to an asymmetric distribution, inverse to a normal distribution;

g) Such as for the values of ε, those of deviations decrease as the volume of the test piece increases, and they present a tendency towards zero when the volume of the test piece tends towards the overall volume of the rock or rock mass;

h) As the size of the test piece decreases, the number of pieces to be tested must augment so that the application of statistics may be realistic and results therefrom be representative of the material.

With the application just presented in a rather generalized way of a deformability study taking the anisotropy and scale effect into account, it was sought to evidence the great interest that studies of this kind may have. The figure exemplifies a possible joint representation of the case dealt with hereinbefore, clearly showing the anisotropy families corresponding to each size of the test piece and the exponential curves of the variation of deviation as a function of the direction and of the size of the test piece.

REFERENCES

DOUGLAS, P.M. and VOIGHT, B (1969) - Anisotropy of granites, a reflection of microscopic fabric; Geotechnique, Vol. XIX No 3.

LEKHNITSKII, S.G. (1963) - Theory of elasticity of an anisotropic elastic body; Holden-Day, San Francisco.

LOUREIRO-PINTO, J. (1966) - Stresses and strains in an anisotropic orthotropic body; Proc. 1st Cong. ISRM, p.625, Lisbon.

LOUREIRO-PINTO, J. (1970) - Deformability of schistous rocks; Proc. 2nd. Cong. ISRM, p. 2-30, Beograd.

PERES-RODRIGUES, F. (1966) - Anisotropy of granites. Modulus of elasticity and ultimate strength ellipsoids, joint systems, slope attitudes, and their correlations; Proc.1st Cong. ISRM, p. 721, Lisbon.

PERES-RODRIGUES, F. (1970) - Anisotropy of rocks: Most probable surfaces of the ultimate stresses and the moduli of elasticity; Proc. 2nd Cong. ISRM, p. 133. Beograd.

PERES-RODRIGUES, F. and AIRES-BARROS, L. (1970)- - Anisotropy of endogenetic rocks; Proc. 2nd Cong. ISRM, p. 161. Beograd.

PERES-RODRIGUES, F. (1974) - Influence of the scale effect over rock mass safety against deformability; 3th Cong. ISRM, p. 202, Denver.

PERES-RODRIGUES, F. and REIS E SOUSA, M.(1982) - - Anisotropy of the thermic characteristics of the rock masses; 4th Cong. IAEG, New Delhy.

ROCHA, M. (1970) et al - Characterization of the deformability of rock masses by dilatometer tests; Proc. 2nd. Cong. ISRM, p. 2-32, Beograd.

UNTERSUCHUNGEN ZUM MECHANISCHEN VERHALTEN GEKLÜFTETEN GEBIRGES UNTER WECHSELLASTEN

Studies on the mechanical behaviour (deformation behaviour) of jointed rock masses under cyclic load

Etudes du comportement mécanique de masses rocheuses fissurées sous charges alternées

Leopold Müller-Salzburg
Honorarprofessor der Universitäten Salzburg (Austria) und Karlsruhe (BRD)

Xiurun Ge
Associate Professor, Institute of Rock and Soil Mechanics, Academia Sinica, Wuhan, China

ZUSAMMENFASSUNG

Wechsellasten, wie sie unter Seilbahnstützen und Staumauern auftreten, verursachen irreversible Verformungen. Gegenwärtig wird die Sicherheit des Felsens nach dem zeitlichen Verlauf irreversibler Deformationen beurteilt, weil Kriterien für eine quantitative Vorausschätzung des Verlaufes fehlen. Die Studie versucht, eine solche Vorraussage aufgrund des Vergleiches von Messungen und Großversuchen in situ zu ermöglichen. Deutlich lassen sich drei Bereiche unterscheiden, welche a) zunehmende Kompaktierung, b) stetige Stabilisierung und c) zunehmende Materialschädigung, an deren Ende der Bruch steht, zugeordnet werden können. Aufgrund der in situ Großversuche an der Kurobe IV Staumauer (Japan) werden konkrete Beispiele durchgerechnet.

SYNOPSIS

Cyclic loads (e.g. under ropeway head masts, concrete dams) create irreversible deformations. The safety of the rock foundation is presently assessed on the basis of the course of irreversible deformations, because criteria for the quantitative prediction of the course are lacking. The present study tries to give such a prediction, based on the comparison between measurements and large-scale in situ tests. A diagram which has been developed allows a clear distinction to be made of three areas, coordinated a) to increasing compaction, b) to steady stabilization, c) to increasing damaging of material resulting in rupture. On the basis of the large-scale tests at Kurobe IV Dam (Japan) actual examples are calculated.

RESUME

Des charges alternées (p.e. sous pylônes téléphériques, barrages en béton) provoquent des déformations irréversibles. Le seuilde rupture des roches de fondation est évalué en fonction de déformations irréversibles, puisqu'on n'a pas de critères pour la prédiction du comportement ultérieur. L'étude présente essaye de rendre possible une telle prédiction au moyen d'une comparaison entre des mesures et des essais à grande échelle "in situ". Un diagramme est développé pour différenicer trois aspects qui peuvent être coordonnés: a) à compaction augmentante, b) à charge stabilisée, et c) à lésion augmentante du matériau qui est suivie de rupture. Des examples concrèts sont calculés sur la base des essais à grande échelle au Kurobe IV (Japon).

1. ZUR EINFÜHRUNG

Bruch- und Deformationskriterien für klüftige Medien sind bisher nur anhand zwei- und dreiachsiger Versuche an Modellkörpern und nur in Bezug auf statische Belastungswirkungen untersucht worden. Die nachfolgend gegebene Auswertung der Messungen an den in den Sechziger Jahren durchgeführten In situ-Großversuchen im Granit der Japanischen Talsperre Kurobe 4 gab Gelegenheit, das Verhalten natürlicher Versuchskörper unter Wechselbelastung zu untersuchen.

Der Unterschied zwischen dem Verhalten unter Wechsellast und dem unter statischer Last dürfte für viele Felsbauten, z.B. für Staumauern und Seilbahnstützen, von Bedeutung sein. Bei Staumauern ereignet sich periodisch wechselnde Be- und Entlastung infolge Füllung und Leerung des Staubeckens sowie infolge von Temperaturunterschieden während ihrer gesamten Funktionsdauer. Dabei wird gegenwärtig die Sicherheit des Gründungsfelsens aufgrund der zeitlichen Entwicklung rückläufiger und unrückläufiger Verschiebungen im Felsuntergrund eingeschätzt. Es fehlen aber Kriterien, welche es erlauben

würden, aufgrund von Messungen in den ersten Betriebsjahren den Verlauf solcher Formänderungen in der Zukunft einzuschätzen.

Die Untersuchung des Materialverhaltens bei den Felsprüfungen unter zwei- und dreiachsigem Druck in situ unter sehr vielen Belastungszyklen gibt die Möglichkeit, einen besseren Einblick in die Brucherscheinungen zu gewinnen, aufgrund deren versucht werden kann, entsprechend Kriterien und einen Weg zur Vorausbestimmung der Tragfähigkeit einer Felsmasse nach wiederholten Belastungen zu finden.

2. KURZE BESCHREIBUNG DER IN SITU-GROSSVERSUCHE BEIM BAU DER STAUMAUER KUROBE 4

Die Materialprüfungen in situ, welche am Beginn der Sechziger Jahre an der Staumauer Kurobe 4 vorgenommen wurden, sind wegen des großen Versuchsmaßstabes und ihrer großzügigen Durchführung weltweit bekanntgeworden.

Nachdem der Aushub für die Staumauer bereits begonnen worden war, bot die damals neue Wissenschaft der Felsbaumechanik die Möglichkeit, die Tragfähigkeit der Felswiderlager aufgrund tatsächlich erhobener Parameter der geklüfteten Felsmasse rechnerisch zu erfassen. Man entschied sich dafür, von dieser Möglichkeit, welche erstmals quantitative Aussagen über die Sicherheit der Felswiderlager versprach, in vollem Umfange Gebrauch zu machen. Ein weiterer Grund für die Durchführung von Großversuchen ergab sich aus der Tatsache, daß die geomechanischen Eigenschaften des Gebirges aufgrund eingehender geologischer, insbesondere gefügemechanischer Untersuchungen offensichtlich weniger günstig waren als ursprünglich angenommen. Deshalb wurde von Müller und Pacher eine geomechanische Widerlagerberechnung unter Berücksichtigung der Diskontinuitäten und der Anisotropie der Felsmasse durchgeführt. Diese Berechnung konnte sich auf die Ergebnisse der Großversuche in situ stützen, welche - nach einem ersten Einzelversuch an der Talsperre Vajont in etwas kleinerem Maßstab - in Kurobe in vorbildlich großzügiger Weise ausgeführt worden sind.

Die Felsprüfungen in Kurobe enthielten Scherversuche unter aufbetonierten Betonblöcken, Scherversuche in der Felsmasse selbst und im Störungsmaterial, dreiachsige Druckversuche an Störungsmaterial sowie einachsige und dreiachsige Druckversuche in klüftigem Felsmaterial (John, 1961; Nose, 1964). An der Talsperre selbst wurden im Gründungsfels sehr eingehende Meßbeobachtungen angestellt (Kansai Electric Power Co. Inc., 1967; Yoshida, 1982). Die Ergebnisse aller dieser Vorkehrungen waren:

- eine grundlegende Abänderung des Bogenmauer-Entwurfes mit dem Ziele, eine Belastung der Widerlager durch Bogenschubkräfte in dem schwachen Fels des oberen Drittels der Einbindung völlig zu vermeiden;

- die unterirdische Ausräumung eines beträchtlichen Volumens von schwachem Fels und dessen Ersatz durch Beton;

- überdies eine umfangreiche Verdübelung etlicher Großkluft- und Störungszonen, ausgeführt in Form von gitterförmig einander kreuzenden und mit Stahlbeton verfüllten Stollen und Schächten.

Heute, nach zwanzig Jahren der Bewährung kann festgestellt werden, daß die Staumauer absolut sicher steht (Yoshida, 1982).

Der gegenständliche Bericht befaßt sich nur mit den Ergebnissen der ein- und dreiachsigen Felsmaterialprüfungen unter Wechsellasten.

Das Volumen der Versuchskörper betrug ca. 11 m^3, ihre Ausmaße waren 2,8 x 2,8 x 1,4 m^3 (Abb. 1).

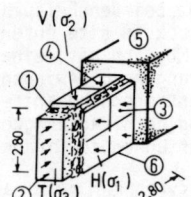

(1) Versuchsblock

(2) 20 Druckpressen in Hauptdruckrichtung

(3) 4 Druckkissenpakete für vertikale Querbelastung H

(4) 2 Druckkissenpakete für horizontale Querbelastung V

(5) Betonverfüllung gegen die Stollenwände

(6) Kontaktfläche der Versuchsblöcke mit dem umgebeden Fels: die Auswirkungen des Umschliessungseffektes am Kontakt wurden durch spannungsoptische Versuche sowie an Gipsmodellen studiert

Abb. 1: Dreiachsige Felsgroßversuche

An jeden Versuchsblock wurden 20 hydraulische Pressen für die σ_3-Belastung sowie Druckkissenbatterien für die σ_1- und σ_2-Belastung installiert ($\sigma_3 > \sigma_2 > \sigma_1$; Druck positiv gerechnet).

Der Fels der Versuchsblöcke war ein in unverwittertem Zustand sehr fester und harter Biotitgranit, der aber bereichsweise in sehr unterschiedlichem Maße durch intensive und räumlich ausgedehnte hydrothermale Einflüsse entfestigt war. Dabei waren die Biotite (durch Baueritisation) zu Chloriten, die Felspäte zu Kaolin umgewandelt worden. Laborversuche zeigten außerordentlich unterschiedliche einachsige Druckfestigkeiten, aber noch viel größere Unterschiede der (ein- und dreiachsigen) Druckfestigkeit ergaben sich aus den in situ-Versuchen im Fels. Überdies waren beide Widerlager von einer Anzahl bedeutender Störungen durchzogen; im rechten Widerlager war der Fels außerdem noch durch eine Schar ungünstig orientierter Großklüfte zerteilt. Diese Gefügeelemente verursachten eine bedeutende Anisotropie und eine beträchtliche Abminderung der Tragfähigkeit in gewissen Richtungen, da sie montmorillonitische Kluftfüllungen mit Reibungswinkel bis herab zu 4° enthielten.

Insgesamt wurden 12 Versuchsblöcke unter ein- und dreiachsigem Druck getestet. Sie waren sämtlich in untertägigen Kammern gelegen. Die Achsrichtung und die Neigung der Blöcke wurden nach folgenden Gesichtspunkten ausgerichtet, und zwar so, daß sie nicht zu sehr von den allgemeinen Druckrichtung der erwarteten Hauptnormalspannungen in den Widerlagern abwichen und daß ein überbetonter ungünstiger Einfluß der geologischen Trennflächen an den Versuchsblöcken vermieden wurde, weil dieser Einfluß in der Berechnung ohnedies gesondert erfaßt werden konnte.

Für drei- und einachsige Druckversuche wurde nach zwei verschiedenen Belastungsprogrammen vorgegangen;

a) bei statischer Belastung gemäß Abb. 2a,

b) bei Wechselbelastung gemäß Abb. 2b.

Die Belastungsstufen betrugen 70 Mp/m^2 (0,7 MN/m^2).

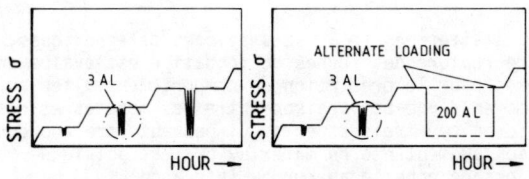

Abb. 2: Belastungs-Zeit-Diagramm
      a) für statische, b) für alternierende Belastung

Sechs der insgesamt 12 Druckversuche (die Blöcke O_2, O_3, Q_2, R_3, P_1 und P_2) wurden unter Wechsellast geprüft; von diesen wiederum drei Blöcke (Q_2, R_3 und P_1) unter dreiachsiger Belastung, und zwar mit σ_1 = 12 Mp/m^2 (0,12 MN/m^2) und σ_2 = 70 Mp/m^2 (0,7 MN/m^2), die anderen drei Blöcke unter einachsigem Druck. Die Versuche wurden bis zum Bruch oder, wenn kein deutlicher "Bruch" eintrat, bis zur offensichtlichen Erschöpfung der Tragfähigkeit ausgedehnt (Nose, 1964). Die Prüfdauer für je einen Block betrug zwischen 4 und 15 Tagen. Ein Beispiel eines Spannungs-Achsialdehnungsdiagramms (für die Block Q_2) zeigt Abb. 2.

Die dreiachsigen Versuche unter zyklischer Belastung wurden in der Absicht disponiert, den Einfluß der häufigen Lastwechsel, welche dem Fels durch wiederholte Füllung und Leerung des Staubeckens zugemutet werden, zu erfassen,

also den Unterschied zwischen der Gebirgsfestigkeit bei
statischer und bei wechselnder Belastung kennenzulernen.
Zugleich erwartete man, daß die Prüfung unter oft wiederholten Lastwechseln in die Lage versetzen würde, das
Kriechverhalten des Gebirges unter Langzeitbelastung,
zumindest intuitiv, einschätzen zu können, nachdem für
wirkliche Kriechversuche über Monate hinweg nicht genü-
gend Zeit verfügbar war. Diese Erwartung gründete sich
auf die Überlegung, daß, ebenso wie unter Wechsellast,
so auch unter Langzeitbelastung eine Ermüdung des Materials an den Korngrenzen und Klüften eine gewisse - mög-
licherweise ähnliche - Rolle spielen würde.

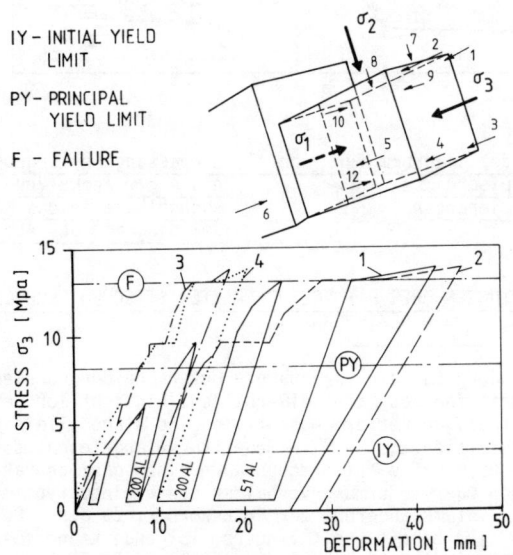

IY – INITIAL YIELD
     LIMIT

PY – PRINCIPAL
     YIELD LIMIT

F – FAILURE

Abb. 3: Spannungs-Dehnungs-Kurve für den Versuchs-
       block Q_2
       IY - erste Fließgrenze; PY - Hauptfließgrenze;
       F - Bruch, Versagen

3. NEUERLICHE AUSWERTUNG DER DREI- UND EINACHSIGEN DRUCKVERSUCHE UNTER WECHSELBELASTUNG

Die Absicht dieser (erst jetzt nachgeholten) Auswertung
ist: Die Beobachtungen des speziellen Verformungsverhaltens der Felsmasse zu analysieren und Beziehungen zwischen dem Charakter dieser Verformungen und dem Bruchverhalten näher kennen zu lernen. In der Felsbaumechanik
erscheint es uns wichtig, sich mehr auf die Verformungen
als auf Spannungen abzustützen und sowohl Bruchkriterien
als auch Kriterien zur Einschätzung von Gefahren, wie bereits bei der Eröffnung des Lissabon-Kongresses vorgeschlagen (Müller, 1966), vorwiegend auf das Verformungsverhalten zu gründen.

Wie bekannt, bestehen große Unterschiede zwischen dem
Verformungsverhalten des Gesteins und des Gebirges. Nach
einer mäßigen Belastung behält die Gesteinssubstanz z.B.
die Fähigkeit elastisch rückzufedern, zeigt also vorwiegend rückläufige (irreversible) Verformungen. Bei der
Entlastung klüftigen Felsens hingegen tritt eine ähnlich
rückläufige Verformung nur in geringem Ausmaß ein, da
sich Formänderungen klüftiger Medien hauptsächlich in
Form von Verschiebungen an den einzelnen Teilkörpern
entlang den Kluftflächen vollziehen (Müller, 1971),
solche, die Reibung mobilisierende Erscheinungen aber
nur eine geringe Tendenz zur Rückverformung haben.

Daraus erhellt, daß dem Studium des Verformungsverhaltens
klüftigen Gebirges hinsichtlich seiner unrückläufigen

Verformungen eine große Bedeutung zukommt. Die Untersuchung dieses weitgehend unrückläufigen Verformungsverhaltens an den Großversuchen von Kurobe unter oft wiederholter wechselnder Belastung zeigt, daß die Größe der unrückläufigen Verformungsanteile eines geklüfteten Mediums
nicht nur von der Einflußgröße n, der Zahl der Zyklen der
Wechselbelastung, abhängt, sondern auch vom Spannungsniveau der größten Hauptdruckspannung σ_3. Deshalb wurde eine
zusätzliche Kenngröße Δs_3 bei der Auswertung der Versuchsdaten eingeführt.

$$\Delta s_3 = \frac{\Delta l_{3\ irr}}{\sigma_3 \cdot n}$$

wobei $\Delta s_{3\ irr}$ die gesamte unrückläufige Verformung in der
Richtung von σ_3 während n Belastungszyklen der Wechsellast unter dem (jeweiligen) Spannungsniveau von σ_3 ist.
Die physikalische Bedeutung von Δs_3 ist klar: es sind die
mittleren unrückläufigen Verformungsanteile in der σ_3-Rich-
tung pro Zyklus und pro Spannungseinheit von σ_3.

Dieser Parameter ermöglicht einen Vergleich des irreversi-
blen Anteils des Deformationsverhaltens eines Blockes auf
verschiedenen Belastungsniveaus, aber auch einen Vergleich
verschiedener Blöcke.

Wenn die Länge des freien Blockes in der σ_3-Richtung mit
l_3 bezeichnet wird, dann kann eine Definition von $\bar{\varepsilon}_{3\ irr}$
gegeben werden:

$$\bar{\varepsilon}_{3\ irr} = \frac{\Delta l_{3\ irr}}{l_3 \cdot n} \ .$$

$\bar{\varepsilon}_{3\ irr}$ ist die durchschnittliche unrückläufige Dehnung
(bzw. Zusammendrückung) in der Richtung von l_3 pro Einheitszyklus der Wechselbelastung auf dem Belastungsniveau σ_3.

In Anlehnung an die Begriffsbestimmung des Verformungs-
moduls können wir definieren

$$D_{irr} = \frac{\sigma_3}{\bar{\varepsilon}_{3\ irr}} \ ,$$

worin D_{irr} der Modul der durchschnittlichen unrückläufigen (=bleibenden) Verformung (bzw. Verformungsanteile)
je Einheitszyklus der Wechselbelastung (für $\sigma_1 = \sigma_2 = 0$)
ist.

Für einachsige Druckversuche gilt dann

$$\Delta s_3 = \frac{l_3}{D_{irr}} \ .$$

Nachdem l_3 für jeden einzelnen Versuchsblock konstant
ist, ist die spannungsbezogene Verkürzung von Δs_3 umge-
kehrt proportional zum Modell D_{irr}. Für dreiachsige Druckversuche hat diese Beziehung gleichfalls eine analoge
Bedeutung.

Daraus kann ein Δs_3-σ_3-Diagramm für jeden dreiachsig
und einachsig getesteten Versuchsblock unter Wechselbelastung gegeben werden. In den Abb. 4 und 5 sind die ent-
sprechenden Kurven Δs_3 in Abhängigkeit von σ_3 für sechs
Versuchsblöcke wiedergegeben.

In Abb. 5 sind auch die Schaulinien für Δs_1 und Δs_2
in Abhängigkeit von σ_3 dargestellt. Δs_1 und Δs_2
sind die Parameter der unrückläufigen Anteile der Querverformung in den Richtungen von σ_1 und σ_2. Ganz offensichtlich wirkt sich in der unterschiedlichen Gestalt der
Diagramme ein gefügebedingt unterschiedliches mechanisches
Verhalten aus. Einige der Versuche erreichten nahezu die
Bruchbedingung, andere hingegen nicht. Abgesehen davon

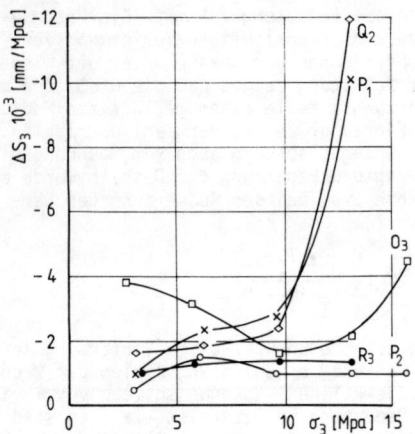

Abb. 4: $\Delta s_3 - \sigma_3$ - Schaulinien für die Versuchsblöcke P_1, P_2, Q_2, O_3 und R_3

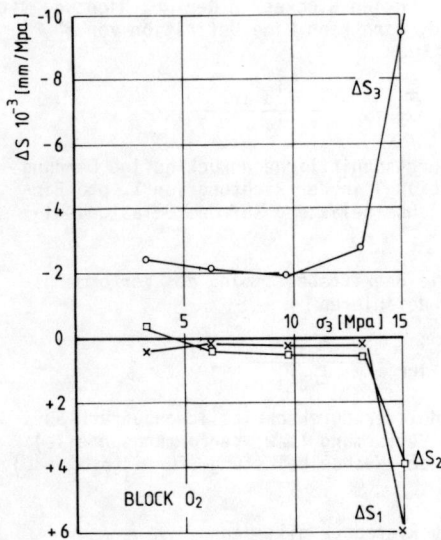

Abb. 5: $\Delta s_3 - \sigma_3$ - Linien für den Block O_2

ist eine allen sechs Diagrammen gemeinsame Regelhaftigkeit festzustellen. Abb. 6 zeigt den Vesuch, diese Regelhaftigkeit des $\Delta s_3 - \sigma_3$ - Zusammenhanges durch eine Modell- bzw. Typen-Kurve zu repräsentieren. Die Diagramme der sechs Versuchsblöcke erscheinen dann als gewisse (unvollständige) Abschnitte dieser allgemeinen vollständigen Modellkurve (Abb. 7).

Demnach zeigt das Materialverhalten der Versuchsblöcke unter oft wiederholten Belastungswechseln drei charakteristische Bereiche, welche bei zunehmendem Belastungsniveau σ_3 durchlaufen werden:

Der Kurvenast I steht nach unserer Meinung in Zusammenhang mit einer gewissen Kompaktion der geklüfteten Felmasse. Der Ast III scheint den Beginn der Bruchprozesse zu charakterisieren (fracture initiation). Der mittlere Kurvenast II könnte im wesentlichen als Bereich relativer (innerer)

Stabilisierung des Gebirges betrachtet werden. Daraus ergeben sich drei charakteristische Punkte der Modellkurve (Abb. 6):

A) Beginn des Prozesses beginnender innerer Stabilisierung,
B) Beginn der Materialschädigung,
C) Bruch.

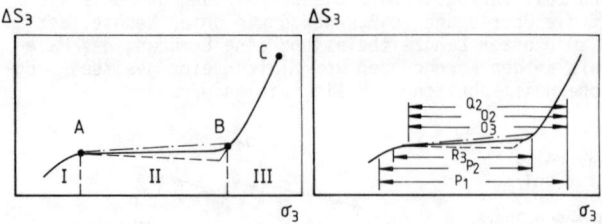

Abb. 6: Modelldiagramm für typische $\Delta s_3 - \sigma_3$-Beziehungen Abb. 7: Einpassung der Ergebnisse der sechs Versuchsblöcke in das Modelldiagramm der Abb. 6.

4. BETRACHTUNGEN ÜBER DIE DREI CHARAKTERISTISCHEN BEREICHE DER $\Delta s_3 - \sigma_3$ - BEZIEHUNG

Eine Begründung für die vorerwähnte Unterscheidung dreier Entfaltungsstufen des Materialverhaltens hinsichtlich seiner unrückläufigen Verformungsanteile sind nicht allein von der Gestalt der $\Delta s_3 - \sigma_3$ - Modellkurve abgelesen, sondern diese Abschnitte entsprechen auch einem ganz charakteristischen Querverformungsverhalten sowie einem typischen Verlauf der Volumenänderung der Probekörper. Zu deren Beschreibung sind bereits seit längerem folgende Kenngrößen in Verwendung (Müller-Salzburg, L., C. Tess, E. Fecker u. K. Müller, 1973):

a) $\quad \psi_1 = \dfrac{-\varepsilon_1}{\varepsilon_3}$; $\quad \psi_2 = \dfrac{-\varepsilon_2}{\varepsilon_3}$.

Diese Kenngrößen sind analog der Poisson-Zahl für Kontinua zu verstehen.

b) $\quad \dfrac{\Delta V}{V} = \varepsilon_1 + \varepsilon_2 + \varepsilon_3$ (negative Werte bedeuten

Zusammendrückung, positive bedeuten Dehnung oder Dilatanz).

c) $\quad \delta_1 = \varepsilon_1 + \varepsilon_3$; $\quad \delta_2 = \varepsilon_2 + \varepsilon_3$

δ_1 entspricht $\dfrac{\Delta V}{V}$, wenn $\varepsilon_2 = 0$ ist;

δ_2 ebenso, wenn $\varepsilon_1 = 0$ ist.

Unter Verwendung dieser Größen kann die Differenz der Veränderung des δ in zwei aufeinander folgenden Richtungen ermittelt werden. Für die Blöcke P_1, P_2 und Q sind die ψ_1-, ψ_2-, $\dfrac{\Delta V}{V}$, δ_1 und δ_2-und die Δs_3-Linien in Abhängigkeit von σ_3 in den Abbildungen 8, 9 und 10 wiedergegeben.

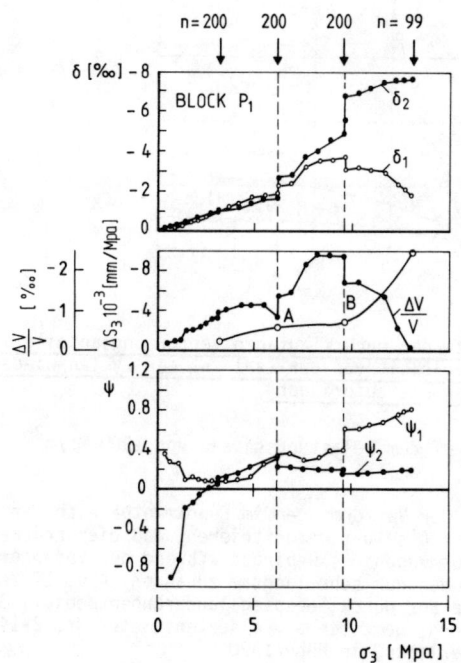

Abb. 8: Diagramm für Δs_3, δ, ψ und $\frac{\Delta V}{V}$ in Abhängig-

keit von σ_3 für den Versuchsblock P_1

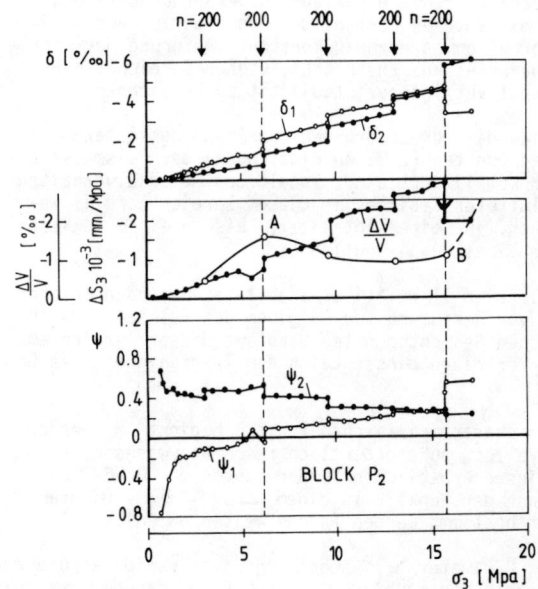

Abb. 9: Kennlinien für Δs_3, δ, ψ und $\frac{\Delta V}{V}$ in Abhängig-

keit von σ_3 für den Versuchsblock P_2

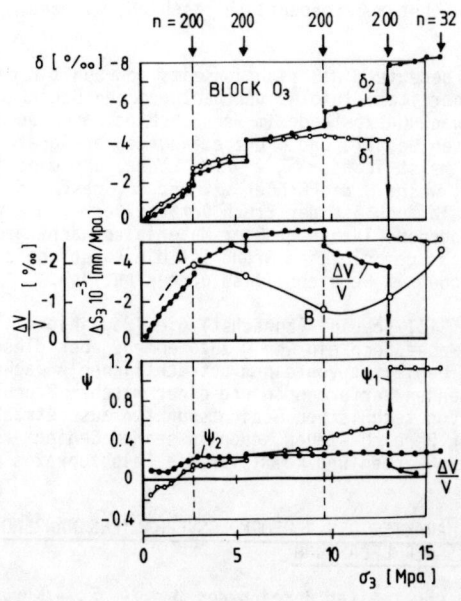

Abb. 10: Kennlinien für Δs_3, δ, ψ und $\frac{\Delta V}{V}$ in Abhängig-

keit von σ_3 für den Versuchsblock O_3

Im folgenden haben wir zu begründen, weshalb wir im ersten Kurvenbereich des Δs_3 - σ_3 - Diagramms ein Stadium anfänglicher Kompaktion zu erkennen glauben:

Aus den Abb. 8, 9 und 10 ist zu erkennen, daß ψ_1 und ψ_2 in diesem Bereich sehr unregelmäßig verlaufen (die Blöcke O_2, Q_2 und R_3 zeigen in diesem Bereich ähnliche Erscheinungen).

In diesem Bereich können ψ_1 und ψ_2 sehr hohe oder ganz niedrige, in einigen Fällen sogar negative Werte haben.

Dieses Verhalten erklärt sich aus dem Wesen der geklüfteten Felsmasse, in welcher bereits vor dem Versuch zahlreiche Fugen und Risse vorhanden sind. Im Verlauf der Freilegung der Blöcke im Stollen können sich viele dieser Klüfte infolge Entspannung des Gebirges öffnen. Am Beginn dieses Bereiches, wo die Belastung noch nicht hoch ist, können etliche der geöffneten Klüfte gewisser Kluftstellungen bei der ersten Zusammendrückung wieder geschlossen werden, andere hingegen können immer noch in geöffnetem Zustande verharren oder sogar noch ein wenig geweitet bzw. in ihrer Erstreckung ausgedehnt werden. Deshalb kann die Querverformung, welche nur an einzelnen Punkten der Probe gemessen wurde, in diesem Bereich sehr unregelmäßig sein und ψ_1 wie ψ_2 können in keiner Weise ähnlich verlaufen wie die Poisson-Koeffizienten kontinuierlicher Medien. Aber diese Unregelmäßigkeit des Verlaufes von ψ_1 und ψ_2 nimmt mit zunehmender Belastung ab. $\frac{\Delta V}{V}$ sowie δ_1 und δ_2 nehmen gleichfalls in diesem Bereich monoton ab.

Im Bereich II nehmen $\frac{\Delta V}{V}$, δ_1 und δ_2 zufolge abgeschlossener Anfangskompaktion und zufolge des hier bereits höheren Belastungsniveaus mit zunehmender Belastung weiterhin regelmäßig ab und die Querverformungskenngrößen ψ_1 und ψ_2 verhalten sich mehr oder weniger stetig. Ihre Werte bewegen sich zwischen 0,2 und 0,4, ganz ähnlich den Werten der Poisson-Zahl für Kontinua.

Ein Vergleich zwischen δ_1 und δ_2 sowie zwischen ψ_1 und ψ_2 zeigt, daß deren Größen sich sehr verschieden entwickeln. Das bedeutet Anisotropie in Bezug auf das Verformungsverhalten, welche aber in diesem Zustand noch keine große Bedeutung hat.

Es sei besonders darauf hingewiesen, daß Δs_3 unter Wechselbelastung in diesem Bereich II nahezu konstant bleibt oder nur ganz wenig ab- bzw. zunimmt, und zwar für einen recht großen Bereich von σ_3. Die Größe der unrückläufigen Verformungen während 200 Lastwechseln ist in diesem Bereich nicht beträchtlich und verläuft ungefähr proportional zu σ_3; mit anderen Worten: der Modul der unrückläufigen Deformation ist nahezu konstant. Aufgrund der Phänomene scheint es uns angebracht, diesen Bereich als Bereich relativer innerer Stabilität zu bezeichnen.

Wenngleich die Formänderungen und Spannungszustände an einzelnen Punkten (z.B. an Kluftenden) der Felsmasse gewiß sehr kompliziert sind, ähnelt das Materialverhalten des geklüfteten Gebirges in diesem Bereich doch im wesentlichen, in seiner Totalität, bis zu einem gewissen Grad dem eines Quasi-Kontinuums.

Dieser Bereich II relativer Stabilität ist nach unserer Meinung von Bedeutung für Aufgaben des Felsbaues. Nach technischen Gesichtspunkten verdient dieser Bereich am meisten Vertrauen hinsichtlich der Tragfähigkeit des Gebirges.

Nach dem charakteristischen Punkt B beginnt im Bereich III der Kurve Δs_3 außerordentlich rasch anzuwachsen. Aufgrund dieser erheblichen Veränderungen der Größe von Δs_3 kann der Punkt B in einem Δs_3 - σ_3 - Diagramm unschwer bestimmt werden (siehe weiter unten).

Auch der Parameter ψ_1 wächst von da ab rapide an und erreicht rasch Werte über 0,4, bis 0,5 und darüber. ψ_2 hingegen verharrt immer noch relativ konstant, ähnlich wie im Bereich II. ΔV wächst nach Erreichen des Punktes B an (was Dilatanz V bedeutet), nimmt aber rapide zu, sobald der Bruch bevorsteht. Die σ_1 - σ_3- und die σ_2- σ_3 -Linien tendieren nach entgegengesetzten Seiten, d.h. σ_1 wächst außerordentlich rasch an, während σ_2 abnimmt.

Das alles bedeutet, daß sich jenseits von Punkt B der Bruch vorbereitet. Infolge der geologischen Bedingungen und der Spannungszustände im Versuchsblock aber auch zufolge dessen Gestalt und Grenzbedingungen ereignet sich der Bruch meist in der σ_3 - σ_1 - Ebene, und dies ist der Grund, weshalb die Größen ψ_1 und σ_1 rasch anwachsen. Deshalb auch ist der Bruch unmittelbar an die vorher vorhandenen Flächen größter Materialschwäche gebunden, und aus dem gleichen Grund ist die Anisotropie der Deformationen im Bereich III so offensichtlich.

Im Stadium III scheint (zunächst) die Tragfähigkeit der gesamten Masse noch ein wenig zuzunehmen, aber diese Zunahme ist begleitet von einem beträchtlichen Anwachsen der bleibenden Verformungen und einer raschen Bruchentfaltung. Von technischen Gesichtspunkten aus betrachtet, muß dieser Bereich - auch schon in seinem Beginn - strenge gemieden werden und kommt für die Felsbaupraxis nicht in Betracht.

5. DER CHARAKTER DES VERFORMUNGSVERHALTENS WÄHREND ZYKLISCHER BELASTUNG

Im ersten und zweiten Bereich der Δs_3 - σ_3 - Kurve unter häufigen Wechselbelastungen haben die Diagramme des unrückläufigen Verformungsanteils in ihrer Abhängigkeit von der Zahl der Lastwechsel n ganz ähnlichen Charakter. Nach den ersten Belastungs- und Entlastungszyklen nehmen die bleibenden Verformungsanteile nur langsam zu und zeigen eine Tendenz zur Stabilisierung an, wie aus Abb. 11 erkannt werden kann, in welcher die Linien für $\Delta l_{3\,irr}$ in Abhängigkeit von n für den Block Q_2 auf den Spannungsniveaus σ_3 = 308 Mp/m² (3,08 MN/m²) und 615 Mp/m² (6,15 MN/m²) wiedergegeben sind.

Die unrückläufigen Formänderungen sind im Bereich III sehr groß. Keinerlei Tendenz zur Stabilisierung ist erkennbar und in etlichen Fällen erfolgt die Zunahme der Formänderungen überproportional zur Zahl der Lastwechsel. Abb. 11 gibt ein Beispiel für den Verlauf der unrückläufigen Formänderungen in Abhängigkeit von n

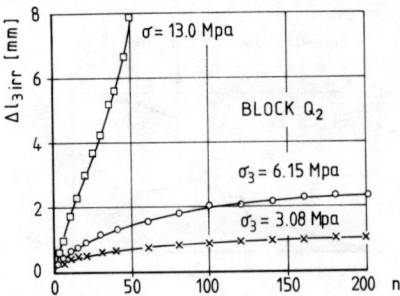

Abb. 11: Linien der unrückläufigen Verformungsanteile in Abhängigkeit von der Zahl n der Zyklen wiederholter Lastaufbringung

für Block Q_2 auf dem Belastungsniveau von 1300 Mp/m² (13 MN/m²).

Die Messungen der Verformungen im Fundamentbereich der Staumauer Kurobe 4 selbst demonstrieren, daß die unrückläufigen Verformungen des Gebirges während der Veränderungen der Stauhöhen nur sehr langsam zunahmen. Abb. 12 zeigt die Ergebnisse der durch Seespiegeländerungen beeinflußten Felsverformungen, gemessen am Felsextensometer Nr. 2-14 im linken Widerlager, in Höhe 1320.
Da die Kraft, die an diesem Widerlager in dieser Höhe angreift, nicht größer ist als 14000 Mp/m (140 MN/m), besteht kein Zweifel darüber, daß die größten Hauptnormalspannungen in diesem Bereich der Felsmasse, welche durch die Wasserlast in das Widerlager der Bogenmauer eingeleitet werden, den Bereich II der Δs_3 - σ_3 - Kurve nicht überschreiten.

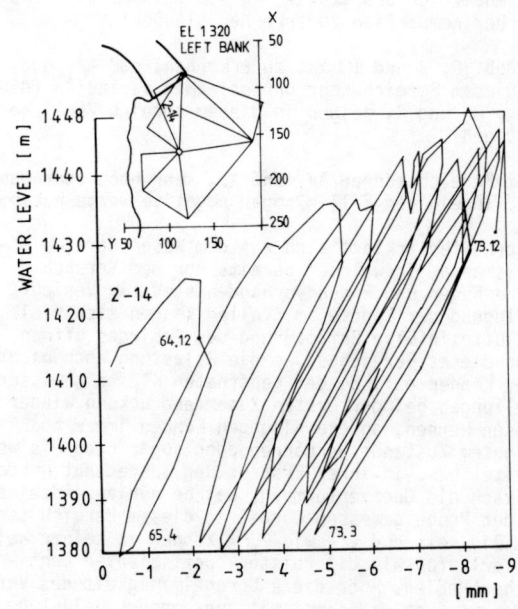

Abb. 12: Felsverformungen, gemessen am Extensometer Nr. 2-14 (nach Unterlagen der Kansai Electric Power Co.)

Die während der Großversuche in situ erhaltenen Diagramme der Verformung unter zyklischer Belastung eröffnen ein Verständnis für die Tatsache, daß die unrückläufigen Verformungen (und Verschiebungen), welche die Felsmeßgeräte während der Füllungen und Leerungen des Staubeckens verzeichnet haben, nach den ersten Jahren der Spiegelschwankungen langsam zunehmen mußten, dann aber die gleiche Tendenz zur Stabilisierung zeigen, wie sie während der Versuche beobachtet worden ist.

In Abb. 13 sind solche im Meßgerät Nr. 2-14 am Bauwerk gemessenen Verformungen in Abhängigkeit von der Zeit für einen Zeitraum von 17 Jahren als Beispiel aufgetragen. Entsprechend den Verformungslinien, die bei Wechselbelastung an den Prüfkörpern erhalten worden waren, kann eine Voraussage über die Größe der bleibenden Verschiebungen gemacht werden, welche in Zukunft auftreten werden. Diese werden in dem genannten Beispiel der Meßstelle 2-14, welche im Jahre 1981 insgesamt -10,7 zeigte, im Jahr 2000 das Maß von -14,0 mm nicht überschreiten wird, wenn man der Ermittlung das zugehörige Verformungsdiagramm für Wechsellast der Abb. 11 zugrundelegt.

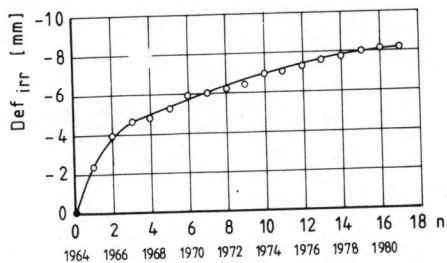

Abb. 13: Unrückläufige Verformungsanteile in der Felsmasse, gemessen am Extensometer 2-14 (revidierte Daten)

Wenn wir in der Lage sind, unsere Schlußfolgerungen auf eine Vergleichung von Messungen an der Bauwerksgründung einerseits mit den Ergebnissen von in situ-Großversuchen in klüftigem Gebirge ähnlicher geologischer Voraussetzungen andererseits zu stützen, können wir neue und wertvolle Aussagen über die Sicherheit unserer Bauwerke gewinnen. Auch für die statische Berechnung der Betonmauer selbst ist es von Wert, zu wissen, wie große bleibende Verformungen des Gründungsfelsens nach einer langen Zeitspanne im Maximum zu erwarten sind. Der in diesem Bericht erläuterte Weg könnte vielleicht eine Ausgangsposition für solche Ermittlungen sein.

6. BESTIMMUNG DES CHARAKTERISTISCHEN KURVENPUNKTES B

Zur Bestimmung des Kurvenpunktes B der Δs_3 - σ_3 - Linie können fünf Phänomene bzw. Kriterien herangezogen werden:

a) der Verlauf der unrückläufigen Verschiebung Δs_3 unter der zyklischen Belastung ändert sich deutlich;

b) die Größe γ_1 (bzw. γ_2) beginnt über den Betrag von 0,4 gegen 0,5 hinauszuwachsen und nimmt rasch zu;

c) δ_1 und δ_2 beginnen sich nach verschiedenen Richtungen zu entwickeln;

d) die Größe der Volumenänderung $\frac{\Delta V}{V}$ (in Abhängigkeit von σ_3 aufgetragen) beginnt anzuwachsen;

e) die unrückläufigen Verformungen (bzw. Verschiebungen), aufgetragen in Abhängigkeit von der Zahl der Lastwechsel n, beginnt rasch anzuwachsen, ohne daß eine Tendenz zu einer Beruhigung zu erkennen wäre.

Wenn obige fünf Kriterien auch nicht immer genau zugleich auftreten, kann dennoch schon aufgrund einiger von ihnen der Punkt B ohne Schwierigkeit bestimmt werden.

Dieser Punkt B der Δs_3 - σ_3 - Linie ist zu verstehen als Grenze der Lastaufnahmefähigkeit des Gebirges unter

Wechsellast. Er darf natürlich nicht ohne weiteres als zulässige Belastungsgrenze des Gründungsfelsens angesehen werden, sondern es muß ein Sicherheitsabstand in Betracht gezogen werden, welcher die Abminderung der tatsächlich zu tolerierenden Belastung gegenüber dieser obersten Grenze der Lastaufnahme angibt.

Abschließend liegt uns daran, festzustellen, daß die Auswertung der Kurobe-Großversuche im Bezug auf Wechselbelastung die große Bedeutung von Analysen des Gebirgsverhaltens auf der Basis von Verformungsbeobachtungen demonstriert, welche uns über die verschiedenen Zustände bzw. Phasen des Materialverhaltens klüftiger Felsmassen zu informieren vermag. Aufgrund von Beobachtungen von Formänderungen und der zwischen ihnen waltenden regelhaften Beziehungen können charakteristische Zustände des Gebirges in Bezug auf Fließen und Bruch erkannt und es kann eine Einschätzung des (jeweils noch nicht in Anspruch genommenen) Reservebereiches der Standsicherheit gegeben werden. Der Weg, welcher in der vorliegenden Mitteilung aufgezeigt wurde, dürfte sich wahrscheinlich auch für die Auswertung anderer Arten von Felsprüfversuchen als brauchbar erweisen.

DANK

Wir sind der Kansai Electric Power Comp. in Osaka, Japan, und ihrem Executive Vice Presidenten, Dr. Minoru Yoshida für die Verfügbarmachung von Unterlagen und für die Zustimmung für die Veröffentlichung unserer Ergebnisse sehr zu Dank verbunden.

Ferner haben wir der Alexander von Humboldt-Stiftung in Bonn, BRD, sehr zu danken, welche einen der beiden Autoren, Professor Xiurun GE, die Befassung mit der Studie über das Berichtthema durch ein Forschungsstipendium an der Universität Karlsruhe ermöglicht hat.

Herrn Professor Dr. Otfried Natau, Leiter des Institutes für Bodenmechanik und Felsmechanik der Universität Karlsruhe und seinem Mitarbeiterstab gebührt unser Dank für Unterstützung und Förderung unserer Arbeit.

LITERATURHINWEISE

John, K.W. (1961)
    Die Praxis der Felsgroßversuche, beschrieben am Beispiel der Arbeiten an der Kurobe-Staumauer in Japan: Geologie u. Bauwesen, Sonderabdruck aus Jg. 27, H. 1

The Construction Department of the Kansai Electric Power Co. Inc. (1967)
    Mechanical behavior of Kurobe IV Dam and its foundation especially the difference from the result of calculation: Proc. 9th ICOLD, Q 34, R. 4

Müller, L. (1966)
    Words by president of the Int. Soc. for Rock Mech., 25th Sept. 1966, Opening Session of the 1st Congress of the ISRM: Proc. 1st Congr. ISRM, Vol. 3

Müller, L. (1971)
    Die mechanischen Eigenschaften der geologischen Körper: Carinthia 11, Sonderh. 28, Festschrift Kahler, Klagenfurt

Müller-Salzburg, L., C. Tess, E. Fecker, K. Müller (1973)
    Kriterien zur Erkennung der Bruchgefahr geklüfteter Medien - Ein Versuch: Rock Mechanics, Suppl. 2

Nose, M. (1964)
    Rock test in situ, conventional tests on rock properties and design of Kurobegawa No. IV Dam based thereon: Proc. 8th ICOLD, Q 28, R. 1

Yoshida, M. (1982)
    Mechanical behavior of Kurobe Dam and its foundation and safety of the dam: Proc. 14th ICOLD, Q 52, R. 2

E

A SITE EXPLORATION TRIAL USING INSTRUMENTED HORIZONTAL DRILLING

Essai de forage instrumenté pour la reconnaissance de terrain

Versuchsbohrung am Ort mit Hilfe von einem mit Messgeräten versehenen Horizontalbohrer

M. V. Barr

Drilling Engineer, Exploration and Production Division, BP Research Centre, Sunbury-on-Thames, U.K.

E. T. Brown

Professor of Rock Mechanics, Imperial College of Science and Technology, London, U.K.

SYNOPSIS

Horizontal drilling may be used in tunnelling side investigations or for probing ahead of the face during tunnelling. The quantity and quality of the information on ground conditions obtained from this or other types of rotary drilling can be improved by continuously recording the drilling variables. An hydraulic diamond drill was fully instrumented and used in horizontal drilling trials in an underground limestone quarry. The results show that important geotechnical data including rock strength indices, changes in rock strength and the presence of open, clay or gouge-filled and water bearing discontinuities, may be recovered using this technique.

RESUME

Le forage horizontal peut être utilisé pour l'investigation d'un emplacement pour un tunnel ou pour sonder à l'avant pendant le percement d'un tunnel. La quantité et la qualité des informations ainsi obtenues peuvent être améliorées par l'enregistrement continu des variables de forage. Un foret hydraulique à mèche de diamant fut entièrement appareillé et utilisé dans des essais de forage horizontal dans une carrière de calcaire souterraine. Les résultats montrent que d'importantes données géotechniques, y compris les indices de la résistance rocheuse, les variations de la résistance rocheuse et la présence de discontinuités ouvertes et remplies, soit d'argile, soit d'argile provenant de frottement, soit d'eau, peuvent être recueillies par l'utilisation de cette technique.

ZUSAMMENFASSUNG

Horizontal-Bohren kann angewendet werden beim Tunnelbohren für Untersuchungen der örtlichen Verhältnisse oder zum Vorsondieren der Oberfläche während des Tunnelbohrens. Der Wert und die Anzahl von Informationen über die Untergrundverhältnisse, die durch diese oder andere Arten von Rotarybohren erbracht werden, können erhöht und verbessert werden durch konstante Messungen der Bohrungs-Parameter. Ein hydraulischer Diamantbohrer wurde mit Messgeräten versehen und in horizontalen Versuchungsbohrungen in einem unterirdischen Kalksteinbruch eingesetzt. Die Ergebnisse haben gezeigt, daß wichtige geotechnische Daten, wie Gesteinsstärke, Veränderungen der Gesteinsstärke und das Vorhandensein von offenen, Lehm- oder Schlammgefüllten und wasserhaltenden Rissen, durch diese Technik gemessen werden können.

INTRODUCTION

Even after a well conducted conventional site investigation, there is usually some uncertainty about ground conditions likely to be encountered during tunnelling. The most certain way of obtaining continuous detailed information about ground and water conditions along the tunnel route is to drill in that direction, more or less horizontally. This is often done from the face or from a side chamber during construction. Such an operation is known as probing ahead. An alternative is to carry out horizontal or sub-horizontal drilling along the proposed tunnel alignment as an advanced stage of the site investigation. However, it must be recognised that, in neither case, will a single drill hole directed along the tunnel axis always detect features that may cause difficulties during tunnelling.

The use of long, horizontal drill holes for site investigations was reviewed by Majtenyi (1976). Carroll and Cunningham (1980) describe an experience of drilling horizontal holes of up to 1125m long ahead of tunnelling. Techniques used in the rotary drilling of long, horizontal holes in rock for other purposes have been described by Thakur and Poundstone (1980) and Woods and Hopley (1980), among others. Horizontal holes with diameters of up to 170mm have been successfully drilled in rock over lengths exceeding one kilometre. Clearly, for such holes, problems can arise with penetration capability, costs and directional accuracy. Sinha, Brown and Green (1982) have considered the influence of drill string flexure and stabilization systems on directional control in horizontal drilling. In probing ahead, hole lengths are usually limited to a few tens of metres, and so directional control is not such a major concern.

In rock, cored drilling will form part of the site investigation, but open-hole water-flush rotary drilling using a non-coring bit is generally used in probing ahead of the tunnel face. In the latter case, the only information directly available on the characteristics of the rock comes from the cuttings carried back in the return water. The driller's observations of the behaviour of the drill can also provide some indication of the nature of the ground ahead. A potentially valuable way of quantifying such information is to instrument the drill rig to monitor the major drilling variables as drilling proceeds. The background to the use of instrumented drilling in tunnelling site investigations and its potential advantages have been fully discussed by Brown and Phillips (1977) and Brown and Barr (1978).

As a result of recommendations made by the Building Research Establishment/Transport and Road Research Laboratory Working Party for Probing Ahead for Tunnels (1975), a programme of laboratory and field research using instrumented horizontal diamond drilling has been carried out. Full details of the instrumented drilling results and of the associated geophysical and television borehole logging have been given by Barr (1982) and West (1980), respectively. The present paper presents and analyses some of the data obtained in an underground horizontal drilling trial.

RELATIONSHIPS BETWEEN ROTARY DRILLING VARIABLES AND ROCK PROPERTIES

The operating variables in rotary drilling can be divided into two groups - underline independent variables that can be independently controlled by the operator, and dependent variables that represent the response of the drill as it penetrates a particular rock mass. The independent variables are usually thrust (F), rotary speed (N) and flushing fluid flow rate; the dependent variables are penetration rate (R), torque generated at the bit-rock interface (T), and flushing fluid pressure. Drilling performance will vary with the type, design and condition of the bit and with the mechanical properties of the rock mass being drilled.

An approach to the determination of the theoretical relationships between the drilling variables used by Teale (1965), Rowlands (1971) and others, is to calculate the specific energy (e) or work done in removing unit volume of rock. Work calculations show that, in terms of the variables defined above and assuming no energy losses, the specific energy is given by

$$e = \frac{F}{A} + \frac{2\pi NT}{AR} \qquad (1)$$

where A is the cross-sectional area of the bit.

The term F/A, (the mean pressure on the bit), has been found to be negligibly small compared with the components of specific energy arising from torque (Rowlands, 1974). The specific energy can then be expressed as

$$e = \frac{2\pi}{A} \cdot \frac{NT}{R} \qquad (2)$$

Laboratory drilling trials carried out by Rowlands (1971, 1974) and others suggest that for ideal conditions in which the rock properties remain constant, there is no bit wear or clogging, and there are no losses due to vibration or friction of the rods on the sides of the hole, e is constant for a given thrust and torque is proportional to thrust. The latter result implies a constant coefficient of 'friction' at the bit-rock interface, μ, defined by $T/r = \mu F$ where r is the radius at which the tangential force may be considered to be applied. In this ideal case, it follows that

$$T \propto F \quad \text{if } \mu \text{ is constant,}$$
$$R \propto T \quad \text{if } N \text{ is constant,}$$
$$R \propto F \quad \text{if } N \text{ and } \mu \text{ are constant,}$$
$$\text{and } R \propto N \quad \text{if } F \text{ is constant.}$$

The applicability of these theoretical relationships has been demonstrated in a variety of rotary drilling trials carried out under controlled laboratory conditions. However, their validity under field conditions has not been so clearly demonstrated. Nevertheless, they do provide a basis on which to interpret recorded relationships between drilling variables. If, for example, thrust and rotary speed are held constant and there is no change in flushing fluid flow rate and pressure, a sudden increase in torque and decrease in penetration rate should indicate that the bit has encountered a rock with a higher resistance to drilling. Similarly, a rapid local increase in penetration rate and a corresponding decrease in torque should indicate a zone of low, or even zero resistance to drilling.

A number of attempts have been made to develop expressions which relate the penetration rate to other drilling variables and some measure of rock strength. An approach that is of particular interest in terms of data collection from site investigation drilling was put forward by Tsoutrelis (1969). In a series of laboratory drilling tests on five rock types using a hard metal rotary bit, Tsoutrelis found that, with rotary speed constant, the initial penetration rate varied linearly with thrust. This suggested that, for a given bit type and design, the thrust (F) and the penetration rate at the commencement of drilling before wear affects bit performance (R_0), are related by an expression of the form

$$R_O = k \ (F - F_O) \qquad (3)$$

where k is the slope of the F - R graph, and F_O is the intercept that this line makes with the F axis.

Data obtained by Tsoutrelis and a number of other investigators show that, up to a certain limiting speed, the parameter $k_O = k/N$ is constant for a given rock and bit design. Tsoutrelis then found that, for a given bit, a unique relationship exists between k_O and the uniaxial compressive strength of the rock (σ_c) such that

$$\sigma_c = \frac{A}{k_O + B} \qquad (4)$$

where A and B are constants for the bit.

This result suggests that it should be possible to determine the in-situ compressive strength of rock by first carrying out a series of controlled laboratory tests to establish the relationship between k_O and σ_c for the type of bit being used, and then determining k_O values from values of R, F and N recorded in the field. A major problem likely to be encountered in attempting to apply Tsoutrelis' method in practice is that wear of the bit in abrasive rocks could make the determination of initial penetration rates difficult. Tsoutrelis (1969) and Rowlands (1971) attempted to take account of this problem by developing special methods of processing data that allow for the influence of bit wear. Other practical problems are likely to arise because of the natural variability of rock properties in situ and mechanical losses occurring down the drill string so that the thrusts recorded at the drill head may not be those applied to the bit.

Despite these obvious difficulties, there is considerable attraction in the possibility of being able to back-calculate the compressive or drilling strength of the strata through which the drill bit passes. Rock strength data may be required for a number of design purposes but, as Brown and Phillips (1977) have shown, the method has particular attraction as a means of acquiring rock strength data for tunnel boring machine design studies. It could eliminate the need for costly continuous coring, since drilling could be carried out with a non-coring rotary bit.

DRILL RIG INSTRUMENTATION

Instrumenting rotary drilling rigs to measure and record the parameters referred to in the previous section is a relatively straight-forward matter. The required technology is readily available and has been applied to a wide spectrum of drilling operations in the past. In the oil industry, drilling variables are recorded as a matter of course. Although oil well instrumentation cannot be used directly in civil engineering site investigations because of the different scales of the two operations, much can be learned from the oil industry's proven field technology.

All modern rotary drills carry some instrumentation and the extra expenditure required to fully instrument them need not be very high. Brown and Phillips (1977) give a wide variety of examples of drill rig instrumentation. Instrumentation of particular site investigation rigs is described by Brown (1975) and by Mouxaux (1978), for example.

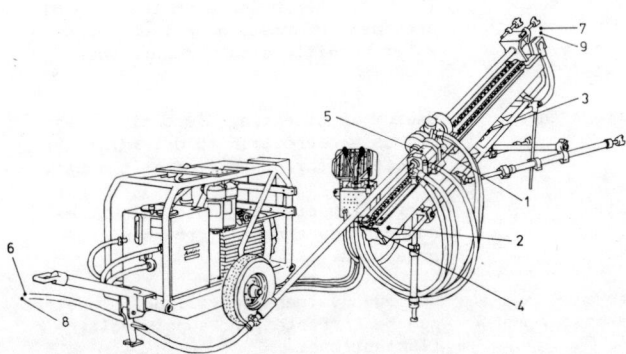

Figure 1. Location of instrumentation on the Atlas Copco Diamec 250 drill.

In the underground drilling trials described herein, an Atlas-Copco Diamec 250 diamond drill (Figure 1) was used. This is a fully hydraulic, lightweight rig that is particularly suited for one man operation in confined underground locations. Rigs of this type have been extensively used in probing ahead operations. Cored horizontal holes 56mm in diameter and up to 150m long can be drilled with this drill using aluminium drill rods.

Table 1 summarises the method of measurement of each of the drilling variables and the use made of each measurement. Of the variables listed, all were periodically sampled via a Mycalex data logger or selected variables

were continuously monitored using a strip chart recorder. At pre-determined intervals, normally from 2 to 10 seconds depending on drilling conditions, the Mycalex logger would scan an array of 20 data channels and transfer recorded values to paper tape for subsequent computer processing. This intermittent sampling of the drilling operation is reflected in the logging system's response to downhole events. Comparison with the continuous strip chart record showed that often Mycalex-recorded events were truncated particularly during rapid penetration by the drill bit. However, these sampling errors were greatly reduced by shortening the interval between data scans. A simple computer program decoded the data, made voltage corrections, applied calibration factors and presented the drilling record in both numerical and graphical form.

By contrast, the chart recorder provided an immediate indication of downhole events by monitoring 5 critical variables, namely: thrust, torque, rotary speed, inlet water pressure and head displacement. This was an invaluable source of information for the driller and provided an ideal diary of the drilling operation.

DRILLING SITE AND PROGRAMME

The initial underground trials of instrumented horizontal diamond drilling were carried out in a disused underground limestone quarry at Corsham, Wiltshire. The Great Oolite limestone known locally as Bath Stone and used as a building stone for several centuries (Hudson, 1971), had been quarried at the test site from galleries approximately 8m wide and 30m below surface. Lithologically the rock is a pale yellow oolitic limestone of Jurassic age (Kellaway and Welch, 1948). It is quite uniform in appearance and mechanical properties, but is intersected by randomly distributed joints which are frequently infilled with clay. Above the old workings is the Bradford clay which acts as a waterproofing horizon thereby providing a relatively dry working environment.

Four 56mm diameter, 19m long horizontal test holes were drilled parallel to a gallery wall by locating the drill rig in a convenient recess in the wall. The arrangement of the drill and the recording equipment at the site is shown in Figure 2. This layout had the additional advantage of affording access to the geological features encountered by the drill bit by simply mapping their exposure in the gallery. Also the boreholes were spaced closely together to better evaluate repeatability of recording downhole events.

Figure 2. Site of the underground drilling trial.

TABLE 1 - INSTRUMENTATION OF ATLAS-COPCO DIAMEC 250 ROTARY DRILL FOR
HORIZONTAL DRILLING TRIALS

Parameters	Method of Measurement	Information Obtained
Rotary speed 1	The output frequency from a tachometer coupled to the drive shaft of the motor is processed by a frequency to voltage converter and transmitted to the data logger as a D.C. voltage proportional to rotary speed.	Important drilling parameter for correlation with rock mass properties. Also used in determination of torque developed at the chuck.
Head displacement 2	A rotary potentiometer monitors the movement of the pulley over which the rotary head drive chain passes. Output voltage is directly proportional to the position of the rotary head.	Used to determine instantaneous penetration rates for correlation with rock mass properties.
Thrust 3	A 0-35 MN/m^2 pressure transducer used in conjunction with an instrumentation amplifier measures the oil pressure applied to the thrust piston.	Useful parameter for controlling drilling performance.
Inlet oil pressure 4	A 0-70 MN/m^2 pressure transducer used in conjunction with an instrumentation amplifier monitors the oil pressure at the control panel.	Used in the estimation of torque presented on the chart recorder (see below).
Swash plate position 5	A rotary potentiometer coupled to the swash plate shaft gives an output voltage proportional to swash plate angle.	Swash plate position and rotary speed are used to numerically determine the torque delivered to the drill chuck from calibration curves obtained using a Heenan-Froude dynamometer.
Torque	An electronic multiplier connected in division mode is used to derive an approximate torque value from inlet oil pressure and rotary speed.	Provides an approximate torque reading to assist the driller during drilling operations.
Water flow rates Inlet 6 Outlet 7	The pressure drop across a venturi nozzle is measured by a differential pressure transducer and the signal processed by a square root extractor to produce a value of flow rate.	The net water balance in the borehole provides information on the groundwater conditions in the rock mass.
Inlet water pressure 8	A 0-1.7 MN/m^2 pressure transducer monitors the water supply to the drill string.	Useful in detecting discontinuities and as a correction to determine the effective thrust applied at the bit.
Outlet water pressure 9	A 0-0.7 MN/m^2 pressure transducer monitors the return water pressure as it passes through a stuffing box mounted on the rock face.	Useful in detecting discontinuities and sensing abnormal groundwater pressures.
Transducer and potentiometer supply voltages	Two precision voltage regulators ensure a stable supply. Any small fluctuations associated with the devices are recorded.	Recorded values of drilling parameters can be corrected for supply voltage fluctuations.

EXAMPLES OF RECORDED OUTPUT

An example of a strip chart record from Borehole 3 is shown in Figure 3. Thrust, torque, rotary speed, head displacement (penetration) and water pressure are continuously plotted against drilling time. The rate of penetration at any depth is given by the slope of the head displacement - time curve, higher penetration rates being characterised by a horizontally-tending slope. Figure 3 shows drilling proceeding uniformly at an approximate penetration rate of 35cm/min, a rotary speed of 2000 rpm and a thrust of 3.1 kN. At 2.30 metres depth, a rapid acceleration in head displacement occurred corresponding to a penetration rate of 117 cm/min over a drilled interval of 8 cm. Momentarily, the rotary speed dropped to 1900 rpm, the torque increased to

17.5 Nm while the thrust and inlet water pressure remained unchanged. Experience has shown this type of drilling response to indicate a discontinuity or cavity.

Thereafter, the drilling parameters tended to their former levels until a second fissure was encountered by the drill at 2.64 metres. A momentary decrease of some 200 rpm was accompanied by a small drop in the thrust and inlet water pressure values as the drill bit traversed a 7 cm. void. Returns were lost to this discontinuity, the flush water discharging from the exposure of this joint in the gallery wall. Beyond this joint, penetration rates gradually decreased from

40 to 15 cm/min. Finally, a smaller fissure was traversed at 2.97 m depth. The recorded drilling parameters display the characteristic signature of this type of geological feature. The record ends at 3.02 m. as the rotary head of the drill has reached the bottom of the drill carriage and a rechuck is required.

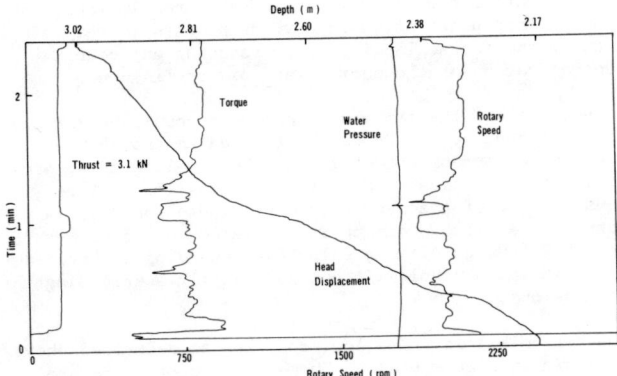

Figure 3. Strip Chart drilling record from Corsham borehole 3 (drilled interval 2.17-3.02m).

The irregular nature of the discontinuities in the Great Oolite Limestone at Corsham was demonstrated by the apparent absence of the latter two joints in an adjacent borehole. This behaviour is confirmed by discontinuities exposed in the gallery. Over very short distances wide clay-filled joints closed to leave only a tight joint trace on the gallery wall.

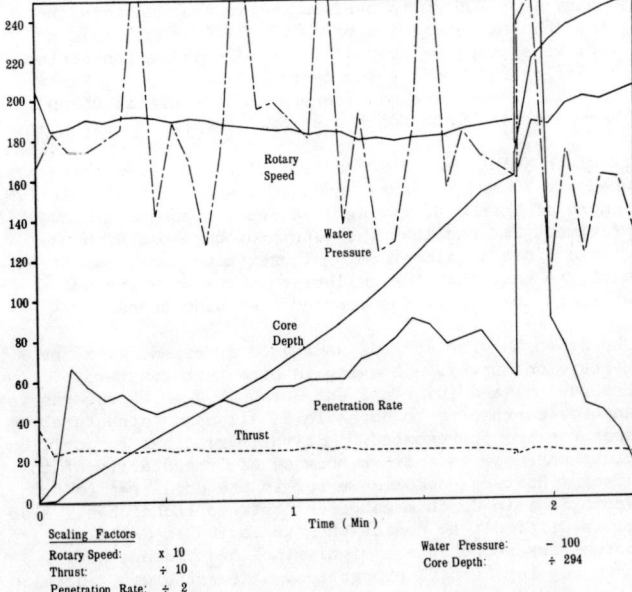

Figure 4. Computed drilling record from Corsham Borehole 2 (drilled interval 16.66-17.51m).

An example of this behaviour is demonstrated by considering Figure 4 which shows the computer-processed drilling record for an 0.85 m interval of Borehole 2. From 16.66 to 17.21 m, drilling progressed smoothly with penetration rate slowly increasing from 25 to 40 cm/min while rotary speed slightly decreased under constant thrust conditions. Without warning the drill string surged forward at a rate of penetration in excess of 240 cm/min through a partly clay-filled zone of some 20 cm width. The surge is marked by characteristic

pulses in the other recorded parameters.

The same feature was encountered in Borehole 3 and is shown in Figure 5. The fissure signature should be recognisable at 17.25 m. In particular, note the response of the rotary speed, water pressure and thrust. All show a marked disturbance when the discontinuity is encountered. The aperture in this instance is only 12 cm in spite of the close proximity of the boreholes.

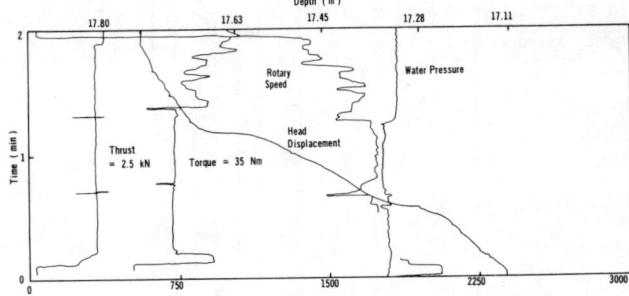

Figure 5. Strip chart drilling record from Corsham Borehole 3 (drilled interval: 17.11-17.80m).

Another 12 cm wide fissure, not seen in the previous hole, was clearly indicated at 17.61 m. This fissure was clay-filled as indicated by increased water pressure resulting from a blocked bit. Erratic torque and rotary speed values caused by snatching of the drill string in the clay zone and high frictional forces at the bit resulted from drilling without circulation. After 0.5 minutes drilling with a dry bit, the drill string stuck in the hole. Early recognition of drilling without circulation in horizontal boreholes is imperative as there is a marked tendency for stuck pipe to occur.

These examples illustrate the ability of an instrumented drilling rig to locate accurately discontinuities and provide early indication of adverse drilling conditions. Figure 6 summarises the results of various downhole discontinuity detection techniques employed during the underground drilling trials. These include the following:

1. Borehole television logging employing a Rees 60 camera as described by West (1980).

2. Impression packer logging as outlined by Barr and Hocking (1976).

3. Analysis of strip chart and computer-processed drilling data.

Borehole 2 was logged with a Rees 60 closed circuit television camera consisting of a 51 mm diameter, 200 mm long miniature television camera with a wide angle lens which can be focused from 20 mm to infinity, a system of illumination, a camera control unit, a monitor, a videotape recorder and a power cable. The camera could be fitted with either a forward-viewing or 45° side-scanning head.

The control unit, the monitor and the recorder were set up at end A of the borehole. A surveyor's measuring tape was attached to the camera which was then advanced down the borehole from A to B using drain rods. Still photographs were taken of the monitor screen as required, and the passage down the hole was recorded on videotape. Runs were made with the camera looking forwards, then the side-scanning head was fitted and particular sections of the borehole wall were examined. Fracture locations were recorded according to depth, fracture inclination, aperture and infill.

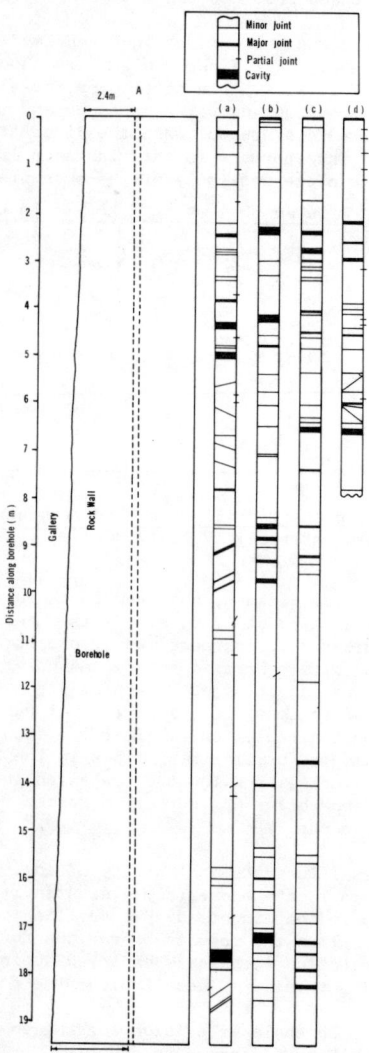

(a) (b) (c) (d)

2.4m A

Distance along borehole (m)

Gallery

Rock Wall

Borehole

3.6 m   B

Figure 6.   Fracture log for Corsham Borehole 2:   a) TV
camera and b) drill record and Borehole 3:   c) drill
record and d) impression packer.

Approximately 8m of Borehole 3 was logged using an
NX-size impression packer.   Because of the packer size,
it was necessary to ream the hole to accommodate the
logging tool.   Due to shortage of stabilisers, reaming
operations were suspended because of severe vibration of
the drill string.   A pneumatically inflatable rubber
packer forms the basis of this simple borehole tool.
The central packer is plugged by stoppers screwed to
receive end guides to centralise the instrument in the
borehole and accept airline terminations.

Overlying the internal packer are two metal shells
suitably curved to conform to the radius of the borehole.
These shells are backed by a resilient foam material which
is in turn wrapped with a thermoplastic film known as
Parafilm 'M'.   When the packer is inflated the stainless
steel shells are forced onto the borehole walls and the
rubber lining with the thermoplastic film forced into any
irregularities which may exist on the wall of the bore-

hole.   After a short period of time the packer is vented,
the plates being mechanically retracted, and the instrument is removed from the borehole.

The action of forcing the resilient foam against the
borehole wall allows the foam to deform and partially
intrude into fissures, vugs and other borehole irregularities.   This penetration into voids results in the deformation of the thermoplastic film, producing an accurate
representation of the particular void or fissure.

Orientation of the structures can be accomplished simply
by using the top of the borehole as datum and overlapping
successive impressions of the borehole wall thereby producing a continuous record of the geologic structure.
Equally, one of the many borehole orientation devices
currently available can be incorporated in the packer
assembly thus permitting selective recording of specific
zones in the borehole without necessarily overlapping the
impressions.

The borehole television log, compiled as described above
is shown in Figure 6(a).   Three kinds of feature are
distinguished: minor joints, major joints and cavities.
These are shown on the log together with joints that
could only be seen in part of the borehole wall.   Minor
joints were recorded when the television image was a thin
white line around the borehole wall, major joints were
recorded when the television image was a thick white line
or when a definite joint aperture could be seen, and
cavities were recorded when a large void could be seen.

West (1980) reported that from the television log a joint
spacing log was derived which showed the mean joint
spacing and the number of joints per metre for each
metre of the borehole.   The mean joint spacing was seen
to vary from 200 mm to over 1m, thus falling into the
joint spacing categories of 'wide' (200 to 600 mm) and
'very wide' (600 mm to 2m) of the classification system of
the Geological Society Engineering Group Working Party
(1977).   There is a section of the borehole in which
there are no joints for almost 5m.

If minor joints are disregarded, it can be seen that there
are only 10 major joints and cavities along the whole 19m
length of borehole, giving an overall mean joint spacing
of almost 2m and that the length of borehole without
joints rises to almost 8m.   These values are consistent
with the fact that the galleries in the underground
workings are mostly unsupported over wide spans.

The other logging methods show good agreement with the
television survey.   Bearing in mind that fracture
records (a) and (b) represent Borehole 2 while records (c)
and (d) correspond to Borehole 3, all major structural
features are represented.   Slight variations in position
occur one hole to another because of irregularity of the
discontinuities, sampling error in the drill record or
variations in depth measurement between techniques.   Also
it is difficult to decide when variation in drilling
parameters represents a highly inclined fissure or
changing intact rock properties.   Nevertheless, analysis
of recorded drilling data provides useful information on
fissure distribution, approximate aperture and infill.

As noted previously, recorded drilling parameters can be
used to derive an estimate of unconfined compressive
strength.   From a series of drilling tests performed at
Atlas Copco and the TRRL, using the Diamec 250 and a TT56
bit, the bit constants defined by Tsoutrelis were $A = 1.8
\times 10^{-6}$ and $B = 1.39 \times 10^{-5}$.

In common with these laboratory tests, a series of drill
tests were performed at Corsham to obtain data from which
to estimate strength.   Rotary speed was maintained at

1500 rpm for various levels of thrust while the penetration rate was recorded. Table 2 lists the results of some strength determination tests in Borehole 3 and the corresponding rock strength values over the drilled interval as determined in the laboratory. The analysis

TABLE 2 - ESTIMATED ROCK STRENGTHS IN BOREHOLE 3, CORSHAM

Depth Interval (m)	Estimated Strength (MPa)	Measured Strength (MPa)
3.41 - 4.26	37.5	30.3
8.17 - 9.02	33.2	34.0
11.96 - 12.81	29.4	25.7
14.13 - 14.98	53.3	49.2

conforms to that outlined by Tsoutrelis except for a slight modification to equation (4) to accommodate a negative thrust intercept on the k_O versus strength plot.

With the exception of the first result, all values are within 15% of the laboratory value. Considering the number of drilling and material parameters over which control must be exercised, the estimated strength values are good and could be used at least at a preliminary design stage for a tunnelling machine.

CONCLUSIONS

Instrumented horizontal diamond drilling can provide useful information in the early stages of tunnelling site investigations or for probing ahead of the working face during tunnel drivage. Field trials with an instrumented Atlas Copco Diamec 250 rotary drill, in conjunction with extensive laboratory investigations, have demonstrated that the following information can be obtained:

1. Fracture frequency, location, aperture and infilling material can be sensed particularly when orientated normal to the drill string.

2. Lithology variation can be detected as 'drilling breaks' by maintaining uniform drilling conditions and monitoring rate of penetration.

3. Acceptable estimates of unconfined compressive strength are obtained following careful calibration of the drill bit.

The interpretation of the recovered data requires a familiarisation period for the drill operator. Once acquired, familiarity with the instrumented drill rig provides the operator with a powerful site investigation tool.

Finally, on the basis of these trials, a similar programme employing a non-coring bit would be a useful extension of this investigation. If it can be demonstrated that open-hole drilling doesn't sacrifice geotechnical data, the reduced necessity to core could be used to reduce project costs or extend the site investigation programme.

ACKNOWLEDGEMENTS

The work described in this paper was carried out under a research contract awarded to Imperial College by the Transport and Road Research Laboratory. The authors gratefully acknowledge the support and encouragement given over an extended period by Mr. M.P. O'Reilly, Head, Tunnels and Underground Pipes Division, TRRL, and the material assistance of members of the Division's staff including Mr. G. West and Mr. P. Johnson. This paper is published by permission of the Director, Transport and Road Research Laboratory and the management of BP Research Centre. The views expressed in the paper are not necessarily those of the Department of the Environment, the Department of Transport or the British Petroleum Company Limited.

REFERENCES

Barr,M.V.(1982). Instrumented diamond drilling for tunnelling site investigation.Ph.D.Thesis,Imperial Col.London.

Barr, M.V. and Hocking, G. (1976). Borehole structural logging employing a pneumatically inflatable impression packer. Proc. of the Symp. on Exploration for Rock Engineering, Johannesburg.

Brown, C.A. (1975). Greater productivity without higher costs objective of new Wesdrill Model 60 drill. The Northern Miner, June 12, 36-38.

Brown, E.T. and Barr, M.V. (1978). Instrumented drilling as an aid to site investigations. Proc. 3rd Int. Congr. Int. Assn. Engng. Geol. Section IV (1), 21-28, Madrid.

Brown, E.T. and Phillips, H.R. (1977). Recording drilling performance for tunnelling site investigations. Technical Note 81. London : Construction Industry Research and Information Association.

BRE/TRRL Working Party on Probing Ahead for Tunnels (1975) Probing ahead for tunnels: a review of present methods and recommendations for research. TRRL Report SR 171 UC Crowthorne: Transport and Road Research Laboratory.

Carroll, R.D. and Cunningham, M.J. (1980). Geophysical investigations in deep horizontal holes drilled ahead of tunnelling. Int. J. Rock Mech. Min. Sci. (17), 2,89-107.

Geological Society Engineering Group Working Party.(1977). The description of rock masses for engineering purposes. Q.Jl.Engng.Geol. (10), 4, 355-388.

Hudson, K. (1971). The Fashionable Stone. Bath: Adams and Dart, 120 pp.

Kellaway, G.A. and Welch, F.B.A. (1948). British Regional Geology: Bristol and Gloucester District, 2nd Edition. London: H.M. Stationery Office.

Majtenyi, S.I. (1976). Horizontal site investigation systems. Proc. 1976 Rapid Excavation and Tunnelling Conf. New York: AIME, 64-79.

Mouxaux, J. (1978). Boring device for recording soil data to avoid pollution due to grouting and to predict unfavourable tunnelling conditions. Tunnelling Under Difficult Conditions, I. Kitamura (ed.). Oxford: Pergamon Press, 257-262.

Rowlands, D. (1971). Some basic aspects of diamond drilling. Proc. 1st. Aust. - N.Z. Conf. Geomech., 222-231, Melbourne.

Rowlands, D. (1974). Diamond drilling with soluble oils Trans. Inst. Min. Metall. (83), A127-A132.

Sinha, K.P., Brown, E.T. and Green, S.J. (1982). Flexural mechanics of horizontal drill-strings. J. Energy Resources Technol., Trans. Am. Soc. Mech.Engrs. (104).

Teale, R. (1965). The concept of specific energy in rock drilling. Int. J. Rock. Mech. Min. Sci. (2), 1 57-73.

Thakur, P.C. and Poundstone, W.N. (1980). Horizontal drilling technology for advance degasification. Mining Engineering (32), 6, 676-680.

Tsoutrelis, C.E. (1969). Determination of the compressive strength of rock in situ or in test Blocks using a diamond drill. Int.J.Rock Mech. Min.Sci. (6), 3, 311-321.

West, G. (1980). Geophysical and television borehole logging for probing ahead of tunnels. TRRL Laboratory Report 932. Crowthorne: Transport and Road Research Laboratory.

Woods, P.J.E. and Hopley, R.J. (1980). Horizontal long hole drilling underground drilling at Boulby Mine, Cleveland Potash Ltd. The Mining Engineer (139), 220, 585-591.

THEORETICAL AND EXPERIMENTAL STUDIES IN BEARING BEHAVIOUR AND CORROSION PROTECTION OF ROCK ANCHORS UP TO A LOAD LIMIT OF ABOUT 4900 kN

Etudes théoriques et expérimentales sur le comportement et la protection contre la corrosion de tirants d'ancrage sous une charge limite allant jusqu'à 4900 kN

Theoretische und experimentelle Untersuchungen zum Tragverhalten und Korrosionsschutz von Felsankern mit Grenzlasten bis zu 4900 kN

O. P. Natau
D. H. Wullschläger
Lehrstuhl für Felsmechanik, Universität Karlsruhe, Bundesrepublik Deutschland

SYNOPSIS

This paper describes the results of a fundamental test of a cement grouted permanent rock anchor with a working load of 2780 kN (load at the yield limit: 4900 kN). The attempt was made to approximate the rock mass conditions around the bond length with a plane analytical calculation and to allocate them to a concreted steel tube. The axial and tangential strains of the steel tube were gauged to determine the course of the load transmission. The large-scale test was simulated by the Finite Element Method. The results show a good conformity with the theoretical basis explained in the beginning of this publication. The assigned measures against corrosion proved to be sufficient.

RESUME

Cette communication donne les résultats d'un essai portant sur un tirant d'ancrage sous charge normale de 2780 kN (charge limite 4900 kN). On a essayé de reproduire les conditions présentes dans la masse rocheuse de la zone de liaison à l'aide d'un modèle de calcul plan, et ensuite de les mettre en rapport avec l'évolution d'un tuyau d'acier dans du béton. On a mesuré l'effort axial et tangentiel du tuyau d'acier afin de déterminer la répartition de la charge. Cet essai à grande échelle a été simulé par la méthode des éléments finis. Les résultats obtenus concordent bien avec la théorie exposée dans l'introduction. Les dispositions prises pour empêcher la corrosion se sont avérées suffisantes.

ZUSAMMENFASSUNG

In diesem Beitrage werden die Resultate der Grundsatzprüfung eines zementmörtelverpreßten Felsankers mit einer Gebrauchs-last von 2780 kN (Grenzlast 4900 kN) vorgestellt. Dabei wurde der Versuch unternommen, die Gebirgsverhältnisse im Haft-streckenbereich unter Hinzuziehung eines ebenen analytischen Verfahrens in einem betonierten Stahlrohr nachzubilden. Zur Ermittlung des Verlaufs der Krafteinleitung wurden die axialen und tangentialen Dehnungen des Stahlrohres gemessen. Der Großversuch wurde mit der Methode der Finiten Elemente simuliert. Die Ergebnisse zeigen gute Übereinstimmung, auch mit den zu Beginn der Arbeit dargelegten theoretischen Grundlagen. Die für Anker der untersuchten Bauart vorgesehenen Kor-rosionsschutzmaßnahmen erweisen sich auch für einen Anker dieser Tragfähigkeit als ausreichend.

1. INTRODUCTION

Today the cement grouted permanent rock anchor is taken to be a nearly perfect and economic construction suited for the use in rock mechanics overground as well as underground. Relative to its capacity no constructive limits seem to exist. Thus rock anchors with an ultimate load capacity of more than 12,5 MN were applied to increase the height of dams. In the Federal Republic of Germany soil and rock anchors must be licensed. For every type of design fundamental tests are prescribed to guarantee high safety standards. The load transmission within the bond length and the constructive measures against corrosion are of special interest in this connec-

tion, because these properties depend on the construction.

2. GENERAL ASPECTS ON CEMENT GROUTED ROCK ANCHORS

2.1 Transmission of prestressing forces

Generally anchors are distinguished in so-called Druckrohr anchors (type A), bar anchors (type B) and strand anchors (type C). An anchor of type A transmits the prestressing at its base subjecting the bond length to compression and shear. Anchors of types B and C have their load maximum at the top of the bond length, which is strained by tension and shear. Bar and strand anchors are objects of the following report.

The prestressing working in the anchor tendon is transmitted to the cement and the surrounding rock mass by bond. The bond stress distribution between grout and tendon can be determined as the first differential quotient of the steel tensile stress distribution. Jirovec (1979) verified this relationship in extensive test series on self-developed measuring anchors and on commercial bar anchors which were prepared with strain gauges. The maximum value of the shear stresses divides the bond length into an adhesion zone and a friction zone, wherein the bond is interrupted by slip (Fig. 1a). According to the increase of force the peak of the bond stress is shifted more and more from the proximal end to the rock side end of the anchor caused by progressive slip. Within the friction zone the cement is ruptured into discs. Nevertheless the anchor can be taken as fully resistant in this zone due to dilatation and wedging of the mortar. Fig. 1b shows the results of a pull-out test on a ribbed anchor tendon of 26,5 mm diameter which verify the relations mentioned above. The maximum of the bond stress is placed in a relative small band between 5 and 7 MN/m^2.

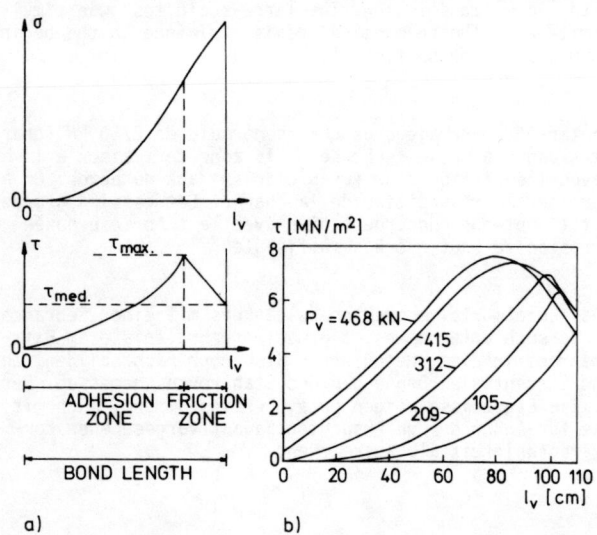

Fig. 1a   Tensile and bond stress distributions (Jirovec, 1979)

Fig. 1b   Grout/tendon bond stress distribution of a single bar anchor (Jirovec, 1979)

It is not possible to get higher shear stresses, the peak becomes flatter and the rear part of the bond length is subjected to strain. The studies made by Jirovec (1979) have shown that prestressing forces up to 500 kN are able to be decreased on less than one meter of bond length as a function of surface conditions and diametre of the tendon. This tendency could be observed on rock anchors with working loads up to 1800 kN in diverse suitability tests. Thus bond lengths of 3 to 8 m mean a high safety factor. Naturally the bond between cement and rock has an effect on the load capacity but in the most cases it has no influence on failure.

2.2 Fissuration of the cement mortar

When the use of cement grouted rock anchors became routine, the opinion existed that rock anchors were not endangered by fissuration of the cement mortar. Studies made by the Karlsruhe Institute of Rock Mechanics have shown the contrary (Jirovec, 1979; Wullschläger, Natau, 1981). fissures are caused by shrinkage of the mortar and by high prestressing forces. Regarding multiple strand anchors there exist tangential fissures, running between the strands and within the cement, radial fissures running outwards starting from the strands or inwards starting from the spaces and axial fissures which divide the bond length into discs (Fig. 2). The last

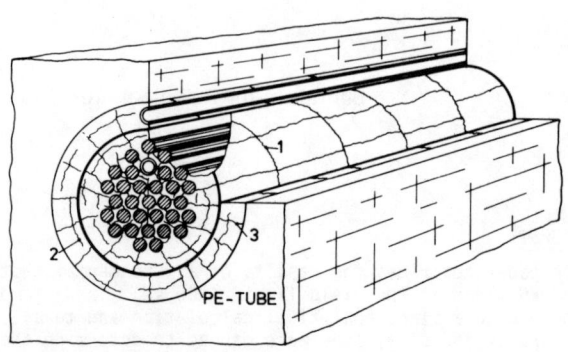

1 AXIAL FISSURES
2 TANGENTIAL (CONICAL) FISSURES
3 RADIAL FISSURES

Fig. 2 Fissuration of the cement grout

named fissures cause the wedge effect. The first fissuration appears, when the bond stress transferred from the steel into the cement gets higher than the tensile strength of the mortar. With increasing loads the fissures will grasp the whole bond length. The volumetric increase respectively the dilatation prevents a further loosening of the anchor. Within the unfissured areas the compound between steel and mortar remains intact.

2.3 Corrosion protection of the bond length

The most important factor for the durability of a rock anchor is its corrosion protection. The analysis of the fissuration led to a simple constructive measure to guarantee the corrosion protection.

The anchor tendon is sheathed in the bond length with a ribbed polyethylene tube, which of course cannot prevent the development of fissures, but stops them at the tube wall (Fig. 2). The high elasticity of the tube makes possible radial and tangential movements. A force transfer of 100% into the external cement cover and the surrounding rock can be realized by the ribbed surface. Many tests have verified the reliability of the sheathing. For this reasons only permanent anchors equiped with ribbed tubes are licensed in the Federal Republic of Germany.

3.   FUNDAMENTAL TEST OF A 4900 KN ROCK ANCHOR UNDER SIMULATED IN-SITU CONDITIONS

3.1 Technical data

The object of the now described studies is a

multiple strand anchor VSL 5 S-31 Losinger system. Important technical data are presented in Table 1:

Table 1.   Technical data of the VSL 5 S-31
rock anchor

number of strands:	n_s	= 31
total cross-section:	A_{st}	= 3100 mm^2
Young's modulus:	E_{st}'	= 195000 N/mm^2
$\sigma_{yield}/\sigma_{ult}$:		1570/1770 N/mm^2
working load:	F_w	= 2780 kN
load at yield limit:	F_y	= 4867 kN
ultimate load:	F_u	= 5487 kN
size of well:	d_B	= 165 mm
polyethylene tube:	diameter: d_{PE} = 115÷125 mm	
	wall thickness: s = 1,3 mm	
	rib spacing: sp = 10 mm	
bond length	l_v	= 4,58 m
free stressing length	l_{fst}	= 3,13 m

3.2 Approximation of natural rock masses

The aim of the described test was to find out the behaviour of the anchor design, particularly its stress-strain behaviour by loading with the 1.5 time working load under in-situ conditions. It is practically impossible to overcore anchors of this magnitude set in natural rock, for uncovering the bond length. Therefore the authors made the attempt to design a concreted steel tube in a way that its stress-strain behaviour is similar to a rock mass with linear elastic material behaviour. The following considerations are based on the theory of the thick wall cylinder (comp. Seeber, 1962; Natau, 1980):

It is assumed that a radial internal pressure p_i works on the borehole wall caused by dilatation of the grout body. Two cases are distinguished (Fig. 3):

case a: $p_i < \sigma_z$; the rock mass does not rupture

case b: $p_i > \sigma_z$; the rockmass ruptures

(σ_z: tensile strength of rock mass)

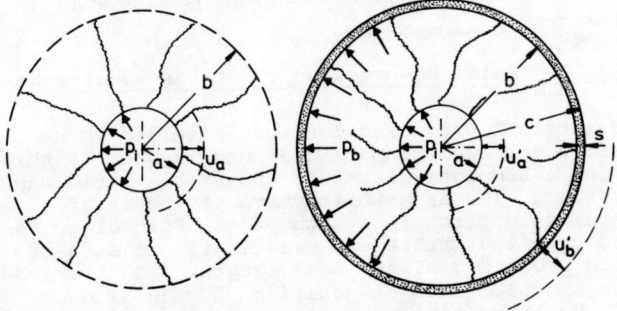

Fig. 3 Displacements u_a of the rock mass (left)
and of the concreted steel tube (right)

From the theory of the thick wall cylinder follows:

case a: $u_a = \dfrac{p_i \cdot a}{E_{rm}} \cdot (1+\nu_{rm})$

case b: $u_a = \dfrac{p_i \cdot a}{E_{rm}} (1+\nu_{rm}+\ln\dfrac{b}{a})$

(E_{rm}, ν_{rm} = rock mass parameters)

In the simulating test the total displacement u_a is composed of the concrete compression u'_a and the steel expansion u'_b:

$$u_a = u'_a + u'_b \quad (u'_b \approx u'_c)$$

case a: $u'_a = \dfrac{p_i \cdot a}{E_c} \cdot (1+\nu_c)$

case b: $u'_a = \dfrac{p_i \cdot a}{E_c} \cdot \ln\dfrac{b}{a}$

For both cases:

$$u'_b = \frac{p_i \cdot b \cdot \nu_{st}}{E_{st}(c^2-b^2)} \left(\frac{1-\nu_{st}}{\nu_{st}} b^2 + \frac{1+\nu_{st}}{\nu_{st}} c^2\right)$$

(E_c, ν_c = concrete parameters)
(E_{st}, ν_{st} = steel parameters)

Using the relation $\qquad p_b = p_i \cdot \dfrac{a}{b}$

the required wall thickness $\quad s = c - b$

can be determined, whereby $\quad c = \sqrt{b^2 \dfrac{A+1}{A-1}}$

with
case a: $A = \dfrac{E_{st}}{E_{rm}}(1+\nu_{rm}) - \dfrac{E_{st}}{E_c}(1+\nu_c) - \nu_{st}$

case b: $A = \dfrac{E_{st}}{E_{rm}}(1+\nu_{rm}) + \ln\dfrac{b}{a}\left(\dfrac{E_{st}}{E_{rm}} - \dfrac{E_{st}}{E_c}\right)$

While planning the test a rockmass with E_{rm} = 5000 MN/m^2 and ν_{rm} = 0,33 (case a) respectively E_{rm} = 10000 MN/m^2, ν_{rm} = 0,33 (case b) were presumed. The parameters for the concreted tube were:

E_c = 30000 MN/m^2, ν_c = 0,2;
E_{st} = 210000 MN/m^2, ν_{st} = 0,3.

The foregoing consideration led to a wall thickness of s = 6,64 mm (inside diameter: d_{si} = 2b = 600 mm) and a borehole diameter d_B = 2a = 165 mm. A tube with insignificant geometrical deviations was used for the test. After uncovering the cylinder it became yet apparent, that the concrete had not cracked under load (case a). Interpreting it must be explained, that the approximation above derived does not consider the shear stresses which are transferred from the tendon into the mortar and the real existing three-dimensional state of stress.

3.3   Set up and method of test

Fig. 4 shows the testing installation, the design and the dimensions of bond length and free stressing length, the single strand measuring system and the arrangement of the strain gauges.

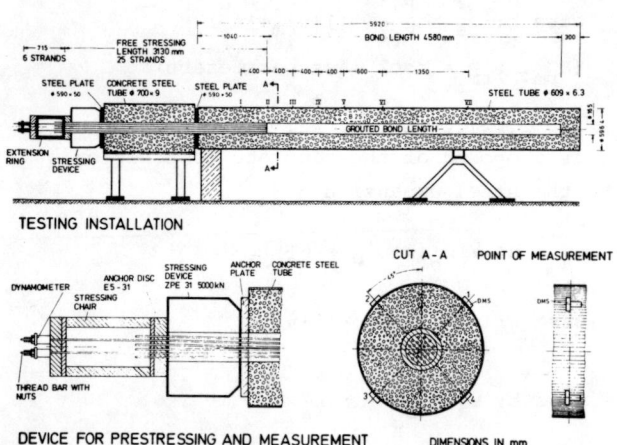

Fig. 4   The VSL 5 S-31 large scale test - setup

The load steps of the tensile test were produced by a 5000 kN hydraulic jack. The displacements of the anchor head were measured with mechanical gauges. The load test and the measurement program were established according to the German standards (DIN 4125). After applying a pre-loading the anchor was loaded in 5 cycles up to the 0.9 time yield force of the steel. The displacement measurements were interpreted in form of load-displacement diagrams, limit line charts and time-displacement curves for determining the creep amount. All those diagrams, not represented here, showed the unobjectionable global behaviour of the anchor under the chosen test loads and given boundary conditions.

3.4   Single strand measurements

In addition to the measurements of normal forces and global strains single force measurements were made on six selected strands to find out possible partial overstressings and to get some informations about the distribution of the anchor force over its cross-sections. The measured strains of the electric dynamometers were monitored by a manual compensator and converted into forces. Fig. 5 shows the positions of the selected strands. From Table 2 it follows that at proof load the outer strands were stressed 6 per cent higher than the inner strands. These deviations are not too significant and do not lead to failure of single strands. Regarding the order of the strand forces the lengthened free stressing length must be considered.

Tab. 2   Strand forces at proof load A_s minus pre-load A_0

number of strand	position	force [kN]
4	inner	106,9
7	inner	105,7
20	outer	112,1
26	outer	113,9
28	outer	112,7
10	---	104,0

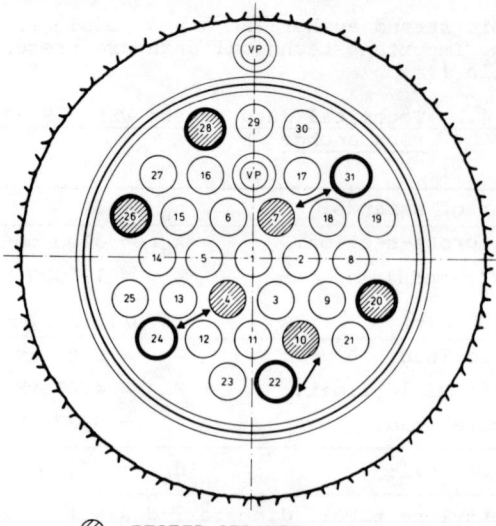

TESTED STRANDS
SPREAD POSITION OF DYNAMOMETER
GROUT INJECTION TUBE

Fig. 5   Position of the measured strands

3.5   Length of force transmission
3.5.1   Strain gauges measurements

To get informations about the course of the prestressing's transmission from the anchor tendon into the cement mortar strain measurements were made on the steel tube. Seven strain gauges stations were installed (Fig. 4). The results of more than 2300 single strain measurements are figured in two diagrams which show the discreet axial compressions (ε_a) and the tangential extensions (ε_t) (Fig. 6 and Fig. 7). The first

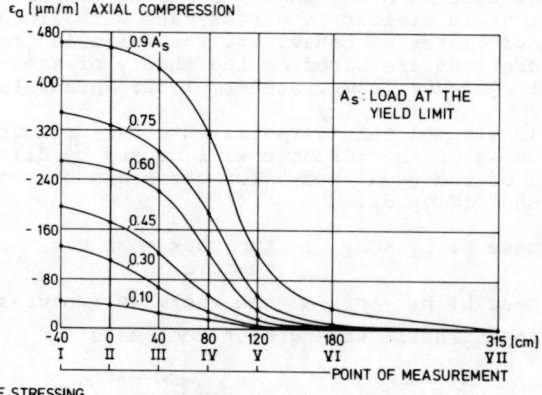

Fig. 6   Axial compression ε_a of the steel tube

measurement cross section was installed 40 cm before the beginning of the bond length. At this point a homogeneous stress and strain field was presumed. The asymptotic curve of the axial strains confirms this assumption. Both diagrams indicate that there are practically no strains from about 2/3 of the bond length, i.e. the load transmission ends at position VI. The course of the tangential extensions has some similarities with the shear stress distribution figured in chapter 2.1. The curves of Fig. 7 show that

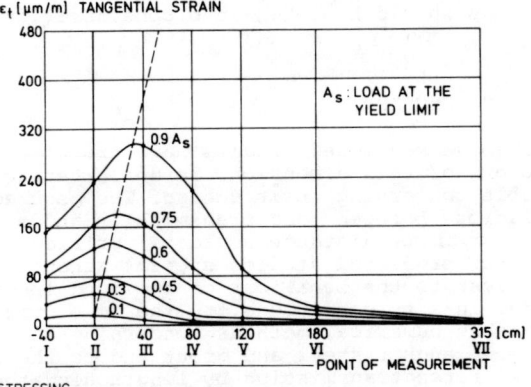

Fig. 7   Tangential expansion ε_t of the steel tube

the position of the maximum local strain ε_t goes to the rock side end of the anchor with increasing forces. Here it is made plain by a dotted line.

3.5.2   Finite element analysis

With a plane cylinder-symmetric finite element calculation it was attempted to get relations between the steel tube deformations and the force input in the bond length. The dimensions of the used mesh correspond to the setup shown in Fig.4, its structure was simplified. The implementation of slip elements was deferred to diminish the arithmetic expense. The used 8-knot-elements with quadratic strain formula are compounded between each other with full bond. The material parameters were chosen according Table 3 or determined in tests, the loading corresponds to the large scale test.

The measured axial respectively radial displacements of the steel tube are drawn together with the calculated values in Fig.8 and Fig.9. The qua-

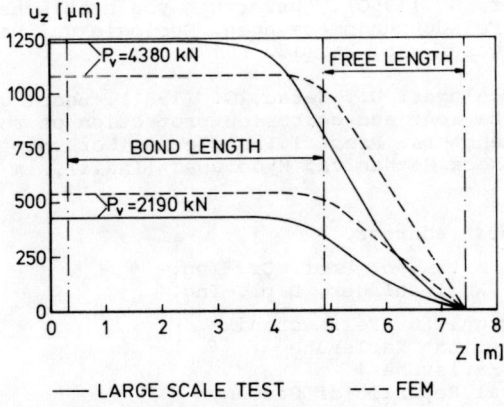

Fig. 8   Axial displacements u_z

litative course shows good conformity, the absolute amounts lie in one dimension with deviations of about 20 %. The calculated axial displacements u_z get much earlier asymptotic than the measured, the radial strains die down earlier. This result is due to the not considered influence of slip between anchor steel and cement grout in the numerical calculation.

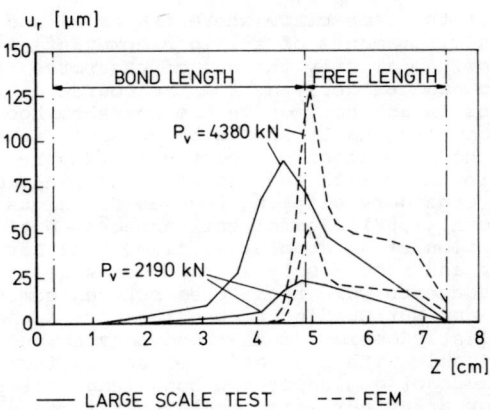

Fig. 9   Radial strains u_r

Tab. 3:   Material parameter for the finite element calculation

material	material model	parameter
anchor steel	linear elastic	$E=195000$ MN/m^2
cement grout	linear elastic/ ideal plastic (Drucker/Prager)	$E=37000$ MN/m^2 $\nu=0,2$ $\sigma_D= 50$ MN/m^2 $\sigma_z= 3$ MN/m^2
concrete	linear elastic/ ideal plastic (Drucker/Prager)	$E=32700$ MN/m^2 (measured value) $\nu=0,2$ $\sigma_D=28,6$ MN/m^2 $\sigma_z=1,7$ MN/m^2
steel tube	linear elastic	$E=210000$ MN/m^2 $\nu=0,269$

In Fig. 10 the distribution of the axial stresses σ_z and of the shear stress τ_{rz} in the cement is shown plotted over the bond length for the highest load step. It is conspicuous that both stresses and asymptotically for all hori-

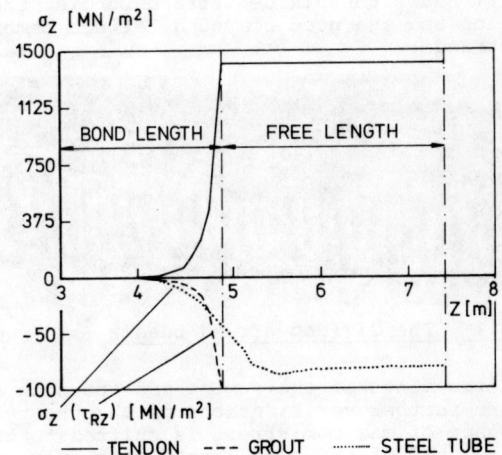

Fig. 10   Axial stresses σ_z and shear stresses τ_{yz} (grout)

zons at the same place where the axial and tangential increments of strain become infinitesimal low. Apparently the end of the force transmission can be derived from the course of the strains in any horizon in the neighbourhood of the bond length. It should be possible to simulate the theoretical respectively measured stress distributions with the aid of a material model which considers slipping such as Cornelius and Mehlhorn (1982) did for soil anchors. With the assumption of an absolutely fixed bond between cement and rock - only in rare cases relative displacements were registered between cement grout and surrounding rock using rock anchors - a realistic design of the bond length would be practicable with indication of safety factors. This mode of design of the bond length is planned for a further large scale test in natural rockmass.

3.6 Condition of the bond length

After the load tests the test body was cut with a diamond saw into 24 pieces and splitted to examine the condition of the whole bond length. The decomposition of the inner grout material delivers the following results:

- No faults or blow holes in the grout; no bond between inner grout and ribbed tube on about 40 degrees of the evolution caused by the injection dipping under 10 degrees.

- Radial and axial fissures along the whole bond length, concentrated in the first third, interrupted by the sheathing; tangential fissures (conical up to $l_v = 1,20$ m (l_v = bond length).

- Till $l_v = 0,8$ m detachments of the cement rips inside and outside of the ribbed tube, caused by shear stresses and partial overloadings at proof load.

- Pulverization in the interfaces of strand and cement ($l_v = 0 \div 0,6$ m); transition of bond into friction.

- The ribbed polyethylene tube is intact over the whole length; this means effective corrosion protection.

Fig. 11 shows the divided strand bundle, the fissuration and the good compound between cement and strands.

Fig. 11   The divided strand bundle

The detachments of the cement and the slip traces are further verifications of the theory that the place of maximum stress is shifted to the rockside end of the bond length with increasing forces. There are no consequences for the design of rock anchors, but the borehole diameter, the diameter of the ribbed tube and the distance of

strands should be taken into consideration.

4.   CONCLUSION

The above mentioned studies have proven that anchors of this dimension are unhesitatingly usable concerning their design. The assumed correlations between load transmission and strains in an optional distance could be verified. In further projected studies special attention will be given to the behaviour of the bond length under in-situ conditions, to the development of suitable numerical methods concerning the slip between anchor steel and grout and to the defined force transmission by lengthening the free stressing length of selected strands.

The authors express appreciation to Dipl.-Ing. R. Porzig and to the Suspa Spannbeton GmbH, Augsburg, for the authority of publication of results from supported and sponsored research work.

5.   REFERENCES

Cornelius, V., Mehlhorn, G. (1982). Tragfähigkeitsuntersuchungen im Verankerungsbereich von Verpreßankern und Pfählen mit kleinem Durchmesser für den Anwendungsbereich Lockergestein. Forschungsberichte aus dem Institut für Massivbau der TH Darmstadt,Nr.46.

Jirovec, P. (1979). Untersuchungen zum Tragverhalten von Felsankern. Veröff. Inst. Bodenmechanik u. Felsmechanik, Universität Karlsruhe, Nr. 79.

Natau, O. (1980). Konzept für die Grundsatzprüfung des VSL Felsankers 5 S-31. Unpublished MS.

Seeber, G. (1960). Auswertung von statischen Felsdehnungsmessungen. Geologie und Bauwesen, Jg. 26, 152 - 176.

Wullschläger, D., Natau, O. (1981). Bearing behaviour and corrosion protection of rock anchors. Proc. 1.Indo German Workshop on Rock Mechanics, Hyderabad (India), in press.

Authors' address:

O. P. Natau, o. Prof. Dr.-Ing.
D. H. Wullschläger, Dipl.-Ing.

Lehrstuhl für Felsmechanik
Universität Karlsruhe
7500 Karlsruhe 1
Federal Republic of Germany

RECENT DEVELOPMENTS OF THE LARGE-SCALE TRIAXIAL TEST

Développements récents des essais triaxiaux à grande échelle

Neuere Entwicklungen des Triaxial-Grossversuches

O. P. Natau
B. O. Fröhlich
Th. O. Mutschler
Lehrstuhl für Felsmechanik, Universität Karlsruhe, Bundesrepublik Deutschland

SYNOPSIS

Recent developments make it possible to determine rock mass parameters in close jointed rock masses by means of large-scale triaxial tests in the laboratory. A sampling technique for large specimens was developed alternatively to expensive triaxial tests in situ. The sampling equipment and technique as well as the testing machine and procedure are described in detail. The boundary conditions of the tests are given. The results for a mixed layer of sandstone and claystone and for a yellow limestone are presented. The comparison to small-scale tests shows in this case that the rock substance parameters cannot be used to calculate a rock construction, because there is no constant relation between both types of parameter.

RESUME

Des développements récents rendent possible la détermination en laboratoire des paramètres d'un massif rocheux à l'aide d'essais triaxiaux à grande échelle. On a développé, comme technique alternative à des essais in situ coûteux et nécessitant une importante mise en œuvre, une technique de carottage permettant d'obtenir des échantillons non remaniés de grandes dimensions. On détaille l'équipement et la technique employée pour le carottage ainsi que les appareils d'essais de laboratoire et le déroulement de ceux-ci. On indique les limites du procédé. On donne comme exemple les résultats d'essais sur un matériau constitué de couches interstratifiés de grès et d'argile compacte ainsi que sur du calcaire jaune. La comparaison avec des essais de petites dimensions montre que les paramètres de la roche ne peuvent pas être employés pour le dimensionnement d'ouvrages en roche dans ce cas précis. Une estimation des paramètres valables pour l'ensemble du massif rocheux avec des essais de petites dimensions est peu sûre car le rapport entre les paramètres de la roche elle-même n'est pas constante.

ZUSAMMENFASSUNG

Neuere Entwicklungen ermöglichen die Bestimmung von Gebirgsparametern mittels großmaßstäblicher Triaxialversuche im Labor. Als Alternative zu aufwendigen Triaxialversuchen in situ wurde eine Entnahmetechnik für ungestörte Großproben entwickelt. Die Entnahme von Großproben, sowie die Versuchseinrichtungen im Labor und die Versuchsdurchführung, werden ausführlich erläutert. Die Versuchsrandbedingungen werden aufgezeigt. Exemplarisch werden die Ergebnisse von Versuchen an einer Wechsellagerung von Sandstein und Tonstein und an einem Kalkstein gegeben. Der Vergleich mit kleinmaßstäblichen Triaxialversuchen zeigt, daß die Gesteinsparameter allein für eine Bemessung von Felsbauwerken nicht angewendet werden können, da deren Relation zu den korrespondierenden Gebirgsparametern nicht konstant ist.

1. INTRODUCTION

The increasing application of modern material laws and calculation methods in rock mechanic problems of underground openings, foundations in rock masses and slopes needs the determination of the stress-strain behaviour of the rock masses more than empirical or half-empirical methods. A great part of the deficit of reliable parameters is caused by incomplete or not existing testing methods. Normally the rock mass parameters are estimated by reduction of the rock substance parameters. This procedure contains the risk that these parameters are either underestimated which

causes an uneconomical construction or overestimated which causes sometimes a collaps. Therefore research activities in testing rock mass behaviour are necessary (Müller et. al., 1977).

The report treats new developments for triaxial tests of narrowly jointed rock samples with dimensions up to 0,6 m x 0,6 m x 1,2 m. The relevant range of applications and technical demands for the tests are shown.

The derivated basic research, financed by the Deutsche Forschungsgemeinschaft, treats a careful sampling of specimens orientated to the

F

joint systems (prismatic and cylindric samples, boring with prestressing, etc) as well as technical conditions of the triaxial cell for testing under defined stress-strain paths (compression and extension tests, multi-step tests under different confining pressure with regard (or not) to the middle principal stress, time dependence of the material behaviour, influence of water-pressure in pores respectively in joints).

2. TESTING POSSIBILITIES AND LIMITS OF THIS TECHNOLOGY

The determination of the rock mass parameters is possible in two different types of investigations. One type is the analytical or numerical procedure (John 1969), (Mühlhaus, Reik, 1978). The rock mass behaviour is calculated on the basis of the rock substance and the joint parameters. This method is not useful if the degree of separation is unknown or the joint orientation is nearly normal or parallel to the principal stresses, especially in close jointed rock masses. The other type is the global determination of the parameters at a jointed specimen as a composite material. This type of investigation includes all effects like those of the rock substance, joints, stratification, degree of seperation etc. Therefore it is necessary to test specimens with a representative size. From sufficient fundamental test series in our institute and elsewhere we know that the parameters of the test specimen are relevant to the considered homogeneous area of rock masses, when 5 to 10 joint planes cut the smallest one way dimension of the test specimen. The limitation of this technology is given by the joint distances. For magnitudes of 10 cm to 20 cm a careful extrapolation is possible (Fig. 15). Larger joint distances are not suitable for this technology. In this case the above mentioned analytical or numerical procedure basing on small scale tests at rock bricks (substance parameters) and direct shear tests at joints (joint parameters) may be an ingenious point of application (Natau, 1979, 1980).

Very often rock mass types are suitable to be tested by the global procedure. These rock masses consisting of limestone, marl, new red sandstone, siltstone, claystone, clay shale or mixed layers are widespread close jointed. The global test procedure can be applicated either in situ or in the laboratory. In-situ-triaxial-tests are much more expensive in comparison to similar laboratory tests. Furthermore the boundary condition, the load path and the reproduction of results are not so exact as in the laboratory tests. Biaxial tests, e. g. double flat jack tests, have not the same efficiency and normally they give only deformation parameters. One principal stress is unknown and the boundary conditions are also not exact. Dynamical methods give the deformation modules only, but they are useful for correlations and interpolations.

A very significant way to get as well the deformation modules as the strength properties is to carry out triaxial tests in the laboratory at large samples. Therefore several special core-drilling and sampling techniques were developed by our research group in Karlsruhe-Sonderforschungsbereich 77. In the beginning the drilling and sampling technique for plastered cores with 60 cm of diameter was applicable in soft rock only (Wichter, 1980). Reik (1978) has tested hard

rock samples of limestone with dimensions of 0,6 m x 0,6 m x 1,2 m with an uniaxial compressive strength of small rock cores up to 100 MN/m^2.

For these prismatic specimens he developed a hole-by-hole bore and sampling equipment with a 100 mm core bit and an air flushing system. Step by step the four faces of the specimen were plastered in combination with special bended sheet steels. After setting of the plaster the positive power compound system was drawn. In this way a careful sampling of very breakable, close jointed or strongly weathered hard rock masses was possible for the first time. A slanting taken sample dismantled in the laboratory is shown in Fig. 1. The prismatic form also allows tests under true triaxial states of stress ($\sigma_1 > \sigma_2 > \sigma_3$) and to study the influence of the intermediate principal stress σ_2.

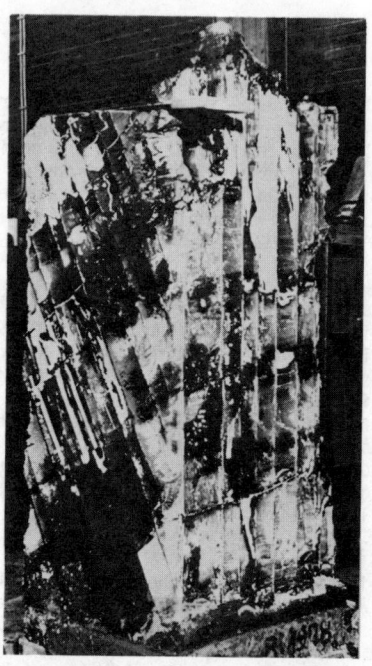

Fig. 1   Rock mass specimen, sampled in definite orientation with hole-by-hole drilling (Reik, 1978, Hesselmann, 1979)

For practical purposes the test with radial symmetric confining pressures ($\sigma_2 = \sigma_3$) is sufficient. Therefore the soft rock drilling and sampling technique with a special large scale barrel in one pass of operation was extended by the authors to a technology, suitable for close jointed and strongly weathered hard rock types, too. During the last three years it was possible to test about 100 samples. Fig. 2 shows an overview of large specimens sampled by drilling in one pass. In nearly all cases we got very good results for the corresponding behaviour of rock masses. Even in such cases where the sampling of small cores was not possible, relevant rock mass parameters could be determined. It was also possible to test mixed layered rock types with a very wide range of substance parameters. The state-of-art of the large core drilling method in one pass and of the sampling technique as well as of the testing procedure of

Fig. 2 a)

Fig. 2 b)

Fig. 2 c)

Fig. 2 d)

Fig. 2 e)

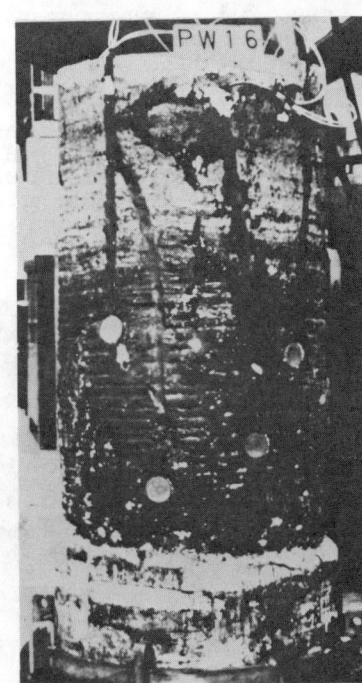

Fig. 2 f)

Fig. 2 Rock mass samples drilled in one pass
    a) grey limestone undulated; b) yellow limestone; c) mixed-layer from hard limestone and
    weathered marl; d) mixed-layer from sandstone and claystone; e) red and green silty clay-
    stones; f) overconsolidated clay with slicken sides

Fig. 3 Starting-point of boring in a close jointed rock

close jointed specimens is described in the following chapters.

3. THE DRILLING AND SAMPLING PROCEDURE

The test of large rock mass specimens and the interpretation of their properties presume that these samples are undisturbed. Therefore we work with our own field crew, provided with a cross country truck, a trailer and a truck-mounted crane, a special developed drilling equipment, a 66 KW diesel-hydraulic power station and all tools for safe preparation and transportation of the samples to our institute (Fig. 4).

Fig. 4 Sampling equipment

The equipment is fitted out with water and air flushing. Normally we use air flushing (7 m^3/min, 6 bar).

The barrel (d = 60 cm, h = 130 cm) is operated with a piston driven carriage and guided by a lunette (Fig. 3). For a good drilling rate and a sufficient flush out of the drillings in different rock types two or four detachable segments of ringbits, tipped with fragments of metal car-

bid in a brazed matrix are developed and proofed in our own workshops. Beside the torsional stress the most important influence upon the stability of the close jointed core in statu nascendi is given by vibrations and knocks at its outer surface. This problem is solved by using ring bits with an inner free cut of more than one cm and by prestressing the core with well sized steel plates, bedded in plaster (Fig. 5). In some cases an additional prestressing with a rock bolt is useful.

Fig. 5 Setting of a dead load for prestressing the specimen

After drilling a cylindric steel casing is put into the annular space and plastered in several steps. When the plaster has hardened the reinforced core is lifted. In all cases the core breaks at the bottom of the borehole.

Before the core is transported to our institute the top and the bottom planes are sealed with plaster, too. The jointed rock sample is now protected against loosening and drainage for two months and more. In weak rock it is also possible and sometimes necessary to use a steel casing with a cutting edge. Beside the core the rock must be excavated step by step while the steel casing is pressed down.

It is known that rock masses even more than rock substances behave anisotropically. Therefore, an orientated sampling is necessary. With the described machine it is possible to traverse the drilling axis from the vertical direction. The maximum of the drilling angle depends on the material itself, because of sample toppling. In both these cases the hole-by-hole drilling and sampling technique has advantages in comparison to the one pass drilling method.

4. THE PREPARATION OF THE SPECIMEN

Still surrounded by the steel casing the sample receives parallel levelled end faces, formed by gypsum or mortar with an accuracy of ± 0,5 mm. After hardening the steel casing and the plaster are removed. In all cases the fabric is documented by several photographs and by recording stratification and discontinuities. For this a steel net with a square mesh size of 5 cm is used (Fig. 6). For triaxial testing the specimens are sealed against the confining fluid with an axial screwed system consisting of two steel plates

specimen can be seen.

Fig. 6 Recording of the discontinuities by means of a square net

and a cylindrical respectively prismatic rubber jacket.

5. THE LARGE SCALE TRIAXIAL TEST CELL

The large scale triaxial test cell is shown in Fig. 7.

- The axial load capacity is 6.4 MN ①. The axial load is controlled by three stiff 2.5 MN-Glötzl-pressure cells ⑤ with electric transducers in the basic frame of the testing machine.
- The confining pressure $\sigma_2 = \sigma_3 \leq 2.2$ MN/m^2 is air-produced and transmitted to the specimen by water ②. For triaxial tests, using prismatic specimens, with $\sigma_1 \neq \sigma_2 \neq \sigma_3$ a special steel frame with hydraulic flat jacks for the intermediate principal stress σ_2 is additionally installed. In all cases σ_2 respectively σ_2 and σ_3 are controlled by manometers ⑥ and electric transducers.
- Axial and lateral deformations are checked by steel wires and vernier scales which can be observed through several windows ③ , ④. The position of the wires for measuring lateral deformation can be chosen according to the fabric of the sample. Additional inductive gauges can be installed.
- The cell is fitted with 10 further outlet connections, e. g. for pore water pressure cells.
- The testing machine is power controlled and allows compression tests in single-step or multi-step variations. Extensiontests $(\sigma_1 \geq \sigma_2 = \sigma_3 > 0, \sigma_1 \rightarrow 0)$ will be possible, too.
- Specimens may have a diameter up to 100 cm and a height up to 170 cm.

In Fig. 8 the basic frame with hydraulic coupling connections for the cell and a well prepared

6. BOUNDARY CONDITIONS, STRESS PATHS AND LOADING PROGRAMMES

The large scale triaxial test has mixed boundary conditions equivalent to the well known small standard triaxial test. At the axial end faces of the specimen a kinematic condition and on the cylindric mantel a static condition is valid. Theorists often want only one type of boundary condition, but this is very hard to carry out and besides this for most practical purposes also

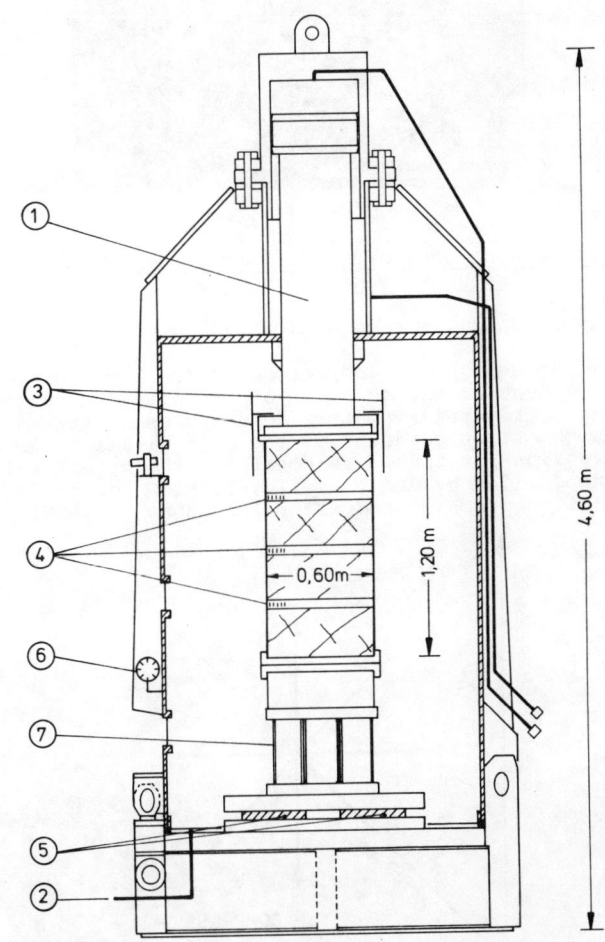

① AXIAL LOAD (MAX. 6,4 MN)
② CONFINING PRESSURE (MAX. 2,2 MN/m²)
③ AXIAL DEFORMATION MEASUREMENT
④ RADIAL DEFORMATION MEASUREMENT
⑤ AXIAL LOAD MEASUREMENT
⑥ CONFINING PRESSURE MEASUREMENT
⑦ ADAPTER

Fig. 7 Outline of the large triaxial cell at Karlsruhe University

Fig. 8 The opened triaxial test cell with a specimen ready for testing

combined boundary conditions are given.

The stress path is understood as the temporary development of the relation of the principal stresses. For a triaxial test only states of stress in the first quadrant of the $\sigma_1 - \sigma_3$-plane (compression positive) are possible. The stress paths are divided by the bisecting-line of the angle into compression and extension paths (Fig. 9).

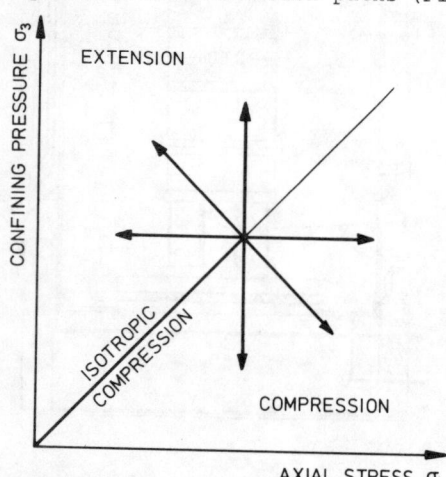

Fig. 9 First quadrant of the $\sigma_1 - \sigma_3$-plane with several stress paths

The developing state of stress is devided in a mean value of principal stresses $(\sigma_1 + 2\sigma_3)/3$ and the deviatoric stress $\sigma_1 - \sigma_3$. The horizontal respectively the vertical stress path keeps one principal stress constant, i. e. both parts are changing. Stress paths vertical to the isotropic compression path cause a constant mean value of principal stresses, i. e. only the deviatoric stress is changing. Other stress paths are suitable for simulating stress transposition in rock masses or a constant relation of the

principal stresses $\lambda = \sigma_3 / \sigma_1$.

In most triaxial tests the compression stress path is used because extension path to realize is much more difficult. The compression test realized with a constant confining pressure is the standard triaxial test to get the yield condition for compression.

Using the well known single-step technique a lot of specimens have to be tested for reliable parameters of the yield condition. The multi-step-technique reduces the number of tests because each specimen allows the determination of the complete yield condition. By one specimen three or four triaxial tests with different confining pressures are carried out with a special loading programme:

- The stress-strain curve is stopped under control when the maximal deviatoric stress is reached.
- The deviatoric stress part is reduced completely and the confining pressure is changed to the next step either increasing (Fig. 11) or decreasing (Fig. 12).
- The deviatoric stress is raised again to the new maximum corresponding to the changed confining pressure.
- Usually this cycle can be repeated three to four times.

Both the above mentioned increasing and decreasing steps of the confining pressure allow the control of independance of the single results on the yield history induced by the multi-step-technique.

7. RESULTS OF LARGE SCALE TRIAXIAL TESTS

The test data:
axial load, confining pressure, axial displacement, change of circumference, change of volume of the cell fluid etc.
are to be interpreted under certain assumptions to get the stress-strain behaviour and the yield condition of the rock mass. The rock mass is to be considered as a quasi-homogeneous composite material, consisting of the rock substance and the discontinuities. The state of stress σ is supposed to be homogeneous. The deformations ε are determined from the average displacements and are assumed to be homogeneous. In the case of constant confining pressure the volume change can be determined from the change of the cell fluid volume.

7.1 Description of the material

The test series described in chap. 7.2-7.3 were carried out with mixed-layer material of sandstone and silty claystone of lower triasic age. Besides the bedded structure of the sedimentary rock mass the discontinuities consisted of two joint systems with a joint spacing of 8 cm to 12 cm. The thickness of the layers varied between 5 cm to 20 cm. Fig. 10a shows a sample before testing.

The normal vectors in the lower hemisphere equal area projection can be seen in Fig. 10b. The joints were closed and partly covered with thin clayey layers. The plane degree of separation was estimated at $\kappa = 0,4$ to $\kappa = 0,7$. The material was tectonically stressed. Today the larger

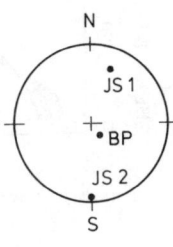

a)   b)

Fig. 10 Mixed-layer material of sandstone and
         silty claystone of lower triasic age (a)
         and joint systems (b)

horizontal stress component in the sampling area
is orientated NNW-SSE (Balthasar, Wenz, 1983).

7.2 Stress-strain behaviour

The stress-strain behaviour is given by curves
$\sigma_1 - \sigma_3$ vs ε_1. Fig. 11 shows a multi-step test
with increasing confining pressure steps. The
test shown in Fig. 12 was carried out with decreasing confining pressure steps.

The stress-strain curve is divided into four
sections.

The first section ⓪ - ① starts from the
isotropic compression with a nearly linear modul
up to a bend at point ① . This point correlates
with the overburden load in situ. The starting
modul is about 150 MN/m² (sample I). The corresponding value of sample II with a rather heigh
confining pressure is 500 MN/m². In both cases
the linear start shows that there are no open
joints in this material. This can only be obtained, if there is no loosening caused by
sampling, transportation and preparation.

The second section ① - ② is also nearly
linear. It gives the deformation modul for initial loading. The values vary from 50 to 135 MN/
m². Up to point ② there is no radial deformation of the specimen.

In the third section ② - ③ the material
yields evidently. The deformation modul decreases rapidly and the lateral deformation increases. The yield condition is reached at point
③ . Sometimes this point is not defined strongly and the engineer has to interpret the internal structure of the sample for determination of
the failure point.

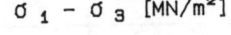

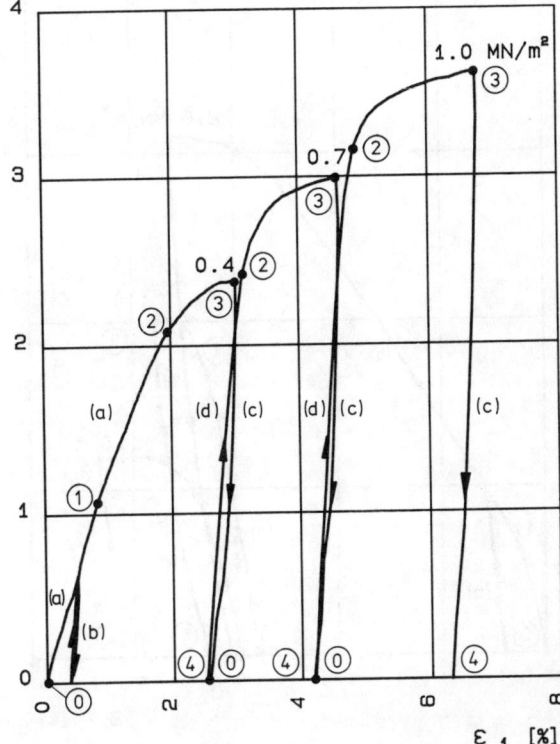

Fig. 11   Multi-step test on sample I with in-
          creasing confining pressure steps
          (mixed-layer of sandstone and clay-
          stone

The fourth section ③ - ④ gives the path of
unloading. The Young modul is obtained from this
section and the axial deformation can be separated in the reversible and irreversible strain.
For the investigated material the Young modul
varies between 440 MN/m² and 650 MN/m². It can be
noticed that the moduls depend upon the stress-
level.

7.3 Strength parameters of rock masses

The yield stresses are figured as Mohr circles
in a τ-σ-diagram.

Fig. 13 shows the circles of the initial confining pressure steps of three samples. Fig. 14
shows the nine circles of all confining pressure
steps of the same three samples.

The Mohr-Coulomb criterion can be applied very
well to this material for the investigated stress
levels using as well the single-step as the multi
step technique.

In Tab. 1 there are listed the values of the angle
of friction, cohesion and uniaxial compressive
strength for each multi-step triaxial test, for
the first confining pressure steps only and for
all confining pressure steps. The values are computer calculated under the condition of linear
regression.

A 71

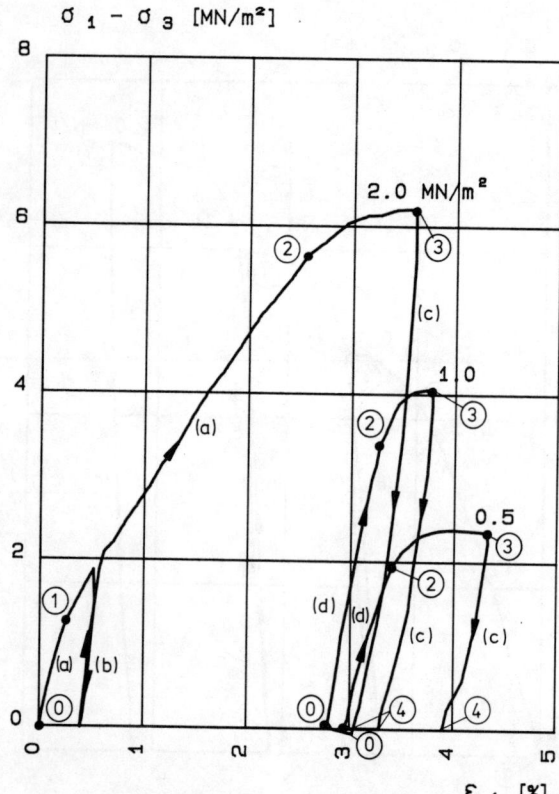

Fig. 12 Multi-step test on sample II with decreasing confining pressure steps
(mixed-layer of sandstone and claystone)

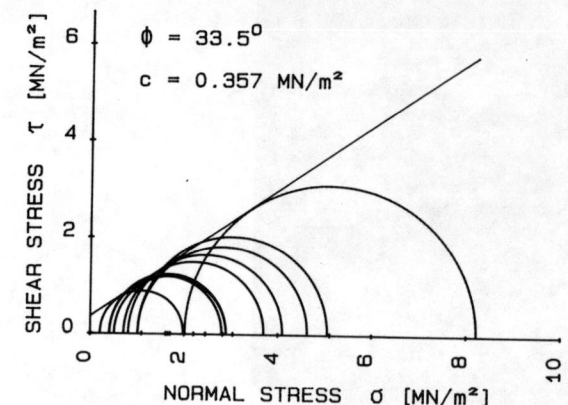

Fig. 14 τ-σ-diagram for all confining pressure
steps of three large scale multi-step
triaxial tests on a mixed-layer of
sandstone and claystone

SAMPLE	ANGLE OF FRICTION φ	COHESION c	UNIAXIAL COMPRESSIVE STRENGTH σ_c
-	deg	MN/m²	MN/m²
I	30.5	0.44	1.54
II	33.8	0.34	1.27
III	33.1	0.37	1.37
I,II,III (INITIAL LOADINGS)	33.3	0.36	1.33
I,II,III (ALL LOADINGS)	33.5	0.36	1.34

Tab. 1: Results of single and multi-step
techniques

For this material there is nearly no difference
between the values of the initial steps and all
multi-steps. Even the values of each particular
multi-step test differ only a little bit. If it
is not possible to mark the yield point exactly
the difference may become larger. Independent
from the single or multi step technique the spread
of the results is very small in comparison to
results of small scale tests, i. e. three to four
large scale multi-step tests are sufficient to
get realistic parameters for a homogeneous area
of close jointed rock masses.

7.4 Strength of rock substance and joints

The parameters of rock substance and joints are
determined attendantly in servo-controlled
testing machines. For many problems the uniaxial
strength σ_C and the Young Modul E of the substance are important. Therefore the result from
small and large scale tests are compared. The
mean values of the substance of the above discussed material are:

	σ_C [MN/m²]	E [MN/m²]
silty claystone	4,8	600
sandstone	10,5	1800

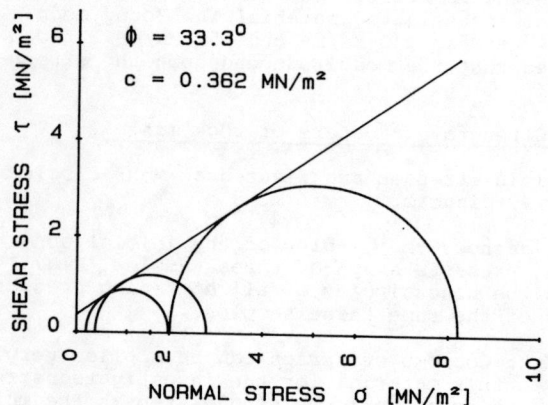

Fig. 13 τ-σ-diagram for the initial confining
pressure steps of three large scale
multi-step triaxial tests on a mixed-
layer of sandstone and claystone

In Fig. 15 the uniaxial compressive strength of close jointed limestone depending on the sample size is drawn. The importance of large scale tests becomes obvious and demonstrates the possibilities of extrapolation. In this case the sample with a diameter of about 60 cm is representative for the rock mass area in situ.

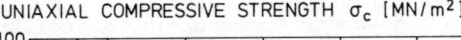
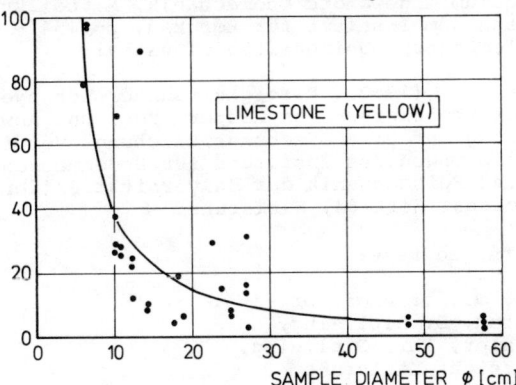

UNIAXIAL COMPRESSIVE STRENGTH σ_c [MN/m²]

LIMESTONE (YELLOW)

SAMPLE DIAMETER ϕ [cm]

Fig. 15 Scale effect of the uniaxial compressive strength for a jointed yellow limestone

7.5 Time dependant material behaviour

The time dependant behaviour of rock masses is influenced by the substance and the joint behaviour and is in general more distinct than the creeping behaviour of the rock substance.

Fig. 16 shows the result for a large scale triaxial test carried out under a constant first stress invariant $(\sigma_1 + 2\sigma_3)/3 = \text{const.}$). The investigated material was a claystone shown in Fig. 2e. It can be seen that creeping stops after a short time until $\sigma_1 - \sigma_3 = 0.425$ MN/m². At $\sigma_1 - \sigma_3 = 0.450$ MN/m² the specimen failed.

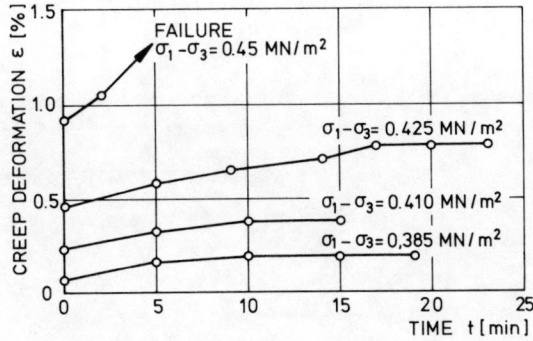

CREEP DEFORMATION ε [%]

FAILURE
$\sigma_1 - \sigma_3 = 0.45$ MN/m²

$\sigma_1 - \sigma_3 = 0.425$ MN/m²

$\sigma_1 - \sigma_3 = 0.410$ MN/m²

$\sigma_1 - \sigma_3 = 0.385$ MN/m²

TIME t [min]

Fig. 16 Creep-curves for claystone from a large triaxial test under constant first stress invariant

7.6 Pressure of pore water

The opinion that in all cases pore water pressure in jointed material has no significant influence on the results of large scale tests was disproved by our investigations carried out with claystone with slicken sides. Without regard of pore water pressure an extremely low angle of friction of about 2° was found. The corresponding value determined by direct shear tests was 11°. Though both values are not directly comparable we started development of a suitable equipment for measuring pore water pressure during the triaxial large scale tests. The determination of realistic pressures of pore water in clay and similar material is a well known problem. At small specimens with diameters of 3 to 5 cm the problem of the end faces with their uncontrolled local states of stress (kinematic boundary condition) are resulting in unsafe values for pore water pressure. Applying the same technology in large scale tests this problem is the same, enlarged by the problem of finishing the end faces with plaster. Therefore the authors developed a modified system consisting of ten pore water pressure cells. These cells were installed in finished flat and plane holes in the mantel area of the specimen. The measuring tubes were installed in small refilled cuts running to connections at the upper face of the specimen (Fig. 2f). Failures in the beginning of this development could be eliminated by using another type of pressure cell. The first successful test with air controlled GLÖTZL-cells (d = 5 cm) in 1981 showed very good balanced values for the pore water pressure. With regard to the effective stresses the angle of friction changed to a realistic order of magnitude. Our present experience is that the realistic pore water pressure in each layer of the specimen can be controlled separately by two to three GLÖTZL cells.

8. CONCLUSIONS

The presented variations of the triaxial large scale test have been successfully applied for research work in tunneling, foundations of power plants and long term stability of slopes in deep open pits. Today the technology is well developed for practical application in all kinds of close jointed rock masses. The drilling technique allows time saving sampling at the surface, in underground openings and in investigation shafts. The sealing with reinforced plaster and steel casings prevents loosening and drainage of the specimens during handling and transportation. Most of the fundamental problems in uniaxial and triaxial large scale testing procedures in the laboratory are solved for a wide range of boundary conditions. The carefully verified results are suitable for statical calculations of all constructions planned in the corresponding rock masses. Our investigations have shown that the multi-step technique is very well applicable for close jointed hard rock, too. In our opinion these results are far more competent than index values estimated by smale scale tests in a more or less sophisticated way. The behaviour of close jointed rockmasses cannot be estimated only by index parameters or by reducing the rock substance parameters.

9. ACKNOWLEDGEMENT

The authors gratefully acknowledge the support of the presented research work by the Deutsche Forschungsgemeinschaft, Bonn-Bad Godesberg, Federal Republic of Germany.

10. REFERENCES

Balthasar, K., Wenz, E. (1983). The determination of the complete state of stress in rock with the flat jack method. 5th Int. Congr. on Rock Mechanics, Melbourne.

Hesselmann, F. J. (1978). Entnahme und Prüfung von Felsproben dünnbankiger Kalksteine. Jahresbericht 1977 des Sonderforschungsbereichs 77 - Felsmechanik -, Universität Karlsruhe.

John, K. W. (1969). Festigkeit und Verformbarkeit von druckfesten, regelmäßig gefügten Diskontinuen. Veröffentlichungen des Instituts für Bodenmechanik und Felsmechanik der Universität Fridericiana, Heft 37, Karlsruhe.

Mühlhaus, H.-B., Reik, G. (1978). Methoden zur Ermittlung mechanischer Eigenschaften von geklüftetem Fels. Felsmechanik Kolloquium Karlsruhe 1978: Grundlagen und Anwendung der Felsmechanik, pp. 67 - 83, Verlag Trans Tech Publications, Clausthal.

Müller, L., Mühlhaus, H.-B., Reik, G., Sharma,B. (1977). Experimental and numerical determination of mechanical behaviour of complex rock formations. Int. Symp. Geotechnics of Structurally Complex Formations, Capri,Italy.

Natau, O., Leichnitz, W., Balthasar, K. (1979). Construction of a computer-controlled direct shear testing machine for investigations on rock discontinuities. 4th Int. Congr. on Rock Mechanics, Vol. 3, pp 241 - 243, Montreux.

Natau, O. (1980). Weiterentwicklungen in der felsmechanischen Versuchstechnik. Kolloquium Angewandte Geomechanik, Mitteilungen aus dem Institut für Bergbau, pp 154 - 172, Technische Universität Clausthal.

Wichter, L. (1980). Festigkeitsuntersuchungen an Großbohrkernen von Keupermergel und Anwendung auf eine Böschungsrutschung. Veröffentlichungen des Instituts für Bodenmechanik und Felsmechanik der Universität Fridericiana, Heft 84, Karlsruhe.

Authors' address:

Natau, O., o. Prof. Dr.-Ing.
Fröhlich, B. Dipl.-Ing.
Mutschler, Th., Dipl.-Ing.
Universität Karlsruhe
7500 Karlsruhe 1
Federal Republic of Germany

ERMITTLUNG DER VERFORMBARKEIT VON ANISOTROPEM FELS AUS DEN ERGEBNISSEN VON FELDVERSUCHEN

Evaluation of the deformability of anisotropic rock masses from the results of field tests

Détermination de la déformabilité des roches anisotropes à partir de résultats d'essais in-situ

J. R. Kiehl
Dipl.Phys.
W. Wittke
Prof. Dr.Ing.
Institut für Grundbau, Bodenmechanik, Felsmechanik und Verkehrswasserbau, Rheinisch-Westfälische Technische Hochschule Aachen, Federal Republic of Germany

SYNOPSIS

The deformability of rock masses is frequently determined with the aid of dilatometer and flat jack tests. However, there are considerable problems in interpreting test results obtained in anisotropic rock conditions since no methods for evaluating the parameters of anisotropic rock masses from these tests are available. In this paper formulae and diagrams for the evaluation of these parameters from the results of dilatometer and flat jack tests based on analytical and numerical methods are presented. An unequivocal determination of individual parameters implies a certain arrangement of tests and a combination of different test results. In-situ triaxial tests on rock specimens largely isolated from the surrounding rock mass by slots make possible the determination of several parameters on one sample if the tests are appropriately carried out.

RESUME

La déformabilité des roches est souvent déterminée au moyen d'essais au dilatomètre et au vérin plat. L'interprétation de ces essais dans le cas de terrains rocheux anisotropes présente pourtant des difficultés considérables, car, jusqu'à présent, il n'existe pas de méthodes d'évaluation pour la détermination des coefficients de roches anisotropes. Dans cet exposé, des procédés correspondants sont traités. Ils ont été développés sur la base de méthodes analytiques et numériques et sont présentés sous forme de diagrammes. Il s'ensuit qu'une détermination univoque des coefficients individuels est possible, seulement si les résultats d'essais différemment disposés sont combinés les uns avec les autres. Les essais triaxiaux sur des éprouvettes, qui sont séparées en grande partie de la roche environnante par des fentes, rendent possible par contre par une mode opératoire correspondante, la détermination indépendante des coefficients de la déformabilité.

ZUSAMMENFASSUNG

Die Verformbarkeit von Fels wird oft mit Dilatometer- und Druckkissenversuchen ermittelt. Bei der Interpretation der Versuchsergebnisse in anisotropen Gebirgsverhältnissen ergeben sich jedoch erhebliche Schwierigkeiten, weil für diese Versuche keine Auswertungsmethoden zur Bestimmung der Kennwerte von anisotropem Fels zur Verfügung stehen. In diesem Beitrag werden auf der Grundlage analytischer und numerischer Methoden entwickelte Formeln und Diagramme zur Ermittlung dieser Parameter aus den Ergebnissen von Dilatometer- und Druckkissenversuchen vorgestellt. Eine eindeutige Ermittlung einzelner Kennwerte ist jedoch nur möglich, wenn die Versuche in bestimmter Weise angeordnet werden und die Ergebnisse mehrerer Versuche miteinander kombiniert werden. In-situ Triaxialversuche an Probekörpern, die durch Schlitze weitgehend vom umgebenden Gebirge getrennt werden, ermöglichen demgegenüber bei entsprechender Versuchsdurchführung die Bestimmung mehrerer Kennwerte an einer Probe.

1. EINLEITUNG

Verschiedene metamorphe Gesteine wie Schiefer und Gneisse aber auch einzelne Sedimentgesteine besitzen infolge eines flächigen Korngefüges eine anisotrope Verformbarkeit, die meist durch fünf Kennwerte hinreichend genau beschrieben werden kann. Dieses transversal isotrope Spannungsverformungsverhalten, bei dem die isotrope Ebene parallel zum flächenhaften Aufbau des Korngefüges angenommen wird, kann auf Felsarten, die sich aus derartigen Gesteinen zusammensetzen, übertragen werden, wenn die Schieferung bzw.

Schichtung das die Verformbarkeit bestimmende Element des Trennflächengefüges darstellt. Das elastische Spannungsverformungsverhalten eines solchen Felsens kann dann durch die in Fig. 1 skizzierten fünf unabhängigen elastischen Kennwerte beschrieben werden (Wittke, 1976).

Weil die Verformbarkeit von Fels darüberhinaus durch den Einfluß des Trennflächengefüges bestimmt wird, ist ihre Ermittlung in der Regel allerdings nur durch Feldversuche möglich, bei

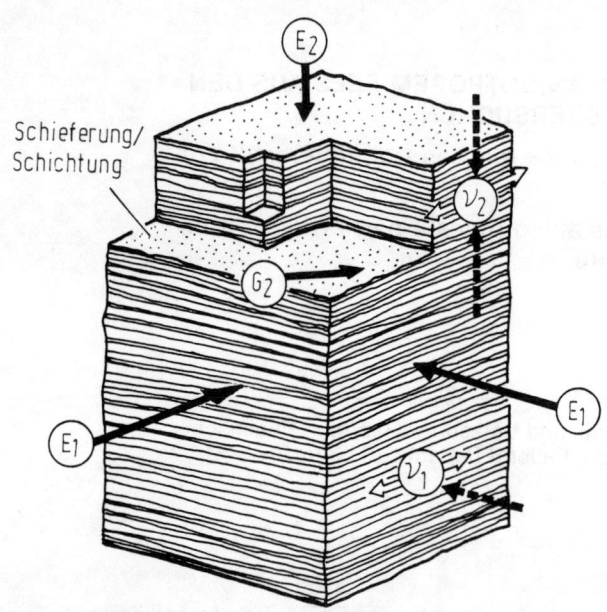

<u>Fig. 1</u>   Elastische Kennwerte von transversal
              isotropem Fels

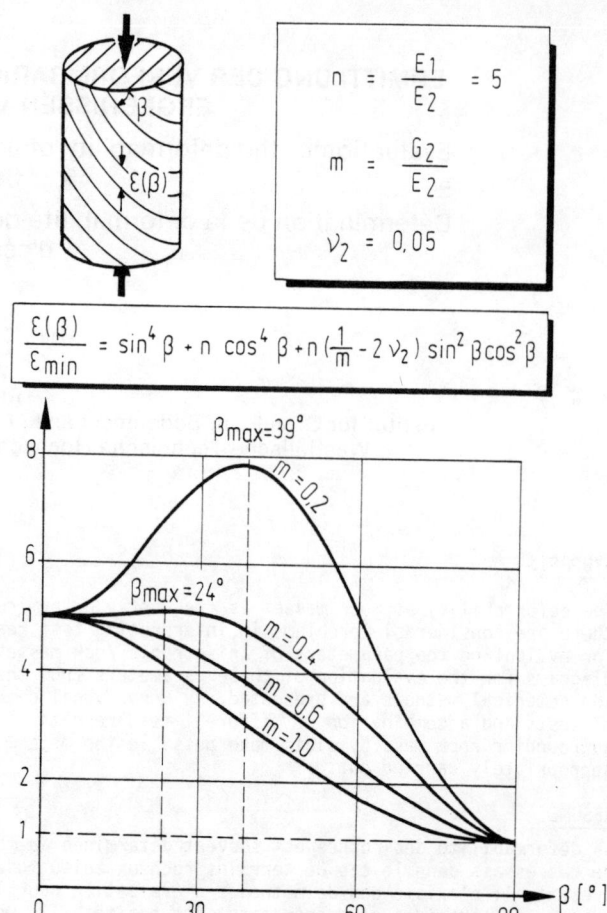

$$\frac{\varepsilon(\beta)}{\varepsilon_{min}} = \sin^4\beta + n\,\cos^4\beta + n\,(\frac{1}{m} - 2\,\nu_2)\,\sin^2\beta\cos^2\beta$$

<u>Fig. 2</u>   Einaxiale Verformbarkeit von transveral
              isotropem Fels in Abhängigkeit von der
              Richtung der Lasteinwirkung und den Mo-
              dulverhältnissen n und m

bei denen größere Volumina als im Labor untersucht werden können. Die Interpretation der Ergebnisse dieser Versuche ist jedoch oft unbefriedigend, weil zwischen dem Entwicklungsstand
numerischer Berechnungsverfahren, die dem Ingenieur heute als Entwurfshilfe zur Verfügung
stehen, und der gängigen Praxis der Auswertung
von Feldversuchen ein großer Unterschied besteht.

Die meisten neueren Berechnungsverfahren ermöglichen eine Berücksichtigung der Verformbarkeits-
anisotropie (Semprich, 1980), während die Auswertung von Feldversuchen in der Regel immer noch
unter der Annahme eines isotropen Spannungsverformungsverhaltens erfolgt. Die Absicht dieses
Beitrags besteht deshalb darin, diese Kluft etwas zu verringern und dem entwerfenden Ingenieur Hinweise zu geben, die einige der von ihm
im Entwurf festgelegten und in Standsicherheitsuntersuchungen eingehenden Kennwerte durch eine
verfeinerte Interpretation von Versuchsergebnissen besser abgesichert werden können.

2. VERFORMBARKEIT VON TRANSVERSAL ISOTROPEM FELS

Die Richtungsabhängigkeit der einaxialen Verformbarkeit von transversal isotropem Fels läßt sich
qualitativ durch die infolge einer einaxialen Belastung in Lastrichtung auftretenden Dehnungen
beschreiben. Um eine von der Belastung unabhängige Größe zu erhalten, kann man, wie in Fig. 2,
die für eine beliebige Lastrichtung nach dem verallgemeinerten Hooke'schen Gesetz erhaltenen Deh-
nungen ε (β) auf den minimalen Wert ε_{min} beziehen. Fig. 2 macht deutlich, daß die einaxiale
Verformbarkeit entscheidend von den Modulverhältnissen

$$n = \frac{E_1}{E_2} \qquad\qquad (1)$$

$$m = \frac{G_2}{E_2} \qquad\qquad (2)$$

abhängig ist. Während das Verhältnis der Elasti-

zitätsmoduln n bei einem Schiefer das Verhältnis
der Verformbarkeiten senkrecht und parallel zur
Schieferung kennzeichnet, wird die Richtung der
größten Verformbarkeit durch das Verhältnis m
bestimmt. Der Einfluß der Poissonzahlen ν_1 und
ν_2 ist, wie die Formel in Fig. 2 zeigt, demgegenüber gering.

3. DILATOMETERVERSUCHE

Eine wenig aufwendige und daher oft angewendete
Versuchstechnik zur Ermittlung der Verformbarkeit von Fels bilden Dilatometerversuche. Das
Versuchsprinzip besteht darin, daß über einen
bestimmten Bohrlochabschnitt ein radialsymme-
trischer Innendruck auf die Bohrlochwand übertragen wird. Die infolge dieses Innendrucks ein-
tretenden Verschiebungen werden als Durchmesseränderungen des Bohrlochs in verschiedenen Richtungen gemessen und im Hinblick auf die Verformbarkeit des umgebenden Gebirges interpretiert.

Fig. 3 zeigt stellvertretend für verschiedene
andere Ausführungen das von Rocha et al. (1966)
in Lissabon entwickelte LNEC Dilatometer, das
i.w. aus einem Stahlzylinder und einer Gummimembrane, die durch Öldruck gegen die Bohrlochwand gepreßt wird, besteht. Am Stahlzylinder

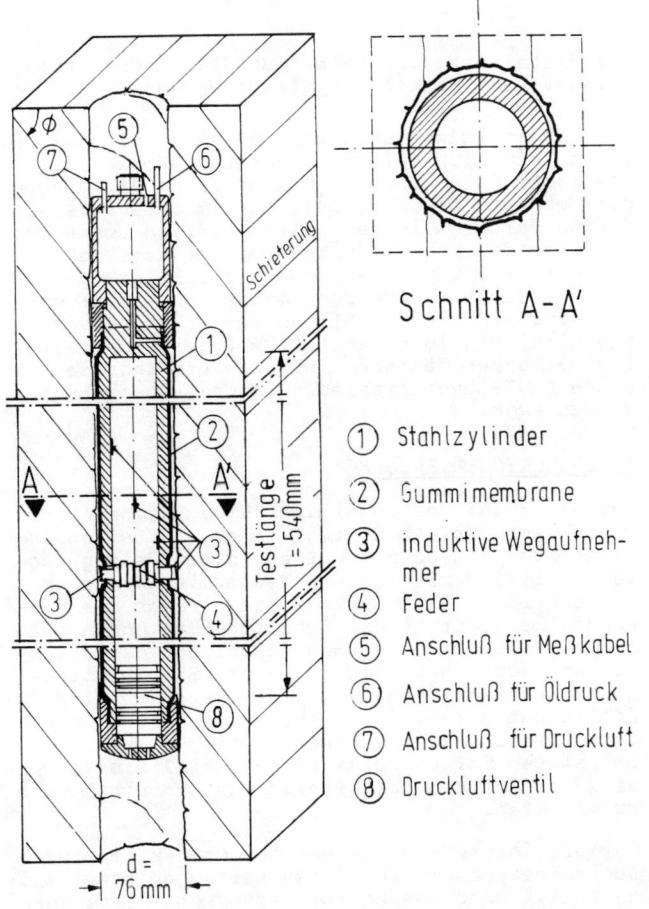

Fig. 3   LNEC Dilatometer nach Rocha et al (1966)

① Stahlzylinder
② Gummimembrane
③ induktive Wegaufneh-
    mer
④ Feder
⑤ Anschluß für Meßkabel
⑥ Anschluß für Öldruck
⑦ Anschluß für Druckluft
⑧ Druckluftventil

sind vier induktive Wegaufnehmer angeordnet,
durch die in vier Richtungen, jeweils um 45° ge-
geneinander versetzt, die Verschiebungen der
Bohrlochwand in radialer Richtung infolge einer
Innendruckänderung Δp als Durchmesseränderungen
Δd gemessen werden (Fig. 4a).

Dabei ergeben sich in den Meßrichtungen in der
Regel unterschiedliche Werte, die etwa durch
örtliche Inhomogenitäten bedingt sein können
wie z.B. das Schließen oder Öffnen einer Kluft
(Fig. 4 b). In diesem Fall ist eine isotrope
Auswertung mit der Formel nach Rocha (1966)

$$E = (1 + \nu)\, d\, \frac{\Delta p}{\Delta d} \qquad (3)$$

im Sinne einer statistischen Erfassung eines Moduls in Meßrichtung gerechtfertigt, wenn sich
der Fels großmaßstäblich isotrop verhält. In (3)
ist E der Verformungsmodul, ν die Poissonzahl
und d der Bohrlochdurchmesser.

Wenn die in den verschiedenen Richtungen gemessenen Durchmesseränderungen jedoch einer ausge-
zeichneten Richtung, z.B. der Streichrichtung
einer Schieferung (Fig. 4c), zugeordnet werden
können, ist eine Auswertung unter Berücksichtigung eines transversal isotropen Spannungsverformungsverhaltens möglich, indem die Meßwerte
nach Kiehl (1980) an eine analytische Funktion
der Form

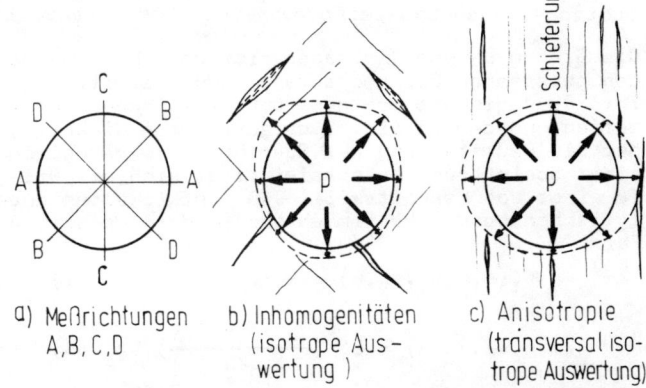

a) Meßrichtungen   b) Inhomogenitäten   c) Anisotropie
   A,B,C,D   (isotrope Aus-   (transversal iso-
                  wertung)   trope Auswertung)

Fig. 4   Interpretation der bei Dilatometerversuchen gemessenen Durchmesseränderungen

$$\frac{\Delta d}{\Delta p \cdot d} = U_o \cos^2 \theta + V_o \sin^2 \theta$$

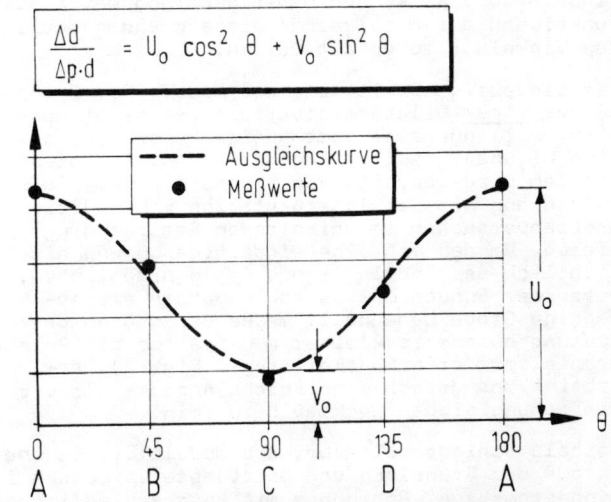

Fig. 5   Anpassung der bei Dilatometerversuchen
in transversal isotropen Gebirgsverhältnissen gemessenen Durchmesseränderungen an eine analytische Funktion
nach Kiehl (1980)

$$\frac{\Delta d}{\Delta p \cdot d} = U_o \cos^2 \theta + V_o \sin^2 \theta \qquad (4)$$

angepaßt werden (Fig. 5). Die Größen U_o und V_o
sind die Extremwerte der relativen Durchmesseränderungen $\Delta d/d$ bezogen auf die Druckänderung Δp. Diese treten senkrecht und parallel zur
Streichrichtung der Schieferung auf, wenn die
elastischen Kennwerte die folgenden Bedingungen
erfüllen:

$$\nu_1 \le 1 - 2\nu_2^2 n \qquad (5)$$

$$m \le m_{max} = \frac{n}{2[\nu_2(1+\nu_1)n + \sqrt{(1-\nu_1^2)(1-\nu_2^2 n)n}\,]} \qquad (6)$$

Die Beziehung 5 folgt aus der Bedingung, daß die
elastische Formänderungsenergie nicht negativ
werden darf, während m_{max} nach Kiehl (1980)
eine obere Grenze für das Modulverhältnis m in

Abhängigkeit von n, ν_1 und ν_2 für transversal isotropes Spannungsverformungsverhalten darstellt.

Die Größen U_o und V_o lassen sich nach Fig. 5 aus den gemessenen Durchmesseränderungen ermitteln. Ihre Abhängigkeit von den fünf elastischen Kennwerten und dem in Fig. 3 dargestellten Winkel ϕ, der die Raumstellung der Bohrlochachse bezüglich einer Schieferung beschreibt, läßt sich mit Hilfe einer von Lekhnitskii (1963) entwickelten analytischen Lösung bestimmen und in der impliziten Form

$$\varkappa_1(m,n,\nu_1,\nu_2,\phi) = E_1 \cdot V_o \qquad (7)$$

$$\varkappa_2(m,n,\nu_1,\nu_2,\phi) = E_1 \cdot \left(\frac{U_o - V_o \cos^2\phi}{\sin^2\phi}\right) \qquad (8)$$

angeben. Parameterstudien von Kiehl (1980), in denen die elastischen Kennwerte variiert wurden, haben gezeigt, daß die Größen U_o und V_o und damit auch die dimensionslosen Funktionen $\varkappa_1$ und $\varkappa_2$ von den Poissonzahlen ν_1 und ν_2 kaum abhängig sind. $\varkappa_1$ und $\varkappa_2$ können daher näherungsweise als Funktionen der Modulverhältnisse m und n sowie des Winkels ϕ aufgefaßt werden.

Die eindeutige Ermittlung der Moduln E_1, E_2 und G_2 aus einem Dilatometerversuch ist nicht möglich, weil nur zwei unabhängige Meßgrößen, nämlich U_o und V_o zur Verfügung stehen. Auf dieses Problem wurde bereits von Kawamoto (1966) im Zusammenhang mit der Interpretation von Radialpressenversuchen im anisotropen Gebirge hingewiesen. Um dennoch eine eindeutige Lösung hinsichtlich der Moduln E_1 und E_2 zu ermöglichen, wurde der Schubmodul G_2 von Kawamoto als abhängige Größe behandelt. Wegen der großen Bedeutung besonders kleiner m Werte für die Anisotropie der Verformbarkeit (vgl. Fig. 2), erscheint uns jedoch eine solche Annahme als Vereinfachung nicht zweckmäßig zu sein.

Deshalb schlagen wir vor, die Moduln E_1, E_2 und G_2 auf der Grundlage von Dilatometerversuchen in mindestens zwei Bohrungen mit unterschiedlichen Raumstellungen zu ermitteln. Aus einer Bohrung mit $\phi = 0°$, d.h. senkrecht zur Schieferung bzw. Schichtung, läßt sich wegen der dann isotropen Verformbarkeit in radialer Richtung der Modul E_1 nach einer zu (3) analogen Formel

$$E_1 = (1 + \nu_1)\, d\, \frac{\Delta p}{\Delta d} \qquad (9)$$

ermitteln, die aus der analytischen Lösung für beliebige Bohrlochorientierungen abgeleitet werden kann. Zur Bestimmung von E_2 und G_2 wird zusätzlich in einer zweiten Bohrung mit $\phi \neq 0°$ eine weitere Versuchsreihe durchgeführt.

Unter den Voraussetzungen, daß beide Bohrungen im gleichen Homogenbereich liegen und in beiden Bohrungen hinreichend viele Versuche durchgeführt werden, kann der Mittelwert von E_1 in die rechten Seiten der Gleichungen 7 und 8 eingesetzt werden. Damit lassen sich dann die Hilfsgrößen $\varkappa_1$ und $\varkappa_2$ bestimmen, wenn gleichzeitig auch für V_o, U_o und ϕ Mittelwerte eingesetzt werden. Mit Hilfe der in Fig. 6 dargestellten Diagramme können durch Interpolation die Modulverhältnisse m und n und damit nach

$$E_2 = \frac{1}{n}\, E_1 \qquad (10)$$

$$G_2 = \frac{m}{n}\, E_1 \qquad (11)$$

die Mittelwerte der beiden anderen Moduln für verschiedene Winkel ϕ bestimmt werden.

Bei dieser Auswertungsmethode ist zu berücksichtigen, daß sich bei Dilatometerversuchen in klüftigem Fels erfahrungsgemäß große Streuungen der Meßwerte ergeben, weil die im Versuch erfaßten Volumina in den meisten Fällen wegen der Klüftung des Felsens nicht repräsentativ sind. Große Streuungen haben jedoch zur Folge, daß für die zuverlässige Bestimmung von Mittelwerten eine größere Anzahl von Versuchen notwendig wird, die im Einzelfall auf der Grundlage statistischer Methoden, wie sie von Charrua-Graca (1979) vorgeschlagen werden, festgelegt werden kann.

4. DRUCKKISSENVERSUCHE

Der von Rocha und da Silva (1970) beschriebene Druckkissenversuch besteht in der Aufweitung von einem oder mehreren schmalen in den Fels gesägten Schlitzen mit flachen hydraulischen Druckkissen (Large Flat Jacks). Die infolge des Kissendrucks auftretenden Schlitzaufweitungen werden an vier verschiedenen Positionen des Druckkissens durch Verformungsmesser (Stahlblattfedern) gemessen (Fig. 7). Diese Versuche sind zur Ermittlung der Verformbarkeit von Fels noch besser geeignet als Dilatometerversuche, weil die belasteten Felsvolumina in der Regel groß genug sind, um für den Fels repräsentative Ergebnisse zu erhalten.

Für die Interpretation der bei Druckkissenversuchen gemessenen Schlitzaufweitungen steht keine analytische Lösung zur Verfügung. Diese Versuche werden deshalb gewöhnlich unter der Annahme eines elastisch isotropen Spannungsverformungsverhaltens nach der Formel

$$E = K\, C_\nu\, \frac{\Delta p}{\Delta s} \qquad (12)$$

mit

$$C_\nu = 1 - \nu^2$$

ausgewertet, wobei E der Verformungsmodul, ν die Poissonzahl und Δs die Schlitzaufweitung infolge einer Änderung des Kissendrucks von Δp ist. Der Faktor K ist abhängig von der Position des Verformungsmessers und der Anzahl der verwendeten Druckkissen. Die K Werte wurden vom LNEC in Lissabon zunächst durch Modellversuche und später durch Berechnungen nach der Methode der Finiten Elemente bestimmt. Für den in Fig. 7 skizzierten Versuch mit zwei Druckkissen wurden von Loureiro-Pinto (1981) die folgenden Werte angegeben:

$$
\begin{aligned}
K_A = K_F &= 150 \text{ cm}\\
K_B = K_E &= 191 \text{ cm}\\
K_C = K_I &= 161 \text{ cm}\\
K_D = K_E &= 215 \text{ cm}
\end{aligned}
\qquad (13)
$$

Zur Interpretation von LFJ Versuchen mit zwei Druckkissen im transversal isotropen Fels haben wir mit einem von Semprich (1980) entwickelten dreidimensionalen Finite Element Programmsystem und dem in Fig. 8 dargestellten Berechnungsausschnitt Parameterstudien durchgeführt, in denen die

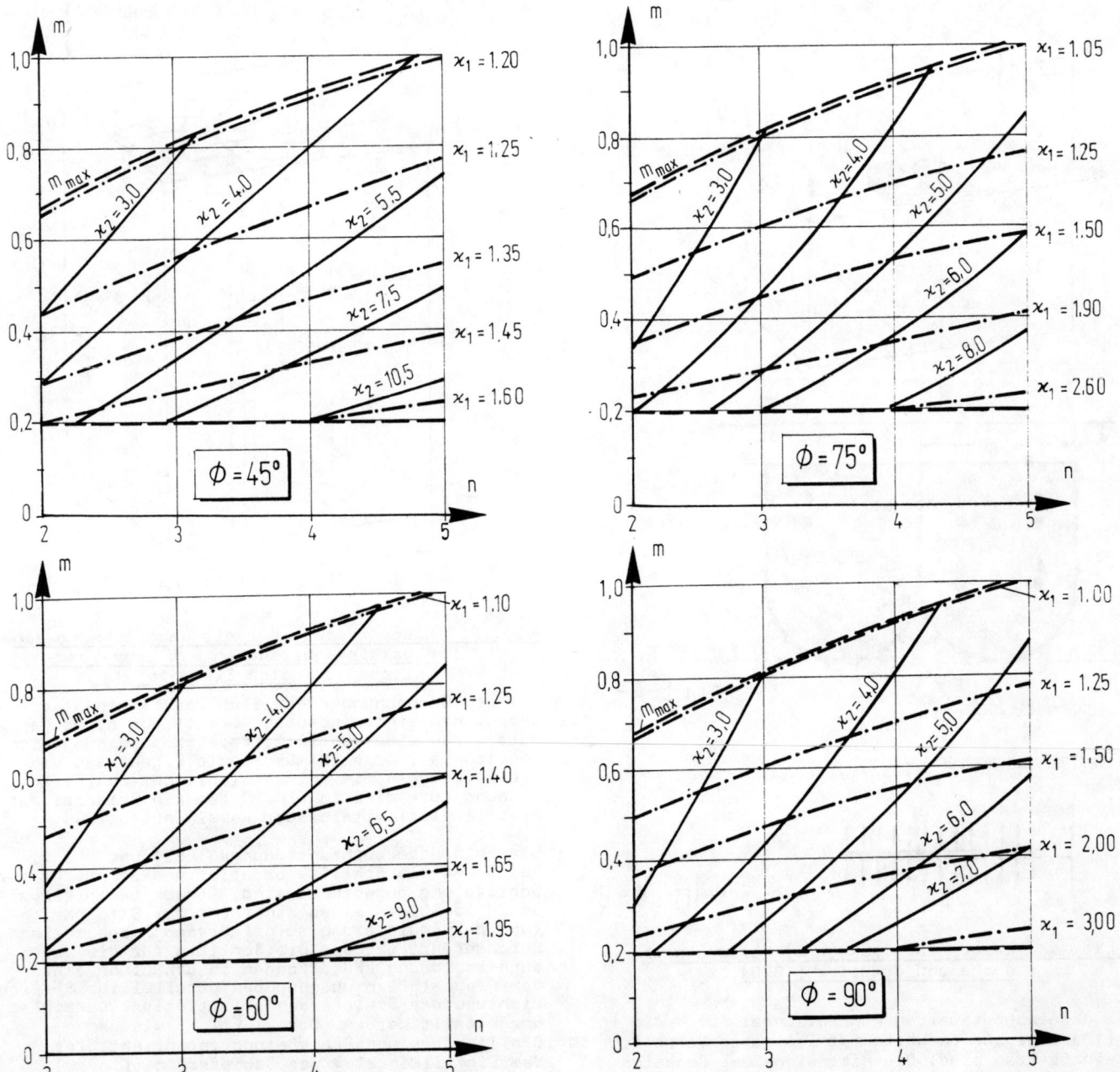

<u>Fig. 6</u>　Ermittlung der Modulverhältnisse m und n aus den Ergebnissen von Dilatometerversuchen im transversal isotropen Fels

elastischen Kennwerte und die Winkel ϕ und θ (s. Fig. 7) variiert wurden. Um den Rechenaufwand zu begrenzen, haben wir uns auf die Fälle $\theta = 0^\circ$ und $\theta = 90^\circ$ beschränkt, für die die Ebene x = 3 m eine Symmetrieebene darstellt (Fig. 8).

In der Ebene y = 0 wurden die Knotenpunkte außerhalb des belasteten Bereichs festgehalten. Diese Vereinfachung bei der Festlegung der Randbedingungen, die zu einer Unterdrückung der Verschiebungen in x- und z-Richtung in dieser

Ebenen führt, beeinflußt jedoch nicht die Verschiebungen in y-Richtung im Bereich der Verformungsmesser, wie durch Vergleichsrechnungen nachgewiesen werden konnte.

Bei den Berechnungen zeigte sich, daß die Schlitzaufweitungen nach der zu (12) analogen Formel

$$\Delta s = K \frac{C_{mn}(\phi, \theta)}{E_2} \Delta p \qquad (14)$$

A 79

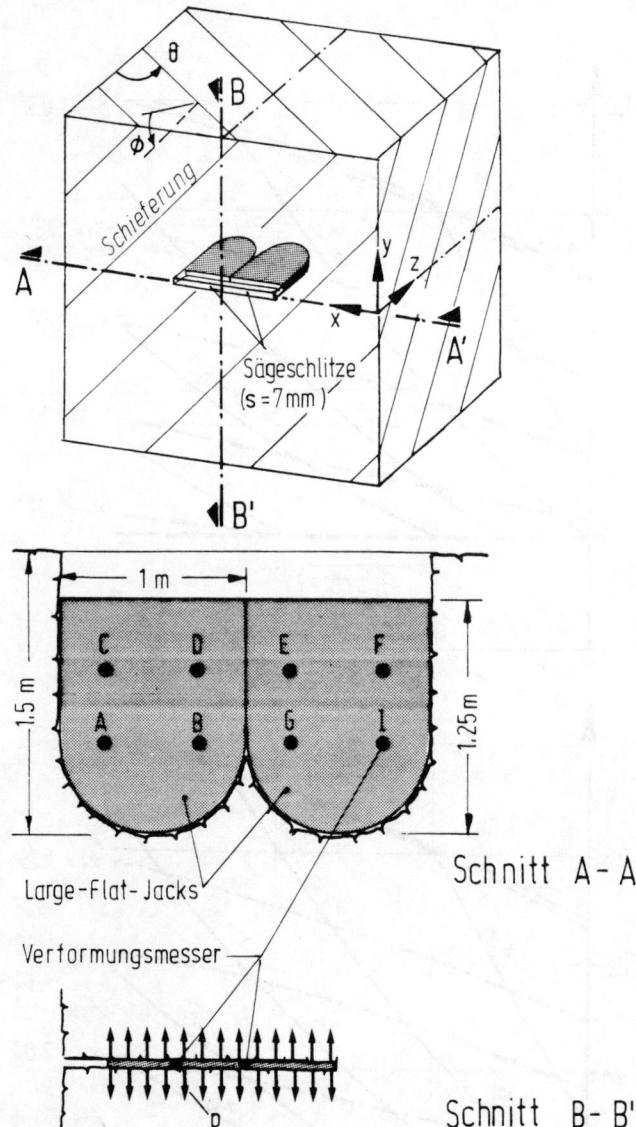

Fig. 7 Large Flat Jack (LFJ) Versuch nach
Rocha und da Silva (1970)

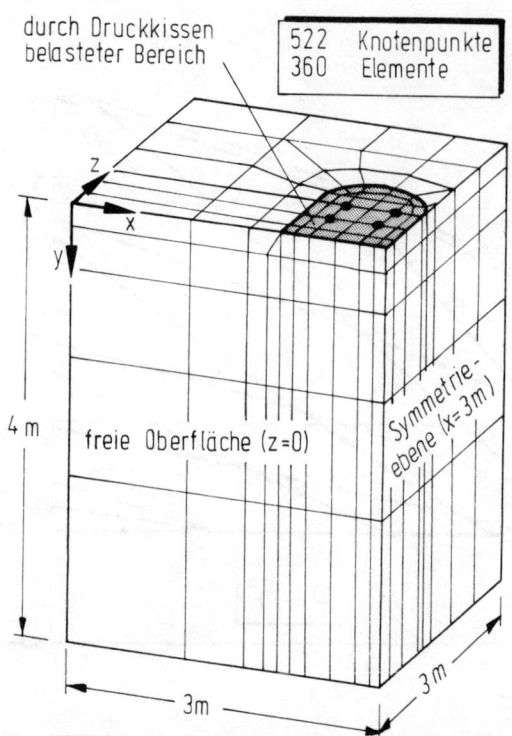

Fig. 8 Elementennetz für die Nachrechnung von
Druckkissenversuchen mit isoparame-
trischen 8-Knoten Elementen

rückgerechnet werden können, wobei für K die in
(13) angegebenen Werte für isotropes Verhalten
einzusetzen sind. Die dimensionslose Funktion
C_{mn} kennzeichnet die Verformbarkeit von transversal isotropem Fels infolge des bei Druckkissenversuchen auftretenden Belastungszustandes.
Der Einfluß der Poissonzahlen ist erwartungsge-
mäß gering, so daß C_{mn} näherungsweise als Funktion von m, n, ϕ und θ aufgefaßt werden kann.

Somit werden die bei einem Druckkissenversuch gemessenen Schlitzaufweitungen i.w. durch die drei
Moduln E_1, E_2 und G_2 und die Orientierung der
Druckkissen zur Schieferung (θ, ϕ) bestimmt. Da
aus einem Versuch mit einer bestimmten Orientierung der Druckkissen immer nur einer der drei
Kennwerte bestimmt werden kann, sind zur Ermittlung aller drei Moduln also mindestens drei Versuche mit unterschiedlicher Schlitzanordnung bezüglich der Schieferung bzw. Schichtung erforderlich.

Aus den Berechnungen hat sich darüberhinaus ergeben, daß eine eindeutige Ermittlung eines Moduls durch eine bestimmte Versuchsanordnung nicht
möglich ist. Deshalb müssen die Ergebnisse von
drei Versuchen mit unterschiedlicher Schlitzanordnung, wie dies in Fig. 9 für ein Beispiel dargestellt ist, miteinander kombiniert werden.

Die beiden Versuchsanordnungen LFJ1 und LFJ 2,
bei denen die Schlitze parallel und senkrecht zur
Schieferung angeordnet sind, können beispielsweise in den Ulmen eines senkrecht zur Streichrichtung der Schieferung aufgefahrenen Probestollens
durchgeführt werden. Die Schlitze für die Versuchsanordnung LFJ 3 können in den Ulmen eines
vom Probestollen ausgehenden parallel zur Streichrichtung der Schieferung orientierten Querschlags
hergestellt werden. Der Winkel ϕ zwischen
Schlitzebene und Schieferung entspricht hier
dem Einfallwinkel β der Schieferung (Fig. 9).

Die Auswertung dieser Versuche im Hinblick auf
die Ermittlung der Moduln E_1, E_2 und G_2 erfolgt
in zwei Schritten: Zunächst werden aus den Ar-
beitslinien der drei Versuche die Hilfsgrößen

$$\varkappa_1(\phi) = \frac{C_1}{C_3(\phi)} = \frac{\Delta s_1}{\Delta p_1} \cdot \frac{\Delta p_3}{\Delta s_3} \qquad (15)$$

$$\varkappa_2(\phi) = \frac{C_2}{C_3(\phi)} = \frac{\Delta s_2}{\Delta p_2} \cdot \frac{\Delta p_3}{\Delta s_3} \qquad (16)$$

ermittelt. Diese Werte ermöglichen eine eindeutige Bestimmung der Modulverhältnisse m und n,
die für die Winkel $\phi = 30^{\circ}$ und 60° durch die in
Fig. 10 dargestellten Diagramme vorgenommen wer-
den kann.

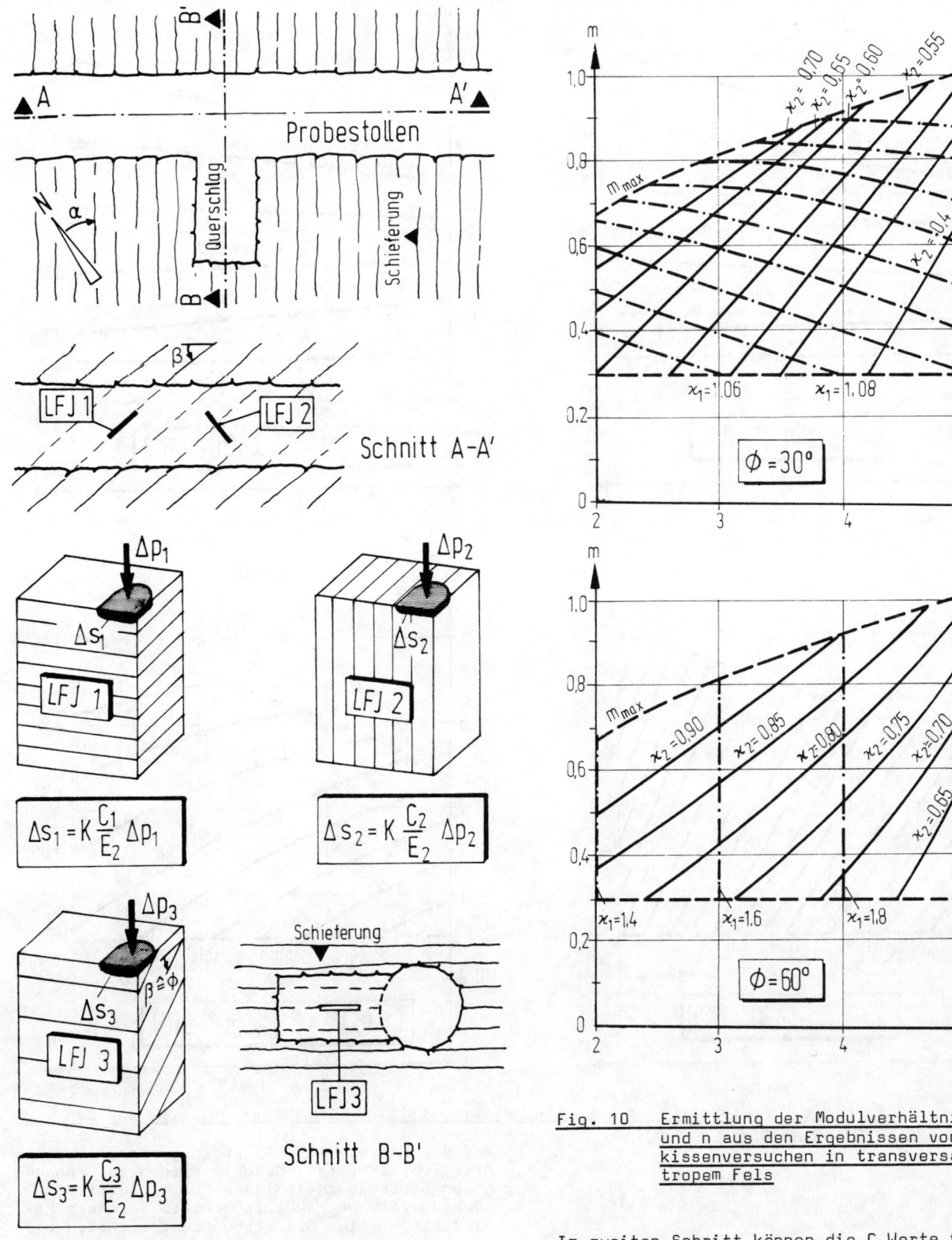

Fig. 10 Ermittlung der Modulverhältnisse m
und n aus den Ergebnissen von Druckkissenversuchen in transversal isotropem Fels

Fig. 9 Vorschlag für die Anordnung von Druck-
kissenversuchen zur Bestimmung der
Moduln E_1, E_2 und G_2 in geschiefertem
Fels

Im zweiten Schritt können die C Werte für die
Versuche LFJ 1 - 3 (vgl. Fig. 9) aus den Diagrammen in Fig. 11 erhalten werden. Die Bestimmung der (mittleren) Moduln erfolgt dann durch
die folgenden Formeln:

G

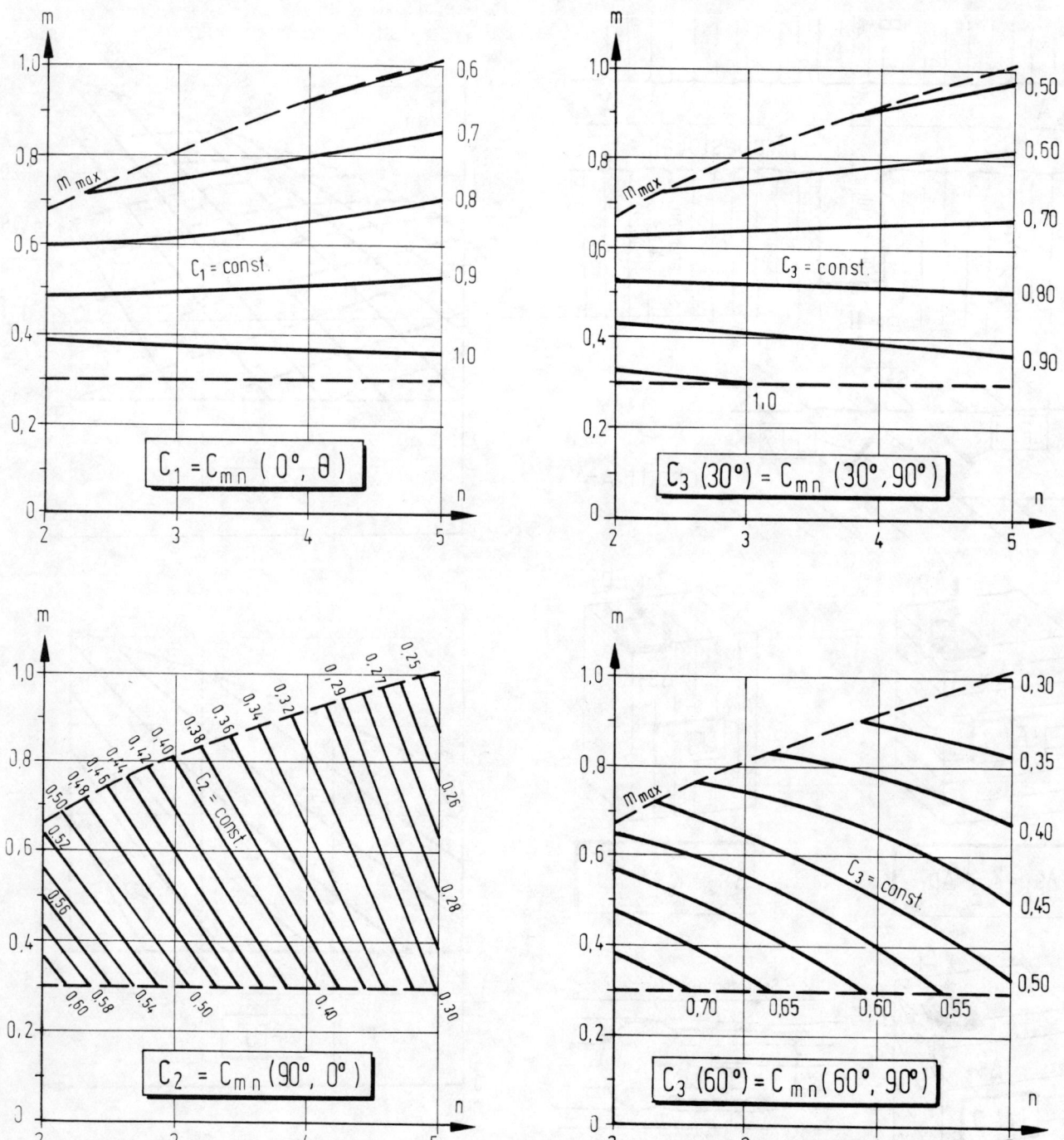

Fig. 11   Ermittlung der C Faktoren für die in Fig. 9 dargestellten Druckkissenversuche aus den Modulverhältnissen m und n

$$E_2 = \frac{1}{3} \sum_{i=1}^{3} K\, C_i\, \frac{\Delta p_i}{\Delta s_i} \qquad (17)$$

$$E_1 = n\, E_2 \qquad (18)$$

$$G_2 = m\, E_2 \qquad (19)$$

Bei dieser Auswertungsmethode ist zu beachten,

daß die Formeln 17 - 19 strenggenommen nur für Mittelwerte gelten. Deshalb werden bei inhomogenen Gebirgsverhältnissen für die drei Versuchsanordnungen jeweils mehrere Versuche erforderlich sein. Die Streuungen, die sich bei Druckkissenversuchen ergeben, sind jedoch erfahrungsgemäß sehr viel geringer als bei Dilatometerversuchen in vergleichbaren Gebirgsverhältnissen, so daß auch die zur Absicherung der Mittelwerte erforderliche Versuchsanzahl geringer ist.

A 82

5. IN-SITU TRIAXIALVERSUCHE

Der in Fig. 12 schematisch dargestellte in-situ Triaxialversuch wurde erstmalig 1972 in einem Flysch des Ghionia-Massivs in Griechenland erprobt. Darüber wurde von Lögters und Voort (1974) berichtet.

Es handelt sich dabei um einen Triaxialversuch, bei dem in der Sohle eines Probestollens ein kubischer Felsprobekörper von ca. 1 m³ mit einer Schlitzbohrvorrichtung aus dem Fels herausgetrennt wird. In horizontaler Richtung erfolgt die Belastung durch vier Druckkissen, die in die Bohrschlitze eingesetzt werden. In vertikaler Richtung wird die Belastung durch Pressen aufgebracht, die durch ein Betonwiderlager gegen die Firste des Stollens abgestützt werden. Die Meßeinrichtungen bestehen aus einem Mehrfachextensometer zur Messung der vertikalen Verformungen des Probekörpers und aus vier Deflektometern zur Messung der Verformungen in horizontaler Richtung (Fig. 12, 13).

Im Unterschied zu allen uns bekannten Feldversuchen zur Ermittlung der Verformbarkeit von Fels bietet dieser Versuch bei entsprechender Durchführung die Möglichkeit <u>alle</u> elastischen Kennwerte von transversal isotropem Fels, einschließlich der Poissonzahlen, unabhängig voneinander in einem Versuch zu bestimmen. Dazu müssen die Bohrschlitze im Falle eines geschieferten oder geschichteten Felsens gemäß Fig. 13 angeordnet werden. Bei dieser Geometrie ergeben sich bei einer Belastung p_y in y-Richtung aus dem verallgemeinerten Hooke'schen Gesetz für transversal isotropes Spannungsverformungsverhalten

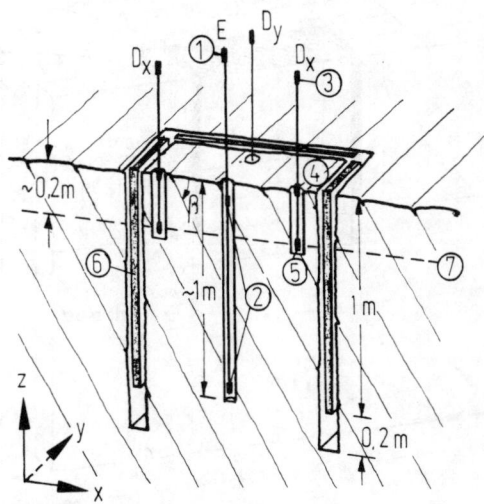

$n = 5$	$\nu_1 = 0,25$	$\beta = 60°$
$m = 0,5$	$\nu_2 = 0,05$	$\sigma_y = 5 \ MN/m^2$

1   Extensometerkopf
2   Extensometermeßpunkte
3   Deflektometerkopf
4   Deflektometermeßglied
5   Deflektometerendstück
6   Druckkissen
7   Meßebene für die Horizontalverschiebungen

<u>Fig. 13</u>   Für die Nachrechnung eines Triaxialversuchs angenommene Kennwerte, Anordnung des Probekörpers zur Schieferung bzw. Schichtung sowie der Instrumente zur Messung der Vertikal- und Horizontalverformungen

für die Kennwerte E_1 und ν_1 die Beziehungen:

$$E_1 = \frac{p_y}{\epsilon_y} \tag{20}$$

$$\nu_1 = \frac{\epsilon_x \cos^2\beta - \epsilon_z \sin^2\beta}{\epsilon_y(\sin^2\beta - \cos^2\beta)} \tag{21}$$

Die Dehnungen ϵ_x, ϵ_y und ϵ_z können aus den Differenzverschiebungen der Extensometermeßpunkte (ϵ_z) bzw. der Deflektometermeßwerte für Horizontalverschiebungen in x-Richtung (D_x) und in y-Richtung (D_y) berechnet werden. In entsprechender Weise ergeben sich bei einer Belastung p_z in z-Richtung die übrigen Kennwerte, wenn die nach (20) und (21) ermittelten Werte für E_1 und ν_1 eingesetzt werden:

$$E_2 = \frac{\cos^4\beta}{\left[\frac{(\epsilon_x+\epsilon_z)}{p_z} - \frac{\sin^2\beta}{E_1}\right]\cos^2\beta - \left(\frac{\nu_1}{E_1}\sin^2\beta + \frac{\epsilon_y}{p_z}\right)} \tag{22}$$

$$G_2 = \frac{\sin^2\beta \cos^2\beta}{\left[\frac{(1+\nu_1)}{E_1}\sin^2\beta + \frac{\epsilon_y}{p_z}\right](\cos^2\beta - \sin^2\beta) + \frac{\epsilon_z}{p_z}\sin^2\beta - \frac{\epsilon_x}{p_z}\cos^2\beta} \tag{23}$$

$$\nu_2 = \frac{\left(\frac{\epsilon_y}{p_z} + \frac{\nu_1}{E_1}\sin^2\beta\right)\cos^2\beta}{\left[\frac{\sin^2\beta}{E_1} - \frac{(\epsilon_x+\epsilon_z)}{p_z}\right]\cos^2\beta + \frac{\nu_1}{E_1}\sin^2\beta + \frac{\epsilon_y}{p_z}} \tag{24}$$

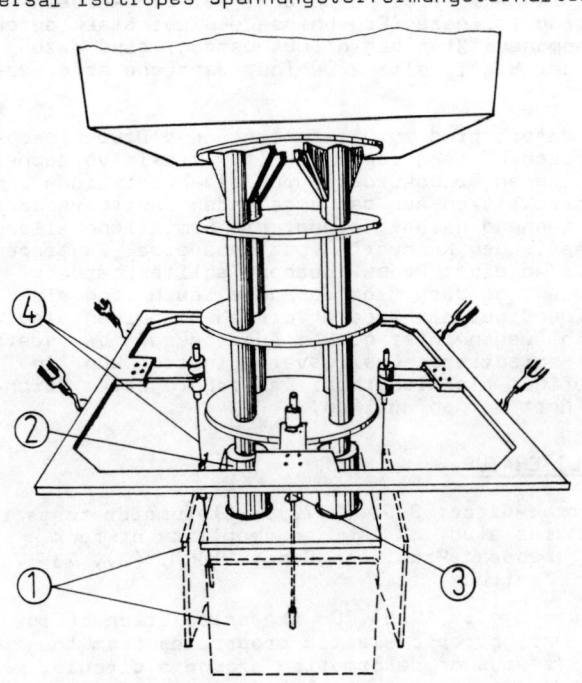

1   Druckkissen
2   Pressen
3   Mehrfachextensometer
4   Deflektometer (an Meßrahmen montiert)

<u>Fig. 12</u>   In-situ Triaxialversuche nach INTERFELS. Belastungs- und Meßeinrichtungen

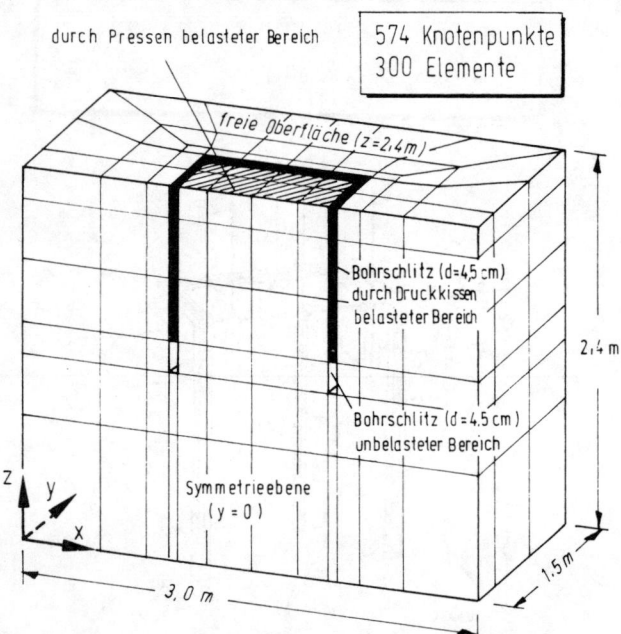

durch Pressen belasteter Bereich

574 Knotenpunkte
300 Elemente

freie Oberfläche (z=2,4 m)

Bohrschlitz (d=4,5 cm)
durch Druckkissen
belasteter Bereich

2,4 m

Bohrschlitz (d=4,5 cm)
unbelasteter Bereich

Symmetrieebene
(y = 0)

z y
x

3,0 m

1,5 m

Fig. 14   Elementennetz für die Nachrechnung
eines Triaxialversuchs mit isoparame-
trischen 8-Knoten Elementen

Die Beziehungen 20 - 24 sind strenggenommen nur
für einen homogenen Spannungszustand im Probekörper, der mit der äußeren Belastung überein-
stimmt, gültig. Der in Fig. 12 und 13 dargestellte Versuchsaufbau zeigt jedoch, daß für den Probekörper die hierfür notwendigen Randbedingungen
nicht genau erfüllt sind, weil er nicht vollständig vom Fels außerhalb des untersuchten Bereiches getrennt ist. Deshalb wurde mit dem in
Fig. 14 dargestellten Berechnungsausschnitt mit
Finiten Elementen ein Triaxialversuch in einem
transversal isotropen Fels mit den in Fig. 13
angegebenen Kennwerten nachgerechnet.

Zunächst wurde mit dieser Rechnung überprüft, ob
sich der Probekörper durch die Herstellung der
Bohrschlitze vollständig entspannt. Dazu wurden
in y-Richtung große horizontale Spannungen σ_y
aufgebracht (Fig. 13). Tatsächlich werden diese
Primärspannungen, wie die Rechnung gezeicht hat,
durch die Herstellung der Schlitze nur in der
oberen Hälfte des Probekörpers vollständig abgebaut.

In zwei weiteren Rechenschritten wurde dann die
Probe horizontal und vertikal belastet. Die
Rückrechnung mit den Formeln 20 - 24 ergab hinsichtlich der Moduln E_1, E_2 und G_2 eine recht
guteÜbereinstimmung mit einem Fehler von weniger als 10 %, wenn man die im oberen Viertel der
Probe berechneten Horizontalverschiebungen zugrundelegt. Das entspricht der in Fig. 13 dargestellten Anordnung der Deflektometer. Für die
Poissonzahlen ergaben sich durch Rückrechnung
allerdings größere Abweichungen gegenüber den
Ausgangswerten (vgl. Fig. 13), die zum einen
durch die geringe absolute Größe von ν_1 und ν_2
und zum anderen durch die Fehlerfortpflanzung
beim Einsetzen der aus (20) und (21) ermittelten
Werte für E_1 und ν_1 in Gleichung 24 bedingt ist.

Dieses Ergebnis zeigt, daß für den betrachteten
Fall bei entsprechender Anordnung der Meßeinrichtungen (vgl. Fig. 13) die Formeln 20, 22 und
23 zur Ermittlung der Kennwerte E_1, E_2 und G_2
durchaus verwendet werden dürfen. Für die Bestimmung der Poissonzahlen ist dagegen eine Mo-
difikation der Formeln 21 und 24 erforderlich.
Entsprechende Korrekturfaktoren können auf der
Grundlage von Parameterstudien festgelegt werden.

6. SCHLUSSFOLGERUNGEN

Für die Interpretation der Ergebnisse von Dila-
tometer- und Druckkissenversuchen, die in anisotropen Gebirgsverhältnissen durchgeführt werden,
ergeben sich besondere Schwierigkeiten. Zum
einen ist eine Nachrechnung dieser Versuche nur
durch aufwendige analytische oder numerische
Rechenmodelle möglich und zum anderen stellt sich
heraus, daß die erhaltenen Versuchsergebnisse in
den meisten Fällen nicht eindeutig im Sinne der
Ermittlung bestimmter Kennwerte aus dem Einzelversuch gedeutet werden können.

Bei transversal isotropen Gebirgsverhältnissen
ist jedoch die Bestimmung der drei Moduln E_1,
E_2 und G_2 durch die Kopplung der Ergebnisse
mehrerer Versuche, die in bestimmter Weise angeordnet werden müssen, grundsätzlich möglich.
Allerdings stellt eine derartige Auswertungsmethode hohe Anforderungen an die Korrelierbar-
keit der Versuchsergebnisse. Diese kann bei Dilatometerversuchen auch bei inhomogenen Gebirgsverhältnissen durch eine entsprechend hohe Versuchsanzahl erreicht werden. Bei Druckkissenversuchen, deren Ergebnisse weniger stark durch
Inhomogenitäten beeinflußt werden, sind dazu
in der Regel weitaus weniger Versuche erforderlich.

Im Unterschied zu Dilatometer- und Druckkissenversuchen ermöglichen in-situ Triaxialversuche,
bei denen Probekörper durch die Herstellung von
Bohrschlitzen aus dem umgebenden Gebirgsverband
weitgehend gelöst werden, die Ermittlung aller
elastischen Kennwerte von transversal isotropem
Fels an einer Probe. Deshalb sollte überdacht
werden, ob derartige Versuche, auch wenn sie
aufwendiger als andere Versuchstechniken sind,
nicht wegen ihrer großen Aussagekraft besonders
bei anisotropen Gebirgsverhältnissen künftig
häufiger als bisher bei Felsbauprojekten durchgeführt werden sollten.

7. LITERATUR

Charrua-Graca, J.G. (1979). Dilatometer tests in
the study of the deformability of rock
masses: Proc. 4th Congr. ISRM, (2), 73 - 76,
Montreux

Kawamoto, T. (1966). On the calculation of the
orthotropic elastic properties from the
states of deformation around a circular
hole subjected to internal pressure in
orthotropic elastic medium: Proc. 1st Congr.
ISRM, (1), 269 - 272, Lisbon

Kiehl, J.R. (1980). Bestimmung elastischer Kennwerte von anisotropem geschiefertem Gebirge
aus den Ergebnissen von Bohrlochaufwei-
tungsversuchen: 4. Nat. Tagung über Felsmechanik, 441 - 473, Aachen

Lekhnitskii, S.G. (1963). Theory of elasticity of an anisotropic elastic body. San Francisco: Holden-Day

Lögters, G., und Voort, H. (1974). In-situ determination of the deformational behaviour of a cubical rock-mass sample under triaxial load: Rock Mechanics (6), 65 - 79

Loureiro-Pinto, J. (1981). Determination of the deformability modulus of weak rock masses by means of large flat jacks (LFJ): Proc. ISRM Symp. on Weak Rock, (2), 423 - 437, Tokyo

Rocha, M., Silveira, A., Grossmann, N., and Oliveira, E. (1966). Determination of the deformability of rock masses along boreholes: Proc. 1st Congr. ISRM, (1), 697 - 704, Lisbon

Rocha, M. and da Silva, J.N. (1970). A new method for the determination of deformability in rock masses: Proc. 2nd Congr. ISRM, (1), 423 - 437, Belgrade

Semprich, S. (1980). Berechnung der Spannungen und Verformungen im Bereich der Ortsbrust von Tunnelbauwerken im Fels: Veröffentlichungen des Institutes für Grundbau, Bodenmechanik, Felsmechanik und Verkehrswasserbau der Technischen Hochschule Aachen, Heft 8

Wittke, W. (1976). A new design concept for underground openings in jointed rock: Publications of the Institute for Foundation Engineering, Soil Mechanics, Rock Mechanics and Water Ways Construction, Technische Hochschule Aachen, Vol. 1, 46 - 117.

Wittke, W. (1977). Interpretation of flat jack tests and field measurements in tunnels by means of Finite Element Analysis: Proc. ISRM Symp. "Field Measurements in Rock Mechanics", (2), 997 - 1018, Zürich

GEOTECHNICAL PROPERTIES OF THE PUERTOLLANO OIL SHALE

Propriétés géotechniques des schistes bitumineux de Puertollano

Geotechnische Eigenschaften der Ölschiefer von Puertollano

Jose L. Berzal
Ing. de Minas, M.Sc. Imp. College, E. N. Adaro, Spain

Carlos S. Oteo
Dr. Ing. de Caminos, Lab. de Carreteras y Geotecnia, Spain

Jose M. Rodriguez-Ortiz
Dr. Ing. de Caminos, E.A.T. Consulting Bureau, Spain

SYNOPSIS

This paper summarizes the geomechanical data — necessarily incomplete — collected from the studies of the oil shale deposit of Puertollano (Spain). The geological features of this deposit and also the basic physico-chemical properties are being described. Special interest has been devoted to the analysis of the shear strength and the deformability of the oil shale samples tested, in order to establish the basic geotechnical data for mining studies.

RESUME

Dans cette communication on décrit l'information géotechnique obtenue pendant les études du bassin aux schistes bitumineux de Puertollano. Les caractéristiques géologiques de ce bassin ont été décrites ainsi que les propriétés physiques et chémiques des schistes. Une attention particulière est prêtée aux propriétés de la resistance à l'effort tangentiel et à la déformabilité des échantillons de roche soumis aux essais, pour établir une information géotechnique fondamentale pour les études d'exploitation.

ZUSAMMENFASSUNG

Die Arbeit beschreibt die Felsmechanik-Forschungsarbeiten der Ölschiefer von Puertollano (Spanien). Die geologischen Kennzeichen und physikalisch-chemischen Gesteinsparameter werden dargelegt. Unter anderem wurden folgende Versuche durchgeführt: Messungen der einaxialen Druckfestigkeit und des Elastizitätsmoduls parallel und normal zur Schichtung. Die fundamentalen geotechnischen Eigenschaften werden dargelegt im Hinblick auf die Abbauverfahren.

1. INTRODUCTION

The increasing cost of oil in the last decade has raised the need for evaluating any potential source of energy. Within this frame, oil shale processing is knowing a rapid expansion all over the world.

In Spain, renewed attention is being paid to the oil shale deposit of Puertollano (in Central Spain), already developed in the 50's. Preliminary research has included a feasibility study comprising technical and economical aspects, as well as the geomechanical problems related to open cast or underground mining.

This paper summarized the geomechanical data, necessarily incomplete, collected in said studies, in the aim to increase the scarce information available on the subject.

2. GEOLOGICAL FEATURES

The Puertollano mining bassin is located 250 Km southwards from Madrid. It is 3,5 Km wide and 12-13 Km long. The ancient coal mine workings have been abandoned and now only open pit coal mining is in operation.

The overall morphology corresponds to a flat valley with geological structure of synclinorium. The Paleozoic materials, folden and diagenized, dip very gentle as a consequence of the Hercinian orogeny.

After the Miocene, the Alpidic orogeny developed in a block tectonics, with vertical faults and the injection of volcanic materials through the longitudinal fractures.

The plan of the deposit appears in fig. 1, along with a typical N-S profil, including the following materials:

a) Quaternary: sub-horizontal deposits of quartzitic and schistose gravels.

b) Miocene: marls and fissured clay marls.

c) Carboniferous, including up to 5 coal seams and the oil shales, as well as layers of schists, sandstones and graywackes.

d) Devorian-Silurian and Cambrian, the sound rock of the Deronian bassin (schists, sandstones and quartrites).

As shown in fig. 1, oil shales appear at a depth between

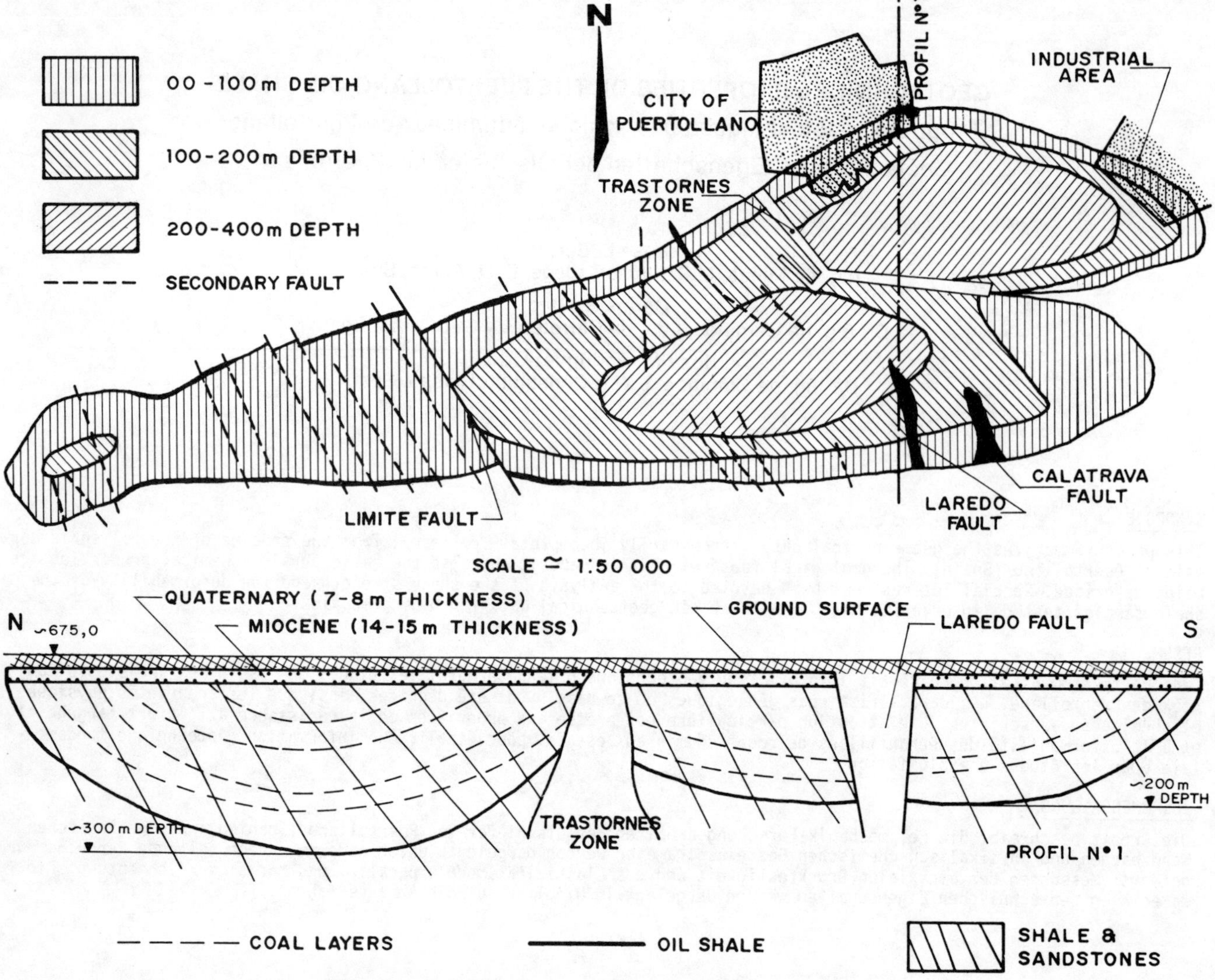

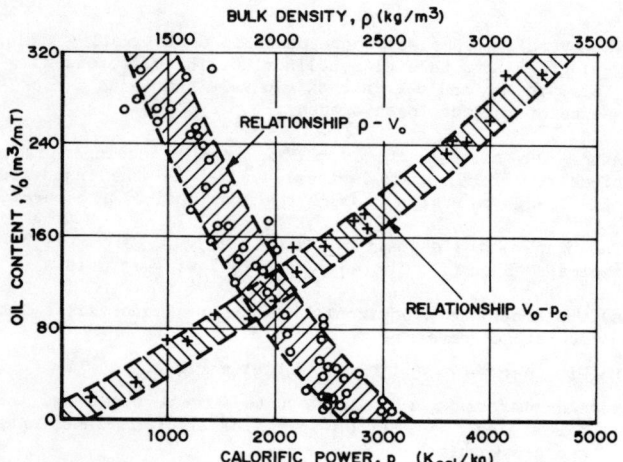

N
00 - 100 m DEPTH
100-200 m DEPTH
200-400 m DEPTH
SECONDARY FAULT
CITY OF PUERTOLLANO
PROFIL N°1
INDUSTRIAL AREA
TRASTORNES ZONE
LIMITE FAULT
LAREDO FAULT
CALATRAVA FAULT
SCALE ≃ 1:50 000
QUATERNARY (7-8 m THICKNESS)
MIOCENE (14-15 m THICKNESS)
GROUND SURFACE
LAREDO FAULT
N
S
~675,0
~300 m DEPTH
~200 m DEPTH
TRASTORNES ZONE
PROFIL N°1
COAL LAYERS
OIL SHALE
SHALE & SANDSTONES

Fig. 1.- Geological scheme of the Puertollano deposit

BULK DENSITY, ρ (kg/m3)
1500
2000
2500
3000
3500
320
240
160
80
0
OIL CONTENT, V0 (m3/mT)
RELATIONSHIP ρ - V0
RELATIONSHIP V0-pc
1000
2000
3000
4000
5000
CALORIFIC POWER, pc (Kcal/kg)

Fig. 2.- Correlationship betwen the oil content, the bulk density and the calorific power

15 and 350 m, in three seams. The upper (A) and lower (C) levels have a mean thickness of 1,5 m, whereas the middle one (B) is 3,5 to 6,0 m thick. The 40% of the deposit dips below 10°, with a gentle increase up to $25\text{-}28^{o}$ at the boundaries of the trough.

Site investigation has comprised a boring campaign, sampling from the a old mine workings and an experimental inclined shaft, 60 m long, for direct appraisal of geomechanical behaviour.

3. PHYSICO-CHEMICAL PROPERTIES

Some 500 samples have been tested in order to assess their mineralogy, bulk density, caloritic power and oil content as established by the Fisher assay.

The main constitutive minerals are mica (40-45%), Kaolinite (20-25) and quartz (15-20%), with minor quantities (0-10%) of calcite, siderite, pyrite, ankerite-dolomite and chlorite.

The Fischer assay has given the following mean values of oil content, V_o , and oil yield, γ_m :

- Seam A: $V_o = 65\ m^3/mT$; $\gamma_m = 8\%$

- Seam B: $V_0 = 145 \ m^3/mT$; $\gamma_m = 13\%$
- Seam C: $V_0 = 125 \ m^3/mT$; $\gamma_m = 14\%$

The oil content V_0, usually taken as a fundamental parameter in oil shale extraction, can be related to the bulk density and calorific power (Fig. 2), without excessive scattering. Oil contents up to 320 m^3/mT have been obtained, corresponding to very low bulk densities, said 1400 Kg/m^3. The mean oil content of the deposit ($\simeq$ 185 m^3/mT) corresponds to a mean bulk density of 1750 Kg/m^3, i.e. a calorific power of 1750 Kcal/Kg. If is believed that said values are sufficiently reliable for estimating the actual oil yields without further testing.

4. GEOMECHANICAL PROPERTIES

4.1. Strength

The following types of tests have been carried out:

- Unconfined compressive strength.

- Point load test

- Indirect tensile strength (Brazilian test)

- Deformability

- Joint shear strength.

Most of the samples have been trimmed out from blocks taken out of the old mine workings or the new inclined shaft. The blocks were mainly plate-shaped cut following the joint system, parallel to the bedding. Many samples disintegrated when trimmed, due to opening of minor cracks.

A preliminary classification of oil shales was attempted following the proposal of Beniawski (1973), and taking into account the oil content. It can be seen in fig. 3 that the tested samples correspond to low or mean strength grades without reflecting the influence of the oil content.

In the fig. 4 the unconfined strength σ_c has been plotted against the oil content V_0. For the most probable values of V_0 (175 to 225 m^3/mT) the unconfined compressive strength varies from 20 to 60 Mn/m^2, without showing a clear variation within a sample being much greater than the range of oil content. The strength is obviously highly influenced by microcracks.

It is interesting to point out that the mean strength of the Puertollano oil shale is clearly below the reported values for other well known shales, as the Colorado ones, as shown in fig. 4 where the results of this research are compared with data of Abel and Hoskins (1978). A lower strength can be justified by a greater oil content.

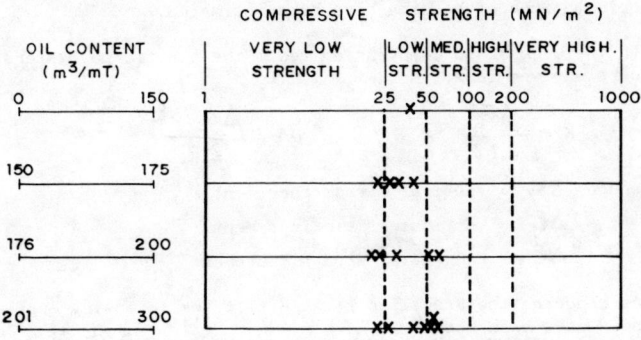

Fig. 3.- Oil shale classification (Bieniawski criteria)

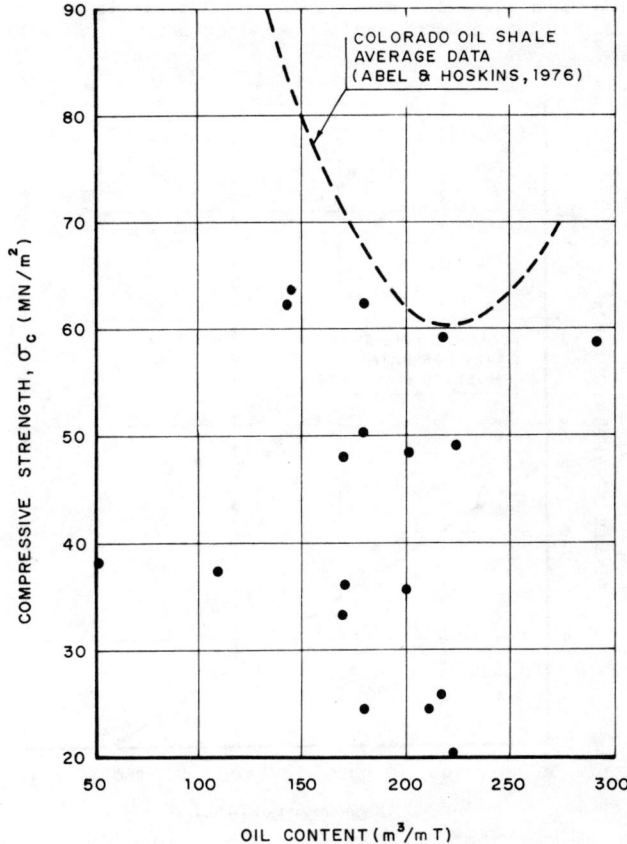

Fig. 4.- Influence of the oil content in the compressive strength

The influence of the orientation is reflected by a mean value of $\sigma_c = 44,5 \ MN/m^2$ (normal to the bedding) against $\sigma_c = 27 \ MN/m^2$ parallel to bedding.

Similar features are observed in the Brazilian tensile tests. The scattering of the values is very high ($\sigma_t = 4$ to 13 MN/m^2) for oil contents between 175 and 240 m^3/mT (fig. 5) the tensile strength has been measured in cylindrical samples trimmed perpendicularly to the bedding, as justified by Chong et al. (1980). The tensile strength may be characterized as mean (Fourmaintraux, 1975) to high (Oteo, 1978).

Again, substantial differences in tensile strength (fig. 5) are encountered when compared with Colorado oil shales (Hoskins et al. 1976), being the Puertollano values considerably higher, although with an important scattering due to the fissures.

The σ_c/σ_t ratio typically ranges from 5 to 10 (fig. 6), well below the usual values. A ratio of 7 may be adopted for preliminary design purposes, but the actual values are highly influenced by fraturing and bedding conditions.

The one-point load test seems very suitable for these materials due to the problems of sample preparation. The relationship between σ_c and the I_s index is shown in fig.7. The mean ratio $\sigma_c/I_s = 25$ is clearly below the value of= 40 is clearly greatest the value of 25.

The influence of the confined stresses in the samples is notable. The fig. 8 the results obtained in a triaxial test, including the average unconfined behaviour, has been plotted. The internal friction angle obtained in this test has 53° and the apparent cohesion is 9 MN/m^2.

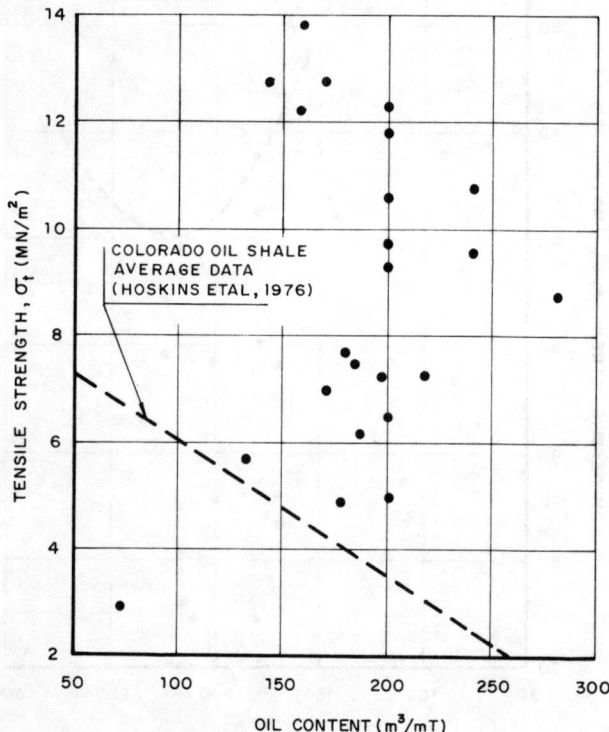

Fig. 5.- Influence of the oil content in the tensile strength

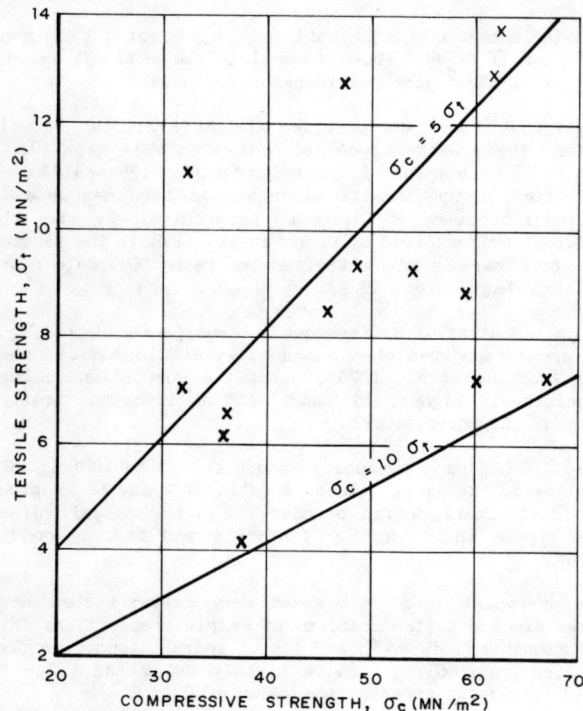

Fig. 6.- Relationship between the compressive and the tensile strength

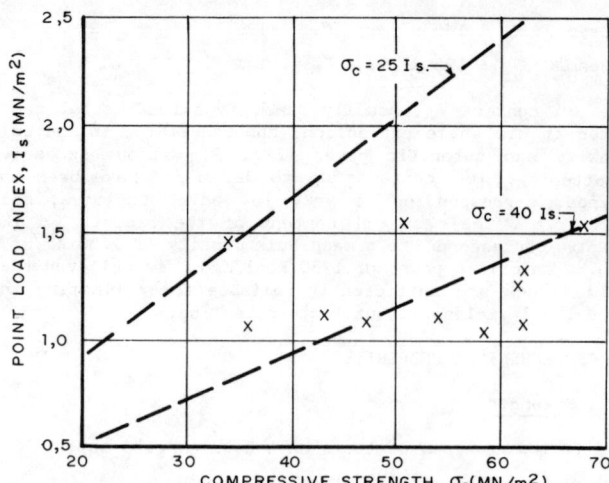

Fig. 7.- Relationship between the compressive strength and the point load test index

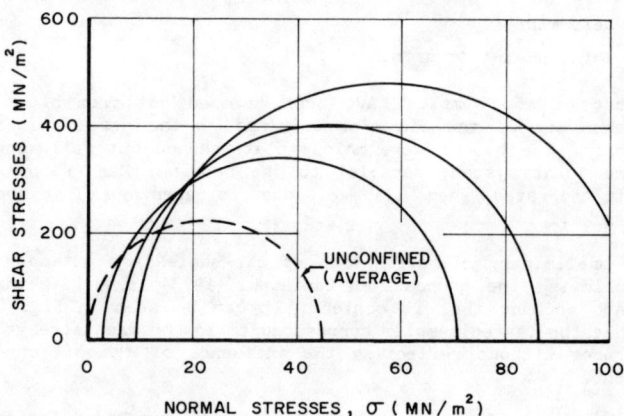

Fig. 8.- Triaxial test

Some test have been also carried out in order to assess the shear strength of joint or bedding planes. The results appear in fig. 9 for tests in dry or wet conditions. Under dry conditions a cohesion intercept of 20 KN/m^2 was found, as well as a friction angle between 20 and 26°. The presence of water causes a decrease of the friction angle to 18-20°, with zero cohesion.

4.2. Deformation characteristics

The testing procedures are schematically shown in fig. 10. Through measurements with strain-gauges and dial gauges the Young moduli and Poisson's ratios for several orientations of bedding have been estimated.

In the unconfined compression tests the elastic parameters were derived from

$$E = \frac{\Delta \sigma_y}{\Delta \varepsilon_y} \qquad , \qquad \nu = \frac{\Delta \varepsilon_c \cdot E}{\Delta \sigma_y}$$

where $\Delta \sigma_y$ = vertical stress increment

$\Delta \varepsilon_y$ = vertical strain increment

$\Delta \varepsilon_c$ = tangential strain increment

As concerns the Brazilian tensile test the following expression have been used

$$E = \frac{16 \, P}{\pi D L (\Delta \varepsilon_1 + 3 \Delta \varepsilon_2)} \qquad , \qquad \nu = \frac{6P + \pi \, DLE \, \Delta \varepsilon_2}{2P}$$

where: P = total load applied to the sample of diameter D
and length L; $\Delta \mathcal{E}_1$ = strain perpendicular to
load; $\Delta \mathcal{E}_2$ = strain parallel to load (fig. 10)

The results of both types of test have been related as
suggested by Deere-Miller (fig. 11). It can be observed
that oil shales fall in the cattegory of low to very low
strength shales, with low relative modulus.

With reference to the elastic moduli the following mean
values have been determined:

$E_1 \simeq 4,6$ GN/m^2 perpendicular to bedding

$E_2 \simeq 5,9$ GN/m^2 parallel to bedding

$E_3 \simeq 10,4$ GN/m^2 sound rock without apparent discontinuities

Poisson's ratio does not appear affected by fissure orientation, with a range of values from 0,20 to 0,45, and
0,30 as mean value.

Creep deformations are not likely under the stresses
expected from underground mining, as checked through
through triaxial testing.

5. CONCLUSIONS

The Puertollano oil shale deposit consists of three seams,
included into an alternance of schists and sandstones,
with an overbunden of Miocene and Plio-Quaternary marly
sediments. Oil shales appear as a bedded material with
spacing of discontinuities from 30 to 50 cm. Micro-cracks
and fracture planes are very close, with spacing of 2
or 3 cm, leading to considerable difficulties in preparing
samples for geomechanical testing.

The compressive strength of this oil shale is low, with
mean values of 44 MN/m when loaded parallel to the bedding
and 27 MN/m^2 normally to the same. The mean tensile strength
is quite high in the order of 9 MN/m^2. This is probably
due to the influence of fissures in the tested samples.

The moduli of deformation are medium to low, with E values
of 6 MN/m^2 parallel to the bedding which reduce to their
80% perpendiculary to the same. Values twice larger were
found for the sound rock without discontinuities as expected, the bedding planes define the orientations of lower
shear strength. On these planes, the dry material snows
a slight cohesion intercept and a friction angle of 24-
-26°, below the strength values of the other schists and
shales of the deposit and due to the presence of oil.

6. ADCKNOWLEDGEMENTS

The authors wish to express their deep gratitude to Empresa Nacional ADARO for your permission to publish this information. Also the colaboration of the Geologist Carlos
Prieto of E.A.T. is reconoced.

7. REFERENCES

Abel, J.F. & Hoskins, W.H. (1976). Confined core pillar
        design for Colorado oil shale. Proc. 9th Oil Shale
        Conf. Vol. 1, 289 p.

Hoskins, W.N., Upadyhyay, R.P., Pills, J. and Sanderg,
        C. (1976). Technical and economic study of candi-
        date underground mining system for deep phick
        oil shale deposit. Proc. 9th Conf. Oil Shale
        Colorado Sch. M. Quat. Vol. 71, nº 4, pp. 199-
        -234.

Oteo, C.S. (1978). Ensayos de laboratorio en la Mecánica
        de Rocas. Boletín de Inf. del Laboratorio del
        Transporte y mecánica del Suelo, Madrid, nº 127,
        pp. 3-33.

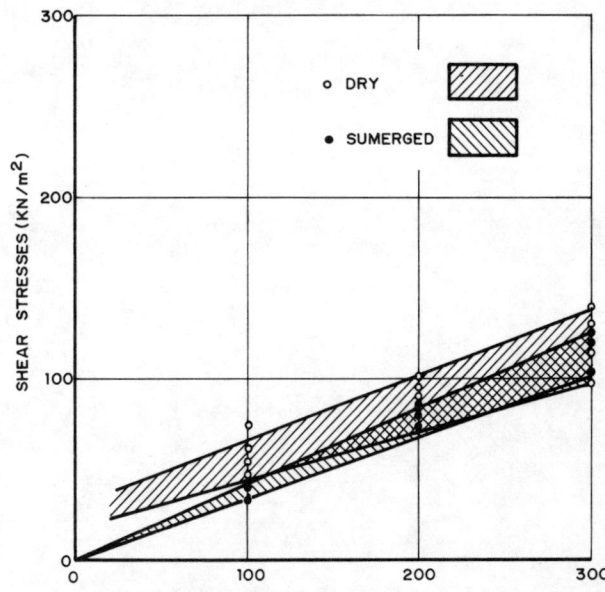

Fig. 9.- Results obtained of the direct shear tests in the
discontinuities

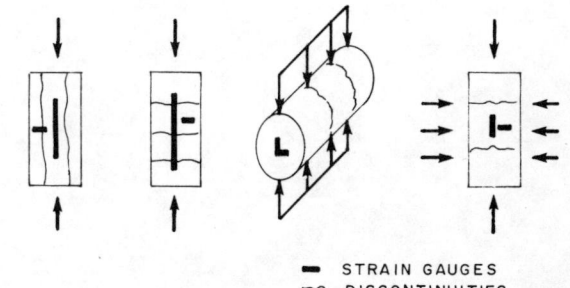

Fig. 10.- Schematic arrangement for measure the deformability

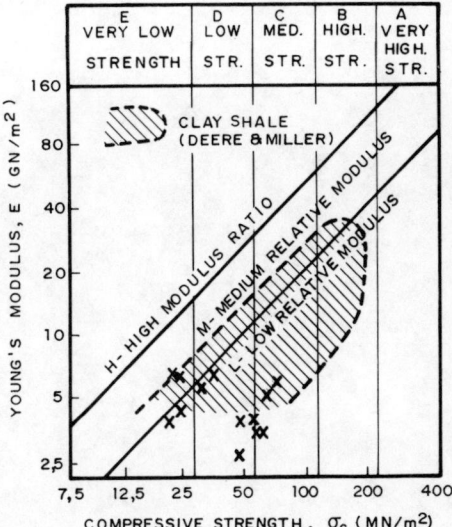

Fig. 11.- Deere-Miller criteria of classifications

IN-SITU DEFORMABILITY vs. WEATHERING OF A GRANITE ROCK IN BARCELONA (SPAIN)

La déformabilité in situ c. l'altération du granite à Barcelone (Espagne)

In-situ Verformungsverhalten und Verwitterungsgrad eines Granits in Barcelona (Spanien)

Manuel Romana
Professor of Geotechnical Engineering, Polytechnical Univ. Valencia, Spain

Davor Simic
Civil Engineer, INTECSA, Spain

SYNOPSIS

Within the work of geotechnical investigation for the design of a new line of the Barcelona Subway, a test shaft 17 m deep was excavated through a weathering profile of the granitic rock which is the substratum of the soil succession. As it was desired to assess the rock deformability as a function of its degree of weathering, several plate bearing tests were carried out at different depths. The total number of tests carried out consisted of 16 load plates in the horizontal and 5 in the vertical direction. This paper describes the testing techniques, displaying the values of the deformation modulus versus the rock weathering and the direction of load application.

RESUME

Pendant les travaux de reconnaissances géotechniques pour le projet d'une nouvelle ligne du Métro de Barcelona on a excavé un puits de 17 m de profondeur à travers une couche de granite météorisé. Le but des travaux étant de lier la déformabilité du rocher avec son degré de météorisation, on a effectué plusieurs essais de plaque de charge à différentes profondeurs. On a fait en tout 16 essais en direction horizontale et 5 dans la verticale. Ce rapport décrit les techniques de réalisation ainsi que les valeurs du module de déformation en fonction de la météorisation du rocher et de la direction d'application de la charge.

ZUSAMMENFASSUNG

Während geotechnischer Bodenuntersuchungen für eine neue U-Bahnlinie in Barcelona wurde ein Testschacht von 17 m Tiefe abgeteuft. Um die Verformungen des Materials (verwitterter Granit) zu bestimmen, wurden verschiedene Plattenbelastungsversuche durchgeführt. Es wurden 16 Versuche in horizontaler und 5 in vertikaler Richtung ausgeführt. Die Versuchstechniken und die Ergebnisse werden dargestellt.

INTRODUCTION

The town of Barcelona in the northeast of Spain has a widespread subway system with a total - length of 49 Km. Additional 23 Km. are currently under construction, so in the short run the complete length under operation of the system will be 72 Km. Middle term plans of extension point towards a subway net 120 Km. long. As the town is seated in a coastal plain, the ground is composed mainly of soils of quaternary age resting directly on shales and granites of the paleozoic age which outcrop in some hills of the town - area. The quaternary soils are series of inter-bedded layers of red gravelly clays, yellow - sandy silts and crusts of limestone which repeat in a somewhat cyclic way.Approaching the mouth - of the two rivers that cross the town - Llobregat and Besós -, deltaic sediments of sands, - silts and clays are found. At its eastern border, the coastal plain is bound by the Catalan coastal batholith of granite, which occupies an intermediate position between the Collcerola - Mts. and the Coastal plain. The granitic formation is composed mainly by granite and granodiorite, with dykes of quartz, aplite and pegmatite.

In such conditions, the majority of tunnels of the Barcelona subway are excavated in the quaternary sediments of the coastal plain. Nevertheless the eastward extension of the line nr. 1 linking the borough of Santa Coloma to Central Barcelona crosses the Besós river entering the geological province of the aforementioned batholith.

In the design of the tunnel, a matter of great - concern was to assess an accurate stratigraphy of the ground and establish the deformability properties of the different strata because the tunnel is to be located near the surface in an urban area where subsidence could be quite damaging. In early stages of the investigation, it was found that from the deformability standpoint there was a variable behaviour, from the granite rock to the grus soil appearing on top. Therefore, a survey to determine the ground deformability was planned.

This paper describes the works of geotechnical investigation that were carried out and discusses the relevant results, aiming at the assessment of the relationship between rock weathe-

EXTENSION OF LINE I OF BARCELONA SUBWAY
SANTA COLOMA TEST SHAFT

FIGURE – 2

HORIZONTAL PLATE LOAD TESTS

FIGURE – I

VERTICAL PLATE LOAD TESTS

ring and its deformability.

Geotecnical surveys for the design of the subway.

The tunnel alignment is wholly included in the granitic formation. Seven boreholes were sunk, down to a depth of 35 m. which allowed enough - penetration within the substratun of fresh granite, and a test shaft 1,20 m wide and 17,30 m deep was excavated up to the layer of weathered granite. About 30 samples were taken at different depths, and 7 borehole permeability tests of the Lefranc and Lugeon type were carried out. Piezometers were located in the boreholes to - find the position of the water table.

Laboratory tests were performed to assess the - geotechnical characteristics of the different - strata.

Plate load tests were carried out at different depths during the shaft sinking. Horizontal and vertical deformability of the strata was found out and related to the degree of weathering of the material. Finally, a survey of the surface vibrations induced by blasting in the bottom of the shaft was performed.

Ground conditions

The soil succession at the site is shown in figure nr. 1 , where the profile of weathering of the granite substratum can be easily identified. However, a neat distinction can be drawn between the materials that have suffered some kind of - transportation and the saprolites that only have suffered local weathering without movement. A - summary of geotechnical characteristics of the - different strata from top to bottom is presented below:

I   - Brown sandy clay and clayey sand, originated from the weathering and transportation of a mixture of the underlying grus and colluvial clays. It belongs to the - CL group of Casagrande classification. The unconfined compression strength ranges - from 0,2 to 0,3 MPa.

II - Reddish clayey sand, originated from the further decomposition and transportation of the grus. It is a soil of the SC group of Casagrande classification, in which - the original rock structure has disappeared. The unconfined compression strength of this material averages 0,2 MPa. The - effective stress parameters obtained from a triaxial c-u test performed on a specimen are c' = 0,07 MPa and ' = 26°

III - Grey grus from the decomposition of the - granite. In this stratum, local weathering of the granite has taken place, without any transportation. This material - belongs to the group SM of Casagrande - classification. However, there is a gradual increase of the proportion of weathered granite fragments with the depth.

IV - Jointed weathered granite, with an average RQD measured in the boreholes of 15%. Its unconfined compression strength is lower than 15 MPa. A slake test performed on a specimen of this material yielded a durability index of 7,4%, which according - with the Gamble criterium corresponds to a very low durability. The Bieniawski - classification of this rock gives a Rock Mass Ratio of 30 indicating a weak formation (class IV).

V   - Jointed fresh granite, with an RQD index between 25 and 50% the unconfined compression strength ranges between 40 and 83 - MPa. The Bieniawski's RMR stands between 50 and 60, indicating a fair rock (class III). This rock constitutes the substratum of the whole formation. Petrological analyses show a medium-grained granite, with 35% of quartz, 25% of K - feldspar, 20% of biotite and 20% of plagioclase. - Measured discontinuities in the test - shaft show three quasi-orthogonal families dipping more than 70° and a near - to - - horizontal set. These joints have soft kaolin filling, which originates from the decomposition of feldspars and plagioclases.

Weathering profile

Although many classifications of rock formations according the degree of weathering have been - proposed, Dearman, Fookes and Franklin (1972); International Society For Rock Mechanics (1978) and so forth, all of them have the common feature of considering a number groups between the - two extreme states: fresh rock and rock disintegrated to soil. Table nr. 1 shows the classification due to the ISRM, which has been employed for this work. It can be seen that this criterium doesn't include the disintegrated rock - that has been transported and therefore has - lost its original structure. Henceforth, the - two uppermost layers of the profile fall outside the range of the ISRM classification.

However, the different strata have been classified from the careful observation of the walls of the test shaft. The grus layer has been subdivided into two groups. The weathering profile is established as follows:

I and II)

Red clay and clayey sand: The material has been transported, and no longer retains the original structure of the rock.

IIIa)

Grey grus from the decomposition of granite: the original structure is still widely   - intact, so it corresponds to group W5.

IIIb)

Grey grus with some blocks of weathered granite: more than a half of the material is - disintegrated to soil, corresponding to group W4.

IV)

Jointed weathered granite: less than a half of the material is disintegrated to soil. - The fresh rock constitutes corestones within the mass. Group W3 is assigned to this stratum.

V)

Although the test shaft has not reached this level, a weathering class of W2 is asigned - to this stratum, from the core recoversed - from the boreholes.
This profile is sketched in figures nr. 2 - and 3.

TABLE NR. 1

WEATHERING CLASSIFICATION

Term	Description	Symbols
Fresh	No visible sign of rock material weathering; perhaps slight discolouration on major discontinuity surfaces.	W1
Slightly weathered	Discolouration indicates weathering of rock material and discontinuity surfaces. All the rock material may be discoloured by weathering and may be somewhat weaker than in its fresh condition.	W2
Moderately weathered	Less than half of the rock material is decomposed and or disintegrated to a soil. Fresh or discoloured rock is present either as a discontinuous framework or as corestones.	W3
Highly weathered	More than half of the rock material is decomposed and/or disintegrated to a soil. Fresh or discoloured rock is present either as a discontinuous framework or as corestones.	W4
Completely weathered	All rock material is decomposed and/or disintegrated to soil. The original mass structure is still largely intact.	W5

Deformability tests

Plate load tests were carried out at different stages during the sinking of the shaft. The tests were performed in the vertical and horizontal direction in order to assess the anisotropy of the soil. In figure nr. 3 it is shown the general arrangement of the equipment for the tests.

Five tests were carried out in the vertical direction, at three different depths, as indicated in figure nr. 1. The general procedure of such tests followed the suggestions of different authors (ISRM, 1978 , Lama et al , 1978). They were performed as follows (see figure 3).

- When the bottom of the shaft reached the selected depth, a horizontal surface was carefully levelled with minimum disturbance of the soil.

- A thin layer of sand/cement was put to ensure proper settling of a square steel plate 30 cm wide and 5 cm thick.

- The reaction support was situated resting on the centre of the plate by means of a spherical hinge. This support consisted of a truss beam with as many spans as necessary screwed to each other up to the top of the shaft.

- On top of the support, a hydraulic jack was located which was to give up to 24 t. for the loading of the plate. The necessary reaction was furnished by means of a loaded dumper that was driven to the top of the shaft.

- The displacements of the plate were measured by means of 4 settlement gauges with a sensitivity of 0.01 mm. Such gauges were fixed on two reference beams which rested on the walls of the shaft.

To measure the deformability in the horizontal direction, 16 plate tests were performed at five different depths, as indicated in figure nr. 2. The general procedure of these tests ran as follows (see figure nr. 3):

- The walls of the shaft were carefully excavated to made two opposite vertical surfaces.

FIGURE 3
EXTENSION OF LINE I OF BARCELONA SUBWAY
SANTA COLOMA TEST SHAFT
PLATE LOAD TESTS

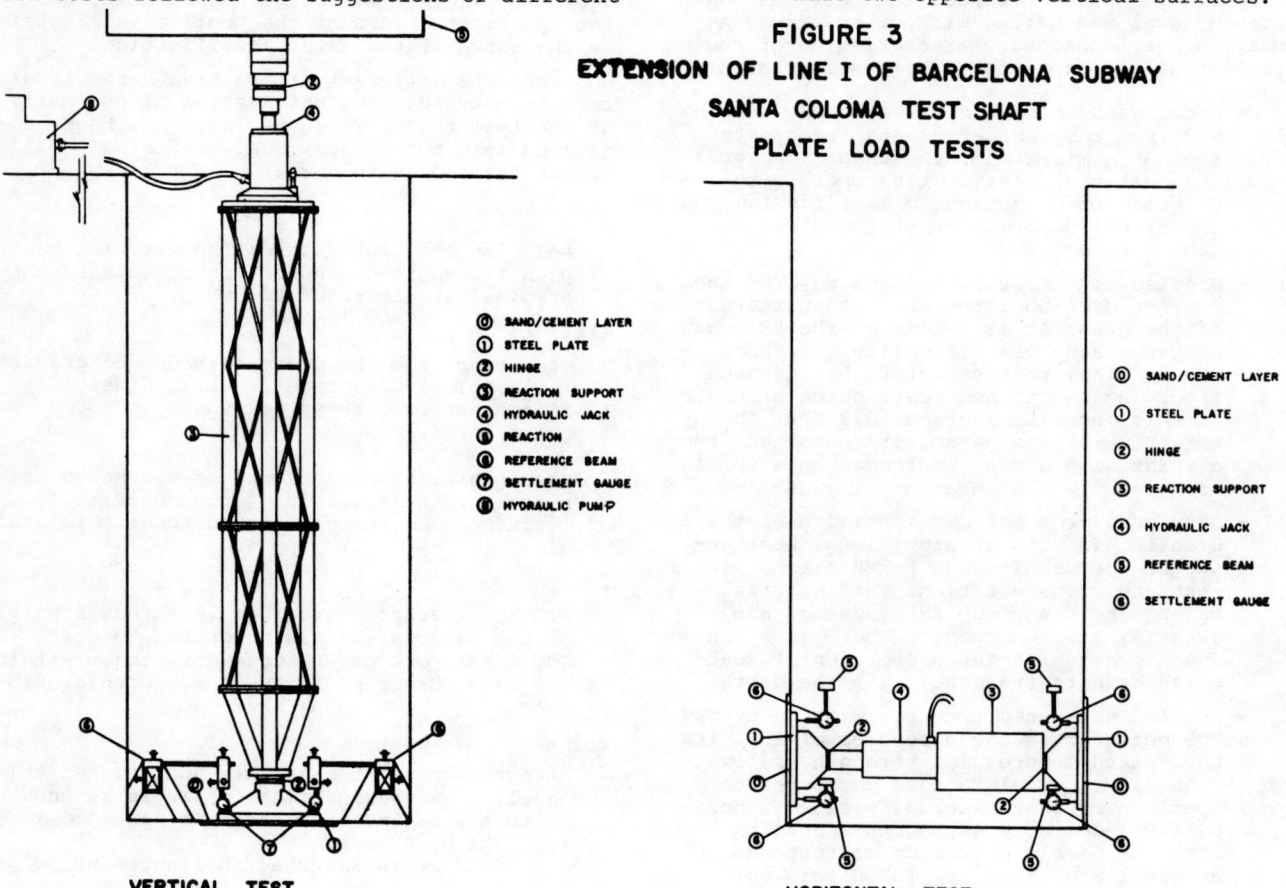

VERTICAL TEST

⓪ SAND/CEMENT LAYER
① STEEL PLATE
② HINGE
③ REACTION SUPPORT
④ HYDRAULIC JACK
⑤ REACTION
⑥ REFERENCE BEAM
⑦ SETTLEMENT GAUGE
⑧ HYDRAULIC PUMP

HORIZONTAL TEST

⓪ SAND/CEMENT LAYER
① STEEL PLATE
② HINGE
③ REACTION SUPPORT
④ HYDRAULIC JACK
⑤ REFERENCE BEAM
⑥ SETTLEMENT GAUGE

TABLE 2.
ELASTICITY MODULI (MPa)

DEPTH (M)	Lithology	HORIZONTAL TESTS									VERTICAL TESTS								
		VIRGIN LOAD			UNLOADING			RELOADING			VIRGIN LOAD			UNLOADING			RELOADING		
		NR. VALUES	AVERAGE	S. DEVIATION	NR. VALUES	AVERAGE	S. DEVIATION	NR. VALUES	AVERAGE	S. DEVIATION	NR. VALUES	AVERAGE	S. DEVIATION	NR. VALUES	AVERAGE	S. DEVIATION	NR. VALUES	AVERAGE	S. DEVIATION
0,00	RED CARBONATE CLAY																		
5,00	BROWN REDDISH CLAYEY SAND	4	26,2	10,7							9	50,4	15,3	8	2.39,4	65,0	6	1.68,2	22,2
	GREY GRUS (W6)	8	55,7	21,5	10	3.09,0	69,4	8	2.66,6	83,8	8	37,8	9,2	9	96,9	11,5	7	86,5	8,6
10,00		4	92,7	33,7	6	3.66,2	85,3	4	3.83,5	99,8	3	46,4	9,3	3	86,5	3,9	3	76,9	3,9
	GRUS WITH SOME BLOCKS OF WEATHERED GRANITE (W4)	6	1.80,8	38,2	17	9.76,5	3.44,2	14	9.52,4	3.13,0									
		6	4.68,4	1.26,2															
15,00	WEATHERED GRANITE (W3)	12	3.91,2	1.52,1	12	14.73,8	4.32,0	8	14.10,8	5.10,9									

TABLE 3   MEASURED MODULI IN GRANITIC ROCKS

REFERENCE	TEST	ELASTICITY MODULUS GPa			
		IN SITU		LABORATORY	
		RANGE	AVERAGE	RANGE	AVERAGE
Tumut(1962)	Plate	1.8 - 52	6.9	41.5-86.1	59.1
	Relaxation	-	11.0		
Gneiss/Granite	Pressure Chambes	13.8-20.6	17.7		
Dworshak(1966)	Plate	3.5-34.5	23.5	-	51.7
Gneiss/massive Granite	Goodman jack	11.6-18.6	14.5		
Turlough Hill (1969) Granite	Flat Jacks	9.6-40.2	29.2	8.0-20.2	15.0
Lake Delio(1970) Gneiss	Plate	7.5-20.4	9.5	15.0-32.4	28.5
	Pressure Chamber	9.7-26.2	18.2		
Churchill Falls (1972) Massive Gneiss	Plate	34.5-48.2	41.5	45.0-75.0	55.0
Project LG-2 (1976) Massive Granite	Plate	38.0-60.9	50.0	-	80.0
Santa Coloma Test shaft(1982)	Vertical Plate	—	0.45	26-52*	39*
	Horizontal Plate	—	1.4		

*Fresh granite

H

- Two square steel plates 30 cm. wide and 5 cm. thick were set against a thin layer of sand/ /cement mortar on the surface.

- The reaction support and the hydraulic jack - were horizontally laid against both plates.

- The displacements of each plate were separately measured by means of settlement gauges - with a sensitivity of 0.01 mm, which were fixed to reference beams set against the walls of the shaft.

In each test, the origin of the displacements - was set at 0,02 MPa to ensure good plate contact. The stress path followed successive steps of - aproximately 0,13 MPa, increasing and decreasing monotonically according to an established pattern of loading up to a maximum load of almost 1,85 MPa. For each loading step, the settlement gauges were read at intervals of 1 minute. The - movements were considered stabilised when the - difference between to succesive readings was - lower than $2,5 \times 10^{-2}$ mm. Once the test was finished, the different elements were taken out and the soil under the plate was excavated to determine its characteristics up to a depth of twice the plate width.

The test paths can be seen in figures 1 and 2. The interpretation of the results involves the determination of the elasticy moduli corresponding to the model of a square rigid loading restings upon an elastic space:

$$E = 0,88 \frac{P}{S} B (1 - \nu^2) \qquad \text{where}$$

E: elasticity modulus
P: stress on the plate
S: settlement of the plate
B: width of the plate, in this case 30 cm
ν: Poisson modulus. In this case the values adopted were 0,3 for the sandy layers and 0,25 - for the weathered granite.

In each test, three different paths have been - distinguished: virgin loading, which includes - the first loading and any successive loading - beyond the maximum stress applied in previous -

steps, unloading and reloading (up to the maximum stress applied previously). Table 2 summarizes the values obtained. It can be seen that:

- Moduli increase with depth

- Horizontal values are, in general, greater - than the vertical ones. In the more representative tests the anisotropy ranges from 3 - to 4.

- For a given test, moduli of the virgin loading are lower than the unloading and reloading - (which are significantly similar). This fact may be due to the closing of fissures, irregular support of the plate and so forth. Also, - greater disparity of values is seen in this - virgin loading values.

- As a consequence, it is suggested that the moduli of the so-called unloading - reloading paths should be taken in consideration to represent the deformability of different strata.

Deformability profile of the ground

The integrated deformability data from in situ and laboratory tests, allow to establish the - following profile:

I) and II). Red clay and sandy clay (weathering class lower than W5). This stratum is the only to show greater vertical than horizontal moduli:

$$E_{vert} = 170 \text{ MPa}$$
$$E_{horiz} = 120 \text{ MPa}$$

IIIa). Grey grus (weathering class W5). Proposed values are:

$$E_{vert} = 90 \text{ MPa}$$
$$E_{horiz} = 300 \text{ MPa}$$

IIIb). Grus with some blocks of weathered granite (weathering class W4). The proposed values - are:

$$E_{vert} = 270 \text{ MPa}$$
$$E_{horiz} = 900 \text{ MPa}$$

IV). Weathered granite (weathering class W3). - The proposed values are:

$$E_{vert} = 450 \text{ MPa}$$
$$E_{horiz} = 1400 \text{ MPa}$$

V). Jointed fresh granite. This stratum has not been reached by the test shaft. Laboratory tests showed moduli values for the rock matrix ranging between 26000 MPa and 52000 MPa which correspond to a high modulus ratio in the chart of - Deere and Miller (1966). To estimate the deformability of the rock mass, the following expression suggested by Bieniawski (1978) has been - employed:

$$E \text{ (GPa)} = 2 \times RMR - 100$$

Entering it with and RMR = 55 atributed to the - jointed fresh granite, a value of E = 10000 MPa is obtained.

Summary and conclusions

Deformability plate loading tests in horizontal and vertical direction have been carried out at differente depths in a soil succession corresponding to the weathering profile of a granite substratum.

Table nr. 2 summarizes the obtained values of Young moduli and shows the corresponding weathering profile, according to ISRM standards. Three parts of the loading curves have been considered virgin load, unloading and reloading. From figure 6 it can be seen that virgin moduli are lower than the unloading an reloading values, which - are significantly similar. Also, horizontal values are greater than vertical ones, with anisotropies ranging from 3 to 4.

The weathering of the rock mass increases significantly its deformability: Estimated Young modulus of 10000 MPa in jointed fresh granite - (W1 class) reduces to 1400 MPa (horizontal direction) in W3 class weathered granite and 900 MPa (horizontal direction) in W4 class grus with - blocks of weathered granite. Finally, Table nr. 3 shows same measured values of elasticity modulus in granitic rocks (Bieniawski, 1978). It can be seen a good fitting of the laboratory and in situ moduli obtained in our investigations.

ACKNOWLEDGEMENTS

The Authors want to thank the following institutions for their help in the investigations that led to the results shown in this paper:

- Servei de Construcció de Transports, Generalitat de Catalunya.

- Intecsa, consulting engineering

BIBLIOGRAPHY

BIENIAWSKI, Z.T. "Determining Rock Mass Deformability: Experience from case Histories". J. Rock Mech. Min. Sci. & Geomech. Abst. Vol 15, pp. 237-247 (1978)

DEARMAN, W.R., FOOKES, P. G and FRANKLIN, J.A., "Some Engineering aspects of weathering with field examples from Dastmoor and elsewere". Quar. Jour, Eng. Geol. 3, 1-24 (1972).

ISRM, "Suggested Methods for the Quantitative - Description of Rock Masses and Discontinuities" Int. J. Rock Mech. Min. Sc. & Geomech. Abstr. 15, 319-368 (1978).

LAMA, R.D., Tutujusi, V.S. "Handbook on Mechanical Properties of Rocks". Volume III. Trans-tech Publications (1978).

THE SHEAR STRENGTH OF CLAY GOUGES IN THE SEDIMENTARY ROCKS OF NATAL

Résistance au cisaillement des failles d'argile dans les roches sédimentaires au Natal

Der Scherwiderstand von tongefüllten Trennflächen in Sedimentgesteinen Natals

P. R. Everitt
Lecturer in Materials, Department of Civil Engineering, University of Natal, Durban, R.S.A.

R. W. S. Goldfinch
Civil Engineer, City Engineer's Department, Durban, R.S.A.

SYNOPSIS

The analysis of the stability of slopes in layered sedimentary rocks requires a knowledge of the shear strength para-meters of the joints and bedding places, including any gouge materials. This paper summarises the results of an investig-ation into the shear strengths of clay gouge in weathered sedimentary rocks found in the Natal Coastal Region. The invest-igation included laboratory testing, back analyses and comparison of conclusions with contemporary references. The res-ults are used in the back analysis and design of stabilization measures for a cutting failure. Further conclusions as to the operating strengths are drawn from this case study.

RESUME

Toute analyse de stabilité des pentes dans une couche de roches sédimentaires exige une connaissance préalable des va-leurs de résistance au cisaillement des diaclases et des plans de stratification y compris tout élément de faille. Nous présentons les résultats d'une étude de la résistance au cisaillement des failles argileuses présentes dans les roches sédimentaires altérées de la côte du Natal. Tests en laboratoire furent inclus, vérifications effectuées et nos con-clusions comparées à celles de la littérature contemporaine. Les résultats obtenus s'appliquent aux vérifications et à la conception des mesures de stabilisation d'une tranchée effondrée. De plus, certaines conclusions relatives aux forces opérantes sont possibles.

ZUSAMMENFASSUNG

Die Stabilitätsanalyse von Böschungen in geschichteten Sedimentgesteinen erfordert die Kenntnis der Scherwiderstandspara-meter von Trennflächen, sowie der darin befindlichen Füllmaterialien. Diese Arbeit umfaßt Ergebnisse einer Untersuchung der Scherwiderstände von tongefüllten Trennflächen in verwitterten, in Natals Küstengebiet befindlichen Sedimentgesteinen. Die Untersuchung umfaßte Labortests, Rückanalysen, und Vergleiche mit anderen, neuzeitlich publizierten Materialien. Die Ergebnisse werden in der Rückanalyse von, und im Entwurf der Stabilisationsmaßnahmen für einen Böschungsbruch angewandt. Weitere Schlußfolgerungen in Bezug auf die Standsicherheit im Betriebszustand werden gezogen.

INTRODUCTION

The Province of Natal of the Republic of South Africa lies on the east coast between the Drakensberg Mountains and the Indian Ocean. The geological structure of the Province comprises a granite basement, overlain by arenaceous and argillaceous sediments which were originally topped by basalt lavas, the remains of which are today found only in the Drakensberg escarpment. Until the late Mesozoic Era, some 120 million years ago the area formed part of a vast inland basin of the Gondwanaland Super-Continent. The break-up of Gondwanaland, which occurred at this time, caused the formation of a hinge structure (de Swart and Bennet (1976)) some distance in from the newly formed coastline, and resulted in a regional eastward dip of the sediments in the coastal area.

During the years 1970 - 1980 a major freeway was constructed parallel to the coastline, which necessitated somewhat deeper cuts than previously excavated, through these dipping sediments. A number of plane failures occurred on clay gouge filled bedding planes, sandwiched between extremely smooth solid rock. In 1975 a research program was begun at the University of Natal with the aim of determining the shear strength of the clay gouges.

SUMMARY OF INVESTIGATION

This investigation included on-site surveys and sampling, laboratory testing to determine peak and residual strengths, back analysis of failures, and the comparison of conclusions with contemporary references. Peak and residual strengths were determined on remoulded drained samples in a 60 mm square Wickham Farrance shear-box modified to incorporate shear reversals. The method followed was based on the recommendations of Cullen et al (1971). The preliminary findings were reported by Knight et al (1977) and these were very briefly:

1. That the residual strengths of the clay gouge were too low to have existed over any large portions of the failure plane before failure.

2. Preliminary back analysis indicated that if the presence of pore water pressures in a tension crack was accepted, failure would be likely to have occurred at a strength close to the remoulded, peak drained strength.

The properties of the clay gouges as determined in the laboratory are given below. (Everitt (1979), Van Wieringen (1977)). Table 1.

Table 1

Slide Location	Ifafa	Ifafa	Mtwalume	Mayat P1	Sea Cow L.
Material	T.M.S.	Dwyka	Dwyka	Ecca	Ecca
Quartz %	15	13	20	36	14
Feldspar %		30			
Kaolin %	16		14		51
Illite %	69	24	27	64	35
Chlorite %		33	39		
Liquid Limit	35	49	49,5	49	55
Plastic Limit	22	27	16,4	26,8	34
Plastic Index	13	22	33	22	21
Sand %	13	33	6	7	8
Silt %	64	45	33	15	34
Clay %	23	22	61	78	58
Activity	0,57	0,99	0,54	0,29	0,26
Peak C'(kPa)	0	0	0	0	0
Peak ϕ'^0	21-25	24-28	23-25	21,3	18,7
Rev. C_r'(kPa)	0	0	0	0	
Rev. $\phi_r'^0$	15	10,5	9,2	9,5-12,5	

Further back analyses using a more refined analysis based on the method of Hoek et al (1974) (Fig. 1) and incorporating sensitivity analyses, and a tension crack, the position of which was varied in relation to the cut face to obtain the critical position, confirmed these findings (Everitt (1979)). The final results may be summarised as follows:

1. In all cases, if initial strengths in the gouge had been at residual values, failure would have been immediate and more widespread. In fact failure was limited to isolated areas and initiated after rainfall.

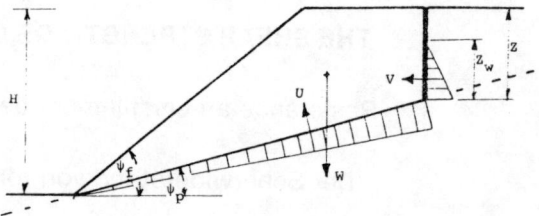

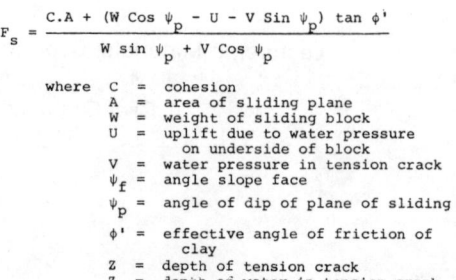

$$F_s = \frac{C.A + (W \cos \psi_p - U - V \sin \psi_p) \tan \phi'}{W \sin \psi_p + V \cos \psi_p}$$

where C = cohesion
A = area of sliding plane
W = weight of sliding block
U = uplift due to water pressure on underside of block
V = water pressure in tension crack
ψ_f = angle slope face
ψ_p = angle of dip of plane of sliding
ϕ' = effective angle of friction of clay
Z = depth of tension crack
Z_w = depth of water in tension crack

FIG. 1 PLANE FAILURE OF A ROCK SLOPE ALONG A LAYER OF CLAY GOUGE

2. In the case of the Table Mountain Sandstone (now known as Natal Sandstone) an initial strength any higher than the remoulded peak strength would have made failure impossible even with a completely filled tension crack.

3. For the remainder; assumption of the remoulded peak strengths gave values of Z_w/Z of between 0,19 and 0,72.

It would thus appear that the remoulded peak drained shear strength may give an upper limit for design shear strengths, whilst the residual strength could be used as a lower limit. These findings coincided with Bishop (1971) who suggested that failure in over-consolidated clays would occur at an interim strength between the (undisturbed) peak and residual strengths. Deen et al (1977) also recommended the use of remoulded peak strengths for temporary slopes or first time failures. In addition Kutter et al (1979) found that, "the residual strength of a filled joint does not fall below the minimum strength of the soil tested alone" which confirms the use of the residual values as a lower limit.

At this stage an effort was also made to correlate laboratory shear strengths with more simply determined parameters. Owing to the mixed mineral nature of the clays the liquid and plastic limits as opposed to mineral composition were used with some apparent success (Fig. 2). As a reference the envelope shown in Mitchell (1976), based on that of Deer, has been shown as well as residual gouge strengths reported by Cording (1977) which appear to show a similar correlation.

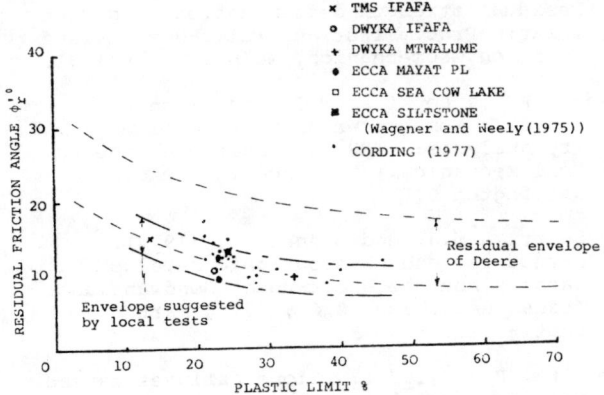

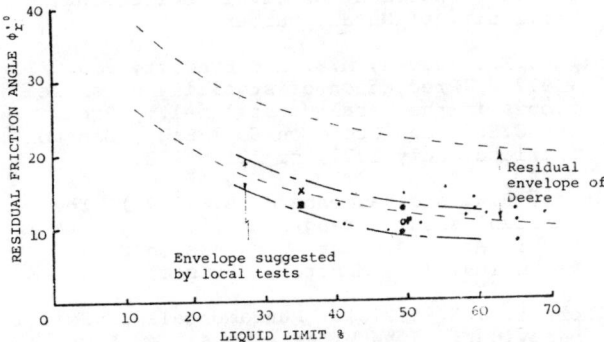

FIG. 2. RELATIONSHIP BETWEEN RESIDUAL FRICTION ϕ_r'
AND ATTERBERG LIMITS (AFTER DEERE (1974))
(REF. MITCHELL (1976))

The Slide at Valley Main Road, Ntuzuma

Subsequent to the investigation, a further
slope cut in extremely weathered yellowish
brown Dwyka Tillite; containing boulders of
hard blue unweathered tillite up to 7 m in
diameter; failed during road construction in
the township of Ntuzuma, north of Durban. This
slope had been cut during October 1976 to a
batter of 1:1 and height 10,0 m, but failed
during autumn rains in 1977. In August 1977
it was trimmed back to 1 vert:2 horiz.(26^0)
but a further failure occurred in November 1977.
At this stage a preliminary geotechnical
investigation was carried out which revealed
that failure was occurring on a layer of pale

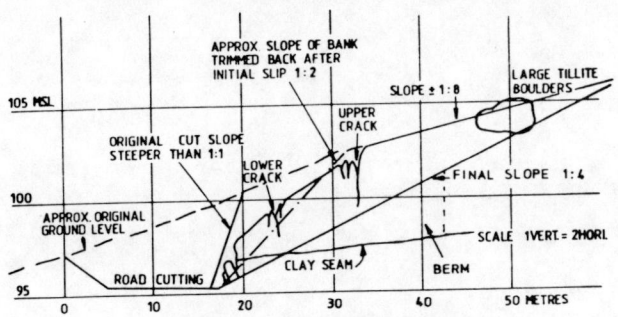

FIG. 3. SECTION THROUGH VALLEY MAIN RD. SLIP
AT THE DEEPEST POINT IN THE CUTTING

grey saturated clay gouge up to 75 mm thick.
This occurred at approximately the level of the
road and dipped in the direction of the road at
only $1\frac{1}{2}^0$. The cut face was then battered back
to 1 vert:4 horiz. (14^0) but failure occurred
again during rain in January 1979. By this
time the head of the failure scarp had retreat-
ed to some 10 metres behind the crest of the
14^0 cut face, onto the natural slope of approximately 5^0 towards the road. The total area
sliding now measured $\pm$ 4 500 m^2.

On 24th January 1979 a back-actor was used to
cut an inspection trench perpendicularly into
the slope in weathered material between boulders. This was taken approximately 18 metres
into the slope. The base of the trench
coincided with the clay gouge failure layer and
it was found to be lying directly onto an
exceptionally smooth and hard layer of blue
tillite. The dip was measured accurately at
$1,35^0$ towards the road. Samples of the clay
which was completely saturated were taken for
laboratory testing and two more similar trenches
at $\pm$ 20 metre spacing were excavated to confirm
the findings. This completed, all trenches were
converted to buttress subsoil drains with a
coarse stone core surrounded by geofabric and
closed by backfilling with weathered tillite.
The drains were connected to a subsoil type
drain along the road shoulder which required
blasting of the base tillite in order to ensure
the fall. The area behind the crest was also
levelled and provided with limited surface
drainage. To date this arrangement has survived
three wet seasons with no further movement.

The results of the laboratory testing were as
follows:

Natural moisture content (in-situ)	34,1%
Liquid Limit	56
Plastic Limit	24
Liquidity Index	0,31

The liquidity index less than 0,36 indicates
overconsolidation (Deen et al (1977)). Remoulded peak drained shear strength was C' = 0,
ϕ' = $19,1^0$. Unfortunately the residual strength
was not ascertained but from Fig. 2 could be
estimated at between 9^0 and 14^0.

A number of back analyses were now performed to
attempt to determine field shear strengths.
Although the material composing the bulk of the
slide was a weathered fissured rock, the same
analysis technique as previously was applied.
This was based on Hoek et al (1974) who state
that the analysis can also be used for "a permeable slope, saturated by heavy rain and
subjected to surface recharge by continued rain
if it is assumed that the tension crack is water
filled".

Analyses of the slope when battered to 14^0 indicated that friction angles ϕ' less than $15,5^0$
would be required for failure. Further analyses
of the slope after drains were installed were
made assuming the drained area to form a gravity
berm. If complete drainage of the 18m berm were
achieved, which is unlikely owing to the drain
spacing, then a value of ϕ' greater than 9^0
would be required for stability even for max
pore pressures in the slope behind. As the
slope was unstable before drainage and appa-

rently stable after drainage these values could be taken as the upper and lower bounds of a residual strength. The fact that these values coincide so closely with the envelope in Fig. 2 appears to be coincidence but nevertheless indicates a good correlation of the most likely values which occur within the boundaries.

Finally analyses of the undrained slope using the remoulded peak value of $\phi' = 19,1^0$ indicated that with maximum pore water pressures failure would occur for slopes cut at angles greater than 17^0. Thus it is feasible that this strength could have been operating at the time of the original 45^0 cut, in which case failure would occur during periods of rain. This is what in fact occurred.

CONCLUSIONS

The work to date then, confirms that of Deen et al (1977), that the remoulded peak drained strength may be used for clay gouges not subjected to previous shear movement.

Secondly the slide at Ntuzuma confirms that with continuing movement residual shear strengths are reached and that these may then be used in analysis.

Finally it would appear that some correlation of residual strength with Atterberg Limits may be possible within fairly close boundaries.

ACKNOWLEDGEMENTS

The authors are indebted to the S.A. Council for Scientific and Industrial Research who sponsored much of the cost of this investigation. The permission of the City Engineer, Durban to publish the latter portion of this paper is also acknowledged.

REFERENCES

Bishop, A.W. (1971). The influence of progressive failure on the choice of the method of stability analysis: Geotechnique (21) 1.

Cording, E.J. (1977). Shear strength on bedding and foliation surfaces: Rock Engineering for Foundations and Slopes, Proc. of Speciality Conference, Univ of Colorado, Aug 15-18 1976. Sponsored by Geotech. Div. ASCE 1977.

Cullen, B.E. and Donald, I.B. (1977). Residual strength determination in direct shear: Proc. First Australia-New Zealand Conf. on Geomechanics, Melbourne. Vol. 1.

Deene, R.C., Hopkins, T.C. and Allen D.L. (1977). Some uncertainties of slope stability analysis: TRR 640, Multiple Aspects of Soil Mechanics. Nat. Acad. Sciences, Washington D.C.

De Swart, A.M.J. and Bennet, G. (1974). Structural and physiographic development of Natal since the break-up of Gondwanaland: Trans. Geol. Soc. S.A., (77), Part 3, Sept - Dec.

Everitt, P.R. (1979). Slope failures caused by clay gouge layers in the sedimentary rocks of the Natal coastal belt: M.Sc. Thesis, Department of Civil Engineering, University of Natal, Durban.

Knight, K., Sugden, M.B. and Everitt, P.R. (1977). Prediction of stability of shale slopes in the Natal coastal belt: Proc. 5th S.E. Asian Conf. on Soil Eng., Bangkok, Thailand, July 1977, pp 201 - 212.

Kutter, H.K. and Rautenberg, A. (1979) The residual shear strength of filled joints in rock: Proc. 4th Int. Congress on Rock Mechanics, (1). Montreux (Suisse).

Mitchell, J.K. (1976). Fundamentals of Soil Behaviour: John Wiley & Sons, New York.

Van Wieringen, M. (1977). Determination of the shear strength, and factors affecting it, of clay layers on the failure planes of slips occurring in the Table Mountain Sandstones, Ecca Shales, and Dwyka Tillite in Natal: M.Sc. Thesis, Department of Civil Engineering, University of Natal, Durban.

Wagener, F. von M., and Neely, W.J. (1975). Stability of a railway cutting in micaceous siltstones: Proc. 6th Regional Conference for Africa, S.M.F.E. (1), Durban.

DEFORMATIONAL AND STRENGTH PROPERTIES OF THE THREE STRUCTURAL VARIETIES OF CARBONIFEROUS SANDSTONES

Caractéristiques de déformation et de rupture des trois variétés structurales de grès carbonifères

Verformungs- und Festigkeitseigenschaften dreier struktureller Varietäten von Karbonsandsteinen

M. Kwasniewski
Senior Research Officer, Silesian Technical University, Faculty of Mining, Gliwice, Poland

SYNOPSIS

The three structural varieties of quartzose carboniferous sandstones were experimentally tested in the conditions of conventional triaxial compression at confining pressures of up to 60 MPa. The effect of the confining pressure on the following deformational and strength parameters of the tested rocks were ascertained: threshold and limit of linearity of longitudinal strain, ultimate strength, Young's modulus, strain at failure, ductility, mode of deformation, type of failure, fracture angle and angle of internal friction. The conditions — pressure — of transition of the tested sandstones from the brittle state via the transitional to the ductile one were likewise determined. Analysing the test results, the granulometric characteristics of rocks were considered as well as their porosity and mineral composition. In theoretical analysis, aiming at determining a suitable strength criterion for the tested rocks, twenty empirical, mathematical and physical criteria, expressed in the convention of principal stresses, were considered.

RESUME

On a soumis trois variétés structurales de grès carbonifères quartzeux aux études expérimentales en compressions triaxiales de révolution sous pressions de confinement jusqu'environ 60 MPa. On a déterminé l'influence de la pression de confinement sur les paramètres mécaniques suivants des roches examinées: le seuil et la limite de la linéarité des déformations longitudinales, la résistance ultime, le module d'Young, la déformation de rupture, la ductilité, le mode de déformation, le type de rupture, l'angle de fracture et l'angle de frottement interne. On a déterminé aussi les conditions — pression — de la transition des grès examinés d'un état fragile à un état transitoire — intermédiaire — vers l'état ductile. En analysant les résultats des recherches on a pris en considération les propriétés granulométriques des roches, leur porosité et leur composition minéralogique. Dans l'analyse théorique tendant à soumettre les grès examinés au critère de rupture convenable, on a considéré vingt critères empiriques, mathématiques et physiques exprimés en convention des contraintes principales.

ZUSAMMENFASSUNG

Drei strukturelle Varietäten von Karbonquarzsandsteinen sind den konventionellen dreiaxialen Druckversuchen bei einem Manteldruck bis zu 60 MPa unterzogen worden. Der Einfluß des Manteldruckes auf die folgenden Verformungs- und Festigkeitsparameter der untersuchten Gesteine wurde ermittelt: Schwelle und Grenze der Linearität der Längsverformungen, Festigkeitsgrenze, Youngscher Modul, Bruchverformung, Dehnbarkeit, Charakter der Deformation, Typ des Bruches, Bruchwinkel und Winkel der inneren Reibung. Die Bedingungen — Druck — des Überganges der untersuchten Sandsteine vom spröden, über den mittleren, bis zum duktilen Zustand, werden ebenfalls beschrieben. Bei der Analyse der Untersuchungsergebnisse wurden die granulometrischen Eigenschaften der Gesteine erwägt, ihre Porosität und ihre Mineralzusammensetzung. Bei der theoretischen Analyse, die hinstrebte, den untersuchten Sandsteinen ein entsprechendes Festigkeitskriterium — Bruchkriterien — unterzuordnen, wurden zwanzig empirische, mathematische und physikalische Kriterien erwägt, die in Konvention der Hauptspannungen bezeichnet werden.

1. INTRODUCTION

Studies on the deformational and strength properties of sedimentary, metamorphic and magmatic rocks in complex states of stress are conducted in the rock mechanics laboratory of the Institute for Design, Construction of Mines and Surface Protection of the Silesian Technical University in order to attend to the needs of mining, geology and geophysics /tectonophysics/. The studies aim at elaborating a general theory of deformation and failure of various kinds of rocks in various conditions of configuration and magnitude of components of triaxial state of stress with special regard to brittle failure, and determination of the conditions of transition of rocks from brittle to ductile state. The purpose is to elaborate a theory of strength and formulate the criteria of failure for various kinds of rocks. In the analysis of test results the aim is to find out the relationships between mechanical properties /characteristics of elasticity, characteristics of plasticity, ductility and the ultimate strength/ and petrographic properties of .

Table 1. Mineralogic and petrographic properties of the three structural varieties of carboniferous sandstones selected for triaxial testing.

Rock type	Mineral composition /vol. %/									Grain-size composition /vol. %/					
						Cement									
	Quartz + fragments of siliceous rocks	Orthoclase	Feldspar	Fragments of effusive rocks	Mica	Argillaceous	Argil-quartzose	Argil-siliceous	Carbonates	up to 0,05 mm	0,05 ÷ 0,1 mm	0,1 ÷ 0,25 mm	0,25 ÷ 0,5 mm	0,5 ÷ 1,0 mm	1,0 ÷ 2,0 mm
1	2	3	4	5	6	7	8	9	10	11	12	13	14	15	16
1. Fine-mediumgrained PNIÓWEK quartzose sandstone	62,1	2,5		4,5	0,7		18,8		11,4			4,9	23,5	71,6	
2. Mediumgrained PNIÓWEK quartzose sandstone	71,1		1,0	4,9	2,3	18,3			2,4	0,6	1,1	8,2	76,7	13,4	
3. Coarse-very coarsegrained JASTRZĘBIE quartzose sandstone	59,9		3,3	8,9	4,8			20,2	2,9		1,5	5,6	11,8	29,6	51,5

Table 2. Basic structural-physical and mechanical properties of the three structural varieties of carboniferous sandstones selected for triaxial testing.

Rock type	Structural-physical parameters				Strength parameters		
	Grain density δ g/cm^3	Bulk density γ g/cm^3	Volume absorption /effective porosity/ n_o %	Weight absorption n_w %	Uniaxial compressive strength σ_C MPa	Uniaxial tensile strength σ_T MPa	Brittleness index $z=\sigma_C/\sigma_T$
1	2	3	4	5	6	7	8
1. Fine-mediumgrained PNIÓWEK quartzose sandstone	2,55	2,35	7,72	3,28	80,8	7,0	11,5
2. Mediumgrained PNIÓWEK quartzose sandstone	2,53	2,36	6,62	2,81	83,9	/7,6/	/11,0/
3. Coarse-very coarsegrained JASTRZĘBIE quartzose sandstone	2,66	2,51	5,41	2,15	98,1	10,5	9,3

rocks, including their textural and structural features /size and shape of grains, bedding, porosity/, and mineral composition.
Below are presented experimental test results on the deformational and strength properties of three structural varieties of carboniferous sandstones from the Rybnik Coal Region /Upper-Silesian Coal Basin/ in the conditions of conventional triaxial compression $/\sigma_1 > \sigma_2 = \sigma_3/$ at confining pressures $/p = \sigma_2 = \sigma_3/$ up to about 60 MPa. Among others was determined the effect of confining pressure on the threshold and limit of linearity of longitudinal strain, ultimate strength, Young's modulus, strain at failure, ductility, mode of deformation, type of failure, fracture angle and angle of internal friction. Also the conditions /pressure/ of transition of the tested

sandstones from brittle into transitional to ductile state were determined. The suitable criteria of failure, expressed in the convention of principal stresses, were found for the tested rocks.

2. EXPERIMENTAL TESTS

2.1 Site and method of taking rock samples

The following three structural varieties of sandstones were tested /cf. Kwaśniewski et al., 1981/:
1. fine - medium-grained /fm/ PNIÓWEK quartzose sandstone form the Orzesze assize /bed 349/3-F - samples 3L/,
2. medium-grained /m/ PNIÓWEK quartzose

sandstone from the Orzesze assize /bed
363-B - samples 3AAA/,
3. coarse - very coarse-grained /cvc/ JASTRZĘBIE
quartzose sandstone from the saddle assize
/bed 508-B - samples 599HH/.

The testing material were 42 mm diameter rock cores from the prospect holes bored by means of a
boring rig TORAM from underground openings of the
XXX-lecia PRL and the Jastrzębie coal mines.
Samples of the PNIÓWEK sandstones /1/ and /2/ come from the Krzyżowice borehole BDK-3/1978,
358,4 m in length, bored /-90°/ from the cross-
heading W-1 on the level 705 of the XXX-lecia PRL
coal mine. The samples of the JASTRZĘBIE sandstone /3/ come from the borehole B 599/1978,
293,2 m in length, bored /-90°/ from the main
drive W-2 in the coal seam 418 at the level -240
of the Jastrzębie coal mine.

2.2 Basic petrographic, structural-physical and mechanical properties of the tested sandstones

The results of mineralogical-petrographic analyses of the tested sandstones are presented in
Table 1. In Table 2 are shown the basic structural-physical properties /grain density, bulk density, porosity, weight absorption/ and mechanical
properties /uniaxial compressive strength, tensile strength /Brazilian method// of the tested
rocks.

2.3 Method of preparing specimens for testing

Specimens were made from cores bored from the
rock mass by means of a double disk /diamond
disks/ rock cutting machine to slenderness ratio
2. Next, the end faces of specimens were ground
by means of a surface grinder with a diamond disk
and, if needed, were hand lapped on glass plate
using fine corundum powder. The accuracy of the
mechanical working and preparation of specimens
for testing met the requirements of the International Bureau of Rock Mechanics /Pforr, 1973/.
Water content in the air-dried specimens prepared for testing amounted to: 0,69 % - /fm/ PNIÓWEK
sandstone, 0,49 % - /m/ PNIÓWEK sandstone, 0,60 %
- /cvc/ JASTRZĘBIE sandstone.

2.4 Description of the triaxial cell KTK-60; testing procedure

Tests on the deformational and strength properties
of the sandstones in the conditions of conventional triaxial compression at confining pressures
up to about 60 MPa were carried out by means of
an apparatus including triaxial cell of von
Kármán's type KTK-60, high-pressure hand pump and
hydraulic testing press EDB-60.
A schematic diagram of the cell KTK-60 is presented in Fig.1. The basic parts of the cell are:
the body /thick-walled cylinder/ 15, cover 2,
piston /plunger/ 1 and clamp nut of the cover 3.
To facilitate measuring of loads acting on the
specimen and of specimen strain by means of electric sensors /e.g., electric resistance strain
gages/ placed inside the chamber, the cover 2
was provided with an electric seal wire 17 /with
a conical element/ with 10 copper leads.
The specimen is loaded in vertical direction by
press EDB-60 /max. load - 600 kN/ through plun-
ger 1.
In order to eliminate high concentration of
stresses occuring usually on the perimeter of the
contact area of the specimen with the elements
transmitting the vertical load /piston, jut on

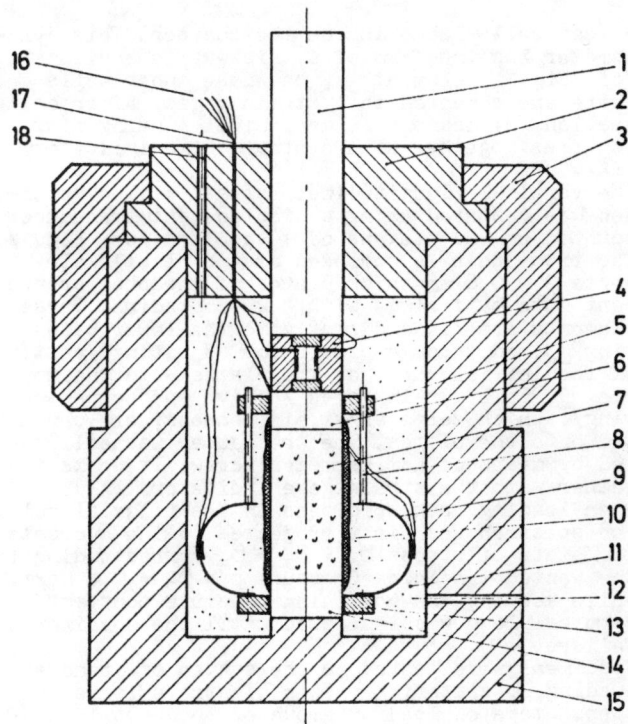

Fig.1 Schematic diagram of triaxial cell KTK-60;
1 - piston /plunger/, 2 - cover, 3 - clamp
nut, 4 - load cell, 5 - upper base ring of
displacement transducer, 6 - upper steel
end platen, 7 - specimen, 8 - pin for adjusting /setting/ initial curve radius of
displacement transducer, 9 - elastic element /plate/ of the transducer with cemented on foil strain gages, 10 - silicone
rubber jacket, 11 - confining medium
/methylsilicone oil/, 12 - to high pressure
pump, 13 - lower base ring of displacement
transducer, 14 - lower steel end platen,
15 - cell cylinder, 16 - to strain indicators, 17 - electric seal wire /10 leads/,
18 - air vent.

the bottom of the chamber/ to the end faces of
specimens disk platens 6 and 14 made of tool steel
heat-treated to hardness of about 55 HRC of the
same diameter as the specimen and thickness equal
to half of the diameter, were cemented on with an
epoxy resin EPIDIAN 5.
The confining pressure is generated by means of
a hand pump. The medium exerting pressure is
methylsilicone oil POLSIL OM-10 - a liquid of
very low viscosity, hydrophobic, with very high
thermal resistance, an excellent dielectric,
physiologically neutral. For measuring the confining pressure the Bourdon's type manometers
of class 1 with a range 250 kp/cm^2 /25 MPa/ -
for pressures lower than 20 MPa, and 1000 kp/cm^2
/100 MPa/ - for higher pressures, are used.
Jackets protecting the specimens against penetration of the oil to the rock, were made of silicone rubber POLASTOSIL M-60 in form of paste
which, shaped at will, was put in a several mil-
limeter layer on the specimen /and the end pla-
tens/. The paste hardened after about one hour
in the process of cold curing, forming an elastic
jacket sufficiently strong and ductile so as not
to be broken /cut/ during brittle fracturing or
ductile barrelling of a specimen.
Measurement of vertical load was made by means of

a load cell placed inside the chamber. This dynamometer has the form of a thick-walled cylinder
/cf. Fig.1 - element 4/, on whose inner walls
there are cemented foil strain gages. After calibration /by means of a mechanical proving ring/
it works together with a strain gage indicator
TSA-4.
The vertical /longitudinal/ strain of the specimen was measured using a displacement transducer
mounted on end platens of a specimen /cf. Fig.1/.
The transducer is composed of two elastic elements 9 - 0,6 mm thick plates of phosphor bronze
bent into half rings with a 25 mm radius. These
elements are fixed directly to the lower base
ring 13 and through pins 8 for adjusting /setting/
of the initial curve of the plates - to the upper
ring 5. Foil strain gages /active and compensa-
ting/, which serve as strain sensors, are cemented onto the plates. The transducer was calibrated by means of a micrometer screw. It works together with the strain gage indicator TSA-4.
When loading rock specimens in the triaxial cell,
the specimen was first subjected to the hydrosta-
tic state of stress $\sigma_1 = \sigma_2 = \sigma_3$ corresponding to
the wanted confining pressure $p = \sigma_2 = \sigma_3$. Next,
while keeping the confining pressure constant,
vertical load was increased until the specimen's
failure.
Specimens were tested in the system of gradated
loading, registering values of strain for each
step. Steps of loading amounted to about 10 kN -
at confining pressures up to about 24 MPa and
about 15 kN - at higher pressures. The rate of
loading /stress rate/ was on the average /cf.
Tables 3 ÷ 5/ equal to about 0,55 MPa/s.
The basis for determining and analyzing the deformational and strength properties of tested
rocks were the plotted on the ground of results
of load and strain measurements, differential
stress - longitudinal /differential/ strain cha-
racteristics $/\sigma_1 - \sigma_3/ = f/\varepsilon_1 - \varepsilon_3/ \overset{def}{=} f/\varepsilon/$.
Individual stages of rock /specimen/ deformation
were separated in those characteristics together
with determination of values of:
- threshold of linearity of longitudinal strain
  $\sigma_{tE} = /\sigma_1 - \sigma_3/_{tE}$,
- limit of linearity of longitudinal strain
  $\sigma_E = /\sigma_1 - \sigma_3/_E$,
- ultimate strength $\sigma_{max} = /\sigma_1 - \sigma_3/_{max}$.
Linear part of the stress - strain characteristic
/corresponding to $\sigma \in [\sigma_{tE}, \sigma_E]$/ was the basis for
determination of the elasticity constant - Young's
modulus:

$$E = \frac{\sum_{i=1}^{n} \Delta\sigma_i / \Delta\varepsilon_i}{n}$$

where n - number of measuring points /observa-
           tions/ between the threshold σ_{tE} and
           limit σ_E of linear strain of rock /spe-
           cimen/.
/NOTICE: It was assumed here that the limit of
linearity is equal to the limit of elasticity of
rock/.
In characteristics $/\sigma_1 - \sigma_3/ = f/\varepsilon/$ the ductility of
the rock /specimen/ ε_η was also determined as a
nonelastic /permanent/ longitudinal strain immediately preceding failure /the difference between
the strain at failure /the total strain/ ε_f and
the strain of compaction ε_c/. As an example,
Fig.2 presents the stress - strain characteristic
of specimen No. 3L-19 of the fine - medium-grained
PNIÓWEK sandstone tested at confining pressure
p = 61,0 MPa.
The value of ductility and the shape of the

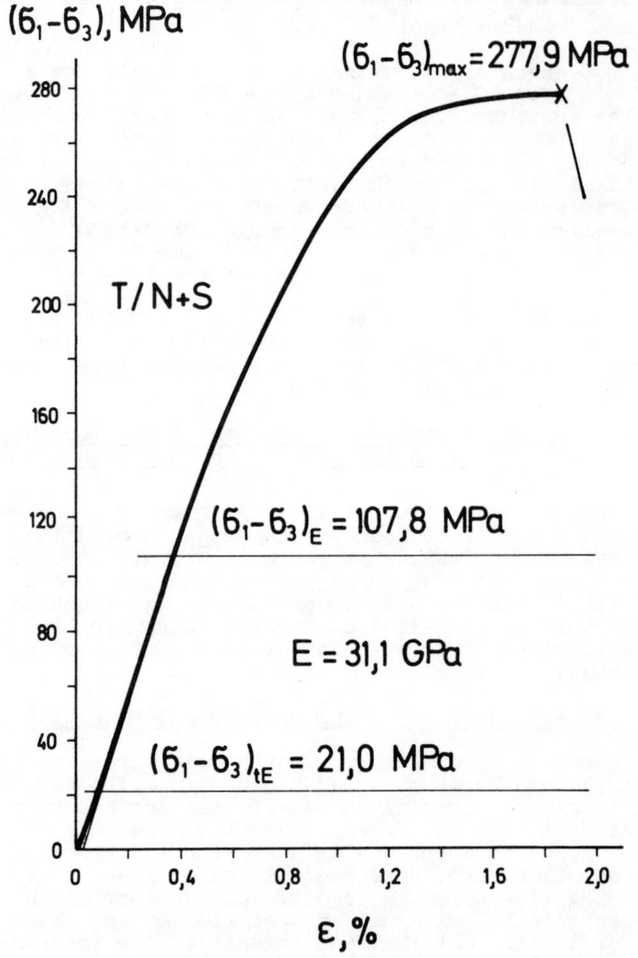

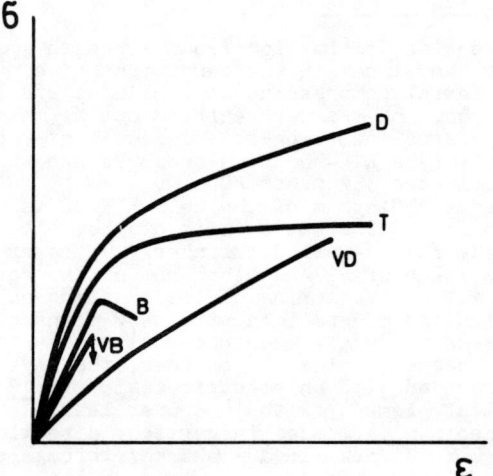

Fig.2   Stress - strain characteristic of specimen
        No. 3L-19 of the fine - medium-grained
        PNIÓWEK sandstone tested at confining
        pressure p = 61,0 MPa.

Fig.3   Model characteristics of rocks deforming
        in a very brittle /VB/, brittle /B/,
        transitional /T/, ductile /D/ and visco-
        ductile /VD/ manner /Hoshino et al., 1972/.

stress - strain characteristic were the basis for determining /cf. Fig.3 and 4/ the mode of rock deformation and failure.

On the basis of appearance of the failed specimen the type of failure was determined /cf. Fig.5/. Also, the fracture angle /the angle between the plane of fracture and the direction of the maximum stress σ_1/ was measured /with an accuracy ± 0,5°/ and, making use of the theory of the envelope of Mohr's circles of limit stresses, the angle of internal friction φ was determined. All test results are collected in Tables 3 ÷ 5 and in Figures 6 ÷ 11. In subsequent chapters these results will be discussed and analyzed in detail in order to define the effect of confining pressure on the deformational and strength properties of the tested sandstones while taking into consideration their granulometric and mineralogical features.

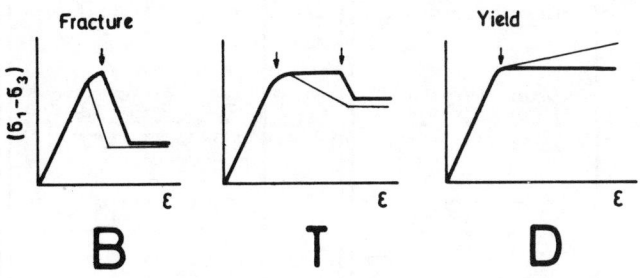

Fig.4 Model characteristics of rocks deforming in a brittle /B/, transitional /T/ and ductile /D/ manner /Mogi, 1972/.

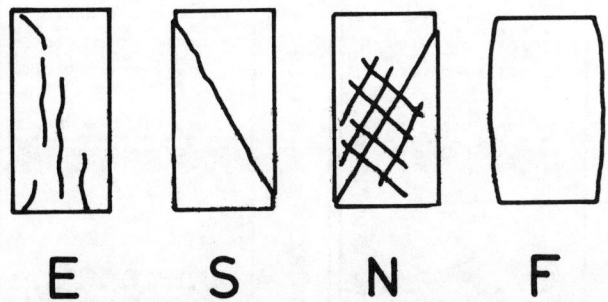

Fig.5 Types of rock /specimen/ failure; E - extension /wedge/ fracture, S - single shear fracture, N - network of shear fractures, F - flow /Hoshino et al., 1972/.

3. DISCUSSION OF TEST RESULTS

3.1 The effect of confining pressure on the deformational properties of the tested sandstones

3.1.1 Characteristics of elasticity

Test results show /cf. column 4 in Tables 3 ÷ 5 and Fig. 6/ that limit of linearity of longitudinal strain of tested fine - medium-grained, medium-grained and coarse - very coarse-grained quartzose sandstones increases with an increase of confining pressure. At the same time, in the case of finer grained /fm/ and /m/ PNIÓWEK

sandstones the rate of increase is lower and the values of the limit of linearity of longitudinal strain settle at a constant level only at pressures higher than about 40 MPa whereas, in the case of coarse - very coarse-grained JASTRZĘBIE sandstone the linearity limit of strain reaches an approximately constant value already at pressures of the order of 20 MPa /cf. Fig.6/. This effect is probably caused by a lower /than that specific to PNIÓWEK sandstones - cf. Table 2 and Fig.6/ porosity of the /cvc/ JASTRZĘBIE sandstone. Since the increase of confining pressure was accompanied at the same time by a drop in linearity threshold of strain /cf. data collected in column 3 of Tables 3 ÷ 5/, the tested sandstones deformed linearly /elastically/ in a respectively widening /increasing/ stress interval $\sigma \in [\sigma_{tE}, \sigma_E]$. As arises from the analysis of data collected in column 5 of Tables 3 ÷ 5 and plots E = f/p/ in Fig.7, to the increase of confining pressure /especially within the range of a dozen or so megapascals/ corresponds an increase of Young's modulus of the tested sandstones. This time though, opposite to the case of linearity limit of strain, the rate of increase of Young's modulus of coarse - very coarse-grained sandstone is lower than that which characterizes finer grained /m/ and /fm/ sandstones and, at the same time, the values of Young's modulus of the latter ones settle at a constant level /equal to about 30,5 GPa for the /fm/ type and about 32,0 GPa for the /m/ type/ already at pressures of the 20 MPa order, while Young's modulus of the /cvc/ sandstone has a constant value E=f/p/=const.=34,5 GPa only at pressures higher than 40 MPa. It should be noted that the increase of Young's modulus of the /fm/ and /m/ sandstones corresponding to the increase in confining pressure from 0 to 20 /40/ MPa, is more significant than that specific to the /cvc/ sandstone; Young's modulus of finer grained sandstones increases on the average by about 45 %, whereas for the coarse - very coarse-grained sandstone by about 35 %. And one more thing - to coarser grained varieties of sandstones correspond higher values of Young's modulus.

3.1.2 Ductility and mode of deformation and failure

Values of ductility of the tested /fm/, /m/ and /cvc/ sandstones are collected in column 9 of Tables 3 ÷ 5. The dependence of ductility on confining pressure is illustrated with plots ε_p=f/p/ in Fig.8. These diagrams presenting a linearly increasing function clearly show the effect of a pronounced increase in ductility, a tendency of the tested sandstones to deform, with an increase of confining pressure. It should be noted that the direction coefficient /gradient/ of the straight line ε_p = a + bp is for the /cvc/ sandstone higher than those of the /m/ and /fm/ sandstones - with an increase of confining pressure the ductility of coarse - very coarse-grained rock increases more strongly than ductility of the finer grained varieties. Within the range of pressures from 0 to about 60 MPa the ductility of the /cvc/ sandstone increased about 5 times on the average; the ductility of the /fm/ sandstone increased only about 3,2 times and in the case of /m/ sandstone the increase of ductility was still lower. High ductility of the coarse - very coarse-grained sandstone is probably also connected /but this hypothesis will have to be verified in further studies/ with a relatively small /equal to about 60 % - cf. Table 1 and Fig.8/ content of poorly deforming quartz; in

RESULTS OF EXPERIMENTAL TESTS ON DEFORMATIONAL AND STRENGTH PROPERTIES OF THE THREE STRUCTURAL VARIETIES OF CARBONIFEROUS SANDSTONES IN THE CONDITIONS OF CONVENTIONAL TRIAXIAL COMPRESSION AT VARIOUS CONFINING PRESSURES p.

Specimen No.	Confining pressure $p=\sigma_3=\sigma_2$ MPa	Threshold of linearity of longitudinal strain $/\sigma_1-\sigma_3/_{tE}$ MPa	Limit of linearity of longitudinal strain $/\sigma_1-\sigma_3/_{E}$ MPa	Young's modulus E GPa	Ultimate strength $/\sigma_1-\sigma_3/_{max}$ MPa	Maximum stress σ_{1max} MPa	Strain at failure ε_f %	Ductility ε_n %	Mode of deformation	Type of failure	Fracture angle Θ	Angle of internal friction φ	Stress rate MPa/s
1	2	3	4	5	6	7	8	9	10	11	12	13	14

Table 3. Fine-mediumgrained PNIÓWEK sandstone

Specimen No.	p	$/\sigma_1-\sigma_3/_{tE}$	$/\sigma_1-\sigma_3/_{E}$	E	$/\sigma_1-\sigma_3/_{max}$	σ_{1max}	ε_f	ε_n	Mode	Type	Θ	φ	Stress rate
3L-4	0,0	17,1	69,0	20,7	80,8	80,8	~0,56	~0,41	B	E	16°	58°	0,22
12	10,5	9,1	65,0	27,6	150,8	161,3	0,93	0,90	B	S	21°	48°	?
13	19,7	9,5	97,2	33,4	191,4	211,3	1,06	1,02	B	S	27°	36°	0,60
18	29,7	0,0	96,3	34,0	228,4	258,1	1,02	1,40	B	S	22,5°	45°	0,51
36	40,5	13,1	78,1	26,1	217,8	258,3	1,41	1,55	B	S	31,5°	27°	0,57
35	44,4	29,2	117,7	27,1	261,4	305,8	1,61	1,46	B	S	30°	30°	0,60
11	49,5	16,4	115,8	30,5	272,8	322,3	1,47	1,83	B	S	33°	24°	0,55
19	61,0	21,0	107,8	31,1	277,9	338,9	1,85		T	N+S	34,5°	21°	0,62

? - time of the experiment was not measured

Table 4. Mediumgrained PNIÓWEK sandstone

Specimen No.	p	$/\sigma_1-\sigma_3/_{tE}$	$/\sigma_1-\sigma_3/_{E}$	E	$/\sigma_1-\sigma_3/_{max}$	σ_{1max}	ε_f	ε_n	Mode	Type	Θ	φ	Stress rate
3AAA-36	0,0	19,1	63,5	23,4	83,9	83,9	0,58	0,50	B	E	15°	60°	0,20
39	10,8	29,8	74,1	26,3	152,0	162,8	0,96	0,95	B	S	28°	34°	0,40
31	20,5	0,0	149,9	35,1	212,3	242,8	0,88	0,88	B	S	28,5°	33°	0,40
28	31,3	x	x	x s	217,3	248,6	x	x	B	N	27,5°	35°	0,61
41	31,6	/0,0/	/68,8/	27,9 s	174,1	205,7	1,33	1,33	D	S	/45°/	/00°/	0,53
34	40,7	0,0	74,8	26,7	201,9	242,6	1,30	1,30	B	S	35°	20°	0,48
40	51,9	0,0	105,7	34,0	256,7	308,6	1,50	1,50	T	S	33°	24°	0,52
33	60,0	0,0	106,3	36,3	267,7	327,7	1,53	1,53	T	S	37°	16°	0,59

x - strain transducer failed during the experiment; s - secant modulus

Table 5. Coarse-very coarsegrained JASTRZĘBIE sandstone

Specimen No.	p	$/\sigma_1-\sigma_3/_{tE}$	$/\sigma_1-\sigma_3/_{E}$	E	$/\sigma_1-\sigma_3/_{max}$	σ_{1max}	ε_f	ε_n	Mode	Type	Θ	φ	Stress rate
599HH-41	0,0	20,8	75,2	30,0	91,8	91,8	0,50	0,42	B	S	17°	56°	0,29
24	11,8	35,5	67,3	27,9	127,7	139,5	0,72	0,69	B	S	29,5°	31°	0,40
39	17,7	0,0	114,5	30,1	165,4	183,1	1,00	1,00	B	S	33°	24°	0,51
26	23,5	24,0	124,1	31,2	194,6	218,1	1,04	0,99	B	S	30°	30°	0,46
38	29,4	0,0	113,6	33,0	193,4	222,8	1,42	1,42	T	N	/45°/	/00°/	0,74
27	35,3	19,9	125,5	29,7	195,0	230,3	1,21	1,19	B	B	39°	12°	0,54
30	41,2	0,0	97,8	35,1	210,8	252,0	1,45	1,45	T	N+S	37°	16°	0,82
33	46,5	0,0	111,2	35,4	222,5	269,0	1,93	1,93	T	N	/45°/	/00°/	0,71
34	52,9	0,0	123,3	32,6	230,2	283,1	1,92	1,92	T	N+S	40°	10°	0,68
35	60,2	0,0	113,7	34,9	229,8	290,0	2,07	2,07	T	N+S	42°	6°	0,70

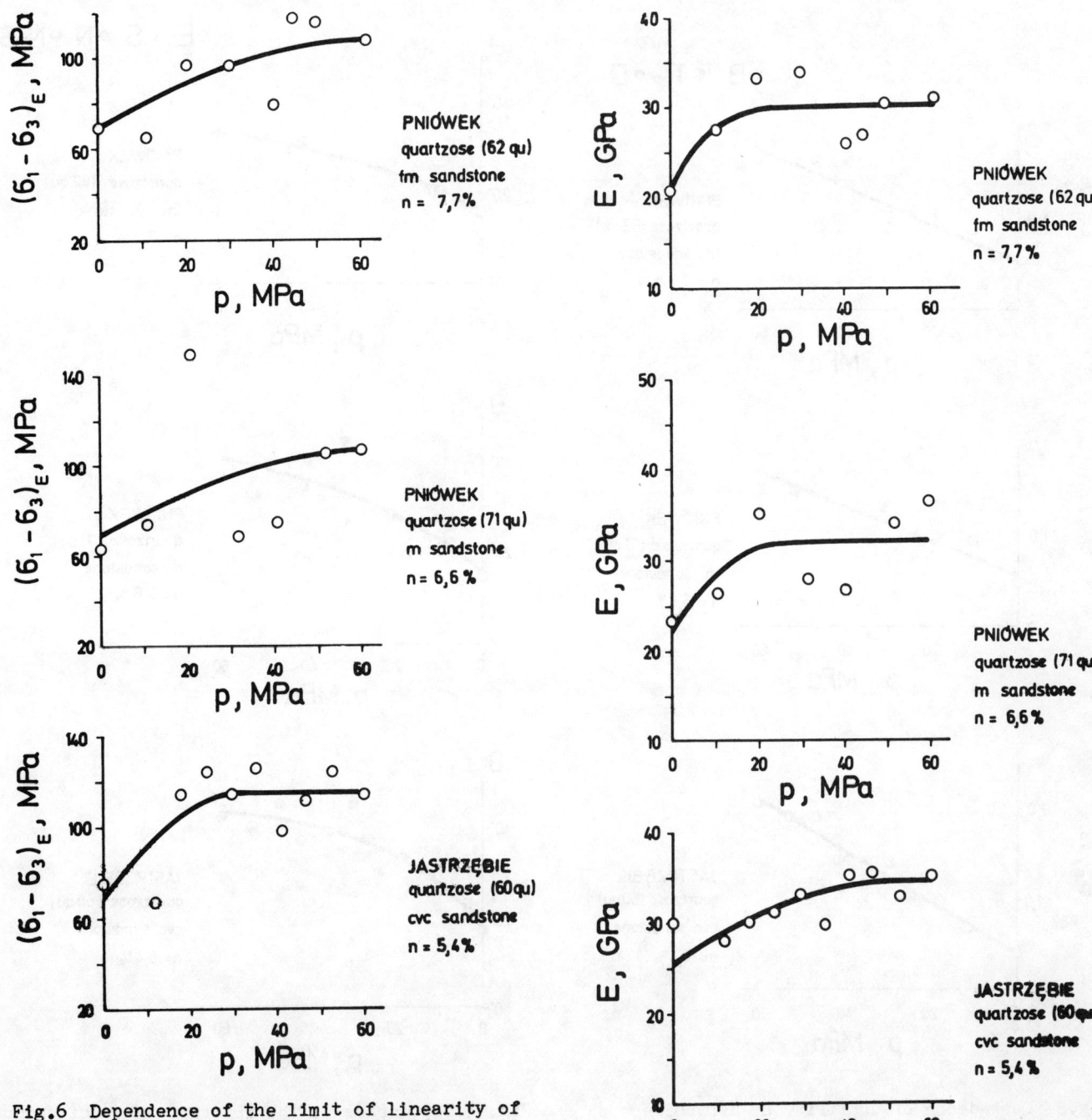

Fig.6 Dependence of the limit of linearity of
longitudinal strain of carboniferous
PNIÓWEK and JASTRZĘBIE sandstones on
confining pressure.

medium-grained sandstone characterized by especially low ductility, quartz occurs in greatest
quantities /about 71 %/.
Interesting information on the deformational properties of the tested sandstones is provided by
an analysis of data collected in column 10 of
Tables 3 ÷ 5. It arises that with an increase of
the grain size of sandstones /passing from the
/fm/ to the /cvc/ type/, the range of pressures
to which corresponds the brittle failure /mode of
deformation and failure B - cf. Fig.4/ becomes
smaller. Specimens of the fine - medium-grained
sandstone failed generally in a brittle manner

Fig.7 Dependence of Young's modulus of carboniferous PNIÓWEK and JASTRZĘBIE sandstones
on confining pressure.

with strains /ε_c/ not greater than 1,61 % /cf.
Table 3/ and their failure at confining pressures
up to 49,5 MPa was accompanied by a loud acoustic
effect.
Medium-grained sandstone failed in a brittle manner /with ε_c ≤ 1,30 %/ at confining pressures now
up to merely 40,7 MPa /an exception here was specimen 3AAA-41 - cf. Table 4 - which at pressure
p= 31,6 MPa deformed in an anomalously ductile

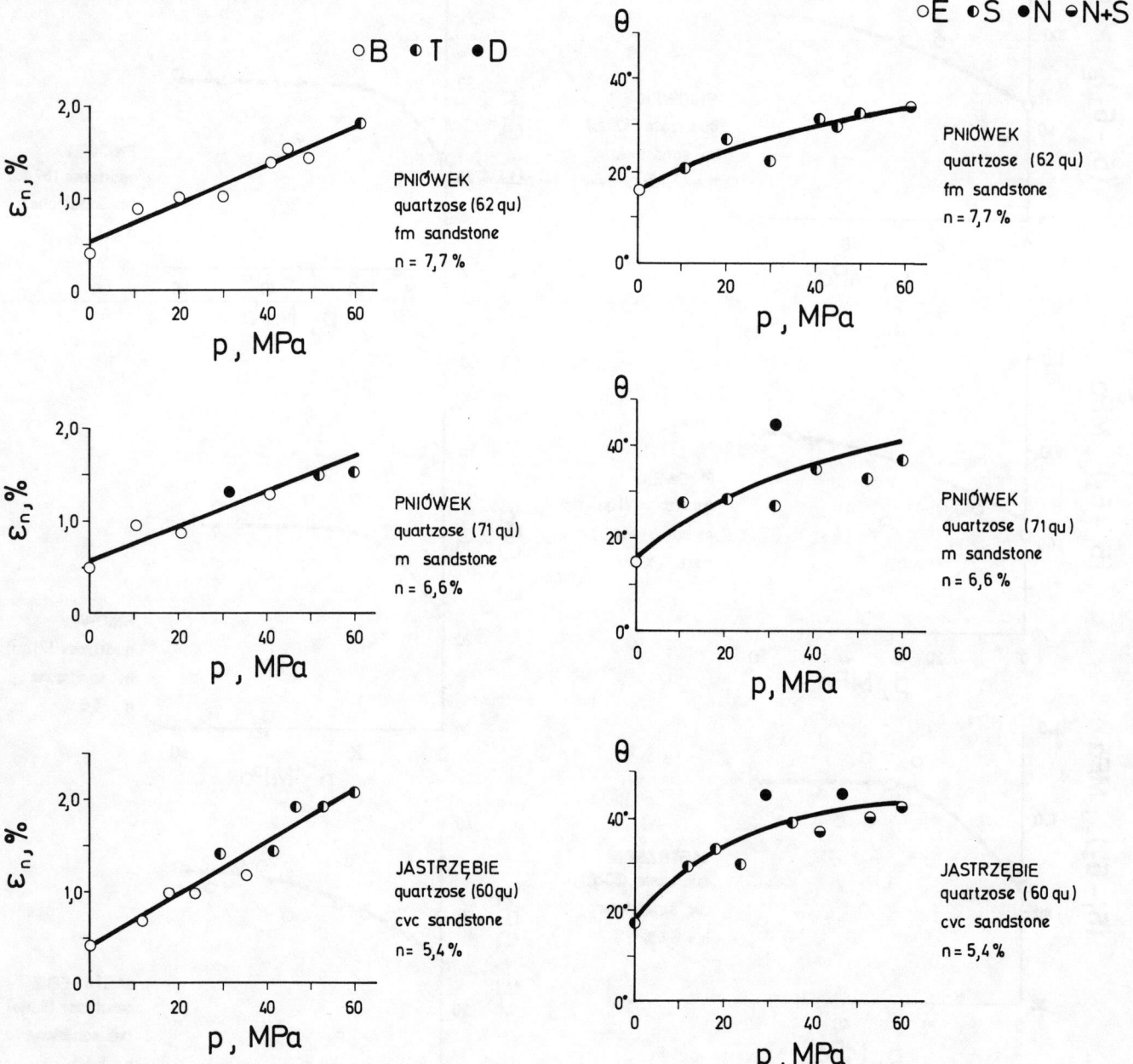

Fig.8 Dependence of ductility of carboniferous PNIÓWEK and JASTRZĘBIE sandstones on confining pressure.

Fig.9 Dependence of the fracture angle of carboniferous PNIÓWEK and JASTRZĘBIE sandstones on confining pressure.

fashion, without the stage of linear strain, showing at the same time low strength. The range of pressures to which corresponds brittle mode of deformation and failure of the coarse - very coarse-grained sandstone is still lower /cf. Table 5/; at pressures equal to 29,4 MPa /specimen No. 599HH-38/ or 41,2 MPa /specimen No. 30/ and higher, the specimens of this sandstone met with certain yielding at stresses close to the limit of strength /characteristics /σ_1-σ_3/=f/ε/ assumed the shape of straight lines almost parallel to the axis ε/, deforming in the stage directly preceding failure in a ductile manner and undergoing failure with a very weak, hollow bang. The mode of deformation

and failure of these specimens was thus intermediate, transitional T between brittle B and ductile D /cf. Fig.4/. Values of ductility were within the range from 1,42 % /specimen No. 38, p=29,4 MPa/ to 2,07 % /specimen No. 35, p=60,2 MPa./.

3.1.3 Fracture angle and angle of internal friction

As it appears /see column 11 in Tables 3 ÷ 5/, dissimilarity of structural features /different grain size/ is also decisive about the fact that types of failure of the three tested sandstones are different. In triaxial tests the specimens

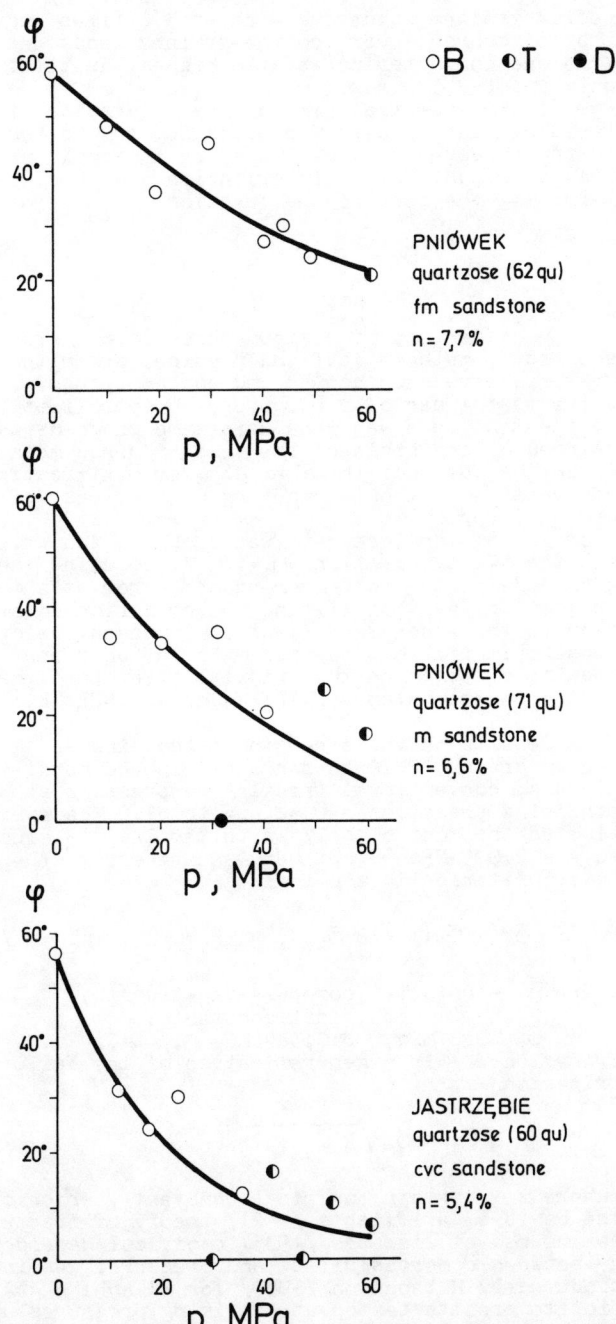

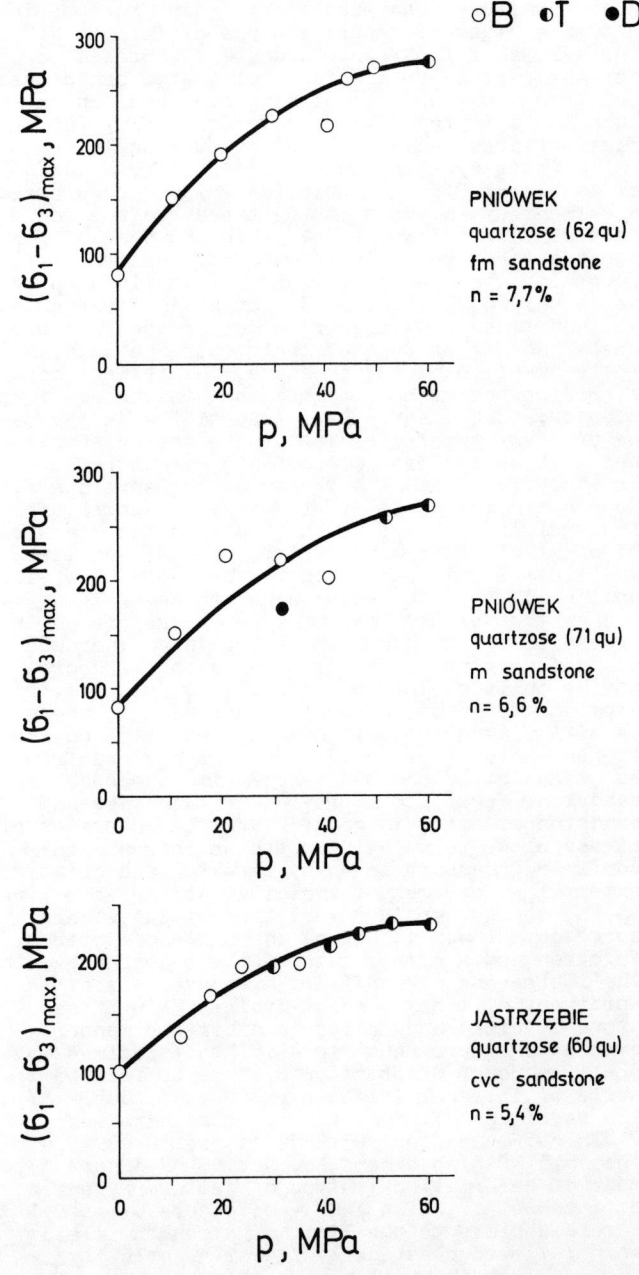

Fig.10 Dependence of the angle of internal friction of carboniferous PNIÓWEK and JASTRZĘBIE sandstones on confining pressure.

Fig.11 Dependence of the ultimate strength of carboniferous PNIÓWEK and JASTRZĘBIE sandstones on confining pressure.

of finer grained varieties /fm/ and /m/ essentially failed /independently of the magnitude of confining pressure/ through shearing in a single plane /again, the only exception here is specimen No. 41 of the /m/ sandstone with anomalous deformational and strength properties - cf. the preceding chapter/. Type of failure S /cf. Fig.5/ was specific to specimens of medium-grained sandstones irrespective of the fact whether they deformed in a brittle B or transitional T manner. However, in the case of coarse - very coarse-

grained sandstone only the specimens deforming in a brittle manner fractured in a single plane of shearing, whereas failure of specimens with intermediate character of deformation occured as a result of displacements, sliding of rock substance along dense, close planes of shearing /inclined to the direction of stress σ_1 at an angle equal or close to 45°/ which did not cause the loss of cohesion of rock material - these specimens /No. 599HH-38, p=29,4 MPa and No. 33, p=46,5 MPa/ assumed the shape of barrels /type of failure N - cf. Fig.5/.
Still more often the specimens of /cvc/ sandstone tested at pressures p ≥ 41,2 MPa assumed the shape of barrel with a clearly marked network of

shear fractures and /secondary/ main fracture in a single plane of shearing /type of failure N+S/. The values of the fracture angle are presented for the particular specimens of tested sandstones /different values of confining pressure/ in column 12 of Tables 3 ÷ 5. The plots $\theta=f/p/$ in Fig.9 clearly show that the fracture angle generally increases from the value equal to a dozen or so degrees for the uniaxially /p=0/ compressed specimens up to about 34° on the average - in the case of /fm/ sandstone, 41° - in the case of /m/ sandstone and 43° - in the case of sandstone /cvc/ for the specimens tested at confining pressures $p \geqslant 60$ MPa. Thus, coarser grained varieties of sandstones show higher values of the fracture angle and higher rate of their increase with an increase of confining pressure. This effect confirms the phenomenon, pointed out in the previous chapter, of a clearly seen /especially in the case of /cvc/ type/ yielding of the tested sandstones with an increase of confining pressure.

In accordance with the theory of the envelope of Mohr's circles of limit stresses, the angle of internal friction φ, which is a specific measure of plasticity /high for brittle rocks, low for ductile, plastic ones - in the boundary case, for an ideally plastic medium equal to zero/ is defined by the formula $\varphi = 90^\circ - 2\theta$. Therefore an increase of the fracture angle θ with an increase of confining pressure corresponds to the decrease of the angle of internal friction φ /cf. Fig.10/ from 58° to about 22° on the average - in the case of /fm/ sandstone, from 60° to about 8° on the average - in the case of /m/ sandstone and from 56° to about 4° on the average - in the case of sandstone /cvc/, thus proving that the tested sandstones, being in general brittle in states of stress close to uniaxial, with an increase in confining pressure show to a greater and greater extent the features of yielding. At the same time to the coarser grained varieties of sandstones correspond lower values of the angle of internal friction and a higher rate of their decrease with the increasing of confining pressure. While the specimens of fine - medium-grained PNIÓWEK sandstone deformed and failed in a brittle manner even at pressure equal to 49,5 MPa /specimen No. 3L-11, mode of deformation B, type of failure S, angle of internal friction $\varphi=24^\circ$ - cf. Table 3/, the specimens of coarse - very coarse-grained JASTRZĘBIE sandstone already at pressures 29,4 and 46,5 MPa /specimens No. No. 599HH-38 and 33, mode of deformation T, type of failure N, angle of internal friction $\varphi=0^\circ$ - cf. Table 5/ at stresses close to the limit of strength /yield point/ flowed plastically /quasi-plastically/.

3.2 Dependence of the ultimate strength on confining pressure; assigning a suitable strength criterion to the tested sandstones

As arises from test results collected in column 6 of Tables 3 ÷ 5 and illustrated with plots $/\sigma_1-\sigma_3/_{max}=f/p/$ in Fig.11, the ultimate strength of the tested sandstones markedly increases with an increase in confining pressure. At the same time it can easily be seen that this effect becomes a little less pronounced with the increase of pressure, i.e. the higher the pressure, the lower is the rate of strength increase corresponding to its further rising. It can also be noted that the larger the grain size of the sandstone, the weaker /lower/ is the increase in its strength. In the tested interval of pressures from 0 to about 60 MPa, the limit of strength of the fine - medium-grained sandstone increased about 3,4 times, of

medium-grained sandstone - about 3,2 times and that of coarse - very coarse-grained sandstone /the one characterized by the highest ductility/ only about 2,3 times.

One of the essential aims of the theoretical analysis of test results was assigning to the tested rocks /assuming that these are isotropic media/ a suitable strength criterion and thus determining the form of the function

$$\sigma_1 = f(\sigma_3) \qquad /1/$$

where $\sigma_3 = p$.

Twenty different, of maximum three parameters, empirical, mathematical and physical strength criteria have been considered and analyzed. While making use of a computer, an approximation of the test data was made using the programs developed on the basis of the least squares method or on the basis of the algorithm for minimization of functions without computing derivatives /cf. Brent, 1973/.

Each set of the test data was approximated with all the twenty equations $\sigma_1=f/\sigma_3/$. Assuming the magnitude of the root-mean-square error as a measure of goodness of fitting the criterion equation to the experimental data, we present below those criteria which approximate the best the results of tests on the triaxial strength of carboniferous sandstones PNIÓWEK and JASTRZĘBIE.

Test results on the strength of the fine - medium-grained PNIÓWEK sandstone in the conditions of conventional triaxial compression at confining pressures $p=\sigma_2=\sigma_3$ up to 61,0 MPa approximated the most closely /with the least - equal to s =12,009 MPa - root-mean-square error/ the power criterion in the form

$$\sigma_1 = \sigma_C(1 + B\,\sigma_3)^C, \quad MPa \qquad /2/$$

where σ_C - uniaxial compressive strength,
    B and C - empirical constants:
    $\sigma_C=80,8$ MPa, $B=0,324$ MPa^{-1}, $C=0,476$.
Criterion /2/ is a generalization of the failure criterion

$$\sigma_1 = \sigma_C\sqrt{k\,\sigma_3 + 1} \qquad /3/$$

/where k - certain empirical constant/, which on the basis of Griffith's /1921/ theory of fracture, making use of Zisman's /1933/ empirical dependence between compressibility and pressure, was introduced by Matsushima /1960/ for KITASHIRAKAWA biotite granite tested at confining pressures up to about 431,5 MPa.

It should be noted here that Matsushima's generalized power criterion /2/ is the same as Balmer's /1952/ type criterion

$$\sigma_1 = a(\sigma_3 + b)^c \qquad /4/$$

for $a = AB^C$, $b = \dfrac{1}{B}$ and $c = C$.
In the considered case of the fine - medium-grained PNIÓWEK sandstone a=47,258, b=3,088 MPa and c=0,476.

Test results on the triaxial strength of medium-grained PNIÓWEK sandstone at confining pressures up to 60,0 MPa were fitted the best /s=23,051 MPa/ by the power criterion of the type

$$\sigma_1 = \sigma_c + B \, \sigma_3^C, \quad \text{MPa} \qquad /5/$$

where σ_c - uniaxial compressive strength,
B and C - empirical constants:
σ_c=83,9 MPa, B=21,178, C=0,580.
The power criterion /5/ corresponds to the one
given by Mogi /1966/:

$$\sigma_1 - \sigma_3 = \sigma_c + B \, \sigma_3^C \qquad /6/$$

and is very frequently used - cf. also Ohnaka
/1973/, Bieniawski /1974/, Borecki et al. /1982/ -
- for the estimation of strength of almost all
kinds of rocks tested in the conditions of conventional triaxial compression at confining pressures as high as 500 MPa.

In the case of coarse - very coarse-grained
JASTRZĘBIE sandstone the relationship between the
maximum σ_1 and minimum σ_3 principal stresses at
failure /for confining pressures, let us remem-
ber, up to 60,2 MPa/ is described the best
/s=8,136 MPa/ by the exponential strength equation in the form

$$\sigma_1 = A \left[1 - e^{-B(\sigma_3 + C)} \right], \quad \text{MPa} \qquad /7/$$

where A, B and C - empirical constants:
A=363,59 MPa, B=0,0222 MPa^{-1}, C=13,105 MPa.
Let us note that the exponential failure criterion
of the type /7/ is identical with the criterion

$$\sigma_1 = a + bc^{\sigma_3} \qquad /8/$$

for a = A, b = -Ae^{-BC} and c = e^{-B}.
Criterion /8/ was mentioned in Franklin's /1971/
paper; however when analyzing studies of various
Authors I haven't noticed even one case where it
would be used for approximating experimental data
$\sigma_1 = f/\sigma_3/$.
For the considered here /cvc/ JASTRZĘBIE sandstone
the empirical constants in criterion /8/ are
equal to: a=363,59 MPa, b=-271,79 MPa and
c=0,978 MPa^{-1}

Among the analyzed twenty different strength criteria expressed in principal stresses, the experimental data were approximated the worst /with
the highest root-mean-square error equal, for the
/fm/, /m/ and /cvc/ sandstones respectively, to
106,7 MPa, 89,3 MPa and 61,8 MPa/ by the original Griffith's /1924/ criterion

$$\sigma_1 = \sigma_3 + 4 \sigma_T + 4\sqrt{\sigma_T^2 + \sigma_T \sigma_3} \qquad /9/$$

where σ_T - tensile strength of material.
This criterion significantly - by 33 to 42 % in
the case of /fm/ sandstone, by 23 to 45 % in the
case of /m/ sandstone and by 12 to 31 % in the
case of /cvc/ sandstone - lowers the values of
maximum stress at failure.

The described above tests on the deformational
and strength properties of the three structural
types of sandstones are at present broadened to
other, in respect to granulometric characteristics and mineral composition, types and varieties of sandstones and mudstones, as well as
silicate magmatic rocks and carbonate sedimentary
and metamorphic rocks.

4. REFERENCES

Balmer, G.G. /1952/. A general analytic solution
for Mohr's envelope. ASTM Proc., Vol.52, 1260.

Bieniawski, Z.T. /1974/. Estimating the strength
of rock materials. The Journal of the South
African Institute of Mining and Metallurgy,
Vol.74, No.8, 312-320.

Borecki, M., Kwaśniewski, M., Oleksy, S., Berszakiewicz, Z., and Pacha, J. /1982/. Odkształceniowe i wytrzymałościowe własności pewnego
piaskowca JASTRZĘBIE w warunkach konwencjonalnego trójosiowego ściskania. In: Metody
i środki eksploatacji na dużych głębokościach
/wybrane zagadnienia/, 55-76, Politechnika
Śląska, Gliwice.

Brent, R. /1973/. Algorithms for Minimization
without Derivatives. 195 pp. Englewood Cliffs,
N.J.: Prentice-Hall Inc.

Franklin, J.A. /1971/. Triaxial strength of rock
materials. Rock Mechanics, Vol.3, 86-98.

Griffith, A.A. /1921/. The phenomena of rupture
and flow in solids. Philosophical Transactions of the Royal Society, London, Series
A, Vol.221, 163-198.

Griffith, A.A. /1924/. Theory of rupture. Proceedings of the First International Congress of
Applied Mechanics, 55-63, Delft.

Hoshino, K., Koide, H., Inami, K., Iwamura, S.,
and Mitsui, S. /1972/. Mechanical properties
of Japanese tertiary sedimentary rocks under
high confining pressures. Geological Survey
of Japan, Report No.244. 200 pp.

Kwaśniewski, M., Pacha, J., Berszakiewicz, Z., and
Oleksy, S. /1981/. Odkształceniowe i wytrzymałościowe własności drobnoziarnistego gra-
nitu STRZELIN i trzech strukturalnych odmian
piaskowców karbońskich PNIÓWEK i JASTRZĘBIE
w warunkach konwencjonalnego trójosiowego
ściskania przy ciśnieniach do 60 MPa. Prace
Instytutu Projektowania, Budowy Kopalń i Och-
rony Powierzchni Politechniki Śląskiej,
MR.I-16/183. 243 pp. Gliwice.

Matsushima, S. /1960/. On the deformation and
fracture of granite under high confining
pressure. Bulletin of the Disaster Prevention
Research Institute, Kyoto University, No.36,
11-20.

Mogi, K. /1966/. Pressure dependence of rock
strength and transition from brittle fracture to ductile flow. Bulletin of the Earthquake Research Institute, University of
Tokyo, Vol.44, 215-232.

Mogi, K. /1972/. Fracture and flow of rocks.
Tectonophysics, Vol.13, 541-568.

Ohnaka, M. /1973/. The quantitative effect of hydrostatic confining pressure on the compressive strength of crystalline rocks. Journal
of Physics of the Earth, Vol.21, 125-140.

Pforr, H. /1973/. IBG-Richtlinien zur Ermittlung
von geomechanischen Kennziffern der Gesteine
und des Gebirgsmassivs. Freiberger Forschungshefte, A 502, Geotechnik und Ingenieurgeologie. 80 pp. Leipzig: VEB Deutscher Verlag
für Grundstoffindustrie.

Zisman, W.A. /1933/. Compressibility and anisotropy of rocks at and near the earth's surface.
Nat. Acad. Sci. Proc., Vol.19, 666-679.

CREEP OF ROCK BASED ON LONG-TERM EXPERIMENTS
Le fluage de roche d'après des expériences à long terme
Kriechen eines Gesteins in einem langfristigen Experiment

H. Ito

Professor in the College of Integrated Arts and Sciences, University of Osaka Prefecture,
Sakai, Osaka, Japan

SYNOPSIS

Nabarro (1948) pointed out theoretically that any polycrystalline solid yields to an applied shear stress to change its shape due to self-diffusion within crystal grains. Since 1957 Kumagai and Itô have carried out the long-term creep experiment on granite beams. Their experimental results over 24 years show that granite flows viscously without a yield stress. It may be essential that a solid has no yield stress. In this chapter the author compiles long-term creep experiments on rock, and discusses the creep, referring to the theory by Nabarro and Herring (1950).

RESUME

Nabarro (1948) a fait remarquer théoriquement que tout solide polycristallin cède à une contrainte de cisaillement appliquée pour se transformer par l'auto-diffusion de grains cristalisés. D'après les expériences à long terme que Kumagai et Itô ont faites sur le poutre en granit pendant 24 ans depuis 1957, le granit se montre fluent et visqueux sans seuil de plasticité. Un solide ne doit, peut-être, pas avoir de seuil de plasticité. Dans ce rapport, les auteurs recueillent les fruits de leurs expériences à long terme sur le fluage de roche et les expliquent en se référant à la théorie établie par Nabarro et Herring (1950).

ZUSAMMENFASSUNG

Nabarro (1948) wies darauf theoretisch hin, daß sich jeder feste polykristallinische Körper infolge seiner Selbstdiffusion innerhalb des Kristallkorns der angelegten Scherspannung ergibt und deformiert. Seit 1957 haben Kumagai und Itô einen langfristigen Kriechenversuch mit Granitbalken gemacht. Ihr über 24 Jahre fortdauernder Versuch zeigt, daß Granit viskos ohne Fließgrenze fließt. Es dürfte sich darum im wesentlichen handeln, daß ein fester Körper keine Fließgrenze hat. In der vorliegenden Arbeit soll das Kriechen, für das Daten aus den langfristigen Versuchen mit Gesteinskriechen gesammelt worden sind, in Verbindung mit der Theorie nach Nabarro und Herring diskutiert werden.

1. YIELD STRESS OF ROCK — EXPERIMENT BY GRIGGS

In a material test, a test-piece is first deformed elastically, then after reaching the yield point, flows plastically. Atoms are fixed at their lattice points in the elastic range, but they are moved in the plastic range. It is well known in a study of metals that the moving mechanism is due to the dislocation within a crystal grain. The yield stress is a strength of material against the plastic flow, while the breaking stress is another strength against rupture.

In a creep test, the lowest stress under which the secondary creep (steady flow) takes place is the yield stress. When only the primary creep occurs but the secondary creep does not, atoms are fixed at lattice points. In this case, when stress is released, the strain is perfectly recovered. Determing the yield stress of material for creep is very laborious, although the value has been known to be smaller than the yield stress obtained by the material test curried out for a short time.

Griggs (1940) tried to determine the yield stress

for creep of rock. The chosen material is alabaster of gypsum. The alabaster was immersed in water chemically saturated by itself. The wet creep tests were carried out at different compressive stresses from 300 to 103 kg/cm^2. The test at 103 kg/cm^2 was continued over 520 days. Strain rates $\dot{\varepsilon}$ of the secondary creep thus obtained are plotted against the compressive stresses σ as shown in Fig.1 (Griggs, 1940, Fig.7 and Table 1), which the author draws using the abscissa ($\dot{\varepsilon}$) scaled ordinarily, though Griggs used a log-scale. The strain rate at 300 kg/cm^2 is too large (23×10^{-9}) to be shown in the figure. The empirical formular got by Griggs is represented by a hyperbolic sine shown in the figure. If the formular were correct, the strain rate would become zero at a stress of 92 kg/cm^2. This is the yield stress of the alabaster under the given condition. However, there is no experimental evidence that the secondary creep does not take place at stresses smaller than 92 kg/cm^2.

2. YIELD STRESS AND SELF-DIFFUSION

Nabarro (1948) pointed out that self-diffusion within grains of a polycrystalline solid can cause the solid to yield to an applied shear

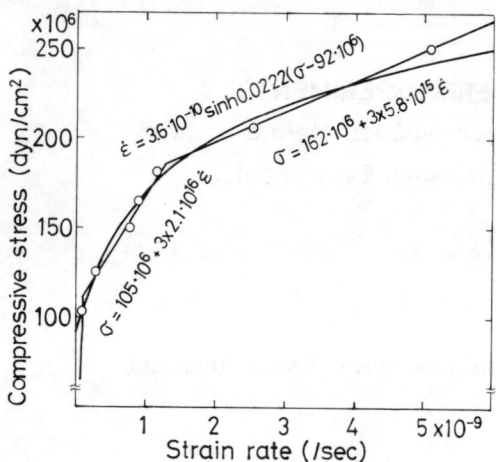

Fig.1 Relation between strain rate and stress on alabaster in wet creep tests by Griggs (1940). The polygonal line shows an interpretation by the author.

stress and the yielding is caused by a diffusional flow of matter within each crystal grain away from boundaries where there is a compression and toward those where there is a tension, as shown in Fig.2. This yielding can cause the solid to behave macroscopically as viscous fluid with an viscosity proportional to the square of the grain size. In actual crystal there are always lattice defects; a lattice vacancy or an interstitial atom is a point defect. The diffusional flow takes place through these point defects. As Herring (1950) perfected the theory, it is called "Nabarro-Herring creep" or the lattice diffusion creep.

Herring (1950) analyzed the atom migration in the crystal made of one type of atom to get the viscosity η for aggregates of quasi-spherical grains with mean diameter d;

$$\eta \propto kTd^2/D\Omega_0 \qquad (1)$$

where k is Boltzman constant, T the absolute temperature, D the self diffusion coefficient and Ω_0 the atomic volume. Here the proportional constant depending on the grain boundary condition is disregarded, because it is not important in this paper.

The self-diffusion coefficient D is given by

$$D \propto N_d \exp(-G_m/kT) \qquad (2)$$

where N_d is the number of point defects and G_m the activation energy necessary that an atom migrates from a lattice point to a point defect. The point defects are created thermally and N_d is given by

$$N_d = N_\ell \exp(-G_d/kT) \qquad (3)$$

where N_ℓ is the number of lattice points and G_d the activation energy necessary to create a point defect. The concentration of point defects N_d/N_ℓ is extremely small in low temperature, accordingly the viscosity η is too large to observe the viscous flow of solid (ref. eqs. 2 and 1).

Rock forming minerals are oxides consisting of an anion (oxigen ion) and cations. In an ionic

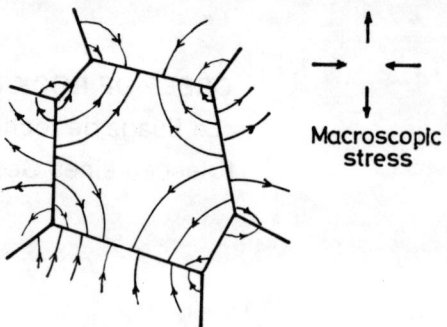

Fig.2 Schematic representation of self-diffusion currents within grains of a polycrystalline solid, when the solid is subjected to a shear stress.

crystal, since the moving of matter must take place keeping electrical neutrality, it is certainly controlled by the oxigen ion with larger volume than the cations. Therefore we might employ the volume and the self-diffusion coefficient of the oxigen ion to Ω_0 and D in eq.(1) in principle. However, in the ionic crystal, the lattice defects can be chemically formed by impurities. Therefore eq.(3) is modified by

$$N_d = N_\ell \exp(-G_d/kT) + N_i \qquad (4)$$

where N_i is the number caused by impurities, being independent of temperature. As the rock forming minerals contain many impurities in general, the lattice defects seem to exist in a large amount. Hence $N_d = N_i$ holds in lower temperature, and the self-diffusion coefficient of eq.(2) is expected to be large. This means that the viscous flow of rock might be observable even in the lower temperature.

On the other hand, creep caused by the dislocation within a crystal grain as mentioned at the top of this paper is called "dislocation creep". The dislocation creep is dominant under higher stress, but the diffusion creep mentioned above is dominant under lower stress.

3. YIELD STRESS OF ROCK — EXPERIMENT BY KUMAGAI AND ITÔ

Since August 7, 1957 Kumagai and Itô have carried out long-term creep tests on granite beams each of 215×12.3×6.8 cm (Kumagai and Itô, 1970; Itô, 1979). One of them, bending under its own weight, is called the unloaded beam, and the other bending under its own weight plus a center-load is called the center-loaded beam. The maximum bending stresses in the unloaded and center-loaded beams are 12.8 and 24.8 kg/cm² respectively. The experimental results are simplified to be shown in Fig.3.

The deflection curve of the beam at the beginning is given by $y = 1/E\ X(x)$, where E (dyn/cm²) is Young's modulus. Since the deflection curve changes with time, it is assumed to be given by $y = T(t)\ X(x)$, where $T(t)$ is a function of time t, having the dimension cm²/dyn, with $T(0)=1/E$. A numerical value of $T(t)$ at the measurement is determined from the most probable deflection curve. $S(t)$ (mm) is the sag of the middle point of beam.

The experiment started in the basement laboratory of Geological and Mineralogical Institute of

Kyoto University. However, because of the reconstruction of the building, the test-pieces
were moved very carefully on October 14, 1967 to
a laboratory of the First Gravity Station of
Kyoto University being close to the university
library. Again, because of the reconstruction
of the library building, the test-pieces were
moved on December 25, 1981 to a basement laboratory of the Fuculty of Science of Kyoto University. The experiment have been similarly carried
out in this third laboratory.

The change of $T(t)$ for the center-loaded beam is
approximated by a straight line except first half
a year as shown in Fig.3. The straight line is
considered to be the secondary creep. The change
of $T(t)$ for the unloaded beam is discontinuous
at the first moving, though the cause of discontinuity is not explained well. However, the
approximate straight lines divided in two parts
show nearly the same inclination. Therefore the
creep rate is considered to be constant over 24
years.

The unloaded beam has the maximum bending stress
of 12.8 kg/cm². Therefore, even if the yield
stress may exist, it must be much smaller than
this value. Thus Kumagai and Itô (1970) concluded, based on the experimental results for 10
ycars, that the granite may flow plastically with
a very small yield stress or viscously without
the yield stress. If the yield stress is zero,
the increasing rate of $T(t)$ is expressed by $1/3\eta$.
The viscosity obtained from Fig.3 is 5.7×10^{20}
poise for the center-loaded beam and 3.2×10^{20}
poise for the unloaded beam. If the beam has
the yield stress, the increasing rate of $T(t)$
cannot be expressed easily with material constants. However, the rate for the unloaded beam
should be considerably smaller than that for the
center-loaded beam. This does not agree with
the experimental results. If the difference of
both viscosities obtained above is meaningful,
the viscosity of granite decreases with a decrease of stress, that is, granite becomes more
fluidal. This is independent of the conclusion
that granite has no yield stress. Thus the experiment by Kumagai and Itô seems to verify the
Nabarro and Herring's prediction.

4. RHEOLOGY MODEL OF ROCK

Materials may be classified rheologically into
elastic (Hooke solid), viscoelastic (Kelvin-
Voigt solid), plastic (Bingham solid), elastoviscous (Maxwell liquid) and viscous (Newtonian
liquid), where the model written in each paren-
thesis is an example. Since a magnitude of the
yield stress is limited, the plastic and elastoviscous (including viscous) materials become an
problem actually; the former has the yiels
stress, but the latter does not.

According to the preceding section, rock is
elastoviscous. Itô (1979) has considered the
rock modeled by the Maxwell liquid for a long
time and drawn the stress-strain diagrams for
different strain rates, where the viscosity $\eta=$
10^{22} poise (ref. next section) and the rigidity
$\mu=10^{11}$ dyn/cm² are given. $\eta/\mu=10^{11}$ sec $=3000$ y
is a relaxation time of the Maxwell liquid. The
stress-strain diagrams suggest that the rock
would behave as the Hooke solid for strain rates
larger than 10^{-13} /sec and would flow as the
Newtonian liquid for times longer than 3000 years
and strain rates smaller than 10^{-14} /sec. It

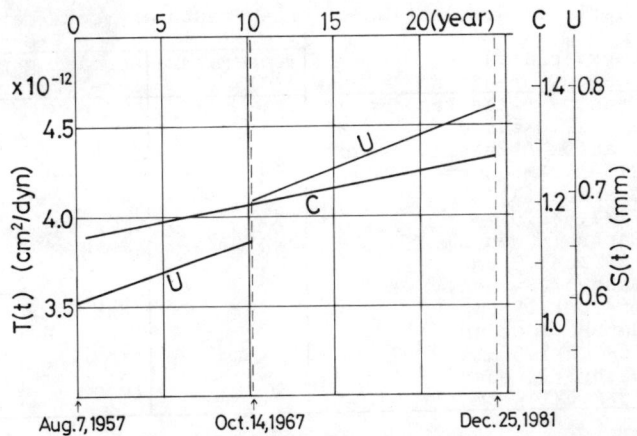

Aug.7,1957   Oct.14,1967   Dec.25,1981

Fig.3 Simplified results of the creep experiment
on granite by Kumagai and Itô, obtained over 24
years. In this figure the data points are omitted. C: center-loaded beam, U: unloaded beam.

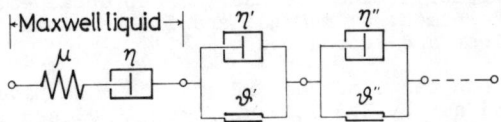

Fig.4 Rheology model of rock for a time as long
as a primary creep ceases, where $\eta>\eta'>\eta''>\cdots$ and
$\theta'<\theta''<\cdots$.

should be noticed that a material test is done
under strain rates larger than 10^{-8} /sec, and
that the geological strain rates are smaller
than 10^{-13} /sec or so.

Behavior of rock under low stresses seems to be
approximated enough by the Maxwell liquid of the
simplest elastoviscous model. To explain the
behavior under somewhat large stresses, the author proposes the model as shown in Fig.4, where
the viscosities are $\eta>\eta'>\eta''>\cdots$ and the yield
stresses $\theta'<\theta''<\cdots$. It may be desirable that
the spring (rigidity μ) is replaced by a visco-
clastic model explaining the primary creep. But
this has no effect on the secondary creep. Employing the model of Fig.4, the experimental results by Griggs (1940) shown in Fig.1 can be represented by the polygonal line as shown in the
figure, though it cannot be drawn for stresses
smaller than 100 kg/cm². If $\theta'=110$ kg/cm² is
assumed, $\theta''=185$ kg/cm² holds, where θ' and θ''
are the yield stress for compression. On the
other hand, the segment of a line for $110<\sigma<185$
kg/cm² is described by the equation shown in the
figure. This is the equation for the Bingham
solid with the yield stress of 105 kg/cm² and
the viscosity of 2.1×10^{16} poise. Also the equation for $\sigma>185$ kg/cm² is for the Bingham solid
with the yield stress of 162 kg/cm² and the viscosity of 5.8×10^{15} poise. These suggest that a
rock would behave as a plastic material for
larger stresses.

The model in Fig.4 may be considered that the
dash-pot η represents the diffusion creep and
the pair of dash-pot η' and slider θ', the pair
of η'' and θ'' etc. do the dislocation creep.

5. VISCOSITY OF ROCK

As mentioned above, an essential property of

Table 1 Compilation of creep tests on rocks having geological strain rates

Rock (place) size of specimen	Experimental condition				Viscosity of secondary creep (poise)
	Temperature (°C)	Confining pressure (kg/cm²)	Maximum bending stress (kg/cm²)	Maximum strain rate (1/sec)	
Granite (Akasaka, Japan) 215×12.3×6.8 cm	24	1	12.8	1.3×10^{-14}	3.2×10^{20}
Granite (Akasaka, Japan) 215×12.3×6.8 cm	24	1	24.8	1.4×10^{-14}	5.7×10^{20}
Granite (Aji, Japan) 21×2.5×2.0 cm	18	1	19.5	6×10^{-14}	1×10^{20}
Gabbro (Sweden) 16×2.0×1.5 cm	18	1	20.4	6.5×10^{-15}	1.0×10^{21}
Gabbro (Sweden) 16×2.0×1.5 cm	25	1000	20	4.2×10^{-14}	1.6×10^{20}
Gabbro (Sweden) 16×2.0×1.5 cm	95	1000	20	3.6×10^{-13}	1.9×10^{19}
Gabbro (Sweden) 16×2.0×1.5 cm	150	1000	20	1.4×10^{-12}	4.8×10^{18}

rock can be observed by an experiment done under the condition that the stress or the strain rate is very small. The creep tests of rock with such condition are compiled in Table 1 to show the viscosities of the secondary creep obtained. The experiment by Kumagai and Itô is shown in the first and second rows.

The creep tests of the 3rd and 4th rows have been carried out by Itô and Sasajima (1980) since August, 1974. In their experiment three granite test-pieces and three gabbro test-pieces are provided, and these deformations are measured by use of interference fringes of light with an accuracy of one-tenth of a wavelength. The viscosities shown here are determined from the experimental results obtained for 7 years.

The tests of the 5th, 6th and 7th rows were performed under 1 kbar confining pressure by Sasajima and Itô (1980). In their experiment a very small deformation of creep was measured also by use of interference fringes of light. For doing the measurements, intermittent breaks of the application of loading, confining pressure and temperature were necessary, and the creep curve was constructed from the intermittent advance of permanent deformation.

As rock undergoes the viscous flow, the earth crust consisting of rock does too. The crust is imagined to float on a more fluidal underlying layer supporting it isostatically. Based on this conception, subsidence of guyots and atolls over geological time has been investigated and the viscosity of the oceanic crust has been found to be 10^{25-26} poise. Analysing the Quaternary crustal movements in the Himalayas and in South-west Japan, the viscosity of the orogenic crust has been estimated to be 10^{22} poise (Itô, 1979).

The viscosities of the crust estimated from geological phenomena are larger than the viscosities of rock obtained by the experiments shown in Table 1 by several orders of magnitude. Rock samples employed in experiments are chosen to be finner in grain size. While most rocks in the crust may be coarser. Therefore, if the Nabarro-Herring creep given in eq.(1) is applicable, the difference in viscosity mentioned above may be explained by the difference in grain size.

From the data of the 5th, 6th and 7th rows in Table 1, the temperature dependence of viscosity is given by

$$\eta = 1.4 \cdot 10^{12} T \, \exp(7600/kT) \qquad (6)$$

where R is the gas constant per mol (Sasajima and Itô, 1980). This equation is based on eqs. (1), (2) and (4) and 7600 cal/mol may correspond to the activation energy G_m.

Comparing the 4th and 5th rows, the gabbro becomes fluidal with increase of confining pressure, as well as ice does. While it is known that metals become hard to flow with confining pressure. It may be first experimental data showing that the creep of rock depends on the confining pressure. The author will discuss the relation between viscosity and confining pressure in other papers.

6. CONCLUSIONS

It is concluded that rock can flow as viscous liquid without any yield stress, based on the experiment by Kumagai and Itô and referring to the theory by Nabarro and Herring. The viscosities obtained experimentally on granite and gabbro are 10^{20-21} poise in room temperature and atmospheric pressure. But the viscosity of natural rock may be greater than those by a few orders of magnitude.

REFERENCES

Griggs, D. (1940). Experimental flow of rocks under conditions favoring recrystallisation. Bull. Geol. Soc. Am., 51, 1001-1022.
Herring, C. (1950). Diffusional viscosity of a polycrystalline solid. J. Appl. Phys., 21, 437-445.
Itô, H. (1979). Rheology of the crust based on long-term creep tests of rocks. Tectonophysics, 52, 629-641.
Itô, H. and Sasajima, S. (1980). Long-term creep experiment of some rocks observed in three years. Tectonophysics, 62, 219-232.
Kumagai, N. and Itô, H. (1970). Creep of granite observed in a laboratory for 10 years. In: S. Onogi (Editor), Proc. 5th Int. Congr. Rheol., 2, 579-590.
Nabarro, F.R.N. (1948). Deformation of crystals by the motion of single ions. Rep. Conf. on the Strength of Solid, Phys. Soc., London, p.75-90.
Sasajima, S. and Itô, H. (1980). Long-term creep experiment of rock with small deviator of stress under high confining pressure and temperature. Tectonophysics, 68, 183-198.

STRENGTH AND DEFORMATION CHARACTERISTICS OF SOFT SEDIMENTARY ROCK UNDER REPEATED AND CREEP LOADING

Caractéristiques de résistance et de déformation de roche sédimentaire poreuse molle sous chargement répété et de fluage

Festigkeits- und Deformationskennzeichen von weichem Sedimentgestein bei wiederholter und kriechender Belastung

Koichi Akai and Yuzo Ohnishi
School of Civil Engineering, Kyoto University, Kyoto, Japan

SUMMARY

The effects of cyclic loading on soft saturated porous rock have been investigated. A typical phenomenon is so called cyclic fatigue in which a material fails at a stress level lower than its static strength. Deformation, strength and behaviour of pore pressure under quasi-static and cyclic loading were studied in undrained test conditions. In addition, creep tests were conducted in order to know the long-term strength. The sample used in the experiments is porous soft tuff whose unconfined compressive strength is about 16 MN/m^2. Laboratory experiments have been performed to know the correlation between creep, cyclic loading and conventional strain-rate constant tests. The concept of complete stress-strain curve was evaluated.

RESUME

Les effets de chargement cyclique sur roche poreuse saturée molle ont été examinés. Un phénomène typique est la soi-disante fatigue cyclique, dans laquelle un matériel se rompt à un niveau de tension plus bas que sa résistance statique. La déformation, la résistance et le comportement de la pression de l'eau interstitielle sous chargement quasi-statique et cyclique ont été visés dans des conditions d'essai non drainé. De plus, des essais de fluage ont été faits pour savoir la résistance à long terme. Les échantillons des expériences sont du tuf poreux mou, dont la résistance de compression sans frottement latéral est environ 16 MN/m^2. Des essais de laboratoire ont été faits pour savoir la corrélation entre le fluage, le chargement cyclique et les essais conventionels de vitesse de déformation constante.

ZUSAMMENFASSUNG

Die Effekte der zyklischen Belastung auf weichen, gesättigten porösen Felsen sind untersucht worden. Ein typisches Phänomen ist die sogenannte zyklische Ermüdung, wobei ein Material bei einem Druck niedriger als die statische Festigkeit bricht. Verformung, Festigkeit und Beträge des Porenwasserdrucks unter quasi-statischen und zyklischen Belastungen unter undrainierten Versuchsumständen sind erforscht worden. Weiterhin sind Kriechversuche zur Erforschung der langfristigen Festigkeit gemacht worden. Die in den Experimenten verwendeten Proben sind poröse, weiche vulkanische Tuffe, deren Kompressionsfestigkeit bei verhinderter Seitendrehung ungefähr 16 MN/m^2 beträgt. Laboratoriumsversuche zur Erforschung der Korrelation zwischen Kriechen, zyklischer Belastung und konventioneller konstanter Verformungsgeschwindigkeit sind angestellt worden.

INTRODUCTION

Foundations of dams, roads and bridges, underground space like tunnels and chambers are subjected to cyclic loading caused by earthquakes, traffics, blasting, etc.. The effects of cyclic loading on several different civil engineering materials such as steel, concrete and soil have been investigated. A typical phenomenon is so called cyclic fatigue in which a material fails at a stress level lower than its static strength. However, little work in this subject have been done in the area of rock mechanics. The influence of combined stresses and pore water pressure have not been investigated.

It is known that the fatigue curve in cyclic loading is similar to the static creep curve. The reason why is not well documented theoretically or experimentally. Scholz and Koczynski (1979) tried to explain these rock behaviors under cyclic loads with hard crystalline rocks. Their conclusion was that three types of

cracking result in dilatancy : stress-induced cracking; stress-corrosion cracking and fatigue cracking. Rock fracture is sensitive to which type is prevalent.

In engineering practice, in relation to cyclic loading on rock, an idea of complete stress-strain curve was presented by Haimson (1974). This idea may be useful to explain phenomenologically the similar behaviors between cyclic and creep loading conditions.

The purpose of this research is to examine a number of features of rock deformation and fracture that are not well observed in more conventional test. Soft saturated porous sedimentary rocks were selected for undrained triaxial tests. Deformation, strength and behavior of pore water pressure under quasi-static and cyclic loading have been investigated. In addition, creep tests were conducted to know the "long-term" strength. The results of these

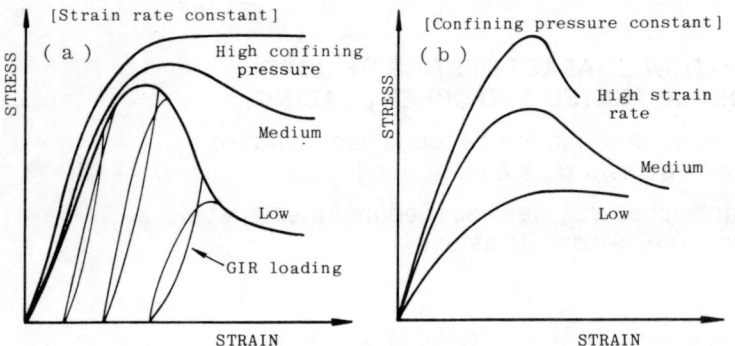

Fig. 1. Complete Stress-Strain Curves

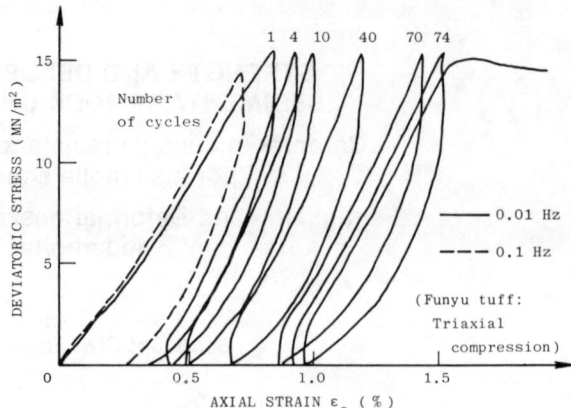

Fig.3. Stress-Strain Relationship

tests are interpreted in view of the complete stress-strain curve.

CONCEPT OF COMPLETE STRESS-STRAIN SURFACE

It is known that the accumulated permanent strain for different upper peak cyclic or static stresses is bounded by the complete stress-strain curve (Haimson, 1974). The envelope of gradually increased repeated (GIR) loading curve is also found to be the complete stress-strain curve (Akai, et. al., 1981) as shown in Fig. 1(a).

However, the concept of complete stress-strain curve is not well established because the curve is dependent upon the strain rate and confinement. The rate of strain in triaxial compression tests has strong effect on the shape of stress-strain curve as in Fig. 1(b). It is recognized that the higher the strain rate is, the stronger is the rock. At the very slow rate of strain test, a rock specimen sometimes does not show even a distinguished peak. The confining pressure in triaxial tests also change the shape of stress-strain curve as depicted in Fig. 1(a). Therefore, the infinite number of complete stress-strain curves may be defined with different strain rates and confinements. This means that a surface of complete stress-strain which bounds a state is possibly established in an adequately defined stress, strain and time space.

If a complete stress-strain curve is a test path on the complete stress-strain surface, it may be used to predict failure of rock as a result of creep and relaxation. As presented in Fig. 2(a), the locus of a creep test in the stress-strain space is a horizontal line and the locus of a relaxation is a vertical line. If the initial shear stress in the rock is close to the peak (point A in the figure),

creep will terminate in rupture at point B after a relatively short time. A creep test initiated at C below critical stress level will approach to point D without rupture after a long time (Goodman, 1980). Definition of critical stress level is not clear yet.

A similar concept applies to cyclic loading beneath the peak load level as shown in Fig. 2(b). Energy is consumed due to microcracking inside the specimen in the process of loading and unloading cycles. Cyclic loading starts at point A will migrate into the complete stress-strain surface and terminate in rupture at point B (Haimson, 1974, Goodman, 1980). Phenomenologically, this may be designated as "cyclic creep loading". It can be estimated that if the initial point of loading is below the critical stress level as indicated above, the increment of accumulation of permanent strain decreases and eventually goes to zero. Hysteresis loop is in equilibrium at this point.

Conclusions obtained in the cyclic creep loading test are also applied to relaxation and corresponding cyclic relaxation loading tests (Fig. 2(c)). Cyclic relaxation loading can be done with controlled amplitude of strain (or displacement), while an amplitude of stress (or load) is controlled in cyclic creep loading.

Examination of three figures (Figs. 2(a), 2(b), 2(c)) suggests that there exists an ultimate complete stress-strain curve which is common to creep, relaxation and cyclic loading (see Fig. 2(a)). The ultimate complete stress-strain curve may be a section of the complete stress-strain surface.

TEST PROCEDURES

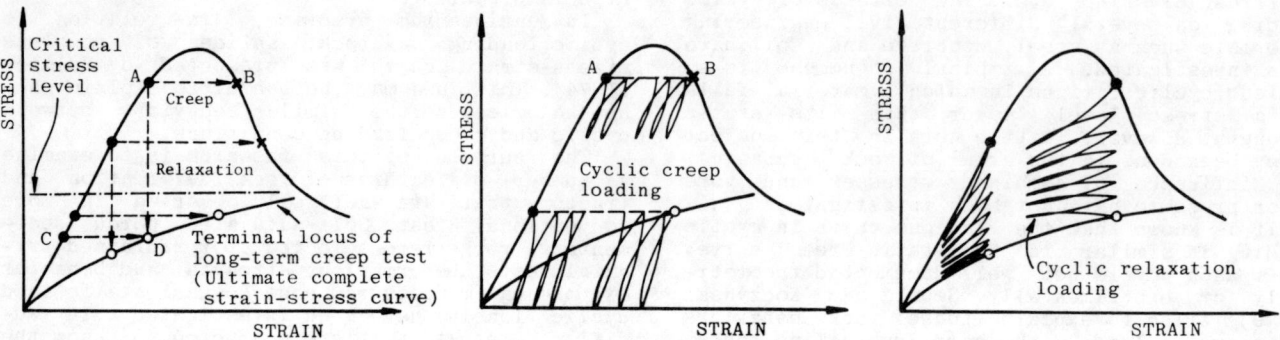

Fig. 2. Concept of Complete Stress-Strain Surface

a sedimentary soft rock called Funyu Tuff (G_S =2.65, n=29 %, γ_d =1.87 g/cm^3, uniaxial compressive strength = 16 18 MN/m^2) is used for a consolidated undrained test in this study. The size of a specimen is 50 mm in diameter, 100 mm in height. The change of pore water pressure in the specimen during compression process was measured by a small trasducer. The consolidation (initial confining) pressure was 1.0 MN/m^2. The standard strain rate controlled test was done by $\dot{\varepsilon}_a$ = 0.12 %/min. In cyclic loading test, various loading patterns were adopted. The loading functions were sawteeth with preset peak levels. The frequencies of cyclic loading were 0.1, 0.01 and 0.001 Hz. The loading test was terminated at 4×10^6 seconds of duration time.

RESULTS AND DISCUSSION

A typical stress-strain curve in cyclic triaxial compression is shown in Fig. 3. The

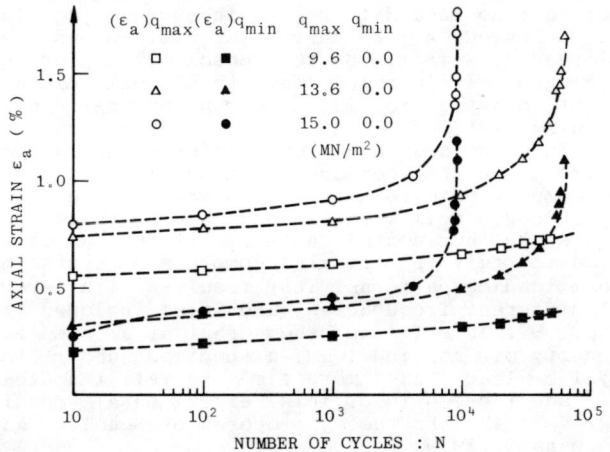

Fig. 4. Max. and Min. Strains at Each Cycle

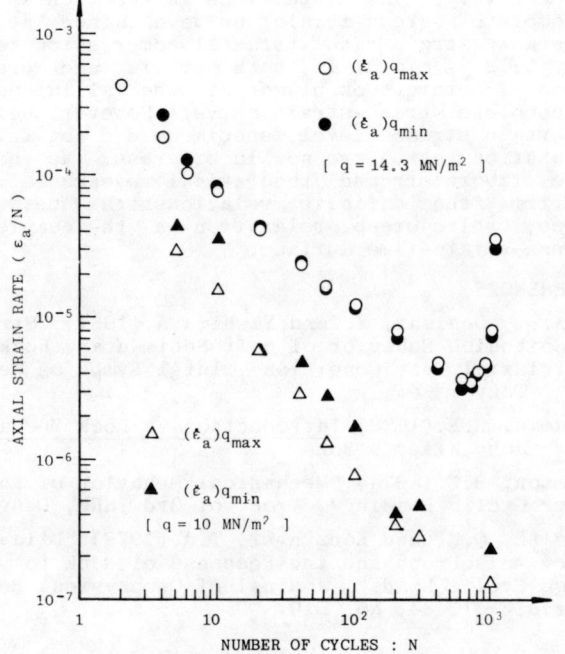

Fig. 5. Axial Strain Rate in Cyclic Loading

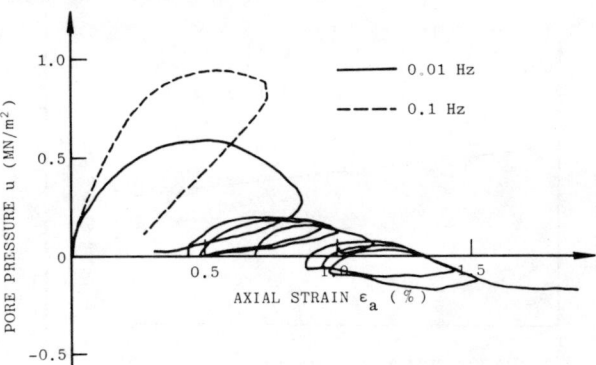

Fig. 6. Pore Pressure-Strain Relationship

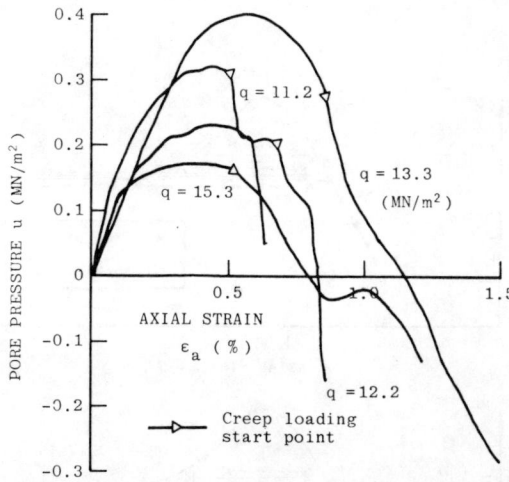

Fig. 7. Induced Pore Pressure in Creep Test

permanent strain increases rapidly at the onset of fracture and, hysteresis loop becomes wider and more inclined. Some of the relations between axial strain and number of cycles are shown in Fig. 4. Where, $(\varepsilon_a)_{qmax}$ is an accumulated total strain at the maximum load and $(\varepsilon_a)_{qmin}$ is an accumulated permanent strain at the minimum load in a preset amplitude. The shape of the curves is very similar to that of creep. The rate of strain per one loading cycle is shown in Fig. 5. This is also similar to the strain rate - time relation of creep. In both cyclic and creep loading the rate of strain drastically increases just before fracture and it constantly decreases when fracture is not expected. This fact implies the existence of critical stress level which was suggested in Fig. 1(a).

Corresponding to the stress-strain curve shown in Fig. 3, the typical strain-pore water pressure in triaxial compression is illustrated in Fig. 6. It shows some characteristics that a large amount of pore pressure is generated at the first cycle and the size of hysteresis loop at the subsequent cycles does not change so much, but the total pore pressure goes into negative region with respect to a initial back pressure. The envelope of the hysteresis loops is almost same to the strain-pore water pressure curve which is obtained in a conventional strain controlled triaxial compression test. The change of pore water pressure at the onset of fracture is not apparent. This suggests the fracture process

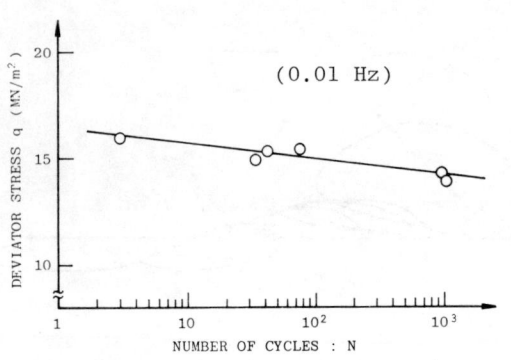

Fig. 8. S-N Curve for Funyu Tuff

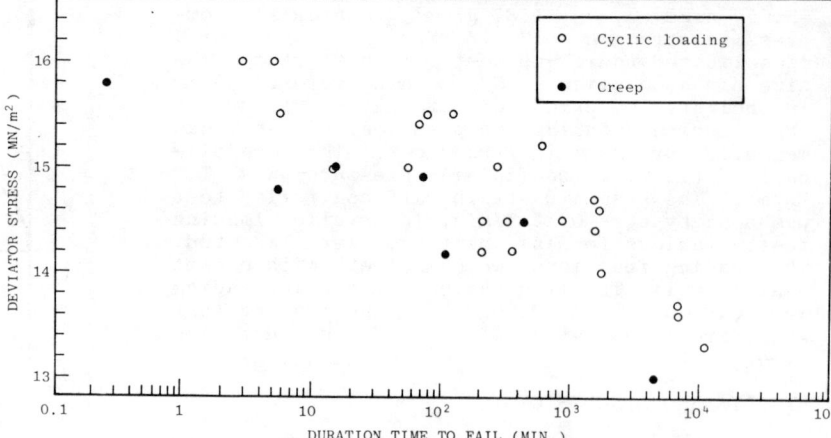

Fig. 9. Time to Fail at Different Stress Level

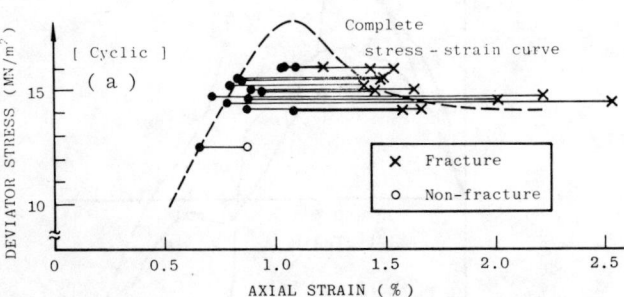

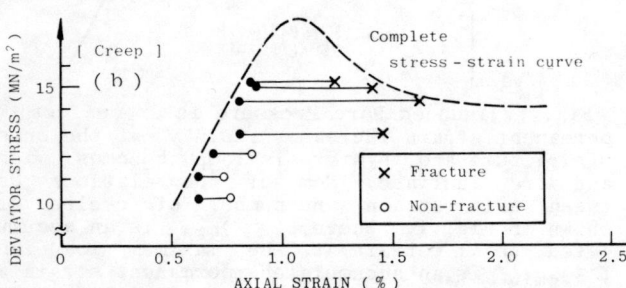

Fig. 10. Cyclic and Creep Test Paths of Soft Rock limited mainly in the vicinity of failure plane (Akai, et. al., 1981).

Strain-pore water pressure relationships of creep tests are shown in Fig. 7. It should be recognized that at the start of creep test (constant stress loading) the initial stress already sheared the specimen and some amount of pore water pressure was generated. As the strain increases with a constant creep load, the pore water pressure changes accordingly. From Fig. 6 and Fig. 7, the pore water pressure created in the triaxial test seems to have a unique relation with strain (axial or volumetric), but more tests have to be done to explore the influences of confining pressure and strain rate. Effective stress paths for these tests were plotted and examined. They were not much different from the total stress paths. This means that we need to test specimen in drained condition and measure the lateral deformation to identify more clearly the dilatancy effect.

Effect of frequency in cyclic loading tests can be seen in Fig. 3 and Fig. 6, especially modulus of deformation and generated pore water

pressure at first cycle. However, frequency did not make so much difference in strength so far as we tested. A S-N curve for 0.01 Hz has been plotted in Fig. 8 to check the fatigue strength. It reveals that this rock did not creat drastic drop in strength during cyclic loading test.

Since we are much more interested in the correlation between creep and cyclic loading, we compare both results in the way of Fig. 9, which could tell when a rock sample fails. The log-scale horizontal axis t_f is the duration time to fail in triaxial compression tests for both loading patterns. Test results with several different frequencies are also included in Fig. 9. This figure shows that at a same intensity of applied load, a specimen under the cyclic load takes more time to fail than that of under creep load. This experimental result agrees with the theory proposed by Scholtz and Koczynski (1979).

Results of cyclic loading and creep loading tests are shown in Fig. 10(a) and Fig. 10(b), respectively. The dotted line in the figures is a complete stress-strain curve obtained at a constant strain rate triaxial compression test ($\dot{\varepsilon}_a$ =0.12 %/min.). In both cyclic and creep tests, fracture took place at the falling part of complete stress-strain curve. However, below a certain stress level, specimens did not fail. Relaxation tests are now in progress. We need more laboratory and theoretical researches to confirm the definite relationships between creep, cyclic creep, relaxation in the complete stress-strain-time surface.

REFERENCES

Akai,K., Ohnishi, Y. and Yashima,A.(1981):"Strain Softening Behavior of Soft Sedimentary Rock in Triaxial Test Condition", Int'l Symp. on Weak Rock, Tokyo.

Goodman, R.E.(1980):Introduction to Rock Mechanics, John Wiley & Sons

Haimson, B.C.(1974):"Mechanical Behavior of Rock under Cyclic Loading", Proc. of 3rd ISRM, Denver

Scholtz, C.H. and Koczynski, T.A.(1979):"Dilatancy Anisotropy and the Response of Rock to Large Cyclic Loads", Journal of Geophysical Research, Vol. 84, No. B10.

A METHOD TO EVALUATE THE DEFORMATION BEHAVIOUR OF DISCONTINUOUS ROCK MASS AND ITS APPLICABILITY

Une méthode pour évaluer le comportement de déformation d'un massif rocheux discontinu et son applicabilité

Methode zur Auswertung des Verformungsverhaltens von discontinuierlichem Fels und ihre Anwendung

Ryunoshin Yoshinaka
Professor of Foundation Engineering, University of Saitama, Urawa, Saitama, Japan
Tadashi Yamabe
Research Associate, University of Saitama, Urawa, Saitama, Japan
Ichiro Sekine
Engineer of Toda Construction Co., Technical Research Center, Matsudo, Chiba, Japan

SYNOPSIS

A method to evaluate the deformability of rock masses is presented. This method considers the geometrical and mechanical properties of discontinuities, and the mechanical properties of intact rock. The constitutive equations to represent normal joint and shear stiffnesses varying against stress-deformation are introduced. The applicability of this method is clarified by means of two-dimension loaded large-scale rock mass models in the laboratory and by numerical analysis.

RESUME

Ce texte présente une méthode pour évaluer la déformabilité d'un massif rocheux. Cette méthode considère comme propriétés géométriques et mécaniques des discontinuités et propriétés mécaniques de la roche intacte. Pour exprimer les diaclases normales et les raideurs de cisaillement, nous avons introduit des équations constitutives variant selon les contraintes normales et les déformations. L'applicabilité de cette méthode est vérifiée par les essais de compression biaxiale au laboratoire utilisant un modèle de massif rocheux de grande dimension et par l'analyse numérique.

ZUSAMMENFASSUNG

Eingeführt wird eine Methode zur Auswertung der Deformierbarkeit von Fels, die die geometrischen und mechanischen Eigenschaften von Diskontinuitäten und die mechanischen Eigenschaften des intakten Felsens berücksichtigt. Weiterhin werden die konstitutiven Gleichungen eingeführt, die die Normal- und Schersteifigkeiten von Klüften im Verhältnis zur Spannungs-Verformung ausdrücken. Die Anwendungsmöglichkeiten dieser Methode werden mit Hilfe von zweidimensional belasteten Felsgroßmodellen im Labor und in numerischen Analysen erläutert.

1. INTRODUCTION

It is well known that the deformation behaviour of rock masses is greatly influenced by the existence of joints. Because of this reason, many studies have been performed about the effects of joints on rock mass. However, almost of them are concerned with the strength of rock masses, and there are few studies to evaluate the deformability of rock mass quantitatively.

In this paper, we take a view points that the deformability of rock mass is the function of fundamental elements such as mechanical properties of intact rock and joints, and geometrical properties of joints. And we present a method to obtain the relation between these elements and deformation behaviour of rock mass by experiments and numerical analyses.

2. MECHANICAL PROPERTIES OF SINGLE JOINT

To obtain joint stiffness, the methods proposed by Goodman (et al.,1968) are adopted basically.

2.1 Compressibility of single joint

Compressibility of single joint is obtained under uniaxial compression and calculated by the difference of deformation between intact and jointed rocks. Physical properties of intact rocks used are represented in Tab.I. Joint normal stiffness is defined by $K_n=(d\sigma/dV)$ where $dV=Vmc-V$. Vmc is the maximum joint closure, and V is joint closure at any stress level. The relation between Kn and (dV/Vmc) for various kinds of joints can be expressed by eq.(1) with reasonable accuracy. Fig.1 shows the example of experimental and calculated results by eq.(1). (Yoshinaka et al.,1980)

$$Kn = m \exp(l(dV/Vmc)) \qquad (1)$$

m,l and Vmc are constant for each joints. And they depend on size and roughness of joint, and wall material. However, if the roughness and material are same, "l" is almost independent on joint size. In general, the greater the area is, the greater the constant Vmc and m is.

Tab.I Physical properties of test rock (tuff)

Gs	γ_{dry}	ω_{sat}	σ_c	σ_t
2.40	1.46 g/cm³	29.2 %	11.2 MPa	1.46 MPa

2.2 Shear stiffness of single joint

Shear deformation of joint was measured by direct shear test equipment as shown in Fig.2. Many measuring points were pasted on the surface of side wall. Shear and dilation deformation were measured with contact gage. Fig.3 shows an example of relation between shear stress τ and shear deformation δ_S, where $\sqrt{A}$ is the average joint length at joint area A. $\tau-\delta_S$ curves for various kinds of joint surfaces, can be expressed accurately by hyperbolic function.(Duncan & Chang 1970) Figs.4(a) and (b) give examples of this approximation. Shear stiffness of single joint can be expressed as follows;

$$Ks = Ksi(1-\tau R_f/\tau_f)^2 \qquad (2)$$

where Ksi is initial shear stiffness at $\tau=0$, and it depends on not only normal stress σ_n but the other condition of joint surfaces. Fig.5 shows the relation between Ksi and σ_n. Ksi can be expressed by eq.(3);

$$Ksi = \alpha Pa(\sigma_n/Pa)^\beta \qquad (3)$$

where α, β and R_f in eqs.(2) and (3) are material constants. τ_f is maximum shear strength and Pa is atmospheric pressure measured in same units with σ_n. β is almost constant for a certain surface, but α depends greatly on joint area.

2.3 Relation between Ksi and Kn

Kn in eq.(1) and Ksi in eq.(3) are also represented as a function of average normal stress σ_n on the joint surface. When dV in eq.(1) is expressed by eq.(4) (Goodman,1976), eq.(5) is obtained by substituting eq.(4) into eq.(1) and eq.(3).

$$dV = -a+b\log\sigma_n \qquad (4)$$

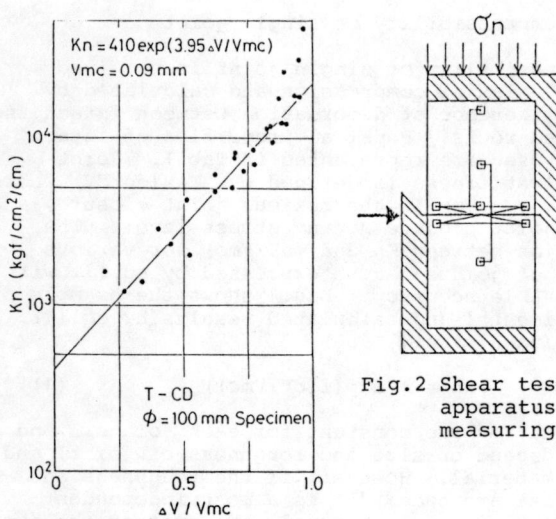

Fig.1 Normal joint stiffness Kn vs. relative joint closure (dV/Vmc)

Fig.2 Shear testing apparatus and measuring points

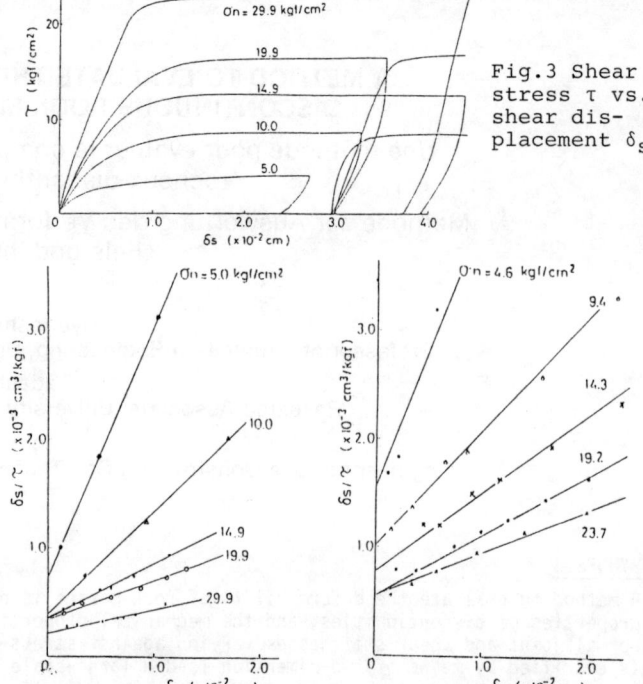

Fig.3 Shear stress τ vs. shear displacement δ_S

Fig.4 Hyperbolic approximation of shear stress-shear displacement curves
(a) $\sqrt{A}=13.7$cm (b) $\sqrt{A}=42.2$cm

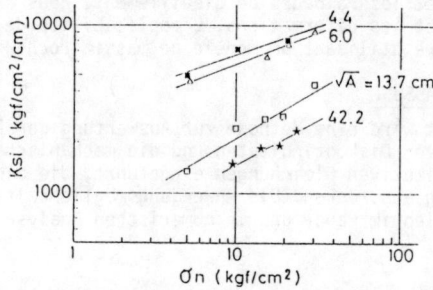

Fig.5 Initial shear stiffness Ksi vs. average normal stress σ_n by eq.(3)

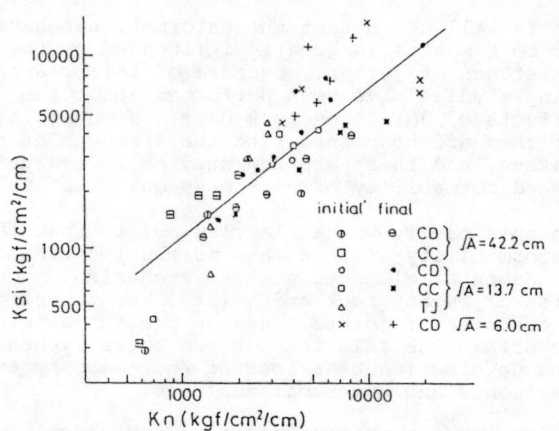

Fig.6 Experimental results of Ksi vs. Kn, and calculated line by eq.(5)
CD:cut by diamond saw,CC:cut by chain saw
TJ:tension joint

Tab.II Mechanical properties of rock element in virgin loading

Intact rock		Joint normal stiffness			Joint shear stiffness				
E_c(kgf/cm^2)	ν_c	m	l	Vmc	α	β	c(kgf/cm^2)	ϕ	R_f
38000	0.25	410	3.95	0.09mm	562	0.582	0.76	37.3°	0.9

$$\log(Ksi) = (\beta/\gamma)\log(Kn) + C \qquad (5)$$

where

$$\gamma = \frac{b\, l\, \log(e)}{Vmc} \quad , \quad C = \frac{\alpha\, Pa^{(1-\beta)}}{\{m\, \exp(-al/Vmc)\}^{\beta/\gamma}}$$

and e is the base of natural logarithm.
In Fig.6, plotted points for average joint length $\sqrt{A}$, represent measured Ksi and Kn, and solid line shows the result calculated by eq.(5) with material constants given in Tab.II. The reasonable agreements are achieved in Fig.6, and this means that eqs.(1) and (2) are effective for the expression of normal and shear stiffnesses.

3. DEFORMATION EQUATION OF JOINTED ROCK MASS

In this paper, it is assumed that deformation of rock mass δ, equates to the summation of the deformations in both intact rock and joints, δ_i and δ_j respectively.

$$\delta = \delta_i + \delta_j \qquad (6)$$

The rock mass model considered is shown in Fig. 7, and strike of each joints is parallel to the axis of minimum principal stress σ_3.

Elastic constants of intact rock are given by Young's modulus E_C and Poisson's ratio ν_C. Joint normal and shear stiffnesses for joint set 1 are Kn_1 and Ks_1, Kn_2 and Ks_2 for joint set 2. Shear strength of joint is represented by cohesion c, and friction angle ϕ of joint surface.

Under these conditions, the equivalent deformation modulus of rock mass is formularized for plane stress of $\sigma_3=0$ as follows;

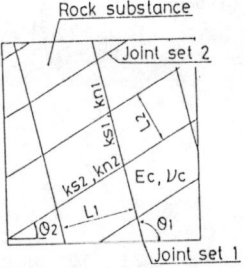

Fig.7 Rock mass model (plan view)

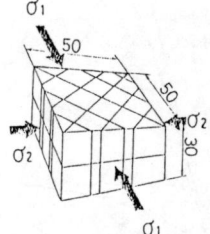

Fig.8 Jointed rock mass model

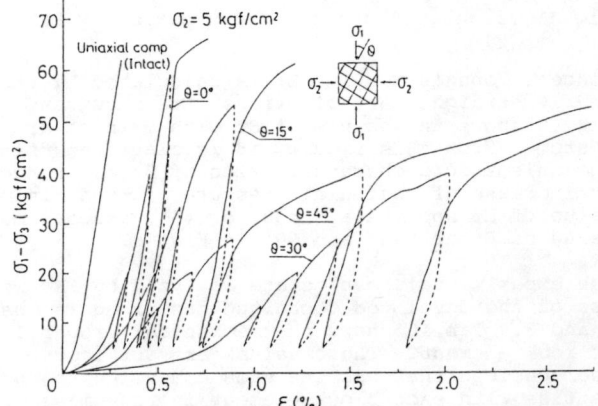

Fig.9 Stress-strain cuves for intact rock and rock mass models with varying orientation

$$\frac{1}{E} = \frac{1}{E_c} + \frac{\cos^2\theta_1}{L_1}\left(\frac{\sin^2\theta_1}{Ks_1} + \frac{\cos^2\theta_1}{Kn_1}\right) + \frac{\cos^2\theta_2}{L_2}\left(\frac{\sin^2\theta_2}{Ks_2} + \frac{\cos^2\theta_2}{Kn_2}\right) \qquad (7)$$

$$\frac{1}{G} = \frac{2(1+\nu_c)}{E_c} + \frac{\cos^2\theta_1}{L_1}\left(\frac{2\sin^2\theta_1 + |\sin2\theta_1|/2}{Ks_1} + \frac{2\cos^2\theta_1 - |\sin2\theta_1|/2}{Kn_1}\right)$$
$$+ \frac{\cos^2\theta_2}{L_2}\left(\frac{2\sin^2\theta_2 + |\sin2\theta_2|/2}{Ks_2} + \frac{2\cos^2\theta_2 - |\sin2\theta_2|/2}{Kn_2}\right) \qquad (8)$$

4. EXPERIMENTAL RESULTS AND ITS CONSIDERATION ON BIAXIAL COMPRESSION TEST RESULTS OF JOINTED ROCK MASS MODEL

Biaxial compression test was performed on the rock mass model shown in Fig.8. Total size of rock mass model is 50cm width, 50cm length and 30cm height, and that of element blocks which consists of the model, is 12.5 12.5 15.0 cm. Joints contained in rock mass, are continuous as shown in Fig.8. Sets of orientation in perpendicularly crossing joints are (0°,90°),(15°,75°), (30°,60°) and (45°,45°). Loading platens are rigid compared with used rocks. And the friction between platen and rock mass model is reduced sufficiently by lubricated teflon sheets. Confining pressure σ_2 is held constant during loading process and the value of σ_2 used are 0.196, 0.49, 0.98 and 1.96 MPa (2, 5, 10 and 20 kgf/cm^2). A typical example of stress-strain curves of rock mass model, is shown in Fig.9.

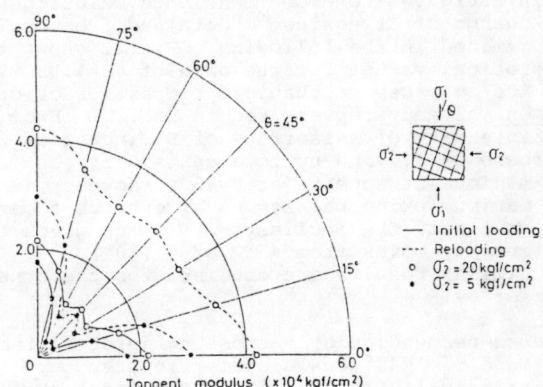

Fig.10 Variation of tangent modulus with joint orientation θ

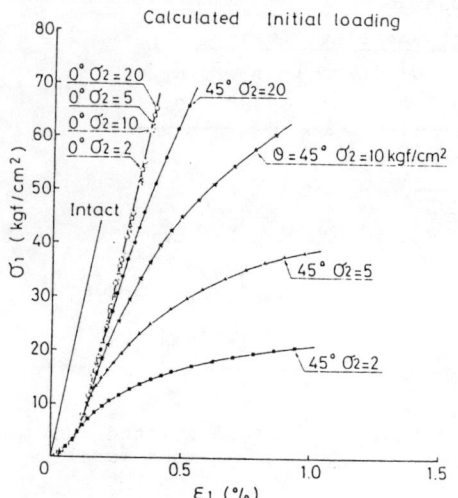

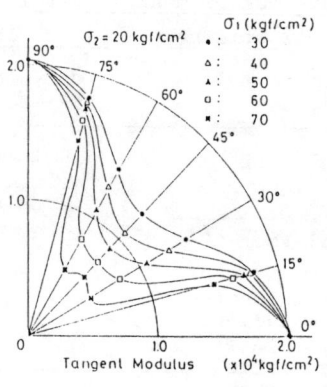

Fig.12 Change of calculated
Et with increase of
σ_1 (σ_2=1.96MPa)

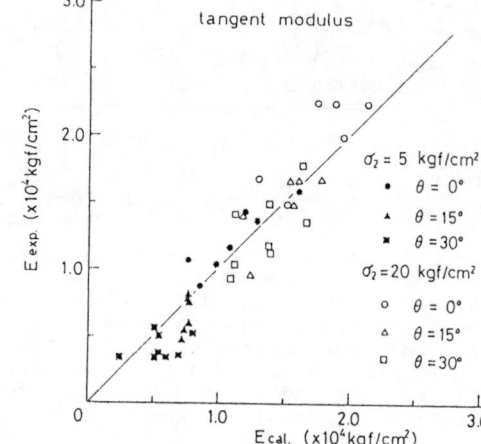

Fig.11 Calculated stress-strain curves by FEM

Fig.13 Comparison of experimental
Et with calculated Et

Tangent Young's modulus Et is calculated by the
nearly straight part of stress-strain curves.
Fig.10 shows the change of Et with θ in polar
system. From this figure, it is clarified that
the anisotropic characteristics of Et are reduced
by increase of confining pressure σ_2, and minimum
value of Et appears at θ=30° or 45°, and maximum
value of Et at θ=0° and 90° in Fig.10.

The experimental results are analyzed by FEM with
use of the developed technique described in Chs.
2 and 3. Tab.II shows the mechanical properties
of rock element. These values are obtained by
the specimen that has the same dimension and properties with each block element in rock mass
model. Fig.11 shows the stress-strain curves by
FE analyses. In this figure, the joint orientation θ is 0° and 45°, and stress-strain curves
for the case of σ_2=0.49 MPa correspond to the experimental results shown in Fig.9. Good agreement between experiment and calculation can be
seen quantitatively for this case and the other
case of σ_2 and θ. However, the calculated strain
is generally smaller than that of experiments for
the same stress level. These differences seem to
be due to the contact condition of joint between
in single joint and model rock mass that has
about 32 block elements.

To investigate the above mensioned relationship,
the change of Et against θ obtained from eq.(7)
is examined in the following. Fig.12 shows the
theoretical values for the case of σ_2=1.96 MPa
(20 kgf/cm²) used mechanical properties of one
element already represented in Tab.II. From Fig.
12, intensity of anisotropy of Et increases with
increase of σ_1, and up to a certain stress level,
the minimum Et appears at θ=45°. However, when
the maximum principal stress σ_1 exceeds a certain
stress limit, the inclination θ which gives the
minimum Et, moves from 45° to 30°(60°).
The similar results are obtained for the other
case of σ_2.

Because perpendicularly crossing joint system for
the case of θ=45° has a symmetric structure to
the principal stress, both of the joint stiffness
Ks and Kn as whole model rock mass can be obtained
by back calculation of experimental results.

Kn and Ks obtained from the case of θ=45°, are
not necessary most suitable to estimate the anisotropic effects on deformability of other
orientation. However, Ks and Kn at θ=45° can be
used to compare the deformability of single
joint with that of rock mass qualitatively.
Fig.13 shows the comparison of the experimental
results with the calculated results obtained
from the above mentioned methods. In this
figure, the plotted points under the same notation correspond to the change of Et with increase of maximum principal stress, and the tendency of deformation behaviour occured in the
model rock mass can be expressed by the above
mentioned methods.

5. CONCLUSION

We present a method to evaluate the deformation
modulus of rock mass by the given properties and
orientations of joints. It is clarified that
the deformation of rock mass can be estimated
by considering the stress dependency, non-
linearity and anisotropy of rock mass. However,
there are many problems to be solved, especially
about modification of joint properties according
to structure of rock mass.

6. REFERENCES

Duncan,J.M. and Chang,C.Y. (1970). Nonlinear
analysis of stress and strain in soils:
Jour. of SM Div.,ASCE (96), 5, 1629-1653

Goodman,R.E. et al. (1968). A model for mechanics
of jointed rocks: Jour. of SM Div.,ASCE(93)
, 3, 637-659

Goodman,R.E. (1976). Methods of Geological
Engineering in Discontinuous Rocks. 173pp.
West Publishing Company

Yoshinaka,R. and Nishimaki,H. (1980) Experimental and numerical studies on bearing
capacity of soft rock foundation: Proc.
of JSCE (304), 113-128

LA DEFORMATION ET LA FATIGUE CYCLIQUE DES ROCHES

Strain and fatigue behaviour of rock

Verformungs- und Ermüdungsverhalten der Gesteine

F. Homand-Etienne
Maître Assistant, Ecole de Géologie, I.N.P.L., Nancy, France

S. Mora
Docteur de 3e cycle, Ecole de Géologie, I.N.P.L., Nancy, France

R. Houpert
Professeur, Ecole De Géologie, I.N.P.L., Nancy, France

RESUME
Le comportement à la fatigue d'un granite et d'un marbre est étudié à partir de chargements cycliques en compression. Les modifications des structures et des déformations sont observées en analysant la courbe en trois phases de fluage cyclique et la géométrie des boucles d'hystérésis. La propagation de la rupture est étroitement liée aux caractéristiques pétrographiques, physiques et mécaniques des roches.

SYNOPSIS
The cyclic fatigue phenomenon in granite and marble has been investigated by cyclic loading in uniaxial compression. General evolution of damages and strains can be observed by means of a three phases cyclic creep and by the evolution of hysteresis loops'geometry. It is concluded that the development of fatigue failure in rocks is closely related to their petrographical, physical and mechanical properties.

ZUSAMMENFASSUNG
Das Ermüdungsverhalten eines Granits und eines Marmors bei wiederholten Drucklastzyklen wird untersucht. Die Entwicklung der Strukturveränderungen und der Verformung kann mittels der drei Phasen der Zykluskriechkurve und der Form der Hysteresis-Schleifen beschrieben werden. Die Bruchausbreitung in Gesteinen ist eng mit ihren petrographischen, physikalischen und mechanischen Eigenschaften verbunden.

1 - INTRODUCTION

Les roches, dans leur emplacement naturel, sont soumises à des charges statiques et également dynamiques. Celles-ci peuvent être d'origine tectonique, lithostatique ou artificielle (activité humaine). Dans ces conditions, les roches subissent un endommagement progressif qui peut se manifester pour des contraintes inférieures à la résistance ultime et même pour des contraintes inférieures à la limite d'élasticité. Cet endommagement, appelé couramment "fatigue", est lié à l'inélasticité de la roche et le mécanisme de son développement est fonction de la structure et des imperfections inhérentes au matériau. Les recherches sur le comportement en fatigue des roches ont commencé il y a une vingtaine d'années. Si pour les métaux, il existe un grand nombre de données expérimentales, ainsi que des théories pour prévoir le comportement en fatigue, les données et les théories concernant les roches sont relativement réduites. De plus, pour les roches, le problème de l'effet d'échelle et de la dispersion des caractéristiques mécaniques se surimpose.

Nous nous proposons de présenter ci-après quelques résultats sur le comportement des roches sous sollicitation cyclique en compression simple.

2 - EXPERIMENTATION

Les roches utilisées dans le cadre de ce travail ont été choisies en fonction de la dispersion relativement faible de leurs propriétés pétrographiques, physiques et mécaniques. Il s'agit du granite de Senones et du marbre de Carrare. Une centaine d'éprouvettes de chaque type de

roche ont été confectionnées et les principaux paramètres physiques ont été mesurés (poids volumique, porosité, perméabilité, célérité des ondes longitudinales). Ces mesures ont permis de sélectionner des lots d'éprouvettes homogènes destinées aux essais mécaniques. Les caractéristiques de ces deux roches sont résumées ci-dessous :

	Granite de Senones	Marbre de Carrare
Poids volumique (kN/m^3)	26,8	27,1
Porosité (%)	0,2	<0,1
Résistance ultime (MPa)	149	93
Limite d'élasticité (MPa)	137	69
Module d'élasticité (MPa)	4,1.10^4	3,8.10^4

Les essais en compression cyclique ont été réalisés avec une presse asservie et un générateur de fonctions (statique-dynamique, force-déformation) a permis d'appliquer des sollicitations cycliques de forme sinusoïdale. La fréquence des cycles choisie pour les expériences présentées est de 0,26 Hz. L'amplitude est de 50 MPa dans le cas du granite de Senones et de 25 MPa dans le cas du marbre de Carrare.

La première phase de l'étude a consisté à déterminer les caractéristiques d'endurance des différentes roches. Les charges cycliques sont appliquées à différents niveaux de contrainte, mais à amplitude constante. Le premier niveau de contrainte maximal choisi se trouve juste au-dessous de la résistance ultime. La durée de vie en fatigue est déterminée en reportant pour chaque échantillon en abscisse le logarithme du nombre de cycles et en ordonnée le taux de sollicitation maximal exprimé par rapport à la contrainte ultime. La courbe obtenue est dite courbe S-N ou courbe de Wöhler. La figure 1 montre la courbe du granite de Senones et celle du marbre de Carrare. De nombreux paramètres, soit intrinsèques à la roche (porosité, degré de fissuration), soit inhérents à l'essai (amplitude, fréquence), influent sur cette courbe et, par suite, sur la détermination de la limite d'endurance σ_E (MORA, 1982). Pour les conditions expérimentales définies sur la figure 1, σ_E est de 85 % pour le granite de Senones et de 83 % pour le marbre de Carrare.

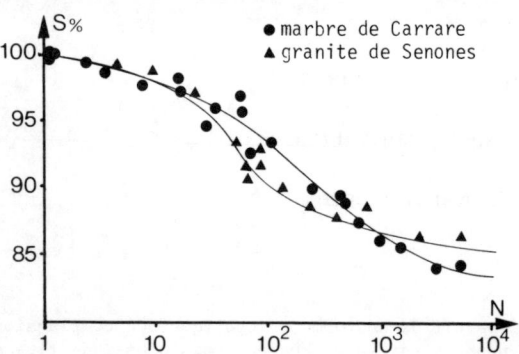

Fig. 1 - Courbes d'endurance S-N du granite de Senones (0,26 Hz ; ± 50 MPa) et du marbre de Carrare (0,26 Hz ; ± 25 MPa).

La limite d'endurance σ_E, replacée sur les courbes contrainte-déformation de compression simple, se situe au voisinage du seuil de dilatance pour le granite de Senones et légèrement au-dessus de ce seuil pour le marbre de Carrare. Des changements de conditions expérimentales, comme la diminution de la fréquence, font diminuer σ_E, les charges maximales étant maintenues plus longtemps.

La détermination de la limite d'endurance n'est qu'un aspect du problème du comportement en fatigue des roches. Les enregistrements de la déformation en fonction de la contrainte (σ-ε) et en fonction du temps (ε-t) permettent d'étudier son évolution tant globale qu'au cours d'un cycle. Les différents aspects de la déformation de l'éprouvette en fatigue sont développés ci-après.

3 - EVOLUTION DE LA DEFORMATION

Bien que les mécanismes de propagation de la rupture sous sollicitations statique et cyclique soient différents, les roches réagissent de façon semblable aux charges cycliques et au fluage sous charge monotone, comme l'ont montré ATTWELL et FARMER (1973), HAIMSON (1974) et HAIMSON et KIM (1972). La figure 2 indique schématiquement la courbe ε-t pour le granite de Senones. Sa forme, rattachée aux caractéristiques minéralogiques et physiques, sera discutée ultérieurement. Il est possible de mettre en évidence, sur cette courbe, des seuils marqués par des changements de pente de la courbe enveloppe des maxima des pics de déformation. Ces seuils permettent de déterminer trois phases qui peuvent être comparées aux phases observées lors d'un essai de fluage statique et qui seront appelées, par analogie, phases de fluage primaire, secondaire et tertiaire. Sur la figure 2, des points marquent respectivement le début de la rupture, la rupture proprement dite et la ruine.

Si, par ailleurs, on tient compte uniquement de la déformation acquise aux pics des maxima des cycles, il est possible de construire une courbe équivalente à la courbe de fluage pour un niveau de contrainte donné. On peut mettre ainsi en évidence des familles de courbes de fluage

à partir des courbes ε-t obtenues par sollicitation cyclique à différents niveaux de contrainte. L'ensemble de ces courbes permet d'observer les variations d'allure et de durée de chaque phase de déformation (fig. 3 et 4).

Les valeurs des déformations sont valables pour les amplitudes utilisées : ± 50 MPa pour le granite de Senones et ± 10 MPa pour le marbre de Carrare.

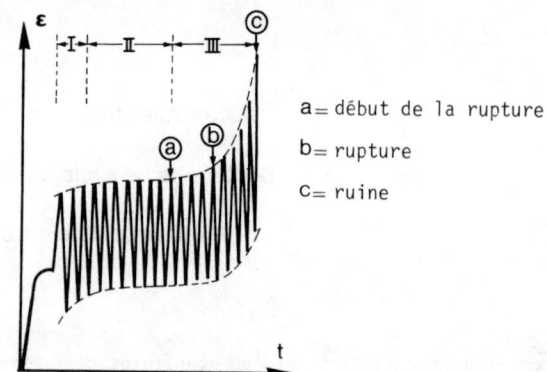

Fig. 2 - Courbe schématique déformation-temps (ε-t) du granite de Senones. Les chiffres I, II, III sont relatifs aux phases de déformation.

3.1 - Première phase

Au cours des premiers cycles, il se produit une déformation assez importante traduisant vraisemblablement un endommagement des bords des fissures inter et intra-granulaires. La vitesse de déformation diminue ensuite et se stabilise suivant le niveau de contrainte de sollicitation. Les déformations atteintes dès ces premiers cycles sont, suivant la roche, de l'ordre de 30 à 50 % de la déformation totale.

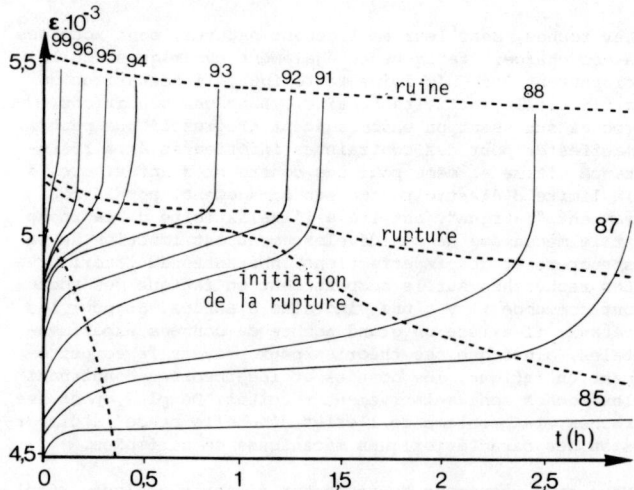

Fig. 3 - Courbes enveloppes des déformations maximales pour le granite de Senones. Les chiffres correspondent au taux de sollicitation S en %.

3.2 - Deuxième phase

Les déformations se produisent à vitesse constante pendant cette phase qui, d'ailleurs, tend à disparaître à de hauts niveaux de contrainte. Lorsque cette période est développée, les boucles d'hystérésis se ferment considérablement du fait de la stabilisation de la fissuration. La déformation

croît linéairement et son importance dépend du type de
roche et de la valeur de la charge appliquée. Pour le
granite, cette phase est bien linéaire. Pour les roches
composées presque exclusivement de cristaux de calcite,
comme le marbre ou les calcaires microcristallins testés
par ailleurs, la progression de la déformation se réalise
de façon légèrement échelonnée, avec des petits paliers.
Ceux-ci peuvent être interprétés comme la manifestation
du caractère plastique des cristaux de calcite. La fissuration ne semble pas jouer un rôle prépondérant dans cette
phase.

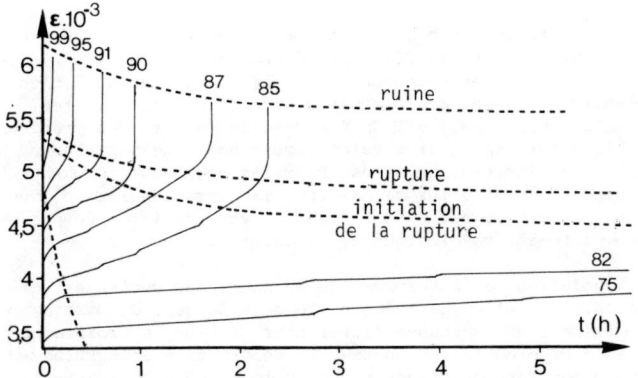

Fig. 4 - Courbes enveloppes des déformations maximales
pour le marbre de Carrare. Les chiffres correspondent au
taux de sollicitation S en %.

3.3 - Troisième phase

La troisième phase commence avec le début du processus
de rupture au moment où la courbe ε-t perd son caractère
linéaire (point a, fig. 2). Cette première inflexion
marque une accélération de la déformation totale qui peut
correspondre à la fin du domaine linéaire de la courbe σ-ε
en compression simple (seuil de libération critique d'énergie). Cette accélération favorise l'accroissement de la
déformation plastique, tandis que la déformation différée
diminue progressivement et la déformation élastique se
dissipe rapidement.

L'accélération de la déformation, à partir du point considéré comme marquant le début de la *rupture*, est la conséquence de la coalescence et de la propagation des fissures
à travers toute l'éprouvette. Elle correspond à la rupture
successive des différents éléments de volume. Généralement, la déformation acquise à ce niveau est équivalente à la
déformation correspondant à la résistance ultime de la
roche.

L'intervalle entre l'initiation de la rupture et la rupture proprement dite varie en fonction du type de roche ;
lorsque l'essai est pratiqué à un haut niveau de contrainte, cet intervalle est très réduit. Dans les granites,
il varie entre 10 et 100 cycles en fonction de la contrainte appliquée, tandis que pour le marbre, il ne dépasse
jamais 50 cycles.

La déformation totale augmente ensuite rapidement,marquant
l'amorce de la *ruine*. Celle-ci est représentée par le dernier pic maximal sur la courbe ε-t et elle provoque la dis-
parition absolue de toute déformation élastique récupérable.

4 - HYSTERESIS

La forme des boucles d'hystérésis varie en fonction des
propriétés rhéologiques des matériaux. Dans le cas des

roches, NISHIMATSU et HEROESEWOJO (1974) ont proposé une
classification de la forme des boucles en trois familles
correspondant à trois comportements rhéologiques. Les
boucles quasi linéaires caractérisent un comportement
élastique, celles de forme elliptique, un comportement
visco-élastique linéaire et celles en croissant,un comportement visco-élastique non linéaire.

La forme générale des boucles reste constante pour une
roche donnée au cours d'un essai de fatigue, mais leur
ouverture change en fonction des différentes phases d'évolution de la déformation. D'autre part, la fréquence et
l'amplitude provoquent également de légères modifications.

Les boucles d'hystérésis du granite de Senones sont quasi-
ment linéaires. La figure 5 montre l'évolution de leur
ouverture au cours d'un essai de fatigue. Pendant une
première phase, elles sont ouvertes, puis elles se referment pendant la deuxième phase, lorsque la déformation
évolue de façon linéaire. Les boucles se rouvrent ensuite,
juste avant la ruine.

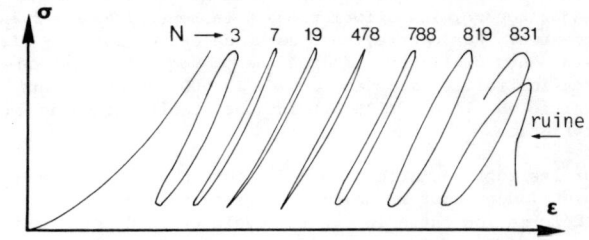

Fig. 5 - Evolution des boucles d'hystérésis pour le granite de Senones.

Le marbre de Carrare présente tout d'abord des boucles de
forme elliptique presque parfaite, ouvertes pendant la
première phase (fig. 6). Au cours de la seconde phase,
elles se referment mais changent d'aspect, leurs extrémi-
tés se rapprochant de la forme d'un losange. Cette caractéristique reste acquise pendant la troisième phase,
malgré la réouverture de la boucle. L'essai de fatigue
est réalisé en faisant varier la charge de façon sinusoïdale. Si la roche a un comportement parfaitement élastique,
l'enregistrement de la déformation en fonction du temps

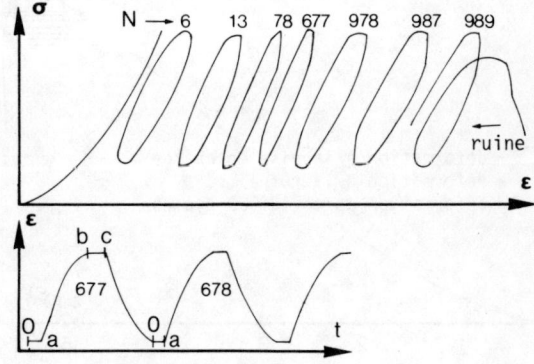

Fig. 6 - Evolution des boucles d'hystérésis pour le marbre
de Carrare. Allure de la sinusoïde ε-t.

doit être de forme sinusoïdale. C'est bien le cas pour le
granite, mais l'observation de la figure 6 montre que la
variation des déformations au cours d'un cycle n'est pas
sinusoïdale. Au début du cycle d'application de la charge,
il y a un retard dans la réponse de la déformation (oa).
Ensuite, la force augmente mais moins rapidement au sommet
de la sinusoïde, tandis que la déformation continue à
croître (ab), ceci étant dû au rattrapage du retard de la

déformation en début de cycle. Après le maximum de la sinusoïde de σ-t, la charge diminue mais la déformation ne
diminue pas en même temps (bc).

D'autres roches calcaires, mais de structure différente,
montrent le même phénomène. L'interprétation de ces retards
de réponse de la déformation par rapport à la force doit
être faite en terme de comportement rhéologique. Ce sont
des manifestations de la viscosité de ces roches.

5 - DEFORMATION TOTALE ET FATIGUE

Quelles que soient les caractéristiques de la contrainte
appliquée (amplitude, fréquence, etc.), il est nécessaire
de dépasser un certain taux de déformation pour que la
rupture d'une éprouvette se produise. La déformation
acquise entre le premier et le dernier cycle avant la
rupture et avant la ruine, augmente considérablement avec
le nombre de cycles nécessaires pour arriver à ces stades.
Au fur et à mesure que la charge diminue vers la limite
d'endurance σ_E, les déformations à la rupture et à la ruine
augmentent. Par contre, en dessous de cette limite σ_E, les
chargements cycliques effectués ne provoquent qu'une déformation faible, même au bout de 10 000 cycles, ce qui
signifie que l'on a bien atteint une stabilisation de la
déformation.

Pour les ruptures atteintes à de hauts niveaux de contrainte, donc pour un nombre de cycles faible, il y a peu
de déformation entre le premier cycle et le dernier. Cependant, la déformation accumulée à la première mise en charge
est plus importante.

Puisque la rupture et la ruine en fatigue cyclique apparaissent lorsqu'un certain taux de déformation a été
atteint, HAIMSON (1974) propose de considérer que les
déformations pour des contraintes au-dessus de σ_E sont
toujours limitées par le tracé de la courbe σ-ε complète
d'une éprouvette vierge en compression simple.

La figure 7 représente, pour le granite de Senones, la
courbe σ-ε "moyenne" obtenue lors d'un essai de compression

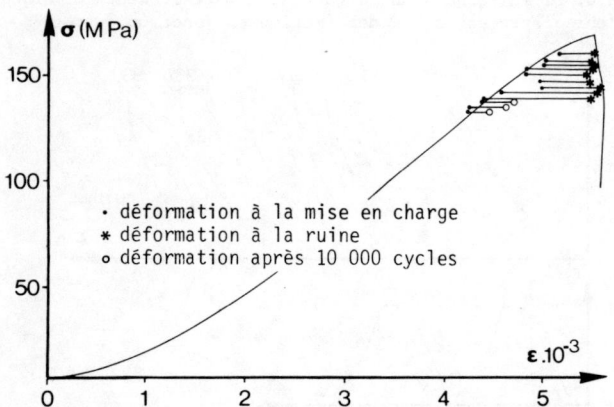

Fig. 7 - Courbe σ-ε du granite de Senones indiquant
l'importance des déformations acquises par les éprouvettes testées en compression cyclique à divers niveaux
de contrainte. Les déformations à la ruine sont limitées
par le tracé de la courbe post-pic.

simple quasi statique. Sur cette courbe ont été portées,
pour un niveau de contrainte donné, la déformation correspondant à la mise en charge, la déformation relative à la
ruine de l'éprouvette et la déformation acquise, après
10 000 cycles, pour les éprouvettes restées intactes. On

remarque que le lieu des points indiquant la ruine est très
proche de la partie post-maximale de la courbe σ-ε. La
courbe σ-ε peut donc être utilisée pour définir, à un niveau de contrainte quelconque situé au-dessus de σ_E, la
charge minimale suffisante pour obtenir un maximum de déformation. Cette courbe de compression simple constitue
une enveloppe définissant les propriétés de déformation
et de longévité en fatigue.

6 - CONCLUSIONS

La microfissuration qui initie et conduit à la rupture en
fatigue cyclique se développe d'une façon identique à celle
se produisant en compression simple quasi statique, c'est-
à-dire au niveau des discontinuités du matériau (pores,
inclusions, fissures). En fatigue, la rupture des granites
a lieu par propagation relativement homogène de fissures
inter et intra-granulaires. Dans les marbres et les roches
composés de cristaux de calcite, la microfissuration joue
un rôle très modeste et la fatigue est conditionnée par le
comportement rhéologique de la calcite.

L'évolution de la déformation au cours des cycles est similaire à celle observée en fluage statique. On distingue
également trois phases différentes du taux de croissance
de la déformation. En outre, la déformation à la ruine est
du même ordre de grandeur que celle obtenue en compression
simple. En effet, en fatigue et pour un niveau de sollicitation donné, les points correspondant à cette déformation
se trouvent au voisinage de la partie post-pic de la courbe
σ-ε de compression simple quasi statique.

En fait, de nombreux paramètres influent sur le comportement en fatigue, comme la fréquence de sollicitation et
l'amplitude des cycles pour une contrainte maximale donnée.
SCHOLZ et KOCZYNSKI (1979) démontrent que l'augmentation
de la déformation durant les charges cycliques est due à
un fluage progressif, augmenté par les dommages causés par
les cycles. Pour de faibles amplitudes, la contrainte
moyenne est élevée et le mode de sollicitation se rapproche
du fluage. Pour de fortes amplitudes, l'hystérésis est
très développée, montrant qu'il se produit de nombreux
dommages, la contrainte moyenne étant pourtant plus faible;
ce mécanisme est celui de la fatigue au sens strict. Pour
des valeurs intermédiaires de l'amplitude, le temps pré-
cédant la rupture serait plus important. Il s'ensuit que
la déformation en fatigue serait le résultat d'une fissuration mixte : celle résultant de la charge maximale
(fluage) et celle produite par la variation cyclique de
la contrainte. La déformation et la rupture sont étroitement liées aux caractéristiques physiques et mécaniques
du matériau, comme le montrent les particularités de comportement des roches composées de cristaux de calcite, le
caractère visqueux de ces cristaux étant, dans ce cas,
prépondérant.

7 - REFERENCES

Attewell, P.B., et Farmer, I.W. (1973). Fatigue behavior
    of rock. Internat. J. Rock Mech. Min. Sci., 10, 1-9.
Haimson, B.C. (1974). Mechanical behavior of rock under
    cyclic loading. C.R. 3e Congr. Internat. Méc. Roches,
    II-A, 373-378, Denver.
Haimson, B.C., et Kim, C.M. (1972). Mechanical behavior
    of rock under fatigue. Proc. 13th Symp. Rock Mech.,
    845-863, Univ. of Illinois, Urbana.
Mora, S. (1982). La fatigue cyclique des roches. Thèse
    Doct. 3e cycle, I.N.P.L., Nancy.
Nishimatsu, Y., et Heroesewojo, R. (1974). Rheological
    properties of rocks under the pulsating loads. C.R. 3e
    Congr. Internat. Méc. Roches, II-A, 385-389, Denver.
Scholz, C.H., et Koczynski, T.A. (1979). Dilatancy aniso-
    tropy and the response of rock to large cyclic loads.
    J. Geophys. Res., 81, 5525-5534.

STUDY OF STRATIFIED ROCK MASSES BY MEANS OF LARGE-SCALE TESTS WITH AN HYDRAULIC PRESSURE CHAMBER

Etudes d'amas rocheux stratifiés par essais in-situ avec chambre à pression hydraulique

Untersuchung geschichteter felsiger Anhäufungen mit in-situ Druckkammerversuchen

G. Oberti
Prof.Ing., ISMES President, Bergamo, Italy

L. Goffi
Prof.Ing., Polytechnic of Turin, Italy

P. P. Rossi
Dr.Ing., ISMES Geomechanical Dept., Bergamo, Italy

SYNOPSIS

This paper discusses the identification of elastic constants of stratified rock through large-scale in-situ tests with an hydraulic pressure chamber. Above all the anisotropy parameteres of a stratified mass have been stated as a function of moduli and thicknesses of the layers. The influence of the stratification is analysed with reference to the equivalent crystalline anisotropy. Experimental measures performed by a special technique in a stratified rock mass consisting of alternating layers of sandstone and marl have been analysed and interpreted in comparison with the results of the theoretical approach. Finally the most recent development of ISMES testing technique is illustrated and discussed.

RESUME

Dans ce rapport on examine le problème de la détermination des constantes élastiques de roches stratifiées au moyen d'essais in-situ à grande échelle avec chambre à pression hydraulique. On rappelle d'abord tous les paramètres d'anisotropie qui caractérisent un amas rocheux stratifié en fonction des modules élastiques et des épaisseurs de couches. L'influence de la stratification est analysée par rapport à l'anisotropie cristalline équivalente. Ensuite on analyse les résultats de mesures effectuées avec instrumentation spéciale dans un amas rocheux stratifié constitué de grès et marne. Les données expérimentales sont comparées aux résultats fournis par le calcul théorique. Enfin, on illustre et discute les plus récents développements de la technique d'essais avec chambre hydraulique mise au point par ISMES.

ZUSAMMENFASSUNG

Berichtet wird über die Untersuchung des Problems der Bestimmung der eleastischen Konstanten geschichteten Felsens durch großmaßstäbliche in-situ Versuche mit einer hydraulischen Druckkammer. Vor allem werden die Anisotropie-Parameter aufgeführt, die geschichteten Fels in Abhängigkeit der elastischen Moduln und der Schichtmächtigkeit charakterisieren. Der Einfluß der Schichtung wird unter Bezug auf die äquivalente kristalline Anisotropie analysiert. Die Meßergebnisse werden fernerhin mit Hilfe einer Spezial-Instrumentierung in einem aus Sandstein und Mergel bestehenden Gebirge überprüft. Die Versuchsergebnisse werden an Hand der sich durch theoretische Berechnung erhaltenen Ergebnisse überprüft. Abschließend werden die neuesten Entwicklungen der von ISMES ausgearbeiteten Versuchstechnik mit einer hydraulischen Druckkammer erläutert.

1. "CRYSTALLINE" ANISOTROPY

The stress-strain relationships for anisotropic elastic continua are recalled:

$$|\varepsilon| = |K| \cdot |\sigma| \qquad (1)$$

or

$$\varepsilon_x = \frac{\sigma_x}{E_1} - \frac{\nu_2}{E_2}\sigma_y - \frac{\nu_1}{E_1}\sigma_z$$

$$\varepsilon_y = -\frac{\nu_2}{E_2}\sigma_x + \frac{\sigma_y}{E_2} - \frac{\nu_2}{E_2}\sigma_z$$

$$\varepsilon_z = -\frac{\nu_1}{E_1}\sigma_x - \frac{\nu_2}{E_2}\sigma_y + \frac{\sigma_z}{E_1}$$

$$\gamma_{xz} = \frac{2(1+\nu_1)}{E_1}\tau_{xz}$$

$$\gamma_{xy} = \frac{\tau_{xy}}{G_2} \qquad \gamma_{yz} = \frac{\tau_{yz}}{G_2}$$

Applying equilibrium and compatibility relationships, and considering problems of plane elasticity (on the plane x, y) in the case of confined transversal strains $\sigma_z = 0$, if we assume

$$\sigma_x - \frac{\partial^2 \phi}{\partial y^2} \qquad \sigma_y - \frac{\partial^2 \phi}{\partial x^2} \qquad \tau_{xy} - \frac{\partial^2 \phi}{\partial x \partial y}$$

A 133

we obtain:

$$\frac{1-\nu_1^2}{E}\cdot\frac{\partial^4\phi}{\partial y^4}+\left[\frac{1}{G_2}-\frac{2\nu_2(1+\nu_1)}{E_2}\right]\frac{\partial^4\phi}{\partial x^2\partial y^2}+\left[\frac{1-n\nu_2^2}{E_2}\right]\frac{\partial^4\phi}{\partial x^4}=0 \quad (2)$$

The problem discussed is that of a circular tunnel of radius R in a indefinite mass of orthotropic elastic rock. The tunnel axis is made to coincide with axis z (Fig. 1). The tunnel is subjected to uniform pressure

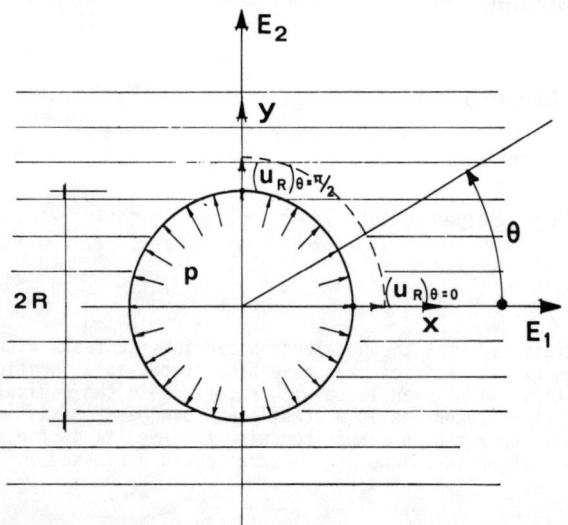

FIG. 1 Scheme of hydraulic pressure chamber test.

p. We propose the calculation of radial displacements at each point of the continuum, but in particular of the points on axes x and y, that is for $\vartheta = 0$ and $\vartheta = \pi/2$ (see Fig. 1).

The solution of the (2), from which values of radial displacement may be derived, exists in a concluded form, however it is remarkably complicated. It has therefore been preferred resorting to a numerical solution. Simple formulae may, on the other hand, be written for displacements of points at the tunnel edge, that is for r = = R

$$(u_R)_{\theta=0}=\frac{pR}{E_1}\left\{(1-\nu_1^2)\left[\sqrt{\frac{1}{1-\nu_1^2}\left[\frac{n}{m}-2n\nu_2(1+\nu_1)\right]+2\sqrt{n}\cdot\sqrt{\frac{1-n\nu_2^2}{1-\nu_1^2}}}-\right.\right.$$
$$\left.\left.-\sqrt{n}\sqrt{\frac{1-n\nu_2^2}{1-\nu_1^2}}\right]+n\nu_2(1+\nu_1)\right\}$$

$$(u_R)_{\theta=\pi/2}=\frac{pR}{E_1}\left\{\sqrt{n}\cdot\sqrt{1-n\nu_2^2}\cdot\sqrt{1-\nu_1^2}\cdot\right.$$
$$\left.\cdot\left[\sqrt{\frac{1}{1-\nu_1^2}\left[\frac{n}{m}-2n\nu_2(1+\nu_1)\right]-2\sqrt{n}\cdot\sqrt{\frac{1-n\nu_2^2}{1-\nu_1^2}}}-1\right]+\nu_2(1+\nu_1)\right\}$$

(3)

where it was assumed

$$n = E_1/E_2 \qquad m = G_2/E_2 .$$

Values of u_R have been calculated and indicated in Table I and diagram of Fig. 2 (in the dimensionless form $\dfrac{u_R}{(p\,R)/E_1}$ versus n and m).

TABLE I

n	m	$\dfrac{(u_R)_{\theta=0}}{p\,R/E_1}$	$\dfrac{(u_R)_{\theta=\pi/2}}{p\,R/E_1}$
1	0.2	1.766	1.766
	0.4	1.224	1.224
	0.6	1.006	1.006
2	0.2	2.502	3.790
	0.4	1.686	2.661
	0.6	1.348	2.193
3	0.2	3.137	5.776
	0.4	2.106	4.068
	0.6	1.672	3.347
4	0.2	3.724	7.697
	0.4	2.511	5.426
	0.6	1.993	4.456

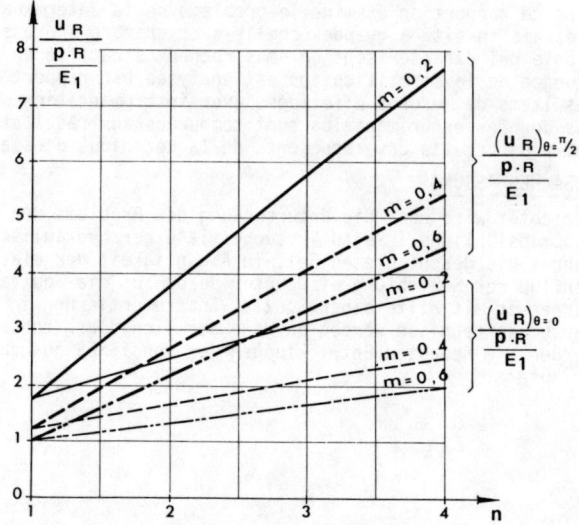

FIG. 2 Radial displacements along principal axes as a function of $n = E_1/E_2$.

The values of the displacements inside the rock mass obtained from the numerical solution of the (2) are reported in Table II still in dimensionless form. ν_1 and ν_2 were assumed equal to 0.2. Values of G_2 were supposed such as to give rise to the following ratios:

$$m = \frac{G_2}{E_2} = 0.2 - 0.4 - 0.6$$

In Fig. 3, the ratio of radial displacement u_r (at a distance r from the tunnel center) to the displacement at the tunnel edge u_R, is plotted versus r/R, using m and n as parameters.

A 134

TABLE II

$n = \dfrac{E_1}{E_2}$	$m = \dfrac{G_2}{E_2}$	$\theta = 0$				$\theta = \pi/2$			
		R	2R	3R	5R	R	2R	3R	5R
4	0.2	3.6266	2.5399	2.0133	1.4066	7.6266	4.9466	3.6199	2.3066
	0.4	2.4266	1.4453	1.0506	0.6733	5.3733	3.0279	2.0906	1.2799
	0.6	1.9199	0.9999	0.6799	0.4133	4.4133	2.2599	1.5186	0.9133
2.6666	0.2	2.8799	1.9599	1.5119	1.0199	5.0799	3.2466	2.3613	1.4999
	0.4	1.9199	1.0999	0.7786	0.4866	3.5733	1.9693	1.3506	0.8199
	0.6	1.5199	0.7599	0.5039	0.2999	2.9466	1.4666	0.9786	0.5886
2	0.2	2.4799	2.6399	1.2426	0.8133	3.7599	2.3799	1.7226	1.0933
	0.25	2.1733	1.3706	1.0093	0.6466	3.3466	2.0266	1.4399	0.8933
	0.30	1.9599	1.1813	0.8506	0.5399	2.9733	1.7199	1.2039	0.7599
	0.40	1.6666	0.9253	0.6426	0.3933	2.6399	1.4293	0.9746	0.5933
1.5384	0.4153	1.4399	0.7639	0.5213	0.3133	1.9466	1.0226	0.6926	0.4199
1.3333	0.2	2.0266	1.2799	0.9413	0.5999	2.4399	1.5133	1.0916	0.6866
	0.4	1.3733	0.7266	0.4906	0.2999	1.6933	0.8866	0.5986	0.3666
1	0.2	1.7599	1.0799	0.7693	0.4866	1.7599	1.0799	0.7706	0.4866
	0.4	1.2239	0.6199	0.4146	0.2466	1.2253	0.6199	0.4146	0.2466
	0.415	1.1999	0.5999	0.3986	0.2399	1.1999	0.5999	0.3896	0.2399

$$\boxed{\dfrac{u_R}{p\cdot R/E_1}}$$

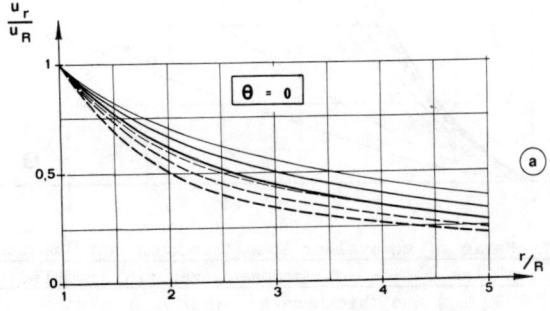

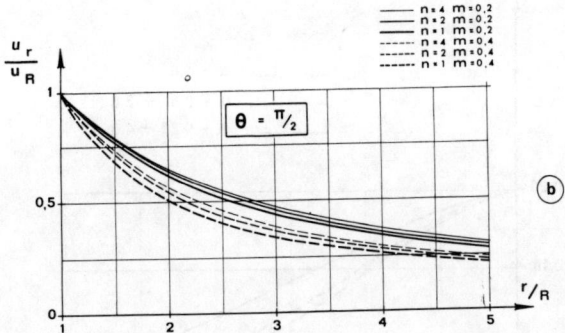

FIG. 3 Ratios between internal and surface radial displacement.

2. "STRATIFIED" ANISOTROPY

Previous conclusions may be applied to the case of stratified rock mass, the orthotropy of which results from
a regular alternation of layers of two isotropic materials A and B of Young's moduli E_A and E_B, Poisson's
ratios ν_A and ν_B and thicknesses s_A and s_B as deduced
from a previous study (Borsetto et al., 1981). This
study starts from Pinto's formulae (1966), which make
it possible to calculate the orthotropy parameters
resulting from the stratification in terms of E_1, E_2,
G_2, ν_1 and ν_2 equivalent for a "crystalline homogeneous
medium":

$$E_1 = \frac{\left[(1+\nu_A) + (1+\nu_B)\frac{s_A}{s_B}\cdot\frac{E_A}{E_B}\right]\left[(1-\nu_A)+(1-\nu_B)\frac{s_A}{s_B}\frac{E_A}{E_B}\right]\cdot E_A}{\left[(1-\nu_A^2) + (1-\nu_B^2)\frac{s_A}{s_B}\cdot\frac{E_A}{E_B}\right]\left(1+\frac{s_A}{s_B}\right)\cdot\frac{E_A}{E_B}}$$

$$E_2 = \frac{\left[(1-\nu_A)+(1-\nu_B)\frac{s_A}{s_B}\cdot\frac{E_A}{E_B}\right]\left(1+\frac{s_A}{s_B}\right)\cdot E_A}{\left[(1-\nu_A)+(1-\nu_B)\frac{s_A}{s_B}\cdot\frac{E_A}{E_B}\right]\left(\frac{s_A}{s_B}+\frac{E_A}{E_B}\right)-2\left(\nu_A-\nu_B\frac{E_A}{E_B}\right)^2\cdot\frac{s_A}{s_B}}$$

$$G_2 = \frac{1+\frac{s_A}{s_B}}{2\left[(1+\nu_A)\frac{s_A}{s_B}+(1+\nu_B)\frac{E_A}{E_B}\right]}\cdot E_A \qquad (4)$$

$$\nu_1 = \frac{(1-\nu_A^2)\nu_B + (1-\nu_B^2)\frac{s_A}{s_B}\cdot\frac{E_A}{E_B}\nu_A}{(1-\nu_A^2)+(1-\nu_B^2)\cdot\frac{s_A}{s_B}\cdot\frac{E_A}{E_B}}$$

$$\nu_2 = \frac{\left[(1-\nu_A)\nu_B+(1-\nu_B)\frac{s_A}{s_B}\cdot\nu_A\right]\left(1+\frac{s_A}{s_B}\right)\frac{E_A}{E_B}}{\left[(1-\nu_A)+(1-\nu_B)\frac{s_A}{s_B}\cdot\frac{E_A}{E_B}\right]\left(\frac{s_A}{s_B}+\frac{E_A}{E_B}\right)-2\left(\nu_A-\nu_B\frac{E_A}{E_B}\nu\right)^2\frac{s_A}{s_B}}$$

If then we assume

$$a = \frac{s_A}{s_B} \qquad E = \frac{E_A}{E_B}$$

and $\nu_A = \nu_B = \nu$,

the aforesaid expressions are simplified in the form

A 135

$$E_1 = \frac{1 + a\,e}{(1+a)\cdot e}\, E_A$$

$$E_2 = \frac{(1 + a\,e)(1+a)}{(1+ae)(a+e) - 2\nu^2\,\frac{(1-e)^2 a}{1-\nu}}\, E_A$$

$$G_2 = \frac{1+a}{2(1+\nu)(a+e)}\, E_A$$

$$\nu_1 = \nu$$

$$\nu_2 = \frac{(1+a)^2\,\nu\cdot e}{(1+ae)(a+e) - \frac{2\nu^2}{1-\nu}(1-e)^2\cdot a}$$

Ratios n and m result therefore from expressions:

$$n = \frac{(1+ae)(a+e) - \frac{2\nu^2}{1-\nu}(1-e)^2 a}{e\,(1+a)^2}$$

$$m = \frac{(a+e)(1+ae) - \frac{2\nu^2}{1-\nu}(1-e)^2 a}{2(1+\nu)(a+e)(1+ae)}$$

and are plotted together with ν_2 versus a and e, in Figs. 4, 5, 6 (*).

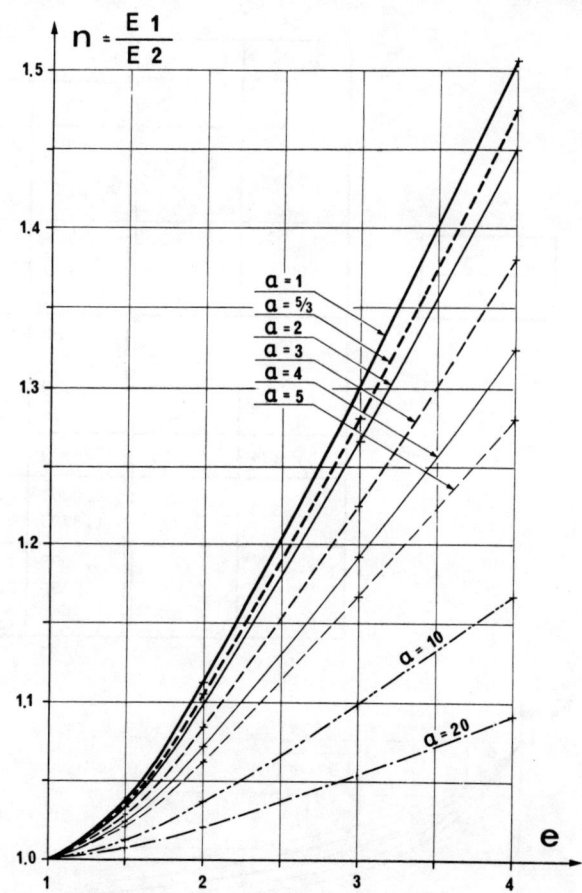

FIG. 4 Ratio of equivalent Young's moduli as a function of the ratios of component material moduli (e= $= E_A/E_B$) and thicknesses (a= s_A/s_B).

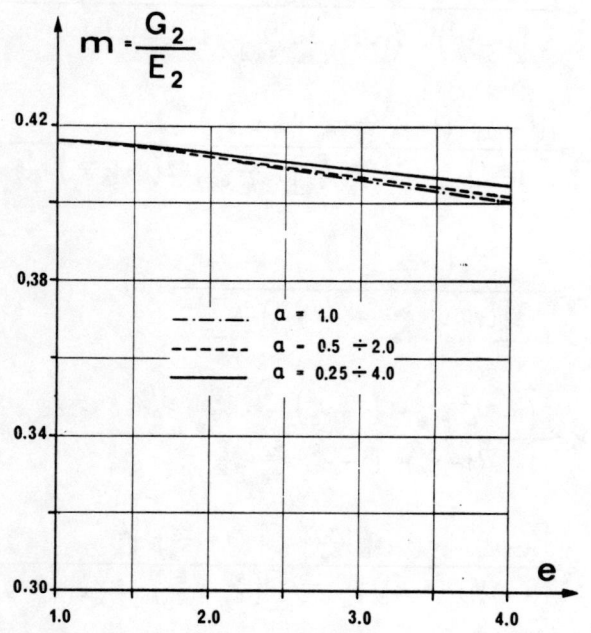

FIG. 5 Ratio between equivalent moduli G_2 and E_2 as a function of e and a.

(*) For semplicity's sake, $e \geqslant 1$ was assumed in the diagrams, that is $E_A \geqslant E_B$; curves for individual values of a are also valid for inverse values, that is for 1/a, given the form of expressions for n and m.

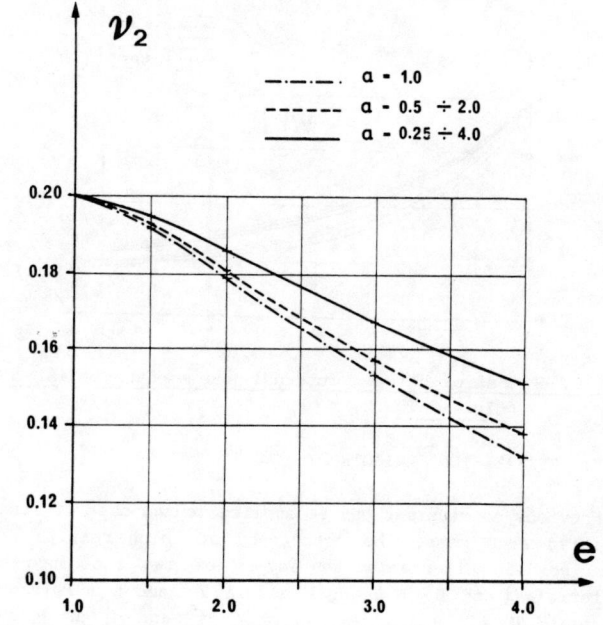

FIG. 6 Equivalent Poisson's ratio ν_2 as a function of e and a.

A 136

As already pointed out, also with high values of the ratio e between moduli E_A and E_B of stratification materials (up to 4) we do not succeed in obtaining, for the ratio n (between equivalent Young's moduli E_1 and E_2) values higher than 1.5. Moreover, the orthotropy deriving from stratification of isotropic materials actually supposes a link between G_2 and E_2. Indeed, as it also clearly appears from the diagram of Fig. 5, $m = G_2/E_2$ is slightly affected by $a = s_A/s_B$ and $e = E_A/E_B$, practically ranging from 0.40 to 0.416. This property also results from the analysis of the formula m which may be written

$$m = \frac{1}{2(1+\nu)} - \frac{\nu^2 (1-e)^2 a}{(1-\nu^2)(a+e)(1+ae)}$$

where the second term is negligible compared with the first one. Therefore G_2 would simply be expressed by

$$G_2 \cong \frac{E_2}{2(1+\nu)}$$

which has the same form of the relationship between G and E for isotropic continua.

The link between G_2 and E_2 may allow to obtain the elastic constants from the simple measure of radial displacements for $\vartheta = 0$ and $\vartheta = \pi/2$, without supplementary tests. The three principal unknowns E_1, E_2 and G_2 are derived from these two measurements, as the third equation is obtained from the aforesaid: $G_2 \cong E_2/\left[2\,(1+\nu)\right]$.

3. SOME EXAMPLES OF F.E.M. CALCULATIONS ON STRATIFIED MEDIA

A study on the actual stratification influence had been undertaken (Borsetto et al., 1981) for a wide case history summarized in Fig. 7. The two cases $e = E_A/E_B = 2$ and 4 had moreover been assumed. Then diametral displacements due to water pressure were calculated for $\vartheta = 0$ and $\vartheta = \pi/2$; the results were compared with those consequent to the assumption of equivalent "crystalline" orthotropy. Results are shown in Table III, and in diagram of Fig. 8. In particular, Fig. 8 shows in abscissa the values of radial displacements, computed in the various assumptions of stratifications, and in ordinate the corresponding values consequent from the assumption of "crystalline" anisotropy. The thickening of points close to a straight line at 45° shows the good correlation between the two assumptions, at least in the 9 cases considered. Therefore we may consider correct the interpretation of experimental results in hydraulic chamber in stratified masses by "crystalline" orthotropy formulae, which therefore may give the equivalent Young's moduli of the rock mass.

4. AN EXAMPLE OF INTERPRETATION OF EXPERIMENTAL RESULTS OBTAINED BY HYDRAULIC PRESSURE CHAMBER

The formulae discussed in the previous sections were applied to the processing of the results of tests carried out at the foundation rock of Ridracoli Dam, located in the province of Forlì. The dam is an arch-gravity structure, max height 120 m, crest length about 400 m; it closes off the River Bidente, forming a reservoir

of about 35 million m^3 capacity.

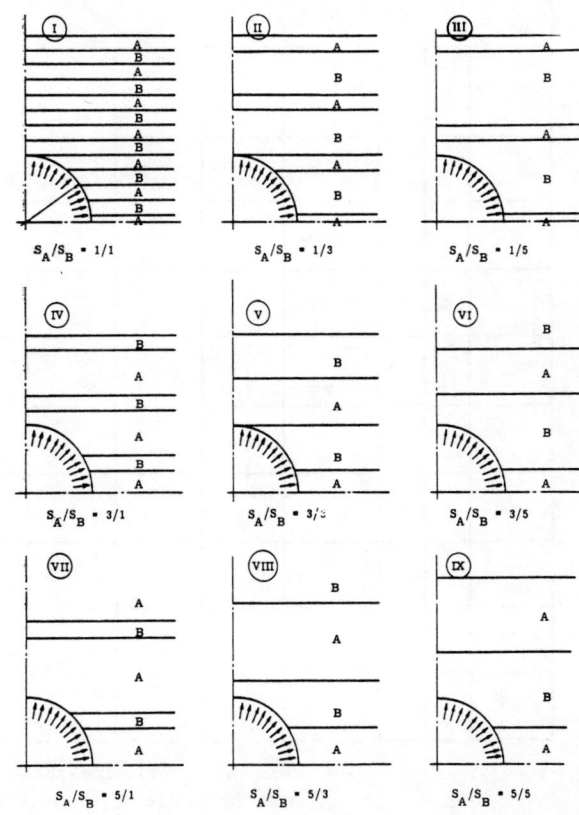

FIG. 7   Alternating schemes of stratifications concerned with displacement computation by F.E.M.

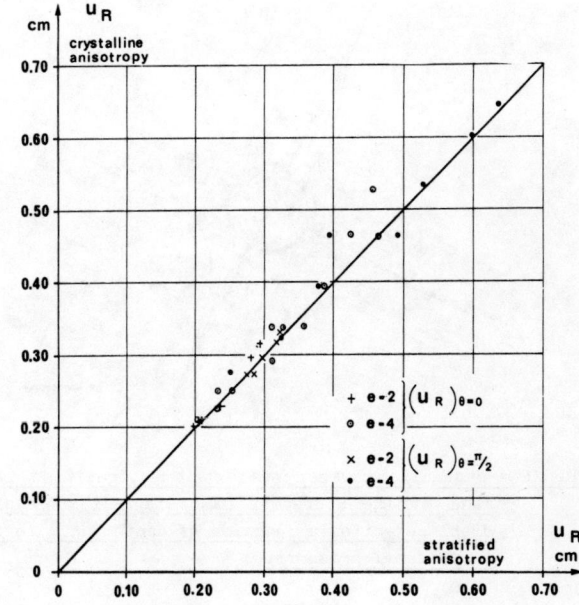

FIG. 8   Comparison between radial displacements in the assumption of crystalline and stratified anisotropy.

A 137

TABLE III (*)

Scheme	S_A	S_B	$a = \dfrac{S_A}{S_B}$	$e = \dfrac{E_A}{E_B}$	$\dfrac{E_1}{\text{Kg cm}^{-2}}$	$\dfrac{E_2}{\text{Kg cm}^{-2}}$	n	Hydraulic chamber stratified anisotropy		Hydraulic chamber crystalline anisotropy	
								$\dfrac{(u_R)_{\theta=0}}{\text{cm}}$	$\dfrac{(u_R)_{\theta=\pi/2}}{\text{cm}}$	$\dfrac{(u_R)_{\theta=0}}{\text{cm}}$	$\dfrac{(u_R)_{\theta=\pi/2}}{\text{cm}}$
I	1	1	1	2	75000	67615	1.1125	0.245	0.273	0.250	0.274
				4	62500	41493	1.506	0.392	0.463	0.3384	0.4644
II	1	3	1/3	2	62499	57636	1.0843	0.280	0.318	0.297	0.3174
				4	43749	31709	1.3791	0.425	0.592	0.4663	0.6021
III	1	5	1/5	2	58333	54901	1.0625	0.293	0.326	0.3159	0.3320
				4	37500	29268	1.2812	0.457	0.617	0.5825	0.665
IV	3	1	3	2	87500	80691	1.08438	0.204	0.229	0.2121	0.2267
				4	81250	58900	1.3795	0.235	0.326	0.2510	0.3241
V	3	3	1	2	75000	67415	1.1125	0.236	0.254	0.250	0.2760
				4	62500	41493	1.506	0.312	0.395	0.3384	0.4666
VI	3	5	3/5	2	68750	62190	1.1056	0.256	0.297	0.2117	0.2952
				4	53125	36026	1.4746	0.386	0.529	0.3951	0.5352
VII	5	1	5	2	91666	86274	1.0625	0.198	0.205	0.2010	0.2113
				4	87500	68292	1.2812	0.226	0.251	0.2265	0.2164
VIII	5	3	5/3	2	81249	73498	1.1056	0.235	0.249	0.2299	0.2498
				4	71874	48741	1.47461	0.312	0.382	0.2920	0.3955
IX	5	5	1	2	75000	67415	1.1125	0.252	0.285	0.250	0.2740
				4	62500	41493	1.506	0.359	0.493	0.3384	0.4644

(*) The values have been calculated following these assumptions: Radius R = 150 cm, Young modulus of material A: $E_A = 100,000$ Kg cm^{-2}, inner hydraulic pressure $p_O = 100,000$ Kg cm^{-2}.

The foundation rock mass consists of alternating sandstone and marl, with layers diping about 28° (Fig. 9).

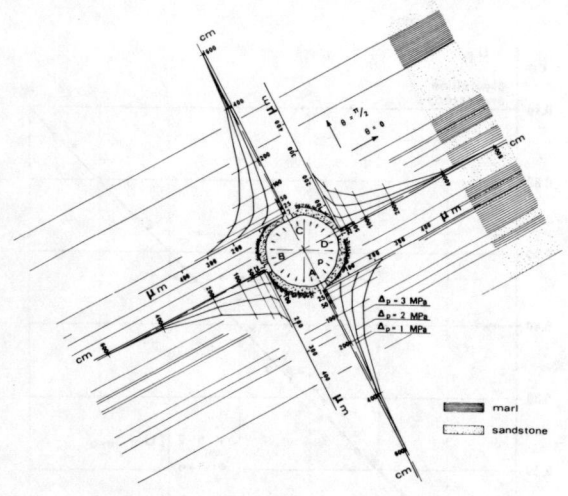

FIG. 9 Hydraulic chamber test on the stratified rock mass at the Ridracoli Dam. Diagrams of radial displacement as a function of depth for 3 levels of internal pressure.

Since layers are 100 to 150 cm thick, large-scale tests were carried out to determine deformability characteristics, using the hydraulic pressure chamber technique. Tests were performed inside two 3 m-diam tunnels, their axis being parallel to stratification planes on the two banks of the dam, at its foundation. The procedures of the tests are described in paragraph 5.

In the comparison of the theoretical and experimental data, we consider, for statical purposes, the radius of the tunnel equal to 157 cm, 7 cm being the depth of the point related to the most superficial measure. In fact we have measured the radial displacements (along the axes perpendicular and parallel to the stratification plane) at different depths as relative to a reference point located at 750 cm from the center of the tunnel (at a depth of two times the diameter of the chamber).

The radial displacements have been measured at the distances of 157 - 175 - 200 - 250 - 350 - 550 cm from the center of the tunnel; the values are the reversible displacements at three different levels of inner pressure (1 - 2 - 3 MPa) starting from a ground pressure of 1.2 MPa (Table IV and Fig. 9). These pressures are relative to the 150 cm radius; therefore, if referred to the "static" radius of 157 cm, the pressures become, according to $\Delta p' = (150/157)\, \Delta p$, 0.955 - 1.911 - 2.866 MPa.

The theoretical and experimental values of the radial displacements have been elaborated considering that the effective radius of the tunnel is 157 cm and the reference point of the displacements is located at 750 cm from the center of the tunnel (Tables V and VI). Experimental values (**) are plotted in Figs. 10 and (**) From average values of the two opposite measurements (A+C)/2 and (B+D)/2 according to Table IV and Fig.9.

TABLE IV   Reversible radial displacements (μm) at the points A,B,C,D (see Fig.9)

Ground pressure 1.2 MPa

Δp (MPa)	depth d (cm)	A	B	C	D
3	7	220	188	295	208
	25	182	149	214	138
	50	146	138	166	115
	100	101	112	120	79
	200	52	61	70	53
	400	7	10	11	10
2	7	129	112	176	125
	25	107	88	125	83
	50	84	81	94	68
	100	56	63	67	48
	200	26	34	36	25
	400	2	5	5	5
1	7	59	48	84	51
	25	50	37	55	35
	50	39	34	38	29
	100	25	26	25	21
	200	12	13	12	10
	400	2	0	0	0

TABLE V $\boxed{\vartheta = 0}$

$n = \dfrac{E_1}{E_2}$	$m = \dfrac{G_2}{E_2}$	$\dfrac{u_r - u_{750}}{p\,\frac{R}{E_1}}$			$\dfrac{u_r - u_{750}}{u_{R'} - u_{750}}$		
		R'	2R'	3R'	R'	2R'	3R'
4	0.2	2.1541	1.0674	0.5408	1	0.4955	0.2510
	0.4	1.7121	0.7308	0.3361	1	0.4269	0.1963
	0.6	1.4775	0.5575	0.2375	1	0.3773	0.1607
2.6666	0.2	1.8065	0.8865	0.4385	1	0.4907	0.2427
	0.4	1.4014	0.5814	0.2601	1	0.4149	0.1856
	0.6	1.1977	0.4377	0.1817	1	0.3655	0.1517
2	0.2	1.6198	0.7798	0.3825	1	0.4814	0.2361
	0.25	1.4871	0.6844	0.3231	1	0.4602	0.2173
	0.3	1.2861	0.6075	0.2768	1	0.4383	0.1997
	0.4	1.2461	0.5048	0.2221	1	0.4051	0.1782
1.5384	0.4153	1.1039	0.4279	0.1853	1	0.3876	0.1679
1.3333	0.2	1.2895	0.6428	0.3042	1	0.4626	0.2189
	0.4	1.0526	0.4059	0.1699	1	0.3856	0.1614
1	0.2	1.2425	0.5625	0.2519	1	0.4527	0.2027
	0.4	0.9589	0.3549	0.1496	1	0.3701	0.1560
	0.415	0.9427	0.3427	0.1414	1	0.3635	0.1500

TABLE VI $\boxed{\vartheta = \pi/2}$

$n = \dfrac{E_1}{E_2}$	$m = \dfrac{G_2}{E_2}$	$\dfrac{u_r - u_{750}}{\Delta p\,\frac{R}{E_1}}$			$\dfrac{u_r - u_{750}}{u_{R'} - u_{750}}$		
		R'	2R'	3R'	R'	2R'	3R'
4	0.2	5.1766	2.4966	1.1699	1	0.4823	0.2260
	0.4	4.0047	1.6593	0.7220	1	0.4144	0.1803
	0.6	3.4338	1.2804	0.5391	1	0.3729	0.1570
2.6666	0.2	3.4859	1.6526	0.7673	1	0.4741	0.2201
	0.4	2.6959	1.0919	0.4732	1	0.4050	0.1755
	0.6	2.3171	0.8371	0.3491	1	0.3613	0.1507
2	0.2	2.5979	1.2179	0.5606	1	0.4688	0.2158
	0.25	2.3936	1.0736	0.4869	1	0.4485	0.2034
	0.3	2.1649	0.9115	0.3955	1	0.4211	0.1827
	0.4	2.0049	0.7943	0.3396	1	0.3962	0.1694
1.5384	0.4153	1.4969	0.5723	0.2429	1	0.3823	0.1623
1.3333	0.2	1.7091	0.7825	0.3608	1	0.4579	0.2111
	0.4	1.3014	0.4947	0.2067	1	0.3801	0.1588
1	0.2	1.2425	0.5625	0.2519	1	0.4527	0.2027
	0.4	0.9589	0.3549	0.1496	1	0.3701	0.1560
	0.415	0.9427	0.3427	0.1414	1	0.3635	0.1500

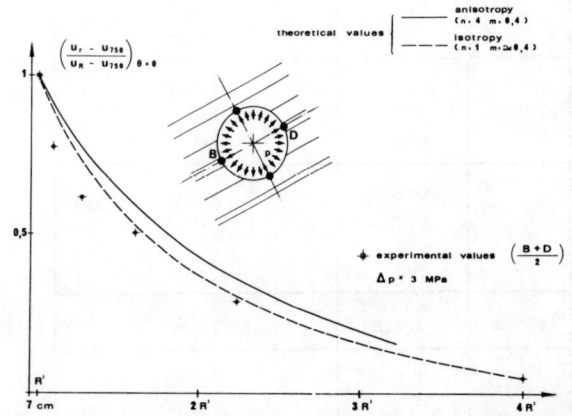

FIG. 10   Comparison between experimental and theoretical values (ratios) of radial displacements along the diameter parallel to the stratification.

11 versus the distance from the tunnel edge as ratios to the relative deformations measured at 7 cm depth; on these diagrams have been plotted also the corresponding theoretical curves for the case of the isotropy and anisotropy (n = 4, m = 0.4).

A good agreement can be observed between the decreasing of experimental and theoretical values of the above ratios. This fact confirms the reliability of the experimental results.

We can interpret the measurements for Δp = 3 MPa referring to the points at 7 cm depth ($u_{R'}$ for R'= 157 cm).

We acknowledge (Table IV) an average radial displacement perpendicular to the stratification (A+C)/2= (220+295)/2= 257 μ m and parallel to the stratification (B+D)/2 = (188+208)/2 = 198 μ m.

The ratio between these displacements gives 257/198 =1.30.

We now compare that value with the ratios between the relative displacements (in the two perpendicular directions) of Table VII.

For m = 0.4 we obtain by interpolation n $\simeq$ 1.43.

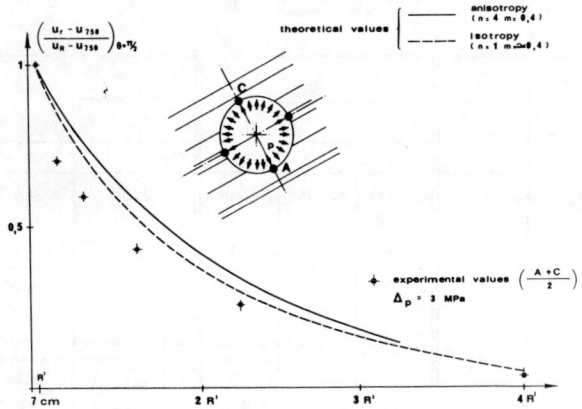

FIG. 11 Comparison between experimental and theoretical values (ratios) of radial displacements along the diameter normal to the stratification.

TABLE VII

$n = \dfrac{E_1}{E_2}$	$m = \dfrac{G_2}{E_2}$	$\dfrac{(u_{R'} - u_{750})\ \theta = \pi/2}{(u_{R'} - u_{750})\ \theta = 0}$
4	0.2	2.403
	0.4	2.340
	0.6	2.324
2.6666	0.2	1.929
	0.4	1.924
	0.6	1.934
2	0.2	1.603
	0.25	1.609
	0.3	1.562
	0.4	1.609
1.5384	0.4153	1.356
1.3333	0.2	1.230
	0.4	1.236
1	0.2	1
	0.4	1
	0.415	1

Always by interpolations, in Table V we find a value

$$\frac{(u_{R'} - u_{750})\ \vartheta = 0}{\Delta p\ \dfrac{R}{E_1}} \simeq 1.08$$

and therefore:

$$E_1 = \frac{1.08 \times \Delta p \times R}{(u_{R'} - u_{750})\ \vartheta = 0} = \frac{1.08 \times 3 \times 150}{0.0198} = 24545 \text{ MPa}$$

$$E_2 = \frac{E_1}{n} = 17165 \text{ MPa}$$

$$G_2 = m\ E_2 = 6866 \text{ MPa} \quad (m = 0.4)$$

5. RECENT IMPROVEMENTS OF HYDRAULIC PRESSURE CHAMBER TESTING TECHNIQUE

The testing technique recently set up by ISMES has been successfully experimented in this investigation. This technique makes it possible to simplify the problem of load application to the rock surface (Fig. 12).

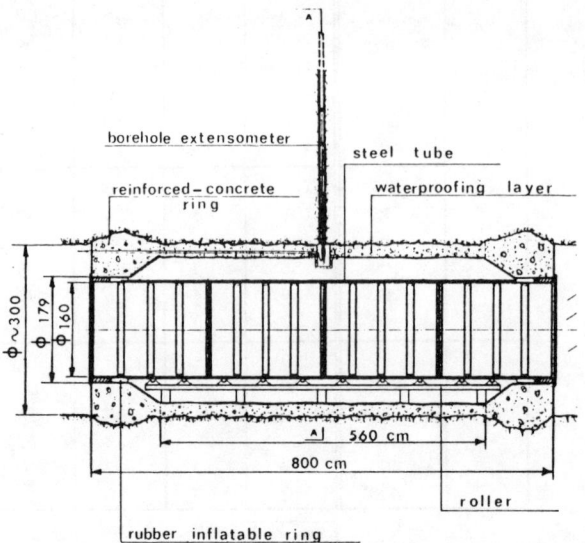

FIG. 12 Scheme of hydraulic chamber showing loading and measuring equipment.

Previous procedure planned the closing of the experimental tunnel by two reinforced concrete plugs and the casting of a concrete lining on the lateral surface of the tunnel. Groutings at subsequent stages ensured the bonding between the concrete lining and the underlying rock. The construction of these expensive works did not allow a frequent use of the hydraulic chamber; therefore, an effort was made to simplify the testing technique.

Now, only two concrete rings at the ends of the chamber are built and the lining of the tunnel lateral surface is realized by a layer of 5 to 10 cm thick spritz-beton, over which a waterproofing lining made with PVC sheets is applied (Fig. 13). The use of spritz-beton considerably reduces the preparation time of pressure application surfaces, making it possible also to exclude the grouting which, by using the previous technique, was necessary to ensure the bond between the concrete lining and the underlying rock. A large-diam central pipe (∅ = 1600 mm), made of composite elements and standing on a base roller, supports the hydrostatic pressure, while the water sealing is ensured by two inflatable rubber rings, inserted between the steel pipe extrados surface and the end rings. The tunnel stretch which is generally involved in the test is 8 m long. Radial holes are bored at the median section; their depth is twice the tunnel diameter. Special multibase strain gauges inside the boreholes make it possible to measure the rock mass displacements at various depths from the loading surface.

The new technique, besides considerably reducing the testing time and costs up to 40%, compared with the

FIG. 13 The inside of hydraulic chamber during the installation of radial borehole extensometers

previous technique, presents additional non-negligible advantages, since it may be used with various-diam tunnels and does not stop the transit through the tunnel during the preparation and performance of the test.

ACKNOWLEDGEMENTS

The authors wish to thank Dr. R. Di Bacco of Mathematical Dept. of ISMES for the valuable collaboration.

REFERENCES

Borsetto, M., Goffi, L., Rossi, P.P. (1981)
    Studio di ammassi rocciosi stratificati riferito a prove di deformabilità in cunicolo. ISMES Bull.no.151

Goffi, L. (1972)
    La determinazione delle costanti elastiche di ammassi rocciosi ortotropi mediante prove in sito. L'Energia Elettrica no. 10

Goffi, L., Rossi, P.P., Borsetto, M. (1978)
    Interpretazione di tecniche sperimentali per la misura dei parametri di deformabilità di ammassi rocciosi. Atti dell' Istituto di Tecnica delle Costruzioni del Politecnico di Torino, no. 112-ISMES Bull. no. 113

Lekhnitskii, S.G. (1963)
    Theory of elasticity of an anistropic elastic body. Holden Day

Oberti, G., Carabelli, E., Goffi, L., Rossi, P.P. (1979)
    Study of an orthotropic rock mass: experimental techniques, comparative analysis of results. Proceedings IV International Congress of ISRM, Montreux

Oberti, G., Goffi, L., Rebaudi, A. (1970)
    Comportement statique des massifs rocheux (calcaires) dans la réalisation de grands ouvrages souterrains. Proceedings II International Congress of ISRM, Beograd

Pinto, J.L. (1966)
    Stresses and strains in an anisotropic-orthotropic body. Proceedings of the I Congress of ISRM, Lisboa

RECENT ADVANCES IN THE INTERPRETATION OF THE FLAT JACK TEST

Développements récents dans l'interprétation des essais au vérin plat

Neuere Fortschritte in der Deutung der Prüfergebnisse mit Druckkissen

M. Borsetto
ISMES, Instituto Sperimentale Modelle Strutture, Bergamo, Italy

G. Giuseppetti
ENEL, (Italian State Electricity Board), Hydraulic and Structural Research Centre, Milan

G. Manfredini
ENEL, Construction Division, Rome

SYNOPSIS

The use of the flat jack test to identify the residual strength of rock masses is discussed. Test interpretation proced-ures developed by means of numerical models of the slot are presented. The same procedures are applicable to the determ-ination of original stresses as an alternative to the "cancellation pressure" method recommended by IRSM Suggested Stan-dards. The use of the flat jack for deformability measurements is also analysed and the limitations of the conventional interpretation formula are outlined.

RESUME

On discute l'emploi des essais au vérin plat pour préciser le résistance résiduelle des amas rocheux. On présente quel-ques procédures d'interprétation de ces essais, développées au moyen de modèles numériques de la coupe. Les même procé-dures peuvent être appliquées pour la détermination des contraintes naturelles comme alternative à la méthode par pres-sion de compensation recommandée par l'ISRM Suggested Standards. On analyse aussi l'emploi des essais au vérin plat pour les mesures de déformabilité, et on souligne les limitations de la formule d'interprétation conventionelle.

ZUSAMMENFASSUNG

Es wird die Benutzung von Druckkissen zur Schätzung der Restfestigkeit von Felsen behandelt, und Verfahren zur Deutung der Prüfergebnisse mit Hilfe rechnerischer Modelle des Schlitzes werden dargestellt. Dieselben Verfahren sind auch zur Bestimmung des ursprünglichen Spannungszustands anwendbar, als Alternative zum "cancellation pressure" Verfahren, das vom IRSM Suggested Standards empfohlen wurde. Es wird auch die Benutzung von Druckkissen für Verformungsmessungen er-örtert, und die Begrenzungen der konventionellen Deutungsformulen werden unterstrichen.

1.INTRODUCTION

The conventional aim of the flat jack test regards rock deformability and original stress determination.

Recently its use in connection to residual strength determination have been proposed (Borsetto, 1980). At the moment the technique appears to be accepted among the set of tests and procedures adopted by ENEL (Ita-lian National Electricity Board) for the geomechanical characterization of rock masses (Dolcetta, 1982).

The positive results obtained by designers of the Edolo underground power station, who first adopted the methodology on the field (Forzano et al., 1980), have motivated the Direction of Construction (DCO) and the Direction of Research (DSR) of ENEL to promote a rese-arch project on the subject.

The activity, carried on at ISMES, is currently inten-ded to improve the theoretical aspects of the interpre-tation and the experimental methodologies.

While it is clear that these two aspects should never be thought undipendent, this paper will mainly be confi-ned on the first as, at the moment of writing, some experimental details are under evolution.

Strenght characteristics of rock masses are very im-portant for the design of underground structures as they heavily condition upon the safety against colla-pse, the deformation development and the interaction with stabilization structures.

The more promising approach to their determination is the direct observation of the behaviour of ancillary openings or experimental tunnels.

Usually displacements are monitored and backanalysis procedures are used in order to identify the strength parameters.

The principal shortcoming of this approach, alone, is that one has to rely upon a preventive determination of many parameters and he still must handle more than one free parameter, consequently a broad variability field for the parameters to be identified is found.

The identification of some strength parameters from the stress measurements is intended to reduce this uncertainty.

During the setting up of a reliable interpretation model of the flat jack test for the afore-mentioned purpose, some attentions have been paied to its conventional uses.

Some limitations of the common interpretation methods, as those of IRSM Suggested Standards, have been found.

2. PRELIMINAR CONSIDERATIONS

In the following some geomechanical aspects of the context where the test is performed and some definitions on what one whould measure are discussed. This provided the guideline for the development of a model of the test and, at the same time, it made clear the actual simplifications one had to accept in order to resort to a conceptually and mathematically treatable matter.

Four main reference states for the variables of the problem are considered:

- undisturbed state

- opening excavation

- slot cutting

- jack pressurization.

2.1 Undisturbed state

The stress state is denoted by σ_o, a six component tensor, which varies from point to point into the rock mass subjected to equilibrium requirements and to boundary conditions. It is assumed to be a continuous function of the position, provided that a sufficiently large observation scale is considered. It is also assumed that this scale is appropriate for describing the behaviour of the rock mass trough the subsequent reference states.

2.2 Opening excavation

A tunnel is usually driven up to the location where the test has to be executed. It is assumed that the excavation of the experimental section is performed under controlled conditions in order to avoid unnecessary disturbance of the rock.

The excavation changes the strain-stress state of the rock mass: this new state will be denoted by $\sigma_1(P)$, $\varepsilon_1(P)$ assuming that reference is made when gross rheological phenomena, are exausted - or at least are so slow to be negligible while the test is executed.

It is clear that any hope to resort to the initial stress σ_o relies upon the capability to describe the relevant phenomena which produce the transformation to the σ_1 state.

If the rock mass behaviour during the excavation is appropriately described by a linear elastic homogeneous isotropic model, modified stresses can be expressed by

$$\sigma_1 = F_1 (G_t, \sigma_o, v) \qquad (1)$$

where G_t denotes the geometry and v the Poisson's ratio. The form of F_1 may be found in textbooks for simple geometries, otherwise it could be obtained analytically

or numerically. The dipendence from the material properties is in this case weak.

Furthermore if the rock mass is isotropic or the axis of the tunnel has an appropriate direction toward the anisotropy principal plane (e.g. for an orthotropic body) some components of the σ_1 stress state uncouple from the original ones so that their determination is feasible with few measurements.

More complex dependence on the characteristics of the rock mass may be found in case of anisotropic rocks (more material constant must be estimated) expecially if the anisotropy is due to some medium-scale dishomogenuities (joints, interlayers, ecc.).

Still a more complicate situation arises if a non-linear stress or strain dependent behaviour takes place.

The setting up of proper F_1 function will in that case depend on the suitable choice of a constitutive stress-strain law, and of related parameters.

In many situations, the most meaningful constitutive model is an elastoplastic one, since the strength of the rock is exceeded nearby the tunnel.

Assuming a perfectly elastoplastic behaviour the Kastner's solution for a circular unlined tunnel gives (Fig. 1) :

$$\sigma_t = c \; cotg\varphi \; (N \; (\frac{r}{a})^{N-1} -1) \; ; \; r < R \qquad (2)$$

with $\qquad N = (1 + sin\varphi)/(1 - sin\varphi)$

and $\qquad R = a \left[(1 - sin\varphi)(1 + S /(c \; cotg\varphi)) \right]^{1/(N-1)}$

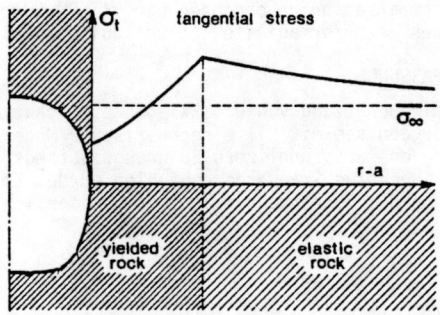

$$\sigma_t = \sigma_t (C_r, \varphi_r, \gamma, geometry)$$

Fig. 1 **Distribution of tangential stresses around a tunnel. Elastoplastic behaviour.**

These results show that the stress at the wall are indipendent from the (isotropic) original stress value S, which only contributes to the width R of the yielded zone.

In this case, it would be impossible to evaluate σ_o from the knowledge of σ_1. This point has been stressed out by Comes in the late sixties (Comes et al., 1969).

The stresses at the wall are function of the tunnel geometry and of the strength values only. This suggests that the stress σ_1 is a measure of the rock strength, that is, the knowledge of σ_1 could give information about the strength. This point will be discussed in section 3.

The change of stress condition may also contribute to the local change of the deformability.

Some non-linear elastic behaviour is some time quite evident at low compressive stresses due to the opening-closing of fissures and microfissures.

Furthermore, if rock undergoes plastic deformations its elastic deformability (that is, the deformability which is evident in a unloading path) can differ from that of the undisturbed state. Lombardi's model for the yielding tunnel, for example, allows for a "residual value" of the deformability (Amberg et al., 1974); a different approach is suggested in (Ribacchi et al., 1977), after the theoretically more consistent model for the elasto-plastic coupling proposed in (Huekel et al., 1977). In conclusion attention must be paid to the fact that the measurement is made in a superficial layer and that its deformability is not always the deformability one is looking for.

At the beginning of this section, mention has been made of deformation ε_1. While it is not strictly connected with the test itself the knowledge of ε_1, at least at level of tunnel convergences, may be very important in order to establish the appropriate constitutive model to be applied at the rock mass during the excavation. This is, in turn, foundamental to establish the meaning of subsequent flat jack measurements.

In addition, it is clear that the tunnel excavation can be regarded as a mere inconvenience for original stress and deformability determinations, while it is a substantial part of the test itself for strength determination.

2.3 Slot cutting

Just before the cutting, some measuring points are marked on the rock. Usually, couples of points simmetrically posed in relation to the slot are used either on the surface of the rock or inside the mass. In the latter case steel pins are usually grouted or mechanically coupled at the bottom of boreholes. The slot is obtained by sawing or drilling. The cutting causes the relief of the normal, $\sigma_n(p)$, and shear components of the σ_1 state of stress acting on the slot surface.

This, in turn, brings to a displacement of the reference points σ_i^c (i being the index of the measuring points). The displacements are function of the position P_i, of the slot and tunnel geometries (G_s, G_t) of the rock mass behaviour during the cutting, and, finally, of the primitive stresses acting on the virtual surface of the slot.

2.4 Jack pressurization

During the pressurization of the jack the reference points will undergo a displacement δ_i^c.

These displacements are function of (P_i), (G_s), (G_t), of the rockmass behaviour during reloading and of the pressure distribution applied by the jack (which may be assumed to be known with good accuracy).

3.MEASURE OF THE RESIDUAL STRENGTH

The residual strength may be defined, for technical purposes, as the strength of the rock after a suitable deformation has occured, and which will remain nearly constant for further deformations of pratical interest.

This implies that, if the strength of the rock is small compared with the original stresses, in an appreciable zone around the experimental tunnel the strength (residual) is not largely dependent on deformation. Thus the stresses in the yield zone around a circular tunnel will only depend on the strength, as in the equation (2).

This observation is valid, provided that the limit equilibrium holds, also for other situations. From a pratical point of view the calculation of stresses in easily performed by simple methods in parametric form which greatly simplifies the backanalysis (Hill, 1950; Borsetto et al. 1978).

Besides, the indipendence from others geomechanical parameters is attractive also from the theoretical point of view because it makes possible an independent extimate.

The residual strenght condition and static determination of stresses (which involve high original stresses, low strength, appropriate geometry and excavation procedures) are the most favorable situation for "strength measurement".

4.SETTING UP OF AN INTERPRETATION MODEL

The identification of strength and that of original stresses are obliviously excluding each other, meanwhile they require that the stress $\sigma_1(p)$, acting on the virtual surface of the slot, is found. In the following the discussion will be confined to some simple forms of the $\sigma_1(p)$ function. The application to some more general situations is easily worked out.

So it is assumed that the slot is cut into the wall of a circular tunnel, in a homogeneous isotropic medium, and that the original stresses oriented in a way that no shear is induced on the future slot surface.

If during the tunnel excavation the rock mass behaviour is elastic, the normal stress on a horizontal slot is given by

$$\sigma_n = \frac{1}{2}(\sigma_v + \sigma_h)(1 + \frac{a^2}{r^2}) + \frac{1}{2}(\sigma_v - \sigma_h)(1 + 3\frac{a^4}{r^4}) \quad (3)$$

where σ_v and σ_h are the vertical and horizontal original stresses. If residual conditions are attained the normal stress is given by:

$$\sigma_n = c_r \, \text{cotg} \varphi_r \, (N_r (\frac{r}{a})^{N_r-1} -1) \quad ; \quad r < R \quad (4)$$

It is worth noticing that if the slot depth (1) is not negligible versus tunnel radius (a), the stress along the virtual cut is quite sensitive to the depth (Fig. 2 and Fig. 3). Usually the recorded displacements are taken as measure of the mean stress σ_n and the stress distribution function, f_n, is neglected ($\sigma_n = \bar{\sigma}_n . f_n$). When by rising the jack pressure, the initial position is restored (avoiding correction factors for an incomplete efficiency of the jack and averaging) it is assumed that:

L

$$\bar{\sigma}_n = \bar{p} \qquad (5)$$

being $\bar{p}$ called "cancellation pressure".

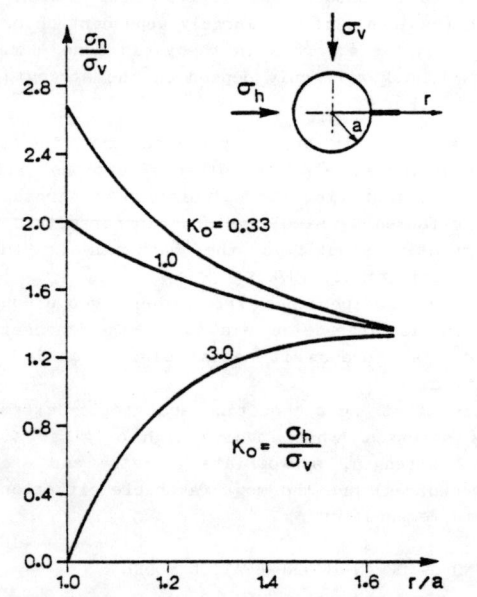

Fig. 2   Normal stress distribution along the slot. Elastic case.

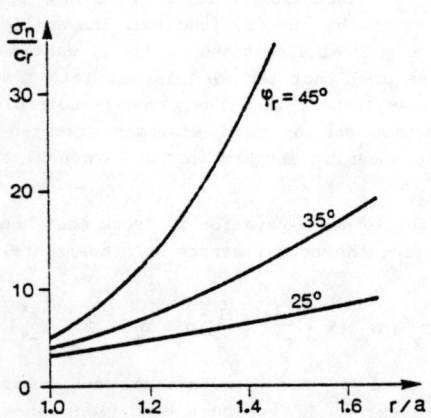

Fig. 3   Normal stress distribution along the slot. Residual conditions.

This eliminates the need to know the actual structural behaviour of the slot. However it is a common experience that various reference points do not come back to their initial position at the same pressure.

With regard to the deformability test the "structure" cannot be disregarded and a simplified structural model is suggested by IRSM recomandations, which derive from Alexander's formula (Alexander, 1960, Jaeger, 1969) for the displacement of a slot of infinite depth subjected to uniform pressure.

A quite general expression for the displacement of the ith reference point during the slot cutting (Fig. 4) can be written as:

$$\delta_i^c = \int_o^1 \sigma_n(x) \ M_i(x) \ dx \qquad (6)$$

where the displacement is found as the contribution of the stress acting on infinitesimal trasversal slices of the slot. The function $M_i(x)$ is an influence function pertinent to the ith measuring point. It depends on the installation characteristics: dimension and shape of the tunnel and of the slot, rock deformability, position of the reference point.

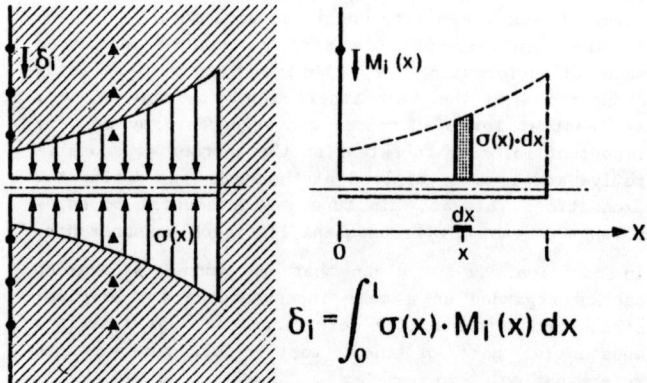

Fig. 4   Calculation of the displacement by means of influence functions.

Similary, the displacement occurring during the pressurization of an ideally efficient flat jack may be written as:

$$\delta_i^p = p \int_o^1 M_i(x) \ dx \qquad (7)$$

According to (6) and (7) the cancellation pressure results:

$$\bar{p}_i = \bar{\sigma}_n \left[\int_o^1 f_n(x) \ M_i(x) \ x \ \Big/ \int_o^1 M_i(x) \ dx \right] \qquad (8)$$

This expression easily shows that if the stress $\sigma_n(x)$ is not constant ($f_n(x) \neq 1$) the pressure required to bring back the measuring points may differ from the mean normal stress and, furthermore, may be point dipendent.

In principle, then, the use of the jack does not eliminate, as assumed in the equation (5), the factors upon which $M_i(x)$ depends. From a pratical point of view is then necessary to investigate the form of the influence functions. The knowledge of the influence functions will obliviously provide also a model for the deformability test.

4.1 Calculation of influence functions

Assuming linear elastic behaviour of the rock the influence function may be easily calculated by means of available numerical methods. In the present work Finite Element method has been used, however alternative methods have been suggested elsewhere (Borsetto, 1981).

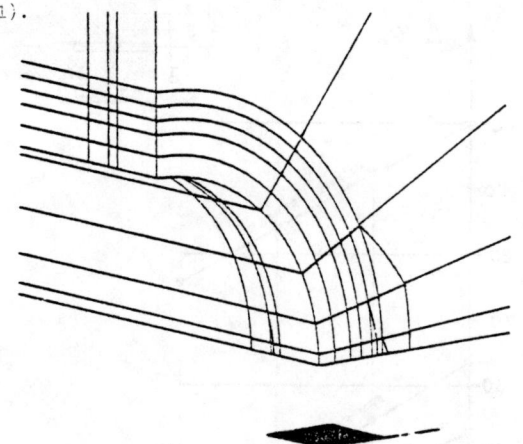

Fig. 5   Mesh of the tunnel.
The trace of the slot is shown.

The mesh shown in Fig. 5 reproduces a tipical installation. Only one eight of the problem is discretized as the influence of a symmetrically placed cut is assumed to be very low.

A zero thickness of the slot has been assumed while other dimensions are the most widely used at ISMES (1.0X.6 m). The numerical model, with the use of the quarter point tecnique and of various polinomial levels for the shape functions allowed by the FIESTA program (Peano et al., 1979,1981) is very accurate.

The displacement have been calculated for eighteen linearly inipendent, load distribution along the slot (six being the rows of elements on it).
In order to resort to an analytical form and produce a qualitative plot, the influence functions have been approximated by polinomials and related coefficients have been found by a linear regression analysis on the aforementioned results (Fig. 6 and 7).The influence functions for a 1 x 1 meter cut are shown in Fig. 8 and 9. These last results however are based on a less accurate - too much stiff - FEM model.
For actual calculations more effective analytical expansions should be worked out.An alternative is to decompose the stress distribution in a sum of the previously said load distribution and then to calculate the displacements with the appropriate combination of the basic displacements.

5. INTERPRETATION OF DEFORMABILITY TESTS

The value of the secant modulus is easily found by the following equation:

$$E(p) = \left(p \cdot \int_0^1 M_i(x) \, dx \right) \cdot E_m \bigg/ \delta_i^p(p) \quad (9)$$

where E_m is the elastic modulus of the F.E. model.
It is of some interest to investigate which discrepancies are found in front of the use of the standard interpretation formula.
The theoretical displacements for different points are given in fig. 10 for the (1 x 0.6 m) slot and an ideally efficient flat jack. It is easily seen that the

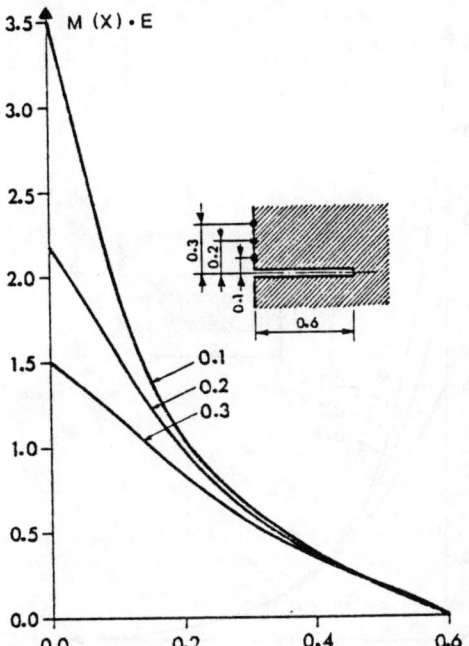

Fig. 6   Influence functions for surface points;
1x0.6 m slot.

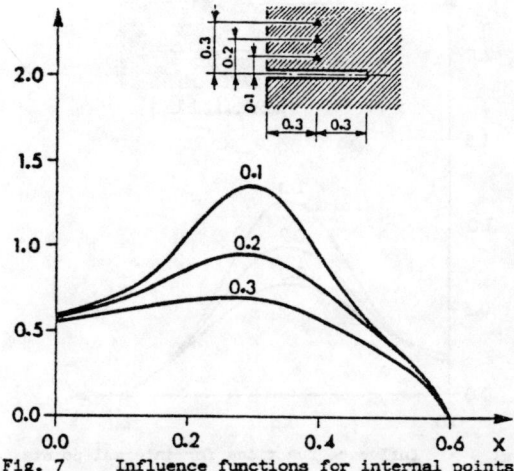

Fig. 7   Influence functions for internal points;
1x0.6 m slot.

standard formula greatly overestimates the displacements and, consequently, the deformability. Similar results are obtained for the 1 x 1 slot, fig. 11 (a Boundary Element solution was been used in this case). The error is not much sensitive to the distance from the slot, d, and it is larger when displacements at depth are considered.

However, in the field, the displacements of internal points are reported to the surface by rigid pins whose small rotations give rise to an additional displacement at the free end of the rod. It is not surprising, then, that, despite the results of fig. 10, the elastic moduli evaluated from deep points were found more reasonable. The spurious displacements caused by the rotation of the pin are of the same order of magnitude of the genuine ones. Furthermore they depend on the free length of the rod and on the distance between the fixed end and the cut.

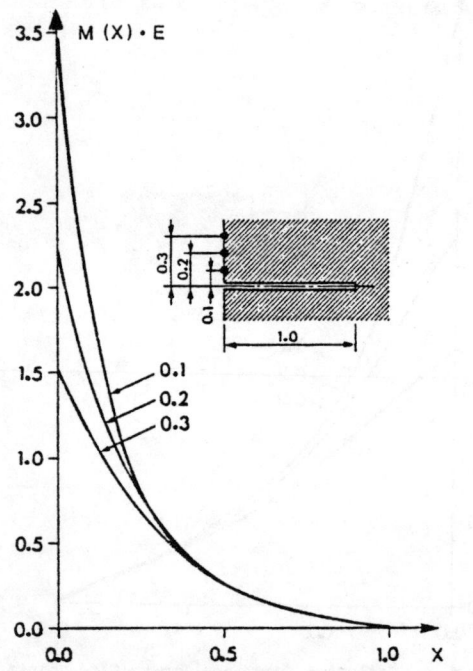

Fig. 8   Influence functions for surface points;
          1x1 m slot.

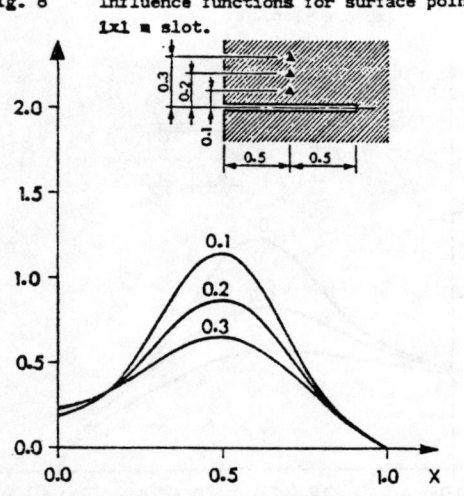

Fig. 9   Influence functions for internal points;
          1x1 m slot.

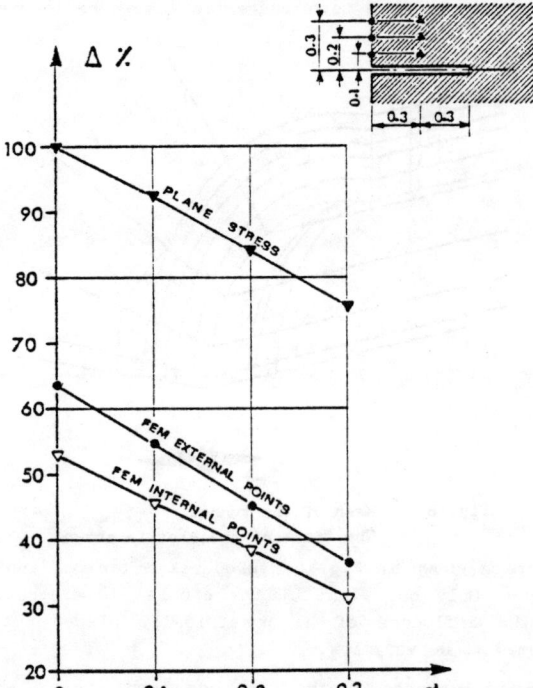

Fig. 10   Displacement under jack pressure.
          Displacement given by Alexander's formu-
          la on the surface of the slot are set
          equal to 100.

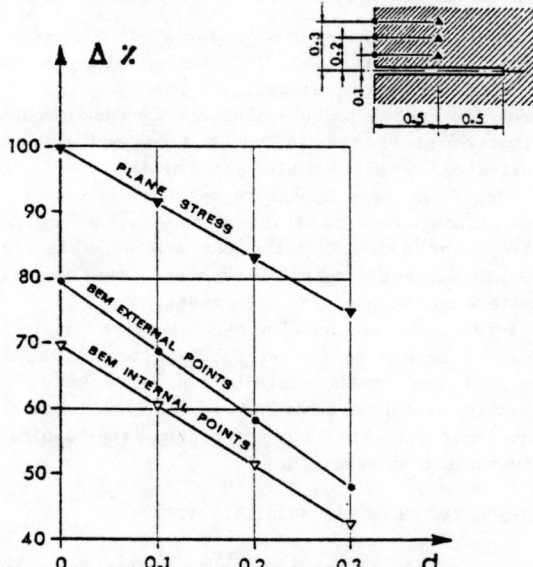

Fig. 11   Displacement under jack pressure.
          Displacement given by Alexander's formu-
          la on the surface of the slot are set
          equal to 100.

In order to illustrate the improvement obtainable in the test interpretation by the use of an accurate numerical model, the results of an in situ test were been selected. The measurements were taken in an experimental tunnel of the Solarino pumped - storage plant; the rock is a non fissured calcarenite (Frassoni et al., 1981). Only deep measurements were taken at this site and the elastic moduli given in Table 1 are evaluated upon the relative displacements of pairs of apposite pins or upon the average of relative displacements of correspondent couples of contiguous pins.

The first columm on the table is evaluated on the basis of the Alexander's formula.The values of the second column have been found according by to the espression (9); in this case the rotations of the pins have been accounted for by assuming their computed values at the midlength of their bonded part.

From the results of the first column a damage of the rock near the slot seems to be likely (as a result of an uncareful application of the formula). No serious damage is evident from results of the second column: here the homogeneity of the rock and the good performance of the test are quite clear.

A 148

E	Alexander's formula	Numerical model
A	19.3	13.8
B	21.9	14.3
C	25.2	14.2
D	11.5	12.5
F	12.8	14.5

TABLE I Solarino site. Flat jack test interpretation. Elastic modulus, GPa.

6. ORIGINAL STRESSES AND STRENGTH EVALUATION

The stress acting on the virtual surface of the slot can be expressed as a function of the parameters to be evaluated by means of a mathematical model of the tunnel excavation. The displacement of the reference point can then be found through the discussed model as

$$\delta_i(k_1, k_2) = \int_o^1 M_i(x) \; \sigma(k_1, k_2) \; dx \quad (10)$$

where k_1, k_2 stand for the appopriate parameters (c_r, φ_1; σ_v, σ_o). Provided that measurements are taken at various points, or even from different tests, it is possible to set up a regression process in order to identify the unknown parameters.

The fact that the model can interpretate displacements at various depth and that it is sensitive to the stress distribution $f_n(x)$ is very important in order to find out if the model used for the tunnel excavation is appropriate.

The effectiveness of the influence functions M_i can be tested, at least partially, on the base of the displacements recorded during the jack pressurization.

In fact, as in this case the load $(p(x))$ is known with good accuracy, the errors z_i:

$$z_i = \delta_i^p(p) - p \int_o^1 M_i(x) \; dx \quad (11)$$

can be due mainly to errors in the M_i functions.

Some measurement points, providing abnormal errors, could be disregarded in the subsequent analysis of the displacements δ_i^c; or a more generalized inconsistency may guide toward a more appropriate model for the calculation of the influence functions. In practice, however, the search for such a "better model" is limited either by economical reasons or by the fact that only simple models can be sufficiently substantiated on the basis of the measurements. Keeping in mind these observations an alternative strategy for the regression analysis can be suggested: $\|e_i \; W_i\| = $ minimum with:

$$\quad (12)$$

with:

$$e_i(k_1, k_2) = \delta_i^c - R_i^p \int_o^1 M_i \; \sigma(k_1, k_2) \; dx$$

Here W_i are penalty values selected on the basis of the error z_i in the reproduction of δ_i^c set of values, evaluated according by to the best available model. In turn, the error is evaluated by scaling down each theoretical influence functions of a factor

$$R_i^p = \delta_i^p / \int_o^1 M_i \; p \; dx,$$

based on the esperimental values.

This procedure, however, cannot eliminate all errors due to a wrong evaluation of the influence functions. In order to emphasize the matter let us write:

$$\sigma_n = \bar{\sigma}_n \cdot f_n = \bar{\sigma}_n + \bar{\sigma}_n \; q_n \quad (13)$$

If the procedure presented at the beginning of the paragraph is used, all the two terms of the right hand side will produce errors. With the second procedure, if an ideally efficient flat jack is used, no errors are induced by the mean normal stress even if an arbitrary influence function is used. The elaboration of the $q_n(x)$ term with approximate influence function is still an error source.

This second procedure is in some way similar to the straight forward method of the cancellation pressure (eq. 5).

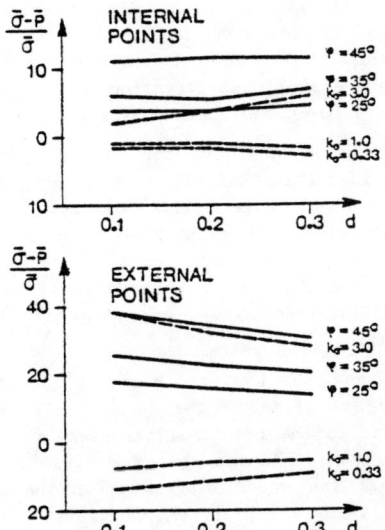

Fig. 12 Comparison between the cancellation pressure $\bar{p}$ and the mean primitive normal stress on the slot $\bar{\sigma}$.

It is interesting to investigate the meaning of the cancellation pressure for the aforementioned simple cases of a circular tunnel. From the results of fig. 12 it is evident that the value of the cancellation pressure for the (1.x0.6 m)slot with ideally efficient jack and deep measurements is, in general, a good approximation of the mean stress σ_n.

If the measurements are taken at the surface (fig. 12) a very poor approximation is found in most cases: that means that these measurements are much more sensitive to the distribution function $f_n(x)$. The differences between reference points at various distances from the slot are larger than in the previous case. Different results are also found for a deeper cut: as an exemple for a 1 m square cut for $\varphi_r = 35°$, the error at a point placed 0.2 m from the slot is 38% at the surface and 8% at depth. Furthermore it must be stressed that previous graphs have been evaluated not taking into account the effects of the pin rotations. It is then evident that the cancellation pressure is only, eventually, a good approximation of the mean normal stress.

By inspection of equations (3) and (4) it is evident that the identification of the strength parameters would require a nonlinear regression analysis, while a linear one is applicable, in case of an elastic behaviour of the excavation, for the identification of original stresses. A sensitivity analysis, however, has shown that the contemporary evaluation of c_r and φ_r is not feasible in practice. In fact, if, as an exemple, a value $\varphi_r = 30°$ is selected, instead of a "true value" of 35°, it is possible, by regressing only the c_r value, to produce a best fitting of the δ_i^c displacements within a 2% scatter (c_r being some 20% greater the true value). Such an error for the friction angle is unacceptable in many situations and a 2% scatter is not significant in real life conditions.

For this reason the friction angle must be evaluated indipendently and then it is possible to resort to a "tuned" c_r value. The sensitivity to φ_r could be increased with a deeper cut (it is more than doubled for the 1m square slot) but attention must be paid that the rock may not be in residual condition.

A similar situation is found for the elastic tunnel case and the identification of σ_v and σ_o when σ_v is the major principal stress and the slot is cutted normally to it. In this case however it is possible to perform tests with different locations of the slot; each one more sensitive to a particular stress component. Various locations are recommanded by IRSM Suggested Standards.

A sensitivity analysis has been performed for the simple case where two cuts on the tunnel wall are used, each one normal to the principal stresses σ_v or σ_o. The results are presented in fig. 13 for "true values" of $k_o = 0.33, 1., 3.$. The upper diagram shows the error in the regressed value of σ_v if an arbitrary k_o value is enforced. The lower diagram shows the corresponding inaccuracy of the fitting of displacement data by means of the ratio between the mean square error and the mean of displacements.
It is clear that if the dispersion caused by the operating condition is not ecceptionally high the regression of the two parameters σ_v and σ_o should be feasible within an acceptable accuracy.

7. CONCLUSION AND RECOMMANDATION FOR FURTHER WORK

The evaluation of the residual stress of rock masses appears to be theoretically feasible when appropriate conditions hold. The results obtained at the Edolo site (Forzano et al., 1980) were encouraging and were successfully (Borsetto et al., 1980) utilized in the definition of design geomechanical parameters. The field experience is anyway still limited.

Current work is intended to ameliorate theoretical and experimental aspects in order to make future test as probant as possible.

In this paper an interpretation procedure has been presented. Moreover it has been found that a refined interpretation model could prevent systematic errors not only in the evaluation of the residual cohesion but also when the test is used for identify original stres-

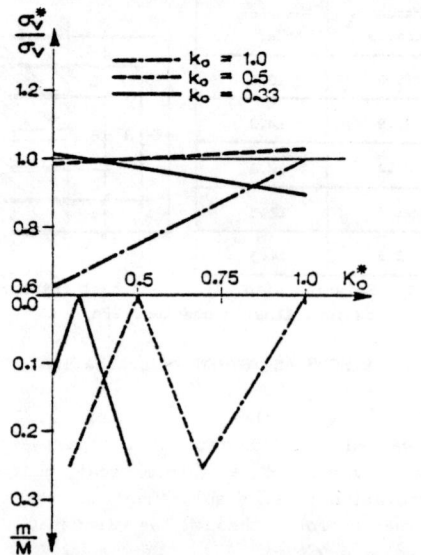

Fig. 13   Sensitivity of the test; K_o^* and σ_v^* are regressed values.

ses and deformability.
The availability of models allowing the interpretation of displacement readings at various depth, in a unified way, will certainly encourage the acquisition of more data in the same test.This, in turn, will improve the significance of a statistical analysis of the data to the benefit of the judgement of the rock mechanician.

Along this line some further developements is possible. As suggested by Rocha (1966), the record of the displacement during the cutting of the slot may give additional informations and control of test performances. The combined use of numerical modelling could enable further strenghtening of the regression capabilities and possibly solve the problem of the identification of residual friction angle.
Another promising tecnique, which is currently discussed for feasibility at the Geomechanical Division of Ismes, could be the application of more than one load distribution by means of special jacks.

As far as the rotation of measuring pins is concerned it is thought that is not a good policy to accept the measurement of the overall displacement without control: whenever in favourable sites the mathematical model could account for this. Then one must either monitor the rotations separately, and with high accuracy, or avoid them by adopting a different measurement tecnique.

ACKNOWLEDGEMENTS

Authors are grateful to Mr. Giorgio Barbieri who carried out the numerical analyses .

REFERENCES

ALEXANDER, L.G. (1960) Field and laboratory tests in rock mechanics, 3rd August - N.Z. Conf. on Soil Mech.

AMBERG, W.A., LOMBARDI G. (1974): Une méthode de calcul élasto-plastique de l'état de tension et de déformation autour d'une cavité souterrain, Proc. 3rd Congr. Int. Soc. Rock Mech., Denver USA, September, V, 2B,

BORSETTO, M., RIBACCHI, R., (1978) Metodi di calcolo per la valutazione dell'efficacia dei tiranti di grandi scavi in sotterraneo. Giornata di studio sulla bullonatura in sotterraneo, Torino.

BORSETTO, M., (1980): Una metodologia per l'identificazione delle caratteristiche di resistenza degli ammassi rocciosi, Proc. XIV Convegno Nazionale di Geotecnica, Firenze, ottobre, V.2.

BORSETTO, M., FORZANO G., VALLINO G. (1980): Verifica in corso d'opera dei parametri geomeccanici di progetto per la centrale in caverna di Edolo, Proc. XIV Convegno Nazionale di Geotecnica, Firenze, ottobre, V.2.

BORSETTO, M. PEANO A.(1981): Numerical simulation of the excavation of large underground openings: an assessment of some past, present and future techniques, Proc. European Simulation Meet., Capri, Italy, September.

COMES, C., LAKSHMANAN J. (1969) La measure des contraintes dans les fondations rocheuses et sur les auvrages en bèton à E.D.F.; 1st Int. Cong. ISRM, Lisbon.

DOLCETTA M., (1982): ENEL's recent experiences in the construction of large underground powerhouses shafts and pressure tunnels, Proc. Int. Symp. on Rock Mechanics related to Caverns and Pressure Shafts, Aachen, Germany, May, in press.

FORZANO G., MORO T., VALLINO G., FRASSONI A., ROSSI P.P., ZANETTI A., (1980): "Scelta e determinazione dei parametri geomeccanici relativi allo studio della stabilità dell'ammasso roccioso interessato dagli scavi della centrale in caverna dell'impianto idroelettrico di Edolo", Atti XIV Conv. AGI, Firenze.

FRASSONI A., MORO T., NOCILLA N., ROSSI P.P., (1981) Physical and Mechanical Characterization of a weak rock involved in the Excavation of an Underground Power-house Int. Symp. on weak rock, Tokio

HILL R. (1950) The mathematical theory of plasticity, Oxford

HUECKEL T. - MAIER G. (1977), Incremental boundary value problems in the presence of coupling on elastic and plastic deformations: a rock mechanics oriented theory, Int. J.Solids and Structures

JAEGER C., COOK N.C.W., (1969) : "Fundamentals of rock mechanics" Metuen & Co. Ltd London.

PEANO A.,PASINI A., (1981) A warning against misuse of quarter-point elements .Int. J. Num. Meth. Engng., 18, 314-320

PEANO A., PASINI A. , RICCIONI R.,SARDELLA L., (1979). "Adaptive approssimations in finite element structural analysis " . Computers and Structures, 10, 333342 .

RIBACCHI M., RICCIONI R., (1977) : Stato di sforzo e di deformazione intorno ad una galleria circolare, Gallerie, V.S., n. 4.

ROCHA M., LOPES J.J.B., DA SILVA J.N., (1966) Proc. Int. IRSM Conf., Lisbon

THE STRUCTURAL EFFECT IN THE MECHANICAL BEHAVIOUR OF CLAY SHALE

L'effet structural dans le comportement mécanique de l'argile schisteuse

Der strukturelle Effekt im mechanischen Verhalten von Tongestein

Guangzhong Sun and Ruiguang Zhou

Institute of Geology, Academia Sinica, PRC

SYNOPSIS

The clay shales of the present study possess rather low mechanical strength and high deformation. They may be categorized as soft rocks. Experimental results show that the mechanical strength and deformation of the rockmass is closely related to the number of the structural body contained in the testing specimen.

RESUME

Les argiles schisteuses étudiées ici présentent une résistance mécanique plutôt faible et une forte tendance à la déformation, de sorte qu'on peut les assimiler aux roches tendres. Les résultats des expériences montrent que la résistance mécanique et la déformabilité de la masse rocheuse sont étroitement liées au nombre d'éléments structuraux dans les échantillons des essais.

ZUSAMMENFASSUNG

Der Ton, den wir untersuchten, besitzt niedrige Festigkeit und hohe Deformation, und ist ein weiches Gebirge. Die Ergebnisse der Experimente zeigten, daß die Festigkeit und der Verformungsmodul des Tones stark mit der Zahl des strukturellen Körpers in der Testsubstanz zusammenhängen.

Rockmasses notably distinguish from continue medium with the existance of diverse structure surfaces in them. Rockmass structure resulted from structure surfaces and rock block cut by them, i.e. so-called structure bodies controls the rock mass mechanical action and mechanical properties. The influence of the rock mass structure exerts on the mechanical behavior is called the structural effect of rock mass on the mechanical behavior. When structure surfaces are evenly distributed in rock masses the structural effect of the rock mass mechanical behaves as a scale effect.

The scale effect of the rock mass mechanical behavior has long been investigated. The results of the previous researches are mainly summarized in the form of the relationship between compressional strength and the modulus of elascity with the scale of samples. The authors had also conducted investigations in this aspect. Attention has not only been drawn to the scale effect, but especially to the structural effect dictated by the extent of the cutting by the structure surfaces. This paper presents some results of such a continuous research project in this field.

GEOLOGICAL FEATURES OF THE EXPERIMENTAL SITE

The clay shales in the experimental site are composed of weathered phyllite hornstone of lower permian series. Mineral composition is mainly kolite, with fewer adlerstein (needle ironstone). The occurrences of the rock strata are

N 10°-5°E/NW∠11°-27°. The thickness of a single layer is about 40-80cm. Joints are well developed in rock masses with normal spacing of 15-20cm. joints are predominantly of closed type, partially cemented with ferrous material. These closed joints are easily subjected to weathering under natural action and manmade perturbation. The natural water content is generally about 13 percent. The dry density of clay shale is 2.0g/cm^3.

EXPERIMENTAL TECHNIQUE AND THE BASIC RESULTS

In order to study the influence of the extent of cutting by the structure surfaces on the rock mass mechanical properties, aspecial series of rock samples has been prepared according to Table 1. Each sample, after having been prepared, is described with eyes on the visible joints, and an estimate of the number of structure bodies in each sample is then made and is listed in Table 1. The experimental resultis indicated that the visible joints play a controlling role in sample failure.

The clay shale layer being studied is of weak rock type with characteristic low strength and remarkable flow deformation. The results of experiment organized in the light of the characteristics of samples flow deformation reveal that is just after 24 hours of constant load than the clay shale layer reaches a stage

Table 1

Serial number of sample	No. 11	No. 10	No. 9	No. 7	No. 6	No. 1-5[*]
Dimension of sample	10x10x10	20x20x20	30x30x30	70x70x70	100x100x100	
Planned number of structure bodies	1	2-4	4-27	64-125	64-216	
Actual number of structure bodies	1	2	13	23	36	
Water content, w %	12.5	13.2	16	16	13.2	14.2
Initial stress of flow deformation (I.S.F.D.) σ_i Kg/cm^2	5.2	3.6	2.1	1.6	1.5	1.55
Normalization factor of I.S.F.D.	1.0	0.69	0.40	0.35	0.29	0.30
Critical failure stress (C.F.S.) σ_c, Kg/cm^2	19.6	12.3	10.87	6.3	5.16	7.8
Normalization factor of C.F.S.	1.0	0.63	0.55	0.32	0.26	0.40
Modulus of elasticity (M.E.) E, Kg/cm^2	5400	2600	600	500	125	1190
Normalization factor of M.E.	1.0	0.52	0.11	0.09	0.023	0.22
Viscosity, $\times 10^{15}$ p	1.2	1.2	0.4	0.3	0.29	0.35
Normalization factor of viscosity	1.0	1.0	0.33	0.25	0.24	0.29

[*]No.1-5 is represented by the average results of uniaxial flow deformation on 5 samples

of full constant rate flow deformation. Therefore, all experiments had been made on each single sample with 24 hours dead load cycle with deformation for each pressure stage, thus the flow deformation of all pressure stages till the failure had been recorded. In such a way each sample is equipped with a deformation history curves.

According to the deformation history curves, following informations can be obtained:
1, Correlation curve of transient strain with stress σ.
2, Correlation curves of the rate of constant velocity flow deformation $\dot{\varepsilon}$ with stress σ.
3, Initial stress σ_i under which flow deformation starts curve (See Table 1).
4, Critical failure stress σ_c for each sample (See Table 1).
In order to ease the date analysis, all experimental results are normalized on the basis of data obtained on sample No. 11. with no explicit joints (that possibly implies with no structure bodies in the sample). Normalized information are also shown in Table 1.

DATA ANALYSIS AND CONCLUSIONS

A comparison of data in Table 1 illustrates that the differences of experimental results are attributed to two important dictating factors: 1) rock mass structure, and 2) water content in different samples ranges from 12.5 to 16 percent. It can markedly affect rockmass mechanical properties of weak rocks. Fig. 1 presents experimental results of above-mentioned influence, obtained by the authors on the clay shales. Since the water content of current experiment changes only in a limited range, namely 12.5-16.0 percent, it does not show apparent influence on the mechanical properties of clay shale. Obviously, the differences shown in Tab. 1 are accounted mainly for the scale of samples or their structural factor.

In order to investigate the influence of the scale of samples or their structural factor on the mechanical properties of clay shales, correlation curves of sample with the mechanical properties (Fig.2) and of the number of structural bodies with the mechanical properties (Fig. 3) of the clay shales have been built, respectively. Figs 2 and 3 indicate:
1, The mechanical properties of clay shales are noted with remarkable structural effect and scale effect; the former effect is more obvious

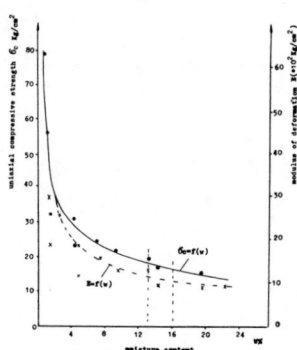

Fig. 1 The influence of water content on the deformation and failure of clay shales.

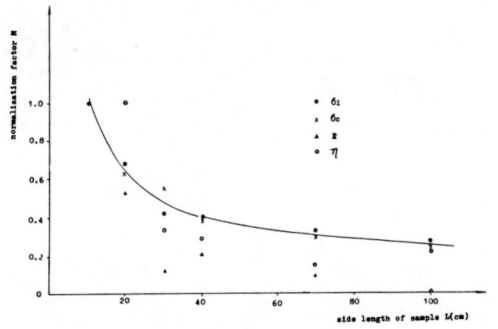

Fig.2 Correlation curve of mechanical properties of clay shales with the sample scale.

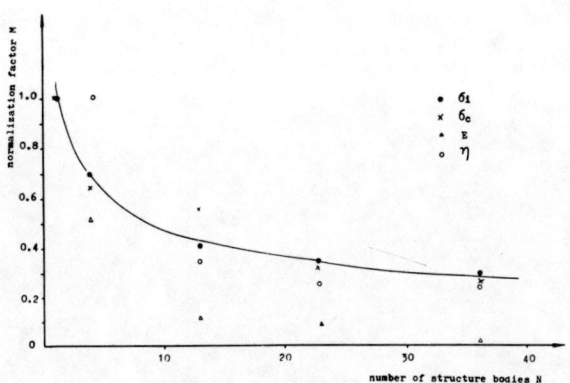

Fig.3 Correlation curve of mechanical properties of clay shales with the number of structure bodies.

than the latter one.

2, The initial stress of flow deformation and the critical stress of clay shales closely related to the number of structure bodies in a form of a curve which transforms to a straight line in a log-log plot. (Fig.4). Fig.4 gives:

1) Relationship between the initial stress of flow deformation and the number of structure bodies in samples (N):

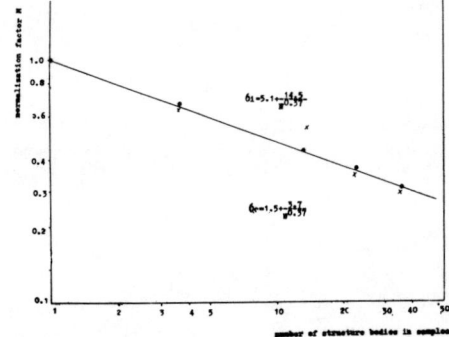

Fig. 4 Correlation curve of failure stress of clay shales with the number of structure bodies.

$$\sigma_i = 1.5 + 3.7 N^{-0.37}$$

2) Relationship between the critical stress of failure and the N:

$$\sigma_c = 5.1 + 14.5 N^{-0.37}$$

3, Neither the date plots of the modulus of elasticity of clay shales (E) and the sample scale, nor the plots of the E and the number of structure bodies (N) are concentrated enough. They appear to be scattered. Nevertheless, some general trends can still be disciminable.

4, There are two types of rate of flow deformation. Samples No.10,11 are of first type, while samples No.6,7,9 are of another type. The viscosity of samples No.10,11 are greater than that of samples No.6,7,9 by a factor of 4.

Based on above analysis we can draw following brief conclusion: The mechanical behavior of clay shales is closely related to the rock mass structure. The influence of rock mass structure is dictated by the mechanical actions and therefore, it is not even for different categories of rock mass mechanical properties. For clay shales as weak rocks with low mechanical strength, the characteristics in combination of structure bodies and surfaces play a dominant role in the course of rock mass failure. Meanwhile, in the course of rock mass deformation, the structure surfaces exert an prominent influence. It results not only in deformation due to the closure of the structure surfaces, but also due to the sliding of the structure surfaces. The latter affects both the transient deformation, and, to a larger extent, the flow deformation with close relation to the occurrences of the structure surfaces.

THE STRENGTH DEFORMATION AND RUPTURE CHARACTERISTICS OF RED SANDSTONE UNDER POLYAXIAL COMPRESSION

Caractéristiques de la résistance, la déformation et la rupture du grés rouge sous compression polyaxiale

Festigkeit, Deformation und Bruch-Charakteristik des roten Sandsteins unter polyaxialer Kompression

Qu-qing Gau, Hong-tai Cheng, and De-pei Zhuo
Southwest Jiaotong University, Sichuan, China

SYNOPSIS

In this paper the characteristics of strength deformation and rupture of red sandstone under polyaxial compression were studied and the feature of the theoretical surface of the strength of rock analysed. The coefficient α, which represents the effect of intermediate principal stress σ_2 on the ultimate compressive strength on the rock, was determined. The relation between the stress-strain curves along the directions of three principal stresses and the lateral pressure was discussed. The rigid and flexible contact modes were adopted during the experiments, and comparisons made between the test results for the two kinds of tests.

RESUME

Les caractéristiques de résistance de déformation et de rupture de grès rouge sous compression polyaxiale sont étudiées et le comportement de la surface théorique de résistance de la roche est analysé dans cet article. Le coefficient α représentant l'effet de la principale contrainte intermédiaire σ_2 sur la résistance à la compression a été déterminé et les relations entre les courbes de contrainte-déformation dans les directions des trois principales contraintes et la contrainte latérale sont discutées. Des contacts rigides et flexibles ont été adoptés durant les expériences et on a effectué des comparaisons des résultats obtenus pour ces deux types d'essai.

ZUSAMMENFASSUNG

In dieser Arbeit wurden die Festigkeit, die Deformation und die Bruch-Charakteristik des roten Sandsteins unter polyaxialer Kompression untersucht. Die Eigenschaft der theoretischen Fläche der Gesteinsfestigkeit wurde analysiert. Der Einflußkoeffizient α, der den Einfluß der mittleren Hauptspannung σ_2 auf die Druckfestigkeit des roten Sandsteins bezeichnet, wurde bestimmt. Die Abhängigkeit zwischen dem Seitendruck und den Spannung-Verformungskurven auf die drei Hauptspannungsrichtungen wurde diskutiert. Beim Versuch wurden die weichen und harten Berührungsmethoden angewandt. Ein Kontrast zwischen beiden Versuchsergebnissen wurde gemacht.

Red sandstone is a sedimentary rock of Cretaceous Period, Mesozoic era. It contains about 40% quartz, 25% feldspar, 15-20% detritus, and small amount of muscovite and chlorite etc. with mainly ferruginous cementing material. The physical indexes are: d=2.57, n=4.95%, w=1.52%, γ =24.33 KN/m³.

The experiments were carried out in a triaxial frame with stress strain curve measuring installations. In this frame 10cm cube specimens were tested under rigid and flexible contact. In rigid contact test thin sheets of PTFE were laid on all the contact surfaces in order to reduce the frictional effect. 99 specimens were tested for their compressive strength and 45 were tested for their deformation. To approach a flexible contact condition emulsified rubber

bags filled with vaseline quartz sand mixture were used. 94 specimens were tested to determine their strength and deformation under this contact condition. A number of tests were also carried out in a conventional ($\sigma_2=\sigma_3$) 500T triaxial machine.

The strength criterion of a rock can be expressed by one of the following equations: $F(\sigma_1, \sigma_2, \sigma_3)=0$, $G(\tau_{oct}, \sigma_{oct}, \lambda)=0$ and $H(R, Q, \lambda)=0$. In these equations $R=(\sigma_1-\sigma_3)/2$, $Q=(\sigma_1+\sigma_3)/2$, $\lambda=(2\sigma_2-\sigma_1-\sigma_3)/(\sigma_1-\sigma_3)$, $-1\leq\lambda\leq1$, and have the relations as shown in Eq(1). When λ=const. Eq. $F(\sigma_1, \sigma_2, \sigma_3)=0$ becomes an equation with two variables σ_1 and σ_3, and in Cartesian coordinates $O\sigma_1\sigma_2\sigma_3$, it represents a curve on the plane $\sigma_2=\sigma_1(1+\lambda)/2+\sigma_3(1-\lambda)/2$. Therefore the space curve surface $F(\sigma_1, \sigma_2, \sigma_3)=0$, after introducing parameter λ, may be looked upon as composed of curves corresponding to different values of λ.

$$\sigma_1 = R + Q = \sigma_{oct} + \frac{(3-\lambda) \cdot \tau_{oct}}{\sqrt{2} \cdot \sqrt{3+\lambda^2}}$$

$$\sigma_2 = \lambda R + Q = \sigma_{oct} + \frac{2\lambda \cdot \tau_{oct}}{\sqrt{2} \cdot \sqrt{3+\lambda^2}} \qquad (1)$$

$$\sigma_3 = -R + Q = \sigma_{oct} - \frac{(3+\lambda) \cdot \tau_{oct}}{\sqrt{2} \cdot \sqrt{3+\lambda^2}}$$

These curves are distributed within the curves with $\lambda = \pm 1$ and intersect at a point of equal tension $\sigma_1 = \sigma_2 = \sigma_3$. For the same reason the space curve surfaces for the other two equations could be looked upon as composed of curves with different values of λ. By cutting these curve surfaces with planes corresponding to different values of σ_3 and λ, the intersection curves will be scattered regularly on the $O\sigma_1\sigma_2$, ORQ and $O\tau_{oct}\sigma_{oct}$ coordinate planes respectively. On the ORQ coordinate plane the curves are so close that shows the characteristics of strength of rock under polyaxial compression. On the other coordinate planes, the concentration of curves is less. So it is convenient for us to utilize the characteristics of curves on ORQ coordinate plane to determine the value of α.

The relations between σ_1 and σ_2 from the test results are as shown in Fig.1 and Fig.2. From the data on Fig.1 and 2 we can obtain the R-Q

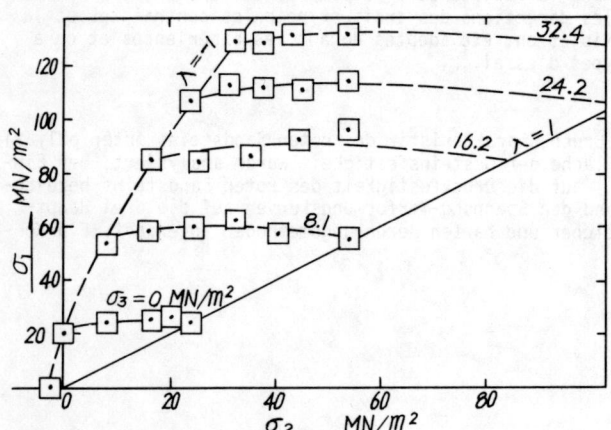

Fig.1 σ_1 versus σ_2 from rigid contact tests

curves corresponding to different values of λ as shown in Fig.3. In Fig.3 these curves have good linearity and the curves from the rigid contact tests lie between the curves with $\lambda = 0$ and -1. With absolute value of λ equal, the curves nearly coincide (e.g. $\lambda = \pm 1$, ± 0.4). With the test data for $\lambda = 0$ and $\lambda = -1$, we obtain the optimum linear lines by the method of least square, namely;

$$R = 0.56Q + 6.77 \qquad (2)$$

$$R = 0.54Q + 5.89 \qquad (3)$$

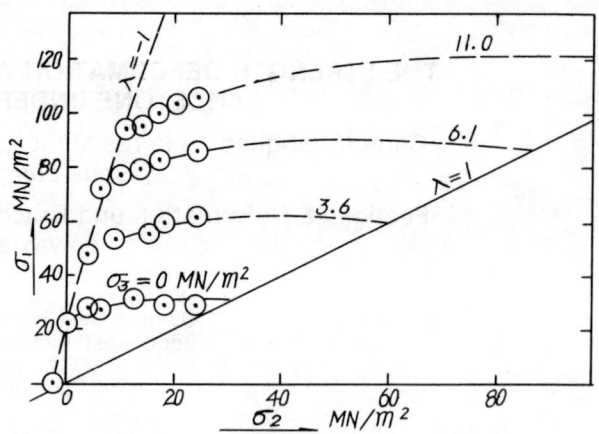

Fig.2 σ_1 versus σ_2 from flexible contact tests

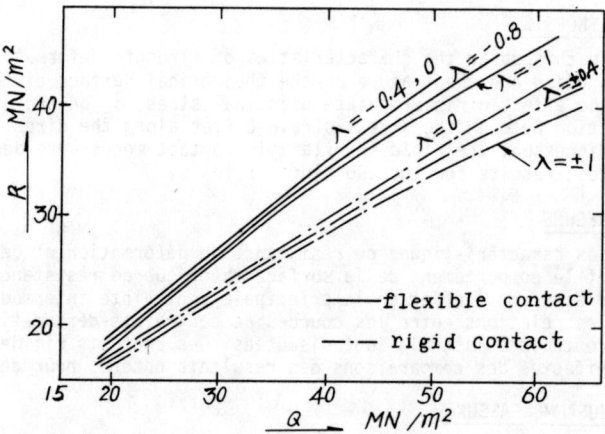

Fig.3 The relation between R and Q

In Eq.(2) and (3) the relative coefficients of linear fitting are r=0.995 and r=0.998. The constant terms differ only slightly and the value of α obtained is 0.04. Therefore the following equation could be used to describe the strength characteristics of red sandstone under polyaxial compression with rigid contact during experiments:

$$R = (1 + 0.04 \cdot \sqrt{1-\lambda^2})(0.54Q + 5.89) \qquad (4)$$

According to the same reason, we can obtain the equation for describing the strength characteristics of red sandstone under polyaxial compression with flexible contact as follow:

$$R = (1 + 0.08 \cdot \sqrt{1+\lambda})(0.74Q + 3.14) \qquad (5)$$

In Eq. (5) r=0.999 for $\lambda = -1$. When $\lambda = -1$ Eq.(4) and (5) when explained according to Coulomb-Mohr theory, we obtain, for rigid contact test: c=6.97 MN/m^2, ϕ =0.57 rad, ß=0.50 rad; and for flexible contact tests: c=4.71 MN/m^2, ϕ =0.83 rad, ß=0.37 rad. From the analysis of the conventional triaxial test results, we obtain the following figures c=6.57 MN/m², ϕ =0.60 rad, ß=0.48 rad and r=0.944. ß is the angle between the normal line on rupture surface and the direction of σ_3.

Fig.4 shows the results from different contact experiments. The figure shows that for the flexible tests, the strength values are highest, and the results from conventional triaxial tests are less, and the test results from rigid contact are the smallest.

Fig.5 shows the curves from the test results and that from calculations according to several strength criterions. The test data are both from rigid and flexible contacts with $\sigma_2=\sigma_3$. In the figure curves 1, 2 and 4 show respectively the linear $\sigma_1/\sigma_c + \sigma_3/\sigma_t$ =1, hyperbolic and parabolic Mohr criterion. Curves 3 and 5 show respectively the revised Griffith criterion and the Griffith criterion. Curves 6 and 7 show lines obtained from equations (4) and (5) with λ =-1.

curve, when σ_1 =17.7 MN/m² = σ_c·83%, θ reaches max. value 0.001. When stress $\sigma_1 \geqslant$ 17.7 MN/m² θ reduces with the increasing of σ_1 and the slope of $\sigma_1-\theta$ curve becomes negative. This signifies that crushing of the rock is accompanied by vol. change. We can estimate the rupture of specimen from the curve $\sigma_1-\theta$.

The stress-strain curves from biaxial and triaxial compressive tests are shown in Fig.6, 7 and 8. In Fig.6 and 7, as σ_2 is increased, ϵ_1 decreases, and curve $\sigma_1-\epsilon_1$ is ascending. For σ_3=0, ϵ_3 is tensile strain from the beginning to the end. As σ_2 is increased, ϵ_3 increases and curve $\sigma_1-\epsilon_3$ is descending. The shape of curve $\sigma_1-\epsilon_2$ differs from that of curves $\sigma_1-\epsilon_1$ and $\sigma_1-\epsilon_3$. ϵ_2 appears to be compressive strain in the beginning stage and the value will increase as σ_2 is increased. On $\sigma_1-\epsilon_2$ curve ϵ_2 will gradually change from compressive to tensile strain. ϵ_2 in the initial stage is larger in Fig.6 than that in Fig.7. This shows that the change of ϵ_2 from compressive to tensile strain is slower in the flexible contact test than in the rigid contact test. On curves $\sigma_1-\epsilon_2$ of the same value of σ_2, value of σ_1 when ϵ_2=0 is larger in Fig.6 that in Fig.(7).

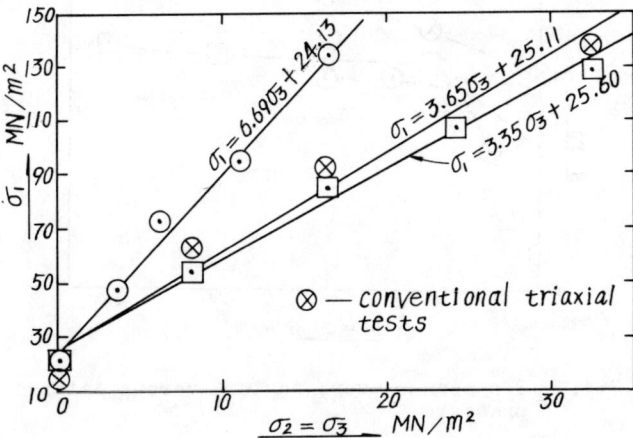

Fig.4 The comparison between the test results

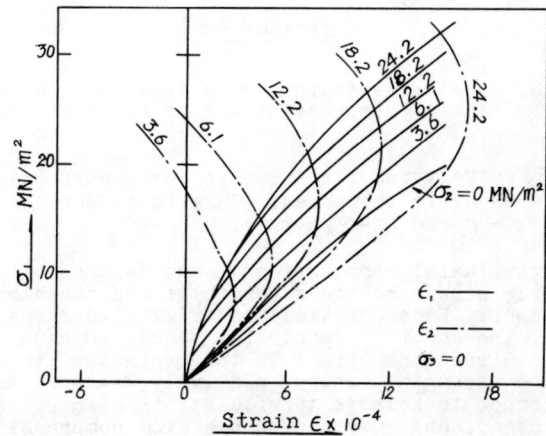

Fig.6 The stress-strain curves from the test of biaxial compression with flexible contact.

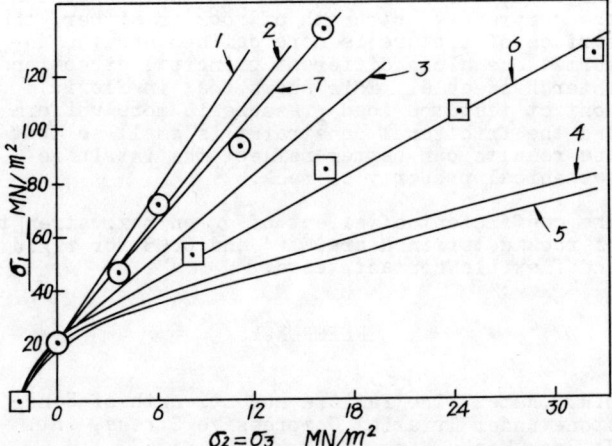

Fig.5 The test results in contrast with the strength criteria.

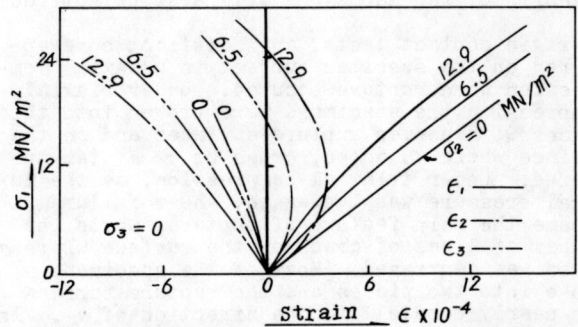

Fig.7 The stress-strain curves from the test of biaxial compression with rigid contact.

The elastic moduli of red sandstone in tension and compression are different. On the $\sigma_1-\theta$

Fig.8 shows triaxial compression stress-strain curves from rigid contact tests, the characteristics of which are similar to that of the biaxial curves as stated above. As $\sigma_3 \neq 0$, in the initial stage ϵ_3 is compressive. When σ_1 is increased, $\sigma_1 - \epsilon_3$ curve turns in direction and is the same shape as that of $\sigma_1 - \epsilon_2$ curve. As σ_3 is the min. principal stress and is maintained constant as it reaches its assigned value, therefore for $\sigma_1 - \epsilon_3$ curves, the curve with bigger value of σ_2 turns to tensile strain first. After

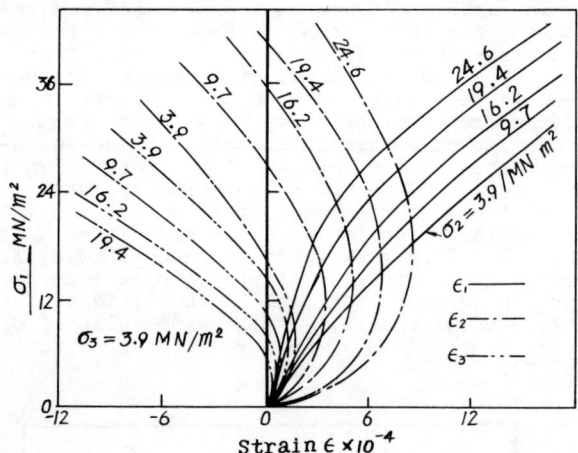

Fig.8 The stress-strain curves from the test of triaxial compression with rigid contact.

$\sigma_1 - \epsilon_3$ curve turns its direction, the curve descends as σ_2 is increased. This is different from $\sigma_1 - \epsilon_1$ and $\sigma_1 - \epsilon_2$ curves.

Under uniaxial compression ϵ_2 and ϵ_3 are both tensile strain so the strength of red sandstone is lower. Under biaxial and triaxial compression, the compressive strains along two and three directions appear in the beginning stage, so the strength becomes relatively higher. But ϵ_3 change to a large tensile strain when σ_2 is increased, and ϵ_2 would change from compressive to tensile strain as σ_1 is increased. As a result, the strength would not increase without limit. Therefore from the stress-strain curves in figure 6, 7 and 8, the effect of σ_2 on the strength of red sandstone will also be noticed.

In rigid contact tests, an indistinct cone appeared on the specimen subject to uniaxial compression when rupture occured. Under biaxial compression the specimens were broken into thick slices with uneven rupture surfaces and on the surface where σ_2 acted, cross crack system appeared. Under triaxial compression, as the lateral pressure was increased, shear failure became the main feature of rupture, while the number of lines of crack on the surface where σ_2 acted was decreased. Most of the specimens broke into two pieces and the rupture surface was nearly paralell to the direction of σ_2. In the case of flexible contact, under uniaxial compression, there was no cone appeared on the fractured specimens, and the rupture surface made a small angle with the direction of σ_1. For biaxial and triaxial tests, most of the specimens broke into two pieces, the surface of rupture was clearly defined and on the surface where σ_2 acted no cross cracks appeared. The rupture surface was nearly paralell to the

direction of σ_2. As the lateral pressure increased, specimens failed mainly by shear.

When $\lambda = -1$, the relation between the average rupture angle $\bar{\beta}$ and lateral pressure is as shown in Fig.9. The figure shows that, for the case of rigid contact tests, when the lateral pressure reaches 8.1 MN/m², $\bar{\beta}$ maintains a constant value. This angle is comparatively close to 0.50 rad obtained from equation (4). In the case of flexible contact tests, when the lateral pressure reaches 6.1 MN/m², $\bar{\beta}$ also maintains a constant value. This value is nearly the same as the value calculated from equation (5) $\beta = 0.37$ rad. In conventional triaxial tests, the relation of $\bar{\beta}$ against the lateral pressure is nearly linear.

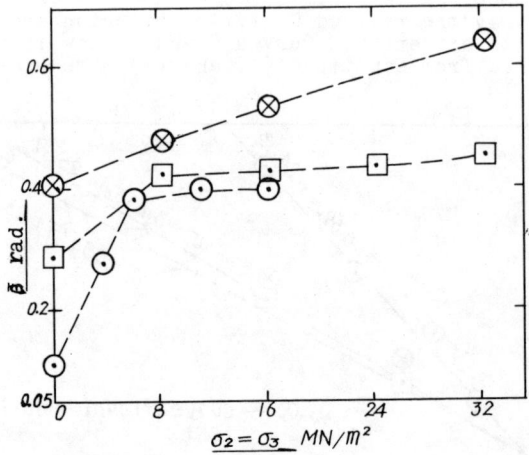

Fig.9 The mean rupture angle $\bar{\beta}$ versus lateral pressure

CONCLUSION

The results of flexible contact tests show that: the compressive strength of rock is higher, the surface of rupture is more defined and the deformations along different principal directions interchangeable. This shows that in flexible contact test the load pressure is more uniform and the frictional constraint is small so that the results can better reflect the intrinsic mechanical property of rock.

The coefficient of effect of σ_2 on max. strength of rock determined are 0.04 and 0.08 for rigid and flexible contact tests.

REFERENCES

M.R.H.Ramez, The Failure and Strength of Sandstone under Triaxial Compressive Stress, Int. J. Rock Mech. Mine. Sci., Vol.4. 1967.
R.D.Lama, S.S.Saluja, V.S.Vutukuri, Hand-book on Mechanical Properties of Rock, Vol.I, 1974 etc.

UN MODELE STATISTIQUE DE RUPTURE MACROSCOPIQUE EN COMPRESSION

A statistical model of macroscopic failure under compression

Ein statistisches Modell des makroskopischen Bruches bei Druckspannungszuständen

Ph. Weber
Professeur, Ecole des Mines d'Alès

P. Saint-Lot
Professeur, Ecole des Mines d'Alès

SYNOPSIS

The behaviour of rocks during compression failure is characterized by the progressive development of axial microcracks. A schematic model in which the heterogeneousness of the microstructure of a rock is simulated by a statistical distribution of fracture energy is proposed. It provides elements for the interpretation of complete stress-strain curves for rocks of Class I.

RESUME

Le comportement à la rupture en compression se caractérise par l'apparition progressive de microfissures axiales localisées. Un modèle schématique dans lequel l'hétérogénéité de la microstructure est simulée par une distribution statistique de l'énergie de rupture a été conçu: il fournit des éléments pour l'interprétation des courbes effort-déformation complètes de classe I.

ZUSAMMENFASSUNG

Das Bruchverhalten bei Druckspannungszuständen ist durch fortschreitende axiale und lokalisierte Mikrorisse gekennzeichnet. Ein schematisches Modell, in dem die Heterogenität der Mikrostruktur bei einer statistischen Verteilung der Bruchkraft simuliert wird, wird vorgestellt. Dieses Modell vermittelt Gründe für die Erklärung der vollständigen Kraft-Formänderungskurven von Klasse I Gesteinen.

INTRODUCTION

La conception d'un modèle purement physique de comportement à la rupture en compression semble illusoire du fait de la complexité de la microstructure de la roche. Les modèles analogiques plastiques à écrouissage négatif, prometteurs dans leurs applications présentent l'inconvénient d'occulter la réalité physique du phénomène. Le modèle proposé constitue une solution intermédiaire entre ces 2 approches : sans chercher à décrire finement le développement du réseau fissural, il fournit des éléments pour l'interprétation des comportements à la rupture en compression, en tenant compte de l'hétérogénéité de la structure.

1. DONNEES EXPERIMENTALES ET HYPOTHESES

Parmi les divers comportements à la rupture en compression, 2 grands pôles se dégagent :

- les roches à microstructure homogène, à haute limite élastique, se caractérisent par le développement quasi-simultané de fissures axiales sur toute la hauteur de l'éprouvette (faciès de rupture en colonnettes). Le comportement est de classe II (figure 1)

- les roches à microstructure hétérogène se caractérisent, par le développement de microfis-sures axiales apparaissant progressivement et se stabilisant après une courte propagation. Ce processus débute avant le maximum de la courbe effort-déformation (basse limite élastique et comportement dilatant) et se poursuit après ce point qui ne coïncide pas avec un évènement particulier de l'évolution du réseau fissural -HOUPERT (1979) -. RÜMMEL (1974) remarque que l'évolution de la vitesse des ondes longitudinales dans la roche ne passe pas par un point remarquable lorsque la contrainte est maximale. Le comportement est de classe I (figure 1). Lorsque cette microfissuration atteint un certain développement, des macrofractures de cisaillement apparaissent.

A partir de ces observations expérimentales, l'hypothèse suivante est formulée :

- le comportement de classe II, associé à une fissuration axiale est intrinsèque des roches à microstructure homogène ; le comportement de classe I, associé dans un premier temps à une microfissuration axiale est la manifestation macroscopique d'une série de micro-ruptures de classe II se développant progressivement.

M

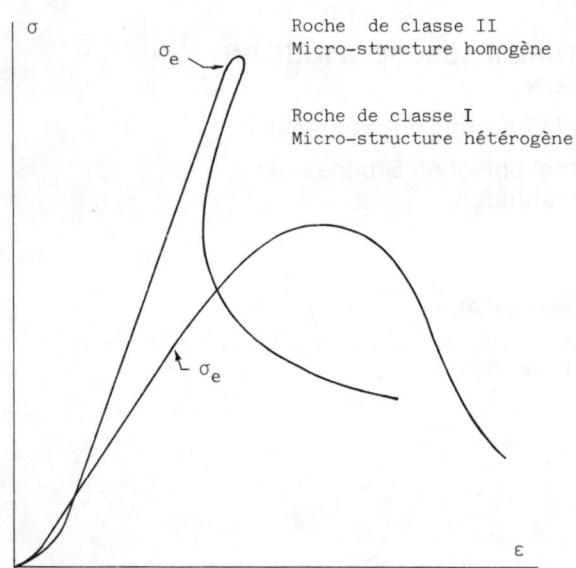

Roche de classe II
Micro-structure homogène

Roche de classe I
Micro-structure hétérogène

Figure 1   2 pôles de comportement à la rupture

2. DESCRIPTION DU MODELE

A partir des hypothèses précédentes est conçu le modèle suivant : il est constitué d'une infinité de "micro-modèles", chargés en parallèle, et simulant l'hétérogénéité de la micro-structure (figure 2). Tous ces micro-modèles sont supposés avoir un comportement de classe II, mais l'ensemble étant sollicité à déformation croissante, leur comportement apparent est de type élasto-fragile avec résistance résiduelle (figure 3). La rupture de chaque élément peut être caractérisée par un seuil de déformation, ε_r, ou par une énergie volumique de rupture :

$$\gamma_v = \frac{1}{2} \left(E.\varepsilon_r^2 - \frac{\sigma_r^2}{E} \right)$$

où E est le module de déformation, ε_r le seuil de rupture et σ_r la résistance résiduelle ; cette énergie est celle qui doit être fournie pour rompre le micro-modèle considéré, ramenée à l'unité de volume. E est supposé inchangé par la rupture.

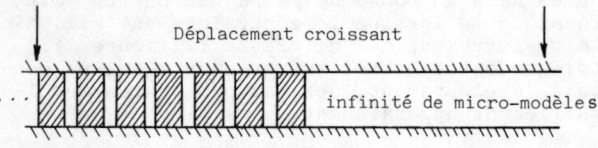

Déplacement croissant

infinité de micro-modèles

Figure 2   Modèle intact

L'influence de l'hétérogénéité de la structure sur l'initiation de la microfissuration est simulée en donnant à ε_r ou à γ_v une certaine distribution dans la population des micro-modèles.

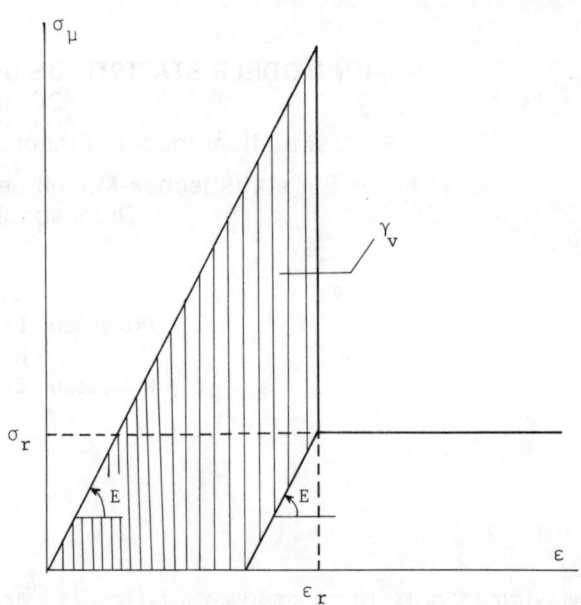

Figure 3   Comportement d'un micro-modèle

Dans un but de simplification, tous les micro-modèles sont dotés du même module de déformation, E, et de la même résistance résiduelle, σ_r. La section S du modèle est égale à la somme des sections dS des micro-modèles ; il en résulte :

$$\sigma_M = (1/S) . \int_{\text{Population}} \sigma_\mu .dS,$$

où σ_μ et σ_M sont respectivement, la contrainte s'exerçant dans les micro-modèles, et dans le modèle.

Soit $P(\varepsilon)$ la fonction de répartition des seuils de rupture dans la population ; pour une déformation ε donnée du modèle, la population est scindée en deux catégories (figure 4) :

- une proportion $P(\varepsilon)$ de la population est en phase résiduelle : $\sigma_\mu = \sigma_r$
- le reste de la population, soit une proportion $(1-P(\varepsilon))$ est encore intact, donc en phase élastique : $\sigma_\mu = E.\varepsilon$

La contrainte macroscopique sur l'ensemble du modèle s'exprime donc sous la forme :

$$\sigma_M = E.\varepsilon.(1-P(\varepsilon)) + (\sigma_r.P(\varepsilon))$$

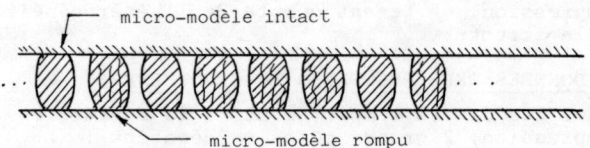

micro-modèle intact

micro-modèle rompu

Figure 4   Modèle en cours de rupture

3. COMPORTEMENT DU MODELE

La figure 5 représente les comportements comparés de modèles pour lesquels la déformation de rupture suit une loi normale de moyenne 10^{-3} et d'écarts-types variables.(E et σ_r sont gardés constants et égaux, respectivement, à 10 000 MPa et 0).

Les courbes ont une allure assez semblable aux courbes expérimentales connues ; en effet, au cours de l'augmentation des déformations, une proportion croissante de micro-modèles perdent leur capacité à supporter une charge, contribuant à l'effondrement progressif du système. Le maximum de la courbe effort-déformation est le point pour lequel un incrément de déformation provoque dans les parties intactes du modèle une augmentation de contrainte qui compense exactement la chute de contrainte par micro-rupture intervenant simultanément.

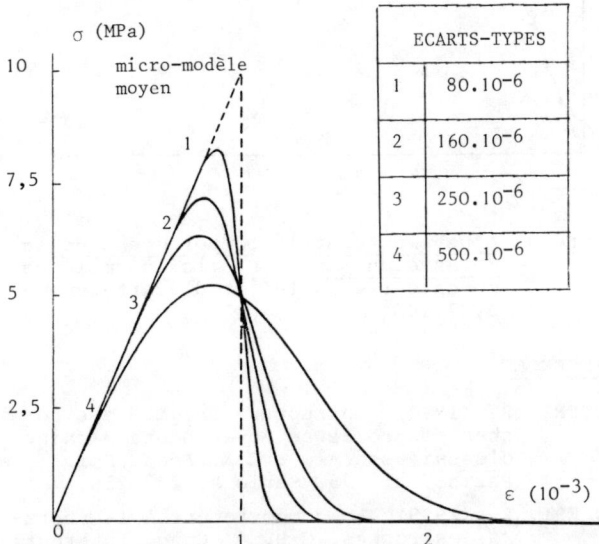

ECARTS-TYPES	
1	80.10^{-6}
2	160.10^{-6}
3	250.10^{-6}
4	500.10^{-6}

Figure 5 Comportement de modèles pour diverses valeurs de l'écart-type

Plus l'écart-type est important, donc plus la structure simulée est hétérogène, plus la résistance en compression σ_c, la limite élastique σ_e, le rapport σ_e/σ_c et la pente de la phase post-maximum sont faibles, plus la courbe "s'arrondit". Sur tous ces points, le modèle est en accord avec les observations expérimentales.

Trois types de distribution ont été étudiés : distribution normale pour ε_r et pour γ_v et log-normale pour γ_v. Dans les 3 cas les observations précédentes sont valables, mais l'évolution de la forme des courbes σ-ε dépend de la loi comme en témoigne la figure 6.

Cette figure représente pour les 3 distributions considérées l'évolution de la limite élastique avec la résistance en compression pour des écarts-types variables, les autres paramètres étant gardés constants.

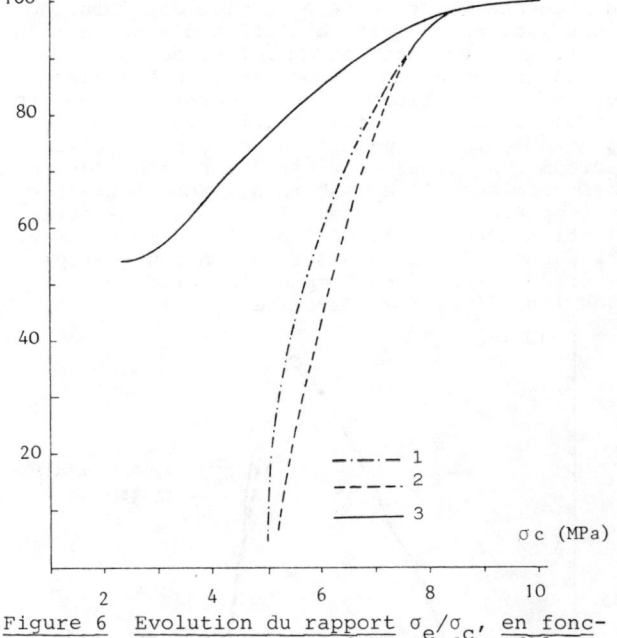

Figure 6 Evolution du rapport σ_e/σ_c, en fonction de σ_c pour 3 types de modèles : ① ε_r : loi normale ; ② γ_v : loi normale ; ③ γ_v = loi log-normale.

4. AJUSTEMENTS

Un échantillon de points $(\sigma_i, \varepsilon_i)$ est extrait d'une courbe expérimentale supposée être celle d'un modèle. La répartition réelle du paramètre choisi $(\varepsilon_r, \gamma_v$ ou $Ln(\gamma_v))$ est déterminée à partir de ces points. Un test de la droite de HENRY permet de la comparer à une distribution normale et fournit, si l'hypothèse est acceptable la moyenne et l'écart-type de la population.

Les figures 7 à 9 présentent les ajustements obtenus par cette méthode sur le minerai de fer de Droitaumont (SAGHAFI (1981)), le marbre de Carrare (HOUPERT (1972)) et la molasse du Gard. Dans les 3 cas on obtient un ajustement assez fin ; cependant pour le minerai de fer et la molasse, il existe un seuil à partir duquel le modèle est inapte à décrire le comportement. Ce point pourrait coïncider avec l'apparition des macro-fractures. Le phénomène évolue alors vers un glissement relatif des fragments découpés par les macro-fractures.

CONCLUSIONS

Le modèle proposé est, certes, très schématique : les "micro-modèles" sont indépendants, leur module de déformation est constant, la phase de macro-fracturation n'est pas prise en compte.

Il fournit cependant des éléments pour l'interprétation des courbes effort-déformation complètes, et présente l'avantage de décrire la rupture en compression comme un processus continu, débutant avant le maximum et se poursuivant au delà de ce point qui ne correspond pas à l'apparition d'un phénomène nouveau dans le développement du réseau fissural. Les comportements pré et post-maximum y sont conjointement liés à un paramètre de dispersion caractérisant l'hétérogénéité de la structure.

Le rôle de l'hétérogénéité de la structure, mis en évidence par ce modèle suscite des interrogations, notamment quant à l'effet d'échelle : en effet, une augmentation du volume de roche sollicité provoque sans aucun doute un élargissement de la distribution de l'énergie de rupture ce qui implique une modification du comportement et notamment une diminution de la pente post-maximum des roches de classe I. De même les roches de classe II au laboratoire ont peut-être un comportement de classe I in situ. La détermination d'une relation liant les dimensions et les paramètres de la distribution pourrait permettre de prendre en compte l'effet d'échelle dans les calculs de stabilité.

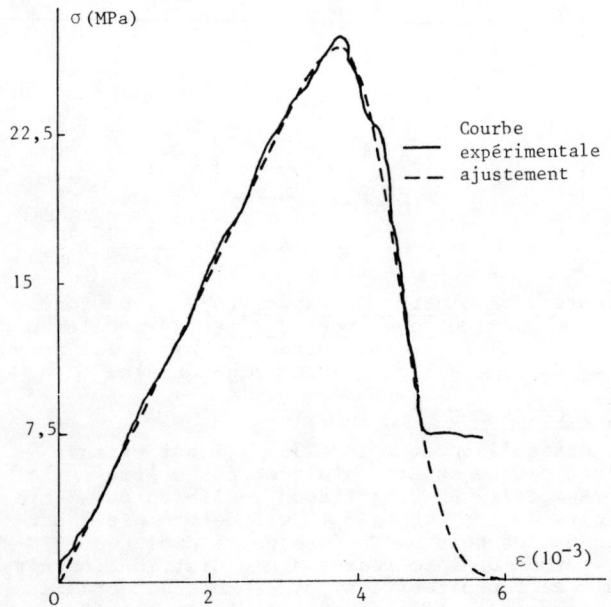

Figure 7   Ajustement sur le comportement du minerai de fer de Droitaumont -ε_r loi normale de moyenne $4,49.10^{-3}$ et d'écart-type $0,49.10^{-3}$

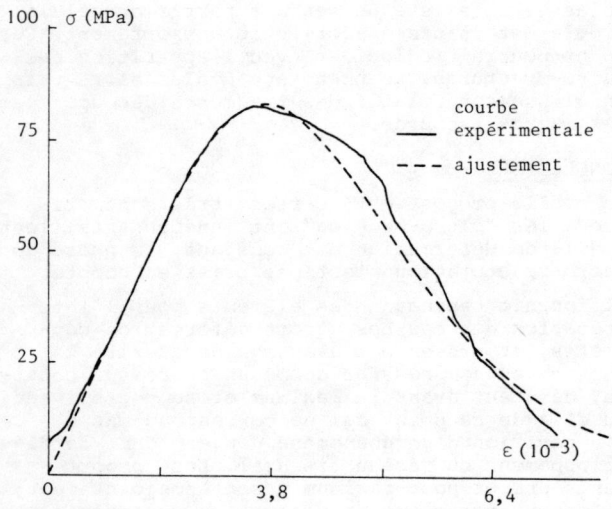

Figure 8 Ajustement sur le comportement du marbre de Carrare-γ_v :loi lognormale de moyenne:$38.10^4 J/m^3$ écart-type:22% de la moyenne des logarithmes.

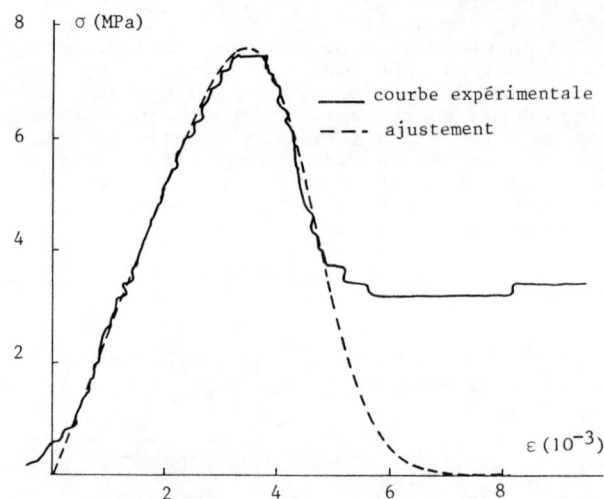

Figure 9   Ajustement sur le comportement de la molasse du Gard - ε_r :loi normale de moyenne : $4,36.10^{-3}$ et d'écart-type $0,87.10^{-3}$

REFERENCES

HOUPERT, R. (1972). La rupture fragile des roches contrôlée au moyen d'une machine d'essai asservie : C.R. Acad. Sci. Paris ; t. 275. Série A. 233-236.

HOUPERT, R. (1979). Le comportement à la rupture des roches. C.R. 4° Cong. Internat. Meca. Roches ; (3), 115-122.

RÜMMEL, F. (1974). Changes in the P-wave velocity with increasing inelastic deformation in rock specimens under compression. In : Advances in Rock Mechanics, C.R. 3° Congr. Internat. Meca. Roches, Denver, Theme 2, 517-523.

SAGHAFI, A. (1981). Etude du comportement mécanique post-rupture du minerai de fer ; Application au dimensionnement des exploitations minières par îlots réduits. Thèse Doct-Ingé., Ecole des Mines de Paris.

PERMEABILITE DU CHARBON DANS DES CONDITIONS TRIAXIALES ET SOUS HAUTES PRESSIONS D'INJECTION

Permeability of coal under triaxial pressure and high injection pressures

Kohlenpermeabilität unter Triaxialdruck und hohem Injektionsdruck

J. Brych
Professeur et Chef de l'Unité de "Constructions MECAROCHES" à la Faculté Polytechnique
de Mons, Ingénieur Conseil en Forages et Mécaroches

P. Defourny
Chercheur à la F.P.Ms — Unité MECAROCHES en 1980/82

J. P. Latour
Chercheur à la F.P.Ms — Unité MECAROCHES en 1980/81

RESUME

Un appareil triaxial permettant d'effectuer les essais de perméabilité radiale et axiale a été dévelopé par L'Unité de Recherche "Constructions MECAROCHES" à la Faculté Polytechnique de Mons. Le diamètre des échantillons peut varier de 40 à 60 mm, la contrainte latérale de 0 à 700 bars et la pression d'injection peut atteindre 500 bars. Une campagne d'essais dans le domaine de perméabilité du charbon a été organisée par l'unité Construction MECAROCHES de la F.P. de Mons et subsidiée à 100% par l'Institut National des Industries Extractives (I.N.I.E.X.) de Liège.

SYNOPSIS

A triaxial device for radial and axial permeability tests has been developed by the Research Unit "Constructions MECA-ROCHES" in the Polytechnical Faculty at Mons. The diameter of specimens may vary between 40 and 60 mm, the lateral constraint from 0 to 700 bars, whilst the injection pressure may reach 500 bars. A series of permeability tests of coal — funded 100% by the National Institute of Extractive Industries (I.N.I.E.X.) at Liège — has been designed by the "Constructions MECAROCHES" unit of the Polytechnical Faculty at Mons.

ZUSAMMENFASSUNG

Die Forschungsstelle "Constructions MECAROCHES" der Polytechnischen Fakultät in Mons hat einen Dreiaxial-Apparat für radiale und axiale Kohlenpermeabilitäts-Untersuchungen entwickelt. Der Probendurchmesser darf zwischen 40 und 60 mm liegen, und der Seitendruck zwischen 0 und 700 Bar, während der Injektionsdruck bis auf 500 Bar ansteigen darf. Eine Reihe von Kohlenpermeabilitätsversuchen, die völlig von dem Nationalinstitut der Bergwerksindustrie finanziert wurde, ist von der "Constructions MECAROCHES" Forschungsstelle der Polytechnischen Fakultät in Mons entworfen worden.

I. INTRODUCTION

En 1976-77, le Fonds National de la Recherche Scientifique de Belgique a subsidié en partie le développement d'une cellule triaxiale (réf. Crédit aux Chercheurs S 2/5 - Gv - E - 35) dont la description a été donnée par J. BRYCH et C. SCHROEDER à l'occasion du dernier congrès de l'I.S.R.M. à Montreux [1].

Cette cellule de base (modèle 700/40-1) a été transformée par la suite en une cellule de perméabilité triaxiale avec la possibilité de mesurer tant la perméabilité linéaire (dans la direction de l'axe de l'échantillon) que radiale (divergente et convergente). La transformation de cette dernière a eu lieu à l'occasion d'un contrat de recherche signé entre l'I.N.I.E.X. Liège et l'Unité de Recherche Constructions MECAROCHES de la Faculté Polytechnique de Mons et concernant une campagne d'essais de laboratoire dans le domaine de perméabilité du charbon. Cette étude devait, en effet, précéder une expérimentation in situ en matière de gazéification souterraine du charbon et avait principalement pour but de mieux comprendre les phénomènes de circulation de différents fluides à travers une couche de charbon en grande profondeur et sous hautes pressions d'injection.

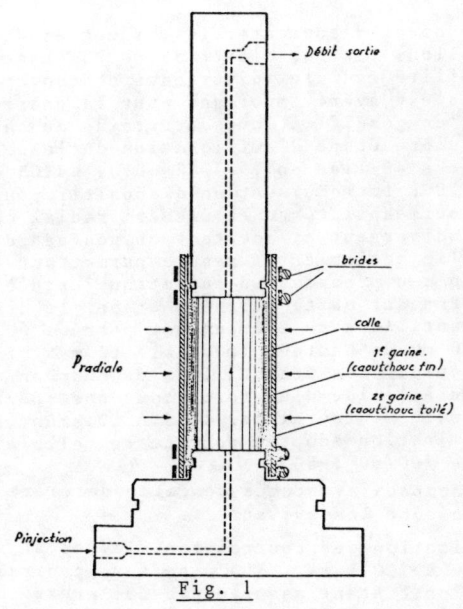

Fig. 1

II. CONCEPTION DU MATERIEL EXPERIMENTAL

Par rapport à la construction initiale [1] de
l'appareil triaxial F.P.Ms - modèle 700/40-1,
les changements suivants, permettant les mesures
de perméabilité en écoulement linéaire ont été
introduits par J. BRYCH et P. DEFOURNY [2],
fig. 1 (réf. Contrat I.N.I.E.X./F.P.Ms MECA-
ROCHES 1980/81).

Le dispositif d'injection et de mesure adopté à
cette période permettrait d'injecter les fluides
liquides ou gazeux à travers l'échantillon et
dans la direction de l'axe de la carotte sous
une pression max. de 200 bars. La cellule ainsi
modifiée porte la dénomination modèle 700/40-2-
200.

Par la suite, étant donné la nécessité d'élargir
l'expérimentation vers d'autres types d'écoulements, P. DEFOURNY et J.P. LATOUR modifient l'enveloppe intérieure de la cellule 700/40-2-200
permettant l'introduction des essais de perméabilité radiale pour les échantillons du 40 mm de
diamètre. Un dispositif amélioré d'injection
permettra d'atteindre les pressions d'injection
de l'ordre de 300 bars. Cette cellule améliorée
portera la dénomination modèle 700/40-3-300.

(réf. contrat INIEX/F.P.Ms MECAROCHES 1981/82)
Fig. 2 [1].

Fig. 4

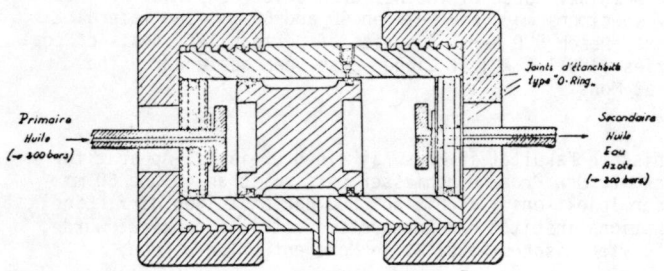

Fig. 2

Dans le souci d'augmenter les diamètres des
échantillons testés, d'introduire les essais de
perméabilité radiale (divergents et convergents)
ce qui s'est avéré important pour la phase du
linking en gazéification souterraine du charbon,
une dernière étape d'amélioration de la cellule
initiale a eu lieu en 1981/82 où J. BRYCH et
J.P. LATOUR introduisent un dispositif pour mesurer la perméabilité en écoulement radial convergent et divergent et adoptent un nouveau dispositif d'injection et de mesure permettant l'utilisation des pressions d'injection jusqu'à 500
bars. Pendant cette dernière étape, le diamètre
des échantillons de charbon est porté à 60,3 mm
et de 80 mm de hauteur (possibilité max. 150 mm
de hauteur). Les échantillons de charbon sont
découpés à l'aide d'un foret couronne spécialement étudié et fourni par la S.A. Diamant Boart.
La dénomination adoptée pour cette cellule est
celle de 700/60-1-500. (Fig. 3, 4).

Les principaux avantages de cette dernière modification sont les suivants :

1. Application des contraintes élevées et allant
   jusqu'à 700 bars. (La mise sous contrainte
   verticale étant assurée par une presse hydraulique de 250 tonnes).

2. Indépendance totale en application et main-
   tien dans le temps des contraintes verticale
   et horizontale.

3. Ecoulement radial effectif, le fluide sor-
   tant (écoulement divergent) ou entrant
   (écoulement convergent) perpendiculairement
   à l'axe de l'échantillon.
4. Possibilité de réaliser un écoulement convergent sous contrainte triaxiale.
5. Possibilité de travailler avec des pressions
   d'injection supérieures à la contrainte radiale.

Un système de régulation de contrepression (à
l'aide d'une vanne pneumatique Kämmer) à la sortie de l'échantillon a été monté sur le banc
d'essais [3] Fig. 5 et 5bis.

Un tel dispositif nous a permis d'effectuer une
série d'essais en matière du comportement hydraulique des échantillons de charbon en écoulement linéaire et radial en fonction de l'augmentation de la pression d'injection des fluides
liquides ou gazeux utilisés et cela sans ou avec
contre-pression à la sortie de l'échantillon.
Le charbon testé était celui du charbonnage
belge de Zolder [4] Tabl. 1 provenant de la couche dans laquelle les essais de perméabilité in
situ ont été réalisés par l'I.N.I.E.X..

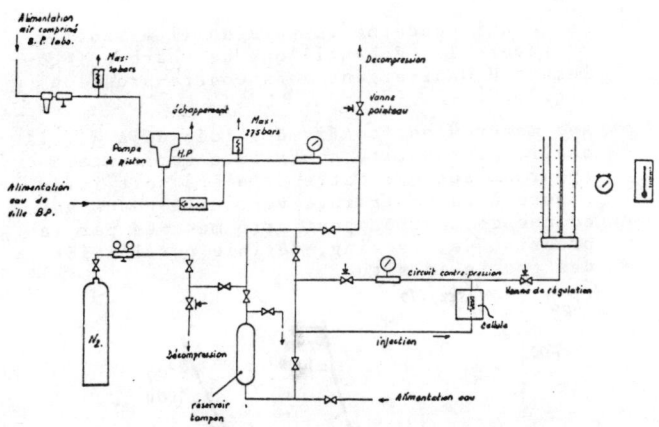

Fig. 5

Fig.5bis

Type d'essai	Résistance R(Kg/cm^2)	Module d'élasticité (Kg/cm^2)
Traction brésilienne	15	
Compression simple	100 - 120	10.000 - 20.000
Compression triaxiale	470 (pour p_{conf}=30b)	28.000
	300-600 (pour p_{conf}=60b)	33.000
	400-700 (pour p_{conf}=100b)	28.000
	(pour p_{conf}=200b)	10.000 à 230.000

Tabl.1

III. TENDANCES DEGAGEES DES ESSAIS

1. En ce qui concerne les essais d'injection d'eau à travers les échantillons de charbon (l'écoulement dans la direction de l'axe de la carotte et sans contre-pression à la sortie).

   En règle générale, les relations débit-pression d'injection et perméabilité-pression d'injection prennent l'allure représentée aux figures 6 et 7. Malgré quelques anomalies plutôt isolées, on a pu constater que les valeurs du débit et de la perméabilité étaient en général assez bien reproductibles. Il est cependant important de noter que nous n'avons jamais constaté d'apparition de fractures (par observation visuelle après essai) parallèles à la direction d'écoulement même lorsque la pression d'injection atteignait 250 bars, c'est-à-dire 70 bars de plus que la contrainte perpendiculaire à l'écoulement.

2. En ce qui concerne l'influence de la contre-pression sur les injections d'eau à travers les échantillons de charbon (écoulement dans la direction de l'axe de la carotte) fig. 8.

   Pour une pression d'injection donnée, l'application d'une contre-pression a pour effet d'augmenter la perméabilité de l'échantillon et de diminuer le débit qui le traverse. Pour une même différence de pression entre les extrêmités d'un échantillon, la perméabilité et le débit augmentent lorsque le niveau de la pression interne dans l'échantillon augmente. Les courbes de la perméabilité en fonction de la pression interne moyenne (fig. 8) ont une allure similaire à celle de la fig. 7 en ce sens qu'il y a également augmentation brusque de la perméabilité lorsque la pression interne dépasse la valeur des con-

En ce qui concerne les contraintes auxquelles les échantillons ont été soumis pendant l'expérimentation, elles étaient correspondantes aux conditions naturelles du site de gazéification souterraine du charbon de Thulin en Belgique, c'est-à-dire contrainte verticale de 180 bars et contrainte horizontale de 100 bars.

Ajoutons que les valeurs de perméabilité déduites des essais ont été calculées à l'aide des formules classiques de Darcy applicables en différentes variantes aux écoulements linéaires ou radiaux des fluides liquides ou gazeux. Ces formules classiques permettent d'obtenir aisément une première idée du comportement du charbon soumis aux divers essais de perméabilité.

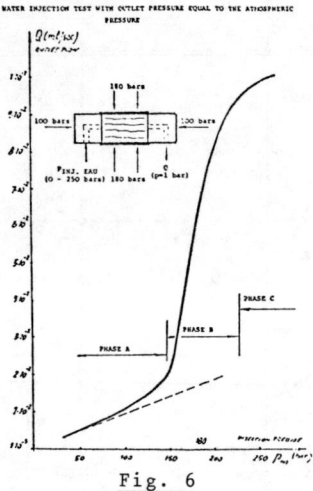

WATER INJECTION TEST WITH OUTLET PRESSURE EQUAL TO THE ATMOSPHERIC PRESSURE

Fig. 6

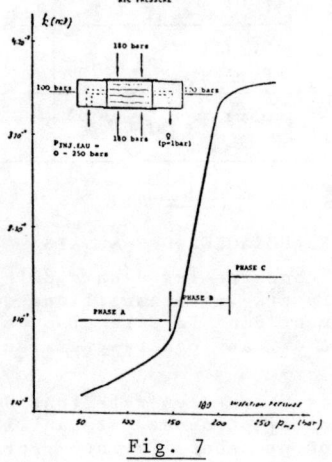

WATER INJECTION TEST WITH OUTLET PRESSURE EQUAL TO THE ATMOSPHERIC PRESSURE

Fig. 7

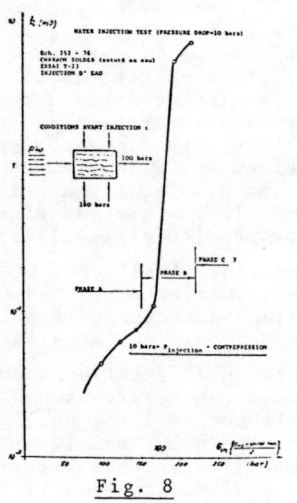

WATER INJECTION TEST (PRESSURE DROP=10 bars)

Fig. 8

traintes perpendiculaires à l'écoulement, sans pour cela qu'il y ait fracturation de l'échantillon.

3. En ce qui concerne la perméabilité radiale à travers les échantillons de charbon (écoulement H_2O divergent sans contre-pression) Fig.9.

Nos mesures ont confirmé l'idée déjà acquise que la plus petite contrainte horizontale (pour autant que cette dernière soit inférieure à la contrainte verticale) dans une couche de charbon peut être mesurée par la pression de fracking, définie par l'INIEX des essais in situ.

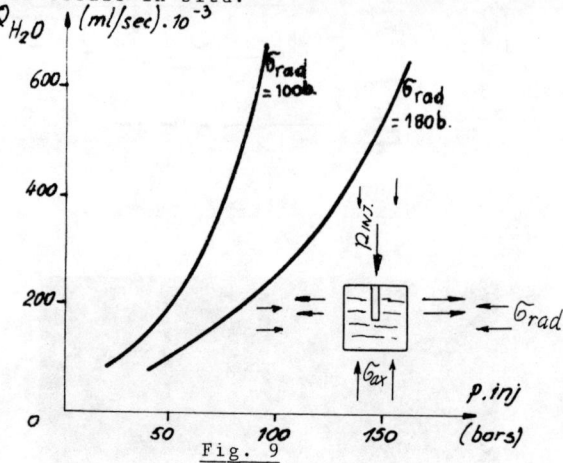

Fig. 9

4. En ce qui concerne la perméabilité radiale à travers les échantillons de charbon (écoulement H_2O convergent sans contre-pression) Fig. 10 et 11.

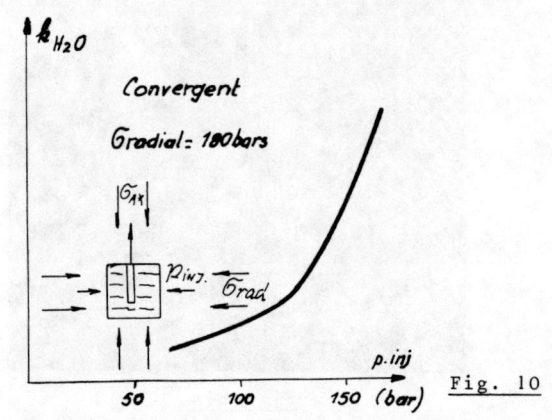

Fig. 10

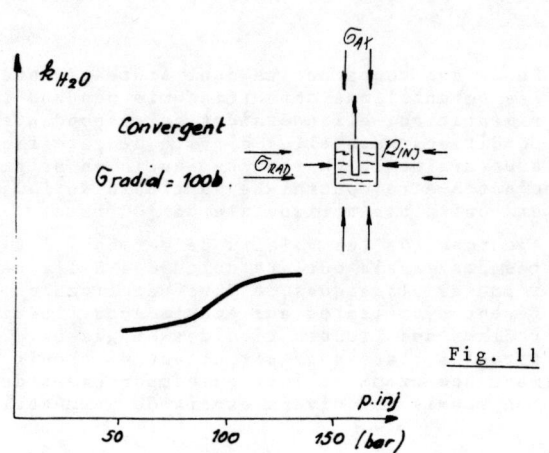

Fig. 11

A 168

On constate que si la pression d'injection
approche la valeur de la contrainte radiale,
la perméabilité augmente rapidement (fig.10).
La fig. 11 indique que si la pression d'injection dépasse la valeur de la contrainte
radiale, la perméabilité tend à devenir constante.

5. En ce qui concerne l'influence d'une fracture en perméabilité radiale (fig. 12 et 13).

Les fig. 12 et 13 montrent que si une seule
fracture existe (verticale ou horizontale)
la contrainte perpendiculaire à la fracture
est déterminante pour l'évolution de k_{H_2O}.

On a constaté ici que la perméabilité du
charbon k_{H_2O} peut être supérieure en cas
d'écoulement convergent par rapport à l'écoulement divergent.   Ce fait a été observé
dans plus de 90 % des essais pour des pressions d'injection inférieures à 100 bars.
Au-delà de cette pression k_{H_2O} croît rapidement et dépasse généralement k_{H_2O} convergent.

6. En ce qui concerne la perméabilité radiale
à travers les échantillons testés avec contre-
pression (écoulement H_2O convergent) Fig. 14.

Contrairement aux observations réalisées en
écoulement axial, les perméabilités observées
avec une contre-pression sont inférieures à
celles observées sans contre-pression.

La perméabilité avec contre-pression augmente cependant plus vite que la perméabilité
sans contre-pression. En ce qui concerne
l'injection d'azote, à l'exception d'un ou
de deux cas où il semble que la turbulence
ait joué un rôle important, les tendances
générales dégagées des essais à l'azote sont
exactement similaires à celles dégagées pour
l'eau.

7. En ce qui concerne l'injection d'azote à travers les échantillons de charbon (écoulement
dans la direction de l'axe des carottes et
sans contre-pression) fig. 15 et fig. 16.

Qualitativement la relation k_{N_2} - P_{inj} est
similaire à la relation k_{H_2O} - P_{inj} déjà observée précédemment.
Une observation plus minutieuse des résultats
a cependant permis de construire une courbe-
type (fig. 16) qui met en évidence quelques
particularités de l'écoulement de l'azote
dans des échantillons testés.

8. En ce qui concerne l'injection d'azote à
travers les échantillons de charbon avec
contre-pression (écoulement dans la direction
de l'axe de la carotte) fig. 17, 18.

Pour des pressions d'injection avoisinant la
pression radiale sur l'échantillon (p_{inj} =
100 bars), l'effet d'une contre-
pression sur l'écoulement d'azote peut être
considéré comme qualitativement similaire à
l'effet constaté sur un écoulement d'eau
(fig. 17 et 18), c'est-à-dire la perméabilité
augmente si la contre-pression augmente et
le débit diminue si la contre-pression augmente.

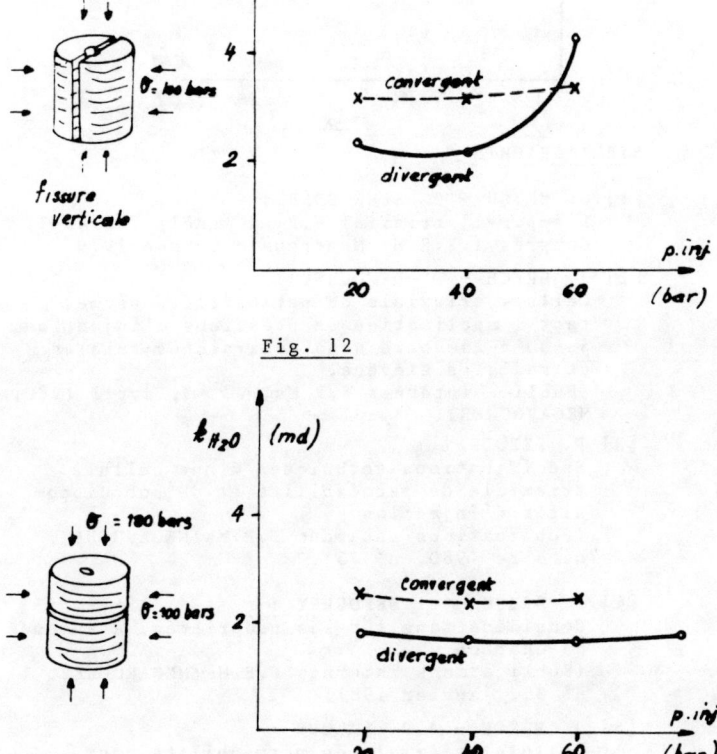

Fig. 12

Fig. 13

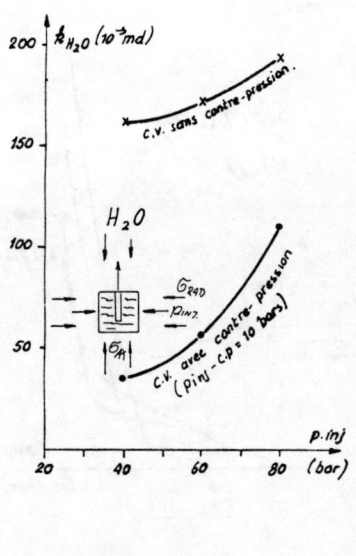

Fig. 14

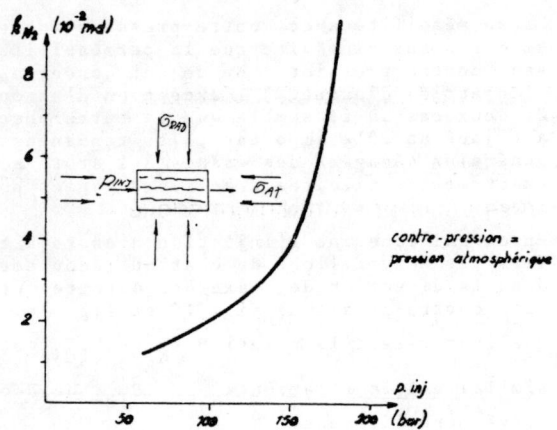

Fig. 15

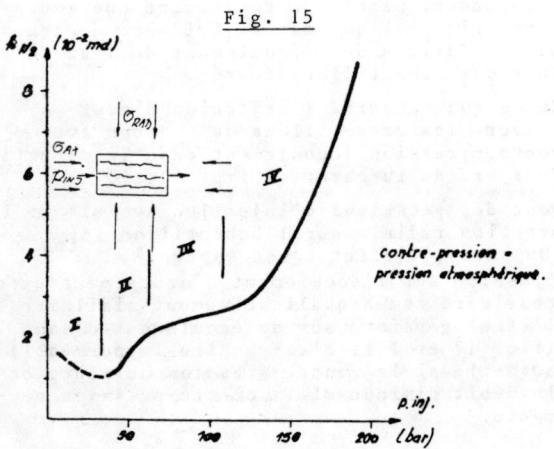

Fig. 16

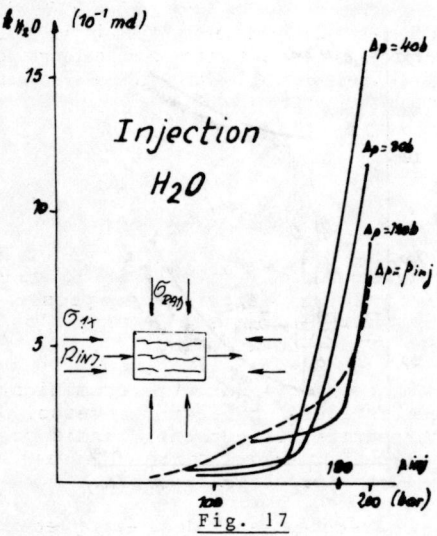

Fig. 17

l'écoulement. Une contre-pression a pour effet d'augmenter k_{gaz}, mais de diminuer le débit.

Pour des pressions d'injection inférieures à 100 bars le cas du gaz est cependant plus complexe que celui de l'eau, vraisemblablement en raison de la très grande compressibilité du gaz et de sa très faible viscosité (lorsque la pression d'injection augmente, l'effet d'ouverture des pores qui accroît la perméabilité du charbon peut être atténuée voire anihilée par l'apparition de la turbulence dans l'écoulement. Ce phénomène a été également constaté pour l'eau, mais à des pressions d'injection de l'ordre de 250 bars. Seules les valeurs très différentes de paramètres physiques de ces deux types de fluides seraient la cause des différences de comportement constatés dans nos essais.

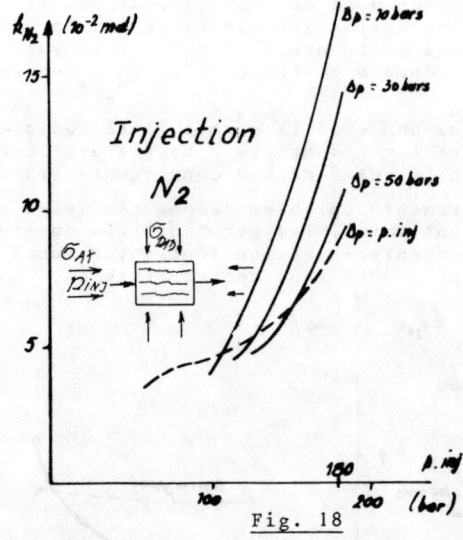

Fig. 18

BIBLIOGRAPHIE

[1] J. BRYCH - C. SCHROEDER :
L'appareil triaxial F.P.Ms modèle 700/40-1.
Congrès I.S.R.M. Montreux - Suisse 1979.

[2] J. BRYCH - P. DEFOURNY :
Cellule triaxiale de perméabilité permettant l'application de pressions d'injection jusqu'à 250 bars sous contraintes axiales et radiales élevées.
(Public. internes F.P.Ms, n° 43, avril 1980, MECAROCHES).

[3] P. DEFOURNY :
Spécifications techniques d'une cellule triaxiale de perméabilité et de son dispositif d'injection.
(Publications internes F.P.Ms/MECAROCHES, octobre 1980, n° 75).

[4] J. BRYCH - P. DEFOURNY :
Considérations sur les propriétés mécaniques du charbon.
(Publications internes F.P.Ms/MECAROCHES, n° 85, janvier 1981).

[5] J. BRYCH - J.P. LATOUR :
Cellule triaxiale de perméabilité pour l'étude des écoulements convergents et divergents.
(Publications F.P.Ms/MECAROCHES, n° 95, mars 1981).

9. Comparaison gaz-liquide

On peut considérer que pour des pressions d'injection supérieures à 100 bars l'écoulement du gaz à travers l'échantillon du charbon est similaire à celui de l'eau, c'est-à-dire k_{gaz} augmente fortement si P_{inj} approche ou dépasse la contrainte perpendiculaire à

NOUVELLE PRESSE TRIAXIALE — ETUDE DE MODELES DISCONTINUS BOULONNES

A novel triaxial press — Study of anchored jointed models
Eine neuartige Dreiaxialpresse — Untersuchung verankerter Kluftkörpermodelle

P. Egger
Chef de Section, Laboratoire de Mécanique des Roches, Ecole Polytechnique Fédérale de Lausanne, Suisse

H. Fernandes
Ingénieur, Laboratoire de Mécanique des Roches, Ecole Polytechnique Fédérale de Lausanne, Suisse

RESUME

L'étude expérimentale de massifs discontinus a donné lieu à la construction d'une presse triaxiale asservi de caractéristiques particulières: dimension des éprouvettes 30 sur 60 cm, rigidité 1,5 GN/m, effort axial maximum 3,5 MN, pression latérale 5 MN/m². Le plateau inférieur de la presse est libre de se déplacer horizontalement, une éventuelle excentricité de l'effort axial étant enregistrée par des capteurs de force disposés en triangle. Une première série d'essais portait sur l'étude d'éprouvettes en béton présentant un joint incliné et boulonné par deux barres d'acier. L'influence de l'angle entre joint et ancrages sur le comportement des éprouvettes est communiquée et discutée.

SYNOPSIS

The experimental study of jointed rocks gave rise to the construction of a servo-controlled triaxial press with particular characteristics: size of samples 30 cm × 60 cm, rigidity 1.5 GN/m, max. axial thrust 3.5 MN, lateral pressure 5 MN/m². The lower platen of the press is free to move in the horizontal plane, excentricities of the axial thrust being measured by load cells arranged in a triangle. A first series of tests dealt with concrete samples crossed by an inclined joint and bolted together by two steel bars. The influence of the angle between bolts and joint on the behaviour of the test samples is presented and discussed.

ZUSAMMENFASSUNG

Zur experimentellen Untersuchung geklüfteter Medien wurde eine servogesteuerte Dreiachsialprese mit besonderen Eigenschaften gebaut: Probengröße 30 cm × 60 cm, Steifigkeit 1,5 GN/m, größte Achsialkraft 3,5 MN, Seitendruck 5 MN/m². Die untere Probenaufnahmeplatte ist horizontal frei beweglich, Exzentrizitäten der Achsialkraft werden mittels im Dreieck angeordneter Kraftmeßdosen festgestellt. Eine erste Versuchsserie wurde an Probenkörpern aus Beton durchgeführt, die eine geneigte Trennfläche aufwiesen und mittels zweier Stahlstangen verankert wurden. Der Einfluß des Winkels zwischen Ankerstangen und Trennfläche auf das Verhalten der Proben wird dargelegt und diskutiert.

1. INTRODUCTION

En mécanique des roches, le comportement mécanique des massifs fissurés continue à constituer un des thèmes principaux de recherche. L'étude expérimentale in situ se heurtant souvent à son prix élevé, l'équipement des laboratoires doit être complété par des presses capables de recevoir des éprouvettes représentatives du massif, de dimensions souvent importantes. Ces considérations ont amené le Laboratoire de Mécanique des Roches de

l'Ecole Polytechnique Fédérale de Lausanne à concevoir et faire construire une presse triaxiale comportant un certain nombre d'aspects particuliers et d'innovations.

Dans la première partie de cette communication, les caractéristiques de cette presse, dite TRIROC sont décrites brièvement, tandis que les résultats d'une première série d'essais sont communiqués dans la deuxième partie.

Ces essais, exécutés sur des éprouvettes en

béton munies d'ancrages passifs, avaient un double but : amélioration des connaissances sur l'effet mécanique d'un boulonnage passif et rodage de la nouvelle presse.

2. CARACTERISTIQUES DE LA PRESSE "TRIROC"

Le choix des dimensions des éprouvettes susceptibles d'être examinées dans la nouvelle presse triaxiale, était guidé par l'abondance de roches stratifiées (flysch, molasse) et de moraines, dans le bassin lémanique, et le coût de construction. Il s'est porté sur 30 cm de diamètre, pour des éprouvettes cylindriques, ou 30 cm de côté, pour des prismes à base carrée, et une hauteur de 60 cm. [Fig. 1].

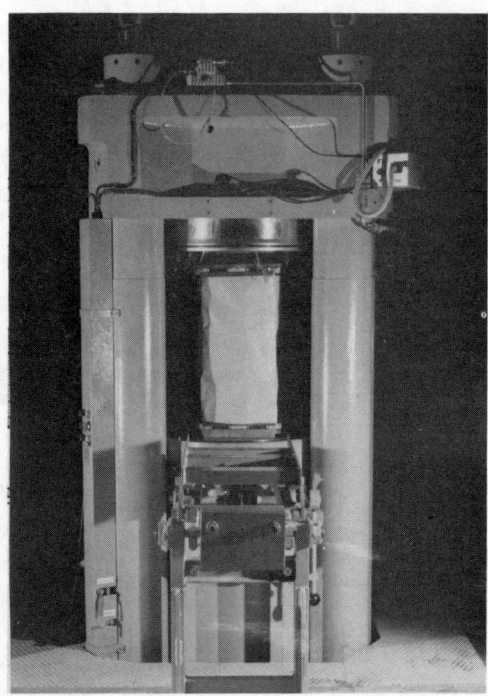

Fig. 1 : Presse TRIROC ouverte avec éprouvette prismatique

Ces dimensions permettent de réaliser des essais sur des roches stratifiées ou fissurées à un intervalle maximum de 5 à 6 cm et sur des sols meubles comportant des grains du même ordre de grandeur.

La capacité de la presse - 3,5 MN d'effort vertical et 5 MN/m^2 de pression latérale - a été choisie en fonction de la nature des éprouvettes décrites ci-dessus. Un soin particulier a été apporté à la possibilité d'étudier le comportement post-rupture des éprouvettes dans les meilleures conditions. Pour ce faire, la presse répond aux trois conditions suivantes :

- Grande rigidité : la rigidité de la presse est de 1,5 GN/m ce qui équivaut à celle d'une

éprouvette ayant un "module de déformation négatif" de 10 GN/m^2 dans le domaine

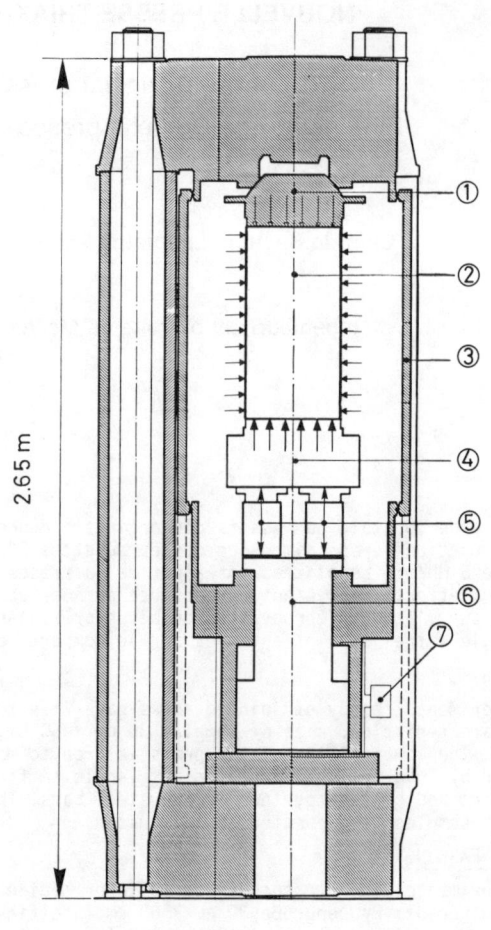

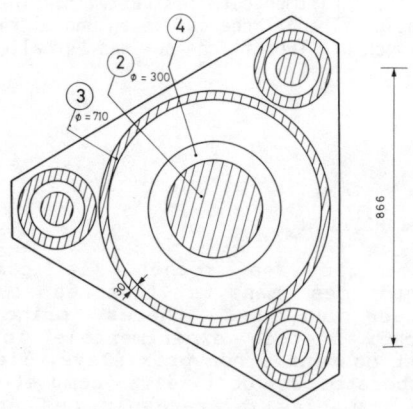

Fig. 2 : Presse TRIROC - Coupes verticale et horizontale
1) plateau supérieur articulé
2) éprouvette (H=60 cm, D=30 cm)
3) jupe mobile
4) plateau inférieur libre horizontalement
5) 3 capteurs de force
6) vérin à double effet
7) servovalve

post-rupture. Dans ce but, l'espacement des trois tirants de la presse a été réduit au minimum en adoptant une jupe coulissant verticalement, pour l'application de la pression latérale, au lieu d'utiliser une cellule indépendante [Fig. 2].

Outre l'augmentation de rigidité, cette disposition permet une économie de place et une simplification de l'exécution des essais.

L'étanchéité est assurée par des O-rings placés entre la jupe coulissante et le corps de la presse. Pour assurer le contact entre O-rings et jupe pour toute valeur de la pression d'eau, les O-rings sont montés sur des supports cylindriques plus déformables que la jupe. La jupe est munie de trois hublots et d'un système d'éclairage à fibres optiques permettant de suivre l'essai pendant son exécution [Fig. 3].

Fig. 3 : Presse fermée avec groupe hydraulique et armoire de commande

- Elimination d'efforts parasites après la formation d'un plan de rupture : dans les presses triaxiales traditionnelles, la continuation de l'essai au-delà de l'apparition d'un plan de rupture incliné entraîne des efforts parasites gênants dans l'éprouvette. Afin d'éviter cet inconvénient, le plateau inférieur de la presse TRIROC repose sur deux niveaux de rouleaux croisés, laissant une liberté de mouvement parfaite dans le plan horizontal et permettant ainsi une translation parfaite sur le plan de rupture de l'éprouvette [Fig. 4].

Les déplacements horizontaux du plateau inférieur sont mesurés à l'aide de deux capteurs inductifs, la force verticale et une éventuelle excentricité de celle-ci à l'aide de trois capteurs de 1,8 MN disposés en triangle.

- Vitesse de déplacement vertical contrôlée : un système d'asservissement comprenant deux servovalves de type MOOG, confère à la presse une grande souplesse d'utilisation, les essais courants se faisant à vitesse imposée

Fig. 4 : Eprouvette de béton homogène après rupture

constante. Le déplacement vertical est mesuré simultanément par deux capteurs, l'un incorporé au vérin et l'autre placé à côté de l'éprouvette, pour garantir un maximum de sécurité lors de l'exécution d'essais.

Toutes les données caractérisant l'essai (déplacements, forces, pressions) peuvent être affichées et enregistrées sur bande magnétique pour un traitement ultérieur.

3. ETUDE D'ANCRAGES PASSIFS

a) Objectifs

Bien que l'ancrage des massifs rocheux par des barres scellées sur toute la longueur ou "boulonnage passif" soit couramment utilisé en pratique, son mode d'action est loin d'être éclairci. Depuis les travaux fondamentaux de BJURSTROEM (1974) et les considérations générales sur la "roche armée" de LONDE et BONAZZI (1974), plusieurs chercheurs, par exemple AZUAR (1977), EGGER (1978), AZUAR et al. (1979), ont contribué à une meilleure connaissance du problème, par des travaux expérimentaux et théoriques. Néanmoins, bien des questions restent ouvertes, en particulier l'influence sur leur effet mécanique de l'orientation des barres par rapport aux discontinuités. C'est à cette dernière question que la série d'essais, exécutée récemment dans la presse TRIROC, essaie de donner des éléments de réponse.

b) Essais réalisés

Dans le but décrit ci-dessus, onze éprouvettes prismatiques en béton, de dimensions 30 x 30 x 60 cm^3 et traversées par un joint lisse à 45°, ont été fabriquées, dans des conditions identiques. Dix éprouvettes ont été armées par deux barres d'acier Ø 6 à surface lisse, d'une résistance à la traction de 14,0 kN et inclinées sous un angle α variant entre 25° et 90° par rapport au joint [Fig. 5].

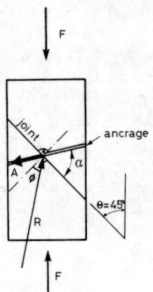

Fig. 5 : <u>Schéma statique de l'éprouvette ancrée</u>

Les éprouvettes ont été sollicitées en compression simple dans la presse TRIROC jusqu'à
la rupture des deux ancrages, le plateau mobile
inférieur de la presse garantissant la vertica-
lité de l'effort appliqué à tout moment de
l'essai.

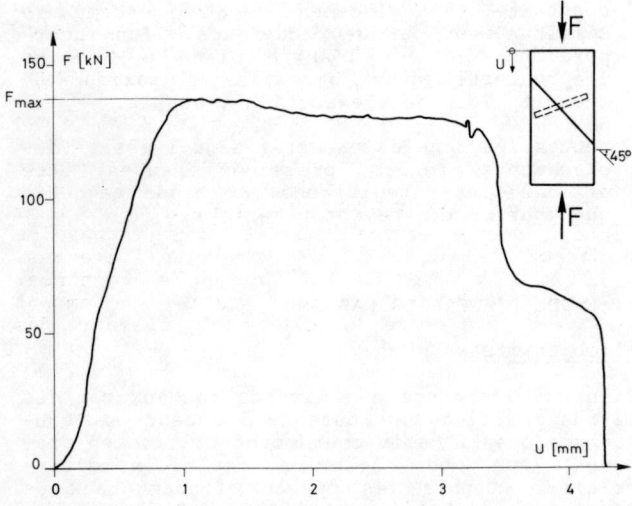

Figure 6 : <u>Allure de la courbe effort-déplace-
ment (éprouvette No 1 : $\alpha = 48^{\circ}$)</u>

La figure 6 montre l'allure typique de la courbe force-déplacement enregistrée pour chaque
essai.

A la fin des essais, les barres d'ancrage ont
été dégagées avec précaution afin de mesurer la
profondeur sur laquelle elles étaient déformées, ainsi que l'angle α* entre le joint et
l'axe des barres, au droit de la zone de rupture.

Par ailleurs, l'évolution de la déformation des
barres pendant l'essai a été suivie sur l'une
des éprouvettes, à l'aide de radiographies
[Fig. 7].

La onzième éprouvette, non armée, a été soumise
à un essai triaxial, pour déterminer l'angle de
frottement sur le joint lisse.

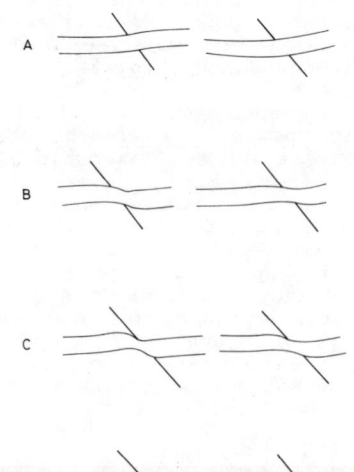

Fig. 7 : <u>Déformation des barres d'ancrage
pendant l'essai (d'après radiographies)</u>. Déplacement vertical
u = <u>A</u>-zéro, <u>B</u>-1,9 mm, <u>C</u>-3,5 mm,
<u>D</u>-4,3 mm.

c) <u>Résultats</u>

Les valeurs maximales de l'effort vertical
F_{max} enregistrées pendant les essais sont
portées dans la figure 8, en fonction de
l'orientation des ancrages. Pour chaque essai,
les valeurs des angles α et α* sont indiquées.
A titre de comparaison, la figure 8 montre
aussi les efforts théoriques exprimés par la
relation

$$F = A \cdot \frac{\cos (\alpha - \phi)}{\cos (\theta + \phi)} \qquad [1]$$

en utilisant la notation de la Fig. 5 et les
résultats de l'essai triaxial (ϕ = 32,5...
...33,5°). Cette relation est basée sur l'hypothèse que les barres d'acier travaillent en
traction simple, à la limite de la résistance.
Le graphique appelle les remarques suivantes :

- En raison d'un incident de fabrication, le
  joint de l'éprouvette No 5 n'était pas
  complètement fermé, au début de l'essai,
  contrairement aux autres éprouvettes. Ceci a
  entraîné une déformation des barres d'ancrage
  jusqu'à l'obtention du contact entre les deux
  lèvres du joint, d'où une valeur initiale
  douteuse de l'angle α.

- L'éprouvette No 7 a été jetée accidentelle-
  ment, après l'essai, mais avant la mesure de
  α*.

On remarque qu'à l'exception de l'éprouvette
No 5, l'angle final α* entre joint et extrémité

A 174

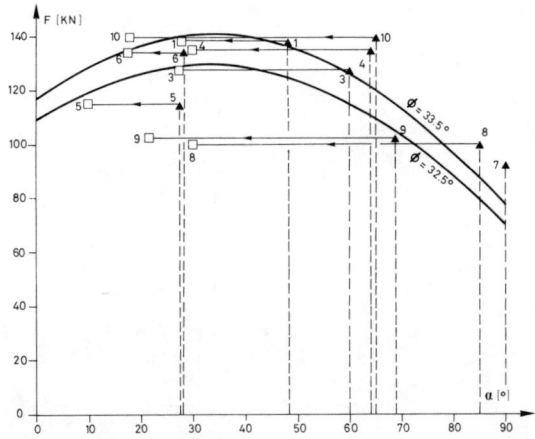

Fig. 8 : Effort vertical maximum en fonction de l'orientation des ancrages (▲ angle initial, □ angle au droit de la zone de rupture).

des barres de toutes les éprouvettes est compris entre 18° et 30° environ, indépendamment de sa valeur initiale α.

La profondeur - perpendiculaire au joint - sur laquelle les barres sont déformées, varie entre 3 et 9 mm environ et augmente avec l'angle α.

La prévision théorique maximale (pour α = φ, voir éq. 1) comprise entre 129 et 140 kN, décrit assez bien les résistances obtenues, pour la plupart des éprouvettes. Font exception, notamment, l'éprouvette No 5 déjà discutée (α corrigé < 30°) et les éprouvettes No 7...9 dont l'angle initial α varie entre 68 et 90°. Pour ces dernières, la résistance mesurée de 93 à 102 kN est comprise entre la prévision donnée ci-dessus, basée sur des ancrages travaillant en traction simple, et la valeur obtenue en cisaillement pur (65... 70 kN, en admettant une résistance au cisaillement égale à 60 % de celle à la traction).

Les déplacements verticaux mesurés (valeurs moyennes des éprouvettes à angle égal) sont environ de 2 à 6 mm jusqu'à la résistance de pic, F_{max}, de 4,5 à 7,5 mm et de 5 à 10 mm lors de la rupture de la première puis de la deuxième barre d'ancrage. Les déplacements les plus faibles ont été observés pour α = 40°...50°. Notons pour mémoire que les déplacements tangentiels le long du joint sont obtenus en multipliant ces valeurs par √2.

d) Conclusions

La série d'essais décrite ci-dessus dans laquelle seule l'orientation des barres

d'ancrage était variable, a permis de tirer les conclusions suivantes :

- Pour des angles d'inclinaison de l'ancrage par rapport au joint α compris entre 30° et 65° environ, la résistance de pic ne dépend pratiquement pas de la valeur de α et correspond sensiblement à la résistance théorique obtenue pour l'orientation optimale (α = φ). Pour α près de 90°, la résistance de pic tombe à 70...75 % de cette valeur, mais elle reste presque de moitié supérieure à la résistance au cisaillement pur.

- Les déplacements nécessaires pour obtenir la résistance de pic, puis la rupture des barres, accusent un minimum prononcé pour α = 40...50° et augmentent fortement pour des angles α < 40° et α > 50°. Ces déplacements paraissent être liés à la profondeur sur laquelle les barres sont déformées. Celle-ci atteint un maximum très net pour des barres perpendiculaires au joint (p_{max} = 1,5 diamètre).

- Dans le cas étudié, l'orientation optimale des ancrages se situe autour de α = φ = 33°; elle combine une résistance élevée avec une grande ductilité, la rupture totale du joint (c'est-à-dire rupture des deux barres d'ancrage) n'intervenant qu'après un déplacement de l'ordre de deux diamètres des barres. D'autres orientations conduisent soit à un joint plus fragile (α ≈ 45°), soit moins résistant (α ≈ 90°). Il fait toutefois rappeler que ces résultats ont été obtenus sur des joints lisses avec des barres de 6 mm de diamètre et ne peuvent être extrapolés directement dans des conditions différentes. C'est pourquoi il est prévu de continuer cette étude en faisant varier notamment l'état de surface du joint et le diamètre des barres d'ancrage.

REFERENCES

AZUAR, J.-J. (1977). Stabilisation des massifs rocheux fissurés par barres d'acier scellées : Rapp. rech. LCPC, No 73.

AZUAR, J.-J. et al. (1979). Le renforcement des massifs rocheux par armatures passives : C.R. 4e Congr. SIMR, 1, Montreux, 23-30.

BJURSTROEM, S. (1974). Shear strength of hard rock joints reinforced by grouted untensioned bolts : C.R. 3e Congr. SIMR, II-B, Denver, 1194-1199.

EGGER, P. (1978). Dimensionnement des ancrages en souterrain : Publ. SSMSR, No 98.

LONDE, P. et BONAZZI, D. (1974). La roche armée. C.R. 3e Congr. SIMR, II-B, Denver, 1208-1211.

NON-DESTRUCTIVE FIELD TEST OF CEMENT-GROUTED BOLTS WITH THE BOLTOMETER

Contrôle non-destructif des boulons à roches ciment-coulés avec le Boltometer

Zerstörungsfreie Kontrolle von Zement-verpressten Felsankern mit Hilfe des Boltometers

Sten G. A. Bergman
Dr. Eng., Consultant, Stocksund, Sweden

Norbert Krauland
Dipl. Ing., Boliden Mineral AB, Boliden, Sweden

Juri Martna
Chief Eng. Geologist, Swedish State Power Board, Vällingby, Sweden

Taisto Paganus
Chief Rock Mechanics, LKAB, Kiruna, Sweden

SYNOPSIS:

The BOLTOMETER is an electronic instrument for non-destructive in situ testing and control of the quality of grouted rock bolts. On the basis of tests with prototype instruments on rock bolts with varying properties a classification system for cement-grouted bolts has been proposed. The paper reports on tests made on 271 cement-grouted bolts and 21 resin-grouted bolts in order to verify the validity of the classification system. The tests on production bolts showed a fairly high percentage of low-quality rock bolts. Some of these bolts have been overcored or tested by hydraulic jacks. The authors conclude that the BOLTOMETER opens up new vistas in the field of tunnel support by bolting.

ZUSAMMENFASSUNG:

Das BOLTOMETER ist ein elektronisches Gerät für die zerstörungsfreie in-situ Kontrolle von Zement-verpressten Felsankern. Aufgrund der Ergebnisse von Feldversuchen mit Prototypen des Instrumentes an Felsankern verschiedener Typen und Güten wurde ein System für die Klassifizierung der Ankergüte entwickelt. Dieser Vortrag berichtet von Versuchen mit 271 Zement- und 21 Kunstharz-verpressten Ankern mit denen das Klassifizierungssystem überprüft wurde. Die Kontrolle von Betriebsankern zeigte einen ziemlich hohen Anteil von Ankern unzulänglicher Qualität. Einige dieser Anker wurden ausgebohrt oder mit Spannpressen überprüft. Die Verfasser folgern daraus, dass das BOLTOMETER neue Möglichkeiten im Bereich dieser Untertageverstärkung mit Hilfe von Ankern eröffnet.

RÉSUMÉ:

Le BOLTOMETER est un instrument électronique destiné au test non destructeur in situ et au contrôle de la qualité des boulons de roche scellés. Un système de classification des boulons scellés au ciment a été proposé sur la base de tests effectués avec des instruments prototypes sur des boulons de roche possédant diverses propriétés. Le texte mentionne des tests effectués sur 271 boulons scellés au ciment et 21 boulons scellés à la résine dans le but de vérifier la valeur du système de classification. Les tests sur les boulons de production ont permis de définir un pourcentage relativement élevé de boulons de roche de mauvaise qualité. Certains de ces boulons ont été surdimensionnés ou testés à l'aide de vérins hydrauliques. Les auteurs concluent que le BOLTOMETER ouvre de nouvelles perspectives dans le domaine du soutènement de tunnels par boulonnage.

1. PRINCIPLE OF MEASUREMENT

At the 4th ISRM Congress at Montreux THURNER (1979) presented a new method for non-destructive testing of the probable loadcarrying capacity of grouted bolts. Since then an instrument has been developed, calibrated and tested and is referred to as the BOLTOMETER.

A specially designed sensor containing piezo-electric crystals is pressed against the free planar end surface of the rock bolt. Compression and flexural elastic waves are transmitted into the metal bolt. When the waves travel along the bolt, some energy is transferred through the grouting into the rock and thus the wave ampli-

tude decreases. At the inner end of the bolt the waves are reflected. The reflected waves are recorded at the outer end of the bolt by means of the piezo-electric crystals. If the grouting surrounds the bolt fully and is of good quality, the amplitude of the reflected wave is damped more than if the grouting is deficient or lacking.

The reflected waves will therefore have a lower amplitude in the case of good grouting, Fig. 1a, than in the case of imperfect grouting, Fig. 1b. The time interval between the excited and the reflected wave gives a possibility to calculate the length of the bolt. The amplitude of the reflected wave or sometimes successive reflec-

N

tions can be analyzed and the probable condition of the rock bolt can be estimated on the basis of calibration tests.

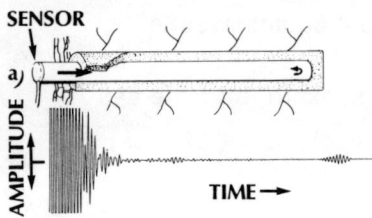

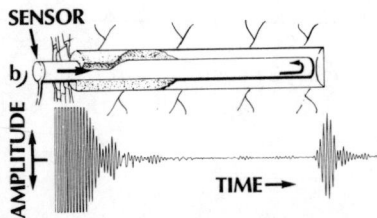

Fig. 1 Examples of registrations obtained
a) for a fully cement-grouted bolt (small amplitude for the reflected wave)
b) for a bolt only partly grouted (large amplitude for the reflected wave)

2. DESIGN AND HANDLING

The BOLTOMETER is composed of the aforementioned sensor connected with an instrument by a flexible cable, Fig. 2. The instrument has a panel (with switches, indicator lamps and a digital display), a microprocessor (for generating input waves and analyzing reflected waves), re-chargeable batteries and connectors for an oscilloscope or a recording unit.

The free end of the bolt has to be cut and/or ground plane. A special contact paste is put on the bolt end in order to obtain a good wave transfer.

When a bolt is tested, the sensor is held against the bolt end, Fig. 2. When good contact is established, four indicator lamps on the instrument panel and on the back side of the sensor light up. Short wave pulses are transmitted into the bolt intermittently and at the same time the reflected waves are analyzed.

The standard version of the BOLTOMETER for cement-grouted bolts - used in the field tests reported here - shows the length of the bolt on a digital display with an accuracy of better than 0.1 m if the quality of the grouting is known or coincident with the quality anticipated. An estimate of the quality of the grouting is shown on indicator lamps of different colours. If a more detailed analysis is required, the time history of the output waves can be displayed on an oscilloscope or a pen recorder unit.

3. PRESENT RANGE OF APPLICATION

In order to get an empirical basis for classification a number of tests have been made on bolts with different but known length and grouting conditions.

Fig. 2 The BOLTOMETER in present design. The sensor is pressed against the bolt end and the result is read on the instrument panel

The present design of the BOLTOMETER is calibrated for control of cement-grouted bolts (Perfo, SN-bolts etc.) or resin-grouted bolts (Celtite etc.).

In the case of perfectly cement-grouted bolts with low water/cement ratio (<0,40) there is a limit bolt length of about 2.3 m, above which no reflection occurs when the grout has hardened. For imperfect or deteriorated grout the limit length increases with increasing degree of imperfection.

Resin grout, for instance polyester, has a much lower acoustic impedance than cement grout. Reflections can therefore be obtained for much greater lengths. The limit length has not been determined experimentally as yet but is estimated to be about 10 m at perfect resin grouting.

4. CLASSIFICATION OF CEMENT-GROUTED BOLTS

THURNER and his co-workers have presented a tentative proposal for practical classification of production bolts. This proposal is based on the aforementioned controlled tests, where the test signal characteristics have been related to the known conditions of the test bolts. The classification system, which has been built into the BOLTOMETER instruments used in the field tests, considers characteristic differences in recorded signals attributable to deviations from optimum performance of a bolt. Naturally, in order to get a practical service system it has been necessary to simplify and generalize the complicated and many-parameter phenomena involved.

The length 2,3 m given in foot-note * to TABLE 1 is valid for hardened cement grout of low water/cement ratio in hard igneous rock. For other grouting qualities and/or ages the limit values must be determined by tests. The properties of the surrounding rock mass may also influence the limit length to some extent.

TABLE I. Tentative BOLTOMETER index for cement-grouted bolts

Class	BOLTOMETER reflection	Estimated bolt performance
A	No reflection	Optimal (up to 2.3 m bolt length)*
B	One small reflection	Reduced
C	One large or several minor reflections	Insufficient
D	Many large reflections	Very poor or non-existant **

* If the bolt is longer than 2.3 m and perfectly grouted in the outer 2.3 m part of the drillhole, no conclusion can be drawn about the grouting quality of the inner part.

** The bolt can probably be pulled out from the drill-hole by means of a hydraulic jack.

The BOLTOMETER index is a rough estimate, which sometimes can be somewhat misleading. For instance, the performance of a grouted bolt depends not only on the amount of grout present but also on which parts of the bolt are grouted. A bolt attached to the rock by grout at both ends may show an inferior BOLTOMETER index but still function satisfactorily - at least temporarily. In the long run the absence of grout may cause deterioration of the bolt by corrosion.

5.   FIELD MEASUREMENTS

During 1980-81 series of tests with the BOLTOMETER have been run at various places in Sweden in order to test the validity of the tentative classification system (TABLE I) on specially prepared test bolts, whose data were unknown to the measuring crew. Furthermore the condition of grouted production bolts of various types and of varying ages were tested. In the latter cases overcoring or pull out tests were carried out on a limited number of bolts.

5.1   Cement-grouted test bolts

At Renström, a Boliden mine, three 2.3 m slack cement-grouted SN-bolts, Fig. 3, were prepared so that the bond in the inner part between bolt and grout was spoiled heavily in two bolts and less in the third. The holes were then filled with low-quality high water/cement ratio grout. The BOLTOMETER gave 1 class B, 2 class C 2.3 m bolts, i.e. good agreement with known facts.

At Laisvall, another Boliden mine, 11 bolts with length 1.7 m were tested. Three of these were cut at various locations and taped together, 3 had their adhesion capacity spoiled to varying degrees and 5 were ordinarily grouted. The bolts were of the slack type, Fig. 3. The BOLTOMETER results were in very good agreement with known facts.

At Kiirunavaara, an LKAB mine, 20 test bolts of the Kiruna type, Fig. 4, with a length of 2.3 m were prepared with varying cement-grouting and pretension. Seventeen of the bolts were later subjected to pull-out tests by hydraulic jacks. The results are given in TABLE II. It can be

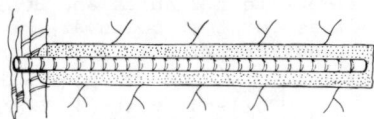

Fig. 3 A slack cement-grouted bolt (rebar). The grout can be placed in the drillhole by various methods - SN, Perfo, Mono-pump, CemBolt and others.

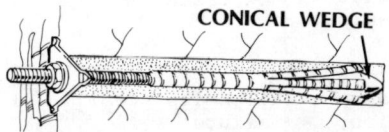

Fig. 4 The Kiruna slot-and-wedge bolt. The drill hole is filled with cement-grout by a hand pump. The bolt is driven through the grout and a few blows of a sledge-hammer cause the conical wedge to lock the bolt in position by splaying out the cross slotted end of the bolt. The flexible washer is tightened by the nut to the required pretension (normally 30-50 kN).

seen that the BOLTOMETER indices gave a fair estimate of the grouting conditions as prepared. The pull-out tests, on the other hand, showed optimum performance for 8 out of 16 bolts with deficient cement-grouting.

Thus, the 34 control experiments on specially prepared test bolts were successful and they indicate (although they are few) that the tentative classification scheme (TABLE I) is reasonable.

5.2   Cement-grouted production bolts

5.2.1 Slack bolts

At Umluspen hydro power plant the Swedish State Power Board (SSPB) has tested 73 bolts of about 3 m length, which were installed 1953-1957 or later. The grouting methods may have varied - probably mostly Perfo and SN, Fig. 3. The BOLTOMETER indicated the following class distribution:

   A 65% - B 3% - C 24% - D 8%

One SN-bolt, Fig. 3, classified B has been successfully overcored. The length was 2.4 m as compared to 2.4 m according to the BOLTOMETER measurement. When the core was split, longitudinal air pockets with a total length of 1.8 m were revealed, Fig. 5.

Two Perfo bolts class A were also overcored but bolts and groutings were damaged in the process. However, air voids and deficiency of grout could be detected.

At the Rinkeby Station of Stockholm Subway the Stockholm Transport Authority, SL, has made 25 SN-bolts, of a length of 1.6-2.4 m, available for BOLTOMETER testing, Fig. 3. The bolts were installed 1976. The following class distribution was obtained:

   A 0% - B 44% - C 56% - D 0%

A 179

TABLE II. Pull-out test results and BOLTOMETER
indices for test bolts at the
Kiirunavaara mine

Number of bolts	Mounting	Full-out test by hydraulic jacks	BOLTOMETER index - Number of bolts			
			A	B	C	D
4	normal pre-tension, normal cement-grout-ing	2 bolts steel failure* not tested	1	1 2		
5	normal pre-tension, partial cement-grout-ing	all bolts steel failure*	1	2	2	
3	normal pre-tension, cement-grout 0.5 m from inner end	all bolts steel failure*				3
4	normal pre-tension, cement-grout 0.5 m near-est to the rock surface	bolts loosened at 165 kN " 80 kN " 60 kN " 30 kN	1	1		1 1
2	normal pre-tension, no grout-ing	bolt loosened at 20 kN not tested			1	1
2	bolts se-cured by a few blows of a sledge-hammer	bolts loosened at 30 kN				2

* Normal bolt fracture at 160-170 kN.

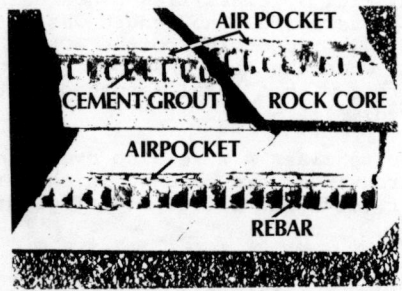

Fig. 5   Part of the overcored SN-bolt class B
from Umluspen. The split core parts show
longitudinal air pockets in the grouting.
(Photo: SSPB.)

No overcoring or mechanical testing has been
planned.

At Laisvall, a Boliden mine, 6 SN-bolts of 2.3 m
length installed 1980 were controlled with the
BOLTOMETER with the following result:

    A 33% (2) - B 17% (1) - C 33% (2) - D 17% (1)

No overcoring or mechanical testing is planned.

At Harsprånget hydro power plant the SSPB has
tested 17 post-grouted simple slot-and-wedge
bolts, Fig. 6, of 4-5 m estimated length. The
bolts were installed 1950. The BOLTOMETER gave
the following class distribution:

    A 24% - B 35% - C 41% - D 0%

One class B bolt was overcored in it's entire
length of 4.6 m by a core drill of 87 mm internal diameter. When the core was split, it was
found that the grout enclosed the bolt in the
75 cm nearest to the rock surface, filled about
half of the hole for the next 75 cm and was then
lacking. Nevertheless, the slot-and-wedge was
still in function and the bolt showed no perceptible signs of corrosion.

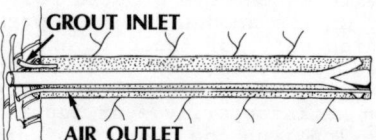

Fig. 6   Post-grouted simple slot-and-wedge
bolt (Harsprånget 1950)

5.2.2 Pre-tensioned bolts

At the Kiirunavaara mine the LKAB Company has
used the BOLTOMETER to test 20 Kiruna-bolts,
Fig. 4, of 2.3 m length installed 1975 and
1980. The class distribution obtained was:

    A 35% - B 25% - C 25% - D 15%

The three class D bolts were later tested by
hydraulic jacks up to 155 kN, thus indicating
full bearing capacity in contradiction to the
BOLTOMETER index.

At Grundfors hydro power station the SSPB has
controlled 96 pretensioned and post-grouted
rock anchors of 8-18 m length with the BOLTO-
METER. The anchors were installed 1955.
The class distribution obtained was:

    A 41% - B 24% - C 25% - D 10%

The results indicate that more than half of
the anchors are insufficiently grouted and thus
possibly subject to corrosion. As regards the
41 % of the bolts indexed class A cf. foot-
note * to TABLE I.

5.3   Slack resin-grouted bolts

Atlas Copco MCT Company has made 21 resin-
grouted (Celtite) production bolts of 2 m
length available for BOLTOMETER control at
Guttusjö. The bolts were installed 1981. The
result was:

A 90% - B 10% - C 0% - D 0%

No overcoring or mechanical testing is planned.

6. DISCUSSION

It is evident that the BOLTOMETER has a considerable potential in discerning bolts with low grouting qualities. For cement-grouted bolts with a length less than 2.3 m it seems justified to conclude that the tentative classification TABLE I will give a fairly correct description of the functional state of a bolt.

For cement-grouted bolts with a length >2.3 m the BOLTOMETER is able to direct attention to bolts with inadequate grouting. The BOLTOMETER can also indicate bolt failures, which have hitherto been undetectable. Fig. 7 shows some types of failures of grouted bolts. Types a and b are visible and commonly observed. Types c and d are invisible and thus more dangerous. The controlled field tests have shown that the latter failure types can be indicated by the BOLTOMETER in terms of reduced performance characteristics, although it is not possible as yet to clearly distinguish the different types of failures.

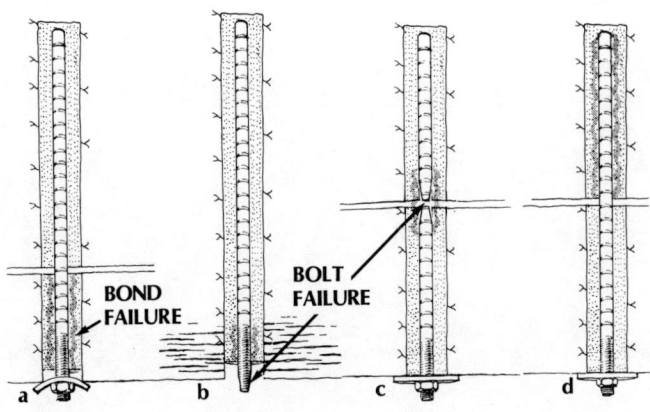

Fig. 7   Some types of rock bolt failures.
a) bond failure near rock surface
b) bond failure followed by bolt failure at rock surface
c) steel bolt failure at depth
d) bond failure at depth

In principle, there are possibilities to extend the reach for fair BOLTOMETER control of grouted bolts to greater lengths than at present. It is also possible to adjust the BOLTOMETER with its present capacity to test other bolt types, for instance non-grouted rock anchors.

The field tests indicate a surprisingly high percentage of bolts with low grouting quality. Thus there seems to be a serious difference between the intended bolt product and the result. If this should be a sign of a more general state-of-affairs the BOLTOMETER already in its present design could

- help us to develop a better production technique to get a higher percentage of good bolts
- help us to control the bolt quality both at an early stage and as a maintenance measure
- help us to monitor changes in bolt performance with time, for instance due to load changes or ageing
- be a tool for a better understanding of the appropriate extent of strengthening (if many bolts now have low-quality performance and nothing happens, we are evidently using more bolts than would be needed if they were all class A).

7. CONCLUSIONS

The BOLTOMETER should be an interesting instrument for owners, designers and contractors, who wish that the effort put into rock mass support by use of grouted bolts should be used rationally.

The present (1982) design of the BOLTOMETER

- seems to be able to give a fair classification according to TABLE I for cement-grouted bolts of a length up to about 2.3 m and low water/cement ratio grout

- seems to be able to pick out also longer cement-grouted bolts with low quality of grouting
- seems to be able to give a fair classification according to TABLE I for resin-grouted bolts of length up to about 10 m.

The BOLTOMETER can be equipped with an oscilloscope or a pen recorder unit for research or development purposes, for instance for tailoring of a special classification system at a certain site or mine.

ACKNOWLEDGEMENTS

The authors want to express their gratitude to Mr. H. Thurner and Mr. Ch. Svensson (Geodynamik AB), LKAB, Boliden AB, the Stockholm Transport Authority and Mr. L. Lindström (SSPB) for their cooperation in the test programme series and their help in the preparation of this paper. The development of the BOLTOMETER has been financially sponsored by the Swedish Work Environmental Fund (ASF) and the Swedish Board for Technical Development (STU).

REFERENCES

THURNER, H.F. (1978): Non-destructive test method for rock bolts, Proc. 3rd Int.Congr. IAEG Madrid, Vol. 10, pp. 309-312.

THURNER, H.F. (1979): Non-destructive test method for rock bolts, Proc. 4th Int.Congr. Rock Mech. Montreux, Vol. 3, pp. 254-255.

DIRECT SHEAR TESTS ON FOLIATION SURFACES

Essais de cisaillement direct sur foliation

Direkte Scherversuche an Schieferungsflächen

Fernando H. Tinoco
Professor of Civil Engineering, Universidad Simón Bolivar and Principal of Sueloproyecto,
Caracas, Venezuela

Daniel A. Salcedo
Professor of Geological Engineering, Universidad Central and President of Ingeotec,
Caracas, Venezuela

SYNOPSIS

Direct shear tests were performed on closed and open foliation surfaces of weathered phyllite samples obtained from slide areas located at the southeast hills of the city of Caracas. The evaluation of the laboratory tests indicated that Coulomb shear parameters failed to match the behaviour of the samples, and the development of the shear strength-displacement theory allowed the interpretation of the direct shear tests. The shear strength envelope, calculated from the aforementioned theory, models the transformation from closed to open foliation during the laboratory test. The results were applied to study the strength of a field sample and compare it to the laboratory shear strength.

RESUME

D'après l'évaluation d'essais de cisaillement direct effectués le long de la foliation, ouverte et fermée, sur plusieurs échantillons de phyllite altérée, prélevés de sites de glissements sur les collines du sud de Caracas, on a pu constater que les paramètres de Coulomb n'étaient pas suffisants pour expliquer le comportement de la roche. D'autre part, c'est l'application de la théorie de la résistance au cisaillement en fonction du déplacement qui a permis l'interprétation de ces éssais. En effet, l'enveloppe des efforts de cisaillement calculée avec cette théorie a permis de modeler le passage de la foliation fermée à ouverte lors d'un essai de laboratoire sur éprouvette. Ce modèle a été appliqué à l'étude du comportement d'un échantillon à plus grande échelle et les résultats ont été comparés avec ceux du laboratoire.

ZUSAMMENFASSUNG

Direkt-Scherversuche wurden in verwitterten Phylliten von Rutschgebieten im Südosten von Caracas mit offenen und geschlossenen Schiefrigkeitsflächen durchgeführt. Die Auswertung der Laborversuche zeigte, daß die Coulombschen Scherparameter zur Beschreibung des Scherverhaltens der Muster ungeeignet sind, und daß die Festigkeit-Verformungstheorie eine befriedigende Darstellung der Versuchsergebnisse ermöglichte. Die Scherfestigkeitshülle, berechnet mittels der obengenannten Theorie, modelliert den Übergang von geschlossenen zu offenen Trennflächen im Laborversuch. Die Ergebnisse wurden zur Beurteilung der Festigkeit der Felsen in-situ und zum Vergleich mit den Laborergebnissen angewendet.

1. INTRODUCTION

The housing needs of an ever growing city have driven the urban developers towards the hills and mountain zones located to the south of the Caracas valley. The characteristics of construction in these zones are the large cuts and fills made in weathered metamorphic rocks. Large slides take place every year that produces important financial losses as well as some life losses. These facts indicate that an adequate knowledge of the properties of weathered phyllites and schists that constitutes the lithology of the metamorphic rocks belonging to Las Brisas and Las Mercedes Formation, Caracas Group. does not exist.

A common factor to the slides of cuts made in weathered phyllite and schists is the amount of time required for failure after the cut has been made. It may take from ten to twenty years or more for the slides to occur and in a few cases

cracking on the crest appears some years before the final event.

The analysis of the slides in weathered phyllite seem to indicate that failure is produced along "closed" foliation surfaces which are herein used to characterize a foliation with unfractured areas in contact with or without roughness. The aforementioned observation led to the direct shear testing of closed foliation surfaces.

The evaluation of the laboratory tests indicated that Coulomb shear parameters failed to match the behavior of the samples of weathered phyllite and, therefore, a modification of them was required to explain the shear strength of the closed foliation surfaces.

The modification did not alter the basic equa

tion of Coulomb theory of shear strength but advantage was taken of the fact that failure of unfractured surfaces and overriding of asperities are very much dependent on displacement to incorporate it in a criteria of shear strength. This also allowed a new definition of shear and normal stresses since the average values did not seem to reveal the phenomena taking place along closed foliation surfaces.

The purposes of this research were to interpret the results of direct shear tests on closed foliation surfaces of weathered phyllite, to incorporate displacement and strain in the shear strength criteria used for evaluation of laboratory tests and to apply the criteria to the evaluation of the shear strength of field foliation surfaces and compare to laboratory tests.

2. DIRECT SHEAR TEST

Weathered phyllite samples were obtained from failure surfaces and they were cut to fit the shear box of a field shear equipment manufactured by Robertson Research Laboratory, London and described by Hoek and Bray (1974).

The test procedure allowed, as its main objective, the determination of the shear stress - displacement curve up to the maximun travel allowed by the equipment (5 mm.). The procedure was divided in three stages:

(1) Shearing of the sample at the selected effective normal pressure up to 5 mm. displacement, release of the applied normal pressure and returning the box to its original porion by means of the application of shear force.

(2) An equal or different normal pressure to the one used in stage (1) is then applied and shearing of the sample up to a 5 mm. displacement is followed by release of the normal pressure and returning of the box to its original position.

(3) Repetition of the steps outlined in stage(2)

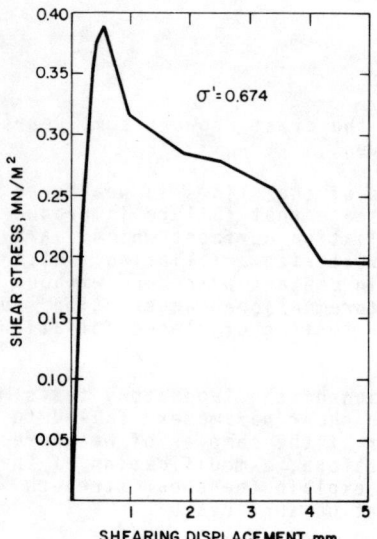

Fig. 1. Shear stress-displacement curve

Fig. 1 shows the shear stress-displacement curve for sample LM-20 obtained as a result of procedure outlined as stage (1). Fig. 2 shows the shear stress-displacement curves obtained by the procedures outlined in stages (1), (2) and (3) for the sample LM-21. They seem to show that the phyllite samples possess a brittle structure that breaks down at peak shear stress and also at lesser values of the shear stress with additional displacement. The behavior is partially brittle (sudden changes of slope) for stage (2) of the test and it is elasto-plastic for stage (3) at reduced normal effective pressure with respect to stage (1).

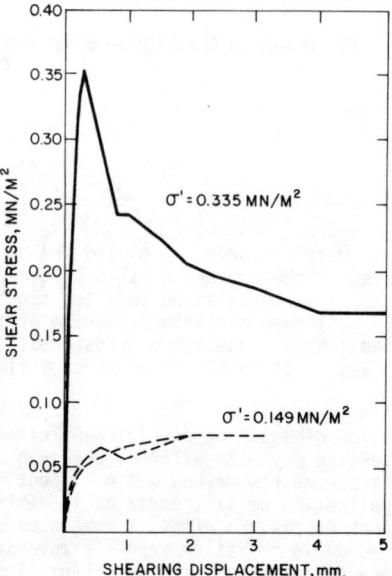

Fig. 2. Shear stress-displacement curves.

The peak values of the shear stresses are plotted in Fig. 3 (a) versus the applied effective normal pressure. It is clear that a unique determination of Coulomb shear parameters c and ϕ, is not possible. Fig. 3 (b) shows the results of stage (2) and (3) for some of the tests. A straight line may be fitted by means of a regresion analysis.

Direct shear tests of samples with "open" foliation surfaces (i.e., by fitting together samples opened along the foliation surfaces) indicated that the minimun value of the angle of internal friction is 12°.

3. STRENGTH-DISPLACEMENT THEORY

The theory proposed by Tinoco and Salcedo (1981) assumes that the process of fracture and sliding in closed foliation surfaces tested in the direct shear apparatus starts as soon as displacement takes place and that shear strength of surfaces is the sum of two components, one is independent and the other is proportional to the effective normal stress acting on the sliding surfaces. Both components are functions of displacement and strain measured along the failure surface. The criterion is described by the following expressions:

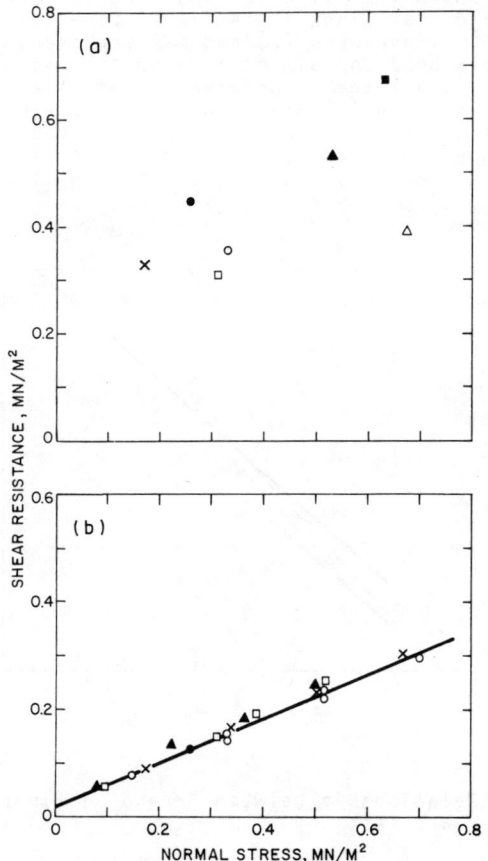

Fig. 3. Results of direct shear tests.

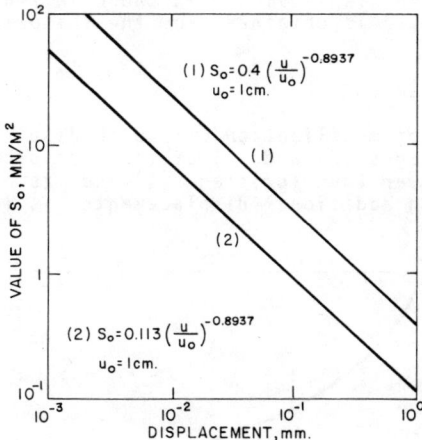

Fig. 4. Relationship S_0 and displacement.

constants appearing in eq. (2) were obtained by a regression analysis and are shown in Fig.4.

The "cohesion" or the shear strength component term that is independent of the effective normal stress and defined by the expression $F_d S_0$

is plotted in Fig. 5 versus displacement.

$$S = F_d S_0 + F_f \sigma' \tan \phi_u \qquad (1)$$

$$S_0 = A (u/u_0)^m \qquad (2)$$

$$F_d = (u/L)^n \qquad (3)$$

where

S = average applied shear stress.
σ' = average effective normal stress
S_0 = strength of unfractured surfaces and asperities.
A = value of S_0 for a given displacement u_0.

u = displacement along the failure surface.
L = length of failure surface.
m = constant.
$n = n_1 + n_2 \log (u/L)$ for pre-peak strains.

n = constant for peak and post-peak strains.
F_f = factor that characterize variation of frictional strength with strain.
$\tan \phi_u$ = coefficient of friction measured at large relative displacements between sliding surfaces.

The variation of the strength of unfractured surfaces and asperities with displacement is shown in Fig. 4. Line (1) represents the results obtained from the application of stage (1) of the test procedure in the direct shear apparatus and line (2) corresponds to the results obtained for stages (2) and (3) of the previously outlined test procedures. The numerical values for the

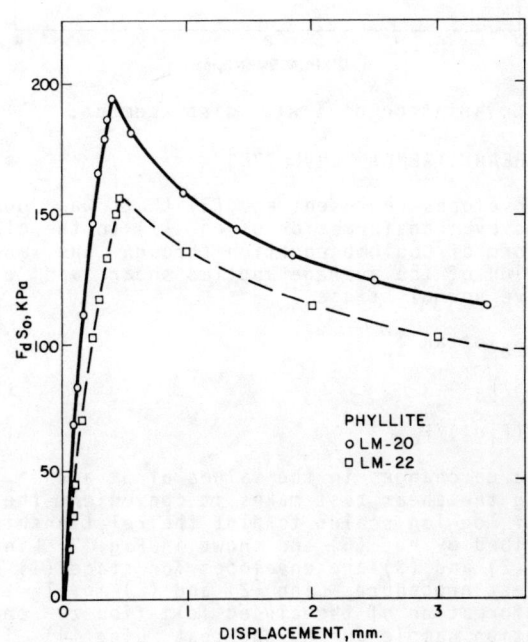

Fig. 5. Variation of $F_d S_0$ with displacement.

It may be observed that very small displacements are required to mobilize the peak strength Additional displacements decrease the shear resistance of the sample. For stages (2) and (3), that involves large displacements, the values of "cohesion" are reduced to less than 20 KPa.

The functional variation of the frictional shear

strength with displacements is shown in Fig. 6. The value of ϕ is obtained from the following expression:

$$\tan \phi = F_f \tan \phi_u \qquad (4)$$

The degree of mobilization of ϕ with displacement for stages (2) and (3) of the test procedures is lower than for stage (1) and its reduction with additional displacements is less marked.

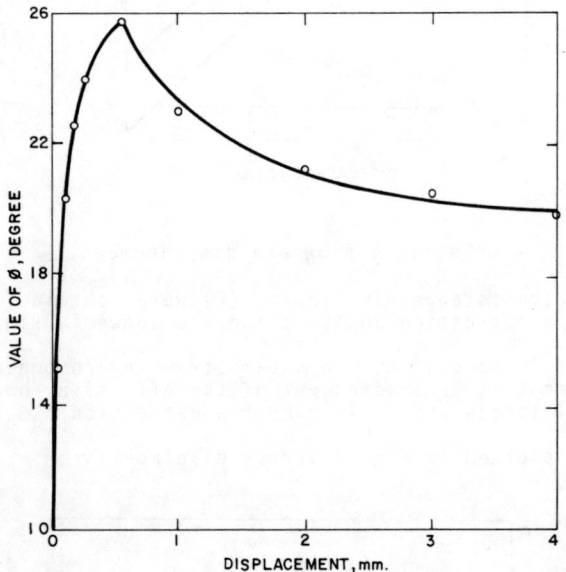

Fig. 6. Variation of ϕ with displacement.

4. SHEAR STRENGTH ENVELOPES

The envelopes represent eq. (5) that was obtained by transformation of eq. (1) to the classic form of Coulomb equation through the modification of the average applied shear and effective normal stress.

$$S^* = S_0 + \sigma^* \tan \phi_u \qquad (5)$$

$$S^* = S/F_d \qquad (6)$$

$$\sigma^* = (F_f \sigma')/F_d \qquad (7)$$

The large changes in the values of S^* and σ^* during the shear test makes it convenient the use of log-log scales to plot the relationship described by eq. (5) and shown in Fig. 7. Line (1), (2) and (3) are envelopes for stage (1) of the test procedure. Line (2) and (3) model the transformation of the closed foliation to open foliation sample during the test. Line (4) is the shear strength envelope obtained from stage (2) and (3) of the test procedure. The mathematical expressions, representing lines(1),(2),(3) and (4), obtained by regression analysis are:

$$S^* = 0.713 \ (\sigma^*)^{0.859} \qquad \text{(line 1)} \qquad (8)$$

$$S^* = 0.378 \ (\sigma^*)^{1.03} \qquad \text{(line 2)} \qquad (9)$$

$$S^* = 0.256 \ (\sigma^*)^{1.03} \qquad \text{(line 3)} \qquad (10)$$

$$S^* = 0.292 \ (\sigma^*)^{0.812} \qquad \text{(line 4)} \qquad (11)$$

Fig. 7 shows that line (2) and line (3) are parallel and that line (4) is almost parallel to line (1). Envelopes (2) and (3) represent the post-peak behavior of the more brittle samples and they model the transformation of closed foliation to open foliation surface represented by envelope (4). It may also be observed that the range of values for σ^* in envelope (4) approach the magnitude of the average applied ef-

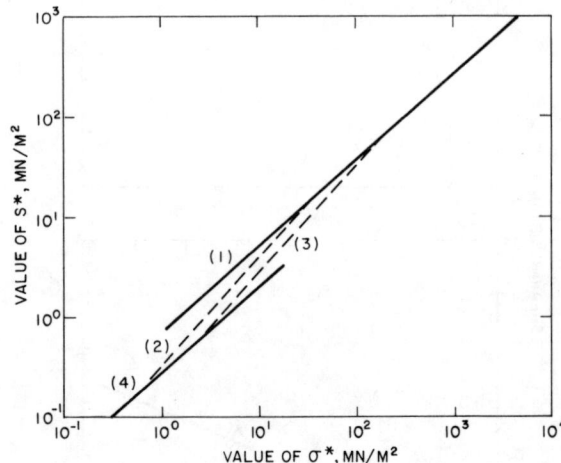

Fig. 7. Relationship between S^* and σ^* for laboratory tests.

fective normal stress.

Since S^* and σ^* are functions of strain, it is necessary to determine the relationship between S^* and the strain along the failure plane. The functional variation of S^* with strain is shown in Fig. 8. Three of the samples tested at different average effective normal stress were selected to show the aforementioned relationship and to point out four noteworthy aspects of the samples behavior:

1. The magnitude of the average applied effective normal stress has no influence in the behavior of the sample for pre-peak and peak strains. The differences shown in Fig.8 are due to variation between samples.

2. Post-peak strains produce a reduction of unfractured areas and the influence of the average applied effective normal stress may be faintly noticed. That is, the process of transformation from closed to open foliation surface starts sometimes after the peak and continues with strain.

3. The influence of the average applied effective normal stress is clearly noted in Fig.8 for stage (2) and (3) of the test procedure.

4. The relationship of S^* and strain corresponding to closed foliation surface fit with the one obtained from peak values of stage (2) and (3) of the test procedure is shown in Fig. 8. This indicates that with a shear apparatus able to allow enough displacement,

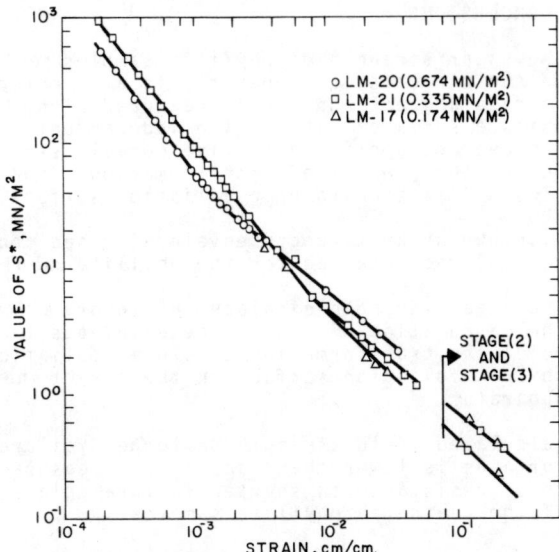

Fig. 8.Relationship between S* and strain.

one may obtain a complete modelling of the transformation from closed to open foliation behavior and to determine the strain required for identification of regions with closed or open foliation behavior.

5. APPLICATION TO A FIELD FOLIATION SURFACE

The prediction of the field shear resistance of weathered phyllite to ascertain the degree of its stability in situ is the main objetive of laboratory tests. Eq.(5)represents the behavior of the samples tested in the laboratory but the modified shear and effective normal stresses are both functions of the strain measured along the sliding surface.

The relationship of $\sigma*$ and strain is shown in Fig.9 for two the samples tested and for stage (1) of the test procedure. The regression analysis yielded the following expressions for samples LM-21 and LM-20, respectively:

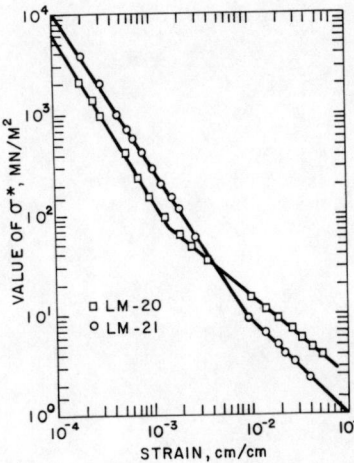

Fig. 9.Relationship between $\sigma*$ and strain.

$$\sigma* = 0.00767 \ (u/L)^{-1.539} \ \text{for} \ \sigma* \geqslant 9MN/m^2 \quad (12)$$

$$\sigma* = 0.10510 \ (u/L)^{-0.967} \ \text{for} \ \sigma* < 9MN/m^2 \quad (13)$$

$$\sigma* = 0.00218 \ (u/L)^{-1.613} \ \text{for} \ \sigma* \geqslant 100MN/m^2 \quad (14)$$

$$\sigma* = 0.33540 \ (u/L)^{-0.840} \ \text{for} \ \sigma* < 100MN/m^2 \quad (15)$$

An example of application of laboratory test results is the calculation of S* for an imaginary field sample of 50 cm. length. The calculation was based on test results from samples(average

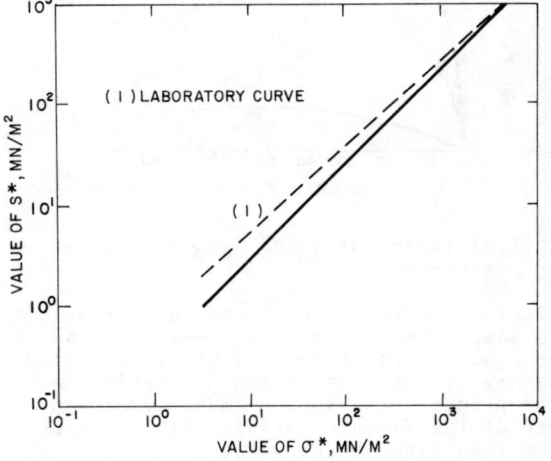

Fig. 10.Relationship between S* and $\sigma*$ for foliation surface.

length, 10 cm.)LM-21 and LM-20 using the following procedure:

1. Assume increasing values of displacement and compute the strain.

2. Use the appropiate eqs. (12) or (13) and/or (14) or (15) to compute value of $\sigma*$.

3. Calculate value of S_o from eq. (2).

4. Compute S* from eq. (5).

The result of the aforementioned procedure is shown by the solid line of Fig. 10. Line (1) shown in Fig. 10 was obtained from laboratory tests and it is the same line (1) of Fig.7 and it has been drawn in Fig. 10 for comparison.The prediction yielded smaller shear resistance S* for the longer sample than for the laboratory shorter samples.

The mobilization of S, F_dS_0 and ϕ with strain is shown in Fig. 11 and Fig.12, for the imaginary field sample of 50 cm. length. The two drawings include also the additionally imaginary shear testing based on sample LM-21 under the pressure of 0.674 MN/m² in order to compare the results with those obtained with sample LM-20.

The value of the average applied shear stress S (see Fig.11) reached a maximun at the same strain and independently of the average applied normal stress for the curves based on two laboratory samples, even though the stress-strain path was very different for each of them. The peak value of S for the long sample is 28% and 39% lower than the peak value obtained for laboratory samples LM-21 and LM-20, respectively.

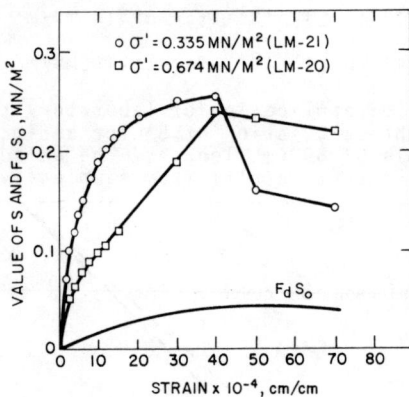

Fig. 11.Variation of S and F_dS_0 for foliation surface.

Similarly, the value of "cohesion" (F_dS_0)of the field sample is 72% and 67% lower than the corresponding values to samples LM-21 and LM-20, respectively. This means that "cohesion" defined by the product F_dS_0 is very much affected by the length of the failure surface. Fig.11 shows a sudden reduction in the value of S at peak strain for the sample based on laboratory test LM-21 and it does model similar laboratory break down.

Evaluation of Fig.12 indicates that mobilization of ϕ with strain depends upon the average applied effective normal stress and it is clearly seen when comparison is made between the results based on the same laboratory test sample LM-21, at two different applied effective normal stress. Even though the path is different for each one of the samples used, the peak values of ϕ at the peak strains are about the same. It may be noted that more strain is required in the longer sample to mobilize the peak value of ϕ than for the shorter laboratory test samples.

6. CONCLUSIONS

1. The shear strength of phyllite samples tested in the direct shear apparatus is a function of displacement and strain measured along the failure surface. It is also independent of the average applied effective normal stress during the process of transformation from closed foliation to open foliation surface.

2. A unique shear strength envelope, based on eq. (5) was obtained for the phyllite samples.

3. The shear strength-displacement theory allows the calculation of resistance envelopes that model the transformation of closed foliation to open foliation surface in the direct shear apparatus.

4. Calculated field strength on longer failure surfaces is lower than laboratory measured shear resistance on shorter failure planes of the weathered phyllite samples.

7. ACKNOWLEDGEMENTS

The writers gratefully acknowledge INGEOTEC and SUELOPROYECT for financial assistance of some phases of this research. Thanks to M. Salcedo G. for preparing the manuscript and V. Callejas for drafting the figures.

8. REFERENCES

Hoek, E. and J. Bray (1974). Rock Slope Engineering. The Institution of Mining and Metallurgy. London.

Tinoco, F. H. and D. Salcedo, (1981). Analysis of slope failures in weathered phyllite, Proc. Int. Symp. on Weak Rock, Tokyo,I:55-62.

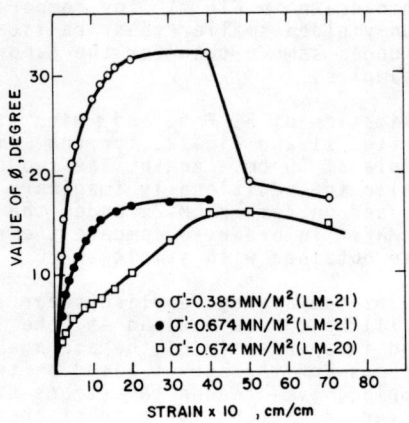

Fig. 12.Relationship of ϕ with strain

ELASTIC CONSTANTS AND TENSILE STRENGTH OF ANISOTROPIC ROCKS
Coefficients d'élasticité et résistance à la traction des roches anisotropes
Elastische Konstanten und Zugfestigkeit für anisotropen Fels

B. Amadei
Assistant Professor, University of Colorado, Boulder, Colorado, U.S.A.

J. D. Rogers
Consultant, Berkeley, California, U.S.A.

R. E. Goodman
Professor, University of California, Berkeley, California, U.S.A.

SYNOPSIS

A closed-form solution is proposed for the distribution of stresses and strains within a thin disc of anisotropic material loaded diametrically by a line or a strip load. This solution is then incorporated into a procedure to measure the elastic constants and the tensile strength of rocks that can be idealized as linearly elastic, homogeneous, orthotropic or transversely isotropic continua.

RESUME
On propose une solution exacte pour la distribution des contraintes et déformations dans un disque constitué d'un matériau anisotrope soumis dans un plan diamétral à une charge linéaire ou répartie. Cette solution est ensuite incorporée dans une méthode de mesure des coefficients d'élasticité et de la résistance à la traction de conditions idéales comme des milieux élastiques linéaires, homogènes, orthothropes ou isotropes dans un plan.

ZUSAMMENFASSUNG

Eine geschlossene Lösung wird hergeleitet für die Verteilung der Spannungen und Dehnungen in einer dünnen Scheibe aus anisotropem Material, belastet mit entgegengesetzt gerichteter Linien- oder Streifenlast. Diese Lösung wird verwendet in einem Verfahren zum Messen der elastischen Konstanten und der Zugfestigkeit für Fels, der idealisiert werden kann als ein linear elastisches, homogenes, orthotropes oder querisotropes Kontinuum.

1. INTRODUCTION

Advanced numerical methods are used extensively in the field of rock mechanics to model the behavior of structures in rock. With these methods, the two-and three-dimensional geometries of rock structures as well as the heterogeneous and anisotropic character of the rock material can be modelled. In addition, non-linear behavior, failure and time-dependent properties of that material can be included. However, despite the sophisticated character of these numerical methods, a proper characterization of rock deformability and strength still remains a difficult task. This is particularly relevant when dealing with rocks whose properties vary with direction, i.e., are anisotropic, such as slates, gneisses, phyllites and other metamorphic rocks.

Several techniques have been proposed in the literature to measure the deformability and strength of anisotropic rocks. A review of these techniques can be found in Amadei (1982). Among them, the diametral compression of thin discs of anisotropic rock has been suggested by Pinto (1979) to measure rock deformability

and by Hobbs (1964) and others to measure rock tensile strength. In both references, however, the rock response to loading was modelled as linearly elastic and the stress distribution in the discs approximated by that for an isotropic material in spite of the anisotropic rock character.

The paper begins with a closed-form solution for the distribution of stresses and strains within a thin disc of anisotropic material loaded diametrically by a line or a strip load. The solution accounts for both the anisotropic character of the material and the orientation of the anisotropy with respect to the loading direction. This solution is then incorporated into a procedure to measure the elastic constants and the tensile strength of a rock that can be idealized as an orthotropic or transversely isotropic, linearly elastic, homogeneous continum.

2. DEFORMABILITY OF ANISOTROPIC ROCKS

Laboratory and in-situ tests conducted on aniso-tropic rocks have shown that their deformability varies with spatial orientation. In order to assess by testing the deformability properties of an anisotropic rock and their directional character, a model is needed to des-cribe the rock behavior in the corresponding test con-figurations. If the response of the rock material to test disturbances can be idealized as linearly elastic, then its constitutive relation can be obtained from those of the theory of linear elasticity for anisotropic media.

2.1 Constitutive Relations for Anisotropic Media

The general constitutive relation for a linearly elastic, homogeneous and continuous material relates the stress and strain tensors by the <u>Generalized Hooke's law</u> as follows:

$$\varepsilon_{ij} = A_{ijkl} \, \tau_{kl} \tag{1}$$

It can be shown that the tensor of compliances A_{ijkl} has, at most, 21 different components based on symmetry properties of the stress and strain tensors and on the existence of a strain energy function (Lekhnitskii, 1957). These components are also related to 21 distinct elastic constants. In matrix form, Equation (1) can be replaced by the following:

$$(\varepsilon)_{xyz} = (A) \, (\sigma)_{xyz} \tag{2}$$

where $(\varepsilon)_{xyz}$ and $(\sigma)_{xyz}$ are respectively (6 x 1) column matrix representations of the strain and stress tensors in an arbitrary x,y,z coordinate system, and (A) is the corresponding compliance matrix with 21 distinct components a_{ij} (i,j = 1 to 6). This number is further reduced if the internal composition of the anisotropic material possesses symmetry of any kind. The number of distinct elastic constants is equal to: 13 if a plane of elastic symmetry exists at each point of the anisotropic material; 9 if the material is orthotropic, i.e., pos-sesses three orthogonal planes of elastic symmetry; 5 if the material is transversely isotropic, i.e., isotropic within a plane; and 2 if the material is isotropic. These numbers apply only when the coordinate system x,y,z is attached to the material symmetry directions. In any other coordinate system, the components of matrix (A) in Equation (2) will depend on these compliances or elastic constants and on the orientation of the x,y,z coordinate system with respect to the symmetry directions.

2.2 Laboratory Testing of Anisotropic Rocks

2.2.1 Measurement of Elastic Constants

If the deformability of an anisotropic rock can be described by the Generalized Hooke's law with one of the elastic symmetries mentioned above, testing is required to measure the corresponding elastic constants. Several examples have been reported in the literature (Dayre, 1969; Masure, 1970; Pinto, 1970; Ko and Gerstle, 1972, 1976; Simonson et al., 1976; Cook et al, 1978; Lerau et al., 1981). Laboratory testing usually consists of cutting samples (cylindrical or prismatic) at different angles to the apparent directions of rock symmetry and testing them in uniaxial, triaxial or multiaxial com-pression. Samples must be cut next to each other in order to eliminate any experimental scatter due to mat-

erial differences. Induced strains and displacements measured during testing are then related to the elastic constants through the constitutive relation of the rock material.

Often the type of rock anisotropy and therefore the constitutive relation are assumed before testing. A common assumption is to describe the rock either as or-thotropic or as transversely isotropic such that planes and/or axes of elastic symmetry coincide with the appar-ent directions of rock symmetry. For example, a plane of transverse isotropy is often associated with folia-tion, schistosity or bedding planes.

Other methods have been suggested to measure the elastic constants or compliances of anisotropic rocks that are idealized either as transversely isotropic or as orthotropic materials. These include dynamic methods and the diametral loading of thin cylinders of rock ("Brazilian test" configuration). The procedure and the analytical solution for measuring the dynamic elastic constants of anisotropic rocks are discussed by Duvall (1965) and others and, therefore, will not be treated here. Instead, the other method is analyzed.

2.2.2 Diametral Compression Tests

Determination of the elastic constants of aniso-tropic rocks by means of diametral compression tests on thin cylinders of circular cross section was suggested by Pinto (1979). The method consists of applying two equal and opposite forces along a diameter of a disc of anisotropic rock. Closed-form solutions were derived that relate the elastic constants of the rock in the disc's cross section to strains measured at its center with a strain gage rosette. The procedure is then repeated for other disc orientations with respect to the overall anisotropy in order to measure all the elastic constants of the rock. In these closed-form solutions, Pinto (1979) assumed that the stress distribution at the center of the disc can always be approximated by that for an isotropic material. Although this is correct when the cross section of the disc is parallel to a plane of transverse isotropy, the stress distribution is affected by the anisotropic character for any other orientation of anisotropy. This can be accounted for by using the analytical solutions derived by Okubo (1952) and Lekhnitskii (1957).

Consider the elastic equilibrium of an homogeneous, continuous, anisotropic body bounded externally by a cylindrical surface of circular cross section with dia-meter D = 2a. Let x,y,z be a cartesian coordinate sys-tem with the z axis defining the longitudinal axis of the body. The latter is assumed to be analogous to a plate or a disc of thickness t, the middle plane of which being taken to coincide with the xOy plane of the coordinate system (Fig. 1). Let X_n, Y_n be the com-ponents in the x,y directions of surface forces per unit area acting along the edge of the plate. These forces are in equilibrium so that the resultant moment and the resultant force are equal to zero. Body forces are assumed to be absent. If, in addition, we assume that the plate is thin, has a plane of elastic symmetry parallel to its middle plane and is loaded by surface forces that vary negligibly across its thickness, then a <u>generalized plane stress formulation</u> (Filon, 1903; Lekhnitskii, 1957) can be used to solve for the distri-butions of stress and strain within the plate. For this formulation, if the equations of equilibrium for stress, the compatibility equations for strain, the strain dis-placement relations, the constitutive relations and the boundary conditions are all expressed in terms of

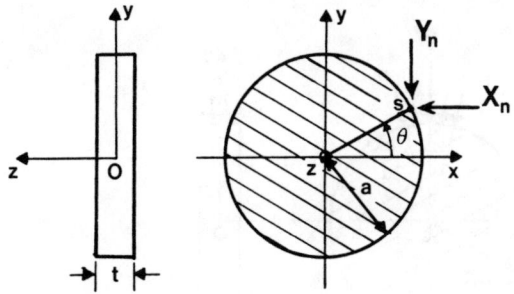

Fig. 1 Geometry of the problem

average values of stress, strain and displacement across the plate thickness, then it can be shown that the mean stress components, $\bar{\sigma}_x$, $\bar{\sigma}_y$, $\bar{\tau}_{xy}$, the mean strain components $\bar{\varepsilon}_x$, $\bar{\varepsilon}_y$, $\bar{\gamma}_{xy}$ and the mean displacement components satisfy the same equations that govern the classical plane strain formulation in the xOy plane.[1] The only difference is that the constitutive relation is replaced by the following:

$$
\begin{bmatrix} \bar{\varepsilon}_x \\ \bar{\varepsilon}_y \\ \bar{\gamma}_{xy} \end{bmatrix} = \begin{bmatrix} a_{11} & a_{12} & a_{16} \\ a_{21} & a_{22} & a_{26} \\ a_{61} & a_{62} & a_{66} \end{bmatrix} \begin{bmatrix} \bar{\sigma}_x \\ \bar{\sigma}_y \\ \bar{\tau}_{xy} \end{bmatrix} \tag{3}
$$

where a_{11}, a_{12}......,a_{66} are the compliance components calculated in the xOy coordinate system. Let F be a stress function such that

$$
\bar{\sigma}_x = \frac{\partial^2 F}{\partial y^2} \quad \bar{\sigma}_y = \frac{\partial^2 F}{\partial x^2} \quad \bar{\tau}_{xy} = -\frac{\partial^2 F}{\partial x \partial y} \tag{4}
$$

Substituting Equations (3) and (4) into the equation of compatibility that $\bar{\varepsilon}_x$, $\bar{\varepsilon}_y$ and $\bar{\gamma}_{xy}$ must satisfy, yields the following differential equation:

$$
a_{22} \frac{\partial^4 F}{\partial x^4} - 2a_{26} \frac{\partial^4 F}{\partial x^3 \partial y} + (2a_{12} + a_{66}) \frac{\partial^4 F}{\partial x^2 \partial y^2} -
$$

$$
2a_{16} \frac{\partial^4 F}{\partial x \partial y^3} + a_{11} \frac{\partial^4 F}{\partial y^4} = 0 \tag{5}
$$

The general solution of this equation depends on the roots, μ_i (i = 1 to 4) of its characteristic equation

$$
a_{11}\mu^4 - 2a_{16}\mu^3 + (2a_{12} + a_{66})\mu^2 - 2a_{66}\mu + a_{22} = 0 \tag{6}
$$

Lekhnitskii (1957) has shown that the roots of this equation are always complex or purely imaginary, two of them being the conjugate of the two others. Let μ_1, μ_2

be those roots and $\bar{\bar{\mu}}_1$, $\bar{\bar{\mu}}_2$ their respective conjugate.[2] As shown by Lekhnitskii, the first derivatives of F with respect to x and y can be expressed as follows:

$$
\frac{\partial F}{\partial x} = 2\text{Re}\ (\phi_1(z_1) + \phi_2(z_2))
$$

$$
\frac{\partial F}{\partial y} = 2\text{Re}\ (\mu_1\phi_1(z_1) + \mu_2\phi_2(z_2)) \tag{7}
$$

where $\phi_k(z_k)$ (k=1,2) are analytic functions of the complex variables $z_k = x + \mu_k y$ and Re is the notation for the real part of the complex expression in the brackets. Combining Equations (4) and (7) we obtain the general expression for the stress components

$$
\bar{\sigma}_x = 2\text{Re}\ (\mu_1^2\phi'_1(z_1) + \mu_2^2\phi'_2(z_2))
$$

$$
\bar{\sigma}_y = 2\text{Re}\ (\phi'_1(z_1) + \phi'_2(z_2))
$$

$$
\bar{\tau}_{xy} = -2\text{Re}\ (\mu_1\phi'_1(z_1) + \mu_2\phi'_2(z_2)) \tag{8}
$$

where $\phi'_k(z_k)$ are the first derivatives of $\phi_k(z_k)$ with respect to z_k. In addition to Equation (5), F must satisfy the boundary conditions along the outer contour of the plate. In terms of $\phi_k(z_k)$ (k=1,2), the boundary conditions take the form

$$
2\text{Re}\ (\mu_1\phi_1 + \mu_2\phi_2) = \int_0^s \bar{X}_n\ ds
$$

$$
2\text{Re}\ (\phi_1 + \phi_2) = -\int_0^s \bar{Y}_n\ ds \tag{9}
$$

where s is the arc length along the contour of the disc (Fig. 1), and $\bar{X}_n$, $\bar{Y}_n$ are the average values of surface force components X_n, Y_n across the plate thickness. Let $\bar{X}_n$ and $\bar{Y}_n$ be expanded as Fourier series in cos mθ and sin mθ where m varies between 1 and an arbitrary number N and θ is an angle defined in Fig. 1. After integration, Equation (9) becomes

$$
2\,\text{Re}(\phi_1 + \phi_2) = \sum_{m=1}^{N} (a_m e^{im\theta} + \bar{\bar{a}}_m e^{-im\theta})
$$

$$
2\,\text{Re}(\mu_1\phi_1 + \mu_2\phi_2) = \sum_{m=1}^{N} (b_m e^{im\theta} + \bar{\bar{b}}_m e^{-im\theta}) \tag{10}
$$

where a_m, b_m and their respective conjugates $\bar{\bar{a}}_m$, $\bar{\bar{b}}_m$ depend on the coefficients of the Fourier series for $\bar{X}_n$ and $\bar{Y}_n$.

General expressions for the functions $\phi_1(z_1)$ and $\phi_2(z_2)$ were proposed by Lekhnitskii (1957) as follows:

---

[1] Indeed, actual values of stress, strain and displacement differ only slightly from the mean values since the plate is thin.

[2] A double bar is used for the conjugate of a complex number.

$$\phi_1(z_1) = A_1 z_1 + \sum_{m=2}^{N} A_m P_{1m}(z_1)$$

$$\phi_2(z_2) = B_1 z_2 + \sum_{m=2}^{N} B_m P_{2m}(z_2) \qquad (11)$$

with

$$P_{km}(z_k) = \frac{-1}{(1-i\mu_k)^m} \left(\left(\frac{z_k}{a} + \sqrt{\left(\frac{z_k}{a}\right)^2 - 1 - \mu_k^2}\right)^m + \right.$$

$$\left. \left(\frac{z_k}{a} - \sqrt{\left(\frac{z_k}{a}\right)^2 - 1 - \mu_k^2}\right)^m \right) \qquad (k = 1,2) . \qquad (12)$$

Coefficients A_m, B_m and their respective conjugates $\bar{\bar{A}}_m$, $\bar{\bar{B}}_m$ can be expressed in terms of a_m, $\bar{\bar{a}}_m$, b_m $\bar{\bar{b}}_m$ $(m=2,N)$ by substituting Equations (11) and (12) into (10) with $z_k/a=\cos\theta+\mu_k\sin\theta$. Then, by combining Equation (8) with the expressions for the derivatives of $\phi_k(z_k)$ in Equation (11), the stress components $\bar{\sigma}_x$, $\bar{\sigma}_y$ and $\bar{\tau}_{xy}$ can be computed at any point (x,y) in the disc. In these components, the contribution of $A_1 z_1$ and $B_1 z_2$ consists of constant quantities that are either function of a_1 and $\bar{\bar{a}}_1$ or of b_1 and $\bar{\bar{b}}_1$.

Consider a uniform radial pressure $\bar{\sigma}_r=p$ applied over the arcs $\pi/2 - \alpha < \theta < \pi/2 + \alpha$ and $3\pi/2-\alpha < \theta < 3\pi/2 + \alpha$ as shown in Fig. 2. This stress distribution can be approximated by the following Fourier series in the angle θ:

$$\bar{\sigma}_r = A_0 + \sum_{n=1}^{N-1} (A_n \cos n\theta + B_n \sin n\theta) \qquad (13)$$

with $\qquad A_0 = 2p\alpha/\pi$

$$A_n = \frac{2p}{\pi} \left(\frac{1 + (-1)^n}{n}\right) \cos \frac{n\pi}{2} \sin n\alpha , \quad B_n = 0.$$

Since there is no shear stress applied along the contour of the disc, the surface force components $\bar{X}_n$, $\bar{Y}_n$ are function of $\bar{\sigma}_r$ only, and can be expressed as Fourier series of $\cos m\theta$ and $\sin m\theta$ with m varying between 1 and N. Using the theory presented above, it can be shown that at any point (x,y) within the disc, the stress field components can be written as follows:

$$\bar{\sigma}_x = \frac{p}{\pi} q'_{xx} \qquad \bar{\sigma}_y = \frac{p}{\pi} q'_{yy} \qquad \bar{\tau}_{xy} = \frac{p}{\pi} q'_{xy} \qquad (14)$$

where q'_{xx}, q'_{yy} and q'_{xy} are <u>stress concentration factors</u> that are functions of the coordinates of the point of interest, the angle α and the compliances a_{11}, a_{22}, a_{12}, a_{16}, a_{26} and a_{66}. Let α be small and p be equal to $W/(\alpha Dt)$ where W is the total force applied on the disc. Substituting this value for p into Equation (14) we obtain

$$\bar{\sigma}_x = \frac{W}{\pi Dt} q_{xx} \qquad \bar{\sigma}_y = \frac{W}{\pi Dt} q_{yy} \qquad \bar{\tau}_{xy} = \frac{W}{\pi Dt} q_{xy} \qquad (15)$$

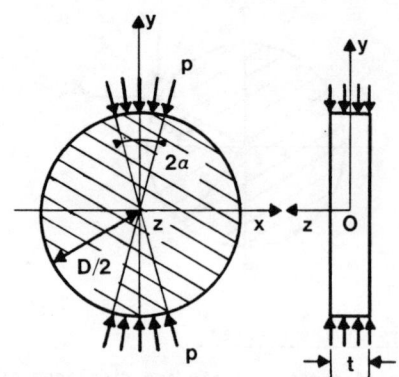

Fig. 2 Diametral compression of a thin disc over an angular width of 2α.

where $q_{xx}=q'_{xx}/\alpha$, $q_{yy}=q'_{yy}/\alpha$ and $q_{xy}=q'_{xy}/\alpha$. Numerical examples for the influence of anisotropy on these factors will be presented in Section 3 in association with the tensile strength of anisotropic rocks.

Considering an <u>orthotropic material</u>, a procedure is now derived to measure its elastic constants or compliances from strain gage measurements taken at the center of discs loaded as shown in Fig. 2. Let x',y',z' be a coordinate system attached to the three planes of elastic symmetry of the material. In this system, nine elastic constants can be defined: three Young's moduli $E_{x'}$, $E_{y'}$, $E_{z'}$; three shear moduli $G_{x'y'}$, $G_{x'z'}$, $G_{y'z'}$; and three Poisson's ratios $\nu_{x'y'}$, $\nu_{x'z'}$, $\nu_{y'z'}$. Therefore, at least nine independent measurements of strain are needed to measure these quantities. Any additional measurement will improve the accuracy of their determination. Consider the three tests shown in Fig. 3. Each test consists of loading a disc whose middle plane is parallel to one of the three planes of elastic symmetry of the material. For each test $j(j=1,2,3)$, strains at the center of the disc are measured in three directions ε_{aj}, ε_{bj}, ε_{cj} using strain gages, at an arbitrary load level W and in the linear elastic range of the material response to loading.[3] To simplify the presentation, we assume that the three discs in Fig. 3 have same geometry, the same strain gages orientation, the same small loading angle α and orientation angle ψ, and that strains are measured at

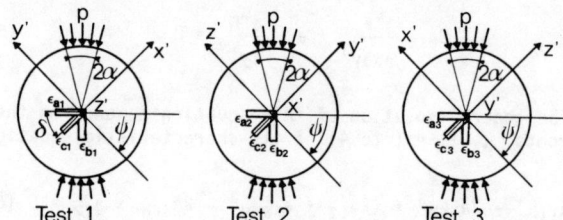

Fig. 3 Measurement of the elastic properties of an orthotropic material using three diametral compression tests.

---

[3] Strain gages are assumed to be small in order to neglect the variation of stress and strain along their length.

the same load level W. In test 1, the stress concentration factors q_{xx1}, q_{yy1}, q_{xy1} are functions of $1/E_x'$, $1/E_y'$, $\nu_{x'y'}/E_x'$ and $1/G_{x'y'}$. Similarly, in test 2, q_{xx2}, q_{yy2}, q_{xy2} are functions of $1/E_y'$, $1/E_z'$, $\nu_{y'z'}/E_y'$ and $1/G_{y'z'}$. In test 3, q_{xx3}, q_{yy3}, q_{xy3} are functions of $1/E_z'$, $1/E_x'$, $\nu_{z'x'}/E_z'$ and $1/G_{z'x'}$. In addition the nine stress concentration factors depend on the angles α and ψ. Combining Equations (3) and (15) for each test j ($j=1,2,3$) and the expressions for ε_{aj}, ε_{bj}, ε_{cj} in terms of $\overline{\varepsilon}_x$, $\overline{\varepsilon}_y$ and $\overline{\gamma}_{xy}$, the nine elastic constants are related to the nine strain measurements as follows:

$$(\varepsilon)_M = \frac{W}{\pi D t}\ (T)(X) \tag{16}$$

where $(\varepsilon)_M$ and (X) are such that

$$(\varepsilon)_M^t = (\varepsilon_{a1}\ \varepsilon_{b1}\ \varepsilon_{c1}\ \varepsilon_{a2}\ \varepsilon_{b2}\ \varepsilon_{c2}\ \varepsilon_{a3}\ \varepsilon_{b3}\ \varepsilon_{c3})$$

$$(X)^t = (1/E_x'\ \ 1/E_y'\ \ 1/E_z'\ \ \nu_{x'y'}/E_x'\ \ \nu_{y'z'}/E_y'\ \ \nu_{z'x'}/E_z'$$

$$1/G_{x'y'}\ \ 1/G_{y'z'}\ \ 1/G_{x'z'})$$

The components of matrix (T) depend on the orientation angles ψ and δ shown in Fig. 3 and on the nine stress concentration factors q_{xxj}, q_{yyj}, q_{xyj} ($j=1,2,3$). The latter are themselves functions of compliance coefficients. Equation (16) can be solved for matrix (X) using the following iterative procedure: first, assume that the nine stress concentration factors are equal to those for an isotropic material[4] and solve Equation (16) for matrix (X), then, calculate the nine stress concentration factors for the new set of elastic properties and repeat the procedure until convergence. It is noteworthy that this procedure applies only when the angle ψ is different from 0 and 90 degrees.

The same procedure applies if the material is <u>transversely</u> isotropic. If the plane of transverse isotropy coincides, for instance, with plane $y'Oz'$, then, the five elastic constants can be measured by conducting tests 1 and 2 in Fig. 3. Finally, if the material is isotropic, the two elastic constants can be measured from one diametral compression test only.

When more than nine, five or two strain measurements are available and the material is respectively orthotropic, transversely isotropic or isotropic, Equation (16) can be solved for the least square estimates of the corresponding elastic constants.

In order to illustrate the procedure presented above consider the following numerical example. A transversely isotropic material is tested as shown in Fig. 3 with $\alpha = 7.5$ degrees. In test 1, the plane of transverse isotropy (plane $y'Oz'$) is inclined at an angle $\psi = 30°$ with respect to horizontal. Measured strains are

such that[5] $(\pi D t/W)\varepsilon_{a1} = -34.54\ 10^{-5}$ MPa^{-1}; $(\pi D t/W)\varepsilon_{b1} = 45.01\ 10^{-5}$MPa^{-1}, and $(\pi D t/W)\varepsilon_{c1} = -7.9\ 10^{-5}$MPa^{-1} with $\delta = 45$ degrees. In test 2, the plane of transverse isotropy is parallel to the middle plane of the disc. Measured strains are such that $(\pi D t/W)\varepsilon_{a2} = -8.75\ 10^{-5}$MPa^{-1} and $(\pi D t/W)\varepsilon_{b2} = 16.25\ 10^{-5}$ MPa^{-1}. Equation (16) is now reduced to a system of five equations with five unknown elastic constants. Among these equations, the two that are associated with test 2 allow the calculation of E_y', and $\nu_{y'z'}$ directly from the strain measurements ε_{a2}, ε_{b2} since the stress concentration factors q_{xx2} and q_{yy2} are equal to those for an isotropic material.

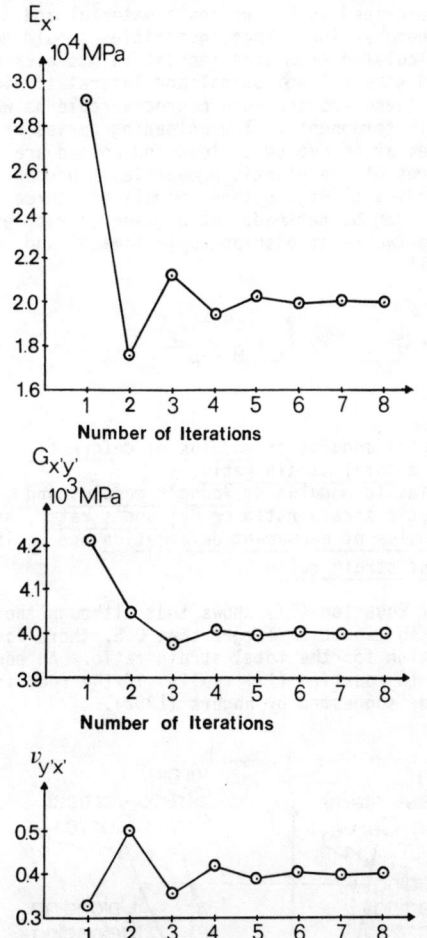

Fig. 4 Variation of E_x', $G_{x'y'}$ and $\nu_{y'x'}$ with the number of iterations

E_y' and $\nu_{y'z'}$ are found to be equal to 4.10^4 MPa and 0.25 respectively. Three equations remain to measure E_x', $G_{x'y'}$ and $\nu_{y'x'}$ using the iterative procedure outlined previously. Fig. 4 shows that for those three elastic constants, convergence is reached after seven iterations.

---

[4]If α is small, the stress concentrations can be approximated by $q_{xxj} = -2$, $q_{yyj} = 6$, $q_{xyj} = 0$.

---

[5]These numbers are theoretical. They were calculated by combining Equations (15) and (3) for each test configuration with $E_x' = 2.10^4$ MPa, $E_y' = 4.10^4$ Mpa, $\nu_{y'z'} = 0.25$, $\nu_{y'x'} = 0.4$, $G_{x'y'} = 4.10^3$ MPa.

2.3 Definition of Elastic Constants

In the testing methods outlined in Section 2.2, the rock response to test disturbances was idealized as linearly elastic. Therefore, strains and displacements that are used in these methods to back-calculate the elastic constants must be measured in the linear elastic ranges of load-deformation or stress-strain curves. Virgin loading curves do not always represent the elastic behavior of tested rock samples.

Fig. 5 shows the difficulty of defining exactly what is meant by Young's modulus E and Poisson's ratio ν for a rock described as an isotropic material and tested in uniaxial compression. These quantities should not simply be calculated from experimental points picked up on the virgin stress-longitudinal and lateral strain curves since these embrace both nonrecoverable as well as recoverable components. The unloading curves or reloading curves after cycles of load and unload are better measures of the elastic properties (Goodman, 1980). According to Fig. 5, three moduli and three strain ratios can be defined. At a given stress level, the following two relationships apply (Amadei and Goodman, 1981):

$$\frac{1}{E_t} = \frac{1}{E} + \frac{1}{M} \qquad \nu_t = \frac{\nu M + \nu_p E}{M + E} \qquad (17)$$

where

E_t is a total modulus or <u>modulus of deformation</u> and ν_t is a <u>total strain ratio</u>,
E is an <u>elastic modulus</u> or <u>Young's modulus</u> and ν is an elastic strain ratio or <u>Poisson's ratio</u>, and,
M is a <u>modulus of permanent deformation</u> and ν_p is a <u>permanent strain ratio</u>

The second of Equation (17) shows that although the Poisson's ratio cannot be larger than 0.5, there is no such restriction for the total strain ratio. An equation similar to Equation (17) applies if the rock is anisotropic as suggested by Rogers (1982).

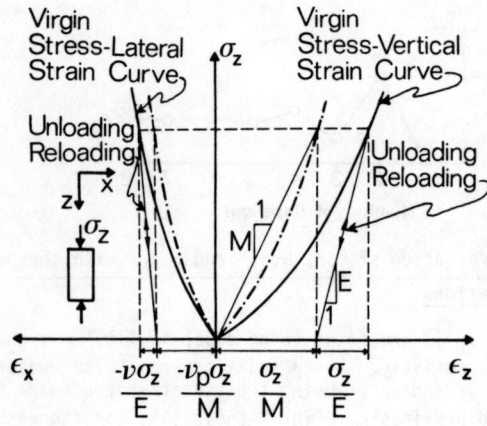

Fig. 5 Definition of elastic properties

A procedure similar to the one for uniaxial compression can be applied for the diametral compression tests described in Section 2.2.2. Strain components that are input in Equation (16) must only account for the elastic response of the rock discs to diametral loading.

3. TENSILE STRENGTH OF ANISOTROPIC ROCKS

Techniques to measure the tensile strength of rocks include the direct uniaxial tensile test and indirect tensile tests. Due to difficulties associated with the former, the latter are more often used. The most popular of these indirect tests is the splitting tension test frequently referred to as the "Brazilian test". It consists of applying diametral compression to a thin disc of rock between the platens of a testing machine. The load can be applied either as a line load or be distributed over a small strip as shown in Fig. 2.

Result of "Brazilian tests" conducted on anisotropic rocks that present a well defined direction of planar anisotropy have been reported by Hobbs (1964), McLamore and Gray (1967), Rogers (1982) and others. In general, tested samples were found to fail along the loaded diameter irrespective of the anisotropy orientation. The reason for this behavior was demonstrated by examining the stress distribution inside a disc of isotropic material under diametral compression by a line load or a strip load. For a line load, the horizontal stress at the center of the disc is tensile with magnitude $2W/(\pi Dt)$ and the vertical stress compressive with magnitude $6W/(\pi Dt)$ where W is the applied force (D and t being defined in Fig. 2). By symmetry, these are principal stresses. For a narrow strip load, the stresses at the center of the disc are approximately the same as above. With a principal stress ratio of 3, failure in the disc ought to result from the application of the tensile stress alone according to the Griffith theory. Therefore, the tensile strength of the rock can be calculated using the expression for the horizontal stress at the center of the disc, W being the load applied on the disc at failure. Experiments showed that the tensile strength of anisotropic rocks calculated that way generally depends on the orientation of the anisotropy with respect to the direction of loading. The minimum tensile strength occurs when the planar anisotropy is perpendicular to the induced tensile stress in the discs while the maximum strength occurs when the anisotropy is parallel to that tensile stress.

The main limitation of the approach summarized previously is that the isotropic solution for the stress distribution within a disc under diametral compression is forced in the interpretation of results from "Brazilian tests" conducted on anisotropic materials. It is preferable to use Equation (15) to account for anisotropy when calculating the tensile strength of anisotropic rock from "Brazilian tests". As an example, consider a disc with the geometry of test 1 in Fig. 3. The rock is assumed to be transversely isotropic in plane $y'Oz'$, its deformability being defined by five elastic constants $E_{x'}$, $E_{y'}$, $G_{x'y'}$, $\nu_{x'y'}$ and $\nu_{y'z'}$. The plane of transverse isotropy is inclined at an angle ϕ with respect to horizontal. If α is small, Equation (15) gives the stress components at any point (x,y) in the disc. The stress concentration factors q_{xx}, q_{yy}, and q_{xy} are functions of the coordinates (x,y), the angles α and ϕ and the compliances $1/E_{x'}$, $1/E_{y'}$, $\nu_{x'y'}/E_{x'}$, $1/G_{x'y'}$. The influence of these compliances on the stress concentration factors can also be expressed in terms of three dimensionless quantities $E_{x'}/E_{y'}$, $E_{x'}/G_{x'y'}$ and $\nu_{x'y'}$. Fig. 6 shows the variation of q_{xx}, q_{yy}, q_{xy} calculated at the center of the disc with the orientation angle ϕ for the following three sets of elastic constants:

	$E_{x'}/E_{y'}$	$E_{x'}/G_{x'y'}$	$\nu_{x'y'}$
Case 1	0.5	5	0.2
Case 2	0.5	1.33	0.2
Case 3	2.	6.06	0.8

The load angle α is equal to 7.5 degrees and N in Equation (13) is set to 15. In comparison to the isotropic solution (e.g., $q_{xx} \approx -2$, $q_{yy} \approx 6$, $q_{xy} \approx 0$)

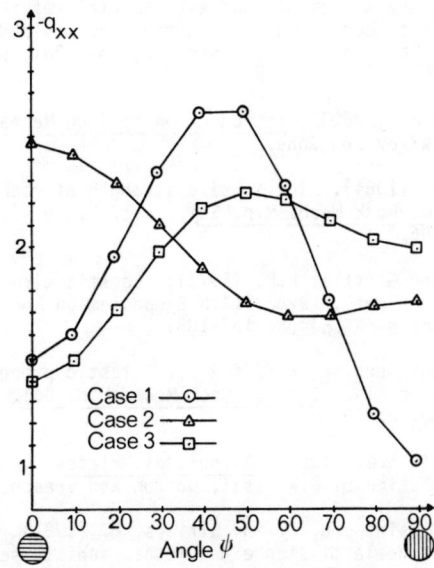

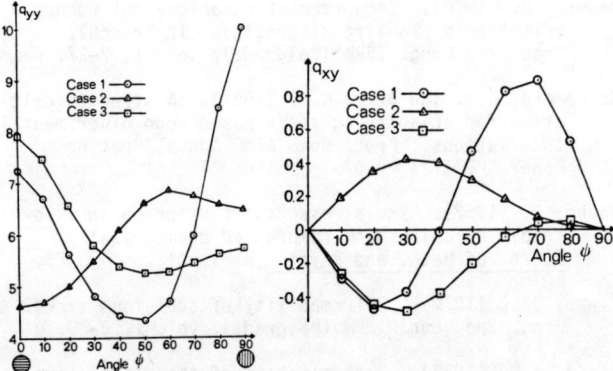

Fig. 6 Variation of stress concentration factors q_{xx}, q_{yy}, and q_{xy} with the orientation angle ψ.

both the anisotropic character of the disc material and the orientation of the anisotropy with respect to the loading direction have a strong influence on the value of the stress concentration factors. For cases 1 and 2, the stress in the x direction is tensile and smaller in magnitude when the plane of transverse isotropy is parallel to the loading direction than when it is at right angle to it, whereas the opposite applies for the stress in the y direction, which is compressive. Case 3 shows conclusions opposite to those previously mentioned since $E_{x'}/E_{y'}$ is now larger than unity. Fig. 7 shows the variation of the stress field along half of the loading diameter (0<2r/D<1) for case 1 and for several values of the orientation angle ψ. The stress field is expressed

in terms of dimensionless quantities $\overline{\sigma}_x/p$, $\overline{\sigma}_y/p$ and $\overline{\tau}_{xy}/p$, where p is the pressure applied on the disc. The stress component perpendicular to the loading diameter is tensile and uniform over 60 percent of the diameter length. At the loading strip contact, $\overline{\sigma}_x$ varies continuously between 1.37 p when $\psi = 0$ degrees and 0.74p when $\psi = 90$ degrees, in comparison to $\overline{\sigma}_x = p$ for the isotropic solution. Due to a non-vanishing shear stress component when ψ is different from 0 and 90 degrees, the principal stress field along the loaded diameter is inclined with respect to the x and y directions. For the numerical example presented in Figs. 6 and 7, the inclination angle does not exceed 7 degrees. If we make the assumption that the tensile strength of the aniso-tropic rock can be calculated from the expression for the horizontal stress at the center of the disc as in the isotropic case, then, by using the first of Equation (15) with q_{xx} as shown in Fig. 6, both the anisotropic character and the anisotropy orientation could be taken into account in the value of the tensile strength.

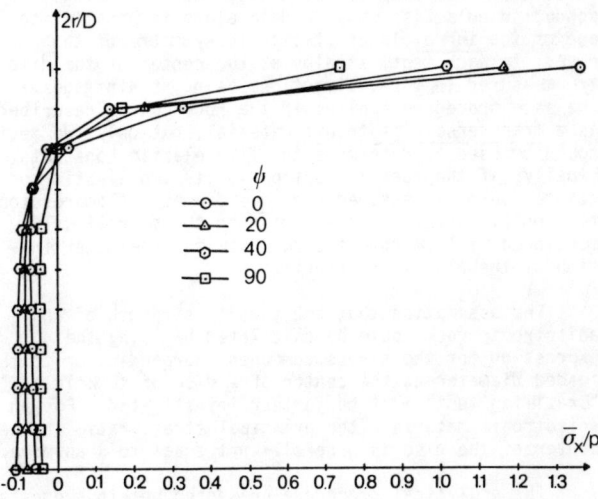

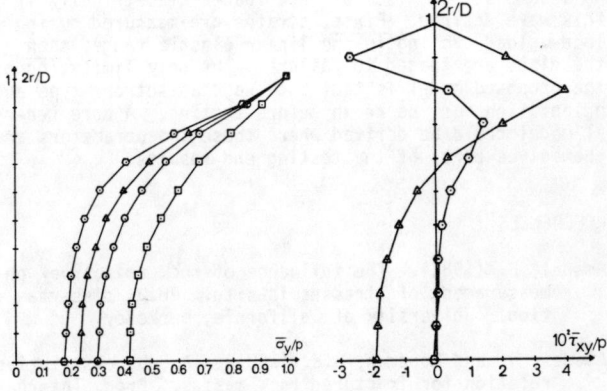

Fig. 7 Variation of the stress field $\overline{\sigma}_x/p$, $\overline{\sigma}_y/p$, $\overline{\tau}_{xy}/p$ along half of the loaded diameter for different values of the orientation angle ψ.

4. CONCLUSIONS

An analytical procedure is proposed in this paper to characterize both the deformability and the tensile strength of an anisotropic rock by diametral compression of thin discs of rock. The rock is modelled as a linearly elastic, orthotropic or transversely isotropic material with one plane of elastic symmetry parallel to the middle plane of the discs.

The stress distribution within a thin disc of anisotropic material loaded diametrically is generally different from that obtained with the isotropic solution. The proposed analytical model shows that this stress distribution depends on the anisotropic character of the material and the orientation of the anisotropy with respect to the direction of loading.

The nine elastic constants of a rock described as an orthotropic material could be back-calculated from three diametral compression tests, each one being conducted on a disc whose middle plane is parallel to one of the three planes of elastic symmetry of the rock. In each test, strains at the center of the disc are measured in three directions using strain gages. The same procedure applies if the rock can be described as a transversely isotropic material, but only two tests could be used to determine the five elastic constants. Finally, if the rock is isotropic, its two elastic constants could be measured from one diametral compression test only. Any test in addition to those previously mentioned will improve the accuracy for the determination of the elastic constants.

The assumption that the tensile strength of an anisotropic rock could be calculated by using the expression for the stress component perpendicular to the loaded diameter at the center of a disc of rock in a "Brazilian test" must be further investigated. For an anisotropic material, the principal stress ratio at the center of the disc is generally not equal to 3 anymore.

The analytical procedure presented herein suggests that both measurements of elastic constants and of tensile strength of an anisotropic rock could be conducted on a same set of discs of rock loaded diametrically if this were desired. First, strains are measured during load-unload cycling in the linear elastic range; then the discs are loaded to failure. The only limitation of the proposed model is that both rock anisotropy type and orientation must be known before testing. A more general model could be derived where these two parameters are themselves parts of the testing end results.

REFERENCES

Amadei, B. (1982). The influence of rock anisotropy on measurement of stresses in-situ. Ph.D. dissertation. University of California, Berkeley.

Amadei, B. and Goodman, R.E. (1981). A 3-D constitutive relation for fractured rock masses. Proc. International Symposium on the Mechanical Behavior of Structured Media, Ottawa, (Selvadurai, A.P.S. ed.), (Elsevier). Part B, pp. 249-268.

Cook, N.E., Ko, H.Y., Gerstle, K.H. (1978). Variability and anisotropy of mechanical properties of the Pittsburgh coal seam. Rock Mechanics 11, pp. 3-18.

Dayre, M. (1969). Anisotropie discontinue d'une formation calcaire (in French). Colloque sur la fissuration des roches. Paris. Numero special de la Revue de l'Industrie Minerale (Juillet).

Duvall, W.I. (1965). The effect of anisotropy on the determination of dynamic elastic constants of rock. Trans. Society of Mining Engineers, December, pp. 309-316.

Filon, L.N.G. (1903). On an approximate solution for the bending of a beam of rectangular cross section under any system of load with special references to points of concentrated or discontinuous loading Proc. Roy. Soc. London, Ser. A., Vol. 201, pp. 65-156.

Goodman, R.E. (1980). Introduction to Rock Mechanics. John Wiley and Sons.

Hobbs, D.W. (1964). The tensile strength of rocks. Int. J. Rock Mech. Min. Sci., vol. 1, No. 3, pp. 385-396.

Ko, H.Y. and Gerstle, K.H. (1972). Constitutive relations of coal. Proc. 14th Symposium on Rock Mechanics (ASCE) pp. 157-188.

Ko. H.Y. and Gerstle, K.H. (1976). Elastic properties of two coals. Int. J. Rock Mech. Min. Sci., vol. 13, pp. 81-90.

Lekhnitskii, S.G. (1957), Anisotropic Plates, English translation by S.W. Tsai, Gordon and Breach, 1968.

Lerau, J., Saint-Leu, C. and Sirieys, P. (1981). Anisotropie de la dilatance des roches schisteuses (in French). Rock Mechanics 13,

Masure, P. (1970). Comportement mecanique des roches a anisotropie planaire discontinue (in French). Proc. 2nd Cong. ISRM (Belgrade), vol. 1, 7-27.

Mc Lamore, R.T. and Gray, K.E. (1967). A strength criterion for anisotropic rocks based upon experimental observations. Proc. 96th AIME Annual Meeting, Paper SPE 1721.

Okubo, H. (1952). The stress distribution in an aelotropic circular disk compressed diametrically. Journ. of Math. and Physics, vol. 31, pp. 75-83.

Pinto, J.L. (1970). Deformability of schistous rocks. Proc. 2nd Cong. ISRM (Belgrade), vol. 1, 2-30.

Pinto, J.L. (1979). Determination of the elastic constants of anisotropic bodies by diametral compression tests. Proc. 4th Cong. ISRM (Montreux), vol. 2, pp. 359-363.

Rogers, J.D. (1982). The genesis properties and significance of fracturing in Colorado Plateau sandstones. Ph.D. dissertation University of California, Berkeley.

Simonson, E.R., Johnson, J.N. and Buchholt, L. (1976). Anisotropic mechanical properties of a moderate and rich kerogen content oil shale. Terratek report TR 76-72 Dec.

STANDARD LABORATORY TESTING FOR COMPACTED SHALES

Essais normalisés pour les schistes compactés

Standardlaborversuche von gepresstem Tonschiefer

C. W. Lovell

Professor, School of Civil Engineering, Purdue University, West Lafayette, Indiana, U.S.A.
47907

SYNOPSIS

Embankments constructed of weak and non-durable rock, like shale, require special design and construction technology. This fact has been underscored in the midwestern U.S. by the unsatifactory performance of hundreds of large shale fills. Research has resulted in a recommended sequence of eight standard laboratory tests. Four tests are required for classification (Atterberg limits, point load strength, and slake durability or cyclic slaking). Tests to define compaction characteristics and design parameters include: moisture-density relations, compaction-degradation, onedimensional compression, and consolidated undrained triaxials on saturated samples. The latter two tests allow prediction of settlement and the assessment of stability of the slopes.

RESUME

Les remblais construits avec des roches de faible résistance et altérables, comme les schistes, demandent une technique de calcul et de construction spéciale. Ceci a été souligné dans le centre ouest des U.S. par le mauvais comportement de centaines de remblais faits de schiste. Un programme de recherche a résulté dans la recommandation d'une séquence de huit essais normalisés. Quatre essais de classification sont nécessaires (limites d'Atterberg, résistance à la pénétration, et durabilité à la désagrégation ou essai cyclique de désagrégation). Les essais pour définir les caractéristiques de compactage et les paramètres de calcul incluent: rapports densité-teneur en eau, compaction-dégradation, compression unidirectionnelle, et essais non drainés sur échantillons saturés reconsolidés à l'appareil triaxial. Ces deux derniers essais permettent la prédiction de tassement et d'évaluation de la stabilité des pentes.

ZUSAMMENFASSUNG

Uferbefestigungen, die aus schwachem, unbeständigem Fels, wie Tonschiefer, errichtet werden, erfordern Spezialplanungs- und Konstruktionstechnologien. Diese Tatsache wurde im Mittwesten der U.S. besonders gut erkennbar wegen des unbefriedigenden Verhaltens von hunderten großer Tonschieferaufschüttungen. Durch Forschung hat sich eine empfehlenswerte Reihenfolge von acht Standardlaborversuchen ergeben. Vier dieser Versuche sind für die Klassifizierung nötig (Atterberg Grenze, Punkttragfähigkeit und Zersetzungswiderstandsfähigkeit oder zyklische Zersetzung). Versuche über Kompaktionskennziffern und Entwurfsparameter enthalten: Feuchtigkeit-Dichte-Verhältnisse, Kompaktions-Degradierung, eindimensionale Kompression und unentwässerte dreiaxial Tests von saturierten Proben. Die letzten zwei Versuche erlauben, die Setzungen vorauszusagen und die Stabilität von Böschungen festzustellen.

1. INTRODUCTION

Shales, siltstones, and similar weak rocks of poor-to-borderline physical durability are abundant rocks in the Interior Plains and Appalachian Highlands of North America. Accordingly, they are frequently excavated and used to build fills such as highway embankments and earth dams. In the eastern parts of the above physiographic divisions, Paleozoic examples of the above rocks often combine relatively high hardness with low durability.

During the somewhat hasty completion of the U.S. Interstate Highway System in the 1960's and 1970's, these nondurable materials were sometimes mistakenly placed essentially in thick lift (rock fill) construction. In the service environment of wetting/drying and freezing/thawing the slaking that took place was evidenced in the following sequence of distress: (a) differential settlement and rough pavement; (b) large total settlements and disruption of drainage facilities in the embankment; and (c) shallow-seated slope failures and loss of guard rail and shoulder. In some sidehill embankments, distress proceeded to deep-seated slope instabilities and loss of service on the highway.

Since there is no proven way to remedy the above interior deterioration (short of rebuilding the embankment properly), highway agencies were very receptive to programs which researched the problem and provided appropriate technology to minimize it in the future. One such program was undertaken over a period of roughly 10 years at the author's institution. Some of the principal findings are summarized in this paper.

2. STUDY OF COMPACTED SHALE CHARACTERISTICS

Shale characteristics important to the economic construction of highway embankments and earth dams are conveniently divided into three groupings: (a) classification, with particular emphasis on physical durability; (b) compaction, i.e., the moisture-density relationships, and degradability in the compaction process; and (c) compressibility and strength in the service environment ...from which settlement and stability against slope failure can be predicted.

Research in each of the above areas led to the development and/or selection of tests which would supply the information needed for design and construction. Experience with these tests in turn led to formulation of standard ways of running the tests and recommended procedures for using the parameters so generated. Since the conclusions reached in such a procedure are peculiar to the population of shales tested, it is important to record that most the author's experience has been with low plastic Paleozoic shales and siltstones from the state of Indiana, USA. All such materials will henceforth be referred to simply as "shales."

Development of a classification system for compacted shales received early research attention (Deo, 1972). Summaries on the various systems developed in North America in the 1970's (Chapman, 1975 and Oakland and Lovell, 1982) show how these systems evolved from (a) specialty tests and (b) discontinuous ratings to (c) common tests and (d) continuous ratings. The Indiana Department of Highways is a case in point. From 1972 until very recently, the 4-level Deo grouping was used, viz., "soil-like", "intermediate 1", "intermediate 2", and "rock-like" shales. The new system is after Franklin (1981) and is illustrated in Figure 1. It uses a continuous "R" rating and a battery of three tests, which will be discussed later.

Since nondurable shales must normally be thoroughly compacted in thin (soil type) lifts in highway embankment and earth dam applications, compaction and compaction-degradation characteristics are of considerable importance. Two approaches are available for compaction control (Bailey, 1976): (1) development of procedural specifications via experimental rolling on test pads, and (2) definition of end results (to be subsequently specified) by varying moisture in standard laboratory soil-type tests. The test pad approach is preferred, particularly when the shale is hard (Oakland and Lovell, 1982), but adds to construction time and expense. The laboratory tests are quick and inexpensive, but must be run on samples scalped of large sizes. Moisture content of shales may be varied and controlled in the laboratory testing by use of curing times (Abeyesekera, 1977 and Witsman and Lovell, 1979). However, in the field it is commonly difficult to change the moisture content of hard shales, requiring that they be compacted at about the natural moisture condition.

As important as prediction of degradation under compaction is, no simple test for accomplishing this existed until very recently. After examining the efforts of Bailey (1976) and others, Hale et al (1981a and 1981b) reported the development of a simple laboratory impact compaction-degradation test. The change in shale gradation produced by the compaction effort is simply represented by the "index of crushing". Hale (1979) shows that this number represents the percentage change in mean aggregate size due to compaction, i.e., the higher the

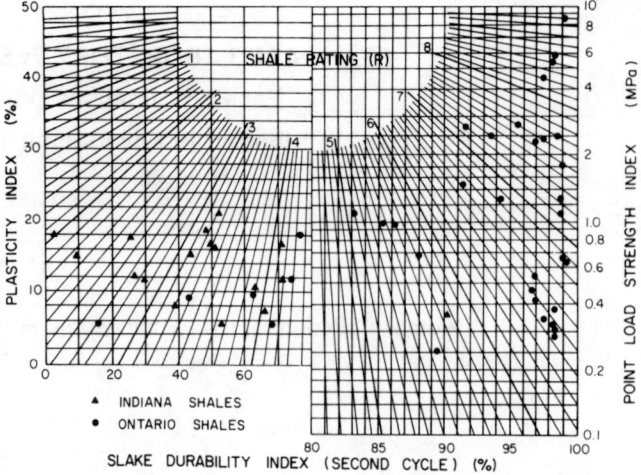

Fig. 1 <u>Typical Indiana and Ontario Shales on a Franklin Chart</u>

index, the more the shale is degraded. At this point in time, the test values are purely relative and uncorrelated with field rolling. Low values of the index for a nondurable shale warn of construction difficulties in achieving a thin dense compacted lift.

Settlement and slope stability predictions involve not only the as-placed responses of the compacted shale, but also a modeling of the compressions (or heaves) and strength changes that occur in the service environment. If laboratory samples are used, they should be as large in size as practicable. Compression of embankment shales under self weight occurs very rapidly (Abeyesekera, 1978 and Witsman and Lovell, 1979), so that this settlement has sensibly occurred by the time that construction is complete. Increases in moisture during service are reflected by embankment heave or (more commonly for the Paleozoic shales) settlement. This process may be modeled in the laboratory by back pressure saturation under simulated embankment confinement. Typical data are shown in Figure 2. Note that at least three pieces of compression information may be obtained from a single compacted sample, viz., the as-compacted compressibility at low stresses, the settlement produced by confined saturation, and the compressibility if loaded after being saturated. Such data may be integrated for the dimensions of the proposed embankment prototype, and a settlement estimate is produced.

Compressibility samples will show a characteristic change in slope at the prestress value produced by the compaction (Lovell and Witsman, 1981). This value is of particular importance to the shearing response of the shale, as explained later in the paper. Instability of slopes should be examined at two times, end of construction and in the long term. The end of construction shear strength is evaluated in confined-undrained tests, using samples which are as large as practicable. The appropriate strength for long term stability assessments may be determined by either drained or undrained testing of saturated samples. Again, the compacted samples should be as large as practicable.

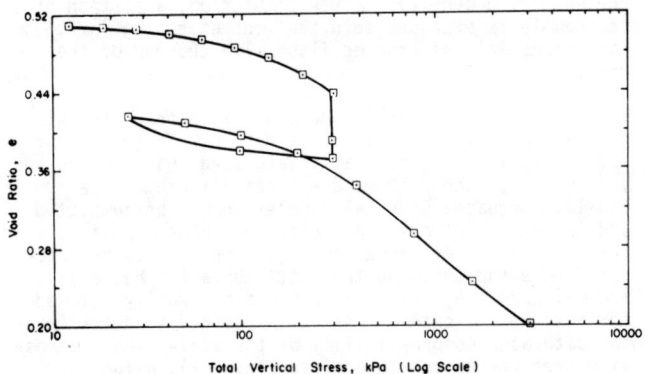

Fig. 2 One-Dimensional Compression Test on the New Providence, Indiana Shale

3. STANDARD TESTS

After about a decade of experimentation, it was concluded that eight tests required standardization...four for purposes of classification, two for compaction, and two for strength and compressibility. Only a few of the required tests have been standardized by ISRM (Brown, 1981), and even these few require modification to fit the needs of hard nondurable shales. All recommended standards are contained in Oakland and Lovell (1982).

3.1 Classification

With the adoption of the Franklin (1981) classification system, three classification tests are needed. Atterberg plasticity limits, point load strength, and slake durability. The Atterberg limits are used for only the less durable shales, which eliminates the need to disaggregate the harder shales in order to run the test. However, the limits are still needed for shales with slake durability indices between 40 and 80, and these materials are quite difficult to reduce to the "ultimate" particulate units required by the limits test. The best technique known to this writer is progressive mixing in a dispersion cup until the liquid limit of the dispersed product becomes constant. As many as 10 successive dispersions may be required, and more research is needed to develop a special version of the Atterberg limits tests for shales.

For the more durable (and harder) shales, a point load strength test is substituted for the Atterberg limits (see Figure 1). The samples should be roughly equidimensional, but need not be trimmed to a geometrical shape. Test values vary considerably into sample size, since planes of weakness are more likely to be included in larger shale pieces. As data such as those in Figure 3 show, the minimum piece dimension should be at least 25 to 30 mm. Correction to a standard size of 50 mm is recommended by the ISRM (1972) and others, but the available correction charts are not generally applicable to shales (Abeyesekera and Lovell, 1982).

Hale (1979) recommended that the point load strength be determined with loading normal to the bedding planes and at the natural moisture content of the shale. Since the stress-strain characteristics in the point load strength test have been successfully correlated with other shale properties (Hale, 1979), these data should be recorded as well.

The slake-durability test has emerged as the premiere durability test. Where the necessary specialized equipment is not available, a five-cycle slaking test, using eight hours of drying and sixteen hours of soaking may be substituted. To parallel the slake-durability results, the slaking index is redefined as the percentage by weight of material _retained_ on a No. 10 sieve. The five-cycle test is more conservative than the normal two-cycle slake durability one (Figure 4), i.e., it gives lower values for the resistance to slaking. It has the further disadvantage of requiring much more testing time.

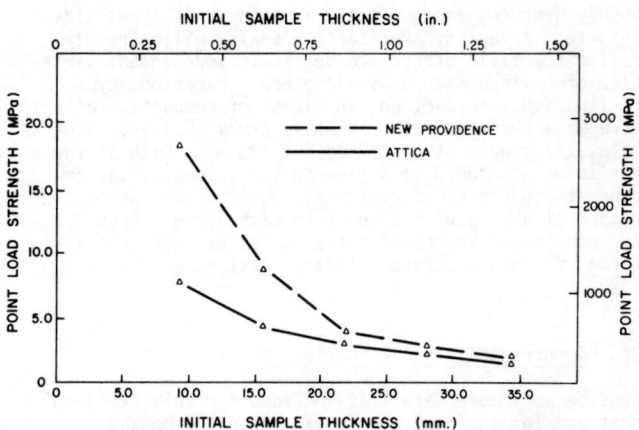

Fig. 3 Size Effect on Point Load Strength Values

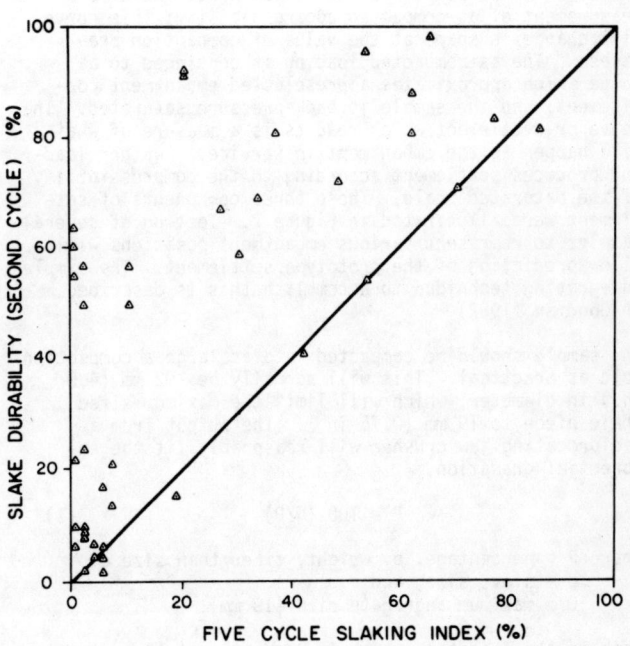

Fig. 4 Slake Durability (2nd Cycle) vs. 5-Cycle Slaking Index.

3.2 Compaction

Whenever possible, compaction control should be generated in a field compacted test pad. If the control must be generated in the laboratory, shale requires several special considerations. The sample should be as coarse as possible, and required curing time for added moisture is 2 days (Bailey, 1976). When such a curing time is allowed, the compaction curve will assume the conventional (soil) shape with a maximum density and optimum moisture content. Practical difficulties with changing the moisture of hard shales in the field may require that the natural moisture content be specified as the control value.

The compaction-degradation test was developed to allow anticipation of a physical problem with hard but nondurable shales. These materials must normally be thoroughly degraded and compacted in thin (soil type) lifts. This test forewarns that certain shales will strongly resist the field efforts to degrade. Independent variables of maximum size of shale piece, gradation, type of laboratory compaction, and level of compactive effort were made the object of extensive study (Bailey, 1976; Hale, 1979; Hale et al, 1981a and 1981b). Each of these variables was found to influence the values of the dependent one (index of crushing), but did not change the rating of shales with respect to each other. Accordingly, convenient values of these variables have been selected for the proposed standard (Oakland and Lovell, 1982).

3.3 Compressibility and Settlement

Considerably more detail is included for this standard test and for the next one (shear strength) because there appears to be almost nothing of this nature in the engineering literature. Full descriptions are available in Oakland and Lovell (1982).

The approach used for compressibility assessment is the conventional one-dimensional consolidation test, conducted in three steps. In the first part, there is a measurement of as-compacted compressibility; this curve will change in shape at the value of compaction prestress. The as-compacted loading is continued to a value which approximates a preselected embankment confinement, and the sample is back-pressure saturated. The heave or settlement which results is a measure of what will happen to the embankment in service. Further loading produces settlement according to the compressibility of the saturated shale. These three components of settlement were illustrated in Figure 2. Testing of several samples to represent various embankment positions will allow prediction of the prototype settlement. The simple integrating technique to accomplish this is described in Goodman (1982).

The sample should be compacted into as large a compaction mold as practical. This will normally be 102 mm (4.0 in.) in diameter, which will limit the maximum sized shale piece to 19 mm (0.75 in.). The output from a reciprocating jaw crusher will reasonably fit the exponential gradation,

$$P = 100 \ (d/D) \qquad (1)$$

where P = percentage, by weight, finer than size d,
  d = sieve size, and
  D = maximum aggregate size (19 mm).

This is the gradation which is compacted at the moisture content which will probably exist in the field embankment. To simulate the field rolling, a kneading type compactor (Oakland and Lovell, 1982) is used to obtain the specified density. After compaction, a portion of the sample is extruded into the oedometer ring (of the same diameter) and trimmed flush with the end of the ring.

The specimen is loaded in the consolidation equipment with a load increment ratio of 0.5 to 0.75 (Holtz and Kovacs, 1981). The partially saturated shale compresses very quickly, and a 10 minute loading increment period is usually adequate. Several samples should be compacted and loaded to approximate various embankment positions in the prototype. At these loadings, the samples are back-pressure saturated, and the settlements (or heaves) are measured. A period of 12 hours is usually allowed for this volume change. If it is desired to determine the saturated compressibility of the shale, the compression test may be continued (see Figure 2), using the conventional procedures for saturated clays (Holtz and Kovacs, 1981).

The settlement prediction (Goodman, 1982), using the test output described above, has not been verified for practical cases, and will likely require empirical corrections. It is, however, a valuable first step in the control of post construction movements in the shale embankment proper.

3.4 Shear Strength and Slope Stability

The laboratory tests recommended are triaxial ones on compacted shale samples, which can produce the necessary input for stability analysis in either a 2-D (Goodman, 1982) or 3-D (Lovell, 1982) model. The sample gradation and preparation are the same as for the compressibility tests described above. The largest sample size practical for most laboratories is a diameter of 102 mm (4.0 in.) and a height of 216 mm (8.5 in.). The maximum size shale piece is again 19.0 mm (0.75 in.).

To approximate the end-of-construction circumstance, the compacted samples may be appropriately confined and sheared undrained in the conventional UU mode (Holtz and Kovacs, 1981). Test results may be interpreted in terms of undrained shear strength as a function of position in the embankment, or as total stress Mohr-Coulomb parameters (c and ϕ). Since the as-compacted state is viewed as less critical than the saturated long term one, these tests are uncommon.

Replicate compacted samples are (a) confined at levels appropriate to various embankment positions, (b) back pressure saturated with volume changes occurring, and (c) sheared. If the saturated samples are sheared undrained, pore pressures should be measured, (CIŪ) test, so that effective stress analysis may be undertaken; the other alternative is to shear them drained, (CID) test, followed by total stress analysis (Holtz and Kovacs, 1981). The saturation is presumed to produce the pessimal strength condition of the shale. It is believed to be a reasonable test condition for the climates of the midwestern U.S.

The Mohr-Coulomb effective stress parameters, c' and ϕ', are more fundamental descriptors of undrained shear behavior than the total stress ones (c,ϕ). For this reason, the CIŪ test procedure has been standardized (Oakland and Lovell, 1982). Typical test data are shown in Figure 5, where ε_a is the axial strain, $(\sigma_1 - \sigma_3)$ is the boundary stress difference which ultimately produced the shear failure, and Δu is the transient pore water pressure induced by the shear. The factor A is after Skempton, and represents the ratio of change in pore pressure to the change in boundary stress on the sample.

Figure 5(d) shows the effective stress development during shear with termination on the effective failure line (not envelope). The values of c' and φ' shown are typical for hard, nondurable Indiana shales. These values for the saturated shale do not vary significantly with the details of laboratory compaction (Abeyesekera, 1977).

The undesirable feature of effective stress analysis is the necessity of predicting the transient pore pressures at failure. This is neither simple nor obvious for the practical problem and for the laboratory data illustrated. The data also indicate another potential problem with an undrained analysis, viz., negative pore pressures and possible overestimation of shear strength. The degree to which negative pore pressures occur in undrained shear depends upon the compactive prestress (Lovell and Witsman, 1981) and its relationship to the effective stress at the start of shear. As the sample is highly prestressed (analogous to highly overconsolidated for an in situ soil), shearing produces volume increase in a drained mode or negative pore pressure in an undrained one.

The two difficulties with applying undrained data, i.e., predicting the excess pore pressure, and possibly employing shear strengths which are greater than drained ones, suggest that drained testing and total stress analysis may be the more appropriate procedure.

5. SUMMARY

After a decade of experimental study of hard but nondurable midwestern U.S. shales, some eight tests for design and construction usage have been identified. Four of these tests are needed to implement the Franklin classification: the Atterberg limits, the point load strength test, and either slake durability or the 5-cycle slaking index. The Atterberg limits are difficult to perform for slake-durability indexes between about 40 and 80, and modification of the basic soil test standard is needed. The 5-cycle slaking test should be used only when the slake-durability device is not available.

Laboratory compaction control curves can be generated for shales in the same general way as for soils...curing time for assimilation of added moisture is critical. It is not generally practical to change the field moisture of hard shales, and the compaction specification may be based on the density achievable at the natural water content. An important part of the compaction process is degradation of the shale pieces so that thin dense lifts may be produced. A standard laboratory compaction-degradation test has been developed which allows the rating of shales relative to their resistance to being degraded during compaction.

Finally, a one-dimensional compressibility test and an undrained triaxial shear procedure have been standardized. The compressibility model may be used to estimate, (a) the settlement of the shale fill under its own weight; (b) the prestress effected by the compactor; (c) the settlement (or heave) to be expected when the fill becomes saturated in service; and (d) the settlement of the fill if loads are added after it becomes saturated. Item (c) is usually the most important movement within the fill. Triaxial tests in either the CIŪ or CID modes on saturated samples may be used to generate the input for long term slope stability analysis. These analyses may be undertaken on either an effective or total stress basis.

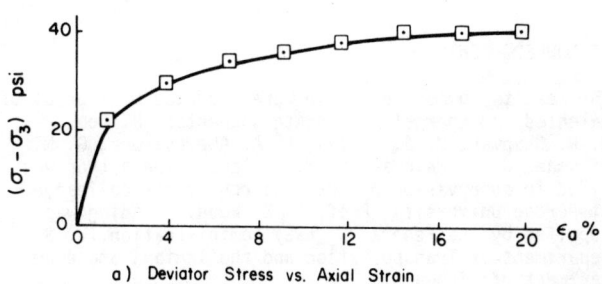

a) Deviator Stress vs. Axial Strain

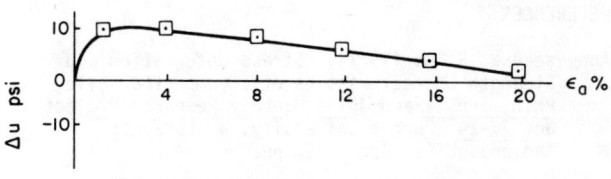

b) Pore Pressure Change vs. Axial Strain

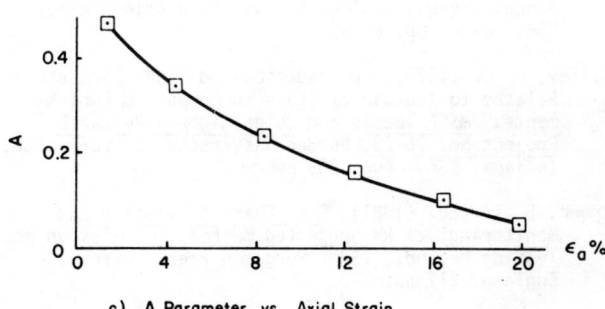

c) A Parameter vs. Axial Strain

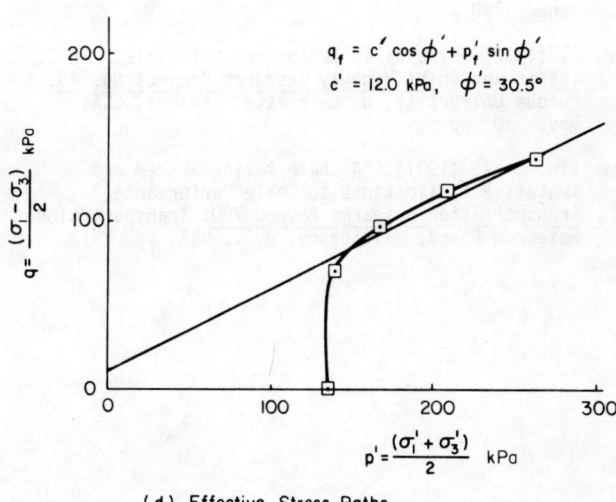

$$q_f = c' \cos \phi' + p_f' \sin \phi'$$
$$c' = 12.0 \text{ kPa}, \quad \phi' = 30.5°$$

(d) Effective Stress Paths

Fig. 5 Isotropically Consolidated Undrained Triaxial Shear Test on New Providence, Indiana Shale

The eight tests and their recommended usage provide a comprehensive guide to the design and construction of embankments using hard but nondurable shales...common constructional materials in the midwestern U.S.

ACKNOWLEDGEMENTS

The results summarized above were produced by a group of talented and energetic graduate students: P. Deo, D. R. Chapman, M. J. Bailey, R. A. Abeyesekera, G. R. Witsman, B. C. Hale and M. W. Oakland. The author was aided in supervision of the research by his colleague at Purdue University, Prof. L. E. Wood. Funding was supplied by the Federal Highway Administration, U. S. Department of Transportation and the Indiana State Department of Highways.

REFERENCES

Abeyesekera, R. A. (1977), "Stress-Deformation and Strength Characteristics of a Compacted Shale," Ph.D. Thesis and Joint Highway Research Project No. 77-24, Purdue University, W. Lafayette, Indiana, USA. Dec. 420 pp.

Abeyesekera, R. A. and Lovell, C. W. (1982), "Characterization of Shales by Plasticity Limits, Point Load Strength and Slake Durability,: Proceedings, 31st Annual Highway Geology Symposium, Austin, Texas, USA. Jan. pp. 65-96.

Bailey, M. J. (1976), "Degradation and Other Parameters Related to the Use of Shale in Compacted Embankments," MSCE Thesis and Joint Highway Research Project No. 76-23, Purdue University, W. Lafayette, Indiana, USA. Aug. 209 pp.

Brown, E. T. (Ed) (1981), Rock Characterization and Monitoring...ISRM Suggested Methods, Commission on Testing Methods, ISRM, Pergamon Press, Oxford, England, 211 pp.

Chapman, David R. (1975), "Shale Classification Tests and Systems: A Comparative Study." MSCE Thesis and Joint Highway Research Project No. 75-11, Purdue University, W. Lafayette, Indiana, USA. June. 90 pp.

Deo, P. (1972), "Shales as Embankment Materials," Ph.D. Thesis and Joint Highway Research Project No. 45, Purdue University, W. Lafayette, Indiana, USA. Dec. 202 pp.

Franklin, J. A. (1981), "A Shale Rating System and Tentative Applications to Shale Performance," Transportation Research Record 790, Transportation Research Board, Washington, D.C., USA. pp. 2-12.

Goodman, Martin (1982), "Design of Compacted Clay Highway Embankments for Improved Statility and Settlement Performance," Joint Highway Research Project, Purdue University, W. Lafayette, Indiana, USA (to be published).

Hale, B. C. (1979), "The Development and Application of a Standard Compaction-Degradation Test for Shales," MSCE Thesis and Joint Highway Research Project No. 79-21, Purdue University, W. Lafayette, Indiana, USA. Dec. 180 pp.

Hale, B. C., Lovell, C. W. , and Wood, L. E. (1981a), "Development of a Compaction-Degradation Test for Shales," Transportation Research Record 790, Transportation Research Board, Washington, D.C., USA. July. pp. 45-52.

Hale, B. C., Lovell, C. W., and Wood, L. E. (1981b), "Factors Affecting Degradation and Density of Compacted Shales," Proceedings, International Symposium on Weak Rocks, Tokyo, Japan, Vol. I, Sept., pp. 321-326.

Holtz, R. D. and Kovacs, W. D. (1981), An Introduction to Geotechnical Engineering, Prentice-Hall, Inc., Englewood Cliffs, New Jersey, USA. 733 pp.

International Society for Rock Mechanics (1972), "Suggested Method for Determining the Point Load Strength Index," Document No. 1, Part 2, Commission on Standardization of Laboratory and Field Tests, Oct.

Lovell, C. W. (1982), "Three Dimensional Slope Stability," Proceedings, 13th Annual Ohio River Valley Soils Seminar, Lexington, Kentucky, USA. October. 9 pp.

Lovell, C. W. and Witsman, G. R. (1981), "Compactive Prestress in Shales," Bulletin, Association of Engineering Geologists. Vol. 18, No. 3, Aug. pp. 297-308.

Oakland, M. W. and Lovell, C. W. (1982), "Classification and other Standard Tests for Shale Embankments," Joint Highway Research Project No. 82-4, Purdue University, W. Lafayette, Indiana, USA. Feb. 171 pp.

Witsman, G. R. and Lovell, C. W. (1979), "The Effect of Compacted Pre-stress on Compacted Shale Compressibility," Joint Highway Research Project No. 79-16, Purdue University, W. Lafayette, Indiana, USA. Sept. 181 pp.

NEEDLESS STRINGENCY IN SAMPLE PREPARATION STANDARDS FOR LABORATORY TESTING OF WEAK ROCKS

La sévérité trop rigoureuse des règles d'analyse appliquées aux tests de roches fragiles en laboratoire

Die unnötige Strenge in den Präparatsnormen für Laboruntersuchungen schwachen Gesteins

P. J. N. Pells
Coffey & Partners Pty Ltd, North Ryde, N.S.W., Australia

M. J. Ferry
Postgraduate Scholar, University of Sydney, Australia

SYNOPSIS

Comparative laboratory testing has been undertaken on samples of Hawkesbury Sandstone in order to evaluate the relative importance of currently recommended procedures for specimen preparation and testing of core samples of weak rocks. The test results indicate that many of the recommendations relating to the measurement of unconfined strength, Young's modulus and Brazilian tensile strength are unnecessarily stringent.

RESUME

Des tests comparatifs ont été enterpris en laboratoire sur des échantillons de grès de la région du Hawkesbury afin de pouvoir évaluer l'importance relative des procédés couramment employés pour la préparation et l'étude des échantillons de roches à structure fragile. Le résultat des analyses indique que nombre de recommandations concernant l'évaluation d'une force unidirectionelle, soit le Module de Young et le test Brésilien d'élasticité d'un échantillon, sont d'une sévérité trop rigoureuse.

ZUSAMMENFASSUNG

Laborvergleichsteste sind an Proben von Hawkesbury Sandstein vorgenommen worden, um die relative Bedeutung der zur Zeit vorhandenen Verfahren zur Probeherstellung und Untersuchung von Kernproben schwachen Gesteins zu bewerten. Die Untersuchungsresultate ergaben, daß viele Angaben bezüglich der Messung der einachsigen Druckfestigkeit, Young's Modulus und indirekter Zugfestigkeit unnötig streng gefaßt sind.

1. INTRODUCTION

Standards for the preparation and testing of rock core samples for unconfined compression, triaxial and Brazilian tests have been published by many authorities such as the International Society for Rock Mechanics (ISRM, 1972), the Canada Centre for Mineral & Energy Technology (CANMET, 1977), the American Society for Testing & Materials (ASTM, 1974) and the Standards Association of Australia (SAA, 1981). Many of the recommendations given by these and other similar authorities clearly come from common sources. For example the Draft Australian Standards are largely based on the ISRM recommendations which in turn reflect the efforts of Bieniawski and Franklin, based on work at the CSIR in Pretoria and Imperial College but also on comparative test data from the US Bureau of Mines (Hoskins & Horino, 1968). The Canadian recommendations are largely based on the ASTM Standards which in turn are also strongly influenced by the work of the US Bureau of Mines.

It is clear that the sample preparation and test procedure recommendations for compression and tensile tests on rock core arose from research organizations largely concerned with mining rock mechanics and, in particular, hard rock mining. Thus most of the rock types used for the comparative test studies had unconfined compressive strengths greater than 50 MPa; in fact the majority greater than 100 MPa. Now in terms of civil engineering rock mechanics, particularly foundation engineering, the majority of rocks of concern have substance strengths of less than 50 MPa. For example in the Australian context,

the Triassic sandstones and shales of the Sydney Basin have unconfined strengths generally in the range 5 to 25 MPa while the Silurian mudstone beneath Melbourne ranges in strength from about 1 MPa to 10 MPa.

In testing such weak rocks the question arises as to the applicability of the specimen preparation and test procedure recommendations given by the various authorities discussed above. In order to investigate this question a programme of testing was undertaken using samples drilled from three blocks of Triassic Hawkesbury Sandstone taken from adjacent to one another in the Summersby quarry north of Sydney. By testing this sandstone in both the oven dry and saturated states it was possible to carry out comparative testing on materials with unconfined strengths of 50 MPa and 20 MPa.

2. PREPARATION & TEST PROCEDURES INVESTIGATED

The specimen preparation parameters and test procedures that were investigated were all related to the recommendations set out by the ISRM because these recommendations have formed the basis for various national standards. The following aspects were investigated:-

Determination of uniaxial compressive strength
. effect of non-parallel ends
. effect of rough ends
. effect of length to diameter ratio less than 2.5
. effect of end capping
. effect of testing rate

Determination of Young's Modulus
. effect of using no dummy strain gauges
. effect of using only a single axial strain gauge
. effect of non-parallel ends
. effect of L/D = 2.0

Determination of Brazilian Tensile strength
. effect of testing directly between test machine platens (i.e. no test jig)
. effect of using cardboard strips against test machine platens (again no test jig)

All testing was carried out in an Instron servo-controlled test machine and all comparisons are made against a data base of samples prepared and tested fully according to the ISRM recommendations. Table 1 lists the number of specimens tested to investigate each of the preparation or test procedure aspects listed above. All specimens were nominally 53mm diameter.

TABLE I
NUMBERS OF SPECIMENS TESTED

UNIAXIAL COMPRESSIVE STRENGTH		
Specimen Preparation and Test Procedure	Moisture Conditon	Number of Specimens
ISRM Standard	Oven-dried	13
ISRM Standard	Saturated	13
Capped	Oven-dried	10
Capped	Saturated	10
L/D = 2.2 nominal	Saturated	11
L/D = 2.0 nominal	Saturated	10
Non-parallel ends	Saturated	15
Non-flat ends	Saturated	15
DETERMINATION OF YOUNG'S MODULUS (Note: all strain gauges 120 ohm, 20mm gauge length)		
ISRM Std + 2 gauges + dummy	Saturated	2
ISRM Std + 2 gauges - no dummy	Saturated	2
ISRM Std + 1 gauge + dummy	Saturated	2
ISRM Std + 1 gauge	Saturated	2
L/D = 2.0 + 2 gauges + dummy	Saturated	2
L/D = 2.0 + 1 gauge + dummy	Saturated	3
L/D = 2.0 + 1 gauge - no dummy	Saturated	3
Non-parallel + 2 gauges + dummy	Saturated	3
Non-parallel + 2 gauges - no dummy	Saturated	2
Use of dial gauge device	Saturated	20
BRAZILIAN TENSILE STRENGTH		
ISRM Standard Jig	Saturated	25
ISRM Standard Jig	Oven-dried	25
Hardboard Strips	Saturated	25
Hardboard Strips	Oven-dried	25
Machine Platens	Saturated	25
Machine Platens	Oven-dried	25

Considering Table 1 it can be seen that a statistically meaningful number of tests was performed for the uniaxial test and Brazilian test comparisons. However, the specimen numbers for the Young's modulus tests are insufficient for statistical evaluation and thus only gross influences are noted.

Space does not permit a detailed explanation of the test procedures adopted in this programme. However, the following notes give some indication of what is meant by the brief description given in Table 1.

i) ISRM Standard Uniaxial Compression Test. The cores were trimmed to a nominal L/D = 2.6 and ground individually using a CSIR type jig described by Benzinger (1974). End parallelness and flatness was better than the ISRM requirements.

ii) Capped Uniaxial Compression Tests. The cores were trimmed to a nominal L/D = 2.6 but with the ends approximately 5° off parallel. Both ends were then capped using standard concrete cylinder sulphur capping. The capped surfaces were flat but deviated

from parallel - as follows:

Oven dried specimens
- at 0°   mean = 0.50°   SD = 0.42°
- at 90°   mean = 0.30°   SD = 0.24°

Saturated specimens
- at 0°   mean = 0.25°   SD = 0.17°
- at 90°   mean = 0.16°   SD = 0.13°

iii) L/D = 2.2 and L/D = 2.0 Uniaxial Compression Tests.
Except for being shorter than required the specimens were prepared to the full ISRM requirements.

iv) Non-parallel Uniaxial Compression Tests.
The specimens were trimmed to a nominal L/D = 2.6 and one end then ground flat and normal to the axis using the CSIR jig. The other end was then cut non-parallel to a mean deviation from parallel of 1.93° (SD = 0.17°).

v) Non-flat Uniaxial Compression Tests.
Specimens were first prepared as per the ISRM Standard (see i above) and then one end was deliberately slightly rounded (10 specimens) or scabbled (5 specimens). No satisfactory measurement of roughness was established but the 'rounded' specimens deviated at least .5mm from flat while the scabbled ends contained pock marks at least .5mm deep.

vi) Measurement of Young's Modulus.
The object of these tests was primarily to investigate whether it was necessary to cement two axial strain gauges to a specimen and also to check on the need for having dummy gauges on a similar piece of rock. The effect of short (L/D=2.0) specimens and non-parallel ends was also evaluated. A simple dial gauge device that clamped across the middle third of the specimen was also tested.

vii) Brazilian Tensile Tests
The recommended ISRM test method involves the use of a special testing jig and for obvious reasons it was worthwhile to investigate the error involved in placing the discs of core directly between the platens of the test machine or inserting a 4mm wide strip of cardboard along the contact lines with the machine platens.

Great care was taken in ensuring that the 'saturated' specimens were at a consistent degree of saturation. This was achieved by repeated application and release of vacuum to the specimens stored under water. This process continued for up to a week and the specimens were then stored under water until testing. This procedure resulted in the uniaxial specimens having calculated degrees of saturation of between 78% and 82% (using SG=2.65). The Brazilian specimens had calculated degrees of saturation of between 94% and 97%. The difficulty in achieving full saturation of the uniaxial specimens was one of the surprises of this work.

3. THE MATERIAL TESTED - HAWKESBURY SANDSTONE

Occupying an area of some 12,500 sq kilometres

in the Sydney Basin, this formation consists dominantly of quite massive sandstone beds up to 20m in thickness. Strong cross bedding is common.

The sandstone is composed primarily of fine to medium, subangular, quartz grains with an argillaceous matrix and some siderite cement. Secondary silica occurs mostly of overgrowths around grains. The degree of overgrowth development is variable and has an important bearing on the strength of the material. Well developed overgrowth of detrital grains results in a strong interlocked structure. Siderite is sometimes found in sufficient quantity to bind grains together and SEM studies have revealed a rather strange potassium aluminium silicate that acts as a true cementing agent between quartz grains. Typical composition of the sandstone is as follows:

Framework quartz	60% - 75%
Matrix clay (70% Kaolinite, 20% Illite)	15% - 35%
Secondary silica, siderite and others	5% - 12%

Three $1m^3$ blocks of sandstone were taken from the Summersby quarry, some 60 km north of Sydney.

The sandstone in these blocks was medium grained with slight to moderate overgrowth development. The porosity varied between 17% and 18%.

As many 53mm core samples as possible were taken from each block. All the uniaxial and Brazilian test specimens were taken from the first two blocks and selection of samples was done in such a way as to avoid specimens in any group coming from the same zone in a particular block. Despite this there was still some difference in the mean dry densities of the samples in the different groups. However, the maximum difference was about 1% and does not appear to have influenced the results. The modulus tests were conducted on specimens from the third block.

4. UNIAXIAL COMPRESSIVE STRENGTH RESULTS

The results of all the uniaxial compressive strength tests are summarised in Table IIa. This table includes the statistical parameters for the specimen dry densities and times to failure - two test variables that could have influenced the compressive strength measurements.

It can be seen from Table IIa that the coefficient of variation on uniaxial strength is very low in all groups except for the rough or rounded and conditions. Excluding these groups the coefficient of variation on uniaxial strength ranges between 3.4% and 7.4%.

Table IIb gives the statistical analysis of significance between the results of the different groups given in Table IIa. From this table it can be seen that on the basis of the t statistic test, at the 1% confidence limit, the following factors had no significant influence on the measured uniaxial compressive strength:

- length to diameter ratio reduced to 2.2 or 2.0

- ends non-parallel by almost 2°

- loading rate 5 x slower than recommended

At the 1% confidence limits the effect of capping both ends of the specimens had a statistically significant effect on the results. However,

TABLE IIa

UNIAXIAL COMPRESSIVE STRENGTH TESTS

(All specimens nominal 53mm diameter)

Rock Condition	Specimen Preparation			Dry Density t/m³		Failure Time min : sec		Compressive Strength MPa		No Specimens
	End Conditions	L/D								
		x	s	x	s	x	s	x	s	
Saturated	ISRM	2.60	0.04	2.178	.018	10:2	0:3	20.5	0.7	13
After oven	ISRM	2.23	0.01	2.180	.028	9:2	0:1	20.9	1.0	11
drying	ISRM	2.04	0.01	2.181	.032	9:0	0:3	20.9	1.0	10
	Non-parallel	2.57	0.05	2.198	.026	10:4	0:4	20.4	1.5	15
	Capped	2.63	0.03	2.187	.028	10:6	0:3	21.9	1.0	10
	Rough	2.64	0.02	2.200	.025	12:14	1:25	15.9	2.0	5
	Rounded	2.61	0.04	2.192	.026	10:41	1:40	9.3	2.4	10
	ISRM	2.60	0.03	2.181	.020	50:17	1:29	19.9	0.8	10
Oven-dried	ISRM	2.61	0.06	2.177	.017	13:55	0:36	50.4	1.8	13
	Capped	2.63	0.03	2.189	.027	13:25	0.31	47.5	2.6	10

TABLE IIb

UNIAXIAL COMPRESSIVE STRENGTH TESTS

STATISTICAL TEST OF SIGNIFICANCE BETWEEN

RECOMMENDED ISRM PROCEDURE AND SIMPLIFIED METHODS

Rock Condition	Test type compared with ISRM procedure	% Difference on mean	t Statistic	Significant difference at 1% Confidence Limit
Saturated	L/D = 2.2	+ 2.0	1.10	No
	L/D = 2.0	+ 2.0	1.08	No
	Non-parallel	- 0.5	0.21	No
	Capped	+ 6.8	3.77	Yes
	Non-flat, rounded	-54.6	15.50	Yes
	Non-flat, roughened	-22.4	6.81	Yes
	Loading rate 5 x slower	- 2.9	1.83	No
Oven-dried	Capped	- 5.8	3.02	Yes

the percentage difference on the mean strength was less than 7% and, interestingly enough, the capped saturated specimens (20 MPa rock) showed an increase in strength while the capped dry specimens (50 MPa rock) showed a decrease in strength. These differences could not be clearly related to any other variables such as degree of saturation or testing rate.

The test results from the groups with non-flat ends were hugely reduced below the 'standard' values and the biggest difference was with the rounded ends.

5. YOUNG'S MODULUS RESULTS

Table III summarises the results of the different methods of modulus measurement in terms of tangent and secant moduli at 50% of the uniaxial compressive strength.

TABLE III

(a) TANGENT MODULUS AT 50% UNIXIAL STRENGTH - GPa

Specimen Type	E1*		E2*		E3*		E4*	
	x	s	x	s	x	s	x	s
ISRM Standard	7.00	.68	7.07	.05	6.72	.11	7.16	.44
L/D = 2.0	7.13	.01	-	-	7.62	1.14	6.91	1.07
Non-parallel	7.22	.53	7.13	.55	-	-	-	-

(b) SECANT MODULUS AT 50% UNIAXIAL STRENGTH - GPa

Specimen Type	E1*		E2*		E3*		E4*	
	x	s	x	s	x	s	x	s
ISRM Standard	3.53	.00	3.55	.04	3.36	.25	3.72	.48
L/D = 2.0	3.31	.29	-	-	4.26	1.34	3.47	.88
Non-parallel	3.78	.22	3.58	.10	-	-	-	-

*Note: E1 - 2 gauges on specimens + 2 gauges on dummy specimen
E2 - 2 gauges on specimens - no dummy gauges
E3 - 1 gauge on specimens + 1 gauge on dummy specimen
E4 - 1 gauge on specimens - no dummy gauge

As shown in Table I there were insufficient specimens tested in each of the modulus groups to allow meaningful statistical analysis of the data given in Table III. However, purely from a engineering judgement viewpoint it appears that with 2 axial gauges on the specimens it did not make any significant difference whether the specimens only had a L/D ratio of 2.0 or were cut with non-parallel (but flat) ends. Using a 240 ohm resistor in place of two strain gauges cemented to a dummy specimen also had no significant effect on the results.

Placing only a single axial strain gauge on the specimens is not satisfactory for accurate determination of modulus. The spread of the modulus measurements using only a single gauge was considerably greater than with the two gauges connected in series. However, using only two or three specimens in each group the maximum error on the mean modulus measurement was only 5% for specimens prepared to ISRM recommendations. For the L/D = 2 specimens this maximum error on mean increased to 28% and for non-parallel specimens it would probably have been worse.

6. BRAZILIAN TENSILE STRENGTH RESULTS

Table IVa summarises the results of the Brazilian tests while Table IVb gives the statistical analysis of significance between the results of the different test groups.

TABLE IVa

BRAZILIAN INDIRECT TENSILE TESTS

(25 Tests conducted in each category)
(- all specimens 53.2mm diameter with)
(thickness to diameter ratio of 0.48)

Rock Condition	Test Method	Dry Density t/m³		Time to Failure sec.		Tensile Strength MPa	
		x	s	x	s	x	s
Saturated	ISRM Jig	2.160	.017	23.3	3.3	1.35	0.11
	Cardboard strips	2.159	.018	27.6	2.8	1.32	0.09
	Platens only	2.160	.017	17.3	2.8	1.17	0.10
Oven-dried	ISRM Jig	2.156	.014	23.3	4.3	4.38	0.17
	Cardboard strips	2.146	.009	67.6	5.9	4.30	0.21
	Platens only	2.165	.028	20.1	2.0	4.08	0.26
Saturated after oven drying	ISRM Jig	2.157	.014	22.5	2.4	1.66	0.09
	Cardboard strips	2.146	.011	28.9	4.5	1.63	0.08
	Platens only	2.168	.027	19.3	2.2	1.40	0.07

TABLE IVb

STATISTICAL TEST OF SIGNIFICANCE BETWEEN STANDARD (ISRM) METHOD AND SIMPLIFIED METHODS

(1% Confidence Limit : t = 2.68)

Rock Condition	Test Method compared with ISRM Method	% Difference on means	t Statistic	Significant difference at 1% confidence limit ?
Saturated	Cardboard strips	- 2.2	1.03	No
	Platens only	-13.3	5.92	Yes
Oven-dried	Cardboard strips	- 1.8	1.45	No
	Platens only	- 6.8	4.73	Yes

These results indicate that there is no significant difference between testing the specimens between cardboard strips as opposed to using the recommended testing jig. Testing the specimens directly between the flat steel platens of the machine gave a statistically significant reduced strength. However, from the engineering viewpoint even this error was not gross.

An interesting aside is to note that all the strength measurements obtained on specimens resaturated after complete oven drying were some 20% greater than those on specimens saturated after coring from the air dried quarry block. This effect was noted by Ballivy, Ladanyi and Gill (1975) and was suggested as being due to 'an irreversible overconsolidation effect, which may be due to a modification in the structure of hydrosilicates, resulting in petrification of bonds'. The present authors have not investigated the physico-chemical nature of this phenomenon and thus cannot comment any further.

7. CONCLUSIONS

The recommendations that have been laid down by the ISRM and other authorities in relation to the specimen preparation and laboratory testing for the determination of uniaxial compressive strength, Young's modulus and Brazilian tensile strength are in a number of aspects unnecessarily stringent for rocks having uniaxial strengths of 50 MPa or less. The test data reported herein indicate that, in the measurement of the uniaxial compressive strength of such rocks, it appears to make no statistically significant difference if:

i) the specimen length to diameter ratio is as low as 2.0

ii) the specimen ends are non-parallel (but flat) by as much as 2°

iii) the loading rate is five times slower than the recommended rate (while true for the sandstone tested here this is probably not true for all rocks)

Capping of specimens with sulphur results in a statistically significantly different result but from the engineering viewpoint the error on means of -6% to +7% is of little consequence.

In the determination of Young's modulus by means of strain gauging compression test specimens it again makes no significant difference if the L/D ratio is as low as 2.0 and if the specimen ends are non-parallel (but flat). Two axial strain gauges connected in series should be adopted but it is not necessary to have a second pair of gauges cemented to a dummy specimen.

Great care should be taken in attempting to determine Young's modulus by means of a dial gauge clamped to the specimen by some sort of jacket device. The device we tested behaved very erratically and yielded modulus values that varied from half to twice the strain gauge values.

All the conclusions given above apply to tests conducted with steel end platens of the same diameter of the test specimen with a spherical seat in the upper platen as recommended by the ISRM.

While it is not important that the ends of the specimens be parallel and at right angles to the specimen axis to the standards recommended by the ISRM, it is critical that they be flat. However, with rocks of uniaxial strength less than 50 MPa this can usually be achieved with a good quality diamond saw arrangement. Lapping of specimen ends using grinding paste should be avoided as this tends to round the ends.

In conducting Brazilian tensile tests on low strength rocks it appears to be satisfactory to simply place cardboard strips, of width equal to 8% to 10% of the sample diameter, along the load application lines. A special testing jig is not necessary.

It appears that many of the laboratory test recommendations that are becoming enshrined in Codes and Standards are based on a rather meagre data base of comparative testing or else are related to the testing of very high strength rocks. Careful laboratory testing is essential but unnecessarily stringent specifications cost money. Further comparative testing of the type described herein is clearly appropriate.

8. REFERENCES

ASTM (1974) - Standard method of test for triaxial compressive strength of undrained rock core specimens without pore pressure measurements. American Society for Testing & Materials; Annual Book of ASTM Standards; Designation D26A-67.

Ballivy, G., Ladanyi, B. & Gill, D.E. (1975) - Effect of water saturation history on the strength of low porosity rocks. Soil Specimen Preparation for Laboratory Testing, ASTM Special Technical Publication 599, pp 4-20.

Benzinger, E.W. (1974) - The preparation of rock material specimens for testing purposes. South African Council for Scientific & Industrial Research, NMERI, Geomechanics Internal Report ME/1173/16, Pretoria.

CANMET (1977) - Pit Slope Manual. Sampling and Specimen Preparation, Supplement 3.5, Canada Centre for Mineral & Energy Technology Report 77-29, 30p.

Hoskins, J.R. and Horino, F.G. (1968) - Effect of end conditions on determining compressive strength of rock samples. Report of Investigations 7171, US Bureau of Mines.

ISRM (1972) - Suggested methods for determining the uniaxial compressive strength of rock materials and the point laod strength index. International Society for Rock Mechanics, Commission on Standardization of Laboratory & Field Tests. Document No. 1.

SAA (1981) - Draft Australian Standard on Method of Testing Rock for Engineering Purposes. DR81227 to DR81239.

THE UNIAXIAL PROPERTIES OF MELBOURNE MUDSTONE

Les propriétés uniaxiales de l'argilite de Melbourne

Die Eigenschaften des Melbourner Tonsteins bei einachsiger Druckbelastung

H. K. Chiu
Geotechnical Engineer, Pilecon Engineering Sdn. Bhd., Kuala Lumpur, Malaysia. Formerly
Postgraduate Scholar, Monash University

I. W. Johnston
Lecturer in Civil Engineering, Monash University, Clayton, Victoria, Australia

SYNOPSIS

The strength and deformation properties of the Melbourne mudstone under uniaxial compression and Brazilian test conditions are examined for the range of weathering characteristics commonly encountered. The relevancy of these simply obtained properties to engineering design are considered with special reference to rock socketed piles. The resultant ratios of uniaxial compressive to tensile strength are examined to demonstrate that while the Brazilian test may provide a reasonable indirect estimation of uniaxial tensile strength for the slightly to moderately weathered mudstone, it may significantly underestimate the tensile strength of the more weathered mudstone.

RESUME

Les propriétés de résistance et de déformation de l'argilite de la région de Melbourne soumise à la compression uniaxiale et à l'essai brésilien ont été étudiées pour toute la gamme des caractéristiques de l'altération dans des conditions normales. La pertinence des données ainsi simplement obtenues pour la conception des ouvrages de génie a été considérée plus particulièrement en ce qui concerne les pieux emboîtés dans de la roche. Les rapports trouvés entre la résistance à la compression uniaxiale et la résistance à la traction ont été examinés. Ils montrent que l'essai brésilien peut fournir une indication indirecte assez valable de la résistance uniaxiale à la traction de l'argilite peu ou modérément altérée, mais dans le cas de l'argilite davantage altérée ce même essai risque de sous-estimer sensiblement la résistance à la traction.

ZUSAMMENFASSUNG

Die Festigkeits- und Verformungskenngrößen des Melbourner Tonsteins unter einachsiger Druckbelastung und unter Brasilianischen Testbedingungen werden für gewöhnlich anzutreffende Witterungsverhältnisse untersucht. Die Relevanz dieser auf einfache Weise erhaltenen Kenngrößen für die ingenieurmäßige Bemessung insbesondere von felsgegründeten Pfählen wird untersucht. Bei Auswertung der erhaltenen Quotienten von einachsigen Zug- und Druckfestigkeiten zeigt sich, daß der Brasilianische Test in der Regel eine relativ gute Abschätzung der einachsigen Zugfestigkeit des leicht bis mittelmäßig verwitterten Tonsteins erlaubt, er jedoch die Zugfestigkeit des stärker verwitterten Tonsteins erheblich unterschätzen kann.

1. INTRODUCTION

In the Melbourne region, a large proportion of construction involves the rocks of the Silurian and Lower Devonian era which form the bedrock of the area. Although this material is predominantly a siltstone, it is known as the Melbourne mudstone and is usually encountered between a relatively hard and brittle slightly weathered condition (uniaxial compressive strength, $\sigma_c \approx 8$ MPa) and a softer more ductile highly weathered condition ($\sigma_c \approx 1$MPa).

For some time the authors have been involved in a detailed research programme aimed at establishing the engineering properties of the mudstone. A wide range of different forms of testing have been used and a full description of the various techniques, their results and interpretation may be found elsewhere (Chiu, 1981; Chiu et al, 1982; Johnston et al, 1980; Johnston and Chiu, 1981, 1982). This paper is however, limited to the uniaxial properties of the Melbourne mudstone; specifically the uniaxial compressive strength and deformation moduli, and the uniaxial tensile strength.

2. TEST PROGRAMME AND TECHNIQUES

The uniaxial tests were applied to mudstone specimens representative of the range of weathering characteristics commonly encounted in construction. However, as the determination of the degree of weathering of any one specimen is a somewhat subjective and qualitative exercise, an alternative quantitative method was used. As discussed by Johnston and Chiu (1982), the saturated water content, w_o, represents a simple and quantifiable parameter which reflects the degree of weathering. The range of water contents concerned in this investigation varied between about 4% (slightly weathered) and 20% (highly weathered).

The methods adopted for the uniaxial compression tests were generally in accordance with the ISRM Suggested Method (Brown, 1981), using specimens of 54mm diameter. Axial loads were measured with an electrical load cell and axial strain was determined from two diametrically opposed displacement transducers clamped to the loading stem of the frame and measuring the relative displacement between the stem and the base platen.

P

The uniaxial compressive strength is simply the maximum axial stress sustainable by each specimen before failure. However, the axial deformation moduli for each specimen usually display some stress dependency and therefore a range of values could be quoted. In this work the definition of axial deformation modulus, E, is in keeping with local practice and is the secant modulus at half the maximum axial stress.

Although the ISRM Suggested Method recommends specimen length to diameter ratios in the range of 2.5 to 3.0, it was considered that a softer rock would permit the use of a smaller ratio. This would enable the preparation of more specimens from the limited supply of high quality rock available for the investigation. As a result of a detailed investigation reported by Chiu et al, (1982), it was concluded that a length to diameter ratio of at least 2.0 was sufficient to yield parameters which were independent of end restraints.

According to the ISRM Suggested Method "load on the specimen shall be applied continuously at a constant stress rate such that failure will occur within 5-10 min of loading, alternatively the stress rate shall be within the limits of 0.5-1.0 MPa/s." While such loading rates may be acceptable for dry hard rocks, for a saturated soft rock such as the mudstone, load applications at the above rates could have a marked influence on induced pore water pressure. As demonstrated by Chiu et al, (1982), these induced pore water pressures can have a significant influence on measured strength and deformation properties. This is particularly true if the alternative rate of loading of the Suggested Method is applied. On the basis of the above investigation and with the aid of analytical methods and mudstone consolidation data (Johnston and Chiu, 1981), it was established that a strain rate of not greater than 3×10^{-4} min^{-1} for the uniaxial compression tests would permit the specimens to drain completely with no excess pore water pressure. Full details of these analyses may be found in Chiu (1981).

Specimens were fully saturated before mounting in the test frame and to maintain saturation, a loose water filled rubber membrane was placed around each specimen with its lower end clamped to the base platen by O-rings.

Poisson's ratio, ν, of a number of selected specimens was determined by a rig described by Sloan (1975). The rig made use of six equally spaced horizontal displacement transducers mounted at the mid-height of the specimens for detailed measurement of radial expansion during loading. The rig also incorporated two vertically orientated displacement transducers for the measurement of axial strains across the middle third of the specimens. These latter transducer results enabled confirmation that the overall displacement measurements provided an accurate determination of axial strain and were not influenced by platen/specimen contact restraints (Chiu et al, 1982).

With regard to the uniaxial tensile strength, a direct determination presents some considerable experimental problems. Therefore an indirect method of determination was required and it was considered that the Brazilian test represented the most practical and potentially reliable method. These tests were conducted in general accordance with the ISRM Suggested Method (Brown, 1981) with the test specimens comprising discs of mudstone of 54mm diameter. Each specimen was fully saturated prior to testing and each test was conducted within a bath of water so that saturation was maintained.

3. RESULTS

Figures 1, 2 and 3 show the results obtained from the uniaxial tests with compressive strength, secant modulus

at half compressive strength and Poisson's ratio respectively plotted against saturated water content.

Figure 4 shows the variation of uniaxial tensile strength as derived from the indirect Brazilian test method. Additional data obtained by Tagell and Tappe (1973) and Williams (1980) have been included in these Figures where relevant.

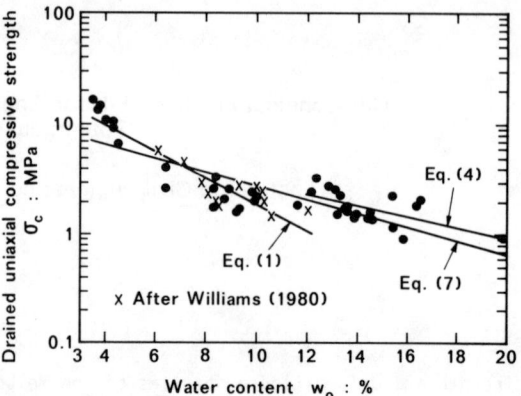

Fig. 1 Variation of drained uniaxial compressive strength with water content.

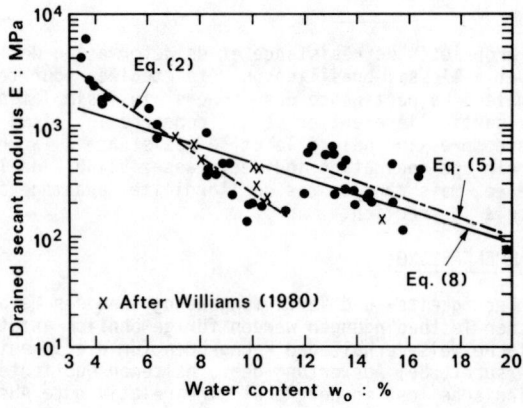

Fig. 2 Variation of drained secant modulus with water content.

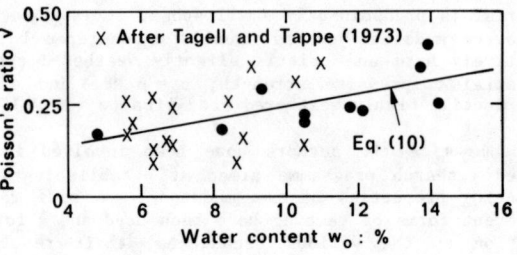

Fig. 3 Variation of Poisson's ratio with water content.

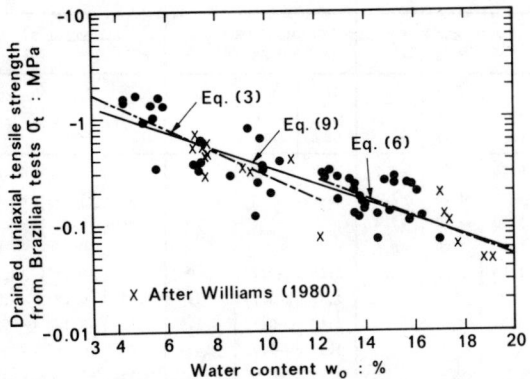

Fig. 4 Variation of drained uniaxial tensile strength
determined by Brazilian test with water content.

A study of Figure 2 will reveal that the secant
modulus, E, generally decreases with increasing water
content. However at about 12% water content there
appears to be a distinct step in the overall trend.
A similar but less marked step can be seen in Figures 1
and 4 but because of the scatter of results, this step is
not apparent in Figure 3. The reason for this anomaly
appears to follow from the general observation that the
mudstone of water content less than 12% was generally
blue-grey in colour, and therefore representative of a
reducing weathering environment. The mudstone of water
content greater than about 12% was generally yellow brown
or occasionally pink in colour and represented mudstone
which had weathered further under oxidising conditions.
The principal difference between the two forms of mudstone is that the oxidised mudstone contains varying
amounts of cementing agents such as iron oxides. It
follows then that the oxidised mudstone may show a
stiffer response to loading than the reduced mudstone,
hence explaining the step in the property variations with
water content.

Least squares analyses were carried out to establish the
respective water content correlations for the reduced
only, oxidised only and all results. For the Poisson's
ratio results, only the correlation for all the results
is presented.

For 4% < w_o < 12% [reduced only]

$$\log \sigma_c \text{ (kPa)} = 4.47 - 0.119\ w_o(\%) \qquad (1)$$

$$\log E \text{ (MPa)} = 4.02 - 0.163\ w_o(\%) \qquad (2)$$

$$\log (-\sigma_t) \text{ (kPa)} = 3.54 - 0.108\ w_o(\%) \qquad (3)$$

For 12% < w_o < 20% [oxidised only]

$$\log \sigma_c \text{ (kPa)} = 3.94 - 0.049\ w_o(\%) \qquad (4)$$

$$\log E \text{ (MPa)} = 3.68 - 0.084\ w_o(\%) \qquad (5)$$

$$\log (-\sigma_t) \text{ (kPa)} = 3.52 - 0.090\ w_o(\%) \qquad (6)$$

For 4% < w_o < 20% [all results]

$$\log \sigma_c \text{ (kPa)} = 4.07 - 0.063\ w_o(\%) \qquad (7)$$

$$\log E \text{ (MPa)} = 3.40 - 0.073\ w_o(\%) \qquad (8)$$

$$\log (-\sigma_t) \text{ (kPa)} = 3.35 - 0.080\ w_o(\%) \qquad (9)$$

$$\nu = 0.08 + 0.017\ w_o(\%) \qquad (10)$$

4. APPLICATION OF UNIAXIAL PROPERTIES TO DESIGN

Although it has been argued that for a weak or soft rock
such as the Melbourne mudstone, the confined properties
are more relevant to engineering design problems (Chiu
and Johnston, 1980; Johnston et al, 1980), the use of
triaxial testing techniques represents a relatively expensive approach when compared with the simplicity of
uniaxial techniques. It follows that if the uniaxial
properties can provide a reasonable indication of performance, then significant economic advantages may be
realised with their use.

Figure 5 presents the back-calculated moduli from a significant number of rock socketed pile and pressuremeter
tests conducted at various sites in the Melbourne region
(Johnston et al, 1980; Williams 1980). The uniaxial
modulus-water content correlations given by Equations (2)
(5) and (8) are also presented. For the reduced mudstone,
Equation (2) forms a lower bound to the in-situ results,
whereas the oxidised mudstone correlation of Equation (5)
provides a reasonable estimation of the average in-situ
moduli. The reason for this observation may be explained
by Figure 6, in which the influence of confining pressure
on modulus is examined for the two weathering environments. For the reduced mudstone (w_o=10±1%), the modulus
shows a rapid initial increase as the confining pressure
increases from zero. For higher confining pressures, the
rate of increase of modulus reduces. For the oxidised
mudstone (w_o=14±1%), the modulus is relatively constant
up to a confining pressure of about 10 MPa. Thereafter
an increase in modulus may be observed. This difference
in behaviour has been attributed to the presence of
cementing agents in the oxidised mudstone. Measurements
of stresses acting on loaded rock socketed piles
(Williams, 1980) indicated that confining pressures of
the order of 1 to 2 MPa are developed. It follows then
that under such confining pressures, the relevant moduli
values for the reduced mudstone are likely to be greater
than the uniaxial moduli, with the latter providing a
lower bound to the in-situ values. For the oxidised mudstone however, as demonstrated in Figure 6, confining
pressures of about 1 to 2 MPa will have little influence
on the relevant moduli, with the result that the uniaxial
moduli provide a reasonable indication of in-situ deformation. The intermediate nature of the overall correlation
of Equation (8) follows with a general correspondence to
the reduced moduli for low water contents and to the
oxidised moduli for high water contents.

In addition to the prediction of deformation, the other
major factor in engineering design involves an assessment
of the stress to cause failure so that an adequate safety

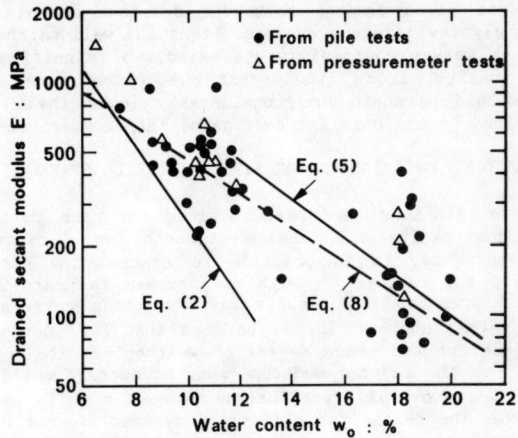

Fig. 5 Comparison of uniaxial compression moduli with
in-situ performance.

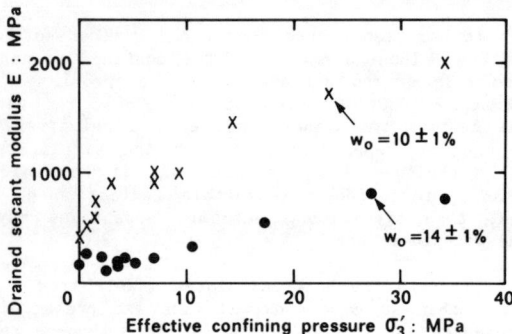

Fig. 6 Influence of confining pressure on uniaxial compressive modulus.

w_o %	REDUCED ONLY 4% < w_o < 12%			OXIDISED ONLY 12% < w_o < 20%			ALL RESULTS 4% < w_o < 20%		
	σ_c Eq(1) (MPa)	σ_t Eq(3) (MPa)	$\dfrac{\sigma_c}{\sigma_t}$	σ_c Eq(4) (MPa)	σ_t Eq(6) (MPa)	$\dfrac{\sigma_c}{\sigma_t}$	σ_c Eq(7) (MPa)	σ_t Eq(9) (MPa)	$\dfrac{\sigma_c}{\sigma_t}$
4	9.86	-1.28	-7.7				6.58	-1.07	-6.1
6	5.70	-0.78	-7.3				4.92	-0.74	-6.6
8	3.30	-0.47	-7.0				3.68	-0.51	-7.2
10	1.91	-0.29	-6.6				2.75	-0.35	-7.9
12	1.10	-0.18	-6.1	2.25	-0.28	-8.0	2.06	-0.25	-8.2
14				1.79	-0.18	-9.9	1.54	-0.17	-9.1
16				1.43	-0.12	-11.9	1.15	-0.12	-9.6
18				1.14	-0.08	-14.3	0.86	-0.08	-10.8
20				0.91	-0.05	-18.2	0.65	-0.06	-10.8

Table I Uniaxial strength data from water content correlations.

margin may be incorporated. However in the case of rock socketed piles, Williams et al. (1980) have shown that performance is generally characterised by work strengthening behaviour. Furthermore, the mechanisms involved in the development of the two components of pile resistance, namely side and base resistance, are not yet fully understood. Therefore their analysis on the basis of detailed confined properties has not been finalised. However, in order to make an assessment of safety in design, the likely minimum failure stresses for both components have been related empirically to the uniaxial compressive strength (Williams et al., 1980). For the side resistance, its ultimate value, f_{su}, is given by the expression

$$f_{su} = \alpha \, \beta \, \sigma_c \qquad (11)$$

where α and β are side resistance reduction factors which are functions of the uniaxial compressive strength of the intact rock and the ratio of mass modulus to intact modulus respectively.

For the base resistance, initial yielding occurred at base stresses of about 3 to 4 times the uniaxial compressive strength, with peak resistances of at least $5\sigma_c$ and often many times greater. It follows that the adoption of $5\sigma_c$ for the minimum peak base resistance provides a conservative design guide. However, because large displacements were required to mobilise the minimum peak base resistances, the designs were usually governed by a displacement criterion.

The above empirical relationships have concerned the design of piles in the Melbourne mudstone wherein the relatively few joints are clean, tight and well matched. For rock masses containing more defects of significantly lower quality, it is likely that the relevant rock mass modulus and strength are considerably lower than those determined by the uniaxial testing of intact specimens.

5. RATIO OF UNIAXIAL COMPRESSIVE TO TENSILE STRENGTH

Since correlations have been obtained for both the uniaxial compressive and tensile strength for a range of water contents, it is possible to examine the general trend of the ratio σ_c/σ_t with an increase in weathering. Table I presents the results for the three correlation groups given above. It may be seen that for increasing water content and hence degree of weathering, the σ_c/σ_t ratios for the reduced mudstone show a decrease while the oxidised and overall correlations produce a quite marked increase. The reason for this trend reversal is not clear and therefore a closer examination is warranted.

For hard brittle rocks, various predictions of the σ_c/σ_t ratio have been made. For example, the plane Griffith Theory (Griffith, 1921) predicts a ratio of -8 while the Extended Griffith Theory (Murrell, 1963) predicts -12 and the Modified Griffith Theory (McClintock and Walsh, 1962) predicts -5.5 to -15 for the coefficient of friction, μ, varying between 0.5 and 1.7 respectively. From a comprehensive summary of test results on a very wide range of different rocks by Lama and Vutukuri (1978), it would appear that the ratio σ_c/σ_t can vary immensely. Although many of the results would have been influenced by anisotropy, it is of interest to note that a crude average of the almost 600 results quoted is about -15, which appears to be in reasonable agreement with the observations of Jaeger and Cook (1976).

For completely weathered rocks and cohesive soils which possess significantly more ductility, the Mohr-Coulomb criterion is usually found to be a reasonable representation of failure stresses, at least in the compressive stress region. In the tensile stress region, because of experimental difficulties, very little work has been reported for these weak materials. As a consequence, engineering designs are generally based on a conservative "no-tension" condition. However, these materials can withstand some tensile stress and therefore must possess a tensile strength. If the Mohr-Coulomb criterion were valid in the tensile stress region, the ratio σ_c/σ_t would vary between -3.0 and -4.6 for the angle of friction, ϕ', varying between 30° and 40° respectively.

The Melbourne mudstone represents a material which is generally more brittle than clays but more ductile than hard rocks. It follows therefore that it would be reasonable to propose that the ratio σ_c/σ_t for the mudstone should reflect its intermediate characteristics with the fresher, less weathered mudstone having properties more typical of harder rocks and the more weathered zones displaying characteristics more in keeping with clays. It follows then that it would be expected to observe a reduction in the σ_c/σ_t ratio as the weathering of the mudstone increases. While this appears to be the case for the reduced mudstone, it is contrary to the results produced by the oxidised mudstone.

In order to resolve this apparent anomaly, it becomes necessary to examine the two components of this ratio. The uniaxial compressive test results give no reason to suspect unusual behaviour because the test itself is a direct determination of the desired property. This conclusion is supported when examining these results alongside the confined triaxial compressive properties which have been presented elsewhere (Johnston and Chiu, 1982).

It would appear therefore that the Brazilian test results may be the source of the observed anomaly.

The basis of the analysis of the Brazilian test is the assumption of the validity of brittle behaviour according to the Griffith failure criteria. Tensile cracking at the specimen centre occurs when the minor principal stress is equal to the tensile strength of the material, which by virtue of the parabolic shape of the Griffith failure envelopes, occurs at the same minor principal stress as would be ideally obtained with a direct uniaxial tensile test on the same brittle material. However, the mudstone, particularly in its more weathered state, is not a brittle material and therefore the Griffith criteria may not be appropriate.

Johnston and Chiu (1982) demonstrate that the compressive strength of the intact mudstone can be adequately described by the non-linear empirical strength criterion

$$\sigma_{1n} = \left[\frac{M}{B}\sigma_{3n} + 1\right]^B, \qquad (12)$$

where σ_{1n} and σ_{3n} are the major and minor principal stresses normalised with respect to the uniaxial compressive strengths given by Equations (1) or (4). The parameters B and M were found to vary with water content according to the relationships

$$B = 0.722 + 0.0093\, w_o(\%) \qquad (13)$$

$$M = 7.80 - 0.224\, w_o(\%) \qquad (14)$$

By assuming the validity of the criterion in the tensile region, Equation (12) predicts that for $\sigma_{1n} = 0$, the ratio σ_c/σ_t is given by

$$\frac{\sigma_c}{\sigma_t} = -\frac{M}{B} \qquad (15)$$

Therefore by using Equations (13), (14) and (15) along with the correlations for uniaxial compressive strength given by Equations (1) and (4), the uniaxial tensile strength and the ratio σ_c/σ_t may be estimated. Table II presents the results of this analysis for water contents between 4% and 20%. The resultant values of the ratio σ_c/σ_t range from -9.1 to -3.6 respectively and these correspond reasonably with the mudstone passing from a more brittle rock-like material at 4% water content to a ductile clay-like material at 20% water content.

information for two selected water contents. This Figure shows that if the empirical criterion of Equation (12) is valid in the tensile stress region, then the Brazilian test does not cause a purely tensile failure but may be initiated by a shear failure as indicated by the point of tangency between the envelope and the Mohr circle. When this condition is reached, the minor principal stress is numerically less than the minor principal stress which would ideally be obtained from a direct uniaxial tensile test. It is of interest to note however, that as the water content becomes smaller, B decreases and M increases to provide failure envelopes which rise more steeply from the normal stress axis and therefore approach more closely the Griffith type criteria of brittle materials. It follows that as the water content decreases, the minor principal stress of the Brazilian test approaches the likely true uniaxial tensile strength as may be seen from Figure 7. This result implies that the Brazilian test estimates of tensile strength may be significantly lower than the true values for high water contents, with the discrepancy becoming less as the water content reduces. Therefore the σ_c/σ_t ratios for the highly weathered mudstone given in Table I may be significantly higher than the likely true values given in Table II. For the less weathered reduced mustone a comparison between the measured and predicted results of Tables I and II respectively show quite acceptable correspondence particularly with regard to the tensile strength. A more detailed discussion of the empirical strength criterion may be found in Johnston and Chiu (1982).

A counter to the above argument arises from observations of the failure mechanisms obtained with the Brazilian tests. Virtually all specimens displayed the classical tensile crack development from their centres with no visible indication of shear development. However, it is not clear whether this failure mechanism and the results obtained were in accordance with the Brazilian test theory

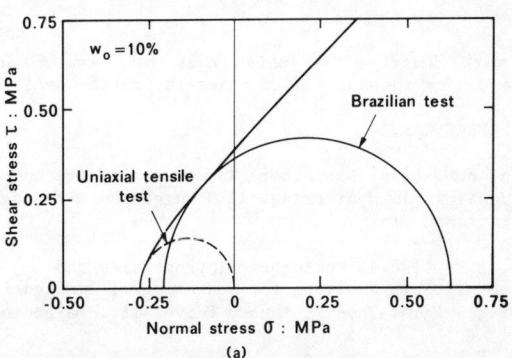

(a)

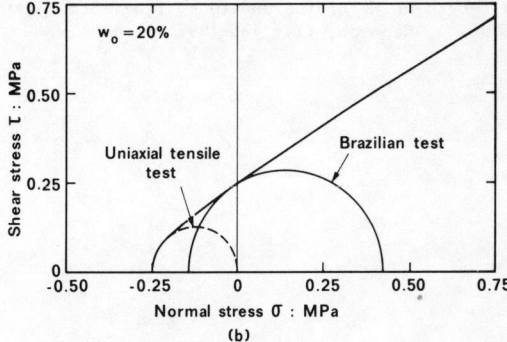

(b)

Fig. 7. Mohr circles at failure for direct uniaxial tensile test and indirect Brazilian test for mudstone according to proposed failure criterion.

w_o %	REDUCED OR OXIDISED MUDSTONE	B Eq(13)	M Eq(14)	$\frac{\sigma_c}{\sigma_t}$ Eq(15)	σ_c Eq(1) or (4) MPa	σ_t MPa
4	R	0.76	6.90	-9.1	9.86	-1.09
6	R	0.78	6.46	-8.3	5.70	-0.69
8	R	0.80	6.01	-7.5	3.30	-0.44
10	R	0.81	5.56	-6.9	1.90	-0.28
12	R	0.83	5.11	-6.2	1.10	-0.18
12	O	0.83	5.11	-6.2	2.25	-0.37
14	O	0.85	4.66	-5.5	1.79	-0.33
16	O	0.87	4.21	-4.8	1.43	-0.30
18	O	0.89	3.76	-4.2	1.14	-0.27
20	O	0.91	3.32	-3.6	0.91	-0.25

Table II Uniaxial strength data from proposed failure criterion.

By transposing the empirical failure criterion into the more conventional $\tau - \sigma$ plane, the Mohr circles at failure for both the direct uniaxial tensile test and the Brazilian test may be examined. Figure 7 presents this

but merely effected by the presence of cementing agents in the oxidised mudstone, or were influenced by stress concentrations and shear development brought about by the increased ductility of the more weathered mudstone.

It follows therefore that the above discussion must at this stage be regarded as speculative. If it were possible to simply obtain reliable and reproducable direct measurements of uniaxial tensile strength, then perhaps some of the issues could be resolved.

6. CONCLUSIONS

The uniaxial properties of the Melbourne mudstone have been examined for the range of weathering characteristics commonly encountered. It has been shown that these properties, in particular the compressive moduli and strength, provide a reasonable indication of the likely in-situ performance of piles socketed into the mudstone. The uniaxial test techniques appear therefore to provide an acceptible alternative to the more expensive confined test techniques.

On examination of the resultant σ_c/σ_t ratios, it appeared that the more weathered oxidised mudstone produced results contrary to expectation. However, in conjunction with an empirically based strength criterion, it is suggested that the Brazilian test may not produce a purely tensile failure but one which is influenced by the development of significant shear stresses. These conditions may lead to an underestimation of the uniaxial tensile strength for the highly weathered mudstone. For the less weathered more brittle mudstone, it seems that the Brazilian test may be more appropriate as an indirect measurement of the uniaxial tensile strength.

7. ACKNOWLEDGEMENT

The work described in this paper has been financially supported by the Australian Research Grants Committee.

8. REFERENCES.

Brown, E.T. (1981). (Editor) Rock Characterization, Testing and Monitoring. ISRM Suggested Methods. Pergamon Press.

Chiu, H.K. (1981). Geotechnical properties and theoretical analyses for socketed pile design in weak rock. Ph.D. Thesis, Monash University, Melbourne.

Chiu, H.K. and Johnston, I.W. (1980). The effects of drainage conditions and confining pressures on the strength of Melbourne mudstone. Proc. 3rd. Aust-N.Z. Conf. on Geomech, (1), 185-189.

Chiu, H.K., Johnston, I.W. and Donald, I.B. (1982). Appropriate techniques for triaxial testing of soft rock. Internal Report, Dept. of Civil Engng., Monash University, Melbourne. To be published.

Griffith, A.A. (1921). The phenomena of rupture and flow in solids. Phil. Trans. Roy. Soc. A 221, 163-198.

Jaeger, J.C. and Cook, N.G.W. (1976). Fundamentals of Rock Mechanics. Science Paperbacks, Chapman and Hall, London.

Johnston, I.W. and Chiu, H.K. (1981). The consolidation properties of a soft rock. Proc. 10th Int. Conf. Soil Mech. and Found., 661-664, Stockholm.

Johnston, I.W. and Chiu, H.K. (1982). The engineering properties of the Melbourne mudstone. Internal Report, Dept. of Civil Engng., Monash University, Melbourne. To be published.

Johnston, I.W., Williams, A.F. and Chiu, H.K. (1980). Properties of soft rock relevant to socketed pile design. Int. Conf. Structural Foundations on Rock, Sydney, A.A. Balkema, 55-64.

Lama, R.D. and Vutukuri, V.S. (1978). Handbook on Mechanical Properties of Rocks, Vol. 2, Trans Tech Publications.

McClintock, F.A. and Walsh, J.B. (1962). Friction on Griffith cracks in rocks under pressure. Proc. 4th U.S. Nat. Cong. Appl. Mech. 2, 1015-1021, Berkeley, California.

Murrell, S.A.F. (1963). A criterion for brittle fracture of rocks and concrete under triaxial stress and the effect of pore pressure on the criterion. Proc. 5th Rock Mech. Symp., Univ. of Minnesota, Pergamon, 563-577.

Sloan, S.W. (1975). Poisson's ratio in rock. Internal Report, Dept. of Civil Engng., Monash University, Melbourne.

Tagell, M. and Tappe, N. (1973). Investigation of Poisson's ratio of Melbourne Mudstone. Internal Report, Dept. of Civil Engng., Monash University, Melbourne.

Williams, A.F. (1980) The design and performance of piles socketed into weak rock. Ph.D. Thesis, Monash University, Melbourne.

Williams, A.F., Johnston, I.W. and Donald, I.B. (1980). The design of socketed piles in weak rock. Int. Conf. Structural Foundations on Rock, Sydney, A.A. Balkema, 327-347.

FIELD MEASUREMENTS FOR THE DESIGN OF THE WASHUZAN TUNNEL IN JAPAN

Mesures in situ pour le projet du tunnel Washuzan au Japon

In-situ Messungen für die Dimensionierung des Washuzan Tunnels in Japan

S. Sakurai

Professor of Civil Engineering, Kobe University, Kobe 657, Japan

SYNOPSIS

The author has proposed a design approach to the dimensioning of underground openings. The proposed method is based on a back analysis of displacements measured at the construction of pilot or exploratory adits which are excavated prior to the construction of main tunnels. This paper shows the proposed method applied to the design of the Washuzan tunnel in Japan. The measurement data of displacements are shown and the results of the back-analysis are discussed in connection with the design of the main tunnel.

RESUME

L'auteur propose une méthode pour la conception des ouvrages souterrains. Cette méthode est basée sur une "analyse en retour" des déformations mesurées pendant la construction d'une galerie pilote où d'exploration excavée avant la construction du tunnel principal. La méthode proposée est discutée en se référant au tunnel de Washuzan au Japon. Les valeurs des déformations mesurées y sont exposées et les résultats calculés à partir de l'analyse en retour sont présentés en rapport avec la conception du tunnel principal.

ZUSAMMENFASSUNG

Eine Methode zur Dimensionierung von Untertagebauten wird vorgeschlagen. Diese Methode beruht auf einer Rückrechnung von Deformationen, welche während des Baus eines vorgängig zum Haupttunnel erstellten Pilot- oder Untersuchungsstollens gemessen wurden. Im vorliegenden Beitrag wird die vorgeschlagenen Methode am Beispiel des Washuzan Tunnels in Japan diskutiert. Die Meßwerte werden zusammengefaßt und die Aussage der damit zurückgerechneten Resultate in Bezug auf die Dimensionierung des Haupttunnels wird dargestellt.

1. INTRODUCTION

The Washuzan tunnel is designed on the basis of new design concept. The new technique was proposed by the author, 1979, and named as "Field Measurement Aided Design Technique(FADT)" in which field measurements play an important role.

The technique requires trial excavations such as pilot and exploratory adits, in which field measurements are carried out during their excavations. Displacement measurements perticularly yield valuable informations about the behaviour of openings. The modulus of elasticity and Poisson's ratio of ground materials and initial stresses existing in the ground prior to excavation can be determined from the measured displacements by means of back-analysis. When we know the material properties and initial stresses, it is possible to analyze the behaviour of main tunnels by using the finite element method for dimensioning support measures and adaptable construction methods. It should be noted, however, that according to the proposed technique those material properties are not necessarily to be real quantities, but to be equivalent values for evaluating the deformational behaviour of openings.

In this paper some details of the proposed technique applied to the Washuzan tunnel are presented. The field measurement and back-analysis results are shown, and discussion is given on the modulus of elasticity and initial stresses in connection with the design of the main tunnel.

2. WASHUZAN TUNNEL

The Washuzan tunnel is located in Okayama prefecture in Japan, on the Kojima-Sakaide rout of highway and railway connecting Honshu(main island) and Shikoku island. The plane view of tunnel is shown in Fig.1. The tunnel consists of two double-track railway tunnels and two double-lane highway tunnels. The railway tunnels are situated under the highway tunnels as shown in Fig.2. Although the length of tunnel is only about 200 meters, this project has many specific features, such that the four large tunnels are constructed very closely each other, and overburden is relatively small.

The work tunnel for carrying construction equipments and materials is constructed in advance of the construction of main tunnel. That is located almost parallel to the main tunnel as shown in Fig.1. In the middle part of the work tunnel the cross section is enlarged for a double-lane traffic.

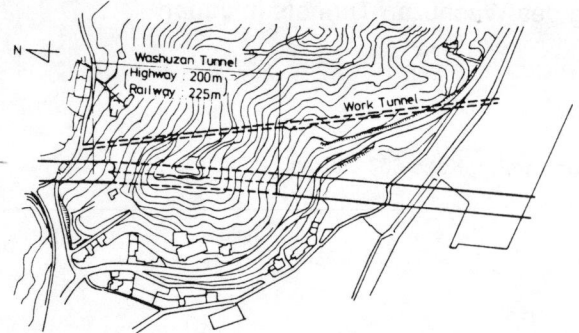

Fig. 1 Plane view of tunnel construction site

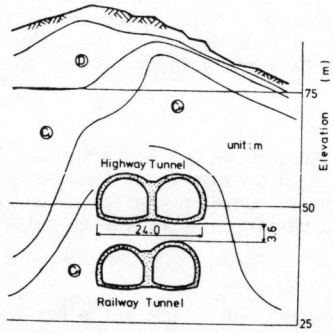

Fig. 2 Cross section of tunnel and geological condition

3. FIELD MEASUREMENT AIDED DESIGN TECHNIQUE (FADT)

The author, 1979, proposed a design technique of dimensioning underground openings. The technique is based on the field measurements, particularly displacement measurements, and named as "Field Measurement Aided Design Technique(FADT)". Since the detail of the technique was presented elsewhere, only the brief summary is described here.

The displacements u appearing around tunnels depend on many different factors such as mechanical properties of ground materials, initially existing stresses in the ground, shape and size of openings, construction method and so on, i.e.,

$$u = f (p, E, \nu, c, \phi, \ \ldots\ldots\ ) \qquad (1)$$

where p, E, ν, c, ϕ are initial stress, modulus of elasticity, Poisson's ratio, cohesion and friction angle, respectively. It should be noted that the evaluation of function f is not an easy task because of complexity of rock mass behaviours. However, the left hand side of Eq.(1), that is, the displacements u can be easily measured at construction period. Therefore, instead of evaluating the function f, the author proposes to use directly the measured displacements for dimensioning the main tunnels.

The proposed method is as follows;

(1) Execute a trial excavation and carry out field measurements, especially displacement measurements of surrounding media.
(2) Assume the media to consist of a homogeneous, isotropic and linear elastic material, even if the behaviour of real media is inelastic. Consequently, the displacements around the trial excavation are expressed as,

$$u^* = g (p^*, E^*, \nu^*) \qquad (2)$$

where E^*, ν^* are the "equivalent modulus of elasticity" and "equivalent Poisson's ratio", respectively. p^* is also equivalent value of initial stresses. Determine the equivalent modulus of elasticity, Poisson's ratio and initial stresses by means of a back-analysis in such a way that $u^* \cong u$ (see Fig.3).
(3) When we know the equivalent modulus of elasticity and initial stresses, the displacements expected around main tunnels can be analyzed by the finite element method for dimensioning support measures and execution methods of excavation. However, the size of main tunnels is generally greater than trial excavations. Therefore, the equivalent modulus of elasticity obtained by back-analysis must be modified considering its reduction with an increase of joint number(see Fig.4).

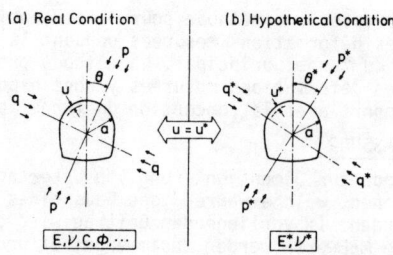

Fig. 3 Relationship between real and hypothetical conditions of tunnel surrounding media

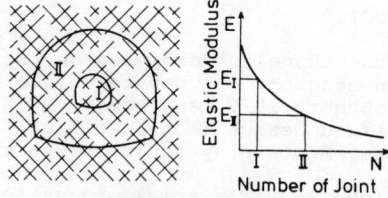

Fig. 4 Influence of joint numbers on modulus of elasticity of rocks

4. FIELD MEASUREMENTS

The field measurements were carried out at the work tunnel during its construction period. The measurement sections are shown in Fig.5, and the type of measurements is also indicated in the figure. The absolute displacements were measured at the measurement section No.5 by inserting the Sliding Micrometer(ISETH) and Inclinometer (SINCO) from the ground surface.

Fig.7 illustrates strain distributions along the borehole B-1 and B-3(see Fig.6). They were obtained by the Sliding Micrometer. It is of interest to know from Fig.7 that the ground being about 10m high at the crown is fairly influenced by excavation. However, the strains are still small enough to keep stability of ground, comparing them with the failure strain of rocks(Sakurai,1981). The strains shown in these figures are of their final values when the tunnel face are far away from the measurement section. The variation of strain with location of tunnel face is illustrated in Fig.8.

The displacements of the ground can be obtained by integrating strains, and they are used as input data for back-analysis to evaluate the equivalent modulus of elasticity and initial stresses.

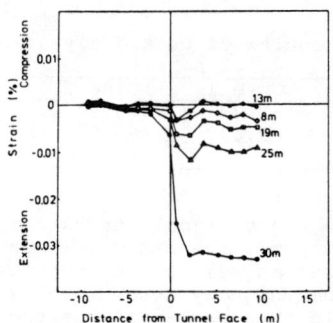

Fig. 8 Variation of strains with distance of tunnel face

5. BACK-ANALYSIS

The equivalent modulus of elasticity and initial stresses are back-analyzed from the results of field measurements. The back-analysis is carried out by a computer program named as "Direct Back-Analysis Program(DBAP)" developed by the author. The program is based on a finite element formulation associated with least square method.Poisson's ratio is assumed in this program. Computer simulations have verified the accuracy and stability of computation of the back-analysis. The detail of the program is not described here because of space limitation. Only the results are presented in the following.

Since the displacement measurements of ground surface were not accurate enough to evaluate the movement of ground due to tunnel excavation, the surface displacements which are needed in the back-analysis are assumed for three different cases as shown in Table I(see Fig.9).

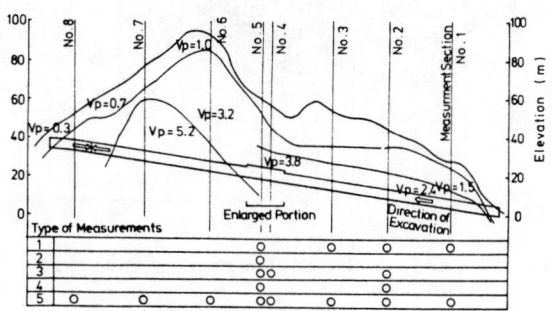

Fig. 5 Measurement sections and type of instrumentations at work tunnel

1 : Measurements of Loosening Zone by Seismic Tests
2 : Stress Measurements of Shotcrete
3 : Measurements of Axial Force of Rock Bolts
4 : Displacement and Subsidence Measurements of the Ground
5 : Convergence Measurements

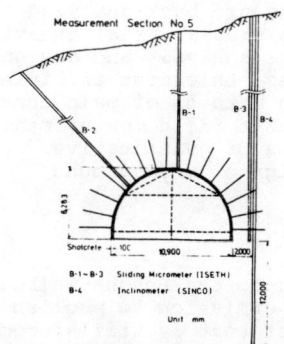

B-1∼B-3 Sliding Micrometer (ISETH)
B-4 Inclinometer (SINCO)

Unit mm

Fig. 6 Displacement measurements along borehole and convergence measurements

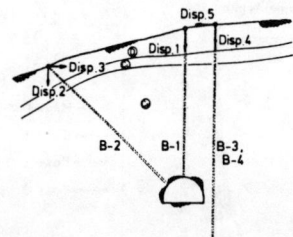

Fig. 9 Displacement vectors at ground surface

Table I Displacement vectors assumed at ground surface

			(mm)
	Case 1	Case 2	Case 3
Disp.1	2.10	1.04	0.0
Disp.2	1.21	0.42	0.0
Disp.3	0.63	0.22	0.0
Disp.4	3.00	0.94	0.0
Disp.5	0.15	0.14	0.0

The results of the back-analysis are shown in Table II, in which the equivalent modulus of elasticity and initial stresses are given as a ratio. In these analysis Poisson's ratio is assumed to be $\nu^* = 0.3$. In order to separate each quantity, we assume the vertical component of initial stress to be approximately equal to overburden pressure.

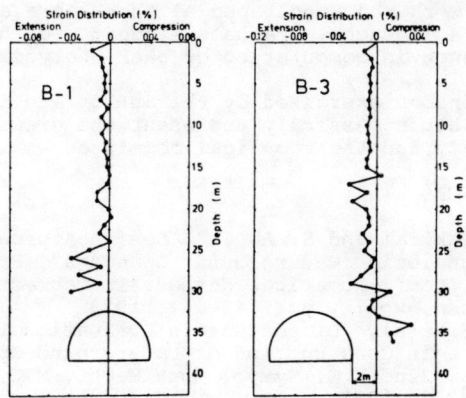

Fig. 7 Strain distributions along borehole

A 217

Table II Results of back analysis by DBAP

	Case 1	Case 2	Case 3
$\sigma_x*/E*$	-7.79	-5.80	-4.36
$\sigma_y*/E*$	-8.00	-5.00	-3.03
$\tau_{xy}*/E*$	-2.48	-1.40	-0.97

$(\times 10^{-4})$

If we assume the vertical stress component to be $\sigma_y* = 10 \text{kgf/cm}^2$, the other components of initial stresses as well as the modulus of elasticity can be uniquely determined as shown in Table III. In this back-analysis the effect of shotcrete was disregarded because the thickness of shotcrete is only 10cm and its effect is insignificant.

Table III Equivalent modulus of elasticity and equivalent initial stresses

(kgf/cm^2)

	Case 1	Case 2	Case 3
$E*$	12500	20000	33003
σ_x*	-9.74	-11.60	-14.39
σ_y*	-10.00	-10.00	-10.00
$\tau_{xy}*$	-3.10	-2.80	-3.20
σ_1*	-6.77	-7.89	-8.32
σ_2*	-12.97	-13.71	-16.08
ψ (deg.)	43.80	52.97	62.22

In order to check the accuracy of this back-analysis, we calculate the displacements by FEM and compare them with the measured displacements. Some of the results are shown in Fig.10. It is noted that a best fit seems to be between Case-1 and Case-2. This means that the modulus of elasticity must be in the range of $E* = 12,500 - 20,000 \text{ kgf/cm}^2$.

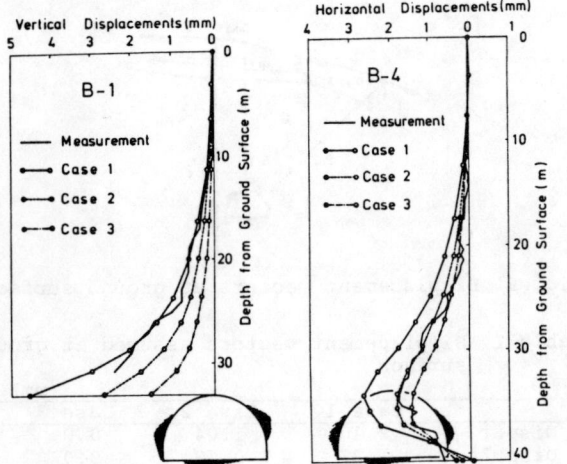

Fig. 10 Comparison between computed and measured displacements

6. DESIGN OF MAIN TUNNEL

Since the equivalent modulus of elasticity and initial stresses have been determined, the main tunnel can be analyzed by the finite element method for dimensioning support measures and construction method. The finite element mesh used

for this purpose is shown in Fig.11. The detail of analysis and design is not presented in this paper because of limitation of space.

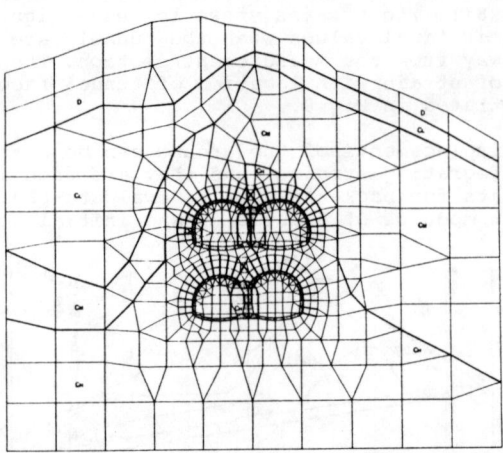

Fig. 11 Finite element mesh for analysis of main tunnel

7. FINAL REMARKS

In this paper the design concept of the Washuzan Tunnel has been briefly described. The method is based on field measurements, especially displacement measurements. That is, the modulus of elasticity and initial stresses are back-analyzed from the displacements measured at trial excavation. They are used for the analysis and design of primary support measures, thickness of lining and adaptable construction methods of main tunnels. It should be noted that field measurements are one of the best sources of quantitative information for final design of underground openings.

ACKNOWLEDGMENTS

The author is much obliged to the Honshu-Shikoku Bridge Authority for the permission to publish some details of the project that is still incomplete. The autor is also grateful to the members of the Technical Committee of Washuzan Tunnel for their valuable contribution to various aspects of the work reported. Special thanks are also due to Mr. K.Takeuchi, graduate student, for his assistance in computation of back-analysis.

The opinions expressed by the author are his own and do not necessarily represent the views of the authority and the technical committee.

REFERENCES

1) S. Sakurai and S. Abe, "A Design Approach to Dimensioning Underground Openings", Proc. 3rd Int. Conf. Numerical Methods in Geomechanics, Aachen, vol.2, pp. 649-661, 1979
2) S. Sakurai, "Direct Strain Evaluation Technique in Construction of Underground Opening", Proc. 22nd U.S. Sympo. Rock Mech., MIT, pp. 278-282, 1981

IN-SITU DETERMINATION OF CREEP PROPERTIES OF ROCK SALT
Détermination sur place des propriétés de fluage du sel gemme
In-situ Bestimmung von Kriecheigenschaften des Steinsalzes

B. Ladanyi
Professor at the Department of Civil Engineering, Ecole Polytechnique, Montréal, Canada
D. E. Gill
Professor at the Department of Mineral Engineering, Ecole Polytechnique, Montréal, Canada

SYNOPSIS

This paper describes the results of short- and long-term borehole dilatometer tests performed in confined blocks of rock salt under simulated in-situ conditions. It is shown that, by using an appropriate interpretation method, such tests make it possible to determine certain basic rheological parameters of rock salt, needed in the design of underground openings.

RESUME

On présente les résultats des essais dilatométriques à court et à long terme, effectués dans des blocs confinés du sel gemme, en conditions quasi-naturelles. En utilisant une méthode d'interprétation appropriée, on peut tirer à partir de tels essais certains paramètres rhéologiques fondamentaux du sel gemme, nécessaires pour la conception de souterrains.

ZUSAMMENFASSUNG

In dieser Arbeit werden die Ergebnisse des kurz- und langzeitigen Bohrlochdilatometerversuche, welche in biaxialbelasteten Steinsalzquadern unter simulierten Feldbedingungen durchgeführt wurden, beschrieben. Es wird gezeigt, daß man mit Hilfe einer dazu geeigneten Bewertungsmethode aus solchen Versuchen einige grundsätzliche rheologische Kennwerte von Steinsalz, die beim Entwurf der unterirdischen Kavernen brauchbar sind, bestimmen kann.

1. INTRODUCTION

The mechanical behaviour of rock salt has for many years been studied essentially in connection with the design and operation of salt mines. In recent years, however, the interest and activity in this field has grown very rapidly because of energy-related projects, such as underground storage of oil, natural gas and radioactive waste.

On the basis of numerous past laboratory studies of the response of rock salt and potash to changes in stress and temperature, a rather complex picture of the behaviour of such materials has been obtained. In addition, it was found that laboratory test results obtained on rock cores may not necessarily well represent the behaviour of the rock mass in situ, because of the effects of scale, destressing and changes in humidity and temperature. For that reason there is presently an increasing interest in developing reliable in-situ testing methods that would be able to furnish a set of most important parameters needed in the design of underground openings.

The paper describes one such method which has recently been developed by the authors. The method uses a high capacity borehole dilatometer, such as the Colorado School of Mines Cell, for performing short-term and long-term borehole loading tests in-situ. A proper evaluation method, proposed by the authors, enables to determine from such a test: the first loading and cyclic loading rigidity moduli, the short term tensile strength, a portion of the stress-strain curve up to the peak strength and the basic creep parameters needed for writing the constitutive creep equation.

The paper describes the method and presents an analysis of borehole dilatometer test results obtained in the rock salt from a salt mine in Eastern Canada.

2. BOREHOLE DILATOMETER TESTING IN ROCK MECHANICS

The borehole dilatometer test has been in a frequent use in rock mechanics since at least two decades, especially after valuable instrumentation developments made Rocha et al. (1966) and by Hustrulid and Hustrulid (1975). These authors have shown that a borehole dilatometer test can furnish reliable in-situ values of short-term deformation properties of the rock mass around a borehole. On the other hand, there has been until recently very little interest in performing long-term hole expansion tests with this instrument for determining the creep properties of rock materials, so that practically no such information can be found in the present rock mechanics literature.

The situation was quite different in other fields, such as in ice and permafrost mechanics, where the long-term borehole dilatometer testing has been in current use since the beginning of the 1970-ies (Ladanyi and Johnston, 1973).

In the recent years, there has been a growing interest in testing short- and long-term mechanical properties of rock salt mainly in connection with the design of underground storage cavities in salt deposits. Although from a large amount of valuable laboratory test results, and from in-situ closure observations in salt mines accumulated through the years, a fairly clear picture of the behaviour of rock salt under changes of stress and temperature has been obtained, it is nevertheless recognized that it would be useful to be able to test the rock salt directly in-situ by using a relatively simple method, such as the borehole dilatometer test.

Use of such a test in rock salt mechanics was mentioned by Passaris (1980) and Albrecht et al. (1980), respectively. The former showed the results of two short-term dilatometer tests to failure, carried out in unconfined thick cylinders of rock salt, using an improvised dila-

tometer device. Albrecht et al. in turn, used a high capacity dilatometer of their own design to determine deformation properties of rock salt in situ. However, although a short creep stage was included in their test,
there is no mention in the paper how this creep information was used.

In this connection it should also be mentined that some
attempts have been made to test creep properties of rock
salt in situ by pressurizing unlined underground cavities
(Passaris, 1979) and boreholes (Nelson and Kocherhans,
1981). This is possible because of relative impermeability of rock salt, but the pressure range in such tests
is limited to relatively low stresses below the hydraulic
fracture level, usually not much higher than the average
ground stress.

Recognizing the need for borehole creep tests in rock
salt and the potential of the method already developed
for frozen soils, the behaviour of which is in many aspects similar to that of the rock salt, the authors have
made an attempt in a previous paper (Ladanyi and Gill,
1981) to show that essentially the same method could be
used in rock salt, provided one has available a dilatometer equipment of a high capacity and sensitivity.
Trial calculations shown in that paper indicated that
one would need for that purpose a dilatometer of at
least 50 MN/m^2 capacity that would be able to record radial strains of the order of 10^{-5}.

3. DILATOMETER TESTS

3.1 General

Although in the previous paper (Ladanyi and Gill, 1981),
a complete testing and interpretation procedure was presented, both for short-term tests and for long-term creep
and relaxation tests, no real test results were shown,
but only a possible behaviour of rock salt in such dilatometer tests was theoretically simulated. As a logical
follow-up, this paper presents the results of such tests
carried out in large blocks of natural rock salt under
simulated field conditions.

Although the possibility existed to make these dilatometer tests underground in a salt mine, the decision was
made to test the equipment and the method first under
controlled laboratory conditions. The experience showed
this to be a right decision, because during the tests
several unforseen modifications to the equipment had to
be made, which could not have been carried out in the
field.

3.2 Equipment

The borehole dilatometer equipment used in the tests was
the CSM Cell, which was originally designed by Panek et
al., 1964, and was further developed by Hustrulid and
Hustrulid, 1975, at the Colorado School of Mines. Its
main intended use was the determination of the modulus
of rigidity of rock. The CSM Cell is a hydraulically
operated system in which the pressure is controlled by
manually activating the piston of a high pressure generator, which also measures the injected fluid volume,
from which the deformation of the borehole is deduced.
The system is rated to 70 MN/m^2, but has been rarely
used up to that level.

The CSM probe consists of a single cylindrical cell,
made of an adiprene membrane, 3.8 cm in diameter and
16.5 cm long, which is mounted on a central steel shaft.
The whole hydraulic system is made of stainless steel
tubing and is relatively rigid.

A detailed description of the equipment and its calibration can be found in the paper by Hustrulid and Hustru-

lid (1975), or in the new ISRM "Suggested Method for Deformability Determination using a Borehole Dilatometer
Test", which is presently under review.

For the purpose of the present investigation, several modifications and additions have been made to the original
equipment. First, a small fluid container was added to
the system to enable to double the volume of the injected
fluid (originally 30 cm^3), because it was found that the
membrane could increase its diameter up to 15% or even
20% without failing. Second, a pressure transducer was
installed in the system close to the cell (Fig. 1),

Fig. 1 Dilatometer probe with pressure transducer,
installed in a confined block of rock salt.

Fig. 2 Complete test set-up and data recording system
for confined borehole expansion tests in salt.

which enabled a continuous recording of applied pressure
to be made. Third, in order to increase the accuracy of
measurement of injected fluid volume, a displacement
transducer was put in contact with the rod of the pressure generator piston, enabling both the applied pressure
and the injected volume to be recorded simultaneously on
a chart moving at a constant speed. Figure 2 shows the
whole equipment used in the tests.

Nevertheless, in spite of all these improvements and additions, the system was found to be difficult to operate, especially when performing long-term creep tests. The main reason for this is the fact that borehole expansion has to be deduced from the volume of injected fluid, which is very sensitive to temperature changes and subject to errors due to leaks. In order to get proper results in creep tests, it was therefore necessary to keep the laboratory temperature strictly constant, and to check possible leakage in the system by performing long-term calibration tests in a steel cylinder before and after each creep test. Similar problems in operating this system, especially under field conditions, have been reported earlier by Patricio and Beus, 1977.

In other words, although the CSM Cell system was used with success in this investigation, it is not considered to be a convenient tool for performing long-term creep tests, because of its hydraulic strain measurement character. For performing in situ creep tests in rock salt, a dilatometer with electronic recording of radial displacements is an absolute necessity.

3.3 Rock salt used in the tests

The rock salt used in the tests was taken from -200 m level of a salt mine located in Eastern Canada. The salt is relatively pure with small quantities of clay, giving it a slightly grey appearance. Its crystal size varies from about 2 to 5 mm in the average. In conventional unconfined compression tests made on five specimens 53 mm in diameter and 106 mm long, the salt showed a non-linear stress-strain curve with an initial tangent Young modulus of about $E = 16$ GN/m^2, and an average uniaxial compression strength of 30 MN/m^2. The Poisson's ratio was about 0.27 in the beginning of the tests, then increased steadily with axial strain, reaching and exceeding the 0.50 level at strains of about 1.5×10^{-3}.

3.4 Test set-up

For the tests, the salt blocks were cut dry by saw into cubes with sides of about 24 cm. The cubes were then put into plywood moulds 30 x 30 x 30 cm and cast in fine-grained concrete. This was necessary in order to get flat sides of the blocks for applying lateral confining pressure to the blocks by a system of jacks (Figs. 1 and 2). For a better confining pressure distribution, plywood boards were inserted between the concrete envelope and the steel platens. The concrete envelope was cut by saw in the middle of each side to avoid its interference with the salt during the test. The confining pressure was controlled by a Structural Behavior Engrg Lab. pressure system, such as normally used for performing triaxial compression tests (Fig. 2). The boreholes for the tests were drilled in the center of unloaded faces of the blocs by using a 38 mm laboratory core drill, normally employed for making rock specimens for triaxial tests. This method produced holes of an average diameter of 4 cm.

3.5 Test results - Short term

In the borehole dilatometer and pressuremeter practice (e.g., Ladanyi and Johnston, 1973, 1978) the results of a short-term borehole expansion test are usually first plotted as a corrected pressure-expansion curve, which represents a relationship of the form $V_m = f(p_c)$, where V_m is the total volume of fluid injected into the measuring cell from the start of pressure application (corrected for the volume change of the hydraulic system, when using the CSM Cell), and p_c is the applied pressure (corrected for the piezometric head and the extension resistance of the unloaded probe).

The "true pressure-expansion curve", used for interpretation, is obtained by shifting the pressure origin of the first curve to p_o, which denotes the average original ground pressure acting normally to the borehole axis. The value of V_m at $p_c = p_o$, denoted by V_{mo}, is read from the plot. The "true pressure-expansion curve" represents then a relationship of the form $\Delta V = f(p)$, where $p = p_c - p_o$ and $\Delta V = V_m - V_{mo}$. It is clear therefrom, that for evaluating a dilatometer test one must know the value of the far-field stress p_o, which has to be determined independently. However, in the present tests, the value of p_o was known, being equal to the constant applied confining pressure.

It is still more convenient for comparison purposes if for plotting the results one uses instead of ΔV, the dimensionless quantity $\Delta V/V$, which represents, in fact, the shear distortion strain at the cavity wall. The value of V, denoting the current borehole volume, is equal to $V = V_o + \Delta V$, where $V_o = V_{in} + V_{mo}$ is the volume of the measuring section of the probe at $p_c = p_o$, and V_{in} is its initial volume, at $p_c = 0$, usually a constant for a given probe.

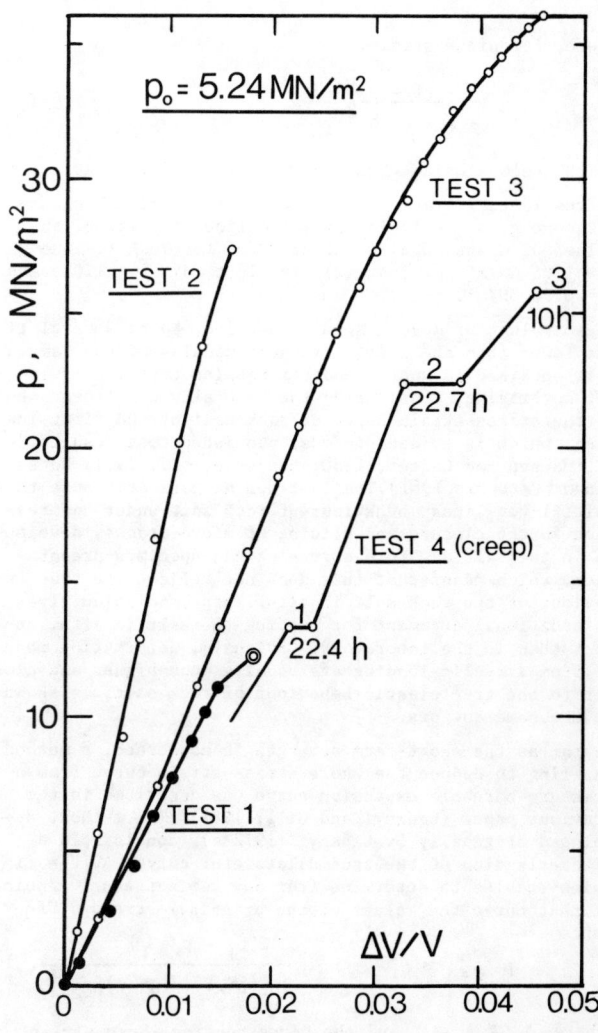

Fig. 3 Pressure-expansion curves for three short-term tests and a stage-loaded creep test.

Figure 3 shows a $\Delta V/V$ vs. p plot of three short-term dilatometer tests (Tests 1, 2, 3) carried out in blocks of rock salt. The forth line in the figure (Test 4) represents the result of a step-loaded test with three creep stages, which will be discussed later. All these tests were carried out under a constant external confining pressure of $p_0 = 5.24$ MN/m².

The geometry and average expansion rate in the three tests were as follows: If a denotes the hole radius, b the distance from the hole center to the nearest side of the block, and $\dot{v} = (\Delta V/V)/dt$ the average rate of hole expansion, then, for

Test 1 : a = 2 cm, b = 12 cm, $\dot{v} = 0.039$ min⁻¹

Test 2 : a = 2 cm, b = 12 cm, $\dot{v} = 0.012$ min⁻¹

Test 3 : a = 2 cm, b = 10 cm, $\dot{v} = 0.023$ min⁻¹

Test 4 : a = 2 cm, b = 10 cm, $\dot{v} =$ variable.

Using the classical theory of an expanding elastic thick cylinder in plane strain, one can calculate the value of the shear modulus from the initial slope of the pressure-expansion curve, as

$$G = \lambda \, \Delta p / \Delta (\Delta V/V) \qquad (1)$$

where, for plane strain,

$$\lambda = \frac{1 - (1 - 2\nu)(a/b)^2}{1 - (a/b)^2} \qquad (2)$$

which becomes unity when $b \to \infty$, as in a field test.

In the Tests (1) and (2), since b/a = 6, and taking $\nu = 0.3$, one gets $\lambda = 1.017$. From the lines in Fig. 3, the values of G are then: $G = 0.76$ GN/m² for Test (1), and $G = 1.85$ GN/m² for Test (2). For Test (3), $\lambda = 1.025$ and $G = 0.98$ GN/m².

These values of moduli seem rather low, being several times lower than the initial tangent modulus of the same salt obtained in unconfined compression tests. This is not surprising, considering the generally non-linear shape of the stress-strain curve of rock salt at the first loading, which is evident in most published test results (e. g., Hansen and Carter, 1980; Hansen et al., 1981; Horseman and Passaris,1981).The last two authors attribute this initial non-linear behaviour of rock salt under compression to the closure and sliding of micro-cracks, developed in the salt due to destressing and specimen preparation, which means that this does not reflect the true behaviour of the rock salt in situ. This conclusion gives an additional argument for testing the salt in situ, rather than in the laboratory. Obviously, deformation moduli from a cyclic loading are usually much higher and closer to the true elastic behaviour of rock salt, as shown by the same authors.

As far as the short-term strength is concerned, a method enabling to deduce the whole stress-strain curve from a pressure-borehole expansion curve was described in the previous paper (Ladanyi and Gill, 1981). The method, developed originally by Ladanyi (1972 a), consists in a discretisation of the true dilatometer curve, $\Delta V/V = f(p)$, which enables to determine from any two consecutive points on that curve the values of the principal stress difference

$$q_{i,i+1} \equiv (\sigma_1 - \sigma_3) = \frac{2(p_i - p_{i+1})}{\ln[(\Delta V/V)_i/(\Delta V/V)_{i+1}]} \qquad (3)$$

where $p_i > p_{i+1}$, and the corresponding shear strain (principal strain difference)

$$\gamma_{i,i+1} \equiv (\varepsilon_1 - \varepsilon_3) = \tfrac{1}{2}[(\Delta V/V)_i + (\Delta V/V)_{i+1}] \qquad (4)$$

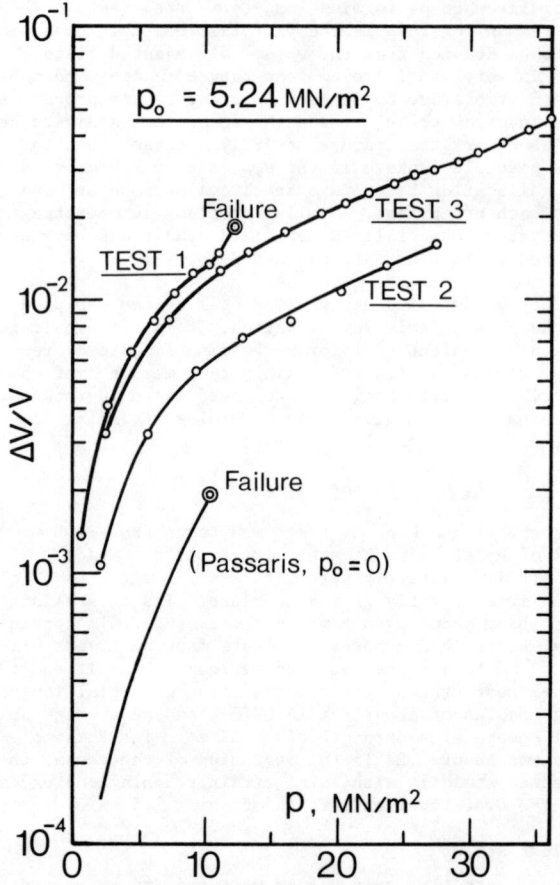

Fig. 4 Pressure-expansion curves from short-term tests, plotted in a semi-log plot, for strength determination.

Because of the lack of space, these curves are not shown for Tests (1) to (3) performed in this investigation. Instead, the dilatometer curves of the three tests were replotted in the semi-log plot, Fig. 4, from which only the peak and the post-peak strengths were determined, using Eq. (3). In such a plot, the peak of the stress-strain curve corresponds to the flattest slope of the line, followed by an increasing slope in the post-peak region. It is noted in Fig. 4 that from the three tests performed, only Test (3) reached the post-peak region, giving the (plane strain) strengths of $q_{peak} = 53.9$ MN/m² and q_{resid} = 38.0 MN/m². Test (2) came close to the peak, with q_{max} = 40.4 MN/m², while Test (1) failed prematurely due to a weakness plane intersecting the block at one end of the borehole. In a block or a thick cylinder, the whole dilatometer curve can only be obtained if an external confining pressure is provided, which was equal to 5.24 MN/m² in these tests. If not, the block will fail by radial cracking, far before reaching the peak compression strength of the salt. This happened in the tests made by Passaris (1980) in unconfined thick cylinders, of which one result is shown in Fig. 4 for comparison.

3.6 Test results-creep

3.6.1 General

For an elastic-non-linear-viscoplastic material, such as the rock salt, the total strain attained after a given time under constant stress can be expressed by

$$\varepsilon_e = \varepsilon_e^{(i)} + \varepsilon_e^{(c)} \qquad (5)$$

where $\varepsilon_e^{(i)}$ is the instantaneous, elastic and plastic, portion of the total strain, and $\varepsilon_e^{(c)}$ is the creep strain, often expressed by a power law expression after Hult (1966):

$$\varepsilon_e = (\dot{\varepsilon}_c/B)^B (\sigma_e/\sigma_c)^n t^B \qquad (6)$$

where the subscript e denotes the von Mises equivalent stress and strain, respectively, σ_c is the reference stress ("creep modulus") corresponding to an arbitrary strain rate $\dot{\varepsilon}_c$, t is the time, and B and n are creep exponents.

In rock salt mechanics most published data show that the exponent n has usually a value between 2 and 5, but the data show less agreement as to the value of the exponent B, which, based on both laboratory tests and field observations of wall closure in salt mines, seems to vary at room temperature between wide limits of 1/3 to 1. (for a review see Ladanyi, 1980).

In the previous paper (Ladanyi and Gill, 1981), it was assumed that B < 1, and a method corresponding to that assumption for finding creep parameters from stage-loaded dilatometer tests was described. However, after performing several creep tests with the salt used in this investigation, it was found that an apparent steady-state creep started rather early in the tests, so that an evaluation method based on steady-state creep assumption seemed more appropriate. The method is considered also to be more pertinent for evaluating long-term in-situ dilatometer tests, where steady-state creep is more often observed than in the laboratory.

For a steady-state creep (B = 1), which follows an instantaneous and a primary creep response, a convenient way of putting stage-loaded creep data into an analytical form is that proposed by Hult (1966) and applied to frozen soils by Ladanyi (1972 b). In that method, all the strains preceding steady-state creep are included into a pseudo-instantaneous strain, which is determined by finding the intersection of the steady-state line with t = 0 axis, as shown in Fig. 5. The total strain attained in a test with N creep stages is then, beyond the primary creep region,

$$\varepsilon = \varepsilon_{pi} + \sum_1^N \dot{\varepsilon}_{ss} \Delta t \qquad (7)$$

where ε_{pi} is the cumulative pseudo-instantaneous strain, $\dot{\varepsilon}_{ss}$ is the steady-state creep rate, and Δt is the length of an individual creep stage. It is usually found (Hult, 1966) that both ε_{pi} and $\dot{\varepsilon}_{ss}$ are related to stress by creep law relationships, but they depend obviously also on temperature, confining pressure and other factors.

Based on the same literature, the following analytical expressions have been adopted for the purpose of dilatometer test evaluation in this paper

$$\varepsilon_{e,pi} = \varepsilon_k (\sigma_e/\sigma_k)^k \qquad (8)$$
$$\dot{\varepsilon}_{e,ss} = \dot{\varepsilon}_c (\sigma_e/\sigma_c)^n \qquad (9)$$

where e stands for "equivalent" as before, σ_k is the reference stress at $\varepsilon_{pi} = \varepsilon_k$, and σ_c is the reference stress at $\dot{\varepsilon}_{ss} = \dot{\varepsilon}_c$. The purpose of a test evaluation is to determine the values of σ_k, σ_c and the two exponents, k, and n.

The effect of temperature on creep rate can be taken into account by means of the temperature function of the form

$$f(T) = \exp(-Q/RT) \qquad (10)$$

generally used in salt mechanics (e.g., Hansen and Carter, 1981; Hunsche, 1981; Munson and Dawson, 1981), where T is the absolute temperature in degrees Kelvin, R = 8.319 J/mol.K is the universal gas constant and $Q \approx$ 54 kJ/mol is the activation energy for rock salt at room temperature, giving $Q/R \approx 6500$ K.

When the creep rate is expressed by Eq. (9), the time function can be made to affect only the value of σ_c, so that, if σ_{c1} and σ_{c2} correspond to absolute temperatures

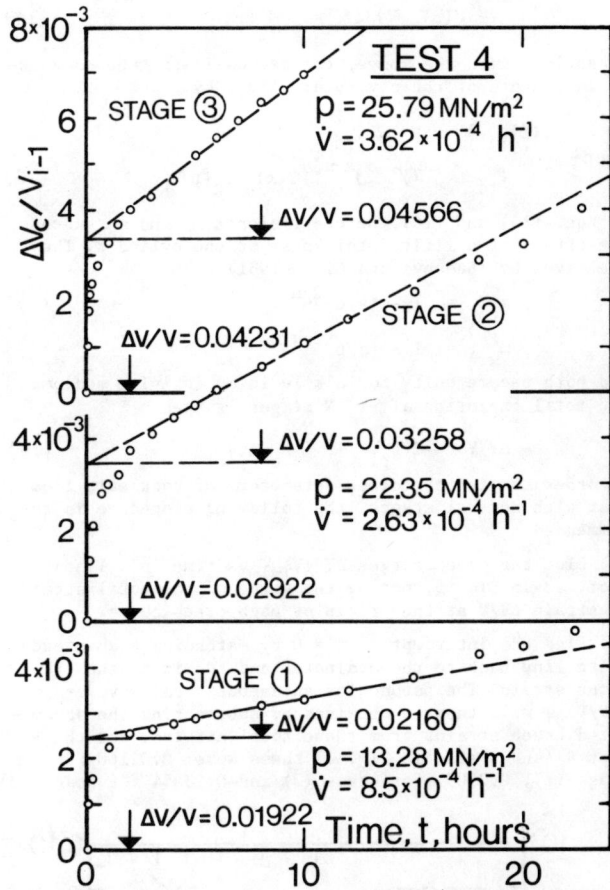

Fig. 5 Creep curves from a stage-loaded borehole expansion test.

T_1 and T_2, their ratio is given by

$$\sigma_{c2}/\sigma_{c1} = [f(T_1)/f(T_2)]^{1/n} \qquad (11)$$

It is noted that the creep formulation adopted here is equivalent in principle to that considered by Hansen and Carter (1981) to be one of the best forms for describing the creep phenomena in rock salt.

3.6.2 Evaluation method

If Eqs. (7) to (9) are adopted for expressing the response of rock salt to loading, the method of evaluation of a stage-loaded dilatometer creep test should be based on the solution of cylindrical cavity expansion in a thick cylinder made of a power law material, as formulated by Ladanyi and Johnston (1973, 1978) and Ladanyi and Gill (1981). The main difference with these previous solutions will be that the exponent B = 1 and that the pseudo-instantaneous response is now also expressed by a power law.

Following Ladanyi and Gill (1981), the creep portion of borehole expansion at a constant stress $p = p_c - p_o$, applied during the stage i, is given by

$$\ln(V/V_{i-1}) = 2 \ln(a/a_{i-1}) \approx (\Delta V_c/V) = 2 F_c t \qquad (12)$$
$$F_c = (\sqrt{3}/2)^{n+1} (2/n)^n \dot{\varepsilon}_c (p/m_c\sigma_c)^n \qquad (13)$$

In Eq. (12), V is the current volume of the cavity, V_{i-1} is its volume at the start of the creep stage and $\Delta V_c = V - V_{i-1}$ is the creep volume increment of the cavity. The volume expansion rate $\dot{v}$ is then from Eq. (12):

$$\dot{v} \;=\; d(\Delta V_c/V)/dt \;=\; 2\,F_c \tag{14}$$

In analogy with the above, the pseudo-instantaneous volume increments of the cavity are given by

$$(\Delta V_{pi}/V) \;=\; 2\,F_{pi} \tag{15}$$

where

$$F_{pi} \;=\; (\sqrt{3}/2)^{k+1}(2/k)^k\,\varepsilon_k(p/m_k\sigma_k)^k \tag{16}$$

In Eqs. (13) and (16), the coefficients m_c and m_k express the effect of a limited thickness of the cylinder. They are given by (Ladanyi and Gill, 1981)

$$m_c \;=\; 1 - (a/b)^{2/n} \tag{17}$$

$$m_k \;=\; 1 - (a/b)^{2/k} \tag{18}$$

and both become unity for a hole in an infinite medium. The total expansion after N stages is then

$$\Delta V/V \;=\; 2(F_{pi} + \sum_1^N F_c\,\Delta t) \tag{19}$$

In order to find the creep parameters of rock salt from a test with several stages, the following procedure is recommended:

(1) Plot the creep curves $\Delta V_c/V_{i-1}$ vs time in a linear plot, as in Fig. 5, noting the value of the total attained strain $\Delta V/V$ at the origin of each creep curve.

(2) Find the intercept at $t = 0$ by extending each steady-state line back to the ordinate, and add it to the previous strain. The pseudo-instantaneous strains $v_{pi} = (\Delta V/V)_{pi}$ will then be obtained by subtracting the accumulated creep strains from these total strains. For the Test 4 (shown also in Fig. 3) these were: 0.02160 for stage (1), 0.03051 for stage (2) and 0.03854 for stage (3).

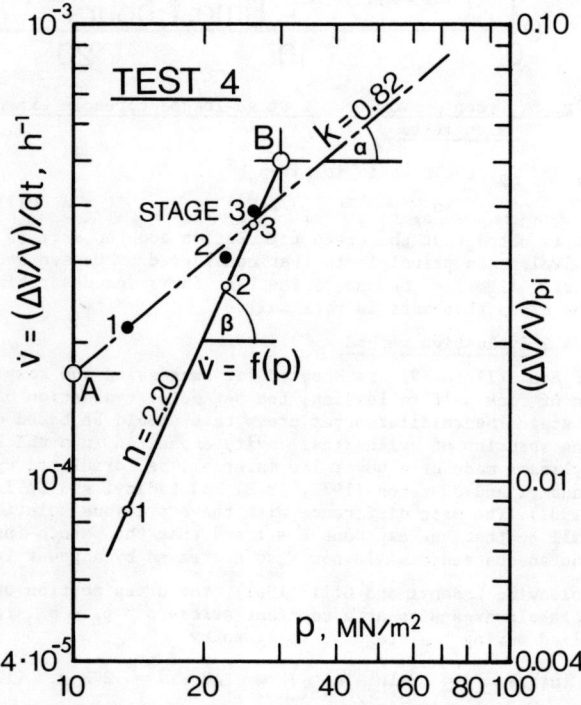

Fig. 6 Pseudo-instantaneous displacements and expansion rates from a stage-loaded borehole expansion test.

(3) Plot these pseudo-instantaneous strains vs applied pressure in a log-log plot, as in Fig. 6. Draw an average straight line across the points and note its slope $k =$

tan α and the coordinates of any point A, (p_A, v_A). For Test (4) it is found that $k = 0.82$ and $v_A = 0.017$ at $p_A = 10$ MN/m^2.

(4) Determine average slopes $\dot{v}$ of the creep curves and plot them in the same log-log plot, Fig. 6, against net pressures p. The slope of a line drawn across these points gives $n = \tan\beta$. For the present test: $n = 2.2$ and for point B: $\dot{v}_B = 5 \times 10^{-4}$h^{-1}, and $p_B = 30$ MN/m^2.

From these data, and for arbitrary values of ε_k and $\dot{\varepsilon}_c$, the values of σ_k and σ_c can be calculated from

$$m_k\sigma_k \;=\; (p_A\sqrt{3}/k)(\varepsilon_k\sqrt{3}/v_A)^{1/k} \tag{20}$$

$$m_c\sigma_c \;=\; (p_B\sqrt{3}/n)(\dot{\varepsilon}_c\sqrt{3}/\dot{v}_B)^{1/n} \tag{21}$$

For the test in question, after adopting $\varepsilon_k = 0.01$ and $\dot{\varepsilon}_c = 10^{-5}$h^{-1}, these expressions give: $m_k\sigma_k = 21.61$ MN/m^2 and $m_c\sigma_c = 5.122$ MN/m^2. Since in that test a/b = 1/5, one gets from Eqs. (17) and (18), $m_k = 0.9803$ and $m_c = 0.7685$, so that $\sigma_k = 22.04$ MN/m^2 and $\sigma_c = 6.665$ MN/m^2.

According to these data, the response of the rock salt to, say, a deviatoric stress increase in a triaxial compression test would be, according to Eqs. (7) to (9):

$$\varepsilon_1 \;=\; 0.01[(\sigma_1 - \sigma_3)/22.04]^{0.82} + 10^{-5}[(\sigma_1 - \sigma_3)/6.665]^{2.2}t$$

with stresses in MN/m^2 and time in hours.

On the other hand, if one wants to simulate the borehole expansion Test (4), the following is obtained from Eqs. (12) to (19)

$$\Delta V/V \;=\; 0.032(p/21.61)^{0.82} + 1.023 \times 10^{-5}\sum_1^N (p/5.122)^{2.2}\Delta t$$

The calculated strains show a good agreement with the values measured in the test.

4. CONCLUSIONS

The main purpose of this paper was to show how the borehole dilatometer test can be used in practice for determining in-situ short-term and creep properties of rock salt. The tests, carried out under simulated in situ conditions, show that the CSM Cell dilatometer can be used for that purpose under laboratory conditions, but an improved dilatometer with electronic strain measurement system will be required for performing such creep tests in situ.

The values of parameters found in the tests agree favorably with the published data, with the exception of the exponent $n = 2.2$, which is lower than expected. This is probable due to a rather short time of each creep stage, during which a true steady state was probably not yet attained. When performing in-situ dilatometer tests, intervals of up to 10 days should be used in each stage in order to get more reliable steady-state creep data.

ACKNOWLEDGEMENTS

This work was partially supported by the Team Grant CRP-297, of the Dept of Education of Quebec. The authors wish also to thank Mr. Roger Lavoie for his valuable assistance in the preparation and performance of all the tests.

REFERENCES

Albrecht et al. (1980). Zur Frage des Standsicherheitsnachweises von Hohlräumen in Salzgesteinen: Proc. 5th Int. Symp. on Salt, Hamburg, (1), 195-211.

Hansen, F.D. and Carter, N.L. (1980). Creep of rock salt at elevated temperatures: Proc. 21st U.S. Symp. on Rock Mech., Rolla, Mo., 217-226.

Hansen, F.D. and Carter, N.L. (1981). Creep of Avery Island rocksalt: Proc. 1st. Conf. on the Mech. Beh. of Salt, Penn State Univ. (in print).

Hansen, F.D., Mellegard, K.D. and Senseny, P.E. (1981). Elasticity and strength of ten natural rock salts: Proc. 1st Conf. on the Mech. Beh. of Salt, Penn. State Univ., (in print).

Horseman, S. and Passaris, E. (1981). Creep tests for storage cavity closure prediction: Proc. 1st Conf. on the Mech. Beh. of Salt, Penn State Univ., (in print).

Hult, J.A.H. (1966). Creep in engineering structures, 115 p. Blaisdell, Waltham, Mass.

Hustrulid, W. and Hustrulid, A. (1975). The CSM Cell – A borehole device for determining the modulus of rigidity of rock: Proc. 15th U.S. Symp. on Rock Mech., Rapid City, S.D., 181-225.

Hunsche, U. (1981). Results and interpretation of creep experiments on rock salt: Proc. 1st Conf. on the Mech. Beh. of Salt, Penn State Univ., (in print).

ISRM (1982). Suggested method for deformability determination using a borehole dilatometer test: 1st draft, 16 p.

Ladanyi, B. (1972 a). In-situ determination of undrained stress-strain behaviour of sensitive clays with the pressuremeter. Canad. Geotech. J. (9), 313-319.

Ladanyi, B. (1972 b). An engineering theory of creep of frozen soils. Canad. Geotech. J., (9), 63-80.

Ladanyi, B. (1980). Direct determination of ground pressure on tunnel lining in a non-linear viscoelastic rock: Proc. 13th Canad. Rock. Mech. Symp., Toronto, CIM Spec. Vol. 22, 126-132.

Ladanyi, B. and Gill, D.E. (1981). Determination of creep parameters of rock salt by means of a borehole dilatometer: Proc. 1st Conf. on the Mech. Beh. of Salt, Penn State Univ., (in print).

Ladanyi, B. and Johnston, G.H. (1973). Evaluation of in-situ creep properties of frozen soils with the pressuremeter. Proc. 2nd Int. Conf. on Permafrost, Yakutsk, North Amer. Contr. Vol., 313-318.

Ladanyi, B. and Johnston, G.H. (1978). Field investigations in frozen ground. Chap. 9 in "Geotech. Engrg for Cold Regions"(O.B. Andersland and D.M. Anderson, Eds), 459-504, McGraw-Hill, New York.

Munson, D.E. and Dawson, P.R. (1981). Salt constitutive modelling using mechanism maps: Proc. 1st Conf. on the Mech. Beh. of Salt, Penn State Univ., (in print).

Nelson, R.A. and Kocherhans, J.G., (1981). In-situ testing of salt in a deep borehole in Utah. Proc. 1st. Conf. on the Mech. Beh. of Salt, Penn State Univ., (in print).

Panek, L.A. et al, (1964). Determination of the modulus of rigidity of rock by expanding a cylindrical pressure cell in a drillhole: Proc. 6th U.S. Symp. on Rock Mech., Rolla, Mo., 427-449.

Passaris, E.K.S. (1979). The rheological behaviour of rocksalt as determined in an in situ pressurized cavity: Proc. 4th ISRM Cong., Montreux (1), 257-264.

Passaris, E.K.S. (1980). The intrinsic anisotropy of rock salt when subjected simultaneously to tensile and compressive stress fields: Proc. 5th Int. Symp. on Salt, Hamburg, (1), 477-485.

Patricio, J.G. and Beus, M.J. (1977). Determination of in-situ modulus of deformation in hard rock mines of the Coeur d'Alene District, Idaho: Proc. 17th U.S. Symp. on Rock Mech., Univ. of Utah, 4B9-1 to 7.

Rocha, M. et al. (1966). Determination of the deformability of rock masses along boreholes: Proc. 1st ISRM Cong., Lisbon, (1), 697-704.

COMPORTEMENT GEOMECANIQUE DES COULIS DE CIMENT INJECTES DANS DES ROCHES

Geomechanical behaviour of cement grouts injected in rock

Das geomechanische Verhalten von Zementinjektionen in Fels

Gérard Ballivy
Professeur agrégé, Laboratoire de mécanique des roches, départment de génie civil,
Université de Sherbrooke, Sherbrooke, Québec, Canada

André Martin
Attaché de recherche, Laboratoire de mécanique des roches, départment de génie civil,
Université de Sherbrooke, Sherbrooke, Québec, Canada

Pierre Niemants
Attaché de recherche, Laboratoire de mécanique des roches, départment de génie civil,
Université de Sherbrooke, Sherbrooke, Québec, Canada

SYNOPSIS

Two kinds of laboratory tests have been developed to determine in which effective stress conditions cement grouts injected in rock are useful. One test concerns the wash-out process of grouted samples using a modified permeameter; the other test is related to bonding strength. It is a pull out test on steel bars grouted in rock core samples.

RESUME

Dans le but d'évaluer le comportement des coulis injectés dans un massif rocheux, il est proposé ici deux types d'essais en laboratoire: des essais de lessivage du coulis de ciment injecté dans des échantillons de roche placés dans des perméamètres modifiés et des essais d'arrachement de tiges d'acier scellées avec du coulis conventionnel. Ces essais sont très utiles pour la conception des projets et ils sont un complément aux essais habituels en chantier.

ZUSAMMENFASSUNG

Zur Auswertung des Verhaltens des injektierten Zementes in einer Felsenmasse werden folgende Laborversuche vorgeschlagen: Versuche von Ausspülung des injektierten Zements in Steinproben in einem umgeformten Durchflüssigkeitsmesser und Ausreißversuche von Stahlstangen, die auf konventionelle Art festgehalten werden. Diese Versuche vervollkommnen die laufenden Versuche an der Baustelle.

INTRODUCTION

La conception de rideaux d'étanchéité injectés dans un massif rocheux ou d'ancrages scellés dans le roc repose sur des critères empiriques ou des modèles théoriques qui ne prennent pas forcément en compte les caractéristiques propres au type de roche et au mode de sollicitation. Les recherches expérimentales présentées ici ont été réalisées dans le but de développer des techniques d'essai en laboratoire qui pourraient aider le projeteur, par exemple, lors de la conception d'ouvrages mettant en oeuvre du coulis de ciment. Il est considéré ici l'étude du comportement des coulis d'étanchéité ainsi que l'étude des coulis de scellement.

1. ETUDE EN LABORATOIRE DU COMPORTEMENT DES COULIS D'ETANCHEITE

1.1 Description du programme expérimental

Les échantillons de roc (tableau 1) sont obtenus par carottage de la roche au diamant. Le diamètre, dans ce cas-ci, est du calibre NQ (49 mm), la longueur de 108 mm. Les échantillons sont perforés axialement puis fracturés, manuellement à l'aide d'un burin, dans la partie centrale et perpendiculairement à l'axe longitudinal. Ils sont injectés sous une pression d'environ 100 kPa grâce à un dispositif, (Niemants, 1981), qui permet de régler l'ouverture de la fissure. La maturation des éprouvettes est faite dans des conditions proches de celles du chantier soit ici à une température de 4°C et à un degré d'humidité

relative de 60% ou sinon avec une immersion complète dans l'eau à 4°C. La durée de mûrissement est comprise entre 24 et 48 heures.

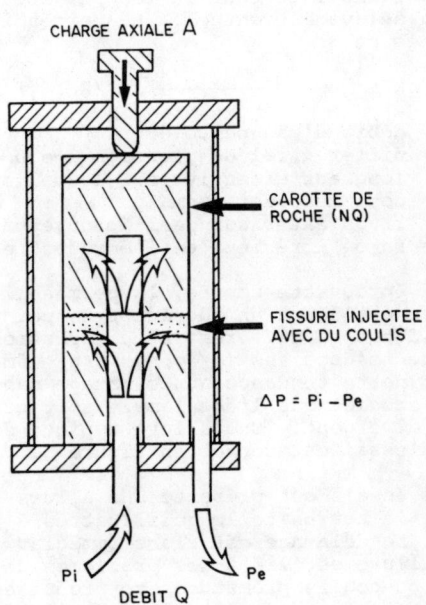

Fig. 1 Schéma du processus de lessivage

Le processus de lessivage schématisé en figure 1 est étudié en disposant l'échantillon injecté dans un perméamètre radial modifié (figure 2), celui-ci permet l'usage de pression d'eau de l'ordre de 2 100 kPa et une charge axiale de 6,7 kN peut être appliquée à l'éprouvette (soit une contrainte normale à la fissure σ_a de 3 700 kPa).

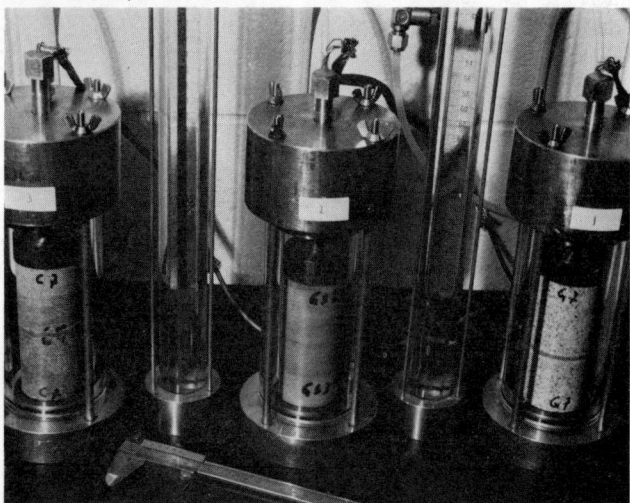

Fig. 2 Perméamètres modifiés
(dispositif du chargement axial)

1.2 Résultats types

Dans une première série d'essais, l'éprouvette injectée est soumise à des valeurs du différentiel de pression d'eau Δp croissantes (jusqu'à 2 100 kPa) pour une charge axiale A constante. L'augmentation du différentiel Δp se fait par paliers de 100 kPa environ pour une durée d'écoulement de 3 minutes à chacun des paliers. L'évolution de la perméabilité de l'éprouvette injectée est suivie tout au long de l'essai jusqu'au délavage éventuel. La perméabilité est donnée par:

$$k = \frac{Q}{2 \pi \cdot \Delta p \cdot L_{eff}} \ln (R_2/R_1)$$

- Q = débit d'eau percolé
- Δp = différentiel de pression d'eau
- L_{eff} = longueur effective sur laquelle l'écoulement est radial
- R_2 = rayon extérieur de l'éprouvette
- R_1 = rayon intérieur de l'éprouvette

Pour une éprouvette donnée, la perméabilité peut être évaluée à un facteur près par le rapport du débit au différentiel de pression $Q/\Delta p$. L'analyse des courbes $(Q/\Delta p)$ vs (Δp) démontre une très nette tendance au colmatage dans 88% des cas étudiés que l'écoulement soit convergent ou divergent. Une allure typique de la courbe d'essai est donnée en figure 3.

Quelques essais ont présenté une allure croissante de la courbe telle qu'illustrée à la figure 4. Le délavage est alors immédiat. Ce type d'allure est lié à des problèmes de décollement roc-coulis prématurés se produisant lors du montage ou lors de l'application du différentiel Δp quand la charge d'assise axiale A est trop faible (A inférieure à 180 N ou contrainte axiale σ_a inférieure à 100 kPa).

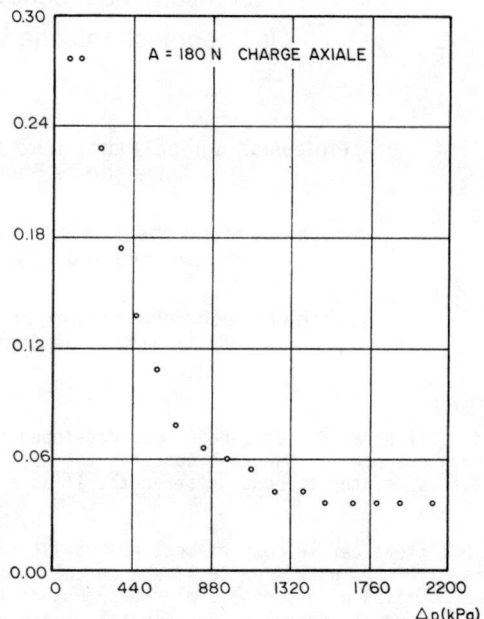

Q = débit
Δp = différentiel de pression

Fig. 3 Colmatage de l'échantillon

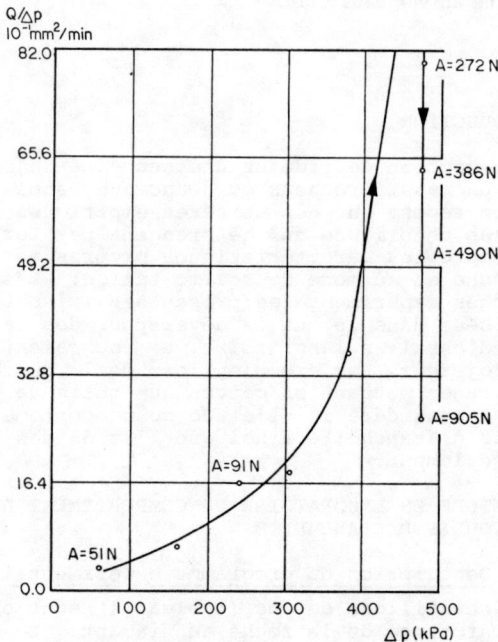

Fig. 4 Délavage de l'échantillon et augmentation de la charge axiale A

Par contre si après le délavage on augmente la charge axiale on observe une très forte réduction de la perméabilité (figure 4). Cet effet est encore plus marqué si on effectue des chargements répétés.

En définitive l'épaisseur de la fissure à injecter, son état de surface, l'état de contrainte variable, le différentiel de pression in-situ, les conditions d'humidité avant et après injection jouent un très grand rôle dans le choix du coulis à injecter dans le massif rocheux. Ces facteurs peuvent être simulés en laboratoire.

2. ETUDE DES COULIS DE SCELLEMENT

L'étude de Dupuis |1| a montré qu'il est possible d'étudier les ancrages injectés grâce à des modèles réduits. Les résultats montrent également que pour les roches dures la répartition des contraintes le long de la tige est du type exponentiel avant la démobilisation partielle (debonding).

D'autres études |3| |4| |5| ont permis de déterminer un modèle de comportement d'un ancrage injecté |8| (figure 5).

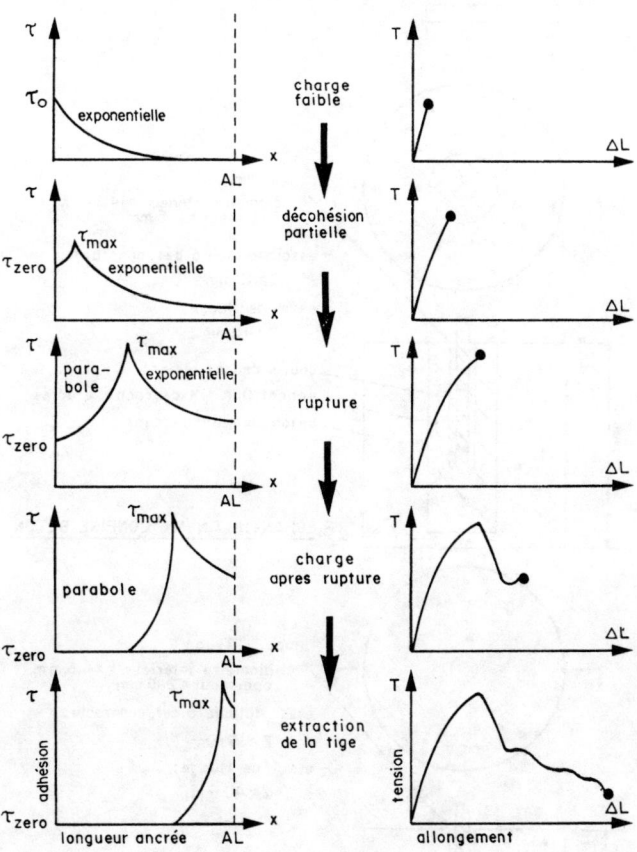

τ = contrainte de cisaillement au niveau de la tige et du coulis de scellement
x = abscisse le long de la partie scellée x=o → tête de l'ancrage
T = tension de la tige d'acier
ΔL = déplacement de la tête de l'ancrage

Fig. 5 Schéma de rupture d'un ancrage scellé dans un massif rocheux

Pour approfondir ces recherches un programme expérimental a été réalisé principalement en laboratoire sur modèles réduits.

L'objet des études présentées ici était de déterminer:
- l'influence des dimensions et du confinement du modèle réduit sur la représentativité des résultats,
- l'influence de la longueur de la tige d'ancrage sur la valeur moyenne de l'adhésion mobilisée,
- le seuil de fluage d'un ancrage injecté en laboratoire ainsi que l'influence de la vitesse de chargement sur la valeur moyenne de l'adhésion mobilisée.

2.1 Description du programme expérimental; résultats

Matériaux utilisés

Quatre matériaux principaux, ayant chacun un rôle différent à jouer dans le système d'ancrage, furent employés.

La roche dans laquelle l'ancrage est fixé. Les modèles réduits d'ancrages (figures 6A à 6F) ont été fabriqués à partir d'échantillons de roc facilement disponibles. Il s'agit de carottes de roc de diamètre NX (54 mm) pouvant provenir d'un forage au diamant et de blocs de roche provenant d'une carrière (Tableau 1).

Type de roc	Compression simple C_O (MPa)	Module de déformation E_R (GPa)
Granite de Winslow (St-Gérard, Québec)	151,6 ± 18,2	24
Calcaire de Trenton (St-Marc des Carrières, Québec)	95,6 ± 1,7	23
Grés de Potsdam (Mirabel, Québec)	177,5 ± 8,0	30

Tableau 1 Caractéristiques des matériaux rocheux utilisés

La tige d'acier chargée de prendre l'effort de tension. Il s'agit d'une tige crénelée telle qu'employée en béton armé:
- diamètre, d = 9,5 mm
- module de déformation, E_A = 200 GPa
- limite d'élasticité déterminée en laboratoire, f_y = 450 MPa.

Le coulis de scellement chargé de répartir l'effort de traction dans la roche grâce à sa capacité à résister au cisaillement. C'est un mélange d'eau et de ciment type 30 (grande résistance initiale). Le rapport en poids eau/ciment est de 0,4. La résistance du coulis en compression simple f'_c est déterminée à partir de petits cubes de 50 mm d'arête conformément aux normes C.S.A. A5-1961 et A.S.T.M. C170-50-1958. Les caractéristiques moyennes du coulis, pour tous les essais réalisés, avec un temps de mûrissement allant de deux à quatre semaines et pour différents types de mûrissement du coulis, sont:

A 229

compression simple, $f_c' = 55,2 \pm 5,3$ MPa
module de déformation, $E_c = 20,7$ GPa

Un béton qui a servi de milieu de scellement ainsi que de milieu de confinement pour les échantillons carottés. La résistance moyenne en compression du béton a été trouvée égale à $C_0 = 62,7$ MPa et son module $E_B = 20$ GPa.

Les techniques de fabrication des éprouvettes ont été décrites ailleurs (Ballivy et Dupuis, 1979). Les types d'éprouvettes utilisées pour cette étude-ci sont décrits à la figure 6.

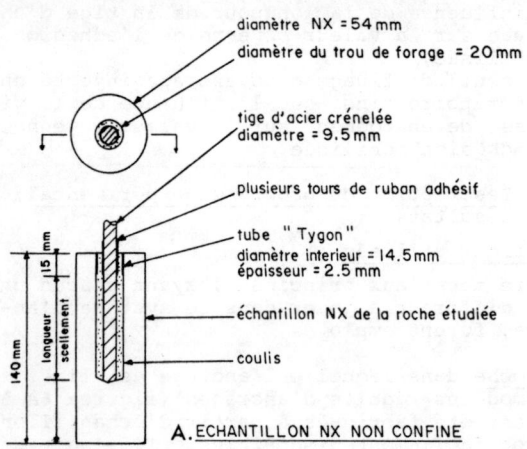

diamètre NX = 54 mm
diamètre du trou de forage = 20 mm
tige d'acier crénelée diamètre = 9.5 mm
plusieurs tours de ruban adhésif
tube "Tygon" diamètre interieur = 14.5 mm épaisseur = 2.5 mm
échantillon NX de la roche étudiée
coulis

A. ECHANTILLON NX NON CONFINE

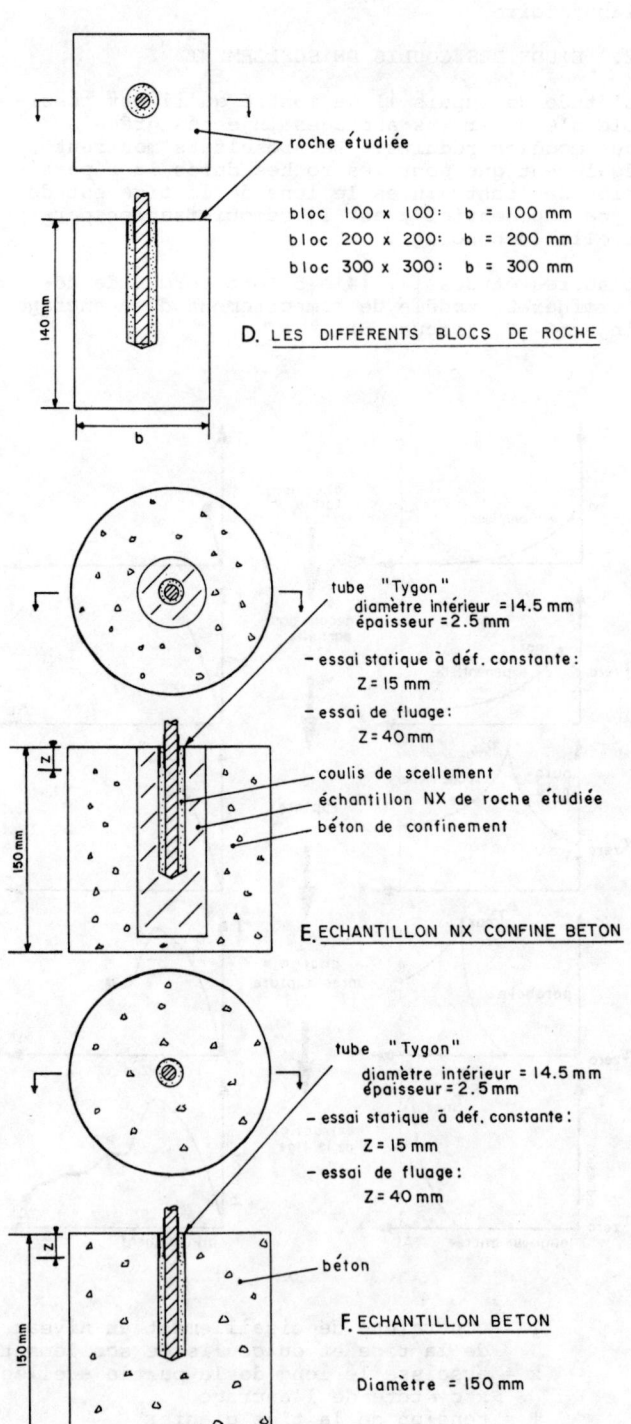

roche étudiée

bloc 100 x 100 : b = 100 mm
bloc 200 x 200 : b = 200 mm
bloc 300 x 300 : b = 300 mm

D. LES DIFFÉRENTS BLOCS DE ROCHE

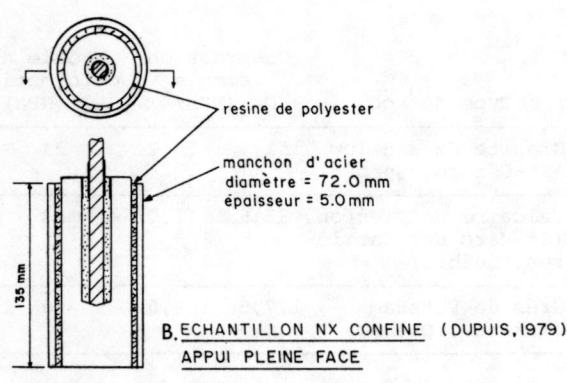

resine de polyester
manchon d'acier diamètre = 72.0 mm épaisseur = 5.0 mm

B. ECHANTILLON NX CONFINE (DUPUIS, 1979) APPUI PLEINE FACE

tube "Tygon" diamètre intérieur = 14.5 mm épaisseur = 2.5 mm
– essai statique à déf. constante : Z = 15 mm
– essai de fluage : Z = 40 mm
coulis de scellement
échantillon NX de roche étudiée
béton de confinement

E. ECHANTILLON NX CONFINE BETON

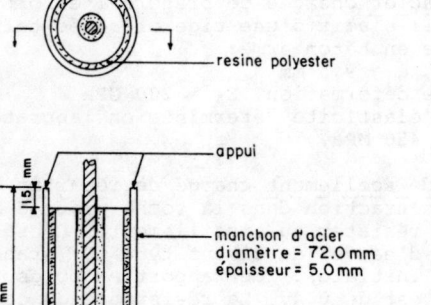

resine polyester
appui
manchon d'acier diamètre = 72.0 mm épaisseur = 5.0 mm

C. ECHANTILLON NX CONFINE APPUI LATERAL

tube "Tygon" diamètre intérieur = 14.5 mm épaisseur = 2.5 mm
– essai statique à déf. constante : Z = 15 mm
– essai de fluage : Z = 40 mm
béton

F. ECHANTILLON BETON

Diamètre = 150 mm

Fig. 6 Divers modes de scellement et de confinement

Essais d'arrachement statique à taux de déformation constant

La résistance de l'ancrage à un taux de déformation constant tel qu'employé au cours de ces essais correspond à la résistance de l'ancrage soumis à un effort relativement bref dans le temps sans être pour autant un effort dynamique: c'est la résistance à court terme.

Trois séries d'essais furent réalisées, les deux premières, avec divers échantillons de calcaire, ne furent différentes que par la nature du conditionnement des échantillons lors du mûrissement du coulis. Ces deux séries d'essais furent menées afin de déterminer et de confirmer les effets sur la résistance de l'ancrage du volume des échantillons, du confinement de la nature du mûrissement. La troisième série d'essais fut réalisée avec des cylindres de béton, plusieurs longueurs d'ancrage furent adoptées afin de déterminer l'influence de la longueur du scellement. On utilisa une presse conventionnelle et un système d'enregistrement automatique |1|, |8|. La rupture du scellement s'est toujours produite au niveau du contact tige-coulis (figures 7 à 9).

L'adhésion à la rupture, ou la contrainte moyenne de cisaillement à la rupture, est définie ainsi:

$$\tau_{moy-rupt} = \frac{\text{charge de rupture}}{\text{surface du contact tige-coulis}}$$

Comparaison avec les essais d'arrachement réalisés en chantier

Les essais en chantier constituent la meilleure référence pour l'appréciation de la représentativité des essais en laboratoire.

Une comparaison a pu être faite entre la valeur

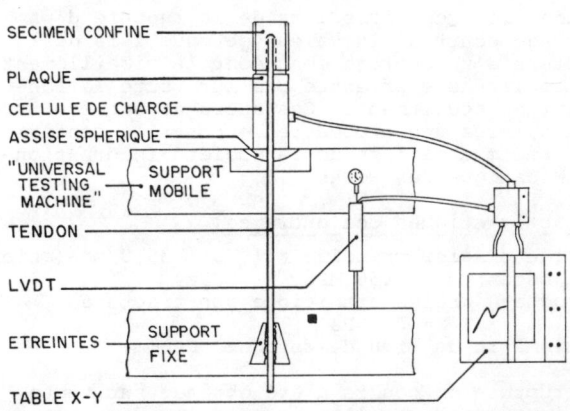

Fig. 7 Schéma du système d'arrachement

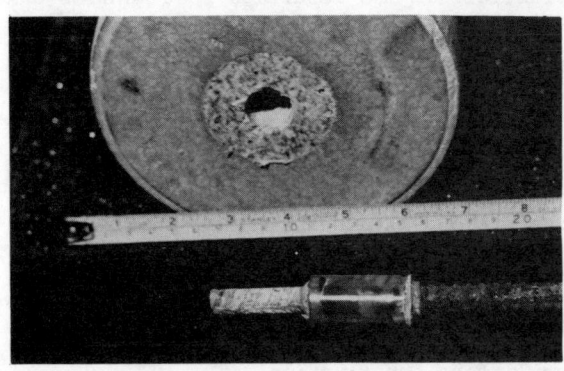

Fig. 8 Rupture au niveau tige-coulis

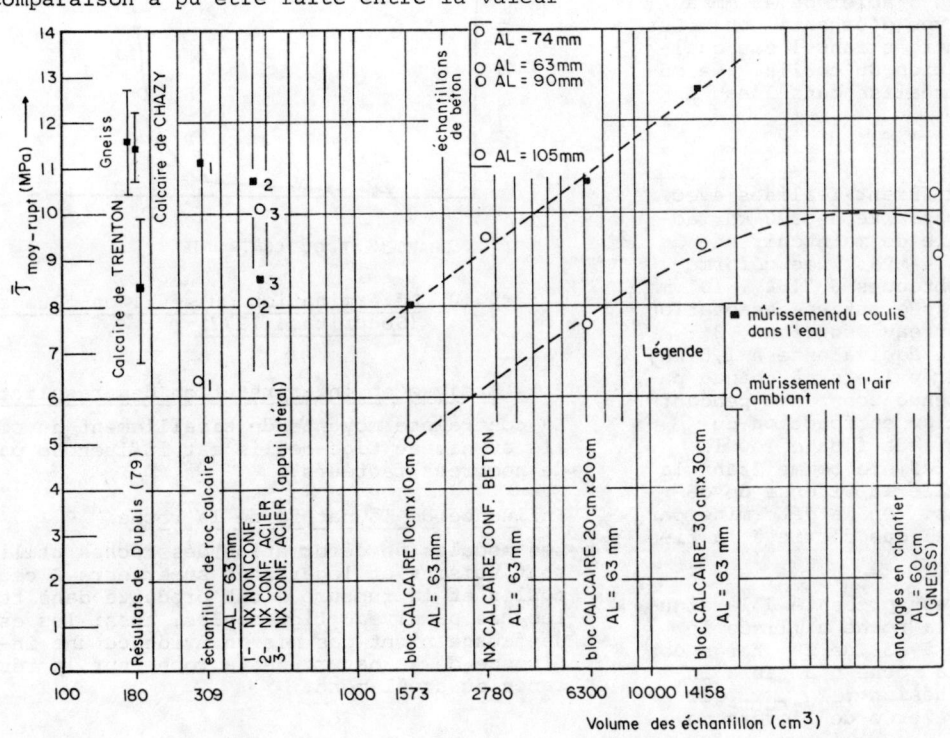

Fig. 9 Résultats des divers essais d'arrachement

d'adhésion mobilisée lors de la rupture d'un ancrage court et la valeur obtenue lors de la rupture d'un ancrage long dont le cisaillement du coulis ne s'effectue pas sur toute la longueur du scellement. Ces ancrages ont été testés d'après une procédure de contrôle conforme aux recommandations de la Société Internationale de Mécanique des Roches |6|.

Caractéristiques_des_ancrages
- tige d'acier crénelée #11, ϕ = 35,8 mm (acier à béton) f_y = 450 MPa, E_A, GPa
- roche: gneiss granitique non fracturé
  E_R = 20 GPa
- diamètre du trou de forage: 76 mm.

La rupture du coulis a été obtenue sur toute la longueur du scellement pour les deux tiges ancrées sur 0,60 m alors que les deux autres tiges d'acier ancrées à 4,50 m ont été mises en plasticité. Les résultats sont montrés aux figures 9 et 12.

Essais d'arrachement statique à charge constange: fluage
Le but principal de ces essais est de déterminer le seuil de fluage d'un ancrage.

Préparation_des_échantillons
L'échantillon "NX CONFINE BETON" (figure 6E) a été choisi pour les essais de fluage parce qu'il présente les caractéristiques suivantes:
- similitude des résistances observées en laboratoire et en chantier
- encombrement acceptable
- facilité de construction.

Des échantillons NX de grès, de granite et de calcaire ont été confinés dans du béton. Des échantillons de béton ont également été préparés, une longueur de tige d'acier de 43 mm a été ancrée dans chaque type d'échantillon. Les échantillons ont été immergés dans l'eau quelques jours avant l'injection du coulis. Le mûrissement du coulis fut réalisé dans l'eau à une température de 21°C.

Matériel_et_procédure
Les essais d'arrachement furent réalisés avec un vérin SIMPLEX d'une capacité de 270 kN, activé par une pompe capable de maintenir une pression constante de 31,4 MPa. Les déformations en palier furent obtenues à 2,54 . 10^{-3}mm près. La procédure utilisée est une adaptation des recommandations du bureau Securitas |3|: après une charge d'assise équivalente à 1,5 kN, on augmente la charge par paliers, à chaque palier la charge est maintenue constante pendant une heure et la déformation en fonction du temps est notée. Le seuil de fluage (ou la charge critique T_c) est définie comme étant la charge à partir de laquelle la vitesse de déformation ne décroît plus. On la détermine par construction graphique tel que décrit à la figure 10.

Un exemple de résultats est présenté à la figure 11. Quelle que soit la roche utilisée le seuil de fluage se situe à 85% de la charge de rupture. La nature de la roche n'a pas d'influence marquée sur l'adhésion $\tau_{moy-rupt}$ qui se situe à une valeur très élevée de l'ordre de 22 MPa.

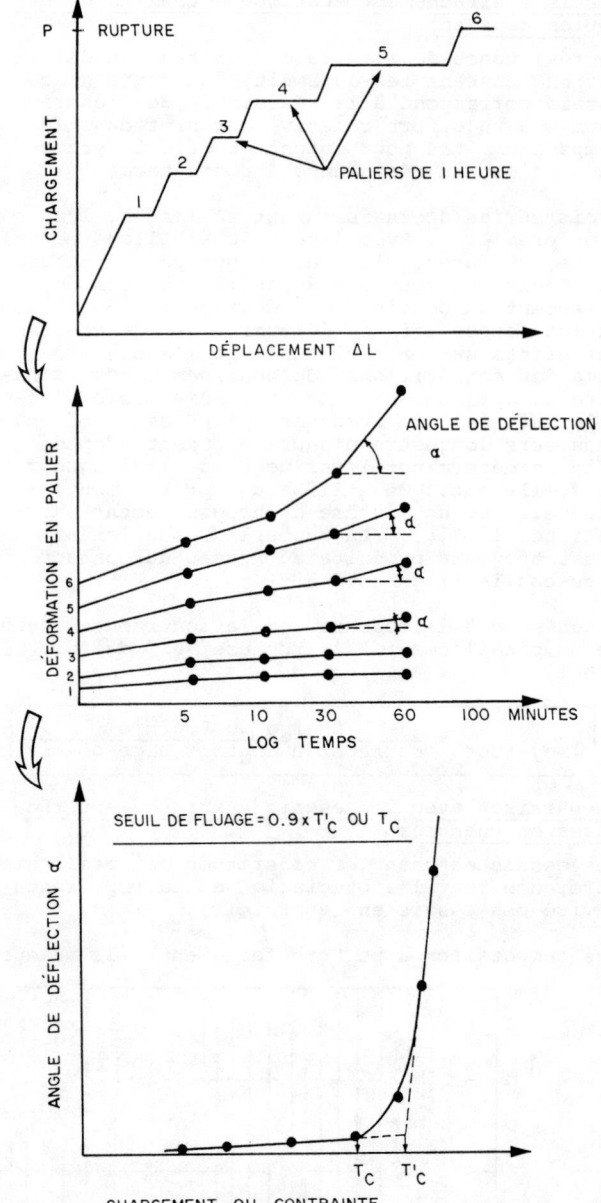

Fig. 10 Détermination du seuil de fluage (Securitas, 1977)

2.1 Analyse et interprétation des résultats

La contrainte moyenne de cisaillement du coulis au niveau tige-coulis est influencée par de nombreux facteurs.

Influence_de_la_nature_de_la_roche

Les modules de déformation des roches utilisées sont voisins et légèrement supérieurs à ceux du coulis et la rupture s'est produite dans tous les cas au niveau tige-coulis; aussi les essais de fluage n'ont pas mis en évidence une influence de la nature de la roche sur la résistance du scellement.

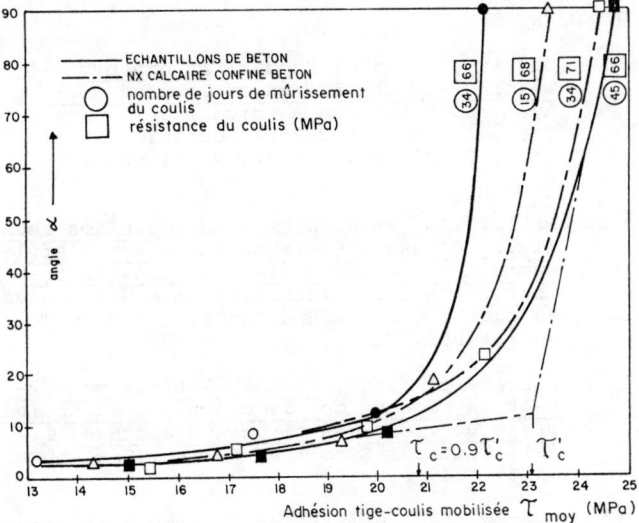

ECHANTILLONS DE BETON
— · — **NX CALCAIRE CONFINE BETON**
○ nombre de jours de mûrissement du coulis
☐ résistance du coulis (MPa)

angle ∝

Adhésion tige-coulis mobilisée τ_{moy} (MPa)

$\tau_c = 0.9\tau'_c$ τ'_c

Fig. 11   Résultats types des essais de fluage

Influence du type de mûrissement du coulis

Le type de mûrissement influence directement la résistance du coulis ainsi que son coefficient de retrait. Une mauvaise hydratation du coulis rend celui-ci moins résistant et le retrait est plus important. Ceci se traduit par une plus faible résistance au cisaillement du coulis pour les échantillons ayant eu un mûrissement à l'air libre (figure 9). D'autre part la résistance du coulis obtenue avec les petits cubes de coulis peut ne pas être représentative de la résistance du coulis injecté si les milieux de prise ainsi que de mûrissement sont trop différents.

Influence du volume des échantillons et du confinement

De nombreuses études théoriques et expérimentales |3| |4| |5| ont montré que lors d'un essai d'arrachement d'une tige scellée il existe des contraintes de traction dans la roche. La roche encaissante limite les déformations du scellement; la roche comprime, par réaction, le coulis: il se crée un serrage |12|. La réaction de la roche est d'autant plus importante que le volume de roche sollicité est grand. L'échantillon optimum est donc ici le bloc 0,30 x 0,30 m ou l'échantillon NX CONFINE beton (figures 6E et 9).

Le confinement par manchon d'acier est tributaire de la résine qui fixe le tubage sur l'échantillon NX. Cette résine peut présenter un retrait plus ou moins important.

Influence de la longueur d'ancrage

Les essais réalisés avec les échantillons de béton ont montré (figure 14) que l'adhésion tige-coulis augmente avec la longueur de l'ancrage jusqu'à la longueur de scellement correspondant à une contrainte de traction dans l'acier équivalente à sa limite élastique. Si on augmente la longueur de scellement, l'adhésion moyenne tige-coulis semble diminuer, ceci est dû à une démobilisation totale d'une certaine longueur de la partie supérieure de l'ancrage (figure 5). D'autre part la résistance

globale de l'ancrage ne semble plus augmenter de façon significative avec la longueur de scellement: c'est la limite élastique de l'acier qui détermine la résistance de l'ancrage.

Influence du taux de chargement; seuil de fluage

Les adhésions obtenues lors des essais de fluage sont nettement supérieures à celles obtenues avec les essais d'arrachement à taux de déformation constant (pratiquement le double); ceci va à l'encontre de ce qu'on a l'habitude d'observer: en général plus le taux de déformation est lent plus la charge de rupture est faible. D'autre part ce phénomène va à l'encontre des résultats trouvés sur l'influence de la longueur de scellement. Ce phénomène contradictoire est à rapprocher de celui observé par Perry et Jundi (1969) en faisant subir un chargement cyclique à des ancrages scellés dans du béton. En effet ils ont montré que la distribution des contraintes le long de l'ancrage n'est pas uniforme et qu'il se produit une redistribution des contraintes après quelques centaines de cycles; mais ils ont montré également que cette redistribution ne va pas toujours dans le sens de la sécurité.

Le seuil de fluage élevé des ancrages testés ici est comparable au seuil de fluage des matériaux utilisés.

Influence de la limite élastique de l'acier et de l'élancement de la partie ancrée

Les essais d'arrachement des tiges scellées dans les cylindres de béton ont montré qu'à partir d'une certaine longueur la résistance moyenne au cisaillement diminue (figure 9). La charge de rupture n'augmente plus et est pratiquement égale à la limite élastique de la tige d'acier. En fait c'est la limite élastique de l'acier qui détermine la rupture.

Ainsi la charge de rupture est:

$$P = f_y \frac{\pi \cdot D^2}{4}$$

f_y = limite élastique de l'acier

l'adhésion mobilisée est:

$$\tau_{moy} = \frac{P}{\pi \cdot D \cdot AL}$$

d'où

$$\tau_{moy} = \frac{f_y}{4} \cdot \frac{1}{\frac{AL}{D}}$$

d'où

$$\boxed{\frac{\tau_{moy}}{f_y} = 0,25 \cdot \frac{1}{AL/D}}$$ figure (12)

Donc ce qui pourrait être interprété comme un effet d'échelle (réduction de l'adhésion tige-coulis en fonction de la surface de contact) (figure 9) n'est que le résultat du mode de rupture (la rupture est induite par l'acier) et de la manière de calculer la valeur de l'adhésion mobilisée (la répartition de l'adhésion n'est pas uniforme le long de la tige) (figure 5).

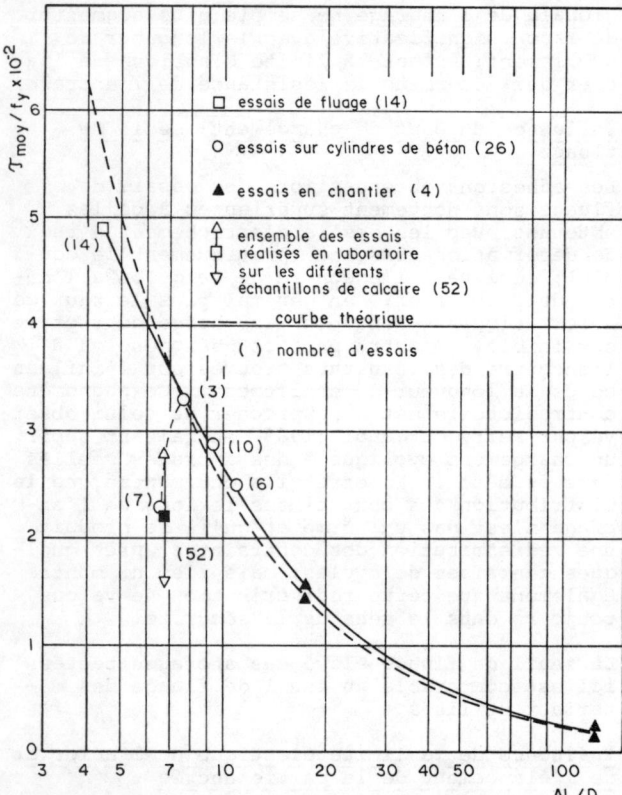

Fig. 12 Résistance moyenne au cisaillement en fonction du rapport AL/D

CONCLUSION

Les présentes études ont mis en évidence les paramètres à prendre en compte pour assurer la représentativité des essais de laboratoire. Les essais sur les coulis d'étanchéité ont permis de déterminer le mécanisme de dégradation d'un rideau étanche injecté dans un massif rocheux.

Le phénomène de la rupture d'un ancrage, constitué d'une tige d'armature crénelée scellée avec du coulis dans une roche dure, a pu être confirmé et documenté.

Les techniques expérimentales décrites précédemment n'ont pas la prétention de remplacer les essais in-situ mais elles permettent de compléter ceux-ci tout en donnant l'ordre de grandeur de la résistance du coulis utilisé en fonction des sollicitations que celui-ci subira in-situ.

REMERCIEMENT

Cette étude a été rendue possible grâce à un octroi du Conseil de Recherches en Sciences Naturelles et en Génie Canada (octroi n° A-42-28) et à la collaboration de Hydro-Québec et du Ministère des Transports du Québec.

REFERENCES

|1| Ballivy, G. et M. Dupuis, Laboratory Testing for the Design of Grouted Rock Anchors, Sixth Pan American Conference, Theme 1 - Soil and Rock Mechanics Problems in Mining, (1979), V. I, pp. 277-291.

|2| Ballivy, G. et P. Niemants, Nouvelles techniques d'essais en laboratoire sur l'efficacité des écrans d'éjection dans le rocher, Commission Internationale des Grands Barrages, Quatorzième Congrès, Rio de Janeiro 1982, V. II, pp. 771-790.

|3| Bureau Securitas, Recommandations concernant la conception, le calcul, l'exécution et le contrôle des tirants d'ancrage (T.A. 77), Editions Eyrolles, Paris, (Septembre 1977), 140 p.

|4| Coates, D.F. et Y.S. Yu. Three Dimensional Stress Distributions Around of Cylindrical Hole and Anchor, Proceedings of the 2nd International Conference on Rock Mechanics, Belgrade, (1970), pp. 175-182.

|5| Hawkes, J.M. et R.H. Evans, Bond Stresses in Reinforced Concrete Columns and Beams, The Journal of the Institution of Structural Engineers, V. XXXIX, #X, (January 1951), pp. 323-327.

|6| Hollingshead, G.W., Stress Distribution in Rock Anchors, Canadian Geotechnical Journal, V. 8, #4, (November 1971), pp.588-592

|7| International Society for Rock Mechanics, Suggested Method for Rock Bolt Testing, Commission on Standardisation of Laboratory and Field Tests, Document #2, (March 1974), 16 p.

|8| Littlejohn, G.S. et D.A. Bruce, Rock Anchors-State of the Art Part 1: Design, Part 2: Construction, Part 3: Stressing and Testing, Ground Engineering, V. 8, #3, 4, 5, 6 (1975), and V. 9, #2, 3, 4 (1976).

|9| Martin, A., Contribution à l'étude de la répartition des contraintes le long d'un ancrage scellé dans le rocher, mémoire de maîtrise de Sciences Appliquées, Génie Civil, Université de Sherbrooke, Qué. (1981), 181 p.

|10| Niemants, P., Etude du comportement des voiles d'étanchéité injectés dans les fondations rocheuses de barrage, mémoire de maîtrise de Sciences Appliquées, Génie Civil, Université de Sherbrooke, Qué (1981).

|11| Perry, E.S. and Jundi, N., Pullout Bond Stress Distribution Under Static and Dynamic Repeated Loading, Journal of the American Concrete Institute, (May 1969), pp. 377-380.

|12| Untrauer, R.E. and Henry, R.L., Influence of Normal Pressure on Bond Strength, Journal of the American Concrete Institute (May 1965), pp. 577-586.

INFLUENCE OF SHEAR VELOCITY ON ROCK JOINT STRENGTH

L'influence de la vitesse de cisaillement sur la résistance des joints d'une roche

Einfluss der Schergeschwindigkeit auf die Festigkeit von Felsfugen

J. H. Curran

Associate Professor of Civil Engineering, University of Toronto, Toronto, Ontario, Canada, M5S 1A4

P. K. Leong

Graduate Student in Civil Engineering, University of California, Berkeley, California, U.S.A.

SYNOPSIS

The influence of the rate of shear displacement on the frictional resistance of rock discontinuities was examined by testing jointed samples of three rock types in a dynamic direct shear machine. The influence of rock type, joint roughness, apparent area of contact, normal stress level and gouge infilling were investigated. The experiments consistently showed that the frictional resistance is dependent on the slip velocity. This rate-dependency was observed within a certain rate of shear velocities below and above which the frictional resistance was essentially independent of the rate of shear displacement. A model based on mechanical instabilities is proposed to explain the observed behaviour.

RESUME

L'influence de la vitesse du déplacement cisaillant sur la résistance à la friction des discontinuités d'une roche a été examinée en effectuant des tests dynamiques de cisaillement direct sur des échantillons de joints dans trois roches différentes. Les influences du type de roche, de la rugosité du joint, de l'aire de contact apparente, du niveau de contrainte normale et du matériel de remplissage ont été étudiées. Les expériences ont clairement montré que la résistance à la friction est fonction de la vitesse de glissement. Cette dépendance a été observée dans une gamme de vitesses de cisaillement, en deça et au delà de laquelle la résistance à la friction est essentiellement indépendante de la vitesse du déplacement. Un modèle basé sur des instabilités mécaniques est proposé pour expliquer le comportement observé.

ZUSAMMENFASSUNG

Der Einfluß der Schubverformungsgeschwindigkeit auf den Reibungswiderstand von Fels-Diskontinuitäten wurde durch Prüfen von Proben dreier Felsarten in einer dynamischen Schubmaschine untersucht. Der Einfluß der Felsart, der Rauhigkeit der Fuge, der Größe der nominalen Kontaktfläche, der Größe der Normalspannung, und des abgeriebenen Materials wurde untersucht. Diese Experimente zeigten in allen Fällen, daß der Reibungswiderstand von der Gleitgeschwindigkeit abhängt. Diese Geschwindigkeitsabhängigkeit wurde innerhalb eines Bereiches von Verschiebungsgeschwindigkeiten beobachtet, über und unter welchem der Reibungswiderstand im wesentlichen unabhängig von der Schubverformung ist. Ein Modell, das auf mechanischer Instabilität aufgebaut ist, wird vorgeschlagen, um das beobachtete Verhalten zu erklären.

1. INTRODUCTION

The performance of engineering structures constructed in rock can be highly influenced by the rate-dependent behaviour of rock and its discontinuities. The stability of these rock structures is of great importance when subjected to dynamic loading. Typical deformation rates vary from very slow creep response to high velocities and accelerations associated with earthquakes and blasting.

Experimental evidence obtained to date shows that the shear resistance of rock discontinuities is dependent on the slip velocity and that many complex factors such as mineralogy, roughness, water content, and normal stress influence this behaviour. Variation in results, the need to establish basic behavioural trends and to formulate constitutive models justifies further investigations.

The influence of the rate of shear displacement on the frictional resistance of rock is investigated in this paper. A series of dynamic direct shear tests were performed in order to examine the rate-effect and the influence of certain factors such as rock type, joint roughness, apparent area of contact of the shear surfaces and gouge infilling on this phenomena. These tests were limited to three rock types with an initial sawn-cut surface of uniform roughness so that the important effects of basic friction and surface roughness could be isolated. A total of twenty-one pairs of blocks were sheared. The displacement rates that were used in the shear tests were two or three orders of magnitude faster than those used by previous investigators.

2. HYPOTHESIS

It is postulated that four major physical processes take place during sliding, namely:

a) Plucking
   This process includes surface damage as a result of the tearing out of grains of minerals which make up the rock mass

b) Shearing of Asperities
   This is the process whereby protuberances are fractured and "cut off" due to high stresses.

c) Ploughing
   When two surfaces of different hardness slide relative to each other, the harder surface will cut grooves in the softer material.

When sliding occurs, a combination of the above processes occur simultaneously though a certain mode may predominate under given circumstances.

In the proposed model it is suggested that when sliding initially takes place on a fresh surface of uniform roughness, surface damage is induced by plucking. At low rates of shear displacement, the shearing of asperities or ploughing dominates depending on the relative hardness of the surfaces in contact. Overriding accompanied by minor shearing of the asperities takes over at higher rates of shear displacement. Justifications of the above model are discussed below.

The typical stress-strain behaviour for an intact rock specimen loaded at different rates is shown in Fig. 1. Generally the apparent stiffness of the material is proportional to the rate of loading. Also, it is found that the failure load increases with a higher rate of loading and vice-versa.

At high shear rates, the asperities come into contact on impact. Unable to fracture the asperity, overriding occurs based on the physical model described previously. The situation when the asperities come into contact on impact is equivalent to a high rate of loading. However it is expected that all materials have a finite stiffness and strength even though the rate of loading may increase indefinitely. Thus, there will come a stage where the asperities are simply fractured as the shear rate is increased beyond a certain shear velocity. Considering an ideal case, a perfectly flat surface will result at higher shear rates. Hence, regardless of the rate at which the shearing of two perfectly flat surfaces takes place, the resulting frictional resistance will be constant.

The shear velocity beyond which the frictional resistance remains constant will be known as the terminal velocity, V_t.

Based on the above rate mechanism for relatively hard rock discontinuities, frictional resistance is independent of the shear rate up to the breakpoint velocity and beyond the terminal velocity. Between these critical velocities, the frictional resistance is rate-dependent.

3. DYNAMIC DIRECT SHEAR MACHINE

The direct shear machine that was used has a design capacity of 1000 kN and 500 kN in the vertical and horizontal directions respectively and is capable of testing artificial as well as natural discontinuities. Two servo-controlled actuators, each with a force capacity of 250 kN and a maximum stroke of 250 mm, react against a braced portal frame to provide the loading in the normal and shear directions.

The horizontal and vertical displacements were measured with LVDT's while the normal and shear forces were measured with load cells located within the actuators. A digital data acquisition system was used to record detailed load-displacement-time histories over the complete testing cycle. Four data sources were scanned at a rate of 23 readings per channel per second.

The reader is referred to Crawford and Curran (1981a) for a full description of the design, construction, testing and instrumention of the direct shear machine.

4. SAMPLE PREPARATION

A test specimen consists of two blocks. The size of the upper surface was either 100*100 mm or 200*200mm and the lower surface was 310*210 mm.

The 100*100 mm surface was obtained by cutting a thin 5 mm thick slice off the original 200*200 mm surface of the upper block such that a square section was left projecting from the the the middle portion of the specimen with its edges sloping at an angle of 45 degrees. The width at the upper surface was about 5 mm less than the lower surface (for the 200*200 mm samples) with its edges rounded off to minimize spalling during sliding. The surfaces were then lapped to a uniform flatness and roughness using silica

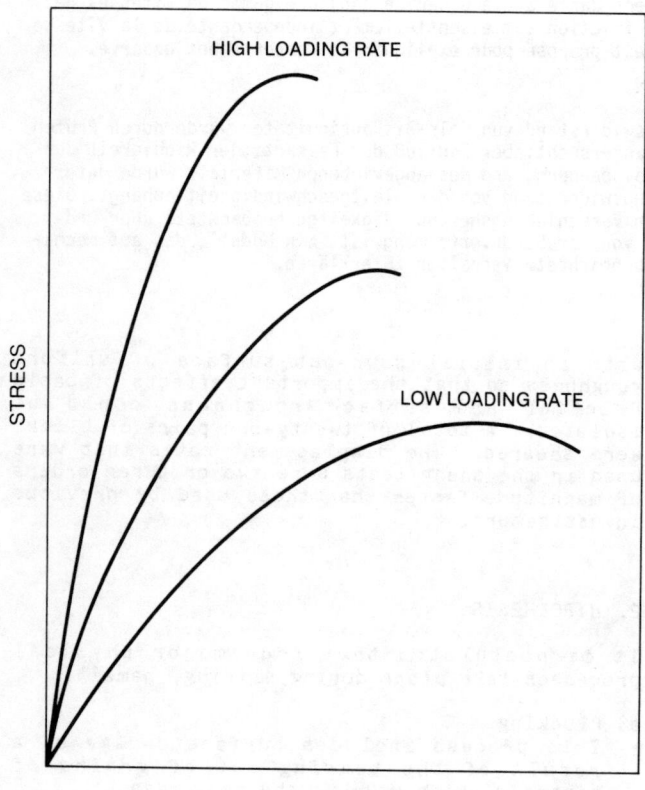

Fig. 1   Typical stress-strain behaviour of rock at different loading rates

carbide grit. After air drying, the test specimens were potted in aluminum boxes with a high strength rapid setting compound consisting of high early strength cement, fine aggregate and a superplastizer.

5. TEST PROCEDURE

As the purpose of this study was to determine the rate-dependent behaviour of rock joints and to ascertain if roughness, rock type, apparent area of contact and gouge infilling have an influence on the rate-effect, the test procedures were designed such that other factors which may influence the dynamic resistance of the rock joints were minimized.

Each pair of surfaces were tested at different shear rates and at different normal load stages as opposed to testing a fresh surface for each normal load and shear rate. The former was preferred as the uniformity of the surface was considered to be far more important than the possible effect of gouge generation. Moreover, the interpretation of the data is further complicated by the introduction of a series of surfaces with varing physical properties. To investigate if roughness has an influence on the rate effect, the initial normal load for each pair of surfaces was different. By conducting the tests in this manner, the surfaces were sheared at different normal loads on a freshly prepared surface of uniform roughness and on surfaces of arbitrary roughnesses due to surface damage created during sliding.

Tests similar to those conducted on the 200*200 mm samples of each rock type were carried out for the 100*100 mm samples of the same rock type (using the same set of normal loads and shear rates) to determine if the apparent area of contact has any effect on the relationship between frictional resistance and shear rate.

The dynamic shear tests on each rock type were divided into two groups; each group consisted of one 200*200 mm and two 100*100 mm samples. Both groups were tested in an identical manner except that one group had the powdery gouge removed from the shear surfaces after every run with a specially designed vacuum cleaner tool so that there was a basis of comparison. Hence it could be determined if the presence of gouge material had an influence on the rate effect.

Except for the different sets of normal loads at which each pair of specimens was sheared, the tests were carried out in a similar manner. Each load stage commenced with a low shear displacement rate which was incremented logarithmically to the most rapid shear rate. This cycle was repeated for the remaining normal loads. The shear rates varied from 0.5 mm/sec to 256 mm/sec. All tests were conducted at room temperature.

6. EXPERIMENTAL RESULTS AND OBSERVATIONS

The results of the black quartz syenite samples and both the granite samples obtained from the dynamic direct shear tests will be presented separately with the aid of summary plots of

shear force versus shear velocity. These plots are obtained by taking the mean of the push-pull frictional resistances.

6.1 Black Quartz Syenite

For all practical purposes i.e. engineering applications, the frictional resistance was independent of the slip velocity for shear displacement rates below 4 mm/sec (breakpoint velocity). For shear velocities beyond 4 mm/sec, there was a decrease in the shear strength. The rate of decrease of frictional resistance was found to increase with increasing normal load though the maximum percentage drop in frictional resistance relative to that corresponding to the breakpoint velocity at a given normal load decreased slightly with higher normal loads. The overall percentage drop in frictional resistance was approximately 30%.

No terminal velocity was obtained for the syenite samples without the gouge removed as the tests were not conducted beyond a shear velocity of 128 mm/sec and at this rate the shear force was still decreasing. This constraint was due to the limited capacity of a single servo-valve on the horizontal actuator during this sequence of tests.

The results were identical for the 100*100 mm and 200*200 mm samples.

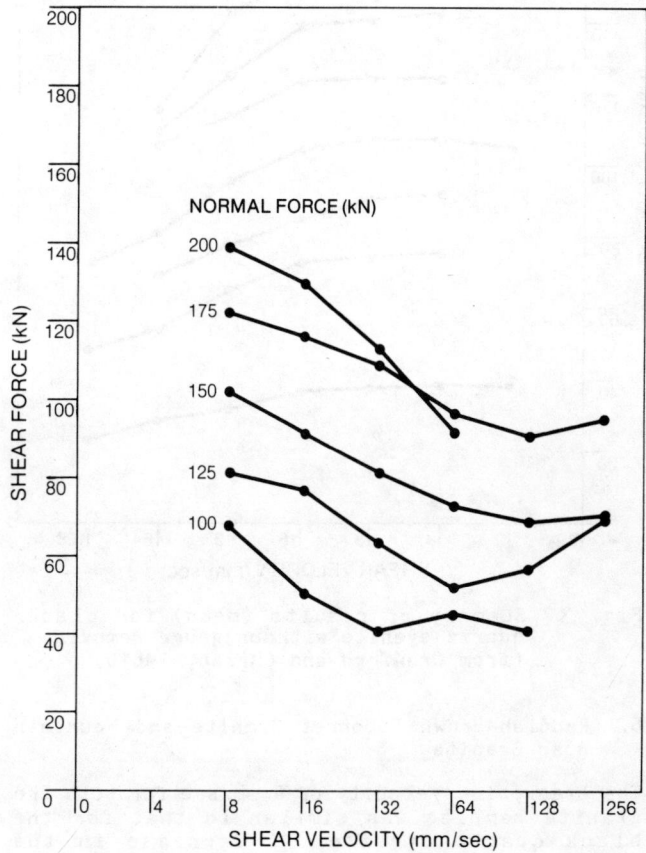

Fig. 2   Summary of results (mean) for black quartz syenite with gouge removed

On completion of the tests fine light grey powdery gouge covered both the shear surfaces together with sheared asperities, striated grooves and cavities. The presence of the latter gives rise to the possibility of the tearing out of grains of minerals besides the shearing of asperities and ploughing during sliding. Spalling at the four edges of the upper surface and the two trailing edges of the lower surface occured even though the edges were rounded off.

Similar results were obtained for the samples with the gouge material removed after every run. Due to installation of an extra servo-valve on the horizontal actuator when this set of tests were conducted, a terminal velocity greater than 128 mm/sec was obtained. Frictional resistance beyond this shear velocity was constant.

A summary of the results for the syenite samples with and without gouge removal is illustrated in Figs. 2 and 3 respectively.

resistance beyond a shear rate of 64 mm/sec while the samples without gouge removal showed a consistent increase in frictional resistance for velocities higher than 64 mm/sec. The rate of decrease in shear resistance per log cycle increase in velocity increased with higher normal loads. The overall mean drop in frictional resistance for both granite samples was 20%.

The overall results for both the granite samples with gouge removal are shown in Figs. 4 and 5.

The sheared surfaces were marked with striated grooves, slickenslides and cavities though these features were not as pronounced as they were for the syenite samples. Fine white to light grey gouge powder covered the damaged surfaces.

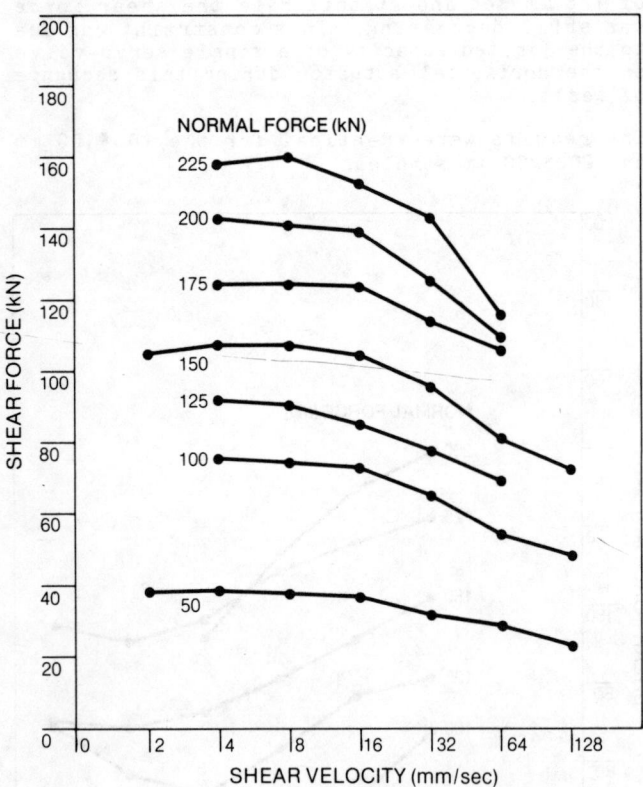

Fig. 3   Summary of results (mean) for black quartz syenite without gouge removed (from Crawford and Curran, 1981b)

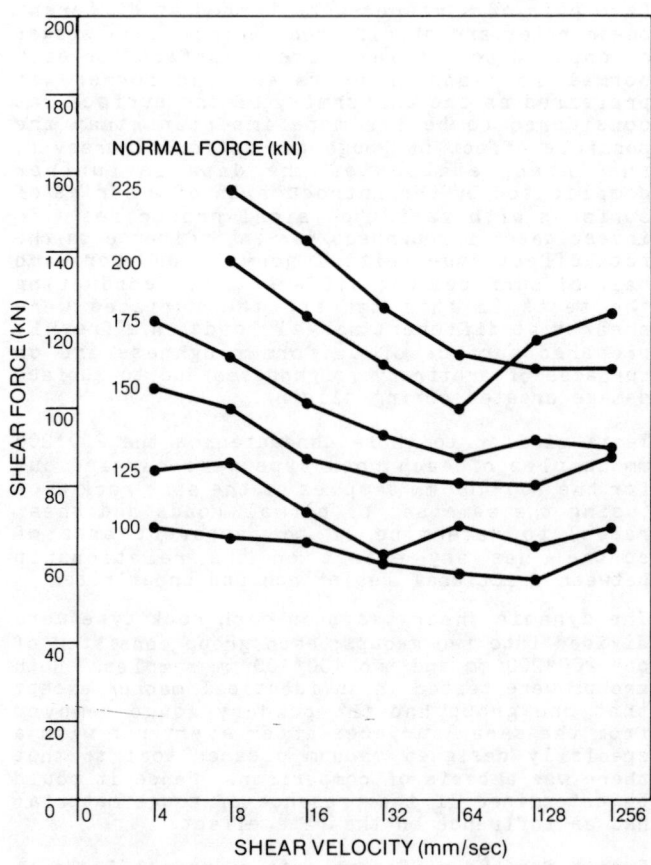

Fig. 4   Summary of results (mean) for reddish-brown dubonnet granite with gouge removed

6.2 Reddish-Brown Dubonnet Granite and Mountain Rose Granite

The breakpoint velocity of 4 mm/sec for both the granite samples was similar to that for the black quartz syenite. A decrease in the frictional resistance was observed for shear rates greater than the breakpoint velocity up to a shear rate of 64 mm/sec. Samples with the powdery gouge removed after every run gave rise to a constant or slightly increasing frictional

7. DISCUSSION AND CONCLUSIONS

These experiments have shown that the frictional resistance to the sliding of rock surfaces of black quartz syenite, mountain rose granite and reddish-brown dubonnet granite is rate-dependent for slip velocities between the breakpoint and the terminal velocity. This phenomena was also observed by Dieterich (1978) for double shear tests on westerly granite. The results obtained

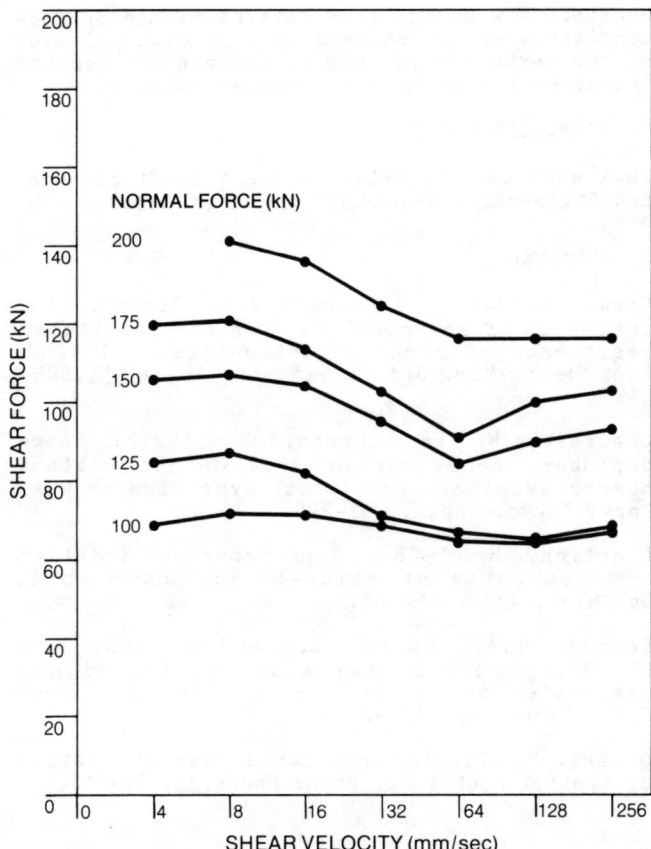

Fig. 5   Summary of results (mean) for mountain
          rose granite with gouge removed

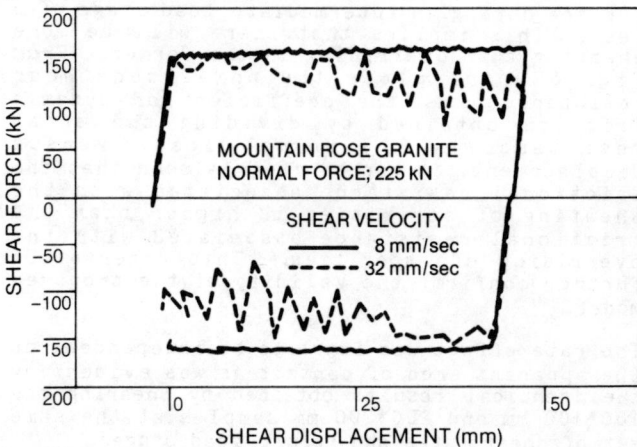

Fig. 6   Typical experimental results

are also consistent with those obtained by Donath et al (1973) while conducting triaxial compression tests on oolitic and lithographic limestone, sandstone and slate. The decrease of frictional resistance was found to be proportional to the logarithmic increase in the rate of shear displacement. A similar, but nevertheless appreciable change in the friction coefficient with loading rate was observed by Ohnaka (1975). This inverse relationship exists whether the mean plus or minus one standard deviation are considered.

The increase in frictional resistance beyond the terminal velocity of 64 mm/sec for the granite samples without gouge removal is attributed to the shearing/ploughing of the compacted powdery gouge within the striated grooves and depressions of the damaged shear surfaces. No shearing or fracturing of the asperities is expected at this stage as the shear surface is considered to be "flat" (under ideal conditions) due the finite stiffness and strength of the material having been reached as explained in section 2. Incomplete removal of gouge, thus leading to the shearing/ ploughing of the compacted material is suggested as a possible explanation for the slight increase in frictional resistance for some of the samples that were tested with gouge removal.

The magnitude of oscillations of frictional resistances was greater the higher the shear

rate (Fig. 6) and can be explained by the proposed model.

At high shear rates, the asperities come into contact on impact so that the asperities have a higher apparent strength than at a lower slip velocity. Unable to fracture the asperity, there will be an increase in force which is required to lift the upper surface as it rides on the asperity. As riding occurs, the contact area of the two asperities decreases, thereby leading to an increased state of stress. A stage will be reached (near the tips of the asperities) where the stress state built up is sufficient to fracture the asperity. The bottom surface will accelerate once this happens. But since the actuator is displacement-controlled, there will be an instantaneous drop in the force exerted by the actuator in order to maintain a constant shear rate.

At lower shear rates, ploughing and/or shearing of asperities occur. There is relatively very little overriding of asperities thereby giving a relatively smaller oscillation of shear resistance. Just as there is an increase in apparent strength with increasing loading rate, a lower shear rate is equivalent to a lower rate of loading associated with the asperities coming into contact. Therefore, a smaller force is required to fracture the material due to a relatively lower apparent strength. It is noted that the above discussion also represents a description of a possible mechanism for stick-slip at high and low shear rates.

It can also be shown that the force required to shear the asperities is higher than the force for overriding to take place.

As the surfaces are being sheared, surface damage occurs which results in a rougher surface mainly due to plucking. As shearing proceeds, it comes to a point where any further shearing of the surfaces will decrease the roughness due to gouge material being compacted in the cavities and depressions of the damaged surfaces. From observations, this transition usually takes place during or after the second loading stage. Hence, the surface at the final load stage will be relatively smoother than a surface during an intermediate load stage of a test. This implies that there will be more shearing than overriding in the former. From Fig. 6 which plots the nomalized shear resistance, i.e. the coefficient of dynamic friction obtained by dividing the shear resistance by the normal load, versus displacement, it can be clearly seen that the frictional resistance associated with the shearing of asperities is higher than the frictional resistance associated with the overriding of asperities. This observation further confirms the validity of the proposed model.

The rate effect was found to be independent of the apparent area of contact as was evident by the identical results obtained by shearing the 100*100 mm and 200*200 mm samples at the same set of shear rates ahd normal load stages.

There was a pronounced reduction of frictional force with an increasing rate of shear displacement. When the shear rate was decreased from the most rapid shear rate to the breakpoint velocity, there was an immediate increase in shear force, indicating that the change in frictional resistance is independent of the surface damage, roughness and the degree of wear, abrasion and gouging of the shear surfaces.

The proposed rate mechanism incorporating mechanical instabilities appears to be consistent with the observations of the surface conditions of the sheared samples on completion of the tests and is able to explain the general trends exhibited by the experimental data.

8. ACKNOWLEDGEMENT

This work was supported by the Natural Sciences and Engineering Research Council of Canada.

9. REFERENCES

Crawford, A.M. and Curran, J.H. (1981a). The influence of shear velocity on the frictional resistance of rock discontinuities: Int. J. Rock Mech. Min. Sci. & Geomech. Abstr., 18, 505-515

Crawford, A.M. and Curran, J.H. (1981b). Rate-dependent behaviour of rock joints - black quartz syenite: Proc. Int. Symposium on Weak Rock, Tokyo, Japan, 291-296.

Diertich, J.H. (1978). Time-dependent friction and mechanics of stick-slip: Pure Appl. Geophys., 116, 115-163.

Donath, F.A., Furth, L.S and Olsson, W.A. (1973). Experimental study of frictional properties of faults: Proc. 14th U.S. Rock Mech. Symposium, 189-222.

Ohnaka, M. (1975). Frictional characteristics of typical rocks: J. Phys. Earth, 23, 87-112.

CRITICAL EVALUATION OF ROCK BEHAVIOR FOR IN-SITU STRESS DETERMINATION USING OVERCORING METHODS

L'évaluation critique du comportement de roc pour la détermination des contraintes in-situ avec les méthodes surcarottées

Eine kritische Auswertung des Verhaltens von Fels bei in-situ Spannungsmessungen mit Hilfe der Überbohrverfahren

Lou P. Gonano
Golder Associates, Inc., Seattle, Washington, U.S.A.

John C. Sharp
Rock Engineering Consultant, Jersey, C.I.

SYNOPSIS

Several significant developments in the interpretation and analysis of in situ stress measurements performed in weak rock using overcoring techniques are described. These developments include the critical appraisal of overcoring strain data, the determination of stress-strain transformation parameters for cross-anisotropic nonlinear rock, and the formulation of a numerical solution for cross-anisotropic rock. By suitable comparisons, it is shown that without these refinements meaningful results to the regional stress field could not have been obtained.

RESUME

On décrit des progrès significatifs dans l'interprétation et l'analyse des mesures de contraintes in situ exécutées dans de la roche faible par les techniques de surcarottage. Ces progrès comprennent l'évaluation critique des données des contraintes de surcarottage, la détermination des paramètres de transformation contrainte-déformation pour de la roche transversale-anisotrope non-linéaire, et la formulation d'une solution numérique pour de la roche transversale aniso-trope. En se basant sur des comparaisons convenables, on montre que sans ces perfectionnements de la technique on n'au-rait pas pu obtenir des résultats valables pour le champ de contrainte régional.

ZUSAMMENFASSUNG

Viele wichtige Entwicklungen in der Darstellung und Analyse von in-situ Spannungsmessungen in schwachem Felsen mit Über-kernungstechnik werden beschrieben. Diese Technik wurde in Verbindung mit Spannungsmeßprogrammen auf drei besonderen Lagen in dem Projektbereich entwickelt. Diese Entwicklungen schließen die kritische Abschätzung von Überkernungsausdeh-nungsdaten, die Bestimmung von Spannungs-Ausdehnungsumwandlungskriterien für kreuz-anisotropische, nichtlineare Felsen, und die Formulierung einer numerischen Lösung für kreuz-anisotropischen Felsen ein. Durch entsprechende Vergleichungen wird bewiesen, daß bedeutungsvolle Ergebnisse für das regionale Spannungsfeld ohne diese Verfeinerungen nicht erlangt werden können.

1.0 INTRODUCTION

Three separate programs of stress measurements were carried out at the site of the Drakensberg Pumped Storage Scheme in South Africa to determine the representative stress field around the major underground caverns and the associated waterways. The design requirements for the large span caverns in weak rock and the concrete-lined pressure tunnels required an accurate assessment of stress conditions. The scheme is located in weak, poten-tially erodable, horizontally bedded sandstones, silt-stones, and mudstones (Figure 1). These difficult geolo-gical conditions also compounded the problem of achieving the required measurement standards.

Three types of refinements in the interpretation and analysis of in situ stress measurements were made:

(1) critical appraisal of the overcoring strain data
(2) determination of the stress-strain transformation parameters
(3) realistic development of a numerical solution to the stress field for cross-anisotropic rock.

The techniques used to improve and interpret the results from in situ stress measurements in weak, non-linear cross-anisotropic rock, e.g., shales, sandstones, schists, are described in this paper.

2.0 BACKGROUND AND APPROACH

2.1 Preamble

Various methods are available for determining the absolute in situ stress field. Because of the weak, bed-ded nature of rock at Drakensberg, borehole overcore methods were considered the most appropriate. Limited additional testing using the hydrofracture techniques was also carried out. Although borehole strain cell devices are generally the most accurate instruments available for measuring in situ stresses, confidence intervals of the order of ±20 percent are generally the limit of accuracy obtainable, even with rock masses that can be described as linear elastic. During the early stages of the inves-tigations, the need for innovations and refinements in the test technique to improve the quality of results for the nonlinear anisotropic rock conditions at Drakensberg was recognized.

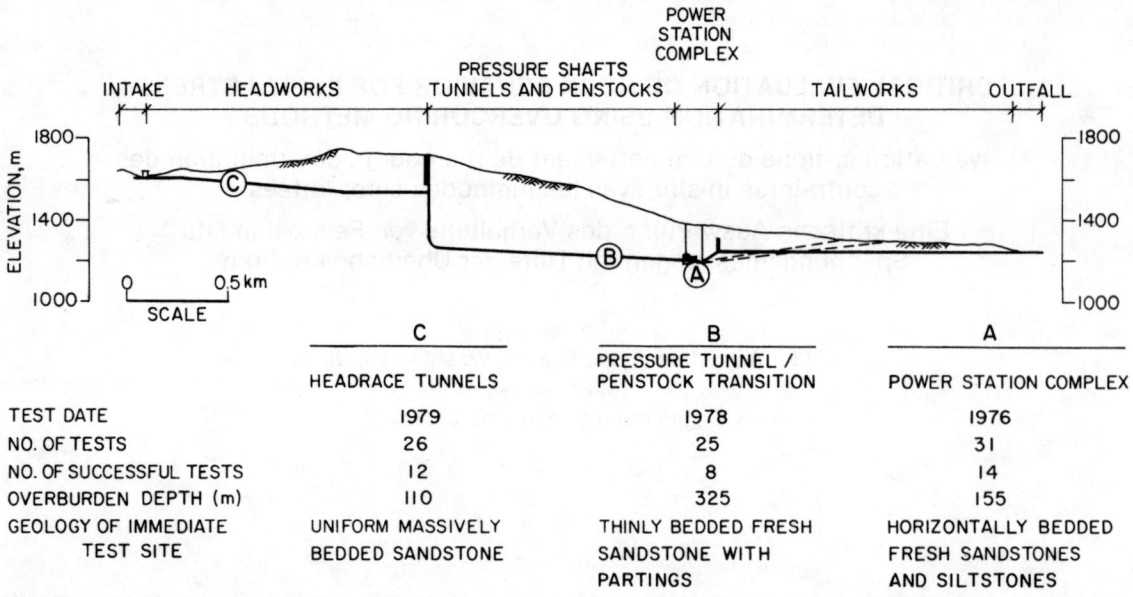

	C	B	A
	HEADRACE TUNNELS	PRESSURE TUNNEL / PENSTOCK TRANSITION	POWER STATION COMPLEX
TEST DATE	1979	1978	1976
NO. OF TESTS	26	25	31
NO. OF SUCCESSFUL TESTS	12	8	14
OVERBURDEN DEPTH (m)	110	325	155
GEOLOGY OF IMMEDIATE TEST SITE	UNIFORM MASSIVELY BEDDED SANDSTONE	THINLY BEDDED FRESH SANDSTONE WITH PARTINGS	HORIZONTALLY BEDDED FRESH SANDSTONES AND SILTSTONES

Figure 1 - Longitudinal Section of Scheme Showing Location of In Situ Test Areas

The detailed procedure for the measurement of in situ stresses using the overcoring principle is adequately described elsewhere (e.g., CSIR, 1973). An EX hole is drilled, the strain cell is inserted into this hole, bonded to the rock and then overcored. The steps involved in the evaluation of the overcoring strain data are the selection of the strain values, the determination of a modulus factor and the calculation of stresses. Corrections may be required for temperature variations, creep of the glue and rock, faulty gages, and stress redistributions caused by proximate openings. Modulus factors E and ν are usually determined from simple tests on the EX core and the computation of stress is generally carried out assuming linear elastic isotropic material. This latter assumption is often stated as a test requirement which severely limits the type of rock mass in which measurements can be made.

Possible sources of major error in this approach are:

(1) nonlinear behavior of the rock (common, particularly in weak rock)
(2) anisotropy of the rock material behavior
(3) inhomogeneities such as bedding, inclusions and grains on the scale of the strain gauges
(4) inhomogeneities on the scale of the overcore resulting in different sets of properties for each actual test
(5) errors, malfunctions and idiosyncrasies of particular gages in the cell
(6) experimental and interpretative misrepresentations of the stress and strain test data.

2.2 Experimental Approach

The basic problem reflected by these potential errors is the indirect nature of the overcoring method; i.e., stresses are not actually measured. Two categories of error sources are recognizable - those which are statistical and experimental in nature, arising from the very problem of representative measurement of the stress field and those which concern the proper representation of the constitutive behavior of the rock material. The traditional concept of the modulus as the "property" relating strain to stress is considered particularly inadequate.

The simplest way of overcoming both types of errors is by making the test as direct as possible; i.e., by measuring a known stress field, in the laboratory, using the same strain gages in the same rock as was used during overcoring. This calibration of stress with strain by back-analysis of representative stress paths eliminates a host of experimental errors. If properly designed, this calibration accommodates in an approximate way a number of secondary factors influencing the constitutive response which otherwise would be extremely difficult to represent explicitly in an analytical model.

The deformability factors obtained from controlled overcore measurements will differ from the traditional constants E and ν, but they will by design describe the stress-strain relationship for the overcored material for the particular set of test, rock and stress conditions.

The question of modulus anisotropy needs special attention. As for the isotropic case, the anisotropic deformability factors should be determinable from direct tests on the overcore. However, in practice, as shown by Ribacchi (1977), a unique set of anisotropic deformability factors cannot be determined, even though enough independent strain measurements are available. This is because individual gage errors cause statistical error "noise" and prevent significant optimization of the deformability matrix to achieve a unique solution from the back-analysis. In order to make the anisotropic solutions tractable, the particular restriction of a measurement borehole in the symmetry plane of the cross-anisotropic material was adopted and the in situ and laboratory testing programs were designed accordingly.

The overall approach adopted in the screening of the data is that, where justified by sufficient understanding, the data and their analyses would be selectively assessed for reliability, corrections, or omission from the "sample." Thus, rather than bias the result, selective treatment of the data would help to make the interpretations less biased by experimental error. This approach to data optimization seeks to remove all data which contain unjustifiable or unaccountable bias or systematic error without censoring random unbiased variations. Recognizing that a larger "sample" size enables a more accurate solution to what is essentially a statistical problem, the need for a compromise in the data screening process using experienced judgement is evident.

In summary, the approach developed for the evaluation and interpretation of in situ stresses is as follows:

(1) Critical appraisal and selection of the overcoring strain data based on the validation of gage and cell functioning and statistical optimization of each strain set.
(2) Determination of the individual anisotropic nonlinear deformability factors for each stress measurement by back-analysis of specially designed laboratory tests.
(3) Explicit evaluation of the significance of factors such as creep, temperature and path-dependent stress-strain behavior.
(4) Development of a stress-strain solution for the overcoring process for the case of a test borehole in the cross-anisotropic plane. The effect of nonlinear behavior is incorporated using iterative solution techniques.
(5) "Weeding out" of in situ stress values using statistical significance testing.

3.0   IN SITU TEST PROGRAM

3.1   Test Program

Three separate suites of tests were conducted at the site of the scheme between 1975 and 1979 at the locations shown on Figure 1. The testing and evaluation techniques adopted were similar apart from minor differences, primarily a result of the developmental process during the duration of the investigations. Only one test suite, namely that at the Pressure Tunnel/Penstock Transition (Site B), has been specifically addressed here to illustrate the interpretative methods used.

The Site B program involved the drilling of 7 bore holes from a small cross-section exploratory adit and the performance of 25 stress measurements using the CSIR cell at borehole depths (8-15m) unaffected by stress concentrations around the test adit. Anisotropic stress solutions were developed for both horizontal and vertical holes. Measurements in holes at both orientations were attempted to minimize any possible directionally controlled systematic errors. Unfortunately, the existence of weak bedding partings prevented useful overcore being obtained from the vertical holes.

3.2   Review of Overcoring Strain Data

The strain relief readings contained anomalies which if not properly evaluated would have significantly affected the overall accuracy of the results. The general validity of the strain relief data was checked using

● redundancy and compatibility checks within rosettes and within a test
● post-overcoring strain behavior
● responses during laboratory tests
● physical examination of the overcore
● overall comparison in terms of calculated stress field with the results of other measurements.

Of the 25 tests, 9 were discarded because of bonding problems or core breakage preventing laboratory testing. Redundancy checks (which test the correspondence of readings within a rosette) and compatibility checks (which test the agreement of each strain value with the overall strain tensor for the measurement) revealed further problems in the bonding. It was found that because of the mechanical design of the cell, only partial bondings

sometimes developed. This situation was not necessarily detected by the intra-rosette redundancy checks. A more useful indicator of the reliability of the strain relief measurement is to calculate the standard deviation from the least squares computation of the strain tensor. Tests with large standard deviations corresponded with physical discrepancies observed in the cell/overcore unit on subsequent examinations. On this basis, five further tests were eliminated. It is concluded that redundancy checks alone are not a good indication of the success of a test; i.e., for accepting a test.

Possible additional sources of error can also be inferred from the typical strain-time response shown in Figure 2. In this example, an overall pattern of dimensional change after overcoring is indicated. Exhaustive laboratory tests were conducted to measure temperature, creep recovery, swelling and humidity effects, and possible electrical problems. Eventually the problem was traced to variations in the temperature and humidity of the ventilation air and the temperature of the drilling water. The rising and falling post-overcoring trends correlated with climatic variations at the project site. The use of a dummy gage does not necessarily nullify the effects of temperature, evaporative cooling and differential thermal expansion. This observation illustrates the extreme importance of cautiously monitoring all in situ test conditions and evaluating the strain data to eliminate or correct all possible interferences.

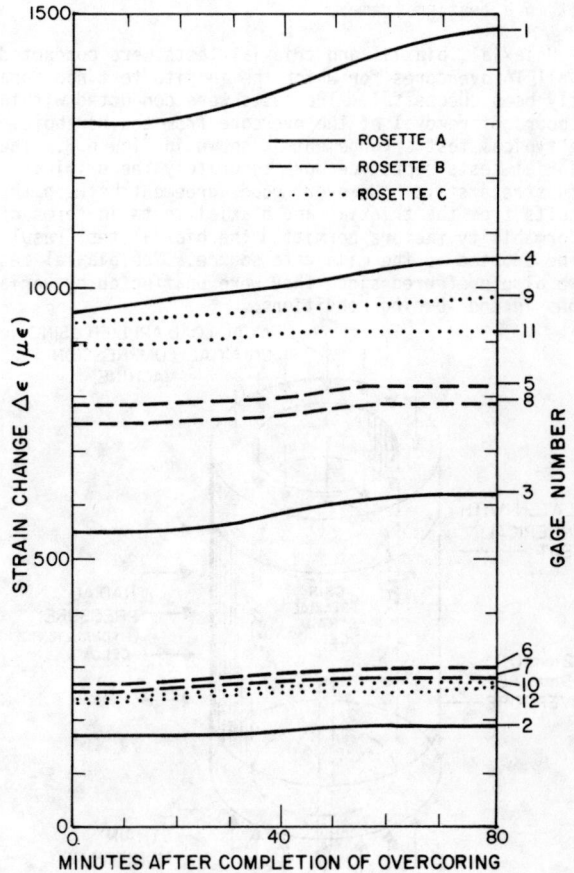

ALL STRAINS ARE RELATIVE TO FINAL DATUM PRIOR TO OVERCORING

Figure 2 - Example of Overcoring Strain Response (Test No. 7, Site B)

The least squares compatibility tests indicated a considerable range in the quality of the data. This variation was recognized in the final evaluation of the average regional field stress tensor, as described later. As for the variations in the quality of individual strain readings, a computer program (supplied by VKE, 1979) was used to selectively omit suspect individual strain values and to test the improvement in the correlation coefficient of the strain tensor for each omission. The redundancy in the strain measurements for any test is only 3, and not 6 as for isotropic rock. The program was thus constrained to eliminate at most only two strain relief values.

The results produced a significant improvement in the quality of the strain tensor for each test and indicated which readings should be omitted in the subsequent stress analyses.

4.0 LABORATORY TEST PROGRAMS

4.1 Test Methods

The objectives of the laboratory test program were to determine the cross-anisotropic deformability parameters for each individual in situ test as accurately as possible and to examine possible sources of experimental errors relating to bonding, temperature and creep effects, as mentioned earlier. To undertake these tests, a special Hoek-type test cell was constructed to load the 72 mm diameter overcore in biaxial compression. Using suitable platens, the core was also loaded in uniaxial and triaxial compression by placing the overcore/cell unit in a testing frame.

Uniaxial, biaxial and triaxial tests were conducted on all 16 overcores for which the in situ test had apparently been successful. The tests were conducted within 24 hours of removal of the overcore from the borehole. The typical test arrangement is shown in Figure 3. The triaxial tests reproduce more accurately the original in situ stress state. However, good agreement between the results from the triaxial and biaxial tests in terms of deformability factors permitted the biaxial test results to be adopted as the main data source. The biaxial tests were also preferred since they were unaffected by variations in end loading conditions.

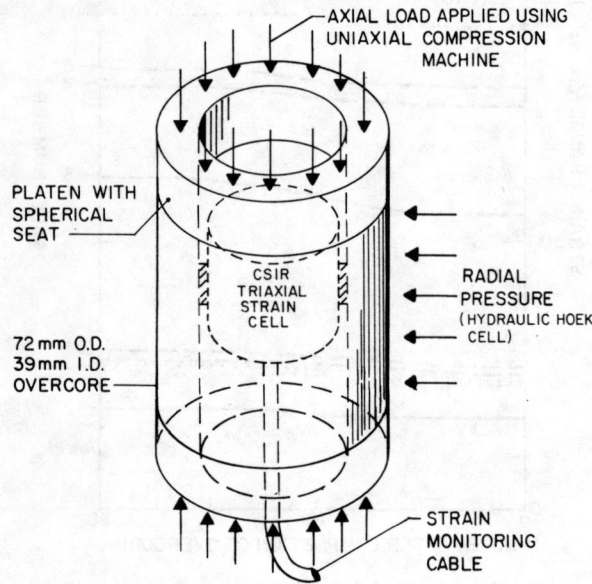

Figure 3 - Configuration of Laboratory Tests on Overcore

To correspond with the unloading phase during the in situ overcoring process, incremental strain readings were taken during the unloading portion of the test cycle. The magnitude of the test pressures and loads were chosen to reproduce as closely as possible the magnitudes of the stresses around the borehole prior to overcoring. The strain cell, still bonded to the overcore, was used to monitor the strain in the overcore during testing.

4.2 Calculation of Deformability Parameters

For the analysis of the laboratory test data, the rock is assumed to be cross-anisotropic with the overcore parallel to the symmetry plane. Five "deformability factors" are required to be calculated as shown in Figure 4. The value of N, the anisotropic shear modulus, cannot be determined from these tests. It was obtained separately, as described below.

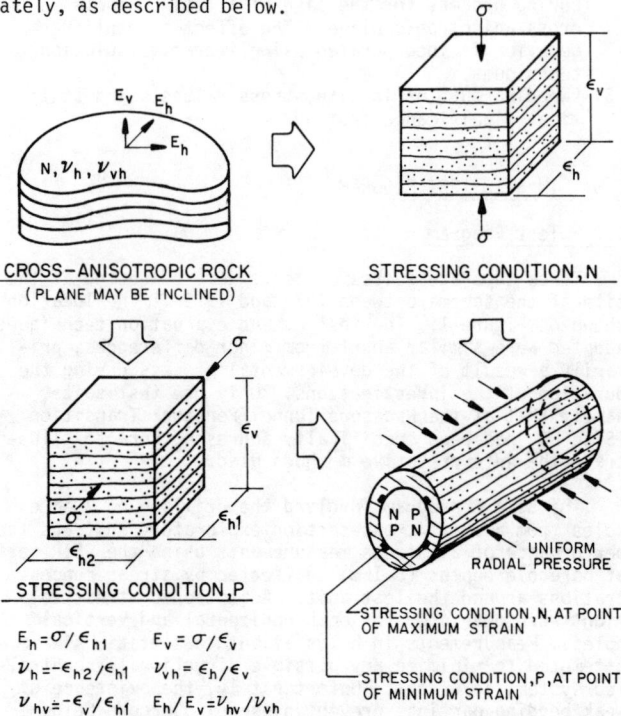

$$E_h = \sigma / \epsilon_{h1} \qquad E_v = \sigma / \epsilon_v$$
$$\nu_h = -\epsilon_{h2}/\epsilon_{h1} \qquad \nu_{vh} = \epsilon_h / \epsilon_v$$
$$\nu_{hv} = -\epsilon_v / \epsilon_{h1} \qquad E_h / E_v = \nu_{hv}/\nu_{vh}$$

Figure 4 - Definition of Deformability Factors in Terms of Biaxial Calibration Test

When a uniform radial load is applied, the variation of tangential strain on the inner surface of the overcore follows a symmetrical sinusoidal pattern with a 180-degree phase. This fact was used to determine the maximum and minimum deformability values; i.e., Eh and Ev, using the strain readings at three points on the inner surface of the overcore. Although this can be done manually, a computer program was written so that all 10, 11 or 12 strain readings from the CSIR cell could be utilized in the evaluation and thereby improve the accuracy of the determination. The program, called ANDLE, is based on the equations for thick cylinder theory (Poulos and Davies, 1974).

Using ANDLE, solutions were obtained for both the magnitudes of Eh, Ev, νh and νvh, and the orientation of the plane of symmetry, angle αc. Typical results for four tests are given in Figure 5. Figure 6 compares the calculated angle of anisotropy, αc, with the value or range of the measured angle of bedding αm. The quality of fit obtained in calculating the values of Eh and Ev was significantly better when αc rather than αm was used in the calculations. There is no *a priori* reason why αc should equal αm. An explanation based on the physical mechanisms of deposition and compaction over geological time would perhaps consider the preferred orientation of the asymmetric sand particles as well as the direction of the cross-bedding. This could explain the proportional rather than constant angular shift observed.

The primary objectives of the above tests and analyses were to establish the individual orientations of the cross-anisotropic plane for each test and to confirm the validity of the assumption of a cross-anisotropic medium. Based on the significance of the anisotropy observed, it was decided to undertake more thorough testing of eight selected overcores using, in addition to the inside strain cell, strain gages bonded to the outside of the overcore. The nonlinear behavior was carefully evaluated in these tests since this characteristic is often evident in anisotropic rocks (Berry *et al*, 1974).

4.3 Stress-Strain Unloading Response

A typical load-strain curve is shown in Figure 7. These curves show a marked similarity with the load-strain response reported by Ribacchi (1977). Based on the loading curve, one would normally classify the material as linear-elastic. However, the unloading response is highly nonlinear. Typical secant values of Eh and Ev as a function of stress level are given in Figure 8. Considerable variations between tests in the values of Eh and Ev are evident. Furthermore, because the rock is much more nonlinear normal to than parallel to the bedding plane, the modulus ratio is also a function of stress level (Figure 9). When assessing the potential significance of anisotropy on the accuracy of calculations for in situ stress measurements, the normal interpretation of the modulus ratio (based on the loading curve) can frequently be misleading. Such high unloading values of Eh/Ev as noted here are probably more common than generally believed.

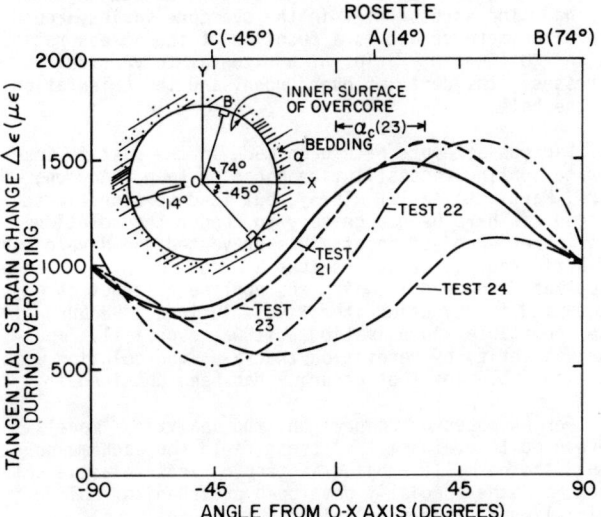

Figure 5 - Examples of Determination of Orientation of Anisotropic Plane (Site B)

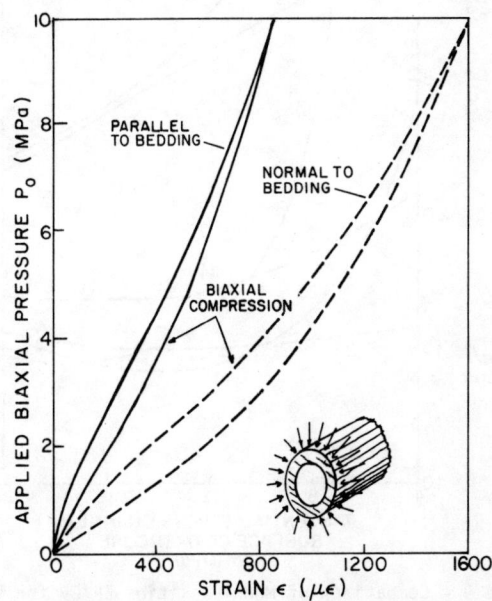

Figure 7 - Typical Results of Overcore Deformability Test (Test No. 21, Site B) Biaxial Test

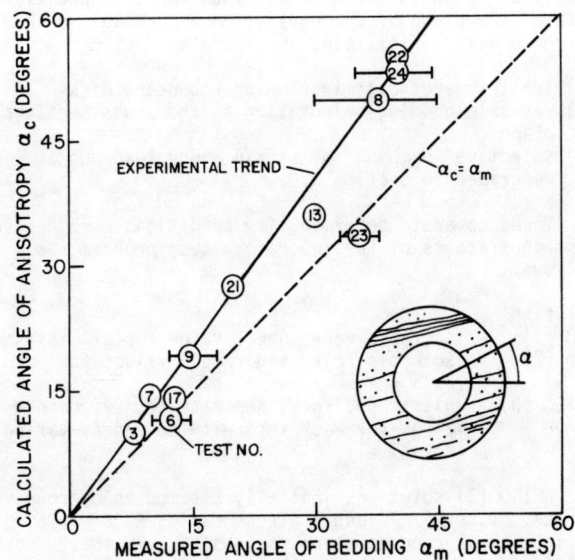

Figure 6 - Comparison of Bedding and Calculated Anisotropy Angles

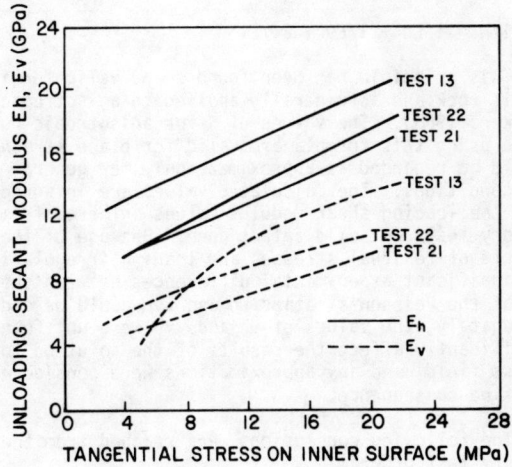

Figure 8 - Typical Values of Secant Deformability Factors Eh and Ev as a Function of Stress Level (Site B)

A 245

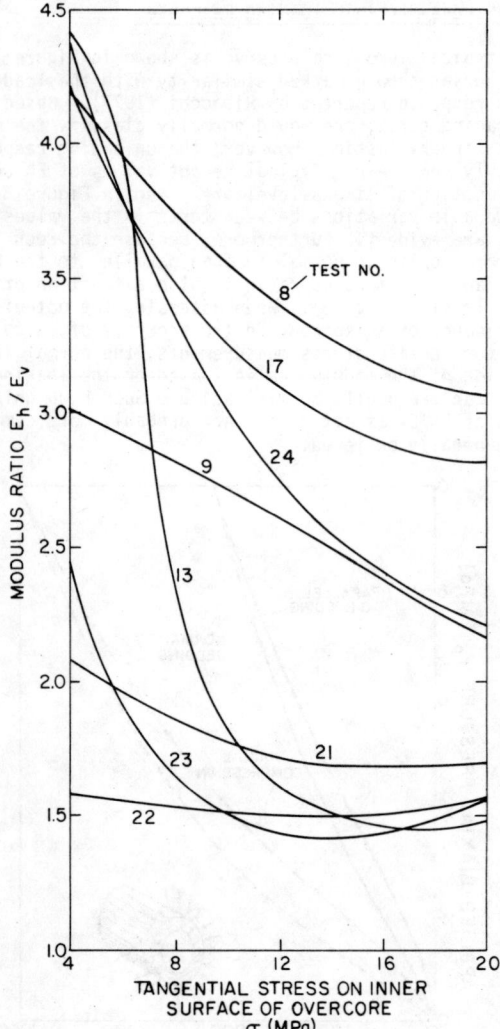

Figure 9 - Comparison of Modulus Ratios Eh/Ev for Various Stress Levels in Overcores (Site B)

Values of N, the shear modulus, are calculated using St. Venant's principle:

$$1/N = 1/Eh + 1/Ev + 2\nu/Ev$$

This principle has been found to be valid for isotropic rock and is generally applied to anisotropic rock (Becker, 1968). The values of N for anisotropic rock deduced using this formula are valid for plane strain but should be regarded as approximate only for general loading conditions. The calculated values are in agreement with the loading shear modulus values inferred from laboratory tests on solid intact core. Because of the effects of residual stresses and loading irregularities, no significant experimental differences between the values of the Poisson's ratios νh and νvh could be deduced. Fortunately, the values of νh and νvh were not found to significantly affect the results of the solution to the stress field, and any approximations were considered to be of no consequence.

The following conclusions were reached from the laboratory test program:

(1) the local variability of rock conditions between test locations is significant.

(2) the rock exhibits a marked anisotropic, stress-dependent, elastic unloading behavior.
(3) the unloading deformational response is distinctly different, normal and parallel to the bedding.
(4) explicit consideration of the deformability factors as functions of the stress level, and the orientation of the anisotropic plane will be necessary to avoid serious errors.
(5) the use of general modulus values to reduce strain data is likely to lead to errors and introduce significant variability in the results.

5.0   CALCULATION OF REGIONAL STRESSES

5.1   Method of Solution

The stress-dependent deformability factors Eh and Ev used in the calculation of stresses need to be selected for the unloading stress change experienced during overcoring. The stress level in the overcore varies around the circumference and is a fucntion of the stress ratios Ka and Kb (the two ratios of horizontal to vertical stresses), the depth of overburden, and the orientation of the hole.

Various attempts have been made in the past to incorporate nonlinear constitutive behavior (e.g., Aggson, 1977; Martinetti *et al*, 1975), but to-date approximate procedures have been necessary to render the solutions tractable. Project constraints prevented the development of a rigorous solution and thus a cross-anisotropic solution was developed in which the nonlinear aspect is accounted for by using iterative solutions. Because of the inevitable approximations, it was especially important to verify by repetition that a unique solution with an acceptable level of accuracy had been obtained.

For purposes of comparison, two analytical models were used to evaluate the stress field for each measurement; the normally adopted isotropic linear elastic solution, and the specially-developed cross-anisotropic quasi-linear (step-wise) elastic solution.

The anisotropic solution is an extensively modified version of the isotropic solution in numerical form (Golder Associates, 1977). The solution is based on the theory of cross-anisotropic continua for the specific case of the plane of symmetry aligned parallel to the borehole axis. Facilities incorporated include:

(1) least squares optimization of redundant data
(2) variable bedding orientation in the cross-sectional plane
(3) selective "dumping" of strain gauge readings as described in Section 3.0.

Three separate deformability conditions were derived from the results of the laboratory test program, as follows:

Solution
ISO (1) - isotropic rock, one E value for all tests
ISO (2) - isotropic rock, separate E values for each test
ANISO (3) - anisotropic rock, separate Eh, Ev, α and ν values for each test with Eh and Ev varied iteratively.

ANISO (3) solutions initially assumed an overburden stress, σv, of 8 MPa and a stress ratio, K, of 1.75, giving stress concentration factors of 1.75 and 2.7 parallel and normal to the borehole axis respectively. These values were successively revised on the basis of the calculated results for the stress field, and then used to re-estimate the values of Eh and Ev. Since in reality only one value of σv and two values of K actually exist, the values of these parameters at any stage in the iteration were applied to all tests. However, the stress levels in the EX borehole inferred for each test were varied according to the orientation of the borehole in space. The iterative solutions were repeated until the changes in the deformability values reached acceptable magnitudes. Generally, three iterations were adequate. The values of the deformability factors adopted for the final solutions are given in Table 1.

TEST NO.		ISO (1)	ISO (2)	ANISO (3)	ν	α
8	Eh Ev N		27.0	32.4 12.2 7.8	0.15	50
9	Eh Ev N		20.8	26.4 7.8 4.5	0.2	19
13	Eh Ev N		16.7	18.7 12.6 6.0	0.2	36
17	Eh Ev N	22.7 22.7 ---	50.0	64.0 24.0 13.4	0.2	14
21	Eh Ev N		14.8	15.0 9.2 3.8	0.3	27
22	Eh Ev N		16.3	16.2 10.6 5.2	0.2	55
23	Eh Ev N		15.1	15.3 8.8 4.4	0.20	34
24	Eh Ev N		18.0	20.7 9.0 5.6	0.15	53
Units		GPa	GPa	GPa	--	degrees

Table 1 -
Deformability Factors Used in Isotropic and Anisotropic Solutions - Site B

5.2 Regional Stress Solutions

The solutions to the regional stress tensors are summarized in Tables 2 and 3, which give the magnitudes and directions of the mean principal stresses, and their variability. The calculation of the mean principal stress magnitude and direction for each solution first requires a separation of the tensor into eigenvalues and eigenvectors.

	ISO (1)	ISO (2)	ANISO (3)
σ v	--------	8,2	--------
σ 1	12,5	11,4	8,69
σ 2	10,60	9,30	7,54
σ 3	8,31	7,14	5,89
Declination/ σ₁	*54/084	*61/084	27/130
Azimuth σ₂	28/306	22/304	*62/324
σ₃	21/205	17/207	07/223
Ka	0,84	0,81	1,15
Kb	0,66	0,63	0,78

*Indicates chosen as sub-vertical principal stress.
All stresses in MPa, angles in degrees.

Table 2 -
Summary Results of Calculated Average Principal Stresses- Site B

Table 3 shows the considerable reduction in the dispersion of stresses effected by the representation of the rock mass as a cross-anisotropic rather than isotropic material. Not only does this significantly modify the magnitude and direction of the mean principal stress, but it also increases the confidence level of the results. The attendant change in the values of each of

Ka and Kb are noteworthy.

Comparing solutions ISO(1) and ISO(2), similar improvements in accuracy and confidence levels are noted when the individual values of the deformability factors are used for each test. While the directions of the principal stresses for ISO(1) and ISO(2) are identical, there has been a substantial correction of the order of 10 to 15 percent to the magnitudes of the mean principal stresses. Table 3 shows the associated reduction in the variability of the individual stress determinations. In this table, variability is expressed as the percentage coefficient of variation; i.e., standard deviation x 100/mean. The variability of the first invariant of the principal stress tensor, I, is also shown as a summary indicator.

		ISO (1)	ISO (2)	ANISO (3)
			Coefficient of Variation	
Principal	σ1	30.0	34.7	28.3
Stress	σ2	31.0	23.3	22.4
Magnitude	σ3	44.0	26.3	16.4
	Av	35.0	28.1	22.4
	I	3.89	3.24	2.04
Principal	σ1	31.4	31.4	31.0
Stress	σ2	35.6	35.6	35.8
Direction	σ3	34.9	34.9	39.0
	Av	33.9	33.9	35.3

Table 3 -
Variability of Calculated In Situ Stress Values

Similar comparisons for the Machine Hall Complex suite of tests (Suite A) indicated a remarkable reduction in the coefficient of variation of stress direction, as shown in Figures 10 and 11 and in Table 4 when an anisotropic solution was used. Again, the change in the estimated mean stress tensor is noteworthy. Not only is the subvertical stress consistent with the theoretical overburden stress, but the inferred K value of 2.5 is in better agreement with that deduced from back-analyses of the Machine Hall Trial Enlargement instrumentation data (Sharp et al, 1979).

		ISOTROPIC		ANISOTROPIC	
		Mean	Variation	Mean	Variation
Overburden		4.0	----	4.0	----
Magnitude	σ 1	12.3	21.7	10.1	21.2
	σ 2	8.0	22.4	7.0	22.9
	σ 3	5.5	36.2	4.0	35.9
Direction	σ 1	11/312	32.6	13/137	21.6
(Declination/	σ 2	29/215	58.6	12/047	21.9
Azimuth)	σ 3	58/061	53.0	78/240	17.2
	Ka	1.5		1.7	
	Kb	2.2		2.5	

Mean stress magnitudes and angles in MPa and degrees respectively.

Variation is - coefficient of variation for magnitude and mean angular deviation for direction.

Table 4 -
Summary Comparison of Machine Hall Complex (Site A) Test Results

A further procedure was used in the refinement of the data. Test results suspected of being inaccurate were selectively omitted from the averaging process. The significance of any reduction in the variability of the "sample" was tested using the Chi-squared test. As a result, test no. 7 was discarded from the set of results, a decision which was supported by the poor redundancy checks on the overcoring strain data mentioned earlier.

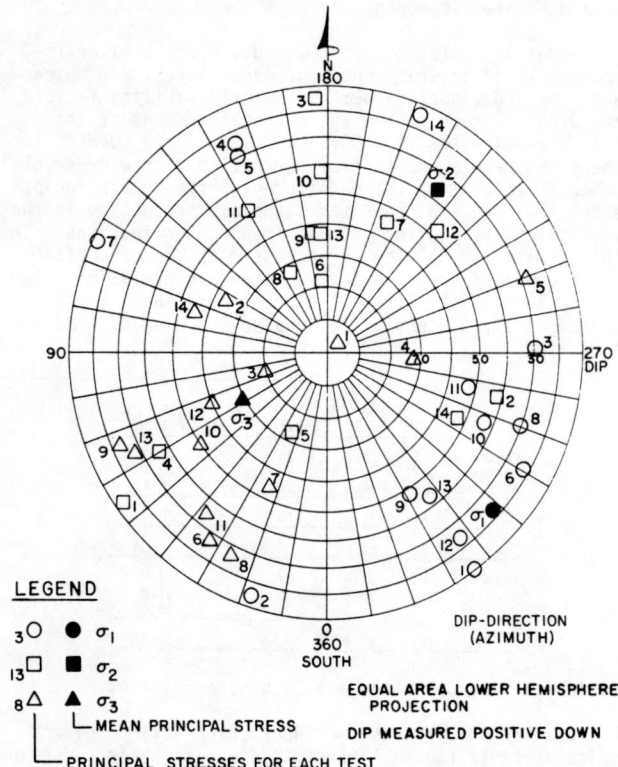

DIP-DIRECTION
(AZIMUTH)

EQUAL AREA LOWER HEMISPHERE
PROJECTION

DIP MEASURED POSITIVE DOWN

Figure 10 - Principal Stress Directions (Site A Tests) -
Isotropic Solution

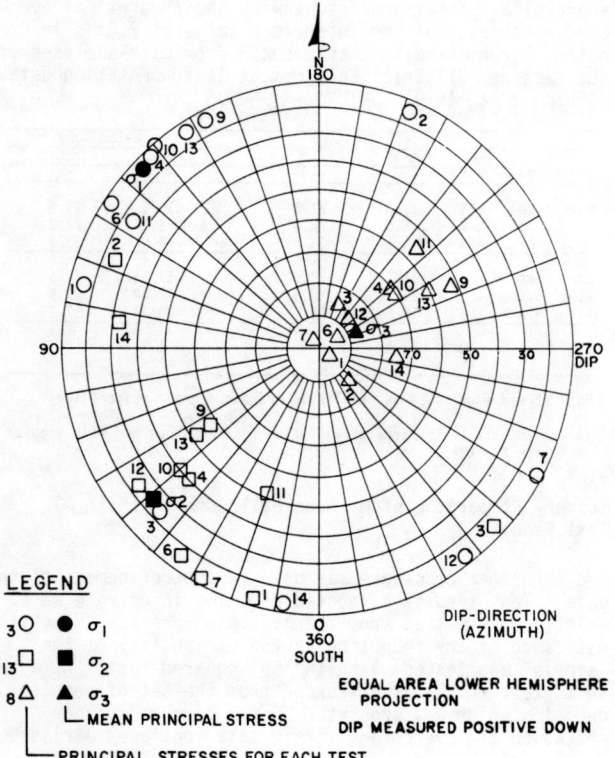

DIP-DIRECTION
(AZIMUTH)

EQUAL AREA LOWER HEMISPHERE
PROJECTION

DIP MEASURED POSITIVE DOWN

Figure 11 - Principal Stress Directions (Site A Tests) -
Anisotropic Solution

The results clearly indicate that for the present geological conditions, acceptable solutions to the in situ stress field could be obtained only by considering the rock as an anisotropic medium. If the nonlinear behavior is not represented in the solution, additional errors of the order of ±10 percent of the stress magnitudes are introduced. It can be seen that using isotropic solutions, the vertical principal stresses are estimated to be between 1.5 and 1.6 times the theoretical overburden pressure, σv. This is not considered realistic for the gently sloping terrain in the test area. For the anisotropic solution, the calculated value of the subvertical stress (7.5 MPa) is bracketed by the values of overburden stresses calculated on the basis of vertical and sloping heights of 8.2 and 7.4 MPa, respectively. Hydrofracture stress measurements at the test site indicated minimum principal stresses in a vertical direction between 7.5 and 8.0 MPa (Golder Associates, 1980).

Assuming a normal distribution, the 95 percent confidence limits for the mean principal stresses in the immediate vicinity of the pressure tunnel/penstock transition area are estimated to be:

Princ. Stress	Mean (MPa)	95% Conf. Limits	Direction
$\sigma 1$	8.69	± 1.70	sub-horizontal
$\sigma 2$	7.54	± 1.17	sub-vertical
$\sigma 3$	5.89	± 0.67	sub-horizontal

5.3 Error Evaluation

Errors in the estimation of the regional stresses may be categorized as either random or systematic. The above calculation of the mean principal stresses implicitly assumes there are no systematic errors. Systematic errors can only be estimated by independent measurements. The comparison with the hydrofracture test results indicates that systematic errors are no greater than between 5 and 10 percent for stress magnitude. Confirmation of accuracy with regard to stress directions is given by the comparison of the results from the three test sites (Figure 12).

Random errors or variations between individual stress determinations are the result of either actual variations between stresses at each test location or experimentally based errors. Actual variations in stresses may arise from residual (lock-in) stresses, as noted by Boch (1979), or by gradual changes in stress as affected by geological conditions. Experimentally based random errors arise in both the in situ testing and laboratory testing phases of the work. The relative significance in the present case of the actual and experimentally induced variations between test results may be gaged from a comparison of the results with those obtained from the Headrace suite (C) of tests. The average standard deviation for the Pressure Tunnel set of results is 1.70 MPa, compared to a value of 2.47 MPa for the results from the Headrace set. The lower value for the Pressure Tunnel set was obtained despite the higher absolute stress levels and the less uniform rock conditions. Thus, it is concluded that the observed random errors are primarily experimental in origin and very much dependent upon the test execution and interpretation techniques. Improper representation of the nonlinear behavior of the rock is regarded as the main factor contributing to any small systematic error that may exist.

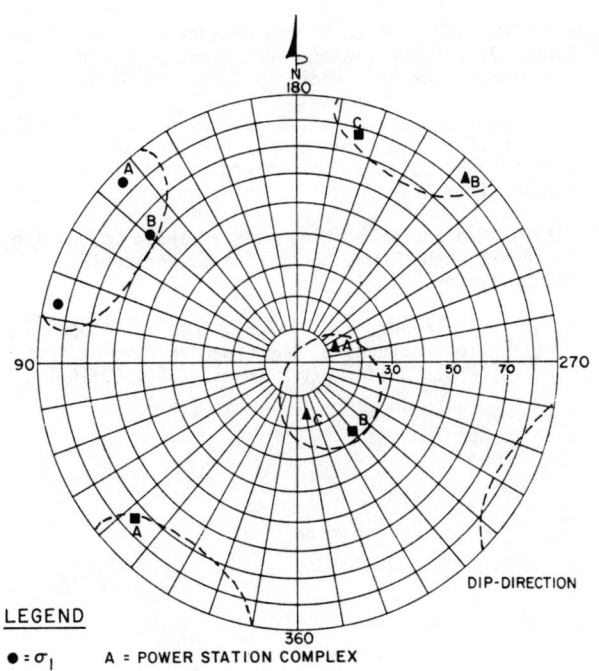

LEGEND

- $\bullet = \sigma_1$ A = POWER STATION COMPLEX
- $\blacksquare = \sigma_2$ B = PRESSURE TUNNELS/PENSTOCK
- $\blacktriangle = \sigma_3$ C = HEADRACE

EQUAL AREA LOWER HEMISPHERE PROJECTION

DIP MEASURED POSITIVE DOWN

Figure 12 - Comparison of In Situ Stresses from the Three Test Sites

6.0   CONCLUSIONS

The overcoring technique using "soft inclusion" cells is the most accurate, albeit commonly problematical, method presently available for measuring the complete state of stress in rock. The description and representation of the deformability of the rock material necessary for converting measured strains to stresses is the major source of errors. The studies described above demonstrate that improvements in the measurement and analysis of the constitutive rock behavior can minimize the major errors. In spite of the complexity of material behavior, reliable estimates of the regional stress field have been obtained.

The particular techniques developed and applied are of general applicability to all types of soft inclusion overcoring measurements. Properly applied, they result in significant and much needed improvements in the precision (accuracy and level of uncertainty) of stress measurements for a wide range of rock conditions, especially at shallow depths where anisotropy is more pronounced.

Of particular significance are the errors introduced by the assumptions of isotropy and linearity in the constitutive response characteristic of the overcore for sedimentary rocks. Extreme caution in the interpretation of stress measurements is suggested since, for these materials, standard laboratory uniaxial tests would indicate relatively insignificant degrees of anisotropy and nonlinearity as normally deduced from loading curves. In addition, temperature and humidity effects need to be carefully assessed.

The results have demonstrated and confirmed that with proper account of anisotropy and nonlinearity, the range of rock conditions for which accurate reliable estimates of in situ stresses can be made can be extended consider-

ably. The need for such refinements can clearly be gaged from the results in the literature for which the calculated minimum principal stresses at shallow depth may be suspect, being two or three times the theoretical overburden pressure in geological environments that do not favor such stress conditions; e.g., Elandsberg Scheme (Bieniawski, 1978), Cruachan Scheme (Young and Faulkiner, 1966). A careful review of results containing such conclusions is warranted since it is the authors' belief that systematic errors are frequently much greater than is generally thought. Clearly, unless one goes to the trouble to understand and incorporate the real behavior of rock, the value of the results is greatly reduced.

The present studies have shown that using a special case rigorous anisotropic stress-strain solution with a reasonable approximation to the nonlinear behavior of the rock, a suitable compromise between accuracy and practicality has been achieved. Systematic errors have been virtually eliminated while random errors, which are largely experimentally derived, have been reduced considerably.

Three types of further improvements in the determination of in situ stresses are recommended:

(1) extension of the anisotropic solution to a more general orientation of the cross-anisotropic medium
(2) more rigorous numerical simulation of the nonlinear behavior
(3) refinement of the triaxial strain cell installed in situ to reduce experimental errors and dispersion of results.

7.0   ACKNOWLEDGEMENTS

The authors wish to thank the Electricity Supply Commission of South Africa for permission to publish this paper, and Gibb Hawkins and Partners and Golder Associates for the material which has been used in its preparation. They also wish to acknowledge the extensive contribution of their many colleagues, particularly Mr. R.J. Pine and Dr. P. Croney, whose contributions in the way of discussions and analyses are reported here.

8.0   REFERENCES

Aggson, J.R. (1977), Testing Procedures for Nonlinearly Elastic Stress-Relief Overcores, *U.S. Bureau of Mines*, RI No. 8251.

Becker, R.M. (1968), An Anisotropic Elastic Solution for Testing Stress Relief Cores, *U.S. Bureau of Mines*, RI No. 7143.

Berry, P., Crea, G., Martino, D. and Ribacchi, R. (1974), The Influence of Fabric on the Deformability of Anisotropic Rocks, *Proc. 3rd Congress*, ISRM, Denver.

Bieniawski, Z.T. (1978), Determining Rock Mass Deformability - Experience from Case Histories, *Int. J. Rock Mech. Min. Sci.* (15), 5, 237-247.

Bock, H. (1979), Experimental Determination of the Residual Stress Field in a Basaltic Column, *Proc. 4th Congress*, ISRM (1), 45-50, Montreaux.

CSIR (1973), Instruction Manual for the Use of the CSIR Triaxial Rock Stress Measuring Equipment, Report MEG ME 1214, *National Mechanical Engineering Research Institute*, Pretoria.

R

Golder Associates (1977), *Drakensberg Main Contract Design Report: In Situ Stress Measurement*, Vol. IX, Unpublished Report.

Golder Associates (1980), *Report on Hydrofracture Testing in the Area of the Pressure Tunnel/Penstock Transition*, Drakensberg Scheme, Main Contract, Unpublished Report.

Martinetti, S., Martino, D., and Ribacchi, R. (1975), Determination of the Original Stress State in an Anisotropic Rock Mass (in Italian), *Revista de Geotecnica* (9), 84-98.

Poulos, H.G., and Davies, I.H. (1974), *Elastic Solutions for Soil and Rock Mechanics*, New York, Wiley.

Ribacchi, R. (1977), Rock Stress Measurements in Anisotropic Rock Masses, *Proc. Int. Symp. on Field Measurements in Rock Mechanics*, Zurich, April.

Sharp, J.C., Pine, R.J., Moy, D. and Byrne, R.J. (1979), The Use of a Trial Enlargement for the Underground Cavern Design of the Drakensberg Pumped Storage Scheme, *Proc. 4th Congress, ISRM*, Montreaux, Vol. 2.

VKE (1979), Van Niekerke, Klein and Edwards, Consulting Engineers, Cape Town, South Africa, Personal Communication.

Young, W. and Falkiner, R.H. (1966), Some Design and Construction Features of the Cruachan Pumped Storage Project, *Proc. Instn. Civil Engrs.* (35), 407-451.

A REAL-TIME INTERPRETATON METHODOLOGY FOR LARGE-SCALE (2 m³)
IN-SITU ROCK SHEAR TEST

Une méthodologie pour l'interprétation en temps réel des essais de cisaillement de rocher in situ et à grande échelle (2 m³)

Eine Methode zur sofortigen Interpretation von in-situ Scherversuchen an 2 m³ grossen Felsproben

Mi. F. Bollo
Civil Engineer, Computer Scientist, Madrid, Spain

E. Herrero
Dr.Engineer, E.N.H.E.R., Barcelona, Spain

J. M. Buil
Civil Engineer, E.N.H.E.R., Barcelona, Spain

SYNOPSIS

We describe a methodology for a large-scale (2 m³) in-situ rock shear test, taking advantage of real time interpretation. Pressure is transmitted by two flat jacks of 2 m² surface each, producing loads over 18 MN. Displacements are measured with a prediction better than 10^{-5} m. The pressure and displacement data are interpreted in real time by means of a portable computer, and the results are used to define the next increases of normal and shear pressure. The early detection of plastification allows the determination of intrinsic resistance with a single test specimen. Test results are related to the rock mass by means of a microseismic elastic survey. A practical case, in the more than 800 m deep Moralets Underground Hydroelectric Power Plant (Spain) in Pirenaic Devonian Schist is included.

RESUME

On décrit une méthodologie pour les essais de cisaillement de rocher in situ et à grande échelle (2 m³) interprétée en temps réel. Les valeurs de pression et de déplacement, introduites dans le programme donnent la définition des incréments successifs de pression normale et tangentielle. La détection précoce de la plastification permet la détermination des paramètres de résistance intrinsèque non-linéaire avec un seul échantillon.

ZUSAMMENFASSUNG

Es wird eine Methode zur sofortigen Interpretation von in-situ Felsscherversuchen an 2 Kubikmeter großen Felsproben beschrieben. Druck- und Verschiebungsdaten bestimmen den Zuwachs des Normal- und Scherdruckes. Ein einziges Probestück genügt für die Feststellung der wesentlichen Widerstandskraft.

1. INTRODUCTION

Rock slope stability analysis are performed daily by rock mechanic Engineers for applications ranging from dam foundations to highways. As a consequence, considerable effort is devoted to develop reliable methods, both for intrinsic resistance determination, and stability analysis. Activating forces may be accurately evaluated in most cases. The main concerns related to intrinsic resistance determination are scale factor (Leichnitz, 1979), decompression and alteration of the rock specimen (Bollo, 1972).

In highways, for instance, the influence of the following factors should be evaluated:
- Excavation quality.
- Drainage system quality and its maintenance.
- Weathering protection and its maintenance.

In dams, grouting quality is essential, in three ways:
- Combined with the drainage system, to guarantee minimal subpressure.
- To produce homogeneous intrinsic resistance in the decompressed foundation surface.
- To clean and fill diaclases.

Microseismic and high density resistivity surveys may be used to check the grouting effect.

Intrinsic resistance data are adquired using a three phase approach:
- Minimal properties may be computed from analysis of stable natural slopes.
- In-situ test. To overcome the scale factor problem, thin flat jacks of 2 m² surface allow rock specimens of 2 m³ volume. (see Figure 1).
- Extension of results to the work area. Correlation with microseismic auscultation data and detailed geological maps are used. (Bollo,1974) Best results are achieved by means of a geological-geotechnical integrated model. (Del Corral, 1982).

The decompression influence is evaluated both in the rock specimen and the work area.

2. INTRINSIC RESISTANCE MODELS

The complex structure of rock, where usually various families of joints with different properties coexist, has determined most authors to search for more realistic formulas than Mohr-Coulomb's model. These formulas are often the result of experimental approaches.

The simplest of them all use two straight lines for describing the (τ,σ) relationship, when indentations in plane faults are considered. (Patton, 1966).

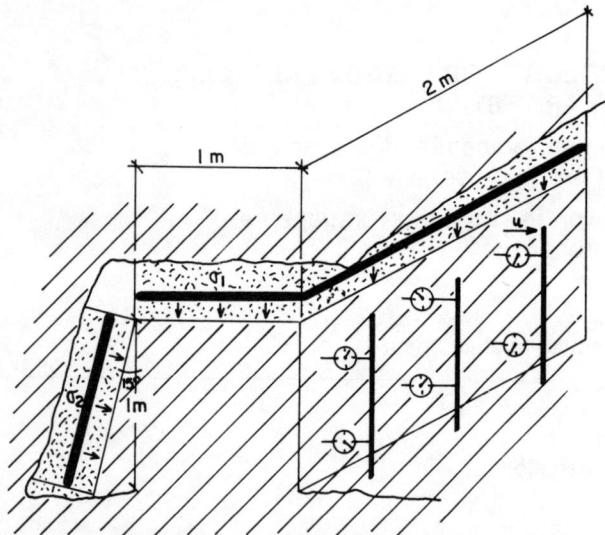

Fig. 1 Large scale rock shear test in the Underground Moralets Hydroelectric Power Plant (Spain) using thin flat jacks.

Non-linear experimental graphs are depicted by Leichnitz (1979), Shiryaev's block model (1979) Fecker (1979), etc.

The general look of these curves may be described as the combination of a linear and logarithmic (τ, σ) relation.

Barton (1974) suggests:

$$\tau = \sigma \cdot \tan \left((JRC) \cdot \log_{10} (JCS/\sigma) + \phi_r \right) \qquad (1)$$

where: τ = shear stress
      c = cohesion
      σ = normal stress
      ϕ_r = basic friction angle
    JRC = joint roughness coefficient
    JCS = efective joint wall compresive
                       strenght

Hayashi (1979) proposes:

$$\tau = c \left(1 - (\sigma/k) \right)^{1/2} \qquad (2)$$

where: k = constant

These models give $\partial \tau / \partial \sigma \to$ undefined and $\partial \tau / \partial \sigma \to 0$ when $\sigma \to \infty$. An obvious improvement, in the second case, would be:

$$\tau = c \left(1 - (\sigma/k) \right)^{1/2} + \sigma \tan \phi_r \qquad (3)$$

A more general failure model for an irregular rock is due to Ladanyi (1969) and has been checked against experimental results by Martin (1974):

$$\tau = \frac{\left(\sigma((1-a_s)(\dot{v}+\tan\phi_\mu) + a_s \tan\phi_r) + c \cdot a_s \right)}{\left(1-(1-a_s)\dot{v} \tan \phi_f \right)} \qquad (4)$$

where:

a_s = shear area ratio = A_s/A
A_s = area over which asperities are being sheared
A = total projected shear area
$\dot{v}$ = dilatance rate at failure
ϕ_μ = friction angle along the planar surface of the teeth
ϕ_f = average friction angle

Bollo (1974) proposes:

$$\tau = c + \sigma \tan \phi_r + \sigma (\tan \phi_i - \tan \phi_r) \Delta \sigma_c^2 / (\sigma^2 + \Delta \sigma_c^2) \qquad (5)$$

where:

ϕ_i = initial friction angle, including the dilatance effect
$\Delta \sigma_c$ = critical stress deviator, to which the limit for dilatance effect corresponds

This offers the advantage of including separately the initial and basic friction angle, like Bertacchi (1974), and introduces the critical stress deviator, somewhat playing a similar role to the clay preconsolidation in Soil Mechanics. (See Fig. 2)

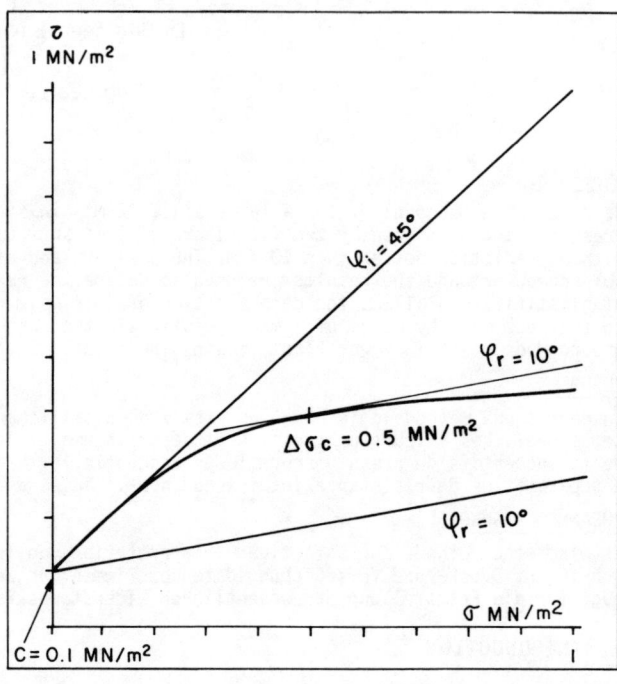

Fig. 2 Bollo's non-linear intrinsic resistance model.

The use of any non-linear intrinsic resistance model places an emphasis on the need for a multiple (τ, σ) definition on a single rock specimen. (Leichnitz, 1979). A regression computation method with a multiple parameter model is only justified if every test specimen has similar geological and geotechnical properties.

In order to get such a multiple (τ, σ) determination with a single rock specimen, a methodology has been defined that allows the early detection of plastification, giving a real-time test interpretation by means of a hand-held computer (HHC).

3. TEST REQUIREMENTS

A convenient size for the rock specimen is 2 m^3. This overcomes the scale effect, in most cases. The excavation should avoid blasting, at least in the vicinity of the final shape, that may be produced with compressed-air hammers.

Two flat jacks of 2 m^2 surface each are used,-
with concrete transmitting pressure to the spe
cimen, up to 18 MN/m^2. The condition of the last
is tested by microseismic means.

In hard rock, displacements should be read with
precission better than 10^{-5} m, which suggests -
the use of capacitive transducers or mechanic -
dial gauges. In soft rock potentiometric transducers are more rugged and therefore adequate.

Shear test galleries are often in a condition -
of humidity, dust and even lack of stable electric power that preclude the use of a normal --
computer.

Mass-storage is limited to rugged digital casse
ttes.However, we aimed at a methodology exclu--
ding the use of military-grade computer equipement.

A minimal configuration may consist of a HHC --
with about 2 Kbytes memory, and dial gauges for
pressure and displacement readings.(See Fig. 3)

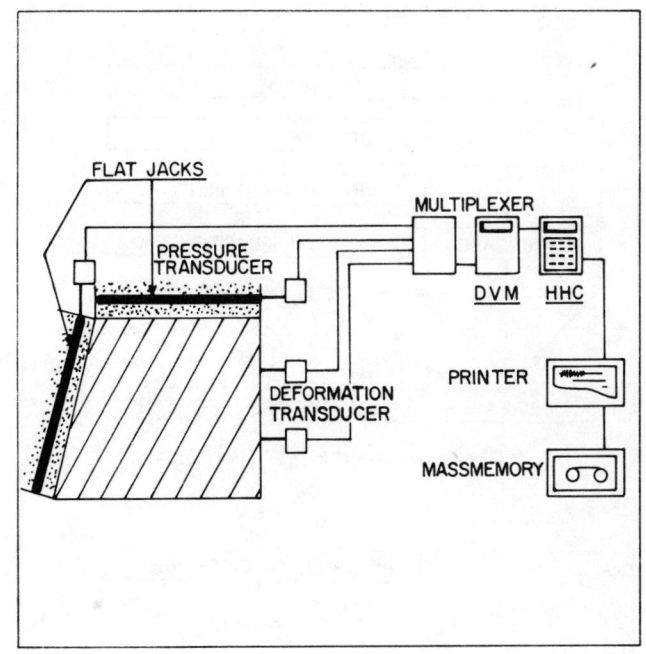

Fig. 4 Optimal test system.

If time intervals are approximatively equal, -
the 4-tuples may be substituted by a succession
of triplets (σ,τ,u).

The possible decisions at any one test stage
are:

- increase τ
- increase σ
- finish test and proceed to discharge

Once a plastification criterion is established,
the decission algorithm becomes quite simple.-
(See Fig. 5).

In order to quantify plastification, displacement increments are decomposed in two terms:

$$\Delta u_{total} = \Delta u_{ellastic} + \Delta u_{plastic} \quad (6)$$

The first is computed by fitting a regression
plane to:

$$\Delta u_{ellastic} = k_1 \cdot \Delta\sigma + k_2 \cdot \Delta\tau \quad (7)$$

where k_1 and k_2 are constants computed at every stage upon the whole (σ,τ,u) succession.

Once k_1 and k_2 are known, solving (6) for Δu -
(plastic) and comparing this value with a preset limit triggers the σ or τ increases. End of
load is decided when σ exceeds another preset
value.

Care should be taken so that $\Delta\tau$ increases in -
steps small enough to prevent the risk of se--
rious plastification, that would invalidate the
consecutive data.

When the limit σ has been reached with minimal
plastification, the discharge may be used to -
explore higher (τ/σ) points.

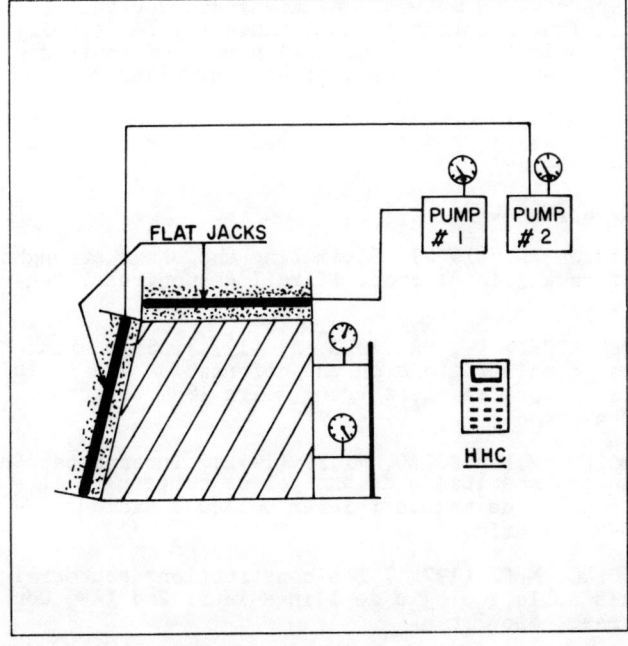

Fig. 3 Minimal test system.

A portable digital cassette interface to the -
HHC may provide mass memory, and a portable --
printer may be used to log the data entered.

In an optimal configuration, pressure and displacement transducers would be conected to a -
multiplexer and this one to a digital voltme--
ter (DVM) interfaced to the controller HHC. -
(See fig. 4).

4. REAL TIME INTERPRETATION

Test evolution may be described as an ordered
series of 4-tuples (σ,τ,u,t) where:

u = average displacement
t = time

A 253

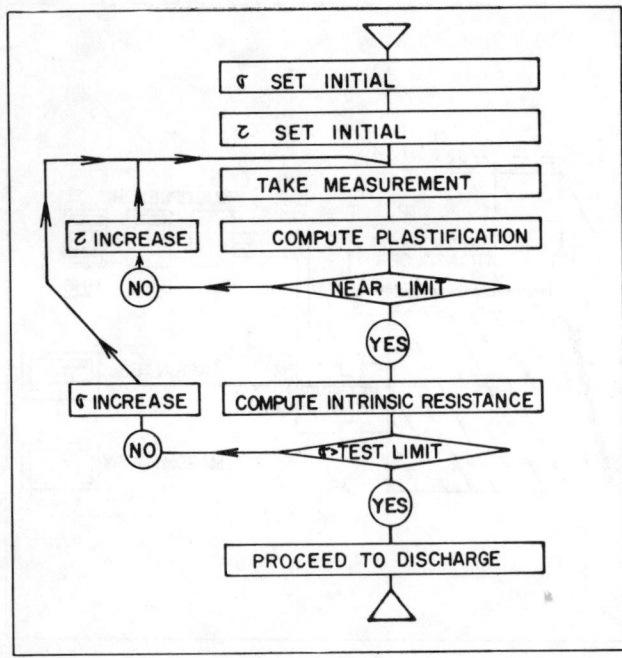

<u>Fig. 5 Decision algorithm based in real time
interpretation.</u>

5. PRACTICAL CASE

Figure 6 shows the application of the descri--
bed methodology to intrinsic resistance determination in Moralets Underground Power House -
(Huesca, Spain).

Initial values for σ,τ are 1.5609 MN/m² and 0,
respectively. The increases of shear pressure
τ are of 0.4963 MN/m². Increases of σ are 0.5
MN/m² when there is no change in τ, and an - -
additional 0.0609 MN/m² when τ grows 0.4963 --
MN/m².

This keeps, in this case, quite constant the -
plastification indicator for the whole range -
of σ : 0 - 9.2187 MN/m², that is about three -
times usual dam foundation loads.

We have represented in the figure by dotted lines the values of plastification indicator - -
$u_{plastic}$, k_1 and k_2. They have been traced relative to the (τ,σ) trajectory, with a tilt of
135°.

One procedure to calculate the non-linear in--
trinsic resistance is as follows:

- the basic friction ϕ_r may be conservatively
computed from the minimal slope of the envelope of shear/normal pressure ratio peaks.
Substracting from the fourth peak τ=3.4739 --
MN/m², σ=4.4265 MN/m² the first τ=0.9925 MN/m²
σ=1.6219 MN/m² we derive ϕ_r=41.5°.

- the initial friction angle may be conservatively estimated substracting from the last shear
/normal pressure ratio peak the first. It re--
sults ϕ_i=49.6°.

- From large scale tensile tests in the same --
rock we know that tensile strenght is about 0.2
MN/m². Therefore, a conservative estimate for -
the cohesion is c=0.2 MN/m².

- the discharge, increasing the τ/σ ratio, produces a rapid grow in the plastification indicator $u_{plastic}$. For σ<5.6 MN/m² the rock specimen
is certainly plastified. It is once more conservative to derive the remaining parameter $\Delta\sigma_c$ --
(critical deviator) from the last charge peak.
The result is $\Delta\sigma_c$ = 10.975 MN/m², in good correspondence with uniaxial compresive strenght data.

So the intrinsic resistance may be computed in
this case by :   (MN/m²)

$$\tau = 0.2 + \sigma \cdot 0.884725 + \sigma \cdot 34.9633/(\sigma^2 + 120.45) \quad (8)$$

which has been also plotted in figure 6.

6. CONCLUSIONS

To overcome the scale effect, large rock specimens are often mandatory. Using thin flat jacks,
microseismic auscultation and the real time interpretation methodology we have described, the
non-linear intrinsic resistance may be accurately estimated with a minimal number of tests dictated only by the geological conditions.

The most important advantage is not the time or
cost saving, but the improved knowledge of the
rock behaviour.

7. REFERENCES

BARTON, N. (1974). Estimating the shear strenght
of rock joints: Proc. IIIrd ISRM Congress, Denver.

BERTACCHI, P., SAMPAOLO, A. (1974). Some critical considerations on the deformation and failure of rock samples: Proc. IIIrd ISRM Congress,
Denver.

BOLLO, M.F., BOLLO, Mi.F. (1972). Interpretación
de los resultados de ensayos de corte para el -
proyecto de taludes: IIIer Coloquio Nacional --
SEMR, Madrid.

BOLLO, M.F. (1974). Les constructions souterraines et la geologie de l'ingenieur: 2nd IAEG Congress, Sao Paulo.

BOLLO M.F., BOLLO, Mi.F. (1974). Etude d'une relation de resistance intrinseque non lineaire -
dans le project des talus: IIIrd ISRM Congress,
Denver.

CAVOUNIDIS, S., SOTIROPOULOS, E. (1979). Strain
softening marly rock: 4th ISRM Congress, Mon- -
treux.

DEL CORRAL, J.M., FERNANDEZ-BOLLO, Mi. (1982).
A systematic approach to the geotechnical con-
sideration of lithoclases in dam design: 14 th
ICOLD, Rio de Janeiro.

FECKER, E., MUELLER, L., REIK, G. (1979). Allgemeine Geotechnische Gesichtspunkte und Grenz-
gleichgewichtsbetrachtungen als erste orientierung bei der planung von Talsperren: 4th ISRM
Congress, Montreux.

LADANYI, B., ARCHAMBAULT, G. (1969). Simulation of shear behaviour of a jointed rock mass: 11th Symposium on Rock Mechanics, California.

LEICHNITZ, W., NATAU, O. (1979). The influence of peak shear strenght determination on the analytical rock slope stability: 4th ISRM Congress Montreux.

MARTIN, G.R., MILLAR, P.J. (1974). Joint - - strenght characteristics of a wheathered rock: IIIrd Int. Congress ISRM, Denver.

PATTON, F.D. (1966). Multiple modes of shear -- failure in rock: Dissertation, University of -- Illinois.

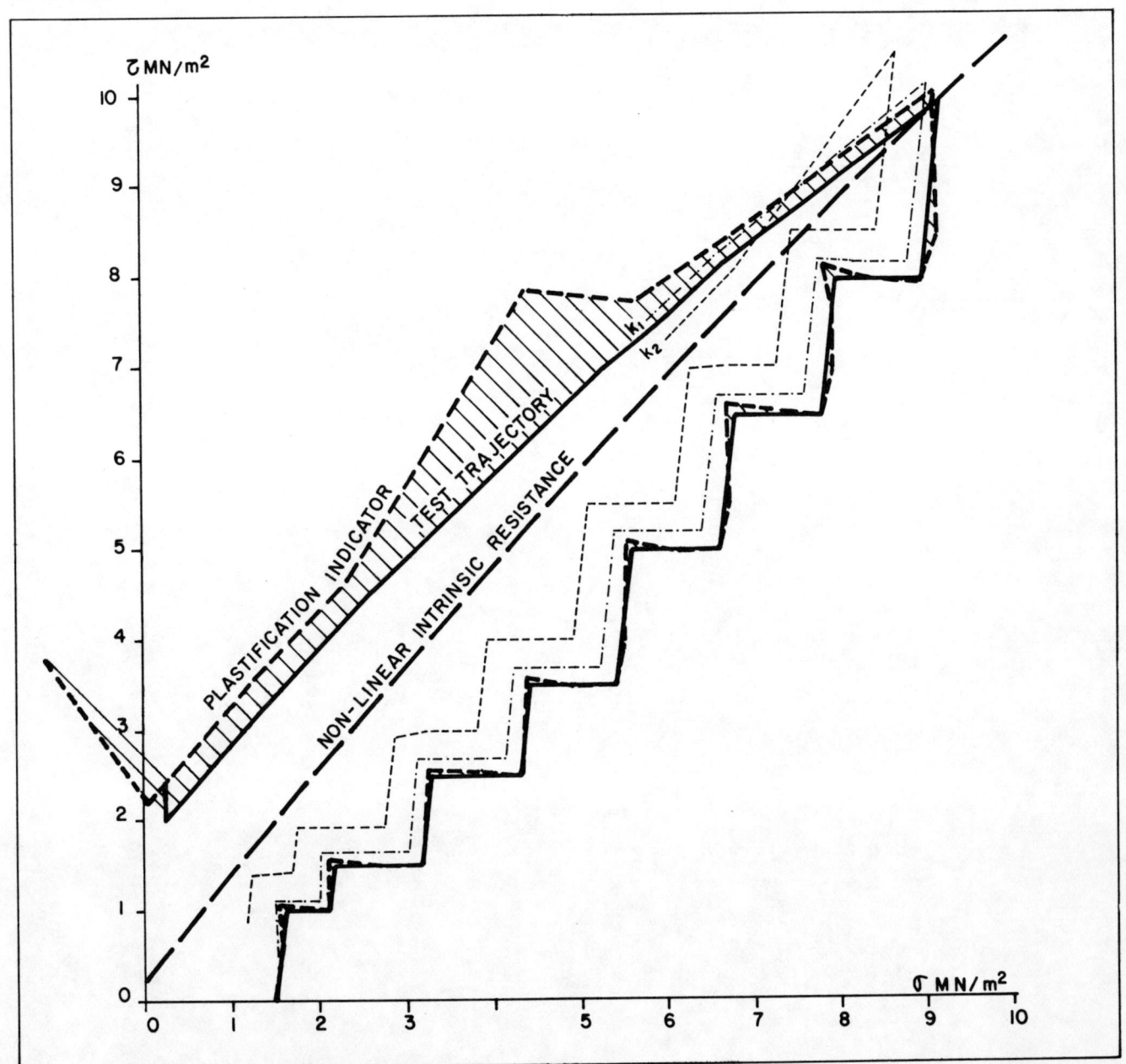

Fig. 6 Application of real time interpretation methodology to the Moralets Underground Hidroelectric Power House (Huesca, Spain)

HAYASHI, M., KANAGAWA, T., HIBINO, S. et al. (1979). Detection of anisotropic Geo-Stresses by accoustic emission and non-linear - trying on Large Excavating Caverns: - Rock Mechanics 4th ISRM, Montreux.

SHIRYAEV, M., KARPOV, N.M., PRIDOROGINA, I.V. (1979). Model studies of the strenght of jointed rock: 4th ISRM Congress, Montreux.

A LOOK AT THE EFFECT OF SCALE ON THE SHEAR EVALUATION OF ROCK JOINTS

Regard sur l'effet "échelle" dans les essais de comportement de cisaillement

Betrachtung der Masstabsgrösseneinflüsse auf das Scherverhalten von Felsklüften

A. G. Paşamehmetoglu, A. Özgenoglu, N. Bölükbaşı, C. Karpuz, A. Bilgin, C. Can
Mining Engineering Department, Middle East Technical University, Ankara, Turkey

SYNOPSIS

The potential influence of joint test-size on measurements if shear strength has often been pointed out and recent experimental studies have shown that there are significant scale effects on both the shear strength and deformation characteristics, especially in the case of rough, undulating joint types. In this paper, in-situ shear tests, conducted at Sır Dam Site, K. Maraş, Turkey, together with small scale shear tests on cores recovered from the same location are explained. Three large-scale and several small scale shear tests are carried out. Blocks 70×70×35 cm are cut in quartzite which is traversed by inclined bedding planes containing filling material. For small-scale tests a portable shear box is employed. The results of both types of test are analysed and the conclusions drawn are given.

RESUME

On a souvent relevé l'influence possible du facteur "échelle" dans les essais de résistance au cisaillement, et des études récentes montrent son effet important, dans de tels essais, sur la détermination des caractéristiques de résistance au cisaillement ainsi que sur celles de la déformation, surtout dans le cas des joints rugueux et accidentés. Cette communication examine les résultats d'essais de cisaillement effectués au chantier du barrage Sır K. Maraş, en Turquie, ainsi que d'essais à petite échelle effectués sur des carottes prélevés au même site. Trois essais à grande échelle et plusieurs essais à petite échelle ont été réalisés. Des blocs de 70×70×35 cm ont été découpés dans du quartzite traversé par des plans de stratification inclinés contenant du matériau de remplissage. Pour les essais à petite échelle on a utilisé une boîte à essais de cisaillement portative. On donne les résultats obtenus pour les deux types d'essai ainsi que les conclusions qu'on en tire.

ZUSAMMENFASSUNG

Die potenziellen Einflüsse des Klufttest-Formats an den Scherfestigkeitsmessungen sind oft hervorgehoben worden, und experimentelle Untersuchungen haben kürzlich gezeigt, daß es besonders im Falle der rauhen und unebenen Kluftarten bedeutende Maßstabeinflüsse an der Scherfestigkeit und dem Deformationsverhalten gibt. In der vorliegenden Arbeit werden die am Sır Staudamm bei K. Maraş (Türkei) durchgeführten Versuche, sowie die unter Verwendung von kleinen Maßstäben an den an gleicher Stelle entnommenen Kernen durchgeführten Scherungstests dargestellt. Insgesamt sind drei großmaßstäbliche und mehrere kleinformatige Scherversuche durchgeführt worden. Blöcke von 70×70×35 cm wurden aus Quarzit herausgeschnitten, der von geneigten und Füllungsmaterial enthaltenden Schichtungsebenen durchsetzt war. Bei den kleinformatigen Versuchen sind tragbare Scherapparate verwendet worden. Die Resultate der beiden Versuchstypen wurden analysiert, und die daraus abgeleiteten Schlußfolgerungen sind beigefügt.

INTRODUCTION

The behaviour of rock masses is generally controlled by discontinuities in the rock mass. Determination of the shear properties of the discontinuities relevant to stability analysis may vary from a large scale test to small scale laboratory direct shear test.

Small scale samples usually represent only a fraction of the natural joint exposures and tests on these samples sometimes yield unrepresentative data. But they are cheap and easy to carry out when it is compared to insitu tests. So, as Wareham and Sherwood (1974) pointed out, the choice of an appropriate joint test-size is generally based on both economic and technical considerations.

The existing data in the literature from small and large scale tests are extremely limited and often inconclusive. Londe (1972) compared the results of small scale laboratory tests with large scale field tests and reported a good agreement between the residual friction angles found by both methods.

Pratt, et. al. (1974) carried out a series of field and laboratory tests on a range of joint sizes in a weathered quartz diorite and showed that the peak shear strength of natural joints decreased by 40% as the sample areas increased from 142 to 5130 cm^2.

Bandis, et. al. (1981) performed tests on different sized and types joint surface replicas obtained by using moulding technique and gave detailed experimental evidence to the positive scale effect as Pratt, et. al. (1974). The results showed significant scale effects on both the shear strength and deformation characteristics. Scale effects are more pronounced in the case of rough, undulating joint types, whereas they are virtually absent for planar joints. Bandis, et. al.(1981) summarised that increasing block size or length of joint leads to:

a) a gradual increase in the peak shear displacement,
b) an apparent transition from a "brittle" to "plastic" mode of shear failure,
c) a decrease of the peak dilatation angle,
d) insignificant scale effects in the case of relatively planar to smooth joint types.

Bandis, et. al. (1981) also stated that negative scale effect met in the literature is due to dissimilar roughness on the small and large scale joints.

On the other hand, in the case of filled joints, as Bandis, et. al. (1981) pointed out no scale effect is to be expected in the cases with a thickness of filling larger than the roughness amplitude.

In this work, shear characteristics of planar to slightly undulated bedding planes with filling material of 0-30 mm thickness are examined with large and small scale tests.

TEST SITE AND TESTING PROCEDURE

This study has been carried out as a part of the geotechnical investigations at Sır dam site of the hydroelectric power plant of Berke project, in K.Maraş, Turkey.

Sır dam site, on Ceyhan river, at 32 km. southwest of K.Maraş is located on regularly dipping hard and fractured metaquartzite. The most critical discontinuity plane of the formation in the area is the bedding plane which dips to the river with angle of 30-40°.

The filling material occuring along the bedding planes varies between 0·30 mm in thickness and contains clay, sand and fine grained angular quartz fragments. This situation is well defined on the left bank of the valley. During tectonic movements, shear failure occured along these bedding planes. This property is obscure or invisible in some bedding planes, which have no filling material and generally located at the right bank of the valley (Sümerman 1979). The surfaces of these presheared bedding planes are slickensided and planar and some of them are planar to slightly undulated (ISRM 1978).

Since the orientation of these planes are unfavourable at the left bank, it is expected to create instability problems.

The shear behaviour of the filled bedding planes at Sır dam site has been determined both by in situ and portable shear box methods.

Three large scale in situ shear tests and seventy small scale (portable shear box) tests were conducted in the area.

The adits at the left bank which have been driven for geological exploration and mapping were utilized for in situ testing while small scale shear tests were carried out on 76 mm diameter core samples which have been recovered by drillings across the bedding planes on both banks. During coring operations, some of the samples recovered as unfilled samples. Totally 30 samples **were** tested in filled condition.

Test blocks of approximately 70x70x35 cm in sizes were tested at normal stresses of 1, 1.5, and 3 MPa in increasing order.

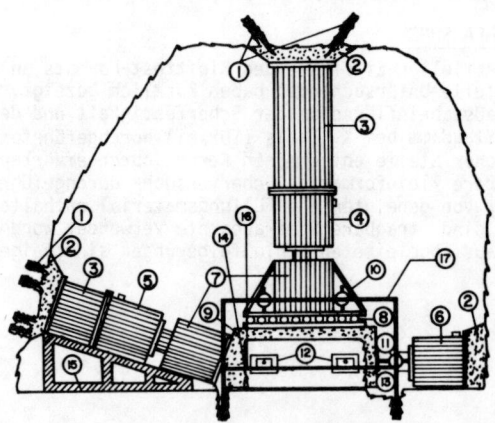

Figure.1 Insitu shear test setup

1-Anchors, 2-Reinforced reaction pads, 3-Reaction columns, 4-Normal load hydraulic jack, 5 Shear load hydraulic jack, 6-Reverse shear hydraulic jack, 7-Load spreaders, 8-Rollers, 9-Reinforced concrete encapsulation, 10-Normal displacement dial gage, 11-Shear displacement dial gage, 12-Lateral displacement dial gage, 13-Glass plate, 14-Filling, 15-Wooden framework, 16-Test block, 17 Dial gages reference beam

The complete test set up, designed at METU (Middle East Technical University), shown in Figure 1 was used for in situ shear tests.

Before testing, consolidation is provided by raising the normal stresses up to their full values in each stress level. The consolidation stage was completed when the rate of change of normal displacement recorded at each of the four vertical gages was less than 0.05 mm in ten minutes (ISRM 1974). The consolidation stages of all tests have approximately been completed in half an hour. Then, the shearing has been carried out in accordance with the ISRM standards (ISRM 1974).

Portable shear box, developed in Imperial College and manufactured by Robertson Research Group was used to test small scale samples, and tests have been carried out in accordance with ISRM standards (1974) at 0.6, 1.2 and 2 MPa normal stresses.

DISCUSSIONS AND CONCLUSIONS

From the analysis of shear stress-shear displacement results of three in situ tests, it is seen that the definite peak shear strengths are absent. They show an accordance with mode 4 shear behaviour as suggested by Wallace (1976) for rock discontinuities. After a certain shear displacement (5 to 20 mm), peak and residual shear strengths are the same (Figure 2a). The reason

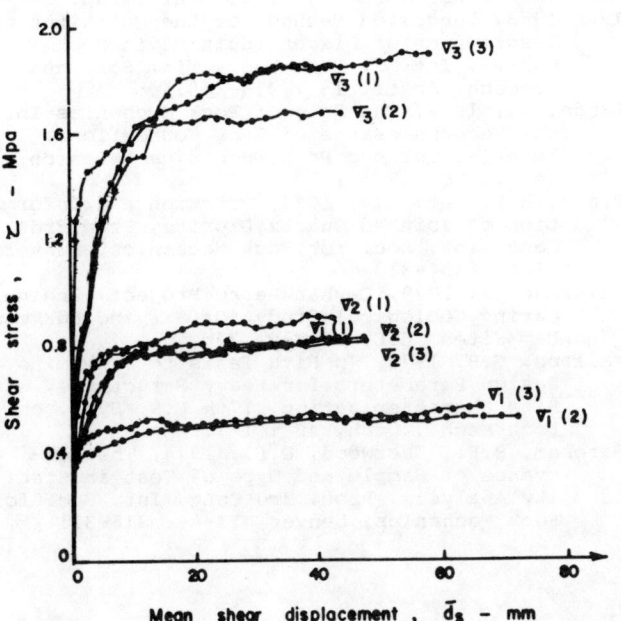

Figure.2a Shear stress-Shear displacement graph for insitu shear tests

for this type of shear behaviour may be due to:

a) Preexisting displacements have been occured along these surfaces (Sümerman 1979),
b) The planar and smooth to slickensided characteristics of the shear surfaces,

c) The existence of filling material whose thickness is generally greater than the amplitude of the joint wall roughness. This situation is also proved having the negative dilatation angles from the tests (Figure 2b). This nega-

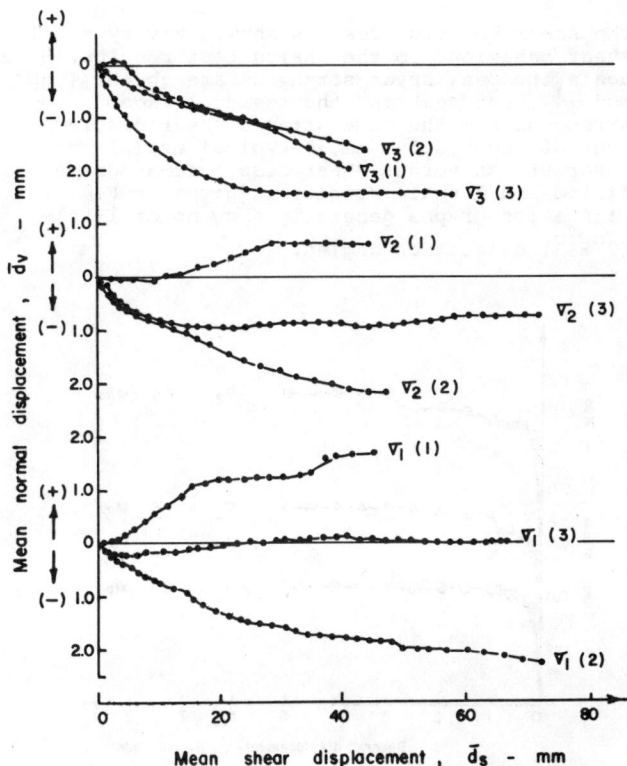

Figure.2b Normal displacement-shear displacement graph for insitu shear tests

tive dilatation was caused by the volume decrease of the filling material, which was dispelled from the sides of the test blocks under applied normal stresses during tests. It was also observed, after the tests, that no rock to rock contact has been occured on the exposed shear surfaces.

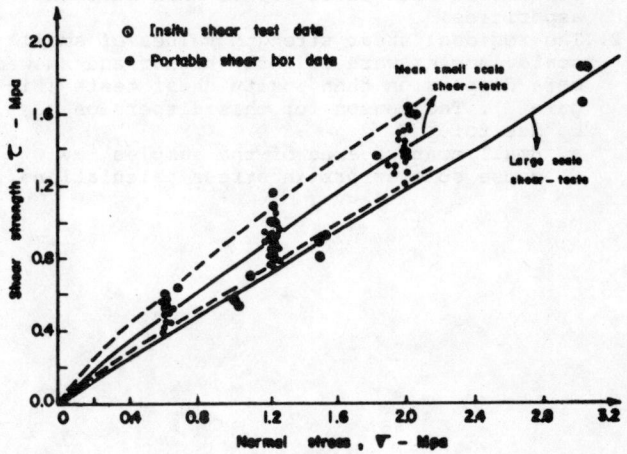

Figure.3 Comparison of shear strength values

A 259

The shear stress versus normal stress graphs generally show linear relationship which is passing through the origin without having no apparent cohesion (Figure 3). This is another indication of the fact that the filling material controls shearing.

The shear box test results showed a very similar shear behaviour to the insitu test results. Here again the peak shear strengths are absent (as mode 4), and peak and the residual shear strengths are the same after a shear displacement of about 0.5 1 mm. A typical example of shear stress versus shear displacement graph for filled small scale samples is given in Figure 4. Dilatation graphs generally show no or little (2^O-3°) dilatation angles.

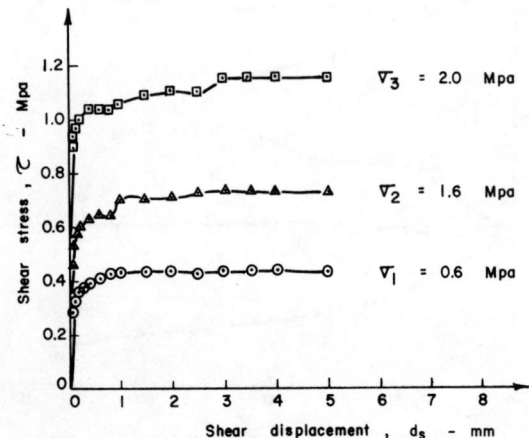

Figure.4 A typical example of shear stress shear displacement graph for shear box tests

When the small scale (shear box) tests, having the shear surface of about $40 cm^2$, are compared with the in situ tests (shear area of $4900 cm^2$), it is found that:

1. The shearing was controlled by filling materials in both tests rather than surface asperities.
2. The residual shear strength values of small scale samples were slightly higher and showed more dispersion than insitu shear tests (Figure 3). The reason for this dispersion may be due to:
   a. Small contact area of the samples may cause some errors in stress calculations,

b. The degree of roughness of discontinuity walls may easily cause an increase or decrease in shear strength in such a small scale,
c. The experimental errors may be greater in shear-box tests than in situ tests.

The mean shear strength envelope of in situ tests show linear relationship passing through the origin, giving the residual friction angle $(\emptyset_r)$ of 30.5^O while the mean shear strength envelope of shear box results show nonlinear relationship at low normal stresses and it becomes roughly parallel to the in situ shear tests envelope at higher normal stresses, giving $\emptyset_r = 31.6^O$ and $c_{app} = 0.15$ MPa.

From this, it can be said that there seems a good agreement between the values of residual shear strengths of small and large scale tests, for filled quartzite bedding planes, at high normal stresses.

REFERENCES

Bandis, S.,et. al.,1981, Experimental Studies of Scale Effects on the Shear Behaviour of Rock Joints, Int. J. Rock Mech. Min. Sci. and Geomech. Abstr. 18 : 1-21.
ISRM 1974, Suggested Methods for Determining Shear Strength, Commission on Standardization of Laboratory and Field Tests, Committee on Field Tests, Final Draft, 23p.
ISRM 1978, Suggested Method for the Quantitative Description of Discontinuities in Rock Masses, Int. J. Rock Mech. Min. Sci. and Geomech. Abstr. 15 : 319-368.
Londe, P., 1972, The Role of Rock Mechanics in the Reconnaissance of Rock Foundations, Imperial College Rock Mechanics Research Report No: 17 : 1-23
Pratt, H.R., et. al., 1974, Friction and Deformation of Jointed Quartz Diorite, Proc 3rd Cong. Int. Soc. for Rock Mechanics, Denver, II-A : 306-310
Sümerman, S. 1979, Ceyhan Berke Project, Engineering Geological Study for Sir and Düzkesme Dam Sites, EIE, Turkey, 79p.
Wallace, G.B. 1976, In Situ Tests to Determine Design Parameters for Heavy Structures, Site Characterization, 17th U.S. Symp. on Rock Mech., Utah, 4B : 1-7.
Wareham, B.F., Sherwood, D.E., 1974, The Relevance of Sample and Type of Test in Stability Analysis, Proc. 3rd Cong. Int. Soc. for Rock Mechanics, Denver, II-A : 316-321.

A STUDY OF ROCK FRACTURES

Analyse de fractures de roche

Analyse von Gesteinsbruchflächen

Xianwei Li, Associate Professor
Yongrui Lan, M.E.
Jungxing Zou, Assistant
China Mining Institute, Xuzhou, China

SYNOPSIS

Fractures resulting from the breaking of rock under different stress conditions can be divided into 10 patterns of tensile fractures and 8 patterns of shear fractures. Photographs of fractures are taken by transmission and scanning electron microscopes. Thus, by the pattern of a fracture it is possible to determine whether it is a tensile or a shear fracture. This might be of value in mining practice.

RESUME

D'après des photographies prises au moyen de microscopes de transmission et de balayage électroniques, les fractures qui résultent de ruptures de roches peuvent se diviser en dix types de rupture par traction, et huit types de rupture par cisaillement. Ainsi, on peut déterminer s'il s'agit d'une fracture par traction ou d'une fracture par cisaillement, ce qui pourrait être utile en matière d'exploitation minière.

ZUSAMMENFASSUNG

Bruchstellen von Gesteinen, die auf verschiedene Beanspruchungen zurückzuführen sind, können in 10 Varianten der Zugbruchstellen und 8 der Scherbruchstellen unterteilt werden. Aufnahmen solcher Bruchstellen werden mittels Transmissions- und Abtastmikroskopen gemacht. Auf diese Weise kan man feststellen, ob es sich um eine Zug- oder Scherbruchstelle handelt, was im Bergbaubetrieb wichtig sein kann.

Materials can be broken under various stresses (such as compression,tension,bend,shear and torsion),exceeding the strength limits of the materials. Fractures appear where the material failed. The characteristics of fractures do not remain the same because the types of stresses are different. Therefore it will be of much help to mining practice to have a better understanding of the properties and features of fractures and to make clear the mechanism of the fracturing process. It would lead to the development of a new field of science and technology which studies the fracturing of materials and the analysis of failure.

Rock is the object of study in the field of rock mechanics. There are only two kinds of rock phenomena in mining, viz. fragmentation and stability. Rock fragmentation includes mechanical fragmentation, high pressure water jet cutting, blasting etc.; rock stability includes pressure of rock strata in mining, slope stability, stability of foundation, geomechanics etc. In mining practice various types of stresses lead to rock fragmentation, then different kinds of fractures occur. If we make beforehand a study of rock fractures under different stresses in order to obtain typical patterns (which is exactly the aim of the present article), the type of fragmentation can be identified and the economical method for rock fragmentation can be determined by comparing the rock fractures which occur in mining practice with the typical patterns.For example,comparison can be made with several cutting tools to find out which one will cause more tensile fractures in rock fragmentation spending less energy.Thus the most favourable cutting tool can be selected.

In this experiment marble(for most cases)and granite are used.Tests are conducted with uniaxial compression(UC),biaxial compression(BC), triaxial compression(TC),diametrical compression of cylinder by wire(DCCW),diametrical compression of cylinder by arc(DCCA),simple tension(ST),point load test(PLT),bending of square beam(BSB),shearing in inclined dies(SID),torsion test(TT)etc.Two types of microscopes have been used:JEM-2oocx type transmission electron microscope with plastic film of carbon twice recovery pattern,projection of chronium and JEE-4x high vacuum film plating unit;Jelo Super-probe 733 type scanning electron microscope with DM-25o(Iv)type vacuum film-plating unit for surface gold plating.

According to observation with transmission and scanning electron microscope patterns of fractures can be divided into two catégories:tensile fractures and shear fractures.

1.TENSILE FRACTURES

1.1Cleavage fracture

Fracture occurs inside the crystals.

1.1.1 Stream pattern(SrP)and step pattern(StP)
When there is vacancy between crystals,stream
pattern is formed with combination of crystal
faces in cleavage along a cluster of parallel
crystal faces on various heights during cleav-
ing (Fig.1). Step pattern is formed if crystal
faces in cleavage do not combine(Fig.2). Stream
pattern is shown in the upper right of Fig.3
and step pattern in the lower left.

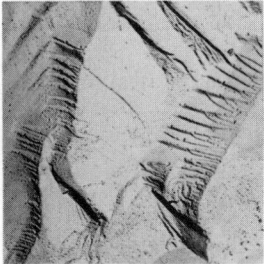
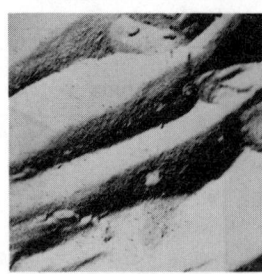

Fig.1 SrP BSB
      TEM 11000x

Fig.2 StP BC
      TEM 22000x

1.1.2 Tongue pattern(TP)
The formation of cleavage tongue is the result
of the main crack intersecting the crystal
twin and development along common crystal face
crystal twin and basic body(Fig.4).

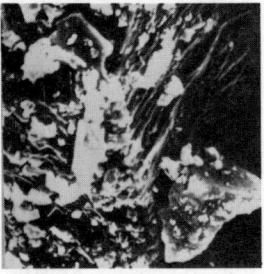
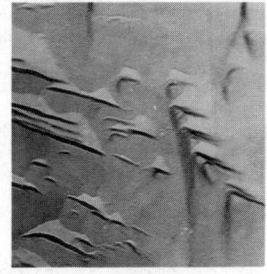

Fig.3 SrP&StP BC
      SEM 1000x

Fig.4 TP BSB
      TEM 66000x

1.1.3 Fishbone pattern(FP)
This pattern is formed when the main cleavage
crack expands forward along crystal twin faces
and basic body faces which form the two sides
of the fishbone but the intersection part of
crystal twin and basic body forms the main bone
at the middle of the fishbone(Fig.5).

1.1.4 Root pattern(RP)
Root pattern can also be found in tensile
stress zone. It appears in conditions of tensile stress and brittle failure(Fig.6).

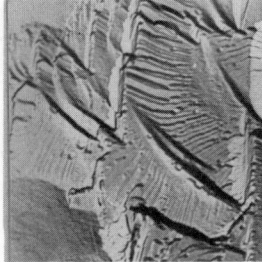

Fig.5 FP BSB
      TEM 15000x

Fig.6 RP PLT
      TEM 15000x

1.1.5 Equiaxial-microhole pattern(EMP)and triangular-microhole pattern(TMP)
Sometimes brittle failure forms equiaxial-microhole and triangular-microhole under tensile
stress(Fig.7 and 8).

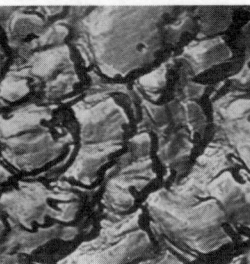
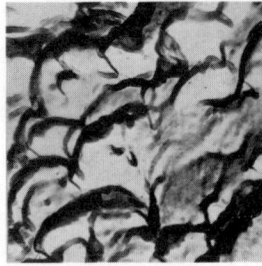

Fig.7 EMP BSB
      TEM 44000x

Fig.8 TMP BSB
      TEM 33000x

1.2 pattern of complete crystalline grain
Fractures are formed either on the crystalline
boundary or in the space between crystals,or in
both places because the strength of crystal is
larger than both the boundary and the space.

1.2.1 Fracture pattern of crystalline boundary
(PCB)
Fracture appears on the face of crystalline
boundary when the strength of crystal and space
between crystals is larger than the strength of
crystalline boundary(Fig.9).

1.2.2 Fracture pattern of crystalline boundary
and space between crystals (PBSC)
Fracture appears on crystalline boundary and
space between crystals because it takes place
on the weakest strength face,when the strength
of crystals is the strongest and the strength
of crystalline boundary nearly equals that of
space between crystals(Fig.10).

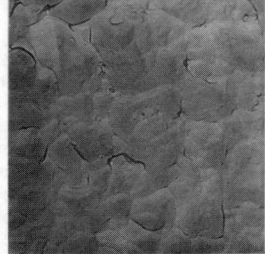

Fig.9 PCB DCCA
      TEM 44000x

Fig.10 PBSC PLT
       TEM 15000x

1.2.3 Fracture pattern of space between cryst-
als(PSC)
Fracture appears in space between crystals when
the strength of crystal and the strength of
crystalline boundary are all larger than the
strength of space between crystals(Fig.11 and
12).

2.SHEAR FRACTURES

2.1 Pattern of parallel slip line(PPSL)
This kind of parallel straight slip line pattern is formed when large single crystal materials are sheared to failure and slip flow
appears(Fig.13).

2.2 Pattern of small grains with linear
arrangement(PGLA)

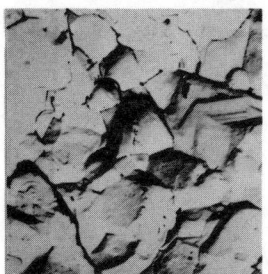

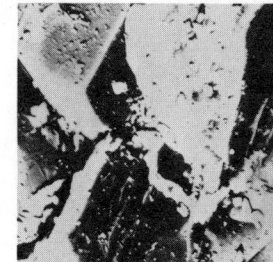

Fig.11 PSC   BC   Fig.12 PSC   ST
      TEM 11ooox   SEM 5oox

These small grains appear when rock is sheared
intensely(Fig.14).

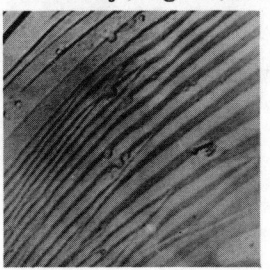

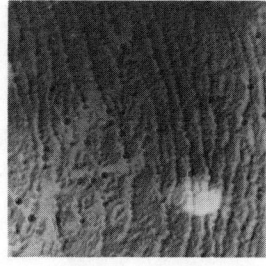

Fig.13 PPSL   PLT   Fig.14 PGLA   TT
      TEM 15ooox   TEM 66ooox

2.3 Stripe pattern(SP)
Stripes may be seen through a lowfold microscope at the fracture of the rock broken by a
shear force(Fig.15).

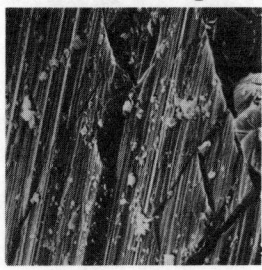

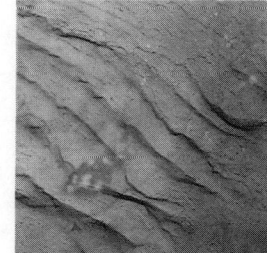

Fig.15 SP SID   Fig.16 SSP TT
      SEM 1ooox   TEM 15ooox

2.4 Snaky-slipping pattern(SSP)
multiple crystal materials slip along many intersecting slipping faces but not slip along a
certain slipping face because of the restriction of crystals with different arrangement.
Therefore snaky-slipping pattern,that is,line-
ar slip crooks,is formed(Fig.16).

2.5 Double-slipping pattern(DSP)
In fig.17 the two slipping systems are shown
with the longer line in one direction and the
shorter one in the other direction.When reso-
lutions of shear stress on the two slipping
systems are higher than critical values,cryst-
al slips in both directions.This is called the
double-slipping pattern.

2.6 Flat-face pattern(FFP)
Flat-face pattern occurs sometimes under
shear stress(Fig.18).

2.7 Tetrahedronal pattern(TP)
Tetrahedronal grains are shown in Fig.19.The

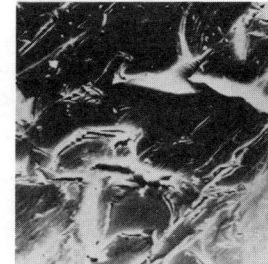

Fig.17 DSP DCCA   Fig.18 FFP   TC
      TEM 44ooox   SEM 5oox

darkest faces are basically parallel.This characteristic is formed by shear effect.

2.8Rectangular-microhole pattern(RMP)
A type of rectangular-microhole pattern is often formed by shear stress on the rock(Fig.2o).

Fig.19 TP   BSB   Fig.2o RMP   TT
      TEM 33ooox   TEM 55ooox
3.CONNECTION OF TENSION AND SHEARING(CTS)
It shows that there can be characteristics of
tension and shearing on one photograph.Fig.21
shows that fracture pattern of crystalline
boundary occur in the upper part on the left
(Fig.9)and double-slipping pattern in the lower part on the right(Fig.17).Fig.22 shows that
fracture pattern of crystalline boundary occurs
on the left(Fig.9)and tetrahedronal pattern
and pattern of linear arrangement of small grains on the right.

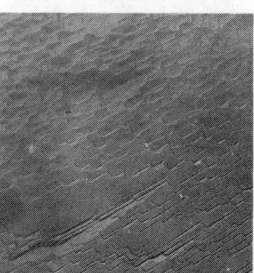

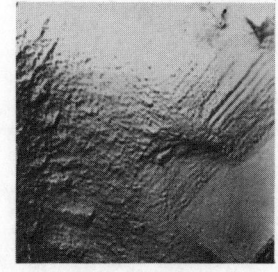

Fig.21 CTS DCCA   Fig.22 CTS DCCW
      TEM 15ooox   TEM 26ooox

Reference:
Editorial group of"Analysis of Metal Fractures"
Shanghai Jiaotong University,"Analysis of Metal Fractures",July,1979,The Publishing House
of Defense.

AN EMPIRICAL FAILURE CRITERION FOR ROCK MASSES

Un critère empirique de rupture pour les masses rocheuses
Ein empirisches Bruchkriterium für Felsgestein

Yudhbir
Professor of Geotechnical Engineering A.I.T., Bangkok, Thailand
Willy Lemanza
Engineer, P.T. Indotec, Bandung, Indonesia
F. Prinzl
Assistant Professor of Geotechnical Engineering A.I.T., Bangkok, Thailand

SYNOPSIS

Results of triaxial tests conducted on a soft rock like material, under intact and disintegrated conditions, are presented. The test data are interpreted to propose an empirical failure criterion for rock masses over the brittle to ductile behaviour range. The parameters necessary for this criterion are evaluated from uniaxial compressive strength obtained from point load test and field data on rock mass quality index. Predictions by the proposed failure criterion are compared with those by the Hoek and Brown relationship.

RESUME

On présente les résultats de tests triaxiaux effectués sur une roche tendre, d'abord intacte, puis desagrégée. Les données des tests sont interprétées de manière à proposer un critère empirique de rupture pour les masses rocheuses dans la gamme de comportement fragile à malléable. Les paramètres nécessaires à ce critère sont évalués à partir de la force compressive uniaxiale obtenue au moyen du test de charge ponctuelle et de données expérimentales relatives à l'indice de qualité de la masse rocheuse. Les prévisions basées sur le critère de rupture proposé sont comparées à celles fournies par la relation de Hoek et Brown.

ZUSAMMENFASSUNG

Die Ergebnisse von Dreiachsialversuchen, die mit unversehrtem und zerbrochenem Fels aehnlichem Material durchgefuehrt wurden, werden gezeigt. Mit der Deutung der Versuchsergebnisse wird ein empirisches Bruchkriterium fuer Fels abgeleitet, das fuer den sproeden und verformbaren Bereich gueltig ist. Die fuer das Kriterium notwendigen Parameter werden mit der einachsialen Druckfestigkeit, bestimmt mit Punkt-Last Versuchen, sowie mit Felsgueteziffern von Feldaufnahmen ermittelt. Aussagen mit Hilfe dieses Bruchkriteriums und aus der Beziehung nach Hoek und Brown werden verglichen.

I. INTRODUCTION

For design of slopes and foundations on rock masses - it is well recognised now that engineers need a failure criterion which is applicable over the stress range from brittle to ductile behaviour of rocks. The criterion should also be able to encompass the whole range of conditions varying from intact rock to highly jointed and disintegrated rock mass. The relationship should preferably be formulated in terms of parameters that are easy to evaluate and are correlated to rock mass quality indices such as Rock Mass Rating, RMR proposed by Bieniawski (1974a, 1976) and/or rock mass Quality, Q proposed by Barton et al (1974). Hoek and Brown (1980) have proposed an empirical criterion which attempts to satisfy some of these requirements. Hoek and Brown criterion was developed from intuitive reasoning and by "trial and error process based upon experience of both theoretical and experimental studies of rock failure".

In this paper results of a laboratory experimental study on a rock-like material simulating soft rock are pre - sented. The main objectives of this study were to examine:

1. the possibility of an engineering failure criterion for rock masses covering the whole range from brittle to ductile behaviour.

2. the stress range of applicability of the Hoek-Brown criterion in the case of soft rock, and

3. the stress range where the transition from brittle to ductile rock behaviour takes place.

II. SELECTION OF MODEL MATERIAL

Stimpson (1970) has discussed in detail the large variety of model materials used in rock mechanics studies. Einstein et al (1970) have discussed the selection criteria for model materials to simulate rock behaviour. In this study a mixture of gypsum and celite was used as a model material. From 20 trial samples, with varying mixture ratios and curing procedures, it was established that a soft rock-like material with combination of brittle behaviour, acceptable strength, lack of bleeding, smooth surface characteristics, and a minimum number of bubble holes was found for a gypsum-celite-water mixture with water/gypsum ratio = 0.55, & water/celite ratio = 32 (celite/gypsum ratio = 1.72 % by weight).

1. Characteristics of model material

The most important material properties and Π-factors of the model material are given in table I. Also shown in table I are the corresponding values for 3 soft rocks.

2. Preparation of model material

Gypsum and celite were dry mixed for 3 minutes and then mixed with water for 1 minute - a duration of mixing that led to practically no bleeding. Then the mix was poured into a mould which was placed on a vibrating table and vibrated for 1-minute to ensure expulsion of much of the air without causing segregation of the mix components.

After 30 minutes when it was at room temperature, the sample was extruded, wrapped in a tight plastic wrap and cured in a humidity room for 3 days.

Table I Material Characteristics & Π-factors Material Characteristics

Property	Model	Shale[1]	Tuff[2]	Limestone[3]
γ, density, t/m^3	1.67	2-2.4	1.84	2.6-2.85
σ_c, compressive strength, MPa	4.1	5-100	4.5	50-200
σ_t, tensile strength MPa	0.7	2-10	0.9	4-7
E, Young's modulus of elasticity, MPa	1.3x10^3	8x10^2-3x10^4	1.5x10^3	5x10^4 - 8x10^4
C_0, shear strength, MPa	1.0	0.3-3.84	1.2	3-7
ν, Poisson's ratio	0.15	0.2	0.29	0.1-0.2
ϕ_u angle of shearing resistance	32	32	30	-

Π-factors

factor	model	shale	tuff	limestone
σ_c/σ_t	6	3-10	5	13-29
E/σ_c	315	160-300	333	1000-400
C_0/E	7x10^{-4}	4x10^{-4}-1x10^{-4}	8x10^{-4}	6x10^{-5}-1x10^{-4}
ν	0.15	0.2	0.29	0.1-0.2
ϕ_u	32	32	30	-

1 = Goodman (1980),2 = Adachi et al (1981), 3 = Szechy (1973)

3. Model samples

A total of 122 model samples were tested during the investigation. Fig. 1 shows types of specimens used to simulate; (a) intact rock, (b) smooth jointed rock, (c) disintegrated rock, and (d) rough jointed rock. Disintegrated rock samples were prepared at two densities, viz 1.65 & 1.25 t/m^3. Intact samples were broken into pieces and disintegrated specimens were composed of 75% coarse particles (-3/8" to + 1/4") and 25% fine particles (-1/4" to + 4/25"). These fractions were mixed uniformly and recompacted to the required value of density (γ). Sonic velocity (before test), water content (after test), and value of density were used to excercise control on the quality of samples moulded throughout the investigation.

III. TYPES OF TESTS

Uniaxial compression, triaxial compression in Hoek-Franklin type triaxial cell (at a rate of 0.0625 cm/minute), direct shear tests (in the shear box commonly used for soils; at a rate of 0.06 cm/minute), and Brazilian tests were carried on during this investigation.

IV. TEST RESULTS

Results of triaxial tests, at different confining pressures, are shown in figs. 2, 3, 4 & 5 for intact, smooth jointed, and disintegrated specimens respectively. In case of intact samples, the stress-strain behaviour is brittle for confining pressures, $\sigma_3 \leq 2$ MPa beyond which a distinct transition to ductile behaviour is exhibited (fig. 2). In case of specimens, with a smooth 30° plane (fig. 3), brittle to ductile transition is evidenced for

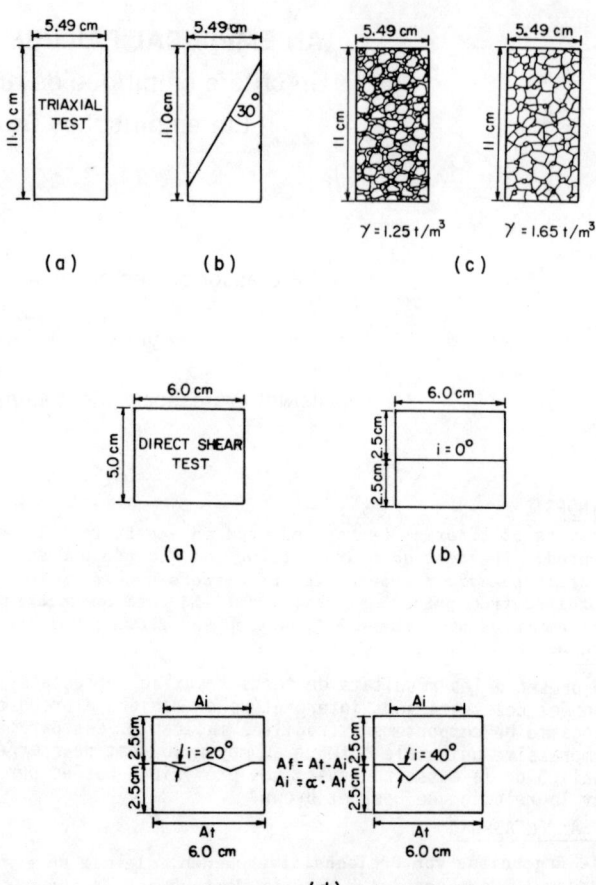

Fig. 1 Types of specimens tested
(a) intact (b) smooth jointed
(c) crushed (d) rough jointed

values of $\sigma_3 \approx 2$ MPa. For $\sigma_3 < 2$ MPa, brittle behaviour is more pronounced and the samples attained peak value of strength (less than the corresponding value for intact samples) at strains less than those for intact specimens. For $\sigma_3 > 2$ MPa both intact and smooth jointed samples give more or less identical stress-strain response. This may be due to the fact that the normal stresses on the joint plane become so large that corresponding shear stresses are sufficient to cause failure through intact material, and little or no sliding takes place along the joint plane. Thus the transition from brittle to ductile response may be coincident with the change in failure mode from sliding to fracture. Einstein et al (1970) observed similar behaviour for their model specimens.

In case of disintegrated specimens, there is continuing yeilding from the start of the test with no distinct transition from brittle to ductile behaviour. In general, for low values of σ_3, samples with higher density gave higher values of strength (fig. 4) compared to those with lower density (fig. 5). At higher values of σ_3 samples at both densities tend to give similar response. At low confining pressures the volume change responses of samples with different densities are expected to be different, and this may explain the observed differences in the stress-strain behaviour. Unfortunately the volume change response could not be recorded in the triaxial set up utilised.

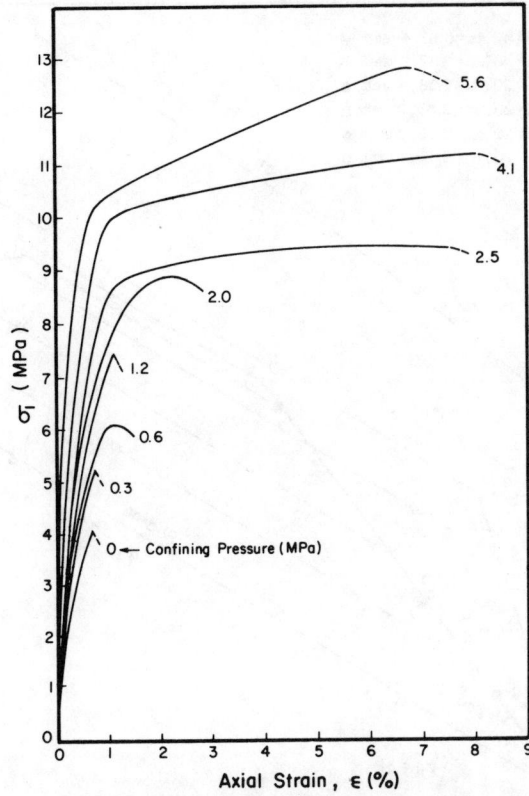

Fig. 2 <u>Stress-strain behaviour of intact speci-mens in triaxial compression</u>

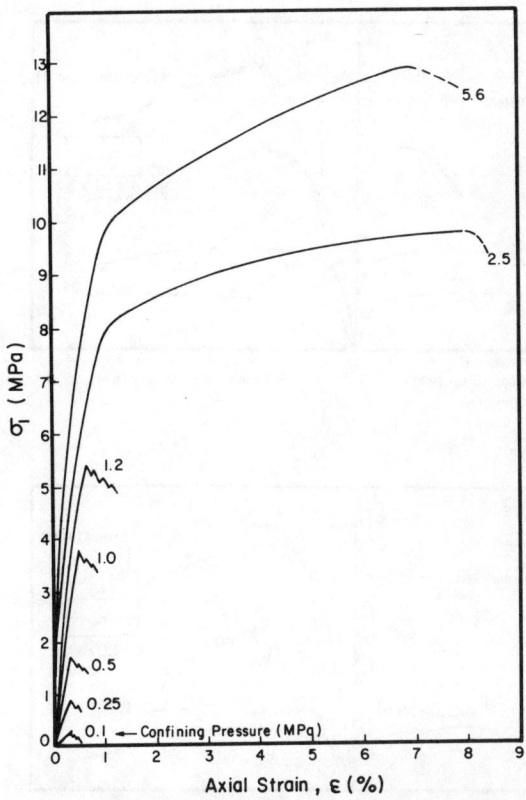

Fig. 3 <u>Stress-strain behaviour of smooth jointed samples in triaxial compression</u>

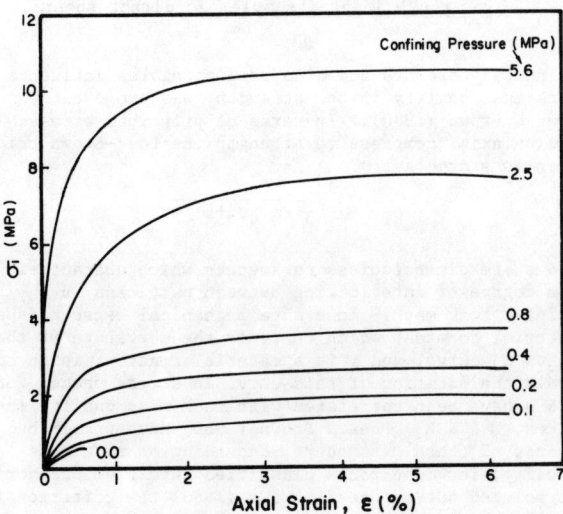

Fig. 4 <u>Stress-strain behaviour of crushed samples (γ = 1.65 t/m^3) in triaxial compression</u>

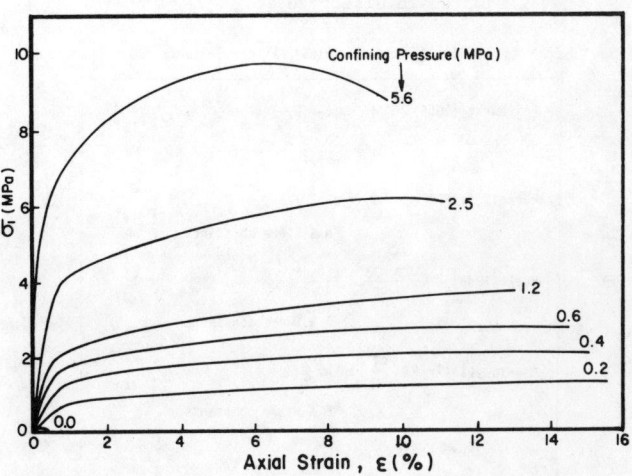

Fig. 5 <u>Stress-strain behaviour of crushed samples (γ = 1.25 t/m^3) in triaxial compression</u>

Figs. 6 & 7 show typical results obtained from direct shear tests on rough jointed specimens. Implications of these test results, especially the influence of the de-gree of joint roughness (ratio of area of asperities to the total area of joint surface, α in fig. 1.d) on the τ vs σ relationships for jointed rockmass as proposed by Ladanyi & Archambault (1970) will be reported elsewhere. It will be noted in fig. 7 that the transition from

sliding to fracture is gradual and not abrupt as postu-lated by a bilinear relationship.

V. EMPIRICAL FAILURE CRITERIA

Attempts to better fit available experimental data has produced some useful criteria for intact rocks. Some of the notable ones are shown in table II. All these cri-

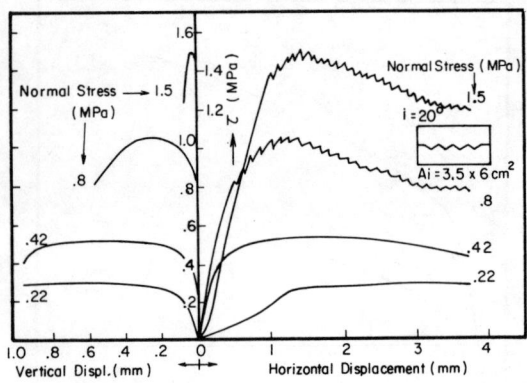

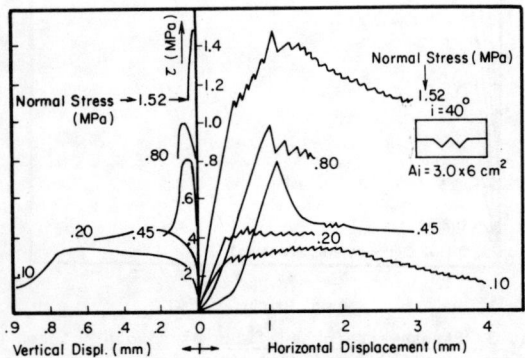

Fig. 6 Stress-horizontal displacement-vertical displacement behaviour of rough jointed samples in direct shear

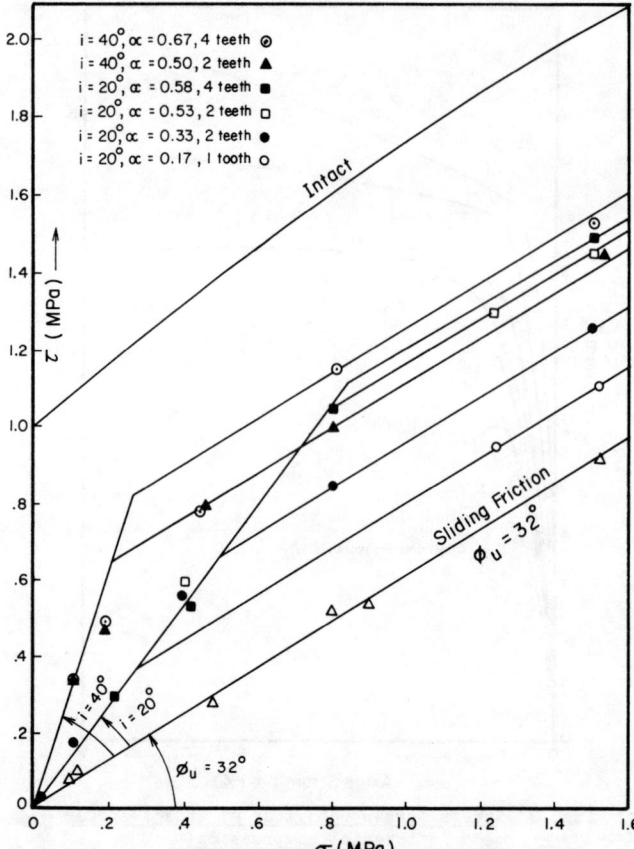

Fig. 7 Shear stress vs normal stress relationships for rough jointed samples in direct shear

Table II Some Empirical Failure Criteria

Fairhurst (1964)	$\tau = \sigma_c \left(\frac{m-1}{n}\right) \left(1+n\frac{\sigma}{\sigma_c}\right)^{\frac{1}{2}}$; $n = \frac{\sigma_c}{-\sigma_t}$, $m = (n+1)^{\frac{1}{2}}$	(1)
Murrell (1965)	$\sigma_1 = F\sigma_3^A + \sigma_c$, A & F are constants	(2)
Hoek (1968)	$\frac{\tau_m-\tau_0}{\sigma_c} = D\left|\frac{\sigma_m}{\sigma_c}\right|^C$; C & D are constants	(3)
Bieniawski (1974)	$\frac{\sigma_1}{\sigma_c} = B\left|\frac{\sigma_3}{\sigma_c}\right|^\alpha + 1$; B & α are constants	(4)

τ = Shear stress

σ = Normal stress

σ_c = Uniaxial compressive strength

σ_t = Uniaxial tensile strength

σ_1 = Major principal stress

σ_3 = Minor principal stress

$\tau_m = \frac{\sigma_1-\sigma_3}{2}$, $\sigma_m = \frac{\sigma_1+\sigma_3}{2}$

teria are applicable in case of intact rocks only. An empirical failure criterion, capable of modelling the highly non-linear relationship between major & minor

principal stresses and also predicting the influence of rock mass quality on the strength, was proposed by Hoek & Brown (1980). In terms of principal stresses and uniaxial compressive strength the Hoek-Brown criterion is stated as:

$$\sigma_1 = \sigma_3 + \sqrt{m\sigma_c\sigma_3 + s\sigma_c^2} \qquad (5)$$

m & s are dimensionless parameters which characterize the degree of interlocking between particles in a jointed rock mass. In a more mechanical sense m is a material constant which controls the curvature of the σ_1 vs σ_3 curve, and s is a material constant which controls the location of this curve in stress space. Both m & s have been correlated with rock mass quality indices, Q & RMR. Hoek & Brown (1980) have prescribed values of these parameters, depending on rock mass quality, for most rocks classified into five categories. As pointed out by Hoek & Brown (1980) the criterion is applicable in the range of brittle rock behaviour, and tends to over predict in the range of ductile behaviour of rocks.

In an attempt to develop an empirical criterion which gives reasonable prediction in the ductile zone also, the authors decided to re-examine the form of criterion discussed by Bieniawski (1974). In a more general form the criterion is stated as:

$$\frac{\sigma_1}{\sigma_c} = A + B \left(\frac{\sigma_3}{\sigma_c}\right)^\alpha \qquad (6)$$

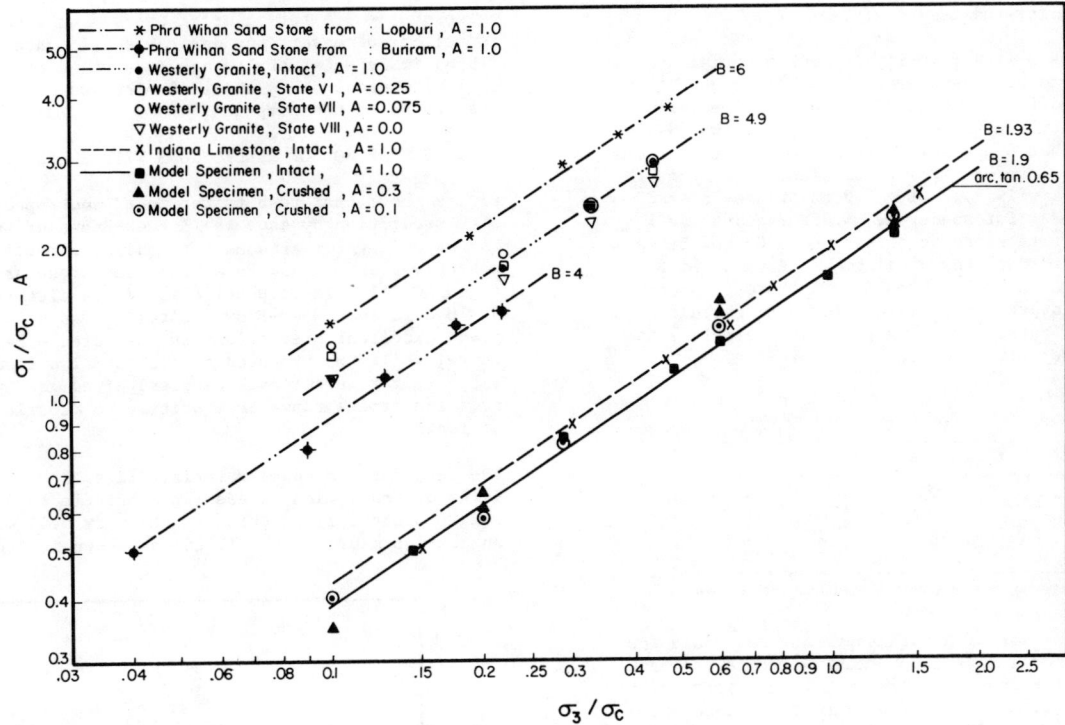

Fig. 8   Evaluation of parameters for the proposed empirical failure criterion

where A — is a dimensionless parameter whose value depends on the rock mass quality. Like Hoek & Brown's s parameter, it controls the position of the σ_1 vs σ_3 curve in stress space & A = 1 for intact rock and A = 0 for completely disintegrated rock. An attempt has been made to correlate A with Q & RMR,

B — is a rock material constant and depends on the rock type, and

α — is a constant and is shown to be independent of rock type and rock mass quality.

In the remaining part of this paper the major differences between the proposed relationship and the one recommended by Bieniawski (1974) will be brought out. Prediction of failure envelopes by the equation (6) will be compared with experimental test results and also with the corresponding predictions by the Hoek-Brown criterion.

VI. EVALUATION OF PARAMETERS α, B and A

The form of equ. (6) is such that on a log-log scale, (σ_1/σ_c-A) as ordinate vs (σ_3/σ_c) as abscissa will give a straight line. Fig. 8 shows triaxial test data for model material under intact and disintegrated conditions. Also shown in fig. 8 are the test results reported in literature (see Hoek & Brown (1980) for Indiana limestone & westerly granite; Jayawardane (1981) for Phra Wihan Sandstone).

Two important observations can be made from the results presented in fig. 8:

(i) within limits of both experimental errors & inevitable scatter, and sparse nature of the data analysed, the general trends in fig. 8. would encourage to make a reasonable assumption that data points for different rocks lie on near parallel lines. This would suggest that the

value of α in equ. (6) may be assumed constant for all rock types (irrespective of the rock mass quality). A close examination of actual data points in fig. 8 will no doubt show deviations from general parallelism, but the overall trends for the rocks examined here have encouraged the authors to assume a constant value of α (= 0.65).

Bieniawski (1974), in an attempt to best fit an empirical relationship to the experimental data, has also attempted to keep same α and varied B values. In this regard it may be stated here that rather than fitting each rock test data separately in σ_1/σ_c vs σ_3/σ_c plot, as suggested by Bieniawski (1974), the mode of presentation in fig. 8 is superior in the sense that the overall trend speaks for itself and presents a more convincing argument in favour of assuming a constant value of α which is independent of both rock type and rock mass quality. While the results in fig. 8 suggest α = 0.65, Bieniawski (1974) showed best fit with experimental data for five rocks with α = 0.75.

Hoek's criterion (Hoek (1968)), as shown by Bieniawski (1974), would also support a constant value of the exponent C in equ. 3 for 5 rock types. Of course the value of C is expected to be different from α because of the fact that Hoek utilised τ_m vs σ_m instead of σ_1 vs σ_3 stress space. The present analysis of limited data would suggest that α is constant for all rocks and its value lies in a close range of 0.65 to 0.75. Triaxial test data on different rocks are being collected to better define the range within which the α value lies so that an average α value could be recommended. The authors are particularly concerned about the lack of sufficient data for rock samples with different rock mass quality indices.

(ii) Parameter B is a property of rock material and
appears to be independent of rock mass quality as
evidenced from test data for the model material
and westerly granite (see fig. 8). It would also
appear that B has low value in case of soft rocks
& high value for hard rocks. Test data from
Bieniawski (1974): B = 3.0 for siltstone and
mudstone, 4 for sandstone, 4.5 for quartzite &
5 for norite supports this trend. On the basis
of limited data now available, the value of B
may be taken from table III. A thorough search
of literature for triaxial test data for different
rocks is currently being conducted to refine the
guidelines given in table III.

Table III Typical Values of Parameter B

B	2	3	4	5
Rock type*	tuff shale limestone	siltstone mudstone	quartzite sandstone dolerite	norite granite quartzdiorite chert

* rock associations have been adapted from Hoek & Brown (1980)

For evaluation of A parameter, the test data for
crushed model rock and westerly granite was back
fitted with equ. (6) for α = 0.65 and appropriate
value of B. The values of A thus obtained are
given in table IV along with the values of m & s
used in Hoek & Brown's criterion. For intact
rocks of course A = 1.0. It is recognised that
these results are not sufficient to generalise
a relationship between A and Q or RMR, however,
a tentative relationship is proposed here:

$$A = 0.0176 \, Q^{\alpha} \qquad (7)$$

$$\alpha = 0.65$$

or utilising the relationship between Q & RMR as
suggested by Bieniawski (1976),

$$A = \exp (0.0765 \, RMR - 7.65) \qquad (8)$$

The main differences between criterion proposed by Hoek
(1968), Bieniawski (1974), and equ. (6) is the fact that
both these criteria were only for intact rocks & A = 1.0,

Table IV A & B Parameters and Hoek & Brown's m & s Parameter

Discription	A	B	m	s
Model specimen:				
- intact	1.0	1.9	5.0	1.0
- crushed (γ = 1.65 t/m^3)	0.3	1.9	2.75	0.07
- crushed (γ = 1.25 t/m^3)	0.1	1.9	1.36	0.017
Indiana limestone	1.0	1.93	3.2	1.0
Westerly granite				
- intact	1.0	4.9	26.96	1.0
- broken:				
.state VI	0.25	4.9	14.31	0.68
.state VII	0.075	4.9	12.77	0.53
.state VIII	0.0	4.9	10.75	0.0
Phra Wihan sandstone				
- Lopburi	1.0	6.0	36.90	1.0
- Buriram	1.0	4.0	23.00	1.0

whereas the proposed criterion can predict for different
conditions of rock mass quality. Furthermore, the values
of B parameter are proposed as different for different
rock types. Bieniawski (1974) recommended an average

value of B = 3.5 for all rock types - an assumption which
is neither necessary nor desirable. In case of Hoek
(1968) it is necessary to assume a value of τ_0 (τ_m at
σ_m = 0) whereas the proposed & Bieniawski's criterion
require no such assumption.

VII. COMPARISON OF PREDICTIONS WITH EXPERIMENTAL DATA

Fig. 9 shows test results for model rock specimens alongwith prediction by equ. (6) & Hoek-Brown criterion.
Corresponding comparisons for Indiana limestone,
westerly granite, and Phra Wihan sandstone are shown in
figs. 10, 11 & 12 respectively. It is clear from figs.
9, 10 & 12 that Hoek-Brown criterian, as stipulated,
gives excellent predictions in the brittle behaviour
range, while the proposed criterion (also Bieniawski's
proposal for intact rock samples) gives good prediction
over the stress range from brittle to ductile behaviour
of rocks.

The test data on model material (figs. 2 & 9) gives
value of transition stress (from brittle to ductile)
equal to about 2 MPa which agrees very well with the
suggestion made by Mogi (1966) as shown in fig. 9.

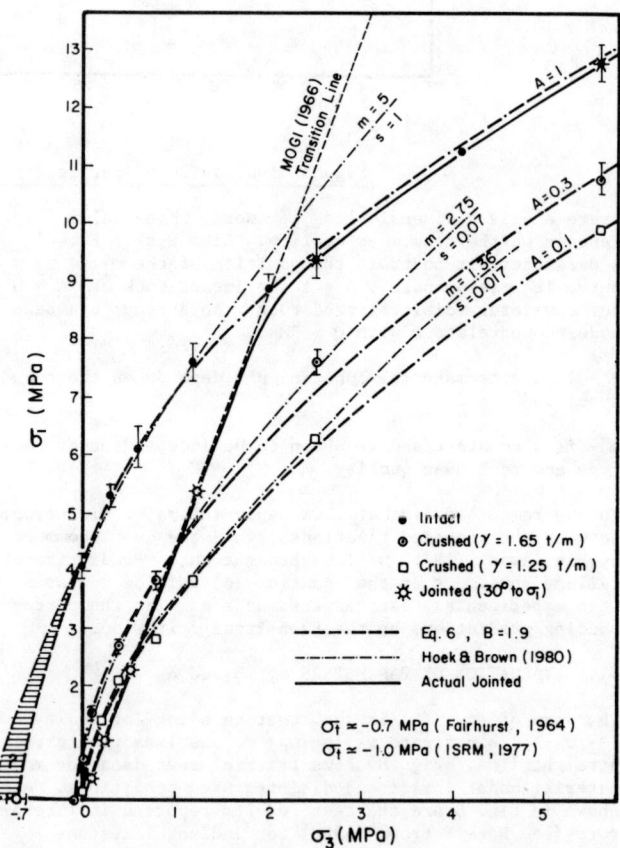

Fig. 9 Principal stress relationships for model
rock samples

VIII. CONCLUSIONS

On the basis of test results presented here it is suggested that an empirical failure criterion for rocks,
over the whole range of brittle to ductile behaviour,
may be stated as:

$$\frac{\sigma_1}{\sigma_c} = A+B \left(\sigma_3/\sigma_c\right)^{\alpha}$$

with α ranging from 0.65 to 0.75, B depending on rock type (see table IV), and A obtained from equs. (7) or (8). Following Bieniawski (1974) the authors would propose that σ_c may be estimated from point load test using the relationship $\sigma_c = 24\ I_s$, where I_s is point load strength index in MPa.

The proposed criterion can thus be evaluated from field results of rock mass classification studies (Q or RMR) where by σ_c & A parameter can be obtained for the rock masses. It is emphasized that guide lines for A & B parameters need to be further refined on the basis of analysis of more triaxial test data on different rock types (particularly for rocks with different rock mass quality).

The Hoek-Brown criterion works for soft rocks also and predicts tests results very well in the brittle rock behaviour range. As already shown by Hoek & Brown (1980) this criterion overpredicts test results at confining stresses greater than transition stress.

Test results on soft rock-like material confirm the validity of Mogi's suggestion to evaluate the magnitude of brittle-ductile transition stress.

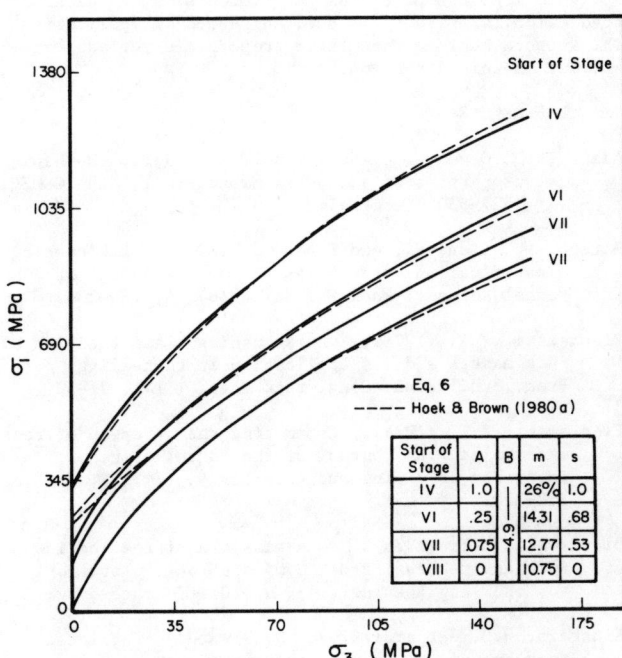

Fig. 11 Triaxial test results for westerly granite

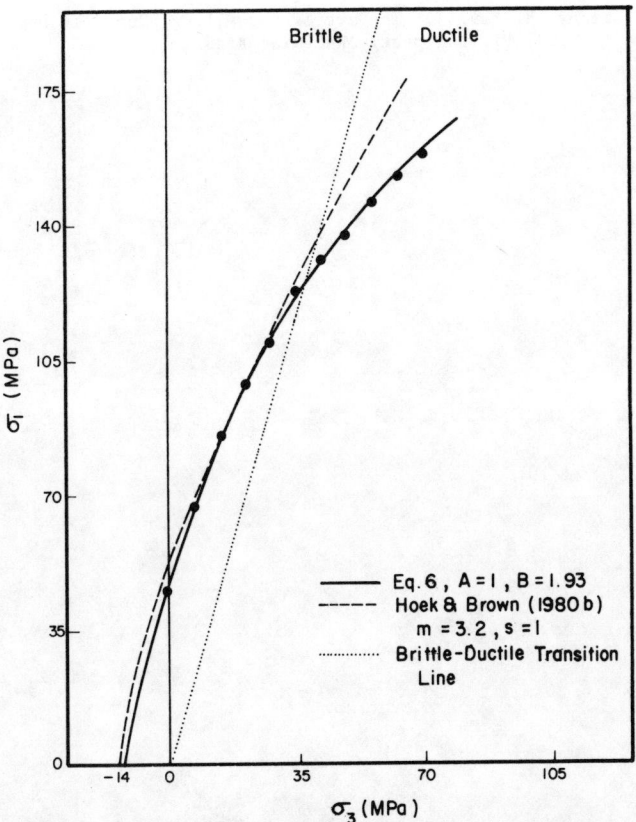

Fig. 10 Triaxial test results for Indiana limestone

It is important to emphasize here that the authors have essentially worked along the general lines set forth by Bieniawski (1974) and Hoek & Brown (1980) in evolving an empirical failure criterion for rocks & rock masses.

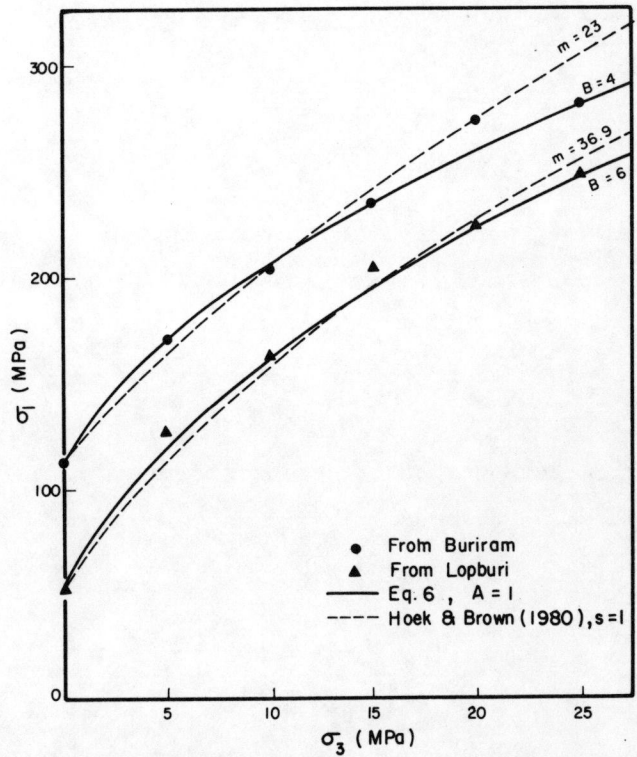

Fig. 12 Triaxial test results for Phra Wihan Sandstone

According to Bieniawski (1974) "such criteria can be selected by fitting a suitable equation into experimental data, and they need not have a theoretical basis. They serve to meet the practical requirements of adequate prediction, simplicity of use, and speed of application". The authors believe that their proposal is yet another attempt in this direction.

IX. REFERENCES

Adachi, T., Ogawa, T., and Bayashi, M. (1981). Mechanical properties of soft rock mass: Proc. 10th ICSMFE, (1), 527-530, Stockholm.

Barton, N., Lien, R., and Lunde, J. (1974). Engineering classification of rock masses for the design of tunnel support: Rock Mechanics (6), 4, 189-236.

Bieniawski, Z.T. (1974a). Geomechanics classification of rock masses and its application in tunnelling. Proc. 3rd. Intnl. Cong. rock mech. (11A), 27-32.

Bieniawski, Z.T. (1974). Estimating the strength of rock materials: The journal of the South African institute of mining and metallurgy, (74), 8, 312-320.

Bieniawski, Z.T. (1976). Rock mass classification in rock engineering: Proc. Symposium on exploration for rock engineering, (1), 97-106, Johannesburg.

Einstein, H.H., et al (1970). Model Study of jointed rock behaviour: Proc. 11th Symposium rock mech., 83-103, Berkely.

Fairhurst, C. (1964). On the validity of Brazilian test for brittle materials: IJRM & MS, (1), 535-546.

Goodman, R.E. (1980). Introduction to Rock Mechanics 480 PP. Chichester: J. Wiley.

Hoek, E. (1968). Brittle fracture of rock : in Rock Mechanics in Engineering Practice, eds. Stagg and Zienkiewicz, 93-124, London: John Wiley & Sons.

Hoek, E., and Brown, E.T. (1980). Underground Excavations in Rock, 527 PP. London: The Institution of mining & metallurgy.

Jayawardane, J. (1981). Triaxial behaviour of Phra Wihan Sandstone, M. Eng. Thesis, A.I.T. Bangkok.

Ladanyi, B., and Archambault, G. (1970). Simulation of shear behaviour of a jointed rock mass: Proc. 11th Symposium rock mechanics, 105-125, Berkely.

Mogi, K. (1966). Pressure dependence of rock strength and transition from brittle fracture to ductile flow: Bulletin, earthquake research institute, Tokyo University, (44), 215-232.

Murrell, S.A.F. (1965). The effect of triaxial stress systems on the strength of rock at atmospheric temperatures: IJRM & MS, (3), 11-43.

Stimpson, B. (1970). Modelling materials for engineering rock mechanics: IJRM & MS, (7), 77-121.

Szechy, K. (1973). The art of tunnelling, 2nd Edition, 1097 PP. Budapest: Akademiac Kiado.

A STUDY ON THE CORRELATION BETWEEN DAM FOUNDATION BEDROCK CLASSIFICATION AND IN SITU SHEARING TEST VALUE

Etude sur la corrélation entre la classification des roches de fondation de barrages et les valeurs des essais de cisaillement in situ

Eine Studie über die Beziehung zwischen der Felsgüte von Staudammgründungen und Ergebnissen von in-situ Scherversuchen

Mitsuaki Mizuno, Tadahiko Fujisawa, and Kozo Saito
Public Works Research Institute, Ministry of Construction

SYNOPSIS

In Japan a bedrock classification method is used to reflect the bedrock mechanical property data obtained from the in situ tests at the design stage. In the present state of the art, however, the results of bedrock tests conducted for the same bedrock classification often vary. This paper analyzes in situ bedrock shearing test data obtained from a number of dam sites in Japan to determine the causes of the variations in the test data and to study the relationships between the bedrock classification of the dam foundation and those strength characteristics. This paper also discusses the procedures for conducting the in situ tests.

RESUME

Au Japon, une méthode de classification des roches utilise les caractéristiques mécaniques des roches. Les données sont obtenues in situ au stade de la conception des travaux. Cependant, jusqu'à présent, les résultats de ces essais. effectués pour une même classification de roche, varient souvent entre eux. Cette communication analyse les données d'essais de cisaillement obtenues in situ à plusieurs emplacements de barrages au Japon afin de préciser les causes des variations des données d'essais et d'étudier les rapports entre la classification de la roche d'une fondation de barrage donnée et les caractéristiques de résistance au cisaillement. On examine également les procédés de réalisation des essais in situ.

ZUSAMMENFASSUNG

In Japan wird eine bestimmte Klassifikationsmethode des Gebirges von Staumauergründungen benützt, in die die mechanischen Eigenschaften des Gründungsfelsens entsprechend den in der Entwurfsphase durchgeführten in-situ Versuchen eingeht. Der gegenwärtige Kenntnisstand ist jedoch noch unbefriedigend, da häufig verschiedenartige in-situ Versuchsergebnisse mit einer gleichen Gebirgsklassifizierungsgüte verbunden sind. In vorliegender Arbeit werden Gebirgsscherversuche an verschiedenen japanischen Dammbaustellen analysiert, um Hintergründe für die Variationen der Versuchsergebnisse zu gewinnen, und um Beziehungen zwischen Gebirgsgüte von Staumauergründungen und Festigkeitscharakteristiken aufzuzeigen. Desweiteren werden Verfahren zur Durchführung der in-situ Versuche diskutiert.

1. INTRODUCTION

In Japan, bedrock is classified according to information obtained by boring, adit observation, so forth, to evaluate the mechanical properties of dam foundation bedrock. The physical properties used to evaluate the mechanical stability of dam foundation bedrock include shear strength and modulus of elasticity.

In Japan, dam foundations are usually classified into several groups according to the bedrock classification method. To improve design accuracy, it is of great importance to improve the correlation between the bedrock classifications and physical properties. At each dam site, in situ tests are made and a review of the bedrock classification and test data is carried out before and after the in situ tests.

Accordingly, the correlation between the bedrock classification and test data is usually good for each dam site. The correlation is not always good between different dams or under different geological conditions, however. Although the bedrock classification is intended as an absolute classification of bedrock independent of geology and petrology, rocks of the same class often show different physical properties where the geography or geology differs.

2. BEDROCK CLASSIFICATION AND IN SITU TEST
2.1 Bedrock Classification for Dam Foundations and Classifying Elements

Dam foundation bedrock is classified into several classes according to combinations of the classsfying elements obtained mainly from observations of adits and boring cores at dam sites, and mechanical properties for each classified bedrock are evaluated.

The classifying elements and their combinations are specified in the Geological survey Methods of Dams published by the Rock Mechanics Committee of the Japan Society of Civil Engineers. There are three classifying elements: rock hardness, spacing of fissures, and state of fissures.

Each element has three levels: A, B, and C for hardness; I, II, and III for spacing of fissures; and a, b, and c for the state of fissures depending on the adhesion, presence of inclusions, and so forth. These are combined to obtain four classes (A, B, C and D) in the overall rating. The relationship between the combination of classifying elements and the four classes is clearly defined.

Class C bedrock is subclassified into C_H. C_M and C_L.

In Japan, Class B and C bedrocks are usually used for dam foundations, and bedrock is selected according to the height of a dam to be constructed.

2.2 In Situ Bedrock Shearing Tests

The in situ bedrocks shearing tests are divided into a block shearing test and rock shearing test (both being practiced in Japan) according to the Guidelines and Exposition for in Situ Bedrock Shearing Tests published by the Japan Society of Civil Engineers in 1979.

The test data discussed in this paper was obtained according to the test methods specified in the Guidelines.

3. RELATIONSHIPS BETWEEN GEOLOGY, BEDROCK CLASSIFICATION, AND MECHANICAL PROPERTIES

3.1 Processing and St dying In Situ Test Data

A considerable number of shear strength test data on C_H, C_M, and C_L bedrocks of recently built concrete dams were processed.

From the in situ test data, a stress VS displacement curve was obtained, and the physical properties of classified bedrocks were determined.

3.2 Necessity of Geological Classification Study

The shearing failure points obtained from the shearing stress VS displacement curve were plotted on the shear strength VS normal stress curve for each bedrock classification to produce Figure 1.

Figure 1 shows the distribution of shear strengths assuming the angle of internal friction to be 45°. Figure 1 shows that the shear strength is greater in the upper class of bedrocks than that in the lower class. As the variations are generally widely spread, it is not proper to marshal the data based on arithmetic mean values. For this reason, the bedrock classification VS shear strength relation ship was studied from the standpoint of geological origins having common features in the physical properties of rock-forming minerals, state and characteristics of fissures, and so forth.

The geology of foundation bedrocks for dams recently built in Japan is classified by origins as follows:

Sedimentary rock - Palaeozoic and Mesozoic sedimentary rocks
- Volcanic sedimentary rocks
Igneous rock - Plutonic rocks
- Volcanic rocks
Regional metamorphic rocks

Table 1 gives a geological classification of the sites where the in situ test data for dam foundation bedrock was collected.

Table 1   Dam site and rock classification of geological origins

Dam No.	Type	Height(m)	Name of rock
1	Palaeozoic and Mesozoic sedimentary rocks		
1.	G	100	Sandstone, slate, chert
5.	G	140	Sandstone, slate
6.	G	96	Schalstein
8.	G	140	Schalstein, sandstone, slate
9.	A	107.5	Sandstone, slate, alternation of state and sandstone strata
12.	G	120	Slate, chert
13.	G	60	Sandstone, slate
22.	G	112	Slate, sandstone, alteration
27.	G	75	Slate, sandstone and alternation of slate and sandstone stara
30.	G	52.5	Sandstone
2	Regional metamorphic rocks		
11.	G	78	Sandy schist, black schist
16.	G	117	Crystalline schist
25.	G	86	Phyllite (slate-like)
26.	G	41.5	Phyllite (schalstein-like)
31.	G	106	Black schist
29.	AG	111	Metamorphic rocks
32.	A	88	Crystalline schist
4.	A	127	Gneiss
3	Igneous plutonic rocks		
1.	G	35	Gneiss-like quartz diorite
2.	A	100	Two-mica granite
4.	A	127	Porphyrite, granite, diabase
10.	G	87	Hornblende black mica granite
12.	G	120	Porphyrite
14.	G	107	Black mica granite
21.	G	97	Quartz diorite
23.	G	84.9	Black mica granite
24.	A	155	Black mica granite
28.	G	72	Granite
29.	AG	111	Peridotite
4	Igneous volcanic rocks		
15.	G	78	Rhyolite
17.	G	100	Basalt
18.	G	112.5	Dacite, quartz porphyry
19.	G	96.5	Basalt
35.	G	55.5	Pyrozene andesite
20.	G	125	Andesite
5	Igneous sedimentary rocks (after the Neogene)		
7.	A	140	Tuff breccia
17.	G	100	Tuff breccia
19.	G	96.5	Tuff breccia
20.	G	125	Tuff breccia
33.	G	65	Tuff breccia
34.	G	155	Igneous gravel tuff

A: Arch dam, G: Gravity dam

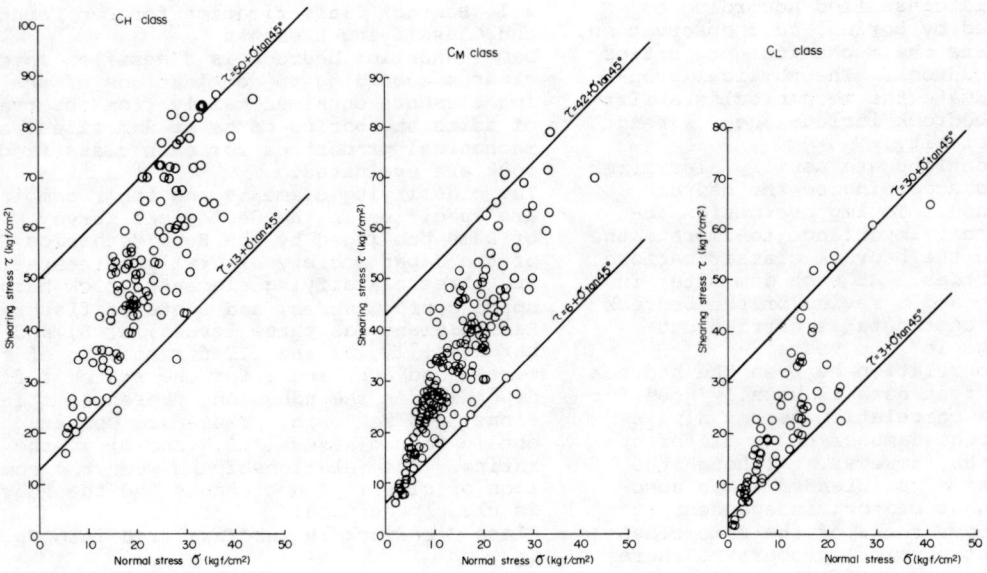

Fig. 1   Relationship between shearing strength and normal stress

4. STUDY OF ROCK CLASSIFICATIONS BY GEOLOGICAL ORIGINS AND SHEAR STRENGTH CHARACTERISTICS

Figure 2 shows a replot of Figure 1 according to geological origins and bedrock classification, and it also shows that the rocks of different geological origins have a different range of strength so that they reduce the data variations.

Table 2 Pure shearing strength and angle of internal friction according to the rock classification by geological origins

Rock classification by geological origins		CH		CM		CL	
		τ_0	ϕ	τ_0	ϕ	τ_0	ϕ
Palaeozoic and Mesozoic sedimentary rocks	Mean	18	55	10	54	6	45
	Upper limit	38	45	30	45	13	45
	Lower limit	13	45	8	45	3	45
Regional metamorphic rocks	Mean	20	58	10	54	5	45
	Upper limit	50	45	27	45	9	45
	Lower limit	17	45	6	45	3	45
Igneous plutonic rocks	Mean	28	60	18	57	10	57
	Upper limit	80	45	42	45	30	45
	Lower limit	24	45	10	45	7	45
Igneous volcanic rocks	Mean	23	55	15	50	6	50
	Upper limit	40	45	28	45	15	45
	Lower limit	16	45	13	45	3	45
Igneous sedimentary rocks (tuff and the like)	Mean	27	50	13	56		
	Upper limit	44	45	26	45		
	Lower limit	22	45	10	45		
General average		28-18	60-50	18-10	57-50	10-6	57-45

Mohr's-Coulomb theory was applied to figures and the shear strength (τ_0) and the angle of internal friction (ϕ) were obtained as listed in Table 2. In addition to mean values, envelopes are drawn for the upper and lower limits except the singular points.

It is found that, for each geological classification, there is correspondence between the bedrock classification and shearing strength. According to Figure 2, however, Palaeozoic and Mesozoic sedimentary rocks, metamorphic rocks and plutonic rocks seem to have a large variation even in the same bedrock classification. The causes of this variation were studied.

The factors of variation to be considered are as follows.

(1) As there was no significant geological difference in the bedrocks, classification was difficult

(2) Differences in test methods.

(3) Effects of the direction of the fissure prevalent in the bedrock against the shearing direction.

The Palaeozoic and Mesozoic sedimentary rocks, metamorphic rocks and plutonic rocks were examined with respect to the three above factors.

As regards factor (1), bedrock whose lower limit of τ_0 largely overlaps with the upper limit of τ_0 of those bedrocks which come a class lower are considered to be difficult to tell apart from the latter. (See Table 2)

The rocks showing the largest overlap are plutonic rocks, most of which consists of granite, and the large dispersion may be due to the lack of clear difference in the rock hardness and the weathering of fissures. As regards factor (2), two shearing test methods are available: The block shearing test method in which the bedrock is tested via a concrete block placed upon it, and the rock shearing test method in which a block is cut out of the bedrock, shaped and

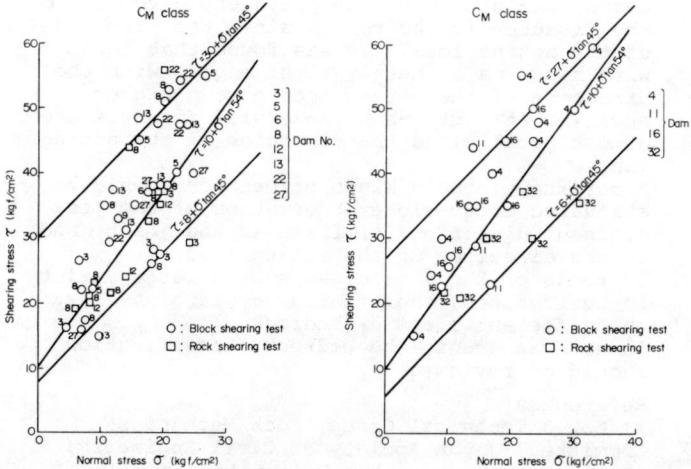

Fig. 2.1 Palaeozoic and Mesozoic sedimentary rocks

Fig. 2.2 Metamorphic rocks

Fig. 2.3 Igneous plutonic rocks

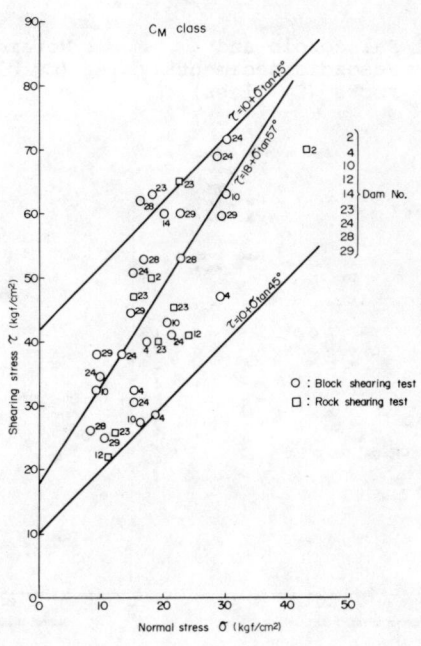

Fig. 2.4 Igneous volcanic rocks

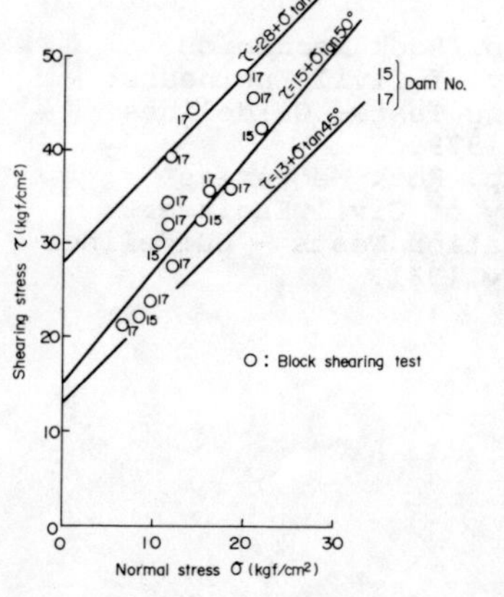

Fig. 2.5 Green tuff sedimentary rocks

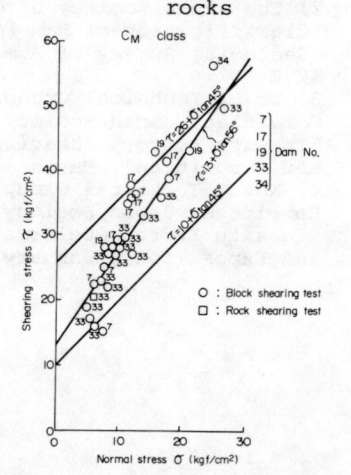

directly tested.   Both methods are widely used.
Figure 2 shows the test data plotted for two
test methods.   According to Figure 2 there
is no definite difference in the results by
each methods. As regards factor (3), the
spacing of fissures and the weathering of fissures are used for classification.   These include the standardization of bedrock classification method.   Among Palaeozoic and Mesozoic
sedimentary rocks with fissures, slate often
have the prevalent in the direction of the fissures.   In the metamorphic rocks, many of
schists and phyllites show similar schistosity.
In addition, they vary greatly in shearing
strength.
Taken altogether, it is inferred that in addition
to the above-mentioned classifying element, the
direction of fissure against the shearing direction has an influence on the shearing strength.
For this reason the direction of the fissure
and schistosity of the Palaeozoic and Mesozoic
sedimentary rocks and metamorphic rocks against
the shearing direction was read from the observed diagram of failured plane, and was classified
into five major categories (figure 3) and
plotted as shown in Figure 4.   Figure 4, shows
that the sedimentary rocks with fissures at

nearly right angles with shearing direction or
intact rocks have a larger shearing strength,
and that those with fissures nearly parallel to
the shearing direction have a smaller shearing
strength.   The metamorphic rocks sometimes show
almost the same tendency as the sedimentary
rocks; but some rocks with fissures at nearly
right angles with the shearing direction show a
small value.   This is probably be due to the
fact that the schists are liable to separate
parallel to the schistosity even when the
fissures are at nearly right angles with the
acting load.

5.   SUMMARY
The bedrock classification method practiced in
Japan is based on observation of adit walls and
boring core samples.   Namely, it is based on
combination of qualitative elements, and is
prone to errors due to differences in personal
evaluations.   So far as the surveyed data is
concerned, the mechanical properties generally
coincide with the bedrock classification.   As
they correspond to the classification of rocks
by geological origins, it is possible to correlate the bedrock classification with the
shearing strength in some degree.   Rocks such
as slate and metamorphic rocks show large
variation in test values even within the same
class of the same rock type.
The shearing strength of these rocks is, in many
cases, influenced by the prevalent direction of
the fissures in the rock against the direction
of the acting load.   It was found that rocks
with fissures at nearly right angles with the
direction of the acting load have a higher
shearing strength than those with fissures are
almost parallel to the direction of the acting
load.
Accordingly, these kinds of bedrock should be
evaluated after close observation of the predominant directions of fissures and joints and
by the direction of the acting load.
If rocks of the same class show a large variety
in test values, shear failured plane may have
the different class of bedrock.
In such an event, the bedrock classification
should be reviewed.

References
1) No. 3 Technical Group, Rock Mechanics
Committee, Japan Society of Civil Engineers:
"A Survey Report on the Deformation and
Shearing Characteristecs of Rocks", Journal of
the Japan Society of Civil Engineers, Vol. 57,
No. 9, 1972.
2) The Japan Society of Civil Engineers,
"Classification of Dam Foundation Bedrock
(Geolocial Survey of Dam Sites)", September
1977.
3) No. 3 Technical Group, Rock Mechanics
Committee, Japan Society of Civil Engineers:
"In situ Bedrock Shearing Tests- Guidelines
and Exposition", March 1979.
4) No. 3 Technical Group, Rock Mechanics
Committee, Japan Society of Civil Engineers:
"In situ Bedrock Deformation Tests - Guidelines
and Exposition", January 1981.

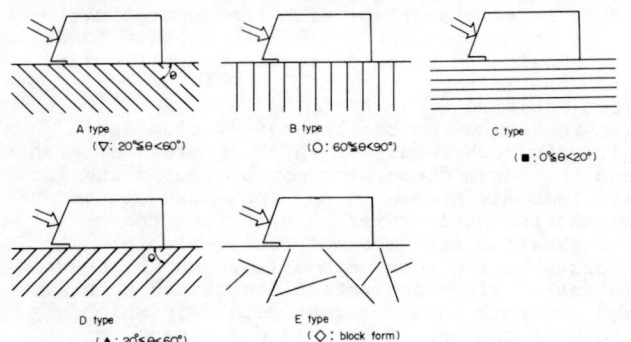

Fig. 3 Models showing fissure angles

4.1 Palaeozoic and   4.2 Metamorphic rocks
    Mesozoic sedimentary   (C_M class)
    rocks (C_M class)

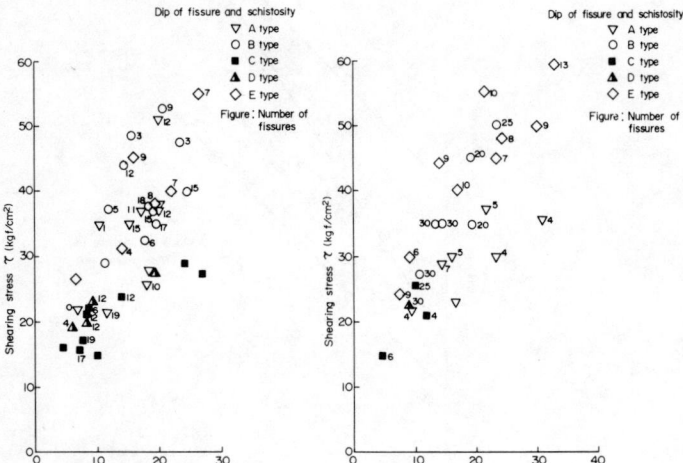

Fig.   4 Relationship between fissures,
        shistosity, and rock strength

STATISTICAL RECONSIDERATION ON THE PARAMETERS FOR GEOMECHANICS CLASSIFICATION

Reprise statistique des paramètres pour la classification géomécanique

Statische Neubetrachtung zu den Parametern der geomechanischen Klassifizierung

K. Nakao
Division Manager, Dr. Eng., Technical Research Institute of Taisei Corporation, Yokohama, Japan

S. Iihoshi and S. Koyama
Senior Research Engineer, Technical Research Institute of Taisei Corporation, Yokohama, Japan

SYNOPSIS

A statistical reconsideration of the parameters for geomechanics classification of tunnel excavation has been carried out to apply it in Japanese geological conditions.
A procedure to analyse the parameters from the data base, collected in conjunction with tunnel deformation and geomechanical rock conditions, was already reported in 1982 at the ISRM symposium at Aachen. In this paper, analysed parameters are considered on the statistical weight. The authors concluded that the conventional parameters are effective for their specific purpose but these items have to be considered in conjunction with their interrelationship.

RESUME

Une reconsidération statistique des paramètres de classification géomécanique des tunnels a été effectuée pour les conditions géologiques rencontrées au Japon. Une méthode d'analyse des paramètres sur la base des données rassemblées sur la déformation des tunnels et les conditions géomécaniques des roches a déjà été présentée en 1982 au Symposium IRSM à Aachen. Dans cette étude les paramètres sont considérés du point de vue statistique. Les auteurs ont conclu que les paramètres conventionnels sont efficaces dans ce but, mais que les rubriques doivent être considérées en corrélation les unes avec les autres.

ZUSAMMENFASSUNG:
Eine statistische Neubetrachtung zu den Parametern für die geomechanische Klassifizierung des Tunnelbaus wurde durchgeführt, um diese auf die geologischen Verhältnisse in Japan anzuwenden. Das Verfahren für die Analyse der Parameter von der Datenbasis, die im Zusammenhang mit der Tunnelverformung und der geomechanischen Gesteinsbeschaffenheit zusammengestellt wurde, wurde bereits auf dem ISRM-Symposium in Aachen vorgetragen.
In dieser Abhandlung werden die untersuchten Parameter nach ihrem statistischen Gewicht betrachtet. Die Verfasser gelangten zu der Schlußfolgerung, daß die herkömmlichen Parameter für ihren jeweiligen Zweck brauchbar sind; jedoch müssen diese Aspekte in Verbindung mit ihrem gegenseitigen Zusammenhang betrachtet werden.

1. INTRODUCTION

Great importance has been attached to new engineering classification of rock mass related to tunnel excavation as a method of assessment. In Japan some types of classification are also applied mainly to the criteria for determining the method of excavation and support system, but most of them are qualitative evaluations.

In the 1970's, techniques were developed in which factors governing rock behaviors after tunnel excavation were analysed from past execution data and quantitative evaluation was made with weighted factors. These techniques included those of Wickham et al. (1972), Bieniawsky et al. (1973) and Barton et al. (1974), and were applied to

actual tunnel excavation for verification purposes by Houghton et al. (1976) and Rutledge et al. (1978).

As a result several problems were posed in the use of their evaluation methods, and the authors already reported the present condition in Japan of these problems at the ISRM Aachen Symposium (1982).

The evaluation methods proposed by above mentioned researchers were developed on geomechanically stable hard rocks and were considered improper, when their methods were to be used for classifying and evaluating rock mass in Japan where tunnels were excavated at soft rock portions having complicated geological structures.

These assessment techniques are composed of the selection, weighting and rating of parameters for the classification. Weighting of these parameters, however, was determined personally by the proposing workers taking into consideration preceding execution of excavation, but mutual correlations between the parameters were not quantitatively determined.

This is considered attributable to difficulties encountered at that time in preparing data to which proper statistical techniques were applicable when the above mentioned determinations were made.

For this reason, the authors collected data concerning the geology, deformation and excavation techniques of tunnels executed by the authors since 1980 and, after statistical processing of the data, reconsidered on the parameters for rock evaluation, in order to prepare a rock classification which was suited for geological structures of Japan and capable of correlating it to the evaluation techniques of above mentioned researchers.

As a result, it is considered that parameters which have been taken into consideration by the previous evaluation method can be effectively applied to the geological condition of Japan, but mutual weighting of parameters should be newly made which reflects the geological conditions of Japan. Thus, when their evaluation techniques are to be applied to geological structures of Japan, it may be necessary to correct the weighting of parameters.

2. METHOD OF STATISTICAL PROCESSING

2.1 Selection of parameters

In order to select parameters required for rock evaluation, necessary survey items at site were set up and rock conditions were classified according to the prescribed procedure. Part of the data sheet used is shown in Table 1.

The above data sheet was prepared by referring to the parameters shown in Table 2 which were previously presented by Wickham, Bieniawsky and Barton.

Table 1. Data sheet

① Name of tunnel				② Excavated date		year		month		day
③ Distance from entrance		k	m	Tunnel dimension	④ Width m	⑤ Height m	⑥ Section m²			
⑦ Rock type ( )	1 Plutonics	2 Metamorphics	3 Sedimentaries	4 Volcanics	5 Others					
⑧ Hardness of rock (MPa)	1 Hard (>80)	2 Medium hard (80 − 20)	3 Soft (20 − 8)	4 Resudials, Clay, Sand, etc.						
⑨ Joint interval d (cm) Horizontal direction on face	1 100 ≤ d	2 50 ≤ d < 100	3 20 ≤ d ≤ 50	4 5 ≤ d < 20	5 0 ≤ d < 5					
⑩ Vertical direction on face	1 100 ≤ d	2 50 ≤ d < 100	3 20 ≤ d < 50	4 5 ≤ d < 20	5 0 ≤ d < 5					
⑪ Axial direction of tunnel	1 100 ≤ d	2 50 ≤ d < 100	3 20 ≤ d < 50	4 5 ≤ d < 20	5 0 ≤ d < 5					
⑫ Joint dip to the direction of drive	1 Favourable	2 Fair	3 Unfavourable							
⑬ Condition of opening	1 No separation, unweathered.	2 Slightly separated and weathered.	3 Partly separated and weathered.	4 Separated and weathered.						
⑭ Water inflow from face	1 None, completely dry	2 Wet condition on surface	3 Partial seepage from joint	4 Seepage from whole joint	5 Partial gushing out from joint					
⑮ Quantity of water inflow (at the entrance)			ℓ/min	6 Gushing out from whole joint	7 Special case, flood					
⑯ Tunnel orientation to the strike of stratum	1 Nearly parallel	2 Nearly right angle	3 Oblique							
⑰ Velocity of P-wave in surrounding rock mass m/sec	⑱ Velocity of P-wave in rock specimen m/sec	⑲ Uni-axial strength Mpa								
⑳ Weight t/m³	㉑ Overburden m	㉒ Ratio of uni-axial strength to the overburden								
㉓ Co-efficient of joint	㉔ Place of observation									
	1 full face	2 arch portion	3 central drift	4 right drift	5 left drift					
㉕ Remarks	1 Yes 0 No									

Table 2 Correlation of parameters

Parameters	Classification			Considered Parameters in Table-1
	Wickham's	Bieniawsky's	Barton's	
	o Rock Type	o Rock Strength	o Geological Condition	o Rock Type
	o Geological Structure	o RQD	o RQD	o Rock Strength
	o Orientation of Joint	o Condition of Joint	o Joint Roughness	o Orientation of Joint
	o Joint Spacing	o Joint Spacing	o Joint Spacing	o Joint Spacing
	o Water Inflow	o Water Inflow	o Joint Alteration	o Joint Alteration
			o Water Inflow	o Water Inflow
			o Thickness of Overburden	

For the deformation measurement after tunnel excavation which was used as reference for selecting parameters, data measured by the latest convergence meter was used and analysis was made by using 152 examples in which the data sheet and measured results showed satisfactory correspondence.

The reasons for not taking RQD into consideration as a parameter were that only a few data of core boring were available at the locations of tunnel deformation measurement and that RQD was considered to be indicated by the spacing and degeneration degree of the joint.

In combination with the above-mentioned parameters, the authors collected data concerning the sectional area of tunnel excavation, excavation techniques, and support techniques and examined the correlation between the above-mentioned parameters and these data items. In obtaining the correlation, statistical processing to be described below was made and the mutual relations were quantitatively calculated.

2.2 Outline of Analytical Procedure

Data which had been investigated at site were given statistical processing on the basis of the flowchart of analysis shown in Fig. 1.

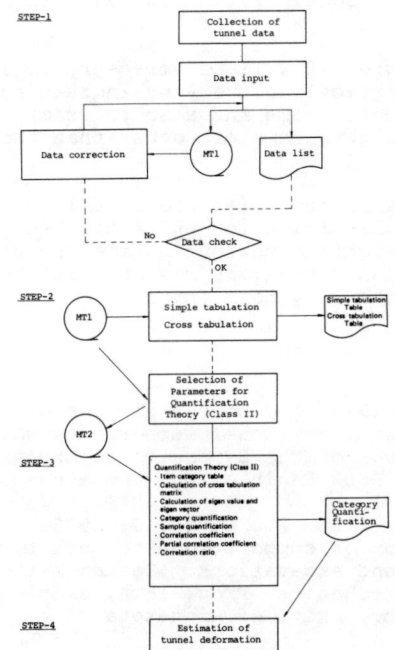

Fig. 1   Flowchart of data processing and analysis

First, collected data items were subjected to simple and cross tabulations at STEP-2 to find out the data characteristics.

Then, the classification range of each category and deviation of sampling were examined.

In STEP-3, all these items were analysed by employing the Qnatification Theory (Class II) and correlation coefficients between various items were investigated. Items affecting the deformation quantity of the tunnel were also selected. This operation was performed on the trial-and-error basis, and the parameter group having the best correlation was selected. If the analysis results of this parameter group are satisfactory, prediction of the deformation quantity of the tunnel to be newly excavated becomes possible at STEP-4.

3.   DISCUSSION ON MUTUAL CORRELATION OF PARAMETERS

Through the analysis by the Quantification Theory (Class II), correlation coefficients between various parameters were calculated and the classification range of categories of various parameters were rearranged.

During this rearrangement, various points mentioned below were examined.

1. Correlation coefficients of joint spacing and tunnel deformations in the horizontal, vertical and axial directions on the tunnel face were of the same degree, but only the joint spacing and tunnel deformation in the axial direction were used because of their high mutual internal relation.

2. The dip of the joint was omitted, because it has high correlations with the cross section and the rock type which have high correlation with tunnel deformation.

3. Water inflow was omitted, because it has a high correlation with the rock type and a low correlation with tunnel deformation.

4. The thickness of overburden has a very high correlation with the sectional area and thus there was no need of adopting it.

5. Rock type was adopted because it had high correlations with water inflow and tunnel deformation.

6. Hardness of rock was adopted because it has high correlations with the orientation of the joint and tunnel deformation.

7. Condition of opening was adopted because it had high correlations with all parameters.

8. Sectional area of excavation of the tunnel was adopted because it had high correlations with most of parameters.

9. Range of tunnel deformation was divided into four as 1mm under 1∿10mm, 10∿50mm and 50mm above.

Results of the quantification analysis made by using the rearranged parameters are shown in Table 3 as simple correlation matrix. The quantity of categories of various parameters are shown in Table 4. The correlation ratio at this time was 0.68.

Table 3.   Simple Correlation Matrix

Parameters	Tunnel section	Rock type	Hardness of rock	Joint spacing	Condition of opening	Tunnel deformation
Tunnel section	1.0000	.3677	.4998	-.0358	.3030	.6267
Rock type	.3677	1.0000	.2393	.3121	.4011	.6047
Hardness of rock	.4998	.2393	1.0000	.3644	.3470	.5298
Joint spacing	-.0358	.3121	.3644	1.0000	.3252	.3896
Condition of opening	.3030	.4011	.3470	.3252	1.0000	.5786
Tunnel deformation	.6267	.6047	.5298	.3896	.5786	1.0000

Table 4.   Category Quantification

Parameter	Category		Quantity	Partial correlation Coefficient	Range
Tunnel section	(1)	0 ∿ 35m²	0.0000	0.49 ①	.7978 ①
	(2)	35 ∿ 60m²	.2898		
	(3)	60m² ∿	-.5080		
Rock type	(1)	Metamorphics	0.0000	0.37 ②	.4340 ③
	(2)	Sesimentaries	.0531		
	(3)	Volcarics	.4340		
Hardness of rock	(1)	Hard	0.0000	0.13 ⑤	.2084 ⑤
	(2)	Medium hard	-.1509		
	(3)	Soft, Clay, Sand	-.2084		
Joint spacing	(1)	20 ≤ d	0.0000	0.29 ④	.3012 ④
	(2)	5 ≤ d < 20	-.3012		
	(3)	0 ≤ d < 5	-.1553		
Condition of opening	(1)	No separation	0.0000	0.35 ③	.7002 ②
	(2)	Slightly separated	.2230		
	(3)	Partly separated and weathered	.0912		
	(4)	Separated and weathered	-.4772		

Frequency distribution in which the sampel quantity was obtained on the basis of this quantity of categories is shown in Fig. 2, which clearly indicates that classification by tunnel deformation is possible.

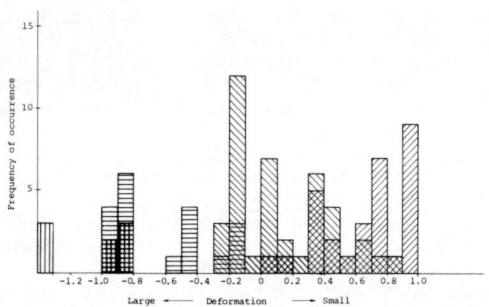

Fig. 2   Histgram of Sample Quantification

The above results clarify that effective contributions were made to classification by the tunnel section, opening condition of the joint, rock type, joint spacing and hardness of rock as factors affecting the tunnel deformation.

4. FINAL COMMENTS

The analysis results of 152 examples of tunnel excavation in the geological condition of Japan so far collected are compared with the weighting of classifications by the three aforesaid researchers in Table 5.

Table 5.   Order of Rating

	Parameters	Weight good ←→ poor		Order of Rating
RSR Concept	o Rock Type o Geological Structure o Orientation of Joint o Joint Spacing o Water Inflow	1 6] 45 25	4 30 9 6	Joint v Geological Structure v Water Inflow v Rock Type
RMR Concept	o Rock Strength o RQD o Condition of Joint o Joint Spacing o Water Inflow	15 20 30 20 15	0 3 0 5 0	Joint Condition v RQD, Joint Spacing v Rock Strength v Water Inflow
Q-System	o Geological Condition o RQD o Joint Roughness o Joint Spacing o Water Inflow	0.5 100 4 0.5 1	20 0 0.5 4(20) 0.05	RQD v Joint Spacing Geological Conditions v Joint Conditions v Water Inflow
Result of Analysis	o Rock Type o Rock Strength o Orientation of Joint o Joint Spacing o Joint Alteration (Opening) o Water Inflow o Thickness of Overburden (Tunnel Section)	0 -0.21 -0.30 -0.48 -0.51	0.43 0 0 0.22 0.29	Tunnel Width or Thickness of Overburden v Joint Condition v Rock Type Joint Spacing v Rock Strength v Water Inflow

As shown in the analysis results, various parameters used in all of the RSR Concept, RMR Concept and Q-System seem to be effectively applicable to the geological classification of Japan. Particularly the rating of the analysis results virtually agrees with that of the RMR Concept.

The only difference lies in that the greatest weight is placed on RQD and the joint condition in their classifications, whereas in the Japanese classification, the thickness of tunnel overburden constitutes the greatest factor.

The above indicates that most of the geology of Japan belong to the so-called "fault zone," that is, the rock body has turned into debris and many joints have occurred, thereby making classification by joint spacing difficult.

For this reason it is necessary to consider, in classifying the rock mass for tunnelling of Japan, that the strength ratio which is determined by the strength of rock specimens comprising the mass and thickness of the overburden has become the factor governing the stability of the tunnel. In connection with this, the size of the tunnel to be excavated also should require full consideration in evaluation of the geology.

In the future, it will be necessary to increase analytical accuracy by increasing the number of data items and also to examine correlations with parameters other than the tunnel deformation.

The results of the trial use of this geomechanics classification which has been prepared by weighting various parameters on the above-mentioned analysis results will be reported in the very near future.

5. REFERENCES

Barton, N. 1976, Recent experiences with the Q System of tunnel support design. Proceedings of the Symposium on Exploration for Rock Engineering.   Johannesburg.

Barton, N., Lien, R., Lunde, J. 1975, Estimation of support requirements for underground excavations. Design Methods in Rock Mechanics, Proc. 16th. Simp. on Rock Mech., Univ. of Minnesota.

Bieniawski, Z.T. 1979, The geomechanics classification in rock engineering applications. I.S.R.M, Montreux.

Houghton, D.A. 1976, The role of rock quality indices in the assessment of rock masses. Proceedings of the Symposium on Exploration for Rock engineering, Johannesburg.

Maeyens, A. 1978, A comparison between the rock mass conditions as predicted and measured after tunnelling. International Association of Engineering Geology III International Congress, Spain.

Rutledge, J.C., Preston, R.L. 1978, New Zealand experience with engineering clas sifications of rock for the prediction of tunnel support. International Tunnel Symposium, Tokyo.

Wickham, G.E., Tiedemann, H.R., Skinner, E.H. 1972, Support determinations based on geologic predictions. North American Rapid Excavation and Tunnelling Conference.

Nakao, K. et al. 1982, Geomechanic classification for assessing rock mass in Japan. Proceeding of the I.S.R.M symposium, Aachen.

A NUMERICAL METHOD FOR THE DEFINITION OF DISCONTINUITY SETS

Une méthode numérique pour la définition des familles de discontinuités

Eine numerische Methode zur Bestimmung von Diskontinuitätsscharen

Nuno F. Grossmann

Research Officer, Rock Foundations Division (Dams Department), Laboratório Nacional de Engenharia Civil, Lisbon, Portugal

SYNOPSIS

The paper describes the numerical method developed at the Laboratório Nacional de Engenharia Civil (LNEC) which allows the complete determination of the existing discontinuity system starting from the total field data collection. The method may be applied to any number of the sets of information obtained and can also take into consideration cases of a non-uniform sampling of the discontinuities. The determination of the existing discontinuity sets is done through different steps, the first one using as a basis the spherical Poisson distribution and the other ones the bivariate normal distribution on the tangent plane at the mean attitude.

RESUME

La communication décrit la méthode numérique développée au Laboratório Nacional de Engenharia Civil (LNEC) qui permet la détermination complète du système de discontinuités existant à partir de la somme globale des données recueillies sur le terrain. La méthode peut être appliquée à un nombre quelconque d'informations obtenues et peut aussi prendre en considération des cas d'échantillonnage non-uniforme des discontinuités. La détermination des familles de discontinuités existantes est faite en plusieurs étapes, la première utilisant comme base la distribution de Poisson sphérique et les autres la distribution normale bivariée sur le plan tangent à l'attitude moyenne.

ZUSAMMENFASSUNG

Das Referat beschreibt die im Laboratório Nacional de Engenharia Civil (LNEC) entwickelte numerische Methode zur vollständigen Bestimmung des existierenden Diskontinuitätengefüges. Die Methode, die von der kompletten Geländedatensammlung ausgeht, kann für irgendeine Anzahl der erhaltenen Informationen angewandt werden und auch Fälle einer ungleichförmigen Diskontinuitätenerfassung berücksichtigen. Die Bestimmung der vorhandenen Diskontinuitätenscharen wird in mehreren Schritten vollzogen, wobei sich der erste der sphärischen Poissonverteilung bedient, und die anderen der bivariierten Normalverteilung auf der zur mittleren Stellung tangenten Ebene.

1. INTRODUCTION

Experience shows that, even on the scale of a dam foundation, it is quite frequent to have a rock mass in which rotation of some parts with respect to the others has taken place. Whenever this fact occurs, the mean attitude of the different discontinuity sets varies from one point to another, and, sometimes, due to the dispersion of the discontinuity sets, discontinuities with the same attitude, but belonging to different sets, may be detected in the rock mass. Because, a priori, occurrence of this situation is always possible (as its non-existance can only be assessed by the study of the discontinuity system), the numerical definition of discontinuity sets is a much more difficult problem than it seems at first sight.

The obvious way to avoid this problem is to study, one by one, small zones of the rock mass for which the assumption may hold that no rotation of one part with respect to the others took place. In a small zone of a rock mass, alas, sometimes the outcropping discontinuities are rather scarce, lying in numbers between 10 and 50. On the other hand, even when it is possible to detect more discontinuities in a given zone, we often do not want to study them all, in order to reduce the total volume of collected discontinuity data.

For the reasons just exposed, the LNEC in 1972 began to develop a numerical method for the definition of discontinuity sets, which may be used whatever the number of discontinuities sampled.

2. GENERAL CASE (UNIFORM SAMPLING)

2.1 - The method begins with the definition of the

c

occurring discontinuity clusters, and considers that a given discontinuity cluster includes all those discontinuities which form with any other discontinuity of that cluster a dihedral angle which is less or equal to a given value α_L .For a pair of discontinuities i and j, this condition may be written as:

$$\arccos\left[\left|\cos\delta_i \cos\delta_j + \sin\delta_i \sin\delta_j \cos(\sigma_i-\sigma_j)\right|\right]\leq\alpha_L \quad (1)$$

σ_k and δ_k being the strike and the dip of the discontinuity k.

Usually, it is not necessary to check condition (1) for all pairs of discontinuities, because, as soon as a discontinuity k gets included in a cluster, there is no longer a need to carry out this test for the pairs formed by the discontinuity k and all the others which have already been included in the same cluster.

As to the limiting angle α_L, its value is obtained with the help of Poisson's law, which describes the probability of occurrence of rare events with a random uniform discrimination.

Let us consider a unit sphere, on which all poles of the detected discontinuities are plotted. According to the definition given, a cluster occurs whenever two poles are nearer than α_L.

Therefore, this value must be chosen in such a way that in a spherical calotte with aperture α_L the occurrence of zero or one poles is a more probable event than the occurrence of two or more poles. If $P(n)$ stands for the probability of occurrence of n poles in the mentioned calotte, this last condition may be stated as:

$$P(0) + P(1) \geq \frac{1}{2} \quad (2)$$

and, assuming $P(n)$ to be given by Poisson's law, also as:

$$e^{-\overline{n}} + \overline{n}\, e^{-\overline{n}} \geq \frac{1}{2} \quad (3)$$

where $\overline{n}$ is the mean number of poles occurring in a spherical calotte of aperture α_L. If N stands for the total number of detected discontinuities, the value of $\overline{n}$ is easily obtained, by multiplying the area of the spherical calotte with aperture α_L by the mean number of poles per unit area of the sphere. This yields:

$$\overline{n} = N (1 - \cos \alpha_L) \quad (4)$$

Introducing (4), condition (3) may be transformed into:

$$\left[1+N(1-\cos \alpha_L)\right] e^{-\left[1+N(1-\cos \alpha_L)\right]} \geq \frac{1}{2e} \quad (5)$$

This inequality, whose left member is of the type $(x\, e^{-x})$, can be solved numerically, giving:

$$x = 1 + N (1 - \cos \alpha_L) \leq 2{,}67834699 \quad (6)$$

and, therefore:

$$\cos \alpha_L \geq 1 - \frac{1{,}67834699}{N} \quad (7)$$

The initial condition (1) becomes, finally:

$$\left|\cos\delta_i \cos\delta_j + \sin\delta_i \sin\delta_j \cos(\sigma_i-\sigma_j)\right| \geq 1 - \frac{1{,}67834699}{N} \quad (8)$$

The analysis of expression (7) shows that the limiting angle α_L decreases with the number of discontinuities considered.

In order to check the proposed method, the LNEC tried two other limiting angles, one about 20% larger than α_L, and the other about 20% smaller. The experience gained (LNEC 1973a, 1973b,1975a, 1975b, and 1976) showed that the smaller angle leads to an excessive number of clusters,whilst the larger one, although being usually satisfactory, sometimes assembles discontinuities that an observation of the corresponding density diagram would immediately separate into two sets.

2.2 - In most cases, the discontinuity sets are identical with the discontinuity clusters just defined.

However, the described cluster definition method does not exclude the existence of situations in which, inside the area of the sphere limited by the convex exterior envelope to the poles of a given cluster, one or more poles occur which have not been included in this cluster. Theoretically, it is even possible that the attitude of one of those non-included poles be equal to the mean attitude of the cluster. Therefore,the proposed method for the definition of the discontinuity sets considers a second stage.

In this stage , the clusters already obtained are grouped according to the criterion that two clusters belong to the same discontinuity set whenever the mean point of one of them falls into the domain of the other.

As can be seen, the cluster grouping criterion is no longer based on the hypothesis of a random uniform distribution of the poles of the discontinuities. It uses, on the contrary,the real distribution of the poles of the different clusters, through the two concepts, mean point and domain.

On the other hand, the LNEC (Grossmann 1977)has verified that the attitudes of the discontinuities of any set may be adequately described by means of a bivariate normal distribution on the plane tangent to the sphere at the point that corresponds to the mean attitude of the set. In order to obtain the corresponding distribution on the sphere, the points on the tangent plane are projected on the sphere, using its center as projection center. With the help of the mentioned bivariate normal model, each discontinuity cluster can, therefore, be characterized by a mean attitude, two principal standard deviations (the maximum and the minimum), and an angle that defines the spatial direction along which the maximum standard deviation occurs (the minimum standard deviation occurs in the direction perpendicular to this one).

The aforesaid mean point of the cluster is the pole corresponding to the mean attitude of the discontinuity cluster.

The domain of the cluster being the area of the sphere which corresponds approximately to the zone where the N_c poles of the discontinuity cluster occur, π was chosen to be the area of the sphere, limited by an equal density of probability of pole occurrence line, where the integral of this density of probability is (1-1/N_c). The equal density of probability of pole occurrence lines are, for the bivariate normal distribution on the tangent plane at the mean attitude, ellipses which are centred with the mean attitude, and whose axes are proportional to the principal standard deviations. For the distribution on the sphere, those lines can be obtained by projecting the corresponding ellipses on the sphere, using the sphere center as projection center.

The formula which allows to check if the mean point of the discontinuity cluster i, having a mean attitude with strike σ_i and dip δ_i, falls into the domain of the discontinuity cluster j, which is characterized by a mean attitude with strike σ_j and dip δ_j, a maximum standard deviation σ_M, a minimum standard deviation σ_m, and an angle ω_M (defining the spatial direction along which the maximum standard deviation occurs), may be written as (Grossmann 1977):

$$\left[\cos\delta_i \cos\delta_j + \sin\delta_i \sin\delta_j \cos(\sigma_i - \sigma_j) \right]^2 \geqslant$$

$$\cfrac{1}{1 + \cfrac{2 \ln(N_c) \; \sigma_M^2}{1 + \sin^2(\omega - \omega_M) \left(\dfrac{\sigma_M^2}{\sigma_m^2} - 1 \right)}} \qquad (9)$$

where N_c is the number of discontinuities which has been included in the cluster j, and ω the angle defining the spatial direction containing the mean attitudes of the two discontinuity clusters i and j, which is given by:

$$\text{tg}\,\omega = \frac{\sin\delta_i \, \sin(\sigma_i - \sigma_j)}{\sin\delta_i \, \cos\delta_j \, \cos(\sigma_i - \sigma_j) - \cos\delta_i \, \sin\delta_j} \qquad (10)$$

If the application of the described cluster grouping criterion reveals the presence of one or more discontinuity cluster groups (each one consisting of two or more discontinuity clusters), the same criterion shall be applied again, after having eliminated all grouped discontinuity clusters, and added one new discontinuity cluster for each one of the above-mentioned groups (which shall include all the discontinuities of the clusters corresponding to that group). This process is repeated until no more grouping is achieved. The remaining discontinuity clusters are, then, identical with the discontinuity sets.

3. NON-UNIFORM SAMPLING

3.1 - The aforesaid method for the definition of discontinuity sets has been elaborated under the assumption that the existing discontinuity data have been obtained through a uniform sampling, that is, that with the sampling performed, and for a random uniform distribution of the discontinuities, the probability of detecting a discontinuity with a given attitude does not depend on this attitude. If this assumption does not hold (e.g., if the discontinuity data are all obtained from vertical boreholes), a modified version of the method must be applied, which uses the auxiliary spherical surface of equal probability presented elsewhere (Grossmann 1980). This spherical surface is characterized by the property that with the sampling performed, and for a random uniform distribution of the discontinuities, the probability of pole occurrence in an elementary area is a constant.

The method for the definition of discontinuity sets in cases of non-uniform sampling performs exactly the same steps with the poles of the sampled discontinuities of the spherical surface of equal probability, as does the method just described for cases of uniform sampling on the usual spherical surface. The only difference is, therefore, that every discontinuity changes its real attitude into an auxiliary attitude, which characterizes the location of the pole on the spherical surface of equal probability.

The practical effect achieved through this change of spherical surfaces is that, in reality, a variable limiting angle α_L is used, which becomes larger for the zones containing the attitudes for which the sampling has not been so good and smaller for the zones containing the attitudes which were better sampled.

3.2 - As has been stated elsewhere (Grossmann 1980), in many cases it is convenient to express the attitude in a coordinate system different from the usual one (x ≡ North; y ≡ East; z ≡ vertical).

If the discontinuity data have been gathered from a borehole, and the attitudes are expressed in the spherical coordinate system (r, ε, ω), whose revolution axis is the borehole axis, the above-mentioned auxiliary attitude of a pole on the spherical surface of equal probability can be obtained by changing its colatitude ε into a new colatitude ε', the length ω remaining unchanged. The new colatitude ε' is given by (Grossmann 1977):

$$\cos\varepsilon' = \frac{A}{B} \qquad (11)$$

where A and B stand for:

$$A = (1 + c_R^2) \, \cos^2\varepsilon + (1 + c_R^2 + \frac{\emptyset}{2\,\bar{R}}) \frac{\emptyset}{\bar{L}} (\frac{\pi}{2} - \varepsilon + \sin\varepsilon \, \cos\varepsilon) +$$

$$+ \frac{4}{\pi} \frac{\emptyset}{\bar{R}} \int_\varepsilon^{\frac{\pi}{2}} E(\varepsilon) \, \sin\varepsilon \; d\varepsilon + \frac{\emptyset^2}{2\,\bar{R}^2} \cos\varepsilon \qquad (12)$$

$$B = (1 + c_R^2) + (1 + c_R^2 + \frac{\emptyset}{2\,\bar{R}}) \frac{\pi\emptyset}{2\,\bar{L}} + \frac{\pi\emptyset}{2\,\bar{R}} + \frac{\emptyset^2}{2\,\bar{R}^2} \qquad (13)$$

$\emptyset$ being the diameter of the borehole, $\bar{L}$ its length, $\bar{R}$ the mean value of the distribution of the equivalent radii of the occurring discontinuities, c_R the coefficient of variation of this distribution, and $E(\varepsilon)$, the complete (between 0 and $\pi/2$) 2nd kind elliptic integral with the modular angle ε. For most practical cases, equation (11) may be simplified into:

$$\cos\varepsilon' = \cos^2\varepsilon \qquad (14)$$

as we may assume that:

$$\frac{\emptyset}{\bar{L}} \simeq 0 \qquad (15)$$

and:

$$\frac{\emptyset}{\bar{R}} \approx 0 \qquad (16)$$

If the discontinuity data have been gathered from a plane circular surface, the attitudes should be expressed in the spherical coordinate system (r, ε, ω), whose revolution axis is perpendicular to the observation plane. In this case, the auxiliary attitude of a pole on the spherical surface of equal probability may again be obtained by changing only its colatitude ε into a new colatitude ε', the length ω remaining unchanged. The new colatitude ε' is given by (Grossmann 1977):

$$\cos\varepsilon' = 1 - \frac{2}{\pi}(\varepsilon - \sin\varepsilon \cos\varepsilon) \qquad (17)$$

If the discontinuity data have been gathered from a rectangular observation surface with the dimensions a and b, the attitudes should be expressed in the spherical coordinate system (r, ε, ω), whose revolution axis is perpendicular to the observation plane, and for which the length ω is measured starting from a direction parallel to the a sides of the rectangle. In this case, the auxiliary attitude of a pole on the spherical surface of equal probability is obtained by changing both its colatitude ε into a new colatitude ε', and its length ω into a new length ω'. The new colatitude ε' is, again, given by equation (17), and the new length ω' by (Grossmann 1977):

$$\omega' = \omega + \frac{\frac{\pi}{2}\left\{a\left[2\ I + (-1)^I \sin\omega\right] + b\left[1 + 2\ J - (-1)^J \cos\omega\right]\right\} - (a+b)\omega}{\frac{a\ b}{\bar{R}\ (1 + c_R^2)} + (a+b)}$$

$$(18)$$

where I and J stand for:

$$I = \text{INT}\left(\frac{\omega}{\pi} + \frac{1}{2}\right) \qquad (19)$$

$$J = \text{INT}\left(\frac{\omega}{\pi}\right) \qquad (20)$$

and INT(x) represents the largest integer which is less than or equal to x.

4. DISCONTINUITY SYSTEM

The analysis of the discontinuity sets obtained for the small zones of the rock mass often shows the presence of large zones of the rock mass where the discontinuity system is approximately the same. The last stage of the method developed for the definition of discontinuity sets has, therefore, the aim of grouping the discontinuity sets determined for the small zones of the rock mass into discontinuity sets which are characteristic of a large zone of the rock mass.

The grouping criterion used is, again, to consider that two small zone discontinuity sets belong to the same large zone discontinuity set whenever the mean point of one of them falls into the domain of the other.

The mean point of the small zone discontinuity set is the pole corresponding to the mean attitude of the discontinuity set.

The domain of the small zone discontinuity set, however, can no longer be defined with the help of the number of poles of the discontinuity set, as this number is a function of the sampling performed (e.g., it depends on the volume of the small zone sampled). In order to have for this case a definition similar to the one of the discontinuity cluster domain, instead of the number N_c, the number N_s given by (Grossmann 1977):

$$N_s = N_T \frac{I_s}{\displaystyle\sum_{i=1}^{i=n_s} I_i} \qquad (21)$$

is used, where N_T stands for the total number of discontinuities sampled in the large zone of the rock mass, I_s for the intensity (the inverse of the spacing) of the small zone discontinuity set s, and n_s for the total number of small zone discontinuity sets determined. With this change, the domain becomes the area of the sphere, limited by an equal density of probability of pole occurrence line, where the integral of this density of probability is $(1 - 1/N_s)$.

In cases in which some of the sampled small zones of the rock mass only show a few discontinuities, the above-mentioned definition of the domain may lead to an area which corresponds to a fraction of the spherical surface larger than (N_s/N_T), the contribution of the intensity I_s of the small zone discontinuity set s to the total joint index (summation $\sum I_i$ over $\bar{i}=1$ to $i=n_s$) of the large zone of the rock mass. In these cases, instead of the number N_s, the number N_s' given by (Grossmann 1977):

$$N_s' = e^{\frac{1}{\sigma_M} \sigma_m \frac{N_s}{N_T}} \qquad (22)$$

should be used. The recommended procedure is to calculate for all cases both N_s and N_s', and to use the smaller one for the definition of the domain.

The formula, which allows to check if the mean point of the small zone discontinuity set i falls into the domain of the small zone discontinuity set j, is identical to equation (9), after having changed N_c for N_s or N_s'.

Here also, the grouping process is repeated, in a similar way to the one described for the grouping of the discontinuity clusters (section 2.2).

5. REFERENCES

GROSSMANN, Nuno Feodor - Contribuição para o Estudo da Compartimentação dos Maciços Rochosos (Contribution to the Study of Jointing in Rock Masses), a research officer thesis - LNEC (Laboratório Nacional de Engenharia Civil), Lisboa PORTUGAL, 1977.

GROSSMANN, Nuno Feodor - Das Gefuegediagramm im Falle einer Ungleichfoermigen Datenerfassung (The Jointing Diagram in the Case of a Non-Uniform Data Sampling), a contribution to the 4th German National Meeting on Rock Mechanics, Aachen GERMANY FR, 1980 May 05-06 - DGEG (Deutsche Gesellschaft fuer Erd- und Grundbau e.V.), Essen GERMANY FR, 1980.

LNEC (Laboratório Nacional de Engenharia Civil)-
- Área de Sines - Estudos pelo Método de
Amostragem Integral (Sines Area - Studies
by the Integral Sampling Method), a report
- LNEC, Lisboa PORTUGAL, 1973a.

LNEC (Laboratório Nacional de Engenharia Civil)-
- Colaboração no Estudo da Compartimentação
do Maciço Rochoso do Local da Barragem de Va
lhelhas (Collaboration in the Study of Joint
ing in the Rock Mass of the Valhelhas Dam Si
te), an internal report - LNEC, Lisboa POR
TUGAL, 1973b.

LNEC (Laboratório Nacional de Engenharia Civil)-
- Aplicação da Amostragem Integral ao Estudo de um Colector em Cascais (Application
of Integral Sampling to the Study of a Col-
lector in Cascais), a report - LNEC, Lisboa
PORTUGAL, 1975a.

LNEC (Laboratório Nacional de Engenharia Civil)-
- Aplicação da Amostragem Integral ao Estudo de Maciços Rochosos (Compartimentação e
Permeabilidade da Fundação da Barragem de
Rebordelo) (Application of Integral Sampling to the Study of Rock Masses (Jointing
and Permeability of the Rebordelo Dam Found
ation)), a research report - LNEC, Lisboa
PORTUGAL, 1975b.

LNEC (Laboratório Nacional de Engenharia Civil)-
- Estudo do Maciço de Fundação da Barragem
do Cabril (Study of the Foundation Rock of
the Cabril Dam), a report - LNEC, Lisboa
PORTUGAL, 1976.

REGIONALISATION DES PROPRIETES MECANIQUES DES ROCHES.
APPROCHE STRUCTURALE GEOSTATISTIQUE

Regionalisation of the mechanical properties of rocks.
A structural geostatistic approach

Regionalisierung der mechanischen Eigenschaften der Gesteine.
Ein geostatistischer, strukturanalytischer Versuch

A. Pineau
Docteur 3ème cycle, Centre de Valorisation des matières premières
E.N.S. Géologie de Nancy-France

A. Thomas
Docteur ès-Sciences, Centre de Valorisation des matières premières
E.N.S. Géologie de Nancy-France

RESUME :

Des résultats issus de 482 essais de compression simple sur des échantillons de minerai de fer révèlent une certaine continuité de la relation entre les propriétés mécaniques et la position spatiale des prélèvements des éprouvettes. Les propriétés mécaniques des roches sont des variables régionalisées justiciables de l'analyse structurale géostatistique. Les variations spatiales importantes des propriétés mécaniques déterminent des hétérogénéités mécaniques liées aux structures géologiques et aux régionalisations géochimiques. Les hétérogénéités mécaniques sont également liées à la régionalisation structurale. Ces liaisons sont évoquées comme outils d'analyse de l'effet d'échelle.

SYNOPSIS

482 uniaxial compression tests on iron-ore samples show a certain interrelationship between the mechanical properties and the spacial position of the samples. The mechanical properties of the rocks are "regionalized" variables which are dependent on structural geostatistical analysis. The important spatial variations of the mechanical properties determine mechanical heterogeneities which are related to the geological structure and to the geochemical and structural "regionalizations" of the examined iron-ore from Lorraine (France).

ZUSAMMENFASSUNG

482 Druckfestigkeitsversuche an Eisenerzproben weisen einen gewissen Zusammenhang der Beziehungen zwischen den mechanischen Eigenschaften und der Probestückaufnahme auf. Die mechanischen Eigenschaften der Gesteine sind regionalisierte Variabeln, die von der geostatistischen Strukturanalyse abhängig sind. Die bedeutenden räumlichen Abweichungen der mechanischen Eigenschaften bestimmen die mechanischen Ungleichartigkeiten, die mit den geologischen Strukturen und geochemischen "Regionalisationen" des studierten Eisenerzes von Lothringen (Frankreich) verbunden sind. Die mechanischen Ungleichartigkeiten sind auch mit struktureller "Regionalisation" der Tektonik verbunden. Diese Beziehungen könnten für die Analyse des Maßstabeffektes von Bedeutung sein.

I - INTRODUCTION

Le terme "géostatistique" désigne l'étude statistique des variables (d'état ou de comportement) relatives au milieu terrestre. De telles variables à distribution spatiale sont appelées "variables régionalisées". Les variations, dans l'espace, d'une variable régionalisée, ne sont généralement pas quelconques, elles sont organisées ou structurées ; on dit que la variable régionalisée présente une structure particulière.

La géostatistique est bien développée et parfaitement codifiée dans le domaine minier pour l'analyse des teneurs (Matheron, 1965 - Journel, 1975 - Maréchal,1970). Depuis quelques années, elle a trouvé beaucoup d'autres applications en Sciences de la Terre ; citons la géophysique (Guillaume, 1977), l'agronomie (Marbeau, 1976), la cartographie sous-marine (Journel, 1969), la météorologie (Regazzaci et al. 1975)....

Il est légitime pour les géomécaniciens de se demander si les propriétés mécaniques des roches s'organisent en variables régionalisées et se prêtent comme telles à l'analyse géostatistique.

II - HYPOTHESE DE REGIONALISATION DES PROPRIETES MECANIQUES.

Dans un volume élémentaire, homogène de roche, on sait que les propriétés mécaniques dépendent de variables représentant d'autres propriétés physiques telles que la composition chimique, la porosité, la fissuration... Une propriété mécanique mesurée sur un volume restreint de roche est souvent considérée comme représentative de la propriété mécanique d'un grand volume, mais cette représentativité dépend de la distribution spatiale des valeurs de la composition chimique de la porosité et donc des hétérogénéités que ces dernières variables présentent.

Ces hétérogénéités peuvent être classées en deux types :

- variations plus ou moins progressives de la nature et de la répartition de la matière (géochimie, porosité matricielle)

- variations plus ou moins progressives de la structure tectonique de la matière (zones de déformations souples, discontinuités ou porosité fissurale).

La géostatistique minière montre que les variables géochimiques sont régionalisées et doivent donc être des "phénomènes structurants" des propriétés mécaniques et par là même leur conférer une régionalisation à caractères analogues. Les hétérogénéités du deuxième type font l'objet de recherches tendant à prouver un caractère de régionalisation certain et elles ont, selon leur échelle, un rôle structurant des propriétés mécaniques.

Il semble que les propriétés mécaniques présentent une structuration spatiale qui résulte de deux régionalisations différentes mais plus ou moins superposées dans les massifs rocheux. Nous n'aborderons ici que brièvement la régionalisation au sens de distribution des déformations continues et discontinues. Notre propos sera essentiellement consacré aux liens existant entre la régionalisation mécanique et la régionalisation de la géochimie et de la porosité matricielle. Notre terrain expérimental est constitué de volumes de minerai de fer prélevés dans la mine de Tucquegnieux, dépourvus de traces macroscopiques de déformations, souples ou cassantes.

III - PRINCIPES DE CARACTERISATION DE LA REGIONALISATION MECANIQUE

Considérons un point $\underline{X}$ dans un massif rocheux et un point $\underline{X} + \underline{H}$ voisin. Les propriétés mécaniques de deux volumes élémentaires de roches prélevés autour des deux points $\underline{X}$ et $\underline{X} + \underline{H}$ ne sont pas indépendantes l'une de l'autre : elles sont auto-corrélées. Cette auto-corrélation dépend :

- du vecteur $\underline{H}$ qui sépare les deux points et
- de la formation géologique particulière.

Notre expérimentation a été menée à l'échelle métrique. Trois blocs de minerai de 2 m de long, 1 m de large et 1 m de haut environ ont systématiquement été carottés en rangées horizontales d'éprouvettes et suivant un maillage régulier. Les éprouvettes sont cylindriques, à axe vertical, de hauteur double du diamètre. Les essais de compression uniaxiale sont réalisés sur éprouvettes entièrement saturées d'eau (Pineau, 1979).

La résistance à la compression uniaxiale est retenue comme propriété mécanique $\sigma_C(\underline{X})$. La figure 1A représente un profil ou variation de $\sigma_C(\underline{X})$ en fonction de la position spatiale $\underline{X}$ de l'éprouvette pour une rangée particulière. Ce profil montre une variation quasi-périodique de $\sigma_C(\underline{X})$; cependant, l'étude directe de la fonction spatiale $\sigma_C(\underline{X})$ n'est pas représentable par une fonction mathématique simple : le phénomène "propriété mécanique $\sigma_C(\underline{X})$" n'est pas parfaitement déterministe. De plus, les valeurs numériques $\sigma_C(\underline{X})$ ne sont pas des réalisations indépendantes de la même variable aléatoire, la statistique pure n'est pas applicable. En effet, une interprétation purement statistique ne tiendrait pas compte de l'auto-corrélation entre deux valeurs $\sigma_C(\underline{X})$ et $\sigma_C(\underline{X} + \underline{H})$ pour deux points voisins.

Cette auto-corrélation est révélée par le demi-variogramme $\gamma(H)$ qui caractérise la variabilité de la variable régionalisée dans une direction de l'espace.

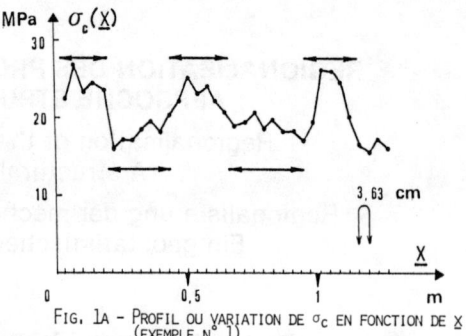

FIG. 1A - PROFIL OU VARIATION DE σ_C EN FONCTION DE $\underline{X}$ (EXEMPLE N° 1)

Son estimateur γ^* est la demi-moyenne arithmétique des différences entre deux mesures expérimentales $\sigma_C(\underline{X})$ et $\sigma_C(\underline{X} + \underline{H})$, différences élevées au carré.

$$\gamma^*(\underline{H}) = \frac{1}{2\,N(\underline{H})} \sum_{i=1}^{i = N(\underline{H})} \left[\sigma_C(\underline{X}) - \sigma_C(\underline{X} + \underline{H}) \right]^2 \text{ où } N(\underline{H})$$

est le nombre total de couples de valeurs

$$\sigma_C(\underline{X}),\ \sigma_C(\underline{X} + \underline{H}).$$

La figure 1B montre le demi-variogramme correspondant à la rangée d'éprouvettes considérée. L'alternance de zones mécaniquement résistantes et de zones moins résistantes se traduit par une croissance non monotone de la courbe ou "effet de trou". Le demi-variogramme permet la mesure, en moyenne, de la périodicité b du phénomène d'alternance et de la largeur a de la zone résistante. La variabilité de σ_C introduite par ces deux types de milieux, résistants et moins résistants, ou, hétérogénéités mécaniques est mesurée par la variance C^*, définie en statistique, lue en ordonnée du demi-variogramme.

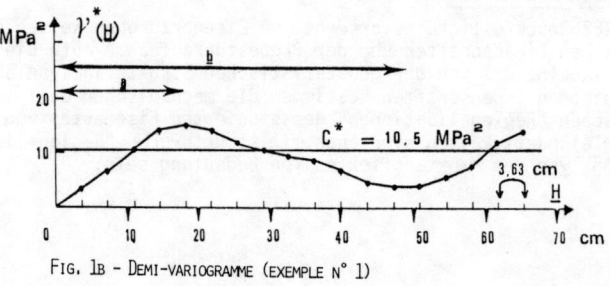

FIG. 1B - DEMI-VARIOGRAMME (EXEMPLE N° 1)

IV - RELATIONS GENERALES ENTRE LA REGIONALISATION MECANIQUE, LA POROSITE MATRICIELLE ET LA REGIONALISATION GEOCHIMIQUE.

La géologie et la géochimie de la formation ferrifère de Lorraine sont bien connues depuis les travaux de Bubenicek, 1968 - Bernard et al. 1960 et Serra, 1968. Nous avons fait appel à ces travaux pour l'interprétation de nos résultats.

Dans l'exemple précédent (fig. 1), les zones résistantes du minerai de fer de Lorraine sont les zones peu poreuses ; elles correspondent :

- soit à un concrétionnement calcaire des oolithes ferrifères, concrétionnement visible à l'oeil nu.

- soit à des zones où les oolithes ferrifères sont liées par un ciment poropelliculaire ou basal de chlorite ou de sidérose, en général non visible à l'oeil nu.

Notre exemple est une illustration quantitative simple de ces liens entre organisation mécanique et structure géologique, mais en général, les hétérogénéités des propriétés mécaniques sont moins nettes. Les concrétionnements calcaires, dans le cas de ce minerai, sont plus ou moins diffus en rognons ou en barres, à bords flous ou soulignant la stratification oblique entrecroisée et ils n'entraînent pas systématiquement une augmentation des propriétés mécaniques.

Il s'avère ainsi que la régionalisation mécanique acquiert une autonomie partielle vis-à-vis des différents facteurs géochimiques et de porosité matricielle. Cette régionalisation est aussi fonction des rapports entre l'échelle des investigations mécaniques et l'échelle des régionalisations pétrographiques. Pour être révélées par le variogramme, il faut que les hétérogénéités aient une taille moyenne comprise plusieurs fois dans le champ d'investigation ainsi qu'une taille suffisamment grande pour être discernée lors des prélèvements d'éprouvettes. Nous nous proposons de détailler quelques exemples d'interprétation de variogrammes.

V - ETUDE DE LA REGIONALISATION MECANIQUE SUR QUELQUES EXEMPLES DU MINERAI DE FER LORRAIN

5.1. - Analyse de variogrammes élémentaires

Une rangée d'éprouvettes prélevées partiellement dans un faciès mixte arénite-lutite (granulométrie d'une arénite ou sable pour les oolithes ferrifères et granulométrie de lutite pour la phase argileuse associée) et partiellement dans la ferri-arénite a fourni des variations assez brusques de σ_c mais qui semblent quelconques (fig. 2A). Le demi-variogramme atteint son palier pour H = 50 cm, ce qui montre l'existence d'une hétérogénéité mécanique d'environ 50 cm (fig. 2B). Dans ce cas particulier, sur des distances inférieures à 50 cm, les résistances à la compression dépendent de la position du prélèvement. Au-delà de cette distance, les mêmes résistances sont des variables aléatoires indépendantes de la position du prélèvement. Le demi-variogramme montre un "effet de pépite" C_o^* qui signifie qu'il existe au moins une hétérogénéité mécanique dont la taille serait inférieure à 4 cm, distance entre deux prélèvements successifs. Il est tentant de supposer que cette hétérogénéité provient de la structure arénitique du minerai de fer car l'étude géochimique a révélé cette même structure (Serra, 1968).

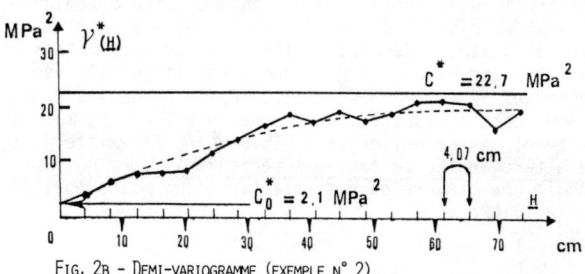

FIG. 2B - DEMI-VARIOGRAMME (EXEMPLE N° 2)

Le profil de la figure 3A obtenu pour une autre série de carottages montre des variations faibles de la résistance à la compression. Le variogramme correspondant (fig. 3B) indique un "effet de pépite pur" ce qui signifie qu'à l'échelle de l'expérimentation, il n'y a pas d'hétérogénéité mécanique. L'hétérogénéité mécanique, dans ce cas particulier est de taille inférieure à 4 cm et elle introduit cette variabilité observée $C^* = 15,7$ MPa^2.

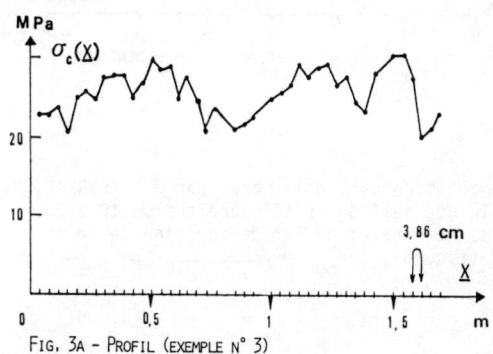

FIG. 3A - PROFIL (EXEMPLE N° 3)

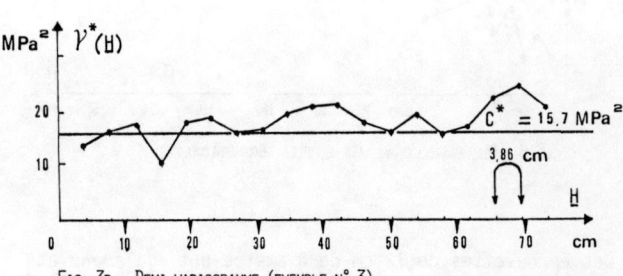

FIG. 3B - DEMI-VARIOGRAMME (EXEMPLE N° 3)

5.2. - Variogrammes moyens : les hétérogénéités mécaniques essentielles à l'échelle métrique.

Le demi-variogramme de la résistance à la compression (fig. 4) tracé à partir de 304 essais sur éprouvettes de 3 cm de diamètre réparties sur 7 rangées révèle les traits essentiels de la régionalisation des propriétés mécaniques à l'échelle métrique. L'effet de pépite C_o^* est peu net : l'hétérogénéité mécanique de taille inférieure à 4 cm est peu importante car elle introduit une faible variabilité des propriétés mécaniques

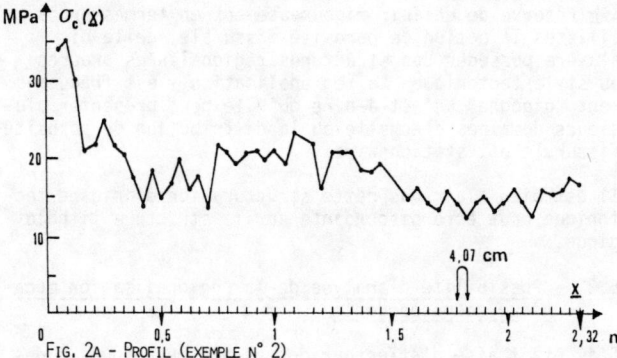

FIG. 2A - PROFIL (EXEMPLE N° 2)

($C_0^* = 2,1$ MPa^2). Il existe une hétérogénéité mécanique essentielle d'environ 20 cm de longueur liée aux migrations et recristallisation de calcite et autres minéraux (chlorite, sidérose). Cette hétérogénéité est essentielle car elle introduit une forte dispersion des propriétés mécaniques ($C_1^* = 21,2$ MPa^2). Le demi-variogramme reste croissant pour H supérieur à 20 cm, mais il n'atteint pas le palier $C^* = 48,8$ MPa^2. Il existe donc une hétérogénéité de taille supérieure à 80 cm qui introduit une dispersion des résistances au moins égale à $C_2^* = 25,5$ MPa^2.

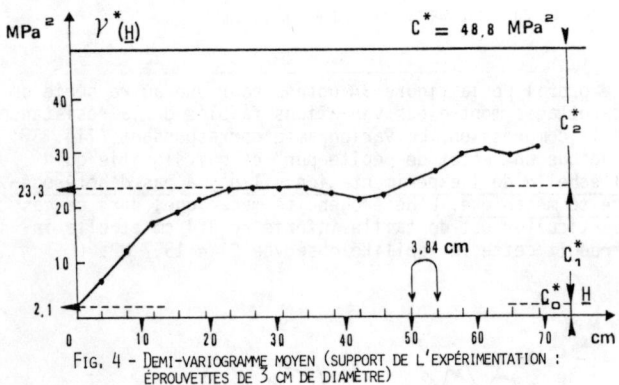

FIG. 4 - DEMI-VARIOGRAMME MOYEN (SUPPORT DE L'EXPÉRIMENTATION : ÉPROUVETTES DE 3 CM DE DIAMÈTRE)

Une expérimentation partielle menée par J.L. PINEAU, 1978, révèle une hétérogénéité mécanique de 10 à 20 m de long dans le minerai de fer de Lorraine de la mine du Paradis (fig. 5).

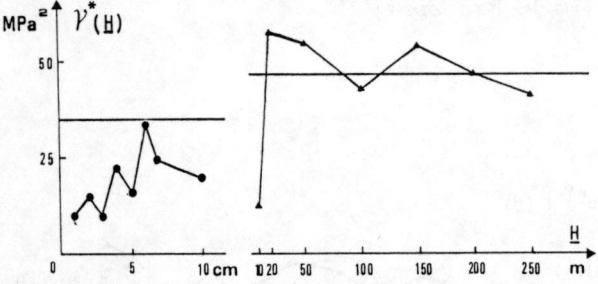

FIG. 5 - DEMI-VARIOGRAMME À L'ÉCHELLE DÉCAMÉTRIQUE

Des éprouvettes de 12 cm de diamètre ont également été prélevées en rangées. Le demi-variogramme moyen (fig.6) montre un effet de pépite important ($C_0^{*'} = 6,3$ MPa^2) : l'hétérogénéité mécanique d'environ 20 cm révélée par l'expérimentation portant sur les éprouvettes de 3 cm de diamètre est déjà suffisamment englobée dans les éprouvettes de 12 cm de diamètre pour ne plus être discernée par l'expérimentation. L'effet de pépite $C_0^{*'}$ présenté par ce demi-variogramme prouve l'existence d'au moins une hétérogénéité mécanique non décelée par l'expérimentation et introduisant une forte variabilité de la résistance à la compression égale à $C_0^{*'}$. Le variogramme moyen (fig. 6) croît lentement sans atteindre le palier $C^{*'} = 30$ MPa^2. L'hétérogénéité mécanique de taille supérieure à 80 cm introduit une dispersion des résistances à la compression au moins égale à

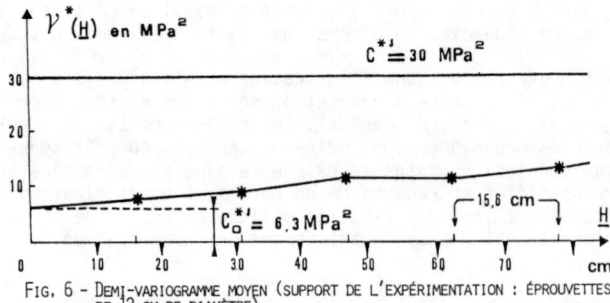

FIG. 6 - DEMI-VARIOGRAMME MOYEN (SUPPORT DE L'EXPÉRIMENTATION : ÉPROUVETTES DE 12 CM DE DIAMÈTRE)

$C_1^{*'} = C^{*'} - C_0^{*'} = 23,7$ MPa^2. Les deux types d'expérimentation portant sur des diamètres différents donnent des résultats dont l'interprétation géostatistique est semblable. Les différences de variance ($C_2^* \neq C_1^{*'}$) sont dues au fait que seulement 37 éprouvettes de 12 cm de diamètre ont été essayées.

Les propriétés mécaniques (résistance à la compression σ_c, module de déformation linéaire E et limite de proportionalité correspondante σ_e) sont parfaitement corrélées entre elles et avec d'autres propriétés physiques telles que la porosité n, ou la célérité des ondes longitudinales (Duffaut P. et al., 1979). Les variogrammes moyens sont d'ailleurs du même type. Cependant, les demi-variogrammes de porosité matricielle et de célérité des ondes longitudinales ne présentent pas d'effet de pépite : l'espace poreux ne présente pas d'hétérogénéité pour des dimensions inférieures à 4 cm ; il est uniformément du même type, porosité matricielle, et ne varie qu'avec les concrétionnements (palier à 20 cm environ). En effet, le prélèvement préférentiel de gros blocs a éliminé les discontinuités de déplacement ou porosité fissurale.

VI - PROBLEMES POSES PAR L'ETUDE DE LA REGIONALISATION DES PROPRIETES MECANIQUES EN FONCTION DE LA REGIONALISATION STRUCTURALE.

6.1. - Caractères de la régionalisation structurale

Les déformations souples ne présentent pas en général de caractères purement aléatoires mais structurent les roches selon un mode déterministe que la tectonique peut analyser.
En revanche, la distribution des discontinuités de déplacement liées aux déformations cassantes obéit à des lois comportant une composante aléatoire et une composante déterministe.

Des travaux récents (Thomas A., Pineau A.) montrent que sous réserve de définir rigoureusement en termes probabilistes la notion de porosité fissurale, celle-ci s'avère posséder des structures régionalisées propres au style tectonique. La régionalisation y est fréquemment "gigogne", c'est-à-dire qu'elle peut présenter plusieurs domaines d'échelle ou la distribution de porosité fissurale est stationnaire.

Il est bien clair que cette structuration d'origine tectonique peut être discordante sur la structure lithologique.

6.2. - Possibilité d'analyse de la régionalisation mécanique correspondante

S'il était aisé d'effectuer des essais mécaniques à des échelles différentes en milieu non fracturé pour étudier les liens avec la pétrographie, la prise en compte de la fracturation suppose des échelles d'investigation qui ne sont guère concevables que *in situ*.

La seule approche rationnelle serait alors la suivante :

- étude régionalisée du champ de fracture avec en particulier détermination des échelles où la distribution est stationnaire

- analyse de déplacements sur un ouvrage réel dans le massif étudié

- mesure *in situ* des contraintes régionales (conditions aux limites)

- constitution d'un modèle mécanique du massif et de l'ouvrage par éléments finis (méthodes des discontinuités de déplacements), le champ de fracture étant simulé aux différentes échelles où il a été reconnu stationnaire

- calcul des déplacements théoriques

- confrontation aux mesures et ajustements des paramètres rhéologiques aux différentes échelles.

De tels programmes d'études qui marqueront le passage de la mécanique des roches à l'échelle du massif sont entrepris aux Etats-Unis et constituent un de nos objectifs.

VII - CONCLUSION

Les propriétés mécaniques sont des variables régionalisées : leur variation spatiale n'est pas quelconque ou aléatoire, elle est au moins partiellement organisée et suit plus ou moins les structures géologiques des dépôts ou les structures tectoniques cassantes ou autres.

Les variations des propriétés mécaniques mesurées s'expliquent clairement par la présence d'hétérogénéités qu'une expérimentation appropriée traitée grâce à l'outil géostatistique peut mettre en évidence.

Les hétérogénéités mécaniques ainsi révélées en taille, en espacement, en variabilité qu'elles introduisent seraient une information précieuse pour aborder rationnellement les problèmes d'effet d'échelle dans les projets miniers et la mécanique des roches en général. Il est certain que l'étude de la structuration spatiale des propriétés mécaniques se posera en termes scientifiques différents car elle se posera à des échelles différentes selon la nature des éléments structurants majeurs, variables pétrographiques ou variables tectoniques.

REFERENCES

Bernard A. et Bubenicek l. (1960) -
Remarques sur les séquences sédimentaires de l'Aalénien de Lorraine.
C.R. Acad. Sci., Paris, t. 250, pp. 3352-3355.

Bubenicek L. (1968) -
Géologie des minerais de fer oolithiques.
Mineralium Deposita, 3 (Berlin), pp. 89-108.

Duffaut P., Wojkowiak F., Josien J.P. et Pineau J.L. (1979) -
Les vides, principal facteur du comportement mécanique des roches.
Congrès international de Mécanique des Roches.
Montreux - pp. 115-121.

Guillaume A. (1977) -
Analyse des variables régionalisées. Traitement du signal en Sciences de la Terre.
Doin, Paris.

Journel A. (1969) -
Etude sur l'estimation d'une variable régionalisée : application à la cartographie sous-marine.
Service hydrographique de la Marine, Paris.

Journel A. (1975) -
Guide pratique de géostatistique minière, CG, Fontainebleau.

Marbeau J.P. (1976) -
Géostatistique forestière.
Doctoral Thesis, CG, Fontainebleau.

Maréchal A. (1970) -
Géostatistique et niveau de reconnaissance.
Applications aux gisements de bauxite métropolitaine.
Doctoral Thesis, CG, Fontainebleau.

Matheron G. (1965) -
Les variables régionalisées et leur estimation.
Masson, Paris.

Pineau A. (1979) -
Effet d'échelle et structures mécaniques du minerai de fer lorrain.
Doctoral Thesis, ENS Géologie, Nancy.

Pineau J.L. (1978) -
Contribution à la caractérisation géomécanique des roches. Application au minerai de fer lorrain.
Doctoral Thesis, ENS de métallurgie et de l'Industrie des Mines. Nancy.

A RELIABILITY ENGINEERING APPROACH TO THE DESIGN OF ROCK SLOPE AND CAVERN IN FRACTURED ROCK

Une technique d'évaluation de la fiabilité des tracés des talus et des cavernes dans de la roche fracturée

Eine Methode zur Abschätzung der Zuverlässigkeit des Entwurfes einer Böschung und einer Kaverne in geklüftetem Fels

Y. Nishimatsu
S. Okubo
Department of Mineral Development Engineering, The University of Tokyo, Tokyo 113, Japan

SYNOPSIS

Taking into account the statistical distribution of spacing and shear strength of the discontinuity in rock mass, the authors give a theoretical formula on the probability of failure of rock structure in fractured rock. The statistical distribution of the shear strength of discontinuity is discussed on the basis of the result of direct shear tests of rock samples. As an example of the application of the theory to the design of rock structure, the stability of a rock slope is evaluated as a function of spacing and shear strength of discontinuity as well as of the height and the gradient of the rock slope.

RESUME

En étudiant la distribution statistique de l'espacement et de la résistance au cisaillement des discontinuités dans les roches, les auteurs formulent une loi de probabilité de rupture de structure dans la roche fracturée. Ils examinent la distribution statistique de la résistance au cisaillement des discontinuités d'après les résultats du test direct de cisaillement effectué sur des échantillons de roches. A titre d'exemple de l'application de cette théorie à la conception de structures taillées dans de la roche, la stabilité du talus rocheux est évaluée en fonction de l'espacement et de la résistance au cisaillement des discontinuités; on en déduit ensuite la hauteur et la pente du talus.

ZUSAMMENFASSUNG

Eine theoretische Gleichung der Wahrscheinlichkeit von der Standfestigkeit von Felsböschungen und -hohlräumen ist unter Berücksichtigung der statistischen Verteilungen des Diskontinuitätsabstandes und der Diskontinuitätsfestigkeit im Gebirgskörper abgeleitet. Die Gesetzmäßigkeit der statistischen Verteilungen der Scherfestigkeiten der Diskontinuität wird am Beispiel des Scherversuches von Bohrkernen betrachtet. Als Beispiel der Anwendung der Theorie wird die Standfestigkeit einer Felsböschung als Funktion des Diskontinuitätsabstandes und der Diskontinuitätsfestigkeit, wie Böschungshöhe und -gradient betrachtet.

1. INTRODUCTION

There are various discontinuities such as joints, faults, bedding planes, and fractures, in most of the rock mass, and it is well known that the stability of rock structure substantially depends upon the strength, orientation, and location of discontinuities which show a widespread fluctuation.

In this situation, the stability of rock structure would not be reasonably designed without some probabilitic approaches, because the mean value of strength is not reliable for design, but the lowest value of measured strength of discontinuity would give an unrealistic design which seems to be too conservative.

The remarkable dispersion of strength, location as well as orientation of discontinuity needs a reliability engineering approach for the reasonable design of rock structure, in which the probability of failure or stability of rock structure is evaluated from the statistical point of view. The design principle of reliability engineering is to keep the probability of failure under a designated small value such as 0.01% or smaller.

In this paper, the authors give a mathematical equation to calculate the probability of failure of rock structure in fractured rock as a function of the state of stress, geometry of rock structure, spacing and shear strength of discontinuity, under the assumption that the spacing of discontinuity follows the exponential distribution.

In order to apply the theory to the design of stability of rock slope, the fluctuation of the result of direct shear test is analysed from the statistical point of view, and as an example of the application of theory to the design of rock slope, the probability of failure of rock slope is calculated.

2. THEORY

2.1 Some Basic Assumptions

For the simplicity, it is assumed that there is a single set of discontinuities. If the presence of a discontinuity does not affect the probability of presence of another discontinuity in its vicinity, i.e. there is no interaction between discontinuities, the each segment of scanline would have an equal probability of presence or occurrence of discontinuity in unit length of the scanline. From this assumption, the probability that the discontinuity does not occur in the interval (0, $x+dx$) is derived as

$$P_0(x+dx)=P_0(x)\cdot(1-\lambda dx) \quad\cdots\cdots\cdots\cdots\cdots\cdots (1)$$

where λ is the transition probability, or the frequency of discontinuity occurrence in unit length.

Considering that the probability of first occurrence of discontinuity in the interval ($x,x+dx$) is given by

$$dP_0(x)=-\lambda dx\cdot P_0(x) \quad\cdots\cdots\cdots\cdots\cdots (2)$$

we have the probability that the interval (0, x) does not include discontinuity, expressed as

$$P_0(x)=e^{-\lambda x} \quad\cdots\cdots\cdots\cdots\cdots\cdots (3)$$

It has been reported that the frequency of discontinuity spacing follows Eq.(3) (Priest, 1976; Kojima, et al, 1981).

In much the same way as described above, the probability that n discontinuities occur in the interval (0, $x+dx$) is given by

$$P_n(x+dx)=P_n(x)\cdot(1-\lambda dx)+P_{n-1}(x)\cdot\lambda dx \quad\cdots\cdots (4)$$

Postulating that the statistical distribution of shear strength of discontinuity is independent of the coordinate x, and the state of stress on the plane of discontinuity depends on its location and gradient, the probability that the discontinuity occurring at the coordinate x does not slide would be expressed as the function of random variable, i.e., coordinate x. This probability function $P_s(x)$ includes the gradient of discontinuity as a parameter.

2.2 Derivation of the formula on the probability of failure of rock structure

Suppose a rock wall with the height h, and let $Q(h)$ be the probability that the failure along the discontinuity does not take place in the rock wall. Then, substituting 1 into n in Eq. (4), we have the probability that a discontinuity occurs in the interval (0, $x+dx$), as

$$P_1(x+dx)=P_1(x)\cdot(1-\lambda dx)+P_0(x)\cdot\lambda dx \quad\cdots\cdots\cdots (5)$$

Combining this equation with the probability $P_s(x)$ that the failure does not take place along the discontinuity, we have the probability that a discontinuity occurs in the interval (0, $x+dx$), but the failure does not take place, expressed as

$$\Pi_1(x+dx)=\Pi_1(x)\cdot(1-\lambda dx)+P_0(x)\cdot P_s(x)\cdot\lambda dx \quad\cdots\cdots\cdots\cdots\cdots (6)$$

Substitute Eq.(2) into Eq.(6), then we have

$$\Pi_1(x)=exp(-\lambda\cdot x)\cdot\int_0^x \lambda P_s(x)\cdot dx \quad\cdots\cdots\cdots\cdots (7)$$

It would not need to be proved that

$$\Pi_0(x)=P_0(x)=exp(-\lambda\cdot x) \quad\cdots\cdots\cdots\cdots\cdots (8)$$

In much the same way as described above, corresponding to Eq.(4), we have the probability that n discontinuities occur in the interval (0, $x+dx$), but the failure does not take place, expressed as

$$\Pi_n(x+dx)=\Pi_n(x)\cdot(1-\lambda\cdot dx)+\Pi_{n-1}(x)\cdot P_s(x)\cdot\lambda dx \quad\cdots\cdots\cdots\cdots\cdots (9)$$

Integrating Eq.(9), we have

$$\Pi_n(x)=exp(-\lambda\cdot x)\cdot\int_0^x \lambda\cdot P_s(x)\cdot\Pi_{n-1}(x)\cdot exp(-\lambda\cdot x)\cdot dx \quad\cdots\cdots\cdots\cdots (10)$$

Denote that

$$\Lambda=\int_0^x \lambda P_s(x)\cdot dx \quad\cdots\cdots\cdots\cdots\cdots (11)$$

then, by means of the mathematical induction, we have

$$\Pi_n(x)=\frac{1}{n!}\Lambda^n\cdot exp(-\lambda x) \quad\cdots\cdots\cdots\cdots\cdots (12)$$

Summing up all of the cases, we have the probability that n discontinuities at its maximum, occur in the interval (0, x) but the failure does not occur along any of them, expressed as

$$Q_n(x)=\sum_{i=0}^n \Pi_j(x)=\sum_{j=0}^n \frac{1}{j!}\cdot\Lambda^j\cdot exp(-\lambda x) \quad\cdots\cdots\cdots (13)$$

Substitute Eq.(11) into Eq.(13), and h into x, respectively, and increase n to infinity, because there does not exist the upper limit of frequency of the occurrence of discontinuity in the rock wall.

Then, we have the formula on the probability that the failure caused from the discontinuity does not occur, expressed as

$$Q(h)=exp[-\lambda(h-\int_0^h P_s(x)\cdot dx)] \quad\cdots\cdots\cdots\cdots (14)$$

3. SATISTICAL DISTRIBUTION OF SHEAR STRENGTH OF DISCONTINUITY

3.1 Experiments

By means of a test rig for the conventional direct shear test, the sliding resistance along the discontinuity,i.e., the shear strength is measured under constant normal stress. A red shale of Cretaceous formation is used as the rock sample to be tested.

In order to model the natural discontinuity, fracture plane is induced by diametrical compression of core sample. Then, the core sample with a diametrical fracture plane is cut into a proper size, and moulded with cement mortar into a die of 50 mm in diameter and 100 mm in height, keeping the axis of cylindrical die on the fracture plane of core sample.

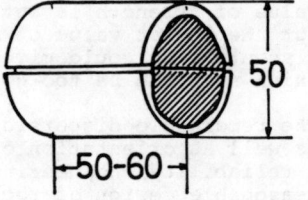

:mm

After the cement mortar has set, the core sample moulded with cement mortar is taken out from the die, and the both ends are cut to get the cylinder of 50 - 60 mm in height, by means of diamond saw.

Fig.1. Cylindrical test piece with slits parallel to the fracture plane

Then, the fracture plane of rock specimen would appear at both ends of the cylindrical test piece. Nextly, narrow radial slits are sawed parallel to the fracture plane from the each opposite side of cylindrical surfaces (see Fig.1).

Then, the test piece with slits is set into the test rig for direct shear test, keeping the fracture plane parallel to the sliding plane of direct shear test. The test result is plotted on the Mohr's stress diagram.

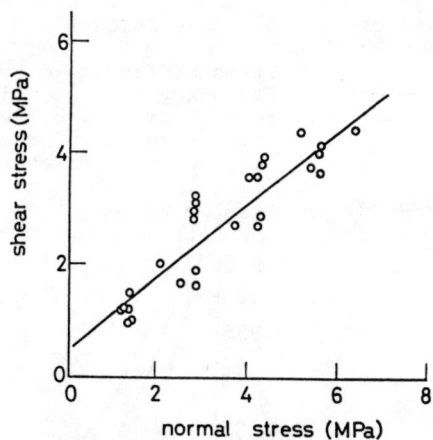

Fig.2. Shear strength of a red shale as a function of normal stress.

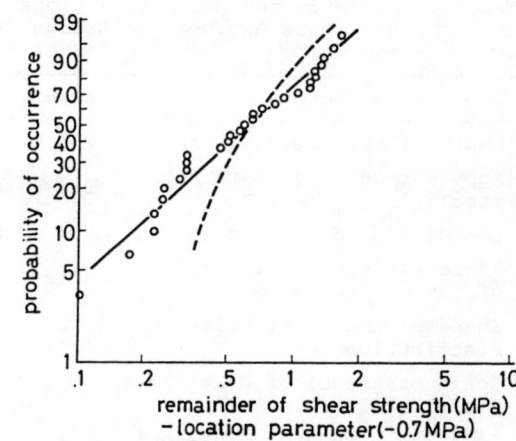

Fig.4. Statistical distribution of shear strength plotted on the probability paper.

Applying Mohr-Coulomb crietrion of fracture

$$\tau_m = \tau_0 + \tan\phi \cdot \sigma \qquad \cdots\cdots\cdots\cdots\cdots\cdots (15)$$

to the test result as shown in Fig. 2, the cohesion τ_0 and coefficient of internal friction $\tan\phi$ are estimated as 0.43 MPa and 0.658, respectively, by the least mean square method.

3.2 Characteristics of Dispersion of Shear Strength of Discontinuity

The remainder or residual $\Delta\tau$ of the fracture shear stress from the mean value is plotted against the normal stress as shown in Fig. 3. It is obvious that the remainder of fracture shear stress is independent of the normal stress. Then, without regard to the normal stress, all of the remainder are plotted on the probability paper of Weibull distribution, because there must be theoretically a finite lowest value of the fracture shear stress. In other words, it is assumed that the remainder of fracture shear stress follows a Weibull distribution, expressed as

$$P_s(\Delta\tau) = exp[-a(\Delta\tau - \Delta\tau_0)^m] \quad \cdots\cdots\cdots\cdots (16)$$

where $P_s(\Delta\tau)$ denotes the probability that the remainder of fracture shear stress is greater than $\Delta\tau$ (Weibull, 1951). The result indicates that the assumption is correct, as shown in Fig.4, where the location parameter $\Delta\tau_0$ has been

determined to give the highest coefficient of correlation on the probability paper. It is calculated that the location parameter $\Delta\tau_0$ = -0.7 MPa, the scale parameter a =1.44, and the shape parameter m =1.5, which give a curve expressed as solid line in Fig. 4.

Considering the definition of the remainder, it is obvious that the fracture shear stress follows a Weibull distribution. Thus, we have the probability that the fracture shear stress is greater than τ, expressed as

$$P_s(\tau) = exp[-a(\tau - \tau_m - \Delta\tau_0)^m] \quad \cdots\cdots\cdots\cdots (17)$$

where τ_m is the mean of fracture shear stress given by Eq.(15), and depends on the normal stress σ.

Putting Eq.(15) into Eq.(17), we have

$$P_s(\tau) = exp[-a(\tau - \tan\phi \cdot \sigma - \tau_0 - \Delta\tau_0)^m] \quad \cdots\cdots\cdots (18)$$

Eq.(18) suggests that the negative fracture shear stress could occur for a small normal stress, when the sum of cohesion and location parameter $\tau_0 + \Delta\tau_0$ is positive. This is the case of the test result of our sample rock, because the cohesion of 0.43 MPa is smaller than the absolute value of location parameter of 0.7 MPa.

In order to avoid this difficulty, the location parameter $\Delta\tau_0$ is assumed as -0.43 MPa, and the scale parameter a is estimated as 3.08 to give the mean value of remainder as zero, for the same shape factor m of 1.5. They give a curve expressed as broken line in Fig.4.

4. APPLICATION OF THE THEORY TO DESIGN OF SLOPE STABILITY

Now, the stability of rock slope would be evaluated by means of the theory described above.

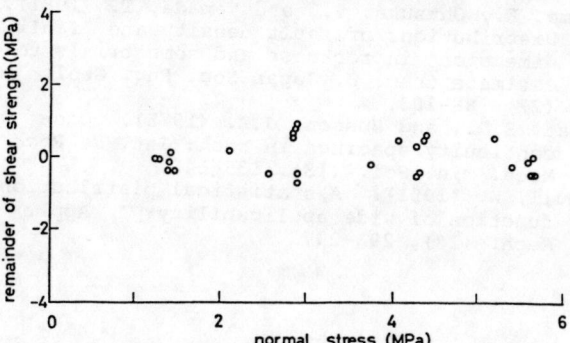

Fig.3. Dispersion of the remainder of shear strength.

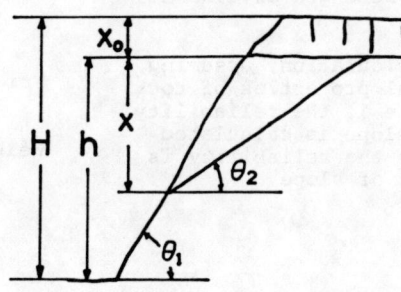

Fig.5. Geometry of rock slope and discontinuity.

Table I Geometry and Mechanical Properties of Rock Mass Assumed for Design Calculation.

Parameters	Assumed Values
Depth of surface soil (x_0)	2.0 m
Apparent density of rock mass (γ)	2.5 x 10^3kg/m^3
Cohesion of discontinuity (τ_0)	0.43 MPa.
Angle of internal friction of discontinuity (ϕ)	33°20'
Shape parameter of Weibull distribution (m)	1.5
Scale parameter of Weibull distribution (a)	3.08
Location parameter of Weibull distribution ($\Delta\tau_0$)	-0.43 MPa.

For the simplicity, it is assumed that there is a single set of discontinuity, the strike of which is parallel to the slope surface, and the hydrological pore pressure is negligible small. The gradient and height of rock slope are given as illustrated in Fig.5. The stress analysis would be conducted in the same way as the limit equilibrium analysis (Hoek, 1970). Then denoting the symbol of geometry of rock slope as shown in Fig.5, the mean normal and shear stress acting on the discontinuity are given by

$$\sigma = \frac{1}{2} \cdot (x + 2x_0) \cdot (\cot\theta_2 - \cot\theta_1) \cdot \gamma \cdot \cos^2\theta_2 \quad \cdots \cdots \quad (19)$$

$$\tau = \sigma \cdot \tan\theta_2 \qquad \cdots \cdots \cdots \cdots \cdots \cdots \cdots \cdots \quad (20)$$

Putting Eq.(20) into Eq.(18), we have the probability that the discontinuity does not slide, expressed as

$$P_s(\sigma) = exp[-a\{(\tan\theta_2 - \tan\phi)\sigma - \tau_0 - \Delta\tau_0\}^m] \cdots \cdots \quad (21)$$

Then, putting Eq.(19) into Eq.(21), we have the probability that the discontinuity occurred at the depth x does not slide, expressed as

$$P_s(x) = exp[-a\{\frac{1}{2}(\tan\theta_2 - \tan\phi) \cdot (\cot\theta_2 - \cot\theta_1) \cdot$$

$$(x + 2x_0) \cdot \gamma \cdot \cos^2\theta_2 - \tau_0 - \Delta\tau_0\}^m] \cdots \cdots \cdots \cdots \cdots \cdots \quad (22)$$

Putting Eq.(22) into Eq.(14), we could calculate the reliability on the stability of rock slope, i.e., the probability that the failure of rock slope does not occur, as a function of height and gradient of the rock slope as well as frequency and gradient of discontinuity, under the given dispersion of shear strength of discontinuity.

As an example of design calculation, assuming the geometry and mechanical properties of rock mass as summarized in Table 1, the reliability on the stability of rock slope is calculated as shown in Fig.6 in which the reliability is plotted against the height of slope.

5. CONCLUSION

It is difficult to evaluate the effect of frequency of discontinuity on the stability of rock structure, because the spacing as well as shear strength of discontinuity show a wide fluctuation. However, it has been reported by a few authors, that the discontinuity spacing follows a exponential distribution. Considering this fact, and assuming that the orientation of discontinuity does not fluctuate, the authors derive a theoretical formula on the probability that the failure of rock structure does not occur.
In order to apply the theory to design of rock structure, it is needed to

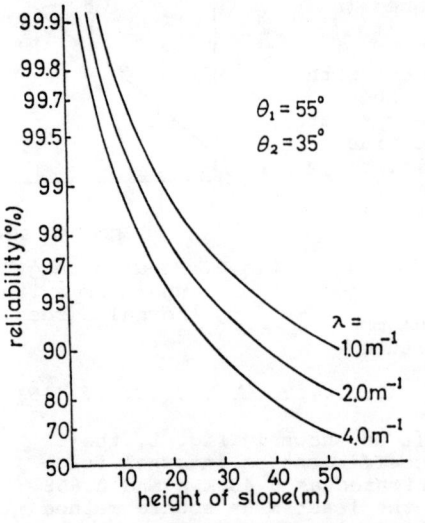

Fig.6. The effect of frequency of discontinuity on the stability of rock slope.

reveal the statistical distribution of fracture shear stress along the discontinuity. Assuming that the fracture shear stress follows Mohr-Coulomb crietrion, the authors show that the dispersion of remainder of fracture shear stress is independent of the normal stress, on the basis of the direct shear test along the discontinuity of core sample. Then, the authors give the equation of the probability that the failure caused from the discontinuity does not occur in the rock structure.

Finally, as an example, by means of a numerical calculation, the authors discussed on the effect of the frequency of discontinuity, height and gradient of rock slope on the stability of rock slope.

References
Hoek, E. (1970). Estimating the stability of excavated slopes in opencut mines: Trans. Instn. Min. Met. (79), A109-132.
Kojima, K., Ohtsuka, Y., and Yamada, T. (1981). Distributions of fault density and "fault dimension" in rockmass and some trials to estimate them: J. Japan Soc. Eng. Geol. (22), 88-103.
Priest, S.D., and Hudson, J.A. (1976). Discontinuity spacings in rock: Int. J. Rock Mech. Min. Sci. (13), 135-148.
Weibull, W. (1951). A statistical distribution function of wide applicability: J. Appl. Mech. (18), 293-297.

STRUCTURE AND MECHANICAL BEHAVIOUR OF A VOLCANIC TUFF

Structure et comportement mécanique d'un tuf volcanique

Struktur und mechanisches Verhalten eines Vulkantuffs

Franco Rippa
Geologist, Istituto di Tecnica delle Fondazioni e Costruzioni in Terra, Università di Napoli,
Italia
Filippo Vinale
Assistant Professor, Istituto di Tecnica delle Fondazioni e Costruzioni in Terra, Università
di Napoli, Italia

SYNOPSIS

A broad investigation is reported on the mechanical behaviour of the volcanic tuff of the Naples area under uniaxial compression. The strength and stress-strain behaviour of this soft rock have been successfully related to genetic and structural features of the tuff, as described by a simple mechanical model in which a matrix, inclusions and cement are distinguished.

RESUME

On présente une étude poussée sur le comportement mécanique du tuf volcanique de la zone de Naples. La résistance à la compression simple et le comportement sous contrainte de déformation ont été expérimentés sur plus de 400 échantillons. Les résultats ont été corrélés avec les caractéristiques génétiques et structurelles du tuf, lesquelles s'intègrent dans un simple modèle mécanique où l'on distingue la matrice, les inclusions et le ciment.

ZUSAMMENFASSUNG

Dies ist ein Bericht über eine im Gebiet Neapel ausgeführten Laboruntersuchung über das mechanische Verhalten des Vulkantuffs unter Zusammendrückung ohne Behinderung der Seitenausdehnung. Die Druckfestigkeit und die Spannung-Dehnung sind bei mehr als 400 Kern-Proben untersucht worden. Die Ergebnisse sind auf genetische und strukturelle Eigenschaften des Tuffs mit Erfolg zurückgeführt worden, wie es sich von einem einfachen mechanischen Modell ergibt, wo Bindemittel, Einschlüsse und Zement unterschieden werden.

1. INTRODUCTION

Wide areas of central and southern Italy, extending from northern Latium to southern Campania, are characterized by the presence of volcanic tuff, that from a rock mechanics viewpoint may be defined as a soft rock.
The experimental investigation herein presented refers to a plain area of Naples, about 300,000 m² wide, where, starting from a depth of about 20 m, two units of volcanic tuff (neapolitan yellow tuff and grey tuff) are subsequently found. The ground surface is about 5 m above sea level, while the water table is 2÷4 meters deep.
The aim is to single out the connexion between genetic, textural and structural features of tuff and, on the other hand, its strength and rheological characteristics under uniaxial compression. For this purpose a mechanical model has been adopted, according to which the tuff consists of an ashy *matrix*, pumice and lithic *inclusions* and a zeolitic *cement*.
The effectiveness of the model in describing the essential features of tuff behaviour has been substantiated by means of mineralogic analysis and Scanning Electron Microscope (SEM) observations. Depending on the different arrangements of matrix, inclusions and cement, different structures have been outlined. Each one is characterized by its own uniaxial compressive strength (σ_c) and stress strain behaviour, both in the region preceeding σ_c and in the post-peak phase until collapse of the specimen.
A wide range of structural and rheological features of the tuff in Naples has been investigated, throughout the whole tuff formation thickness.

2. MAIN ASPECTS OF TUFF GENESIS

In the area around Naples tuffs originated from volcanism of the Phlegrean Fields. The major event of this volcanic activity is represented by the emplacement of large volumes of pyroclastic products. According to literature, the pyroclastic materials were deposited with two different mecha

D

nisms: either a pyroclastic flow, as a result of a fissure activity, or a pyroclastic fall, subsequent to a typical volcanic explosion.

The volcanic deposits of Phlegrean Fields are relative to three main periods of activity, the first and the second of which are characterized respectively by the Grey Campanian Tuff (Campanian Ignimbrite) (28,000÷35,000 years B.P.) and the Yellow Neapolitan Tuff (10,000÷12,000 years B.P.) (Barberi et al. 1978; Lucini, Tongiorgi 1959; Lirer, Munno 1975).

The grey campanian tuff has an areal distribution of about 7,000 km^2, while the neapolitan yellow one outcropps in a smaller area, which stretches from Naples to the Phlegrean Fields.

The loose pyroclastic materials (ash), that gave rise to tuff, are formed mainly by volcanic glass particles, and secondarily by crystal and lava fragments. The grains are irregular shaped and their size is that of a sandy silt. They have a vitreous texture which is spongy and frothy. The particles exhibit high porosity, but only a limited number of pores are in communication with one another and with the atmosphere.

In the ashy matrix pumice and lithic fragments are chaotically arranged. Their size may range from few mm to several cm. Pumices are generally more numerous and larger than lithic fragments.

Lithification of the complex of loose materials (ash + pumice and lithic fragments = pozzolana) is a consequence of a diagenetic process that, under favourable conditions, yielded to the development of zeolites.

Zeolites are neo-formed hydrate minerals of Al and Si, grown as a result of unwelded volcanic glass modification under the action of fluids in subaereal environment.

The zeolitic minerals generally found in neapolitan tuffs are mainly phillipsite, and chabazite; they appear as variously arranged crystals.

Duration of zeolitization process has been determined in 4,000÷5,000 years (Capaldi et al. 1971). More recent researches, however, indicate also much shorter duration (de' Gennaro et al. 1982). Depending on several factors (water alkalinity, solid/liquid ratio, temperature, ...) the distribution of zeolites in the tuff may be variable.

Quantitative evaluations have revealed that the high percentages of zeolites are to be related to the high amount of fine vitreous material in the tuff, which are responsible for high reactivity (Colella et al. 1973).

The content of zeolites in the lithified tuff can be up to 60% in weight, although its average value is about 30% (Scherillo et al. 1980).

It is interesting to note how a central core of zeolitized tuff surrounded by unzeolitized material is often found in situ.

A transition phase, in which zeolitization is not complete exists between pozzolana and tuff, and is commonly called "mappamonte".

3. EXPERIMENTAL PROGRAMME

In the investigated area two tuff units have been found, either directly lying on each other or with loose pyroclastic soil layers interbedded.

The upper unit is made by yellow neapolitan tuff whose thickness is considerably less than in other areas of the town, as a result of the larger distance from the volcanic center.

Beneath the upper unit of y.n.t., a grey tuff unit is found. No geological data are available at present to determine its actual date, which anyway can be attributed either to the I or to the II Phlegrean period. In this research, the determination of the absolute date of some peat levels immediately below grey tuff formation has started and it is being carried out.

Structural features of this unit appear of interest, as related to the characteristics of grey campanian tuff. Furthermore, the data provided by this investigation may be of interest considering the lack of data on mechanical behaviour of grey campanian tuff.

In the investigated area tuff formations have been crossed by means of 19 boreholes. Thicknesses of the two units are about 15 m each.

Experimental programme involved the carrying out of more than 400 uniaxial compression tests on cylindrical specimens (diameter = 80÷100 mm; height/diameter ratio = 2÷2.5).

In order to reduce the end effect, a thin layer of vulcanized rubber was inserted between the specimens and the testing machine plates.

Failure tests have been carried out both in the natural fully saturated condition (~300 samples) and at water content very close to that of tuff over the water table. The latter is a widespread situation in Naples.

For about 200 tests, stress-strain curves have been completely described, also in the post-peak region until the collapse of the specimen. In this case the axial shortening has been measured, over a length of about 15 cm, between two pairs of points placed directly on opposite sides of the sample. In addition displacement has been measured externally too.

The ability to follow the complete load-deformation curve, even in the post-peak region, is related to the use of a very stiff testing machine as compared to tuff stiffness. In fact, the stiffness of the machine corresponds to a slope that is steeper than the steepest descending part of the load displacement curve of the samples.

A constant average strain rate of $2 \cdot 10^{-7}$ s^{-1} has been used.

In order to point out structural features that can have an influence on tuff behaviour, various observations have been performed both in situ and in laboratory.

The length of each core sample has been measured, even if shorter than 10 cm.

More than 1100 specimens were cut for laboratory tests. Each one has been photographed and recorded with remarks on the main structural features (appearance; matrix sharpness; size, frequency, arrangement of inclusions; alteration degree of

pumices; average diameters of the 3 largest pumi
ce and lithic inclusions). On collapsed samples
(>400) these features have been checked upon the
failures surfaces and the inner core.

4. TUFF STRUCTURES IN THE INVESTIGATED AREA

Field and laboratory observations have made possi
ble to classify four main structures, identical
within each tuff unit.
The first structure (A) is homogeneous, i.e. it
is characterized by a limited presence of inclu
sions, that in any case are small sized (fig.1).
Pumices generally have a size less than 5 mm, whi
le lithics, even more infrequent, have a size
less than 3 mm.
Within other structures (B, C, D) inclusions play
a very significant role, which becomes more and
more important as their number and size increase.
As a general rule, pumice inclusions prevail by
number and size; they are round shaped and with
sub-rounded edges.
Pumices appear in different ways; unwelded, com
pletely welded or welded in the outer zone and
unwelded inside. In the third case it is possible
to enucleate them from the material matrix. As a
consequence of pumice welding, more or less large
voids may be present.
The arrangement of pumices is nearly always chao
tic; they rarely tend to arrange in sub-horizon
tal beds.
Within structure B (fig.1) pumice inclusions have
a maximum size of about 15 mm while lithics size
is < 8 mm. Inclusions are never in contact with
one another.
Structure C differs from B as inclusions, equal
in size to those in B, become more abundant.
Finally, structure D (fig.1), essentially present
within the grey tuff unit, is characterized by the

abundance of large sized pumice inclusions (<50mm)
very often in direct contact with one another.
Within the grey tuff unit some peculiarities, com
mon to each structure, have been observed, that
distinguish it from yellow tuff. In the ashy ma
trix a large number of well formed k-feldspar (sa
nidine) crystals are visible; furthermore the
grain size of the inclusions is well sorted, compa
red to the grain size of the inclusions in the yellow tuff.

The previous distinction of 4 structures is consi
stent with dry unit weight (γ_d) determinations.
For instance, cumulative frequency curves related
to structures A, B, C in the yellow tuff are plot
ted in figure 2. As it was to be expected, γ_d decre
ases as the material becomes rich in pumices.

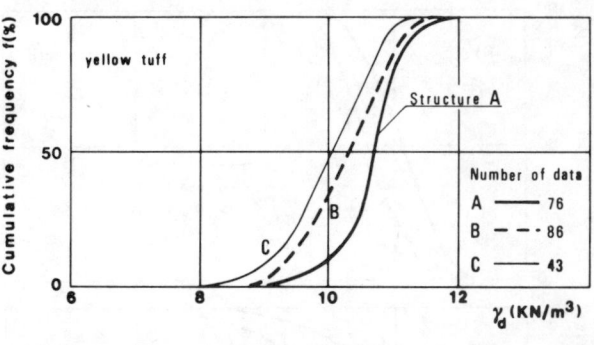

Figure 2

Structures A, B, C, D are not equally distributed
within each tuff unit. As a broad exemplification,
it is worthwhile to report that examinations of
the samples show approximately that structure A
represents 35% on the total number of yellow tuff
samples, structure B 40%, C only 20%. Within grey
tuff, on the contrary, structure D prevails (≈40%)
on structures B and C (B+C≈50%) and A.
The spatial distribution of the structures within
the yellow tuff is chaotic along each borehole.
Consequently, it is impossible to draw stratigra
phic correlations between boreholes. This had to
be expected, bearing in mind the characteristic
features of the yellow tuff formation elsewhere
in Naples.
Even within the grey tuff unit, structures are ir
regularly arranged, although horizontal heteroge
neity is limited. As a matter of fact, it is so
metimes possible to make out the presence of so
me layers very abundant in pumices, across boreholes.

5. UNIAXIAL COMPRESSIVE STRENGTH

Uniaxial compressive strength (σ_c) of yellow nea
politan tuff has been extensively investigated
(Pellegrino 1967, 1968, 1970). Tests have been
carried out on specimens taken in different areas
of Naples, at a water content far from saturation
(degree of saturation S<0.1); it has been found
out that σ_c varies in a rather wide range, bet
ween 2 and 8 MPa about.

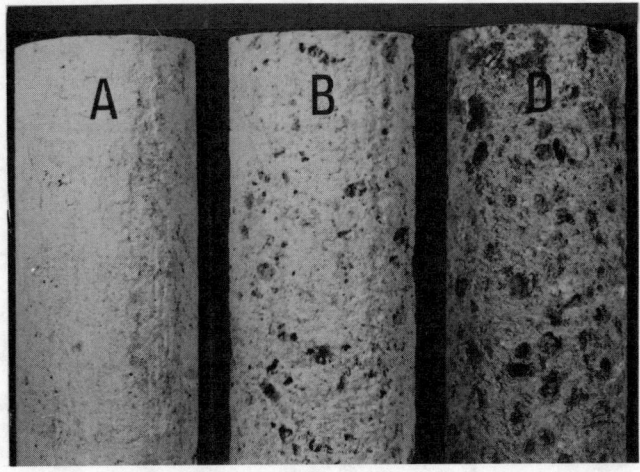

Figure 1

The study of yellow tuff behaviour, when it was saturated in laboratory starting from an initial natural condition of S<<1, has been recently undertaken (Evangelista 1980). A reduction of 25% in σ_c, with reference to unsaturated material has been pointed out.

The experimental data herein reported mainly refer to samples in their natural fully saturated condition. Furthermore, several tests have been carried out on wet air dried specimens (S<0.1).

In both conditions results show good agreement with previous data (figure 3a). In fact, the results obtained approach the lower limit of the previous data, as the greater distance from the emission centre justifies.

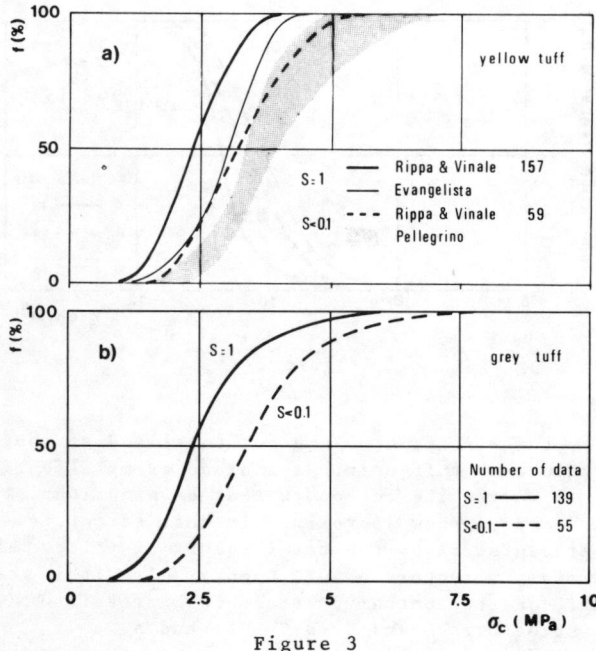

Figure 3

As regards the grey tuff, tests have been carried out both on saturated and unsaturated specimens. A graphical synthesis of the results is plotted in figure 3b; there are not reference curves as previous data are lacking. Uniaxial compressive strength σ_c of saturated samples varies between 1 and 6 MPa about, thus showing a range larger than that related to yellow tuff. Saturation causes a σ_c reduction of about 30%.

The data above discussed are summarized in figure 4, that showsthat they are mostly placed in the high modulus ratio region.

In order to achieve the main aim of this research, i.e. to single out a connexion between tuff uniaxial compressive strength and deformational behaviour (see par.6), and the genetic and structural characteristics of the material, the results have been separated in accordance with structures described at par.4 (figure 5). Plotted data are related to yellow tuff structures A, B and C. It is evident that moving from an homogeneous struc-

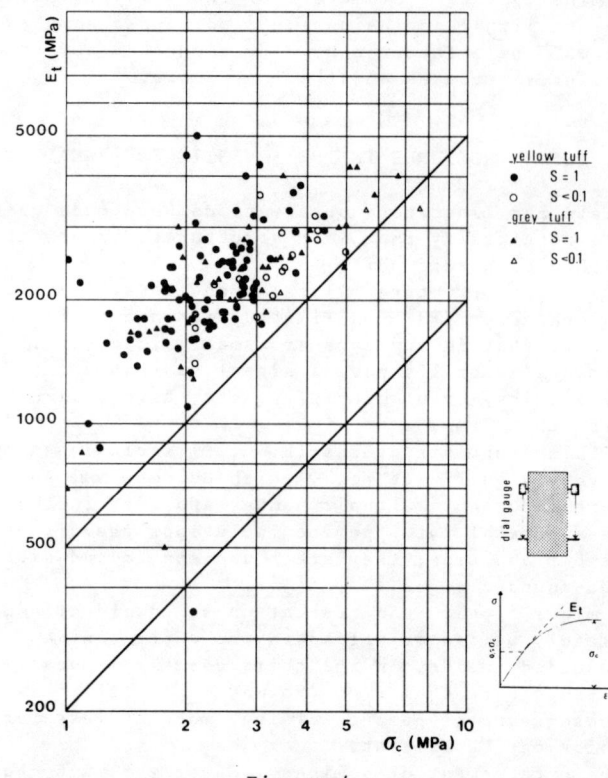

Figure 4

ture (A) towards structures with higher content of pumices, σ_c appreciably decreases.

It is worthwhile to note that σ_c distribution curve is clearly more uniform within structure B and C than in structure A, although the latter is apparently homogeneous. As it will be pointed out at par.6, this apparent discrepancy may be ascribed to the variability of the degree of cementation within structure A and to the essential role that the presence of zeolitic minerals plays at a microstructural scale. It is true that even within other structures zeolitic cement is variable, but its influence is at least partially obscured as a result of the overlap of macrostructural features.

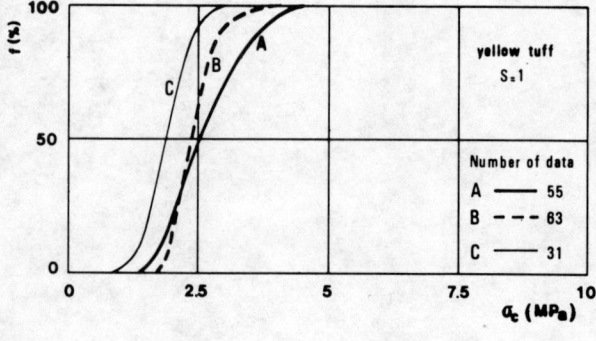

Figure 5

B 36

As concerns the influence of the type of structure on σ_c in conditions far from saturation, available data are not sufficient to draw conclusions. However, the results collected for structures A and B, allow to foresee a different effect of the water content on the 2 structures (figure 6).
A possible interpretation of this experimental behaviour may be in the rising of water capillarity tensions within an unsaturated medium (Pellegrino 1968; Evangelista 1980). As a matter of fact, it seems reasonable to suppose that the absolute tension field is lower within structure B than A, where pumices inclusions are rare and small sized.

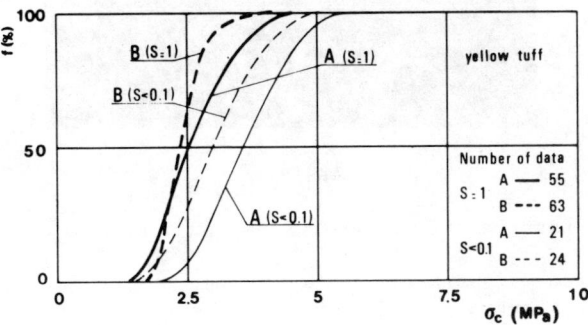

Figure 6

As it was previously said, this experimental work covered the two tuff units on their whole thickness. Thus, it has been also possible to examine the variability of σ_c, separating the results among the different boreholes, both within yellow tuff and grey tuff (figure 7).
As concern yellow tuff, fig.7a shows that the re

sults relative to four different boreholes lead to clearly different curves, characterized by a uniformity degree higher than the overall curve of fig.3a. On the whole, these observations suggest that within the yellow tuff the total variability is a consequence of both an horizontal and a vertical heterogeneity.
With reference to grey tuff figure 7b shows, on the contrary, that within this material the total heterogeneity is of the same order than the vertical one.
This difference between yellow and grey tuff could be a confirm of the marked horizontal heterogeneity of yellow tuff formation, compared with grey tuff, as just pointed out at par.4.

6. DEFORMATIONAL BEHAVIOUR OF TUFF UNDER UNIAXIAL COMPRESSION

In order to investigate the deformational behaviour of tuff a suitable mechanical model has been adopted, according to which tuff consists of an ashy *matrix*, *inclusions* and zeolitic *cement*.
With the aim of testing the mechanical reliability of this model, several SEM observations have been made.
It has been pointed out that the grains aggregate that forms the matrix is not continuous It includes surfaces of discontinuity (grains or grains aggregates boundaries and microcracks) as well as roughly spherical or elongated voids.
Surfaces of discontinuity occur also in the zeolitic cement. In fact, the cement appears as crystalls aggregates and microcracks are sometimes present (figure 8).

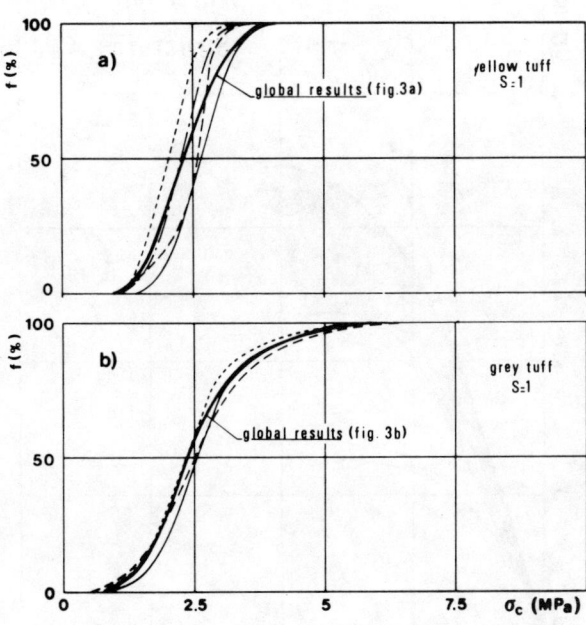

Figure 7

Figure 8

As zeolites grow welding the outer zone of vitre
ous particles, they originate from the grain boun
daries and partly occupy voids creating a sort of
intergranular bridges (figure 9).

Figure 9

The rheological behaviour of a tuff without signi
ficant inclusions (structure A) actually depends
on these microstructural features.
Otherwise, when inclusions increase either in num
ber or in size, their presence plays a significant
role on the rheological behaviour of tuff, so that
attention has to be devoted to "macrostructural"
features as well.
Lithic inclusions do not partecipate to zeolitization process and therefore they have a weak con
nexion with the ashy matrix.
On the contrary, pumice inclusions actively parte
cipate to zeolites growth and pumice-matrix bonds
become more and more significant as their size de
creases. However, as pumice inclusions have always
a significant size as compared with matrix grains,
the cementation degree between pumices and matrix
is lower than inside the matrix. As consequence,
discontinuity surfaces are present at the bounda
ries of the inclusions (figure 10).
Furthermore, the internal pores of the pumices
add to these discontinuities. They generally ha
ve an elongated shape and appreciable size.

The model herein adopted allows to understand de
formational behaviour under uniaxial compression,
as it has been singled out within different stru
ctures. The results deduced for the yellow tuff
will be examined in the following pages. Similar
results have been determined for the grey tuff,

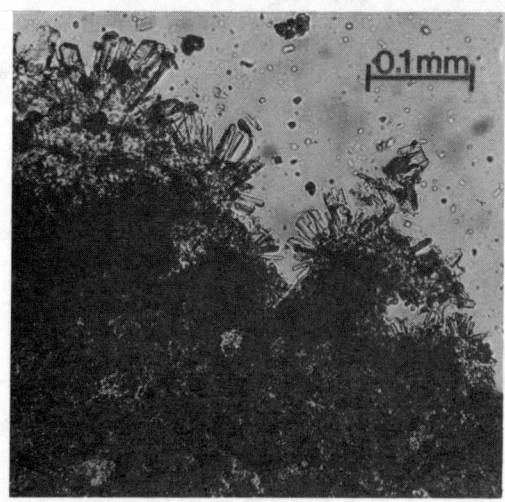

Figure 10

so that identical remarks may be done for it.

Three main types of $\sigma-\varepsilon$ behaviour may be outlined
for structure A, for which the curves in figure
11 show the upper and lower bounds.
With reference to the usual distinction of 4 stages
in a $\sigma-\varepsilon$ curve (Paterson 1978), fig.11 shows that
no initial squeezing phase is observed.
Subsequent phases are quite different within the
three main types classified. Particularly the near
ly perfect linear region extends almost up to σ_c wi
thin the more resistant type, while it becomes
proportionally shortened as σ_c decreases and conse
quently the non-linear region becomes more pronounced.

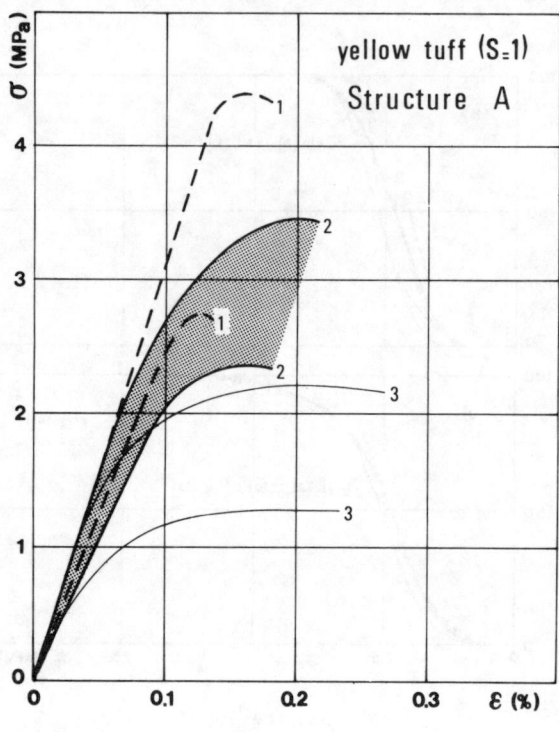

Figure 11

It is interesting to note that the specimens of
the three types do not differ in number and dimen
sions of inclusions.
Differences in σ_c and in $\sigma-\varepsilon$ behaviours may be ex
plained looking at the microstructural features
previously outlined. Microstructural features relati
ve to the cement vary in a wider range than those
relative to the matrix. As a consequence, while struc
ture A may be considered homogeneous as far as tho
se latter features are taken into account, on the o
ther hand it has to be considered heterogeneous, regards
the former ones. Therefore, the differences poin
ted out are related to different zeolitic cement
content. For this purpose thermo-gravimetric ana
lysis have been carried out; the results obtained con
firm the expected differences in cement contents, although
further investigation is recommended (de'Gennaro et al 1982).

Within structures other than A, as number and si
ze of the inclusions increase, macrostructural fea
tures have more and more prominence, until they beco
me the prevailing factors and obscure the effects
of microstructure on mechanical behaviour of tuff.
This can be shown plotting $\sigma_c=f(\gamma_d)$ in a diagram,
and assuming γ_d as a rough estimate of the presen
ce of inclusions (figure 12). It can be observed
that in structure A, in which microstructural fea
tures prevail, data appear to be random, while in
structure B and, ever more, in structure C, a cor
relation between σ_c and γ_d may be observed.

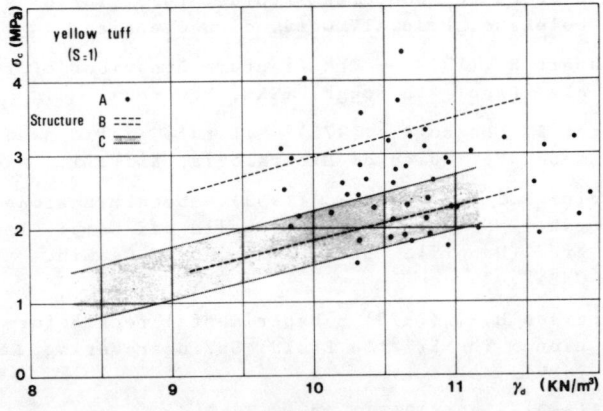

Figure 12

The influence of the inclusions on $\sigma-\varepsilon$ curves is
shown in figure 13. The curves are always charac
terized by a pronounced non-linear region, which
indicates a major presence of structural defects
(Houpert 1979).
Within structure C, which is characterized by a
larger number of inclusions, the non linear region
is even more pronounced; as a consequence the li
near phase is significantly shortened.

The $\sigma-\varepsilon$ behaviours just described substantially
modify when samples are far from saturation. In
fact, while σ_c increases, on the other hand there
is a remarkable extension of the linear region,
independent of the structure type.
This is clearly shown in figure 14, where the ran
ges of $\sigma-\varepsilon$ curves of saturated and wet specimens

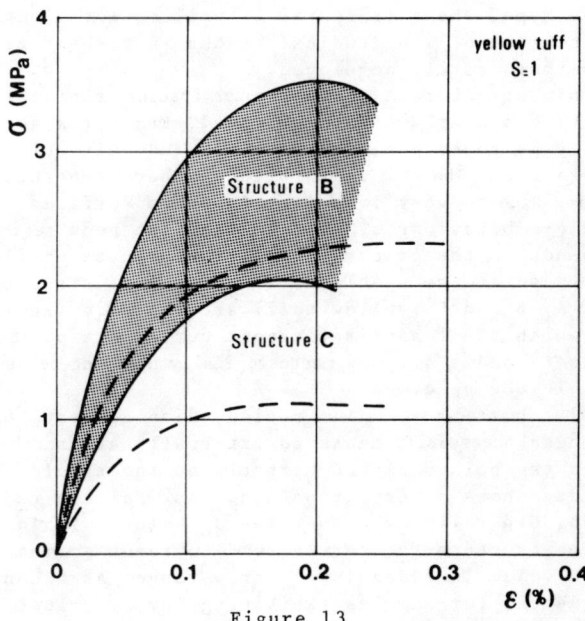

Figure 13

are plotted.
Compared results are relative to samples taken
from the same borehole. For each structure, the
same number of tests has been examined in both
groups.

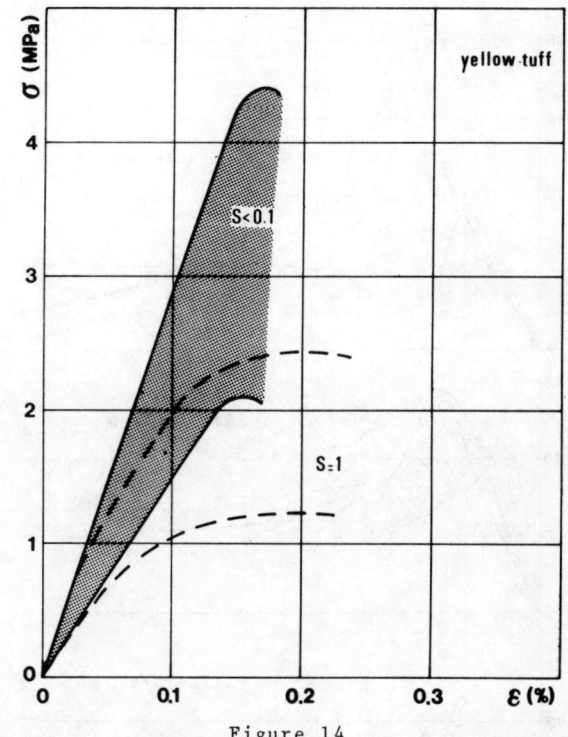

Figure 14

During the phase up to peak-load it has been ge
nerally observed the onset of longitudinal cracks,
fairly uniformly distributed within the specimen.
In this phase there is no growth of macroscopic
fractures.
After the peak load, cracks join together to form
macroscopic fractures, mainly of the axial split

ting type. shear fractures as well as local con centration of longitudinal cracks in a shear zone have been rarely developed.

Within structure A rough regular fracture surfaces have been nearly always observed. Fractures gene rally go round pumice inclusions and only seldom cross them. The same occurs within other structures giving rise to very irregular fracture surfaces.

The σ-ε behaviour of tuff in the post peak region depends on the cracks propagation process until collapse of the specimen. Behaviours of structu res A, B and C (yellow tuff) are shown in fig.15, in which the load-displacement curves are plotted, since σ and ε are now macroscopically nonhomogeneous within the specimen.

In the advanced post-failure region, displacements ha ve been necessarily measured externally and not bet ween the points placed directly on the sample. Figure shows a very pronounced post failure re gion. Generally, whatever the σ$_c$ value, within each structure the curves tend to reach a final load value included in a narrow range. As a con sequence, for samples exhibiting low σ$_c$ values, load reduction is moderate leading to a less brit tle behaviour.

The results reported in previous figure are re presentative samples of more than 200 curves.

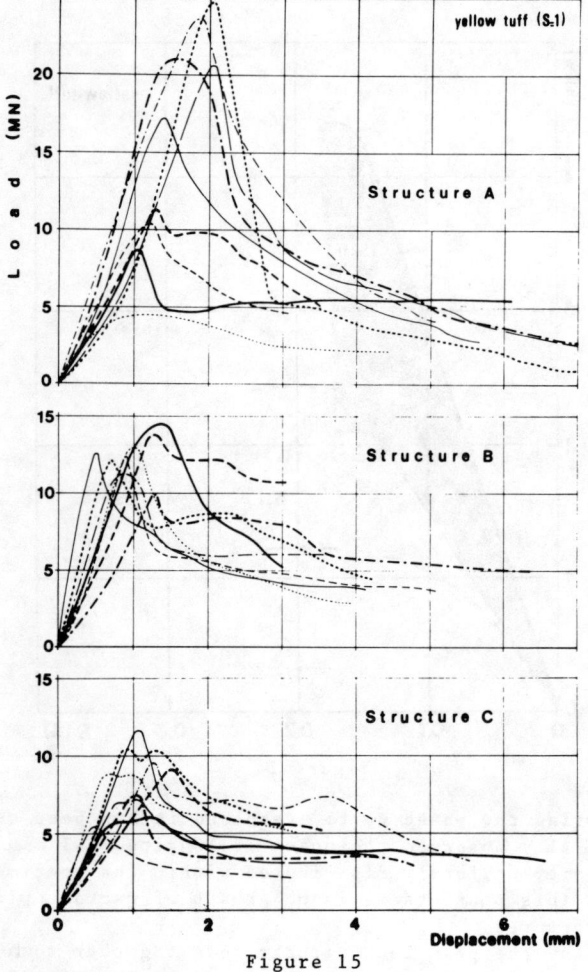

Figure 15

7. ACKNOWLEDGEMENT

The research work going on since several years on tuff genesis and mechanical behaviour at Mineralo gy and Soil Mechanics Institutes of Naples Univer sity provided a valuable framework for this study. The authors would like to express their gratitude to prof. de'Gennaro and prof. Lirer, of the Mine ralogy Institute of Naples, for the stimulating discussions and for making available SEM.

8. REFERENCES

Barberi F., Innocenti F., Lirer L., Munno R., Pe scatore T., Santacroce R. (1978) - The Campa nian Ignimbrite: a major eruption in the Nea politan area (Italy). Bull.Vulcanol. V.41.

Capaldi G., Civetta L., Gasparini P. (1971) - Fra ctionation of U^{238} decay series in the zeoliti zation of volcanic ashes. Geoc.Cosm.Acta, 35

Colella C., Aiello R., Sersale R. (1973) - Gene sis, occurrence and properties of zeolitic tuff. Rend.Soc.Ital.Min.Petr., 29.

de'Gennaro M., Franco E., Langella A., Mirra P., Morra V. (1982) - Le phillipsiti nei tufi gialli del Napoletano. Periodico di Mineralogia (in press).

Evangelista A. (1980) - Influenza del contenuto d'acqua sul comportamento del tufo giallo na poletano. Proc.XIV Conv. di Geotecnica.

Houpert R. (1979) - The fracture behaviour of ro cks. Proc. 4th Congr. ISRM, Montreux.

Lirer L., Munno R. (1975) - Il tufo giallo napole tano. Periodico di Mineralogia, XLIV, 1.

Lucini P., Tongiorgi E. (1959) - Determinazione con il C^{14} di un legno fossile nei Campi Fle grei (Napoli). "Studi e Ric. Div. Geomin. CNEN", 2.

Paterson M.S. (1978) - Experimental rock deforma tion - The Brittle Field. Springer-Verlag, New York.

Pellegrino A. (1967) - Proprietà fisico-meccani che dei terreni vulcanici del napoletano. Proc. VIII Convegno di Geotecnica, Cagliari.

Pellegrino A. (1968) - Compressibilità e resisten za a rottura del tufo giallo napoletano. Proc. IX Convegno di Geotecnica, Genova.

Pellegrino (1970) - Mechanical behaviour of soft rocks under high stresses. Proc. 2nd Congr. ISRM, Beograd.

Scherillo A., Porcelli C., Franco E., de' Genna ro M. (1980) - Guide to the tuff deposits in the Neapolitan and Roman volcanic areas. 5th International Conference on Zeolites, Naples.

THE APPLICATION OF THE RESEARCH ON ROCK PROPERTIES AND MICRO-STRUCTURE TO COAL MINING ENGINEERING

L'étude des propriétés et de la microstructure des roches et son application dans les travaux miniers

Anwendung der Forschungsergebnisse über die Eigenschaften und Mikrostruktur von Gesteinen im Kohlenbergbau

Pan Quinglian
Xin Yumei
Wang Shukun
Niu Xizhuo
The Central Coal Mining Research Institute, Beijing, China

SYNOPSIS

By means of some examples this paper describes the dependence of the solution of many coal mining problems on the investigation of the properties and the microstructure of rocks. As regards a solid and thick roof, the microstructure, the mechanical and physical properties, as well as the weaking effect of water injection under high pressure, and also the factors of magnitude and the mechanism are described in detail.

RESUME

Dans ce texte, on explique, en citant des exemples concrets, comment on peut combiner l'étude des propriétés et de la microstructure des roches avec les travaux miniers. On analyse également la microstructure et les propriétés physico-mécaniques de toits de grès épais et dur, ainsi que les effets des facteurs d'influence et le mécanisme de l'amollissement de la roche par l'injection d'eau.

ZUSAMMENFASSUNG

Im Beitrag wird vor allem durch einige Beispiele die Abhängigkeit der Lösung mancher Probleme im Kohlenbergbau von der Erforschung der Eigenschaften und Mikrostruktur von Gesteinen behandelt, wobei in Bezug auf festes und mächtiges Hangendes sowohl die Mikrostruktur, die mechanischen und physikalischen Eigenschaften, sowie die Wirksamkeit der Schwächung mit Hockdruckwasserinjektion, als auch die Einflußgrößen und der Mechanismus ausführlicher beschrieben werden.

How to closely combine rock mechanical research with mining practice and to solve specific engineering problems has been an important subject for coal mining research and also a matter of interest to whole rock mechanical circles. In recent years, some studies have been carrying out by our Institute on hard roof control, bump, support of the roadway in soft strata and roof classification.

1. THE MICRO-STRUCTURE AND PHYSICO-MECHANICAL PROPERTIES OF SANDSTONE

The investigation and study on rock property and micro-structure of sandstone roof of seam No. 2, Datong Coal Area had been completed so as to explore a safe and economic way for hard sandstone roof control.

1.1 Rock micro-structure

Based on rock composition, the sandstone roof of seam No. 2 can basically be divided into two categories: clay cemented arkose and carbonate cemented arkose, with the majority being the former. The sandstone is compact, well cemented and greyish-white in colour. The clay in clay cemented arkose amounts to 13-19 %, being mainly Kaolinite (Fig. 1). With some pyrite, well-developed joints and micro-fissures perpendicular to these joints, the rock mainly consists of

quartz, feldspar and other minerals. The amount of the feldspar can be up to 25%. Feldspar, mainly orthosyenite and microcline with developed joints and micro-fissures, is liable to be eroded and turned into clay minerals such as gaolin, illite and secondary quartz along the joints and fissures. As a result, the joints and fissures are filled with clay minerals or become openings upon wetted (Fig. 2).

Fig.1 25 202 4516   Fig.2 25 661 5108

Fig.1 The cement of the sandstone is irregular arranged Kaolinite crystal.

Fig.2 The joints and fissures of the feldspar and filled clay.

In addition to these mineral fragments, small amount of fragments of aposandstone, pelitic sandstone and clay rock are found in the sandstone. There are also fissures filled with clay mineral in quartz.

Since a lot of clay minerals remain in the sandstone in form of cement material, joints and fissures are well-developed and filled with clay mineral and coal line exists in part of the rock, the roof of seam No.2 has an affinity for liquid, and is advantageous to be softened by water injection.

1.2 Result and mechanism of rock being softened by water immersion

Based on the affinity of rock for liquid mentioned above, some experiments have been carried out in our lab (Table I).

From the Table we can see that the strength of clay cemented arkose, the principal part of the roof strata reduced 26-49% after water immersion. The experiments show that this kind of roof can be softened by water injection.

Table I

Rock type	Compressive strength before immersion (kg/cm^2)	Dry volume weight (g/cm^3)	Specific gravity	Porosity (%)	Water absorption (%)	Strength reduction after being immersed in water for 6-9 days (%)
Carbonate cemented arkose	975-1352	2.56-2.60	2.64-2.75	2.3-6.9	1.29-1.50	10.3-29 an average of 20
Clay cemented arkose	490-1243	2.35-2.50	2.63-2.69	6.3-11.5	1.62-3.76	26-49 an average of 38

Observing original sandstone sample, sandstone samples immersed by tap water and by acid mine water under an electromicroscope, we found that the size of the mineral was becoming smaller successively.

Having the grains of mineral enlarged and examining them again, micro-fissures about 0.1 μ in width can be seen all over the surface of the grains, After immersed by water, the width of the micro-fissures got smaller and the number of fissures increased (Fig. 3,4).

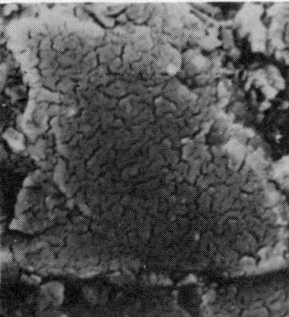 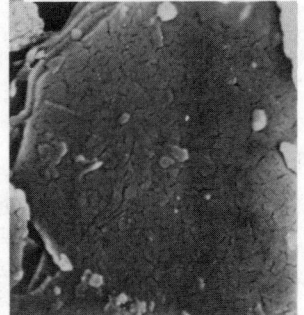

37002 20kv 0.5u   37003 20kv 0.5u
   Fig. 3   Fig. 4

Fig. 3 Fissures on clay mineral of sandstone before immersion

Fig. 4 Fissures on clay mineral of sandstone after immersion

It is believed that because of absorption, wedge effect and dissolution, the size of the mineral in rock would become smaller after water immersion. Meanwhile, as water moleculae enter in-between mineral layers, the mineral would swell up, which makes micro-fissures narrow. When more water moleculae enter in-between layers, new micro-fissures would be produced. All these result in the reduction of binding force between minerals, and thus the strength of rock. And because there are a great amount of clay minerals (0.005 mm in size) in rock, the rock has a very large specific surface area. As a result, large amount of water moleculae are adhered and absorbed into the structure between mineral

layers by intensive binding force. Moreover, because water moleculae often exist in dipole, and the grains of clay mineral have a large specific surface area and is negative charged, the dipole water moleculae are directional arranged around the grains of clay mineral, forming electrified hydrated moleculae. Having the same electric charge, these particles repel each other, which in turn destroy the binding force between clay minerals and greatly reduce rock strength.

So, adhesion and absorption, wedge effect, hydration and dissolution are factors reducing sandstone strength after water immersion.

In addition to above mentioned factors, in conducting high-pressure water injection in thick strata, joints, beddings and fissures will be enlarged by high-pressure water, thus reducing the binding force between layers, destroying integrity and stability of strata, and increasing cavability. It is especially so when there is a coal seam embedded in rock strata; e.g. the strength of rock sample containing coal line can be reduced by over 50 % after water immersion.

1.3 Main factors affecting rock softening

(1) The influence of rock texture and composition

As shown in Table I, the strength of clay cemented arkose decreases an average of 38 % after being saturated with water, but that of carbonate cemented arkose only decreases 20 %, about half as many as the former. The reason for that is because clay mineral is flaky grain in shape. These flaky grains are irregularly arranged, forming large pores between grains (Fig. 1) and possessing higher adhesiveness and absorptivity to water. The carbonate cement is grate in texture. It is aphroid and compact between grains. The porosity and absorptivity is low. In addition, carbonate is insoluble in water, so it is unfavourable for carbonate cemented arkose to be softened by water immersion.

Although there are 8-10 % of clay mineral contained in this kind of rock, they do not present as cement material, but as aggregate of clay mineral or grains of pseudo-arkose distributed in carbonate cement texture. For these reasons lab study on the type of rock, occurrence and micro-structure of rock minerals should be carried out before appling water injection for rock softening.

(2) The influence of water quality

a. Effect of water containing oxygen on rock softening

Comparative tests on immersing clay cemented arkose in water dissolved with air and without air have been done in our lab. The result shows that the effectiveness of rock being softened in the former is about 10 % higher than that in the later case. The reason is that oxido-hydrolysis takes place when pyrite in sandstone run into water, producing sulphuric-acid. The sulphuric-acid may dissolve part of carbonate, thus promoting the softening process and greatly reducing the rock strength. In areas where roof strata have higher pyrite content, it is advantageous to inject moderate amount of air at the same time with water injection.

b. The influence of acid water on rock softening

In order to improve poor result of softening carbonate cemented sandstone, acid water immersion has been tested. The solute used is pyrite from the roof, with PH value being 3. The strength of carbonate cemented arkose reduced by 64 % after being immersed in the solution. This indicates the method of increasing acidity of injected water is feasible in dealing with this kind of rock.

(3) The influence of immersion time

Fig. 5 shows the relationship between absorption rate and immersion time and between softening coefficient and immersion time for sandstone.

The sandstone is saturated after being immersed in water for four days. But it should take nine days for the strength of the sandstone to become stable. That means to achieve best results, a period of time is still needed after the rock is saturated with water. The experiments have shown that water injection process should be set for about ten days.

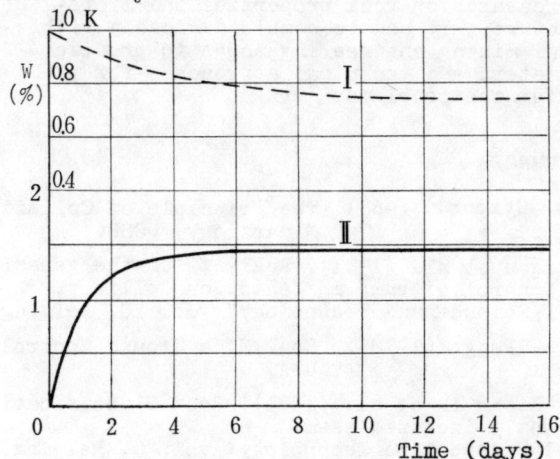

Fig.5 I. Softening coefficient (K) versus immersion time
II. Absorption rate (W) versus immersion time

(4) The influence of rock porosity

It can be seen from Fig. 6 that rock porosity has great influence on the result of rock softening by water injection. Therefore, a general sampling and measuring of rock porosity of roof should be done before conducting water injection.

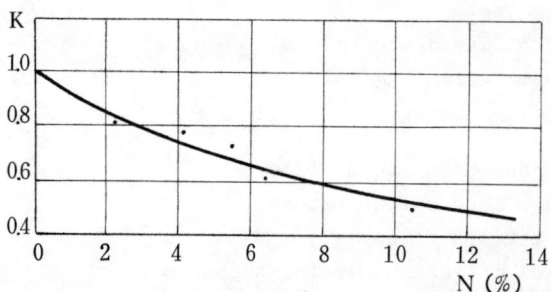

Fig. 6 The porosity of clay cemented arkose (N) versus softening coefficient (K)

A conclusion can therefore be drawn that the rock composition, micro and macro-structure, texture, porosity, water quality and length of injection time all exert great influence on the result of rock softening. Prior to the underground tests, lab studies should be done first so as to gain necessary basis and parameters.

2. THE APPLICATION OF THE RESEARCH ON ROCK PROPERTIES AND MICRO-STRUCTURE TO MINING ENGINEERING

Mining practice has already proved that, by combining the study on rock properties and micro-structure with geological and field measured data, the following items can be predicted:

a. roof condition of the longwall face, desired roof control method, type of the support and supporting density.

b. deformation characteristic of surrounding rock of the roadway and the desired type of roadway support and layout of the roadway.

c. the possibility and affected range of bumping, and the desired preventive measures.

For example, through experiments, it is extimated that hard and thick sandstone roof of seam No. 2, Datong Mining Area can be controled by water injection. According to this extimation, underground tests have been conducted in three faces with thick sandstone roof in Colliery No.4, Datong, and good results have been obtained. Table II and III show the main parameters used and techno-economical results of water injection.

Table II

hole spacing	30 m
hole diameter	60-62 mm
horizontal swing angle	70°
elevation angle	17°-20°
hole depth	70-80 % of the face length
duration of water injection	10 days
amount of mine water injected	300-400 m³/hole

Table III

Techno-economic index	Non-water injection area	Water injection area
hanging roof over the goaf	more than 10 % over 5 m	3.67 % over 5 m
caving height	79.3 % over 5 m	99.1 % over 5 m
caving angle	40° in average	70° in average
sized of caved blocks	8 - 20 m	caved in layer, small blocks
periodic weighting interval	39.6 m in average	9.3 m in average
intensity of periodic weighting	impact loading greater than 600 kg/cm^2	even loading less than 600 kg/cm^2
dust concentration at face	100 %	54.56 %
average face output per day	100 %	105.2 %
cost per ton of coal	100 %	81.82 %
cost of the damage to the supports	100 %	26.3 %
cost of forced roof caving	100 %	14.3 %

There are seven collieries in Beijing Mining Area having serious hazard of bump, of which Mentougou Colliery is the worst one. Using the method proposed by General Mining Institute of Poland, we measured, before and after water injection, the energy index of coal from seam No. 8 where having bump problem. The test showed that the energy index of coal after water injection was reduced from 8 to 3.5, so the possibility of bumping was minimized. The mini-fissures on coal after water immersion was increased remarkabley (Fig.7,8), resulting in the increase of plasticity of coal and decrease of energy index.

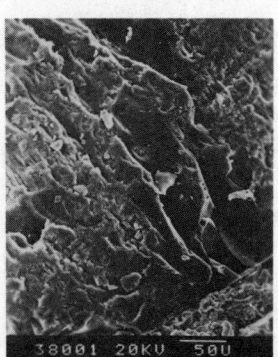 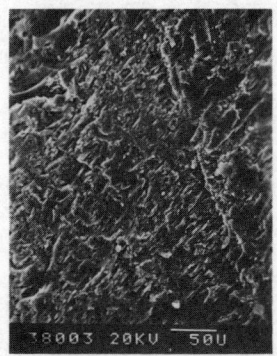

Fig. 7   Fig. 8

Fig. 7 Mini-fissures on coal before water immersion
Fig. 8 Mini-fissures on coal after water immersion

In the light of the lab tests, in-seam water injection was carried out in seam No. 8, Mentougou Colliery when mining coal pillar. The result was encouraging with the number of bumping over 2.5 in Richard scale in water injection area being 79 % less than that in non-injection area. In-seam water injection has now been one of the measures for bump prevention in this mine. The coal formation in Shenyang Mining Area belongs to Tertiary soft strata. According to lab analysis, we learned that both the roof and floor of the seam are argillaceous shale with

easily swellable montmorillonite. The average compressive strength of the shale is 33.8 kg/cm^2, nature water content 18%. Under this rock conditions, it is very difficult to maintain the roadway by using the normal supporting method. In this case, a full-circle yielding support was tested underground and turned out to be effictive. Although it needs higher initial investment, the use of the support is still reasonable in the view of the whole maintenance cost. In addition to this support, the combination of shotcreting and bolting and U type metal support was used in some mining areas, which also resulted in good techno-economic results.

In order to help to select the right type of power support, support density and roof control method in line with the specific geological and production conditions in coal mines in China, "Roof classification of gently inclined seam in China" has been worked out. In which, the immediate roof is classified as four categories according mainly to compressive strength of the rock. The appropriate type of support, support density and roof control method is also proposed.

The above examples indicate the importance of the research on rock properties and microstructure. It can be combined closely with actual mining engineering so as to provide parameters and technical approaches for solving the problems.

REFERENCES

Ф. В. Чухров. (1965). The Principle of Colloid-Mineralogy, USSR.

Lu Zigan et al. (1981). Analysis on the genesis and control of bump in Mentougou Colliery, "Coal Science and Technology" vol. 10, Beijing.

Syd S. Peng. (1978). Coal Mine Ground Control, USA.

Wang Xianglin et al. (1981). Roof Classification of gently inclined seam. "Coal Science and Technology" vol. 9, Beijing.

MESSGERÄT MIT DIREKTER MECHANISCHER ANZEIGE FÜR ANKERKRÄFTE UND ANDERE LASTEN

Measuring device with a direct mechanical read-out for anchor forces and other loads

Cellule de mesure avec indicateur direct et mécanique pour forces d'ancrage et autres charges

H. Habenicht
Bergbau-Ingenieur, Eichenweg 1, A-9220 Velden, Österreich

E. Behensky
Betrieb für Maschinenbau, Ursulinenplatz 5, A-5020 Salzburg, Österreich

ZUSAMMENFASSUNG

Die Prinzipien der Funktion und des Aufbaus einer hydraulischen Zelle werden erläutert. Diese Zelle enthält eine Anzeigeeinheit hoher Qualität. Die Anzeigeeinheit besteht aus einem Federsystem, welches durch den hydraulischen Druck deformiert wird. Die Verformung kann an einem Anzeigestift abgelesen werden. Der Stift wird durch die Verformung des Federsystems in axialer Richtung verschoben, wobei die Verschiebungsbeträge dem hydraulischen Druck proportional sind. Die Vorteile dieses Meßsystems werden vorgestellt.

SYNOPSIS

The operating principles and the design of an hydraulic cell are explained which incorporates a high-quality read-out unit. The latter consists of a spring system which is compressed and deflected by the hydraulic pressure so that the deflection can be read from an indicating rod. The rod is being shifted in an axial direction at a rate proportional to the hydraulic pressure. The advantages of this measuring system are being stated.

RESUME

On décrit la fonction et la composition d'une cellule hydraulique comportant un dispositif de lecture de haute qualité. Ce dispositif incorpore un système à ressort qui se comprime et se dévie en fonction de la pression hydraulique, permettant ainsi la lecture de la déflexion sur une tige indicatrice. La tige se déplace en direction axiale à une allure proportionnelle à la pression hydraulique. Les avantages de ce système de mesure sont expliqués.

1. Einleitung

Für die Messung von Kräften oder Drücken im Bereich der Bautechnik, des Bergbaus, oder der Gebirgsmechanik gibt es derzeit eine Vielzahl von Instrumenten. Ihre Bauweise richtet sich nach der Erfordernis, daß die Kräfte oder Drücke vom Instrument aufgenommen werden müssen, wofür meist eine weitgehend steife und hochfeste Kapsel oder Zelle vorgesehen wird. Die Größe der Kraft oder des Drucks wird durch ein Übertragungssystem in eine physikalische Meßgröße umgewandelt und zu einem Anzeigesystem weitergeleitet, an dem der Betrag des Meßwertes abgelesen werden kann. Die Aufgaben der Übertragung und der Anzeige werden durch verschiedene Lösungsformen bewältigt. Hiezu bedient man sich elastomechanischer, hydraulischer, elektrischer und photoelastischer Wechselwirkungen (1, 2, 3).

Solche Instrumente haben einen hohen Grad der Reife erreicht, so daß sie infolge ihrer Genauigkeit, Zuverlässigkeit, Robustheit, Beständigkeit und Wirtschaftlichkeit in vielen Projekten mitverwendet werden. Die Erfordernisse der Praxis drängen jedoch - wie allgemein in der technischen Entwicklung - nach Verbesserungen. Dazu gehören jene bezüglich der Einfachheit, Energiefreiheit, Einsetzbarkeit unter Schlagwetter-Bedingungen, Verbindbarkeit mit vielerlei Ablese-, Registrier-, Schreib- oder Warnsystemen auch nach erfolgtem Einbau, Unempfindlichkeit gegenüber Temperaturschwankungen, erhöhtem Meßbereich, erhöhter Ablesegenauigkeit und günstigen Kosten.

Ein Schritt in dieser Richtung wurde mit der Neuentwicklung einer patentrechtlich geschützten Zelle gemacht, welche als Übertragungssystem und Anzeigesystem eine Kombination von hydraulischen und elastomechanischen Elementen benützt. Diese Kombination wird als Hydromechanisches Anzeige-System (HAS) bezeichnet. Der

Aufbau der Zelle und die Wirkungsweise des HAS,
sowie einige grundlegende theoretische Betrach-
tungen, die darauf beruhenden Vorteile, und erste
Einsatzerfahrungen werden nachstehend erläutert.

2. Aufbau und Wirkungsweise der HAS-Zellen
In Abb. 1 sind beispielsweise zwei Zellentypen
gezeigt, welche das HAS zur Grundlage haben (4).

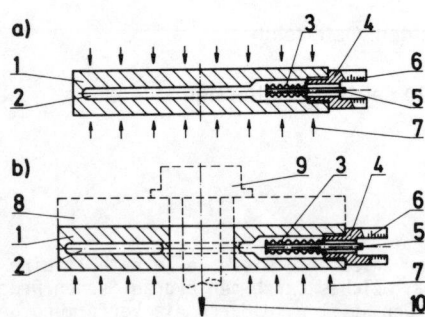

Abb. 1
Schematische Darstellung des Bauprinzips von
HAS-Zellen (Schnitt normal auf die Zellenebene).
a) Druckmesszelle (DHAS) für Einbettung in Gebirge, Beton, Mörtel oder Schüttgut. b) Ankerkraft-Meßzelle (AHAS) für die Messung von
Kräften an Gebirgsankern oder Zuggliedern in der
Bautechnik.
1... Zellenkörper, 2... Schlitz mit Druckflüssigkeit, 3... Federbalg, 4... Verschlußzylinder,
5... Anzeigestift, 6... Ableseskala, 7... Außen-
druck bzw. Last, 8... Druckausgleichplatte,
9... Ankerkopf, 1o... Ankerkraft

Die Zellen bestehen aus dem Zellenkörper, der
aus Blechen oder Platten durch Schweißen hergestellt wird und dem Federbalg (evtl. Kolben)
mit Verschlußzylinder. Zwischen den Blechen oder
Platten ist ein Schlitz zur Aufnahme der Druckflüssigkeit freigehalten. Dieser Schlitz ist an
einer Stelle zu einer Zylinderform ausgeweitet,
in welche der Federbalg eingesetzt wird.

Bei Wirksamwerden eines Außendrucks oder einer
Last auf die Zelle überträgt die Druckflüssigkeit diesen auf den Federbalg, der sich nur entsprechend seiner Federkonstante in der Richtung
seiner Längsachse verformen kann. An der Stirnfläche des Balgs sitzt ein nach ausserhalb der
Zelle führender Anzeigestift. Sein freies Ende

ist somit ausserhalb der Zelle sichtbar gemacht.
Die Stellung des freien Endes des Anzeigestiftes
gegenüber einer im Verschlußzylinder eingekerbten Skala kann visuell abgelesen werden.

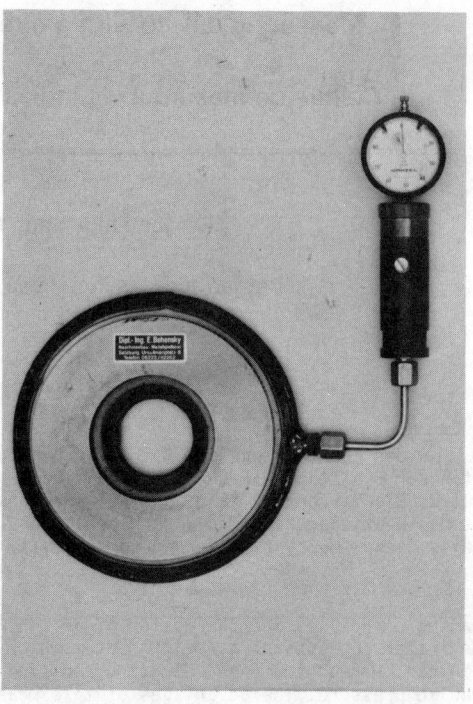

Abb. 2
Ansicht einer Ankerkraft-Meßzelle mit außerhalb des Zellenkörpers angeordnetem Anzeige-
system. Die mechanische Feinmeßuhr ist in den
HAS-Zylinder (rechts) eingesetzt.
Verschiedene Grade der Ablesegenauigkeit sind
wahlweise verfügbar, da die Ablesung entweder
mit freiem Auge gegenüber der eingekerbten
Skala erfolgen kann oder mit einer Schiebelehre,
oder mit einer mechanischen Feinmeßuhr. Auch
können mechanische Schreiber angeschraubt werden
oder induktive Weggeber mit über Kabel gehender
Fernregistrierung angeschlossen werden. Damit
ist im Bereich der Meßwerterfassung jede Möglichkeit gegeben und insbesondere die aufwand-
freie Direktablesung mit freiem Auge durchführ-
bar.

Die geschützte Unterbringung des Anzeigestiftes
im Schlitz des Verschlußzylinders verhindert Be-
schädigungen oder Störungen durch Schlag, Staub,
Wasser usw. Da der äussere Teil des Verschlußzylinders drehbar und der Schlitz damit um seine
Längsachse schwenkbar ist, kann die Ablesung von
verschiedenen Blickrichtungen aus erfolgen.

Die achsiale Verschiebung des Anzeigestiftes

entspricht exakt dem Einstauchweg des Federbalges
(oder eines manchmal eingesetzten Kolbens). Bei
Annahme einer völlig inkompressiblen Druckflüssigkeit entspricht der Einstauchweg des Balges
dem Betrag der Deflexion der Zellenplatten mul-
tipliziert mit dem Verhältnis von Schlitzfläche
zu Stirnfläche des Balges. Bei einem solchen
Verhältnis von z.B. 4oo : 1 kann das HAS somit
eine Vergrößerung der Zellendeflexion um das
4oo-fache anzeigen.

Die Größe der Zellendeflexion würde dabei dem
Widerstand entsprechend verlaufen, welchen die
Zelle als Gesamtsystem der von außen wirkenden
Last entgegenstellt. Bei inkompressibler Druckflüssigkeit enspricht dieser Widerstand einer
kombinierten Wirkung der Balgsteifigkeit und der
Steifigkeit des Zellenkörpers. Wird die Druckflüssigkeit kompressibel so wirkt sie nicht mehr
als vollkommenes Übertragungselement sondern sie
dämpft die flächenproportionale Verschiebung des
Balges. Näheres hiezu wird im folgenden Abschnitt
behandelt.

Die Benutzung eines mechanischen Federbalges
führt zur weitgehenden Vermeidung von Hysterese-
wirkungen, welche bei hydraulischen Systemen mit
Manometeranzeige üblich sind. Dies bedeutet auch
eine Verringerung der Ansprechträgheit. Dadurch
wird ein erhöhtes Maß an Genauigkeit erzielt.

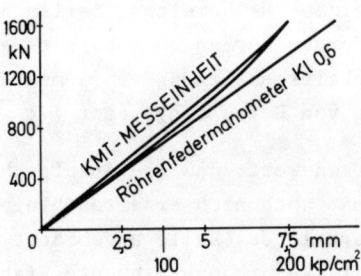

Abb. 3
Diagramm einer Belastungs- und Anzeige-Prüfung
an einer Ankerkraft-Meßzelle. Die obere Kurve
mit der Hystereseschleife stammt aus der Ablesung mittels eines Manometers (Skala der Abszisse in kp/cm^2). Die untere Kurve entspricht
einer Geraden erhalten aus der Ablesung mittels
des HAS und der Feinmeßuhr (Skala der Abszisse
in mm) und weist keine Hysterese auf. Die
Ordinate ist in Einheiten der Druckkraft der
Prüfpresse geteilt.

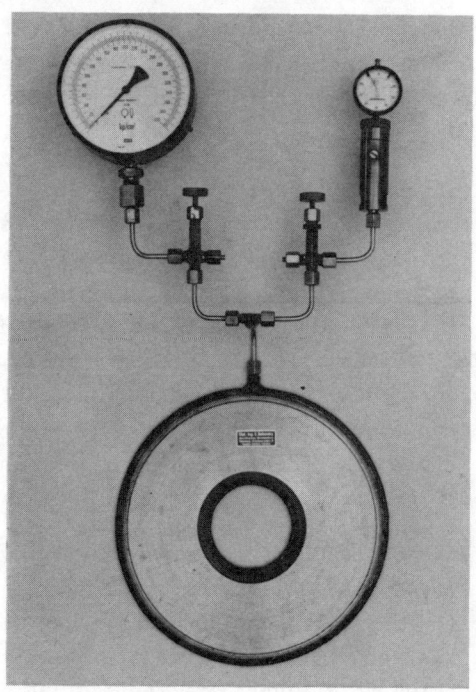

Abb. 4
Ansicht der für die Kurven der Abb. 3 eingesetzten Ankerkraft-Meßzelle, an welche ein
außenliegendes HAS und ein außenliegendes Manometer angeschlossen sind.

Das HAS hat des weiteren den Vorteil, daß mit
der Anzeige in Form einer Längenangabe und mit
dem flächenproportionalen Übertragungsprinzip
praktisch auch eine Ablesung der Zellenverfor-
mung erfolgt, so daß zweierlei physikalische
Meßgrößen aus der Ablesung hervorgehen. Dies hat
zur Folge, daß auch eine Verformungsmessung
möglich wird.

Eine weitere bemerkenswerte Verbesserung, welche
das HAS mit sich bringt, ist auch die Möglichkeit einer temperaturunabhängigen Anzeige. Diese
ergibt sich durch die Bauart eines temperatur-
kompensierenden Balges.

Abb. 2 zeigt eine Zellenbauform von geringer
Dicke, bei welcher das HAS außenliegend angeordnet ist.

3. Fragen der Zellensteifigkeit.
Die Steifigkeit der Zellen ist vor allem bei
eingebetteten Zellen (Beton, Gebirge), welche
den Druck messen sollen, von Bedeutung. Für
Zellen, die äußere Lasten (konzentrierte Kräfte
wie Stempellasten, Ankerkräfte) messen sollen,
ist die Steifigkeit bzw. die Nachgiebigkeit oft

von untergeordneter Bedeutung, wenn die Verformung gegenüber jener des kraftausübenden Elements nur vernachlässigbar kleine Bruchteile ausmacht.

Bei den eingebetteten Zellen wäre der Idealfall dann gegeben, wenn die Steifigkeit der Zelle gleich jener des umgebenden Mediums wäre. In diesem Fall gäbe es keine Spannungskonzentrationen und die Zelle würde direkt den Zustand des umgebenden Mediums erleiden, diesen also am besten wiedergeben. Dies erscheint für den Einbau in Beton oder Spritzbeton durchaus sinngemäß, weil der Druck des Mediums sich auch erst ab dem Einbau entwickelt.

Für den Einbau ins Gebirge oder in bereits vorbelastete Bauteile (Betonstrukturen) hat die Gleichheit der Steifigkeit nur dann einen Sinn, wenn man imstande ist, beim Einbau den Kraft- und Formschluß völlig herzustellen, so daß die Zelle sofort und ohne Verzerrung anspricht. Kann man dies nicht, so bestehen Probleme für die Interpretierbarkeit der Meßwerte. Einen Ausweg aus der Situation der Probleme mit ungleichen Steifigkeiten und mit ungenügendem Kontakt zwischen Zelle und Medium hat POTTS (5) gesucht. Die von ihm vorgeschlagene Maßnahme war, eine Zellensteifigkeit zu wählen, welche mindestens 3 mal so groß ist wie diejenige des Mediums, weil dann die Spannungskonzentration in der Zelle nicht höher als auf den Wert 1,5 ansteigt (bei den meisten praktischen Gebirgsarten) und somit der Meßwert um den bekannten Faktor von 1,5 auf den Betrag des Mediums reduziert werden kann. Dennoch ist zu diesem Zweck eine Vorspannung der Zelle beim Einbau erforderlich, welche zusammen mit möglichen Unvollständigkeiten des Kontakts die Reaktion der Zelle einigermaßen verzerrt.

Für die Abstimmung der Steifigkeit der Zellen auf jene des Mediums bestehen jedoch vom Entwurf her einige Möglichkeiten. Wie auch PRAGER (6) zeigt, kann die Steifigkeit einer Zelle einerseits durch die Steifigkeit des Zellenkörpers und andererseits durch die Steifigkeit der hydraulischen Flüssigkeit angepasst werden, wobei auch der Anteil an Flüssigkeitsvolumen gegenüber dem Stahlvolumen (Dickenverhältnisse zwischen Schlitz und Blech) noch eine Rolle spielt. Im HAS kommt noch eine weitere Möglichkeit hinzu, nämlich die

Steuerbarkeit durch die Steifigkeit des Federbalges.

Von allen Möglichkeiten, welche somit zur Anpassung zur Verfügung stehen, soll hier nur jene mittels des Federbalgs näher erläutert werden. Wird nämlich die Schlitzweite und damit das Volumen der Druckflüssigkeit verhältnismäßig klein gehalten gegenüber der lastbedingten Deflexion der Zellen-Membranen, so wird auch der Einfluß der Kompressibilität der Flüssigkeit herabgesetzt, so daß die Volumskompensation durch die Stauchung des Balges in den Vordergrund tritt.

4. Ergebnisse von Laboruntersuchungen.

Prüfungen des HAS im Labor haben die hohe Empfindlichkeit, den linearen Anzeigeverlauf und das beinahe völlige Fehlen einer Hysterese nachgewiesen. Eine Kennlinie einer HAS-Ankerkraft-Maßzelle ist in Abb. 3 dargestellt. Sie zeigt den Zusammenhang zwischen Druckkraft der Presse und dem Einstauchweg des Federbalges. Der ansteigende und der absteigende Ast des Lastspiels fallen so zusammen, daß eine Hysterese nicht erkennbar ist. Demgegenüber zeigt die im selben Diagramm dargestellte Kurve der Ablesewerte eines Präzisionsmanometers einen gekrümmten Verlauf und eine Hysterese.

Die Anzeigecharakteristik des HAS erlaubt daher Messungen größerer Genauigkeit trotz robuster Bauart und großen Meßbereichs. Zellen dieser Ausführungsformen wurden auch bereits erfolgreich zur Inhaltsbestimmung von Zement-Silos und zur Mischung von Beton eingesetzt (7).

Es ist bemerkenswert, daß die Anzeigecharakteristik des HAS auch noch erhalten bleibt, wenn das HAS außerhalb der Zelle angeordnet wird, wobei bisher als Verbindungselemente Stahlrohre eingesetzt worden sind. Die Längen der Verbindungsleitungen haben bis 15 m betragen.

Abb. 4 zeigt die Anordnung von HAS und Manometer als außenliegende Einheiten mit der für das Prüfdiagramm der Abb. 3 benutzten Zelle. Die Anzeigeeinheiten wurden jeweils einzeln zur Aufnahme der Kennlinie hinzugeschaltet.

Literaturstellen

1 HABENICHT, H.: Anker und Ankerungen zur
        Stabilisierung des Gebirges. Springer-
        Verlag, Wien New York (1976) 196 S.

2 HABENICHT, H.: Meßtechnische Vorkehrungen
        für Fels- und Bodenanker. Taschenbuch
        Tunnelbau 1982, S. 323-368, Deutsche
        Gesellschaft für Erd- und Grundbau e.V,
        Verlag Glückauf, Essen (1981)

3 HABENICHT, H.: Gebirgsdruckmessung mit
        passiven hydraulischen Flachzellen.
        Berg- und Hüttenmännische Monatshefte
        122 (1977) H. 12, S. 563-568, Springer-
        Verlag, Wien New York

4 BEHENSKY, E.: Patentschrift

5 POTTS, E.L.J.: Underground Instrumentation.
        Quarterly of the Colorado School of
        Mines, Vol. 52, Nr. 3, July 1957,
        p. 137-182

6 PRAGER, R.: La mesure des contraintes dans
        les sols et la cellule Glötzl de
        pression totale. These No. A.O. 923o,
        L'Universite Scientifique et Medicale
        de Grenoble, Grenoble 1974, 157 pp.

7 BEHENSKY, E.: Betriebsunterlagen

E

ROCK MASS CLASSIFICATION FOR BLOCK CAVING MINE DRIFT SUPPORT

La classification des massifs rocheux pour le soutènement des mines à foudroyage en masse des galeries

Die Klassifizierung des Gebirges für den Grubenbau im Blockbruchbau und Abbaubruchbau

F. S. Kendorski
R. A. Cummings
Engineers International, Inc., Westmont, Il., U.S.A.

Z. T. Bieniawski
The Pennsylvania State University, State College, Pa., U.S.A.

E. H. Skinner
U.S. Bureau of Mines, Spokane, Wa., U.S.A.

SYNOPSIS

A rock mass classification system has been developed that enables the planner or operator of block or panel caving mines to arrive at support recommendations for production drifts. This system has been named the MBR (standing for Modified Basic RMR) System, since it follows closely the Geomachanics System of Bieniawski and incorporates many of the ideas of Laubscher, who considered mine drift support in particular. The MBR itself is the first of three sequential ratings generated by the system, and is a pure description of geological conditions. The MBR is adjusted to obtain a rating useful for estimating temporary, or development, support. In the third step, the adjusted MBR (AMBR) is again adjusted for final, production support (FMBR) ratings. The MBR system is based on an in-depth field study of experience at several caving mines in the western United States.

RESUME

On a développé un système pour classifier les massifs rocheux qui permet au planificateur ou à l'opérateur de mines à foudroyage, en masse ou en panneaux, de deviner le soutènement nécessaire aux galeries de production. Le système fut appelé le système MBR (Modified Basic RMR) parce qu'il suit rigoureusement le système Geomechanics (RMR) de Bieniawski et incorpore plusieurs des idées de Laubscher qui considéra le soutènement des galeries en mines de foudroyage en particulier. MBR devient le premier de trois classements successifs produits par le système et constitue tout simplement une description des conditions géologiques. On ajuste le MBR afin d'obtenir, dans un deuxième temps, un classement utile pour estimer le soutènement temporaire des galeries de développement, par exemple. Troisièmement, le MBR ajusté (AMBR) est encore modifié afin de déterminer le classement ultime pour la production (FMBR). Le système MBR est basé sur une étude détaillée des expériences à quelques mines à foudroyage des Etats-Unis occidentaux.

ZUSAMMENFASSUNG

Ein Gebirgsklassifikationssystem wurde entwickelt, das dem Entwerfer oder Bergwerksbetriebsführer in Blockbruchbau- oder Abbaubruchbaubergwerken die Gelegenheit bietet, den Grubenbau der Produktionsstrecken zu bestimmen. Dieses System wurde MBR (Modified Basic RMR) genannt, weil es Bieniawskis Geomechanics System (RMR) genau folgt und viele Ideen von Laubscher beinhaltet, der sich insbesondere mit Streckenausbau in Bruchbaubergwerken befaßt. MBR ist die erste von drei aufeinanderfolgenden, von dem System erzeugten Berechnungen und ist eine reine Beschreibung der geologischen Bedingungen. MBR wird abgeändert, um ein Berechnungssystem zu erhalten, durch welches zeitweilige oder Entwicklungsunterlagen abgeschätzt werden können. Während der dritten Stufe wird das abgeänderte MBR (AMBR) nochmals zum Zwecke der endgültigen Produktionsunterlagen (FMBR) Bewertung abgeändert. Das MBR System beruht auf einer eingehenden Feldforschung der bei verschiedenen Bruchbaubergwerken im Westen der Vereinigten Staaten gewonnenen Erfahrungen.

INTRODUCTION

A rock mass classification system has been developed (Engineers International, Inc., 1982) for use in estimating support requirements in mines using caving methods. The system is based in part on the RMR system of Bieniawski (1973, 1976, 1979a, 1979b) and uses many of the same input parameters with additions necessary due to the mining nature of the application. The system has been termed the "Modified Basic RMR" or MBR.

Purpose and Scope

The MBR system is intended to aid planners, developers and operators of caving mining systems by providing a rational recommendation for drift support. The method is useful for advance planning during exploration, predevelopment, and expansion, for estimating support costs during feasibility studies, and for re-examining existing support philosphies in light of new geologic or engineering information. However, it should be very clear that the support recommendations made by this system are semiquantitative and should not be used as the sole basis for evaluating the effectiveness of a support system that is in use. Also, the MBR system is not a substitute for detailed design.

The MBR system was adapted from existing systems in use for civil tunneling. These adaptations were necessary because the development of a caving mine is radically different from driving a tunnel. In developing the modifications and adaptations for the MBR system, use was made of the current support philosophy and observed support effectiveness at a number of United States mines using block, panel, or mass caving. These mines encompass a spectrum of geological conditions and engineering factors. Data were collected for horizontal drifts in these mines. Thus, the MBR system is not necessarily valid for non-horizontal workings (inclines, raises, shafts) or for other mining methods. It is hoped that the MBR system will eventually be extended in these areas, and also to considerations of cavability, drawpoint spacing, and undercut requirement. It is hoped that with continued use, the MBR system can be improved in generality and predictive capability.

Using the Results

In using this system, it is necessary to distinguish between cave initiation deformations, and deformations due to production by caving. The former are unloading and abutment phenomena, which depend directly on rock mass conditions and geometry of the mine layout. The latter are very strongly dependent on operational factors which are (primarily) the continuity of production, and (secondarily) the spatial distribution of draw rates. Although production effects may be overriding factors in determining support behavior, there is no good way to allow for these in any classification system. It has therefore been assumed that draw control practices will be adequate to keep peak rock mass strains during production at levels below those occurring during abutment loading. This translates into "good draw control" and "effective undercutting." The user of the MBR system should always be aware of problem areas. Stress concentrations at corners and re-entrants in the mine plan may be significant, as well as drawpoints and drop-points (loading cutouts), intersections, and other workings of increased span.

The supports recommended are not as conservative as would be specified for civil projects, in recognition of the temporary life of most mine openings. The system has been made conservative enough to withstand small reductions in rating within a given rating class, because in driving a mine drift, it is seldom feasible to "fine-tune" supports for each small rock competence change.

OVERVIEW OF THE METHOD

Derivation

The MBR System was developed under a contract with the U. S. Bureau of Mines, to examine how a ground classification approach could be fruitfully used in planning caving mine drift supports. The Geomechanics System of Bieniawski (Bieniawski, 1979a) and modified by Laubscher (1976) was found to have the greatest potential.

In the Geomechanics System, numerical ratings are given to various geological aspects: geometry and condition of rock fracturing, intact rock strength, and water condition. The sum of these ratings is the RMR, which varies between 0 and 100, and which can be related to the rock mass cohesion, the deformation modulus, stand-up time, and support recommendations. The method has been widely accepted for tunnels, as a means to describe rock mass competence, and to estimate support requirements.

Laubscher (1976) developed a technique for adjusting the RMR value to more accurately reflect mining circumstances. The adjusted classes were to define cave angle and cavability as well as drift support. For additional information, the reader is referred to Laubscher (1977, 1981) and Laubscher and Taylor (1976).

In developing the MBR System, basic required data were collected for numerous block caving mines in the U. S. Original field data were generated wherever possible. Key factors affecting drift performance were identified as adjustment factors, and a rating scale was attached to each, based on field observations, theoretical analyses in the absence of field data, or both.

In deriving the numerical adjustments, much use was made of the conceptualization of rock mass strains that develop with removal of broken rock from a growing underground volume. The tendency of the rock mass to strain is resisted by the rock mass which surrounds the tunnel supports, which in turn develop stresses in the process. By constructing typical strain history curves for several mines having different MBR values, and applying strain-related failure criteria from structural texts and field observations, a relationship was developed between FMBR (final rating) and support requirement. The ratings and adjustments were assigned initial values, based on expectations of the rock mass strain distributions and related stress for various mining geometries and geologic settings. These were cross-checked against the field data, and changes were made, where necessary, to make the system agree with field observations.

The ratings were then computed for the studied mines, and the support recommendations were adjusted to reflect the prevailing support practice. Appropriate consideration was given to apparently overdesigned or underdesigned support.

Rock masses in the data base range in MBR from about 20 to almost 70. Development in rock of MBR less than 20 is ordinarily avoided and is never very extensive, and plus-70 rock would be very difficult to cave, so the data base is fairly representative of likely mining conditions. Depths ranged from about 213 m (700 ft) to well over 610 m (2,000 ft), although mining experience at depths greater than 762 m (2,500 ft) is limited.

Description

The MBR (Modified Basic RMR) (Fig. 1) system makes use of the RMR approach and uses some of the concepts of Laubscher. Key differences lie in the arrangement of the initial rating terms and in the adjustment sequence. In the MBR system, the inputs are selected

and arranged so that a rational rating is still possible using very preliminary drill-hole geotechnical information. The MBR is also a multi-stage adjustment; the output at each stage can be related to support for various mining conditions.

The MBR rating is the result of the initial stage, and is the simple sum of the raw ratings.

The MBR is an indicator of rock mass competence, without regard to the type of opening constructed in it. This MBR value is used in the same fashion as the RMR and other systems for determining support requirements, by consulting support charts or tables. The MBR recommendation is for isolated single tunnels that are not in areas geologically different (in a structural sense) from production areas.

The second stage is the assignment of numerical adjustments to the MBR that adapt it to the ore block development process. With regard to support, the principal differences between production drifts and civil tunnels (in development only) are the excavation techniques and the need for multiple, parallel openings. Unfavorable fracture orientations may also strongly influence stability. Input parameters relate to excavation (blasting) practice, geometry (closeness, size, and orientation of openings), depth, and fracturing orientation. The adjustment values are obtained from tables and charts, and the MBR is multiplied by the decimal adjustments to

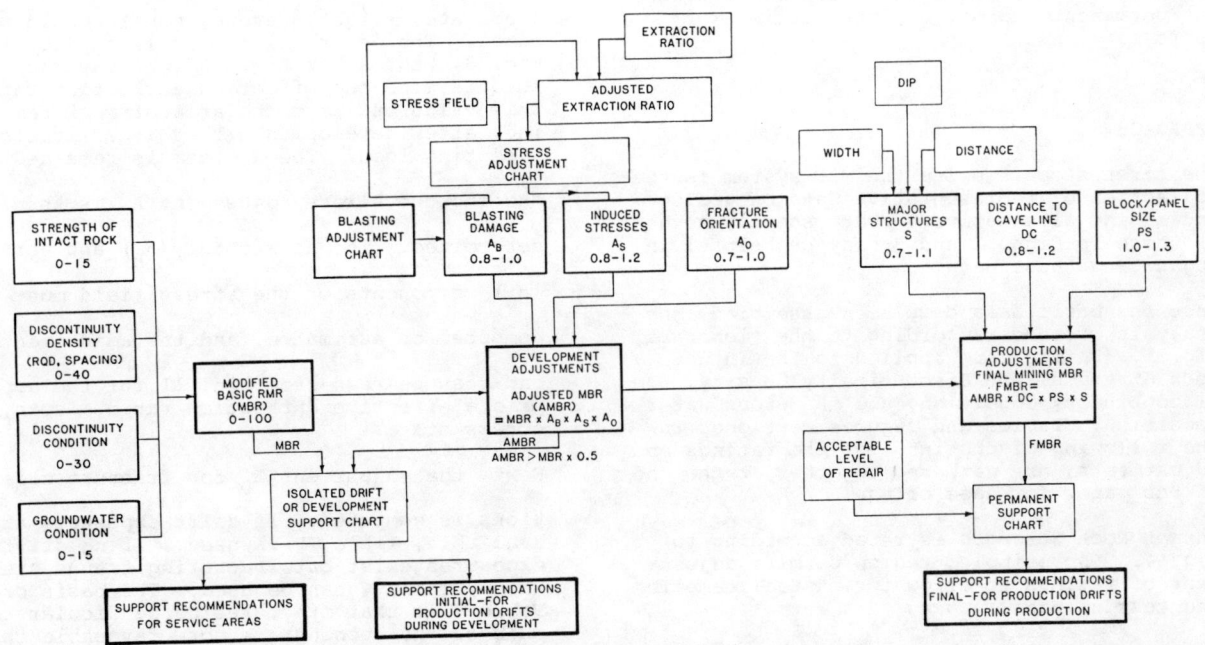

Fig. 1 Organization of the MBR System. The general flow begins at left. Intermediate data inputs originate at the top, and outputs are at the bottom. The multi-stage approach is evident in the groupings of the adjustments.

B 53

obtain the AMBR (Adjusted MBR). Drift support charts are again consulted to give a range of supports for drift development (initial support). The user may select support according to the performance period desired, since lighter support will be adequate in some rock for shorter periods. The objective is to initially stabilize the opening during development so that permanent support may use its full capacity to resist the abutment loading increment.

The third and last classification stage deals with the additional deformations due to abutment loadings. As stated before, caving deformations will also be accounted for if proper undercutting and draw control practices are followed. The most significant identified factors influencing abutment load, as experienced by a drift support element, are the location and orientation of the drift with respect to the caved volume, the size of the caved volume, the ability of the rock mass to withstand stress (rock mass competence--MBR), the tendency of the lining to attract stress, and the role of any major structural trends that may serve to localize or transfer the abutment deformations. Input variables relate to block or panel size, undercutting sequence, level layout, MBR, and general structural geology of the area. The adjustment values are obtained from tables and graphs and are used as multipliers to the AMBR, and yield the FMBR (Final MBR). This value, together with an assessment of repair acceptability (depending on the type of opening) is correlated with recommendations for permanent support at intersections or drift sections.

APPROACH

The first step in using the MBR system is the collection of representative data on geology and mining alternatives. Data sheets, such as those in Figs. 2 and 3, may be helpful in organizing these data.

Once the basic data have been assembled, the analysis proceeds according to the flowchart, Fig. 1. Ratings are applied to the Intact Rock Strength, the Discontinuity Density, the Discontinuity Condition, and the Groundwater Condition. Tables and figures mentioned in the following discussion, for the ratings and adjustments, are gathered together at the end of the paper for ease of use.

Intact Rock Strength is rated according to Fig. 4. The stippled region permits adjustment of ratings to allow for natural sampling and testing bias.

The Discontinuity Density is related to blockiness and is the sum of ratings for RQD and joint spacing (Fig. 5). If either type of data is lacking, it can be estimated through the use of Fig. 6.

Table I is used for rating the Discontinuity

Condition. The most representative conditions are assessed for this step. The degree or type of alteration can be a useful index for this as well.

The Groundwater Condition rating is simple and made by the use of Table II.

To obtain the MBR, the four ratings mentioned above are summed. The ranges of the input parameters are given in Fig. 1. At this point, the Development Support Chart, Fig. 7, gives support for service areas away from production areas, but in rock still considered within the chosen interval.

Having thus obtained the MBR and the applicable support recommendations, the AMBR is computed for development adjustments.

First, the extraction ratio is computed for the mining layouts under study. For single drifts with multiple intersections or that are otherwise affected by other openings, the extraction ratio may depend on the extent of the area considered. Only in such instances is the convention adopted that all openings within 1.5 drift diameters of each rib are considered in computing the extraction ratio. The ratio is computed at springline and therefore includes the horizontal planimetric area of the finger or transfer raises.

Next, blasting damage is assessed according to the criteria of Table III. Both the blasting damage adjustment A_B and the descriptive term (moderate, slight, severe, none) should be noted. The descriptive term is applied to Fig. 8, Fig. 9, or Fig. 10 (whichever applies) to determine the effective extraction ratio. This value reflects the area of rock remaining, after development, that is effective in accepting load. The A_B term is retained as is.

The induced stress adjustment A_S is then determined. The horizontal (σ_h) and vertical (σ_v) components of the stress field must be computed or estimated, and the adjustment A_S can then be read from Fig. 11 for the appropriate effective extraction ratio, depth, and stress state.

Next, the adjustment A_O for fracture orientations is computed. If drift exposures are available, Table IV is used. If no drift exposures exist but fracturing trends are known, Table V can be used. The basis of Table V is that fractures perpendicular to the axis of the opening are more favorable than fractures parallel to it, that both development and support are facilitated by fractures that dip away from the heading rather than towards it, and that steep dips are preferable to shallow dips.

If fracturing trends are not known but core is available for examination, fully interlocking

Project Name _____ Site of Survey _____ By _____ Date _____
_____

1. Geologic Region: _____ Rock Type _____ Location _____

2. Compressive Strength: Average _____ Range _____ Method _____ Comment _____

3. Core Recovery: Interval _____ Average _____ Range _____

4. RQD: Interval _____ Average _____ Range _____

5. Discontinuity Spacing: Average _____ Range _____ Comment _____

6. Discontinuity Condition Wall Roughness   Wall Separation   Joint Filling   Wall Weathering
        Most Common   _____   _____   _____   _____
        Intermediate   _____   _____   _____   _____
        Least Common   _____   _____   _____   _____
        Consensus   _____   _____   _____   _____

7. Water Condition   Dry   Damp   Wet   Dripping   Flowing

8. Fracture Orientations   Set 1   Set 2   Set 3   Set 4   Set 5
        Strike   _____   _____   _____   _____   _____
        Dip/Dir   _____   _____   _____   _____   _____
        Rank   _____   _____   _____   _____   _____
_____

9. Major Structures   Strike   Dip   Dip Dir.   Width   Location/Comment
   Name:_____   _____ ____ _____   _____   _____
   Name:_____   _____ ____ _____   _____   _____
   Name:_____   _____ ____ _____   _____   _____

10. Stress Field σ_1: Direction _____ Magnitude _____ Measured? _____
                  σ_3: Direction _____ Magnitude _____ Measured? _____

11. Source of Geological Data _____
    _____
    _____

Fig. 2 MBR Input Data Sheet - Geological Data

core can be examined for the number of groups of joints of similar inclinations in the core. This requires considerable geologic judgement and skill, but a rating can be derived from Table V if the relative numbers of steep sets, shallow sets, and so on, can be estimated.

The adjustment A_O can be difficult to make if the data are not well-suited, but it has a significant bearing on drift stability, and the best estimate should be made.

The three adjustments, A_S, A_B, and A_O, are multiplied, yielding for most situations a decimal value between 0.45 and 1.0. The MBR is multiplied by this value or 0.50, whichever is greater, to yield the AMBR.

The Development Support Chart, Fig. 7, is then again consulted for support recommendations. It should be decided what degree of support reliability is desired for development. It is recommended that the development support be selected so as to stabilize the opening for as long as it will take to bring the block into production. For long lead times (1 year or more) a more conservative support should be used.

Finally, the FMBR is computed. In this third stage, the role of abutment loadings is accounted for. This is addressed through considerations of structural geology and mining geometry ("production adjustments").

Faulting and sheeted or shattered zones disrupt the mining-induced stress pattern and are dealt with through the adjustment for major structures, S (Table VI). Although any zone of significantly less competence is theoretically eligible for adjustment, it is suggested that only the larger, nearby features are worthy of consideration. The limiting width-distance relationship will become clear for each mining property. Where information is too sparse or preliminary to be geographically specific, it may be possible to characterize blocks of ground according to an expected or typical distribution of weakness zones.

The adjustment for the proximity to the cave line, DC, is computed from Fig. 12. This rating refers to the point of closest approach of the caved area. In some cases, this means the vertical distance, and in others, horizontal. The term reflects the dissipation of abutment load away from the point of application.

Project Name _____ Site of Survey _____ By _____ Date _____
_____

1. Type of Drift(s) _____ 2. Orientation(s) _____ 3. Design Life _____

4. Design Dimensions Width _____ Width variation _____
                       Height _____ Height variation _____

5. Drift Spacing (Horizontal) _____

    Other Openings   Type _____ Size _____ Spacing _____

6. Extraction Ratio
        Multiple Openings: Excavated Area _____ Unexcavated _____ e$_r$ _____

        Single Opening: 1.5 (width) _____ Excavated _____ Unexcavated _____ e$_r$ _____

7. Distance below undercut - drift floor to undercut floor _____
                              drift crown to undercut floor _____

8. Method of Excavation: Machine bored   Controlled D & B   Conventional D & B

9. Excavation conditions:
    Perimeter Hole Traces _____
    Rib or Crown Looseness _____
    New or Existing Cracks _____
    Overbreak & Barring-Down _____
    Other Criteria _____

10. Intersections, turnouts: Type _____ Location _____ Max. Span _____

11. Block Dimensions: Side _____ Orientation _____ End _____ Orientation _____

12. Cave Line   Direction _____ Direction of Progress _____

13. Drift Location (in block, with respect to major structures and their dips, with respect to cave) _____
    _____
    _____

Fig. 3 MBR Input Data Sheet - Engineering

The block or panel size adjustment, PS, (Fig. 13) reflects the relationship between magnitude of abutment stress and size of caved volume. Smaller panel or block sizes are associated with lower abutment load levels because the caved volume is smaller. The effect is negligible for blocks larger than 61 m (200 ft) or so, so these larger blocks as well as level-wide (mass) caving systems receive an adjustment of 1.0. PS may also be applied to blocks that are partially undercut.

These three adjustment values, S, DC, and PS, are multiplied together and then multiplied by the AMBR rating to yield the Final MBR (FMBR), which is used to obtain permanent drift support recommendations. The range of values for the product of these adjustment ratings is 0.56 to 1.7; there are no other restrictions on this range. In practice, the high end of this range will seldom be reached, because small caving blocks are uncommon in present practice.

The recommended support is then arrived at for drift sections or intersections, through Fig. 14. For spans of more than 6 m (20 ft) the rating scale for intersections is used.

The degree of acceptable repair refers to the occurrence of cracking, spalling, slabbing, or other unacceptable deformation of the lining, that requires a production interruption while repairs are made. Repairs required because of damage resulting from excessive secondary blasting, wear, or poor undercutting or draw control were not addressed in developing the support chart. A higher incidence of repair is tolerated in slusher or grizzly drifts than in fringe drifts or haulageways.

In selecting a support type based on FMBR, the user should have in mind the level of conservatism that was applied in picking the development support. A high degree of support reliability in development will permit up to one repair category lighter support in production than might otherwise have been selected. For lower FMBR values, the support charts indicate a range of supports. This reflects the variability in conservatism among mine operators. Generally, the support used in such cases is the lightest in the range, although this will depend on the reliability desired.

CLOSURE

The MBR system is proposed as an aid to planners and operators of block caving mines. It is a ground classification approach and as such, will benefit from further use in a wider range of mining and geological conditions. The system draws from the prior work of Laubscher, Bieniawski, and others.

The MBR system allows the user to consider various mining layouts and procedures to obtain the lowest-cost workable support system. Ratings are improved by the following, which are within the control of mine operators:

- controlled blasting or improved blasting practice,

- widest possible separation of openings, both laterally and vertically, given the constraints of drawpoint spacing and ore column height,

- driving openings perpendicular to major structural trends,

- keeping critical openings (haulage and fringe drifts) well away from the caved volume,

- minimal undercut area; and, clearly,

- avoidance of weakness zones and major discontinuities for all openings, where possible.

ACKNOWLEDGEMENTS

The MBR System was developed in support of Contract No. JO100103 with the U. S. Bureau of Mines. The authors appreciate the cooperation of the Bureau in connection with the preparation of this paper. Thanks are also due to the Anaconda Copper Company, Magma Copper Company, Climax Molybdenum Company and Noranda Lakeshore, Inc., for the particular cooperation and assistance offered by these corporations and mine personnel. Finally, the cooperation and financial support of Engineers International, Inc. is gratefully acknowledged in the preparation of this paper.

REFERENCES

Bieniawski, Z. T. (1973). "Engineering Classification of Jointed Rock Masses," Transactions, South African Institution of Civil Engineers, (15), 12, pp. 335-344.

Bieniawski, Z. T. (1976). "Rock Mass Classifications in Rock Engineering," Proceedings of the Symposium for Rock Engineering, Johannesburg, A. A. Balkema, (1), pp. 97-106.

Bieniawski, Z. T. (1979a). "The Geomechanics Classification in Rock Engineering Applications," Proceedings, 4th International Congress on Rock Mechanics, ISRM, Montreux, Switzerland.

Bieniawski, Z. T. (1979b). "Engineering Rock Mass Classifications--New Aid for Engineers and Geologists," Bulletin of Earth and Mineral Sciences, The Pennsylvania State University, (49), 1, September, pp. 1-5.

Engineers International, Inc. (1982). "Caving Mine Rock Mass Classification and Support Estimation," (Final Report on U. S. BuMines Contract No. JO100103), Westmont, Illinois, September.

Laubscher, D. H. and H. W. Taylor (1976). "The Importance of Geomechanics Classification of Jointed Rock Masses in Mining Operations," Proceedings of the Symposium for Rock Engineering, Johannesburg, November, pp. 119-128.

Laubscher, D. H. (1977). "Geomechanics Classification of Jointed Rock Masses - Mining Applications," Trans. Institute of Mining and Metallurgy, Section A, (86), pp. A1-A7.

Laubscher, D. H. (1981). "Selection of Mass Underground Mining Methods," pp. 23-38 in Stewart, D. R., ed., Design and Operation of Caving and Sublevel Stoping Mines, Society of Mining Engineers of AIME, New York, 843 p.

MBR SYSTEM-RATINGS AND ADJUSTMENTS TABLES AND FIGURES

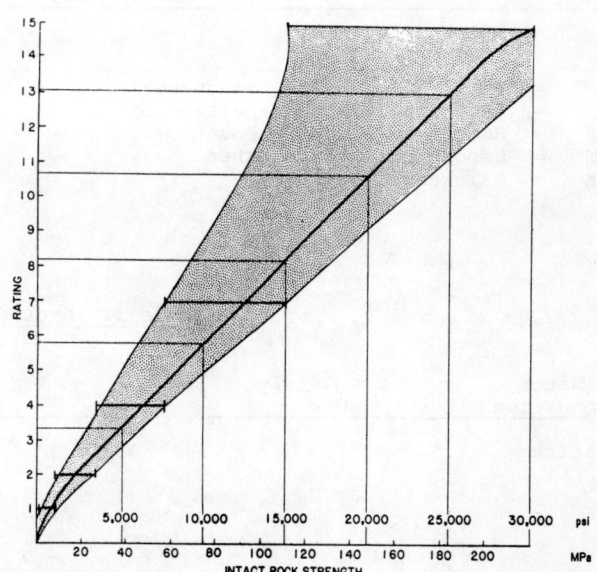

Fig. 4 Rating for Intact Rock Strength. Intended source of data is point load testing.

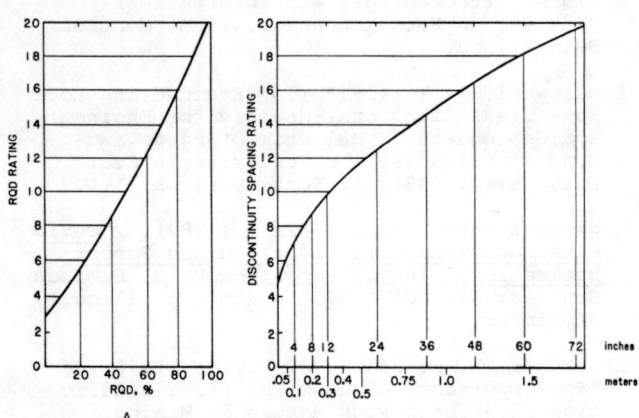

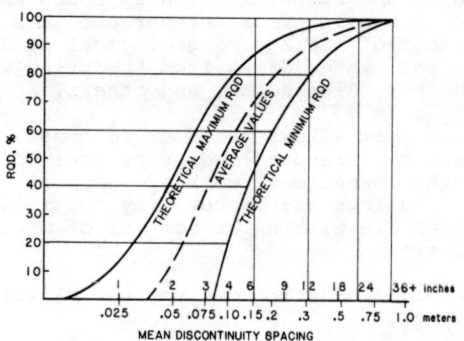

Fig. 5 Ratings for Discontinuity Density.
A - RQD rating.
B - Discontinuity spacing ratings

Fig. 6 Theoretical Relationship Between RQD and Discontinuity Spacing (after Bieniawski, 1979b).

Table I. Discontinuity Condition Ratings

		Description of Discontinuity			
Wall Roughness	VR	R-SR	SR	SM-SK	SM
Wall Separation	None	Hairline	Hairline to 2mm	2-6mm	6mm
Joint Filling	None	None	Minor Clay	Stiff Clay Gouge	Soft Clay Gouge
Wall Weathering	F	SL	SO	SO	VS
Rating	30	25	20	10	0

VR = Very rough (coarse sandpaper)
R = Rough (medium or fine sandpaper)
SR = Smooth to slightly rough
SM = Smooth but not polished
SK = Slickensided, shiny

F = Hard, unweathered, fresh
SL = Hard, slightly weathered
SO = Softened, strongly weathered
VS = Very soft or decomposed

Table II Groundwater Condition Rating

Water Condition	Completely Dry	Damp	Wet	Dripping	Flowing
Rating	15	10	7	4	0

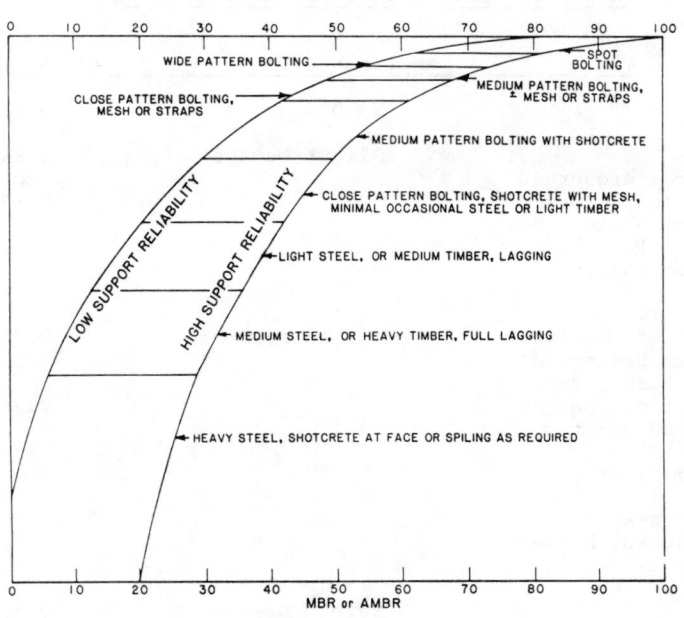

Fig. 7 Development Support Chart. Explanation of the support types shown are given in the Explanation, following.

Explanation for Fig. 7.

a. Spot Bolting: Bolting to restrain limited areas or individual blocks of loose rock, primarily for safety.

b. Wide pattern bolting: Bolt spacing on 1.5 m (5 ft) to 1.8 m (6 ft) centers, or wider in very large openings.

c. Medium pattern bolting, with or without mesh or straps: Bolts on 0.9 m (3 ft) to 1.5 m (5 ft) centers, 0.23 m (9 in.)-wide steel straps or 0.1 m (4 in.) welded wire mesh.

d. Close pattern bolting, mesh, or straps: Bolt spacing less than 0.9 m (3 ft), 0.1 m (4 in.) welded wire mesh or 0.30 m (12 in.) straps, or chain link if fracture spacings are low enough that ravelling may be a problem.

e. Medium pattern bolting with shotcrete: Bolts on 0.9 m (3 ft) to 1.5 m (5 ft) centers, and 0.8 m (4 in.) (nominal) of shotcrete. Light mesh for wet rock to alleviate shotcrete adherence problems.

f. Close pattern bolting, shotcrete with mesh, minimal occasional steel or light timber: Bolt spacing less than 0.9 m (3 ft), with 0.1 m (4 in.) welded wire mesh or chain link throughout, and nominal 0.1 m (4 in.) of shotcrete. Localized conditions may require light [0.3 m by 9 kg (6 in. by 20 lb)] wide flange steel sets or 0.2 m (8 in.) timber sets in some areas, lagged at the crown, as necessary.

g. Light steel, medium timber, lagging: Bolting as required for safety at the face--full contact (grouted or Split Set) bolts only. Light [0.2 m by 16 kg (8 in. by 35 lb)] wide flange sets or 0.25 m (10 in.) timber sets on 1.5 m (5 ft) centers, full crown lagging with rib lagging in squeezing areas.

h. Medium steel, heavy timber, full lagging: Medium [0.25 m by 20 kg (10 in. by 45 lb)] wide-flange steel sets or 0.3 m (12 in.) timber sets on 1.5 m (5 ft) centers, fully lagged across the crown and ribs. Support should be installed as close to the face as possible.

i. Heavy steel, shotcrete at face or spiling as required: Heavy [0.3 m by 33 kg (12 in. by 72 lb)] wide-flange steel sets on 1.2 m (4 ft) centers, fully lagged on crown and ribs, carried directly to face. Spiling, or shotcreting of face, as necessary.

j. General:
    Bolting: bolts in a. through f. are considered to be .019 m (3/4 in.), fully grouted or resin-anchored standard rockbolts; mechanical anchors are acceptable in material of MBR >60. Some success is reported for Split Sets at present but their ability to create a rock arch in supporting the drift has not been demonstrated with engineering or numerical backup. Split Set use is at the discretion of the operator.

Table III   Blasting Damage Adjustment A_B

Conditions/Method	Applicable Term	Adjustment A_B
1. Machine Boring	No Damage	1.0

2. Controlled Blasting
    a. Practically all hole traces preserved
    b. No loosened blocks or opened joints
    c. Overbreak: mostly less than .15m (6 in.) always less than (1 ft)
    d. No or very minor new inter-joint cracking
	Slight Damage	0.94
		to
		0.97

3. Good conventional blasting
    a. Some perimeter hole traces preserved.
    b. Some loosened blocks and slabs, some barring down necessary but not enough to impede production. Some joints opened.
    c. Overbreak: commonly .30m (1 ft), locally higher.
    d. Noticeable tendency for cracks to develop in intact rock blocks, between joints, even in harder areas.
	Moderate Damage	0.90
		to
		0.94

4. Poor conventional blasting
    a. At best, only a few perimeter hole traces preserved; most are partially complete.
    b. Many loosened blocks in crown and also in the ribs. Moderate to extensive barring down required, may impede production in the worst cases. Many joints loosened so that blocks are free to fall out.
    c. Overbreak: almost always greater than .30m (1 ft) locally reaches 1m (3 ft) or more.
	Severe Damage	0.90 (best)
		to
		0.80 (worst)

d. No experience whatsoever in this rock	Moderate Damage	0.90 (nominal)

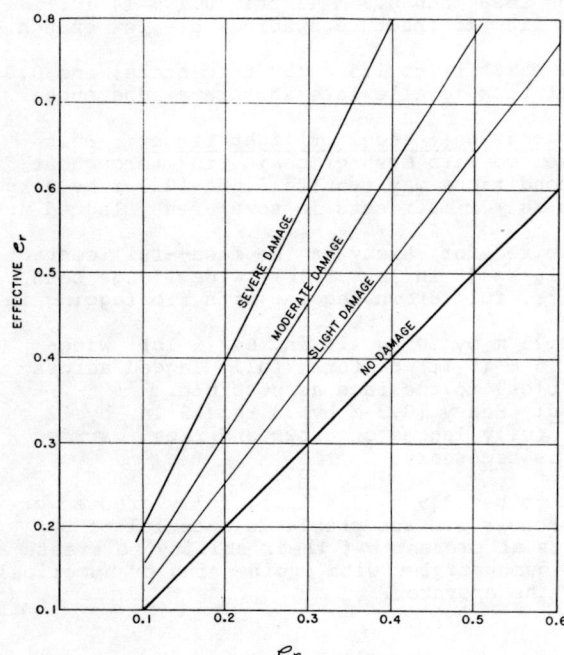

Fig. 8   Effective Extraction Ratios for 2.1 m (7 ft)-Wide Drifts, for Various Degrees of Blasting Damage.

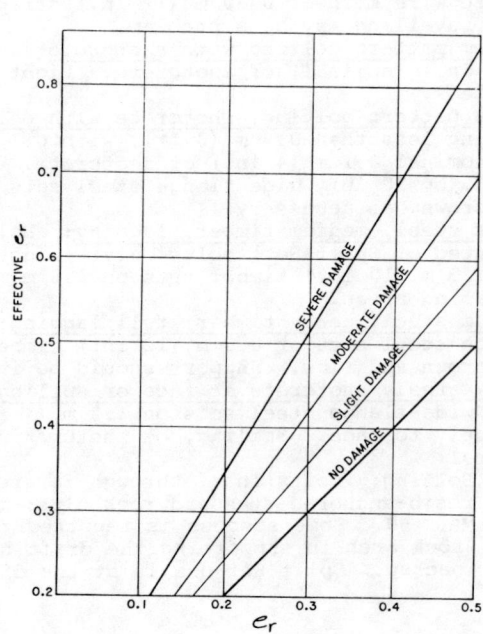

Fig. 9   Effective Extraction Ratios for 3.0 m (10 ft)-Wide Drifts for Various Degrees of Blasting Damage.

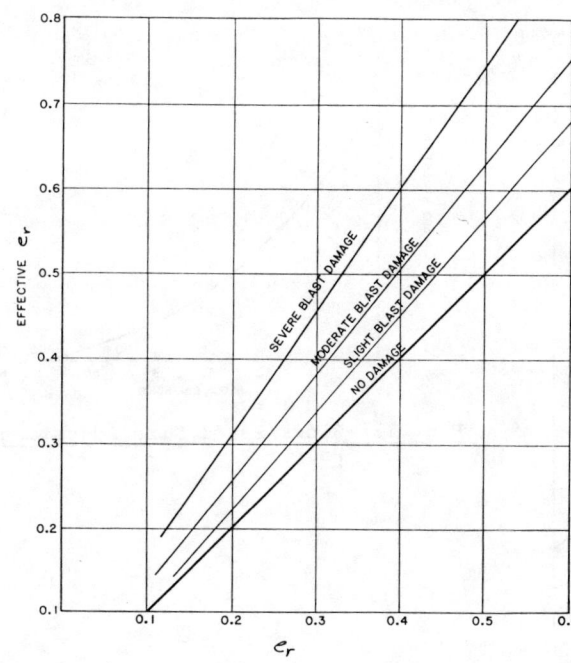

Fig. 10 (shown left) Effective Extraction Ratios for 4.6 m (15 ft)-Wide Drifts for Various Degrees of Blasting Damage.

Fig. 11 (shown below) Adjustment A_s for Induced Stresses Due to Multiple Openings.

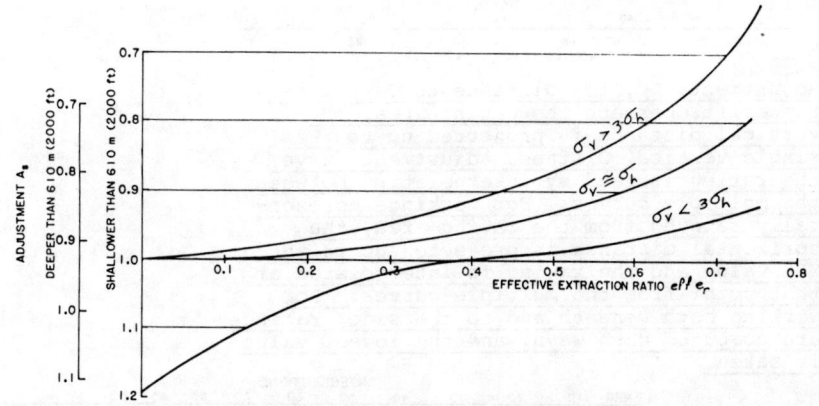

Table IV Fracture Orientation Rating A_O Based on Direct Observation in Drift

Number of Fractures Defining Block	Number of Non-Vertical Faces					
	1	2	3	4	5	6
3	–	0.95	0.80	–	–	–
4	–	0.95	0.85	0.80	–	–
5	1.0	0.95	0.90	0.85	0.80	–
6	1.0	1.0	0.95	0.90	0.85	0.80

Table V Fracture Orientation Rating A_O Based on Indirect

Observation of Fracture Statistics

Strike Heading Direction	Perpendicular				Parallel		(Flat dip)
	With dip		Against		–		–
Dip amount	45-90	20-45	45-90	20-45	45-90	20-45	0-20
	1.0	0.95	0.90	0.85	0.80	0.90	0.85

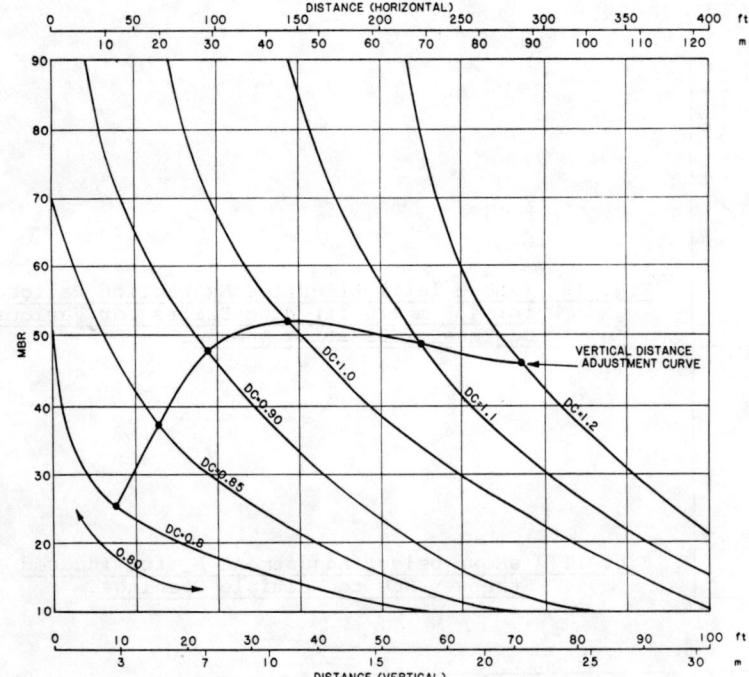

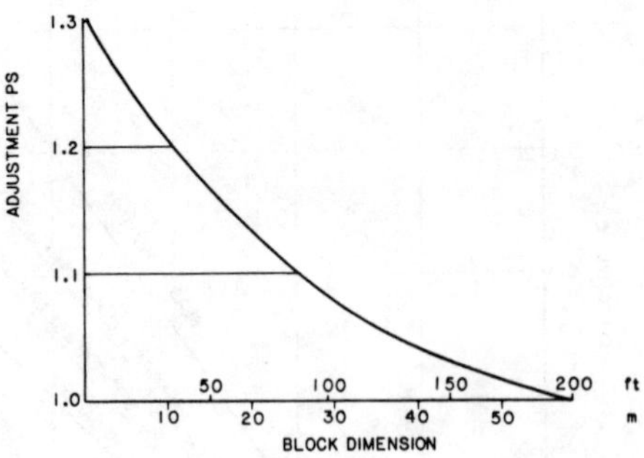

DISTANCE (HORIZONTAL)

VERTICAL DISTANCE
ADJUSTMENT CURVE

Fig. 12 Adjustment, DC, for Distance to Cave Line.
For drifts beneath the caving area, the
vertical distance is projected up to the
single Vertical Distance Adjustment Curve;
the rating is read by interpolating between
the multiple curves. For workings horizontally removed from the caving area, the
horizontal distance is projected up to the
MBR value and the rating is interpolated at
that point from the multiple curves. For
working both beneath and to the side, ratings
are computed both ways, and the lowest value
is taken.

Fig. 13 Block/Panel Size Adjustments
PS.

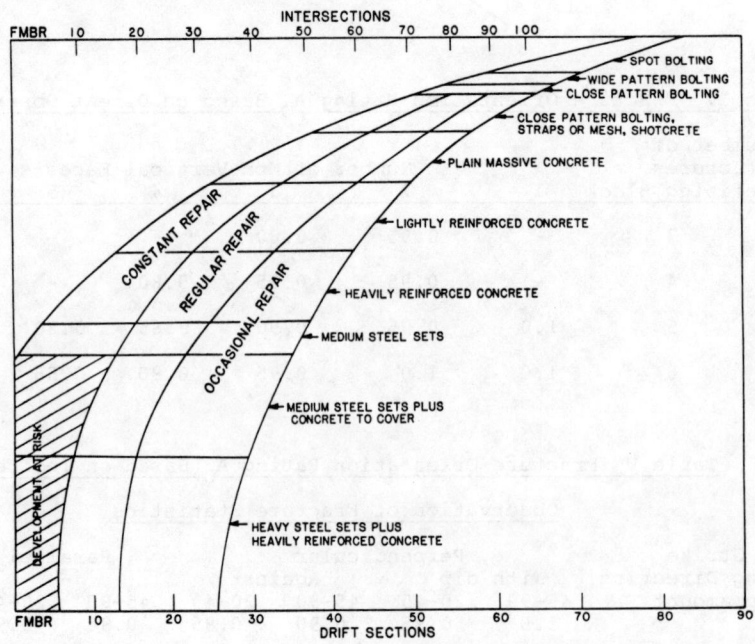

Fig. 14 Permanent Support Chart. Gives recommendations for permanent support of production
area drifts. Further explanation of the support types are given in the Explanation,
following.

B 62

a. Spot bolting: Bolting to restrain limited areas or individual blocks of loose rock, primarily for safety.

b. Wide pattern bolting: Bolts on 1.2 m (4 ft) to 1.8 m (6 ft) centers. May be wider in very large openings, when longer bolts are used.

c. Close pattern bolting: Bolts on centers less than 0.9 m to 1.2 m (3 to 4 ft), practical limit 0.6 m (2 ft).

d. Close pattern bolting, mesh or straps, shotcrete: Close pattern bolts with welded wire mesh or chain link in ravelling ground, and nominal 0.1 m (4 in.) of shotcrete.

e. Plain massive concrete: Cast-in-place massive concrete lining, 0.30 m to 0.46 m (12 in. to 18 in.) thick, may be applied over bolts or bolts and mesh when necessary. Prior shotcrete, if not damaged , may be considered part of this concrete thickness. Concrete should have a minimum as-placed 28-day compressive strength of 20.7 MPa (3,000 psi). Concrete must be placed so as to be in rock contact around the entire perimeter, especially the crown.

f. Lightly reinforced concrete: Massive, cast-in-place concrete lining 0.30 m to 0.46 m (12 in. to 18 in.) thick as above, lightly reinforced with .025 m (No. 8) rebar on 0.6 m (2 ft) centers or continuous heavy chain link. Reinforcement mainly in brows, crown, corners, and intersections. Prior light steel, if regularly placed with tiebars, may be considered reinforcement.

g. Heavily reinforced concrete: Massive, cast-in-place concrete lining as above, heavily reinforced with .025 m rebar on 0.6 m (2 ft) centers or less, on ribs and crown, throughout the interval. Prior light steel with tie rods may be considered adequate reinforcement.

h. Medium steel sets: Medium [0.25 m by 20 kg (10 in. by 45 lb)] wide-flange steel sets on 1.2 m (4 ft) centers, fully lagged on the crown and ribs.

i. Medium steel sets plus plain concrete to cover: Medium [0.25 m by 20 kg (10 in. by 45 lb)] wide-flange steel sets on 1.2 m (4 ft) centers, with plain, cast-in-place (3,000 psi) (minimum) concrete of sufficient thickness to cover the sets.

j. Heavy steel sets plus heavily reinforced concrete: Heavy [0.3 m by 33 kg (12 in. by 72 lb)] wide-flange steel sets on 1.2 m (4 ft) centers, with minimum 0.3 m (12 in.) thick heavily reinforced (as above) concrete throughout.

k. General:
   1. Concrete: it is assumed that proper concrete practice is observed: negligible aggregate segregation, full rock-concrete contact, adequate curing time, few blockouts.
   2. Chain link or steel sets, from development support, are considered reinforcement if the concrete between the sets is also reinforced in some way, either by tie rods or supplementary reinforcing bar.
   3. Repair in at least some locations is unavoidable in very poor rock. Development in rock of less than FMBR = 10 should be avoided if at all possible.

Table VI Major Structures Adjustment S

Adjustments are >1 if abutment stresses tend to be carried away from the excavation; <1 if stresses are concentrated.

Fault strike versus heading direction	Most nearly perpendicular						Most nearly parallel*					
Fault dip direction with respect to works	Towards			Away			Towards			Away		
Amount of Dip	Sh	M	St	Sh	M	St	Sh	M	St	Sh	M	St
Distance to nearest fault zone (W = Zone Width)												
<0.1 W	0.8	0.75	0.75	–	0.9	0.95	0.75	0.7	0.7	–	0.9	0.95
0.1 W to 1.0W	0.85	0.8	0.8	0.90	0.9	1.0	0.8	0.75	0.9	0.85	1.0	0.95
1.0 W to 10.0W	0.95	0.85	0.9	0.95	1.05	1.05	0.9	0.9	0.95	0.90	1.05	1.10
10.0 W to 50.0W	1.0	0.95	1.0	1.05	1.05	1.0	1.0	1.0	1.0	1.0	1.10	1.0
>50W	1.0	1.0	1.0	1.0	1.0	1.0	1.0	1.0	1.0	1.5	1.0	1.0
Within Zone	0.85	0.80	0.90	0.85	0.80	0.90	0.80	0.75	0.85	0.80	0.75	0.85

Sh = Shallow (<30°)
M = Moderate (30° to 60°)
St = Steep (>60°)

<0.1 W factor not to be applied for W<3.0 m (W<10 ft)
>50W factor not to be applied for W>3.0 m (W>10 ft)

*For those workings not "screened" from fault effect by caved volume. If "screening" exists, use 1.0.

EXPERIMENTAL INVESTIGATIONS AND FORECAST OF DEFORMATIONS AND FAILURE MECHANISM IN ROCK MASSES

Etudes expérimentales et prédiction des déformations et du mécanisme de la rupture dans les massifs rocheux

Experimentelle Untersuchungen und Prognose der Verformungen und des Bruchmechanismus im Gebirge

I. V. Baklashov
Professor, The Moskow Mining Institut, USSR

K. A. Ardashev
U. M. Kartashov
Leningrad, USSR

A. D. Alekseiev
Doneck, USSR

SYNOPSIS

The paper deals with the results of experimental studies of mechanical processes of the deformation and fracture of rocks. The methods of carrying out the research and the results of investigations of mechanisms of post-failure deformations and fracture of rocks under uni- and tri-axial compressive loading are discussed in detail. The results of the research are used for the prediction of load-bearing layers in rock masses and for mining technology development.

RESUME

Ce rapport considère les études expérimentales du processus mécanique de la déformation et de la rupture des roches. On examine, en détail, les méthodes de réalisation et les résultats des études portant sur le mécanisme post-rupture des déformations et cassures de roches soumises à des charges compressives mono-axiales et tri-axiales. On en utilise les résultats pour évaluer la force portante des strates soumises à ces charges dans des massifs rocheux, ainsi que pour le perfectionnement de la technique minière.

ZUSAMMENFASSUNG

Der Beitrag behandelt experimentelle Untersuchungen über mechanische Verformungs- und Bruchvorgänge im Festgestein. Dabei wird die Methodik der Untersuchungen über den Verformungs- und Bruchmechanismus des Festgesteins im Post-Bruchbereich in dem ein- und dreiachsigen Spannungszustand ausführlich behandelt. Die Ergebnisse von den Untersuchungen werden zur Voraussage der Standsicherheit der lasttragenden Gebirgsschichten und zur Vervollkommnung der Abbautechnik angewandt.

The most vital problem of rock mechanics is investgation of mechanism of rock masses deformation and deterioration. The mining production activities always involve the mechanical processes of rock deformation and destruction. Therefore, investigation of these processes yields the basis for forecasting the stability of rock masses where mining is carried out, for ensuring the safety of mining operations.

Today we can state with assurance that when studying the mechanical processes in rock masses it is preferable to investigate the deformations and not the stressed condition of the masses.This statement is based on the following considerations.First, in the course of experimental studies deformations are measured in the mass, and stresses are determined by calculations involving the measured deformation values, i.e. measuring instruments serve as deformation meters. Second, the deformations provide more data for estimation of mechanical processes - take, for example, the methods of rock mass stability estimation by the values of observed deformations. Third, the mechanical process develop in conditions of limited deformations and the level of rock mass stressed condition is determined by the deformed

state. Therefore, main efforts of the investigators shoull be concentrated on the studies of deformation mechanism in rocks and rock masses.

It is proved by experiments that the most important factor in the rock deformation mechanism is fracture. For instance, the amount of deformation is determined mainly by increase of rock volume after fracture. It follows, that the primary task for investigators is the study of fracture mechanism in rocks and rock masses, though this subdivision into deformation mechanism and fracture mechanism is quite arbitrary as the two processes are interconnected very closely.

When undertaking the studies of deformation and fracture of rocks and rock masses, it is necessary to base them on certain facts established experimentally by many investigators:
(1)Deterioration of rocks usually results from emergence of microcracks propagating, as a rule, along the mineral grain boundaries;
(2)The cracking begins at a loading level below the ultimate strength and corresponding to the level of long-term strength of the rocks;
(3) As acoustical observation data show, the

F

crack propagation process is developing like an avalanche, ending with destruction of the rocks; duration of this process depends on the level of acting loads ranging from long-term to instantaneous strength of the rocks;
(4)If the tests are carried out on the basis of preset deformations, the ultimate strength reached will indicate the maximum supporting capacity only and not the full exhaustion of supporting capacity;as the deformations develop beyond the ultimate strength level, the supporting capacity is decreased down to the residual strength level which is clearly illustrated by the full strain diagrams;
(5)Besides the commonly alopted notions of rock specimen strength and rock mass strength, one more term is needed - rock material strength which does not depend on the scale factor and differs from both above notions;
(6)In a properly arranged compression test,i.e. excluding any friction at the specimen bult faces, the fracture will be of columnar nature, with direction of cracks coinciding with the direction of maximum main stress; similar type of fracture is observed in rock masses;
(7)It tensile tests are carried out at preset levels of deformations, full strain diagrams can be plotted, just as in compression tests, including the deformations beyond the ultimate strength;
(8)Destruction of load-bearing elements (pillars, ceilings, high walls, masses around working, etc.) in rock masses can occur in the form of plastia flow (along the flat line) or brittle fracture (along the sloping- down line in the beyond- ultimate-strength section of the full strain diagram);
(9)Brittle fracture of rock masses can occur in the form of main traverse crack or, more often, in the form of areas filled with broken rock (henceforth referred to as"fracture areas");
(10)Boundaries of fracture areas have a fixed location, and changes of these boundaries in space and time will determine the supporting capacity and durability of load-bearing elements in rock masses.

Having analyzed the above experimental findings, the task of investigation of the deformation and fracture mechanism can be formulated as follows: to develop, on the basis of the experimental studies, a general theory of rock deterioration, starting from microfractures due to defects in mineral structure and ending with macrofracture of rock mass load- bearing elements; the theory should take into account the temporal kinetic nature of fractures and their dependence on the deformed state, viewed in the light of modern concept of beyond-ultimate-strengh deformation of rocks. The rock stregth criterium within the framework of such theory could be expressed as follows:

$$f(\sigma_1,\sigma_2,\sigma_3,\varepsilon_1,\varepsilon_2,\varepsilon_3,t,T)=0, \quad (1)$$

where $\sigma_1,\sigma_2,\sigma_3$ are components of main stresses in a rock mass;
$\varepsilon_1,\varepsilon_2,\varepsilon_3$ are the corresponding components of main strains;
t is time;
T is temperature.

The specific novel feature of this task formulation consists in taking into account of rheological and deformational aspects of fracture: the rock mass fracture is considered here as a process and not as a single-stage, single-time action, contrary to existing conceptions. Such conceptions, with their inherent deficiencies, are reflected in mechanical (e.g., Moores) and microdefect (e.g., Griffitt's) strength theories which gained wide recognition in the rock mechanics. According to them, the rock mass strength is determined by the stressed state level only:

$$f(\sigma_1,\sigma_2,\sigma_3)=0. \quad (2)$$

The recent years saw successful emergence of deformational strength theories (e.g., Lin'kov theory) taking into account the beyond-ultimate- strength deformation of rocks. They are characterized by the following strength criterium:

$$f(\sigma_1,\sigma_2,\sigma_3,\varepsilon_1,\varepsilon_2,\varepsilon_3)=0. \quad (3)$$

Well recognized now are kinetic strength theories (e.g., Zhurkov theory) taking into account temporal and thermal aspects of rock deterioration, as reflected in the new strength criterium:

$$f(\sigma_1,\sigma_2,\sigma_3,t,T)=0. \quad (4)$$

The above brief analysis of the accepted strength theories shows that they cover only individual aspects of the complex procrss of rock deformation and deterioration. It is obvious that scientifically sound forecasting of the rock mass stability is impossible without comprehensive study and estimation of all specific features of rock deformation and deterioration, as was stated under item (1).

To solve this problem, the authors carried out a series of laboratory and analitical studies. The studies of deformation and fracture mechanism at the micro-level, i.e. at the level of mineral structure defects, resulted in creation of a crystalline rock model consisting of a system of heterogeneous elements (mineral grains) built into a matrix and oriented arbitrarily. The tensor of grain moduli of elasticity is taken to be known, and the tensor of the system moduli of elasticity under impact of outer forces is to be determined taking into account the mutual influence of the grains. The effective elastic properties thus determined are characteristics of the material which differ considerably from the elastic properties defined experimentally on specimens, which was illustrated by comparing such characteristics of salt-carrying rocks.

The subsequent study of stress concentrations on the boundaries between grains and matrix proved the emergence of tensile stresses under outer compressive loads. It is interesting to note that the level of compressive loads at which tensile stresses reach the value of rock ultimate tensile strength and cause the appearance of tensile cracks, corresponds to long-term compressive strength of the rock. Propagation paths of microcracks converging into main traverse cracks were studied, and it was found out that their orientation corresponds to the direction of maximum compressive stresses. When there is no friction on the specimen bult faces, the main cracks reach these faces without interruption, which fact explains the columnar mechanism of rock specimen fracture. In case of friction, compressive stress zones are formed in areas adjacent the bult faces and prevent straight-line propagation of main cracks; in this case the specimen fracture is of conical pattern.

To investigate the kinetic character of rock specimen fracture in laboratory conditions, the acoustic emission from the specimens under compression was measured, and the beginning of this emission corresponded to the long-term strength level, and the maximum intensity to the level of instantaneous strength of the rocks. For description of kinetic rock fracture mechanism it is recommended to use the well-known Kachanov's equation where the continuity function is employed as measure of fracture.

The authors completed a series of rock compression tests in various stressed conditions of specimens and stress-strain diagrams were plotted, also called "full strain diagrams". Generally, in such tests rigid loading devices are used, or loading devices with a tracing system. In both cases the test is carried out with present deformation levels, corresponds to actual deformation processes in rock masses and differs essentially from conventional tests at present loads.

The most promising is the test method employing "rigid" attachments to ordinary hydraulic or mechanical presses. The aothors made use of an installation incorporating a compression chamber with "rigid" loading device and providing independent application of axial and lateral loads to the specimen. Tested in this installation were cylindrical specimens of various rocks: salt rocks, oil shale, marble, limestone, hard coal, and others. The full strain diagrams were plotted, with the beyond-ultimate-strength branch, and data were recorded on the changes in rock volume, development of longitudinal and lateral strains, loosening factor, strain modulus before and after the ultimate strength level, also complete strength charts. The beyond-ultimate-strength strain modulus determines the slope of beyond-ultimate deformation curve and is called the modulus of brittleness in some papers and the modulus of decay in others.

For instance, the modulus of elasticity defined for marble (of Koelga site) was $2,5 \cdot 10^4$ MN/m^2, the modulus of brittleness was $3 \cdot 10^4$ MN/m^2, for sulphide ore (Norilsk site) these moduli were $7 \cdot 10^4$ MN/m^2 and $7,7 \cdot 10^4$ MN/m^2 respectively. By the moment of complete breakdown their volume increas (dilatation) reached 10% or more.

Another installation employed by the authors for the purpose of plotting the full strain diagrams provided non-uniform ($\sigma_1 > \sigma_2 > \sigma_3$) three-axis compression of the specimens until final destruction. The test procedure made it possible to observe the complete process of deformation and transition of the rocks from before-ultimate to ultimate strenght, beyond-ultimate and destruction stages.

Subjected to testing were specimens of the rocks containing coal seams: fissile shales, sand shales, limestones and sandstones, all having different mechanical characteristics (modulus of elasticity, ultimate strenght of single-axis compression) and physical properties (porosity, volume weight). At first the specimens were loaded by single-axis compression, normal and parallel to direction of the layers, in order to determine strenght anisotropy. The latter was best represented in alevrolites: 32 MN/m^2 normal to layer direction, 15 MN/m^2 parallel to it.

Testing of specimens in spase field of compressive stresses was carried out according to program $\sigma_1 > \sigma_2 > \sigma_3$ where the components of compressive stresses were selected to correspond the field of compressive stresses in the walls of a working. The loading process resulted in a stressed state correponding to mine depth of 1000-1500 m.

The full diagrams of rock deformation plotted at volumetric loading, like in the first experiment revealed the following characteristic areas of deformation: before-ultimate quasi-elastic deformation; ultimate quasi-plastic deformation at slight reduction in resistance approaching the ultimate strength; beyond-ultimate deformation, with sharp reduction in resistance and fracture; deformation with almost unchanging residual rock strength after fracture.

At the before-ultimate deformation area the slope of the upward curve in the full diagram is identical for all types of rocks, wich is in compliance with the generalized Gook law concerning the complex stressed state. When the specimens are subjected to maximum loads, a plastic deformation zone is formed. Its special feature is the growth of strains, with the load unchanging, and sharp loss of density. The highest plasticity is displayed by fissile and sandy shales, less plastic are limestones, immune to plastic deformation are sandstones.

Of special interest in the above experiment were the studies of deformation process in the rock after the latter reached the maximum resistance level. Deformation of specimens after crossing the ultimate strength limit is obtained by unloading them along the axis. This process models deformation of the rocks in the destruction zone around workings. The zone can be identified by a sharp fall in supporting capacity, rapid growth of tensile strains, intense volume expansion. An important feature of the deformation process in the beyond-ultimate area is a certain stabilization of specimens resistance, or so-called residual strength. This value defines the deformational and supporting capacity of the rock after it lost its cohesion and allows to determine the production parameters necessary for maintenance of the workings.

Full strain diagrams at single-axis compression and temperatures down to -120°C were plotted for specimens taken of salt-rock masses employed for construction of underground cavities for storage of liquified gases. It was found that the diagrams change substantially with decrease of temperature: the modulus of elasticity and modulus of brittleness of salts both go up.

For plotting of experimental full strain diagrams analytical linear-increment approximations are proposed: with four areas (before-ultimate, ultimate, beyond-ultimate and residual strength areas) or three areas (before-ultimate, beyond-ultimate and residual strength). The latter case is more convenient for practical purposes.

On the basis of experimental studies of rocks under compression a model was developed explaining the inner mechanism of strength falling, volume deformation, side pressure effect and time factor in the beyond- ultimate strength deformation area.

Plotting of full strain diagrams under tension is complicated by certain technical problems. The only correctly arranged experiments known to us so far are those of Peng. The authors have developed an original experimental procedure for plotting of such diagrams with deformational treatment of Griffith's energy presentations, where the deterioration energy is to be estimated using experimentally- determined coefficient of stress intensity.

The above experimental studies and models revealing the deformational and kinetic nature of rock fracture mechanism were employed for analysis of deformation and deterioration in load--bearing elements of rock masses. In this case the deterioration of rock masses is studied in the light of main cracks propagation or formation of fracture zones with boundaries, also called the brittle fracture fronts, changing with time and advancing from the vacated space limits inside the rock mass. As a measure of rock mass deterioration, the notion of "rock mass supporting capacity" is introduced and the deterioration process is viewed as reduction of this capacity. Also introduced is another notion - endurance of load-bearing. elements of rock masses, which is expressed by the time internal between the start of utilization and full expenditure of supporting capacity or reaching the residual supporting capacity. It shoud be kept in mind that the rock mass endurance, as well as the supporting capacity, is a primary parameter in mining design.

The main travers cracks concept was employed for estimation of supporting capacity of salt pillars in study of their deterioration note was taken of salt rock cristalline structure non-homogeneity and of the possibility of tensile cracks initiation along the boundaries of crystals. The main crack trajectories were studied for various geological structures and geometrical shapes of the salt pillars: in case of strong salt contacts between the pillar and rock; in case of clay layers at the contacts and in the middle part of the pillar; in case of band-type and columnar pillars with straight walles and with bellows at contacts with the roof. It was found that the supporting capacity of the pillars with strong salt contacts is 20-30 % higher than that with clay layers, and the bellows at contacts with the roof promote further increase in the pillar supporting capacity.

The fracture zone concept was employed for estimation of the supporting capacity of ore and coal pillars. The deformation and deterioration of the pillars was analyzed following two patterns: by preset loads and by preset strains. The preset load pattern is suitable for stall-and-brest system of ore mining, when the rigidity of pillars and ceiling is comparable, and the kinetic character of deterioration is the key factor. The following aspects were investigated in this pattern: starting time of pillar deterioration after this uncovering, position

of the brittle fracture front and speed of its displacement inwards, supporting capacity and endurance of the pillars. A mining system is proposed, with pillars of variable cross-section, within one panel which are of equal strength from the point of view of endurance and so reduce the percentage of ore losses.

The second pattern (by preset strains) was employed for analysis of operational condition of protective coal pillars in seams with hard-collapsible roof, where the rigidity of pillars is inferior to that of the roof the studies resulted in plotting of strain diagrams and establishing the maximum and residual supporting capacity, as well as endurance of coal pillars. Recommendations were formulated for control of operational condition of the pillars based on measuring the displasements of pillar walls and roof, also recommendations for pillarless stopping of coal.

New studies using deformational concepts of rock deterioration under compression were carried out for permanent workings and underground construction projects. Axial assymetry problem was solved for a horizontal round-section working excavated in a rock mass with non-hydrostatic initial stresses state. It was proved that taking into account the beyond-ultimate-strength deformation of rock it is possible to level out in a large degree the ellipcity of fracture zone boundaries around the working, and the respective axial symmetry problem solution can be applied for a great variety of practical tasks.

To analyze mechanical processes of deformation and deterioration around a mine face, a volumetric axial symmetry problem was solved by method of displacement. The following widely discussed items were also investigated: stress field, strained condition of rock contour, boundaries of fracture zones, which are used for estimation of workings stability and loads acting on supports.

It was proved that stability of workings depends mainly on mechanical processes of deformation and deterioration of rock masses around the working's face. It is here that the qualitative features of mechanical processes are formed, and with further transfer of the face the processes merely undergo quantitative changes.

The deformational concept taking into account the time factor and low temperature field was employed for analysis of stressed/strained condition and for estimation of stability of a salt mass incorporating a spherical underground cavity filled with a liquified gas under pressure. It was shown that at the beginning of utilization the strength criterium is of primery importance and at the end of utilization the deformational criterium, i.e. the level of ultimate strains, takes this part.

Thus, experimental studies of the mechanism of rock deformation and deterioration allow to obtain reliable forecasts of rock mass stability and to formulate recommendations for more efficient mining operations.

MAIN PRINCIPLES AND METHODS OF INVESTIGATIONS OF IN-SITU ROCK MASSES

Principes et méthodes de l'étude in situ des caractéristiques mécaniques des massifs rocheux

Allgemeine Konzeption und Untersuchungsverfahren für die Durchführung von geomechanischen Untersuchungen des Gebirges in situ

Yu. A. Fishman
Cand.Sc. (Tech.), Chief of Rock Foundations Department, Hydroproject, USSR

S. B. Ukhov
Dr.Sc. (Tech.), Prof., Civil Engineering Institute, Moscow, USSR

A. B. Fadeev
Dr.Sc. (Tech.), Prof., Civil Engineering Institute, Leningrad, USSR

SYNOPSIS

The in-situ determination of rock mass characteristics (test procedure and interpretation of results) must be closely associated with the engineering problem encountered (type of structure, foundation and the methods of calculation being used) and must produce the necessary valid information. The present article gives a descriptive account of the general approach to field investigations of rock masses. Some problems connected with test procedure, equipment used and the processing of the results, taking into account specific features of fissured rock masses, are discussed.

RESUME

La détermination in situ des caractéristiques des massifs rocheux (essais et interprétation des données) doit être étroitement liée au problème technique posé (type d'ouvrage, de fondation et méthode de calcul utilisée) et doit assurer l'information initiale nécessaire et valable. On examine le procédé relatif aux essais in situ des massifs rocheux. Quelques problèmes concernant les méthodes de réalisation des essais, le matériel utilisé et l'interprétation des données obtenues, compte tenu des particularités des massifs rocheux fissurés sont également étudiés.

ZUSAMMENFASSUNG

Die Bestimmung der Kennwerte von Gebirge in situ (Durchführung der Versuche und Definition der Ergebnisse) muß mit der gestellten Aufgabe (unter Berücksichtigung von Bauwerkstyp, Gründung und Berechnungsverfahren) abgestimmt werden, um die erforderliche aussagekräftige Ausgangsinformation zu liefern. Im Beitrag wird die allgemeine Konzeption der Durchführung von Großversuchen im Gebirge behandelt. Es werden auch einige Fragen der Methodik für die Durchführung der Versuche, der bei Versuchen eingesetzten Einrichtungen, sowie die Auswertung der ermittelten Ergebnisse, unter Berücksichtigung von besonderen Eigenschaften des Felskörperverbandes betrachtet.

1. INTRODUCTION

Development of numerical methods of calculations of rock foundations and the finite elements method in a non-linear statement in particular enlarges the scope of the problems to be solved but at the same time places heavy demands on the scope and quality of geomechanical characteristics. Aside from traditional studies of shear resistance parameters and deformation modulus of the rock mass a demand was created for determination of strength and deformation properties in tension, compression and triaxial loading, the value and orientation of natural stresses, permeability factor as a function of a stressed state of a rock mass, etc.

In-situ investigations are known as the most valid methods of determination of characteristics of the rock mass. But at the same time high costs and labour content of these efforts demand a proper staging of experiments and maximum data acquisition from in-situ investigations at a limited number of tests.

2. PRINCIPLES

One of the main principles of in-situ investigations consists in the procedure of experiments closely associated with the solution of the raised engineering problem: with consideration for the type of the foundation, design of the structure, behaviour of the structure integrally with the foundation and methods of calculation. Thus modulus of deformation may be determined in different ways, e.g. by loading of a plane surface of the foundation (method of plane loading plates), loading of the surface of a cylindrical opening(method of radial plates or pressure chamber), triaxial loading, etc. In each case different results will be obtained because rock masses as contrasted to the ideal elastic media are deformed differently depending on the type of a stressed state. It is obvious that the first method of determination of deformation modulus is the most suitable for calculations of structure settlements and the second one is suited for calculations of lining of hydropower tunnels and the third one is adequate for calculations of a stress-strained state by numerical methods, etc. The same is true to determination of other geomechanical characteristics and particularly shear resistance parameters. In this case not onty the level of stresses but the pattern of their distribution is of importance which is governed by the mode of application of loads to the block. Shears of blocks with moment and momentless patterns of their loading produce different parameters of shear resistance (Fishman, 1976, 1979). For stability analysis of concrete retaining structures (dams, abutments, etc.) the loads on which create the moment with respect to the center of gravity of the footing section the first method of shear tests is evidently the most suitable one which will be described below in detail. Thus the methods of in-situ investigations should to a certain extent simulate the conditions of foundation functioning which they are intended for (Fishman, 1976).

The second principle of in-situ investigations following from the first one consists in consideration of changes in properties of rock masses which occur during construction and operation of the structure.

Unloading of the rock mass during rock excavation, decompaction and disturbances due to explosions and the following compaction through loading by the structure, water saturation, grouting, instantaneous (seismic) and continuous power effects, etc. - these are the factors which as a rule result in drastic changes in properties and conditions of the rock mass compared with the conditions of its natural occurrence. In connection with that in-situ investigations whereever possible should be conducted under conditions simulating construction and operation effects upon the rock. It seems expedient to reveal their properties both in unloaded and compacted conditions before and after grouting under conditions of a "dry" and water-saturated rock mass at instantaneous and continuous effects (Ukhov, 1970, 1975; Fishman, 1974, et al.).

The third principle following from the first one consists in consideration of the scale factor. The universally adopted idea now is that fissured rocks in a rock mass exhibit a pronounced scale effect. In a general case the curve of the scale effect for actual rocks is determined by numerous factors and is of the complicated nature (Ukhov, 1976, 1979). Extrapolation of values determined by small volumes of rocks to large sections of the rock mass appears invalid. Therefore it is necessary to determine characteristics of mechanical properties of fissured rocks in a rock mass on the basis of the scale effect. Such experiments are conducted in the field conditions. As for interpretation of results of in-situ tests a body of mathematics of mechanics of the continuous medium is used it is necessary that the volume of the rock under study should satisfy quasi-dense conditions. According to Ukhov, 1975, 1979, it is necessary that the linear size of the considered volume of rock

should be at least 5-7 times higher than the linear size of the non-uniformity element.

The fourth principle is that the in-situ tests should be accompanied by auxiliary investigations: rock sample tests, geophysical experiments, studies of fracturing, porosity, density, etc. Such complex of investigations allows proper interpretation of in-situ test results and makes it possible to study the effect of the scale factor within a wide range of changes in the studied volume of the rock mass and to obtain the results to be extended into the whole construction site and to be an analogy for designing of other projects.

Finally the fifth principle consists in revealing of importance of studied characteristics for determination of the required detailing of their investigations. High costs of in-situ tests demand determination of the most optimum composition, scope and procedure of experiments. The similar scope and accuracy of investigations are not always necessary for one and the same structures. Everything depends on particular conditions of construction. The quest for maximum accuracy in all cases may result in expensive field tests even though it is not necessary and on the other hand the application of only "rough" and approximate methods of investigations may have negative effects on reliability of the structure.

The principle of optimization of investigations of rock foundations which nowadays advances in investigations in the field of hydropower engineering (Fishman, 1982) is based on the method of "sounding" of geomechanical characteristics, analysis of their effect on the design solutions. Let us illustrate the main distinctive features of this procedure by the following particular example.

During planning of in-situ investigations of the rock mass at the site of the Tashlykskaya pumped storage power plant in the Ukraine the question of the necessary scope and details of studies of deformation modulus for the designing of pressure tunnels was raised. According to preliminary evaluations made by geophysical methods with the use of analogues the values of deformation modulus were: in zone I of sound rocks - 6000 - 7500 MPa and in zone II of tectonic disturbances - 300 - 500 MPa (shown as shaded sections in Fig. 1). As the analysis of the relationship between [*] width of opening of cracks in tunnel lining and modulus of deformation demonstrated that variation of the value of modulus of deformation has an effect on opening of cracks in the lining only at $D < 5000$ MPa for concrete and at $D < 2500$ MPa for reinforced concrete linings and with relation to the permissible opening of cracks $a_{lim} = 0.15$ mm of particular geological conditions of the project at $D < 3200$ MPa and $D < 1200$ MPa accordingly. At higher values of deformation modulus its variation has no effect on parameters of the lining; in this case it is important to avoid considerable errors in evaluation of its lower value. As shown in Fig. 1 for the considered zone of

_____

[*]- Allowable width of opening of cracks a_{lim} is the value governing reinforcing and cost of lining accordingly

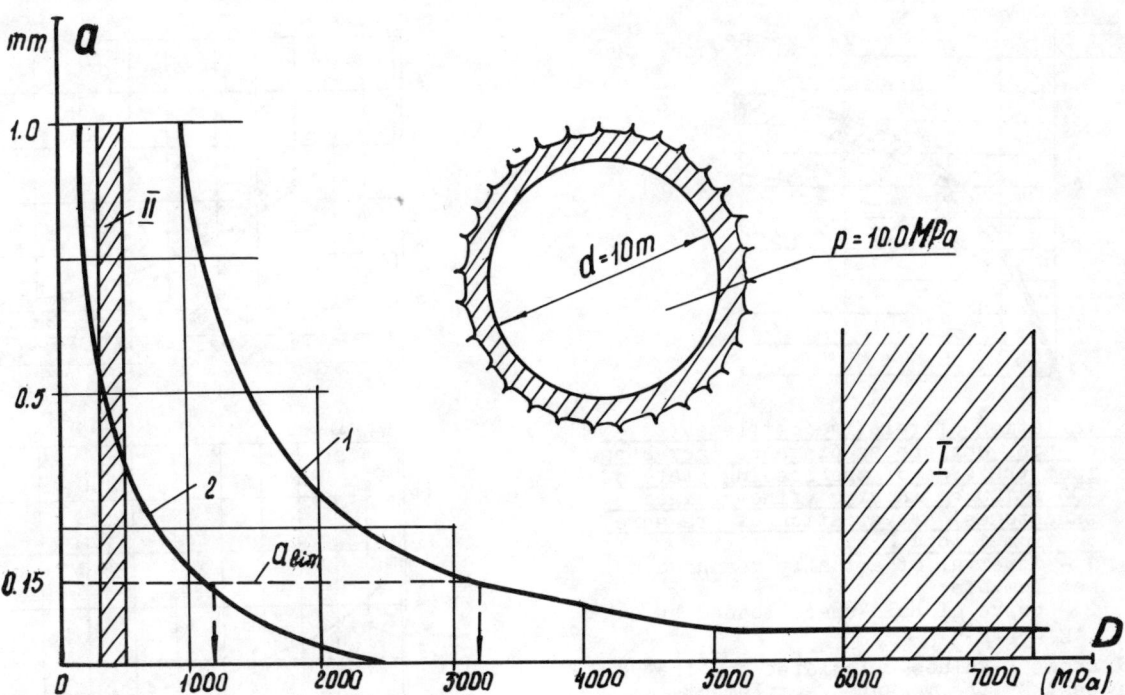

Fig. 1. Curve of dependence of opening of cracks in tunnel lining on modulus of deformation of surrounding rock mass (Tashlykskaya pumped storage power plant).
a_{lim} - permissible value of crack opening for local conditions.

sound rocks even if the error is of a double value modulus of deformation is beyond the range of its variation effect on the lining structure. Thus in the zone of sound rocks there is no need to carry out large-scale in-situ investigations and it is quite enough to limit ourselves to geophysical tests and other proximate methods. At the same time the values of deformation modulus in the zone of tectonic disturbances are within the range of their pronounced effect on parameters of tunnel linings. Variation of their values even within 100 MPa results in drastic changes in reinforcement and cost of the lining. Consequently in the zone of tectonic disturbances it is necessary to conduct more precise analysis of values of deformation modulus. This accuracy may be reached by sufficiently representative in-situ tests, e.g. by the method of radial loading plates or pressure chambers. However at small thickness of tectonic zones these expensive experiments may appear economically inexpedient which really happened at the Tashlykskaya hydropower plant and which allowed under given conditions to reject large-scale in-situ investigations.

The similar analysis and selection of the optimum composition, scope and methods of investigations may be carried out for all geomechanical characteristics utilized in designing. The given procedure of optimization is presented in more detail in Fishman's article, 1982.

Now let us describe some questions concerning

the methods of in-situ investigations.

3. METHODS

3.1 Investigation of deformation properties

For calculation of rock foundations and underground structures the following main characteristics are used now: D - modulus of deformation, K_o - coefficient of specific rebound and ν - Poisson's ratio.

3.1.1. Modulus of deformation. Calculation of the value of the modulus of deformation on the basis of results of loading of the foundation surface by plane loading plates is usually made by Schleicher formula derived for the design diagram of half-space. The results of experiments demonstrate that deformations of the rock foundation surface around the loading plate form a smaller settlement depression than in the problem of the theory of elasticity of loading of half-space by a rigid or flexible load over the area of the loading plate (Fig. 2). One of the reasons of the given effect consists in non-uniformity of the rock mass under the loading plate resulted both from rock excavation (decompaction and explosion effect) and natural processes of unloading and weathering.

For interpretation of results of loading plate tests Tsytovich, Ukhov et al., 1970, proposed a calculating scheme of a limited thickness layer, the formula and the Table for calculation of D and settlement S of the foundation surface at different distances from the loading plate. Having constructed theoretical curves of relative settlements at different values of thickness of the compressed layer H according to the proposed method and having compared them with the test curve it appears possible to find the actual value H and the required value D by coincidence of these curves.

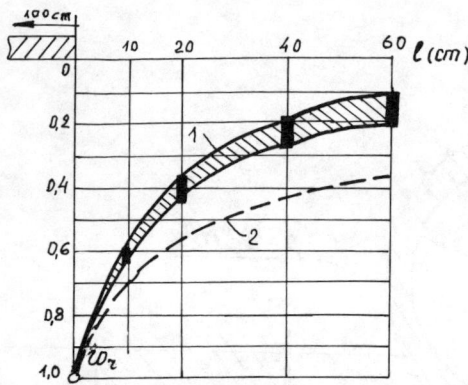

Fig. 2. Variation of relative settlement of rock surface as distance increases from the edge of the loading plate of 100 x 100 cm in size (fine-grained sandstones) at variation of pressure from 1 to 4 MPa.
1 - interval of actually measured settlements;
2 - curve of half-space loaded by a rigid plate.

Table I shows D-values calculated for the particular test by the measured settlements at different values H/b (where b - halfwidth of the loading plate).

Table I

Relative thickness of layer H/b	D-values, MPa, for points located at a distance of (cm) from the edge of the loading plate				
	0	10	20	40	60
1.0	6900	5050	4100	2350	–
1.5	8680	8970	9170	9600	9600
2.0	9750	11500	12900	16500	25900
∞	14200	22000	29700	59800	144000

As illustrated in Table I at a relative thickness of compressed layer H/b = 1.5 the values D calculated for the points located at different distances from the loading plate are about the same. These values are the actual values of modulus of deformation.

As stated above to comply with the principle of conformity of the test and in-situ conditions very often it is necessary to hold the loading plate under each step of loading during a certain period of time. The data presented in Fig. 3 (Ukhov) give some ideas of the foregoing. If we assume the required accuracy of D as e.g. 20%, then in the first cycle of loading the time of holding of the rock under the load at each step shall be in the order of 5-6 hours and in the second cycle of loading it shall be 3-4 hours.

For determination of modulus of deformation values of large areas of the rock mass it is expedient to use seismoacoustic surveys. Correlation relationships between D and D_d on the basis of in-situ tests are presented in the article, Ukhov, 1979.

3.1.2 Coefficient of rebound. Another deformation characteristic employed for calculation

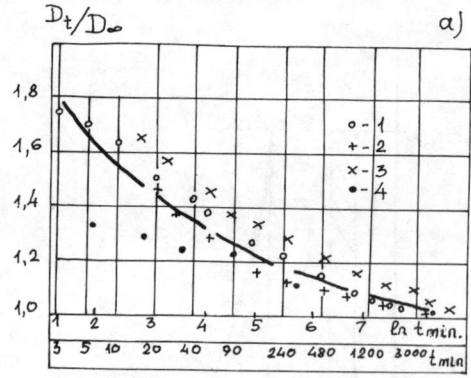

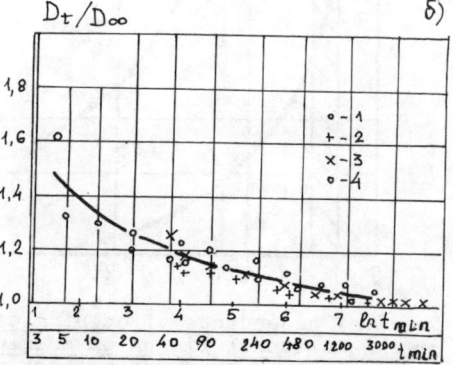

Fig. 3 Variation of $\frac{D_t}{D}$ with t.

D_t - modulus of deformation at holding under load during t; D - modulus of deformation at full stabilization of settlements.
a) Ist cycle of loading; b) IInd cycle of loading.
1 - metamorphic schists, loading plate 100 x 100 cm in size, variation of pressure $\Delta\sigma$ from 3 to 4 MPa;
2 - limestones, loading plate 80 cm in diameter, $\Delta\sigma$ from 2 to 4 MPa;
3 - ditto, $\Delta\sigma$ from 4 to 7 MPa;
4 - fine-grained sandstones, loading plate 100 x 100 cm, $\Delta\sigma$ from 1 to 2 MPa.

of linings of underground structures is coefficient of rebound determined by the method of loading of the surface of circular excavations by radial loading plates or pressure chamber. Coefficient of rebound is the value similar to the bedding coefficient used for calculation of slabs and beams by the Winkler method and therefore depends not only on compliance of the rock but on the radius of excavation. For convenience of comparison between coefficients of rebound obtained in different experiments the concept of coefficient of specific rebound was introduced which is the value of coefficient of rebound in the first radius of the tunnel and determined by the following formula:

$$K_o = \psi \frac{qr}{100 U_r} \qquad (1)$$

in which q - intensity of uniformly distributed load on the surface of circular excavation, r - radius of excavation, U_r - radial movements of excavation walls, ψ - coefficient taking into account the length of the loading section and the line of measurements (Talobre, 1957).

Frequently coefficient of specific rebound is recalculated through modulus of deformation D obtained in experiments with plane loading plates according to the following relationship of the theory of elasticity (Galerkin formula):

$$K_0 = \frac{D}{100(1+\nu)} \qquad (2)$$

or in contrast modulus of deformation D is determined by the given formula recalculating it through K_0. Since in experiments with plane and radial loading plates a considerably different stressed state occurs (compression of the rock mass in the first case and compression in the radial direction and tension in the tangential direction in the second case), the values of coefficient of rebound calculated by the formula (1) and recalculated by the formula (2) should differ from each other. As investigations carried out by the Institute of Hydroproject showed the values K_0 or D determined by the tests with radial loading plates are at least 1.5 times lower than by the tests with plane loading plates.

3.1.3 Poisson's ratio. The third deformation characteristic of fissured rocks is Poisson's ratio which presents considerable difficulties. Reliable methods of direct determination of this value in-situ conditions have not been developed. There is no correlation between ν and ν_d - dynamic coefficient of lateral deformation even in the tests with samples. Ternovsky and Ukhov, 1977, proposed an indirect method of determination of ν of rocks under in-situ conditions at vertical and horizontal loading of loading plates. It is interesting to note that in 5 tests carried out for 4 types of rocks the value varied from 0.08 to 0.11.

3.2 Investigations of strength properties.

These include: studies of compression strength, tensile strength and shear strength of fissured rocks.

3.2.1 Compression strength. Much attention is being given to investigation of problems of strength properties of rock masses under conditions of unconfined compression in mining industry because of necessity to make stability analysis of rock blocks left during excavation of underground chambers. Usually the tests are carried out by the method of crushing of rock prisms. Dimensions of loading sections as a rule do not exceed 0.6 x 0.6 m because of difficulties encountered in making the prisms and heavy loads required to crush them. In this case the obtained strength parameter of the rock mass in unconfined compression R_m varies from 1-2% to 40-60% of strength of the rock sample depending on the method of rock block delimitation and character of fracturing. The minimum values are related to blocks delimited by perforators and the maximum ones are related to blocks delimited by non-percussive methods (rotary drilling or saw cutting with diamond disk wheels).

Generalization of results of determination of

strength characteristics of the rock mass in unconfined compression accumulated by mining experience allows to recommend (Fadeev) the following Table of coefficients of structural weakening $\lambda = R_m/R_s$.

Table II

α (degree) \ M	4	4-8	8-12	12-14	14
0-15	0.6	0.4	0.36	0.34	0.3
20-35	0.4	0.3	0.27	0.24	0.20
40-60	0.35	0.25	0.22	0.20	0.18
65-90	0.55	0.40	0.36	0.32	0.28

M - modulus of fracturing, i.e. number of fractures for one l.m.; α - angle between the plane of the main system of fractures and the direction of the compressive load.

Disturbance of the natural structure of rocks by explosions (Fadeev, 1972) results in a decrease of strength properties of rocks. In this case strength of sound rocks decreases higher compared with soft rocks. The Table III is demonstrated for tentative evaluation of the coefficient of structural weakening λ according to strength of the sample in unconfined compression R_s:

Table III

R_s (MPa)	2	2-10	10-40	40
λ	0.1	0.08	0.06	0.08

3.2.2 Tensile strength. Investigations of tensile strength of rock masses in the USSR began not long ago (Ukhov, 1975). At the beginning investigations were carried out using the method of tearing away of the loading plate concreted to the rock foundation. Since the breaking surface as a rule is not at the contact between concrete and the rock but occurs at nearby cracks, the obtained characteristic of tensile strength may be related to the whole rock mass. As a matter of fact in this case the edge effect shall be taken into account, i.e. squeezing of the rock mass at edges of the loading plate. Experiments of the Moscow Civil Engineering Institute (Ukhov, 1975) conducted on sedimentary rocks (sandstones, argillites, microschists) produced the values of ultimate tensile strength R_T = 0.01-0.7 MPa. In similar experiments of Hydroproject conducted on hard porphyrites the values R_T = 0.1 - 0.5 MPa were obtained. At the same time the tests of tearing away blocks of the rock mass (for the purpose of which a delimiting trench was provided around the loading plate down to 10-15 cm) conducted on the same rocks and the same chamber produced lower values R_T = 0.05 - 0.25 MPa which is explained by the absence of the effect of lateral stresses (restraining of the rock mass). It should be noted that in all experiments due to nonuniformity of the rock mass a turn of the loading plate takes place at forcing out with respect to the axis which as a rule coincides with one of the edges. Therefore the obtained values R_T should be considered as the low value of strength characteristics in unconfined tension.

Though the ability of rock masses to accept tensile stresses is questioned investigations conducted by the Moscow Civil Engineering Institute and Hydroproject demonstrated that rock masses possess certain tensile strength consideration of which in calculations should be beneficial for raising the efficiency of design solutions.

3.2.3 _Triaxial compression strength._ For modern calculation methods it is necessary to have strength characteristics of rock masses under conditions of triaxial compression. However in-situ triaxial compression tests are conducted very rarely because of a high labour content and a complicated procedure. Separate in-situ tests and laboratory studies as well make it possible to recommend only an approximate method of construction of the Coulomb-Mohr envelope curve by unconfined compression and tensile strength characteristics of the rock mass.

3.2.4 _Shear strength._ Methods of analysis of shear strength of concrete and rock blocks are well-known. Usually the results of experiments with all plates involved are plotted on coordinates τ_t and σ (where τ_t and σ are ultimate tangential and normal stresses) and the Coulomb linear relationship $\tau_t = f(\sigma)$ is constructed parameters $tg\,\psi$ and C of which are shear strength characteristics:

$$\tau_t = \sigma tg\,\psi + C \qquad (3)$$

Tsytovich, Ukhov and Burlakov, 1970, proposed a theoretical body and method of determination of ultimate and critical values of shear strength of rocks on the basis of data of each separate shear block test:

$$\tau_t = \sigma tg(\psi_o + \alpha) + C_o \qquad (4)$$

$$\tau_{cr} = \sigma tg\,\psi_o + C_o(1 - tg\,\psi_o tg\,\alpha) \qquad (5)$$

where σ - normal pressure (stress), $tg\,\psi$ and C - real friction and cohesion coefficients, $tg\,\alpha = \frac{\Delta U}{\Delta V}$ - coefficient of lift of the block in shear, ΔU and ΔV - average increments of horizontal and vertical displacements at the shear block.

At least two shear tests by one and the same shear block are made to obtain the values $tg\,\psi_o$ and C_o. In this case τ_t is related to the first cycle characterized by the failure of the shear block - rock contact and τ_{cr} is related to the second and succeeding cycles - along the failed contact.

It was found by in-situ tests that the value $tg\,\psi_o$ does not depend on the value σ practically does not change during duplicate tests and does not change with enlargement of the shear block area. The maximum change interval of the value $tg\,\psi_o$ for the tested types of rocks varied from 0.50 to 0.82. It should be noted that at approximation of shear block tests by the Coulomb relationship the parameter $tg\,\psi$ ranges from 0.3 to 2.7 which presents difficulties for physical interpretation of this value.

The parameter $tg\,\alpha$ does not depend practically on the area of the loading plate but varies regularly with variation of :

$$\tau_\alpha = \left(ctg\,\psi_o + \frac{\sigma}{2\beta}\right) - \sqrt{\left(ctg\,\psi_o + \frac{\sigma}{2\beta}\right)^2 - ctg\,\psi_o} \qquad (6)$$

in which $\beta = \dfrac{\sigma tg\,\psi_o}{(ctg\,\psi_o - tg\,\alpha)^2}$ -

- parameter determined by the results of the test with σ = const.

The value C_o does not depend on normal pressure but changes considerably with conditions of rocks (fracturing, silicification, etc.).

Having determined values $tg\,\alpha$, $tg\,\psi_o$, C_o on the basis of the given test at σ = const it seems possible using formulae (4) and (5) to calculate τ_t and τ_{cr} for other values of σ , i.e. to obtain the relationship $\tau_t = f(\sigma)$ and $\tau_{cr} = f(\sigma)$ for the rock of the given shear block area. These relationships are materially non-linear.

Fishman, 1979, proposed the following non-linear relationship $\tau_t = f(\sigma)$.

$$\tau_t = \sqrt{(nR_c)^2 + 2m\sigma R_c - \sigma^2} - nR_c \qquad (7)$$

where $n = \frac{h}{b}$ and $m = \frac{e}{b}$ - coefficients characterizing arms of forces applied to the shear block (Fig. 4), R_c - parameter of crushing strength of the rock mass determined by the formula (Fishman, 1979):

$$R_c = \frac{\sigma^2 + \tau_t^2}{2(m\sigma - n\tau_t)} \qquad (8)$$

The relationship $\tau_t = f(\sigma)$ was obtained theoretically and confirmed by experiments for the system "shear block (dam) - foundation", characterized by high strength of the contact and the contact zone. In this case the failure in contrast to the classical scheme of shear is not accompanied by sliding but by a turn of the shear block with its separation from the pressure side and crushing from the downstream side.

Under real conditions different strength of the contact zone may take place and consequently both forms of failure may take place: shear and turn and the curves $\tau_t = f(\sigma)$ thereof. Criteria characterizing the boundary between them may be found from the simultaneous solution of relationships (3) and (7). The critical value of the normal stress:

$$\sigma_{cr} = \sigma_a + \sqrt{\sigma_a^2 - \sigma_\beta^2} \qquad (9)$$

where $\sigma_a = \dfrac{(m - ntg\,\psi)R_c - tg\,\psi C}{1 + tg^2\psi}$; $\sigma_\beta^2 = \dfrac{C^2 + 2nCR_c}{1 + tg^2\psi}$

The critical value of shear coefficient:

$$tg\,\psi_{cr} = \sqrt{\left(n\frac{R_c}{\sigma}\right)^2 + 2m\frac{R_c}{\sigma} - 1} - n\frac{R_c}{\sigma} \qquad (10)$$

where $tg\,\psi = tg\,\psi + C/\sigma$.

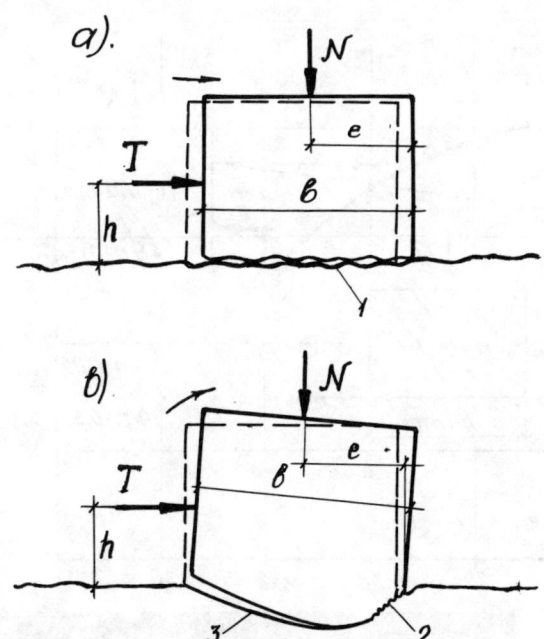

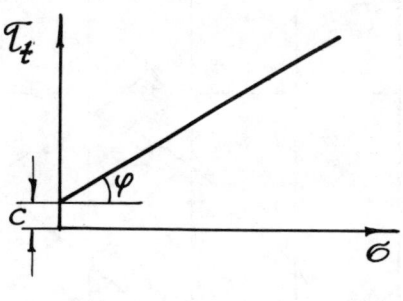

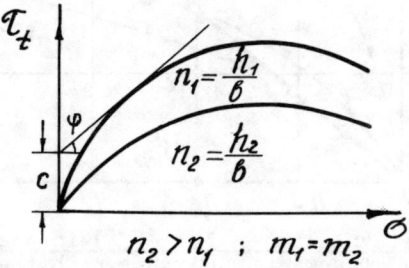

$$n_2 > n_1 \; ; \; m_1 = m_2$$

Fig. 4 Two typical schemes of failure of the
"shear block foundation systems" and
corresponding curves $\tau_t = f(\sigma)$.
 a) shear with sliding;
 b) turn with foundation crushing.

At $\sigma < \sigma_a$ and $tg\,\psi$ $tg\phi_{cr}$ shear takes place and
at $\sigma > \sigma_a$ and $tg\,\psi$ $tg\phi_{cr}$ a turn happens.

Special model tests with shear of blocks of
the different strength of their contact with
the foundation were made to verify these criteria. In particular the tests with the
shear block isolated completely from the foundation (for this purpose a steel shear block
is placed on a gypsum foundation without cementing) demonstrated the following: at small values of σ the failure always happened in the
form of sliding of the shear block on the foundation, i.e. according to the classical scheme
of sliding independent of the moment created
by the force T with relation to the center of
gravity of the footing section. In this case
the relationship $\tau_t = f(\sigma)$ is rigorously linear, i.e. Coulomb's law with parameters $tg\,\psi =$
0.8 and C = 0 (Fig. 5). However beginning from
a certain critical value of σ the failure
scheme changed drastically. The shear block
did not slide on the foundation but separated
from the pressure side and was pressed in the
foundation from the downstream side loosing
the bearing capacity at this moment. The relationship $\tau_t = f(\sigma)$ deflected sharply from
the linear form; the higher is the point of
application of the force T (the bigger is the
moment it creates), the earlier (i.e. at lower
values of σ) this deflection is observed.

Thus it is confirmed that at low values σ and
parameters $tg\,\psi$ and C a shear pattern takes
place and the linear relationship $\tau_t = f(\sigma)$
is established; at higher values a turn pattern and the non-linear relationship $\tau_t = f(\sigma)$
take place. For the turn pattern the parameters $tg\,\psi$ and C are of no physical sense because they are not constants of the material

(or the contact of two materials) and vary
widely depending on the value of the normal
stress σ . For this pattern of failure Fishman,
1979, proposed the method of calculation of
stability of concrete dams with the use of the
parameter R_c which is a constant of the material and does not depend on the value σ .
However for the purpose of maintaining the
single method of calculation of stability of
dam we may use to a cerain extent of conditio-
nality the parameters $tg\,\psi$ and C considering
them as mathematical parameters of a tangent
line and curve $\tau_t = f(\sigma)$ (Fig.4) and defining for each value of σ by the following
formula:

$$tg\,\psi = \frac{d\tau_t}{d\sigma} = \frac{mR_c - \sigma}{\sqrt{(nR_c)^2 + 2m\sigma R_c - \sigma^2}} \qquad (11)$$

$$c = \tau_t - \sigma\,tg\,\psi \qquad (12)$$

These and other experiments on different models
and in-situ demonstrated the dependence of
parameters $tg\,\psi$ and C from the loading pattern (coefficients n and m or the moment created by exterior forces (Fig. 4)). The recently adopted momentless pattern of loading of
blocks in most cases does not comply with the
loading pattern of retaining structures and
as a rule results in higher values of shear
strength parameters compared with the moment
pattern (Fishman, 1979).

In connection with the above mentioned one of
the main requirements for the procedure of in-
situ shear tests of concrete shear blocks
(Fishman, 1976) is to meet the conditions of
similarity of the tests and reality with regard to the pattern of load application. Experience in designing points to the fact that
the loading pattern of certain types of structures is adopted similar, e.g. in gravity dams
coefficients characterizing the arms of horizontal and vertical forces are $m \approx 0.7$ and
$n \approx 0.45$ which allows to standardize the experi-

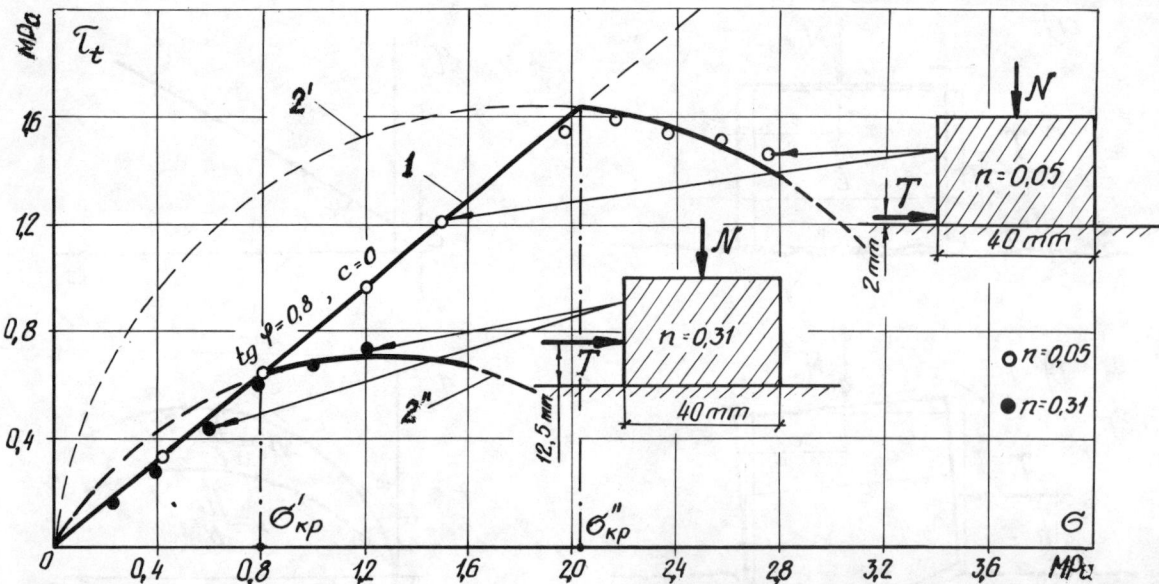

Fig. 5 Dependence curve $\tau_t = f(\sigma)$ in shear of
the block without bonds with the
foundation.
1 - linear relationship in shear;
2', 2" - non-linear relationship at turn.

ments for one and the same types of structures.
The scheme of this test carried out at the
Krapivinsky hydrodevelopment is shown in Fig.6.
If the type of the structure is unknown or
results of experiments are to be used for stability analysis of structures of different
type,,the tests shall be carried out according
to one of the schemes and then parameters $tg\psi$
and C are recalculated by formulae (11 and 12)
in which values m and n are substituted which
correspond to the loading of the structure
under design. For instance, if shears of the
shear blocks are induced according to the
momentless pattern (n = 0, m = 0.5) and at
σ =1.0 MPa the value τ_t = 3.0 MPa is obtained,
then strength parameters are:

$$R_c = \frac{\sigma^2 + \tau_t^2}{\sigma} = \frac{1^2 + 3^2}{1} = 10 МПа$$

$$tg\psi = \frac{mR_c - \sigma^2}{\sqrt{\sigma R_c - \sigma^2}} = \frac{0.5 \cdot 10 - 1}{\sqrt{1 \cdot 10 - 1^2}} = 1.33$$

$$C = \tau_t - \sigma tg\psi = 3 - 1 \cdot 1.33 = 1.67 MPa$$

For the gravity dam of 100 m in height ($\sigma \cong$
1.0 MPa) at m = 0.7 and n = 0.45 we obtain:

$$tg\psi = \frac{mR_c - \sigma}{\sqrt{(nR_c)^2 + 2m\sigma R_c - \sigma^2}} = \frac{0.7 \cdot 10 - 1}{\sqrt{(0.45 \cdot 10)^2 + 2 \cdot 0.7 \cdot 1 \cdot 10 - 1^2}} = 1.04$$

$$C = \sqrt{(nR_c)^2 + 2m\sigma R_c - \sigma^2} - nR_c - \sigma tg\psi =$$

$$= \sqrt{(0.45 \cdot 10)^2 + 2 \cdot 0.7 \cdot 1 \cdot 10 - 1^2} - 0.45 \cdot 10 - 1 \cdot 1.04 = 0.21 MPa$$

Fig. 6 Shear of the block loaded similarly
to the loading pattern of the dam
(Krapivinsky hydrodevelopment).

These are standardized or test values of
strength parameters.

The design values $tg\psi$ and C may be obtained

by introduction (according to the construction code, 1977) of the safety factor K_φ = 1.15 and K_c = 1.8 (or in case of a sufficiently large number tests they may be obtained by statistic treatment):

$$\text{tg}\,\varphi = 0.9 \text{ and } C = 1.2 \text{ kg/cm}^2$$

As may be seen from above the given procedure allows to obtain strength parameters by the results of shear of each separate shear block with the following statistic treatment.

REFERENCES

Construction Code (SNIP II - 16-76), 1979, "Foundations of hydraulic structures", Moscow.

Fadeev A.B. (1972). Crushing and seismic effects of explosions in borrow pits. Nedra, Moscow.

Fishman Yu.A. (1974). Development of mechanics of rock masses and investigations of rock foundations of hydraulic structures in the USSR, "Transactions of Hydroproject", N° 33, Moscow.

Fishman Yu.A. (1976). Investigation procedure of shear resistance of concrete blocks on rock foundations. "Transactions of Hydroproject", N° 50, Moscow.

Fishman Yu.A. (1979). Investigations into the mechanism of the failure of concrete dams rock foundations and their stability analysis. Proc. of the IV Congress of the ISRM, vol. II, Montre.

Fishman Yu.A. (1982). Optimization of investigations on the basis of analysis of design solutions. "Gidrotechnicheskoye stroitelstvo", N° 2, Moscow.

Talobre J. (1957). La Mécanique des Roches, Paris.

Ternovsky I.N., Ukhov S.B. (1977). Method of determination of coefficient of lateral deformation of fissured rocks. "Transactions of Hydroproject", Issue 50, Moscow.

Tsytovich N.A., Ukhov S.B. (1970), et al. To the problem of determining the modulus of deformation of in-situ fissured rock masses. Proc. Of the II Congress of the ISRM, Beograd.

Tsytovich N.A., Ukhov S.B., Burlakov V.N. (1970). Failure mechanism of fissured rock dam upon displacement of loading plate. Proc. of the II Int. Congress on Rock Mechanics. Beograd.

Ukhov S.B., Kubetsky V.L., Fishman Yu.A., Lapin L.V. (1972). In-situ investigations of bearing capacity of rock foundations. Transactions of coordination meeting on hydraulic engineering, VNIIG, Issue 77, Leningrad.

Ukhov S.B. (1975). Rock foundations of hydraulic structures, "Energy" Publishers, Moscow.

Ukhov S.B. (1979). Principles used in developing geomechanical models of rock masses for solving engineering problems. Proc. of the IV Congress of the ISRM, vol. II, Montreaux.

ESSAIS GEOSTATIQUES SUR UN MODELE DE TUNNEL DANS UN MILIEU DISCONTINU

Geomechanical model tunnel testing in discontinuous media

Geotechnische Modelluntersuchungen eines Tunnels im Diskontinuum

Z. LANGOF

Z. Langof
Institut de Géotechnique et de Fondations, Sarajevo, Yougoslavia

RESUME

Grâce aux modèles surface plane, nous avons effectué des essais portant sur l'inclinaison des couches, sur la réparti-tion des pressions souterraines, sur le revêtement de tunnel, sur les contraintes ainsi que sur la déformation d'un mas-sif rocheux. Le modèle du massif rocheux présentait les dimensions de 210 × 156 × 12 cm et était composé de blocs de 2 × 2 × 12 cm. Le modèle reconstitue un tunnel percé à une assez grande profondeur et dont le poids des couches au-dessus du tunnel est remplacé par des charges correspondantes.

SYNOPSIS

The influence of the slope on layers, using geomechanical models, has been tested in terms of the underground pressure distribution on tunnel linings as well as stresses and deformations of rock masses. The rock mass model was 210 × 156 × 12 cm and was reproduced using 2 × 2 × 12 cm blocks. The model represents a case of a tunnel found at a sufficiently great depth where the weight of overhead material was replaced by corresponding loads.

ZUSAMMENFASSUNG

Mittels geotechnischer Modelle in der Ebene wurde die Untersuchung der Einwirkung der Schichtgefälle auf die Verteilung von den unterirdischen Drücken, auf die Tunnelauskleidungen, sowie auf die Spannungen und die Verformungen von Fels durchgeführt. Das Modell der Felsmasse hatte die Abmessungen 210 × 156 × 12 cm und wurde mit Blöcken in der Grösse von 2 × 2 × 12 cm reproduziert. Das Modell stellt einen Fall eines sich in größerer Tiefe befindlichen Tunnels dar, bei dem das Gewicht der Überlagerung durch entsprechende Belastungen ersetzt wurde.

1. INTRODUCTION

Le problème de la détermination des pressions sou-terraines sur le revêtement de tunnel ainsi que les phénomènes apparaissant autour de l'ouverture du tunnel ont été traités sur modèles géostatiques. Nous avons effectué des modèles - surface plane, où le massif rocheux était reproduit comme un milieu dis-continu anisotrope.

Nous avons examiné une série de quatre modèles dont chacun présentait une inclinaison des couches différe-nte /0°, 30°, 60° et 90°/.

En ce qui concerne les modalités et conditions des essais, les quatre modèles étaient identiques, les différences n'apparaissant qu'en ce qui concernait l'inclinaison des couches. En reproduisant les modè-les de cette manière-là il nous était possible de ré-aliser le but principal des essais: examen portant sur l'influence de l'inclinaison des couches sur la répartition des pressions souterraines sur le revête-ment de tunnel ainsi que sur les déplacements et les

contraintes dans le massif rocheux.

La conception des essais sur modèles consistait en établissement préalable de l'état de contrainte pri ma-ire dans le modèle, après quoi nous relâchions le revêtement de tunnel pour pouvoir, de cette maniè re--là, obtenir les déplacements et les états de contra-intes secondaires dans le massif rocheux ainsi que les pressions sur le revêtement de tunnel.

2. DISPOSITION DU MODÈLE

Nous avons examiné le cas du tunnel construit assez profondément où la charge appuyant sur le haut du tunnel a été remplacée par une charge adéquate. Le modèle a été construit dans une surface verticale et situé dans un châssis d'acier dur, présentant les di-mensions de 210 x 160 x 120 cm. Le châssis du mo-dèle était ouvert du côté avant, et fermé du côté arrière par une tôle en acier.

Fig. 1. Vue générale sur le modèle pendant les essais

Nous avons obtenu l'état de contrainte primaire grâce à deux systèmes indépendants de presses hydrauliques. La transmission de la charge sur le modèle a été opérée grâce à un système de poutres, de rouleaux et d'une épaisse couche de caoutchouc perforé. Pour assurer l'appui et l'orientation de la transmission de la charge nous nous sommes servis d'un cadre particulier, situé sur le pourtour du modèle /fig. 1./.

Le modèle du tunnel présentait une forme circulaire /diamètre de 45 cm/ et était composé de six parties, mutuellement liées et constituant un tout. Certaines de ces six parties peuvent indépendamment des autres se déplacer vers le centre du tunnel, ce qui en réduit le rayon /fig. 2./.

Le modèle du tunnel s'effectuait simultanément avec la simulation du massif rocheux. L'excavation du tunnel dans le massif rocheux a été reproduite grâce au déplacement graduel du revêtement vers le centre du tunnel.

Pour que le tunnel ne représente pas un corps étranger dans le massif rocheux, pendant l'établissement de l'état de contrainte primaire la déformabilité du tunnel correspondait approximativement à celle du massif rocheux.

Le modèle du massif rocheux était fait en blocs de 2 x 2 x 16 cm, qui dans le sens longitudinal étaient armés au moyen des fibres "de glace". En vue de réduire le frottement nous avons mis entre les couches une feuille en matière plastique.

Fig.2. Modèle du revêtement de tunnel et du massif rocheux après les essais

Nous avons effectué sur modèle les mesures suivantes:
- déplacement du massif rocheux - 52 endroits,
- déformations spécifiques du massif rocheux - 32 endroits, en vue d'obtenir les contraintes secondaires,
- contraintes secondaires dans le massif rocheux - au moyen de 30 cellules de contraintes,
- pressions radiales sur le revêtement de tunnel - au moyen de 16 cellules de contraintes,
- réduction du diamètre du revêtement de tunnel.

3. DÉTERMINATION DES CARACTÉRISTIQUES MÉCANIQUES DU MODÈLE DU MASSIF ROCHEUX

Les caractéristiques de déformations ont été obtenues grâce aux "prélèvements" du massif rocheux examinés /dimensions: 30 x 30 x 12 cm/ avec une charge biaxiale.

Le principe de base de ces essais était d'assurer simultanément une charge sur les "prélèvements" et sur le modèle. A cause des inclinaisons des couches différentes les procédés de la charge dans le sens vertical et parallèle aux couches présentaient certaines différences chez les modèles particuliers.

Il importe de souligner que le point de départ était toujours l'état de contrainte primaire que l'on obtenait toujours de la même manière.

Les essais ont été effectués en trois phases. Lors des deux premières phases nous n'avons pas dépassé les contraintes primaires, ces deux phases ayant pour but "l'entraînement" du modèle, tandis que pendant la troisième phase la contrainte a dépassé les contraintes primaires.

Le module de déformations /D/ est obtenu grâce aux charges dépassant celles primaires et le module d'élasticité grâce aux "décharges" /E/ /fig. 3./.

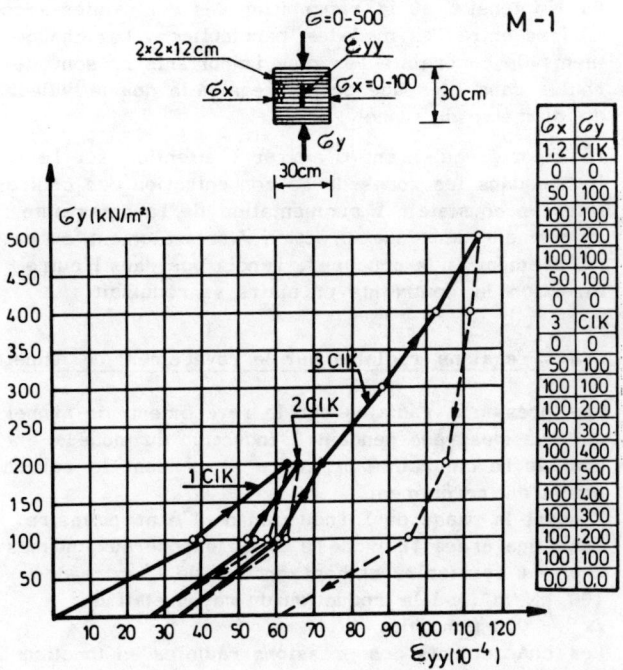

Fig. 3. Examen des caractéristiques de déformations sur "le prélèvement" pris dans le massif rocheux

La résistance au cisaillement a été examinée également sur les "prélèvements" /dimensions: 15x30x12 cm/ qui présentaient les inclinaisons des couches différentes. Les essais ont démontré que la résistance ne dépendait pas uniquement de l'inclinaison des couches mais aussi de la direction où le cisaillement se produit.

4. MÉTHODOLOGIE DES ESSAIS SUR MODÈLES

Une fois le modèle construit et les instruments de mesure installés, le modèle était induit en état de contrainte primaire en trois phases. L'objectif des deux premières phases était uniquement de provoquer dans le modèle l'état de comportement quasi-élastique. Les trois phases ont été identiques et correspondaient aux essais effectués sur "les prélèvements" /fig. 3/.

Les données offertes par la troisième phase nous ont permis d'obtenir les caractéristiques de déformations pour chacun des endroits de mesure sur modèle. Outre cela, pendant la troisième phase nous avons simultanément effectué les mesures des cellules de contraintes "in situ", ce qui nous a permis d'obtenir les correspondants coefficients de correction pour chacune des cellules.

Tous les modèles présentaient les mêmes relations des contraintes primaires /λ = 0,5/, c'est-à-dire la contrainte horizontale était de 100 kN/m^2 et celle verticale de 200 kN/m^2.

La relation des contraintes primaires de l'ordre de 0,5 a été adoptée en vue de mettre plus en jour l'influence de l'inclinaison des couches. Ayant obtenu la contrainte primaire de la troisième phase nous avons procédé au relâchement graduel du revêtement de tunnel, c'est-à-dire à la réduction du rayon du tunnel /R/, en étapes de 1,0 mm /0,44%/

5. RÉSULTATS DES ESSAIS

5.1. Déplacements du massif rocheux

L'"excavation" de l'ouverture du tunnel, c'est-à-dire la réduction du rayon du revêtement du tunnel (ΔR) en étapes de 1,2,3,4 et 5 mm, a provoqué les déplacements du massif rocheux. Le caractère et l'importance des déplacements du massif rocheux ont été bien différents chez les modèles particuliers /fig. 4./.

Si l'on fait l'analyse comparative des résultats des déplacements alors on peut conclure qu'il existe une limite en ce qui concerne l'inclinaison des couches, où l'on constate de considérables changements qualitatifs du caractère des déplacements.

Chez les deux premiers modèles, présentant l'inclinaison des couches de 0° et 30°, les vecteurs principaux des déplacements étaient approximativement orientés verticalement par rapport aux couches, et chez les deux autres modèles /inclinaison des couches de 60° et 90°/ les vecteurs principaux des déplacements étaient approximativement orientés parallèlement à l'inclinaison des couches.

De là dé découle que la limite mentionnée se trouve

G

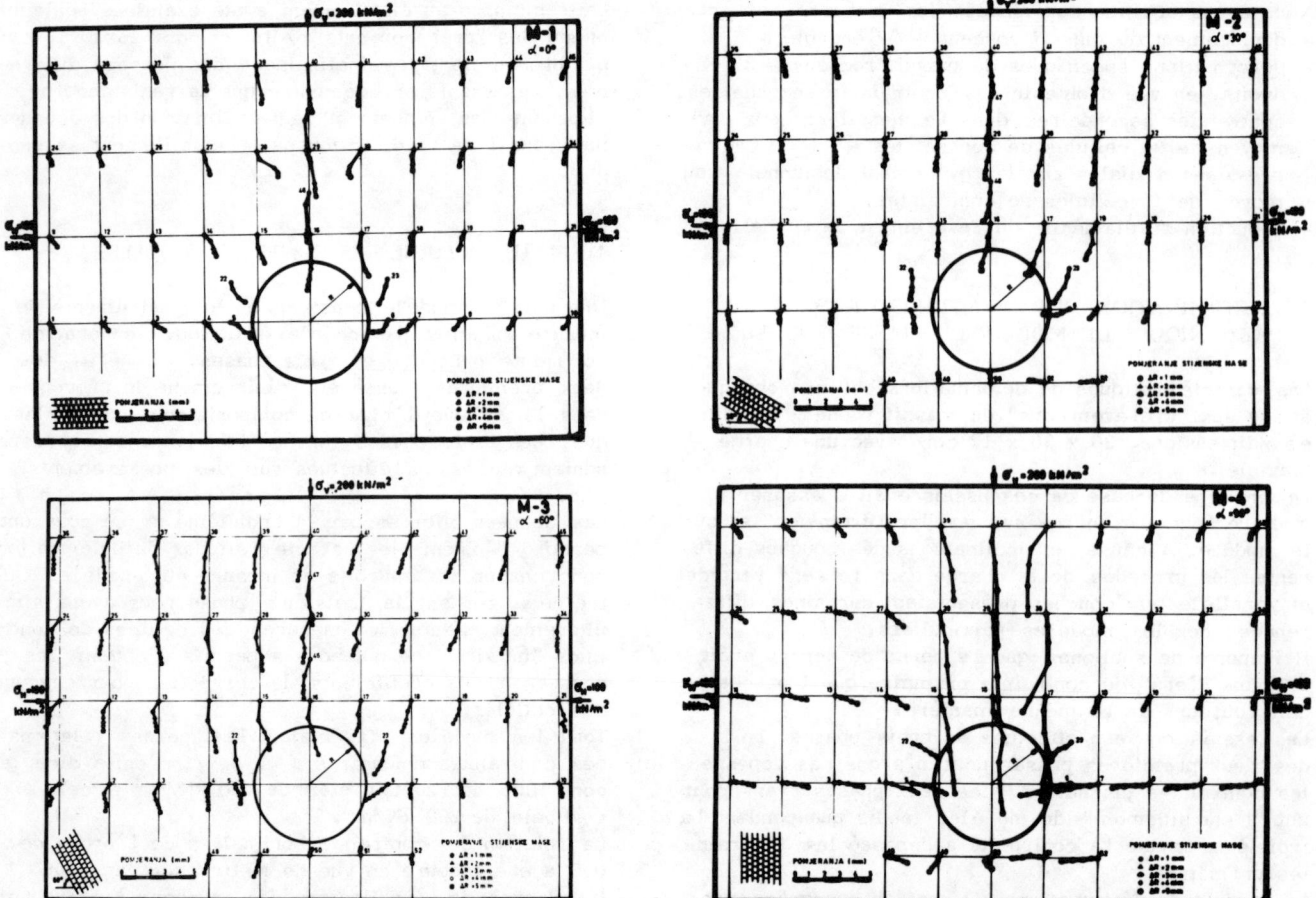

Fig. 4. Déplacement du massif rocheux en fonction de la réduction du rayon du revêtement de tunnel

chez les modèles examinés entre 30° et 60°.
Ici il importe également de souligner que l'on a enregistré chez les quatre modèles d'importants déplacements dans le sens de l'axe horizontal du tunnel.

5.2. Etats de contraintes secondaires dans le massif rocheux

Pour des raisons quinematique et à cause du glissement le long des couches il nous était impossible /partant des relations bien connues entre les contraintes et les déformations que l'on utilise quand il s'agit des milieux continus /d'obtenir les valeurs réelles des contraintes principales pour les milieux discontinus, tels qu'ils étaient dans les modèles. Nous n'avons pu l'obtenir ni au moyen des cellules de contraintes. C'est pourquoi nous avons examiné les contraintes componentales dans deux directions mutuellement verticales, qui correspondent aux axes principaux de symétrie /parallèlement ou verticalement par rapport aux couches/.
En vue de permettre la comparaison des contraintes secondaires dans les modèles particuliers nous avons représenté sur la figure 5 les contraintes secondaires componentales dans les surfaces horizontales et verticales.

Il est évident qu'il existe de considérables différences

vu l'intensité et la répartition des contraintes secondaires entre les modèles particuliers. Les changements de contrainte les plus importants se sont déroulés dans l'espace qui est égal à la double valeur du diamètre du tunnel /2 R/.

Il importe également d'attirer l'attention sur le fait que dans les zones de la concentration des contraintes on constatait l'augmentation de la contrainte uniquement dans une direction /verticalement ou parallèlement aux couches/, tandis que dans l'autre direction la contrainte primaire se réduisait.

5.3. Pressions radiales sur le revêtement de tunnel

Les pressions radiales sur le revêtement de tunnel ont été mesurées pendant l'induction du modèle en état de la contrainte primaire et pendant le relâchement du revêtement.
Pendant la phase de l'induction de l'état primaire, dans une étape le modèle a été exposé aux mêmes charges verticales et horizontales de l'ordre de 100 kN/m², où la réduction du rayon était de $\Delta R = 0,5$ mm /fig. 6/.
Les changements des pressions radiales en fonction de la réduction du rayon sont représentés sur la figure 7. Les pressions radiales différaient considérablement d'un modèle à l'autre et dépendaient largement

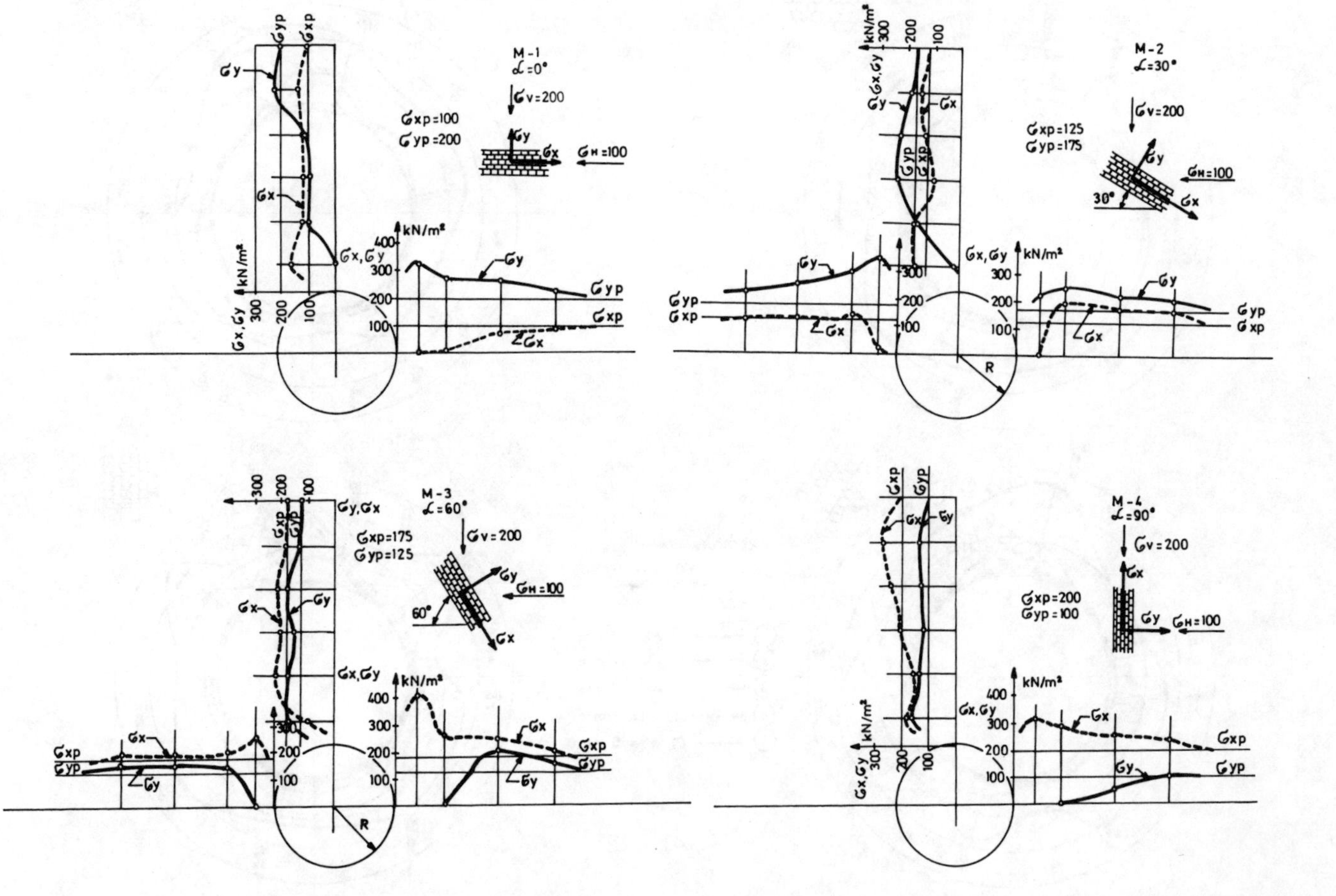

Fig. 5. Contraintes secondaires dans le massif rocheux dans les directions verticale et parallèle aux couches

de l`importance de la réduction du rayon. En ce qui concerne la répartition des pressions sur le revête- ment on peut soutenir que l`on a constaté chez tous les modèles de considérables pressions latérales, et que ces répartitions étaient asymétriques chez les modèles présentant des couches inclinées.

D`une manière générale, les pressions sur le revête- ment présentaient des intensités relativement petites et se réduisaient rapidement avec la réduction du ray- on. L`unique exception est présentée par le modèle aux couches verticales où la pression sur le revête- ment s`est maintenue même après la réduction du rayon de 2,22% /fig. 8/.

5.4. Formes et importance des zones de la "déchar- ge" et concentration des contraintes autor de l`ouverture du tunnel

Partant de tous les résultats des essais portant sur l`influence de "l`excavation" du tunnel dans le massif rocheux nous avons fait des diagrammes sommaires - fig. 9. Les données de la figure 9 peuvent servir du point de départ permettant de résumer le compo- rtement du massif rocheux des modèles particuliers et de procéder à une comparaison mutuelle.

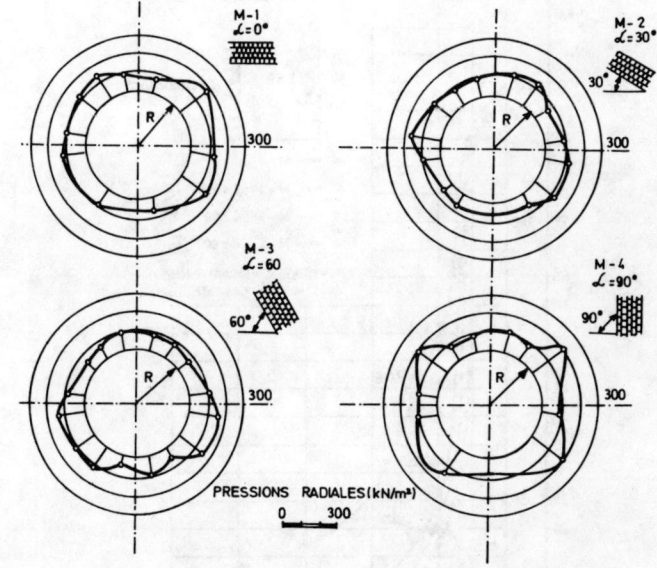

Fig. 6. Pressions radiales sur le revêtement pour les états de $\sigma_v = 100$ kN/m^2 et $\sigma_h = 100$ kN/m^2, et pour la réduction du rayon du revêtement $\Delta R = 0,5$ mm /$\Delta R = 0,22\%$/

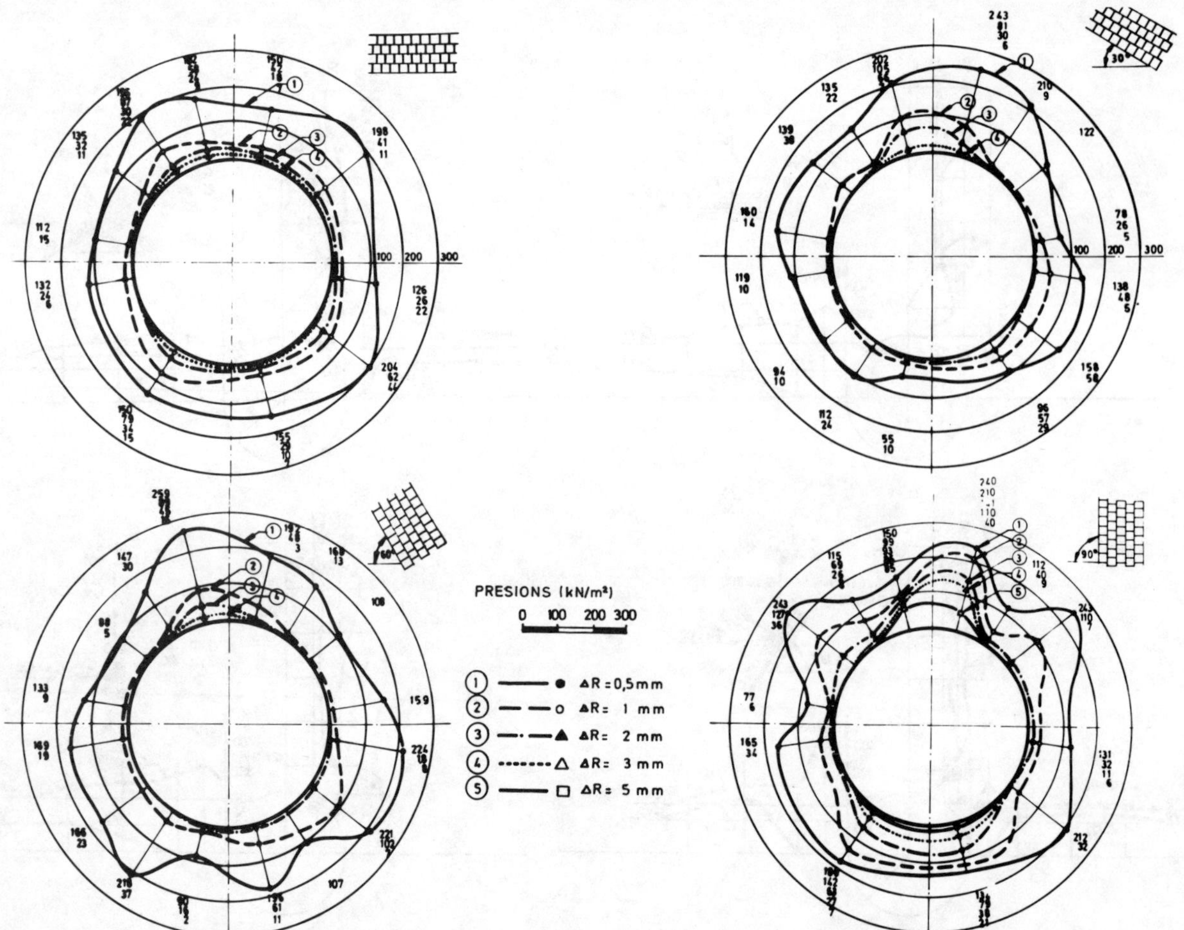

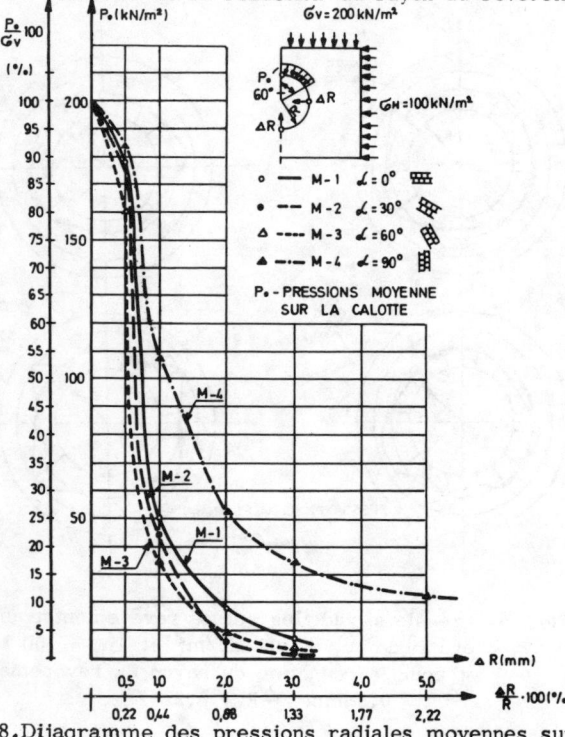

Fig. 7. Pressions radiales sur le revêtement pour l'état de σ_v = 100 kN/m^2 et σ_h = 200 kN/m^2, en fonction de la réduction du rayon du revêtement de tunnel

PRESIONS (kN/m^2)

0 100 200 300

① ——— ● ΔR = 0,5 mm
② — — — ○ ΔR = 1 mm
③ —·—·— ▲ ΔR = 2 mm
④ ········· △ ΔR = 3 mm
⑤ ——— □ ΔR = 5 mm

Fig.8.Dijagramme des pressions radiales moyennes sur la calotte en fonction de la réduction du revêtement

P_o - PRESSIONS MOYENNE SUR LA CALOTTE

○ ——— M-1 $\angle$ = 0°
● — — — M-2 $\angle$ = 30°
△ ---- M-3 $\angle$ = 60°
● —·— M-4 $\angle$ = 90°

Chez les modèles à la composition symétrique /M-1, M-4/ l'augmentation des contraintes primaires était maximalement de 100%, et chez les modèles aux couches inclinées inclinées /M-2, M-3/ il y avait des zones où les contraintes primaires ont augmenté de plus de 100%.

Les valeurs minima des zones de la "décharge" ont été constatées chez le modèle M-2, et les plus importantes chez le modèle M-4.

Chez les modèles aux couches inclinées nous avons obtenu des répartitions particulièrement asymétriques. Les zones des grands écroulements dans la région de la calotte sont similaires chez les modèles particuliers, bien que le modèle M-4 ait présenté des déplacements considérablement plus importants.

Le trait caractéristique commun des quatre modèles c'est que les changements les plus importants, aussi bien en ce qui concerne la "décharge" que les zones de concentration des contraintes, se sont déroulés dans l'espace présentant le rayon de 2 R.

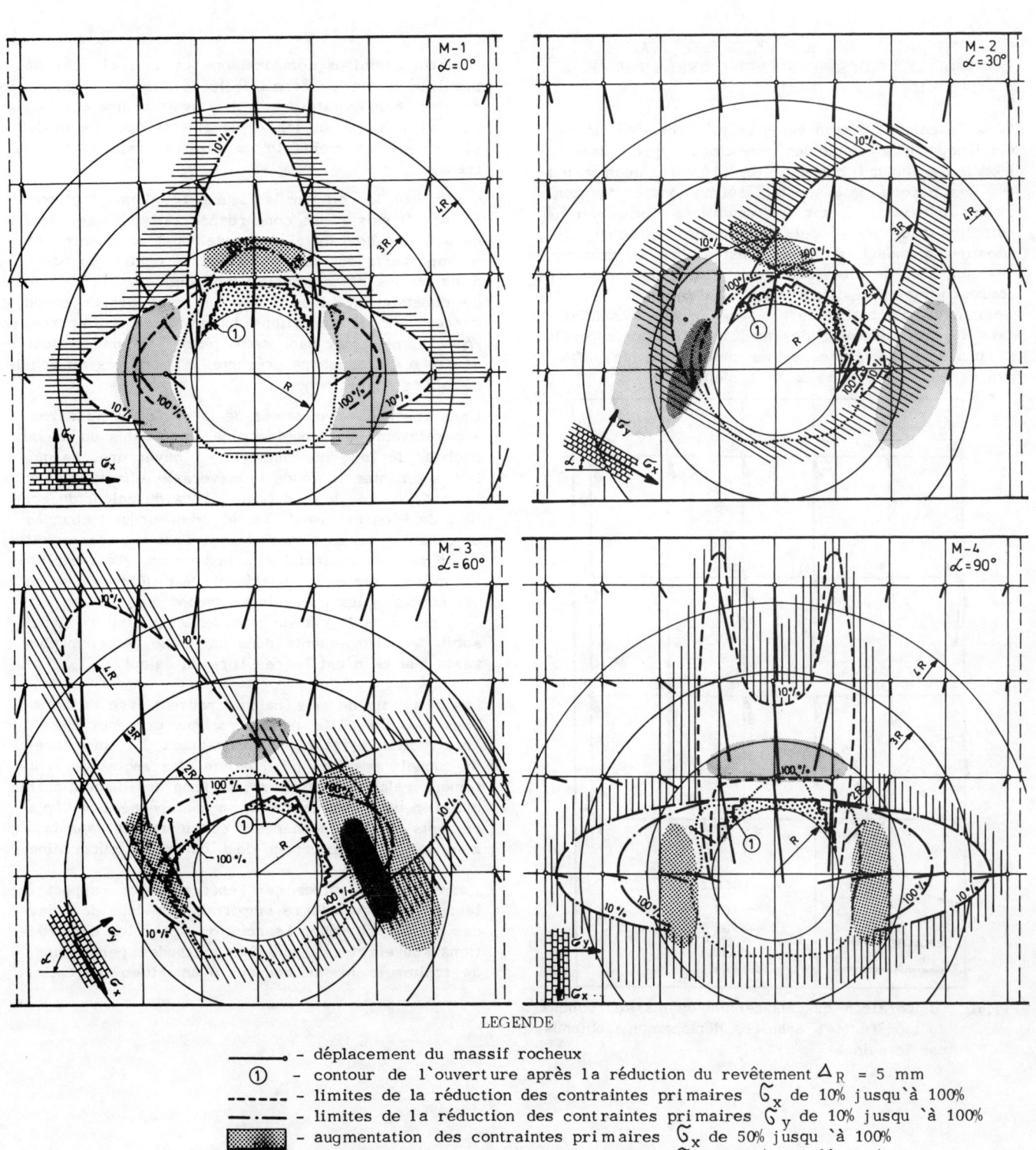

LEGENDE

_____ - déplacement du massif rocheux

① - contour de l'ouverture après la réduction du revêtement Δ_R = 5 mm

— — — — - limites de la réduction des contraintes primaires σ_x de 10% jusqu'à 100%

— · — · — - limites de la réduction des contraintes primaires σ_y de 10% jusqu'à 100%

▨ - augmentation des contraintes primaires σ_x de 50% jusqu'à 100%

▨ - augmentation des contraintes primaires σ_y de 50% jusqu'à 100%

■ - augmentation des contraintes primaires de plus de 100%

▨ - zones du libre écroulement des blocs

·········· - pressions radiales sur le revêtement.

Fig. 9. Graphique des zones de la "décharge", de la concentration des contraintes, des zones du libre écroulement des blocs, des pressions radiales sur le revêtement et des déplacements du massif rocheux.

B 85

6. COMPARAISON DES RÉSULTATS DES ESSAIS SUR MODÈLES ET DES RÉSULTATS OBTENUS PAR LE CALCUL

Dans le calcul effectué selon la méthode des éléments finaux on a simulé les conditions régnant dans le modèle en ce qui concerne les éléments géométriques, les caractéristiques de déformations et des conditions d'essais. Toutes les conditions citées ont été remplies sauf une et cela essentielle: caractère discontinu du massif rocheux, c'est-à-dire la présence des surfaces de glissement potentielles présentant les correspondantes résistances au cisaillement.
Comme les discontinuités ont joué un rôle essentiel dans le comportement des modèles, il était superflu de procéder aux comparaisons mutuelles, les différences ayant été grandes et évidentes.

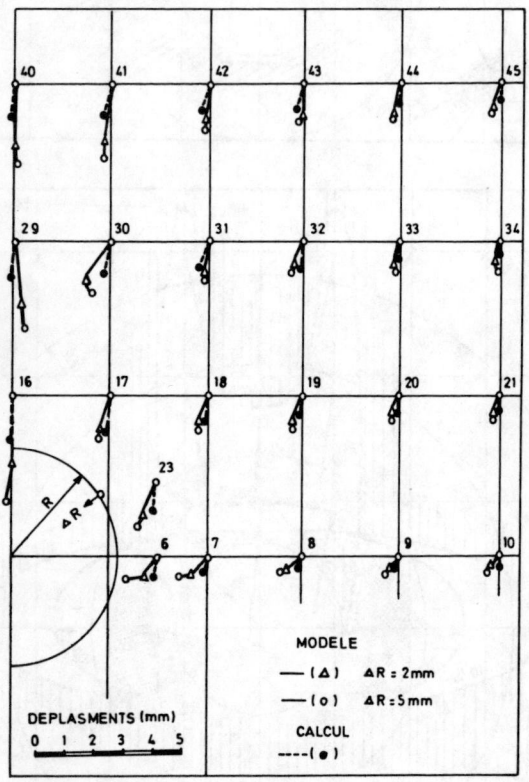

DEPLASMENTS (mm)
0 1 2 3 4 5

MODELE

— (△)   △R : 2mm
— (o)   △R : 5mm

CALCUL
----(●)

Fig.10. Comparaison du déplacement du massif rocheux du modèle M-1 avec les déplacements obtenus par le calcul

Pourtant certaines comparaisons et la recherche des similitudes ont pu être effectuées dans le modèle aux couches horizontales, car il y avait là une composition symétrique et les déplacements chez ce modèle-là ont été les moins importants par rapport aux autres.

Comme on le voit sur la figure 10, dans la plupart points il existe une concordance satisfaisante, sauf dans l'axe horizontal du tunnel où l'on a enregistré d'importants déplacements vers le centre du tunnel à cause des discontinuités horizontales. Il importe de constater que l'on trouve une plus grande concordance chez les réductions du rayon moins importantes /△R= 2 mm/, car dans cette phase le contour de l'ouverture du tunnel ne présente toujours pas de considérables changements.

Dans les phases suivantes de la réduction du rayon apparaissent des troubles plus importants du massif rocheux le long du contour de l'ouverture, ce qui fait augmenter la coupe transversale effective dans le modèle. Les axes opposes, lors du calcul du contour de l'ouverture du tunnel, demeurent inchangés et capables de supporter les contraintes tangentielles ainsi que les contraintes à la tension. Cela explique en même temps de considérables différences en ce qui concerne les contraintes secondaires. Il est évident que dans le cas du modèle le massif rocheux subit des changements dans une zone beaucoup plus vaste que ce n'est le cas lors du calcul.

Les résultats de tels calculs peuvent être utilisés dans les cas où le massif rocheux présente une construction symétrique par rapport à l'ouverture du tunnel, ensuite, où il existe des angles de frottement relativement élevés le long des discontinuités ou bien dans les cas où l'on a empêché des déplacements et des "décharges" considérables dans la zone autour du pourtour de l'ouverture suterraine.

Dans tous les autres cas les écarts par rapport à la réalité peuvent être importants et dans de tels cas il faut chercher la solution, pour les constructions souterraines, dans les méthodes permettant de traiter le massif rocheux d'un milieu discontinu.

AN APPROACH TO ROCK MASS CLASSIFICATION FOR UNDERGROUND WORKS

Essai de classification des masses rocheuses pour la construction d'installations souterraines

Der Versuch einer Gebirgsklassifikation für Untertagebauten

P. Lokin
Faculty of Mining and Geology, Belgrade University, Yugoslavia

R. Nijajilović
Centroprojekt Design Company, Belgrade, Yugoslavia

M. Vasić
Faculty of Technical Sciences, Novi Sad University, Yugoslavia

SYNOPSIS

The paper first analyses existing rock mass classifications. The authors propose a classification which is developed in stages, similar to the evolution of the design. Classification should start right from the choice of site and be developed all the way through to the final completion of the work. All the stages of classification together must constitute an integral system. In each successive stage the classification is filled in with new, more reliable and specific data, with more and more quantitative parameters, as produced by current investigations and required by the current stage of design. Thus its character changes from the most general in the initial stage to become completely specific at the end.

RESUME

Cette communication commence par analyser les classification existantes. Ensuite les auteurs proposent une classification en plusieurs phases. Elle devrait être commencée dès le choix de la location des installations et finir à la fin de la construction. Toutes les phases de la classification doivent présenter un système unique. A chaque nouvelle phase on doit ajouter de nouvelles données plus sûres et plus précises avec des paramètres de plus en plus quantitatifs: autant que les recherches en cours le permettent et que les besoins progressifs du projet le demandent. De cette manière le caractère de la classification se changerait en devenant de plus en plus précis au fur et à mesure de l'avancement du projet.

ZUSAMMENFASSUNG

In dieser Arbeit werden die bestehenden Methoden der Gebirgsklassifikation analysiert. Die Autoren schlagen vor, daß die Klassifikation in Phasen gemacht wird, sodaß sie sich dem Entwurf ähnlich entwickelt. Mit der Klassifikation sollte schon bei der Wahl des Bauortes begonnen werden, und am Ende des Bauens sollte sie fertig sein. Alle Phasen der Klassifikation müssen ein System darstellen. In jeder folgenden Phase wird die Klassifikation mit immer mehr quantitativen Parametern ergänzt, sodaß der Charakter der Klassifikation dem konkreten Objekt immer näher kommt.

1. INTRODUCTION

Underground construction engineering has for some time now been a rapidly expanding field. Apart from the ever increasing need for road and water tunnels ensuing from the ramification of highway networks and the rapid multiplication of hydroelectric plants, many other works are now built underground: underground railways, subways warehouses, shelters, powerhouses, recreation facilities, etc. Serious consideration is already being given to the construction of entire underground communities (F.Moreland, 1977).

The increasing scope and diversity of underground engineering projects has elicited advances both in investigation and desing methods and in excavation and building technology. A specific feature of underground works compared with most other projects is their very great dependence on geological conditions, so that the investigation and forecasting of these conditions is especially important.

State-of-the-art desing practices for underground works are based shiefly on classifications of rock masses, the results of site investigations, including in situ tests, and mathematical stress/strain models. These three groups of methods in fact characterize the three main design approaches. Howeren, none of them by itself can provide suficiently reliable data at the right time for design purposes. In practice they are therefore usually combined and integrated. This may also be seen as the most likely trend in the future development of investigation and design methodology.

This paper critically reviews some of the most recent rock mass classifications, and presents a new approach which in our judgement would greatly augment the applicability of site investigation data.

2. A CRITICAL REVIEW OF EXISTING ROCK MASS CLASSIFICATIONS

An analysis of various classifications which have found wide application in underground engineering projects reveals that despite their indisputable value they all have certain shortcomings which limit their usefulness.

One of the oldest classifications, widely used especially in the USA, is that of Terzaghi (1946). It is used mostly in the case of steel arch supporting, but is not applicable with more modern excavation methods. More recent classifications are adapted to the so-called New Austrian Method of tunneling, using shotcrete, iron mesh and rock bolting.

The classifications of Rabcewicz (1957), Lauffer (1958), and Pacher, Rabcewicz and Gosler (1974) introduce the excavation stand-up time into the classification scheme.

In 1970 Deer proposed a classification for application in underground engineering projects in which rock mass quality is estimated from the length of core segments yielded by drilling (RQD). This parameter has been included in the great majority of later classifications.

In 1972 Wickham proposed a classification based on a number of parameters, to each of which he assigned numerical values. He recommends the kind of support to be used in different classes of rock mass.

Important innovations in the classification system were introduced by Bieniawski (geomechanical classification) and by Barton, Lien and Lunde (Q-system). These two are also the corrently most comprehensive classifications, derived for the most part from analysis of ground conditions during the construction of a large number of tunnels. They use quantitative criteria of rock mass properties. Bieniawski's criteria are the following: uniaxial strength, core quality (RQD) joint density, joint orientation properties of joint surfaces, water inflow into the excavation. The classification of Barton et al. (Q-system) uses quantitative criteria of the following properties: core quality (RQD), number of joint sets, roughness index of joints, index of joint weathering, reduction factor for the influence of water in joints, stress reduction factor.

The first step in applying one of these classifications (either geomechanical or Q-system) is the visual differentiation of quasi-homogeneous zones with approximately the same quality of rock mass. The length of quasi-homogeneous zones has to be decided with reference to the variability of rock mass properties, but also to the economical use of appropriate supporting.

After a critical examination of Barton et al.'s classification, Bulichev (1977) proposed a very similar one based on eight indices, of strength, jointing, groundwater, and the dimensions and orientation of the excavation.

Table I

Elements of Classification - Properties of the Rock Mass (Ground)			TERZAGHI 1946	RABCEVICZ 1957 LAUFFER 1958 PACHER, RABCEWICZ, GOSLET 1974	DEER 1970	BIENIAWSKI 1973 (geomch.)	BARTON, LIEN, LUNDE 1974 (Q-system)	LOUIS 1974	BULICHEV 1977
DATA ON ROCK MASS	Jointing	1 Overral quality estimate	+	+					
		2 Number of joint sets					+		+
		3 Joint density				+			
		4 Rock quality (RQD)			+	+	+	+	
		5 Joint orientation				+			+
		6 Joint roughness				+	+		+
		7 Gap width, filling, weathering				+	+		+
	Mechanical Properties	8 Uniaxial strength				+	+		
		9 Uniaxial stren.und.point load						+	
		10 General strength index (f)							+
	Stress States	11 Initial stresses					+		
		12 Secondary stres.due to excav.	+				+		
		13 Swelling stresses					+	+	
	Groundwater	14 Permeability						+	
		15 Groundwater inflow				+	+		+
		16 Groundwater pressure				+	+		
DATA ON THE WORK		17 Type and purpose					+		
		18 Shape and dim.of cross section	+	+					+
		19 Stand-up time		+					
		20 Unsupported span		+					
APPLICATIONS		21 Estimating stab.of the excav.	+	+	+	+	+	+	+
		22 Prediction of unsuppor.span				+	+		
		23 Prediction of stand-up time				+	+		
		24 Choice of excavation techn.						+	
		25 Choice of permanent supporting	+	+	+	+	+	+	+

Apart from the RQD quality index, Louis also introduces a strength index from which the mechanical behaviour of the rock mass can be forecast and the excavation technology chosen. To estimate the influence of groundwater on rock mass behaviour the following properties are taken into account: permeability, water pressure and swelling.

The above classifications, especially those of Barton et al. (Q-system) and Bieniawski, were developed upon experience gained in the building of a large number of tunnels and other underground works. Their application is generally very simple and quick, since the parameters involved are easily determined and quantitatively expressed. In using the classifications of Rabcewicz and Lauffer, considerable experience is needed to be able to predict the time for which an excavation of given dimensions will he stand up. Barton s classification, in contrast to the majority of the others, is developed in detail for a large number of possible cases, and the basic classes are subdivided into corresponding subclasses.

The application of these classifications yields data of great importance for design decision --making, as for instance: whether or not iron mesh has to be used; the thickness of shotcrete needed; the type, density and length of rock bolts; the thickness of concrete lining required; the maximum unsupported span; the time for which the excavation will stand up, and with Louis s classification, what mode of excavation to use.

Table 1 shows what elements enter into the different classifications and for what purposes they are appropriate.

A careful analysis reveals that most of the classifications neverteless have certain shortcomings which limit their applicability. The principle shortcomings are:

- Most of them do not take into account specific features of the work, except the Q-system classification which takes into account the type and purpose of the work in terms of the ESR factor. This factor has not, however, been sufficiently verified in practice and will need to be adapted and improved as more experience is gained with it.

- Some classifications have emerged from engineering experience under specific geological conditions, making them less applicable to cases where the geology is essentially different. In other words, the specific features of the ground within which the work is to be built are not given sufficient weight.

The small number of parameters which enter into some of the classifications, e. g. Deer s, are not able to fully express all the rock mass properties relevant to the engineering task. Little use is made of in situ testing, as for instance of: rock mass deformability parameters convergence and extensometer measurements, shear strength parameters, natural stresses, etc.

- The classifications are not appropriate to special excavation techniques, such as smooth blasting, mechanical without blasting, etc.

- They do not offer solutions for specific problems, e.g. large unstable blocks. Some of them have only been developed for compact rock.

- Most of the classifications consider only the stability of the excavation. To take into account other geotechnical factors they would have to be modified. For example, Barton remarks that for a general classification the

first four parameters of the Q-system could be retained, while the remainder would have to be amended. Louis s classification does take into account the mode of excavation, but only in a rather general way.

- The classifications are not adapted to the sequence of desing stages: they are generalized and based mostly on data obtained from mapping, boring and only the simplest field and laboratory tests. Considering the reference data for the classifications and the kind of results they can yield, one may say that they are appropriate roughly to the phase of preliminary desing.

3. A NEW APPROACH TO THE CLASSIFICATION OF ROCK MASSES

3.1 Basic Principles

In the construction of underground works the basic factors related to the ground as an engineering environment are as follows:

- stability of the excavation
- stability of the work
- excavation conditions
- drainage and ventilation conditions
- the possibility of interaction between the rock and groundwater and materials built into the work or stored in it
- the influence of construction on the surrounding ground and other structures.

All these factors, usually grouped under the common heading of "geotechnical factors", are the subject of appropriate site investigations. A rock mass classification should be able to give the necessary assistance in solving problems relating to these factors. Classifications used at the present time can be considered as a synthesis of all investigation results. Hence the procedure classification, and the utilization of the results it yields, should be viewed in terms of the entirety of all the site investigations. We therefore take as our point of departure the basic principles of site investigation. In order to yield appropriate results these investigation must be planned and executed in accordance with the following principles: investigation stage-by-stage, completemess, uniformity, economy.

The stages of investigation should be synchronized with the stages of design and construction. It is useful to have relevant classification data on hand throughout design andconstruction, and not only at certain stages of design as has so far been the case. At each stage of design and construction the classification should be adapted to the data then available and to the engineering requirements at that stage. The classification at each stage should represent a synthesis of all the investigations performed up to that time.

The principle of completeness requires that the classification should be able to provide information not only on the stability of the excavation (opening) but also of other geotechnical factors (e.g.: resistance of the rock mass to excavation, drainage, conditions, interactions, etc.). This means that in shoosing the parameters to be used for the classification, all the properties of the rock mass which can influence these factors must be taken into account.

In the overall evalution of all the elements

of a classification their relative importance, that is to say their influence on the geotechnical factors should be taken into account in each specific case. Similarly, specific features of the local geology must be given due consideration. This is important because, for instance, the same amount of groundwater can have a different effect in different rock masses, or joints of the same characteristics can differently affect the strength and deformability of different rocks.

A classification of rock masses for an underground engineering project must cover the entire area relevant to the given stage of investigation. Preliminary investigations must differentiate this area into zones which are roughly homogeneous in geotechnical factors. For practical reasons it is convenient to carry out this zoning so that each quasi-homogeneous zone corresponds to one rock class. Only after this, for assigning an overall rating to each zone and determining the corresponding classes, do quantitative parameters become predominaut (in situ measurements, sample tests, etc.). The zoning procedure is theorefore of key importance. To be meaningful in must cover the entire investigated volume to approximately the same level of reliability.

Since the classification will be used in designing and building various works, it should be able to take account of their specific features. It is also useful if it takes into account the specifics of different excavation techniques which might be applied.

The principle of economy implies that the cost of the investigations be kept as low as feasible, i.e. that at each stage only the data essential to the corresponding stage of the project be determined,, while making maximum use of all the results of earlier investigations.

3.2 Stage-wise Design and Investigation

As already noted, the planning, design and construction of a work are divided into a number of sequential stages. The number of stages, and the principal problems which have to be solved in each of them, vary somewhat from country to country, but there is a general overall similarity. Neglecting some details and minor discrepancies, the following stages of planning and design may be recognized:

- initial project conception, choice of site
- investment program (a Yugoslav term; corresponds roughly to "feasibility study")
- preliminary design
- detailed design
- design of adjustments to meet actual site conditions - during construction
- design of work as executed
- design of monitoring, and of demolition if necessary

It is a spesific feature of underground works that the zone of interaction between the work and the ground cannot be thoroughly investigated before excavation, so that design and investigation usually continue during construction, and the design is often modified in the light of conditions actually encountered.

Each stage of design has its specific targets, which in turn determine the targets of investigation and hence also their type and scope.

Table II

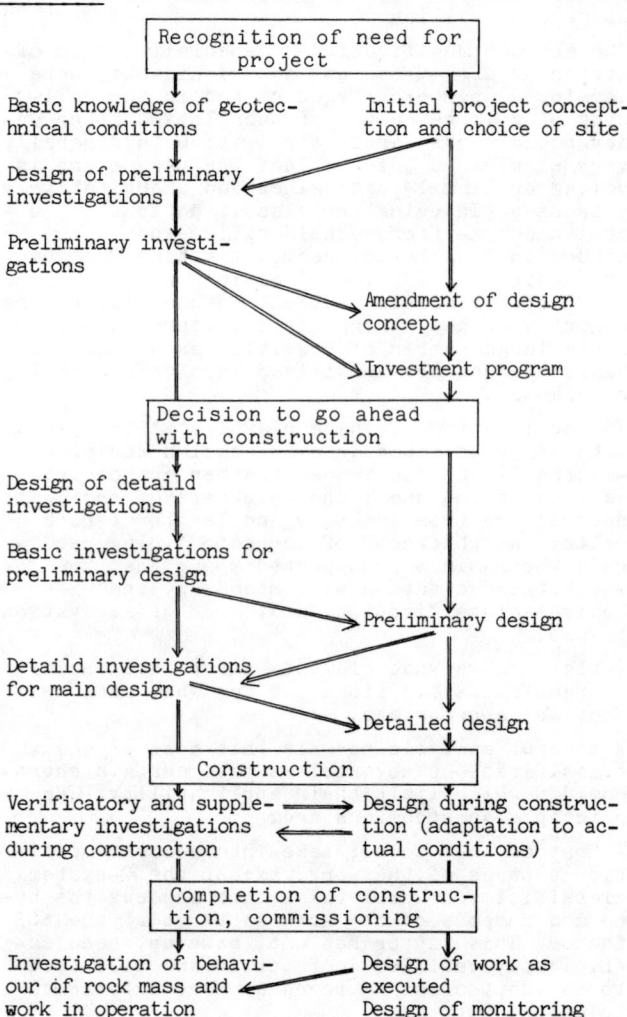

The problems resolved by investigations at any given stage should be resolved to a level of reliability such that they will not be contradicted by a later stage of investigation.

Table II shows the relationship between stages of investigation and design, and the interchange of information. The links between investigation and design make it evident that they in fact constitute an integral process. It follows that the investigator ought to be a member of the design team from the very start of the project right through to its completion.

3.3 Principles and Content of the Rock Mass Classification at Different Stages of Investigation

The idea of a classification as a process which begins at the time of the shoice of location and continues until construction is completed, is essentially very close to the concept od the New Austrian Method, and to Kujundžić s formulations (1974, 1982) of the evolution and modification of engineering-geological sections and models, or of mathematical models. The classification gradually evolves through a number of stages, like the design itself. However, each stage must emerge from the one preceding it so that all together they constitute an integral system which develops in accordance with the needs of design.

The classification has its most general form at the initial stage, gradually becoming more and more in subsequent stages, as it becomes more directly dedicated to the specific work and site. The data on which it is based become more and more numerous and reliable, with an ever increasing proportion of quantitative data.

The system of classification undergoes essential changes when data obtained chiefly by observation (mapping, boring) give way to data from laboratory and field measurements, and when data from surface investigations are succeeded by investigations in the excavation and in sity measurements.

Table III shows which data are used for the classification at which stages of design and construction.

Table III

DATA FOR CLASSIFICATION AND APPLICATION OF RESULTS			STAGE OF DESIGN OR CONSTRUCTION	INITIAL PROJECT CONCEPTION (CHOICE OF LOCATION)	FEASIBILITY STUDY	DESIGN PRELIMINARY	DETAILED	DURING CONSTRUCTION	DESIGN OF WORK AS EXECUTED
DATA FOR CLASSIFICATION	Data on the ground	Geology	Lithological composition	1(2)	2(3)	3,5	3,5(4)	4(5)	
			Tectonic fabric	1(2)	2(3)	3	3 (4)	4	
			Neotectonic activity	1	1	2	2 (4)	4	
			Weathering	1(2)	2(3)	3,5	3,5(4)	4(5)	
			Karstification	1(2)	2(3)	3	3 (4)	4	
		Discontinuity of ground	Number of joint sets	(2)	2(3)	3	3 (4)	4	
		Condition of joints	Joint density	(2)	2(3)	3	3 (4)	4	
			Strike/dip	(2)	2(3)	3	3 (4)	4	
			Length and continuity	(2)	2(3)	3	3	(4)	4	
			Width (Gap and filling)					(4)	4	
			Characteric.of filling					(5)	5	
			Roughness					(4)	4	
			Degree of disc.(RQD,block size etc)	1	1	3	3	(4)	4	
			Fissure porosity	1	1	1		(4)	4	
		Physical and mechanical properties	Bulk density			5	5	(5)	
			Compressive strength			5	5	(5)	
			Abrasion			5	5	(5)	
			Deformability	1	1	5(3)	5(3,6)	6	
			Shear strength	1	1	5	5 (6)	6	
			Permeability	1	1	3	3 (6)	6	
			Temperature				(6)	6	
		Initial stress state		1	1	1	(6)	6	
		Groundwater	Occurrences and levels	(2)	2(3)	3	3 (4)	4	
			Direction and rate of flow			3	3 (4)	4	
			Quantity of water		2(3)	3	3 (4)	4	
			Shemistry and temperature		5(3)	3,5	3,5	(5)	
		Processes in the ground	Chemical		(3,5)	3,5	3,5(4)	4	
			Swelling			5	5 (6)	6	
			Piping				(4)	4	
	Data on the work and phenomena appearing during construction	Work	Type and purpose	7	7	8	8	8	
			Gross section dimensions	7	7	8	9	9	10
			Shape of cross section	7	7	8	9	9	10
			Position and orientation	7	7	8	9	9	10
			Type of strcture	7	7	8	9	9	10
		Construction	Excavation tech.and method					10	10
			Secondary stresses					10	10
			Stand-up time					10	10
			Unsuported span					10	11
			Occurences of instability					10	11
			Water inflows and pressures occurrences of gases and heating					10	11

Key

1 – forecasting data derived from analysis of available documentation or analogy with similar cases

2 – forecasting data obtained by reconnaissance

3 – data obtained by detailed mapping, exploratory boring or similar investigations

4 – data obtained by in situ observations in an excavation

5 – data yielded by laboratory tests

6 – data yielded by in situ measurements in an excavation

7 – outline data from the project conception

8 – data from the investment program (on the grounds of which constr.has been approved)

9 – data from design documentation

10 – data from observations during construction

11 – data obtained by analyzing the behaviour of the work

For the purposes of the initial project conception and shoice of site the classification should provide a reference for an evaluation of the surrounding area az a whole in terms of construction of the given project, and allow a comparison and optimum choice among various alternative locations. It is based chiefly on qualitative data about the rock masses, above all the lithological composition and tectonic fabric, and the groundwater situation. These data are obtained by analysis of any existing documentattion and field reconnaissance. Other data on rock properties are deduced from analysis of available documentation and analogy with similar grounds elsewhere. In these analyses the type, purpose and size of the work must also be taken into account.

At the stage of the investment program (feasibility study) the classification should anable an approximate evaluation of the engineering conditions at the shosen location and approximate cost of the project. Apart from the data from the preceding stage, the classification now calls for basic data on the density and orientation oj joints and on weathering of the rock obtained by engineering-geological mapping and a limited amount of geophysical investigation, exploaratory boring and auxiliary investigations.

At the preliminary design stage decisions should in principle be taken on all the crucial aspects of construction and operacion of the future works,as: shape and size of cross section, its position and orientation, method of excavation, type and rating of temporary and permanent supporting, etc. This calls for investigations of cinsiderable scope, usually more than any other stage: detailed engineering-geological mapping, exploratory boring, geophysical investigations, laboratory tests, etc. They yield data which enable a level of classification corresponding roughly to the classifications of Bieniawsky or Barton et al. A serious limitation in the making and application of the classification at this stage is the impossibility of sufficiently reliable and precise differentiation of quasi-homogeneous zones in the ground. Thid practically means that extreme geotechnical conditions can be predicted, but not the diversity of these conditions over the whole investigated volume.

At the detailed design stage, if there are as yet no data from exploratory excavations or excavations on the work itself, the classification is based on more or less the same data as in the preceding stage, except that they are now more numerous and more reliable, so that a more detailed zoning of the ground and a more complete classification are possible.

If investigatory excavations are made for the detailed design, then the classification can be brougth up to a level close to the conditions which will be encountered in actual construction, when the design is adapted to the actual state of the ground (design during construction).

Underground excavations, be they exploratory or constructional, can yield a large amount of very reliable data on the rock mass, as, for example, all possible data on rock jointing, actual amounts and influence of groundwater, etc. Then also, in situ tests can be carried out, yielding data on the strngngth

and deformability characteristics of the rock masses and the stress state in them. It is especially important that it is then possible to observe the behaviour of the ground under excavation and to try out various methods of improving its engineering properties. All in all, the information obtained at this stage allows a detailed zoning of the ground and an estimate of geotechnical conditions within each quasi--homogeneous zone so identified.

In ramified underground works (shelters, mines, underground railways, etc.), the orientation, purpose and other features should be taken into account for each part of the work, not just its cross-section dimensions. At this stage of investigation the classification should be completely dedicated to the actual work to be built.

At the stage of designing the work as executed, information on the behaviour of the ground during escavation and in coaction with temporary and permanent supporting enables further elaboration of the classification. Thid classification can be reffered to should need arise for any alterations to the work, as in a change of function, extention of space, etc. It can also be useful in investigations and design of similar projects under similar geotechnical conditions.

4. CONCLUSION

This paper only presents the philosopy of a different approach to rock mass classfication for underground works. The authors are now working on a detailed procedure of classification and its application to a number of specific projects now being designed and built. The results of these endeavors will be published at a later date.

REFERENCES (BIBLIOGRAPHIE) LITERATUR

Attewell,B.P. Farmer, W.I., — Principles of Engineering Geologu , London, 1979.

Barton, N Lien, R. Lunde, J., — Engineering Classification of Rock Masses for the Design of Tunnel Support, Rock Mechanics, Vol.6, No 4, 1974.

Bieniawski,Z.T., — Geomechanics Classification of Rock Masses and its Application in Tunneling, Advances in Rock Mechanics, Proceedings of the Third ISRM Congress,Volume II, Denver, 1974.

Bulichev S.N., — Ocenka ustojchivosti treshcinovatih skalnih porod pri provedenii gornyh vyrabotok, Ustojchivost i krepljenie gornyh vyrabotok, Mežvuzovskij zb. vyp.1, izd.LGI, 1977.

Deer, D.U., — Technical Description of Rock Cores for Engineering Purposes Rock Mechanics and Engineering Geology, Vol.1, No 1, 1963.

Kujundžić, B., — Methods of modelling in engineering geology and geo-engineering, Advances in Rock Mechanics, Denver, 1974, Volume II, Part A, 60-65

Lauffer, H., — Das Innkraftwerk Prutz-Imst. Ost.Wasserwirtsch,7 (1955) H. 5/6

PETROPHYSICS: THE PETROGRAPHIC INTERPRETATION OF THE PHYSICAL PROPERTIES OF ROCKS

La pétrophysique: Interprétation pétrographique des propriétés physiques des roches

Petrophysik: Gesteinskundliche Interpretation der physikalischen Eigenschaften von Fels

Modesto Montoto
Prof. of Petrology, University of Oviedo, Oviedo, Spain

SYNOPSIS

Petrophysics, the knowledge of the physical properties of rocks and their interpretation in terms of petrographic considerations, is briefly discussed, focussing on those rock-forming components that clearly influence the geomechanical behaviour of the intact rock: pores, cracks, textures and minerals. Microscope techniques and instrumental procedures for their observation and automatic discrimination, mapping and quantification are described.

RESUME

La pétrophysique, c'est à dire la connaissance des propriétés physiques des roches et leur interprétation en termes de considérations pétrographiques est revisée brièvement. On met l'accent sur les composants de roches qui influent directement sur le comportement géoméchanique de la matrice rocheuse: les pores, les fentes, la texture et la minéralogie. On décrit des techniques microscopiques et des procédés instrumentaux pour leur observation, discrimination automatique, représentation graphique et quantification.

ZUSAMMENFASSUNG

Die Petrophysik, die Kenntnis der physikalischen Eigenschaften der Gesteine und ihre Darstellung unter dem Gesichtspunkt der Gesteinskunde, wird kurz dargestellt, unter besonderer Berücksichtigung der Eigenschaften der Bestandteile, die das geomechanische Verhalten der Festgesteine beeinflussen (Poren, Risse, Struktur und Mineralogie). Es werden mikroskopische und Instrumentalverfahren für ihre Beobachtung und automatische Unterscheidung, genaue Festlegung und Messung beschrieben.

1. INTRODUCTION

Petrophysics is an interdisciplinary branch of Science concerned with the physical properties of rocks and the petrographic interpretation of these properties. The basic principles of this discipline are mainly apported by Physics, Materials Science, Engineering Geology and Petrography.

From the specific point of view of Rock Mechanics it has been classically disputed that different petrographic factors influence the geomechanic properties of the rock masses. Among these intrinsic factors are mineralogy (and its state of alteration), fabric, and a wide spectrum of discontinuities which range from faults and joints in the field to pores and microcracks only observable under electron microscopy.

This paper focuses only on those rock-forming components, whose influence on the geomechanical behaviour of the intact rock has been positively demonstrated. The influence of the physicochemical nature of the environment, where the rock is placed, is not considered. As we will show the rock-forming components can generally be listed, in decreasing importance, as: voids (pores and cracks), textures and minerals.

Consequently a petrographic description, on a microscope scale, of the different rock types involved in any Rock Mechanics problem is, in our opinion, more than necessary. The aforementioned rock-forming components should always be described in detail and quantified. The way of presenting this information, to be of petrophysical interest, is tentatively set forth in the next chapter.

Further studies on a smaller scale will require specific crystallographic considerations related to the symmetry, physical properties, and chemistry of the crystalline network of the rock-forming minerals (including crystalline de fects, dislocations, etc.).

The ability of petrographic studies in helping to satisfactorily interpret the present geomechanic behaviour of rocks and the potential to predict future behaviour explain the continuous growth of Petrophysics. Today, in fact, this scientific approach is of prime interest in applied Geosciences and new problems of unprecedented complexity must be solved.

For instance, the storage of high-level radioactive wastes in crystalline rock-massifs at depth is a controversial subject which is currently under study. The requisites of long-term stability and isolation imposed on the rock-massif, selected as a "natural barrier" for the stored radionuclides, must be guaranteed for periods of time of unprecedented length (tens to hundreds of thousands of years). Among the petrophysical studies undertaken in this area, those related to permeability, deformability and diffusivity stand out; all of them must take into consideration the geographic and geologic environment of the reservoir and its possible variations along the mentioned periods of time. Undoubtedly, this kind of problem constitutes a new challenge to the Geosciences scientific community and furthermore the obtained results are of immediate social impact.

2. ROCK-FORMING COMPONENTS OF PETROPHYSICAL SIGNIFICANCE

The mechanical behaviour of an intact rock is generally evaluated on the theoretical assumption of an ideal material. Such material is assumed to be homogeneous, isotropic, and free of discontinuities and voids throughout its mass. An example of such an idealization is met in Griffith's theory, developed to interpret the initiation and propagation of cracks in glassy materials, which has been basically accepted in Geomechanics.

But, as any petrographer knows, this is an unrealistic view of the rock interior; observations under the microscope corroborate that rocks are inhomogeneous materials, built up of several different crystalline phases (minerals); these are generally anisotropic, with a crystallograpic structure more or less distorted and variable density of local defects, sometimes with compositional zoning and variable state of alteration, and important local discontinuities such as cleavage planes, cracks, pores, mineral inclusions, etc. Besides the mentioned inhomogeneties and discontinuities affecting the rock-forming minerals, the rock, as a whole, is full of different textural discontinuities, openings and grain borders being the most significant. These different rock components, and in particular the internal rock discontinuities must be unavoidably considered when trying to interpret the physical properties of the rock, especially its transport properties.

Accordingly, all the rock-forming components contribute to the departure of rocks from an ideal material. For this reason they must be carefully identified, quantified, and mapped, to provide a basis for proper petrophysical interpretations. Among those components, void spaces (pores and cracks), texture, and minerals clearly stand out.

A general commentary on each of them, related to some case histories follows.

The rock-forming minerals are not the more influential petrographic components on the physical properties of the rock, as intuition might suggest; rather, the lack of minerals, that is, the presence of internal voids in the rock provides the greater influence. So Franklin (1974) clearly stated: "When giving a mineralogical description, intended for use in an engineering study, one should remember that pores, voids and microfissures are probably the most important "mineral" to be observed".

A similar statement can be cited from Duffaut et al (1979): "In the rocks, as in any other material, that that is not is more important than that that is. So, it is highly advisable to all geomechanicists, to start a rock description for that of its voids and to present, as a first physical characteristic, the porosity value".

The most immediate dependence establishes is between the bulk volume that is not occupied by solid matter (porosity, $\emptyset$) and ultimate strength, σ_s. This relationship has been frequently reported in most types of composite and natural materials (metals, ceramics, wood, rocks,...), under the form of simple exponential equations. A more precise equation of the form $\emptyset = Ae^{-b\sigma_s}$, where A and b are constants that take into consideration the effective confining pressure and significant petrophysical aspects such as mineral alteration and composition, granulometry, petrofabric, etc., has been presented by Hoshino (1974).

More aspects related to porosity and its influence on the mechanical behaviour of rocks have been reviewed by Friedman (1976). The review also covers other aspects of interest in Rock Mechanics such as permeability, k, and the prediction of $\emptyset$ and k at depth, as well as mechanisms of porosity reduction (compaction), collapse of pore space at depth, and fluid flow through a fractured rock mass. The mentioned prediction of $\emptyset$ and k at depth is, at present, of basic interest on account of the growing industrial use of the subsurface space.

Porosity has also been used in other different geomechanic applications; to quantitatively express the weathering stage of a rock (which controls its deformability, strength, etc.), and to interpret the acoustic emission/microseismic activity character, AE, for instance.

Granites, covering a wide range of different petrofabric and geomechanic characteristics (σ_c in the 46-156 MPa range, Young modulus in the .7-4 MPa.10^4 range and total porosity from 1.26 to. 5.21 %) were loaded in the laboratory under uniaxial compressive stress having recorded the associated AE, Montoto, M. et al (in press). Three different AE stages were registered in all of them. What is more, these geomechanically different granites could be grouped in three main types, based on their unique AE characteristics; the petrophysical authors' interpretation is based on porosity and microfractographic considerations, and the state of weathering of these rocks.

In addition to the amount of void space contained in a rock, the shape of the pores can also have a remarkable influence. Some efforts to quantify this (or the crack and pore aspect ratio), by means of measurements of the velocities of longitudinal and transversal elastic waves through the rock, can be found in the literature.

Saito (1981) emphasized the influence of the character void spaces on some different physical properties in igneous rocks: longitudinal wave velocity, uniaxial compressive strength, rebound value by Schmidt test hammer and the progress of alteration. Saito's study demonstrated that there is a remarkable difference in these properties between volcanic and plutonic rocks. He attributed this mainly to the difference in pore shape between the two rock types.

A specific type of opening is of basic interest: fractures. These discontinuities range from cracks, observed by means of microscopic procedures, to faults more or less easily recognized in field geology. Considering only the intact rock scale, microcracks play a very important role in the physical properties of rocks.

This importance has been extensivelly demonstrated and quoted: even if their percentage is only 0,1 %, Feves et al (1977). Friedman (1975): "fracture is perhaps the most important topic in Rock Mechanics today", and Atkinson (1981), who stated: "strength, transport, elastic and inelastic characteristics are all highly dependent upon the size, shape, number and distribution of microcracks. These physical properties change dramatically as cracks characteristics are modified under stress". A review of these aspects can be found in the aforementioned references. Unfortunately our present state of knowledge in establishing the relationships between microfractography and in-situ physical properties must be considered as poor.

As a consequence of the growing industrial use of the underground in such politically or economically strategic fields as nuclear waste storage and geothermal reservoirs, a new research subject has been widely developed in Rock Mechanics: the thermophysical study of rock masses. In particular, thermally induced microcracks must be carefully controlled and understood.

So, in heated rock masses, the development of significant intergranular thermal stresses can be petrophysically explained by the marked differences among the thermal expansion coefficients of the main rock-forming minerals and the thermal anisotropy of some of them (for instance, calcite, quartz, K-feldspar),

On a textural scale the differential volumetric variability caused by thermal stresses gives rise to intracrystalline defects and significant textural discontinuities such as intergranular cracks, which modify the mechanical properties of the intact rock, Houpert and Homand-Etienne (1979), Johnson et al (1978). These last authors review other interesting petrophysical contributions related to rock-heating as follows: "Thermally-induced microcracks, which are usually of grain-size dimensions or smaller, decrease elastic moduli and fracture strength and increase porosity and permeability. In addi-

tion, fracture toughness, thermal expansion and diffusivity coefficient change". Therefore, the development of these microcracks must be avoided and the rock interior must be continuously monitored to ensure that the microfissuration threshold (experimentally determined) is not reached.

Although the internal void spaces are significant, the importance of rock-forming minerals must not be diminished. Many relationships can be found in the literature comparing mineral content with strength properties while the other petrographic factors remain constant. Most of the relationships were formulated in regard to important economic mining properties such as drillability or crushability. The most common mineral content variations exemplified were ore component, carbonate, clay, quartz, and feldspar.

Occasionally these relationships have been used to define practical indexes. For instance, abrasiveness has been used as an index since this factor is considered to have an influence on the economics of mechanised tunnel excavation. This relationship can be measured by combining the quantitative mineralogical composition with the hardness of each rock-forming mineral. In studies of this type West (1981) obtained abrasiveness values of 6.25 in granites, 6.87 in quartz-sandstones, and 3.08 in limestones.

One of the clearest examples of the geotechnical applicability of rock composition investigation is reported by Shevkun (1976). In an USSR iron deposit in ferruginous quartzites, Shevkun determined important relations between the Fe content and drilling and blasting operations:

$$E = 0.005 (Fe)^2 - 0.662 (Fe) + 22.34$$
$$V_{max} = 0.028 (Fe)^2 - 2.27 (Fe) + 49.12$$
$$V_p = 9.88 (Fe)^2 - 1228.9 (Fe) + 38910$$

Where E is the specific energy expenditure of roller-bit drilling in KW-h/m; V_{max} is the crushability of the ore in cm^3; V_p is the velocity of longitudinal waves in Km/s and (Fe) is the quantity of basic iron minerals in the deposit (martite and hematite) expressed as the percentage of total iron content.

The influence of rock texture, in particular petrofabric anisotropies, has been extensively studied in Petrophysics. In this sense it is essential to remember that "the simplifying assumption of elastic isotropic behaviour of rocks may lead to systematic errors", Duellman and Heitfeld (1978).

A simple example about textural interpretation of some geomechanical properties is presented by the turkish Pasamehmetoglu et al (1981). They compared the degree of weathering of the Ankara andesites and their triaxial compressive strength. According to their results the finely crystalline facies shown a higher cohesive strength than the porphyritic types, but lower internal friction angles for all the weathering grades of that facies. They texturally argued "that in fine-grained materials, the grain boundaries are strongly interlocked and hence giving a higher cohesive strength. Internal friction of grain on grain, plus the resistance due to interlocking of grains, is smaller in that facies because of the smaller surface roughness due to its finely crystalline nature".

According to all the previous considerations the different aspects that, in our opinion, must essentially be studied and quantified in a rock description of petrophysical applicability are:

Pores. Their percentage, size distribution, shapes and their location in relation to the rock texture and mineralogy (intra- , or inter-granulars, associated or not to zones of mineral alteration, etc.). Their possible interconnectivity character, which can be recognized by means of SEM observations on flat specimens. Similarly their possible aligned orientation, according to preferred directions, that, under future stresses could represent weakness planes and consequently further oriented cracks.

Cracks. A mapping and quantification of the microfractographic network, in relation to the rock texture (inter-, intra-, or trans-granular character of cracks) and mineralogy (autophasic, or heterophasic location of cracks), and its density and orientation. Analysis of the nature of cracks (open, mineralized, healed,...).

Texture. Granulometric analysis(distribution of size and shape of the essential rock-forming minerals).

The isotropic behaviour of the most frequent igneous rocks is a common presumption, based on the apparent lack of visible preferred petrographic orientations in them; so, this aspect must necessarily be elucidated by means of a petrostructural analysis (orientation diagrams) of: cracks, grain elongation (especially in highly anisotropic minerals, e.g. micas), and any other type of detectable internal discontinuities in the rock.

These measurements, as well as the analysis of the crystallographic preferred orientations, and the determination of any type of symmetry, are essential in the study of metamorphic rocks.

Analysis of the grain-interlocking character and percentages of autophasic and heterophasic contacts, between the essential rock-forming minerals.

In the sedimentary rocks the nature and composition of the matrix and the cement.

Minerals. Modal analysis (percentage in volume) of the essential rock-forming minerals, but singularizing the possible different "classes" of each mineral according to various levels of alteration; (for instance, more significant than the single datum: feldspars - 24%, would be the data: sound feldspars (say, alteration affecting less of 20% of the grain volume) - 4%, medium-altered feldspars (alteration affecting 20 - 50%) - 12%, core-altered zoned feldspar - 8%). A tentative classification in "classes" of each rock-forming mineral in granites, according to their physical and/or chemical alteration, can be found in Ordaz et al (1978).

Chemical composition (major oxides percentage) of essential and accesory minerals.

3. PETROGRAPHIC PROCEDURES

As previously stated, the aforementioned rock-forming components need to be observed, quantified, and mapped, to visualize their mutual spatial relationship.

Accordingly, specific microscope petrographic procedures need to be applied. A delicate preparation procedure is required since the specimens must be capable of being observed and quantified under very different optic and electronic microscope techniques.

3.1 Sample preparation.

Pores and cracks, as internal rock discontinuities of a remarkable geomechanical influence, can be better observed and discriminated if specimens in the form of flat rock surfaces are prepared for the microscope studies.

These specimens are substantially different from the conventional ones used in the standard petrographic studies. In fact they are "fluoresceine-impregnated, Au-Pd metallized, polished, thin section", Montoto, M. et al (1981). The preparation of these specimens usually requires careful procedures to guarantee that no new structures ("artifacts") have been developed during the stages of sawing, grinding and polishing. Ion-thinning procedures must be applied occasionally, as a final stage procedure to remove morphological artifacts that could be developed in the form of cracking, pitting, or plucking. The presence of artifacts could give rise to misleading interpretations.

3.2 Observation.

Light-transmitted polarizing microscopy provides the most classical view of rocks under study. Very valuable information can be obtained about rock texture, mineralogy, and state of alteration. Furthermore, when this technique is simultaneously combined with light-reflected fluorescence microscopy, the rock microfractography can be clearly related to the rock texture and mineralogy for mapping purposses, Montoto, M. et al (1981).

This petrographic information, obtained under optic microscopy, can be complemented with observations under Scanning Electron Microscopy, SEM. The use of a secondary electrons detector allows the best topographical view of the internal discontinuities in the rock, such as pores and cracks. With a detector for backscattered electrons, the different rock-forming minerals can be discriminated by the ability of this technique to show various mineral phases formed by different atomic numbers under different emission levels. Finally, if an X-ray detector is used, chemical information of any mineral can be obtained, even a mapping of each constituent element.

A new microscope technique, acoustic microscopy, opens new possibilities in petrophysical studies; with particular reference to the deformability (elastic modulus) of the rock components, internal discontinuities and inhomogeneties, and any other parameters controlling the propagation of ultrasonics in materials.

All the mentioned optic and electronic microsco

pe techniques can be sequentially applied to each rock-specimen to obtain a set of different images, or a "multi-image". These sets of images will be the basis for the automatic mapping and quantification procedures now described.

Other methods to observe, decorate and interpret cracks in rock materials have been considered by Simmons and Richter (1976).

3.3 Quantification.

Petrophysical interpretations need the support of reliable and appropriate petrographic data obtained through different microscope techniques, as previously described.

Nowadays the petrographic information that can be measured through the available procedures on quantitative microscopy, is restricted to very conventional data; basically: mineral content in volume, crack density, granulometry, crystallographic orientation,... Moreover the information is obtained through manual, time-consuming, tedious, and error-subjected methods. Among them, the old "point-counter" and the analogic image analysis systems, stand out.

These analogic systems (commercialized from two decades ago) essentially consist of a TV-camera connected to an optic or electronic microscope, and a minicomputer. The image is displayed on a b&w TV-tube, and the components in the specimen under observation can be discriminated if displayed under different grey levels. But these systems very helpful in many metallurgical, biological, etc. applications, find a very poor applicability in Petrography. This is due to their inability to discriminate, under polarizing microscopy (the most used and standarized procedure in the study of rocks), the rock-forming minerals.

Consequently, on account of the large amount of rock specimens to be analyzed in this type of studies, and the different microscope techniques implied, the present situation is clearly unsatisfactory and new quantitative microscopy procedures are, without doubt, needed in Petrography.

Accordingly, this Department of Petrology and the IBM Scientific Center in Madrid, have entered upon a collaborative study in this field under the title: "Digital image processing in microscope Petrography". The different steps implied in the developed procedure and the partial results up to now obtained follow.

a) Image digitization. Micrographies on 35 mm film, are recorded and digitized with a flat-bed microdensitometer Perkin Elmer 1010 MP. The size of the digitized images is usually 460 x 700 pixels.

b) Image registration. The different images, which form the multi-image set (as described in 3.2 Observation), are subjected to an image registration process. As a result of this process information contained in two, or more, different images can be subjected to a pixel to pixel comparison.

This is of great interest in Petrophysics, allowing one to establish very significant relationships among cracks, pores, texture, mineralogy, altered zones, etc. For example, the mapping of the crack network and the quantification of the crack density in terms of mineral and textural locations.

c) Image-analysis of selected rock-forming components. Each digitized image is subjected to segmentation and feature extraction, which allows a proper classification and quantification of the petrographic information it contains.

The algorithms employed in the method are written in PL/1, and this interactive image analysis system, runs under VM/CMS in an IBM 370/158 computer, using a Ramtek RM 9351 display terminal.

Details about the different steps included in the described procedure, such as: sampling interval resolution of the digital images, pixel size, transformation algorithm and number and selection of control points used during image registration, etc., and its application to discriminate, integrate and quantify rock-forming components, can be found in Montoto, L. (1982), Montoto, L. et al (1978), and Montoto, M. et al (1981).

Among the main results obtained up to now, those related to discrimination of the grain borders (avoiding errors due to the presence of twinned minerals), mapping of pores and cracks, granulometric and microfractographic analysis, stand out. Present efforts are devoted to a better discrimination of the rock-forming minerals, determination of percentage of contacts between grains of mineral A and grains of mineral B, elimination (by processing and shape recognition procedures) of artifacts developed during sample preparation, and to find more petrophysical applications of this method.

In addition, an automatic procedure which consists of a drive-computer microscopy system, able to discriminate, integrate and quantify different type of petrographic information, under the form of digitized multi-image sets, is at present under development in collaboration with the IBM Scientific Center in Madrid.

Such an image analysis system is being developed based on an IBM S/1 computer with 512K on storage. The computer will be connected to an optic microscope Universal-Pol Zeiss (equiped with light-reflected fluorescence) and to a scanning electron microscope Philips PSEM-500, for image input purposes, and to a Ramtek RM 9350 for image display.

4. CONCLUSIONS

Openings (pores and cracks), texture and minerals, are, from the petrographic point of view, the main rock-forming components governing the geomechanic behaviour of the intact rock.

Accordingly, they must be quantitatively analized and mapped as a support for the petrophysical interpretations. In general the following petrographic aspects require, in our opinion, a more accurate quantification: a) character, orientation and location (in relation to minerals and texture) of openings, b) character of grain interlocking, c) contacts between the essential rock-forming minerals (in particular between those presenting higher contrast in the physical property which, to the rock-scale, is under study), and d) sound and altered minerals.

The development of computarized procedures in petrographic microscopy, based on the "multi-image" concept, allows the obtention and processing of a large amount of petrophysically significant information, of the above mentioned type, and moreover, new in this field, the integration of all the available information.

These procedures would obviate: a) the inadequate petrographic analyses (in general of no petrophysical significance) which support most of the present Rock Mechanics studies, and b) the tedious, time-consuming and error-subjected character of those analyses.

5.   ACKNOWLEDGMENTS

Most of this work has been supported by project 4447.79 of the "Comisión Asesora de Investigación Científica y Tecnológica" of the Spanish government. The described procedures on quantitative petrographic microscopy are part of a collaborative study with the IBM Madrid Scientific Center; the author acknowledges the significant contribution of Dr. L. Montoto, of that Center, in developing the ideas and methods here presented.

6.   REFERENCES

Atkinson, B. (1981)
Cracks in rocks under stress: Nature, (290), April 23, 632.

Duellman, H. and Heitfeld, K.H. (1978)
Influence of grain fabric anisotropy on the elastic properties of rocks: Proc. 3rd Cong., Int. Ass. Engng-Geology, Sect. II (1), 150-162, Madrid.

Duffaut, P. et al (1979)
Les vides, principal facteur du comportement mécanique des roches: Proc 4th Cong. Int. Soc. Rock Mech., 115-121, Montreaux.

Feves, M., Simmons, G. and Siegfried, R. (1977)
Microcracks in crustal igneous rocks: Physical properties: in, The Earth's crust: its nature and physical properties, J.G. Heacock, Ed., Geophys. Ser. (20), Monogr.

Franklin, J.A. (1974)
Rock quality in relation to the quarrying and performance of rock construction materials: Proc. 2nd Cong., Int. Ass. Engng. Geology, IV-PC-2.1/11. Sao Paulo.

Friedman, M. (1975)
Fracture in rock: Rev. Geophys. and Space Physics (13), 352-389.

Friedman, M. (1976)
Porosity, permeability, and Rock Mechanics A review: 17th U.S. Rock Mechanics Symp. 2A1-1-17.

Hoshino, K. (1974)
Effect of porosity on the strength of the clastic sedimentary rocks: Proc. 3rd Cong. Int. Soc. Rock Mech., (II A), 511-516. Denver, CO, USA.

Houpert, R. and Hommand-Etienne, F. (1979)
Influence de la temperature sur le comportement mecanique des roches: Proc. 4th Cong., Int. Soc. Rock Mech., (1), 177-180, Montreux.

Johnson, B., Gangi, A.F. and Handin, J. (1978)
Thermal cracking of rock subjected to slow, uniform temperature changes: Proc. 19th U.S. Symp. on Rock Mech., 259-267.

Montoto, L. (1982)
Digital multi-image analysis: application to the quantification of rock microfractography: IBM J. Res & Develop., Special issue "Image processing & Pattern recognition". (26), 6, Nov.

Montoto, L., Montoto, M. and Bel-Lan, A. (1978)
Amethod to measure pores and fissures in geologic materials under SEM by digital image processing: Proc. 9th Int. Cong. on Elect. Microscopy. (1), 212-213, Toronto.

Montoto, M., Montoto, L., Roshöff, K. and Leijon, B. (1981)
Microfractographic study of heated and nonheated Stripa granite: Subsurface Space, Proc. Int. Symp. Rockstore-80, Pergamon Press, (3), 1357-1368, Stockholm.

Montoto, M. Suárez del Río, L.M., Khair, A.W. and Hardy, H.R. (in press)
Acoustic emission in uniaxially loaded granitic rocks in relation to their petrographic character: 3rd Conf. on Acoustic emission/Microseismic activity in Geologic Structures and Materials. The Pennsylvania State University, Oct. 1981, Clausthal, Trans Tech Publ.

Ordaz, J., Esbert, R.M. and Suárez del Río, L.M. (1978)
A proposed petrographical index to define mineral and rock deterioration in granitic rocks: Proc. Int. Symp. on Deterioration and Protection of Stone Monuments, (1), 2.6, 16 p., París.

Paşamehmetoğlu, A.G. et al (1981)
The weathering characteristics of Ankara andesites from the rock mechanics point of view: Proc. Int. Symp. on Weak Rock, 185-190, Tokyo.

Saito, T. (1981)
Variation of physical properties of igneous rocks in weathering: Proc. Int. Symp. on Weak Rock, 191-196, Tokyo.

Shevkun, E.V. (1976)
The influence of mineral composition on the properties of rocks: Fiziko-Tekhnicheskie Problemy Razrabotki Poleznykh, (12), 1, 125-127.

Simmons, G. and Richter, D. (1976)
Microcracks in rocks: Physics and Chemistry of Rocks. Strens, R.G.J. (Ed.), Wiley, 105-137.

West, G. (1981)
A review of rock abrasiveness testing for tunnelling: Proc. Int. Symp. on Weak Rock, 585-594, Tokyo.

THE SIGNIFICANCE OF THE PIEZOMETER IN ROCK ENGINEERING

L'importance du piézomètre in mécanique des roches

Bedeutung der Piezometer in der Felsmechanik

Jachen Huder
Prof. Dr.sc.tech., o.Prof.ETH, Institute of Foundation Engineering and Soil Mechanics,
Federal Institute of Technology Zurich, Zurich, Switzerland

Gian Amberg
Dipl.Ing. ETH, Institute of Foundation Engineering and Soil Mechanics, Federal Institute of
Technology Zurich, Zurich, Switzerland

SYNOPSIS

This article deals with the importance of a knowledge of the groundwater conditions, piezometer measurements and the associated measuring systems. Useful measures to relieve pressure in the groundwater can only be applied when based on piezometer measurements. In this way dangers can be directly averted or the safety can be increased. Three examples of groundwater effects are briefly presented and the successful relief of pressure in the groundwater by means of boreholes is described. Experience with different measuring systems over a period of many years permits a critical overview, and attention is drawn to several typical defects.

RESUME

L'article traite de l'importance de la connaissance de la nappe phréatique et des mesures du niveau piézométrique ainsi que des différents systèmes de mesure. C'est seulement sur la base de contrôles du niveau piézométrique que l'on peut efficacement rabattre la nappe éloignant ainsi le danger en augmentant la sécurité. On présente brièvement trois exemples de l'influence de la nappe phréatique et on décrit son rabattement couronné de succès à l'aide de forages. Une longue expérience permet une considération critique des différents systèmes de mesure et quelques défauts typiques sont indiqués.

ZUSAMMENFASSUNG

Im vorliegende Artikel werden die Bedeutung der Kenntnis der Grundwasserverhältnisse, der Piezometermessungen und ihre Meß-Systeme behandelt. Erst aufgrund von Piezometermessungen können dienliche Maßnahmen zur Entspannung des Grundwassers getroffen werden und dann direkt zur Abwendung von Gefahren oder zur Erhöhung der Sicherheit führen. Drei Beispiele mit Grundwassereinflüssen werden kurz vorgestellt, und die erfolgreiche Entspannung des Grundwassers durch Bohrungen wird beschrieben. Eine lanjährige Erfahrung mit verschiedenen Meß-Systemen läßt eine kritische Betrachtung zu; außerdem wird auf einige typische Mängel hingewiesen.

In the field of soil mechanics a knowledge of the porewater pressures acting in the ground is of prime importance in dealing with foundation engineering problems. Several different piezometer measuring systems have been developed for the very varied conditions often encountered (Terzaghi, 1967). The concept and installation of a piezometer must be given great attention as well as the measurements themselves (Londe, 1982). It is all too easy for error to creep in resulting in wrong interpretations. In soil mechanics today it is possible to look back over many years of practical experience in the determination of porewater pressures (Chauvineau 1972, Huder 1976). Rock engineering can profit from this experience, for the importance of porewater pressures is also very important in this field, although this fact is often overlooked. Although much data is available for water inflow into tunnels and galleries often, besides data on the quantity of flow, data on the magnitude of the porewater pressures in

the rock is lacking. Even when the flow of water is very small or almost imperceptible due to the low permeability of the rock mass the measurement of porewater pressures should never be neglected.

In the recent years different projects, as for instance underground storage or disposal systems have again drawn attention to the significance of the groundwater conditions and rock permeability in rock masses (Johansen, 1979). The fundamentals concerning this problem have been known for a long time (Theis 1935, Gibson 1963). Whereas earlier in the case of hydro-electric plants one was content to rely on the data of Lugeon (1932) for permeability values, today for such projects more extensive data is necessary. For example, the exact position of the groundwater table or a knowledge of the porewater pressures and their time variation are considered important. These, however, can only be determined with the aid of piezometer measurements. The permeability can be reliably determined by means of boreholes (Chalmers 1979, Carlson A. and Olsson T. 1982). This method, however, does not permit determination of the variation of the water pressure conditions with time. If the system of jointing or the fracturing and fissuring of the rock is very intensive so that the material is like a well compacted granular medium, as in the case of a mylonite zone, then the comparison with soil mechanics is established.

The assumption that groundwater flow in an anisotropic rock obeys the laws for a homogeneous medium is certainly not completely true (Wittke, 1970). However, for the engineer this model can be valuable as a first approximation for dealing with groundwater problems, whether for measurement programmes or to estimate the effects of constructional works. Here, the term groundwater is taken to apply to the water filling all the voids in the rock mass that are continuously connected and which is subject to the action of gravity (flow potential). Thus a rock borehole penetrating the groundwater represents a potential line and depending on the system it may become a drain. In an isotropic porous medium the streamlines are orthogonal to the potential lines. Along the streamline the potential reduces in the form of a pressure drop. Boreholes serving the purpose of drainage influence the potential field such that exit hydraulic gradients at large underground openings no longer present any danger. An intelligent use of such boreholes in rock masses can reduce seepage forces and eliminate damage. In the following a brief report is given of constructional problems in three typical zones of disturbance involving groundwater effects.

Slope instability at Disentis (The Grisons)

The cause of a scree instability could be traced back to dammed up water. A several metre thick mylonite rock zone due to its relative imperviousness dammed up the water in the jointed rock mass so that the overlying scree became saturated and springs appeared at the surface.
Three boreholes showed that the rock was heavily jointed to a considerable vertical depth. The deepest borehole (30 m) was sunk entirely in the mylonite layer. This rock consisted of fine calcareous fragments with a varying content of the silt-sand fraction. The grain size distribution curve for the material is shown in Fig. 1. The liquid limit was 21%, the plastic limit 14% and the permeability was found to be $k \simeq 10^{-6}$ cm/s. In some boreholes from which good quality cores were extracted the original rock (i.e. gneiss) structure was still recognizable. The rock was classified as a sericite-muscovite-plagioclase gneiss. The original rock, however, as the gradation curve (Fig.1) shows, was strongly weathered. Where the silt content predominated the gneiss structure could no longer be identified and the material was like a pulverized grey to white talc. The dammed up water in the rock joints, caused by the relatively impermeable mylonite zone, could be clearly identified by means of the piezometers installed at various locations.

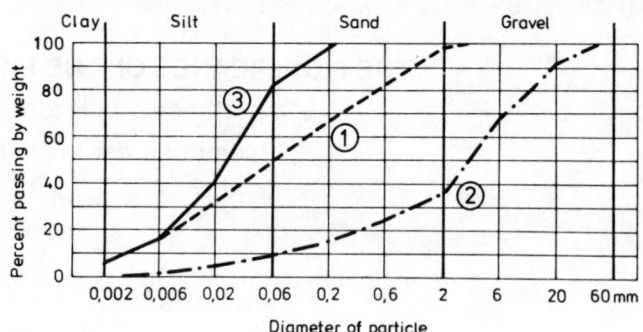

Fig. 1   Grain size distribution curves (percentage weight summations)

1   Material from Disentis

2   Material from the gallery Pradella / Schuls

3   Material from the tunnel Isla Bella - Rothenbrunnen

Feed gallery for the Pradella hydro-electric plant in the Engadine

The free surface flow gallery was driven in the Lower Engadine dolomite rock. With increasing overburden the degree of fissuring of the rock increased. Although the rock could still be regarded as stable - the distance excavated per day amounted to about 8 m - it was necessary to keep increasing the thickness of the lining. The working face showed signs of more and more moisture without the water inflow becoming noticeably greater. However, after a change of shift one Monday granular material over a distance of about 30 m was found in the tunnel. A typical grain size distribution curve for this material is shown in Fig. 1. The gallery was gradually filled with a heap of debris with an angle of repose of about 6°. Failure had occurred in the strongly fissured rock. After carefully removing the material, boreholes were made in the working face. Due to high water pressures the first borehole could not be extended beyond 5 m length. After the fourth borehole, however, the water pressure was slowly relieved so that the depth of boreholes in the direction of the tunnel could be correspondingly increased.

In order to reduce further the water pressure (the measurements gave values of about 40 bars) drainage boreholes were made, which fanned out around the working face to depths of about 40 m. In this way a reduction of water pressure was achieved and at the same time the direction of the seepage forces could be changed such that they were no longer towards the working face of the tunnel.

The decision to use the boreholes was made only after the conventional methods had been tried out and the rate of advance was only 30 cm a day even with hydraulically jacked pile sheets. After the pressure relief was obtained using the boreholes the rate of excavation could be accelerated and as a result the excavation proceeded at a rate of 3 m a day in the fissured zone.

Isla Bella Tunnel, Rothenbrunnen (The Grisons)

The Isla Bella Tunnel traverses Bündner schist of various facies over a length of 2447 m. Driving could only be carried out from the south end with an upward gradient of approximately 12%. Full-face attack with an area of 100 m² at the working face was employed. To avoid interrupting of the excavation work due to the rearrangement of the various service conduits, it was decided after 600 m to drive a top pilot heading as far as the north portal, which after completion would serve for drainage and ventilation. However, at km 0,80 the first difficulties were encountered. The rock pressure became so great that the tunnelling machine had to be removed under difficult conditions. Exploratory boreholes showed that the length of fissured section to be driven through was about 30 m long. It was decided to complete this section using traditional mining techniques. However, the conditions of tunnel excavation even with this arrangement became increasingly difficult. The rock pressure increased so much that the material in the tectonically highly disturbed Bündner schist (see gradation curve in Fig. 1) was squeezed like a paste out of the gaps in the lining. The cause, here too, was to be sought in the very high water pressures in the fissured rock. Measurements of water pressure could not be carried out due to time limitations, though they may have been as great as 30 bar (3 MPa) considering the topography. The geological section showing the disturbed zone is shown in Fig. 2.

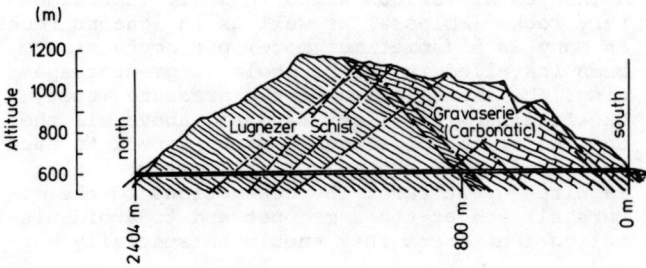

Fig. 2 Geological section, Tunnel Isla Bella, with the mylonite zone at km 0,80

Due to the low permeability of the material the water inflow was small. However, the water pressure in the vicinity of the pilot tunnel had, above all, to be reduced. The first relief boreholes - sunk with the tunnel driving machine - were placed normal to the top heading axis at the roof of the tunnel. Their average length was about 5 m. This first stage of pressure relief is shown by the dotted line in Fig. 3. Later on, boreholes were sunk using a bigger machine, mainly normal to the tunnel axis, and having an average length of 15 m (i.e. 7,5 - 28 m, total length 164 m - shown as a dashed line in Fig. 3). By reducing the water pressure the conditions in the rock were noticeably improved, despite the fact that the total amount of water flow was very small, amounting to a total of 6 - 8 litres/s for all drainage boreholes. After relieving the pressure in the water the excavation of the top heading proceeded quickly.

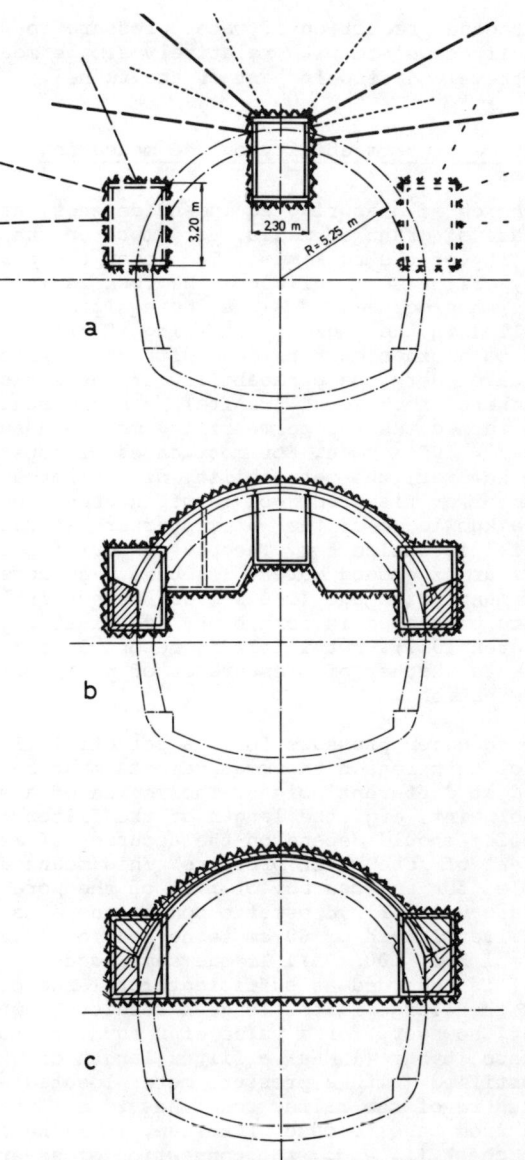

Fig. 3 Stages of excavation in the mylonite zone, sketch of constructional procedure with relief boreholes, first stage l ≈ 5 m (dotted line), second stage l ≈ 15 m (dashed line)

a) Excavation of the top heading gallery with the drainage boreholes, followed by the excavation of the two side drifts with drainage boreholes

b) Excavation of the top heading and concreting the abutments

c) Temporary support of the walls

To be on the safe side two drifts were subsequently constructed, which were also provided with normally directed boreholes (see Fig. 3). Afterwards the roof section of the tunnel could be excavated without any important problems arising. Thus, in this case also, it was possible by changing the direction of flow and a simultaneous reduction of water pressure to successfully complete with relatively simple means an extremely difficult part of the tunnel.

Influence of permeability on the measuring systems

The choice of measuring equipment depends, as in the case of granular media, very much on the permeability of the rock mass. The lower the permeability or the more deficient the rock is in joints, the more sensitive is the equipment to be installed in the rock and shifting it from one place to another must be done with the corresponding care. Here the permeability of the intact unweathered rock (e.g. granite) is not discussed, as in such cases the permeability may be assumed to be $k \simeq 10^{-12}$ cm/s. For most cases of construction, however, the permeability of a jointed system or of fissured rock is of interest, and thus a knowledge of the water pressure in such material is called for. Theories describing the behaviour of ground water flow have been developed many years ago (Theis 1935, Horner 1951) and are presented in text books (de Wiest 1965, Cedergren 1977), but little is reported of measurements whether of pressure or of permeability determinations.

The pore water pressure forms a potential field and for this reason the measurements must be performed at different points. The region of a measuring point, e.g. the length of the filter in a borehole, should depend on the accuracy of measurement of pore water pressure. This means, for example, for an accuracy of ± 3% of the pore pressure u under hydrostatic conditions a maximum filter length of 60 cm is needed for a value of u = 1 bar (100 kPa). If the same accuracy (± 3%) is regarded as sufficient for large pressures then large filter lengths result. To what extent, however, for a value of u equal to for instance 10 bar (1 MPa) a filter length of 6 m is justified (with a pressure meter located at the centre of the filter zone) has to be investigated from case to case. Likewise, it is necessary to check if, e.g., the connection of several jointed systems is desirable, or if this should be prevented by having shorter measuring lengths.

The section over the length of the filter zone is a potential line. In order not to invalidate this assumption the free part of the borehole must be sealed with impermeable material. This may be accomplished by a grout mixture (cement - clay - bentonite - water) or by ramming balls of clay to form a plug.

If the measuring position in the open borehole is only activated periodically, e.g. by introducing packers on both sides of this location, the potential line is given by this open connection and the results must be interpreted with great care since there is a danger of flow around the packers.

In order to get as much information as possible from long boreholes it is recommended that several piezometers are placed at various layers in a borehole as shown schematically in Fig. 4.

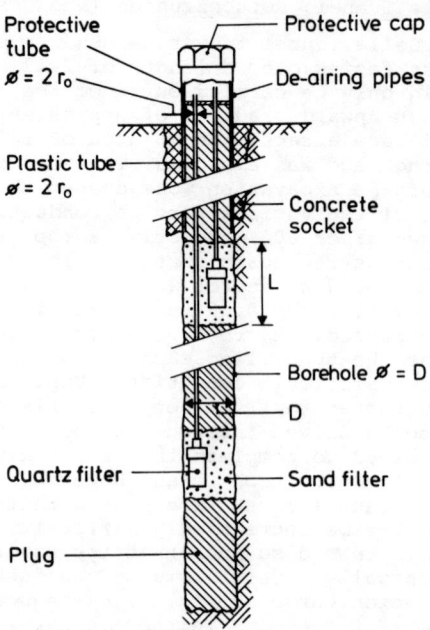

Fig. 4 Sketch of piezometer in a borehole in different layers separated by impermeable sealing material

This technique has been successfully applied by our Institute of Foundation Engineering and Soil Mechanics at various sites in soil, in sedimentary rocks (Molasse) as well as in igneous rocks. As many as 5 (sometimes more) piezometers have been installed in the borehole at greater spacing. The installation of the pressure meter, backfilling with filter sand and above all the sealing of the piezometers with respect to each other and along the measuring tubes or cables requires great care. The connections to the meters all end at the same spot and to avoid mistaking the meters they should be specially marked.

The order of magnitude of the permeability should be known beforehand if a system to measure the pore pressure has to be planned and designed, especially if the effect of a rapid change of loading has to be observed (e.g. variation of pressure in the rock base of a dam due to changes in the water level in the reservoir). With a reference time of t_{90}, i.e. with the time interval for a sudden change in water pressure Δu at the measuring location to be recorded as $0,9 \Delta u$ in the measuring system, it can be shown from the equation of continuity (de Wiest 1965, Cedegren 1977) that:

$$Q = 4 \pi r_s H k$$

whereby Q is the quantity of water $\frac{dV}{dt}$ flowing from an aquifer into (or out of) a spherical surface of radius r_s with a pressure head difference H to the new pressure head H_o at the given permeability k.

According to Hvorslev r_s is given by

$$r_s = \frac{L}{2 \ln \left[L/D + \sqrt{1 + (L/D)^2} \right]}$$

in which L is the length of sand filter and D is the diameter of the borehole. In the above the compressibility of the water and the rock has been neglected.

$Q = \dfrac{dV}{dt}$ is the quantity of water that caused by an external pressure flows into or out of the piezometer tube, the volume being given by

$$dV = r_o^2 \pi \, dH$$

in which r_o is the radius of the piezometer tube; refer to Fig. 4.

It follows that

$$4 \, r_s \, H \, k \, \pi \, dt = r_o^2 \, \pi \, dH$$

and by integrating

$$(1 - H/H_o) = 1 - \exp(-4\pi \, k \, r_s \, t/\pi \, r_o^2)$$

For $H/H_o = x$ the reference time $t_{(1-x)}$ becomes

$$t_{(1-x)} = -\ln (H/H_o) \frac{r_o^2}{4 r_s k}$$

so that for $H/H_o = 0,1$

$$t_{90} = 2,3 \frac{r_o^2}{4 r_s k}$$

For example, for L = 200 cm, D = 10 cm, (r_s = 27,10 cm) k = 10^{-7} cm/s, r_o = 0,4 cm the value of t_{90} becomes

$$t_{90} = 2 \cdot 3 \frac{(0 \cdot 4)^2}{4 \cdot 27,10 \cdot 10^{-7}}$$

$$= 33'948 \text{ sec} = \underline{9,4 \text{ hours}}$$

This result illustrates the significance of the volume change or the time in piezometer measurements. Many measuring systems attempt to reduce the volume change to a minimum and are thereby largely independent of the permeability of the ground.

Measuring systems

Besides the permeability and the required accuracy of measurement also of importance regarding the choice of the measuring system is the position of reading of the pressure meter. Roughly speaking there are two types: the open system and the closed system. By "open system" is understood the measurement of the free water surface in an open tube by means of a probe. In the "closed system", as a rule, the water pressure is measured at the desired location itself using a membrane and the signal is transmitted electrically, pneumatically or hydraulically to the rea-

ding unit. If the reading position is located below the measuring position, or if the piezometric head is greater than the elevation of the reading position then the membrane transducer can be determined at the end of the tube using a pressure meter, e.g. manometer.

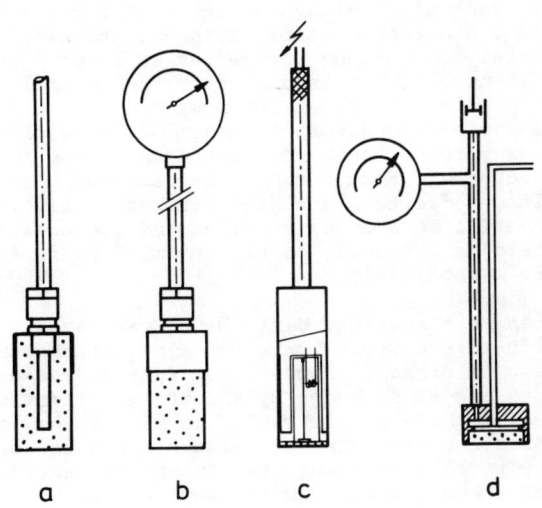

Fig. 5   Different types of piezometer

a) and b) open systems (b) with manometer measurements)

c) and d) closed systems: c) electrical
d) pneumatic

Open systems are simple and operationally reliable, but they have the disadvantage that rapid changes in pressure cannot be measured if the ground has a low permeability. The tube diameter of the measuring system should be kept to a minimum in order to obtain as small a reference time as possible. The tube diameter, however, is limited due to the diameter of the probe, which for constructional reasons cannot be made arbitrarily small. Our institute uses a tube of internal diameter 8 mm and, for greater depths, 12 mm. The measuring tube must be placed in such a way that the probe can be inserted and lowered to the water surface without any effort, i.e. the tube must not develop any bends and must lie vertically above the position of measurement. The main advantage of the system, besides its simplicity and low cost, is the possibility to check the proper functioning of the piezometer at any time. The pressure level can be changed by a chosen amount by adding or taking away water in the measuring tube so that the return to the original pressure level can be controlled.

The "closed systems", i.e. pressure transducers, which measure the pressure directly at the measuring point, have the advantage that the measuring tubes or cables can be arbitrarily brought out. The readings can be made far from the measuring position. Further, pressure variations can be measured with a small reference time. As already mentioned there are various systems, which differ in the manner of transmitting the membrane pressure. Common to all systems is the separation of the water from the transducer by means of a membrane. The membrane deformation can, for example, be directly measured electrically or balanced out using pneumatic or hydraulic pressure. The closed systems, however, are much more expensive than the open systems, though they can be installed in practically every situation. A direct control of the functioning of the interaction between pore water - filter - membrane is no longer possible after the installation of the pressure meter for most electrically operated closed systems. In addition, in alpine regions purely electrical systems have the big disadvantage that they may be damaged by excess voltages due to lightening strikes or electric potential in storms. An all-round protective shield is extremely costly and must be installed very methodically.

Often an electrical disturbance can cause a zero drift in the transducer, whose magnitude must be known beforehand, however, if further measurements are to be made. By inserting a plastic tube in the electric cable, or vice versa, it is possible by applying stepwise a known backpressure to calibrate the transducer and thus determine the new zero reading, saving thereby the measuring station. With the additional plastic tube the proper functioning of the measuring system can be checked by applying backpressures.

The pneumatic or hydraulic systems, in our experience, are more suitable. With these systems the membrane is lifted in its measuring position like a valve by backpressure produced to make a measurement. Since usually two to three relatively stiff tubes are used for the purpose of measurement the installation of the pressure meter is more difficult than in the case of shielded electrical cables. There is also a certain danger of the tube bending and even developing a leak, which is not a problem with cables.

References

Carlsson A., Olsson T. (1982); Characterization of Deep-Seated Rock Masses by Means of Borehole Investigation.
The Swedish State Power Board, Final Report

Cedergren H.R. (1977); Seepage, Drainage, and Flow Nets.
(2nd Edition), Wiley & Sons New York

Chauvineau M. (1972); Sonde de pression interstitielle à contre-pression pneumatique et débit contrôlé.
Bull. Liaison Labo. P. et Ch. 58

Chalmers A. et al (1979); A Modified Form of Aquifer Depletion/Recovery Test for Assessing Potential Water Makes into Deep Excavations.
Proc. ISRM Vol.II pp. 67-72, Montreux

de Wiest R.J.M. (1965); Geohydrology.
Wiley & Sons New York

Gibson R.E. (1965); An Analysis of System Flexibility and its Effect on Time-Log in Porewater Pressure Measurements.
Géomechanique vol.XIII

Horner D.R. (1951); Pressure Build-up in Wells.
Proc. World Pet. Cong. Leiden II, pp. 503

Huder J. (1976); Erkundung der Grundwasserverhältnisse.
Schw. Bauzeitung, Heft 37

Hvorslev M.J. (1951); Time Lag and Soil Permeability in Ground-water Observation.
Waterway Experiment Station, Vicksburg, Bull. no. 36

Johansen P.M. et al (1979); The Performance of a High Pressure Propane Storage Cavern in Unlined Rock, Rafnes, Norway.
Proc. ISRM , Montreux

Londe P. (1982); Concepts and Instruments for Improved Monitoring.
ASCE, SM GT6

Lugeon M. (1979, repr. from 1932); Barrages et Géologie.
ISRM - Montreux 1979

Terzaghi K., Peck R.B. (1967); Soil Mechanics in Engineering Practice.
(2nd Edition), Wiley & Sons New York

Theis Ch.V. (1935); The Relation between the Lowering of the Piezometric Surface and the Rate and Duration of Discharge of a Well using Ground-water Storage.
U.S. Geological Survey

Wittke W. (1970); Rechnerische und elektroanaloge Lösung dreidimensionaler Aufgaben der Durchströmung von klüftigem Fels.
ISRM Vol.III, pp. 253, Beograd

ASSESSMENT OF THE GROUTABILITY OF DISINTEGRATED GRANITES BY MEANS OF HYDROGEOLOGICAL, ENGINEERING GEOLOGICAL AND GEOPHYSICAL FIELD INVESTIGATIONS

L'évaluation in situ de l'injectabilité des granits fissurés par des essais hydrogéologiques, géologiques et géophysiques

Beurteilung der Injizierbarkeit aufgelockerter Granite aufgrund hydrogeologischer, ingenieurgeologischer und geophysikalischer Felduntersuchungen

A. Blinde
J.-P. Koenzen
F. Metzler
Institute of Soil and Rock Mechanics, University of Karlsruhe, Kaiserstrasse 12, D-7500 Karlsruhe, FRG

H. Hötzl
G. P. Merkler
Department of Applied Geology, University of Karlsruhe, Kaiserstrasse 12, D-7500 Karlsruhe, FRG

SYNOPSIS

Hydrogeological, engineering geological and geophysical field experiments have been carried out to assess the groutability of disintegrated granites. The studies included in-situ investigations of fabrics, optical soundings in bore holes, joint tracing techniques and water pressure tests (LUGEON) as well as geoelectrical self-potential measurements and seismical experiments. Up to now results from packer tests have been used as the most important criterion for sealing measures. By means of the above-mentioned research methods additional parameters have been determined to characterize the rock mass with special attention being paid to the detection of the preferred directions of the water paths, thus leading to a more precise and more economical judgment of necessary grouting measures.

RESUME

Des mesures hydrologiques, géotechniques et géophysiques ont été effectuées pour contrôler la perméabilité de masses de granite altéré. Les études ont comporté des investigations in situ concernant la structure des roches, le sondage optique en forage, le reconnaissance des joints par injections d'eau colorée et des tests de charge (LUGEON) ainsi que des mesures de la polarisation naturelle et des mesures sismiques. Les résultats des essais de charge pour tester la perméabilité à l'eau des massifs rocheux ont été utilisés jusqu'à maintenant presque en exclusivité pour la détermination d'injectabilité des roches fissurées. Par des études appliquées on a obtenu des paramètres supplémentaires pour caractériser les massifs rocheux spécialement en ce qui concerne la direction préférentielle de la migration de l'eau. En tenant compte des résultats des expériences exécutées on a pu formuler des possibilités plus précieuses et plus économiques pour l'exécution de l'injection.

ZUSAMMENFASSUNG

Für die Beurteilung der Injizierbarkeit aufgelockerter Granite wurde ein kombiniertes Verfahrensprogramm von hydrogeologischen, ingenieurgeologischen und geophysikalischen in-situ Untersuchungen gewählt. Sie umfaßten Methoden der Gefügestatistik, optische Bohrlochsondierungen, Kluftmarkierungen und WD-Versuche (LUGEON) sowie dem Einsatz der geoelektrischen Eigenpotentialmethode und seismischer Messungen. Die bisher nahezu ausschließlich für Injektionen als Kriterium verwendeten Ergebnisse aus WD-Versuchen mit ihren pauschalen Wasseraufnahmewerten wurden durch die verwendeten Untersuchungsverfahren wesentlich erweitert. So wurden zusätzliche gebirgsspezifische Parameter ermittelt und insbezondere die Richtungen bevorzugter Wasserwegigkeit festgestellt. Die Berücksichtigung dieser Angaben ermöglicht einen gezielteren und wirtschaftlicheren Einsatz notwendiger Injektionsmaßnahmen.

1. INTRODUCTION

For the judgement of the necessity of sealing measures for the improvement of the underground the investigation of the permeability behaviour of the concerned rock mass is of great importance. In the compass of a research program the essential parameters within a selected granitic deposit have been determined thus allowing a characterization of the permeability properties in view of grouting purposes. The investigations have been executed by a combination of different in-situ measurement methods of engineering geology, hydrogeology and geophysics. The area to be investigated lies within an upper-Carboniferous granitic complex in Southwest Germany showing all transitional stages from solid rock to granitic grus. The investigations have been sponsored by a grant of the DFG (German Research Council) in Bonn, to whom we want to thank for the supportance.

2. FIELD INVESTIGATION METHODS

2.1. Engineering geological mapping

At the beginning of the research program outcrops within a defined granitic areal of different stages of rock loosening have been selected and have been treated mineralogically-petrographically. At the same time researches for the characterizing of the chemical weathering have been made. The point of issue regarding the selection of material has been the possibility of technical execution of the in-situ experiments. Quarries in operation have been preferred as not only the provision installations could be used but also control investigations could be done in the territory of worked quarries. The detailed fabric - statistical studies in defined homogeneous zones have been done in view of a most complete description of the essential joint features being responsible for the permeability and thus for the grouting possibilities. Besides of the orientated statistical analysis of the measured fissures together with the observed joint films resp. joint fillings the spacings of the individual system of joints, the degree of separation as well as the spatial extent of the discontinuities (Müller, 1963) has been investigated for the standardization of the rock loosening. Joint parameters essential for flow calculations and grouting measurements (selection of injection liquid etc.), width of joints and asperity could be measured on the surface only inexactly. By using an optical probe it was, however, possible to get more exact indications regarding the width of joints and the joint specific data received by surface work could be confirmed essentially. Further important points regarding the water routing of individual directions of fissures could be deduced by the joint genesis.

2.2   Hydrogeological in-situ testing

2.2.1 Joint tracing

For the determination of the discontinuities used by the water, the method of joint tracing after Hötzl et.al., 1982, has been applied. The injected colour medium has been infiltrated at constant pressure head in bore holes resp. small test pits and thus marked the water paths flown through by the adsorptive binding of the tracer (e.g. Rhodamin, Methylenblau, Astra-Diamantgrün) to the rock. The coloured joint planes could be mapped and analysed in an orientation statistical way after having set free the tested granitic territory. Besides, the coloured fracture-bordered rock bodies rendered possible the demarcation of hydraulically effective joint-bordered rock bodies after Meier, 1978.

2.2.2 Water pressure tests

A main point within the research program were the water pressure tests (LUGEON), which have been up to now nearly the only criterion for the assessment of sealing measures. For the execution of the investigations the latest methodical knowledge has been considered e.g. Ewert, 1979 and 1981, Heitfeld, 1979, Heitfeld and Krapp, 1981, Houlsby, 1976, Pearson and Money, 1977, Schetelig et al., 1978. The packer tests have been carried out with a fully electronical testing installation (COMDRILL System) by use of 1.0 m long, pneumatically expandable packers in 1.0 m steps from bottom to top. A direct seepage around the packers could almost be excluded at the performed single and double packer tests. injection pressure has been measured by a piezoresistive pressure gauge directly in the test section and been transfered analogically on a recorder. At the same time on a second trace of the writing unit the corresponding flow quanti-

ties had been registered chronologically. Therefore different measurement ranges could be selected. In order to examinate pressure losses in the pressure pipe system additional casing-head measurements had been done and also were recorded synchronically. The packer tests have been performed in increasing and decreasing pressure steps in view of steady state flow conditions and long-time effects (Blinde et al., 1981a, Hötzl et al., 1981). In order to define the essential flow ranges as well as to record the possible anisotropical permeability behaviour of the tested area, additional level measurements have been carried out. The process of the water propagation has been observed by the aid of geoelectrical self-potential measurements during the pressure test. For the analysis of the test data and for the examination of the reproductionability the results have been described in P,Q,t-diagrams with level changes as well as in P/Q-diagrams of steady state flow phases. The presentation of the water absorption capacity of the individual test sections has been done by the LUGEON-values (1 LUGEON = 1 $l \cdot min^{-1} \cdot m^{-1}$ at 10 bar).

2.3   Engineering geophysical measurements

2.3.1 Geoelectrical self-potential measurements

The movement of water solutions in porous or jointed rock mass causes an electrical filtration potential, which can be measured at the surface (Blinde et al., 1981b, Merkler et al., 1970). Different hydraulic gradients resp. flow processes thus influence the self potential field and lead to its measurable changement By using the SP-method measurements have been carried out at the same time as the water pressure tests. They rendered possible the determination of the preferred directions of the water paths. Furthermore, the results illustrated the flow processes and could be compared with the level changes as well as with the joint statistical measurements. Adding of a concentrated sodium chloride solution caused a supplementary artificial change in the potential field and thus we got more significant results. The SP-measurements have been analysed by the potential resp. measuring data differences for chronologically different stages of the test. The individual local distribution of the differences has been represented in isoline maps. In order to objectify the representation of the measurement results and to work out statistically reliable trends, the isoline maps have been analysed by isanomalic diagrams (Neumann, 1954, Buchheim and Lauterbach, 1954).

2.3.2 Seismical measurements

The aim of the seismical measurements was to register the elastical anisotropy of the granites caused by the joints in dependance of direction and depth by the aid of the propagation of seismical waves. A further point was the examination of the mechanical-elastical improvement after cement grouting tests based on Vp-wave velocities. The applied in-situ measurement methods in bore holes included seismical down-hole, up-hole and cross-hole measurements (Stokoe and Woods, 1972). On the terrain surface seismical rotary soundings and short hammer-blow refraction profiling have been carried out. For the registration of the measurement values a seismical recording apparatus with 12 channels with enhancement (Mc.Seis 1300, OYO) has been used.

3. TEST RESULTS

The results of the joint statistical studies yield indications about the main joint sets and their frequency. A spatial limitation of the hydraulically effective joints being essential for grouting experiments could be obtained by using joint tracing techniques with coloured water. We found out that only distinct joint sets showing typical oxide coatings were preferred by the water. The relationship between the medium width of joints to the joints used by the water was revealed by optical sounding in bore holes with measured widths between 0.2 and 2.5 mm. They thus showed to be suitable for cement grouting. Characteristical medium values of asperity after DIN 4762 (German norms) for the individual directions of fissures could only be determined sporadically (e.g. joint set A: 2.08 (±1.29) mm or joint set B: 1.59 (±1.04) mm). However, a qualitative specification of joint plane characteristics for the different types was possible. The packer tests executed in three test fields gave characteristic values for the water absorption capacity, thus rendering possible a standardization of rock loosening regarding the pressure dependant water absorption. Fig. 1 shows average LUGEON-values within the different test sections. Example 1a) reveals heterogenous characteristics in the water absorption values caused by some specially hydraulically effective joints with great widths.

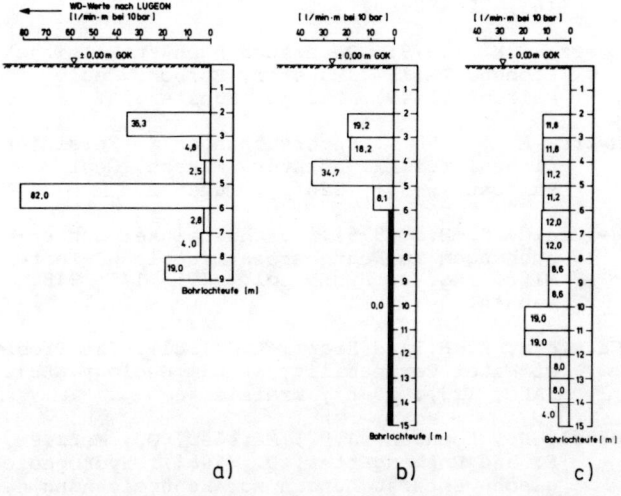

a)   b)   c)

Fig. 1 Distribution of average LUGEON-values of three typical test fields

Example 1b) brought forth a more homogenous distribution of the LUGEON-values within a granite with a higher closeness of fissures. These results depend on a smaller but uniform gap width of 0.2 - 0.5 mm of the discontinuities. In example 1c) an almost homogenous distribution of the water absorption is shown all over the test range with the influence of a stronger water-bearing fault zone near the hole bottom. The most remarkable feature of all three analysed granitic types is the considerable influence of fine-grained fillings of hydraulically effective joint systems due to the position close to the surface. In almost 50 % of the 128 executed water pressure tests the erosion of those infillings was clearly shown. It was proved by the

flush out of suspended matter in piezometer holes during unsteady state flow conditions as well as in P/Q-variations of the registrated diagrams. Geodetical high precisely measurements during the packer tests did not show any significant dislocation of the concerned rock mass even at relatively high pressures. A comparision between the results of LUGEON tests and seismical up-hole measurements (Fig. 2) shows good conformity regarding the depth-dependant rock loosening. Referring to the measured seismical velocities of the rock mass a "seismical jointing coefficient K (%)" could be calculated beeing typical for the rock loosening.

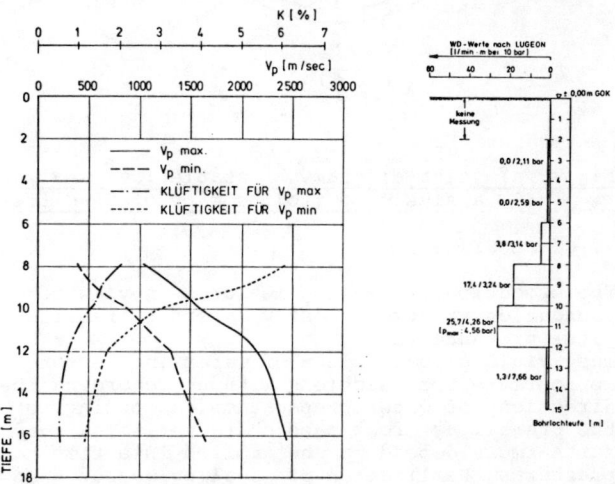

Fig. 2 Results of seismical up-hole measurements with calculated "seismical jointing coefficient K" (left) and packer tests (right)

Results of level observations and SP-measurements have been elaborated by statistical analysation of the direction of level differences measurements and SP-measurement value differences of same test intervals. They show a clear conformity compared to the joint diagram and the results obtained from the joint tracing experiments (fig. 3).

The method of SP-measurements reveals a preferred water flow in NE-direction (fig. 3c). In the perpendicularly orientated main joint direction (NW/SE) follows a pressure rise in the piezometers but the waterflow, however, is low in this direction. The effect of cement grouting experiments has been examined by engineering geological mapping of the grouted joints after blasting of the tested area. The results showed a clear reference to the water-bearing joint system. Thereby the width of joints of the preferred direction of propagation could be determined with an average value of 2.0 mm. The filling of just these discontinuities with cement could be confirmed by seismical measurements. The Vp-wave velocity before grouting amounts 1500 $m \cdot s^{-1}$ and after injection 2860 $m \cdot s^{-1}$ measured perpendicular to the main joint set, thus proving also a mechanical-elastical improvement of the grouted granite of about 60 %. Results from water pressure tests also coincide with the grouting experiments.

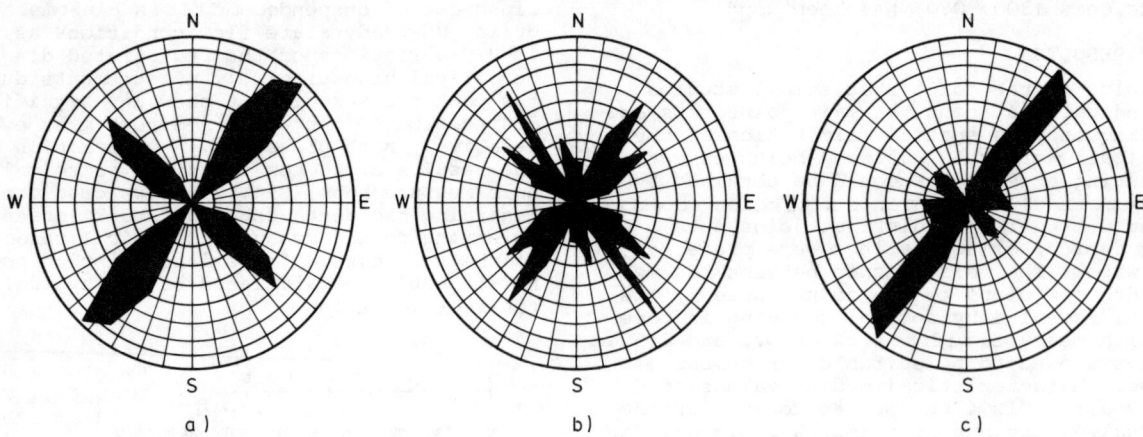

Fig. 3 a) joint diagram, b) statistical analysis of water level changes in bore holes, c) statistical analysis of isoline maps of SP-differences values

4. CONCLUSIONS

The execution of sealing measures in view of foundation treatment mostly refers to results of water pressure tests.
They yield global values of water intake for appropriate test sections without regarding the directions of water propagation. According to the prevailing rock mass characteristics they furthermore depend on the applied injection pressures. Realization of sealing measures however requires more detailed information about the nature of rock loosening and about the direction dependant water routing. This report shows, that the combination of hydrogeological, engineering geological and geophysical field investigation methods render possible a representative characterization of the required properties. In the presence of convenient conditions of exposure which can be expected regarding large-scale foundations preliminary exploration by using the above-mentioned research methods can give specific information of rock characteristics for the assessment of grouting purposes. Results of packer tests before and after grouting measures can be used for the quantification of the amendment of the underground. Geoelectrical SP-measurements inform about the directionally propagation of the injection liquid. It can be expected that the detection of the depths of penetration will be possible by further improvement of this method. Datas of seismical measurements can be used not only for preliminary site investigations but also for the verification of sealing results especially in view of the determination of the anisotropical propagation behaviour after grouting. Regarding the necessity of sealing measures, the application of the described research methods help to a technical and economical optimization of foundation treatment.

5. REFERENCES

Blinde, A., Hötzl, H., Koenzen, J.P. and Metzler, F. (1981a). Quantifizierung von Durchlässigkeitseigenschaften aufgelockerter Granite. Ber. 3. Nat. Tag. Ing.-Geol., 109 - 118, Ansbach.

Blinde, A., Hötzl, H. and Merkler, G. (1981b). Ingenieurgeophysikalische Untersuchungen der Auflockerungsanisotropie von Gesteinskörpern. Ber. 3. Nat. Tag. Ing.-Geol., 201 - 207, Ansbach.

Buchheim, W. and Lauterbach, R. (1954). Isoanomalenrichtungsstatistik als Hilfsmittel tektonischer Analyse. Gerl. Beiträge Geophys. (63). 2, 88-89.

Ewert, F.K. (1979). Untersuchungen zu Felsinjektionen, Teil 1. Münster. Forsch. Geol. Paläont., (49), 292 pp. Münster.

Ewert, F.K. (1981). Untersuchungen zu Felsinjektionen, Teil 2. Münster. Forsch. Geol. Paläont., (53), 326 pp. Münster.

Heitfeld, K.-H. (1979). Durchlässigkeitsuntersuchungen im Festgestein mittels WD-Testen. Mitt. Ing.-u. Hydrogeol., (9), 175 - 218, Aachen.

Heitfeld, K.-H. and Krapp, L. (1981). The Problem of Water Permeability in Dam Geology. Bull. IAEG, (23), 79-83, Krefeld.

Hötzl, H., Koenzen, J.P., Merkler, G., Metzler, F. and Rothengatter, P. (1981). Hydrogeologische Untersuchungen zur Kennzeichnung der Durchlässigkeit von klüftigen Festgesteinen. Veröff. Inst. Boden- u. Felsmechanik, (87), 143-179, Universität Karlsruhe.

Hötzl, H., Metzler, F. and Rothengatter, P. (1982). Die Kluftmarkierung - Eine Anwendung der Markierungstechnik zur Ermittlung von Durchlässigkeitseigenschaften klüftiger Gesteine. Beitr. z. Geologie der Schweiz - Hydrologie, (28 II), 381-393, Bern.

Houlsby, A.C. (1976). Routine Interpretation of the Lugeon Water Test. Q. Jl. Engng. Geol. (9), 303-313, Belfast.

Meier, G. (1978). Zum Problem der hydrodynamisch wirksamen Trennflächen. Freiberger Forschungshefte A 597, 1 - 57, Berlin.

Merkler, G. and Moldoveanu, T. (1970). Einige Beispiele zur Anwendbarkeit geophysikalischer Messungen bei Baugrunduntersuchungen. Proc. 2nd Congr. ISRM, (1/18), Beograd.

Müller, L. (1963). Der Felsbau. 1. Bd., Theoret. Teil. Felsbau über Tage. 1. Teil, 624 pp. Ferdinand-Enke-Verlag. Stuttgart.

Neumann, W. (1954). Praktische Untersuchungen zur Isanomalen-Richtungsstatistik. Freiberger Forschungshefte C 13 (Geophysik). Berlin.

Pearson, R. and Money, M.S. (1977). Improvements in the Lugeon or Packer Permeability Test. Q.Jl Engng. Geol., (10), 221 - 239, London.

Schetelig, K., Schenk, V. and Heyberger, W. (1978). Neues Meßverfahren für die Durchführung von Wasserabpressungen. Veröff. 3. Nat. Tag. Felsmechanik, Aachen.

Stokoe, K.H. and Woods, R.D. (1972). In Situ Shear Wave Velocity by Cross-Hole Method. Journal of the Soil Mechanics and Foundations Division, ASCE, Vol. 98, No. SM5, Proc. Paper May, 444 - 460.

POST-CONSTRUCTION SEEPAGE TOWARDS TUNNELS IN VARIABLE HEAD AQUIFERS

La percolation des eaux vers des tunnels creusés dans des nappes phréatiques à charge variable

Wassereindrang in fertiggestellten Tunneln bei schwankenden Grundwasserspiegeln

Arturo A. Bello-Maldonado, C.E.
General Manager, Geosistemas S.A., Mexico
President of the Mexican Society for Rock Mechanics, 1981-1983

SYNOPSIS

Ground water flow into tunnels driven in aquifers with variable head is evaluated for the construction period as well as for the time life of a tunnel to which ground water is allowed to enter, as is the case in tunnels driven with the associated purpose of water reclamation. The time-dependent behaviour of water inflow from the transitory construction stage up to the permanent state after completion is likewise established for tunnels driven in constant head aquifers.

RESUME

On évalue le flux d'eau dans les tunnels excavés dans des nappes phréatiques à charge variable, aussi bien pour la période de construction que pour la durée de vie utile d'un tunnel dans lequel on permet l'entrée de l'eau de sous-sol après sa construction, tel est le cas des tunnels excavés ayant pour but additionnel de procurer de l'eau potable. On établit aussi le comportement dans le temps, le débit qui pénètre dans le tunnel, depuis l'état transitoire de la construction jusqu'à l'état stationnaire après la construction pour les tunnels excavés dans des nappes phréatiques à charge hydraulique constante.

ZUSAMMENFASSUNG

Der Grundwassereindrang in Tunnel, die in wasserführende Schichten mit schwankenden Grundwasserspiegeln getrieben wurden, wird bewertet, sowohl für die Bauperiode als auch die aktive Zeit, in der der Grundwassereindrang in den Tunnel erlaubt ist, wie es der Fall von Tunnel mit der gleichzeitigen Zielsetzung der Wassergewinnung ist. Ebenso wird das zeitabhängige Verhalten des Wassereindrangs von der vorübergehenden Bauperiode bis zum stationären Zustande nach Fertigstellung für Tunnel, die in konstanten Grundwasserspiegeln getrieben wurden, festgestellt.

INTRODUCTION

Seepage towards tunnels being driven in constant head aquifers can be predicted for the under construction transitory stage, as close as the permeability and effective porosity of the subsoil mass are known, BELLO, 1974. In many instances the after construction water inflow has none or little interest, because the tunnel lining is impermeable in the general case. However, for water reclamation projects involving tunnels along the waterways, the possibility of allowing groundwater to get into the tunnel, to increase the volume reclaimed acquires relevance, mostly because this additional water can be gained without increasing the construction cost of the tunnel. Furthermore, the construction of tunnels driven through aquifers with the sole purpose of water reclamation could be a mediate and economical way to solve this problem in some very populated locations.

BELLO, 1974, developed a procedure to estimate the amount of water inflow for both the transitory and the stationary stages for a tunnel driven at a constant rate in a constant head aquifer. This paper deals with the relationship between time and water inflow after the end of the excavation for the above mentioned case and also for that one in which the tunnel is driven in a variable head aquifer. For both cases the transitory stage is the most important for the under construction period, because during this time takes place the larger amount of seepage and it must be handled to allow tunnel driving, whereas the stationary stage is of most importance when it becomes necessary to know the water inflow after construction, to base on it the fi--

nancial characteristics of a water reclamation project.

AFTER CONSTRUCTION SEEPAGE IN CONSTANT HEAD AQUIFERS

The results from the mathematical model for water inflow prediction for a tunnel driven at a constant rate in a constant head aquifer, are summarized in Fig. No. 1. From them, the after construction water seepage into the tunnel already excavated to its final length L_t, can be obtained as follows:

As long as the tunnel is being driven at a constant rate $r = dL/dt$, the water inflow per unit length in each section is given by:

$$q = k \cdot H \cdot \frac{T + 1}{\sqrt{T(T+2)}}$$

where $T = \dfrac{Kt}{H}$ and $K = \dfrac{3k}{2n_e}$

as it is shown in Fig. No. 1.

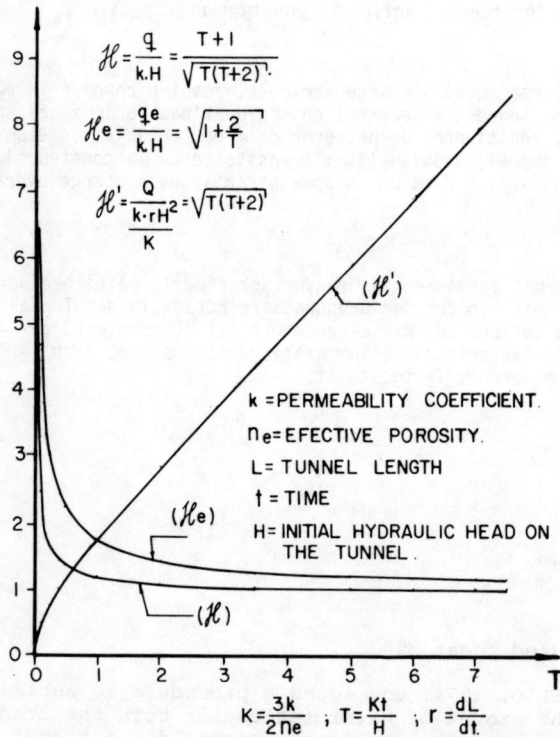

$$\mathcal{H} = \frac{q}{k \cdot H} = \frac{T+1}{\sqrt{T(T+2)}}$$

$$\mathcal{H}_e = \frac{qe}{k \cdot H} = \sqrt{1 + \frac{2}{T}}$$

$$\mathcal{H}' = \frac{Q}{k \cdot r H^2} = \sqrt{T(T+2)}$$

k = PERMEABILITY COEFFICIENT.
n_e = EFECTIVE POROSITY.
L = TUNNEL LENGTH
t = TIME
H = INITIAL HYDRAULIC HEAD ON THE TUNNEL.

$K = \dfrac{3k}{2n_e}$; $T = \dfrac{Kt}{H}$; $r = \dfrac{dL}{dt}$

FIGURE 1.- VARIATION OF UNIT DISCHARGE AND TOTAL WATER INFLOW WITH TIME AND LENGTH OF THE TUNNEL UNDER CONSTANT HYDRAULIC HEAD. (REF. Nº 1).

Total water inflow up to the time t before the final length is reached, is obtained by means of:

$$Q = \int_o^L q \cdot dL = r \int_o^t q \cdot dt$$

Let ζ be the time at which final length L_t of

tunnel is reached, or the excavation is stoped, afterwards, the total after construction discharge can be obtained by integration of:

$$Q_p = r \int_{t-\zeta}^t q \cdot dt$$

as $dt = \frac{H}{k} dT$, last equation can be writen:

$$Q_p = r \cdot \frac{k \cdot H^2}{K} \cdot \int_{T-T_F}^T \frac{T+1}{\sqrt{T(T+2)}} \, dT$$

where $T_F = \dfrac{K\zeta}{H}$

After integration, it results:

$$Q_p = \frac{r \cdot k \cdot H^2}{K} \left[\sqrt{T^2 + 2T} - \sqrt{(T-T_F)^2 + 2(T-T_F)} \right]$$

It is convenient to define the dimensionless parameter $\mathcal{H}''$, proportional to Q_p:

$$\mathcal{H}'' = \frac{Q_p}{r \cdot H^2} \frac{K}{k}$$

given by:

$$\mathcal{H}'' = \sqrt{T(T+2)} - \sqrt{(T-T_F)^2 + 2(T-T_F)}$$

or:

$$\mathcal{H}'' = T\sqrt{1 + \frac{2}{T}} - (T-T_F)\sqrt{1 + \frac{2}{T-T_F}}$$

From this last equation it can be easily noted that for a very long time t, that is, for a great value of T, the value of $\mathcal{H}''$ approaches the value of T_F, so: $\mathcal{H}''_F = T_F$ that is, if Q_F is the final total inflow to the tunnel:

$$\frac{Q_F K}{k \cdot H^2 \cdot r} = \frac{K\zeta}{H}$$

or:

$$Q_F = k \cdot H \cdot L_t$$

as it is expected for the long term.

Fig. 2 shows the different variations of $\mathcal{H}''$, that is of Q_p, for different times ζ at which the tunnel reaches its final lenght and a maximun value of $\mathcal{H}'$, that is of Q, is present, with the value:

$$\mathcal{H}'_M = \sqrt{T_F(T_F+2)}$$

Relationship between the maximun total discharge and stationary total discharge in the post-construction stage, is given by:

$$\frac{\mathcal{H}'_M}{\mathcal{H}''_F} = \frac{\sqrt{T_F^2 + 2T_F}}{T_F} = \sqrt{1 + \frac{2}{T_F}}$$

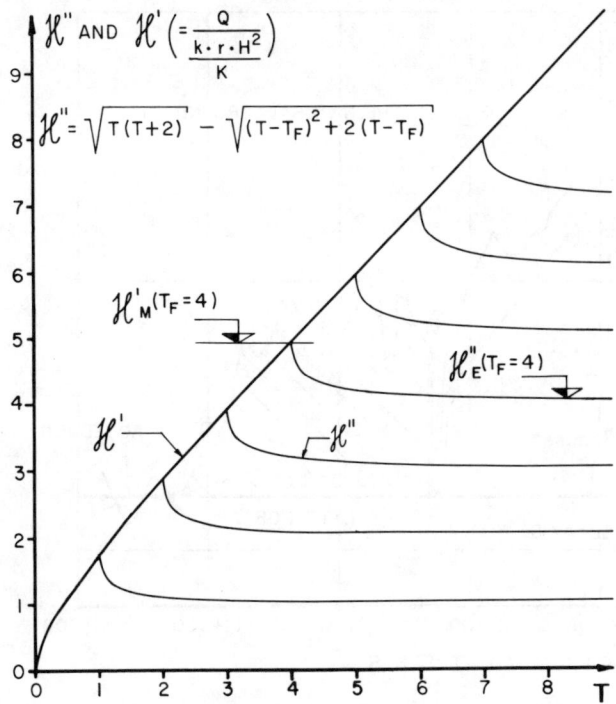

$\mathcal{H}''$ AND $\mathcal{H}'\left(=\dfrac{Q}{k\cdot r\cdot H^2}\right)$

$\mathcal{H}'' = \sqrt{T(T+2)} - \sqrt{(T-T_F)^2 + 2(T-T_F)}$

$\mathcal{H}'_M(T_F=4)$

$\mathcal{H}''_E(T_F=4)$

$\mathcal{H}'$

$\mathcal{H}''$

FIGURE 2.- VARIATION OF AFTER-CONSTRUCTION TOTAL DISCHARGE INTO A TUNNEL UNDER CONSTANT HYDRAULIC HEAD.

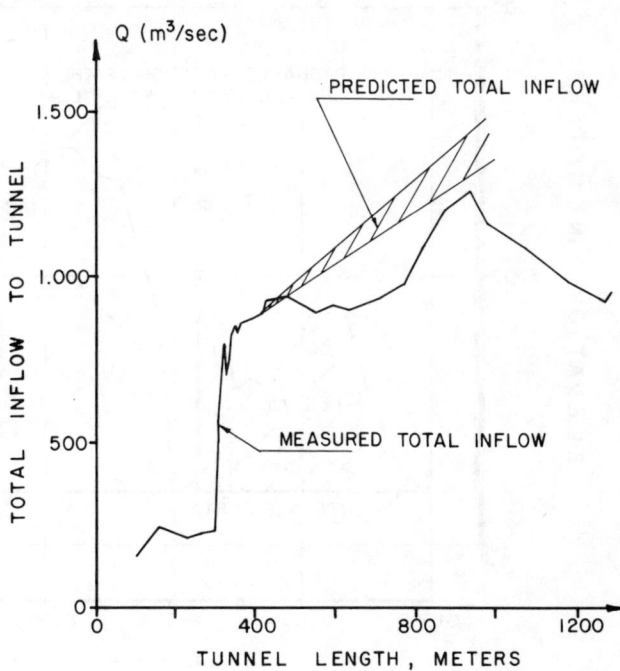

Q (m³/sec)

PREDICTED TOTAL INFLOW

MEASURED TOTAL INFLOW

TUNNEL LENGTH, METERS

FIGURE 3.- PREDICTED AND MEASURED TOTAL INFLOW TO THE EMISOR CENTRAL TUNNEL, PART - 14-14.A, FROM AN AQUIFER 640 m. IN THICKNESS.

This relation is of great help to analyze the - variation of pumping equipment required for maximum and final total inflow discharges to a tunnel; as it has the same form of the equation - defining the average unit discharge $\mathcal{H}_e$, its - value can be read from the corresponding curve in figure No. 1, once the value of T_F is known.

The theoretical behavior of total water inflow established above, has been compared with the - discharge actually measured in a part of the -- Emisor Central tunnel, driven north of Mexico - City as part of the deep sewage system of the - metropolitan area. The total discharge shown - in relation to time corresponds to a length of n, made tributary to a shaft for seepage -- handling to outside the tunnel. As can be seen in Fig. 3, actual and predicted discharge values are in good agreement; the subsoil conditions assumed in the development of the mathematical model were well acomplished in this part of the tunnel, as the aquifers are recharged not only by annual rain, but also by an underground wa-- ter flow coming from the near highland areas.

SEEPAGE TOWARDS TUNNELS DRIVEN IN VARIABLE HEAD AQUIFERS

The mathematical model developed for constant - head aquifers, has been extended to cover the - variable head aquifers case, which is present - in a tunnel under construction across the Las - Cruces mountain range, which runs in a north -- -south direction in the western part of Mexico City. The aquifers in the mountainous area are recharged by rain in annual periods, as it has been proved by the comparison of annual rain --

volume in the area and the water inflow getting into another tunnel operating in the same area from 25 years ago, and in which provitions for - groundwater penetration were made. This old tunnel was used to calibrate the extended mathematical model and also, it was used to prove -- that the water table close to it remains lowe-- red down to the tunnel elevation, as predicted by the model. Permeabilities of subsoil formations along the new tunnel were obtained by -- means of field testing with the Lugeon procedure, made on sampling borings used to confirm -- subsoil characteristics and to know about the - elevation of the water table. Fig. No. 4 shows general geological formations, the position of water table and the values of permeability coefficients obtained.

Assuming that the hydraulic head varies linearly from Zero at the origin of the excavation, - to H_M when the tunnel reaches the length L_M, -- and denoting the constant rate of driving by -- $r = dL/dt$, it can be established that for each section of unit lenght located at a distance L_x from the origin, the water inflow is given by:

$$q_x = k \cdot H_x \frac{T+1}{\sqrt{T^2+2T}}$$

where

$$T = \frac{K\mathcal{z}}{H_x} \quad \text{and} \quad K = \frac{3k}{2n_e}$$

$\mathcal{z}$ being the time after that particular section of the tunnel was reached with the excavation.

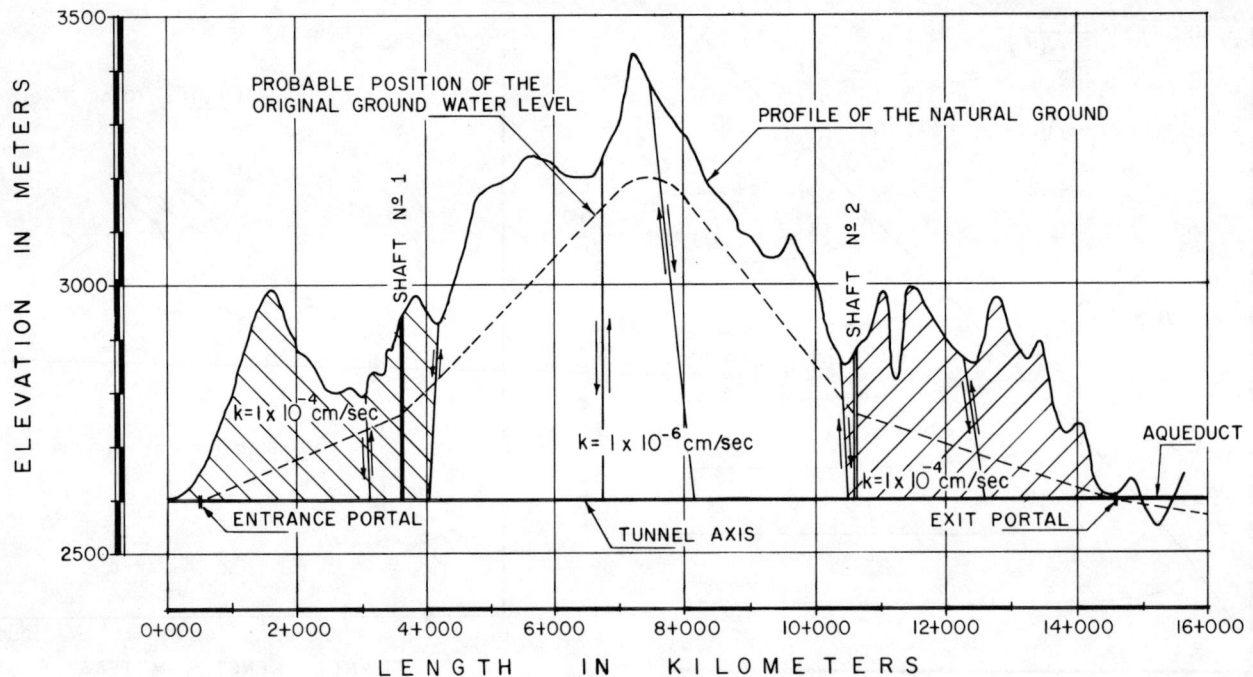

FIGURE 4.- GENERAL GEOHYDROLOGICAL CONDITIONS FOR A TUNNEL DRIVEN
THROUGH AQUIFERS WITH VARIABLE HYDRAULIC HEAD.

Total discharge into the tunnel at any time t_x and length $L_x = r \cdot tx$ is obtained from:

$$Q = \int_0^{L_x} q_x \cdot dL_x = r \int_0^{t_x} q_x \cdot dt_x$$

t_x being the time measured from the very ins--- tant the tunnel passed through the origin o.

If $\beta = \dfrac{H_M}{L_M} = \dfrac{H_x}{L_x}$ is the slope of

the hydraulic head, then:

$$T = \frac{K \mathcal{Z}}{H_x} = \frac{K}{\beta} \frac{\mathcal{Z}}{L_x} = \frac{K \cdot \mathcal{Z}}{\beta \cdot r \cdot t_x}$$

Denoting by $\lambda = \dfrac{K}{\beta \cdot r}$

it results: $T = \lambda \dfrac{\mathcal{Z}}{t_x}$

so: $q_x = k \cdot \beta \cdot r \cdot t_x \dfrac{\lambda \dfrac{\mathcal{Z}}{t_x} + 1}{\sqrt{\lambda \dfrac{\mathcal{Z}}{t_x} \left(\lambda \dfrac{\mathcal{Z}}{t_x} + 2 \right)}}$

and

$$Q = k \cdot \beta \cdot r^2 \int_0^{t_x} t_x \frac{\lambda \dfrac{\mathcal{Z}}{t_x} + 1}{\sqrt{\lambda \dfrac{\mathcal{Z}}{t_x} \left(\lambda \dfrac{\mathcal{Z}}{t_x} + 2 \right)}} dt_x$$

Calling t the time at which the total inflow into the tunnel is to be evaluated, it is established that $\mathcal{Z} = t - t_x$, because for small va-- lues of t_x, that is, in those sections close - to the origin, elapsed time $\mathcal{Z}$ to evaluate q_x is close to t and on the other extreme, for va- lues of t_x close to t, that is, for those -- sections close to the front of the excavation, the elapsed time $\mathcal{Z}$ to evaluate q_x is close to zero. Therefore:

$$Q(t) = k \cdot \beta \cdot r^2 \int_0^t t_x \frac{\lambda \dfrac{t - t_x}{t_x} + 1}{\sqrt{\lambda \dfrac{t - t_x}{t_x} \left(\lambda \dfrac{t - t_x}{t_x} + 2 \right)}} dt_x$$

Integration leads to the following expression for the total water inflow during tunnel exca vation progress:

$$Q(t) = k \cdot \frac{\beta \, rt}{2} \cdot rt \cdot \emptyset(\lambda)$$

Noting that $rt = L_x$ and $\beta \, rt = H_x$

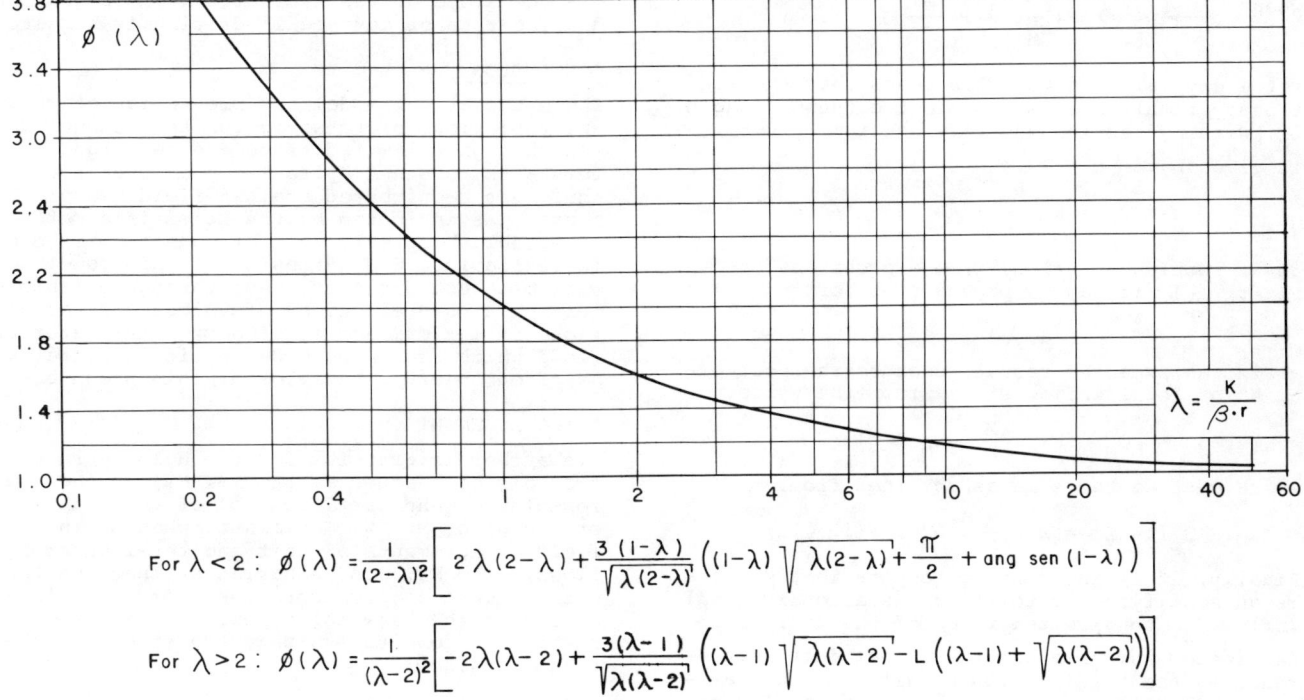

For $\lambda < 2$: $\phi(\lambda) = \dfrac{1}{(2-\lambda)^2}\left[\, 2\lambda(2-\lambda) + \dfrac{3(1-\lambda)}{\sqrt{\lambda(2-\lambda)}}\left((1-\lambda)\sqrt{\lambda(2-\lambda)} + \dfrac{\pi}{2} + \text{ang sen}\,(1-\lambda)\right)\right]$

For $\lambda > 2$: $\phi(\lambda) = \dfrac{1}{(\lambda-2)^2}\left[\, -2\lambda(\lambda-2) + \dfrac{3(\lambda-1)}{\sqrt{\lambda(\lambda-2)}}\left((\lambda-1)\sqrt{\lambda(\lambda-2)} - L\left((\lambda-1)+\sqrt{\lambda(\lambda-2)}\right)\right)\right]$

FIGURE 5.- AMPLIFYING FACTOR OF THE STATIONARY TOTAL INFLOW TO GIVE MAXIMUM TOTAL INFLOW
FOR TUNNELS DRIVEN IN INCREASING HEAD AQUIFERS.

$$Q(t) = k \cdot \frac{H_x}{2} \cdot L_x \cdot \phi(\lambda)$$

$\phi(\lambda)$ is a function with a dual expression depending on the values of λ; both expressions and their graphical representation are shown in figure No. 5.

As $\dfrac{H_x}{2} = H_m$ is the mean hydraulic head along the length L_x, last equation can be written:

$$\frac{Q(t)}{k \cdot H_m \cdot L_x} = \phi(\lambda)$$

which is a dimentionless parameter.

When the total tunnel length L_M with the increasing hydraulic head is reached at the time t_M, the maximum total inflow Q_M is present, with the value:

$$Q_M = Q(t_M) = k \cdot \frac{H_M}{2} \cdot L_M \cdot \phi(\lambda)$$

Thereafter the value of $Q(t)$ for that portion of the tunnel will show a reduction with time, as on every section of the tunnel, the unit discharge q_x is being reduced down to its stationary value. After excavation total inflow to that part of the tunnel is given by:

$$Q_p(t) = k \cdot \beta \cdot r^2 \int_0^{t_M} {}_{t_x} \frac{\lambda \frac{t-t_x}{t_x} + 1}{\sqrt{\lambda \frac{t-t_x}{t_x}\left(\lambda \frac{t-t_x}{t_x} + 2\right)}}\, dt_x$$

It can be noted that the integration extends only over the interval $0 \leq t_x \leq t_M$ thus considering the increase of $H_x = \beta r t_x$ in the interval $0 \leq H_x \leq H_M$ while the elapsed time $\mathcal{Z} = t - t_x$, for evaluation of the unit discharge q_x at every section of the tunnel, varies from $\mathcal{Z} = t$ corresponding to the section at the origin, to the value $\mathcal{Z} = t - t_M$ corresponding to the section at the distance L_M from that point.

Integration of the last equation leads to the following expression for the post-excavation total inflow:

$$Q_p(t) = k \cdot \frac{\beta r t_M}{2} \cdot r t_M \cdot F\left(\frac{t}{t_M} \cdot \lambda\right)$$

or:

$$Q_p(t) = k \cdot \frac{H_M}{2} \cdot L_M \cdot F\left(\frac{t}{t_M} \cdot \lambda\right)$$

$F\left(\dfrac{t}{t_M}, \lambda\right)$ is a function of the time t, in relation with the excavation elapsed time t_M, and of λ; its graphical representation is shown in figure No. 6 and its dual expression, depending on values of λ, is given at the end of the paper as an appendix.

From figure No. 6 can be noted that in the long term, that is:

when $t \longrightarrow \infty \Rightarrow F(\frac{t}{t_M}, \lambda) \longrightarrow 1$

for every value of λ. This means that the stationary total water inflow to the tunnel length L_M, with hydraulic head varying from 0 to H_M, - is given by:

$$Q_E = k. \frac{H_M}{2} . L_M$$

as it should be. Also, from expressions of $F(t/t_M, \lambda)$ it can be proved that for $t = t_M$:

$$F(1, \lambda) = \emptyset(\lambda)$$

again, an expected result, because the post-excavation total inflow starts from the value:

$$Q_M = k. \frac{H_M}{2} . L_M . \emptyset(\lambda)$$

which must be the same as obtained from:

$$Q_p(t=t_M) = k. \frac{H_M}{2} . L_M . F(1, \lambda)$$

Finally, it is interesting to note that $\emptyset(\lambda)$ - is an amplifying factor of the stationary total inflow Q_E to give the value of the maximum total inflow which enters the tunnel just when -- reaching the total length L_M with the increasing hydraulic head. This ^{M} amplifying factor decreases with increasing values $\lambda = \frac{K}{\beta \cdot r}$ that - is, the amplification will be greater if λ is - smaller, wich ocurrs when:

a.- The permeability coeficent is smaller,
b.- The efective porosity is greater,
c.- The slope of the hydraulic head is greater.
d.- The rate of advance in driving is higher.

CONCLUSION.

The mathematical model for prediction of values and variations of total inflow to a tunnel driven through unlimited, free and recharged aquifers with constant hydraulic head over the tunnel, has been tested against field measurements. As indicated by the Scientific Method - of establishing this type of models, this one - is realiable. The extension of this model to variable head aquifers, made throughout this paper leads to predictions of values and variations of maximum and stationary total water inflows which are in accordance with expectatives based on observed behavior of this phenomena.

ACKNOLEDGEMENTS.

The author is grateful for the help he received from his colleagues at GEOSISTEMAS during the - formulation and certification of the models described; from Mr. V. Gaxiola S. when making the field measurements of unit and total water inflows, from Mr. J.A. Albarrán at checking the - mathematical developments and from several --- others in the drawing, typewriting and the numerical work done to evaluate the several functions obtained.

REFERENCES.

BELLO-MALDONADO, A.A. 1974. Seepage towards tunnels. Proceedings 2nd. International Congress of the IAEG. Sao Paulo, Brazil, paper VII-5.

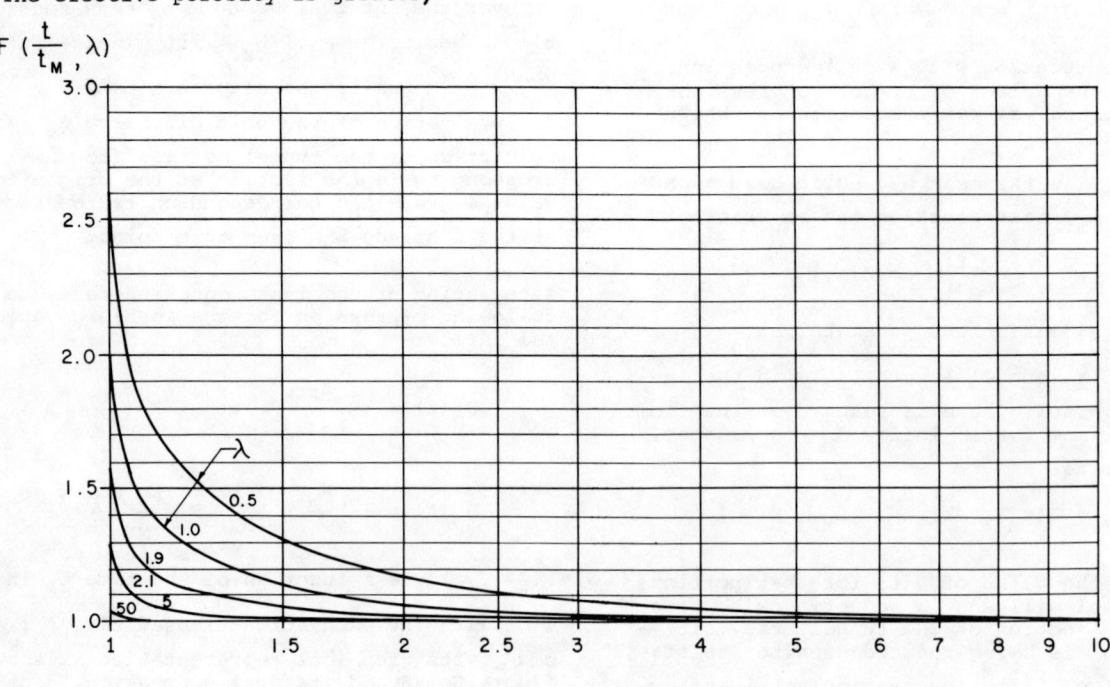

FIGURE 6.- VARIATION OF MAXIMUM TO STATIONARY TOTAL INFLOW
FOR TUNNELS DRIVEN IN INCREASING HEAD AQUIFERS.

APPENDIX:

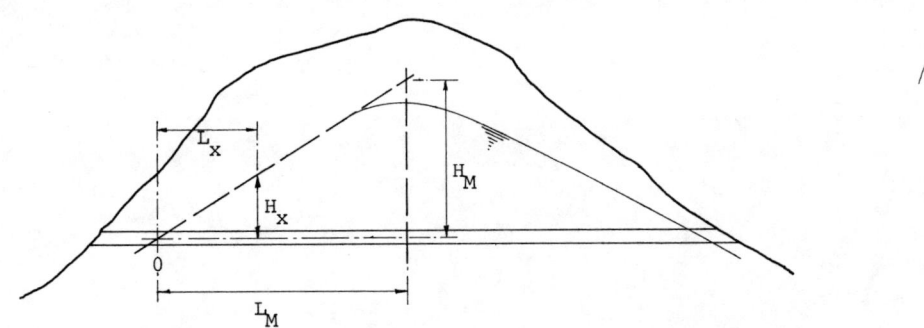

$$\beta = \frac{H_x}{L_x} = \frac{H_M}{L_M}$$

NOTATIONS FOR THE CASE OF INCREASING HYDRAULIC HEAD ALONG THE TUNNEL

EXPRESSIONS FOR THE FUNCTION $F\left(\frac{t}{t_M}, \lambda\right)$:

For $\lambda < 2$:

$$F\left(\frac{t}{t_M}, \lambda\right) = \frac{2}{\lambda}\left\{ -\sqrt{\lambda^2\left(1 - \frac{t_M}{t}\right)^2 + 2\lambda\frac{t_M}{t}\left(1 - \frac{t_M}{t}\right)} + \frac{t^2}{t_M^2}\frac{\lambda^2}{2-\lambda}\left[1 - \left(1 - \frac{t_M^2}{t^2}\frac{2-\lambda}{\lambda} + 2\frac{t_M}{t}\frac{1-\lambda}{\lambda}\right)^{3/2}\right] + \right.$$

$$+ \frac{t^2}{t_M^2}\cdot\frac{3}{2}\frac{(1-\lambda)\sqrt{\lambda(2-\lambda)}}{(2-\lambda)^3}\left[(1-\lambda)\sqrt{\lambda(2-\lambda)} - \left[(1-\lambda) - \frac{t_M}{t}(2-\lambda)\right]\sqrt{1 - \left[(1-\lambda) - \frac{t_M}{t}(2-\lambda)\right]^2}\right.$$

$$\left.\left. + \left(\text{ang sen}(1-\lambda) - \text{ang sen}\left[(1-\lambda) - \frac{t_M}{t}(2-\lambda)\right]\right)\right]\right\}$$

For $\lambda > 2$:

$$F\left(\frac{t}{t_M}, \lambda\right) = \frac{2}{\lambda}\left\{ -\sqrt{\lambda^2\left(1 - \frac{t_M}{t}\right)^2 + 2\lambda\frac{t_M}{t}\left(1 - \frac{t_M}{t}\right)} - \right.$$

$$- \frac{t^2}{t_M^2}\frac{\lambda^2}{\lambda-2}\left[1 - \left(1 + \frac{t_M^2}{t^2}\frac{\lambda-2}{\lambda} - 2\frac{t_M}{t}\frac{\lambda-1}{\lambda}\right)^{3/2}\right] +$$

$$+ \frac{t^2}{t_M^2}\cdot\frac{3}{2}\frac{(\lambda-1)\sqrt{\lambda(\lambda-2)}}{(\lambda-2)^3}\left[(\lambda-1)\sqrt{\lambda(\lambda-2)} - \left((\lambda-1) - \frac{t_M}{t}(\lambda-2)\right)\sqrt{\left[(\lambda-1) - \frac{t_M}{t}(\lambda-2)\right]^2 - 1} - \right.$$

$$\left.\left. - L\frac{(\lambda-1) + \sqrt{\lambda(\lambda-2)}}{\left((\lambda-1) - \frac{t_M}{t}(\lambda-2)\right) + \sqrt{\left[(\lambda-1) - \frac{t_M}{t}(\lambda-2)\right]^2 - 1}}\right]\right\}$$

J

ANALYSIS OF WATER LOSSES IN BASALTIC ROCK JOINTS

Analyse des pertes en eau à travers les fissures dans les roches basaltiques

Analyse der Wasserverluste in Klüften basaltischer Gesteine

P. T. Da Cruz
D.Sc. Engineer, Professor at Universidade de São Paulo-Brasil, Consulting Engineer,
Institute de Pesquisas Tecnológicas do Est. de São Paulo — IPT

E. F. De Quadros
M.Sc. Engineer, Researcher at The Division of Mines and Engineering Geology, Instituto de
Pesquisas Tecnológicas do Est. de São Paulo — IPT

SYNOPSIS

Modified water loss tests are proposed to improve the knowledge of water flow in rock discontinuities. Laboratory tests were performed in three types of rock fractures with varying surface roughness to analyse the nature of flow in rock discontinuities. Field and laboratory test results were compared. The concept of "equivalent opening" is introduced for flow prediction and analysis.

RESUME

L'étude de la percolation d'eau dans des massifs rocheux nous a amenés à la proposition d'une amélioration des essais du type Lugeon. Des essais en laboratoire ont été effectués sur trois types de fractures avec des rugosités différentes. La comparaison des résultats des essais en laboratoire avec les résultats des essais in situ a permis de formuler le concept d'"ouverture équivalente".

ZUSAMMENFASSUNG

Wir schlagen verschiedene Wasserverlustteste vor, um das Verständnis von Wasserfluß in Felsbrüchen zu verbessern. Laboratorium-Tests wurden an drei verschiedenen Arten von Felsbrüchen mit verschiedener Oberflächenrauhheit durchgeführt, um die Art des Flusses im Felsbruch zu bestimmen. Feld- und Laboratorium-Testergebnisse wurden verglichen. Wir führen den Begriff der "Aquivalenten Durchlässigkeit" für Flußvorhersage und -analyse ein.

1- MODIFIED WATER PRESSURE TEST

To investigate flow characteristics in a rock discontinuity at the site of Nova Avanhandava dam, in São Paulo, Brasil, six tests were performed using a modified procedure that inclu des several stages of pressure both on loading and unloading cycles. At the begining tests were performed with very low pressure stages, in cluding infiltration tests to make possible the analysis of a complete curve relating flow with effective head.

Typical tests results are shown in Figs. 1 and 2. The same results can be reploted as in Fig. 3 and 4 relating flow over effective head versus effective head.

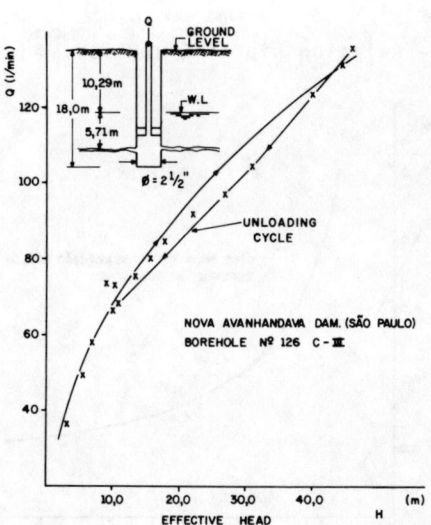

Fig. 1- Relationship of flow Q versus effective head H (field test).

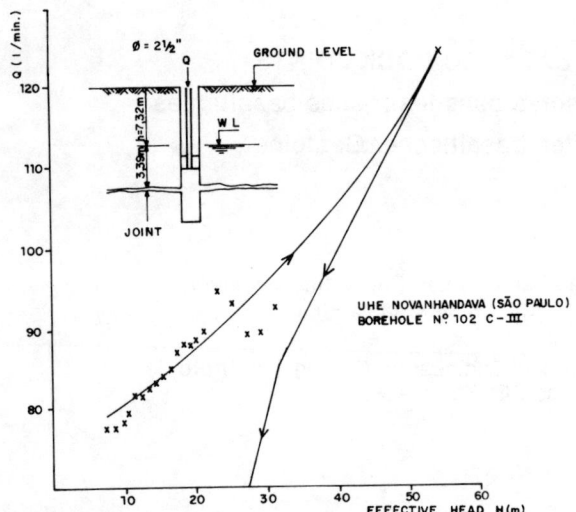

Fig. 2- Relationship of flow Q versus effective head H (field test)

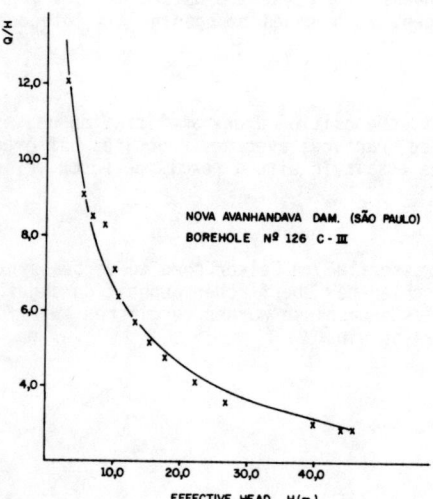

Fig. 3- Relationship of Q/H versus H (field test)

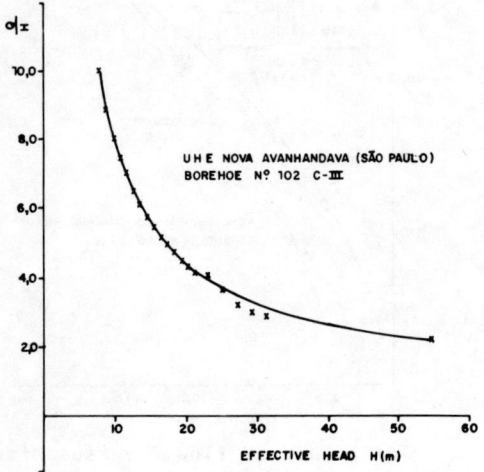

Fig. 4- Relationship of Q/H versus H (field test)

The non linearity of relation of flow to effective head is indicative of flow regimen, that should start as linear for very low pressures passing by a transitory regimen and leading to a turbulent flow.

Tables 1 and 2 shows the traditional water loss coefficient for laminar flow and the modified water loss coefficient for turbulent flow. These tables already includes an evaluation of opening of the joint using concepts of laminar and turbulent flow. The relationship of water loss coeficcient (WLC) to equivalent opening could be obtained from Fig. 5, considering that only one joint was involved within the tests and using conventional well formulae (Cruz, 1979).

TABLE 1 - Computation of the water loss coefficient (WLC) and opening for Borehole Nº 126 C-III

EFFECTIVE HEAD Ho(m)	WATER LOSS COEF. FOR LAM. FLOW(L/min x x m x atm)	WATER LOSS COEF. FOR TURB. FLOW(L/min x x m x atm)	LAMINAR OPENING e (cm)	TURBULENT OPENING e' (cm)
10,64	16,25	16,76	0,044	0,080
13,24	14,42	16,60	0,043	0,079
15,44	13,11	16,30	0,042	0,079
17,64	12,10	16,02	0,041	0,078
22,14	10,50	15,63	0,040	0,076
26,74	9,18	15,01	0,038	0,075
31,54	8,02	14,25	0,036	0,073
39,74	7,82	15,60	0,036	0,076
43,94	7,53	15,79	0,035	0,077
45,74	7,47	16,29	0,034	0,079

TABLE 2 - Computation of the water loss coefficient (WLC) and opening for Borehole Nº 102 C-III.

EFFECTIVE HEAD Ho(m)	WATER LOSS COEF. FOR LAM. FLOW(L/min x x m x atm)	WATER LOSS COEF. FOR TURB. FLOW(L/min x x m x atm)	LAMINAR OPENING e (cm)	TURBULENT OPENING e' (cm)
7,72	25,22	22,16	0,052	0,098
8,72	22,30	20,83	0,050	0,092
9,72	20,22	19,93	0,048	0,090
10,62	18,79	19,37	0,046	0,086
12,52	16,32	18,26	0,044	0,084
13,42	15,40	17,84	0,035	0,083
14,42	14,47	17,37	0,043	0,082
15,32	13,74	17,01	0,042	0,080
16,32	13,06	16,69	0,042	0,079
19,22	11,46	15,80	0,041	0,078
22,92	10,36	15,09	0,040	0,076
25,12	9,30	14,74	0,038	0,074
27,12	8,26	13,60	0,037	0,072
29,12	7,71	13,16	0,036	0,070
31,12	7,45	13,15	0,032	0,070
54,52	5,72	13,15	0,032	0,073

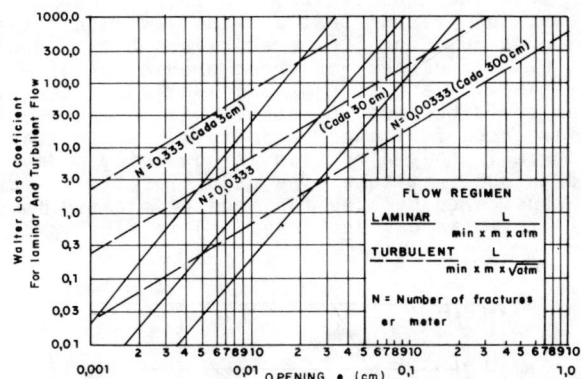

Fig. 5- Relationship of WLC (laminar and turbulent) to equivalent opening.

2- LABORATORY TESTS

In order to understand the nature of flow in rock discontinuities, hydraulic conductivity tests were performed on a very simple device using 2" samples, 30 cm in lenght and varying both the opening as well as the rugosity of the rock surfaces. Typical test results are shown in Figs.6, 7 and 8, relating flow with gradient (Quadros, 1982). The same non linearity of flow versus pressure (gradient) is obtained.

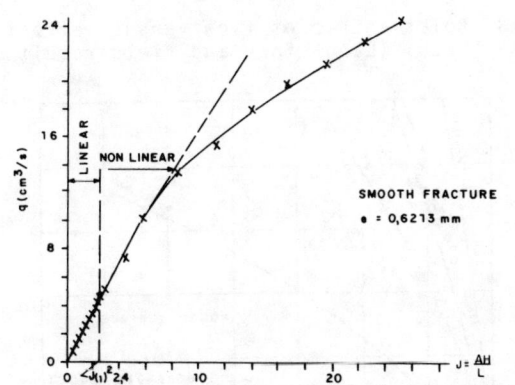

Fig. 6- Relationship of unit flow q versus gradient J.

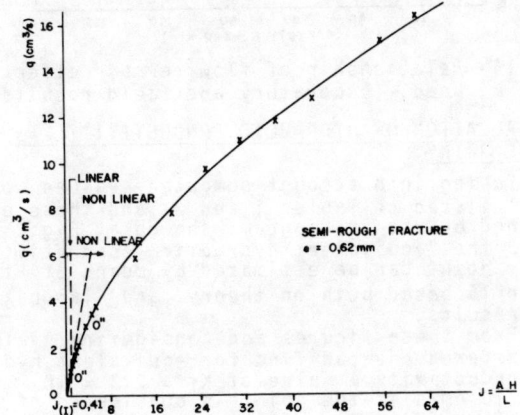

Fig. 7- Relationship of unit flow q versus gradient J.

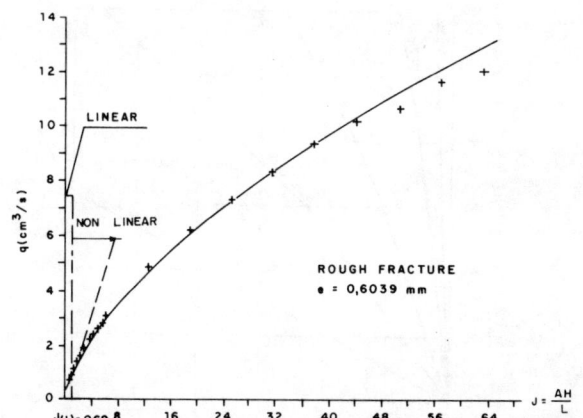

Fig. 8- Relationship of unit flow q versus gradient J.

In order to simulate a field water pressure test, laboratory test results had to be transformed in relations of flow versus effective head.

In the laboratory test flow develops on a rectangular section of aproximately constant width, as shown in Fig. 9 (Quadros, 1982).

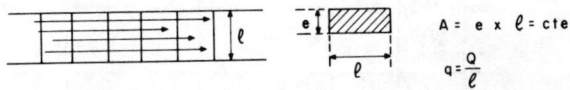

Fig. 9- Single joint with a constant rectangular section.

In a field water pressure test the flow is radial and in increasing areas. Fig. 10 shows the assumed radial distribution of flow.

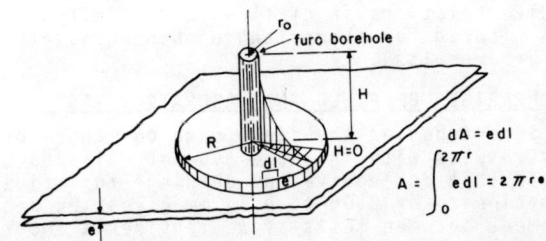

Fig. 10- Radial distribution of flow (Rissler, 1978).

Considering the above model and the empirical equations derived from regression analysis relating unit flow (q), gradient (J), opening (e) and absolute rugosity (k), diagrams relating flow with effective pressure were prepared for the three types of rock surfaces tested on the laboratory.

Fig. 11 shows test results for the smooth surface and Fig. 12 the test results for the three rock surfaces tested at the same opening (0,7 mm).

It is evident from the last figure the influence of the rock surface rugosity on the amount of flow for the same gradient (or the effective head).

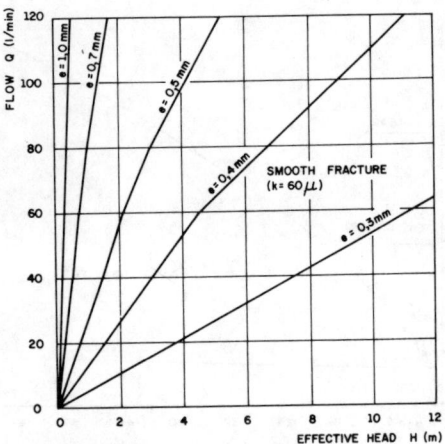

Fig. 11- Relationship of flow versus effective head for smooth surface. (Laboratory results).

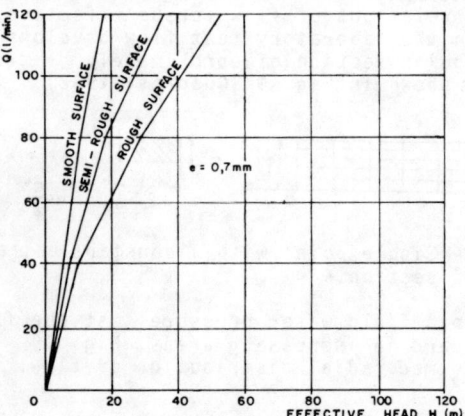

Fig. 12- Relationship of flow versus effective head for a constant opening.(Laboratory results).

3- COMPARISON OF FIELD AND LABORATORY TESTS

Superimposing the field data on laboratory tests results with the same type of relation we can see that an "equivalent opening" for field discontinuity of 0,60 to 0,70 mm gives the best agreement between tests if we consider a smooth surface but for a rough surface the best agreement is reached with an "equivalent opening" of about 0,80 to 1,0 mm. The test results shows a non linear relation of flow versus effective head, suggesting a transitory to turbulent flow.

A comparison between this value and those ones listed on Tables 1 and 2 shows that for test in Borehole nº 126 C-III an "equivalent opening" of 0,70 to 0,80 mm is obtained and for borehole nº 102 C-III a value of e varying from 0,70 to 0,10 mm for turbulent flow (see Fig. 5) corresponds to the best agreement.

It is important to observe that the "equivalent opening" for laminar flow is much smaller.

Rugosity is an important factor, because the head losses are increased. As a result the flow velocities are reduced, and consequently the hydraulic conductivity (Figs. 13 and 14).

The "equivalent opening" reflects the effective area where the water flows. In practice the fracture width measured directly (in cuts

on shafts) can be some orders of magnitude larger than the area (voids) in which the water flows. The present test and analysis are then quite usefull for flow predictions but should not be used either for grout prediction or compressibility and strenght evaluation. Correlation could be however, stablished in each particular case, for other parameters. A research program in this direction is now being developed in our Institute. The nature of flow must be considered because the indiscriminated use of well formulae can be misleading in flow predictions.

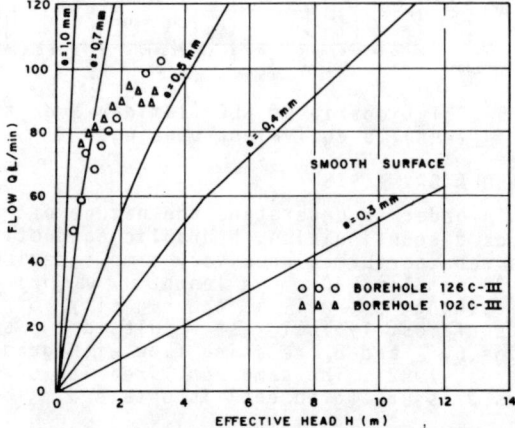

Fig. 13- Relationship of flow versus effective head. (Laboratory and field results).

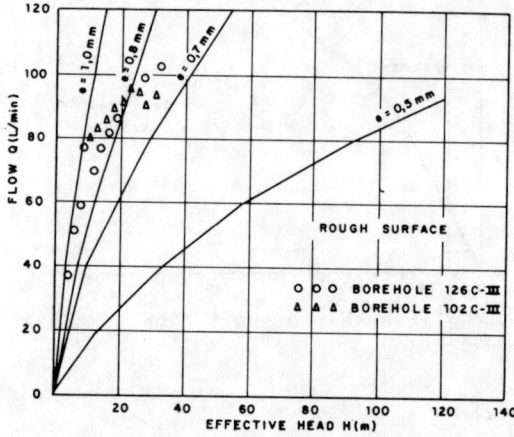

Fig. 14- Relationship of flow versus effective head. (Laboratory and field results).

4- EVALUATION OF HYDRAULIC CONDUCTIVITY K_f OF THE JOINT

Taking into account computed values of e and e' listed on Tables 1 and 2, and those ones obtained by use of diagrams showed at Figs. 12 and 13 the "equivalent hydraulic conductivity" of the joint can be estimated by means of Figs. 14 and 15 based both on theory and laboratory test results.

From these figures and considering values of e refered one can find for equivalent hydraulic conductivity a value of $K_f = 2,3 \times 10$ to $2,5 \times 10$ cm/s if the joint is considered of smooth surface (Fig. 14) and $K_f = 2,2 \times 10$ to $3,0 \times 10$ cm/s for a rough surface fracture (Fig. 15).

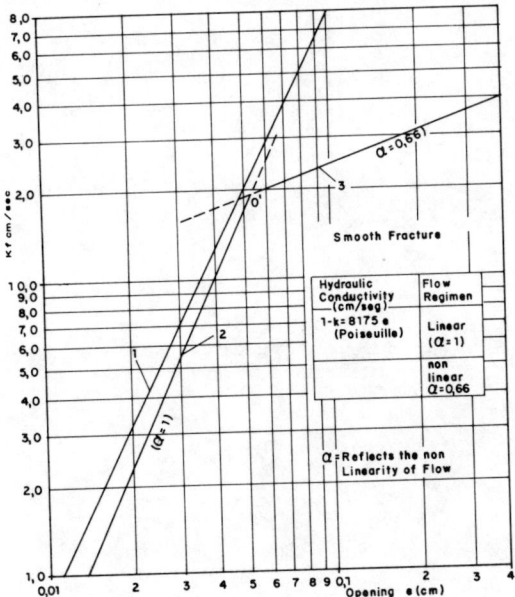

Fig. 15- Hydraulic Conductivity K_f versus equi
valent opening $\underline{e}$

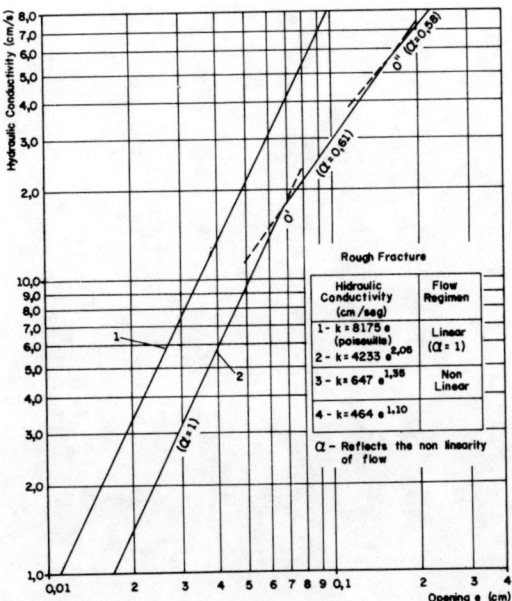

Fig. 16- Hydraulic Conductivity K_f versus equi
valent opening $\underline{e}$

5- CONCLUSIONS

Multiple stages water pressure test is an advance towards the understanding of flow of wa ter in rock fractures.

This test makes possible the knowledge of flow regimen involved and improve predictions of seepage into large escavations or into drains into the foundation of concrete structures which can be underestimated if the flow regimen is not correctly evaluated.

Laboratory and field tests shows that tran sitional or turbulent flow establishes with ve ry low pressure which means that care should be taken when applying laminar theory to explain results of water pressure test conducted in rock joints.

Results of tests for Nova Avanhandava dam shows that flow in practical problems of seepa ge trought rock fractures is most of the time of non linear nature (transitory or turbulent flow).

AKNOWLEDGMENTS

Authors want to express the aknowledgments for the Insituto de Pesquisas Tecnológicas do Estado de São Paulo-IPT and Companhia Energéti ca de São Paulo-CESP for the opportunity of pre senting this paper.

REFERENCES

ASSOCIAÇÃO BRASILEIRA DE GEOLOGIA DE ENGENHARIA (ABGE)-1975 - Ensaio de perda d'água sob pres são; diretrizes. São Paulo. 16p. (ABGE, Bo letim 02).

CRUZ, P.T. - 1979 - Contribuição ao estudo do fluxo da água em meios contínuos. São Paulo, IPT (pre-print.).

CRUZ, P.T. and all - 1982 - Evaluation of ope ning and hydraulic conductivity of rock dis continuities. In: American Symposium on Rock Mechanics, Berkeley, 1982.

LOUIS, C. - 1969 - Étude des ecoulements d'eau dans les roche fissurées et de leurs influen ces sur la stabilité des massifs rocheurs.Bul letin de la Direction des Études et recher ches. Serie A (3): 5-132. (These presentee à l'Université de Karlsruhe).

QUADROS, E.F. - 1982 - Determinação das caracte rísticas do fluxo de água em fraturas de ro chas. São Paulo. (Dissertação de Mestrado, Es cola Politécnica - Universidade de São Paulo).

RISSLER, J.C. - 1978 - Determination of the wa ter permeability of jointed rock. Aachen,Ins titute for Foundation Engineering, Soil Me chanics, Rock Mechanics and Water Ways Cons truction.

EIN ÜBERARBEITETES KONZEPT DES WASSERABPRESSVERSUCHS

A revised concept of the water pressure test
Une conception revisée de l'essai de Lugeon

N. J. Schneider

Dr.Ing. Dipl.Geol., Fachbereichsleiter Untertagespeicher Fels Kavernen Bau- und Betriebs-
GmbH Hannover, Bundesrepublik Deutschland

ZUSAMMENFASSUNG

Ein überarbeitetes Konzept des Wasserabpreßversuches wird vorgestellt. Änderungen zum herkömmlichen Verfahren bestehen bei der Versuchsapparatur, der Versuchsdurchführung und der Versuchsauswertung. Verbesserungen der Meßgenauigkeit ergeben sich durch die Verwendung elektrischer Druckaufnehmer im Verpreßsegment des Bohrloches und induktiver Durchfluß-mengenmesser sowie analoger Meßdatenerfassung oder gleichwertiger digitaler Verfahren. Die Versuchssteuerung über eine konstante Verpreßmenge mit zeitäquivalenten Pausen zwischen den Verpreßstufen ermöglicht die Durchlässigkeitsbestimmung mit Verfahren, die den tatsächlich vorliegenden instationären Strömungsbedingungen entsprechen.

SYNOPSIS

A revised concept of the water pressure test is being presented. Compared with the conventional procedure, the testing apparatus, procedure and assessment have all been altered. The accuracy of measurement is improved by the use of electrical pressure sensors in the pressure segment of the drilling hole, the use of inductive flow-meters, as well as the recording of measuring data by analog or digital processes of identical accuracy. The control of the experiment by means of a constant quantity under pressure and intervals of equal time between the various stages of the pressure test make it possible to determine the porosity through processes which correspond in fact to the actual flow conditions.

RESUME

On présente un concept modifié de l'essai de Lugeon. Le dispositif et le protocole des tests ainsi que l'évaluation des résultats ont tous été modifiés par rapport au procédé courant. La précision des mesures est améliorée grâce à l'emploi des sondes de pression électriques insérées dans le segment de pression du trou de forage et d'appareils de mesure d'écoulement inductifs ainsi que par l'enregistrement de données de mesure par des calculateurs analogues ou numériques de précision identique. Le contrôle des expériences est rendu possible grâce à une quantité d'eau constante sous pression appliquée à des intervalles de temps égaux aux divers stades de l'essai, ce qui permet de préciser la porosité par des processus qui correspondent aux conditions d'écoulement réelles.

1. EINFÜHRUNG

Der Wasserabpreßversuch gehört im Felsbau immer noch zu den gebräuchlichsten Versuchsverfahren, da er einfach in der Durchführung ist, keine langen Versuchszeiten benötigt und keine aufwendige Versuchsapparatur beansprucht. Diese Versuchsmethode wird heute sowohl zur Durchlässigkeitsprüfung des Gebirges als auch zur Planung und Prüfung von Injektionsarbeiten verwendet. In seiner herkömmlichen Form birgt der Wasserabpreßversuch eine Reihe von systematischen Fehlern, die zu groben Fehlinterpretationen der Ergebnisse führen. Diese lassen sich durch eine Änderung des apparativen Aufbaus, der Durchführung und der Auswertung des Versuchs korrigieren.

Aufgrund der Komplexität der Strömungsvorgänge im Fels ergeben sich außerdem eine Reihe von Problemen, die bei der Durchführung und Auswertung besonders zu berücksichtigen sind. Die Wasserführung im Fels findet in der Regel in drei Systemen statt - dem Gestein, den Klüften und Störungen. Diese unterscheiden sich in ihren Durchlässigkeitseigenschaften stark voneinander (Abb. 1). Einzelne Singularitäten wie Großklüfte oder Störungszonen können die Durchlässigkeit ganzer Gebirgsbereiche dominieren.

Bei der begrenzten Reichweite des Wasserabpreß-versuchs kann weiterhin der Fall eintreten, daß eine statistische Regelung des Kluftgebirges in dem untersuchten Gebirgsbereich nicht vorliegt und somit das Versuchsergebnis nur die Durch-lässigkeit einzelner Kluftindividuen weitergibt. Die Kenntnis über die Beschaffenheit des Kluft-gefüges ist deshalb eine wesentliche Voraus-setzung für eine folgerichtige Durchführung und Auswertung der Versuche. Die Ermittlung der Durchlässigkeit kann sowohl für einzelne Klüfte als auch für quasi homogenes Gebirge nach dem diskontinuierlichen bzw. nach dem kontinuierli-chen Durchlässigkeitsmodell erfolgen (WITTKE 1970). Da jedoch analytische Lösungen für die diskontinuierliche Durchlässigkeit auf einfache Fälle beschränkt sind (RISSLER 1977), muß die Versuchsauswertung ohne numerische Hilfsmittel überwiegend nach dem kontinuierlichen Modell erfolgen.

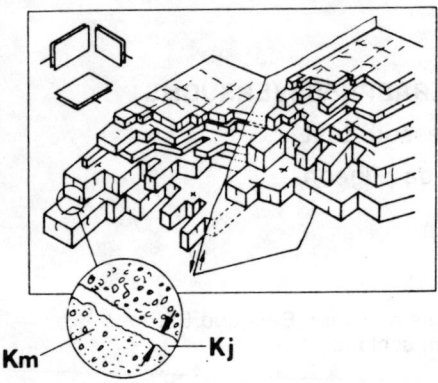

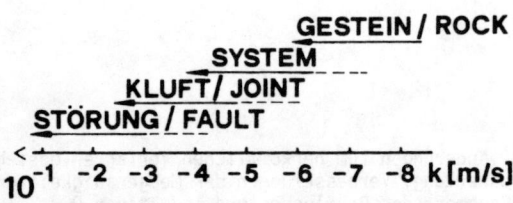

Abb. 1: Durchlässigkeits- und Wasser-
führungssysteme in Fels

Das nachfolgend vorgestellte Versuchskonzept
des Wasserabpreßversuches dient zur Bestimmung
der durchschnittlichen Durchlässigkeit und zur
Prüfung der Injizierbarkeit von Fels. Der Versuch kann sowohl als Einbohrlochversuch wie als
Mehrbohrlochversuch durchgeführt werden. Zur
Ermittlung der Durchlässigkeit wird vorausge-
setzt, daß die Verpreßstelle unterhalb des
Grundwasserspiegels liegt, die Störung recht-
winklig zum Bohrloch verläuft, der Aquifer ge-
spannt ist und instationäre Strömungsbedingungen vorliegen.

2. VERSUCHSAPPARATUR

Die Versuchsapparatur besteht aus einer Wasser-
pumpe mit Tank, einem Durchflußmengenmeßgerät,
einem Absperrschieber am Bohrlochkopf, einem
Doppelpackersystem und einer Meßsonde. Im Mehrbohrlochversuch sind weitere Packersysteme und
Sonden erforderlich (Abb. 2).

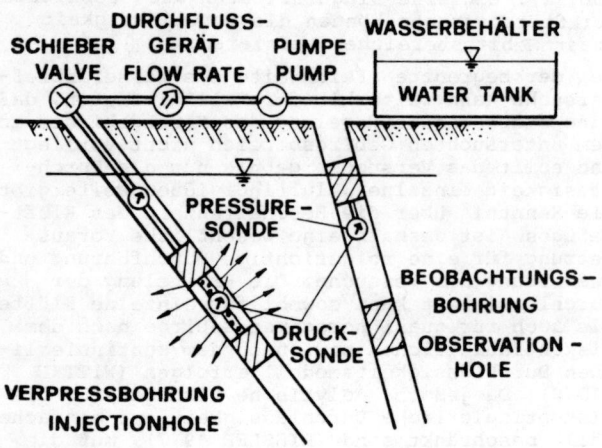

Abb. 2: Apparatur des Wasserabpreßversuches

Die Dichtung des Verpreßsegments erfolgt mit
einem Doppel- oder Mehrfachpacker. Pneumatische
Packer haben sich zur Dichtung des Verpreßsegments in der Injektionsbohrung durch ihre Fähigkeit, sich den Unebenheiten der Bohrlochwandung
anzupassen, besonders bewährt. Die Manschetten-
länge der Packer sollte 0,6 m nicht unterschrei-
ten, größere Längen bis zu 1,5 m gewähren grös-
sere Dichtungserfolge. Die Abdichtung der Seg-
mente in den Beobachtungsbohrungen erfolgt in
derselben Weise. Im Verpreßsegment und in den
Beobachtungssegmenten ist eine Meßsonde zur Beobachtung des Verpreßdruckes bzw. Bergwasser-
drucks erforderlich. Die Messung erfolgt mit
elektrischen Druckaufnehmern. Über dem Packersy-
stem wird noch die freie Wassersäule registriert.
Zur Bestimmung der Verpreßmenge wird ein induk-
tives Durchflußmengenmeßgerät verwendet. Die
Verpressung wird mit einer mechanisch regelbaren
Pumpe vorgenommen, die eine konstante Mengenför-
derung im Bereich von ca. 1 - 150 l/min erlaubt.
Am Bohrlochkopf ist eine Schließvorrichtung mittels Schieber erforderlich. Die Versuchsdaten
sind kontinuierlich und lückenlos zu erfassen.
Hierzu kann entweder eine analoge Aufzeichnung
über einen Mehrkanalschreiber mit Parameter-Zeit-
achsenregistrierung oder eine elektronische Da-
tenerfassungsanlage mit digitaler Registrierung
verwendet werden. Bei letzterer sollten die Zeit-
abstände zwischen den einzelnen Messungen 0,5
Sekunden nicht überschreiten.

3. VERSUCHSVERFAHREN

Die Bohrungen sind entsprechend dem Kluftgefüge
normal zu den einzelnen Kluftscharen auszurich-
ten. Vor der Durchführung der Versuche ist im
Bohrloch das Kluftgefüge mit Hilfe einer opti-
schen Bohrlochsonde, einer Bohrlochfernsehsonde
oder einem Televiewer einzumessen. Die geologische Abfolge der Schichten ist an den Bohrkernen
zu ermitteln. Im Mehrbohrlochversuch sind die Beobachtungsbohrungen in drei Richtungen in einem
Abstand von 2 - 20 m und die Verpreßbohrung an-
zuordnen. Die Position und Länge des Verpreßsegments ist entsprechend den Ergebnissen der
Bohrkernaufnahme und den Gefügeaufnahmen im Bohrloch festzulegen. Es ist dabei darauf zu achten,
daß einzelne Schichten gleicher Petrographie und
gleicher Klüftigkeit gesondert untersucht werden,
damit die hydrologische Charakteristik der be-
treffenden Schicht als ungespannter, halbgespann-
ter oder gespannter Aquifer bei der Auswertung
berücksichtigt werden kann. Großklüfte, z.B.
Schichtflächen oder Störungszonen sollten als
mögliche Hauptwasserwege gesondert untersucht
werden. Die Länge des Verpreßsegments kann je
nach Problemstellung, ob eine Durchlässigkeitsprüfung für größere Gebirgsbereiche oder für
einzelne Klüfte gewünscht wird, von ca. 50 m
bis 0,2 m schwanken.

Für die im Felsbau in der Regel angetroffenen
Gebirgstypen reicht eine Verpreßlänge von 5 m
aus. Durch eine Verlängerung des Verpreßsegments
kann die Versuchsgenauigkeit - insbesondere bei
gering permeablen Gesteinen - gesteigert werden.

In den Versuchen wird die Verpressung in einzelnen Mengenstufen in ansteigender und absteigender Reihenfolge vorgenommen (Abb. 3). Die Verpreßmenge wird pro Stufe konstant gehalten. Die
Verpreßdauer pro Stufe beträgt i.a. 10 Minuten,
kann jedoch, sofern eine größere Reichweite des
Versuches erwünscht ist, auf eine Stunde und

mehr erhöht werden. Vor der Verpressung wird die
gewünschte Fördermenge an der Pumpe reguliert
und über einen By-Pass abgeführt. Die Verpressung beginnt durch schnelles Öffnen des Schie-
bers am Bohrlochkopf. Am Ende der Stufe wird
die Wasserzufuhr zum Verpreßsegment mittels Ab-
sperrventil ebenso rasch unterbrochen. Zwischen
den einzelnen Verpreßstufen sind Pausen einzu-
legen, die mindestens das 1- bis 3-fache der
Verpreßzeit betragen. Die Druckmessung ist sowohl während der Verpressung als auch in den
Pausen vorzunehmen.

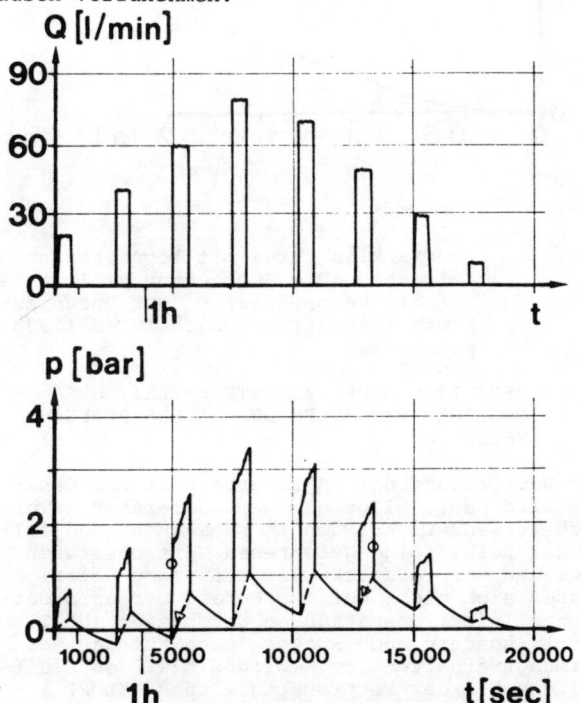

Abb. 3:   Mehrstufen-Wasserabpreßversuchs-
          Diagramme nach dem überarbeiteten
          Versuchskonzept mit Pausen zwischen
          den einzelnen Verpreßstufen
          Verpreßmengen (Q)-Zeit Diagramm
          Verpreßdruck (P)-Zeit Diagramm

4.   BESTIMMUNG DER DURCHLÄSSIGKEIT UND SPEICHERKAPAZITÄT

Die Strömung beim Wasserabpreßversuch ist i.a.
instationär, d.h., bei konstanter Verpreßmenge
steigt der Verpreßdruck bzw. der Bergwasserdruck
im Gebirge stetig an (Abb. 4). Liegt ein gespannter Aquifer vor, so kann die Durchlässigkeit nach COOPER & JACOB (1946) aus der zeitlichen Druckänderung ermittelt werden. Hierzu ist
die Druckentwicklung auf logarithmischer Zeitachse aufzutragen (Abb. 5). Aus dem linearen
Druckanstieg Δ h pro logarithmischer Einheit,
der Verpreßmenge Q und der Länge des Verpreßsegments L resultiert die durchschnittliche
Durchlässigkeit wie folgt:

$$K = \frac{Q}{4 \pi \Delta h L}$$

Aus der Druckmessung in den Beobachtungsbohrungen kann ferner der Speicherkoeffizient S des
Gebirges ermittelt werden. Hierzu wird die line-
are Druckanstiegskurve bis zur logarithmischen
Zeitachse verlängert und die Zeit t_0 bei Druck
O bestimmt (Abb. 5). Der Speicherkoeffizient S
errechnet sich daraus wie folgt:

$$S = \frac{2.25 \, KL \, t_0}{r^2}$$

K = Durchlässigkeitsbeiwert
L = Länge des Verpreßsegments
r = Abstand zwischen Verpreßbohrung und Beob-
    achtungsbohrungen

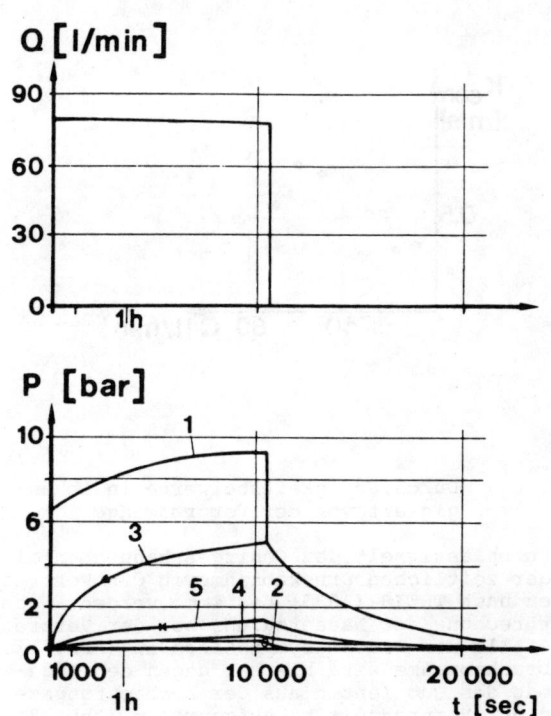

Abb. 4:   Druck-Zeit-Diagramm eines Wasserab-
          preßversuchs mit konstanter Verpreß-
          menge und Druckabklingkurve in der
          Verpreßpause, Druck im Verpreßsegment
          (1), Druck im Gebirge in 20 m (2),
          10 m (3), 15 m (4), 5 m (5) Entfer-
          nung von der Injektionsstelle

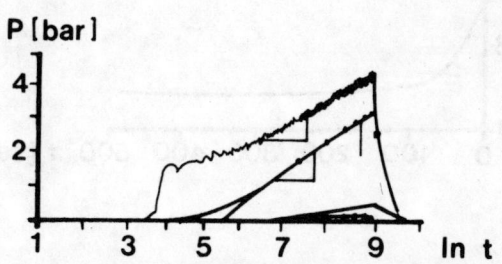

Abb. 5:   Derselbe Versuch wie in Abb. 4 mit
          logarithmischer Zeitachse zur Durch-
          lässigkeitsermittlung nach Cooper &
          Jacob (1946)

Die Durchlässigkeitsbeiwerte K sind vom Verpreßdruck bzw. von der Verpreßmenge abhängig,
da das Hohlraumvolumen des Kluftgefüges durch
druckabhängige Kluftaufweitung eine Veränderung
erfährt (Abb. 6).

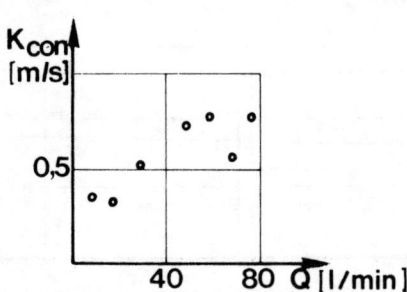

Abb. 6:   Durchlässigkeitsbeiwerte in Abhän-
           gigkeit von der Verpreßmenge

Die Durchlässigkeit des Gebirges kann ebenfalls
aus der zeitlichen Druckabnahme in den Verpreß-
pausen nach THEIS (1935) bestimmt werden. Nach
Unterbrechung der Wasserzufuhr bei der Verpressung fällt der Druck asymptotisch ab (Abb. 7).
Die Druckabnahme wird hierbei gegen den Loga-
rithmus des Quotienten aus der Beobachtungszeit
t und der Verpreßzeit t_o aufgetragen (Abb. 8).
Aus der Druckminderung Δh pro logarithmischer
Einheit, der vorangegangenen Verpreßmenge und
der Länge des Verpreßsegments L ergibt sich die
Durchlässigkeit K wie folgt:

$$K = \frac{Q}{4 \pi L \Delta h}$$

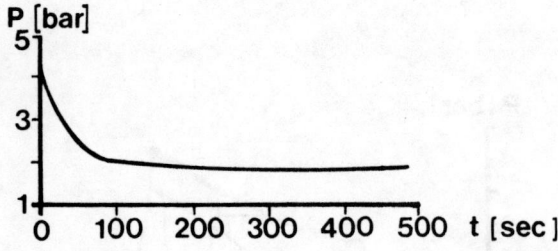

Abb. 7:   Durckabklingkurve in der Verpreß-
           pause

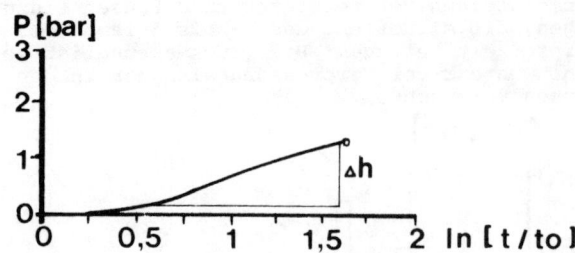

Abb. 8:   Druckabklingkurve mit logarithmischer
           Zeitachse. Die Beobachtungszeit t ist
           auf die Verpreßzeit t_o zur Durchläs-
           sigkeitsermittlung nach THEIS (1935)
           normiert.

5.   BEURTEILUNG DER INJIZIERBARKEIT DES
      GEBIRGES MIT HILFE DES WASSERABPRESS-
      VERSUCHS

Zur Beurteilung der Injizierbarkeit des Gebirges wird heute allgemein der Wasserabpressver-
such verwendet. Aus den in steigender und fal-
lender Reihenfolge gefahrenen Versuchsstufen
bzw. den Verpreßdruck-/Verpreßmengen-Relationen
lassen sich nicht nur der Umfang der erforder-
lichen Injektionsmaßnahmen abschätzen (HEITFELD
1965), sondern auch Rückschlüsse auf das Verformungsverhalten des Gebirges bzw. des Kluft-
gefüges bei der Verpressung ziehen (KLOPP &
SCHIMMER 1977) sowie geeignete Verpreßdrücke
festlegen. Die Form der Kennlinie im Verpreß-
druck-/Verpreßmengen-Diagramm gibt an, welches
Fließregime - laminar, turbulent - vorliegt
(LOUIS 1974), welche Verformungen - reversibel,
irreversibel - des Gebirges auftreten und/oder
ob Neubrüche im Gebirge durch Hydraulic-Frac-
turing gebildet werden.

Im herkömmlichen Wasserabpreßversuch (Abb. 9)
werden die relativen Druckwerte in den einzelnen Stufen jeweils auf den Ausgangsdruck zu
Beginn des Versuchs bezogen, was zu Fehlinterpretationen führt, da sich die hydraulische Aus-
gangssituation im Bohrloch und dem umgebenden
Gebirge ständig ändert. Diese Verfahrensweise
führt im Druckmengen-Diagramm i.a. dazu, daß
bei absteigender Stufenfolge die Wasseraufnahme
des Gebirges infolge des gestiegenen Bergwasser-
druckes generell kleiner ist, als bei ansteigender Stufenfolge (Abb. 10). Werden jedoch zwischen den einzelnen Versuchsstufen Pausen einge-
schaltet (Abb. 3), so wird dem Wasserdruck im
Verpreßsegment und im Gebirge die Möglichkeit
gegeben, sich zu entspannen. Hiermit lassen sich
für jede einzelne Versuchsstufe nahezu dieselben eindeutig bestimmbaren Ausgangsbedingungen
erzielen. Wie der Vergleich des Mehrstufenver-
suchs, der nach der herkömmlichen (Abb. 10) und
nach der neuen (Abb. 11) Versuchsmethode durchgeführt wurde, zeigt, wird beim herkömml.Verfah-
ren durch den falschen Bezug auf den Ausgangsdruck zu Beginn des Versuchs eine irreversible

Gebirgsverformung vorgetäuscht. Die Ergebnisse
nach dem neuen Verfahren zeigen, daß die Vor-
gänge der Kluftaufweitung infolge der Verpressung tatsächlich reversibel sind. Der Verlauf
der Verpreßdruck-/Verpreßmengen-Relation im
herkömmlichen WD-Versuch ist deshalb eindeutig
auf die Versuchsbedingungen zurückzuführen und
spiegelt daher   keine Gebirgscharakteristik
wieder.

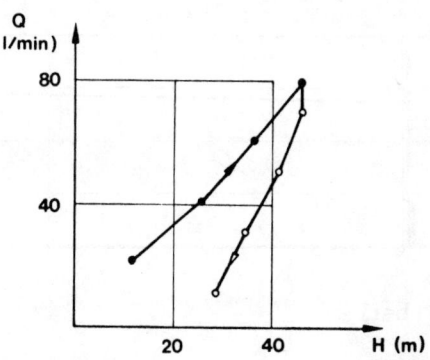

Abb. 10:   Verpreßmengen-Verpreßdruck-Kurve
            eines Mehrstufenversuchs nach her-
            kömmlicher Versuchsdurchführung
            ohne Pausen (derselbe Versuch wie
            Abb. 9), ansteigende Stufenfolge (o)
            absteigende Stufenfolge (O)

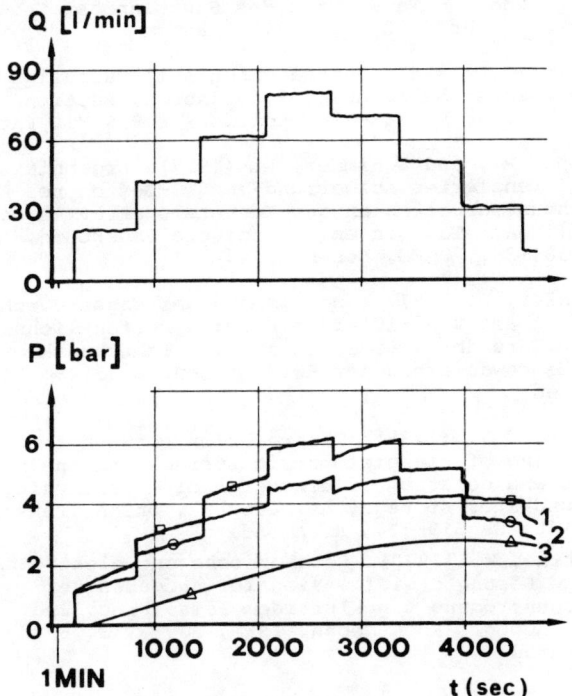

Abb. 9:   Mehrstufen-Wasserabpreßversuchs-
            Diagramme nach dem herkömmlichen
            Versuchskonzept
            Verpreßmengen (Q)-Zeit Diagramm
            Druckverlauf (P)-Zeit Diagramm
            aus Pumpenabgang (1), im Verpreß-
            segment (2), im Gebirge (3) in 10 m
            Entfernung von der Verpreßstelle

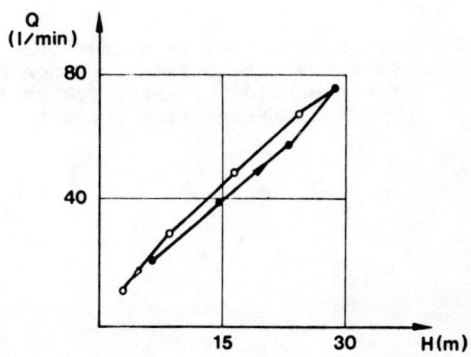

Abb. 11:   Verpreßmengen-Verpreßdruck-Kurve
            eines Mehrstufenversuchs nach über-
            arbeiteter Versuchsanordnung, der
            an demselben Ort wie der Versuch in
            Abb. 10 durchgeführt wurde
            (derselbe Versuch wie Abb. 3)

In Versuchen mit konstanter Verpreßmenge können
Neubrüche eindeutig am Druckabfall bei zunehmender Verpreßdauer erkannt werden (Abb. 12).

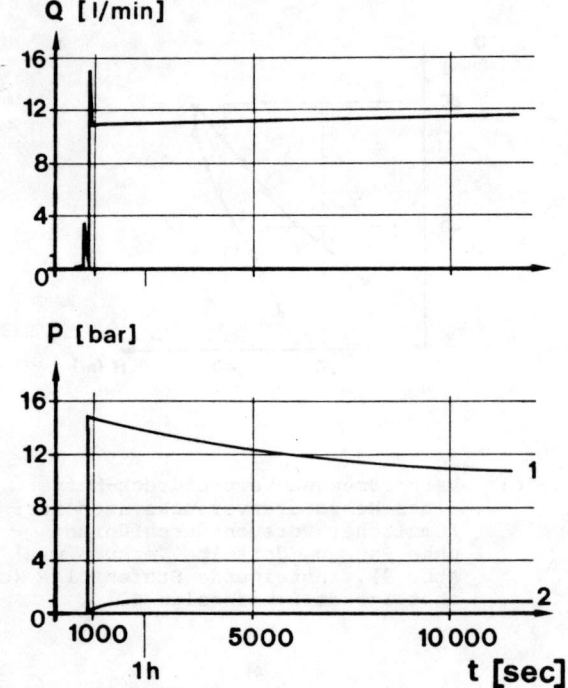

Abb. 12: Druckverlauf eines Wasserabpreß-
versuchs mit konstanter Menge bei
Kluftneubildung, vergleichbar dem
Hydraulic Fracturing Versuch

LITERATUR

Cooper, H.H., und Jacob, C.E. (1946).
A generalized graphical method for evaluating
formation constants and summarizing well-field
history: Trans. Am. Geophys. Union, Nr. 27,
S. 526-534, Richmond

Heitfeld, K.H. (1965). Hydro- und baugeologische
Untersuchungen über die Durchlässigkeit des Un-
tergrundes an Talsperren des Sauerlandes: Geol.
Mitt., 5, Nr. 1/2, S. 1-210, Aachen

Louis, C. (1974). Introduction à l'hydraulique
des roches: Bull. BRGM, 2ième série, Section
III, Nr. 4, S. 283-356, Editions B.R.G.M., Paris

Klopp, R., und Schimmer, R. (1977). Ergebnisse
differenzierter Auswertung von WD-Testen bei Ab-
dichtungsarbeiten an der Möhnetalsperre: Berichte 1. Nat. Tag. Ingenieurgeologie Paderborn,
S. 381-392, DGEG, Essen

Rissler, P. (1977). Bestimmung der Wasserdurch-
lässigkeit von klüftigem Fels: Veröffentlichung
Inst. Grundbau, Bodenmechanik, Felsmechanik und
Verkehrswasserbau der RWTH Aachen, H. 5, S. 144,
Aachen

Theis, C.V. (1935). The relation between the
lowering of the piezometric surface and the
rate and duration of discharge of a well using
groundwater storage: Am. Geophys. Union Trans.,
Nr. 16, S. 519-524, Richmond

Wittke, W. (1970). Rechnerische und elektroana-
loge Lösung dreidimensionaler Aufgaben der
Durchströmung von klüftigem Fels: Proc. 2nd
Int. Congr. Rock Mechn. ISRM, Bd. 3, S. 6-18,
Belgrad

EXPERIMENTAL STUDY ON ROCK SLOPE STABILITY BY THE USE OF A CENTRIFUGE

Etude expérimentale sur la stabilité des talus rocheux au moyen du centrifugeur

Experimentelle Untersuchung zur Böschungsstabilität unter Verwendung einer Zentrifuge

Katsuhiko Sugawara
Associate Professor, Dept. of Mining Eng., Kumamoto Univ., Kumamoto, Japan

Masatsugu Akimoto
Associate Professor, Dept. of Civil Eng., Kyushu-Tokai Univ., Kumamoto, Japan

Katsuhiko Kaneko
Research Associate, Dept. of Mining Eng., Kumamoto Univ., Kumamoto, Japan

Hiroshi Okamura
Professor, Dept. of Mining Eng., Kumamoto Univ., Kumamoto, Japan

SYNOPSIS

A series of physical model experiments with the use of a centrifuge are carried out to make clear the progressive failure process of rock slopes and the influence of discontinuities. The initiation of the progressive failure process, the stable-unstable transition and the limiting equilibrium have been discussed by monitoring the strain on the upper surface of the slope. It is pointed out that the stable-unstable transition may present a designing method of rock slope by the peak strength of the rock mass.

RESUME

On a effectué une série d'expériences sur modèle réduit à l'aide d'un centrifugeur afin d'éclairer le processus de déformation progressive des pentes de talus rocheux de même que l'influence des discontinuités. En se basant sur le contrôle de la contrainte exercée dans la partie supérieure de la pente, on analyse l'amorce du processus de ruptures progressif, la transition stable-instable et les équilibres limites. On signale que la transition stable-instable peut permettre de formuler une méthode pour le tracé des talus à l'aide des données de résistance maximum de la roche.

ZUSAMMENFASSUNG

Es wird eine Serie von Modellversuchen unter Verwendung einer Zentrifuge durchgeführt, um den progressiven Versagensverlauf von Felshängen und den Einfluß von Diskontinuitäten zu klären. Die Einleitung des progressiven Versagensverlaufs, der Übergang vom stabilen zum instabilen Zustand und das Grenzgleichgewicht werden an Hand der Beobachtung der Dehnung an der oberen horizontalen Fläche des Abhanges diskutiert. Es wird darauf hingewiesen, daß der Übergang vom stabilen zum instabilen Zustand eine Entwurfsmethode für Felshänge unter Verwendung der maximalen Festigkeit der Felsmasse geben kann.

1. Introduction

The fact that movements of the rock mass forming a slope can occur for many years before the slope finally collapses suggests that the failure process is progressive rather than instantaneous as is assumed in most forms of stability analyses. Understanding of the progressive failure mechanism is very important in designing a slope. Timedepending phenomena such as weathering and creep play an important role in this process but it is considered that the initiation of this process must be concerned with the stress state and the strength of the potential failure surface. In this paper, by analyzing the results of physical model experiments with use of a centrifuge, the progressive failure process of rock slopes will be discussed. A circular failure, which will occur when the material forming the slope is homogeneous, and the influence of discontinuities will be analyzed.

2. Experimental apparatus

The present model experiment with use of a centrifuge is based on the principle that if in the small model, made in every part of the same materials as its prototype, the pull of gravity on each part can be increased in the same proportion as the linear scale is decreased, then the unit stresses at similar points in the model and prototype will be the same, and the displacement of any point in the model will represent to scale that of the corresponding point in the prototype.

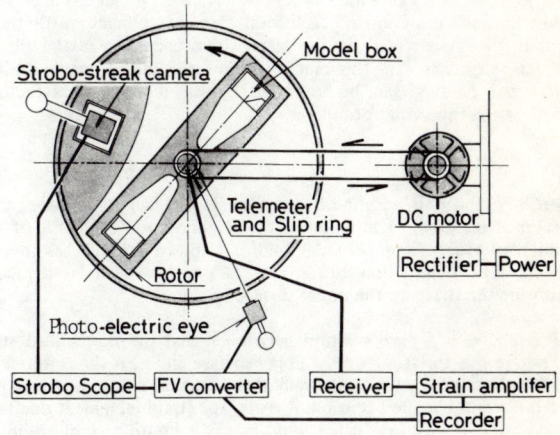

Fig.1 Schematic diagram of centrifuge and measuring apparatus.

Fig. 1 shows the experimental equipment. The rotor, 128cm long, is symmetric with respect to the axis and has two chambers to set the model boxes at both ends. The D-C motor, 22kw and 2500r.p.m. in maximum, is controlled by the Thyristor-Leonard system and the rotating speed of the axis is measured by a photo-electric transducer. For the strain measurement of the model, a bridge circut is formed with a strain gauge affixed on the model surface and three standard resistances in the rotor and output signals feed into the recorder through a telemeter. The model box is used to perform the plane-strain condition in the model. Its inner space is 12cm high, 17.8cm wide and 3cm thick, and it is made by aluminum and coated with chromium to decrease the frictional resistance on the lateral surfaces.

In the present centrifuge, the radius of rotor is not sufficiently large compared to the dimensions of the model, then it may be difficult to secure the similarity in the strict sense. Namely the model is slightly loaded in the horizontal direction and also the overburden pressure in the model is slightly greater than that in the prototype. The ratio of the overburden pressure is given by $\beta = 1 + 0.5h/r$, in which h: the depth; and r: the radius of rotation at the upper surface of the model. In the present experiment, $\beta = 1.133$ at the bottom of the model.

3. Physical properties of materials

The models of rock mass are made by pouring the mixture of plaster, slaked lime, standard sand and water into the aluminum model box. After leaving for 30 minutes and drying at 80°C for 48 hours, they are tested. Table 1 shows the physical properties of three kinds of materials which are classified by the cohesion of peak strength.

Table 1 Physical properties of materials

Model	γ	E	cp	ϕp	cr	ϕr	E/cp
A	140	80.5	137.0	30.0	34.0	29.0	588
B	142	60.8	97.9	30.0	31.0	29.0	621
C	145	40.3	71.3	30.0	34.0	29.0	565

γ: Unit weight(N/m³); E: Young's modulus(MN/m²); cp: cohesion of peak strength (KN/m²); ϕp: friction angle of peak strength(degree); cr: cohesion of residual strength(KN/m²); ϕr: friction angle of residual strength(degree).

4. Circular failure process

Fig.2 shows a circular failure of homogeneous slope, analyzed by FEM giving the gravitational load incrementally. The rock mass is represented by triangular elastic elements which are connected each other with elasto-plastic joint elements of which directions have been pre-determined in the manner that the direction of each joint element coincides with that of potential slip line defined by the mean stress condition of the adjacent triangular elements. In this analysis, the generalized magunitude of gravitational loading can be indicated by the dimensionless factor n, defined by the following formula:

$$n = \gamma H/c \qquad (1)$$

in which γ: the unit weight of material; H: the height of slope; c: the cohesion of material. The failure in the neighborhood of the toe of slope has initiated at $n = ni$ and a circular failure, as shown in Fig.2, has appeared at $n = nc$. The completion of circular failure has been confirmed by monitoring the strain on the upper surface of slope.

Fig.3(a) shows a typical relation between n and the normalized strain, ϵ/ϵ_o, where ϵ is the strain on the upper surface and ϵ_o is the material constant which is the elastic axial strain at the uniaxial compressive strength. After the strain becomes tension at $n = nc$, the strain increment due to the increase of n is infinite. Then $n = nc$ means a limiting equilibrium. At $n = nt$, the strain increment turns into tension. Such a change is caused by the progress of sliding on the slip lines. It is allowable to say that $n = nt$ represents a stable-unstable transition, because the progressive failure process will be clearly detected at $n > nt$. As a matter of course, ni, nt and

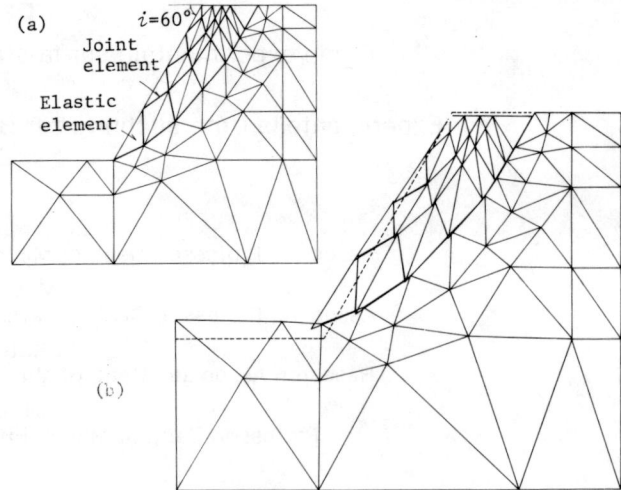

Fig.2 Deformation of homogeneous slope analyzed by FEM, (a) mesh diagram (b) deformation of slope.

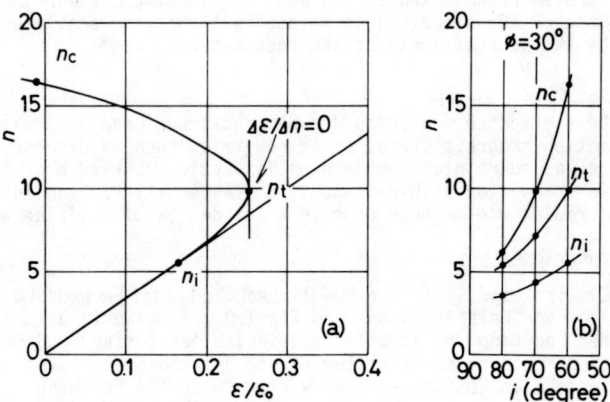

Fig.3 Results obtained by FEM analyses, (a) a typical relation between n and the normalized strain on the upper surface, ϵ/ϵ_o, (b) Relations among nc, nt, ni and dip angle of slope, i.

nc depend on the friction angle of material and also the dip angle of slope as illustrated in Fig.3(b). Such a failure process should be examined by the following experiments.

Fig.4 shows the geometry of model. The dip angle of slope surface, i, ranges from 50 degrees to 80 degrees and the three kinds of material in Table 1 are used. Fig.5 and Fig.6 show the final features of model slope. In the case of $i = 80$ degrees, the failure always occurs rapidly and it consists of a circular failure with a vertical tension crack in the upper surface, of which length is about $0.5H$. In the case of $i \leqslant 70$ degrees, the sliding block is generally divided into several pieces by parallel tension cracks in the slope surface.

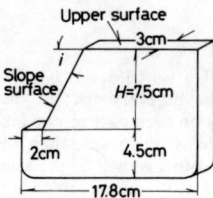

Fig.4 Geometry of homogeneous slope model.

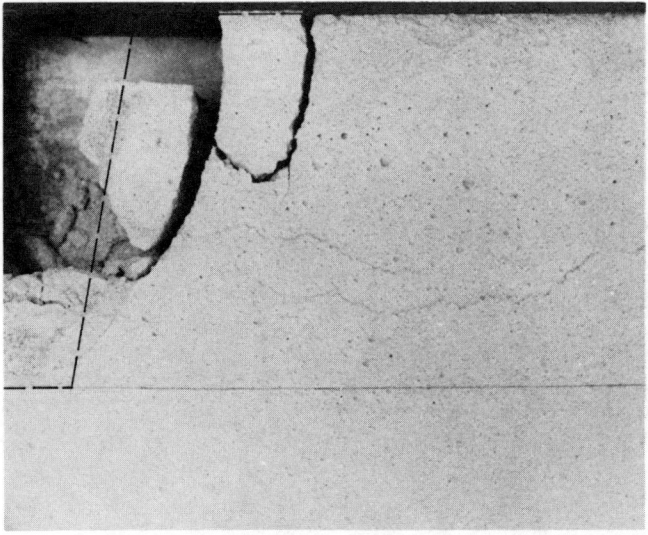

Fig.5 Collapse of steep slope model, $i=80$ degrees.

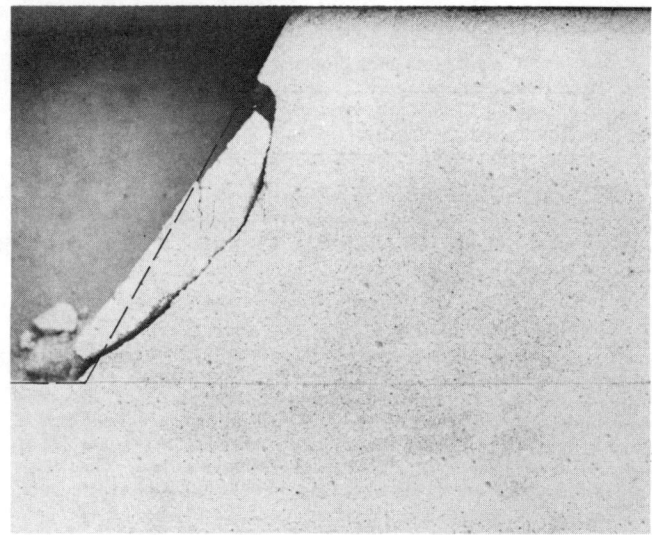

Fig.6 A typical failure of slope model, $i=60$degrees.

Fig.7 illustrates the relationship between n and ϵ/ϵ_o, where n is defined by eq.(1) using the cohesion of peak strength as c, ϵ is the strain on the upper surface measured by the strain gauge of 5cm in length and ϵ_o is the elastic axial strain at the peak strength in the uniaxial compression test. Curves of n vs. ϵ/ϵ_o are similar to that in Fig.3(a), then nt can be estimated in the manner previously described. Fig.8 shows the relation between nt and i.

It is suggested that nt is approximately linear to i and that it does not depend on the brittleness of material. Fig.9 shows the relation between nc and i. It is concretely shown that nc depends on not only i but also the post failure behaviors of material. Otherwise, the limiting equilibrium condition will be able to indicate by n calculated using the cohesion of residual strength as c in eq.(1). But Fig.10 shows conclusively that it can not be defined only by the residual strength of material.

The experimental results suggest that in designing a homogeneous rock slope, if one attempts to calculate the limiting height of slope on the basis of the conventional limit analysis with the peak shear strength of rock mass, an unrealistic design is obtained, since the strain softening phenomena is underestimated, and that on the other hand, if one uses the residual strength, a design obtained may be pessimistic in many cases. It is confirmed that the analysis of the stable-unstable transition of rock slope may present a more reasonable way of designing and that the stable-unstable transition can be concretely estimated by monitoring the strain increment on the upper surface.

5. Influence of discontinuity

Fig.11 shows the failure model which assumes the onset of sliding on two planes simultaneously, namely one is a potential slip plane BG in the intact rock and the other is a joint surface DG. The force acts on BG is represented by R, of which direction is with an angle θ to the horizontal and also with an angle α to BG. It can be assumed that α is equal to the

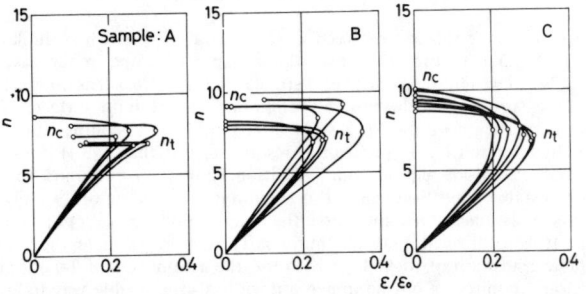

Fgi.7 Relations between n and ϵ/ϵ_o at $i=60$degrees.

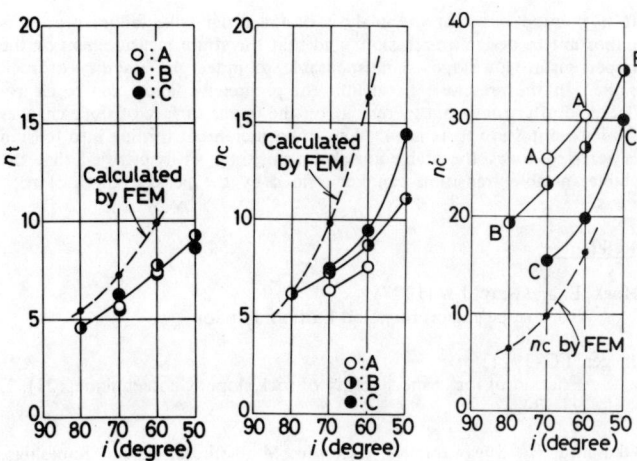

Fig.8 Relations between nt and i.

Fig.9 Relations between nc and i.

Fig.10 Relations between $n'c$ and i.

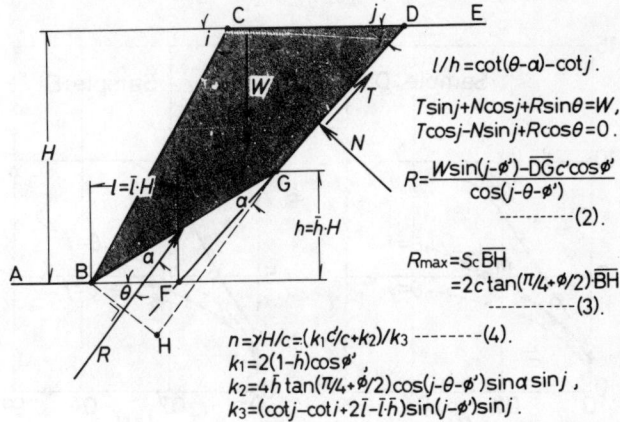

$l/h = \cot(\theta-\alpha) - \cot j$.

$T\sin j + N\cos j + R\sin\theta = W$,
$T\cos j - N\sin j + R\cos\theta = 0$.

$R = \dfrac{W\sin(j-\phi') - \overline{DG}c'\cos\phi'}{\cos(j-\theta-\phi')}$
$\qquad\qquad\qquad$ ---------- (2).

$R_{max} = S_c\overline{BH}$
$\qquad = 2c\tan(\pi/4+\phi/2)\cdot\overline{BH}$
$\qquad\qquad\qquad$ ---------- (3).

$n = \gamma H/c = (k_1 c'/c + k_2)/k_3$ -------- (4).
$k_1 = 2(1-\bar{h})\cos\phi'$,
$k_2 = 4\bar{h}\tan(\pi/4+\phi/2)\cos(j-\theta-\phi')\sin\alpha\sin j$,
$k_3 = (\cot j - \cot i + 2\bar{l}-\bar{l}\cdot\bar{h})\sin(j-\phi')\sin j$.

Fig.11 A failure model of slope with a discontinuity.

Table 2 Mechanical properties of models

Model	Intact rock material				Joint surface	
	c_p	ϕ_p	c_r	ϕ_r	c'	ϕ'
D	138.0	30.0	30.0	29.0	5.0	22.0
E	62.0	30.0	33.0	29.0	10.0	24.0

c': Cohesion of joint surface(KN/m^2); ϕ': friction angle of joint surface(degree).

angle of fracture of the intact rock, so that $\alpha=\pi/4-\phi/2$, in which ϕ is the friction angle of intact rock. The equilibrium condition is given by eq.(2) in Fig.11, in which W: the weight of sliding block; T: the shear force on DG; N: the normal force on DG. If the maximum value of R is given by eq.(3), the limiting value of n is calculated by eq.(4), in which c: the cohesion of intact rock; c': the cohesion of joint surface. The failure model discussed above should be examined by the following experiments.

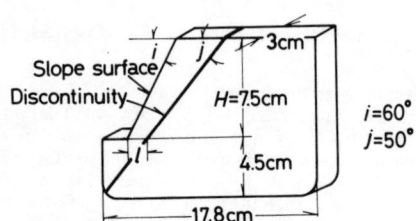

Fig.12 Geometry of slope model with a discontinuity.

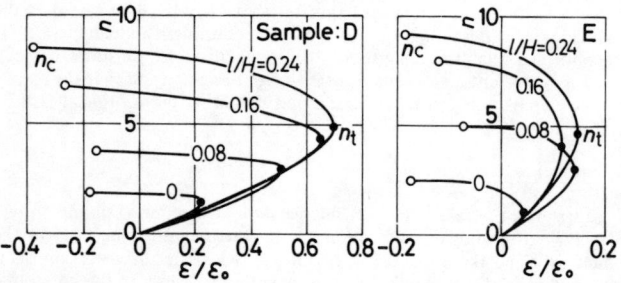

Fig.13 Relations between n and ϵ/ϵ_o at $0 < l/H < 0.24$.

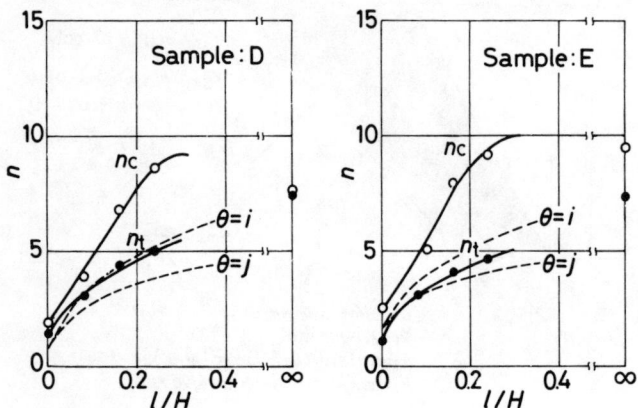

Fig.15 Relations among n_c, n_t and l/H. Dotted lines are calculated by eq.(4).

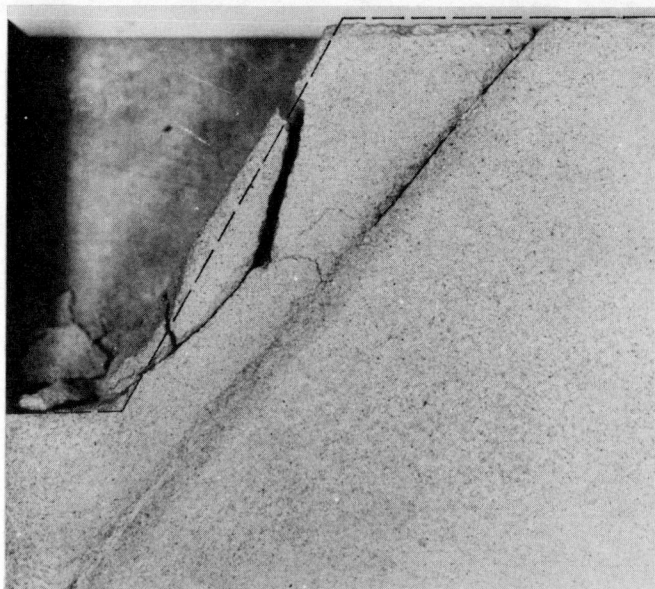

Fig.14 A typical failure of slope model with a discontinuity, l/H=0.24.

Fig.12 illustrates the geometry of model. The horizontal distance, l, between the toe of slope and the joint surface, ranges from 0 to 0.24H. The model of joint is made by placing a thin seam of grease between the intact materials. The mechanical properties are summarized in Table 2.

Fig.13 shows the relation between $n=\gamma H/c_p$ and ϵ/ϵ_o, which is similar to that in Fig.3 (a). Fig. 14 shows the collapse of slope in the case of l/H=0.24. The failure consists of a circular failure with a tension crack in the slope surface and the progress of sliding on the joint surface. The relation among n_t, n_c and l/H is summarized in Fig.15 as compared with n calculated by eq.(4). Experimental results suggest that eq.(4) does not define the limiting equilibrium condition, but it corresponds to the stable-unstable transition, since the simultaneous sliding on two planes previously assumed does not satisfy the kinetic condition except its initiation. It is confirmed that the more safty way to estimate the stable-unstable transition of rock slope is to use the assumption of $\theta=j$ and that the strain monitoring on the upper surface is also a reliable way to judge the rock slope stability in such a complex case.

6. Conclusion

It is pointed out that the analysis of the progressive failure process is important to design a rock slope and that the strain measurement on the upper surface of slope is indispensable to judge the stability of rock slope. In the present experiments, the progressive failure can be clearly detected after the strain increment on the upper surface of slope changes from compression to tension. The strain increment turning into tension is referred to as the stable-unstable transition. It is clarified that the stable-unstable transition can be defined by the peak strength of rock mass.

References

Hoek, E. and Bray, J.W.(1977).
    Rock slope engineering. 2nd Edition. London.

Jaeger, J.C.(1971).
    Friction of rocks and stability of rock slope: Geotechnique, (21), 2, 97-134.

Okamura, H., Sugawara, K., Akimoto, M., Kubota, A. and Kaneshige, O.(1979).
    Experimental study on rock slope stability by the use of a centrifuge: J. of Mining and Metallurgical Institute of Japan, (95), 1091, 7-14.

ROCK SLOPE STABILITY AT THE MARQUESADO MINE

Stabilité des talus rocheux de la mine "Marquesado"

Böschungsstabilität im Marquesado Tagebau

Evariston Portillo (Geologist)
INTECSA, Madrid, España

Manuel Romana (Professor)
Universidad Politecnica de Valencia (España)

SYNOPSIS

The MARQUESADO Mine is the biggest Spanish open pit with a depth of about 200 m and an annual production of 2,500,000 tons of iron ore. The rock is a very complex formation of karstic limestone and dolomite with an alluvial overburden of more than 100 m. The design of rock slopes has included three steps:
- Tectonic evaluation of the pit zone
- Classification and geotechnical characterisation of the rock mass and joints
- Evaluation of rock stability
A review of work done and results is presented.

RESUME

La mine du MARQUESADO est la plus grande mine à ciel ouvert en Espagne avec une profondeur de 200 m et une production annuelle de 2.500.000 t. La roche est une formation complexe de calcaire karstifié et de dolomites avec une couverture alluviale de plus de 100 m. Le projet de talus comprend 3 opérations:
- estimation des efforts tectoniques.
- classification et caractérisation géotechnique du massif rocheux et des joints.
- évaluation de la stabilité du massif.

ZUSAMMENFASSUNG

Das MARQUESADO Bergwerk ist der größte Tagebau Spaniens mit etwa 200 m Tiefe. Der Fels ist Kalkstein und Dolomit mit mehr als 100 m alluvialer Überlagerung. Der Entwurf der Böschungen ist in drei Stufen gemacht worden:
- Tektonische Erkundung des Tagebaus
- Geotechnische Charakterisierung des Gebirges und der Klüfte
- Standsicherheitberechnung
Es werden die Ergebnisse vorgestellt.

1. INTRODUCTION

The deposit is located north of the Sierra Neva da, east of Granada. The mine originally underground, began to be exploited open-cast by the Compañia Andaluza de Minas (C.A.M.) from 1967 on, and at present produces over 2500,000 tons/year, with a open pit of 1000 x x 800 metres and a depth of 200 metres (Fig. 1).

2. GEOLOGY

The deposit is located in the Dollar Unit of the Mulhacén Mantle, which is included in the Nevado-Filabride complex of the Betica S. str. Units (PUGA et al, 1974) and includes:

- A base Paleozoic micaschists and quartzites

- A thrusted covering of limestone and marbles with a complex range of lithological and mine ralogical facies over 100 m. thick.

- A deep deposit of plioquaternary alluvions

The deposit is made up of irregular hematite and goetite masses included in the metamorphised limestones. In the relation between the mi ne and the enclosing rock, there is generally a steep step, either with rectilinear contacts (diaclase, fractures) or with an irregular sha pe. The upper and lower limits are usually fairly parallel with the stratification, but this is not the case of the side limits, which usually cut the structures sharply (TORRES, P. 1980) (Fig. 2).

The limestone formations are two triassic rock masses shided on paleozoic basis. The mine is lo cated in a synclinorium with layer dipping around 3° Contacts with paleozoic schists are very irregular (Fig. 3).

About 1,200 measurements of diaclases and stra tification planes were taken, and computer pro cessed to obtain the frequency diagrams, which showed five important families (Fig. 4).

A survey of 60 faults planes with slickensides was carried out applying the M method (ARTHAUD, 1969) and showed, statistically, the direction of principal axis of strain ellipseis (Fig. 5 and 6) supporting the conclusions of joints study.

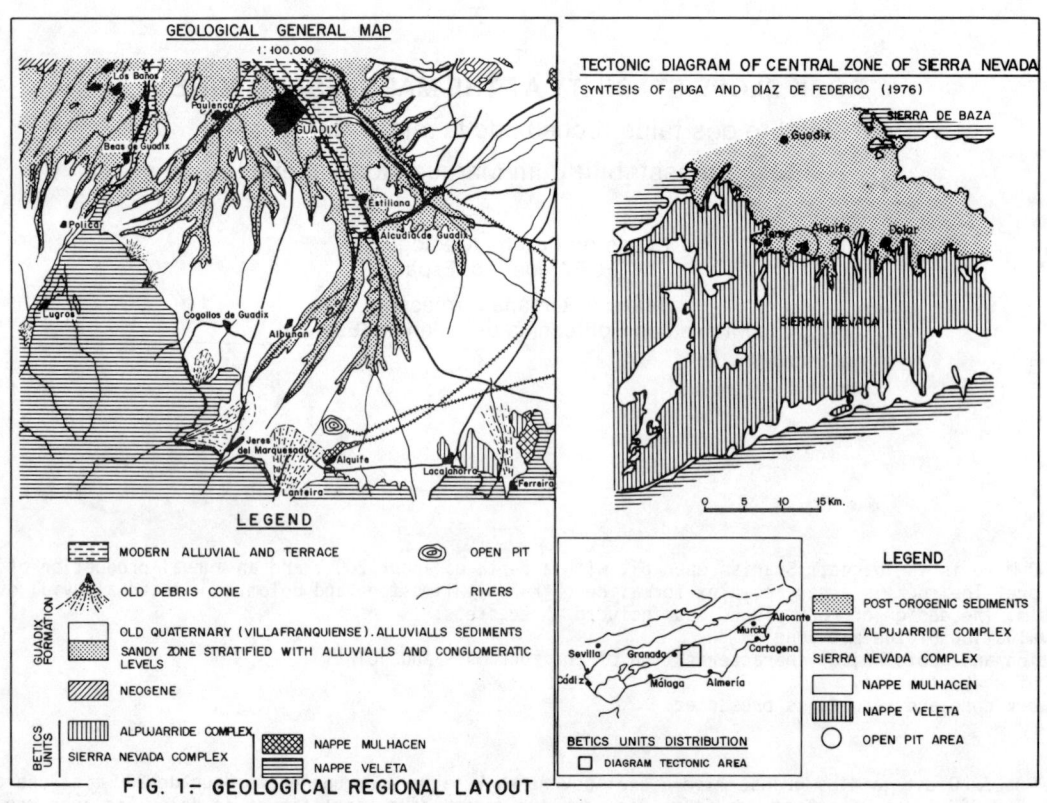

GEOLOGICAL GENERAL MAP
1:100.000

TECTONIC DIAGRAM OF CENTRAL ZONE OF SIERRA NEVADA

SYNTESIS OF PUGA AND DIAZ DE FEDERICO (1976)

LEGEND

MODERN ALLUVIAL AND TERRACE   OPEN PIT

OLD DEBRIS CONE   RIVERS

GUADIX FORMATION
OLD QUATERNARY (VILLAFRANQUIENSE). ALLUVIALLS SEDIMENTS

SANDY ZONE STRATIFIED WITH ALLUVIALLS AND CONGLOMERATIC LEVELS

NEOGENE

BETICS UNITS
ALPUJARRIDE COMPLEX   NAPPE MULHACEN
SIERRA NEVADA COMPLEX   NAPPE VELETA

BETICS UNITS DISTRIBUTION
DIAGRAM TECTONIC AREA

LEGEND

POST-OROGENIC SEDIMENTS

ALPUJARRIDE COMPLEX

SIERRA NEVADA COMPLEX

NAPPE MULHACEN

NAPPE VELETA

OPEN PIT AREA

FIG. 1.- GEOLOGICAL REGIONAL LAYOUT

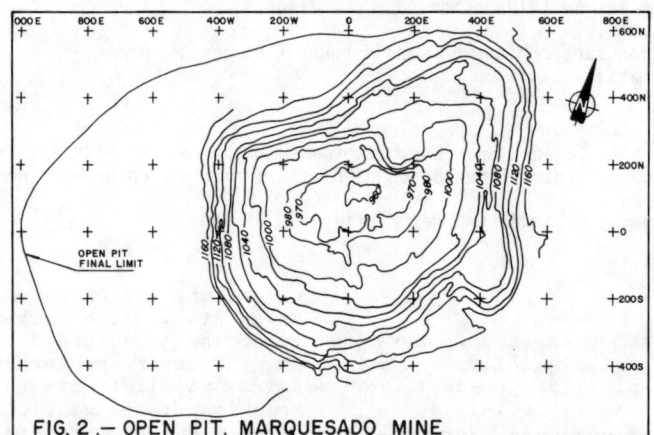

FIG. 2.- OPEN PIT. MARQUESADO MINE

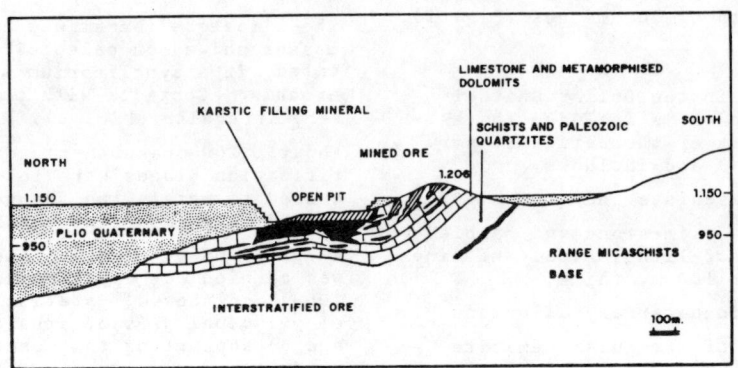

FIG. 3.- GEOLOGICAL SECTION OF THE MARQUESADO DEPOSIT

C 6

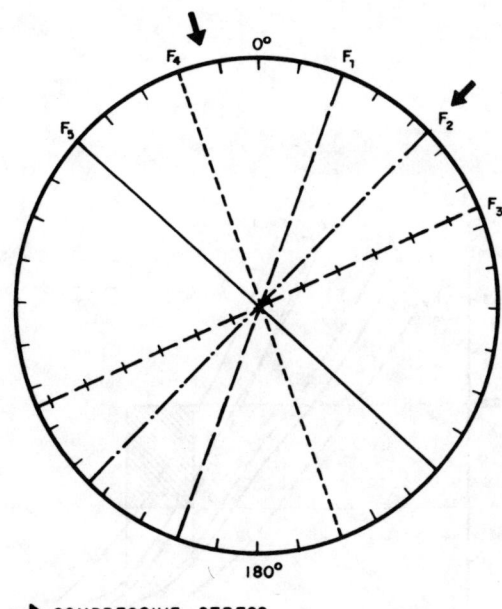

COMPRESSIVE STRESS

FIG. 4.- DIRECTIONS OF PRINCIPAL
JOINT FAMILIES

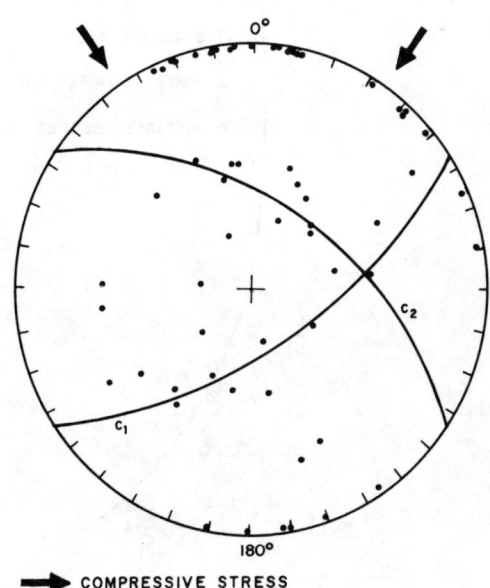

COMPRESSIVE STRESS

FIG. 5.- PLANES M. REPRESENTATION

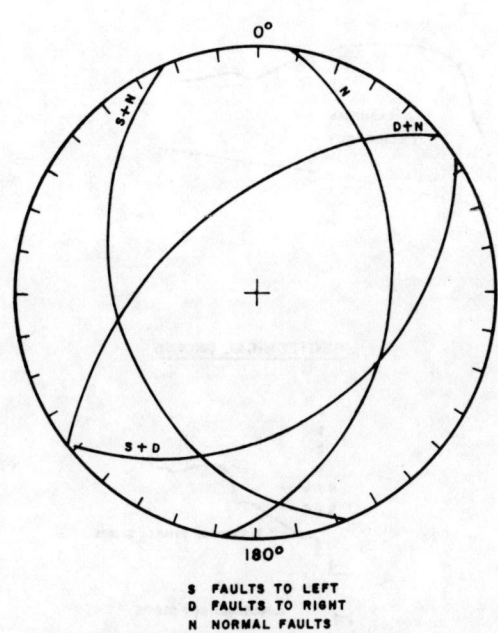

S   FAULTS TO LEFT
D   FAULTS TO RIGHT
N   NORMAL FAULTS

FIG. 6.- MOST FRECUENT FAULT
PLANES REPRESENTATION

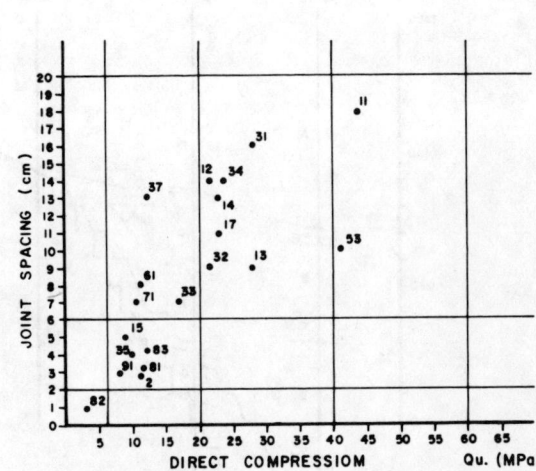

FIG. 7.- AVERAGE PROPERTIES OF THE
DIFFERENT LITHOLOGICAL FACIES

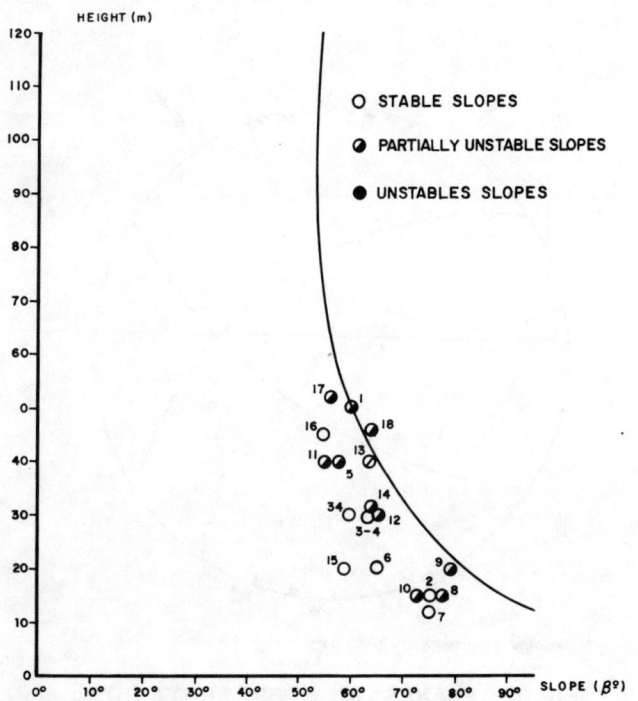

FIG. 8.- BACK-ANALYSIS OF ACTUAL ROCK SLOPES

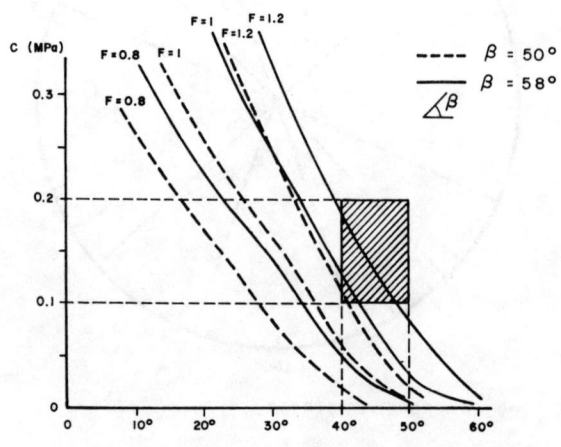

FIG. 10.- FACTORS OF SAFETY CURVES FOR CIRCULAR SLIDE (ZONE-2)

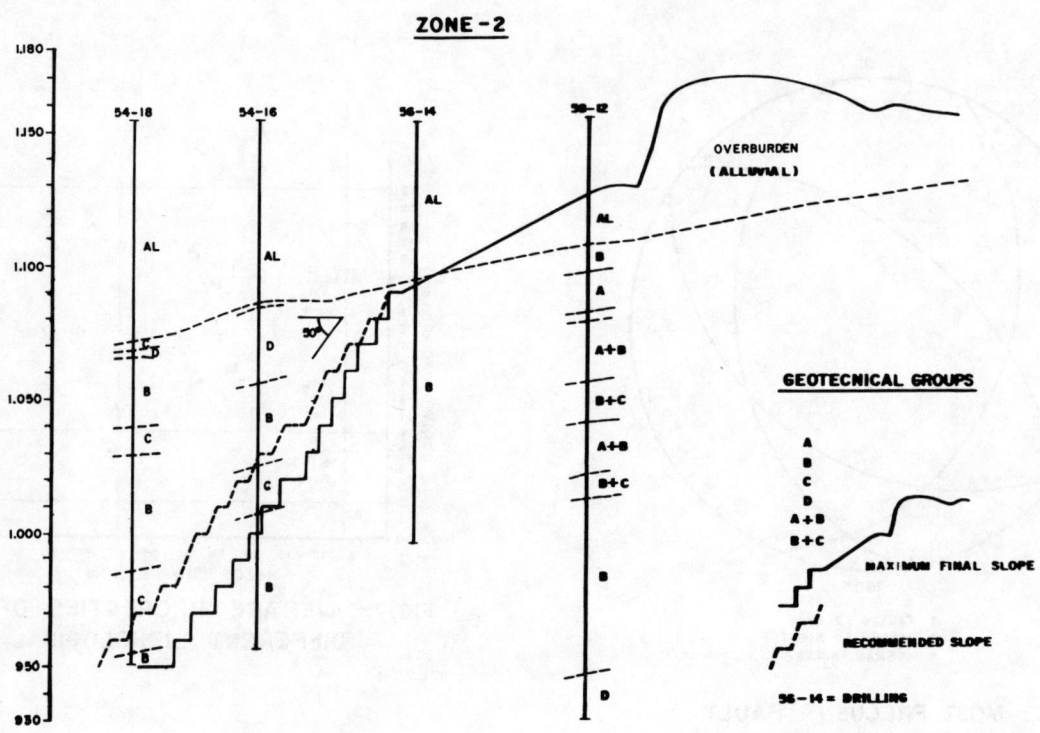

FIG. 11.- RECOMMENDED SLOPE

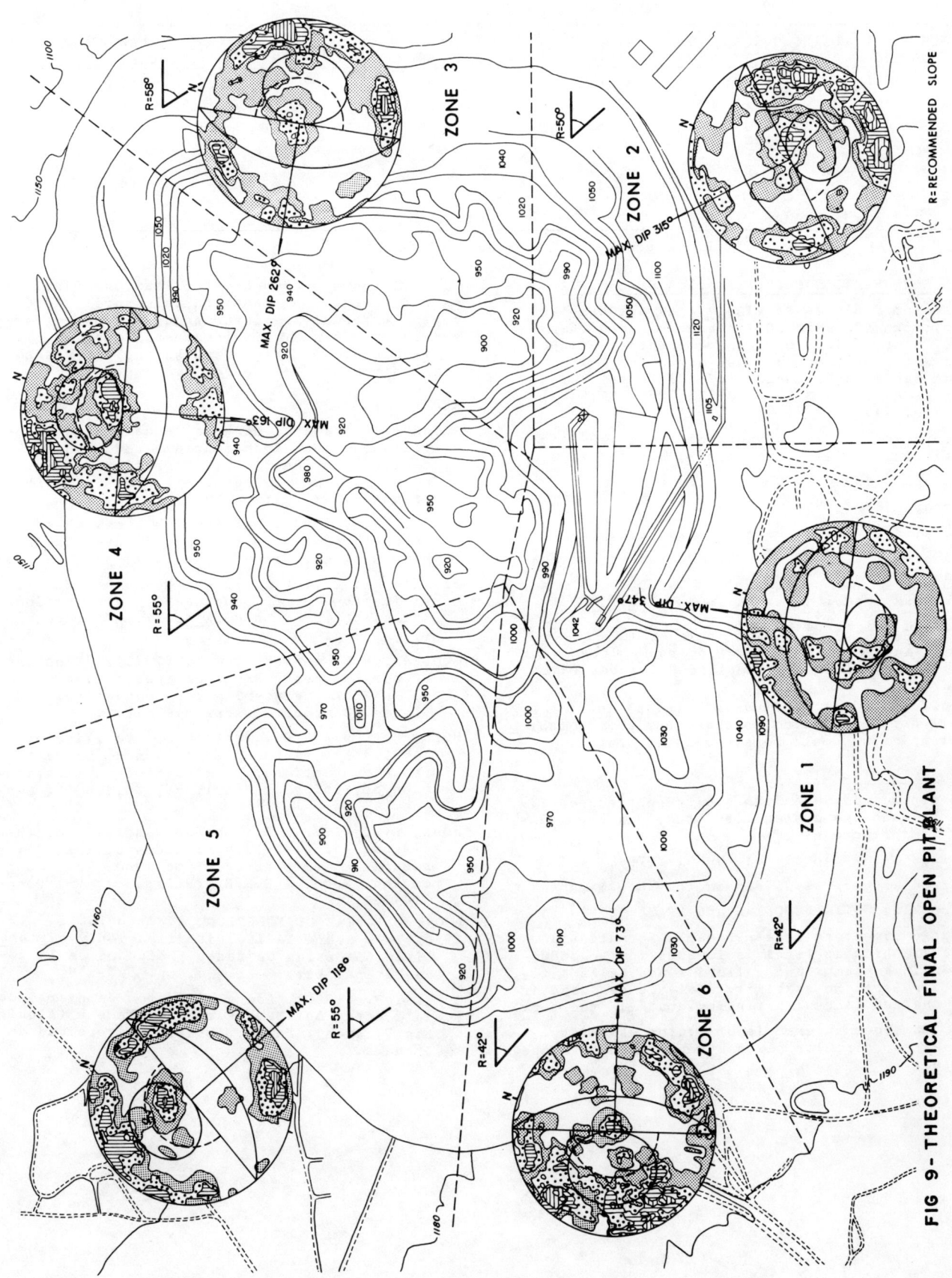

FIG 9 - THEORETICAL FINAL OPEN PIT PLANT

R = RECOMMENDED SLOPE

C 9

GEOTECNIC GROUP	LITHOLOGICAL FACIES	Freq. %	R.Q.D %	Joint Spacing (cm) CORES	OUTCROPS	Q_u(MPa)	I_s(MPa)	BIENAWSKI CLASS
A	11,53	9,5	39-50	10-19	60	40	2,3-2,6	III
B	12,13,14,17, 31,32,33,34, 37	44,7	28-50	9-17	20-60	20-30	1,1-2,6	III-IV
C	61,71,15,35 91,2,81,83	33	7-20	3-9	20	7-12	1-1,5	IV
D	82	9	0	0	0	<3	-	V

TABLE 1   Summary of geomechanical data

3. GEOMECHANICAL CLASSIFICATION OF MATERIALS

778 point loads tests (I_s), 78 direct compression tests (Q_u) and RQD logging of more than - 8.000 m of boring cores gave more than 20.000 data which were computer adjusted to 20 lithological facies and finally to 4 different -- Bieniawski classes (BIENIAWSKI,1979).(Table 1 - and fig. 7).

Observation and measurement of about 1200 discontinuities with BARTON (1973) methods gave:

- joints   JRC = 8-10   ≃ 37°
- strata   JRC = 4-6   ≃ 33°

Some slides were observed on planes dipping 37°. An average value of   = 35°was assumed for analysis.

Rock-mass cohesion was assumed 0.1-0.2 MPa.

Specific weight is 26 KN/m^3.

A back-analysis of 18 existing rock slopes was carried out with Hoek simplified methods (HOEK, 1970).

Results were not conclusive (Fig. 8) due to - small heigth of actual slopes, but showed that real friction could be at 40° - 45° range.

4. SLOPE STABILITY

The pit was divided into 6 different zones (Fig. 9). In each one a stability study was carried - out taking account of:

- plane and wedge slides (JOHN, 1978).
- circular general slides in the rock mass
- toppling falls (GOODMAN and BRAY, 1976)

Karstic limestone is a permeable formation and general phreatic level is lowered in the zone - by drainage sumps and galleries and wells lateral barriers. So stability analysis was done in the drained case with tension crack in the top.

Fig. 10 shows an example of factor of safety - curves for circular slides (zone 2) and 2 slope angles.

Fig. 11 shows recommended and maximum final slope for zone 2 on a geotechnical profil. As different material are interbedded and not clearly separated the analysis is general and some local falls can happen. Fig. 9 shows the pit plans - with recommended slope for every zone.

REFERENCES

ARTHAUD, F. (1969). Méthode de détermination - graphique des directions des raccourcissement,d'allongement et intermediarie d'une - population de failles Bull. Soc. Geol. de - France (7) XI, 1; pp 729-737

BARTON, N.R. (1973). Review for a new shear - strength criterion for rock joints. Engineering Geology, Elsevier, Vol. 7, pag. 287-332.

BIENIAWSKI, Z.T. (1979). The Geomechanics Classification in rock engineering applications. 4st. Int. Cong. on Rock Mech., Montreux. T.2, pp. 41-48.

GOODMAN, R.E. and BRAY, J.W. (1976), "Toppling of rock slopes". Proc. of specialty conference on Rock Engineering for Foundations and Slopes. Boulder, Colorado. Vol. 2.

HOEK, E. (1970). Estimating the stability of excavated slopes in opencast mines. Extract - from Transactions. Section A of the Institution of Mining du Metallugi, Volume 79, London.

JOHN, K.W. (1978). Engineering analysis of three dimensional stability problems utilising the reference hemisphere". Proc. 2nd. Int. Cong. on Rock Mech. I.S.R.M. (Belgrado). Vol 2, pp. 314-321.

PUGA, E.; DIAZ DE FEDERICO, A. y FONTBOTE, J.M. (1974). Sobre la individualización y sistematización de las Unidades profundas de la zona Bética. Est. Geol. 30, pp. 543-548.

TORRES, P. (1980). Los yacimientos de hierro de la comarca del Marquesado del Zenete (Alquife, Las Piletas). Ph. D. Thesis. Universidad de Granada.

GEOTECHNICAL PROCEDURES FOR OPEN-PIT COAL MINE DESIGN

Procédés géotechniques pour la conception des houillères à ciel ouvert

Geotechnische Verfahren für die Projektierung des Tagebergbaus von Kohle

Luis Gonzalez De Vallejo
Dr. en Geología Económica. Empresa Nacional ADARO. Spain

Carlos S. Oteo
Dr. Ing. de Caminos. Lab. de Carreteras y Geotecnia. Spain

SYNOPSIS

In this paper a geotechnical procedure for the study of six open pit coal mines in Spain is presented. The experience gained in the methods of investigation, engineering geological surveys, in situ and laboratory testing as well as back analysis of the shale rock formations is included.

RESUME

Cette communication présente une méthode pour l'étude géotechnique de six houillères à ciel ouvert en Espagne. Elle traite notamment des méthodes de reconnaissance de site, des essais in situ et en laboratoire et de l'analyse en retour des formations de schistes argileux.

ZUSAMMENFASSUNG

Die Arbeit beschreibt die geotechnischen Methoden für die Projektierung von 6 Tagbergbauen von Kohle in Spanien. Es werden die Erfahrungen dargelegt, die bei in-situ Untersuchungen, Ingenieurgeologie, Laboratoriums-Versuchen, Rückanalysen und Scherversuch in Diskontinuitäten von Felschiefer gemacht wurden.

1. INTRODUCTION

The energy crisis of the early 70's due to the increase of the oil prices has given rise to a reactivation of the open pit coal mining in Spain. Other contributory factors were the modern excavation techniques with highly mechanicised systems, as well as the substantial financial support by the Spanish Government. New coal basins have been explored since them mainly are composed by low grade coals.

This circunstances have developped a new situation for the spanish coal mining which has had to face the exploitation of low quality coals under unfavorable geological conditions. Feasible economic ratios has reached so high values as much as 15 to 1 and up to 20 to 1 (cubic meters of overburden to Tn of coal).

Before the last decade few attention has been paid to the geotechnical aspects related with the open pit coal mining in Spain, but due to the reasons previously described, particularly the high ratios values, geotechnical input because one of the fundamental factors considered for coal mining feasibility and development. The main geotechnical problems which have had to cope could be summarized as follows:

- Geotechnical investigations over large areas ranging from 2 to 12 Km long and 150 to 300 m in depth.

- Specific geotechnical problems derived from the great depths to be reached over highly fractures rocks, of weak matrix and low durability.

- Incidence of environmental problems, hydrogeological constraints and old underground mining works.

- Critical geotechnical conditions to be reached in order to optimised the economic feasibility of the mines.

- Relationships between geotechnical recommendations for mining design and other related factors, e.g. geometry, explotaition development, ripability, drainage, spoil mining recovery, etc.

In this paper spanish experiences on six large open pit coal mines: Lloseta (Majorca), Juliana-Albardado (Cordoba), Cervantes Este (Cordoba), San Ricardo (Cordoba), La Castellana (Cordoba) and Puertollano, are discussed. Methods of investigations and geotechnical problems, and its solutions, are presented, pointing out those aspects of more generalized nature where general conclusions can be applied for any of the studied cases.

2. METHODOLOGY

The large excavations to be faced needed to develop new working procedures according with the geotechnical pro-

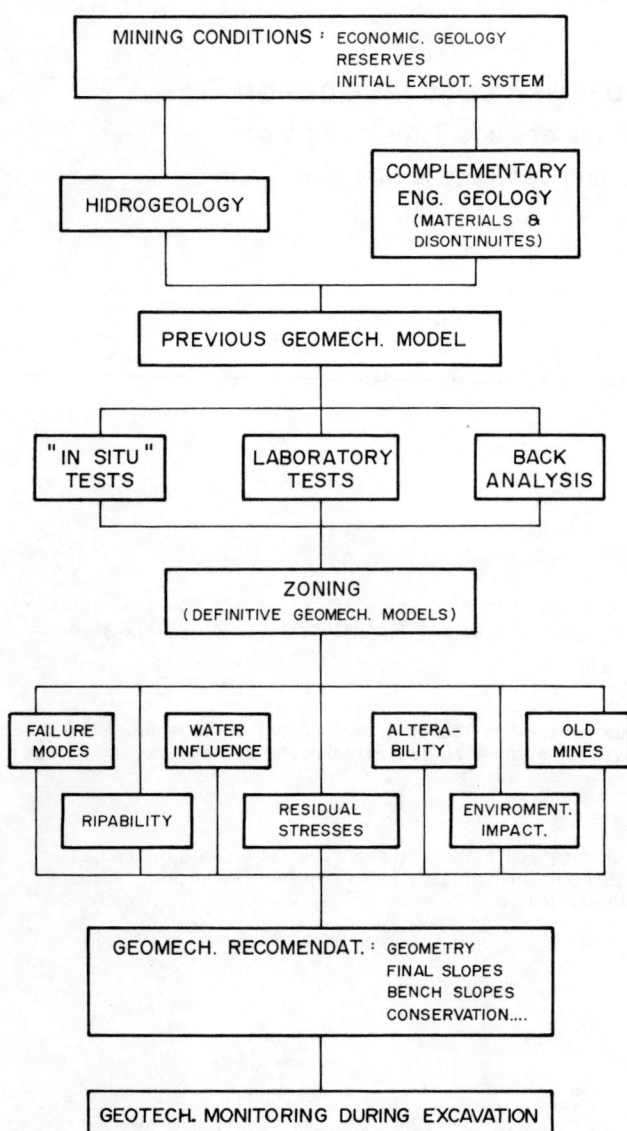

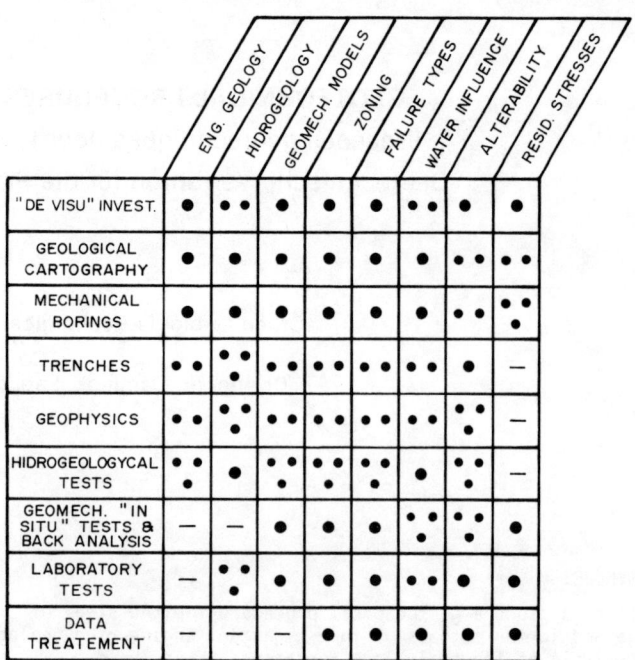

Fig. 2.- Geomechanical procedures used in open pit mines

● NECESARY ●● RECOMENDABLE ⋮ COMPLEMENTARY

Fig. 1.- Organigram of work method

blems described in Section 1. Very deep excavations over great lengths as well as other related geomechanical and environmental factors contributed to such new working procedures. The last objective was to stablish the final slopes angles, drainage conditions, geometry of the open pit and reccomendations for the explotaition and recovery.

A working compromise between theoretical principles and practical engineering has had to be achieved. Thus, it has been tried to get the maximum geomechanical information from economic geological and mining studies. For this purpose a good coordination among that fields and geotechnical works is needed. After some previous experiences it has been possible to establish a working methodology which has been applied to 6 Spanish mines (Section 1). Such methodology is synthesised in fig. 1. The procedures has the following three major stages:

1 St. Stage: Once Economic and mining geological investigations are carried out Engineering geology and Hydrogeology works are proceeded. The main objective of these studies is to establish the previous Geomechani-

cal Model (P.G.M.) as described in Section 3.

2nd Stage: Once the P.G.M. in established, a full site investigation program is carried ont based on in situ and laboratorory testing. Back analysis of failed slopes also performed. Special attention is paid to groundwater flow, discontinuities and the three physical properties coming into stability calculations: Specific gravity (γ), angle of internal friction (φ) and cohesion (c). Finally the open pit is divided in different sectors accordingly with its own geomechanical properties defining in such away the Definitive Geomechanical Model (D.G.M.).

3rd Stage: On each DGM following factors are analysed: a) Possible modes of slope failures; b) Influence of the groundwater conditions; c) Influence of the residual stresses; d) Weatherability; e) Influence of the old underground mining works; f) Explotaition aspects including mine geometry and ripability; g) Spoil Piles; h) Environmental impacts e.g. induced seismicity, mining recoveryand, land reclamation, etc; i) Geotechnical monitoring during the excavation.

This procedure has been carried out accordingly with the works described in fig. 2 where ranking values have been applied to point out the relationships between the procedure and their objectives. This figure is based on the experience gained on the mines mentioned in Section 1.

3. GEOLOGICAL STUDIES

Geotechnical studies must be based on detailed geological investigations of the future mined area. In most of the cases and Economic Geological Study is available before the geotechnical surveys are carried out. This Economic Geological Study provides information on the coal reserves, geometry of the deposits and properties of steam coals for calorific power .However, many geological aspects relevant for engineering purposes are not induced. On the other hand, the major emphasis is devoted on the

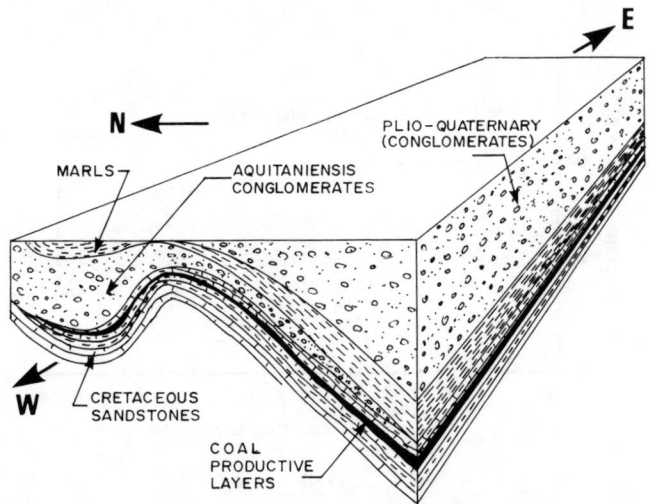

Fig. 3.- Geologycal scheme of Lloseta open pit mine (not scale)

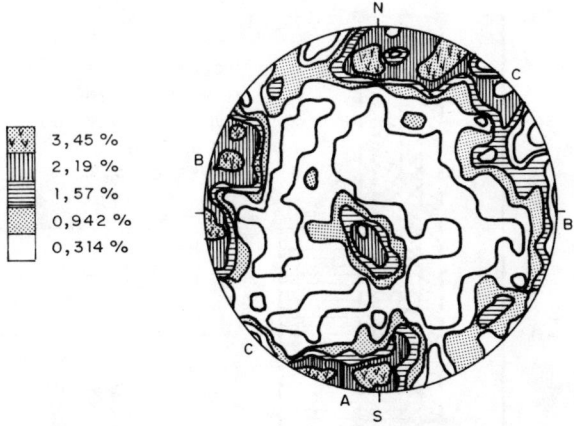

▒	3,45 %
☰	2,19 %
▤	1,57 %
▨	0,942 %
☐	0,314 %

Fig. 4. Schmidt's stereographic representation of discontinuities at Lloseta open pit mine

area closer to the coal deposits, being unsufficiently studied the upper zones of the basin where most of the excavation processes should be taken place. Besides, a large number of borehole logs are generally available although few geotechnical data can be obtained from these boreholes which have been drilled only for mining economic evaluation.

From the geological studies the stratigraphy and tectonic of the area, as well as Geological maps and cross sections, are obtained. Isopacs and isohipses of the coal deposits are also included in these studies which are important to define the coal thickness and the Overburden materials, and therefore the depths of the slopes.

An example of the general structure of a coal basin in Lloseta (Majorca Island) is shown in figure 3. Engineering geological studies start from these previous economics geological works. Photogeological and structural analysis of discontinuities are proceeded as one of the most important stages of the investigation. Detailed structural surveys and logging of boreholes and open pits are carried out. Systematic analysis of discontinuities in terms of orientation, spacing, continuity, rugosity, filling materials, etc, are described. Stereographic projections are extensively used to represent orientation and genetic studies of discontinuities. An example of the main discontinuities present in Lloseta coal basin is shown in fig. 4. As it can be seen these studies are complementary of the general structure showing fig. 3. The Lloseta case provided a good example of complex geological structure -asociated with synclinoriums- where extensive structural studies were needed to carried out cinematic analysis for slope stability calculations.

During this stage new borehole drilling and open pits are carried out. Usually they are made accompassing mining engineering geology and hidrogeology. This is due to the high cost of these boreholes where great depths are reached. From the point of view of the engineering geology, a detailed logging are proceded. Core orientation, RQD determination, and representative samples are obtained.

Besides that boreholes can be ready for further tests, e.g. well logging, seismic geophysics, permeability tests, etc, as they are described later (fig. 2). Geomorphological surveys of natural and certificial slopes are carried out studying the lithology, structure slope angles and slope heights, as well as their relationships.

Engineering hydrogeological studies range from groundwater and hydrological surveys, permeability and pumping tests in boreholes or wells, to the investigation of the groundwater flow nets before and during the excavation. These studies are conduced to assess the discharge, seepage forces on the slopes and to provide drainage measures.

From the Engineering Geology and Hydrogeology a Previous Geomechanical Model is obtained (The main objective of these studies). Such a model represents a selective synthesis of the those geological aspects relevant for engineering purposes. The model should includes: the main lithological units, principal discontinuities and weakness zones, boundary conditions of seepage and an assesstmant of the possible failures types related to natural and artificial slopes. Graphical representation of the model should included cross profiles, block diagrams and plans, as different means to explain a three dimension engineering geological model.

4. GEOMECHANICAL CHARACTERITATION

One of the most important steps during the geotechnical investigations is to find the representative geomechanics properties of the materials. For this purposes three types of works are generally carried out.

- Field testing including any "in situ" test which allows to find information on the rock mass strength and deformability.

- Laboratory tests: Detail identifications and classification of rock specimens is carried out as well as strength and deformability properties of cores and small discontinuities.

- Back analysis of failed slopes. Theoretical analysis on natural and artificial slopes are carried out in order to find $c-\phi$ relationships and to compare them with Laboratory results.

Besides of engineering geological logging described in Section 3 the following tests are carried out inside the boreholes and trenches:

a) Well logging techniques, density log, sonic velocity, gamma-gamma, neutron, etc.

b) Down-hole or up-hole seismic geophysics to assess V_p and V_s velocities.

c) Permeability Tests, packer tests and Lugeon test.

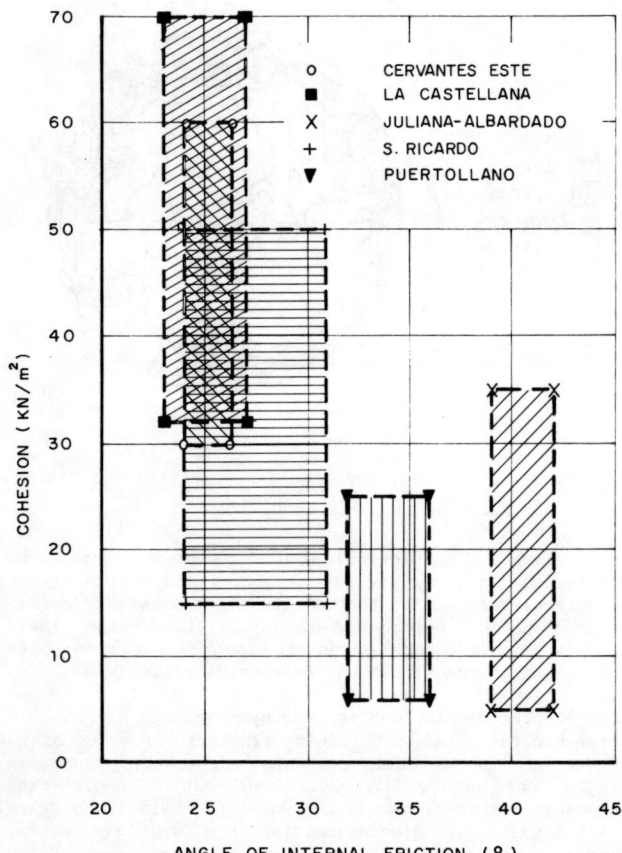

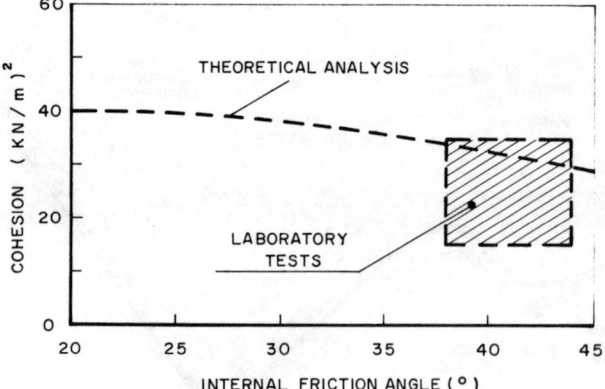

Fig. 6.- Back analysis carried out at Juliana-Albardado
open pit mine

Fig. 5. Relationship between the apparent cohesion and the
internal friction angle in the discontinuities of
some Spanish shale rocks

d) Core orientations

e) In situ shear strength test in trenches.

Laboratory test are mainly carried out on core samples
and block samples of maximum size about 40 x 40 cm. Typical classification tests consists in unconfined compressive strength, point load tests, Brazilian test, density, porosity, sonic velocity, etc. This tests trie to
identify and classify possible zones of different rock
qualities.

Mineralogical identification as well as fabric analysis
both determined by optical and electronic scanning microscopes and X-Ray analysis have proved to be an important
way to study the influence of shales and coal seams on
geotechnical properties such as durability and shear
strength. Durability is also studied using the slake
durability test and other cyclic tests.

Small and large box shear strength tests on discontinuities are extensively used. Shear strength values in terms
of cohesion (c) and internal friction (φ) are determined
on natural and induced discontinuities. Results from a
large number of tests are presented in figure 5. All the
tested material came from 5 open pit mines where the
overburden were shales with intercalations of coal seams.
As it can be seen from fig. 5 some relationship could
be suggested between c-φ values. Generally the highest
φ values correspond to shales with a sandy matrix. Besides them rocks have an RQD ranging from 60 to 80. Low φ

values correspond to shales with large quantities of
clays in its composition and less than 40 RQD values.

From the actual studied cases a clear relationship between c-φ values and other properties, such as weathering
or RQD, can not be establised although the results obtained in the present investigations are promising for
future studies on this field.

Back analysis calculation of failed slopes are fundamental to determine c-φ relations on in situ large discontinuities. In most of the cases back analysis results
and those c-φ values measured in the Laboratories shown
a acceptable aggreement (fig. 6). Other important application of back analysis investigations is to observe
modes of failures. In this way a buckling failure could
be identify in ones of the studied open pit mines (San
Ricardo).

5. GEOMECHANICAL ZONING

All the open pit mines mentioned in Section 1 where located over a wide range of geological and geomechanical
properties, extended on several Kms. of length and up
to 300 m of depth. These circunstances obviously affected
the geotechnical behaviour of the rock masses which were
remarkable different from one sector to another of the
open pit mines. Therefore a rock mass classification of
the diferent sectors were carried out interms of a so
called Geomechanical zoning (D.G.M.).The main criteria
followed were based on the following aspects:

- The P.G.M., including Hidrogeology.

- Fracturing of the rock masses

- Rock mass quality evaluation

- Assigment of appropiate geotechnical properties to
each sector.

The P.G.M. described in Section 3 was complemented with
an assessment of the state of fracturing of the different
rock masses affected by the open pit mine. A three dimensional model in developed using the data of Section 3,
RQD values, relationship between seismic waves velocity
(V_p) obtained in down-hole or seismic geophysics, and
V_p obtained from sonic tests in rock cores, as well as
other possible relationship between them.

Any other fators than could affect the state of fracturation such as old underground mining works was also
taken into account. From this data, a rock mass classi-
ficastion was developed getting together zones with the

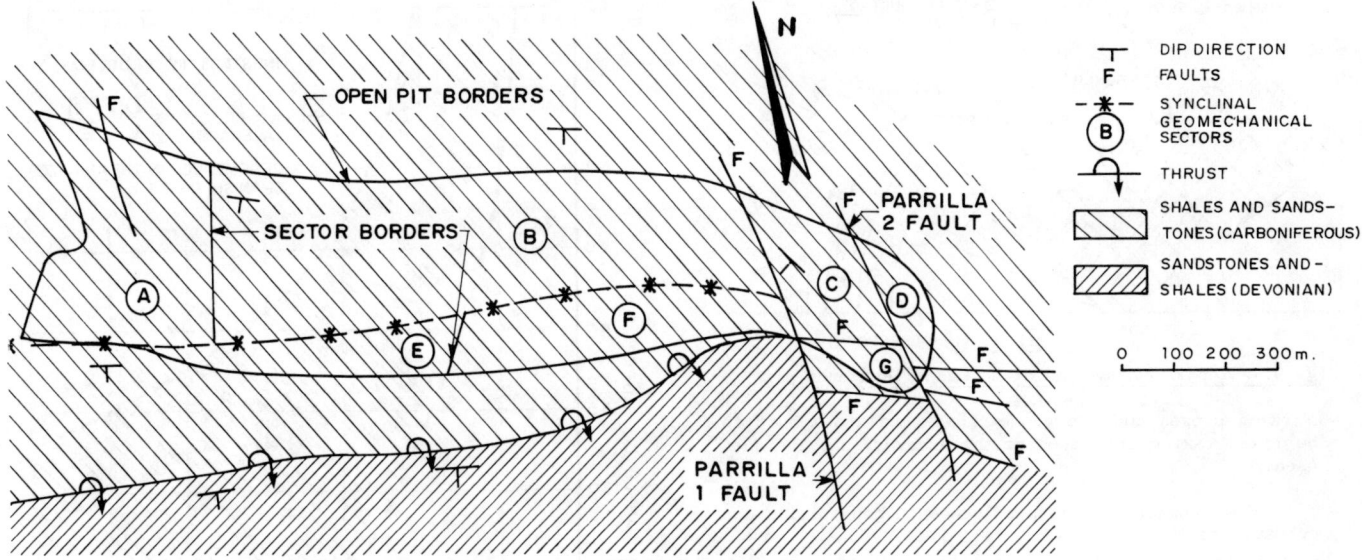

Fig. 7.- Zoning of La Castellana area

same range of rock mass values.

The Geotechnical properties described in Section 4 were assigned to the different zones defined by the rock mass values unconfined compressive strength values (σ_c) and its changes with depth were used as well as in situ deformation moduli (E) evaluated by geophysics , or the ratio E/σ_c . Shear strength values or ranges of them were also assigned to each sector. It is important to point out that for these exercises a significant amount of subjective criteria is used. Obviously as much as the geomechanical data is available as better matching can be done between sectors described in terms of rock mass classification and its strength properties.

Other useful properties such as average density values, durability or weathering grades and ripability indices were provided for each geomechanical sector. The geomechanical zoning concept constitute an essential criteria for slope stability calculations and mining geometry design. The type of zoning carried out in the open pit mine of La Castellana (Peñarroya, Córdoba) is shown in fig. 7.

6. FAILURE MODES

The definition of the possible failure slope modes is one of the essential stages in the previously described procedure. Some of the aspects that has to be considered are the following:

a) Block failures, in terms of planes, and wedges mainly. Graphycal analysis using K. John and Hoek and Bray's methods have been applied. Probabilistic analysis have been also used in some cases following the Castillo and Serrano (1973) method.

b) Total slope failure are assessed using plane failure analysis with or without tension cracks in the slope top. When the upper part of the slope is severely wathered at and its behavies as a soil-like a circular or poligonal analysis in carried out.

c) Specific problems arised from the productive wall slope where coal seams are in contact or very close to it. In this case the possible models shown in fig.8 should be taken into account. The height of the benches strongly influence the final slope angle.

d) Specific problems avised from the highwall slope. The geological structure of the studied cases consisted on a more or less complex folded structure (fig. 3), where the toppling was the most characteristic type of slope failure. This type of failure influences the width of the berms as well the angle of the benches in this highwall slope.

Typical safety factors used were 1.3 for final slopes and 1.1 to 1.2 for working slopes. To reach this coefficients an efficient drainage of the slopes were needed.

7. FINAL CONSIDERATIONS AND CONCLUSIONS

The previously described procedure has been proved to be efficient to achive the following objectives:

- Final slopes design in terms of slope angle, and geometry of berms and benches. Fig. 9 presented an example of the La Castellana open pit mine. The geomechanical sectors of this mine were described in fig. 7.

- The final slope should be ajusted using not only the results of a complete investigation but also the experienced obtained from similar cases especially if great depth has to be reached. Relationships between slope angle and slope height is presented in fig. 10 from spanish and elsewhere open pit mines.

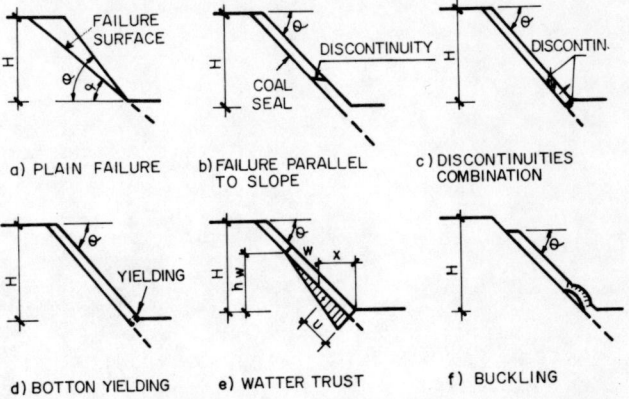

Fig. 8.- Failure types wen a coal seal is near the slope
(Piteau and Martin, 1981)

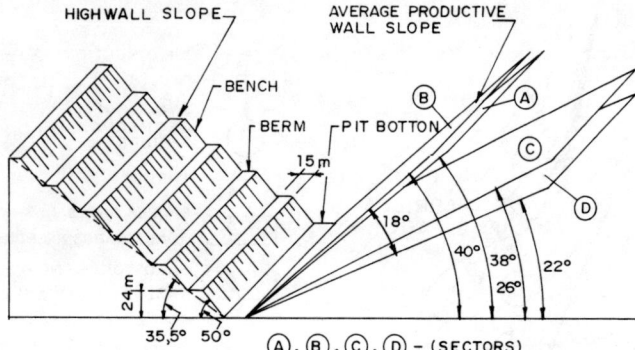

Fig. 9.- Slopes recommended at La Castellana open pit

- Working slopes and its geometrical conditions must to be fixed taking into account the excavation techniques used.

- Plan view geometry and its influence on convexities and concavities shapes.

- Excavation techniques, and drainage works

- Environmental impacts, induced seismicity, influence of waste piles, land reclamation, pollution, etc.

- Because the length and depth of studied mines the design must be checked and complemented during the explotation trhough out a geotechnical monitoring program mainly consisting in field instrumentation of the slopes and systematic surveying.

8. ADCKNOWLEDGEMENTS

The authors wish express their gratitude to mining organitationfor cooperation and permission to used some of the information here described: Encasur and Lignitos, S.A. Also, the cooperation of our colleges who running the field and laboratory test is gratefully acknowledged.

9. REFERENCES

Castillo, E. and Serrano, A.A. (1973). "Análisis probabilístico de la estabilidad de taludes rocosos". Boletín de Información del Laboratorio del Transporte y Mecánica del Suelo, Nov-Dic. Spain

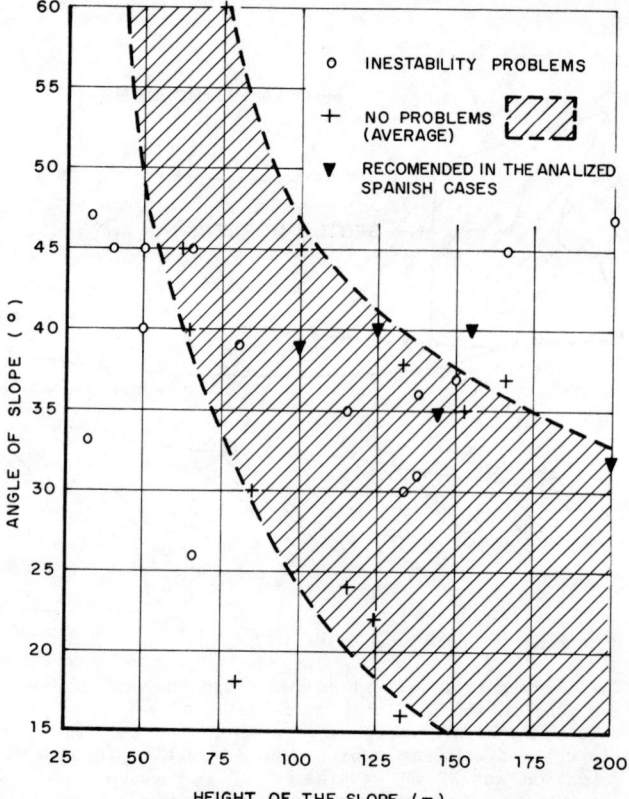

Fig. 10.- Relationship between the height and the angle of slope in shale materials (Data of Kley and Lutton)

Kley, R.J. and Lutton, R.J. (1967). "A study of selected rock excavations as related to large nuclear conditions". U.S. Army Corps. of Engs. PNE-5010.

Piteau, C. and Martin, J. (1981). "Mechanics of rock slope failure". Proc. 3rd Int. Conf. on Stability in Surface minin. Vancouver.

ROCK SLOPE ENGINEERING IN HONG KONG
Confortement des talus rocheux à Hong Kong
Untersuchung und Gestaltung von Felshängen in Hong Kong

E. W. Brand, S. R. Hencher and D. G. Youdan
Geotechnical Control Office, Engineering Development Department, Hong Kong

SYNOPSIS

Slope failures and boulder falls are common in Hong Kong, and preventive and remedial works are constantly being under-taken. The paper reviews the history of rock slope engineering in Hong Kong, outlines the present investigation and design practice, and describes the methods employed to stabilise rock slopes and boulders.

RESUME

Les glissements de terrain et les chutes de caillasses sont courants à Hong Kong et l'on entreprend constamment des travaux de confortement et de réparation. Cet article donne un aperçu de l'historique des études de stabilisation des talus rocheux à Hong Kong, esquisse la pratique courante en matière de reconnaissance et de conception de travaux et décrit les méthodes de confortement des coteaux et des caillasses.

ZUSAMMENFASSUNG

Böschungsbrüche und Felsstürze kommen in Hong Kong häufig vor und vorbeugende und Ausbesserungsarbeiten sind ständig im Gange. Die nachfolgende Abhandlung gibt einen Überblick über die Untersuchung und Gestaltung von Felshängen in Hong Kong sowohl in der Vergangenheit als auch nach dem gegenwärtigen Stand, und beschreibt die zur Sicherung von Böschungen und Felsblöcken angewandten Methoden.

1. INTRODUCTION

The object of this Paper is to provide a general review of rock engineering practice in Hong Kong as related to the design and performance of slopes. No such review has been made to-date, although there are several publications which provide useful information.

The Territory of Hong Kong, with a land area of only 1 050 sq. km, is situated on the southern coastline of China (Fig.1). The population of more than 5 million is concentrated largely in the urban areas of Hong Kong Island and Kowloon, and increasingly in the new towns now under construction in the New Territories. The economical development and use of land is of prime importance, and land values are very high.

The terrain of Hong Kong is extremely hilly, with the land rising to over 500 m on Hong Kong Island in a distance of less than 2 km from the sea. Most of the Kowloon peninsular has now been levelled to provide fill for reclamation, but isolated hills of up to 100 m still eixst. Little low lying land exists in the New Territories, with peaks of 550 m being common. Natural slopes throughout the Territory are steep, typically with upper slopes greater than 35 degrees, midslopes of 25 to 30 degrees and footslopes of about 15 degrees. Undeveloped slopes are often densely vegetated.

The majority of the slope failures in Hong Kong are caused by rainfall, which averages about 2 200 mm annually, the majority of which falls between May and September. Intensities of 50 mm per hour are not uncommon, and intensities of more than 100 mm per hour are sometimes recorded. Daily rainfalls can exceed 400 mm in extreme cases. Many slope failures therefore occur each year, both in natural slopes and cut slopes, and some of these have been disastrous (So, 1971; Lumb, 1975).

The rainstorm that occurred on 28 to 31 May 1982 will serve to illustrate the extreme conditions that can prevail in Hong Kong. During the four-day period, the rainfall recorded at the Royal Observatory was 654 mm, of which 437 mm fell on the first two days. The maximum intensities recorded at particular locations were 111 mm per hour and 394 mm per 24 hours. Widespread landslides and flooding were caused throughout the Territory, with a loss of 28 lives. The Geotechnical Control Office gave advice and assistance at 533 landslide 'incidents' that resulted from the rains, but a quick review of aerial photographs taken after the event revealed that well in excess of 1 000 separate slope failures had occurred. Although the majority of the failures were in 'soil' slopes, there were some notable 'rock' slope failures and boulder falls. Examples of typical rock slope failures are given in Figs.2 and 3.

Fig. 1.   Territory of Hong Kong showing geology

1 HONG KONG ISLAND   RECLAMATION
2 KOWLOON   GRANITE
3 NEW TERRITORIES   VOLCANICS
4 LANTAU ISLAND   SEDIMENTARY

Fig. 2.   Sheeting joint failure in granite

It should be mentioned that the climate of Hong Kong is such that frost is virtually unknown and plays no part in engineering design considerations. Neither are earthquake forces considered, since the Hong Kong area is of low seismicity.

Fig. 3.   Complex failure in weak highly-jointed granite

2.   GEOLOGICAL CONDITIONS

The geology of Hong Kong has been described by Ruxton (1960) and Allen & Stephens (1971). The predominant rock types are granite and volcanic rocks (Fig.1), the main difference between the two from a rock slope stability viewpoint being their joint patterns. The granite joint spacing is typically 2 to 10 m, while the volcanics usually have a blocky structure as a result of joint spacings of only 0.2 to 1 m. Extensive sheeting joints can occur in the granite and more rarely in the volcanics, and these often govern the engineering behaviour of the rock mass (see below).

The parent rocks are often weathered to considerable depth, resulting in places in the existence of mantles of residual material (saprolite) up to 50 m thick over the granite (Ruxton & Berry, 1957) and up to 20 m thick over the volcanics. These zones of highly and completely weathered rock commonly contain corestones of less weathered material, and are often overlain by colluvium.

The designer of cut slopes in Hong Kong is rarely confronted with a single 'grade' of rock material with which to work. The importance of the weathering profile concept has been emphasized by Brand (1982), and a description and classification system has been adopted for Hong Kong conditions in the Geotechnical Manual for Slopes (Geotechnical Control Office, 1979), which tends to govern Hong Kong practice. This uses a descriptive system of *zones* based on work by Ruxton & Berry (1957) and *grades* based on work by Moye (1955). A more sophisticated system of grade identification has recently been introduced in Hong Kong by Hencher & Martin (1982), and this will probably be used in a revised version of the Geotechnical Manual for Slopes which is now under preparation. It is also likely that a six-zone classification system will be adopted much along the lines of that recommended in the British Code of Practice for Site Investigation (British Standards Institution, 1981).

3. HISTORY OF SLOPE DESIGN IN HONG KONG

In 1979, an exercise was completed by the newly established Geotechnical Control Office to cata-logue all the existing cut slopes and retaining structures in Hong Kong, and to place them into a ranked order of priority for the purposes of investigation with a view to establishing the need for preventive works. A computer programme was devised to sort and rank the slopes and walls on the basis of a number of simple parameters, such as height, slope angle, geology, proximity to a building, slope condition, etc. Of the 6 000 cut slopes catalogued, about 4 000 were described as being 'rock' and 'rock/soil' slopes, and 89 of these featured in the highest ranked 100 slopes. The highest ranked rock cut slope is 70 m high at an angle of 75 degrees, and there is a dwelling immediately adjacent to the toe. Figure 4 shows the cut slope which is ranked as no. 692; the toe of this 40 m high granite slope is only 20 m from the new high-rise housing block.

It should be realized that the majority of the cut slopes that exist in Hong Kong were never designed on engineering principles. Prior to 1950, there was no great pressure to develop the midslopes and upper slopes of natural hillsides. Steep slopes were only cut for roads or access tracks, and cut angles were typically 75 degrees. Between 1950 and 1963, slopes were cut at a 10 vertical to 6 horizontal profile (59 degrees), regardless of the slope height or the nature of the soil or rock. The importance of slope height was acknowledged in 1963, from which time cut-tings exceeding 9 m high were provided with 0.9 m wide berms, thus reducing the overall slope angle to about 56 degrees. From 1968, a profile of 59 degrees (10 on 6) was only considered acceptable if the slope was not more than 7.6 m high. Slopes exceeding this required 1.5 m wide berms at 7.6 m intervals, giving an overall slope angle of 52 degrees. At about the same time, a few private firms started to produce slope designs on the basis of rudimentary cal-culations based on data from proper site investi-gations.

In 1972, after the disastrous Po Shan landslide (Government of Hong Kong, 1972), the government required a comprehensive geotechnical report to support any proposal for a major cutting. Despite this, the old rule-of-thumb (10 on 6) was often used as a justification for submissions. It was not until a Guide to Site Investigation and Earthworks was produced by the government in 1973 that there was a significant shift to a soundly based geotechnical engineering design procedure.

In 1979, the Geotechnical Manual for Slopes (Geo-technical Control Office, 1979) was produced

under the aegis of a Steering Committee. This Manual presently dictates the standards of safety required of slopes in Hong Kong, and it provides general guidance for slope design. The recom-mended approach to stability assessment is that of 'classical' limit equilibrium analysis, by means of methods of analysis for non-circular surfaces for soil (e.g. Janbu, 1954), and by means of the methods recommended by Hoek & Bray (1977) for plane, wedge and toppling failures in rock. The Manual specifies that a new slope must have an acceptable theoretical factor of

Fig. 4.   High-rise building close to old quarry slope

safety related to the risk category of the slope, assessed in terms of the likelihood of loss of life should the slope fail. *High risk* slopes are required to have F-values of 1.4, *signifi-cant risk* slopes 1.3, and *low risk* slopes 1.2.

In recognition of the difficulties inherent in applying 'classical' methods of analysis to deeply weathered rock materials, the Geotechnical Control Office has recently completed the first phase of an attempt to establish a semi-empiri-cal design method for Hong Kong slopes (Brand & Hudson, 1982). Two series of Geotechnical Area Studies, based largely upon terrain evaluation from aerial photographs, have also been embarked upon to provide adequate geotechnical data for planning and land use management purposes (Brand, Styles & Burnett, 1982).

In the context of 'rock' slopes, special mention must be made of the use of old quarry sites for building development. In a situation of acute land shortage, it is not surprising that disused quarries have been found to be obvious sites on which to build high-rise low cost housing, since suitably flat platforms were left by the quarrying operations. Many of these sites, however, were left with unsatisfactory rock slopes as a result of overblasting and because no geotechnical engineering methods were used in their 'design'

The slope shown in Fig.4 was formed as part of a quarry, and extensive preventive works were subsequently carried out to bring it to its present satisfactory state.

4. INVESTIGATION METHODS

For the design of a rock cut slope in Hong Kong, the site investigation typically commences with the mapping of rock exposures, which are generally plentiful, followed by a programme of drilling, and by the installation of piezometers to measure insitu water pressures.

The present site investigation practice in Hong Kong is described in detail by Brand & Phillipson (1982) and sampling techniques are the subject of a paper by Brenner & Phillipson (1979). In 'soft' material, rotary-wash boring is used, the hole being advanced by surging and drilling the casing down with water flush; sampling is by means of open-drive samplers or, on better jobs, by means of triple-tube Mazier retractable core barrels to give 73 mm samples. Rotary drilling is used in rock, double-tube core barrels being used that give 40 to 92 mm diameter cores; on rare occasions, triple-tube barrels are used.

In addition to full visual descriptions, rock cores obtained from Government site investigations are described in terms of Total Core Recovery, Rock Quality Designation and Fracture Index; full-page colour photographs are also taken of core boxes. As well as providing valuable information for engineering assessment, full core records are essential in Hong Kong, because of the early disposal of cores as a result of the high cost of storage.

Because of the importance of joint orientation in rock stability analyses, more frequent use is now being made of the Triefus Borehole Impression Packer Device, but its use is still rare. Reliance is more commonly placed on surface joint surveys before and after excavation.

Piezometers are very commonly placed in completed drillholes, either at the hole bottom or at a level of change in weathering zone, sometimes with two piezometers in one hole. Although piezometer tips are usually 300 mm long, the surrounding sand filter is usually extended both above and below the tip to give a length of 1 to 2 m. This is to ensure interception of groundwater from preferential drainage paths such as joints or relict joints. The piezometers are provided with standpipes (usually 25 mm ID) and, unfortunately, measurements are commonly made by dipmeters, although more frequent use is now being made of devices that record the maximum transient pressure that occurs (see below).

5. DESIGN OF ROCK CUT SLOPES

The methods of rock slope engineering applied in Hong Kong are based largely on those summarized by Hoek & Bray (1977). Specific applications of these to Hong Kong conditions have been described by Beattie & Lam (1977), Starr & Finn (1977) and Starr, Stiles & Nisbet (1981).

The general stability of a rock slope is governed by the jointing pattern, by the shear strength along the joints, and by the distribution of water pressures in the slope. Whereas there are adequate methods available for a fairly accurate assessment of the true jointing pattern, joint shear strength and critical water pressures usually pose problems of interpretation for the designer.

Preliminary stability analysis in Hong Kong is usually carried out by plotting joint measurements stereographically and then using the well established methods for assessing kinematic capability of wedges to fail (Hoek & Bray, 1977). As in many other parts of the world, however, there is a tendency for many workers to rely too much on such techniques to give complete answers. In particular, the statistical sorting of data can and does lead to dangerous practice by eliminating critical joints from consideration. A useful method to avoid this is to number each joint in the field during the survey. After the data has been contoured to arrive at pole concentrations, any original data points not incorporated in such concentrations and yet which fall in the unstable zone should be reassessed in the field. It should also be noted that a preliminary joint survey is rarely adequate to characterise the nature of critical joints for design. A single measurement taken with a geological compass will not represent the true orientation of a natural discontinuity. To assess orientation and roughness of critical joints sensibly, grid surveys should be carried out on joint surfaces using plates of various diameters.

For persistent joints in rock, the field strength at relatively low stress levels, is given by :

$$\gamma = \sigma \tan (\phi_b + i) \quad . \quad . \quad . \quad (1)$$

$$\text{or} \quad \gamma = a\sigma^b + \sigma \tan i \quad . \quad . \quad . \quad (2)$$

where σ is the effective normal stress, ϕ_b is the basic angle of friction for planar surfaces, i is the roughness angle relevant for the field stress level, and a & b are parameters determined from laboratory shear tests for non-linear strength envelopes corrected for dilation.

Additional strength is often available due to the impersistence of joints and is one of the most important yet most difficult factors to assess in the field. One of the best ways to calculate this additional strength is by backanalysis of failed and stable slopes following assessment of the strength of persistent joints.

Despite the appreciable number of laboratory investigations that have been carried out on the granites and volcanics of Hong Kong, their joint shear strengths are not well known, and there is a lack of good published data. For this reason, well documented published and unpublished data known to the Authors is sum-

Table I. Nature and shear strength of discontinuities

Rock Type	Discontinuity Pattern	Joint Characteristics	Roughness ($i°$)	Shear Strength of Planar Surface	Reference
Granite (Undifferentiated)	Widely spaced cooling joints and tectonic joints	Rough textured, grade II Clean, grade III-V Heavily iron-stained/clay infilled Heavy staining (manganese) Kaolin infilled	Not measured but probably 0 - 5°	$\sigma \tan 39°$ $\sigma \tan 38°$ $\sigma \tan 31°$ $\sigma \tan 26°$ $\sigma \tan 30°$	Report to Hong Kong Government, Golder Associates (1974) Hencher & Richards (1982)
Granite (Hong Kong)	Sheeting joints parallel to natural slopes	Rough, wavy, persistant, grade II-IV	15°	$\sigma \tan 40°$	Hencher & Richards (1982) Richards & Cowland (1982)
Monzonite	Tectonic, varied attitude	Chlorite coated Iron-oxide coated	9° 9°	$0.2\sigma^{1.17}$ $\sigma \tan 38°$	Internal Government Report (Slope Failure at Yip Kan Street, 1982)
Volcanic (Undifferentiated)	Tectonic, blocky	Smooth, stained/weathered Clay filled Relict in grade V (manganese) Iron-oxide coated	Not measured but probably 0 - 5°	$\sigma \tan 30°$ $\sigma \tan 26°$ $\sigma \tan 20°$ $\sigma \tan 32°$	Report to Hong Kong Government, Golder Associates (1974) Koo (1982) Internal Government Report (Island Road Improvement, 1982)
	Sheeting joints	Rough textured, iron-oxide coated	Not measured but probably up to 15°	$\sigma \tan 38°$	
Phyllite	Close foliations	Micaceous, weak rock	---	$\sigma \tan 28°$	Internal Government Report (Table Hill Reservoir, 1982)

marized in Table I for the major rock types and joint materials encountered in Hong Kong. It should be noted that the weathering grades referred to are as defined by the Geotechnical Control Office (1979) and in more quantifiable terms by Hencher & Martin (1982).

Although the basic friction angles are reasonably well established for the main Hong Kong rock types, the Authors consider that there is still a need for further study of the typical characteristics of joints, of the effects of weathering and infill on shear strengths, and of apparent cohesion due to impersistence in different quality rock masses. In addition, the quality of rock testing in Hong Kong is not good, and there is a need for improved standards in laboratories to follow the methods recommended by the International Society for Rock Mechanics (1974) and by the Canadian Pit Slope Manual (Gyenge & Herget, 1977).

The shear strength data given in Table I are extremely valuable for the preliminary design of rock slopes in Hong Kong. For safe and economical design of large slopes, however, a specific programme of sampling and shear testing is called for. The testing procedure for this was recently described by Hencher & Richards (1982) for Hong Kong granite, and Richards & Cowland (1982) discussed the application of these strength measurements to the study of the stability of large slopes containing sheeting joints in the North Point area of Hong Kong Island.

The third major factor controlling the stability of rock slopes, water pressure, is usually critical to design in Hong Kong where the majority of slope failures are associated with heavy rainfall. For design purposes, therefore, water pressures are required to be monitored throughout an entire wet season to allow the prediction of the conditions that are likely to exist for storm return periods of 10 years and 1 000 years.

During storms, peak water levels several metres above normal are commonly attained, and the monitoring system must therefore be capable of recording the *maximum* pressures, which might exist for only very short periods. A simple device, patented by Sir William Halcrow & Partners, that is used increasingly in Hong Kong comprises a series of plastic 'buckets' suspended at close intervals down a standpipe. These are closed, apart from small inlet and outlet ports, and each contains a small float. The maximum water level during a storm is indicated by the highest 'bucket' that contains water.

Other 'automatic' recording systems incorporating hydraulic or pneumatic piezometers have been adopted for major projects in Hong Kong in recent years and are considered to have produced reliable data. In many cases, however, there are inadequate data on water pressures, and the designer can only guess the peak pressures that might develop along critical joints. This is not, of course, a design problem unique to Hong Kong.

6. PREVENTIVE AND REMEDIAL WORKS

On the basis of the catalogue of ranked slopes and retaining structures mentioned in section 3 above, the Geotechnical Control Office initiated a programme of stability investigations in 1979 to rectify a backlog of unsatisfactory slopes and walls. Slopes which are found to have unacceptably low factors of safety are generally put into the ongoing Lanslip Preventive Measures programme for rectification. In addition to these *preventive* works, *remedial* works are carried out in situations where failures have occurred.

Where preventive works are found to be required for hard rock slopes, the type of work necessitated is usually simple and straightforward.

Rock anchors are quite extensively used, but more commonly other support techniques are employed, along with surface protection, scaling and relief drains.

For many of the slopes in Hong Kong, bad drilling and blasting have left very uneven faces with overhangs and obvious blast damage to the cut faces. Preventive works in these situations is usually based on an experienced visual assessment aided by joint surveys and simple back analyses of exposed faces. Frequently, the appearance of a blast damaged face is considered from a cosmetic viewpoint and more material is removed than would be necessitated by consideration of the local factor of safety. Such conservative measures are considered economic as they reduce the need for future maintenance, and the relative costs of such works are low.

Mass concrete buttresses are commonly used as a preventive measure in Hong Kong, sometimes with stone pitching. These are usually introduced below small overhangs, and to provide lateral support to dangerous wedges and planes. Simple mechanical support is often provided by means of rock bolts or dowels. Although bolts usually rely on a grout envelope and a concrete head cap for corrosion protection, increased protection is now being obtaibed with proprietry shrink-wrap bolts with stainless steel heads.

Rock anchors were employed fairly extensively on rock slopes in Hong Kong until about 1979; a typical example of their use is shown in Fig. 5. Since that time, anchors have been less popular for permanent support, because local practice is such that the standards of installed anchors have often not conformed to the best international practice as far as corrosion protection is concerned. The Geotechnical Control Office has recently produced a draft General Specification for Prestressed Ground Anchors, and it is hoped that this will result in better practice in the industry. The central core of the Specification is the adoption of a 'prior approval' procedure for proprietry ground anchor systems, but it will still be some time before this procedure can be fully operational.

Chunam (1 cement : 3 lime : 5 sand) and sprayed concrete are commonly used as a surface protection for poor quality rock to limit infiltration and erosion. Where substantial support is required, sprayed concrete is sometimes reinforced. In all cases, adequate measures must be taken to prevent adverse water pressures developing. Occasionally, surface meshing is used where there is inadequate room for a toe fence, but this generally has a short design life because of corrosion.

To prevent damage from minor rock-falls, a toe barrier (fall-zone and fence) can be a very economical solution. Where there is no space for a proper barrier, a chain link fence is very effective as long as it is properly maintained. Toe barriers, however, have their limitations in Hong Kong, since space is limited and boulders from natural slopes have been known to land as far as 10 m from the toe of a cut.

Inadequate drainage is the key factor in most failures. For drainage to be efficient, special attention must be given to high transient ground-

Fig. 5. Example of extensive use of rock anchors

water pressures and to the considerable volumes of water flow, but these important aspects of preventive measures have received little attention in Hong Kong to-date. Raking drains are used occasionally, but problems can result from blockage due to poor design and lack of maintenance. Surface drains are usually constructed to intercept up-slope runoff, but they often similarly suffer from poor design and lack of maintenance.

Frequently, vegetation which is aesthetically beneficial can mask dangers and can actively reduce stability due to the penetration of roots. Many failures are caused by trees being blown down in cyclones, resulting in blocks of rock being dislodged.

At present, many emergency slope incidents occur each year during times of heavy rainfall. Some of the most important slope failures occur on highways that need to be reopened as soon as possible because no alternative road links are available. In these cases, repair works are often carried out with half the road opened after the immediate danger has been overcome. In order to expedite the repairs, a remedial works design must often be based on a professional visual assessment with limited access and without the benefit of proper site investigations. After an initial amount of remedial work to permit road opening, circumstances often justify the design of permanent remedial works to be carried out at a later date; the full range of solutions used for preventive works can then be employed.

7. ROCK CLIFFS

Because of urban development around the harbour, many buildings are situated where they may be threatened by rock falls from the steep natural cliffs that flank many of the major peaks of Hong Kong Island and Kowloon. The debris from an ancient major natural rock cliff failure is illustrated in Fig.6.

The preventive works to support degrading natural rock slopes, which often extend over heights of more than 100 m at steep angles, would be a major engineering task. Even investigation is extremely difficult because of the scale of the problem and the lack of easy access. Pilot schemes are continuing to identify and stabilise the most dangerous areas. These studies have indicated that, in many cases, removal of the features or provision of effective barriers are not viable. Massive anchored buttresses appear to be the only reasonable solution.

8. BOULDER FALLS

The natural degradation of the hillsides of Hong Kong on a geological time-scale has resulted in the deposition of many hundreds of thousands of boulders on the natural slopes. These vary greatly in size ranging up to more than 1 000 cu m, but boulders of about 10 cu m are perhaps the most common. They occur as isolated boulders or as extensive boulder fields. The manner in which individual boulders are supported on the slopes ranges from boulder-to-rock contact to almost complete embedment in soil. Many of the boulders are perched precariously, or appear to be so (Figs.6 and 7), but there is often no evidence to indicate recent movements.

In many locations throughout Hong Kong, boulders pose apparent threats to urban developments at lower elevations. Boulder falls, although not common, do occur from time to time, and deaths and injuries have been caused in this way. As with slope failures, boulder falls occur almost entirely during times of heavy rainfall, largely because the supporting soil is washed away.

The problem of boulder falls is largely an intractable one in Hong Kong. The large number of boulders and the often densely vegetated and inaccessible slopes make it exceptionally difficult to carry out any meaningful investigations on boulder stability. Where small slope areas have been studied, no rational method has been devised for assessing boulder stability, and subjective judgement has been used to decide on which boulders should be removed or stabilized. Removal of boulders is usually effected by reducing the size by mechanical and chemical splitting, but explosives are sometimes used for very large boulders and for emergency remedial situations. Stabilization works most commonly take the form of concrete buttressing, which is largely maintenance free but sometimes aesthetically unsatisfactory.

It is presently accepted in Hong Kong that an extensive programme of Territory wide boulder stabilization is impractical and that some element of risk from boulder falls must be accepted. Present government policy dictates that preventive works are carried out only where a boulder poses an immediate and obvious danger to life.

9. CONCLUSIONS

The majority of slope problems in Hong Kong are

Fig. 6.   Debris from a major rock cliff failure

concerned with the stability of weak, decomposed materials, and rock engineering has been largely overshadowed as a result.

A considerable number of rock slopes in Hong Kong require preventive works because of past unsatisfactory rock engineering practices. It will be some years before this backlog is cleared. The present state of rock engineering is unsophisticated, and there are a number of fields where improvements can be made.

There is a need for an improved understanding of the behaviour of rock slopes in Hong Kong and this must be reflected in improved quality investigations and shear strength determinations. The characteristic nature of joints in the common rock types also requires extensive investigation.

Fig. 7.   Typical boulders on a Hong Kong slope

It is clear from some failures that occurred in the rainstorms of 1982 that normally adopted methods for determining water pressures are not satisfactory, and this applies to both soil and rock slopes. Piezometers capable of recording maximum water pressures are necessary for rational design decisions to be made, and the installation details must be specifically designed for individual cases.

It is hoped that the publication of a specification for ground anchors in Hong Kong, with particular emphasis on corrosion protection, will allow this much needed support system to be used with confidence.

ACKNOWLEDGEMENTS

This Paper is published with the permission of the Director of Engineering Development of the Hong Kong Government.

REFERENCES

Allen, P.M. and Stephens, E.A. (1971). Report on the Geological Survey of Hong Kong. 107 pp. plus 2 maps. Hong Kong: Government Printer.

Beattie, A.A. and Lam, C.L. (1977). Rock slope failures - their prediction and prevention: Hong Kong Engr, (5), 7, 37-40.

Brand, E.W. (1982). Analysis and design in residual soils: Proc. ASCE Spec. Conf. Engineering and Construction in Tropical and Residual Soils, 89-143, Honolulu.

Brand, E.W. and Hudson, R.R. (1982). CHASE: Empirical approach to the design of cut slopes in Hong Kong: Proc. 7th S.E.Asian Geotech. Conf., (in press), Hong Kong.

Brand, E.W. and Phillipson, H.B. (1982). Site investigation practice in Hong Kong: Site Investigation Manual, International Society for Soil Mechanics and Foundation Engineering (in press). Tokyo: ISSMFE.

Brand, E.W., Styles, K.A. and Burnett, A.D. (1982). Geotechnical land use maps for planning in Hong Kong: Proc. 4th Congress Int. Assoc. Eng. Geol., (in press), New Delhi.

Brenner, R.P. and Phillipson, H.B. (1979). Sampling of residual soils in Hong Kong: Proc. Int. Symp. Soil Sampling, 104-120, Singapore.

British Standards Institution (1981). Code of Practice for Site Investigation, B.S. 5930. 147 pp. London: British Standards Inst.

Geotechnical Control Office (1979). Geotechnical Manual for Slopes. 242 pp. Hong Kong: Government Printer.

Government of Hong Kong (1972). Final Report of the Commission of Inquiry into the Rainstorm Disasters, 1972. 91 pp. Hong Kong: Government Printer.

Gyenge, M. and Herget, G. (1977). Laboratory tests for design parameters: Pit Slope Manual, Supplement 3.2 (Canmet Report 77-26). 74 pp. Ottawa: Canmet.

Hencher, S.R. and Martin, R.P. (1982). Description and classification of weathered rocks in Hong Kong for engineering purposes: Proc. 7th S.E.Asian Geotech. Conf., (in press), Hong Kong.

Hencher, S.R. and Richards, L.R. (1982). The basic frictional resistance of sheeting joints in Hong Kong granite: Hong Kong Engr, (11), 2, 21-25.

Hoek, E. and Bray, J.W. (1977). Rock Slope Engineering. 2nd Edition. 402 pp. London: Inst. Min. Met.

International Society for Rock Mechanics (1974). Suggested Methods for Determining Shear Strength. Committee on Field Tests, Document No. 1. 23 pp. Lisbon: ISRM.

Janbu, N. (1954). Application of composite slip surface for stability analysis: Proc. Europ. Conf. Stability of Earth Slopes, (3), 43-49, Stockholm.

Koo, Y.C. (1982). Relict joints in completely decomposed volcanics in Hong Kong: Canad. Geotech J., (19), 117-123.

Lumb, P. (1975). Slope failures in Hong Kong: Q.J. Eng. Geol., (8), 31-65.

Moye, D.G. (1955). Engineering geology for the Snowy Mountains Scheme: J. Inst. Engrs Australia, (27), 287-298.

Richards, L.R. and Cowland, J.W. (1982). The effect of surface roughness on the field shear strength of sheeting joints in Hong Kong granite: Hong Kong Engr, (in press).

Ruxton, B.P. (1960). The geology of Hong Kong: Q.J. Geol. Soc. London, (115), 233-260.

Ruxton, B.P. and Berry L. (1957). Weathering of granite and associated erosional features in Hong Kong: Bull. Geol. Soc. Amer., (68), 1263-1292.

So, C.L. (1971). Mass movements associated with the rainstorm of June 1966 in Hong Kong. Trans. Inst. Brit. Geographers, No. 53, 55-65.

Starr, D.C. and Finn, P.S. (1979). Practical aspects of rock slopes stability assessment in Hong Kong: Hong Kong Engr, (7), 10, 49-56.

Starr, D.C., Stiles, A.P. and Nisbet, R.M. (1981). Rock slope stability and remedial measures for a residential development at Tsuen Wan Hong Kong: Q.J. Eng. Geol., (14), 175-193.

CONFORTEMENT DES TALUS ROCHEUX DE LA CARRIERE CHENAL DE MONTEZIC

Consolidation of rock slopes of the quarry "Chenal de Montezic"

Felsböschungsbewehrung des Steinbruchs "Chenal de Montezic"

G. Colombet
Département Géotechnique — Coyne et Bellier — France

M. Glories
Région d'Equipement Alpes Marseille — EDF — France

RESUME

L'exploitation de la retenue de Montézic et principalement les vidanges hebdomadaires vont créer des sollicitations importantes dans les talus rocheux de la carrière chenal qui peuvent provoquer leur rupture. Le massif granitique a donc été armé par câbles d'acier scellés sur toute leur longueur. Une auscultation appropriée permet le contrôle de la stabilité.

SUMMARY

The operation of the Montézic reservoir and particularly the weekly discharges will create significant stresses in the rock slopes of the quarry channel which may cause their failure. The granite rock mass was reinforced by steel cables anchored over their entire length. A suitable measurement system makes it possible to check the stability.

ZUSAMMENFASSUNG

Durch die Nutzung des Montézic Stausees, und insbesondere dessen wöchentliche Entleerung, wird die Felsböschung vom Steinbruch "Chenal de Montézic" hohen Beanspruchungen ausgesetzt, sodaß ein Gleitflächenbruch entstehen kann. Das Granitmassiv wurde mit über der ganzen Länge verankerten Stahlseilen bewehrt. Eine geeignete Prüfuntersuchung soll der Kontrolle der Stabilität dienen.

1. INTRODUCTION

L'aménagement hydroélectrique de Montézic a été réalisé entre 1979 et 1982 par Electricité de France dans la région d'Aurillac (France). Le projet prévoit la création d'une installation de transfert d'énergie par pompage entre le réservoir supérieur et la retenue existante de la Couesque.

La carrière chenal, objet du présent article, relie le réservoir supérieur aux prises d'eau. Les déblais extraits ont fourni les enrochements nécessaires à la construction de deux barrages bordant ce réservoir.

Pendant les travaux, le substratum granitique est apparu plus fissuré que prévu. Les talus rocheux du chenal présentaient un risque d'instabilité en exploitation, le réservoir faisant alors l'objet d'une vidange hebdomadaire sur 23 m de dénivelée, à la vitesse moyenne de 0,16 m/heure.

Cet article décrit les principes originaux retenus dans le confortement de ces talus.

2. ETABLISSEMENT DU PROJET DE CONFORTEMENT

2.1. Description sommaire de l'ouvrage

La figure 1 représente schématiquement la coupe transversale de ce chenal long de 200 m.

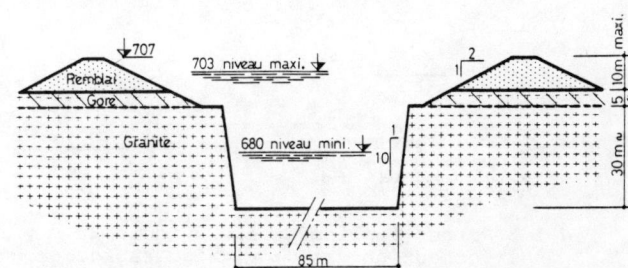

Fig.1 : Coupe transversale schématique de la carrière chenal.

Le chenal est excavé dans un granite à gros grains, compact et massif en profondeur. Il est altéré en surface et prend alors l'aspect d'un gore. La fracturation est dans l'ensemble peu marquée (espacement de 2 à 3 mètres). Les études géologiques réalisées sur le site n'ont pas montré de direction de fracturation privilégiée.

Les fissures sont fermées. La perméabilité en grand du massif est faible (quelques unités Lugeon). Dans ces conditions, il est probable que l'exploitation de la retenue -principalement les vidanges entre les cotes 703 et 680- va développer des sous-pressions importantes dans les talus qui peuvent provoquer leur rupture. Il a donc été estimé nécessaire de consolider ces talus. Le choix s'est porté sur un confortement à base d'ancrages passifs, solution rapide à mettre en oeuvre et économique.

Une difficulté résidait néanmoins dans le dimensionnement des aciers. Les calculs ont été conduits avec plusieurs hypothèses permettant de chiffrer l'influence des paramètres essentiels vis-à-vis de la stabilité (direction des plans de fracturation, caractéristiques mécaniques, sous-pressions).

2.2 Prédimensionnement des aciers

On a recherché la quantité d'acier nécessaire pour que les sollicitations en vidange rapide soient sensiblement les mêmes que celles après ouverture des fouilles. On a retenu un modèle simple ; les lignes de rupture dans le rocher sont planes et on fait varier leur inclinaison. On a considéré le cas pessimiste de la pleine sous-pression (fig.2).

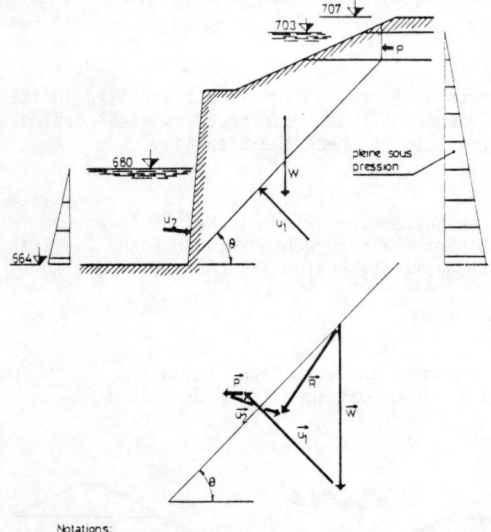

Notations:

W = poids d'un coin rocheux découpé par une fracture plane inclinée
d'un angle θ sur l'horizontale.

u1 & u2 = poussées hydrostatiques.

P = poussée horizontale dans le remblai

$\vec{R} = \vec{W} \cdot \vec{u_1} \cdot \vec{u_2} \cdot \vec{P}$

Fig.2 : Forces appliquées sur un coin
rocheux en cas de vidange rapide.

L'hypothèse retenue dans le calcul des ancrages est la mise en traction pure de ceux-ci par suite de la dilatance des joints, quelle que soit leur inclinaison. Ainsi si l'on appelle α l'inclinaison d'un ancrage sur le plan de discontinuité et t_a sa tension, l'apport de l'acier est donné par l'expression :

$$B = t_a \cos \alpha + t_a \sin \alpha \, \mathrm{tg}\phi$$

avec ϕ angle de frottement limite du joint.

Avec ce mode de calcul, la direction de fracturation la plus dangereuse correspond aux plans

inclinés à 55° sur l'horizontale. La force à appliquer pour assurer la stabilité des talus en vidange rapide est de 280 t/ml. Cette valeur a été retenue dans le projet d'exécution des ancrages passifs. Il a été admis de plus que les aciers travaillaient aux deux tiers de leur limite élastique.

2.3 Dispositif d'auscultation associé

Le dimensionnement des ancrages dépend largement des hypothèses de calcul retenues. La solution définie précédemment est apparue raisonnable, à condition de lui associer un programme d'auscultation permettant de suivre le comportement réel du massif pendant les premières mises en eau, et de contrôler la stabilité des talus en exploitation.

Le projet prévoyait dans ce but (fig.3) :

- des mesures de déformation par Distofor (extensomètres électromagnétiques TELEMAC) introduits dans des forages de 15 mètres de longueur,

- des mesures de pression interstitielle par cellules de pression TELEMAC placées dans des forages subverticaux de 25 et 30 m de profondeur.

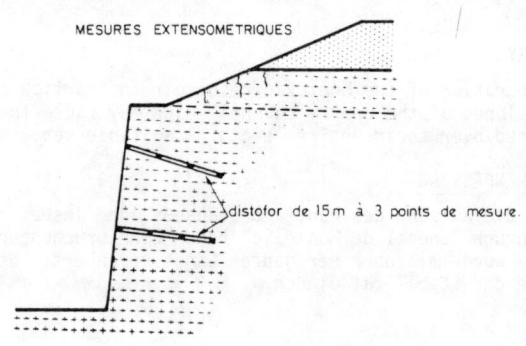

MESURES EXTENSOMETRIQUES

distofor de 15 m à 3 points de mesure.

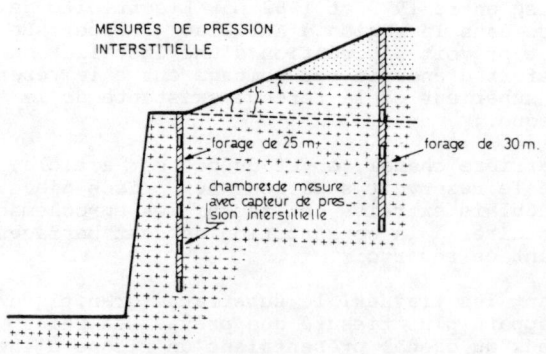

MESURES DE PRESSION INTERSTITIELLE

forage de 25 m·   forage de 30 m.

chambres de mesure avec capteur de pression interstitielle

Fig.3 : Dispositif retenu pour l'auscultation des talus.

3. CHOIX D'UN TYPE D'ANCRAGE PASSIF ET MISE EN OEUVRE

L'Entreprise SOLETANCHE, adjudicataire du marché, a proposé de placer des câbles en acier à haute résistance constitués de 12 torons de 15 mm de diamètre (12 T 15) dans la partie inférieure du talus et des câbles de 6 torons (6 T 15) dans la partie supérieure. La protection des aciers contre la corrosion est assurée d'une part par le mortier de scellement et d'autre part par une gaine nervurée PVC. Des écarteurs permettent de centrer le câble dans la gaine.

La coupe type du talus consolidé est représentée sur la figure 4. Les ancrages sont placés en quinconce ; sur chaque rangée, les ancrages sont espacés de 4 m. Leur longueur permet de coudre les plans potentiels de glissement passant en pied de talus et inclinés de 45°.

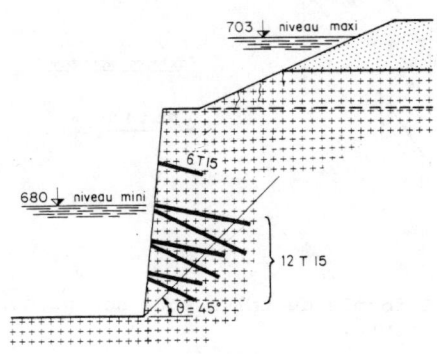

COUPE TYPE DU TALUS CONFORTE PAR CABLES
6 T 15 & 12 T 15

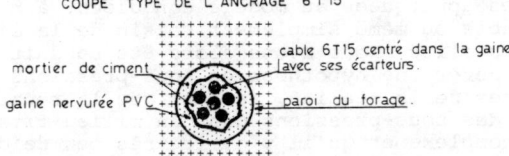

COUPE TYPE DE L'ANCRAGE 6 T 15

Fig.4 : Principe adopté pour le confortement des talus.

4.   CALCUL DU TALUS PAR UN MODELE DE BLOCS

Pour nous aider dans la compréhension de la cinématique du massif en roche armée et dans l'interprétation des mesures de déformation définies en 2.3., nous avons réalisé un calcul du talus par un modèle de milieu fissuré. Le massif est considéré comme un ensemble de blocs indéformables se déplaçant les uns par rapport aux autres selon les lois rhéologiques des contacts. Ce type de modèle représente mieux le comportement réel du massif que celui utilisé dans le calcul des ancrages (§2.2). Par contre, il se prête moins à l'analyse paramétrique nécessaire pour leur dimensionnement.

4.1. Programme utilisé

Le programme utilisé est le programme BLOC de COYNE et BELLIER. Ce programme utilise une méthode originale, parente de la méthode des éléments finis qui permet de modéliser les milieux hétérogènes fissurés. Des articles présentant cette méthode seront publiés courant 1983.

Le modèle est bidimensionnel, chaque bloc possédant 3 degrés de liberté (deux degrés de trans-

lation et un degré de rotation). Les blocs sont simplement en contact les uns des autres. Le comportement tangentiel des joints entre blocs est du type élastoplastique avec frottement limite de Coulomb.

Les blocs peuvent être reliés par des ancrages schématisés dans le modèle comme des ressorts.

Comportement des joints

Lorsque le joint est ouvert, l'effort normal N et l'effort tranchant T sont nuls.
Lorsque le joint est fermé ou partiellement ouvert (fig.5) :

- la contrainte normale σ est proportionnelle à l'enfoncement relatif des deux faces :

$$\sigma = k_n \cdot u$$

avec k_n rigidité normale du joint,
     u enfoncement relatif des deux faces du joint.

- la contrainte de cisaillement τ est proportionnelle au déplacement relatif tangentiel :

$$\tau = k_t \cdot v$$

avec k_t rigidité tangentielle du joint
     v déplacement relatif tangentiel.

On en déduit)

. effort normal   : $N = \int \sigma(x) \cdot dx$
. effort tranchant   : $T = \int \tau(x) \cdot dx$
. moment au centre du joint   : $M = \int \sigma(x) \cdot x \cdot dx$

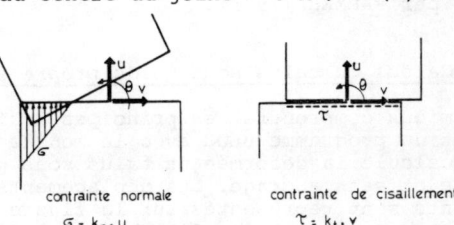

contrainte normale   contrainte de cisaillement
$\sigma = k_n \cdot u$ $\tau = k_t \cdot v$

Fig.5 : Comportement des joints fermés ou partiellement ouverts dans le programme BLOC.

4.2. Modèle retenu

Deux familles de joints ont été considérées dans la définition du modèle numérique :

- une première, inclinée à 55° sur l'horizontale; les calculs précédents montraient qu'il s'agissait là de la direction potentielle la plus dangereuse en exploitation ;

- une deuxième, verticale, schématisant une décompression possible du massif.

L'espacement des fractures est compris entre 2 et 3 mètres. Le modèle (figure 6) comprend ainsi 17 blocs susceptibles de glisser reposant sur 7 blocs fixes.

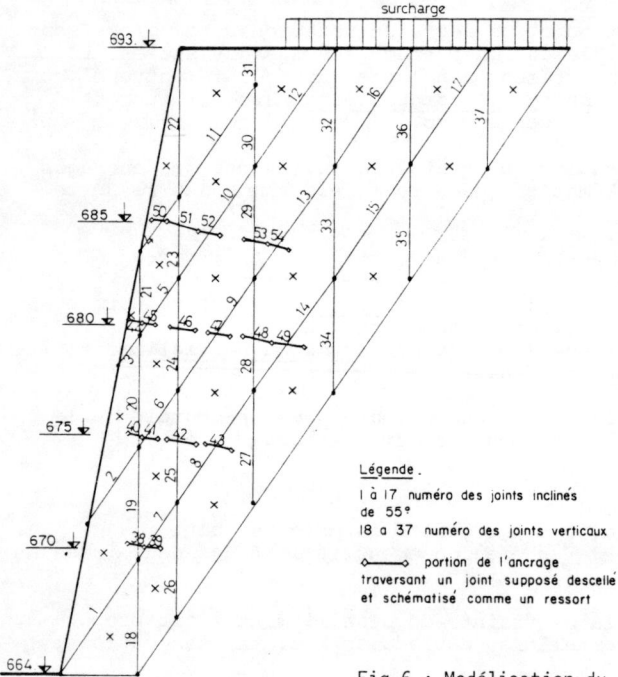

Fig.6 : Modélisation du
talus de Montézic
utilisé avec le
programme BLOC.

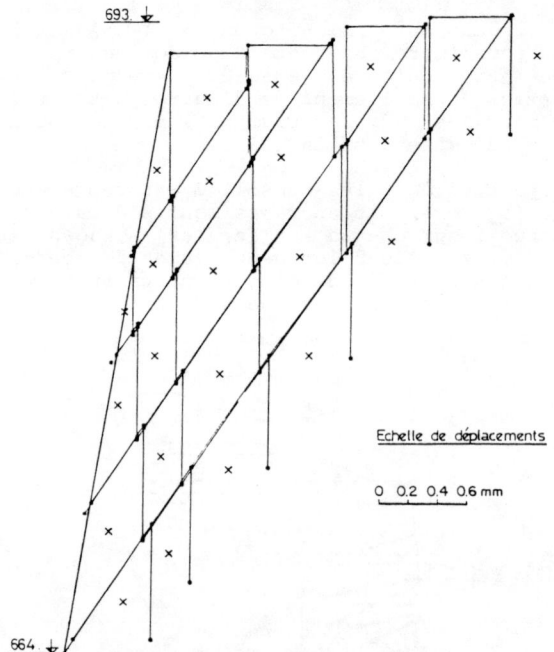

Fig.7 : Déformée du talus sous son poids
propre.

Les ancrages ou ressorts sont figurés à leur emplacement réel. Leur comportement est élastique
sur toute leur longueur "libre" supposée égale à
50 cm de part et d'autre de la fissure couturée.
Cette valeur peut être déduite des formules indiquées par FARMER (1975).

4.3. Calcul du talus sous poids propre seul

Pour mieux comprendre les principes de fonction-
nement du programme BLOC avec le modèle proposé,
on a calculé la déformée du talus sous poids
propre et sans ancrage. Les déplacements correspondants sont représentés sur la figure 7. On
constate que les blocs glissent parallèlement
aux joints inclinés à 55°.

A l'intérieur du massif, les joints verticaux
s'ouvrent et ne transmettent aucun effort. Dans
cette zone, le déplacement des blocs est propor-
tionnel à la rigidité tangentielle des joints
inclinés.

A proximité du versant, les joints verticaux
sont fermés et transmettent une poussée aux
blocs situés en aval. L'équilibre est atteint
lorsque la relation T=N.tgφ est vérifiée au niveau de chaque joint incliné (avec φ angle de
frottement limite). Le déplacement des blocs à
proximité du versant ne dépend plus de la rigidité tangentielle des joints, mais de la valeur
de l'angle de frottement limite φ.

4.4. Calcul du talus en vidange rapide

Le programme BLOC a permis une analyse du comportement du talus et de son confortement sous
l'effet de la vidange rapide.

On s'est aperçu que le programme ne convergeait
pas avec l'hypothèse pessimiste de la pleine
sous-pression (fig.2) -ceci apparaît lorsque les
forces appliquées au modèle conduisent à un état
instable ou même simplement voisin de la limite
de stabilité- Nous avons alors été conduit à
considérer une hypothèse de sous-pressions plus
proches de la réalité, sachant que la répartition des sous-pressions dans un milieu fissuré
est complexe et qu'il y avait très peu de données à ce sujet dans le granite de Montézic.

La cinématique des mouvements en vidange rapide
est indiquée sur la figure 8. Les colonnes verticales découpées par les joints verticaux ont
tendance à basculer vers le vide. La valeur des
déplacements est essentiellement fonction des
hypothèses de sous-pressions, et non pas de la
valeur de la rigidité tangentielle des joints.

Les aciers limitent les déplacements. Ils sont
d'autant plus tendus que les joints qu'ils traversent ont tendance à plus s'ouvrir. La figure 9, donnée à titre d'exemple, indique la traction des aciers avec une hypothèse de sous-pression semblant réaliste.

Par suite de l'effet de basculement des colonnes
verticales, la portion d'acier la plus tendue
correspond à la traversée du premier joint vertical par la rangée d'ancrages supérieure
(49 tonnes par mètre linéaire de talus soit 100t
environ par ancrage). D'une manière générale,
les aciers sont d'autant moins tendus qu'ils sont
plus bas et que l'on pénètre à l'intérieur du
massif.

On peut en déduire qu'avec le modèle mathématique utilisé dans le programme BLOC, les blocs
rocheux dont la stabilité est la moins assurée
sont les blocs superficiels en haut du talus.
Ce résultat n'est pas a priori étonnant puisque

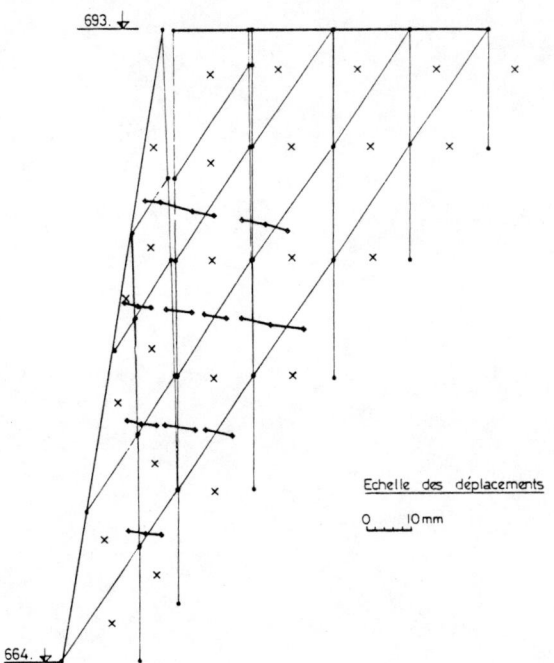

Fig.8 : Déformée du talus en vidange rapide.

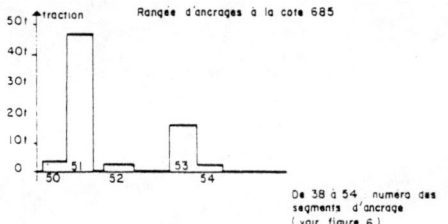

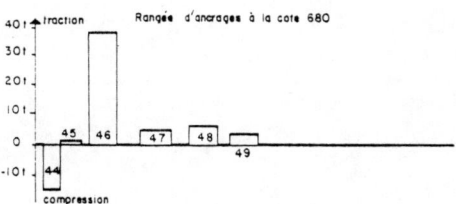

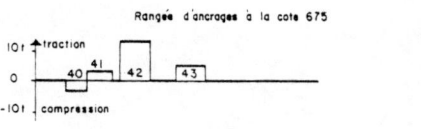

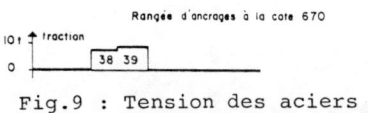

Fig.9 : Tension des aciers en
vidange rapide.

le projet de confortement a cherché principalement à assurer la stabilité des coins rocheux de
grand volume, peu d'ancrages étant placés dans
la partie supérieure du talus.

5. CONCLUSIONS

Le renforcement des talus rocheux de la Carrière
Chenal de Montézic par câbles d'acier passifs
scellés sur toute leur longueur constitue une
solution économique. Cependant, le dimensionnement des aciers garde un caractère quelque peu
empirique par suite des conditions locales du
site (structure géologique complexe et effet de
la vidange rapide difficile à quantifier).

C'est pourquoi nous avons cherché à analyser le
comportement du massif en roche armée avec le
programme BLOC. Enfin, une auscultation était
indispensable pour vérifier les hypothèses de
calcul et contrôler la stabilité des talus pendant l'exploitation de l'ouvrage.

REFERENCES

FARMER J.W. (1975). Stress Distribution along a
Resin Grounted Rock Anchor : Int. J. Rock
Mecha.Min.Sci.(12).

GROUPE FRANCAIS (1979). Le renforcement des massifs rocheux par armatures passives :
4ème congrès international de la SIMR,
Montreux.

HABIB P. (1978). Behaviour and Reinforcement of
Rock Masses : Rapport, octobre 1978.

HOEK E, BRAY S.W. (1977). Rock Slope Engineering
The institution of mining and metallurgy,
London.

LONDE P., HOEK E.Y. (1974). Travaux de surface
au rocher : Rapport général, thème 3, 3ème
Congrès International de la SIMR, Denver.

PANET M.(1978). Stabilisation et renforcement
des massifs rocheux par aciers précontraints
et aciers passifs : Séminaire de l'ATTI,
Stresa.

THE STABILITY OF THE ORVIETO ROCK

La stabilité du rocher d'Orvieto

Über die Stabilität des Felsens von Orvieto

C. Cestelli-Guidi, G. Croci, P. Ventura
University of Rome, Italy

SYNOPSIS

The town of Orvieto is built on a tuffaceous platform having subvertical sides surrounded by talus debris and lying on a base formation of over-consolidated clays, marine in origin and dating back to the Pliocene era. The rock has had a number of landslides, and marginal portions of it have collapsed, this being due both to natural causes resulting from developments taking place on a geological time-scale, and to a number of anthropic causes that have developed over a period measured in decades. The various causes will be analyzed in this paper, and the operations performed to contain or eliminate the present phenomena will be described. In order to check on the development of these phenomena, monitoring systems have been provided in the areas mostly affected.

RESUME

La cité d'Orvieto est construite sur un plateau tufacée à bords subverticaux, entourées de talus d'éboulis et située sur une base d'argile très surconsolidée d'origine marine du cycle pliocène. Le rocher a connu de nombreux écroulements sur les côtes et de nombreux glissements de pente soit par causes naturelles au cours des temps géologiques, soit par causes humaines, survenues dans l'échelle décennale. On analyse toutes ces causes et les interventions relatives afin de contenir et d'éliminer les phénomènes en cours. Dans le but de contrôler l'évolution de ces phénomènes on a installé des appareils enregistreurs in situ dans les zones les plus intéressées par les déformations.

ZUSAMMENFASSUNG

Die Stadt Orvieto ist auf einer tuffsteinartigen Platform mit subvertikalen Wänden aufgebaut, die von Anschwemmungsschichten umgeben sind und die auf einer überkonsolidierten Tonbasisablagerung liegen, deren Ursprung ins pliozäne Zeitalter zurückgeht. Der Felsen war vielen Wandeinstürzen unterworfen, mit Gehängeschutt, der sich im Ablauf der geologischen Zeiten entwickelte, und war auch das Objekt vieler menschlischer Fehler gewesen, die sich im Laufe der Jahrzehnte entfaltet haben. Alle diese Ursachen sowie auch die diesbezüglichen Eingriffe, um die aktuellen Erscheinungen zu beschränken und zu beseitigen, werden analysiert. Um die Entwicklung dieser Erscheinungen zu kontrollieren, hat man in den Punkten, die am meisten von der Verrüttung betroffen sind, ein Monitorsystem vorgesehen.

1. INTRODUCTION

The artistic patrimony of the Orvieto is characterized by vestiges of ancient Etruscan constructions and by the famous Duomo built to commemorate the miracle of Bolsena.The cliff has been threatened for time by both natural and man-causes.

To perform the recovery operations, saving the town, was given the proper weight to the numerous technical and economical aspects.Therefore the works are executed by the Consortium of Companies GEO SONDA S.p.A. of Rome,GRASSETTO COSTRUZIONI S.p.A. of Padova and SO.GE.STRA. S.p.A. of Rome; the general project was givenbirth by the collaboration of the Authors with the following:

-Eng: P. De Carolis: Geomechanical and stress-state analyses of the rock.
-The Geosonda Technical Department: Engrs. L. Diamanti and C. Soccodato, Dr. G. Ghiarini: General aspects and project coordination.

-The Land System Engineering Offices: Prof. V. Barberis, Eng. A. Banella: Project of the aqueduct and sewage system.
-The Ecosuolo Engineering Offices: Profs. R. Coltro, V. Gualdi, and Dr. F. De Santis: Hydraulic andforestation setting of the slopes.
-The Paolo Marconi Offices: Prof. P. Marconi: Architectonical Restoration.

2. THE GEOLOGY

Figure 1 shows a typical view of Orvieto's tuff rock bench, lying to the west of the Tiber river and the Paglia its tributary, the city's geomorphology being common to many cities in the Tuscia interland (Middle Italy). The tuff platform at an altitude above mean sea level of 300 m,has subvertical sides surrounded by detritus slopes, the platform's morphology, approximately eliptical with semiaxes of about 800 m and 1600 m, having been surveyed aerophotogrammetrically.

Fig. 1 - Tipical view of the Orvieto tuff cliff before the works.

The platform lies on an extensive formation of "grey base clays", marine in origin, dating to the Pliocene, with a thickness of some hundreds of meters, very overconsolidated since sediments were eroded to a thickness of around a hundred meters.

Deposits on the base clays, are continental sediments forming the "Albornoz formation", comprising lacustrine and river conglomerates, gravels, sands, and silts, and strata of scoria, pumice and ash, in thicknesses of from 5 to 15 m, coming from the Vulsino volcanic system in which the present Lake Bolsena lies.

Deposited on these formation is then the "black-scoria lithoid tuff", having two main lithofacies: a grey "pozzolanic" and a yellowish "scoriaceous". A geologic section, such as that in fig. 2, through the tuff platform shows an average thickness of 50 m, with free portions of from 20 to 35 m in thickness, and with a bed everywhere connected to the base clays by means of the aforementioned Albornoz formation (Conversini, Lupi, Martini, Pialli, Sabatini, 1978).

Other lacustrine-river deposits were then made, being mainly characterized by deposits of travertine, present now in the area, where St. Patrizio's well was dug. The piroclastic deposits were then eroded away, untill the present inversion of the ground's configuration was created together with the typical tuff platform. The foot of the rock bench, finally, is characterized by talus debris formed of sands, gravels, pebbles and tuff boulders in an ashstone and pozzolanic matrix produced by the breaking up of the rock itself. This detritus has been the seat of landslides that mainly developed along the 5 dit-

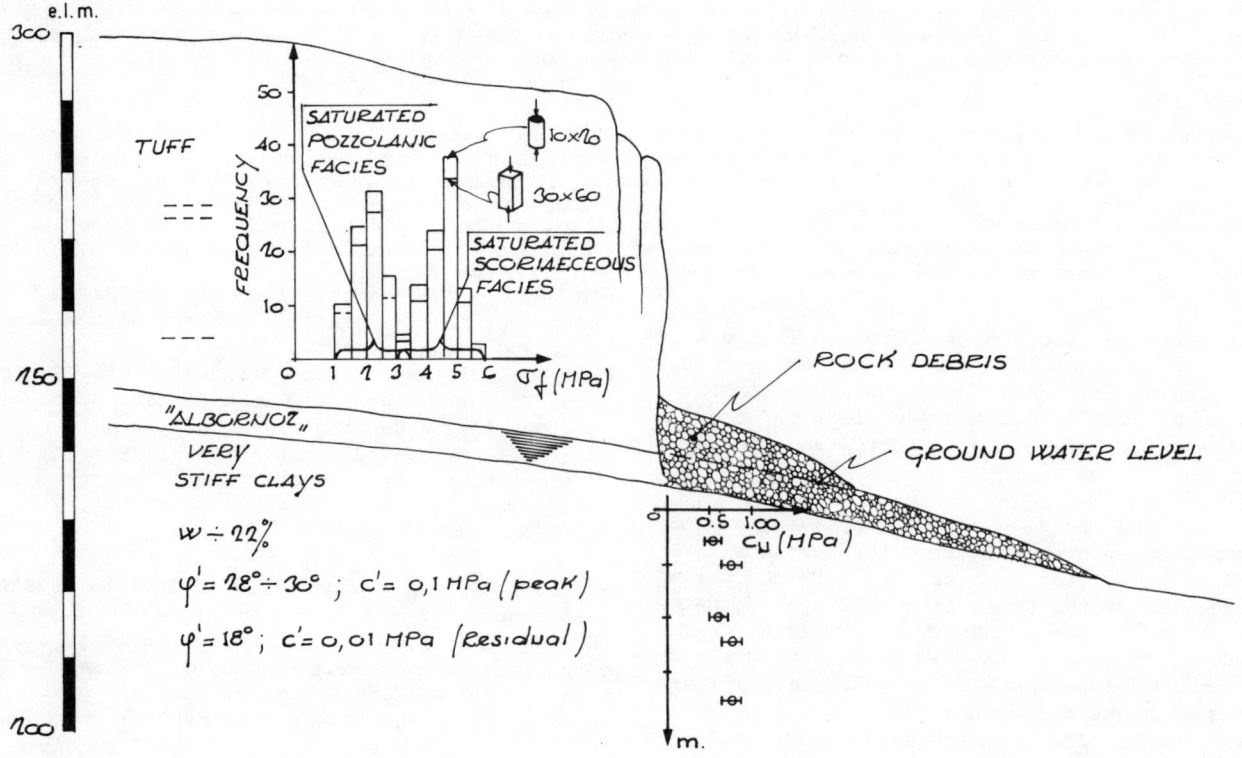

Fig. 2 - The geological section and strength parameters measured in Laboratory

C 32

ches radiating from the tuff platform, as will be described hereafter.

3. HYDROGEOLOGY

The ground water, as piezometers show, mainly invests the permeable Albornoz formation and the bordering tuffaceous talus detritus, these too being very permeable. The bed (fig.2) of the acquifer is the base clays, and water was drawn from it when the town was besieged through the famous, 62 meters deep, well of St. Patrizio, which is 13 meters wide and is driven through the entire tuff platform down to the Albornoz formation.

Rainwater, an average of 900 mm for year, especially percolates through the perimetral band of the tuff formation, to supply this acquifer, which shows up, especially in about 30 springs, in the form of both perched and contact springs.Percolation is increased, moreover, by leaks from the town's acqueduct and, in particular,by leaks from the sewer system, as pollution surveys made on the springs show. Water consumption has,furthermore, quadrupled during the past century,and the broken down sewer system has continued to discharge especially into the 5 ditches mentioned before.

4. THE MECHANICAL AND STRUCTURAL CHARACTERISTICS OF THE FORMATIONS

4.1 The tuffaceous platform:

The unconfined compressive strength of the tuff is summarized in fig. 2, where the strengths of the scoriaceous facies and the pozzolanic facies are distinguished. The measurements were made on satured samples, both cylindrical samples taken

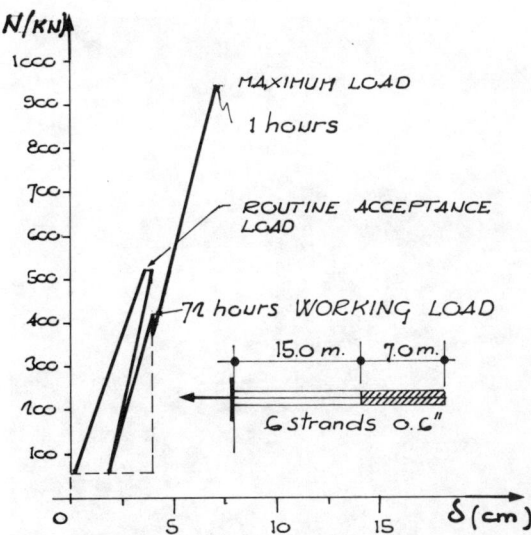

Fig. 3 - Suitability test on anchor in tuff according to a draf of italian specification

from drill holes and prismatic ones taken by hand from existing grottoes.Fracture is of the fragile type under relatively low isotropic stresses (Pellegrino,1970).

The pozzolanic facies, subjected to higher values of the stress state as induced by its own dead weight, therefore leaves the stronger scoriaceous facies in cantilevers 1 to 2 m long. This phenomenon is accentuated by the weathering processes,especially that due to saturation varying with the climate; this involves however only a surface exfoliation of the softer tufface-

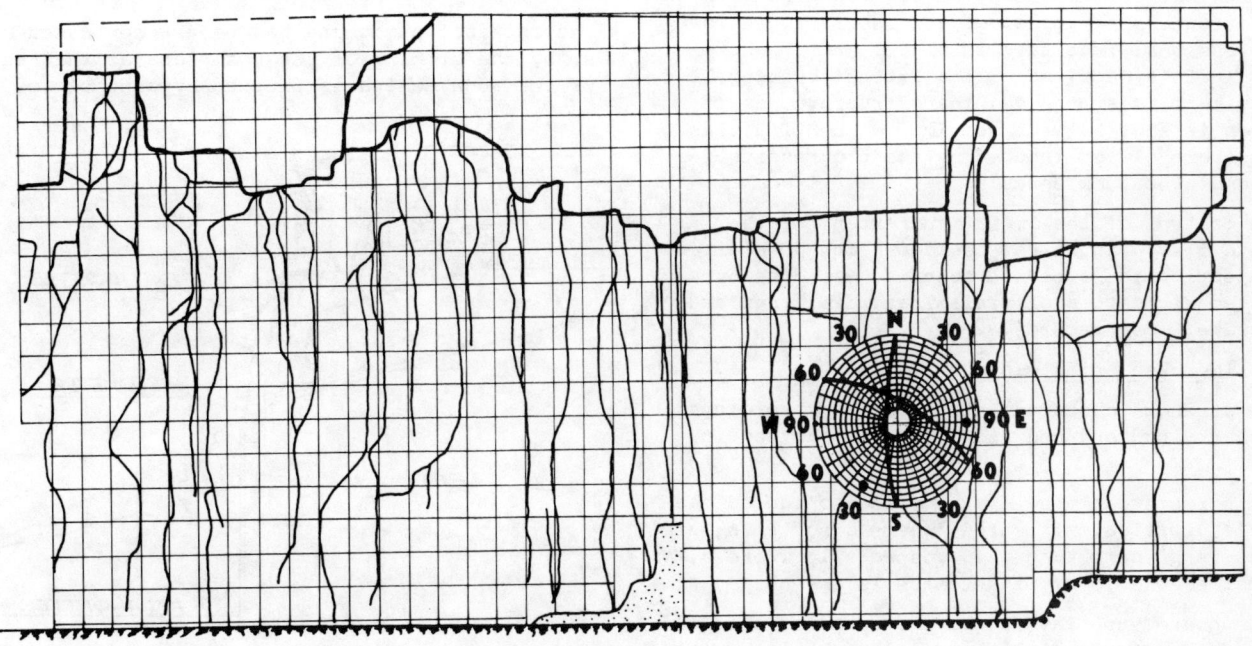

Fig. 4 - Discontinuities and joints in the S.Chiara Cliff,directly surveyed from a scaffolding (1,8 x 2,0 m) and typical Schmidt contouring diagrams.

Fig. 5 - Typical columnar slices

ous part.

The bond and shear allowable stresses of the in situ tuff was controlled by preliminary direct test on anchors, as shown in fig.3, according to a draft of italian specifications.

The tuff platform is structurally characterized by two main series of discontinuities and joints subvertical.The discontinuities and the joints, directly surveyed a scaffolding, are shown in fig. 4, regarding the Santa Chiara zone (see also fig. 7) with even morphology of the cliff. In the figure is also shown   centers of Schmidt contouring diagrams of two principal sets of joints. Others two sets, at 45° compared with the front, are typical of the "spur" zones (see Professional Institute in fig. 1 and 7) where the tuff is deprived of the lateral bearing.It's also particular the columnar structure, as is shown in fig. 5, with vertical discontinuities, those are always induced by the progressive failure of the rock bench.

The control of the amount and extent of the in depth discontinuities, besides being done direct inspection in caves and tunnels, was done by a systematic series of preliminary drillings and by water loss records.

4.2 The base stiff clays

The strength characteristics of the base clays, both drained and undrained, are summarized in fig. 2.

The structure of the very overconsolidated clays is "cortical" fissured, where a softening process is present (Esu 1976),especially where a considerable cover of detritus is absent.

This phenomenon favors the progressive instability of the slopes (5.2),especially where the large deformation existed previously,so that the strength tends to take on residual values   (Manfredini, Martinetti, Ribacchi, Sciotti, 1980).

5 ANALYSIS OF THE MECHANISM AND CAUSES OF THE FAILURES

5.1 Rockfalls in the tuff cliff:

The failure mechanism typical of the tuff cliff is that already described, closely correlated with the discontinuities, with joints that propagate for 10 to 15 meters inside the rock,as a function of the height of the rock front.This mechanism is favored by the progressive decrease in strength of the part of the rock having a pozzolanic facies,especially in surface, where acts the weathering and the tuff is subjected to a high deviator component of the stress tensor for the stresses corresponding to the dead loads, and very low horizontal stresses. The upper stronger tuff has therefore remained "cantilever-wise" over the pozzolanic facies.

The afore-mentioned deviator component similarly acts on the base clays,inducing creep process in them. However the rock debris,15 to 25 m in thickness below the front of the cliff, have been deposited according to the equilibrium profil and geostatic stresses in the base clays.Considering the geological time that has already passed,it is very probable that these process, if not troubled by man,have exhausted their capacities. Though prestresses-states have been left in the tuff platform, predisposing it to the cracking mentioned.

When the aforesaid pozzolanic facies is more extensive, megablock type collapses have taken place, in such a way as to form conoid detritus,restoring a new equilibrium profile with the cliff (Vinassa de Regny, 1904; Verri, 1905), and with the slope more downhill. (fig.7).

The rockfall phenomena has been accentuated also by the stoping of the pozzolana at the base of the rock, which is pierced by a number of

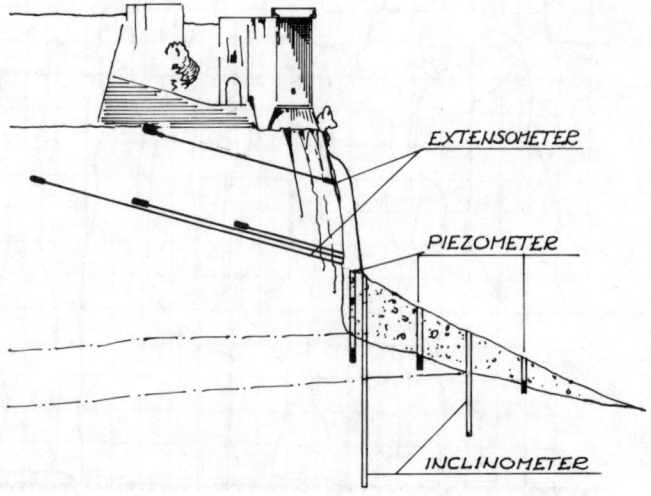

Fig. 6 - Typical instrumented section.

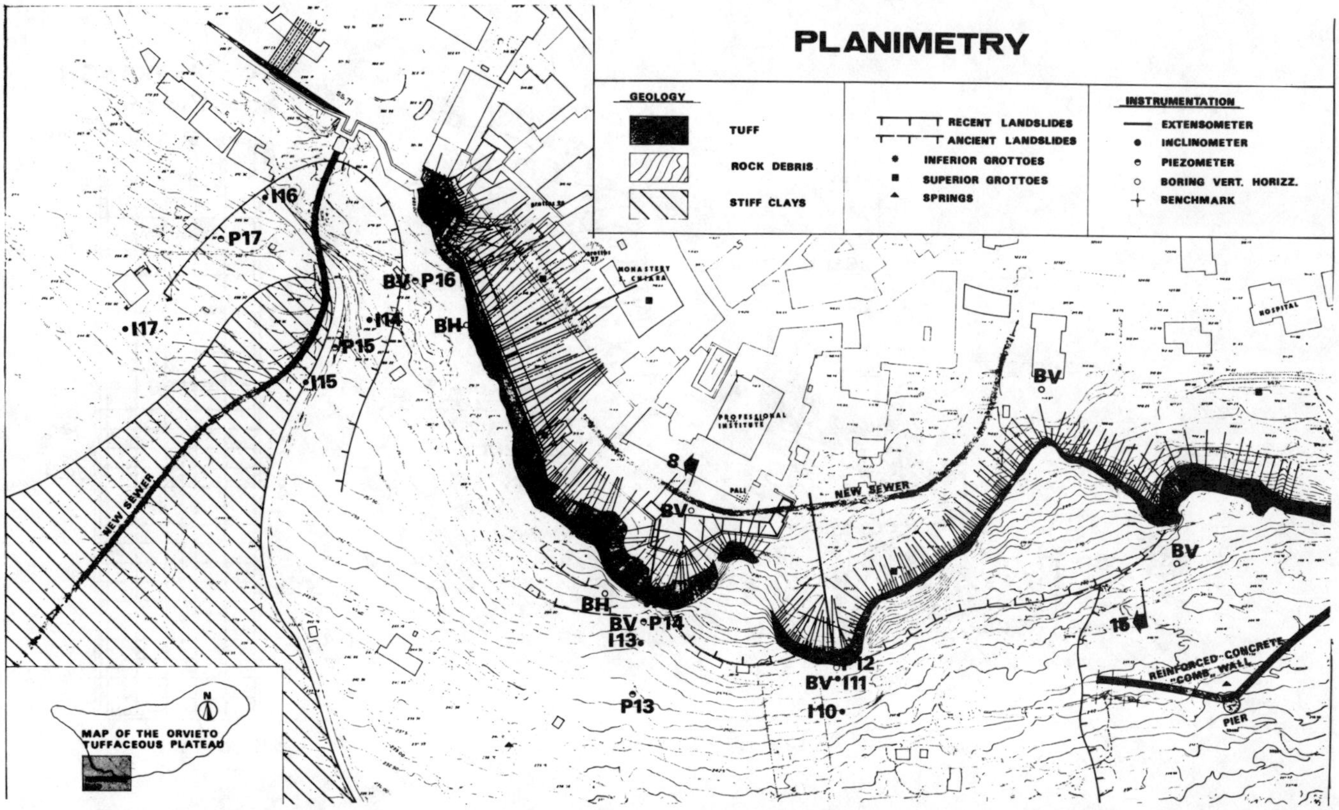

GEOLOGY

TUFF

ROCK DEBRIS

STIFF CLAYS

RECENT LANDSLIDES
ANCIENT LANDSLIDES
INFERIOR GROTTOES
SUPERIOR GROTTOES
SPRINGS

INSTRUMENTATION

EXTENSOMETER
INCLINOMETER
PIEZOMETER
BORING VERT. HORIZZ.
BENCHMARK

MAP OF THE ORVIETO
TUFFACEOUS PLATEAU

Fig. 7 - Geological map; Strengthening and recovery operation; Instrumentation for in situ monitoring

tunnels, these have been carefully surveyed so as to stabilize them.

Landfills by enbankment have also been made at the summit, so as to increase the exploitable area, increasing the gravitational stress state in the cliff.

5.2 Landslides of the slopes:

The Orvieto hill are characterized by numerous landslides in the talus debris,in movement on the base clays. The horizontal stresses at the foot of the rock bench are reduced in this manner, interacting with the aforementioned rockfalls.

The landslides are seen to be especially located near the town's sewer discharge in ditches, these having flows of 10 m^3/s on the average in connection to the habited surface of the plateau, that has been almost entirely waterproofed by the roofs and asphalted streets.

The piezometric level, measured in particular in the landslides,is shown in fig. 2 and 6 for a typical vertical section radial to the rock, and is subtended by the aforementioned ditches in the circumferential direction.The water-table oscillations are thus a function of the waters discharged into the ditches, as was found in-situ during the rainfalls.

Such landslides moreover are monitored by inclinometers,as shown in the figure 6 scheme,and by

geodetic references(fig.7). Movement of the detritus is also favored by the general causes due to the natural evolution of the hill. However, these causes are acting on time-scales that are much longer than those due to man's interventions.

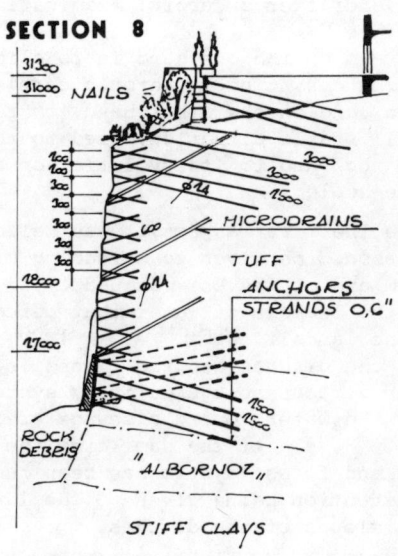

Fig. 8 - Rockbolts and anchors in the most failured cliff.

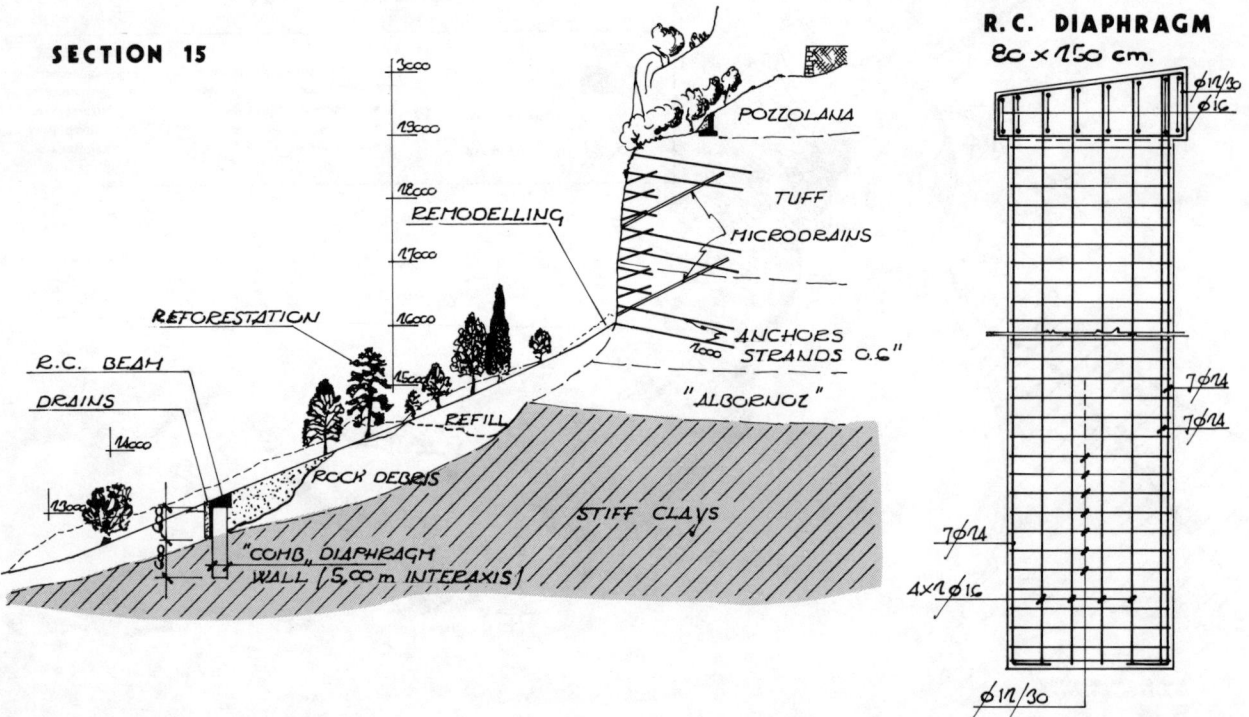

SECTION 15

3000
19000
18000
17000
16000
15000
14000
13000

REMODELLING
REFORESTATION
R.C. BEAM
DRAINS
REFILL
ROCK DEBRIS
"COMB" DIAPHRAGM
WALL (5.00 m INTERAXIS)

POZZOLANA
TUFF
MICRODRAINS
ANCHORS
STRANDS O.C"
"ALBORNOZ"
STIFF CLAYS

R.C. DIAPHRAGM
80 x 150 cm.

ϕ12/30
ϕ16
7ϕ14
7ϕ14
7ϕ14
4 x 1 ϕ16
ϕ12/30

Fig. 9 - Landslides improvments: "comb" diaphragm wall, drains and reforestation.

6. STRENGTHENING AND RECOVERY OPERATION

6.1 General criteria:

The operational criteria had their source in the interpretation of the phenomena described above, in light of aerophotogrammetric and photogrammetric surveys of the rock to scales of 1:500 and 1:200, and after a careful examination of the in situ state.

Thus, the aim on the one hand is to eliminate the disturbances, tied to chronic carelessness of the man, and on the other hand, to recovery the overall static situation, tied to the nature of the rock and to its evolution on a geological time-scale.

As regards the first aspect, it was believed that priority should be given to a redoing of the sewer system and of the town acqueduct, so as to eliminate the leakage of water that percolates down to the landslides. For this end are provided the diking of the ditches and the channeling in pipes of the various sewage delivery system, so as to convey the water toward a sewage treatment system.   At the same time the entire hydraulic and forest system has been recovered; special attention being given to the landslides along the slopes of the ditches.

Regarding the second aspect, the main operational criteria concern both the rock itself and the landslides. Where the rock displays the greatest failures near constructions, and a greater cliff height (30÷35m), a series of anchors have been predisposed. In particular have been executed the prestressed anchors at the foot of the cliff reinforced by rockbolts and by the general injections of grouting. In this way, the aim is to impress the stress-state of confinement of the foot of the rock bench, and contain the static fatigue phenomena. The anchors, 30 meters long on the average, in fact are headed on the concrete walls especially where the tuff is weaker; thus completing the masonry walls built at the foot in the past strengthening. The rock-boltings of 6 to 8 m of the most highly fractured tuff have been moreover predisposed, so as to form an in-situ reinforced tuff "rock wall". This "rock wall" is anchored, besides as already described at the foot, and at the summit by long anchors, not prestressed however, so that the wall is doubly constrained. Finally the above-mentioned general improvement of the landslides has been integrated by remodelling the slopes. In particular were the landslides interact with spur-type tuff morphology, to stabilize the area, diaphragm walls arranged combwise have been built, as shown in fig. 7.

The quantitative definition of the operations was estimated principally on the basis of preliminary in-situ surveys and investigation, already indicated, continually verified during the execution. The random nature of the parameters requires, in fact, that the theoretical outlined analysis is used as a qualitative guide.

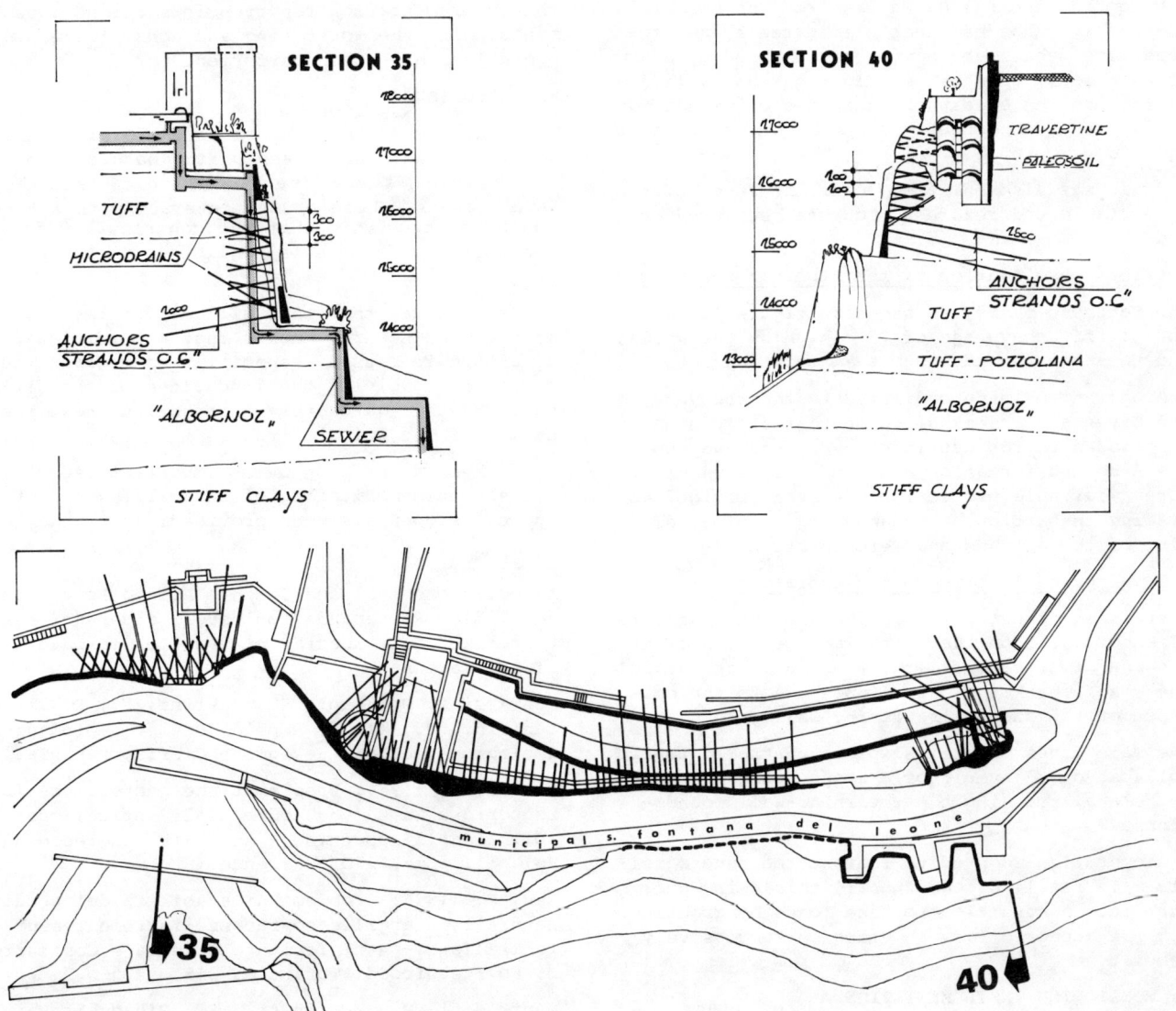

Fig. 10 - Strengthening of the Albornoz fortress and the main Orvieto gate

6.2 Consolidation at the Santa Chiara Monastery and at the Professional Institute

The front of the cliff, in this area in situ verified, has its greatest height (36 m max), and is characterized, figure 7, by an only slightly off-vertical surface, being subtended, for around 80 m, between the spurs, with a morphology that is unique over the entire perimenter of the tuff platform (about 4300 m). The summit of the rock is characterized by a 9 m to 12 m high wall, fig. 4, built of tufa fragments, having a 1.5 m thickness at its base. This wall, built as a retaining wall for an orchard, has collapsed over a 25 m stretch, thus creating in the 6 m pozzolanic embankment a typical amphitheatre-type detachment.

The need to eliminate the thrusts on the wall and to lighten the brow of the cliff, which is subject principally to a plane stress state (vertical and normal to the front) suggested that the collapsed wall not be rebuilt. The summit of the rock has been reinforced by an anchored sheathing, this being in tune as well with the picture of the discontinuities and joints as described. The operation is coupled, as fig. 7 and 8 indicates, with the rockbolts in 6 ÷ 8 m long Ø 24 mm steel bars, these being held to the tuff in depth, by anchors formed of three 0.6" strands. The anchors have been sheathed to protect them against corrosion, grouted with mortar and prestressed on the reinforced tuff.

Finally drain holes have been drilled into the tuff face, to avoid "lamina" effects caused by the percolation of rainwater into the discontinuities that were not consolidated by the grouting.

6.3 The consolidation of the Cannicella landslides

The landslides that began in the talus debris are

due to the discarge of waters from the Hospital, a discharge that had been eliminated by new sewers. The landslides have been contained by a comb of bulkheads, as shown in figures 7 and 9, so as to include the area where the base clays has been discovered, subsequently remodelled according to equilibrium profil.

A drain was also built to lower the level of the water table and reforestation has been predisposed.

6.4 The consolidation of the Albornoz fortress:

The Fortress bears, as shown in figure 10, on tuff cliff an average of 25 m high on the whole, and non-prestressed anchors are provided for.

The operation in subordinated to the recovery of the Civetta ditches, since in that strip area landlisdes in the detritus have taken away both the Albornoz formation and the base clays. There fore a suitable protection has been provided con sisting in particular: channeling in pipe, diking of the ditches and reforestation.

6.5 The consolidation on the Confaloniera:

A special strengthening is the one that has esta blished itself on the north face of Orvieto, in the Confaloniera locality, where a large part of the road embankment has slipped due to the collapse of the ancient retaining wall.

The main cause of this slip can be traced to the collapse of the vault of a stope tunnel in the pozzolana, reaching the retaining-wall foundations.

The operation adopted to restore the zone consists in refill of the tunnel, this being put back into a condition to take loads by grouting it ; the anchored r.c. sheathing is successively rebuilt.

7. MONITORING INSTRUMENTATION

The checking of the effectiveness of the strenghtening described, and the acquiring of data for back analysis, is done by an extensive monitoring network.

The preceding figure 6 and 7 showed the typical instruments in the slopes these being inclinometers, piezometers, and geodetic reference points in the landlisdes.

At the same time, the tuff cliff is controlled by extensometers with triple invar rods (60,40 and 20 m) at the foot of the rock, and a single rod (40 m) at the top.

The instrumentation report points out no movements where the monitoring and consolidation operations have been performed.

8 CONCLUSIONS

The analysis of the failure of the Orvieto hill points up that, superposed on the natural "long-term" causes, there are anthropic "short-temp" causes. These causes interact with a time-scale relationship that the contribution of the non-antropic factors is negligible in the planning.

It is believed that this will arrest the development of the natural causes that are initiated by anthropic causes, especially as regards the phenomena involving the landslides of the ditches and as regards the constructions next the summit of the failured rock.

So as to keep a check on these important aspects, a systematic monitoring of the cliff and of the typical slopes has been provided.

REFERENCES

Associazione Italiana Cemento Armato Precompresso (1981) - Raccomandazioni sugli ancoraggi nei terreni e nelle rocce - Atti delle Giornate AICAP, Ravenna

Conversini, P., Lupi, S., Martini, E., Pialli,G., Sabatini, P., (1978) Rupe di Orvieto - Indagini geologico-tecniche - Regione Umbria, Perugia.

D'Elia, B. (1981) Problematiche geotecniche in rapporto alla salvaguardia degli antichi centri abitati dell'Appennino - Atti XIV Congresso AGI Vol. III pp. 63-70 - Firenze

Esu, F. (1976) Problemi di stabilità dei pendii naturali in argille sovraconsolidate e fessurate italiane - Atti Istituto Scienza e Costruzioni Politecnico di Torino n. 315.

Manfredini, G., Martinetti, S., Ribacchi, R., Sciotti (1980) - Problemi di stabilità della rupe di Orvieto - Atti del XIV Convegno di Geotecnica Firenze Ottobre Vol. II pp. 231-246.

Meigh, A.C., Wolski, W. (1979) - Design parameters for weak rocks - VII ECSMFE Brighton

Pellegrino, A. (1970) Mechanical Behaviour of soft rocks under hight stress - Proc. 2nd Congr. Int. Soc. Rock Mech, pp. 3-25 Beograd

Vinassa de Regny (1904) Le frane di Orvieto - Giornale di Geologia pratica pp. 110-130

Verri, A. (1905) - Le frane di Orvieto - Boll. Soc. Geologica Italiana - Vol. 24 - pp. XXXI - XXXII.

PRINCIPAL GEOLOGICAL MODELS OF DEFORMATION AND FAILURE OF ROCK SLOPES IN CHINA

Principaux modèles géologiques de la déformation et de la rupture des talus rocheux en Chine

Prinzipielle geologische Modelle für die Deformation und den Bruch einer Felsböschung in China

Sun Yuke
Associate Professor, Institute of Geology, Academia Sinica, Peking, China
Yao Baokui
Assistant Professor, Institute of Geology, Academia Sinica, Peking, China

SYNOPSIS

This paper describes a geological model of five typical large-scale landslides. The character of the deformation and the failure pattern which reflect the basic character of the geological model of the slope are analysed, and the principal dynamic factors which affected the stability of the slpe are outlined.

RESUME

Cet article présente le contenu d'un modèle géologique dans cinq types de glissement de terrain à grande échelle. Le caractère des déformations et l'aspect des ruptures qui reproduisent les éléments de base du modèle géologique des pentes sont analysés. On indique les facteurs dynamiques principaux qui influent sur la stabilité des pentes.

ZUSAMMENFASSUNG

Dieser Beitrag bschreibt ein geologisches Modell von fünf typischen großen Erdrutschen. Der Charakter der Deformation und die Bruchsweise, die das geologische Modell des Hanges kennzeichnet, werden analysiert, und die wesentlichen Faktoren, die einen Einfluß auf die Standsicherheit des Hanges hatten, werden angegeben.

1. PRINCIPAL CONTENT OF THE GEOLOGIC MODEL ON THE ROCK SLOPE

The geologic model of the slope is conprehensive expression of the various factors which affect the slope stability,in general,the geologic conditions are the basic factors,on which deformation character and failure pattern depend and concentrically reflecte basic characteristics of the geologic model.Therefore,the geologic model of the slope,in general,include the following principal content:
1)the basic geologic conditions of the slope(the background of the regional geology,the medium structure and structure characteristics of the rock mass)and mechanical properties of rock mass,
2)the structure of the slope(the dip of the rock layer and orientation of slope surface are same or opposite),
3)principal artificial and natural dynamicfactors affecting stability of the slope,the former include the underground mining,the excavation of the slope toe,blast vibration etc.,and the latter include rainfall,condition of underground water and characteristics of the tectonic stress field and so on,
4)the developing process and characteristics of the rock mass deformation on the slope,
5)the failure pattern of the slope.

2. THE ENGINEERING CASE-HISTORIES

The five typical large-scale landslides in China have been described by authors(the Jin Chung open mine pit;the Ge Zhou Ba 2# river foundation pit; the Tang Yan Guang landslide;the Yan Chi He landfall and the Bai Hui Chang large-scale landslide).Their basic geologic condition,the structure of the slope,principal dynamic-factors,deforma-

tion characteristics and failure pattern of the slope are shown in table I.

3. THE DEFORMATION CHARACTERISTICS AND FAILURE PATTERN OF THE SLOPE

The shetch curves of deformation and their failure pattern of the abovementioned slopes are shown in Fig.1.It may be seen that abovementioned slopes all have evident creep characteristics.In the Fig.1,the solid lines are the real deformation curves,the dashed lines are the tendance deformation curves obtained by approximation.The complete process of creep deformation appeared in the Jin Chung open slope(primary or transient creep,secondary or steady state creep,tertiary or accelerated).For foundation slope of the Ge Zhou Ba nonaccelerated creep appeared,since after completion of the foundation pit excavation through one year or so,the stress of the rock mass in the slope are redistributed and reach to a new stable equiponderant state.The Tang Yan Guang slope and the Bai Hui Chang slopes practically all are in process of steady state creep shown by the dashed line,deformation rapidly increase or after the increase again rapidly decrease respectively because of heavy rainfall and drain of the seepage water.The Yan Chi He hillbody befor the landfall also was in secondary creep state,the heavy rainfall accelerates development process of the deformation,to appear tertiary creep resulted in the landfall.It should be pointed out that deformation of the slope in general is comprehensive effect of various deformation factors,but as the deformation of the slope often is controlled by the weak rock mass which has evident property of creep deformation, therefore characteristics of creep deformation can be found from the deformation curves of the slopes.

Table I. the engineering case-histories on five typical large-scale landslides

slope	geologic condition	factor affected stability of the slope	deformation characteristics of the slope	failure pattern of the slope	stability evaluation
the Jin Chung open mine slope	violent tectonic deformation area,in the maximum excavated depth tectonic stresses 100kg/cm² or so,the Sinian metamorphic rocks,single dip ,dip angle 70° or so,numerous faults,hard-soft interlayered medium structure,maximum depth of excavation 250 m.	1.hard-soft interlayered medium structure, 2.the dip of rock strata towards interior of the slope, 3.loose effect of blast, two large blasts(charge ton or so)and everyday, 4.the more horizontal tectonic stress which is perpendicular to the slope.	evident creep characteristics,the maximum horizontal and vertical displacement all more than 10m, discontinuous deformation is principal(slide along the bedding plane and roll).	typical toppling failure,there is not the evident entire sliding plane,the rock mass of the slope can be divided 3 deformation area: 1.violent toppling area(15-20m), 2.toppling deformation area(20m), 3.microtoppling deformation area.	toppling failure often occured in the slope with steep inversely dipping layer,in general,rapid large-scale slide of the whole slope can not be occured
the Ge Zhou Ba foundation pit slope	violenter tectonic deformation region,in depth of excavation tectonic stresses 30kg/cm²,the Cretaceous period sedimentary rocks(siltstone,siltstone contained clay and clay rock),formation are gentle,dip angle 6° or so,there are several weak interlayer,angle of the slope about 17°,vertical excavation depth 60-65m.	1.combinative medium structure contained the weak interlayers, 2.rebound effect after excavation, 3.the tectonic stress is approximately perpendicular to the slope, 4.several weak interlayer play control role to deformation of the slope.	1.shear slide along the weak interlayer with antidip characteristics (1),displacement 9cm, 2.continuative rebound deformation except on bedding plane of the weak interlayer, 3.after excavation one year, deformation no longer developed, 4.deformation of upper reaches and lower slope are of symmetry.	failure of the slope mainly appear as entire slide along weak interlayers.	because the tectonic stress and excavation depth of the slope are all less,dip angle of slope is gentle, after excavation one year or so new equilibrium is reached,deformation of the slope developed no longer.
the Tan Yan Guang landslide	the slope of the reservoir, the Sinian period slight metamorphic rocks(sandstones), dip angle of the bedding plane and the angle of the slope are same, about 30-35°, the rock of slide surface is the thin layered sand slate with clay on bedding plane, the volume of the landslide is about 100×10m.	1.dip of the slope and the bedding plane are approximately agreement, 2.the slide surface is weak rock containning the clay, 3.excavation of the slope toe, 4.rainfall,storage water of the reservoir, 5.cut of the steep dip faults on two slides.	befor large-scale slide deformation of the slope is no evident,once the slope loses stability,the slope rapidly slided.	failure of the slope are controlled by the bedding plane and faults,rapid slide along bedding plane,highness Of surge wave is 21m.	the landslide body seems to be the rigid block mass, befor landslide, there are not evident premonition, slide failure often taken place in violent way to produce catastrophic result.
the Yan Chi He landfall	violenter tectonic deformation region,the valley topography,the Sinian period metamorphic rocks(dolomite rock,shale and so on),dip angle 15° or so,there are the more vertical joints and unloading cracks,slide bed is the dolomitic mudstone, the volume of the landfall approximately 100×10m.	1.the hillbody is of 3 free face, 2.the hard-soft alterlayered medium structure, 3.underground mining, 4.heavy rainfall(80mm), 5.tectonic condition (joints, unloading cracks).	1.underground mining directly results in the surface fracture, 2.the hillbody is of characteristics of whole deformation, 3.displacement of the fracture early mainly are horizontal,lately mainly are vertical, all having evident creep characteristics.	landfall with slide toppling characteristics,the vertical fracture extended to cut the hillbody,the slide surface develop along dip direction from lower to upper,hillbody first was sliding and later toppling.	the landfall are of evident premonition phenomena, in general,landfall took place in violent way to produce catastrophic hazard.
the Bai Hui Chang landslide	the Jurassic period coal series stratum,dip angle is gentle(6° or so),slide surface is the clay parting which have very low strength(φ=8°, c=0.04kg/cm²), under slide surface about 120m the coal seam had been mined,the ancient landslide , the volume about 500×10m.	1.the slide bed is the clay rock having very low strength, 2.the coal seam had been mined under slide body, 3.rainfall, 4.excavation of the slope toe.	1.having evident creep characteristics, 2.the active slidebody mainly gives vertical displacement,the passive slidebody mainly gives horizontal displacement, 3.front edge of the slope slide out 10cm perday, 4.the velocity of deformation is relevant to rainfall.	leveled pushing slide with the collapse character,the slidebody is divided into two parts:the active slidebody and the passive slidebody.	because the dip angle of clay parting is very gentle,althrough having the more deformation velocite ,but in general the large scale instantaneous rapid landslide can not produce.

The deformation of the slope rock mass can be divided two category:the discontinuous deformation and continuous deformation.Continuous deformation include elastic deformation(rebound after the excavation),plastic deformation,creep deformation and swell deformation of the weak rock mass.Discontinuous deformation include landslide,turn or roll of the rock masses,shear slide along structure surface and the whole landslide or toppling of the hillbody.The rock masses in fact all are discontinuous medium,the distinction only is to have different degree of separation,therefore,before the whole slope loses stability or is destroied,althrough as a whole deformation of the slope mainly is various continuous deformation, but may be affected by different discontinuous deformation factor.The various slope not only are of different deformation type and failure pattern,but also are of different deformation properties.For example,the Ge Zhou Ba slope of the foundation pit mainly gives the continuous deformation,besides sliding appeared on the weak bedding plane,the Jin Chung open pit slope mainly gives discontinuous deformation characteristics.The different deformation characteristics of the slope result in different instability or failure pattern,the failure pattern of the abovementioned slopes model may be summarized as follows:
1)the Jin Chung model,toppling failure on the slope with layer dipping towards interior of the slope,

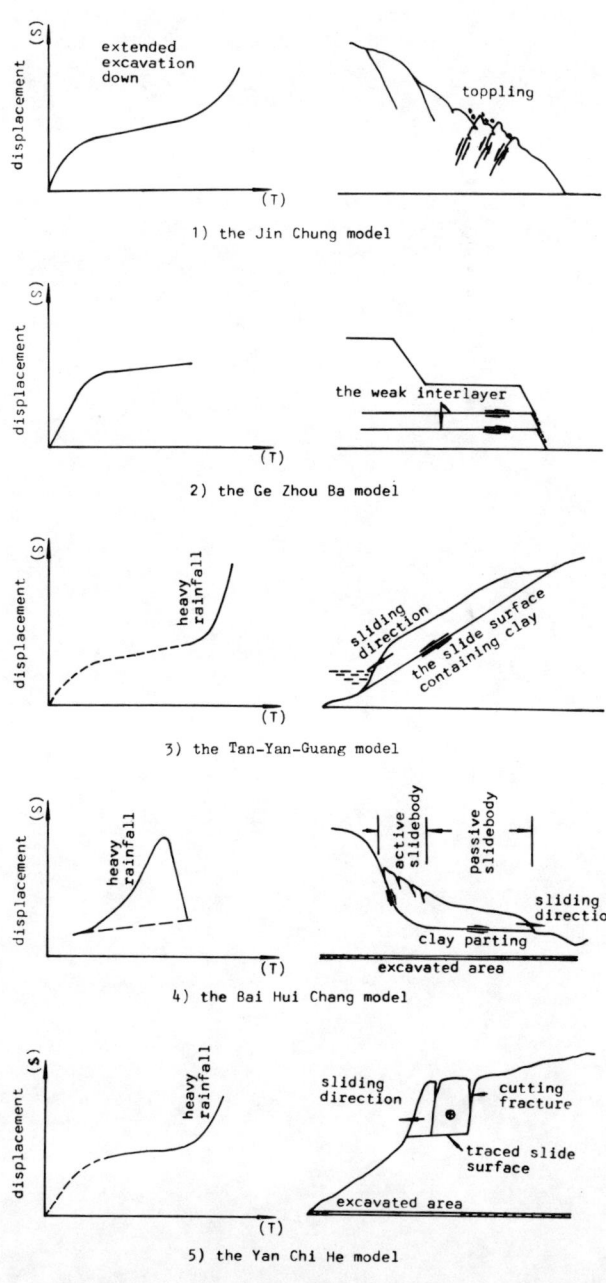

1) the Jin Chung model

2) the Ge Zhou Ba model

3) the Tan-Yan-Guang model

4) the Bai Hui Chang model

5) the Yan Chi He model

Fig.1. deformation characteristics and failure pattern of five typical slope modeles

───── real deformation curve
----- approximated deformation curve

2)the Ge Zhou Ba model,entire slide along the horizontal weak intercalations,
3)the Tan Yan Guang model, rapid slide along the layered surface,
4)the Bai Hui Chang model, leveled pushing slide with the collapse character,
5)the Yan Chi He model, landfall with slide-toppling characteristics.
The stability of the slope depends on the deformation characteristics and the failure, which has obvious different to the various slope modeles.Althrough Jin Chung slope gives obvious toppling deformation,accumulative horizontal disp-

lacement and the vertical displacement are all more than 10m,but because there are no the evident entire slidesurface,therefore,the large-scale slide failure is not produced and can not be produced.The slide velocity of the Bai Hui Chang slope is larger,althrough the front margin of the slope are of slide velocity of 10cm per day, but because the dip angle of the slide surface(claystone)is very gentle,the large-scale catastrophic rapid slide also can not be produced.The deformation of the Ge Zhou Ba slope mainly is rebound effect after excavation,as the tectonic stresses are lesser, therefore,althrough the slide along the weak parting is more than 8cm,after stresses rearrangement,deformation of the slope developes no longer.It should point out that the slope having toppling failure characteristics from slope surface into the interior rock mass according to deformation degree in general can be divided into 3 deformation area:the violent toppling deformation area,the toppling deformation area and the microtoppling area.in general depth of the violent toppling area and the toppling area are all not more than 20m,because in the slope there are not the evident entire slide-surface,this is principal reason that the Jin Chung slope having toppling deformation characteristics can not produce large-scale slide.

4. PRINCIPAL DYNAMIC FACTORS AFFECTED THE STABILITY OF THE SLOPE

The stability of the slope is affected by the various factors which not only include geologic background of the area and geologic condition of the slope,but also include various artificial and natural dynamic-factors.Rainfall,storage water of the reservoir and the tectonic stress in the rock mass are several main nature dynamic-factor, but the underground mining or excavation are the best important artificial dynamic-factor.The underground mining disturbed or damaged overburden rock mass,not only to reduce strength of the rock mass ,but also to increase permeability of the rock mass.In fact the landfall or slide of the Yan Chi He hilltop and the Bai Hui Chang slope all directly are related to the underground mining,the cut-excavation of the slope toe also change stability condition of antislide on the slide surface,play very unadvantageous effect.It is evident that frequent blast also are of the important effect to stability of the slope,accelerate process of deformation and failure of the slope.

5. CONCLUSION

The geologic modeles of the slope are of extensive content,and are comprehensive expression of various factors affecting stability of the slope ,while deformation property and failure pattern of the slope concentrically reflect basic characteristics of the geologic model.The geologic modeles of the abovementioned slopes are of the certain universal significance in some degree ,therefore in the engineering practice having certain reference value to the stability evaluation and the engineering-analogy analysis of the slope.

REFERENCE

Evans.R.,Stability analysis of a rock slope against toppling failure,International Symposium on Weak Rock,theme 3,62-68,1981.

GEOTECHNICAL INVESTIGATIONS FOR THE STABILITY OF A MINE SLOPE

Etudes géotechniques sur la stabilité du talus d'une mine

Geotechnische Untersuchung zur Standsicherheit einer Tagebauböschung

E. Shanmukha Rao, R. K. Jain, J. Bhagwan and D. Mukherjee
Geotechnical Engineering Subdiscipline, Central Road Research Institute, New Delhi, India

SYNOPSIS

The paper deals with the geological and geotechnical investigations carried out for a landslide so as to understand its mechanism and to recommend suitable corrective measures. A typical wedge failure has been identified based on principles of rock mechanics and interpretation of Schmidt's Stereonet. A slope stability analysis is also carried out from the soil mechanics angle to arrive at a profile which will be safe during the uninterrupted progress of the mining operations.

RESUME

Cet article présente les études géologiques et géotechniques effectuées sur un glissement de terrain en vue de déterminer son mécanisme et de prescrire les mesures correctives à prende. Un cas caractéristique de la rupture de coin a été identifié en tenant compte des principes de la mécanique des roches ainsi que par l'interprétation de la projection de Lambert. Une analyse de la stabilité de talus a été faite du point de vue de la mécanique des sols afin de développer un profil sûr permettant les exploitations minières ininterrompues.

ZUSAMMENFASSUNG

Der Bericht befaßt sich mit geologischen und geotechnischen Untersuchungen, die für einen Erdrutsch durchgeführt wurden, um dessen Mechanismus zu begreifen und mögliche Verbesserung vorzuschlagen. Auf felsmechanischen Prinzipien und der Auswertung mit Hilfe des Schmidtschen Netzes aufbauend, wurde ein typischer keilförmiger Bruch identifiziert. Eine Standsicherheitsanalyse nach bodenmechanischen Prinzipien wurde ebenfalls durchgeführt, um ein Böschungsprofil zu erreichen, das während der fortlaufenden Bergwerkstätigkeit sicher ist.

INTRODUCTION

The paper presents the case history of a landslide (Photo 1) in an open-cast mine and the geotechnical investigations that followed to find a solution for the uninterrupted progress of mining operations. The type of the rocks available in this area are iron bearing sedimentary rocks of Proterozoic era. The outcrops consist of lateritic ore, soft laminated ore and blue dust with occasional bands of haematite quartzite. Structurally, the deposit lies on the western limb of the synclinorium of Dunn (1937) complicated by subsequent generation of folds, open and cross types. On the eastern side of the hill, the dip is towards the west whereas along the western slope, i.e. the affected slope, the dip is towards the east. The area experiences rainfall of about 200 cm per annum between June and September. The area is thickly vegetated and ranks among one of the world's best Sal forests. The hill slopes mildly from El.918 m at an angle of 70° upto El.904 m which further tapers at an angle of 18° upto El.892 m where it attains a slightly steeper gradient of 26° upto El.880 m. Beyond this point on the downhill side, the mining and plant construction operations changed the hill slope bringing it from an original gradient of about 35° to the present average slope of about 42° upto the primary crusher point. There are also about 11 benches built by the project authorities of 3-5 m width and height in order to serve as relief terraces to increase the stability of slope (fig.1).

Photo 1. A panoramic view of the landslide area with relief benches above the crown.

A major landslide occured on the western slope of the mine area on the 8th July 1980. The records of rainfall indicated that the area experienced a cumulative rainfall of about 55 cms during the 30 days preceding the landslide. Further, the affected slope of the hill was completely denuded to accommodate mining operations and thus lost the vegetative cover against the surface run-off and direct rainfall.

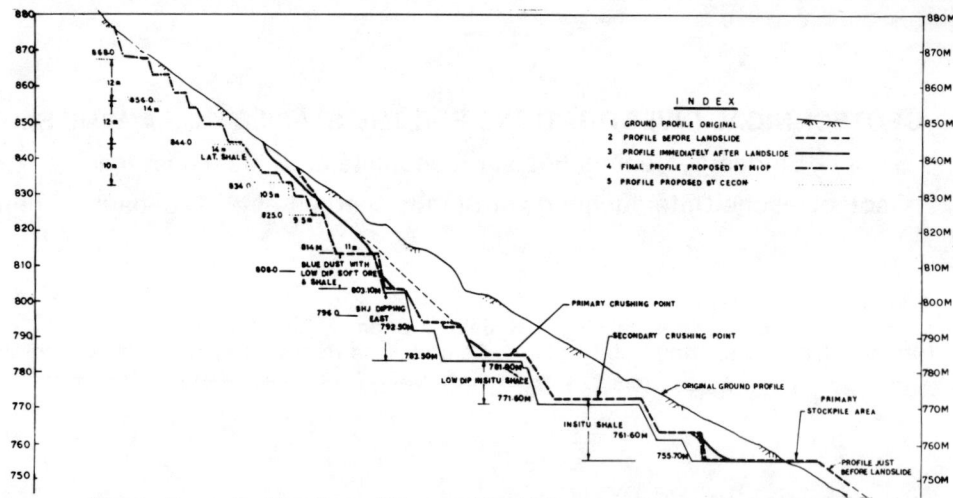

Fig.1. Cross Sectional profiles of the hill slope at various stages.

It was reported that the slope was subjected to blasting as a normal part of mining operations. Coupled with this, the hill was made up of rocks which were geologically interesting (fig.2) but not competent from the engineering point of view. The three broad categories of rocks were banded haematite quartzite (BHQ), blue dust and shales which have widely varying properties in composition, rock quality, compressive and shear strengths. The dips, strikes and joint pattern were sharply varying at short intervals and the ferruginous shales, which could further be categorised as soft, medium hard and hard varieties, were heavily fractured, jointed and fissured.

The hill had a thin mantle of 2-3 m of lateritic soil with about 60% of more than 60μ size indicating its highly permeable nature. This slope was brought to an average inclination of about 42° to facilitate founding of primary and secondary crushing plants. It was also reported that at the time occurrence of the landslide near prependicular uphill slope of about 25 m high was existing and that the construction activity along with drilling and blasting operations was in full flow. There were no tension cracks noticed on the uphill slope before the slide occurred excepting for the fact that the entire slope was completely saturated and became slushy and inaccessible.

The landslide occurred suddenly and in a matter of seconds, about 25,000 cu m of debris consisting of boulders of all sizes mixed with saturated soil mass tumbled down and covered the lower reaches. The landslide, as the investigations revealed, was not a deep-seated one and might be classified as a 'debris avalanche'.

To sum up, the landslide appeared to have resulted as a consequence of the combination of (1) Deforestation of the hill slope to facilitate mining operations, (2) Geology of the area, (3) Heavy rainfall preceding the occurrence of landslide, and (4) Undercutting the hill slope for the foundations of the primary and secondary crushing plants.

LABORATORY AND FIELD INVESTIGATIONS

The field investigations included reconnaissance of the slide area; study of the original, present and future profiles of the hill slope from the stability point of view; detailed geological survey; recovery of rock samples through four boreholes exclusively made for the purpose; conducting field permeability tests; performance of in-situ direct shear tests and collection of disturbed and undisturbed soil and rock samples for laboratory testing. The laboratory studies cover the consistency tests for soil and shale samples, grain size distribution, direct shear tests on the soil and rock samples, etc.

The geological survey, the study of fracture pattern and the fracture frequency, the rock quality designation (RQD) based on modified core-recovery ratio (Table I), the condition of the cores and the field permeability testified to the effect that the area was geologically unstable. The only favourable note was that the rock dips into the hill (towards east) ruling out the possibility of a planar failure. Four boreholes CBH-1 to CBH-4 (fig.2)

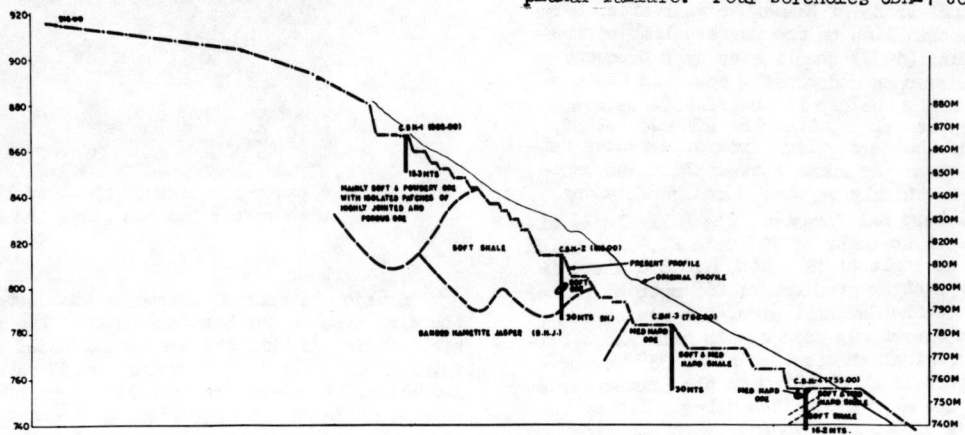

Fig.2. Geological section showing borehole location.

TABLE I: BORELOG OF CBH-1 RL + 867.67 m

Depth in metres	Description	Approx. core length (mm)	No. of fragments >10 mm	RQD	Fracture frequency per m.	K X10^{-5} cm/sec
0 - 1	Brown powdery ore with pieces of limonitic haematite.	-	Powder	0	-	2.054
1 - 2	Soft powdery ore with pieces laterite	-	Powder	0	-	0.076
2 - 3.5	Soft powdery ore with fragments of medium hard porous laminated ore.	20	2	2.66	-	1.53
3.5 - 4.5	Powdery ore (brown) with small fragments	-	Powder	-		
4.5 - 6.0	Powdery ore (brown) with small fragments		Powder	-	-	2.038
6 - 7	Medium hard **highly** porous and jointed haematite ore	165	22	17.5	22	0.898
7 - 7.5	Medium hard highly porous and jointed haematite ore	20	3	4	13	0.697
7.5 - 9	Powdery ore(blue) with small fragments of laterite	60	13	4	13	0
9 - 9.5	Blue dust with a few fragments of medium hard porous laminated ore	-	Powder	0	-	1.231
9.5-10.2	Blue dust with small fragments	-	Powder	-	-	
10.2 - 11	Limonitised ore (brown) with small fragments	-	Powder	-	-	2.04
11.0-11.5	Limonitised ore (brown) with small fragments		Powder			
11.5-12.2	Limonitised ore (brown) with small fragments	20	2	2.35	8	106.18
12.2-12.7	Fragments of medium hard porous laminated ore	-	Powder	-	-	
12.7-13.5	Blue dust with small fragments	-	Powder	-	-	
13.5-13.9	Powdery ore (blue)					144.9
13.9-15.3	Porous limonitised ore cones and fragments of varying length	295	17	21.07	17	120.23

located along the central line of crushing plants give an insight into the complexity of substrata formation. A typical borelog (CBH-1) is presented in Table I.

The field permeability of about 1×10^{-5} cm/sec upto a depth of 10 m in CBH-1 was indicative that the BHJ was more compact than the underlying shale. The values of 'K' presented in the borelog represent only average values considering the varying depth as the water level went down fast and should not be misconstrued as exact values at a particular depth. The results of field permeability tests at CBH-2, CBH-3 and CBH-4 were baffling for the simple reason that water could never be seen inspite of pouring 7000, 5500 and 4000 litres of water respectively in each of these boreholes. Some field permeability test results by variable head method are presented in fig.3. It thus established the highly jointed nature of the shales and their vulnerability to absorb water leading to softening and further disintegration. It was, thus the need to trap the subsoil water was felt in the interest of the stability of the slope.

In situ direct shear tests were conducted at four different locations on soft ferruginous shale samples of 30 cm x 30 cm size, carefully trimmed in miniature caves, specially dug for the purpose. The normal and lateral loads were given through hydraulic jacks (Photo 2). The test results gave C' = 19 KN/m^2 and $\emptyset$' = 32.2° (fig.4) which agreed well with the published literature on intact rocks. Block samples of rock were also subjected to the laboratory direct shear tests under consolidated drained conditions, and the plane of failure was chosen to be along the bedding planes. The results gave C' = 40 KN/m^2 and $\emptyset$' = 22° giving the least possible strength. Powdered soft shale samples were also subjected to direct shear

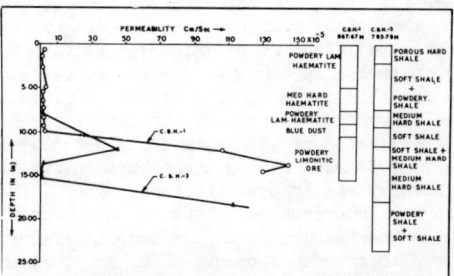

Fig.3: In situ permeability test results by variable head method.

tests and C' = 0 and $\emptyset$' = 24° were obtained. The Mohr's envelopes based on direct shear block samples of shale are presented in fig.5. The residual shear strength parameters gave a C' = 0 and $\emptyset'_r$ = 18.5°. A last set of tests simulating the field conditions with the same dip orientation, gave C' = 55 KN/m^2 and $\emptyset$' = 41°.

The plasticity indices of the overlying lateritic soil, the blue dust and the yellowish soft shales were found to be 4.0, 5.5 and 27.17 respectively. The wet and dry densities of the BHJ, hard ferruginous shale and soft shale are 3.34 and 3.22 gm/cc, 3.20 and 2.95 gm/cc and 1.70 and 1.50 gm/cc respectively.

DISCUSSION

Graphical Representation and Interpretation:

About 100 readings on the attitudes of the rocks and 142 readings on the joint sets were taken in and around the landslide area as a part of the geological investigations. The object was to arrive at the most favourable

C 45

plane of discontinuity through a stereoplot so that the mode of failure, whether it is a planar, wedge, toppling or circular type, could be identified.

The polar diagram using Schmidt's stereonet for the dip and joint sets is presented in fig.6. It can clearly be seen that there are two bedding plane concentrations 21% (D1) and 10%(D2) in the north-west quadrant of the stereoplot. This is indicative of the existence of a local asymmetric fold, with each of the limbs dipping in the same south-east direction but at different inclinations of 66° and 41°. The strike is generally N 37°E.

The stereoplot for the joint sets unfolds three

Photo 2: In-situ direct shear test in progress in a miniature cave.

vertical joint sets J1, J2 and J3, J1 being the most dominant with a pole concentration of 18% and a strike direction of N60° W-S60°E. Another joint set J4 with a pole concentration of 6% dipping 62° SE and the prependicularly intersecting J2 and J3 are minor joint sets.

By drawing the plane of the slope face dipping at 35° NW and the friction circle corresponding to the in-situ direct shear Ø' value of 32.2°(fig.4) for the interbedding soft shales in the stereo diagram, we can delineate the unstable zones and the mode of failure. From the stereo diagram, we find that the dip planes 'D1' and 'D2' are intersected by the major joint set 'J1' at two places. Wedge formation does occur because two planes of discontinuity are intersecting each other. The plunges of the two intersection points is SE, which is same direction as the favourable dip of bedding planes. Thus, failure by sliding cannot occur. Similarly, J2 and J3 also intersect D1 and D2 but cannot disturb the stability.

But on the other hand, J1 and J3 intersect the slope face in the north-west quadrant in a region where Ø f> Ø, Ø f and Ø being slope face angle and angle of shearing resistance of the soft shale, which being the weakest rock is most likely to contribute to failure. The plunges of intersecting planes are greater than Ø' and also in the same NW direction as that of the slope face, establishing the wedge failure beyond doubt. The unstable zone is shown by the shaded portion (fig.6). The evidence of wedge formation, through stable non-disturbing type due to J1, J2 and J4 at the microlevel and the plunging unstable type due to J1 and J3 in the direction of slope at microlevel, are depicted in photographs 2 and 3 respectively.

Stability Analysis of Slope:

Based on field and laboratory investigations, the stability analysis of the western slope of the hill was attempted. Thus it was categorised into three sections El.916 m to El.868 m. El.868 m to El.814 m and El.814 m to the primary stock pile (fig.1) and analysed separately Finally, analysis of the entire slope taking all the th-

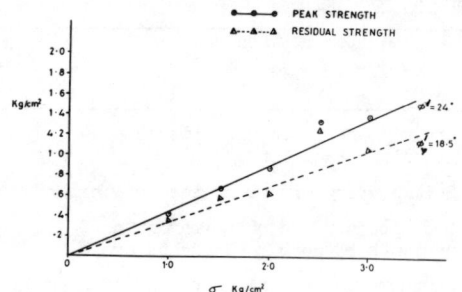

Fig.5: Laboratory shear strength parameters.

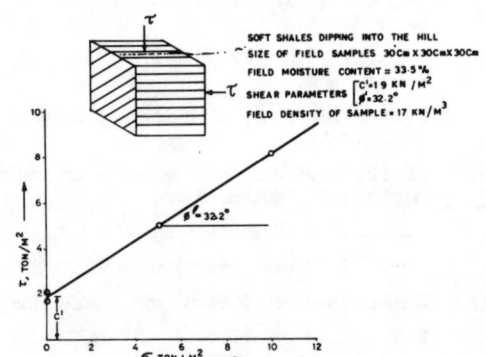

Fig.4: Shear strength parameters from in situ direct shear test.

ree above sections together was done. The unit weights for BHJ and soft shale were taken as 3.34 and 2 t/m3 respectively. The C' and Ø' in terms of effective stress were respectively as 20 KN/m2 and 30°. It was decided to have a minimum bench width of 7 m, a bench height of 7 m or 12m coinciding with the future mining benches and a bench face of 70° as far as possible. The geology of the area, the high tensile stresses that were likely to develop near the bench crests leading to tension cracks and overhangs the safety of men and machinery, the mechanics of rolling boulders were some factors which influenced the above design parameters.

Having decided upon the methodology, it was now required to arrive at a stable, economical and workable profile through different trials. First of all, the profile as existed after the landslide occured alongwith 11 relief benches was analysed and it gave a factor of safety 0.8. A second attempt was made by providing six benches between El.868 m and El.796 m conforming to the future mining benches but it yielded a factor of safety of 0.78, which is not only unsafe but involved extensive earthwork. Yet another profile having three relief benches between El.868 m and El.784 m, by combining or widening or increasing the heights of existing benches, was analysed separately in three sections and as a whole but resulted in factors of safety ranging from 0.8 to 0.9.

All this led to the fact that minor modifications do not serve any purpose unless some drastic changes are made in the cross section to make it safe. Such a profile was arrived at (fig.7) taking into account all possible modes of failure. The stability calculations gave F.S. values of 1.2 to 1.8, the lower & upper limits being for the saturated and dry conditions. Thus, it was once again shown that interception of subsoil water through horizontal drains to eliminate development of excess porewater pressures would go a long way in the long term stability of the slope. The use of circular arc analysis is debatable for a slope involving rock and soil but in view of the fact that the area was highly disturbed and

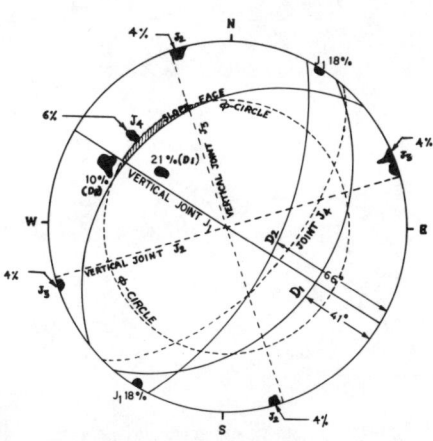

Fig.6: Stereoplot depicting polar concentrations, corresponding planes and failure mechanism.

Photos 3 & 4: Wedge failure because of the joints sets at micro & macro level respectively.

fractured, the analysis is not considered unreasonable.

CONCLUSIONS

1. Stereoplot of an area under investigation is a useful tool to understand the failure mechanism and to predict rock slope instability.

2. Subsurface drainage through horizontal drains would help in increasing rock slope stability, especially in highly disturbed and discontinuous rock masses.

3. Each bench should have an inward transverse slope to lead the surface run-off into a lined side drain from where it can be taken far away from the disturbed area.

4. Special attention is required during blasting operations as uncontrolled blasting operation can result in uneven contours, overbreak, overhangs, excessive

shattering and tension crack development.

ACKNOWLEDGEMENT

The paper is published with the kind permission of the Director, Central Road Research Institute, New Delhi. The encouragement and total freedom given by Sh.T.K.Natarajan, Deputy Director and Head, Geotechnical Engineering Subdiscipline is gratefully acknowledged. Thanks are due to Sh.G.N. Tilak, General Manager, NIOP for his keen interest and close cooperation. The authors are also thankful for the help and cooperation extended by Shri O. Mascarenhas during the field investigations.

REFERENCES

1. Billings, M.P. (1974) "Structural Geology", Prentice Hall Inc.

2. CRRI (1981), Unpublished Report on "Geotechnical Investigations for the Stability of a Hill Slope at Meghahatuburu", New Delhi.

3. Dunn, J.A. (1937) "Mineral Deposits of Eastern Singhbhum", Mem. Geological Survey of India,

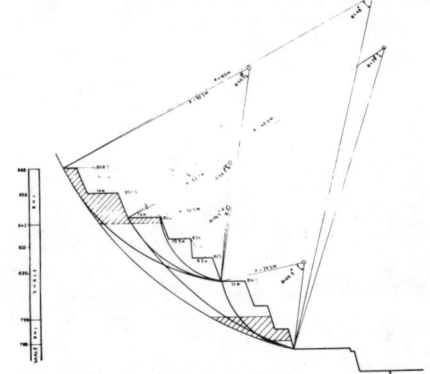

Fig.7: Stability analysis of the slope by circular arc method.

vol.LXIX Pt.1.

4. Hoek, E. and Bray, J. (1977) "Rock Slope Engineering" Institute of Mining and Metallurgy, London.

5. Jones, H.C. (1933) "Iron Ore Deposits of Bihar and Orissa", Mem. Geological Survey of India, Vol.LXIII, Pt.2.

6. Lambe, and Whitman, (1969) "Soil Mechanics", John Willey & Sons, Inc. New York, pp.553.

7. Phillips, F.C. "The use of stereographic projections in structural geology", Edward Arnold, London.

8. Ragan D.M. (1968) "Structural Geology - An introduction to Geometrical techniques", John Wiley & Sons, New York.

PROBLEMS IN DESIGNING AND SECURING ROCK EMBANKMENTS

Problèmes de conception et de consolidation des talus rocheux
Probleme bei der Gestaltung und Sicherung von Felsböschungen

Dipl.-Ing. Dr. Franz Pacher, Dipl.-Ing. Nejat Ayaydin
Büro für Fels- und Tunnelbau, Franz Josef-Strasse 3, A-5020 Salzburg, Austria/Europe

SYNOPSIS

Actual examples are used to demonstrate certain problems in designing and securing rock embankments. In a later section the conclusions to be drawn are discussed.

RESUME

A l'aide d'exemples donnés, on examine certains problèmes surgissant lors de la formation et de la consolidation des talus rocheux. Dans une partie on discute des conclusions à en tirer.

ZUSAMMENFASSUNG

Anhand ausgeführter Beispiele werden bestimmte Probleme aufgezeigt, die bei der Gestaltung und Sicherung von Felsbö-schungen auftreten. In einem weiteren Teil werden die daraus zu ziehenden Folgerungen besprochen.

1. GENERAL

Rock embankments vary in gradient and in the steps needed to create them far more widely than loose stone (earthen) embankments. Besides, a wide variety of problems can arise which signi-ficantly influence their design and stabiliza-tion. Light is thrown on these problems in the following examples, from which conclusions are drawn and discussed.

2. EXAMPLES

2.1 Case 1 (see fig. 1)

For a bridge exit, a rock spine had to be cut out to a height of 35 m. Beneath the humus layer and weathering crust, weathered granulitic gneiss cropped out, whose quality and stratification pattern was not exactly known at the time of planning. The embankment point and gradient were decided on under the assumption that the firm-ness of the material would increase rather than decrease as the depth got lower. But this was by no means guaranteed. Because of the height of the existing cut, a later flattening of the gradient was out of the question.
For these reasons, steps had to be planned from the outset to deal with a possible deterioration of conditions.
As a safety precaution, a horizontal, anchored concrete beam was laid in at the halfway point of the cut, which was intended to secure the area above it. This step proved very effective and prevented a large scale embankment collapse. Large, mylonitic deposits which had not been expected came to light and could be cleared away with an excavator. The steepened embankment in the lower third was secured with the help of an anchored wall.

2.2 Case 2

The 15-20 m high cut for widening a lakeside road had already been made at the time of further planning.The embankment gradient was 70-80°. According to the geologist, the slope, except for certain isolated areas, was stable. The problem here was to increase the long range stability and also to secure the oversteep em-bankment against further weathering and erosion. An additional requirement was that the embank-ment not be secured in the usual way with anchor-ed pillars, beams, or a wall, but in a land-scaped style so unobtrusive that the view both from the lake and from the road would not be disturbed.

The problem was solved this way:
Instead of a projecting support construction, the lowest area of the slope cut was sufficient-ly stabilized by injection and anchoring to form a supporting dam (see fig. 2). This was done with a network of anchors at intervals of 2.5 x 2.5 m with a depth of 8 - 12 m.
The surface in areas that were not stable for the long run was stabilized either by gunite (shotcrete) or by covering nets, etc. (see sig. 3).

2.3 Case 3 (fig. 4)

The intended embankment is located in mottled sandstone consisting of stronger and weaker sandstone deposits with intervening deposits of argillaceous (clay) rock. The stratification is basically horizontal. Numerous vertical rips and cracks in cm and dm widths through the sand-stone beds aggravate the situation and decrease stability. The embankment has a maximum height of 40 m.

E

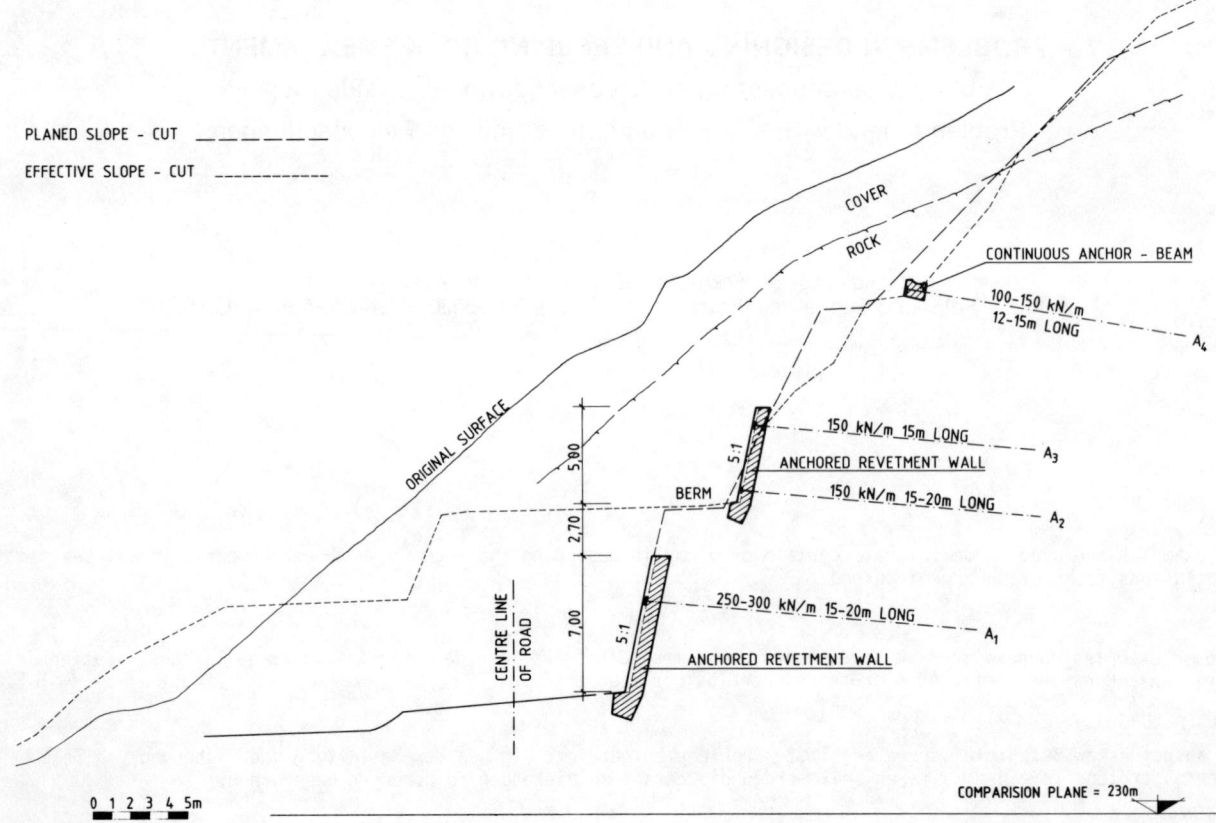

PLANED SLOPE - CUT — — — —
EFFECTIVE SLOPE - CUT — — — — —

COVER
ROCK

ORIGINAL SURFACE

CONTINUOUS ANCHOR - BEAM
100-150 kN/m
12-15m LONG
A₄

150 kN/m 15m LONG
A₃
ANCHORED REVETMENT WALL
BERM
150 kN/m 15-20m LONG
A₂

5:1

5.90

2.70

CENTRE LINE OF ROAD

7.00

5:1

250-300 kN/m 15-20m LONG
A₁
ANCHORED REVETMENT WALL

0 1 2 3 4 5m

COMPARISION PLANE = 230m

Fig. 1

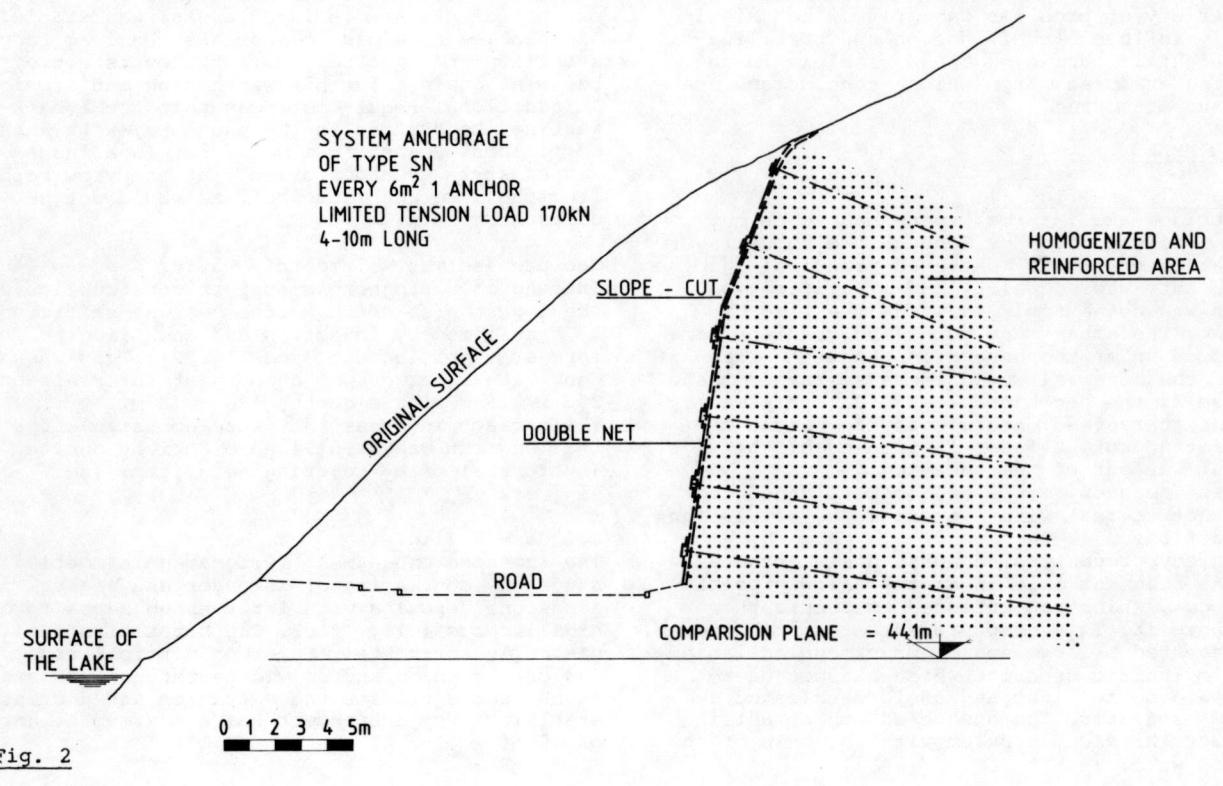

SYSTEM ANCHORAGE
OF TYPE SN
EVERY 6m² 1 ANCHOR
LIMITED TENSION LOAD 170kN
4-10m LONG

HOMOGENIZED AND
REINFORCED AREA

SLOPE - CUT

ORIGINAL SURFACE

DOUBLE NET

ROAD

SURFACE OF
THE LAKE

COMPARISION PLANE = 441m

0 1 2 3 4 5m

Fig. 2

C 50

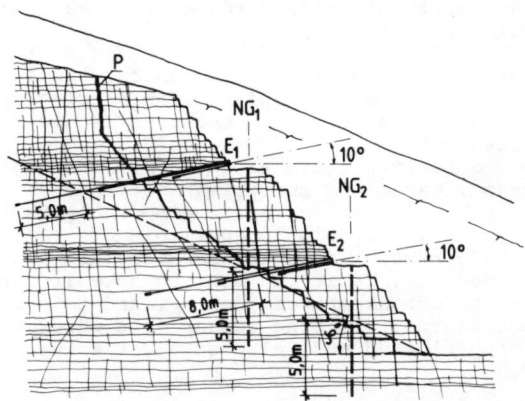

E.........EXTENSOMETER
NG......SLOPE - INDICATOR
P.........POTENTIAL SLIDING SURFACE

Fig. 3

Conditions are investigated by borings and exploratory digging. The mountain water table
lies deep (see fig. 5).

Fig. 5

Further requirements are:
- The extent of the stabilizing measures should
  be kept as small as possible.
- The embankment should require as little future
  upkeep as possible, as it was a railway cut.
- Clearing out the bottom and work on the slope
  should be done from several berms.
Here too, the rock conditions, in particular
the quality and decomposition of the rock layers
deeper down, could not be so thoroughly explored as to enable a definitive plan, especially
as the slope cut is very long and divided up
by different disturbances into differing homogeneous areas.

Fig. 4

Consequently, the job is done with large steps
and individual berms; as required, the bottom
area is stabilized by a 5-8 m high foot wall.
Other securing steps such as anchoring, consolidation, and drainage are decided upon and taken
depending on the particular area.

3.   CONCLUSIONS FOR PLANNING AND EXECUTION

3.1   Shape
With regard to the shape of the cross section,
the embankment gradient cannot be generally
planned but depends on the conditions (rock
quality, structure, water, and much else). In
particular, the influence of the height of the
cut must be taken into account. Respect for
stress flow lines makes a flattening out (roun-
ding out) of the lowest area unavoidable.
In addition, deeper cuts at the foot of the embankment for water ditches or cable trenches are
unsiutable, as they undercut the embankment and
break up the rock bond, even if done sectionally.

It should further be noted that a concave cross
section makes a better embankment than a convex
one, as the former shape creates a horizontal,
supporting arch.

3.2   Preliminary Investigations
It must be acknowledged that the preliminary investigation cannot usually be carried out to the
point where all problems are revealed and the
planner is protected from surprises. In certain
circumstances, borings alone will give no sufficient information. Deep exploratory slot cuts
combined with borings and geophysical methods
have shown the best results.

The plan has to allow for all possible circumstances. Among the main considerations are:
- the stability of the embankment itself;
- later upkeep;

C 51

- landscape preservation;
- the safety of traffic routes (e.g. against falling rock).

In order to find the best technical and economical solution, several plans of differing gradient, shape, and securing measures should be carried out.
Where embankments are affected by mountain streams, early and sufficient drainage must absolutely be planned, as it is almost always the most economical solution possible
The safe removal of surface water must also be provided for. It is the best to have it run over the embankment divided up, and to avoid a concentrated fall. The danger of a watercourse getting blocked must also be dealt with.

3.4 Security Measures
These apply to both overall security and the security of the partial embankments.
The condition of the open surface must be taken into account. Depending on the material and its stability, there are various possibilities. For weathered rock, gradients must absolutely be selected that can be biologically stabilized.

Traffic routes are to be protected from falling rock by covering nets, intercepting fences, and catching areas.

3.5 Construction
Removal work can only be done with protective methods, whether it involves blasting or cutting. This aspect must be checked continuously. In par-

ticular, weak areas must be identified early on and incorporated into the general planning concept.

Between planning and construction, a concept must exist that allows the adjustment of the steps being taken to the acutal conditions. In this respect we must distinguish between immediate steps that secure the stability of the embankment and so-called cosmetic steps which are not urgent and can be applied later.

3.6 Technical Measurement Supervision
(see too fig. 4)
In principle, higher embankments should be kept under technical measurement supervision, and this not just once they are finished, but also while under construction.

Proven devices to this end are:
- Slope indicator, which can be used relatively early to check for deformation;
- Extensometer, which can only be installed once the instrument location point has been reached;
- Measurement bolts;
- Piezometer in drilled holes to monitor the mountain water table.

For the completed embankment, especially if it contains walls or beams, an occasional geodetic inspection is also applicable.

3.7 Maintenance
The operating embankment hat to be maintained and checked for eventual changes in conditions.

INTERACTIVE GRAPHICS IN THE BACK-ANALYSIS OF SLOPE FAILURES

Calcul graphique interactif dans l'analyse post rupture des talus

Graphisch-Interaktive Berechnungen zur Rückanalyse von Hangrutschungen

Dr. C. Dinis da Gama
Assistant Director, DMGA — I.P.T., P.O. Box 7141, São Paulo, Brazil

SYNOPSIS

In order to perform a reliable back-analysis of slope failures, an understanding of the mechanisms of sliding is required and adequate computation methods are needed. Appropriate consideration of the geomechanical factors influencing slope failures (such as slope and slide geometry, water levels and acting forces at the moment of rupture) is essential; on the other hand, the available methods to analyse limit equilibrium situations should be adaptable to complex slope geometries and water levels, as well as to the heterogeneous composition of most rock masses. Interactive graphics show great potential to fulfil those requirements, as is illustrated by means of back-analysis studies conducted by the author.

RESUME

Pour exécuter des analyses post-rupture de talus représentatives, il est nécessaire de comprendre les mécanismes de glissement, ainsi que les méthodes d'analyse compatibles. Il est indispensable d'opérer une choix appropriée des facteurs géomécaniques qui influencent la rupture des talus (tels que la géométrie des pentes et du glissement, les niveaux d'eau et les forces qui jouent un rôle important au moment de la rupture); par contre, les méthodes disponibles pour analyser les situations d'équilibre limite doivent être ajustées aux complexités de la géométrie, du niveau d'eau ainsi qu'aux hétérogénéités des massifs rocheux. Des exemples développés par l'auteur montrent la grande utilité du calcul graphique interactif pour la résolution de ces questions.

ZUSAMMENFASSUNG

Um zuverlässige Rückanalysen von Hangrutschungen durchzuführen, benötigt man geeignete Analysemethoden und Kenntnisse der Rutschungsvorgänge. Eine zweckdienliche Betrachtung der geomechanischen Faktoren, welche die Hangrutschung beeinflussen (bzw. Hang- und Rutschungsgeometrie, Grundwasserstand und wirkende Kräfte im Augenblick des Bruches) ist wesentlich. Andererseits müssen sich Analysemethoden für Grenzgleichgewichtssituationen auf die Komplexität der Geometrie und des Grundwasserspiegels sowie auf die heterogene Zusammensetzung der meisten Felshänge anpassen lassen. Die graphisch-interaktive Berechnung besitzt ein großes Potential zur Erfüllung dieser Ansprüche, wie der Verfasser hier durch Rückanalyseberechnungen veranschaulicht.

1 INTRODUCTION

Design and utilization of rock slopes cannot satisfactorily be accomplished without adequate information on the shear strength properties of the geologic features that usually control their stability, as well as knowledge of the most probable mechanisms of failure that are likely to pose a risk to their safety. Both these aspects of rock slope engineering are difficult to evaluate without field data, together with appropriate observation and judgment.

On the other hand, laboratory determination of cohesion and friction angles of joints in a certain rock mass is a risky method of characteriza

tion due to the lack of representativity of the shear testing in terms of a large volume where variability is rule rather than exception, and also because these parameters should be obtained from samples taken along the failure surface, which is not known in advance, usually. Thus it is virtually impossible to reproduce either in the laboratory or "in situ" the real constraints that may cause a slope to fail, upon a certain stress or displacement history, taking into consideration all material heterogeneities and corresponding strengths.

In his Rankine Lecture, Prof. Jaeger (1971) pointed out that measurement of sliding in rock surfaces has been made under different conditions and for various purposes, giving results that are lumped together, and figures quoted without real appreciation of the complexity of the situation.

Therefore, extreme care must be exercised when using a pair of parameters c and Ø to establish the safe geometry of a slope. Not only these parameters have to represent the strength of the rock mass under a certain type of external loads, but also the equivalent situation which may cause failure in the future. Thus, only field studies conducted on similar slopes and analogous geometries are able to lead to reliable design of new slopes.

As these pre-requisites are very difficult to get in practice, the designer must at least base his evaluations on the study of case histories regarding failure of slopes excavated in the same type of rock mass, and/or under equivalent loadings.

When a certain slope rupture is studied, back-calculation of its stability is the only way to quantitatively explain the phenomenon and the best method for estimating the shear parameters of the weakest link which caused failure to occur.

Several assumptions are needed in order to perform back-analysis (Sancio , 1981):
- Information on the geometry and geologic composition of the failed slope.
- Data on water pressures and other forces acting at the moment of failure.
- Rock mass physical properties.
- Formulation of a failure model, involving the most probable mechanism of sliding.
- Utilization of an adequate method of analysis for the calculation of the strength parameters along the failure surface.

It is obvious that two types of back-analysis may exist: with or without previous slope instrumentation and monitoring. Thus, if the event is predicted in advance, chances to gather data on the slope mechanism of failure are greater than if no information is obtained before the sliding occurred.

The differences in the quality of the back-analysis (and in the reliability of its results) are so great that it is becoming common to excavate experimental slopes (or trial pits, as they are known in surface mining) to determine with greater accuracy the shear strength of rock masses.

Gama (1981) described the work done in an uranium open pit where a very steep slope 250 m long and 48 m deep was monitored through several types of instruments in order to study its stability.

In the course of this work a number of slope failures was detected and the corresponding back-calculations done in order to determine shear strength of the various joint types and infilling materials, which were considered responsible for those slidings. As a result of this research, reliable parameters were incorporated in the stability analysis of the ultimate pit, thus improving its design, and reaching greater reliability in the solutions to be adopted in practise. This example of what may be called "induced back-analysis" together with similar experiences, seems to be a trend indicating the growing importance of this methodology.

2 INTERACTIVE GRAPHICS CAPABILITIES

For the purpose of conducting realiable back-calculations of slope failures, methods of analysis must be as realistic as possible with respect to the constraints and all other aspects of the actual problem under consideration. Approximations should be limited to a minimum and be reserved for essential data not collected before slope failure, rather than calculation methods or the analytical procedure.

Because of the sensitivity of shear strength parameters in relation to the approximations introduced in the calculations, it is essential to use mathematical tools that won't distort reality. Also, it requires a permanent participation of the analyst during the computation procedure, in order to effectively communicate with the computer as the calculation is being carried out.

Interactive computer graphics seems to solve most of those difficulties, for it is simultaneously a process for incorporating complex geometries (slope shapes, water lever surfaces and irregular rock contacts) into the models, as well as to allow the participation of the user to modify algorithm assumptions and to orient the solution for the basic purpose of back-analysis.

Furthermore, this process provides inovations on the presentation of the outputs, thus improving the sought results, which may be fed-back for subsequent analysis.

Referring to the advantages of interactive computer graphics, Ingraffea et al. (1981) mention the reduction in tedium while increasing the accuracy obtained through this technique, which leads to greater cost-effectiveness in geotechnical structural analysis.

The use of interactive graphics for the purpose of inputting finite element meshes (both 2-D or 3-D) and for displaying results of various types (including color representations and perspectives), is also a great improvement in this field.

As far as back-analysis studies are concerned, interactive graphics have proven to be an excellent technique and has allowed considerable reliance be placed on back-calculations of slope failures by the author.

In the examples that follow, ilustrations are provited on this subject.

2.1 Graphic input by means of photos

In order to represent the actual geometry of a slope, the shape of the underground water surface, and the structure of the rock mass, we need adequate techniques. Obviously, the back-calculation method should accept actual data through a series of coordinates of points in stead of geometrical abstractions required by the limit equilibrium algorithm.

For this purpose, interactive graphics allows to input data by means of digitalization made on detailed maps or photos of the slope. The photos should be taken at positions that do not distort the image and they should contain a linear scale, like a ruler or other object which indicates a direct conversion factor between the real distances and the units of distance in the photo. The latter are put on the digitizing table and thus they correspond to G.D.U. (graphic display units), easily converted into real scale units, with which calculations of distances, inclinations, areas, weights and water pressures are done.

Existence of differences in the vertical and horizontal scales are permitted when necessary. Fig. 1 shows the digitalization operation of a slope, for supplying data to be stored in the computer memory for subsequent slope stability analysis or back-calculations.

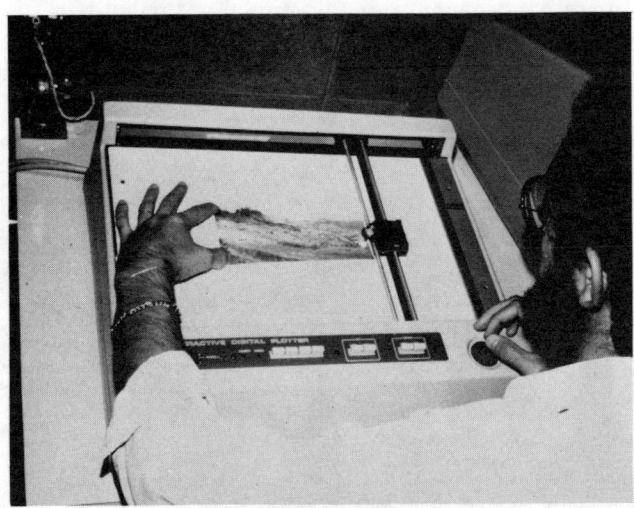

Fig. 1

2.2 Dynamic representation of slope movements

The interactive computer graphics capacity of representing successive images of a certain phenomenon makes possible to simulate movements of slope during its failure.

Using a pre-established sliding surface and the center of rotation of the movement, it is easy to represent various phases of the slide, as shown in Fig. 2.

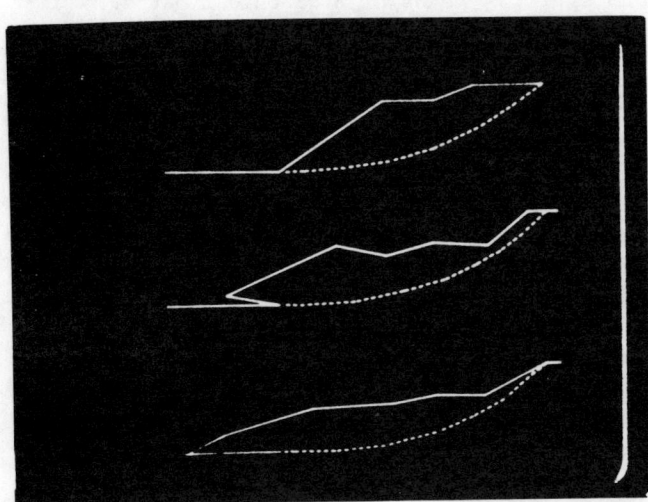

Fig. 2

This picture resulted from a program based on the rotation of a constant mass of ground, allowing the change of its shape according to the influence of gravity when kinematic instability occurs. The three images concern the case of slope failure that will be discussed in 4, where the first one indicates the original geometry, the central image explains an intermediate situation and the third one represents the final geometry upon sliding.

2.3 Multi-block movements

When the constitution of a rock slope can be considered as a system of semi-rigid blocks, its displacements or rotations can be computed upon the interaction of the various blocks that come in contact with each other.

The technique developed by Cundall (1974) allows the representation of block movements at successive instants thus simulating a progressive slope failure, after selecting the lowest stability mode of failure. This technique incorporates realistic friction laws between blocks, in accordance with the properties of jointed rock masses.

Fig. 3 shows the progressive displacements of rock blocks forming a steep slope which suffered a horizontal acceleration of magnitude 0.1 g in the instant t=0. The block movements are represented as time proceeds, and in a further occasion their positions are represented in Fig. 3. This method has great potential but its application for back-analysis requires further developments.

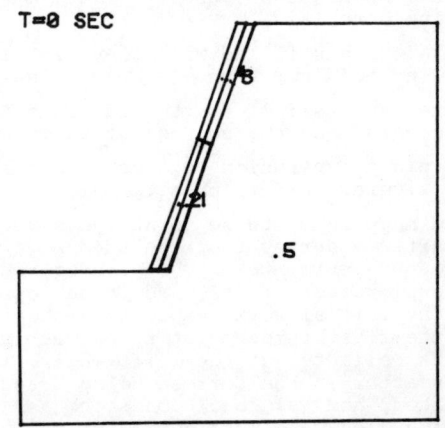

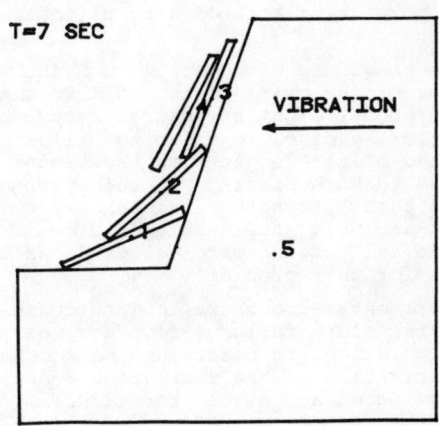

Fig. 3

3 BACK-ANALYSIS OF FAILED SLOPES IN AN URANIUM MINE

Slope stability studies have been conducted in the Cercado uranium mine, near Caldas,MG (Brazil) in the past six years, since the pre-feasibility phase of the project. Early work was described elsewhere (Gama and Silva, 1978) , and subsequent research was developed in order to determine the shape of the open pit which has to incorporate the steepest possible slopes due to the very high mining strip ratio. Because of the need to use low safety factors, careful design of the mine slopes was accomplished by means of obtaining shear strength parameters from back-analysis studies of various failures that occurred in the pit preparatory work, and upon the excavation of an instrumented experimental trench.

These activities were described by Gama (1981) and these have proceeded recently with observation and monitoring work that goes on with pit excavation in the production phase of the mine. As predicted, most bench failures result from slidings along discontinuities of the rock mass, according to their spatial frequency and attitude. Data for the back-analysis are obtained not only by permanent observation of the slopes, but also by measuring several types of instruments installed in benches that form the ultimate geometry of the pit.

Classification of the join types existing in the ore-body was established by means of four distinct groups:

a) Natural open joints, usually flat, and without in-filling materials.

b) Natural joints filled with clay minerals and showing no alteration of the rock surfaces.

c) Natural joints filled with clay minerals and presenting wheathered rock at their walls.

d) Long planar intrusions of basic rocks showing altered in-filling materials.

Detailed mapping of these joint types was accomplished after a survey that included more than 7,000 measurements, taken both underground (along exploration drifts) and in the open pit. During the initial work, laboratory shear testing of the first three joint types was performed on blocks collected at the drift walls. Considerable scatter in the corresponding results was noticed, and values of cohesion and friction angles were considered not reliable for the purpose of establishing ultimate pit slope angles.

Upon the knowledge of location, orientation and attitudes of the main joints and discontinuities of the orebody, eight structural sectors were defined in the mine, in order to allow the prediction of safe angles for the slopes to be excavated in each domain. For this purpose reliable shear strength parameters were required and back-analysis studies rather than additional geomechanical testing were selected as the solution for this problem.

The experimental trench was opened consequently and several slope failures recorded, permitting back-calculations to determine the shear strength characterization of the four joint types, which were considered acceptable for final pit design because they represent weighted mean values along those joints.

Continuation of this research lead to the systematic observation of slope behaviour, with special attention given to slope failures in order to back-analyse them and obtain new values of cohesion and friction angles for those joints. The method of calculation improved due to the recent utilization of interactive computer graphics, as may be seen in the following examples.

3.1 Slope failure in zone 8-A

A very regular rock slide occurred in the 8-A zone of the Cercado open pit, located in a 16 m high bench at elevation 1372 m. Fig. 4 is a photo of the rock face after the sliding, showing that rupture occurred along a planar surface, without lateral constraints.

Fig. 4

Because no water pressures were present at the moment of failure (the local water level was below bench floor, as determined in a nearby borehole) the situation for the slip along a joint type c) (filled with clay and having weathered walls) was back-analysed int the computer system, by means of a graphical method, using the well known polygon of forces (Fig. 5).

This equilibrium of forces results from application of the Coulomb friction law at the limit equilibrium situation of that slope.

Each polygon of forces corresponds to a pair of c, Ø values, and in Fig. 5 six cases are represented, one of which is impossible because it gives a negative friction angle.

An interesting conclusion taken from the interpretation of the various polygons of forces results from the fact that it is possible to evaluate the relative importance of the cohesion and of the friction terms in the particular case under analysis. This circumstance helps to decide which pair (c,Ø) is selected, taking into consideration that cohesion has greater influence on the rupture of steep slopes, while friction angle affects higher slopes. This criterion is useful when no other information on slips in the same joint type is available, thus helping the analyst to choose the most probable combination of c and Ø that characterizes the joint material under analysis.

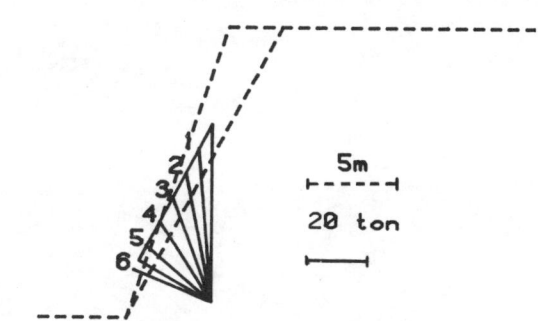

NO.	COHESION (kN/m2)	FRICTION ANGLE (degrees)
1	5	55.98
2	10	49.21
3	15	39.90
4	20	27.18
5	25	10.80
6	30	-7.52

Fig. 5

3.2 Slope failure in zone 9

Another 16 m high rock bench slipped parallel to the face at zone 9 in the Cercado open pit, at elevation 1364 m. Data on the failure indicated that it happened after blasting of a nearby bench, and no water pressures were acting.

Fig. 6 shows the bench under consideration, where a block is partially suspended at the top.

Fig. 6

The joint along which this slip occurred was type c), and again using the graphical representation of the limit equilibrium relation of the Coulomb failure criterion the solutions contained in Fig. 7 were obtained.

It must be pointed out that the above analysis was based on the calculation of a pseudo-static safety factor given by:

$$F_s = \frac{(W\cos\alpha - KW\sin\alpha)\ \tan\emptyset + cL}{W\sin\alpha + KW\cos\alpha}$$

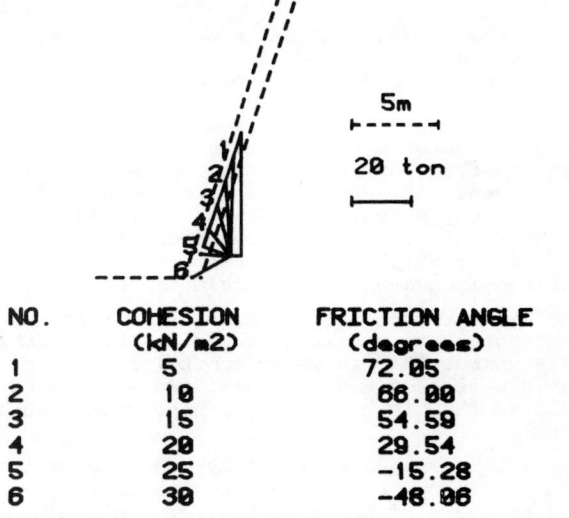

NO.	COHESION (kN/m2)	FRICTION ANGLE (degrees)
1	5	72.05
2	10	66.00
3	15	54.59
4	20	29.54
5	25	-15.28
6	30	-48.06

Fig. 7

Where:
W - weight of the failed rock mass
α - inclination of failure plane
K - fraction of g (seismic coefficient)
L - length of failure plane
c,∅- cohesion and friction angle of joint

The value of K was selected to be 0.1 (vibration acceleration of 10% of g) which was estimated by comparison with previous blasts that were recorded in the mine.

Although the two reported slips occurred in different circumstances, distance from each other was only about 200 m and hence the successive (c,∅) pairs were put together in the same graph (Fig. 8) thus yielding only one solution corresponding to the point of intersection of the two straight lines. This information of c=21 KN/m2 and ∅=25.1º is in accordance with previous determinations for the shear strength parameter of the type c) joint in the Cercado uranium mine.

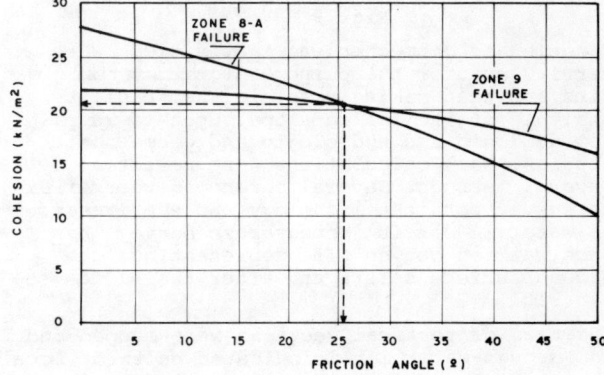

Fig. 8

4 BACK-ANALYSIS OF A FAILED SLOPE IN A MANGANESE MINE

The Serra do Navio manganese mines are located in the Amapá Territory, northern Brazil. They form a group of open pits producing 2,3 million tons of manganese are (1979 figures), half of which is exported, mined from sedimentary deposits covered by vegetal soils and weathered schistose rocks. Excavation is done by mechanical means and does not require the use of explosives. Benches are 7,5 m high, and each group of five have a general slope angle of 45°, after which a douple berm is left, thus forming an overall slope angle of about 39°.

The deepest open pit was mine A-12, reaching a depth of 80 m, and it suffered a severe failure in November 1980 in which about 1.1 million cubic meters of ground moved down-slope and covered the pit bottom containing high grade ore.

The author was asked to study the problem in order to propose new slope angles allowing the recovery of the remaining ore, and minimizing the removal of the slipped terrain.

Fig. 9 shows a general view of the sliding in mine A-12.

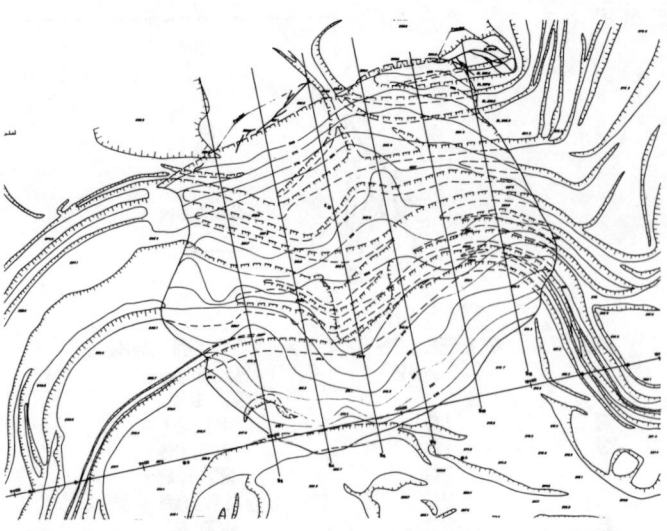

Fig. 10

4.1 Bi-dimensional back-analysis

Sections S-2 to S-12 permitted to develop back-calculations with the use of computer graphics. Each section was submitted to digitalization to transfer its geometry and ground water position to the computer memory and then use the limit equilibrium method of slices to relate c and ∅. The center of rotation is determined after a search to locate by iterative means the one that corresponds to a minimum safety factor.

Fig. 9

A series of geomechanical activities were carried out for the purpose of characterizing the site and then making possible the stability analysis of the failed slope. Upon topographic, geologic and hydrologic and geomechanic studies, back-calculations were performed by several methods. Several boreholes were drilled to survey both the lithology and the presence of water. A careful topographic survey has been carried out on site representing the ground surface before and after the slide (see Fig. 10).

A series of vertical sections were mapped and the ground-water level indicated on them. Local observation and surface evidence led to the hypothesis of a circular type failure, from which the limit-equilibrium analysis was applied in the back-calculations.

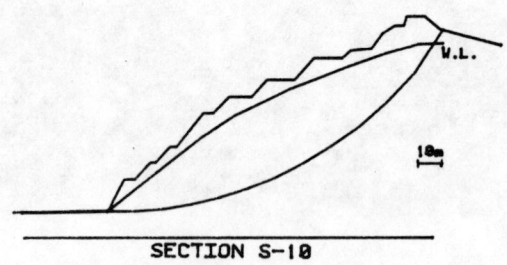

Fig. 11

Fig. 11 shows the example of one vertical section, in which the back-analysis result was expressed by means of an equation relating c and ∅. In the six sections these equations are:

Section S-2: $c + 42.01 \tan ∅ = 22.75$
Section S-4: $c + 39.49 \tan ∅ = 22.39$
Section S-6: $c + 41.71 \tan ∅ = 22.19$
Section S-8: $c + 41.52 \tan ∅ = 24.59$

Section S-10: c + 40.14 tan ∅ = 27.20
Section S-12: c + 35.61 tan ∅ = 23.37

To determine the two parameters that would represent average values for the overall slope, the following hypothesis was adopted: the friction angle is equivalent to the angle of repose of the failed material, which by coincidence equals the original slope angle before mining.

That angle is 19° and the average of the resulting six values for cohesion gives

$$c = 10 \ t/m^2 = 100 \ KN/m^2$$

These values of c and ∅ were utilized in the bi-dimensional slope stability analysis for the re-excavation of the mine.

4.2 Three-dimensional back-analysis

The existing data for the A-12 mine sliding were so detailed that a three-dimensional stability analysis was performed.

Using the method proposed by Hovland (1977) and incorporating water pressures in his equations, a back-calculation was tried. As a first step data on the geometry of three surfaces was supplied: points given by its (x,y,z) coordinates were furnished for the original ground surface, for the ground water level surface, and for the probable failure surface.

A perspective of the slope as given by the graphic system is shown in Fig. 12.

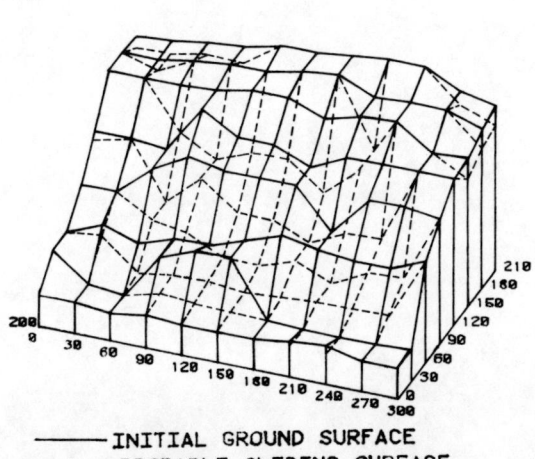

—— INITIAL GROUND SURFACE
– – – – PROBABLE SLIDING SURFACE

Fig. 12

The method of calculation proceeded by interpolating, for the vertices of a regular square mesh, the values of the elevations for the three surfaces. By this process, elementary prismatic blocks were defined to represent the ground volume, and also uplift water pressures were computed for each block. Thus, application of Hovland expression of the three-dimensional slope safety factor was possible:

$$F_3 = \frac{\sum_x \sum_y \left[c \ A_i + (W_i \cos D_i - U_i) \tan \emptyset \right]}{\sum_x \sum_y W_i \sin \alpha_{yz_i}}$$

Where:
W_i is the weight of elementary prism i
A_i is the area of the base of elementary prism i
D_i is the angle of dip for the base of prism i
U_i is the water pressure acting on prism i
α_{yz_i} is the inclination of prism i base along slip direction

In the three-dimensional back-analysis conducted for mine A-12, the failed ground was divided in 88 columns (or prismatic blocks) with a section of 30 m x 30 m. The equation resulting from this back-analysis was:

$$c + 22.29 \ tg \ \emptyset = 11.01$$

By comparison with similar equations obtained in the bi-dimensional calculation, it must be pointed out that 3-D analysis gives lower values of c and ∅, indicating that lateral constraints provide greater stability to the slope.

Fig. 13 represents the variation c vs. ∅ obtained both from 2-D and 3-D back-calculations, showing the different ranges of application for the values.

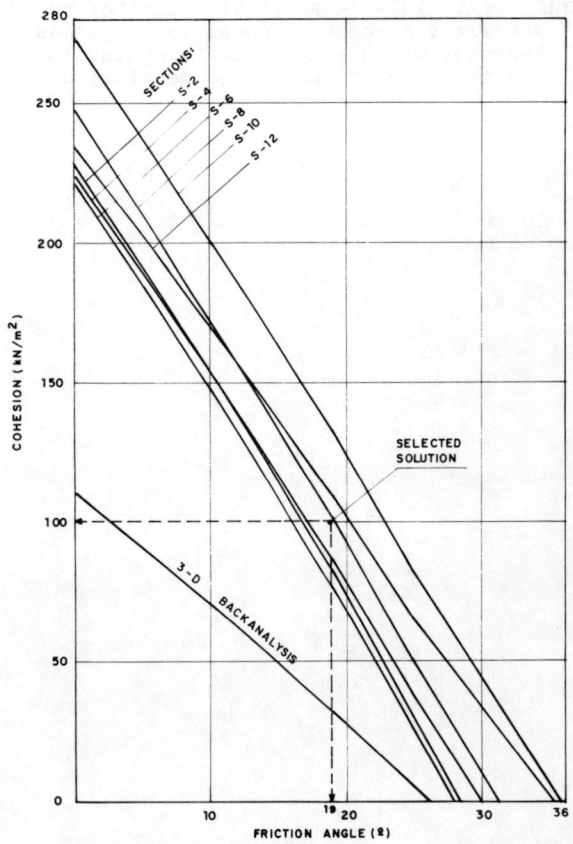

Fig. 13

Utilization of 2-D or 3-D results depends on the further application of c and ∅. If these will be used for 2-D slope stability analysis then the corresponding values from back-calculations should be taken, and similarly for the 3-D parameters.

In the specific case fo mine A-12 because the design of new slopes was done by 2-D methods the average c and ∅ values were selected, such as Fig. 13 represents.

5 CONCLUSIONS

The importance of back analysis to determine the shear strength parameters of joints and rock masses requires the utilization of realistic input data, and the application of appropriate calculation methods and techniques. Interactive computer graphics plays an expressive role in obtaining greater reliability in the back-analysis of failed slopes because it uses simultaneously the input and output facilities of those machines, permits the utilization of graphical methods of evaluating stability, and allows the human interference to conduct computations in accordance to the engineering judgment of the analyst. This type of computer utilization enhances the engineer's confidence in the back-analysis results, thus providing more reliable parameters for subsequent slope design.

6 REFERENCES

BARTON, N. R. (1971) - Estimation of in situ shear strength from back-analysis of failed rock slopes. Proc ISRM Symp. on Rock Fracture, II-27. Nancy.

CUNDALL, P.A. (1974) - Rational design of tunnel supports: a computer model for rock mass behavior using interactive graphics for the input and output of geometrical data. U.S. Army Corps of Engineers Technical Report MRD-2-74.

GAMA, C.D. and SILVA, R.F. (1978)- Engineering geological studies for the Cercado uranium mine. Proceedings 3rd Int. Cong. I.A.E.G., Vol. 2, 239-250, Madrid.

GAMA, C.D. (1981) - Back-analysis of slope failures in the Cercado uranium mine (Brazil). Proc. 3rd. Int. Conference on Stability in Surface Mining. Ed. AIME, 745-772. Vancouver.

HOVLAND, H.J. (1977) - Three-dimensional slope stability analysis method. J. Geotechnical Engineering Division, Proc. ASCE, 103 GT 9, Sept., 971-986.

INGRAFFEA, A.R.; KULHAWY, F.H. and ABEL, J.F. (1981) - Interactive computer graphics for analysis of geotechnical structures. Proc. Int. Conf. on Computers in Civil Engineering. Ed. ASCE, 864-875. New York.

JAEGER, J.C. (1971) - Friction of rocks and stability of rock slopes. 11th Rankine Lecture. Geotechnique (21), No. 2, 97-134.

SANCIO, R.T. (1981) - The use of back-calculations to obtain the shear and tensile strength of weathered rocks. Proc. Int. Symp. on Weak Rock, 647-652. Tokyo.

MULTI-STAGE DIRECT SHEAR TESTS IN SITU ON SCHIST FOR SLOPE STABILITY STUDIES (CAUÊ MINE, ITABIRA, BRAZIL)

Essais de cisaillement in situ et à plusieurs niveaux exécutés sur un schiste pour des études de stabilité de talus (Mine de Cauê, Itabira, Brésil)

In mehreren Laststufen durchgeführter Scherversuch in situ an Schiefer zur Bestimmung der Hangstabilität (Cauê Bergwerk, Itabira, Brasilien)

Engº Prof. A. J. da Costa Nunes
President of Tecnosolo S.A.

S. S. Sandroni
Consultant

J. M. Souza Ramos
Geologist of Cia.Vale do Rio Doce

SYNOPSIS

A vast geotechnical investigation program has been in progress in the last 13 years as part of the Stability Studies of the Cauê Open-Pit Iron Mine. After a brief description of the site and of the several materials and investigation activities which have been and are being carried out, the paper concentrates on the laboratory and in situ direct shear tests executed in the hard fractured Nova Lima Schist. It is shown that a multi-stage in situ direct shear testing technique, in which care was taken not to fail the rock in each stage, can be utilized to determine an engineering estimate of the in situ strength of the fractured schist.

RESUME

Un vaste programme d'investigations géotechniques est en cours de réalisation depuis 13 ans pour les études de stabilité de la mine de fer à ciel ouvert du Cauê. Après une brève description du site et des divers types de matériaux ainsi que des investigations passées et en cours, cet article se concentre sur les essais de cisaillement en laboratoire et in situ exécutés sur le schiste "Nova Lima" dur et fracturé. On montre que la technique des essais de cisaillement in situ à plusieurs paliers où l'on prend soin à chaque palier de ne pas rompre la roche peut être utilisé pour évaluer la résistance in situ du schiste fracturé.

ZUSAMMENFASSUNG

Ein umfangreiches geotechnisches Untersuchungsprogramm wurde in den vergangenen 13 Jahren ausgeführt als Teil der Stabilitätsuntersuchungen der offenen Eisenerzmine Cauê. Nach einer kurzen Beschreibung des Abbaugebietes und der zahlreichen Materialen und Untersuchungsaktivitäten, die ausgeführt wurden und werden, konzentriert sich dieser Beitrag auf die direkten Scherversuche, ausgeführt im Laboratorium und "in situ", am harten, zerklüfteten "NOVA LIMA" Schiefer. Es wird gezeigt, daß direkte "in situ" Scherversuche in mehreren Laststufen, wobei der Bruch des Gesteins sorgfältig vermieden wurde, benutzt werden können, um eine Festigkeit des zerklüfteten Schiefers ingenieursmäßig abzuschätzen.

1. INTRODUCTION

The present paper discusses the results of laboratory and in situ direct shear tests on a hard schist carried out as part of the stability studies of the 500 to 600 m-deep Cauê Open-Pit Iron Mine, owned by Vale do Rio Doce Company, Itabira, Minas Gerais, Brazil. The geology is rather complex consisting structurally of a assymmetric syncline with an E-W axis plunging some 30ºE. Rocks are pre-Cambrian in age and correlate with the Minas and Rio das Velhas series which form the so-called Iron Quadrangle of Minas Gerais State. The local stratigraphic column is as follows in the Table.

As a result, a variety of slope stability situations, characterized by various ground water conditions and different sequences of strata and dip of layers in relation to the walls of the ellypse-shaped pit, have to be dealt with.

Efforts with the aim at determing the most economic stable pit have required the attention of several engineers and engineering organizations in the last 13 years or so.

The Piracicaba Quartzite, the Caraça Quartzite and weathered near surface layers of the Nova Lima rocks

TABLE

SERIES	GROUP	LOCAL DESIGNATION
Minas	Piracicaba	Piracicaba Quartzite
	Itabira	Itabirite and Hematite
	Caraça	Caraça Quartzite
Rio das Velhas	Nova Lima	Nova Lima Schist
		Nova Lima Quartzite

which are soft, hand-excavatable materials, have been submitted to an extensive program of soil laboratory tests coupled with back-analysis of several failures which occurred as exploration progressed. The Iron bearing rocks of the Itabira group, which are also soft materials with occasional hard spots, have been recently submitted to a, as yet in progress, program of laboratory and field tests linked with the construction of an experimental slope some 45 meters in height. The hard unweathered rocks of the Nova Lima Group have been tested in direct shear both in the laboratory and in situ.

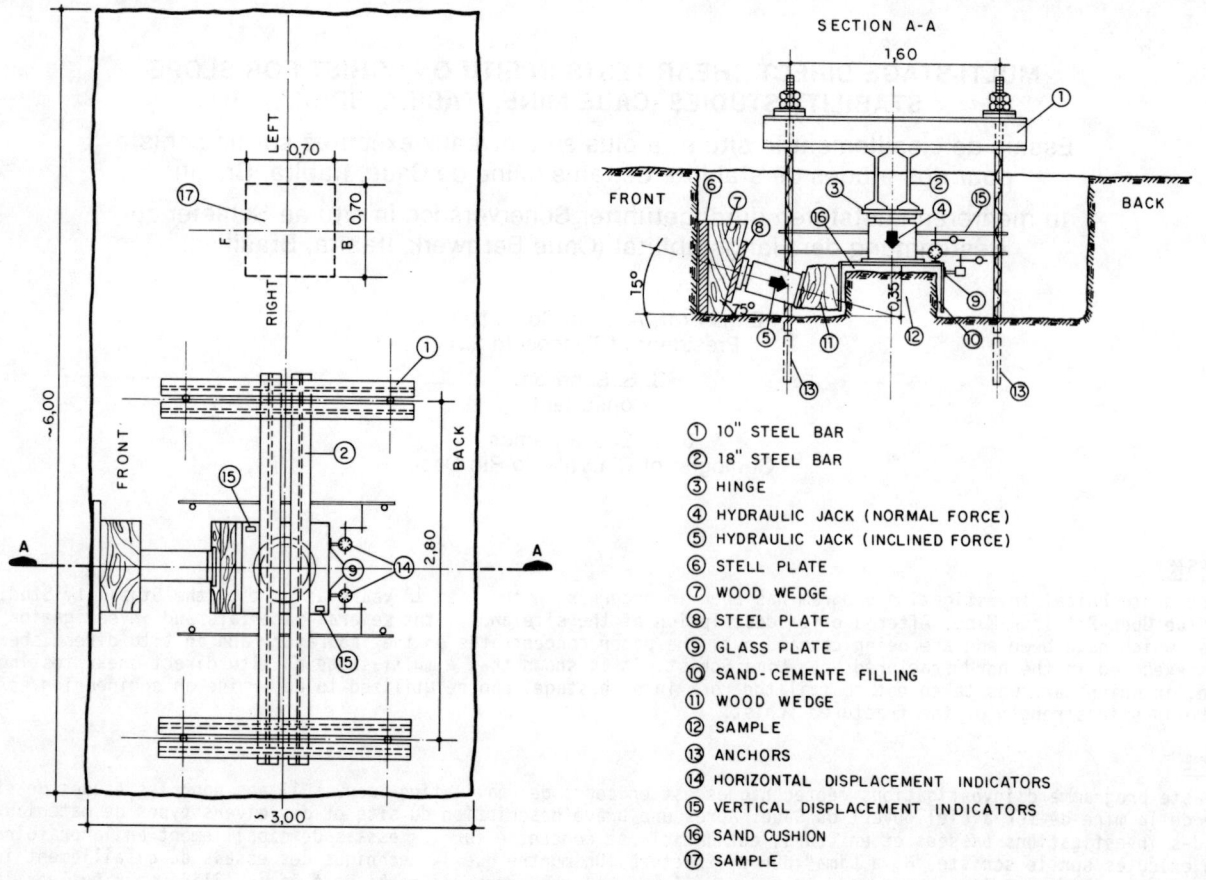

SECTION A-A

1,60

FRONT

15°

BACK

① 10" STEEL BAR
② 18" STEEL BAR
③ HINGE
④ HYDRAULIC JACK (NORMAL FORCE)
⑤ HYDRAULIC JACK (INCLINED FORCE)
⑥ STELL PLATE
⑦ WOOD WEDGE
⑧ STEEL PLATE
⑨ GLASS PLATE
⑩ SAND-CEMENTE FILLING
⑪ WOOD WEDGE
⑫ SAMPLE
⑬ ANCHORS
⑭ HORIZONTAL DISPLACEMENT INDICATORS
⑮ VERTICAL DISPLACEMENT INDICATORS
⑯ SAND CUSHION
⑰ SAMPLE

NOTE: DIMENSIONS IN METRES

Fig. 1   In situ direct shear test set up

Some geotechnical aspects of this vast engineering enterprise have been brought to print (e.g. Ramos et at, 1974) but most of the accumulated results are still awaiting divulgation.

In what follows, the authors report on the laboratory and the large scale in situ direct shear tests carried out in the hard Nova Lima Schist in the hope that, appart from adding one more drop to the ocean of information on rock behaviour, this will encourage further publications on the numerous geotechnical aspects of this fascinating job. Without exploring details of interpretation, the paper reports a sucessfull attempt into obtaining a reasonably representative "en masse" strength from in situ multi stage direct shear tests.

2. DESCRIPTION OF THE TESTS

The tests herein described have been executed in the hard unwheathered fractured Nova Lima Schist exposed during the progress of exploration of the mine. Fractures are locally parallel to schistosity both dipping some 45o SE. The failure planes induced by the in situ tests were horizontal therefore cutting across fractures and schistosity. The laboratory tests, carried out in intact samples, had failure plane parallel to the schistosity.

The laboratory direct shear tests, performed by Geotecnica S.A., have been carried out in square samples saw-cut from intact blocks of the rock. The conventional procedure was followed of applying a normal load and failing the sample by progressively increasing the shear load. After failure along a clearly defined surface, the ultimate or residual shear strength was measured for different values of the applied normal stress.

The in situ direct shear test have been carried out in carefully formed square blocks with 0,55m sides and 0,35m height. The set up is shown in figure 1. Due to economic as well as time schedule restrictions a multi--stage test procedure was adopted: nominal normal stress stages of 5,0; 10,0; 15,0 and 20,0 kgf/cm^2 were applied and, in each stage, the shear force was gradually increased until a tendency towards a sharp change in horizontal displacements was noted. Before each normal stress increment the applied shear force was fully released. In all four tests of this type, i.e. sixteen loading stages, have been carried out. Full apreciation was given to the fact that, if care was not taken, only the first stage could be considered as representative, the remaining ones corresponding (in the case of excessive displacement in the first stage) to a "pre-failed" (and, therefore conservative) situation.

The results, plotted in figure 2 together with the results of the laboratory tests suggest that the in situ test program has been successfull in obtaining strength values intermediat between the intact sample strength and the failed or residual situation. This strength can, therefore, be used on an engineering basis to represent the strength characteristics of the rock mass.

C 62

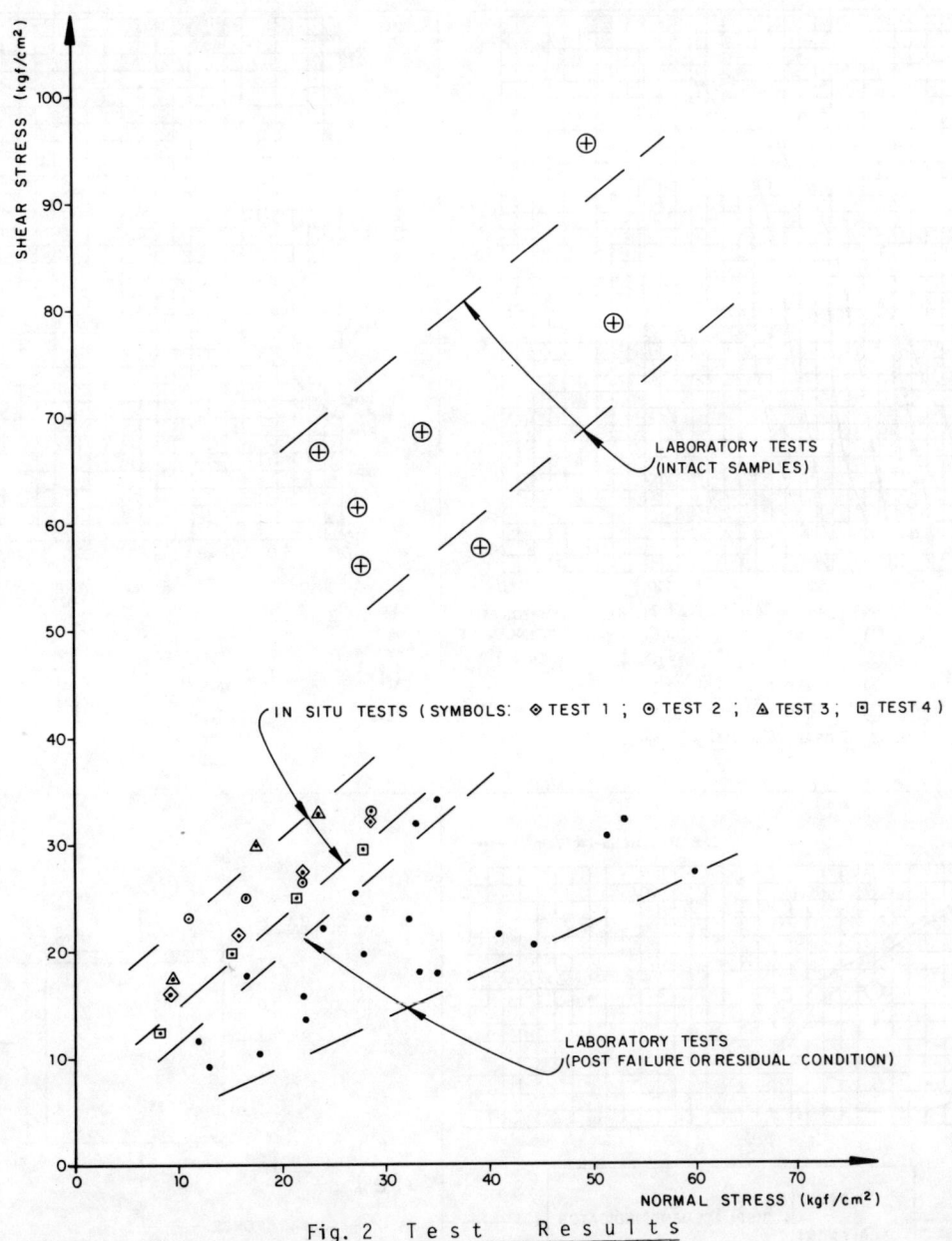

Fig. 2 Test Results

3. COMMENTS ON EXECUTION OF THE IN SITU TESTS

Detailed results of one of the in situ direct shear tests (test No. 1, which is typical) are shown in figure 3. The horizontal displacement vs. shear stress curves are given in 3(a) as an indication of the criterion utilized to interrupt each stage. In 3(b) the resulting apparent strength envelope is shown. Vertical displacements are given in 3 (c) and 3 (d) for the first and the last stages of the test and the shape of the failure surface is depicted in 3 (e).

4. CONCLUSION

A multi-stage in situ direct shear testing technique in which care was taken not to fail the rock in each stage was utilized to determine engineering values for in situ strength of a massif of fractured schist.

5. BIBLIOGRAPH

RAMOS, J.M.S.; GUIMARÃES, P.F. and BRÃO, P.C. (1974)
Slope Stability of CVRD Cauê Mine Itabira - Brazil.
Soc. Mining Engineers, AIME, Fall Meeting Acapulco,
México.

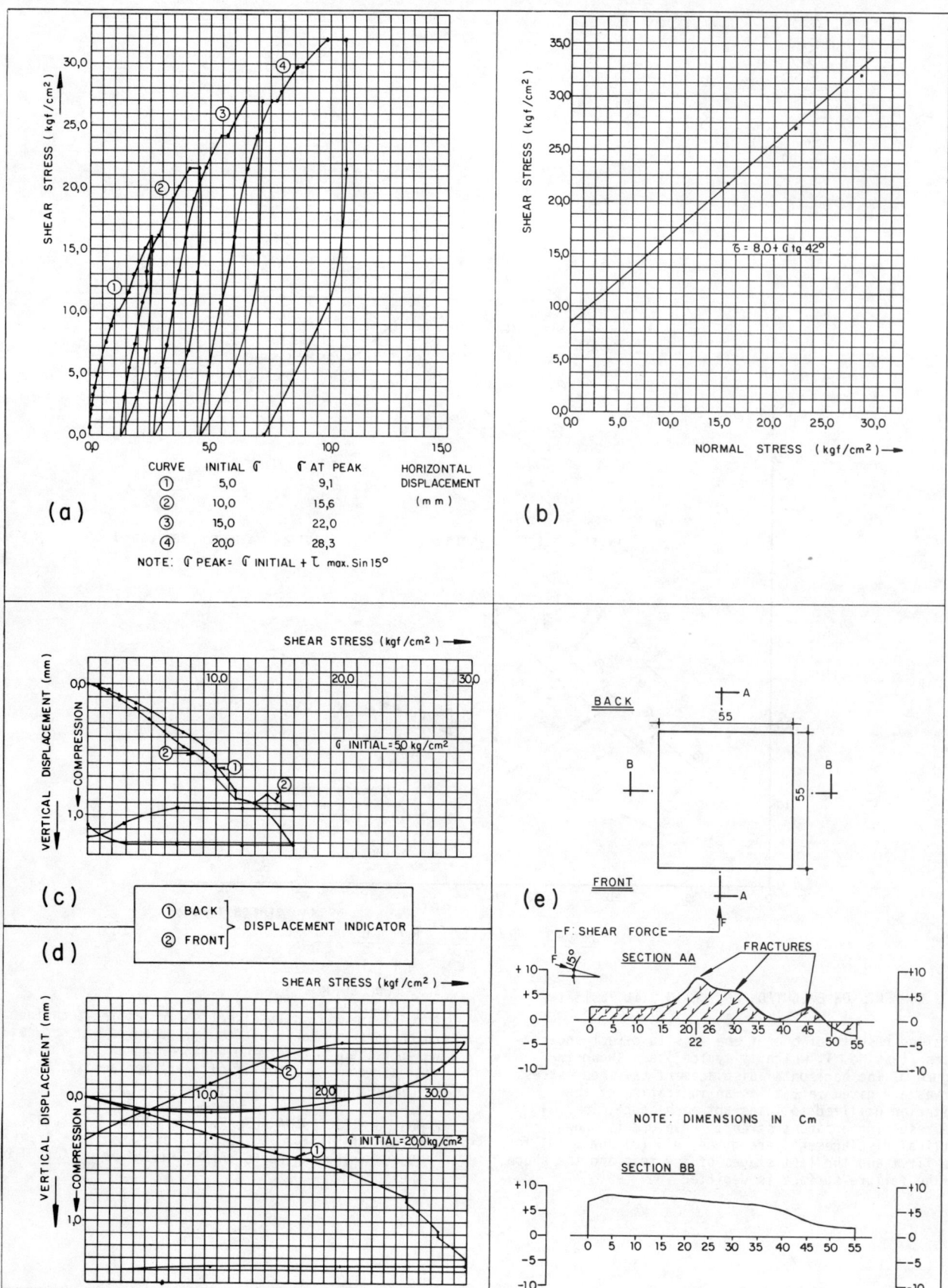

Fig. 3   Results of a typical in situ direc shear test

GRAPHICAL METHOD FOR THE ANALYSIS OF ROCK SLOPES IN URBAN AREAS

Méthode graphique pour l'analyse de talus rocheux en zone urbaine

Graphische Methode für die Analyse von Felsböschungen in Stadtgebieten

Gianfranco Perri A.
Professor of Rock Mechanics, Universidad Central, Apartado 47351, Caracas 1041-A,
Venezuela

SYNOPSIS

The purpose of the paper is to apply a three-dimensional graphical method to the slope analysis in weathered schists that constitute the hills surrounding Caracas, the capital of Venezuela. The method was applied to a particular urban zone with an area of one million square meters. The result was used to draw a map showing areas with different degrees of stability related to the three-dimensional arrangement of discontinuities within the topographic constraints.

ZUSAMMENFASSUNG

Die Absicht dieser Arbeit ist es, für die dreidimensionalen Analysen der Böschungs-Stabilität der verwitterten Schiefer, aus denen die Hügel in der Umgebung Caracas, der Hauptstadt Venezuelas, bestehen, eine graphische Methode zu verwenden. Die Methode wurde an einer spezifischen städtischen Zone mit einer Fläche von einer Million Quadratmeter angewendet. Das Ergebnis wurde benützt, um eine Stabilitätskarte zu zeichnen. In dieser werden Flächen mit verschiedenen Stabilitäts-abstufungen gezeigt, die sich auf die dreidimensionelle Lage zwischen den geologischen Ungleichförmigkeiten und topographischen Oberflächen beziehen.

RESUME

Cette communication concerne l'application d'une méthodologie graphique pour l'analyse tridimensionelle de la stabilité de talus dans les schistes altérés, qui constituent les collines autour de Caracas, capitale du Vénézuéla. Cette méthodologie a été appliquée à une zone spécifique urbaine sur une superficie d'un million de mètres carrés. Les résultats ont été utilisés afin de tracer une carte montrant les zones dont la stabilité varie en rapport avec la répartition tridimensionnelle des discontinuités à l'intérieur des contraintes topographiques.

INTRODUCTION

In the rocky areas characterized by a highly anisotropic mechanical behavior, owing to the presence of surfaces of lesser resistance, systematically arranged in the rock mass, the parameters which decisively control the stability of the slopes, naturaly or artificially formed, are of two principal orders:

- On one hand, the parameters of shear strength (cohesion and friction which may develop along the length of the different surfaces of lesser resistance or planes of geological discontinuity.

- On the other hand, the reciprocal location between these planes of discontinuity and the planes of the free surfaces of the rock mass: the faces of the slopes.

In particular, concerning the problem of the stability of the slopes in rock masses, at least three possibly characteristic situations can be defined:

a) *Stable kinematic conditions*. This is the relative location in the space between the planes involved in the problem, which does not give rise to the existance of any freedom for all the structure. These, consequently, result stable, independently of the characteristics of shear strength that may develop.

b) *Conditions of kinematic instability and of mechanical stability*. When the geometric conditions of kinematic instability are present, the shear strength that can develop along the surface of the different planes of discontinuities involved, intervenes to prevent the movement. When these resistant forces are greater than the unstabling forces, the slope will be stable.

F

c) *Conditions of kinematic and mechanical instability.*
If besides verifyin situations of kinematic instability, it happens that the unstabilizing forces are
superior in intensity to those of resistance, then
the slope will be unstable.

It is known (Hoek and Bray, 1974; Goodman, 1976) that
the techniques of hemispheric projection form an extrem
ely convenient and simple instrument for the represen -
tation in a plan, based on a network, of the complicated
phenomena of geometric interaction in the space between
the structures involved in the problem of the three-di-
mentional analysis of slopes in rock masses. At the same
time, it is also convenient and simple to introduce,by
use of a graphical method, the effects of the shear -
strength, in its frictional and even cohesive component
and finally, the most advanced methods permit the complete
quantification of the problem, even determining the
numeric values of the safety factors and of the effects
of eventual non-gravitational external forces.

DESCRIPTION OF THE METHOD

The analysis procedure presented here is especially effective when applied to relatively vast areas, on which
a territorial zoning is desired, based on the stability
of the slopes that compose them.

The topographic scales used are generally small to medium
(specifically, the author has used the method, with sat
isfactory results, on scales of 1:1000, 1:2000 and 1:5000),
and the approach of the procedure of the elaboration and
analysis of the situation, was carried out on statistical
and probabilistical bases. At the same time, the results
are given in terms of different degrees of geological-
ricks of instability.

The inmediate consequence of aforementioned is the
efficiency and reliability of the results of such a study,
which are closely linked to the abundance and the repre
sentativity of the data to be used as inputs of the problem.

These inputs, in the initial phase of the analysis, are
the location in the space and the geologic-geotechnical
nature of the structural discontinuities present in the
rock formations of the area (the first family of planes)
and, the location in the space of the planes that form
the topographic gradients of the same area (the second
family of planes).

The fundamental stage of the procedure is to analyse the
geometric interaction existing between the different -
planes of the first family and then between these and
the second familiy of planes. This analysis allows certain interesting aspects of the problem to be made evi
dent and can lead to the quantification and topographic
location of the phenomena in its integrity.

In order to illustrate the method of the analysis, direct
reference must be made when it is applied in the zoning
of risks, in an area of approximatelly one million square
meters, on which a residential urban zone is to be devel
oped.

Namely, an area located in the hills in the south-east
of the valley of the city of Caracas, Venezuela, whose
geologic ambient corresponds with the *Las Mercedes Forma
tion of the Caracas Group.* This was originally described
by Aguerrevere and Zuloaga (1937) as a series of calcareous schists with zones of graphite, locally micaceous,
which outcrop in a fairly weathered state in the area
under study.

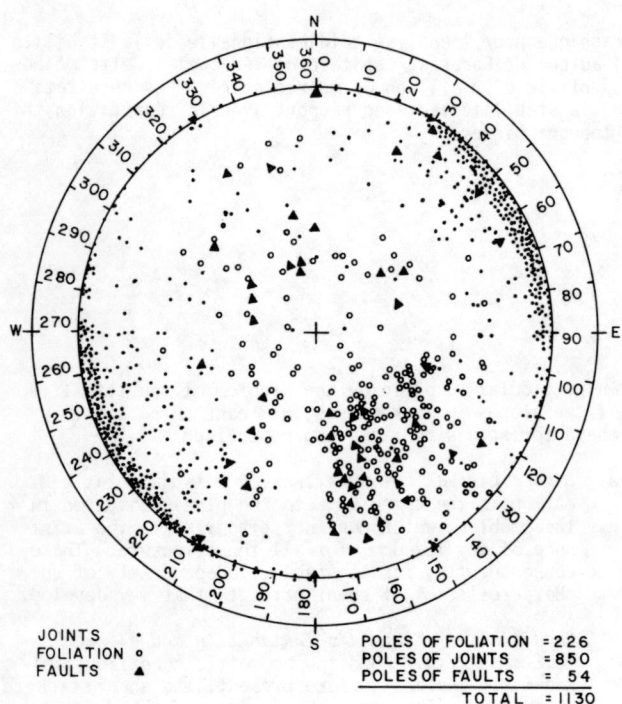

JOINTS   ·
FOLIATION o
FAULTS   ▲

POLES OF FOLIATION = 226
POLES OF JOINTS   = 850
POLES OF FAULTS   = 54
TOTAL   = 1130

Fig. 1 *Plot of poles of discontinuities. Equatorial
equal-area projection.*

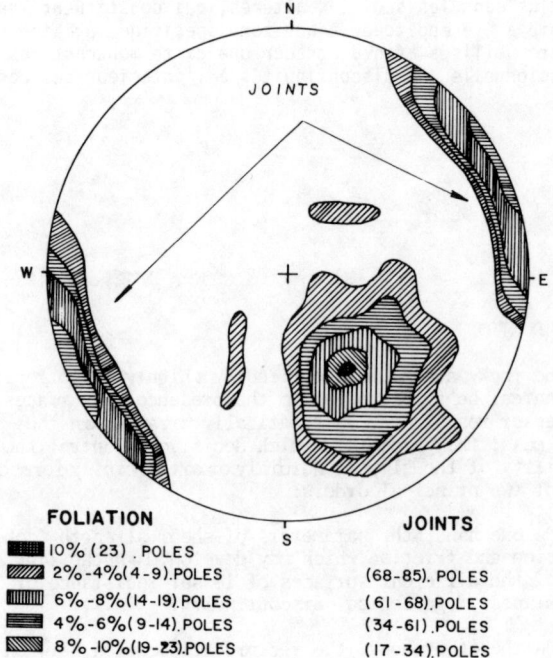

FOLIATION   **JOINTS**

▓ 10% (23) . POLES
▨ 2%-4%(4-9).POLES   (68-85). POLES
▥ 6%-8%(14-19).POLES   (61-68).POLES
▤ 4%-6%(9-14).POLES   (34-61).POLES
▧ 8%-10%(19-23).POLES   (17-34). POLES

Fig. 2 *Contour diagram of discontinuities: joint,
foliation and faults.*

The rocks present in all the area belong to one litho-logical gruop, thus allowing all the structural data obtained in the geologic survey of the field to be treated as one entity in this example.

This information consists of the location and geomorphologic description of the main structural accidents existing which, for succesive elaboration, are diferentiated in foliation joints and minor faults. Figure 1 presents the diagram of the concentration of poles, based on Lambert's equal-area network, of all the observed discontinuities (1130 poles).

Figure 2 presents the corresponding countour diagram which permits to observe the existence of two distinct patterns of discontinuities: the first corresponds to the planes of foliation and the second to the planes of joints; the minor faults follow a pattern similar to that of the foliation.

As a complement, useful for future consideration, in figure 3 and 4 respectively, the envelope of the great circles of the foliation planes and of the planes of joints, are reported. Also outlined, in both diagrams, is the common area of the two envelopes, which represent the zone of the network in which the intersections between the two families of planes of discontinuities area located, or in other words, the zone in which the lines of all the possible structural wedges present in the area under study, fall.

Until now, has been analyzed and arraged the first geometric parameter: the structural aspect. It now remains to analyze and arrange the second geometric parameter of the problem: the topographic aspect.

In order to do this, it is convenient to imagine the hills which form the area, as a collection of a great number - of small slopes, each one of which represents a small portion of the side, crest or base of the relief.

Each elementary plane thus established, is characterised by a well defined value of dip and an equally well defined value of strike; this is the bearing of each elementary area of the gradients.

At this point, follows the quantification of the problem, or rather, of the definition and localization of the corresponding geometric characteristics of strike and dip, direction for each one of these ideal slopes which are generally quite numerous.

For this purpose, we proceed to elaborate a map of the gradients which, besides being characterized by the values of dip of each area, contains the information relative to the bearing of the gradients, or rather, the dip direction of each elementary plane.

In this example which is being illustrated, four classes of dip and eight classes of dip direction were used. All of these are shown in figures 5 and 6 where by means of diagrams, the quantified results of this analysis of the area being studied, are given. For example, there may be observed a dip direction of the gradients, predominatelly towards the N-NE and an almost complete lack of gradients towards the W. With respect to the angle of dip, we note that those of the category 28°-35° predominate. The flat areas are excluded from the calculations of the percentages in the diagram of dip, while they are taken into consideration in the percentage calculations in the diagram of dip direction allowing the latter to give us an exact idea of the portion of non-level territory in the area being studied.

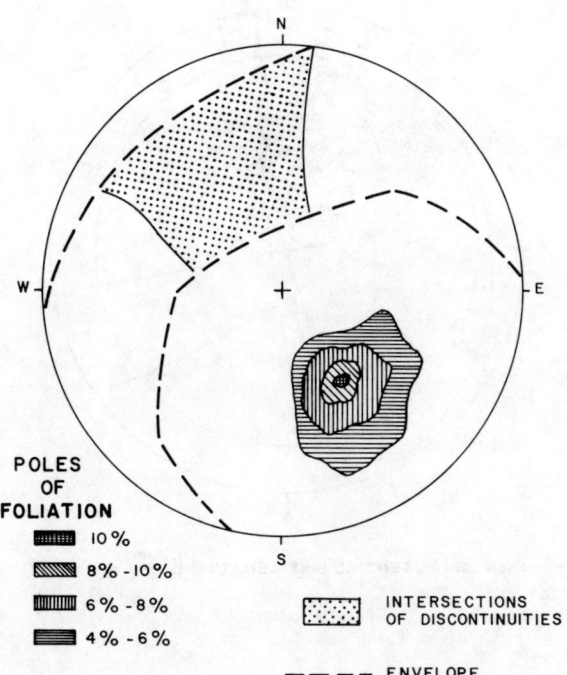

Fig. 3 Envelope of great circles of foliation and area of intersections foliation-joints.

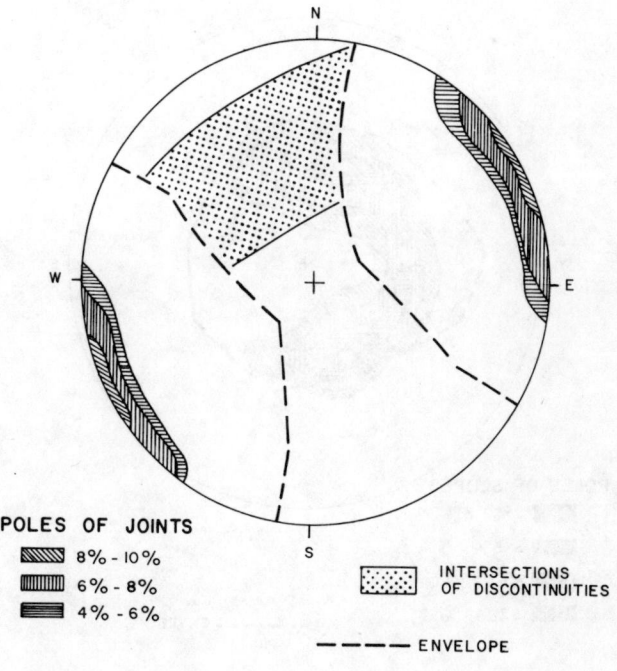

Fig. 4 Envelope of great circles of joints and area of intersections joints-foliation.

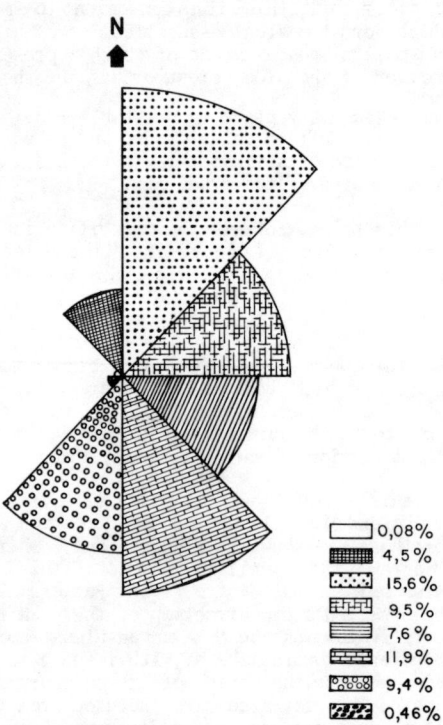

Fig. 5. *Distribution of dip directions or dip azimuth of topographic gradients*

☐	0,08%
▦	4,5%
⊙	15,6%
⊞	9,5%
▤	7,6%
▩	11,9%
⊚	9,4%
▧	0,46%

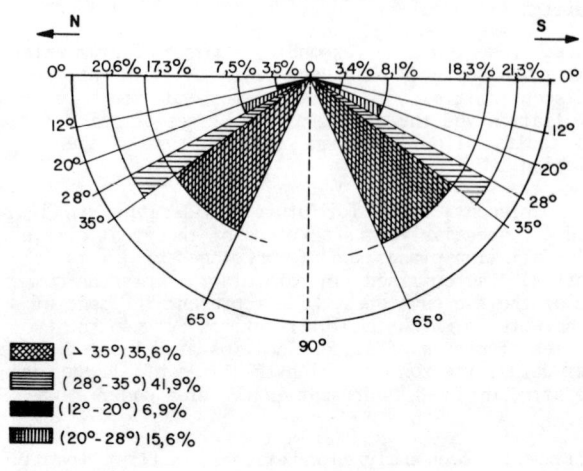

Fig. 6. *Distribution of true dip of topographic gradients.*

▨	(⩾ 35°) 35,6%
▤	(28°-35°) 41,9%
▪	(12°-20°) 6,9%
▥	(20°-28°) 15,6%

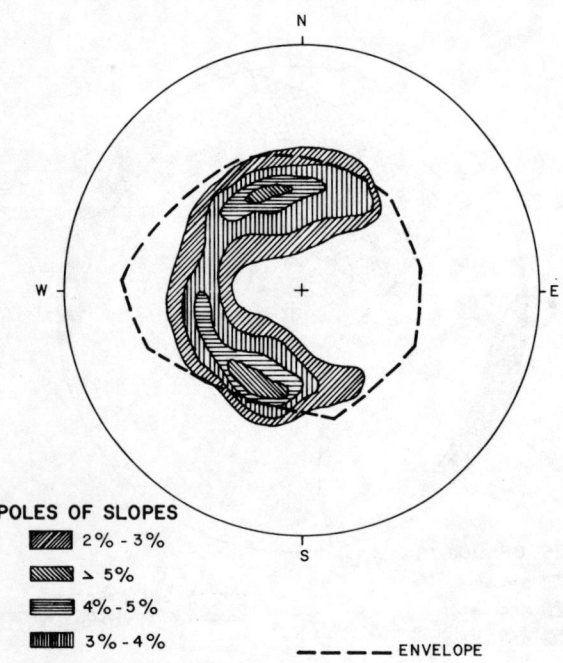

POLES OF SLOPES

▨	2% - 3%
▧	⩾ 5%
▤	4% - 5%
▥	3% - 4%
– – – –	ENVELOPE

Fig. 7. *Contour diagram and envelope of great circles of topographic gradients.*

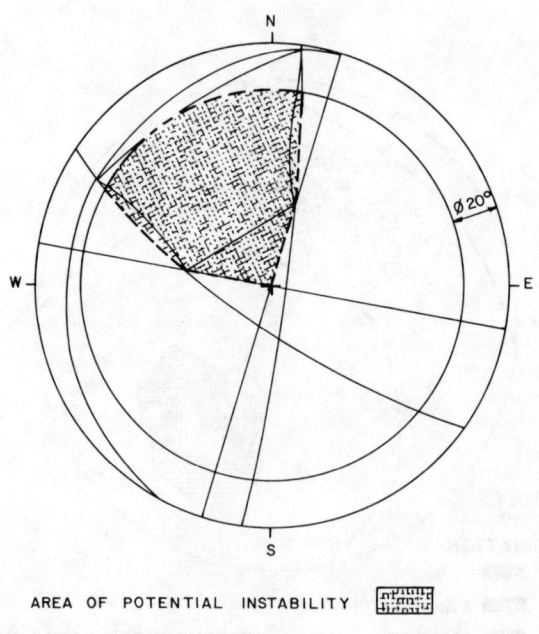

AREA OF POTENTIAL INSTABILITY ▦

Fig. 8. *Stereo plot of great circles representing limit stable slopes.*

C 68

It can now be used in an original form, the idea by which the relief of the rock has been associated with a group, made up of small slopes which, geometrically, are no more than planes, whose bearing is known with suf ficient precision.

On the other hand, this group of planes can be shown in the hemispheric projection by means of a simple diagram of the concentration of poles, which, in turn, allows the making the corresponding diagram of countour of poles of the gradients an the envelope of the corresponding great circles. These same elements are reported in fig ure 7 for the example that is being illustrated.

At this point, follows the confrontation of all the ge ometric factors involved in the problem and quantified in the aforementioned stages.

Referring to the three (a.b.c.) characteristic situa tions possible in the stability of rock slopes, it is worth while, on one hand, to individualize the stable kinematic conditions, so as not to bother with them further and, on the other hand, to individualize which, among the unstable kinematic conditions, can and can not constitute mechanical instability.

In effect, it is possible in a phase of general zoning as in that which is being analyzed, to take to into con sideration a minimal value of the frictional parameter of the shear strength along the different surfaces of discontinuity analyzed and thus also to set aside the slopes that become stable, with the mere contributions of a minimal friction, reducing even more, the cases to be analyzed in more detail, for being potentially and effectively unstable.

In order to make this step, by means of the sequence of hemispheric projections, it is sufficient to trace, in the diagram, a concentric circle with the network which marks the value of the angle of friction available, from the periphery towards the centre.

In figure 8, the area must not be intercected by the great circles of the stable slopes, is outlinded in the manner thus described. Successively the new simple mechanical elaboration which is detailed, prosecutes:

- Superimpose figures 7 and 8.
- Trace the figure outlined by the envelope of the planes of the gradients and by the area of the potentially unstable slopes.
- Outline the section of contour diagram of the poles of the gradient corresponding only to those poles whose great circles form part of the outlined fig ure in the previous step. In other words, among all the existing gradients, individualize the poles of those planes whose great circles intercept the figure mentioned in the previous step, along its entire extention.

The result is figure 9, where we may observe that:

a) The area of the location of the intersections(folia tion-joint) which are potentially unstable (because they possess an angle of dip lesser than that of the slopes) represents around 50% of the total area of the location of these intersections.

b) The area of the contour of poles of the gradients which are potentially unstable, (as they possess an angle of dip greater than that of the lines of wedges) represents almost 22% of total area of the contour of the gradients poles.

c) As a result of (a) and (b) the *probably* unstable

slopes (for which the two unfavourable conditions area verified simultaneously) will be around 11% of the total existing in all the area studied, (obviously only of all those reported in the coun tour diagram and corresponding to inclinations su perior to those of 20°).

At this point, the procedure of analysis, which has been illustrated with reference to the stability of the struc tural wedges, must be repeated, also in order to analyze the stability relative to the two systems of existing discontinuities, considered as planes in a separate form, following exactly all the same steps already described.

In this particular case, these other analisis is of seconda ry importance, owing to the fact that, first of all, the joints area pseudo-vertical and second, the foliation planes cause unstable situations coincidents with those already individualized by the wedges. Consequently, the situations shown in the previous point (c) will not suffer important variations, neither in quality nor quantity.

Once the problem is quantifid in the form described, there remains simply its placement in the area, in order to achieve a greater benefit in its application. This last operation is inmediate, considering that we already posses all the necessary information. The results are reported in the plan of figure 10, called *The Map of Stability Analysis*, which summarizes all the principal aspects involved in the problem. (The figure represents only a limited portion of the plan which, in the example in question, was originally elaborated on a scale of 1:2000).

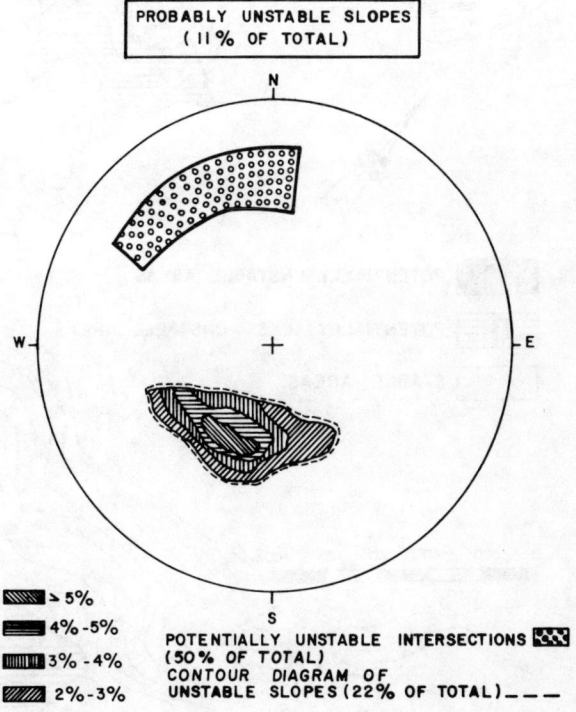

Fig. 9 *Contour diagram of poles of potencially unstable slopes and area of potentially unstable intersec tions joint-foliation.*

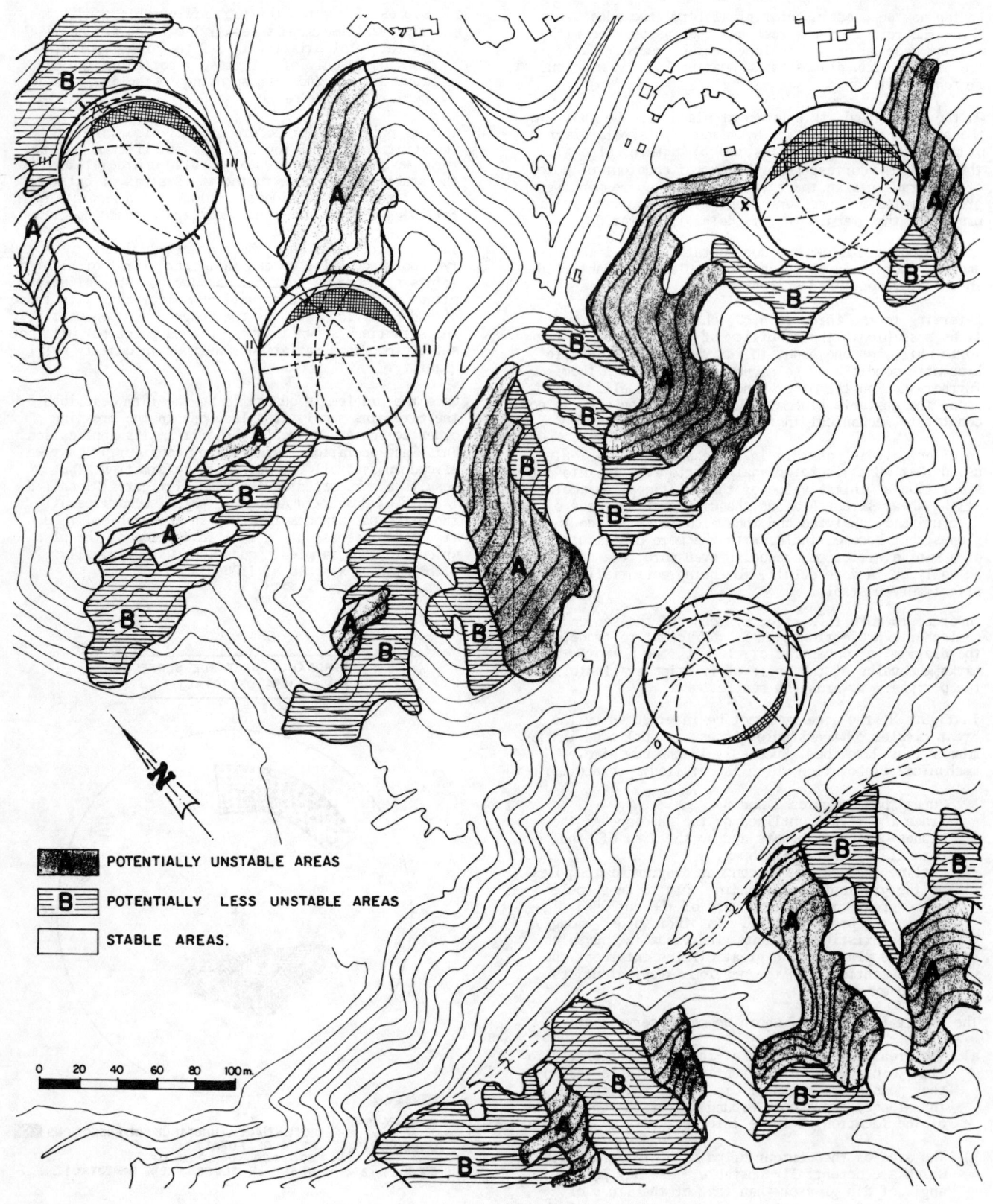

POTENTIALLY UNSTABLE AREAS

B POTENTIALLY LESS UNSTABLE AREAS

STABLE AREAS.

0 20 40 60 80 100 m.

Fig. 10. Partial representation of the "The Maps of Stability Analysis". The original map was elaborated on a scale of 1:2000. The total area was approximately one million square meters.

In spite of having been able to realize a more detailed zonification, or rather, individualizing a great number of possible categories as the large amount of information available would have permited, it was preferable to use a zoning of only three categories, considering the degree of precision available and above all, considering the practical purpose of the use of such an analysis. In other words, a more detailed subdivision at this stage and at this scale of study would perhaps result in pure academic speculation.

For the example illustrated, the following categories of zonification have been used:

a) *Zones of potential geometric-structural instability;* which comprise slopes with strike-sub-parallel to that of the foliation structures and dipping at an angle greater than 28° in the same direction as the discontinuities.

At the same time, these zones contain slopes with gradients greater than 28°, on whose faces outcrop intersection lines of discontinuities with a direction pseudo-perpendicular to the strike of the slopes; similarly, slopes with gradients greater than 35°, on the face of which outcrop intersection lines of discontinuities with a non-perpendicular direction to the strike of the slope.

b) *Zones of potentially reduced geometric-structural instability;* these comprise slopes with gradients greater than 35° and on the face of which enentually may outcrop intersection lines of discontinuities whit a pseudo-parallel strike to the slope. Also included, at the same time, are slopes with gradients greater than 28° and on the face of which outcrop discontinuities with a non-parallel strike of the slope.

c) *Zones of geometric-structural stability;* which com prise slopes with gradients lesser than 28° and slopes with steeper gradients which are situated in such a way as not to permit kinematic conditions of instability, owing to the favorable location of the geological structures.

Once the zonification of the area has been completed in the described manner, it may pass to the more detailed phase of analysis, in which it will proceed to carry out localized calculations of stability for the zones and the specific slopes belonging to the areas with a greater potential of instability.

CONCLUSION:

A method for the zonification of urban areas based on the geological risks has been presented.

The result is the elaboration of a "Map of stability analysis" in which the total area is separated in three zones: potentially unstable areas, potentially less unstable areas and stable areas.

The method is based on the use of the techniques of hemispheric projections for graphical three dimensional analysis of the stability of rock slopes characterized by surfaces of lesser mechanical resistance and a frankly anisotropic behaviour.

In fact, in those geotechnical conditions,the fundamental factor for the control of stability is the number, location and characteristics of the discontinuities, and it is possible and easy to introduce in the proposed analysis, the effects of other factors that can concur to define the stability conditions: hidrology, external and seismic loads and others.

On the other hand, in certain cases, the stability of the area depend essentially on factors that escape from a analysis such as the proposed in this paper. For example, when the processe of external geodynamics are very active: erosion, surface runoff, regional landslides, etc.

In those cases, however, the result produced by the method described can be considered as an important contribution to the complete definition of geological risks in urban areas.
It is not the purpose of this study to described the aforementioned detailed analyses, which were carried out using the routine methods of the graphical techiques of hemispheric projections, taking into account for each specific geometry of the slope, the corresponding existing discontinuities,with their respective parameter or shear strength (friction and or cohesion), which may develop. (Perri, 1979).

REFERENCES:

AGUERREVERE, S.E. y ZULUOAGA, G. (1937) "Observaciones Geológicas en la parte Central de la Cordillera de la Costa", Bol. Geol. y Min., Caracas, Tomo I, Nos. 2 y 4.

GOODMAN, R. (1976) "Methods of geological engineering in discontinous rocks" West Publishing Co., New York.

HOEK, E. y BRAY, J.W. (1977) "Rock slope engineering", Institution of mining and metallurgy, London.

PERRI, G. (1979) "La cohesión en el análisis estereográfico de estabilidad de taludes", Revista Latinoamericana de Geotécnia, Caracas, Vol. V, N° 2.

A ROCK SLIDE IN AN URBAN AREA: A CASE HISTORY

Un glissement de roche dans une région urbaine: Etude d'un cas spécifique

Felsrutschung im Wohngebiet: Eine Fallstudie

D. A. Salcedo
Professor of Geological Engineering, Universidad Central and President of Ingeotec,
Caracas, Venezuela

F. H. Tinoco
Professor of Civil Engineering, Universidad Simón Bolívar and Principal of Sueloproyecto,
Caracas, Venezuela

SYNOPSIS

A slide occurred in weathered rock near residential buildings located in Caracas, Venezuela. The purpose of the paper is to describe by means of this case history the technical problems and decisions that engineers have had to face to arrive at a reasonably weighted distribution of factors associated with the occurrence of slides. Factors such as orientation of discontinuities, fill weight and water pressure influence were evaluated in terms of Mohr-Coulomb and strength displacement theories. It is concluded that it is possible to arrive at a weighted distribution of factors inducing failure. However, this distribution is strongly dependent on subjective engineering judgment.

RESUME

Un glissement s'est produit dans des roches altérées près d'édifices résidentiels au sud-est de Caracas, au Vénézuéla. Cet article a pour but de décrire, en se rapportant à ce cas spécifique, les problèmes que les ingénieurs ont à confronter et les décisions techniques qu'ils ont à prendre afin d'arriver à une détermination pondérée valable des facteurs liés au déclenchement des glissements. Des facteurs tels que l'orientation des discontinuités et l'influence du poids de remblais et de la pression d'eau ont été évalués d'après la théorie de Mohr-Coulomb et celles du déplacement de la résistance. On en tire la conclusion qu'il est possible d'arriver à une distribution pondérée des facteurs qui déclenchent un glissement — néanmoins cette distribution dépend en grande partie du jugement subjectif de l'ingénieur.

ZUSAMMENFASSUNG

Ein Hangrutsch hat in verwitterten Felsen in der Nähe von Wohnhäusern in Caracas, Venezuela, stattgefunden. Die Absicht dieses Aufsatzes ist es, durch diesen Fall zu einer richtig bewerteten Verteilung aller mit der Rutschung verbundenen Faktoren zu gelangen. Streichen und Fallen der Trennflächen und Einfluß der Auffüllung und des Grundwassers wurden nach Mohr-Coulomb und Festigkeitsverformungskriterien ausgewertet. Man kommt zur Folgerung, daß es zwar möglich ist, allen Faktoren Werte zuzuordnen, aber die Verteilung ist stark von subjektiver Ingenieurbeurteilung abhängig.

Urban developments in hilly areas of southern Caracas, Venezuela, are every year subjected to rockslides which cause importance loss of properties, and expensive repair work. It also originates lawsuits between owners, owners and builders, owners and real state agencies.

The urgent need for the repair work to be accomplished, makes impossible to wait for a legal verdict. This verdict usually takes many years due to the complexity of determining legal responsibilities of the parties involved, such as urban developers, earthwork contractors, builders, geotechnical engineers, real state agencies and county or city authorities. Due to this fact affected owners and contractors have search for a way to solve the problem of determining responsibilities in order to avoid sending the case to court. It has been therefore a recent practice to establish an agreement between parties in which they agree to hire an independent board of consulting engineers with the following purpose: "to determine causes of failure and express by means of a percentage the influence of each factor in the occurrence of the rockslide".

The results of the technical report would be the base to determine responsibilities and to distribute the

The purpose of this paper is to describe by means of a case history, the technical problems and decisions that engineers have to face in order to arrive at a reasonable weighted distribution of factors causing the rockslide.

2. SITE DESCRIPTION

Natural slopes are composed of weathered quartz-micaceous schists and sericite quartz-phyllites. Foliation planes are very well developed with a strike parallel to the strike of the slope and dip of 30 - 40° towards the free surface. Joints strike approximately normal and parallel to the slope close to a 90° dip, representing released surfaces for the slide that occurred along foliation surfaces.

Reinforced concrete buildings were built between 1973 and 1975 on foundations embedded in weathered rock. Above the weathered rock existed a 10 m high fill that was placed approximately thirteen years prior to the construction of the buildings.

Results of subsoil investigation made in 1973 indicated

that the fill has N values, determined from Standard Penetration Tests, ranging from 7 to 20.

Once all the apartments were sold out, owners decided to build new water tanks locating all necessary pipes on the surface in order to keep control of water leakages. Cracking in walls, broken pipes and fill settlements were observed and repaired from 1975 to 1981. The slide occurred on July 25, 1981 severely damaging parking lots and backyard areas of two residential buildings. Fig. 1 shows geometry of slope before failure.

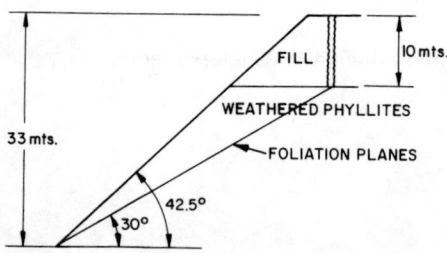

Fig.1. Geometry of slope before failure.

3. SAMPLING AND TESTING

Oriented monolithic samples of the weathered sericite-quartz-phyllites of the failure surface were taken to be tested in a direct shear box. Foliation planes are usually closed but during the geological investigation it was noticed that most of the samples broke apart very easily exhibiting a very low tensile strength in the direction normal to foliation surfaces. Some samples were tested as open foliation surfaces fitting the two halves together as a typical open joint. Other samples were tested as closed foliation surfaces. Results of shear tests indicated an ultimate friction angle of 20°. The unit weight of the weathered phyllite is typically 2.2 KN/m^3.

4. EVALUATION OF FACTORS ASSOCIATED TO THE OCCURRENCE OF FAILURE

The following factors were evaluated:

4.1 Geological and topographical conditions.

    a. Regional geological characteristics.
    b. Geometry of original slope.
    c. Orientation of foliation and joint planes.
    d. Type of rock and shear strength characteristics.

4.2 Effect of fill

    a. Fill as surcharge.
    b. Fill settlements.
    c. Decrease of shear strength parameters with time.

4.3 Effect of water

    a. Water from rain.
    b. Water from broken pipes and sewers.
    c. Water due to the excess of irrigation.

In order to evaluate the aforementioned factors, the following activities were accomplished:

a. Review of existing information. Earthwork, soil investigations, construction records and drawings, water bills, etc.
b. Analysis and interpretation of existing aerial photographs of the area from 1951 to 1978.
c. Detailed geological investigation of the area.
d. Excavation of trenches along existing pipes and tank areas to determine importance of damages.
e. Stability analysis of original slope and original slope plus fill.

5. BACK ANALYSIS

A back analysis was made assuming a rigid block sliding on an average plane foliation surface. (Hoek & Bray, 1974). The tension crack location was calculated assuming that it was developed in dry condition. Fig. 2 shows back analysis of original slope for dry condition, tension crack and foliation surface full of water, and water pressure in the lower half of the tension crack.

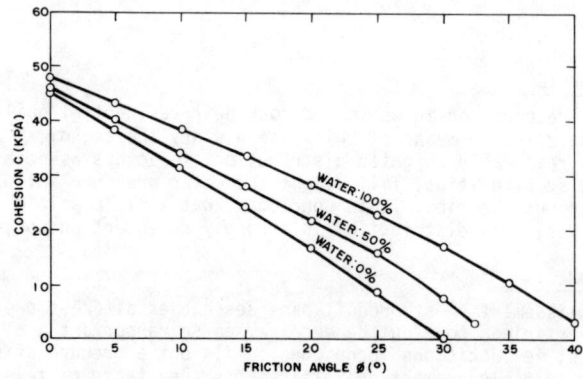

Fig.2. Back analysis results. Original slope.

Results of back analysis of the original slope assuming that tension crack was at the same location that the crack that had developed in the slide, are shown in Fig. 3.

The same type of back analysis was made for the natural slope plus the fill. In this case according to field observations made after the failure, it was assumed sliding along foliation surfaces and a tension crack developed in the fill. Fig. 4 shows the result of this back analysis.

Based on a 20° ultimate friction angle, the variation of the factor of safety with height of water in the tension crack was calculated for different values of cohesion. This analysis was made for both, the original slope and the original slope plus fill. Results are shown in Fig. 5, 6 and 7.

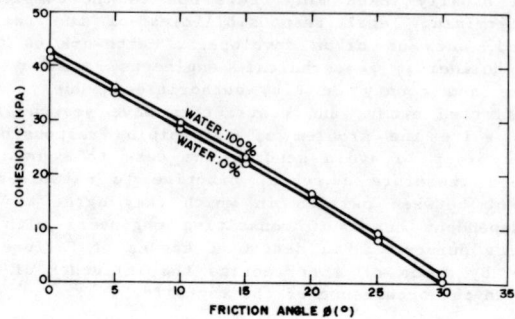

Fig.3. Back analysis results. Original slope.

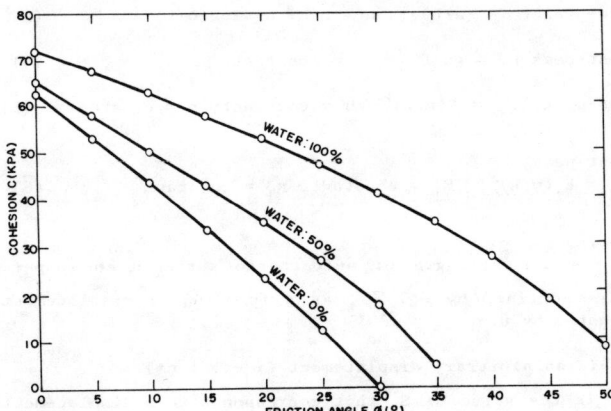

Fig.4. Back analysis results. Original slope plus fill.

Results of back analysis lead us to the following conclusions:

a. Back analysis of original slope, assuming critical tension crack position for a dry slope, shows that for a 20° friction angle the cohesion required at equilibrium varies from 16.5 KPA to 28.5 KPA, depending on the water pressure conditions. These values agree with similar results obtained in other back analysis of slides that have occurred in the same type of rock. Experience in the shear behavior of these materials has indicated that for design purposes, a reasonable value of cohesion along foliation surfaces may be taken as 25 KPA.

b. The relative effect of water pressures in the factor of safety of the original slope, assuming tension crack has developed during dry conditions, can be evaluated from Fig. 5. From this figure it can be seen that the factor of safety has a relative decrease of approximately 23° from dry to full water pressure conditions. For c = 25 KPA it is necessary a reduction of 19% to reach limiting equilibrium which corresponds to a water pressure condition of $z_w/z = 0.86$.

c. Back analysis of original slope (Fig. 3 and 6) assuming tension crack occurring at the same location that the real crack developed in original slope plus fill, shows that for $\phi_u = 20°$ the cohesion required at equilibrium is equal to 15 KPA, below the values of cohesion calculated from back analysis of similar slope failures. The fact that the natural slope did not fail before the fill was placed, indicates that the cohesion along foliation surfaces must have been higher than the calculated from limiting equilibrium. It can also be concluded that under this hypothesis, water pressure would have not significantly influence the stability of the original slope. (See Fig. 6).

d. The effect of water pressures in the factor of safety of the original slope plus fill, can be evaluated from Fig.7. This figure shows that the factor of safety has a relative decrease of

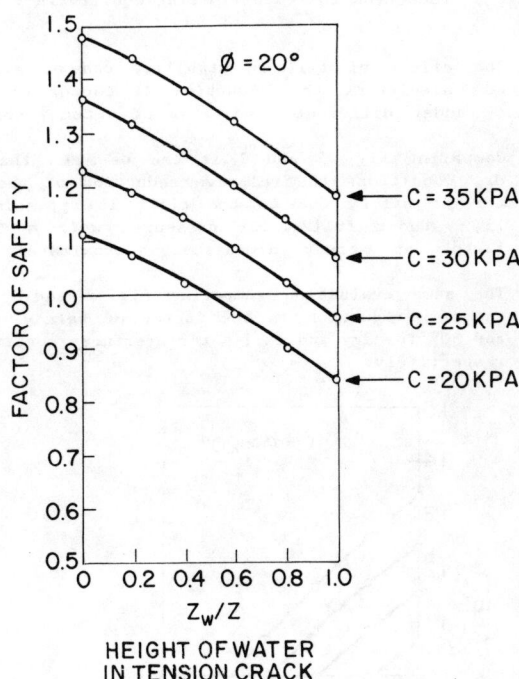

Fig. 5. Influence of water pressures in factor of safety. Original slope. Tension crack developed in dry conditions.

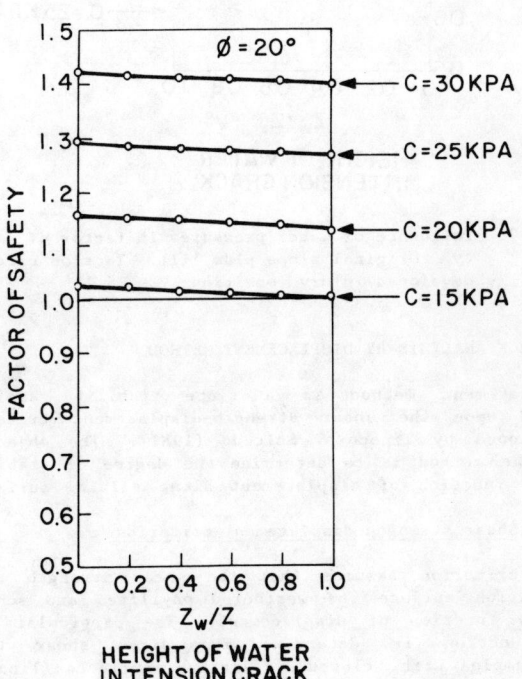

Fig. 6. Influence of water pressures in factor of safety Original slope.

approximately 43% from dry to full water pressures conditions. For c = 25 KPA it was necessary only a 6.5% reduction to reach limiting equilibrium.

e. The effect of fill in stability can be evaluated calculating the reduction of factor of safety under different conditions of water pressures.

Comparing Fig. 5 and 7 it can be seen that for dry conditions the relative reduction of the factor of safety due to the fill, is approximately 15%. Under full water pressures $z_w/z = 1$, the factor of safety diminishes approximately 35%.

The same evaluation comparing Fig. 6 and 7 indicates a reduction in the factor of safety of 18% and 50% for dry and full water pressures conditions, respectively.

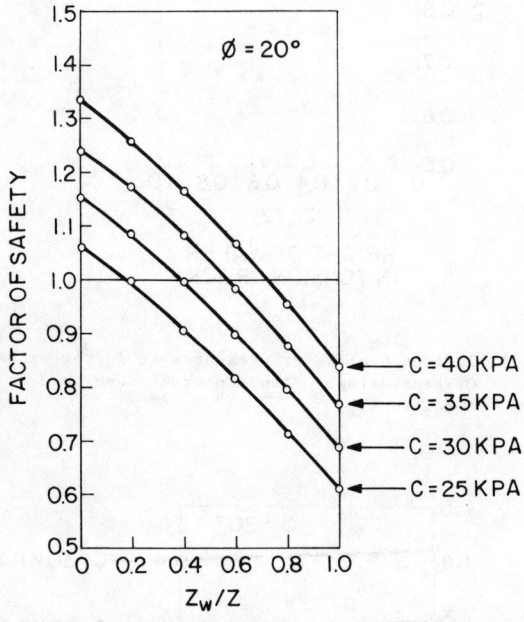

Fig. 7. Influence of water pressures in factor of safety. Original slope plus fill. Tension crack developed in dry conditions.

6. BACK ANALYSIS BY DISPLACEMENT METHOD

Displacement method is a slope stability analysis based upon the shear strength-displacement criterion developed by Tinoco & Salcedo (1981). The objective of the method is to determine the degree of stability as a function of displacement along sliding surfaces.

6.1. Shear strength-displacement criterion.

The criterion assumes that the shear strength along foliation surfaces of weathered phyllites and schists is a function of displacement. The particular type of function is determined from direct shear tests on samples with "closed" foliation surfaces (Tinoco & Salcedo, 1981) and (Tinoco, 1982). The criterion is described by the following equations:

$$S = F_d S_o + F_f \sigma_a \tan \phi_u \qquad (1)$$

$$S_o = A (u/u_o)^m \qquad (2)$$

Pre-peak $F_d = (u/L)^n$; $n = n_1 + n_2 \log(u/L)$ (3)

Post-peak $F_d = (u/L)^n$; $n = $ constant (4)

Pre-peak, $F_f = B(u/L)^p$; B = constant, p = constant (5)

Post-peak,
$F_f = B_1(u/L)^p$; $B_1 = $ constant, $p_1 = $ constant (6)

where:
S_o is the strength of unfractured surfaces and asperities defined by eq. (2) as a function of displacement denoted by u.

u_o is an arbitrary displacement (i.e., 1 cm).

A is the value of S_o that corresponds to a displacement u_o

F_d, defined by equation (4) for pre-peak and eq. (5) for post-peak behavior, is a power function of the strain along the sheared surface.

L is the length of the failure plane or foliation surface.

F_f, defined by eq. (5) for pre-peak and eq. (6) for post-peak behavior, is a function of the strain along the failure surface and it is also a function of the applied effective normal stress.

σ_a' is the effective normal stress.

$\sigma_a' \tan \phi_u$ denotes the ultimate frictional resistance as a result of wear and gouging of the sliding surfaces.

The interpretation of eq. (1) in terms of Mohr-Coulomb parameters is:

$$c = F_d S_o = A(u/u_o)^m (u/L)^n \qquad (7)$$

$$\tan \phi_{mob} = F_f \tan \phi_u = B_1 (u/L)^{p_1} \tan \phi_u \qquad (8)$$

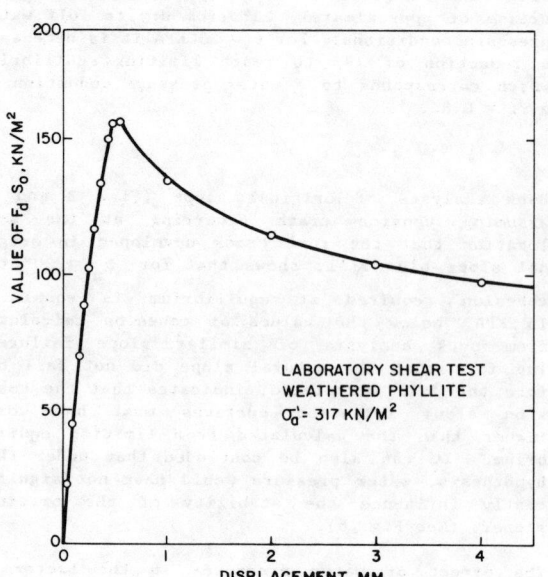

Fig. 8. "Cohesion" (eq. 7) as a function of displacement.

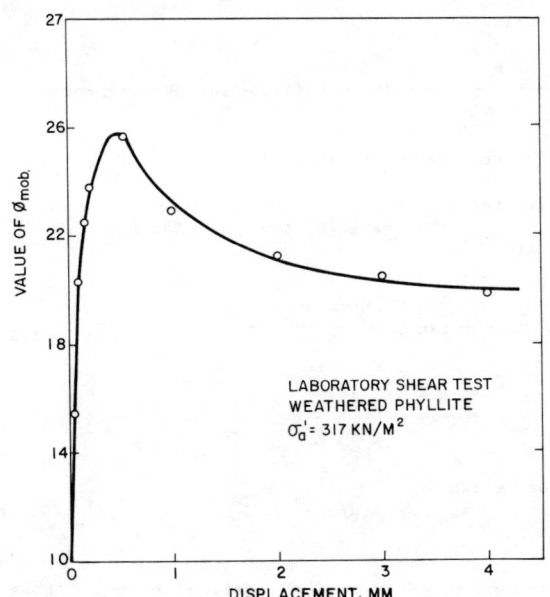

Fig. 9. Mobilization of friction angle with displacement.

The values of c and ϕ_{mob} calculated from eqs. (7) and (8) are shown in Fig. 8 and Fig. 9, respectively. The results shown in those figures were obtained from a direct shear test on "closed" foliation surface of a sample of weathered phyllite. In order to demonstrate the use of equations and values of the parameters used in the back analysis, the calculations for the laboratory sample tested as well as for a field foliation surface (i.e., length = 45 cm) are shown in Tables I and II, respectively.

For this particular case, the field strength of closed foliation surfaces is 79% lower than the laboratory strength for the same amount of displacement.

6.2 Method of analysis.

The objective of the back analysis of the rockslide was to determine the weight of each of the following factors in producing the slope failure: weight of the in situ rock, weight of the added fill and water pressures.

The calculation of the influence of each factor becomes easier if their action are additive and assumed to be independent of each other. The standard calculation of a factor of safety does not meet the aforementioned criteria. The concept of unstable excess proposed by Barton (1972) was used to evaluate the independent influence of each factor at states of stability and at limiting equilibrium.

$$P_n = (W_n + ql_n - U_n)(\tan\phi - \tan\beta) + cl_n \qquad (9)$$

and
$$\Sigma P_n = 0 \qquad (10)$$

Where P_n is the unstable or stable excess; W_n is the weight of the slice; q is the surcharge; U_n is the force exerted by water pressure along length of slice l_n; c is the cohesion; ϕ is the friction angle and β is the angle of inclination of the failure plane.

Eq. (9) allows to separate the influence of the weight of the slice, the pressure due to fill and the water pressure.

The equations used in the back analysis were:

Table I. Shear strength-displacement parameters.

A kPa	ϕ_u deg.	u_o cm	m	n_1	n_2	$n^{(1)}$	$B^{(2)}$	$B_1^{(3)}$	$P^{(4)}$	$P_1^{(5)}$
400	12	1	-0.8937	-0.154	-0.339	0.6557	59.89	1.188	0.504	-0.105

(1) Value of n for peak and post-peak behavior. (2) Value of B for $u/L \leq 1.75 \times 10^{-3}$; (3) Value of B_1 for $u/L > 1.75 \times 10^{-3}$; (4) Value of p for $u/L \leq 1.75 \times 10^{-3}$; (5) Value of p_1 for $u/L > 1.75 \times 10^{-3}$.

Table II. Results of calculations (eqs. 2, 3, 4, 5, 6, 7 and 8)

u cm	l_f cm	n	$F_d \times 10^{-3}$	S_o kPa	F_f	ϕ_{mob} deg.	σ_a' kPa	$F_d S_o$ kPa	S kPa
0.02	12	0.7878	6.48	13195.6	2.383	26.87	317	85.47	246.0
0.2	12	0.6557	68.24	1685.5	1.826	21.21	317	115.00	239.0
0.02	45	0.9824	0.51	13195.6	1.224	14.59	317	6.72	89.2
0.2	45	0.6557	28.69	1685.5	2.098	24.00	317	48.35	189.7
0.8	45	0.6557	71.19	488.3	1.815	21.10	317	34.76	157.1

upper triangle:

$$P_{1-N} = \left[\frac{2 N_s - 1}{N^2}\right] \frac{\gamma H^2}{2} (1 - \cot \alpha \tan \beta)$$

$$\cos \beta \ (\cot \beta - \cot \alpha)(F_f \tan \phi_u - \tan \beta)$$

$$+ \frac{q H}{N} (\cot \beta - \cot \alpha)(F_f \tan \phi_u - \tan \beta)$$

$$- \gamma_w Z_w \frac{H(1 - \cot \alpha \tan \beta)}{N \sin \beta} \left[1 - \frac{1}{2}(1 - \cot \alpha \tan \beta)\frac{N_s}{N}\right]$$

$$\cos \beta \ F_f \tan \phi_u .$$

$$+ \frac{H(1 - \cot \alpha \tan \beta)}{N \sin \beta} A \left(u/u_o\right)^m \left(u/L_f\right)^n \qquad (11a)$$

lower triangle:

$$P_{1-N} = \left[\frac{2 N_s - 1}{N^2}\right] \frac{\gamma H^2}{2} (1 - \cot \alpha \tan \beta) \cot \alpha \cos \beta$$

$$(F_f \tan \phi_u - \tan \beta)$$

$$+ \frac{q H \cot \alpha \tan \beta}{N \sin \beta} \cos \beta \ (F_f \tan \phi u - \tan \beta)$$

$$- \gamma_w Z_w \cot \alpha \tan \beta \left(\frac{H \cot \alpha \tan \beta}{N \sin \beta}\right) \cos \beta \ F_f \tan \phi_u$$

$$\left[\frac{N - (N_s - 1)}{2N}\right]$$

$$+ \frac{H \cot \alpha \tan \beta}{N \sin \beta} A(u/uo)^m \ (u/L_f)^n \qquad (11 \ b)$$

The development of "cohesion", defined by eq. (7) as a function of displacement for the slope, is shown in Fig. 10. The values were calculated on the assumption that foliation surfaces had an average length of 0,45m. The selection of this average length was explained by Tinoco & Salcedo (1981). Fig.11 shows the mobilization of the friction angle ϕ with displacement along the failure plane.

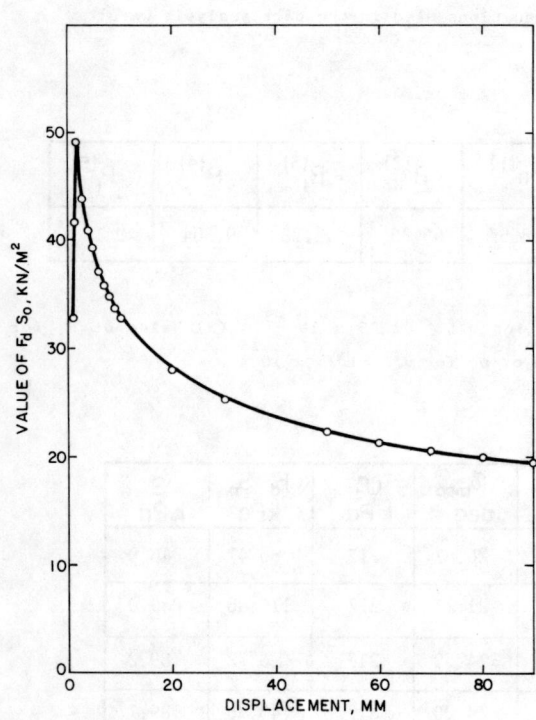

Fig. 10. Developed "cohesion" (eq. 7) with displacement along sliding surface.

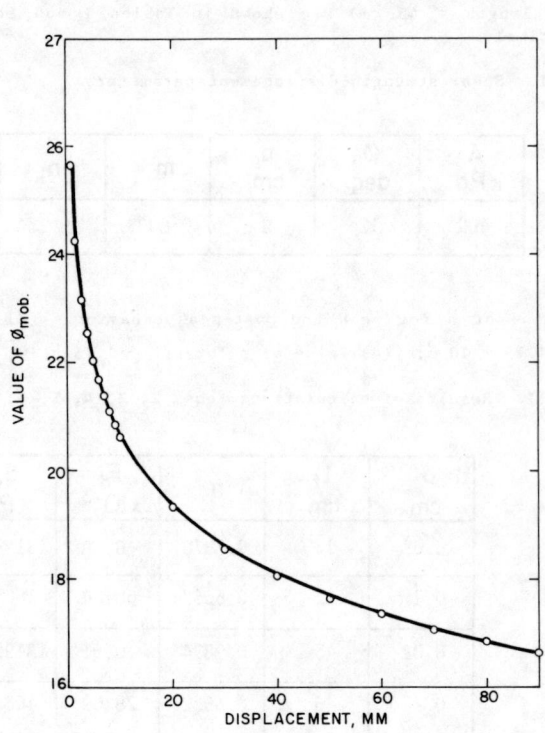

Fig. 11. Mobilized friction angle with displacement along slidiing surface.

6.3 Evaluation of results.

The results of the back analysis with parameters $A = 400 \ KN/M^2$ and $u_o = 1 \ cm$, are shown in Fig. 12. The ordinate is the ratio λ_T of active forces to the sum of "cohesive" forces. The acting forces are due to the weight of the slope above the failure plane, the weight of the fill as a surcharge and different values of water pressures assuming a triangular distribution along the failure plane. That is, maximum water pressure at the top of the plane and zero at the toe of the slope. The ratio λ_T is defined by:

$$\lambda_T = \sum_1^n (S_w + S_f + U) \bigg/ \sum_1^n F_d S_o l_n \qquad (12)$$

and

$$\lambda_f = \sum_1^n S_f \bigg/ \sum_1^n F_d S_o l_n \qquad (13)$$

where:

S_w = shear force due to weight of in situ rock.

S_f = shear force induced by the weight of fill as a surcharge.

U = resultant force of water pressure distribution along failure plane

l_n = length of failure plane for slice n.

$F_d S_o$ = "cohesion" as defined by eq. (7)

Values of water pressures of 0, 10, 20 and 30 KN/M^2 were used to calculate the sum of shear forces. Fig. 12 shows that less displacement is required for failure at the higher values of water pressure. It also shows that the ratio λ_f defined by eq. (13), increases with displacement, and at failure its value is 0.57, indicating the great influence that the weight of fill as a surcharge, had in producing the rockslide.

The extreme case of maximum possible water pressure value of 100 KN/M^2 was also studied. Limiting equilibrium was reached at a displacement of little more than 3 mm and in this case, the weight of this factor is 0,375 of the total "cohesive" force, and it is also equal to the weight of the factor due to fill surcharge.

The behavior of the slope without the weight of fill was studied by the same procedure and the results are shown in Fig. 13, where the ratio λ'_T is defined as:

$$\lambda'_T = \sum_1^n (S_w + U) \bigg/ \sum_1^n F_d S_o l_n \qquad (14)$$

Values of water pressure of 0, 2, and 10 KN/M^2 were used to calculate λ'_T. Fig. 13 shows that a large amount of displacement (90 mm) is required to reach limiting equilibrium. It may also be noticed that when failure is approached, a small increase in water pressure may accelerate the process although its weight as percent of total "cohesive" force is small for the water pressure values shown in Fig.13.

The back analysis of the slope was also made with a

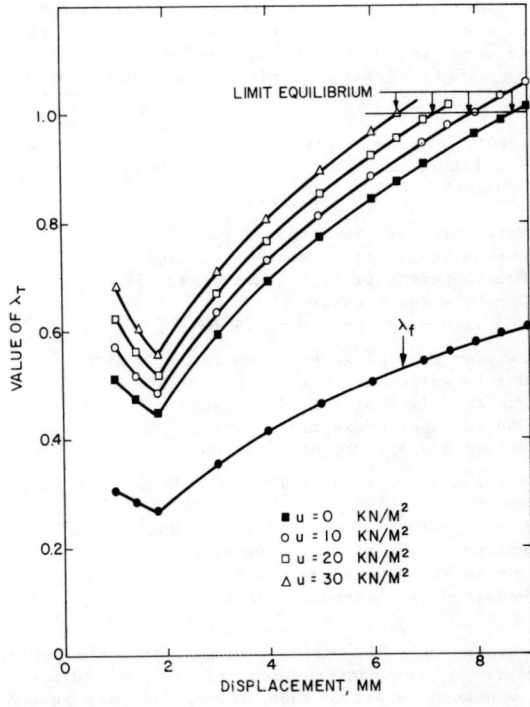

Fig. 12 Relationship of λ_T and λ_f with displacement.

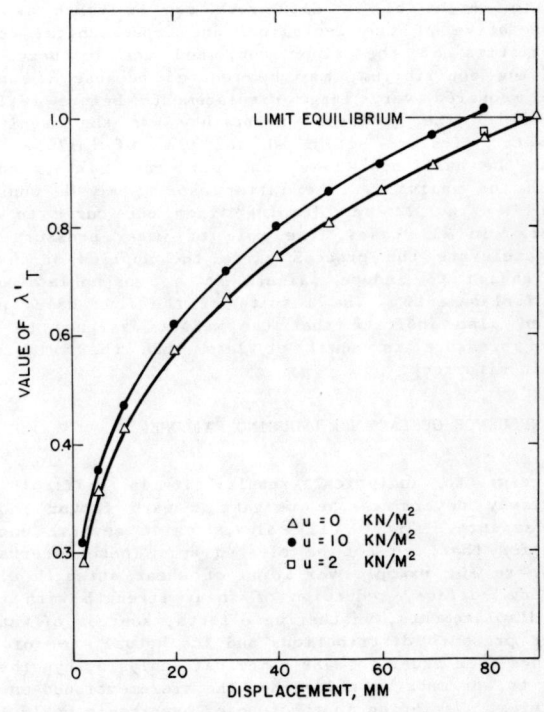

Fig. 13. Relationship of λ'_T with displacement.

higher value of A, 1.5 times the value used in the previous calculations, equal to 600 KN/M^2. That is, the resistance of the foliation surface to displacement is now larger. The calculations showed that, at limiting equilibrium, the weight that must be assigned to each factor studied is approximately the same. The only difference between the two strength calculations is reflected in the amount of displacement required to obtain limiting equilibrium condition.

The result of the analysis indicate that the weights to be assigned to each factor at limiting equilibrium is as follows:

a. Water pressure less than 30 KN/M^2
    Due to weight of in situ rock: 36%
    Due to weight of fill (surcharge): 54%
    Due to water pressure: 10%
    Required displacement: 6.5 mm

b. Extreme case of water pressure: 100 KN/M^2
    Due to weight of in situ rock: 25%
    Due to weight of fill (surcharge): 37.5%
    Due to water pressure: 37.5%
    Required displacement: 3.2 mm.

d. Water pressure of 100 KN/M^2, higher strength parameter
    Due to weight of in situ rock: 29%
    Due to weight of fill (surcharge): 42%
    Due to water pressure: 29%
    Required displacement: 10 mm

Considering that the slide occurred approximately seven years after construction of the buildings, and that movements started soon after, one may reasonably consider that the extreme case of water pressure assumed (100 KN/M^2) did not occur. If that is an admisible reasoning, according to cases a and c the weight of the fill should carry the greater percentage of the three aforementioned factors. The factor due to weight of in situ rock can be taken as representative of the geological and topographical characteristics of the slope cut, and its influence at limiting equilibrium, may be reduced because it would have required very large displacements before failure occurred. The greatest uncertainty was the magnitude of water pressure acting at the time of failure, and also, the particular time that its action is introduced in the analysis. The latter aspect, may be considered in Fig. 12 by "jumping" from one curve to another. In all cases, the role of water pressure was to accelerate the process since the applied surcharge was enough to induce failure at a reasonable amount of displacement. The results of the four cases presented also indicate that the weight assigned to the water pressure is equal or less than the weight of the in situ rock.

7. INFLUENCE OF FACTORS INDUCING FAILURE

According to analytical results it is difficult to precisely determine the weight of each factor inducing failure. There will always exist several uncertainties that can not be cleared up. These uncertainties are for example variation of shear strength along sliding surface, reduction of shear strength with time and displacements, weathering effects, sources of water, water pressure distributions and its actual time of influence. In order to make numerical analysis, hypotheses have to be made considering the aforementioned uncertainties. Variation in the type of hypothesis would lead to different results.

Based on the above considerations it is evident that any estimation of the weight of each factor inducing failure, must necessarily be partially technical and partially subjective, both strongly dependent on engineering judgment.

Subjective analysis based on analytical results lead us to classify the main factors producing failure in three groups: A) Geological and original topographical conditions. B) Effect of fill and C) Effect of water.

As an initial hypothesis, in absence of any data, it is reasonable to assume that each group has the same influence, in other words, approximately 33%.

Results showed that original slope could be considered stable on dry conditions and that very severe water pressures and large displacements would have been necessary to reach limiting equilibrium. This fact suggests that its percentage value should be much less than 33%. Considering the unfavorable orientation of foliation surfaces, low shear strength parameters of the rock and its degree of weathering, it was decided that a 20% influence was a reasonable percentage ·for the geological and original conditions.

The effect of the fill construction on the overall stability has been considered as the most important factor (60%) based on the following facts:

a. The effect of fill as an active surcharge reduced the factor of safety close to limiting equilibrium, even at dry conditions.

b. Fill settlements originated breakage of pipes resting on it, allowing water infiltration.

c. Based on the hypothesis that shear strength on foliation surfaces decreases with displacement (Tinoco and Salcedo, 1981), the fill as a surcharge has contributed to those displacements and to the loss of shear strength along foliation surfaces.

The effect of water in overall stability can be summarized as follows:

a. Increase of fill weight due to saturation.

b. Water pressures along foliation surfaces and in tension crack developed in the fill.

c. Decrease in shear strength along foliation surfaces.

Consequently, water was considered to have a weighted influence of 20% in the occurrence of failure. It is interesting to note that even though water pressures could be considered as the factor that triggered the slide, it was not selected as the main factor influencing stability.

Based on the above discussion and on the analysis of existing information and field observations, the following percentage values were selected:

1. Geological and original topographical conditions: 20%.

2. Effect of fill: 60%, distributed as follows:
    a. Surcharge due to fill constructions: 45%
    b. Fill deformations due to insufficient compaction and progressive saturation: 15%

3. Influence of water: 20%, distributed as follows:

    a. Rainwater infiltration due to lack of an adequate
       drainage system: 5%
    b. Water infiltration from broken pipes: 2%
    c. Water infiltration from excess of irrigation: 1%
    d. Water infiltration coming from a completely bro-
       ken sewer located at the crest of the slide: 12%.

The interrelationship between some of the aforementioned factors is clearly evident, however, for a numerical estimation it was necessary to assume that
they were independent of each other. An example of
this relationship is the following: fill settlements
make possible breakage of pipes and sewers and, as a
result, water leakages increase the degree of saturation and fill settlements.

8. CONCLUSIONS

It is possible to arrive at a reasonable weighted
distribution of factors associated to the occurrence
of slides. However, this distribution may vary according to the hypotheses that have to be made by engineers. Therefore, results are strongly dependent
on subjective engineering judgment.

Due to the fact that results will be inmediately used
to determine responsibilities, as stated in the a-
greement between parties involved, further research
is needed to develop techniques and procedures, o-
riented to define the weighted distribution of factors associated to failures. These procedures should
be directed to minimize the subjective judgment.

ACKNOWLEDGEMENTS

The writers gratefully acknowledge INGEOTEC and SUE-
LOPROYECTO for financial assistance of some phases
of this research. Particular gratitude to M. Salcedo
G. for preparing the manuscript and to V. Callejas
for drafting the figures.

REFERENCES

Barton, N. (1972). Progressive failure of excavated
rock slopes: Proc. 13th Symp. on Rock Mechanics, pp.
139-170.

Hoek, E., and Bray, J. (1974). Rock Slope Engineering. London. Institution of Mining and Metallurgy.

Tinoco, F. (1982). Teoría de resistencia-desplaza-
miento y la estabilidad de taludes: I Seminario Sur-
americano de Mecánica de Rocas. Bogotá, Colombia.

Tinoco, F., and Salcedo, D. (1981). Analysis of
slopes failures in weathered phyllite: Proc. International Symp. on weak rock, pp. 55-62, Tokyo, Japan.

G

EXPERIENCE IN ENGINEERING EVALUATION OF ROCK SLOPE STABILITY IN THE USSR

Expérience en matière d'évaluation de la stabilité des talus rocheux en URSS

Erfahrungen in der Einschätzung der Felsböschungsstandsicherheit in der UdSSR

E. G. Gaziev, Dr.Sc., G. L. Fisenko, Dr.Sc., S. S. Grigorian, Dr.Sc., J. M. Kazikaev, Dr.Sc.,
V. I. Rechitski, Dr.Sc., V. G. Zoteev, Dr.Sc.
Commission on Stability and Strengthening of Rock Masses,
Soviet Committee for Rock Mechanics

SYNOPSIS

The report drawn up by a group of Soviet engineers generalizes experience in the analysis and strengthening of rock slopes in the USSR and the observed forms of failures. The report gives a descriptive account of approaches to stability calculation, criteria, different procedures of stability analysis including calculation of rock failure-flows, mathematical statistics methods for the analysis of jointing, and the method of the probabilistic evaluation of the stability of rock masses.

RESUME

Ce rapport, rédigé par une équipe d'ingénieurs soviétiques, présente une somme d'expérience en matière d'analyse et de consolidation des talus rocheux en Union Soviétique, et énumère les formes de rupture observées. On donne un compte rendu des façons d'envisager les calculs de stabilité, les critères qui y entrent, les différents procédés d'analyse de la stabilité y compris le calcul de rupture de roches, la méthode statistique pour l'analyse des fissures et l'évaluation de la stabilité des massifs rocheux.

ZUSAMMENFASSUNG

Erfahrungen in der Felsböschungssicherung in der UdSSR, im rechnerischen Nachweis der Böschungsstandsicherheit und beobachtete Formen von Böschungsbrüchen werden beschrieben und verallgemeinert. Grundsätze des rechnerischen Nachweises der Böschungsstandsicherheit, verschiedene Berechnungsverfahren (einschließlich Berechnung von fließartigem Felsbruch), Methoden der statistischen Analyse der Klüftigkeit und der Wahrscheinlichkeitseinschätzung der Standsicherheit von Gebirge werden erläutert und zur Diskussion gestellt.

1. GENERAL

1.1 Basic approaches to stability analysis

Engineering evaluation of stability of rock slopes is realized in several stages:
a) Calculation of a stress-strained state of the rock mass and identification of potential failure surfaces;
b) Stability analysis of definite rock masses by the methods of limit equilibrium;
c) Evaluation of stability and permissible parameters of slopes through observed deformations during excavation and operation.

Development of numerical methods of computer-aided calculation at the present time has opened up strong possibilities for analysis of a heterogeneous anisotropic jointy media, e.g. rock masses. Finite elements method (FEM) and development of kinematic stability programs (Fairhurst, 1977) provide strong possibilities for the given analysis. Undoubtedly the most complicated and responsible problems of stability of rock masses shall be solved on the basis of a detail analysis of their stress-strained state.

However there are some considerations limiting a wide application of these methods, i.e.:
a) The analysis of a stress-strained state requires much information on deformability and strength of rock blocks and joints separating them in a complicated stressed state which takes into account non-linearity of deformation characteristics, dilatancy and self-strengthening of joints in case of displacements;
b) When identifying the process of formation of the slip surface in a rock mass not only the values of natural initial stresses in the rock mass are of high and sometimes decisive importance but history of their formation is very much important as well. Variation of these factors may bring about appreciably different final outcomes and result in significantly erroneous evaluations of rock mass stability;
c) Limited possibilities of the present computer facilities makes it difficult to use these methods widely.

Moreover, the experience demonstrates that frequently it appears possible to locate the potential slip surface on the schematical structural model of the rock mass without any analysis of its stress-strained state.

Stability calculation of rock masses with the located potential slip surface is made by the methods of limit equilibrium taking into account the following conditions:

- The sliding rock masses are not absolutely rigid bodies and consist of rock blocks or slices interacting with each other during displacement. This just interaction that determines the process and mechanism of displacement.

- Establishment of a state of limit equilibrium in any section of the potential slip surface does not mean the loss of stability of the rock mass which is determined by interaction between these unstable blocks or slices with the below located stable sections of the rock mass.

- Analysis of stability of rock slopes consists in determination of stability deficit of separate rock slices and the rock mass as a whole.

- The shear strength of a rock joint or a weakened zone is a curvilinear function (Gaziev, 1977) which for the purpose of simplification of mathematical calculations is approximated at the given interval of normal stresses by the linear (Coulomb) expression.

- Tensile strength of a rock mass in joints as a rule is equal to zero.

- Stability analysis of rock slopes should not consist in determination of the absolute value of stability criterium because nature is always more complicated and varied than inevitably simplified models that may be considered in analytical calculations. The probabilistic method of stability analysis only makes it possible to evaluate reliability of the obtained solution taking into account the level of validity of initial information introduced into the calculation.

1.2 Stability criteria

Safety factor

$$K = \frac{B}{A} \geqslant 1 \qquad (1)$$

is used most extensively in engineering experience for evaluation of the permissible value of loads and strength, where: A - the resultant of all acting forces or acting stress in a critical region, and B - the resultant of shear forces or mobilized strength of the material in a critical region.

However, the use of this parameter in rock mechanics is not always convenient and correct (Rocha, 1978; Habib, 1979).

Safety or stability factor is valid as applied only to a definite calculation scheme and a definite effect which may disturb the rock mass from an equilibrium condition. Taken apart the idea of safety factor is of no sense.

Safety factor may be used only for comparison

of different solutions obtained for one and the same rock mass.

The more convenient criterion is to determine the very value of stability reserve (or strength reserve) or the reversed sign value of stability deficit (or strength deficit):

$$S = A - B \leqslant 0 \qquad (2)$$

Prababilistic character of strength and deformation properties of rock masses, their geometric parameters and acting loads define expediency of use of the probabilistic criterion of safety which is the concept of reliability already widely used in engineering practice.

1.3 Models

The basis of construction of any model is a geological model of the rock mass showing its structure, composition, and allowing to forecast failures, rock-slides, processes of unloading, weathering and similar natural phenomena.

The complex of specialized models: strength, deformation, structural, filtration is usually called a geological engineering model of the rock mass.

A design (geomechanical) model is constructed on the basis of the geological model for the solution of the definite problem by definite methods. The difference between the design model and the geological one is that apart from geological factors the design model should take into account the mechanism of the process it is designed for and the used methods of analysis as well.

The design model includes a structural model of the rock mass, a hypothesis of a possible character of displacement or deformation and the necessary strength and deformation parameters of the rock mass in conformity with the adopted hypothesis and the method of analysis.

1.4 The most widespread forms of failure, methods of stability analysis and methods of strengthening of rock slopes

The Commission on the stability and strengthening of rock masses of the Soviet Committee for Rock Mechanics questioned a number of designing, research and scientific institutes and construction organizations in the Soviet Union dealing with construction of quarries and pits, and performing the analyses of stability of rock slopes. The results of this questioning are shown in Table 1.

Fig. 1 represents examples of possible forms of failures of different rock slopes. In engineering experience in the USSR the most widespread failures were of type E, H, A, and D, i.e. sliding along one joint (E) or along a system of joints (A) or along two systems of joints across the edge of intersection (D) or along it (H). Large rock-slides of F type also occur.

All questioned organizations believe that the structure of the rock mass is a determining factor for analysis of its stability and one of the most important parameters is orientation of joints and shear strength along them.

Table I.

Questioned organizations	Structural particularities	Joint orientation	Shear strength of joints	Modulus of deformation	Natural stressed state	Wedge analysis	Algebraic summation	Prof. G.L.Fisenko	Stability deficit	Prof. A.L.Mozhevitinov	Finite Elements Method	Other methods	A	B	C	D	E	F	G	H	I	Drainage	Excavation	Anchors	Steel grid	Spray concrete	Buttresses	Retaining walls	Filtration flow	Earthquakes or blasts	External structures	Necessity of probabilistic approach
VNIITSVETMET (Ust-Kamenogorsk)	*		*				*	*												*					*	*	*	*				
VNIMI Urals Branch (Sverdlovsk)	*	*	*			*	*								*					*										*		*
HYDROPROJECT (Moscow)	*	*	*						*	*		*	*		*	*	*	*		*		*	*	*						*	*	*
LENHYDROPROJECT (Leningrad)	*			*		*			*	*	*	*	*		*	*	*		*	*		*	*	*			*			*	*	*
GIPRORUDA (Leningrad)	*	*	*			*		*							*					*			*		*		*	*		*		*
Mining Institute (Sverdlovsk)	*	*	*	*	*	*		*					*		*	*				*		*	*	*	*					*		*
Polytechn.Institute (Karaganda)	*	*	*			*	*	*					*		*					*		*									*	*
KRIVBASSPROJECT (Krivoy Rog)	*														*					*		*					*	*				*
Mining Institute (Moscow)	*							*						*						*	*	*		*			*		*		*	*
UNIPROMED (Sverdlovsk)	*	*	*			*	*	*							*		*		*	*	*	*	*	*			*	*	*	*		*
YOUZHGIPRORUDA (Kharkiv)	*	*	*										*	*				*	*				*				*	*				

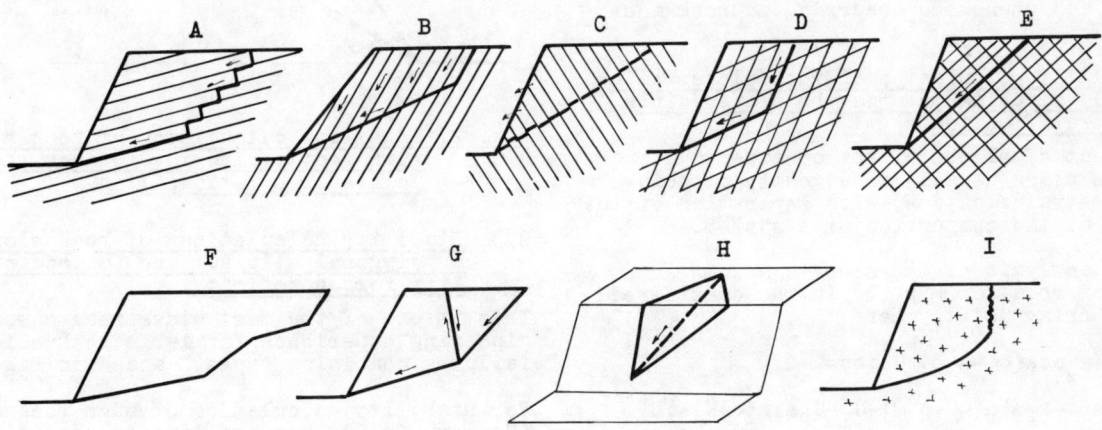

Fig. 1. Possible forms of rock slopes failure

The most widespread strengthening facilities are drainage, buttresses and anchors.

The overwhelming majority of questioned organizations believes it necessary to make probabilistic calculations of stability of rock masses for evaluation of their reliability.

2. STABILITY CALCULATIONS

2.1 Calculation of a stress-strained state of rock masses by the finite elements method

These calculations are used widely for the elastic or non-linear problems (Efimov et al.,1970; Rozin, 1971; Zolotarev, 1973; Amusin, Fadeev, 1975). The program has been developed which takes into account dilatancy and selfstrengthening of rock joints in shear, and non-linear character of deformation properties of the rock and joints with variation of normal stresses (Gaziev, Arkhippova, 1976; Arkhippova et al., 1976).

For representation and consideration of a complicated stressed state of rocks and structures resulted from a combined effect of structural-tectonic non-uniformity of rock masses, hydrodynamic and tectonic fields of stresses and other factors, the procedure of assigning the boundary conditions based on iterative principles has been worked out (Kazikaev et al.,1979). The program taking into account a block structure of the rock mass and its initial stressed state has been developed which provides possibilities to determine parameters of the unloading zones during driving (Zoteev, Nozhin, 1979).

For the solution of axial symmetric and three-dimensional elastic problems the programs were developed at the Leningrad All-Union Research and Scientific Institute of Mining Geomechamics and Survey of the USSR Coal Ministry (VNIMI) (Amusin, Fadeev, 1975) and in the Institute of Applied Mathematics, the KazakhSSR Academy of Sciences (Erzhanov et al., 1975).

For a short period of time the finite elements method has found wide use among specialists engaged in rock mechanics and favoured development of a qualitatively new stage of studies of processes and phenomena occurring in rock masses.

2.2 Stability calculation of rock slopes with gently pitching towards the slope joints or stratification

Displacement of slopes of the considered structure takes place usually in a gently pitching joint (or stratification) with separation of the rock mass at the conjugate joint system.

Stability analysis of the slope consists in checking of relationship (2)$_*$in the considered gently pitching joint, where[*]:

$$A=G\left[\sin\alpha +k_s\cos(\alpha - \delta)\right]+U_2\cos\alpha -Q$$
$$B=\mu\left\{G\left[\cos\alpha -k_s\sin(\alpha - \delta)\right]-U_1-U_2\sin\alpha +N\right\}+cL\Bigg\}(3)$$

---

[*] - All letter parameters are explained in Section 6 under the title "Notation".

Stability of the slope may be considered ensured if at all possible values of L the condition (2) is satisfied.

For rock slopes of a simple form (benches of quarries and pits) which are exposed only to dead weight it appears possible to determine the depth of a tension crack (Fig. 2):

$$h_*=c\left[\gamma\cos^2\alpha (tg\alpha -\mu)\right]^{-1} \qquad (4)$$

its distance from the edge of the bench:

$$b=h_*\sqrt{ctg\alpha -ctg\beta} , \qquad (5)$$

and the maximum permissible height of the bench:

$$H_*=h_*(ctg\alpha +\sqrt{ctg\alpha\cdot ctg\beta })/(ctg\alpha -ctg\beta). \qquad (6)$$

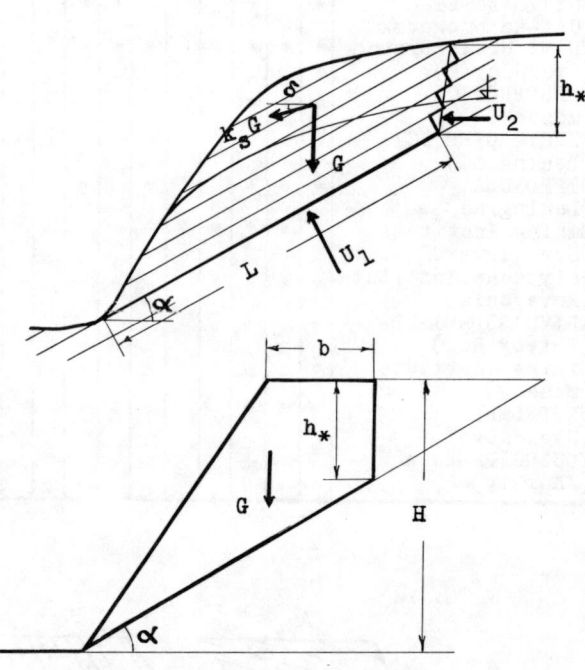

Fig. 2. Stability calculation of rock slopes with gently-pitching towards the slope joints or stratification

2.3 Stability calculations of rock slopes at a polygonal slip surface (method of stability forces deficit)

This is one of the most widespread cases in engineering experience for large rock slides and failures combining types D and F in Fig. 1.

For stability calculation of such rock masses (Gaziev, 1979) they are divided conditionally by vertical planes into slices located at appropriate sections of a polygonal slip surface (Fig. 3). For each slice (beginning from the

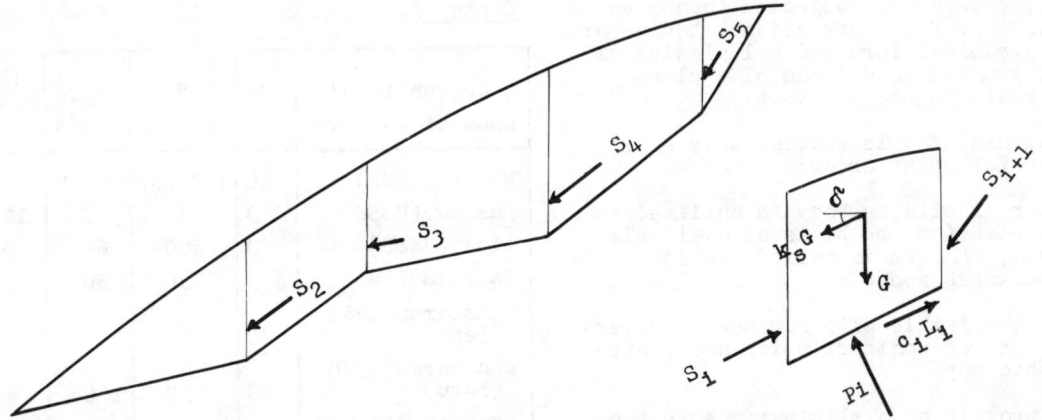

Fig. 3. Stability calculation of rock slopes at a polygonal slip surface (method of stability forces deficit)

upper one) stability deficits S_i are calculated which represent inner forces of interaction between slices (at separation planes limit equilibrium is not necessary):

$$S_i = G_i \left\{ \left[\sin\alpha_i + k_s \cos(\alpha_i - \delta) \right] - \right.$$
$$- \mu \left[\cos\alpha_i - k_s \sin(\alpha_i - \delta) \right] \right\} +$$
$$+ S_{i+1} \left[\cos(\alpha_{i+1} - \alpha_i) - \mu \sin(\alpha_{i+1} - \alpha_i) \right] -$$
$$- \mu \left[(U_{21} - U_{31}) \sin\alpha_i - U_{1i} + N_i \right] + Q_i - c_i L_i , \qquad (7)$$

where i - numbering of slices in upward direction, S_{i+1} - stability deficit of an overlying slice transferred to the considered slice.

Direction of forces S_i is supposed to be coincided with a dip direction of the appropriate slip plane as it is shown in Fig. 3.

If the value S_i calculated by (7) is negative, then the value $S_i = 0$ has to be introduced into calculation of the below located blocks.

Stability criterion is the condition (2) written for lowest slice, i.e. $S_1 \leqslant 0$.

2.4 Stability calculation of rock slopes at a polygonal slip surface and presence of cross-cut joints (method of wedge analysis)

A traditional method of wedge analysis is used under the assumption that limit equilibrium is reached in all dividing planes of joints (Shahuniants, 1941; Fisenko, 1965; Mgalobelov, 1970; Freiberg and Kaufman, 1981).

2.5 Stability calculation of rock slopes in the absence of well-defined systems of joints

Under the guidance of Dr. G.L. Fisenko the complex of methods for stability calculation of slopes of benches and dumps of opencast collieries and openpits was developed at VNIMI Institute in Leningrad (Fisenko, 1956, 1965; VNIMI, 1972; Fisenko et al., 1974).

Basic propositions on which the methods of calculation are developed are the following:

- in the absence of unfavourably located planes of weakness in the slope the slip surface is monotonic and close to circular in form;

- in the presence of unfavourably orientated weakness planes the slip surface coincides completely or partially with them.

The form and location of the slip surface in a non-weakened rock mass adjoining the slope is determined by the main propositions of the theory of limit equilibrium of a loose medium.

The value of cohesion in the rock mass is determined by:

$$c = c_0 / \left[1 + a \cdot \ln(H/b) \right] \qquad (8)$$

where c_0 - cohesion determined on rock samples; H - total height of the slope; b - average linear size of blocks separated by adjoining fractures; a - coefficient depending on strength of rocks in a block, degree and character of jointing (may vary from a = 0.5 for sandy-clay formations to a = 10 for hard igneous rocks with a developed cross-cut jointing).

2.6 Stability calculations of three-dimensional blocks

The most widespread case of failure of three-dimensional rock blocks is a slip along a dihedral angle formed by two intersecting joints (type H in Fig. 1). In this case calculation is effected by analytical (Kovari & Fritz, 1976; Komarov, 1979) or by graphical (Gaziev, 1973, 1977) methods. In the presence of a great number of planes dividing the block and sequence of interacting of three-dimensional blocks a graphical method is more suitable for analysis because it allows to consider both cohesion forces in slip planes and interaction of blocks in sequence.

2.7 Stability calculation of large rock slides

In mountainous regions large rock slides occur frequently which may come into motion when they

are under cut, watered or subjected to any engineering activity. Such rock sliding bodies are
usually of a pyramidal form and calculation of
their stability under conditions of a plane
problem may result in heavy errors.

Stability analysis of this sliding body shall
be made in the following order:

a) A possible rock sliding body is outlined in
a topographic plan (on the basis of available
geological data, changes in relief, joints, outcropping of weakness zones);

b) A shape of a possible slip surface is determined in the form of contours which are plotted
on a topographic map;

c) A typical profile of a slip surface in the
direction of the most probable movement of the
rock slide is drawn up, curvilinear slip surface is approximated by polygonal one divided
into a series of sections with different dip
angles;

d) The volume of the rock mass located at the
appropriate sections of the slip surface is calculated for which a topographic plan with contours of the day surface and the slip surface
is used;

e) Stability calculation of the rock sliding
body by the method of stability force deficit
described in item 2.3 is carried out.

2.8 Calculation of rock failure-falls

A number of large-scale failures of rocks are
known for which a strong "anomalous" mobility
and a large distance of the cone of the failured mass (Table 2) are typical. For explanation
of the mechanism of these rock slides Dr. S.S.
Grigorian (1979) proposed a new law of dry friction which lies in the fact that the value of
tangential stresses τ acting on the contact
of the crushed rock flow with the foundation is
limited by shear strength of a weaker contacted
material (τ_*) and at $\tau < \tau_*$ the Coulomb law
functions, i.e.

$$\tau = \begin{cases} \mu\sigma, & \mu\sigma \leqslant \tau_*, \\ \tau_*, & \mu\sigma > \tau_*, \end{cases} \qquad (9)$$
$$\mu, \ \tau_* = \text{const.}$$

For qualitative description of such failures a
corresponding design mathematical model was proposed. According to this model movement of the
rock flow in a rough one-dimensional "hydraulic"
approximation may be described by differential
equations:

$$\frac{\partial v}{\partial t} + v\frac{\partial v}{\partial s} = g \sin\alpha - (\frac{\tau}{\rho h} + \frac{k_v v^2}{2h})\text{sign } v - g\frac{\partial h}{\partial s},$$
$$\frac{\partial h}{\partial t} + \frac{\partial (vh)}{\partial s} = 0, \qquad (10)$$
$$\sigma = \rho \, gh \cos\alpha.$$

under appropriate initial and boundary conditions.

Table 2.

Conventional name of failure	s km	V mln.m^3	v_{max} m/s	τ_{*1}	τ_{*2}
				MPa	
Khait (USSR)	10	300–500	25–30	20–30	3
Raslak (USSR)	3	6	20	15–30	2,5–3
Aini (USSR)	3	100	60	30	7,5
Baipaza (USSR)	1,5	30	50	15	15
Huascaran 1962 (Peru)	20	13	100	–	10
Huascaran 1970 (Peru)	20	50	180	–	24
Sherman-Glasher (USA)	3–5	23	80	–	1
Black Hauk (USA)	8	320	70		7,5
Tsiolkovsky (Moon)	100	–	–	–	0,1–1

Back analyses for the given model showed that
strength τ_* assumes two quite different values: one of them τ_{*1} characterizes movement
along an initial steep slope and in this case
the failured material is the material of the
rock flow; the second value (τ_{*2}) characterizes movement along a gentle path of the flow
(valley, canyon) covered with deposits and in
this case the failured material is the underlying material (deposits). We can say that
$\tau_{*1} \gg \tau_{*2}$.

Data presented in Table 2 demonstrate that inspite of roughness of the design model, inaccuracy of initial data and a wide range of variation of different-scale phenomena the values
τ_{*1} and τ_{*2} are sufficiently stable which
undoubtedly confirms applicability of the law
(9) for qualitative description of large-scale
failures. The exceptions are the values of τ_{*2}
for failures of Sherman-Glasher and near the
Tsiolkovsky crater on the Moon. This is explained by the difference of the material of the
underlying surface; in the first case it was ice
(the fallen rock moved on the surface of the
glacier, and in the second case it was a soft
moon regolith.

Utilization of the proposed law of friction and
the mathematical model of type (10) may appear
useful for prediction of possible spontaneous
failures in potentially dangerous particular regions and for evaluation of mobility of the flow
of the rock mass fallen down by explosions during construction of dams in mountainous gorges.

The proposed law of friction is also applicable
when the flow consists of some other materials:
the flow of ice fragments, snow slide, ash flow
in volcanoes, etc. In these cases the values τ_*
will be considerably lower than the shown above.
The proposed method of calculation is applicable
for a quantitative description of failures on
planets and on the Moon; this application allows
to propose a new interpretation of nature and
origin of a number of morphologic features of
the surface of these celestial bodies (canyons

and "lava" streams on the Mars, rock slides and failures, etc.). As an example of the described application the last line of Table 2 shows the results of calculation of τ_{*2} for a gigantic rock slide near the Tsiolkovsky crater on the reverse side of the Moon (Grigorian, 1979).

On the basis of the proposed law of friciton methods of physical simulation of large-scale failures have been developed.

2.9 Other employed methods

Apart from the above mentioned methods a number of some other methods and procedures and most frequently stability calculation on the one plane of displacement (Type E, Fig. 1) are employed in the USSR. Among them there are the methods developed by Prof. A.L. Mozhevitinov (Mozhevitinov, Shintemirov, 1970), methods developed by V.G. Zoteev (1982), R.P. Okatov (Popov, Okatov, 1980).

All of them are characterized by different adopted hypotheses about interaction between separate slices (direction of forces, points of their application, their values, etc.). Besides there are methods which take into account more complicated mechanisms of stability disturbance of the rock slope (Gaziev, Rechitsky, 1974; Freiberg, Kaufman, 1981; Kazikaev, 1981).

3. PROBABILISTIC STABILITY EVALUATION OF ROCK MASSES

3.1 Genral

In engineering experience we always operate with the systems in which something is unknown to us. This idea becomes evident especially during designing and building of structures on natural foundations where the main parameters are characterized by uncertainty and are azimuths and angles of dip of joints, parameters of shear strength in joints and force actions resulted from earthquakes and floods. In these conditions it appears natural to employ probabilistic methods of evaluation and calculation.

Probabilistic analysis is presupposed to take into consideration not only average values of parameters but their dispersions that allow to obtain the final result of calculations (e.g. stability deficit) in the form of a random variable with a definite dispersion. This makes it possible to determine probability and evaluate reliability of the obtained result.

For simplification of probabilistic analysis the important condition is that the law of distribution of both angles of dip of joints and parameters of shear strength in joints may be taken as normal (Vistelius, 1958; McMahon, 1971, 1974; Maranhão, 1974; Mirtshulava, 1974; Rocha, 1974); in this case all these parameters with the exception of shear strength parameters in the same slip surface may be considered as independent.

3.2 Method of statistic analysis of parameters of joint orientation

For evaluation of reliability of initial information about jointing of the rock mass a probabilistic method of determination of joint system parameters was developed (Gaziev & Tiden, 1979) which allows to reveal systems of joints and to determine average values and dispersions of their azimuths and angles of dip which are necessary for probabilistic evaluation of reliability. The determination of the existence of the sets of joints is performed by χ^2-criterion. Then assuming the normal law of distribution for azimuths and angles of dip in each set of joints it is possible to determine the average values and dispersions of joint azimuths and dips. The corresponding computer program was developed at the Hydroproject Institute (Moscow).

3.3 Method of stability evaluation

The method of probabilistic analysis of reliability as applied to evaluation of stability of rock slopes was developed at the Hydroproject Institute (Gaziev & Rechitsky, 1979).

Assuming the condition (2) as criterion of stability it may be noted that functions A and B in a general case may be functions of all random variables which are angles of dip of failure surfaces and shear strength in them.

Determination of numerical characteristics of values A and B may be effected by any available method of stability calculation with the use of the linearization method which as calculations demonstrate introduces errors not exceeding some per cent.

If we apply the linearization method to functions A and B, we'll obtain function S linear as well. Taking into account that the law of distribution of linear functions of random variables each of which is distributed by the normal law, is also normal, the condition (2) at the taken value of reliability (P) may be written:

$$S_P = \overline{S} + \Psi(P)\sqrt{D[S]} \leqslant 0, \qquad (11)$$

where $\Psi(P)$ - function inverse to Gauss-Laplace; $\overline{S}$ - average value of stability deficit; $D[S]$ - dispersion of stability deficit:

$$\overline{S} = \overline{A} - \overline{B}; \qquad (12)$$
$$D[S] = D[A] + D[B] - 2 \, \text{cov}\,[AB]; \qquad (13)$$

where $\text{cov}[AB]$ is covariance between interdependent functions A and B.

At reliability P = 95% the condition (11) is written:

$$S_{0,95} = \overline{S} + 1,645\sqrt{D[S]}. \qquad (14)$$

The probabilistic analysis allow to find weak points of the design, to reveal the strongest factors determining reliability which in its turn makes it possible to outline the optimum composition and scope of surveys.

4. EVALUATION OF STABILITY AND PERMISSIBLE PARAMETERS OF SLOPES ON THE BASIS OF DEFORMATIONS OBSERVED DURING CONSTRUCTION AND OPERATION

Sometimes during mining or construction operations associated with quarries, cutting of slopes or creation of reservoirs in mountainous regions movements of rock masses resulted from variation of a stress-strained state, unloading of the rock mass or disturbance of its stability take place. The character of displacements and deformations may be different and may depend on their nature, structure of the rock mass and character of external action. For instance, during deformation of slopes of type C (Fig. 1) of more than 100 m in height the values of displacements may reach some meters at the crest but at slopes of types A, D and E displacements up to the moment of the failure may be of some centimeters only. Experience shows that in slopes of type D strong movements may take place in a steeply dipping joint but in a gently pitching joint no movements will be observed and the slope will be in a stable state. Besides it should be noted that displacements preceding the failures frequently are comparable with unloading deformations.

Displacements in rock masses are realized first of all through opening of existing joints and slides on their surfaces as a result of which orientation of a full vector of displacement gives usually a clear idea of sets of joints determining stability of the considered rock mass.

On the basis of the diagrams of the measured displacements, velocities and accelerations are calculated at different points of the rock mass.

During development of deformation of a rockslide or a failure type the characteristic features are a rapid growth of displacements with excavations, and relatively slow attenuation of deceleration of displacement (even acceleration) during laying-up of the slope and cyclic acceleration of displacements during heavy rains and snowmelt.

After mechanism of failure is revealed back calculations of stability are made (proceeding from a limit state of the rock mass at the moment of displacement), on the basis of which possible values of shear strength parameters of the slip surface are determined.

Analysis of the diagram of displacement together with stability calculations allow to provide the most effective measures for stabilization of the rock mass (Gaziev, 1978).

5. ACKNOWLEDGEMENT

The authors believe their duty to express appreciation for fruitful cooperation and valuable remarks and ideas expressed by: I.V. Baklashov, Dr. Sc., and V.M. Varichuk (Moscow Mining Institute); Yu.S. Bogoslovski (YOUZHGIPRORUDA); V.P. Budkov, Dr.Sc. (VIOGEM); K.P. Katin, Dr.Sc., and A.M.Pavlik (VNIITSVETMET); B.V. Mezhevykh (GIPRORUDA); R.P. Okatov, Dr.Sc. (Kar.PI); B.I. Reznikova and N.K. Popova (HYDROPROJECT); V.T. Sapozhnikov,Dr. Sc. (VNIMI, Urals Branch); A.A. Sorokin, Dr.Sc. (LENHYDROPROJECT); V.P. Storozhuk (KRIVBASS-PROJECT); M.V. Vasiliev (Sverdlovsk Mining Institute); V.I. Zobnin, Dr.Sc. (UNIPROMED).

6. NOTATION

A - sum of active shear forces,
B - sum of retaining forces,
D - dispersion,
G - weight of considered block or slice,
g - gravitational acceleration,
H,h - height of slope, depth of joint,
K - safety factor,
k_s - seismicity coefficient,
k_y - coefficient of hydraulic resistance,
L - length of slip surface,
N - sum of projections on the normal to the plane of displacement of external forces applied to the slope,
P - probability of occurrence,
Q - sum of projections on the direction of displacement of external forces applied to the slope,
S - deficit of stability forces,
s - path coordinates,
t - time,
U - forces of hydrostatic pressure of water filtrating in joints to the foot of the rock block or slice (U_1) and to the upstream (U_2) and downstream (U_3) faces as well,
V - volume of failure,
v - displacement velocity,
$\propto$ - angle of joint dip,
β - angle of slope,
γ - dry density of rock mass,
δ - angle of seismic acceleration vector to the horizon,
μ, C - parameters of linear (Coulomb) relationship of shear strength,
ρ - density,
τ - tangential stress, shear strength,
σ - normal stress.

7. REFERENCES

Amusin, B.Z., Fadeev, A.B. (1975). Finite elements method in solving problems of rock mechanics. Moscow: Nedra.

Arkhipova, E.K., Gaziev, E.G., and Suponitsky, L.I. (1976). Some problems in calculation of jointy physical non-linear rock foundations of hydraulic structures by finite elements method: III Research and Scientific Conference of Hydroproject, Theses of reports and communications, 215-216, Moscow.

Erzhanov, Zh.S., Karimbaiev, T.D. (1975). Finite elements method in problems of rock mechanics. A.-Ata: Nauka, Kazakh SSR.

Efimov, Y.P., Sapozhnikov, L.B., and Troitsky, A.P. (1970). Realization of finite elements method in electronic computer for solution of plane problems in theory of elasticity. VNIIG, v. 93, 81-101, Leningrad.

Fairhurst, C. (1977). Le bilan énergetique en mécanique des roches: Revue française de géotecnique, N°1, 17-36.

Fedorovsky, V.G., Freiberg, E.A., and Vasiljev,

I.M. (1981). Three methods of slope stability analysis: 10th International Conference on Soil Mechanics and Foundation Engineering, 401-404, Stockholm, Sweden.

Fisenko, G.L. (1956). Stability of slopes of coal quarries: Ugletechizdat.

Fisenko, G.L. (1965). Slope stability of quarries and dumps. Moscow: Nedra.

Fisenko, G.L., Revazov, M.A., and Galustyan, E.L. (1974). Strengthening of slopes in quarries. Moscow: Nedra.

Freiberg, E.A., Kaufman, M.D. (1981). Stability calculations of rock slopes: Izv. VNIIG, v. 147, 114-123, Leningrad.

Gaziev, E.G. (1973). Rock mechanics in civil engineering. Moscow: Strojizdat.

Gaziev, E.G. (1977). Stability of rock masses and methods of their strengthening. Moscow: Stroyizdat.

Gaziev, E.G. (1978). Analysis of storage reservoir slope stability: 3rd Int. Congress of the IAEG, Section 3, v.1, 127-132, Madrid.

Gaziev, E.G. (1979). Principles and methods of stability analysis of rock slopes of different structure: Geological engineering, N° 1, 60-69, Moscow.

Gaziev, E.G., Rechitski, V.I. (1974). Stability of stratified rock slopes: 3rd Congress of the ISRM, Denver (USA), v. 11-B, 736-791.

Gaziev, E.G., Arkhipova, E.K. (1976). Behaviour of a joint in calculation of rock masses by the finite elements method: Geological engineering of rock masses, 148-151, Moscow: Nauka.

Gaziev, E.G., Rechitski, V.I. (1979). Method of probabilistic analysis of rock slopes stability: 4th Congress of the ISRM, 637-643, Montreux (Suisse).

Gaziev, E.G., Tiden, E.N. (1979). Probabilistic approach to the study of jointing in the rock masses: Bulletin of the IAEG, N°20, 178-181.

Grigorian, S.S. (1979). New law of friction and mechanisms of large-scale rock failures and slides: Reports of USSR Academy of Sciences. Mechanics, v. 244, N° 4, 846-849.

Habib, P. (1979). Le coefficient de sécurité dans les ouvrages au rocher: 4th Congress of the ISRM, v. 3 , 18-22, Montreux (Suisse).

Kazikaev, I.M., Surzhin, G.G., Fomin, B.A., and Bondarenko, O.I. (1979). Strength and stability of pillars in the field of tectonic stresses: Rock pressure, methods of control. Frunze: Ilim.

Kazikaev, I.M. (1981). Geomechanical processes during ore mining. Moscow: Nedra.

Komarov, V.V. (1978). Solution of three-dimensional problem of stability of slopes in jointy rocks: Transactions IGD MCHM USSR, issue 57, 44-53, Sverdlovsk.

Kovari, K., Fritz, P. (1976). Stabilitätsberechnung ebener und räumlicher Felsböschungen: Rock Mechanics, v. 8, N° 1, 73-113.

Maranhão,N. (1974). Geometrical characterization of jointing of rock masses: 2nd Congress of the IAEG, V1-3, 1-10, São Paulo (Brazil).

McMahon, B.K. (1971). A statistical method for the design of rock slopes: Ist Australia - New Zealand Conference on Geomechanics, v. 1, 314-321, Melbourne.

McMahon, B.K. (1974). Design of rock slopes against sliding on pre-existant fractures: 3rd Congress of the ISRM, v. 11-B, 803-808, Denver (USA).

Mgalobelov, Y.B., Khalina, L.I. (1970). Analytical method of stability calculation of rock slopes under conditions of a plane problem and determination of anchor force for raising of stability: Gidrotechnicheskoye stroitel'stvo, N° 9.

Mirtsukhulava, Ts.E. (1974). Reliability of irrigation works. Moscow: Kolos.

Mozhevitinov, A.L., Shintemirov, M. (1970). General method of stability calculation of slopes of earth structures: Izvestiya VNIIG, v. 92, 11-22.

Popov, I.I., Okatov, A.P. (1980). Control of rock slides in quarries, Moscow: Nedra.

Rocha, M. (1974). Present possibilities of studying foundations of concrete dams: 3rd Congress of the ISRM, v. 1-A, 879-897, Denver (USA).

Rocha, M. (1978). Analysis and design of the foundations of concrete dams: Int.Symp. on Rock Mech. Related to Dam Foundations, v.2, III, 11-70, Rio de Janeiro (Brazil).

Rozin, L.A. (1971). Calculation of hydraulic structures by the finite elements method with the use of electronic computer. Moscow: Energia.

Shahunyanz, G.M. (1941). Determination of stability conditions of landslides: Railway building and track facilities, N° 2, 48-51.

Vistelius, A.B. (1958). Structural diagrams. Academy of Sciences, Moscow-Leningrad.

VNIMI (1971). Instructions of procedure of determination of inclination of slopes, edges, benches and dumps of quarries under construction and in operation. Leningrad: VNIMI.

Zolotarev, G.S. (1973). Experience of stability evaluation of slopes of a complicated geological structure by finite elements method and by model experiments. Moscow:MGU.

Zoteev, V.G. (1982). Stability calculation of rock slopes of deep quarries: Transactions IGD MCHM USSR, issue 69, 29-35,Sverdlovsk.

PIT SLOPE DESIGN METHODS: BOUGAINVILLE COPPER LIMITED OPEN CUT

Méthode d'étude de talus de puits à la mine de cuivre à ciel ouvert, Bougainville

Entwurfsmethoden für Grubenböschungen: Tagebau der Bougainville Copper Limited

J. R. L. Read and G. N. Lye
Senior Engineering Geologist and Rock Mechanics Engineer, Bougainville Copper Limited, Panguna, Bougainville, P.N.G.

SYNOPSIS

The Bougainville Copper open pit will be about 2700 m in diameter with wall heights varying between 350 and 950 m. This paper describes how pit slope stability is evaluated using limit equilibrium principles with deterministic and stochastic input parameters gathered from investigations of rock mass strength, geological structures, groundwater and regional seismicity. The methods of investigation and special analytical techniques are outlined, and measures taken to improve bench design and steeper overall slopes by using dozers rather than shovels to cut final slopes are described.

RESUME

La mine de cuivre à ciel ouvert de Bougainville sera approximativement de 2700 m de diamètre avec des murs d'une hauteur comprise entre 350 et 950 m. Cet article décrit comment on évalue la stabilité du talus de la carrière en utilisant les principes des équilibres limites avec des paramètres d'entrée déterministiques et stochastiques obtenus à partir d'études de la résistance des masses rocheuses, des structures géologiques, de la nappe superficielle et de la séismicité régionale. De plus, les méthodes d'étude et les techniques analytiques spéciales sont esquissées, et on décrit les mesures prises afin de perfectionner les études de gradins et de talus plus raides en utilisants des bull-dozers plutôt que des pelleteuses pour couper les talus définitifs.

ZUSAMMENFASSUNG

Der Bougainville Tagebau wird einen Durchmesser von 2700 m haben mit Wandhöhen, die zwischen 350 und 950 m liegen. Die Arbeit beschreibt, wie die Stabilität der Grubenböschung bestimmt wurde und wie dazu Gleichgewichtsansätze zusammen mit deterministischen und stochastischen Daten verwendet wurden, die aus Untersuchungen der Gebirgsfestigkeit, der geologischen Struktur, des Grundwassers und der lokalen seismischen Aktivität hervorgehen. Die Untersuchungsmethoden und spezielle analytische Verfahren werden beschrieben, wie auch Maßnahmen zur Verbesserung der Gestaltung der Bermen und zur Anwendung allgemein steilerer Neigungen mit Hilfe von Bulldozers anstelle von Schippen, für die Formung der endgültigen Böschungen.

1. INTRODUCTION

Bougainville Copper Limited is mining a porphyry copper deposit by means of an open pit at Panguna, Bougainville Island, Papua New Guinea (Fig. 1).

Politically Bougainville Island is part of the North Solomons Province of Papua New Guinea, although geographically it is part of the Solomon Islands which are situated immediately to the southeast. The island is 200 km long and 30 to 60 km wide with mountainous terrain, dense tropical vegetation and an annual rainfall of 4500mm spread evenly throughout the year. Several volcanoes occur on the Island. Most are extinct, but two are dormant and Mt. Bagana, situated 40 km northwest of the mine site, is active. The island is also located in one of the world's most active seismic zones.

2. GEOLOGY

The orebody is a typical island arc porphyry copper deposit with dioritic and granodioritic

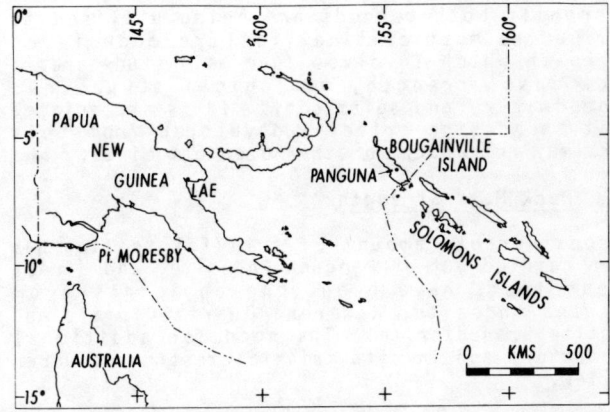

Fig. 1 Location of Panguna Mine.

intrusions into andesitic volcanic rocks.

The andesitic volcanic rocks comprise the Panguna Andesite Member (Baumer & Fraser, 1975) consisting of lavas, agglomerates and tuffs. The sequence is 1000m thick and dips 5 to 20° to the southeast. The intrusions include, in order of their emplacement, Kawerong Quartz Diorite, Biotite Diorite, Biotite Granodiorite, Leucocratic Quartz Diorite and Biuro Granodiorite. Mineralisation is mainly in quartz veins, in joints and in relatively minor sulphide veins within the Biotite Diorite, Biotite Granodiorite, Leucocratic Quartz Diorite and the Panguna Andesite.

Mineralised breccias occur in nearly all rock types. These are mainly pipelike but some form linear belts along the intrusive contacts. Northeasterly striking pebble dykes, consisting of pebbles of all rock types in a chloritic matrix, intersect all rock types. Overlying the whole sequence are areas of boulder alluvium and up to 30m of unconsolidated volcanic ash.

Ore body reserves are estimated at 800 million tonnes of grade 0.40% copper and 0.46 grams gold per tonne, at a waste:ore ratio of 0.66. Operations at the mine commenced in 1972 and are expected to continue until at least 2000.

3. DESIGN INVESTIGATIONS

There have been two previous phases of slope design work at Panguna.

Initially the walls were designed using circular failure criteria on the assumption that the rock mass was randomly fractured (Pentz, 1969). The decision to proceed in this manner was guided by the evidence available from the limited number of surface exposures and from two exploratory adits excavated during the feasibility studies.

As excavation of the mine proceeded, it became apparent that although the rock mass was highly and very closely fractured geological structures were not randomly orientated. Accordingly detailed structural geological analyses were carried out and design procedures were re-assessed using a probabalistic approach (McMahon, 1979).

Currently both methods are being utilised to define the most critical failure mode in an approach which involves four main study areas; rock mass strength, geological structure, groundwater, and seismicity. It is appreciated that the design criteria developed from this work may not apply to other open cut mines.

3.1 Rock Mass Strength

A considerable amount of triaxial testing has been carried out on Panguna Andesite, and direct shear tests have been done on joints from Panguna Andesite, Kawerong Quartz Diorite and Biotite Granodiorite. The need for additional laboratory and in-situ triaxial testing is under review.

3.1.1 Triaxial tests

The strength of jointed Panguna Andesite has been assessed on the basis of 6-inch triaxial tests carried out on:

. Undisturbed core samples obtained using 6-inch triple tube diamond drilling equipment. These tests were carried out at the

Australian National University, Canberra (Jaeger, 1969).

Recompacted graded samples taken from face rills in the pit and tested in the mine laboratory at Panguna. The grading used was scaled down to 100% -19mm by using a curve parallel to that obtained from grading analysis of the first run sample. The regraded samples were then recompacted to as close to in-situ density as possible (intact rock = 2.55 tonne/m^3). Values of approximately 2.5 tonne/m^3 were obtained.

The triaxial test results are shown in Figure 2.

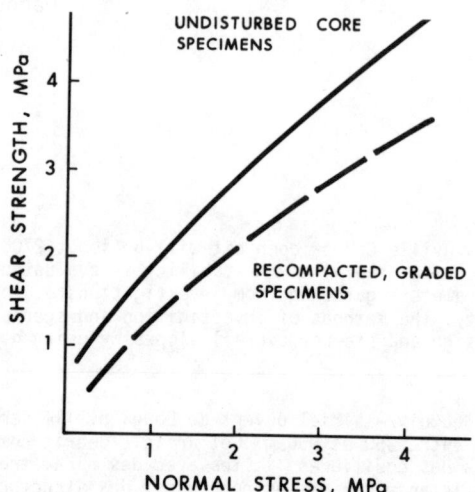

Fig. 2 Triaxial test results, Panguna Andesite.

Initially the strength was assessed on an assumption that there were a sufficiently large number of closely spaced joints to permit the formation of "kink" bands associated with particle rotation and overall rock mass dilation at low effective normal stress levels (Hoek, 1980). On this basis, and assuming a circular failure mode, the tests on intact core were taken as an upper bound and those on recompacted graded material as a lower bound; design strength was taken as the average between these two limits.

Detailed examination of the pit walls, particularly where trial batters have been shaped by dozers rather than with shovels, has led to the conclusion that there is no rational basis for postulating a "kink" band failure mechanism. Rather than representing an upper bound, it is now believed that the tests on intact core provide a more realistic estimate of the rock mass strength than the average value previously used.

3.1.2 Direct shear tests

Several hundred direct shear tests have been carried out in the mine laboratory at Panguna. Tests were carried out on a variety of combinations of joint surfaces and coatings, and on saw-cut surfaces.

The peak friction angle and cohesion values for surfaces tested in Panguna Andesite and Kawerong Quartz Diorite are about 42° and 1.8 MPa; residual friction angles are about 26°. Sixty percent of the ultimate pit wall will be in Panguna Andesite with the remainder being in Kawerong Quartz Diorite.

3.1.3 Future tests

Although Panguna Andesite has been extensively tested there have been no rock mass strength tests done on Kawerong Quartz Diorite. The original exploration adits and much of the early mining were done in Panguna Andesite. Until now exposure of Kawerong Quartz Diorite has been limited.

The Kawerong Quartz Diorite is less closely jointed and by inspection is considered to be stronger than the Panguna Andesite. However, although it may be thought rational to accept the Panguna Andesite strength as a conservative estimate of Kawerong Quartz Diorite strength, consideration is being given to a programme of Kawerong Quartz Diorite triaxial testing.

3.2 Geological Structure

Structural information is gathered in two stages. Initially fault and joint orientations are recorded during daily routine mapping of the mining faces and committed along with all other basic geological information to a general purpose computer data base. Subsequently specific structural information including joint orientation, spacing, length, roughness, infilling and moisture condition is recorded at selected sites and committed to a rock mechanics computer data base.

Overall the geological structure of the ore-body is complex. However, quantitative analysis of fault and joint orientations has revealed clear structural patterns within the pit. Initially four domains were recognised (Furstner, 1979), but as the pit has expanded and more information has become available seven domains have been delineated.

Joints in the ultimate wall rocks are very closely spaced with spacings ranging from 20mm in Panguna Andesite to 35mm in Kawerong Quartz Diorite. Despite this very close spacing specifically orientated joint sets can be distinguished within and between structural domains. There is, however, a dominance of steeply dipping joints. As shown in Figure 3, 72% of all joints in Panguna Andesite and 88% of all joints in Kawerong Quartz Diorite are steeper than 50°.

3.3 Groundwater

The groundwater regime is complex with wide variations in groundwater levels over short distances. Characteristically, clay-filled faults form impermeable barriers and impart a cellular pattern to the regime. The rock mass within each cell may be permeable, but drainage from a cell does not occur until the impermeable barrier is penetrated.

Groundwater levels are monitored by a network of piezometers installed both beyond the ultimate pit perimeter and within the pit. Those beyond the perimeter are pneumatic piezometers installed as permanent reference points in diamond drilled holes up to 450m deep. The piezometers within the pit are open standpipe

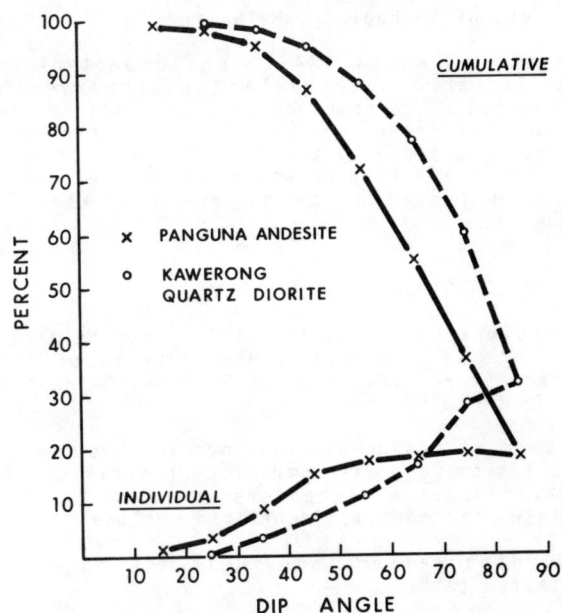

Fig. 3 Distribution of joint dip angles in Panguna Andesite and Kawerong Quartz Diorite.

piezometers installed in down-the-hole hammer drilled holes up to 150m deep. These piezometers are replaced as they are mined out so that changes in the groundwater regime can be monitored as mining proceeds.

3.4 Seismicity

Bougainville Island is located in one of the world's most seismically active regions. Most of this activity originates from a subduction zone at the boundary of the Solomon Sea and South Pacific crustal plates. This zone is defined by the 5000m to 9000m deep North Solomons trench, situated about 80 km west of Bougainville Island, and dips steeply to the northeast between the trench and the west coast of the island.

From historical records the estimated return period for an event of Richter Magnitude 6.5 is two years. For events of Magnitutde 7.2, 7.5, and 8.0 the return periods are 10, 25 and 100 years. A Magnitude 7.2 event, which had an estimated site felt intensity of MM VII, occurred in 1975.

However, until 1982 the monitoring system used was not sensitive enough to determine whether any active faults occur on the island. To remedy this situation the system was upgraded and now consists of three Sprengnether S6000 siesmometers linked to digital recorders at remote outstations and one Spregnether L4C seismometer coupled to a Teledyne Model RV-301B helicorder at the original Panguna station.

Digital tape playback is controlled by a Spectrum Model SS23 minicomputer which also utilises an interactive computer programme to locate the earthquakes and determine risk statistics. The digital recording and playback equipment, and the computer software, were assembled by the Seismology Centre, Phillip

Institute of Technology, Melbourne.

Kinemetrics Model SMA-1 accelerometers are installed across the island to observe the attenuation of seismic energy with distance from the source. Fresh bedrock sites are operated at Panguna, and close to the east and west coasts. Additionally, sites on weathered rock and soil, and on mine waste dumps, are operated at Panguna.

4. DESIGN STUDIES

The ultimate pit diameter will be approximately 2700m with wall heights ranging from about 350m on the western side to about 950m on the eastern side of the pit.

The current design studies involve checking the overall slope and its components for failure by whatever failure mechanisms are kinematically possible. Mechanisms identified include:

. Sliding failure on single fractures or stepped paths.

. Sliding failure on wedge combinations of two intersecting fractures or stepped paths.

. Sliding failure on two block combinations of active and passive wedges for both single fracture and stepped path combinations.

. Rotational slip-circle failure.

. Sliding failure on single fractures or stepped paths with failure through the rock mass in the toe area.

. Toppling failure.

Stability analyses are carried out using limit equilibrium principles with both deterministic and stochastic input parameters.

Special development work has been undertaken with respect to rock mass strength, the simulation of failure paths, pit slope drainage and earthquake loading.

4.1 Rock Mass Strength

The test data obtained from the triaxial tests on Panguna Andesite was used in the development of the Hoek-Brown non-linear failure criterion equation (Hoek & Brown, 1980) which was first used in the design studies in 1978.

The equation has since been modified according to the following solution obtained by Dr. John Bray of Imperial College (Hoek, 1982a):

$$\tau = (\cot \phi_i \cos \phi_i) \frac{m\sigma_c}{8}$$

where the instantaneous friction angle ϕ_i is given by :

$$\phi_i = \text{Arctan} \frac{1}{\sqrt{4h\cos^2\theta - 1}}$$

where $h = 1 + \frac{16(m\sigma + s\sigma_c)}{3m^2 \sigma_c}$

and $\theta = \frac{1}{3} (90 + \text{Arctan} \frac{1}{\sqrt{h^3 - 1}})$

The instantaneous cohesive strength c_i is given by : -

$$c_i = \tau - \sigma \text{ Tan } \phi_i$$

All analytical programmes which require the use of rock mass strength parameters have been modified to use the new equation by:

. Estimating the normal effective stress range appropriate to the failure surface (or part thereof).

. Determining appropriate srength parameters from the non-linear envelope.

. Calculating safety factor.

. Iterating the above steps until a convergence in safety factor is obtained.

4.2 Failure Path Simulation.

Monte Carlo simulation is used in two different ways:

. To determine stepped failure paths using statistical distributions of joint set orientations, lengths, spacings and strengths. The simulation uses a joint termination index and probability of joint set occurrence to determine how much of the failure path occurs through the rock mass rather than along joints.

. To determine the distribution of safety factors in an active-passive wedge failure analysis using the average strength and orientation distributions determined by the stepped path simulation. The probability of failure is then calculated as a probability that the safety factor is less than a design value.

Rosenbleuth's method (Rosenbleuth, 1975) is also used to calculate the mean and standard deviation of safety factor, but when calculating the probability of failure a normal distribution is assumed.

4.3 Pit Slope Drainage

Earlier studies (Furstner, 1979) have always assumed that a gallery would be required to drain the groundwater behind the highest pit slopes and reduce the groundwater pressures acting on potential failure surfaces to an acceptable minimum. Current work has indicated that it is likely that horizontal drain holes will provide adequate drainage in those zones where the groundwater table is high.

The drilling machine used for drilling piezometer holes within the pit features a drill carriage capable of vertical, angle or horizontal orientation and is capable of drilling 200m horizontal drain holes. Initial trials using horizontal drain holes have been successful. If future trials are successful, horizontal drain holes will be used for slope drainage in preference to a drainage gallery.

4.4 Seismic Stability

In the past pseudo-static methods were used to analyse seismic stability. Originally a peak

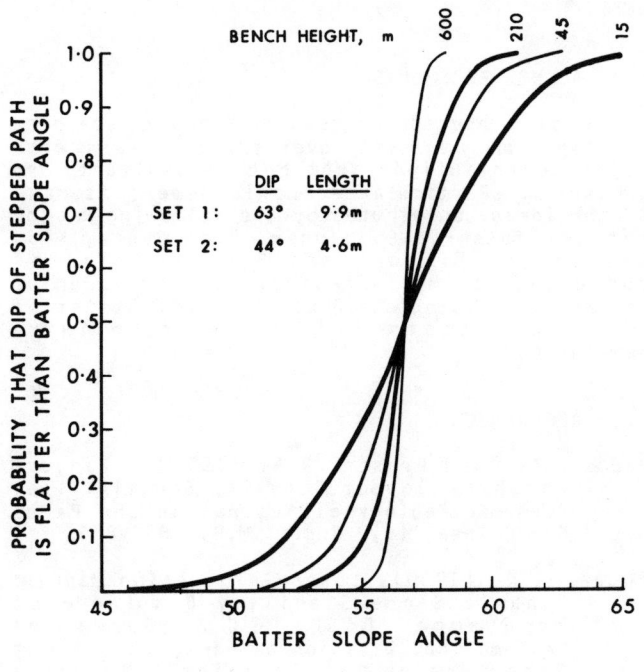

Fig. 4a

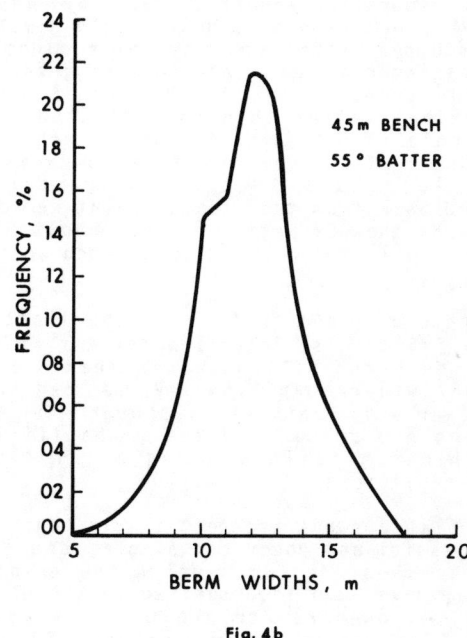

Fig. 4b

Fig. 4 Statistical representation of stepped path and berm width data (after McMahon, 1979).

horizontal ground acceleration value of 0.13g was used, but this was reduced to 0.06g. It is now recognised that the pseudo-static analysis bears little relation to the performance of the pit wall under rapid cyclic loading conditions and that the use of horizontally applied seismic vectors in pseudo-static analysis could lead to an unconservative allowance for earthquake loading.

An additional constraint is that the early site accelerometer data was inconclusive. The design value of 0.13g represented the maximum acceleration recorded on fresh bedrock with values in the range 0.3 to 0.4g being recorded on weathered rock and soil. However, no accelerometers had been operating during the ten year event which occurred in 1975. Depending on the approach adopted, accelerations up to 0.3g horizontal can be inferred for that event which raises uncertainties as to the appropriate average seismic co-efficient to use in the pseudo-static analysis. In view of these uncertainties a new approach was initiated. As already described the seismometer and accelerometer network was upgraded, and consultants were commissioned to examine the feasibility of adapting existing dynamic methods of analysis for pit slope design.

As a result of the consultant's studies (Hoek, 1982b) a Newmark (1965) displacement analysis programme (Chugh, 1980) formulated to use horizontal or combined horizontal-vertical earthquake acceleration time histories was adapted for design use. Site work still in hand includes studies of the dynamic properties of the rock mass in the ultimate pit walls and additional dynamic analytical programmes.

5.   ULTIMATE SLOPE DESIGN

5.1 Bench Design

Operational experience has shown that the optimum pit slope cannot be shovel dug. From early in the mine life 15m high benches were excavated at a spacing of 15m crest to crest with a theoretical batter slope of 56° and berm width of 5m. In practice the undercutting action of the shovel induced joint toppling and sliding failures which both reduced the width of the berm above and covered the working berm with rill. Under these conditions access and clean-up was difficult and dangerous. Double benches were then introduced as a means of increasing berm width to 10m, and containing berm failures and rill formation without decreasing overall slopes. This proved operationally acceptable and is the current procedure. However, the problem of berm failure and rill clean-up remained making it impossible to consider any steepening of the overall slope using this method. Alternatives were therefore examined.

The first alternative accepted for a field trial was a dozer cut, 45m high triple bench incorporating a 55° batter slope and 10m wide berm. The choice of this configuration, which has an overall slope of 47°, followed detailed structural reappraisal of the shovel dug benches in Panguna Andesite. It was noted in this study that the shovel dug benches tended to fall back to a batter slope of about 57°, representing a stepped failure path formed by two dominant joint sets dipping out of the batter at angles of about 63° and 44°.

The stochastic parameters used to design the

trial configuration are illustrated by Figure 4. Figure 4a, determined by Monte Carlo simulation of the stepped path formed by the two dominant joint sets over a number of slope heights, shows that the probability of the stepped path dip angle being flatter than the selected batter slope angle of 55° for a 45m high bench is 0.2. Figure 4b, plotted from the data summarised by Figure 4a, shows that if the batters are excavated back from the steeper joint set dip of 63° towards the 55° batter slope, about 85% of the berms will be wider than 10m and about 95% wider than 7m.

In the case of plane failure on the flatter of the two joint sets, joint lengths are such that 50% of the berms will be wider than 8.5m and about 95% wider than 6.5m. Values can also be derived for wedge failures although when joint strengths are considered the probability of a 45m high bench failure along a joint wedge becomes negligible.

The initial and a subsequent triple bench trial was successful and dozer cut single, double and triple benches, designed using the principles outlined, have been incorporated in the ultimate pit wall. However, triple benches are not regarded as the optimum design. Figure 4a illustrates that the probability of the stepped path dip being flatter than any selected batter slope angle decreases with increasing bench height: other things being equal, the optimum overall slope could in fact be obtained by totally eliminating benches. Consequently bench heights greater than 45m are being evaluated and a trial 60m high dozer cut bench is to be constructed.

5.2 Overall Slopes

In the proposed ultimate pit triple benches extend between haul roads for a maximum height of 275m. In the deepest section of the pit triple benching is combined with haul roads over a maximum height of 590m; this reduces the intermediate slope of 47° to an overall slope of 43°. When the flatter slopes in weathered rock and soil leading to the pit perimeter are included the overall slope at the proposed maximum height of 950m decreases to 38°.

Possible failure mechanisms identified in the intermediate and overall slopes are:

. Sliding failure on two block combinations of active and passive wedges for stepped path combinations.

. Rotational slip-circle failure.

. Sliding on single fractures or stepped paths with failure through the rock mass in the toe area.

When analysed statically the slopes are stable for each of these mechanisms with minimum drainage, that is, with the groundwater maintained about 50m behind the pit wall. With earthquake loading better drainage is necessary. The analyses carried out so far indicate that under what are considered the worst foreseeable conditions the groundwater table must be drawn back to a distance of at least 150m behind the wall.

6. ACKNOWLEDGEMENTS

The design procedures used at Panguna have been developed progressively over several years with major contributions from both consultants and BCL staff. Particular acknowledgement is made of the ideas contributed by the following : the late Professor J.C. Jaeger, D.L. Pentz, K.J. Rosengren, E. Hoek and B.K. McMahon as consultants; M.J. Furstner, formerly Senior Engineering Geologist, BCL; and H.J. Stoter of CRA Consultants for computer software developments.

7. REFERENCES

Baumer, A., and Fraser, R.B. (1975). Panguna Porphyry Copper Deposit, Bougainville. Economic Geology of Australian and Papua New Guinea, (1), Aus. I.M.M., 855-866.

Chugh, A.K. (1980). User Information Manual: Dynamic Slope Stability Displacement Programme "DISP". U.S. Bureau of Reclamation, Division of Dams, Embankment Dams Section, Engineering & Research Centre, Denver.

Furstner, J.M.M. (1979). Tunnelling at the Panguna Open Pit Copper Mine. Proc. 3rd ICIAEG (III), 2, 45-51, Madrid.

Hoek, E. (1980). Report to Bougainville Copper Limited on shear strength assessment and slope stability analysis for Pan Hill. Golder Associates Report, October, 1980, Unpubl.

Hoek, E. (1982a) Report to Bougainville Copper Limited on Hoek-Brown non-linear failure criteria equation. Golder Associates Report, June, 1982, Unpubl.

Hoek, E. (1982b). Report to Bougainville Copper Limited on seismic slope stability analysis. Golder Associates Report, June, 1982, unpubl.

Hoek, E., and Brown, E.T. (1980). Underground Excavations in Rock. T.I.M.M., London, 137-177.

Jaeger, J.C. (1969). Report on mechanical properties of Bougainville joints. Unpubl.

McMahon, B.K. (1979). Report to Bougainville Copper Limited on slope design studies, Pan Hill. McMahon, Burgess and Yeates Report, July, 1979, Unpubl.

Newmark, N.M. (1965). Effects of earthquakes on dams and embankments. Geotechnique (15), 2, 139-160.

Pentz, D.L. (1969). Report on slope stability of Bougainville Copper Open Pit. Unpubl.

Rosenbleuth, E. (1975). Point estimates for probability moments. Proc. Nat. Acad.Sci. U.S.A., (72), 10, 3812-3814.

AN INVESTIGATION INTO THE STABILITY OF A SLOPE IN AN OPEN-CUT MINE

Investigation sur la stabilité du versant d'une mine à ciel ouvert

Untersuchung der Standsicherheit einer Tagebauböschung

R. H. T. Cox

Experimental Officer, CSIRO, Division of Applied Geomechanics, Mount Waverley, Australia

SYNOPSIS

Deformation and moisture measuring instruments were installed at a potentially unstable highwall of an open-cut coal mine. These were monitored for several months during which time two failures, encompassing more than 600 metres of highwall, occurred. In both cases a deep-seated, double-wedge mechanism was identified. The mechanism was modelled on a computer using the method of slices which, together with the results of an intensive material testing program, confirmed that a critical overburden depth-to-strength ratio could be reached. This situation was exacerbated in the presence of moisture, increased slope batter angle, and coal and overburden blasting.

RESUME

Des instruments de mesure de déformation et d'humidité ont été installés sur le haut front d'abattage d'une mine de charbon à ciel ouvert. Ces instruments ont été contrôlés pendant plusieurs mois au cours desquels deux écroulements se sont produits sur une distance de plus de 600 mètres le long du front d'abattage. Dans les deux cas un mécanisme double-cale profonde a été identifié. Ce mécanisme a été modelé sur un calculateur électronique, selon la méthode des tranches, ce qui conjointement avec les résultats d'un programme intensif de contrôle des matériaux, a confirmé qu'un rapport profondeur-puissance pour le terrain de couverture peut être atteint. Cette situation s'acerbe à cause de l'humidité, de l'augmentation de l'angle d'inclinaison de l'escarpement et à cause du foudroyage du charbon et du terrain de couverture.

ZUSAMMENFASSUNG

Geräte für das Messen von Verformung und Feuchtigkeit wurden an einer potentiell instabilen Abbaufront eines Kohlentagebaus eingerichtet. Diese Geräte wurden mehrere Monate lang überwacht, und während dieser Zeit kamen zwei Böschungsbrüche vor, die sich über mehr als 600 Meter der Abbaufront erstreckten. In beiden Fällen wurde ein tiefsitzender Doppelkeil-Mechanismus festgestellt. Dieser Mechanismus wurde mit Computer nachgebildet, wobei Gebrauch von der Schnittmethode gemacht wurde, welche, zusammen mit den Ergebnissen einer intensiven Materialprüfung, bestätigt hat, daß ein kritisches Abraumtiefe-Festigkeits-Verhältnis erreicht werden konnte. Diese Umstände wurden in der Gegenwart von Feuchtigkeit, und mit zunehmendem Böschungswinkel, als auch beim Sprengen von Kohle und Abraum verschlechtert.

1. INTRODUCTION

With the gradual deepening of the open-cut coal mining operation at Goonyella Mine in central Queensland, pit-wall failures were becoming frequent and the consequences of these failures more severe. In an endeavour to understand the causes and mechanisms of these failures, the CSIRO, in collaboration with the Utah Development Company, initiated an investigation.

It was evident from previous work by Fuller and Cox (1978) and Wooltorton (1978, confidential report) that a biplanar boundary was associated with most of the slope failures monitored in the area. Furthermore, it was clear that the following parameters had the greatest influence on highwall stability;
  a)   slope angle and depth
  b)   blasting practice
  c)   geological structure of the highwall
  d)   moisture content of the overburden
While slope angle is kept as steep as possible to improve the mining efficiency in handling

overburden, the depth is fixed by dragline reach. Thus a bench is created in the highwall so that overburden depth is kept constant. Overburden blasting had already been modified from in-line to diagonal as discussed by Hagen et al. (1978) and this appeared to reduce back break considerably. Buffer blasting had also been discontinued at the time of the investigation.

The overburden at the mine is a sequence of siltstones, sandstones and mudstones of Tertiary and Permian age which overlays coal of Permian age. Tectonic deformation has been minimal at the northern pit, with little folding or faulting (Golder, Brawner and Associates, 1972). However, a flat dipping black clay shear plane was observed just above the coal seam in the Cleanskin pit by Godfrey (1977, confidential report). It appeared the highwall movement was taking place on this plane.

This paper describes part of an investigation which concentrated on a study of highwall behaviour in the Cleanskin pit opposite access Ramp 11. A line of subsurface instruments was installed perpendicular to the pit on the bench so that this ground movement and overburden moisture conditions could be measured. This information was recorded at regular intervals over several months using a 100-channel data-acquisition system. The data stored on cassette tapes was posted to the Division of Applied Geomechanics in Melbourne for processing.

2.   HISTORY OF RAMP 11 HIGHWALL BEHAVIOUR

Highwall blasting followed the removal of coal from cut #5. This fractured the overburden so that the dragline could excavate the material without difficulty. This it did by creating a 'bridge' across the pit and transferring over-burden material from the highwall to the spoil pile. (Figure 1.) The coal seam was uncovered as the bridge moved north.

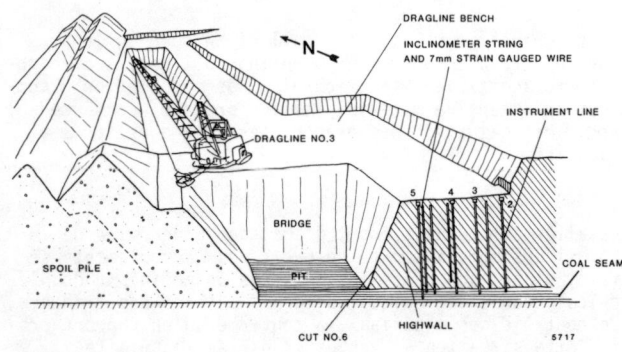

Fig. 1. Mining cut #6 after installation of

instruments

Coal blasting followed pit excavation. Sections about 100 m long were fired sequentially and the fractured coal transferred to trucks using a coal shovel.

On 28th October, 1977 a section of the highwall, approximately 450 m in length, failed by moving into the pit. (Figure 2).

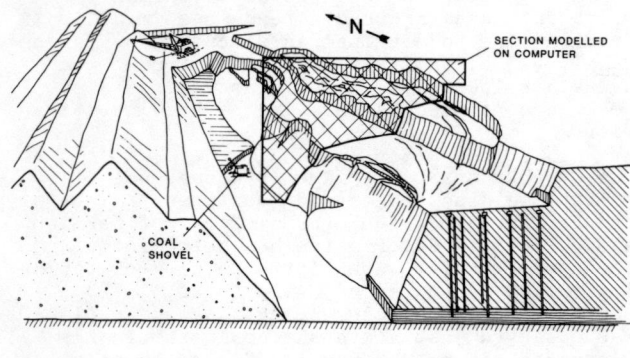

Fig. 2. Highwall failure #1 on 28/10/1977

The coal-shovel operator noticed movements

starting in the north at 2 a.m. He had time to move the shovel close to the spoil pile before the failure had extended far enough south to cut his power cable. It appeared that the move-out of the overburden and the generation of the scarp east of the failure occurred simultaneous-ly. Moveout continued for a number of hours, slowing progressively with time. This was accompanied by the toppling forward of blocks on the slope face.

By the end of November, 1977, mining had ceased in the pit and the summer rains were increasing in frequency. On 1st January, 1978 a 100 m section of the bench moved into the pit, taking with it the instruments closest to the edge. Figure 3 shows the extent of failure relative to the first. Within the next four weeks the failure boundaries had extended eastwards - due, it appears, to exceptionally heavy rainfall. However, no further mining operations were disrupted.

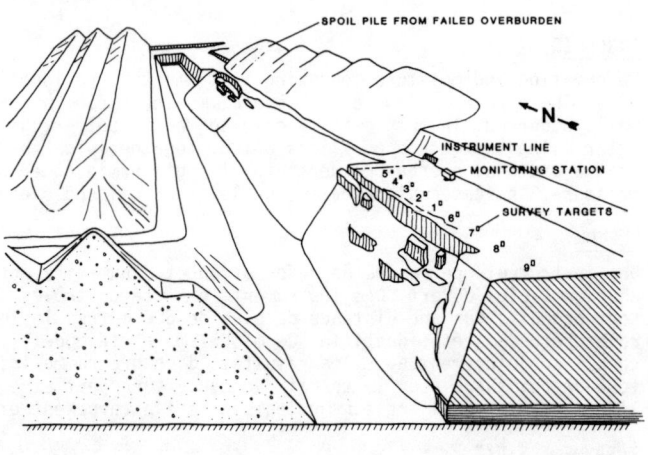

Fig. 3. Highwall failure #2 on 1/1/1978

3.   RESULTS OF THE MONITORING EXERCISE

With the dragline to the south, the following instruments were installed at four locations in the highwall prior to excavation of cut #6:
a) inclinometer string - 15 tiltmeters at 2 m intervals were used to detect highwall profile changes at site 5 (Figure 1)
b) strain gauge wire - 30 strain gauges were cemented to a 7 mm diameter steel wire located at site 5 to complement the inclinometer string
c) vertical extensometers - to determine extension or compression in both the upper and lower sections of the overburden at each of the four sites
d) horizontal extensometers - to establish surface displacement at each of the four sites. Their output was complemented with nine survey targets located to the south of the instrument site (Figure 3)
e) piezometers - to measure ground-water levels in the zone just above coal at each of the four sites
f) psychrometers - to measure negative pore water pressure. Twelve instruments were

spaced at 6 m intervals at sites 2 and 4.

The overburden removal, each of the coal blasts, and the highwall failures all caused significant changes in the output of the subsurface instruments. Four of the inclinometer outputs clearly show this in Figure 4, for example.

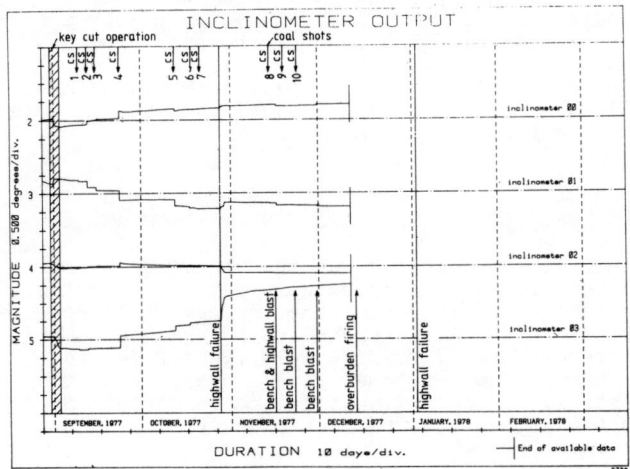

Fig. 4. Sample output from the four lowest inclinometers

The changes that these outputs represent are graphically illustrated in Figures 5 and 6 which show the ground profile development as defined by the inclinometers and strain gauges respectively. The inclinometer profile is shown as straight lines connecting the midpoints between tiltmeters. The strain gauge profile is determined by radii, each tangential to the next. In both cases, the profiles have been plotted so that movement is always towards the pit. The accuracy of the representation of the profiles suggested by these figures is reduced by the limited number of instruments used. This would account for the discrepancy in horizontal scales. However, a number of very interesting features can be observed.

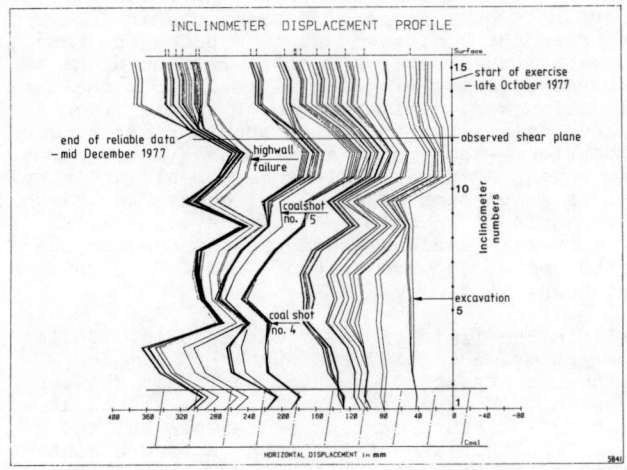

Fig. 5. Ground profile defined by inclinometers

First, overburden excavation, coal blasting and highwall failure #1 contribute to most of the movement towards the pit. (Coal blasts 4 and 5 were the most significant, being opposite the instrument site). The movement as a result of creep was very small in comparison.

Secondly, it is apparent that movement must have occurred on a number of shear planes, the most dominant of which was located a few metres above coal. The second obvious shear plane was about 7 m below the surface, verified by observation down an open borehole.

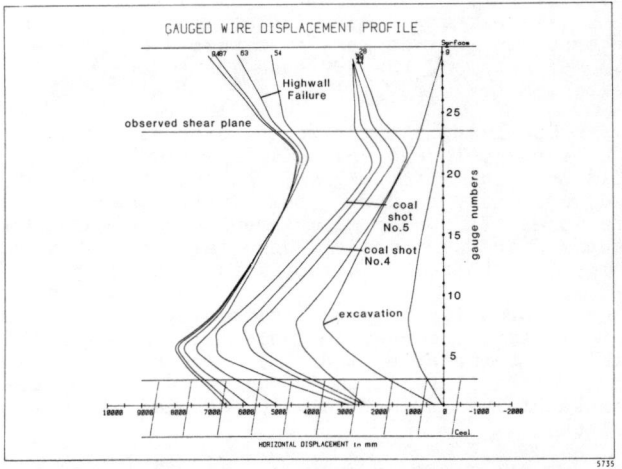

Fig. 6. Ground profile defined by strain gauge 7 mm diameter wire

For comparison, the results of the vertical and horizontal extensometers are listed in Table I.

Table I. Results of extensometer changes

INSTRUMENT T=TOP B=BOTTOM		EXTENSION CHANGE (mm)			
		COAL FIRING		HIGHWALL FAILURE	
		#4	#5	#1	#2
ANGLE	SITE	23/9/77	12/10/77	28/10/77	1/1/78
Vert.	2 T	-1.00	0.35	0.00	2.18
	B	-0.49	-0.58	0.03	0.38
	3 T	-2.37	-0.88	-0.65	-2.29
	B	-1.45	-0.65	-2.11	4.70
	4 T	-1.38	-0.39	-4.73	-2.18
	B	2.83	-0.26	-0.20	4.75
	5 T	-2.86	-0.35	-0.54	-
	B	-2.80	-0.99	-0.22	-
Horiz.	2	-0.73	-0.19	+1.03	0
	3	-0.03	-0.08	+1.18	+0.27
	4	-0.23	-0.01	+0.22	-0.15
	5	+0.08	+0.19	+0.27	-

The readings are the changes in extension as a result of coal firings 4 and 5 and the two highwall failures. It is interesting to note that compaction of the highwall is indicated in almost every event throughout the overburden except for the second failure. In this case, the lower extensometers indicated extension. It is reasonable to assume that this extension was caused by shear movement in the lower part of the overburden during this failure.

The extent of compaction is only very small in all cases. It must be remembered that this area had been bench blasted eight months, and buffer blasted two years previously. Therefore it would be expected that further settling in disrupted bedding planes would occur. The lack of extension especially in the lower overburden (except during the second highwall failure) implies that any shear displacements within the overburden must be kept to within 4 or 5 mm - the diameter of the PVC tube containing the extensometer wire.

By comparison, the horizontal extensometers on the surface showed very little change. A 1 mm extension at sites 2 and 3 was the most recorded, and that was during the first highwall failure. Virtually no movement was recorded on the surface during the second failure although site 5 in this case disappeared into the pit.

The information is substantiated by the results of the target survey in Table II. (See locations of targets in Figure 3). During the second highwall failure on 1st January, 1978, displacements of less than 2 mm were recorded at these sites.

At the time of instrument installation about 4 m of water lay above the coal seam. However, after the excavation of the overburden, and as a result of the coal blasting in the pit, this drained away and no water was again observed until the end of the monitoring exercise.

Table II.   Results of survey - displacements towards pit (mm)

Target point	9/11/77	16/11/77	30/11/77	7/12/77	4/1/78
1	0	1	0	2	3
2	0	0	0	2	4
3	0	0	0	2	2
4	0	1	1	3	4
5	1	1	1	3	3
6	0	1	0	3	1
7	1	1	1	1	1
8	1	1	1	1	1
9	1	1	1	2	1

The psychrometers were equally distributed in holes 2 and 4. The two lowest instruments, at site 2, were saturated during installation and rendered inoperative. The rest settled down to show a reasonably constant soil suction pressure of about -3500 kPa. The first summer rains started about mid-November and a psychrometer at site 2, located about mid-way through the over-burden, reached saturation and ceased functioning.

At site 4 the second lowest psychrometer (positioned close to the shear zone identified by the displacement instruments) showed increasing moisture preceding the second highwall failure. Over 100 mm of rain had fallen in a few days after mid-December and this may have contributed to the increase in moisture at this location. Further evidence of the penetration of moisture to the base of the overburden was provided by the lowest psychrometer at this site which rapidly reached saturation with the very heavy rainfall late in January (over 350 mm fell in four days). The rest of the psychrometers showed very little change for the duration of the monitoring period, implying that the rain did not soak through the overburden but followed well defined channels.

4.   RESULTS OF MATERIAL TESTING PROGRAM

In order to help analyse the behaviour of the highwall, cored samples were removed from four boreholes located near the instrument line and a total of 84 direct shear and 18 triaxial tests were performed. Godfrey (1977, confidential report) also performed direct shear tests in both saturated and unsaturated conditions on material removed from the black clay shear plane.

The resistance acting on a shear plane may be defined by the Mohr-Coulomb equation:

$$\sigma_s = C + \sigma_n \tan \phi$$

where σ_s = shearing stress

σ_n = normal stress

C = cohesion

ϕ = angle of shearing resistance

The results of Godfrey's tests on the black clay shear plane showed that at a natural moisture content of about 17%, the C value was 40 kPa and ϕ about 12°. However, when the material was saturated (about 21% moisture content) these values dropped to a C value of 0 to 10 kPa and ϕ to about 4 to 10°.

In contrast, the direct shear and triaxial tests performed on a range of material taken from between the coal seam and 12 m below the bench, showed much greater strength. Although the characteristics of material taken from the same location were similar for both testing techniques used, this strength varied considerably between locations - some only a few metres apart. The average peak shear resistance for all the samples gives a cohesion value, C, of about 740 kPa while the angle of shearing resistance, Ø, is 19°. The average residual shear resistance, on the other hand, shows a drop of C to 290 kPa with Ø at about 16°.

Within the limitations of the material testing programme, a significant number of samples showed a strength dependancy on their moisture content. Most of the samples were tested at a natural moisture content of between 10% and 16%. An increase of 1% in the moisture content reduced the cohesion value by up to 200 kPa in one case.

5. HIGHWALL FAILURE MECHANISMS

The horizontal stress, which increases with depth within country rock, alters to accommodate the excavation of a trench. This gives rise to the development of shear stresses which reach a maximum at the toe of the highwall face (Stacey, 1970). If the shear strength of the material is exceeded, relative movement across the developing shear plane will occur. This movement will cease when the resisting force generated by the plane balances the components of horizontal ground force and gravitational force along that plane.

5.1 Primary shear plane

It was clear from observation and subsurface instrumentation that movement was taking place on a shear plane inclined at about 8° into the pit just above the coal seam, as illustrated in Figure 7. Movement on this plane was inevitable because it showed much lower strength characteristics than the surrounding rock (Section 4). After the initial response to excavation of cut 6#, most of the move-out could be attributed to coal blasts in the pit. This resulted in a reduction of the horizontal confining stresses above this plane and the subsequent opening up of joints at the surface.

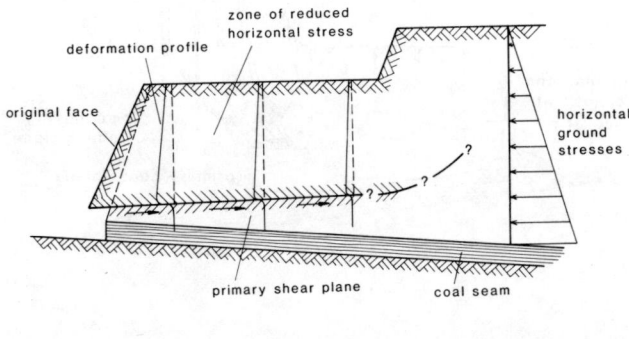

Fig. 7. Overburden deformation as a result of movement on the primary shear plane

5.2 Secondary shear plane

The development of a steeply inclined secondary plane was apparent in both cases of highwall failure. Movement of material down this plane provided the driving force required to overcome resistance on the primary plane. In the case of highwall failure #1, this plane appeared to intersect the primary plane at some point beneath the top of the bench.

The results of direct shear tests on material taken from three locations between 29 and 36 metres below the bench have been plotted in Figure 8. For a vertical stress assumed proportional to overburden depth at the point of intersection of the planes, and a shearing resistance $\tau_n = 24$ kPa $+ P_n \tan 8°$ acting on the primary plane, then Mohr's circle, as shown in Figure 8, can be constructed. If it is to intersect the line one standard deviation from the average rupture line, then the horizontal confining pressure must drop to less than 1/4 of the vertical stress.

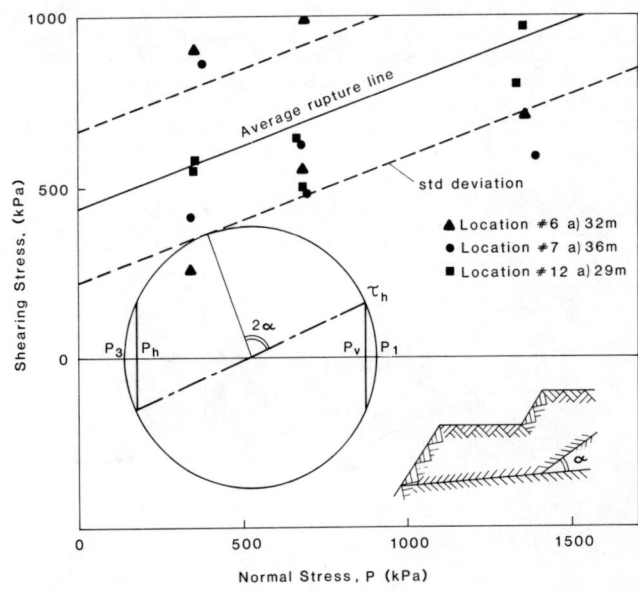

Fig. 8. Representation of critical stresses for rupture of material using Mohr's Circle

Clearly, if lower strength material is present, particularly on steeply inclined bedding planes, then the development of the secondary plane is assured when the horizontal confining stress drops to a low value. Since the field study further geological evidence has been collected which indicates the existence of bedding planes which gradually steepen in dip towards the pit away to the east of the highwall. It is quite possible that the secondary plane coincides with these bedding planes and that this secondary bedding plane is merely an extension of the primary plane.

5.3 Static stability analysis

Instability is not necessarily implied by the development of the secondary shear plane. The behaviour of the slope is determined by the residual frictional properties on both the primary and secondary planes as movement begins.

This system can be readily modelled on computer by dividing the overburden, within the failure boundaries, into a number of slices and comparing the gravitational components of force down the slope relative to frictional resistance on the shear planes. The force transmission between slices is assumed to act parallel to the slip plane on which the slice rests.

Factor of safety is defined as $\dfrac{\Sigma F \text{ resistance}}{\Sigma F \text{ driving force}}$. If the factor of safety is less than one, then failure is predicted.

A cross-section of the slope through the centre of highwall failure #1 is represented in Figure 9(a). The inclination of the primary and secondary shear planes have been selected for the most accurate representation of the real failure. However, it can be shown that a variation of the secondary plane angle between 35° to 65° will not influence the factor of safety greatly. Alterations in the residual strength properties of the shear plane material are very significant, as shown in Figure 9(b).

(a)

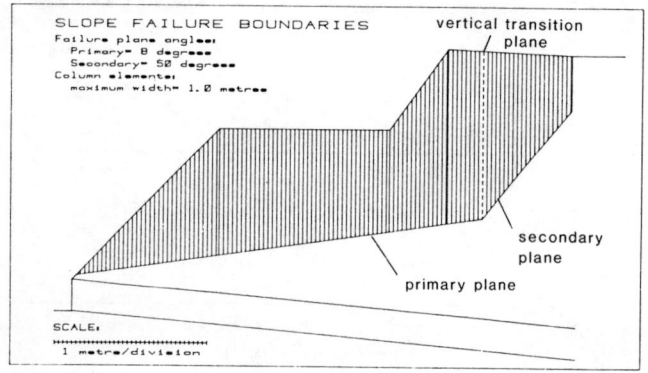

(b)

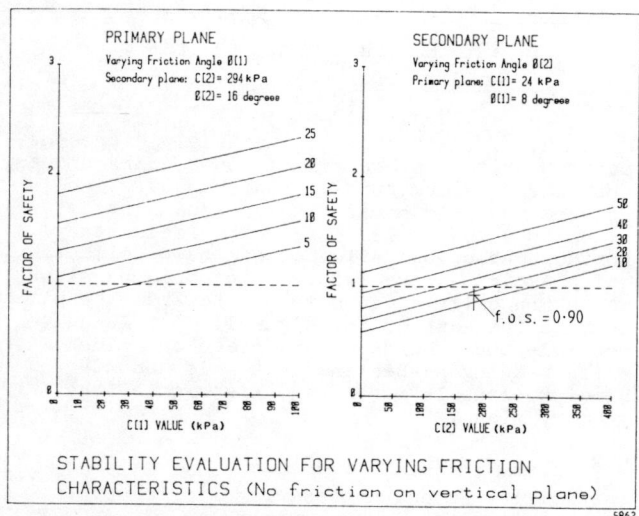

Fig. 9. Establishing the factor of safety of the slope for differing friction values in the primary and secondary planes

5.4 Dynamic stability analysis

Highwall failures cannot be accurately modelled using a simple 2-dimensional approach only. The failure records and post-failure photographic evidence suggest that instability and collapse, particularly in the case of the first failure, did not occur simultaneously along its length. It appears that the reduction in confinement of the overburden after the removal of the dragline bridge caused the development of the secondary plane from beneath the crest of the highwall. The collapse of this material close to the edge caused a further reduction in confinement in

intact overburden. This resulted in an eastwards extension of the failure boundary.

The energy released by this movement of the mass towards the pit was transmitted to the overburden in the south which followed in sympathy. Eventually movement ceased when 450 m of the highwall had failed.

The restabilization of the overburden can best be described in Figure 10. The surface geometry of a stylized cross-section through failure is shown in Figure 10(a). A computer model of the 2-dimensional dynamic behaviour of the slope is shown in Figures 10(b) and 10(c). Stability returned as the mass of the driving wedge diminished.

If the average residual shearing resistance (Section 4) is used for the secondary plane, then Figure 9(b) gives a factor of safety of 1.1, which implies stability. However, if the cohesion value is dropped to 170 kPa, the factor of safety drops to 0.9 and movement of the overburden commences. These values have been used in Figure 10(b) to produce displacements as shown.

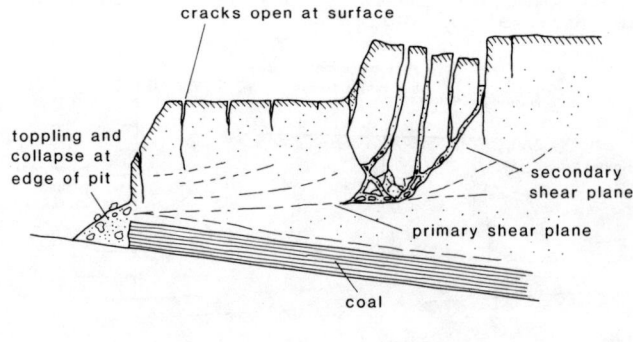

Fig. 10(a). Postulated failure boundaries based on surface observations

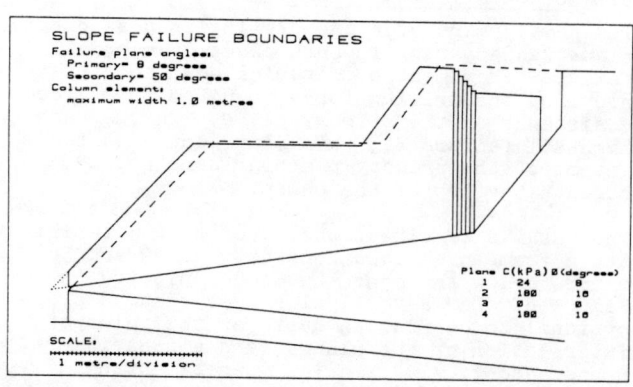

Fig. 10(b). Computer model failure boundaries based on the bi-planar mode of failure

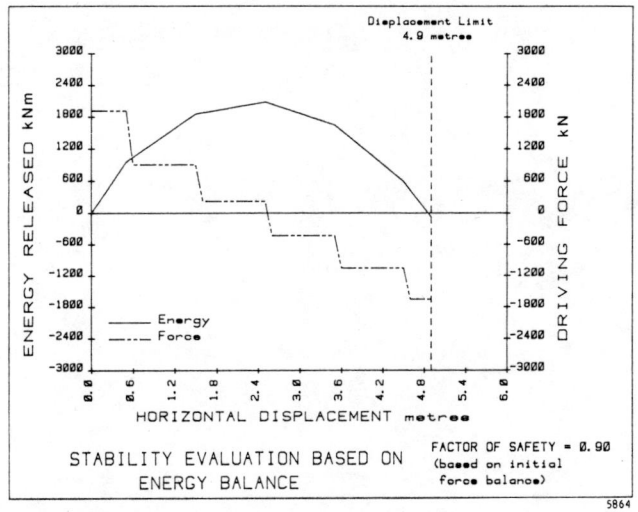

Fig. 10(c). Graph of energy released and driving force

6. CONCLUSIONS

Two highwall failures were recorded in a northern pit at Goonyella during the course of this study. Although they differ in magnitude, the mechanisms governing both were the same.

A redistribution of the horizontal confining stress within the overburden follows a creation of a trench to expose the coal seam. If a shallow dipping plane of low shear strength exists near the toe of the wall then it is highly probable that the shear stress generated by the redistribution will be dissipated by movement on this plane. The result is a reduction in the horizontal confining stress, and it is this factor that determines the instability and ultimate failure of the wall.

The shear strength of the intact overburden could be exceeded when the horizontal confining stress drops well below the vertical stress. This is most likely to occur at some point in the material just above the primary plane. The following factors contribute to this condition:
1)   Increased slope batter angle
2)   Moisture penetration to the shallow dipping plane leading to a reduction in resistance to movement
3)   Buffer blasting, which would both reduce the shear resistance on the plane and provide vertical cracks for the ingress of moisture
4)   Coal blasting shock wave, which provides additonal force to overcome the frictional resistance to movement on the plane.

The failure of the intact overburden is dependent on development of a steeply-inclined secondary plane. The wedge bounded by this plane moves, driving the overburden towards the pit. This wedge must then split up into segments, each undergoing a directional transition where the secondary and primary planes meet. This results in a reduction of the driving force, the cessation of movement, and return to stability.

It is also clear from evidence that failure does not occur simultaneously along its length. The initial instability usually encompasses only a small section of the highwall. The energy released by this movement can be enough to extend the failure boundaries and so include sections of the highwall that would otherwise have remained stable.

7. REFERENCES

Fuller, P.G. and Cox, R.H.T. (1970). Highwall and spoil pile failure processes identified by sub-surface instrumentation, Goonyella Mine - phase 1. CSIRO Aust. Division of Applied Geomechanics, Technical Report No. 80.

Golder Brawner and Associates (1972). Report to Utah Development Company on Slope stability at the Goonyella Mine, Queensland, Australia, 4 Volumes.

Hagen, T.N., McIntyre, J.S. and Boyd, G.L. (1978). The influence of blasting on mine stability. First International Symposium on Stability.

Stacey, T.R. (1970). The stresses surrounding open-pit mine slopes. Planning Open Pit Mines. A.A. Balkema, Amsterdam pp. 199-207.

OPEN-CUT SLOPE DESIGN USING PROBABILISTIC METHODS

Profile d'exploitation à ciel ouvert utilisant les méthodes de probabilité

Der Entwurf einer Tagebergbauböschung unter Anwendung von
Wahrscheinlichkeitsmethoden

P. Morriss
Senior Rock Mechanics and Mining Engineer, Golder Associates, Melbourne, Australia
H. J. Stoter
Technical Consultant, CRA Consultants, Melbourne, Australia

SYNOPSIS

Hamersley Iron operates open cut mines at Tom Price and Paraburdoo in Western Australia. Over a period of four years probabilistic methods have been developed to optimize the pit slopes from economic and safety viewpoints. The reasons for using such techniques are tied directly to the behaviour of the rock mass. Some aspects of the collection and analysis of data to determine structure and strength properties are given. Computers were used to assist with the analysis and design. A new computer related technique for sampling bivariate structural data is discussed. This allows Monte Carlo sampling of structural and strength parameters to derive a probabilistic model of pit slope sections. The strengths and weaknesses of this approach are discussed.

RESUME

Hamersley Iron opère des mines à ciel ouvert à Tom Price et Paraburdoo en Australie occidentale. En l'espace de quatre ans des méthodes de probabilité ont été développées pour améliorer l'utilisation des pentes de l'exploitation minière, du point de vue économique et de sécurité. Les raisons qui ont permis d'utiliser ces techniques sont directement liées au comportement de la substance du rocher. L'article présente quelques aspects de d'échantillonage et de la méthode d'analyse utilisés pour déterminer les paramètres de stratification et de résistance. Des ordinateurs ont été utilisés dans cette analyse et dans la réalisation du projet. Une nouvelle technique informatique pour échantillonner les données stratigraphiques bivariées est discutée. Cela permet l'échantillonage Monte-Carlo des paramètres de stratigraphie et de résistance pour dériver un Modèle de Probabilité des sections des pentes du puits de la mine. Les avantages et les inconvénients de cette méthode sont aussi considérés.

ZUSAMMENFASSUNG

Die Fa. Hamersley Iron betreibt Tagebaubergwerke in Tom Price und Paraburdoo in Westaustralien. Über eine 4-jährige Zeitspanne wurden Wahrscheinlichkeitsmethoden entwickelt, um die wirtschaftlich als auch sicherheitsmäßig bestmögliche Böschungsneigung zu ermitteln. Die Gründe für die Anwendung solcher Methoden sind direkt auf das Verhalten des Gebirges zurückzuführen. Es werden einige Angaben zur Datenerhebung und -analyse zur Bestimmung der Struktur und der Festigkeit gemacht. Zur Analyse und zum Entwurf wurden Elektronenrechner eingesetzt. Eine neue, auf Elektronenrechner beruhende Methode zur Stichprobenentnahme struktureller Daten auf zwei normal zueinander stehenden Ebenen wird beschrieben. Dies macht die Zufallsprobenentnahme von Struktur- und Festigkeitsparametern möglich, um daraus ein Wahrscheinlichkeitsmodell der Tagabauböschungsprofile herzuleiten. Die Vor- und Nachteile dieses Verfahrens werden besprochen.

1. INTRODUCTION

During the feasibility study stage of an open cut mining project, the amount of geotechnical data available, from which slope design parameters must be deduced, is usually very limited There is a temptation to use statistical techniques to evaluate this limited data because of its variability. However, unless it can be determined that such data is truly representative of certain 'domains' of the proposed pit walls, there is no basis for this approach.

In an operating mine the opportunity exists to collect a great deal more data relating to the rock structure and discontinuities which are relevant to the design of final wall slopes. In this situation, understanding and allowing for the variability in the orientation, spatial distribution and shear strength of relevant discontinuities is of prime importance. In this way it is possible to design a slope where the likelihood of failure of that slope for a given safety factor is known.

The basic concepts of designing slopes using probabilistic techniques are well documented (Refs. 1-6). This paper highlights some of the difficulties and technicalities encountered using these techniques at a large open-cut mine.

2. BACKGROUND

Hamersley Iron operates two open cut Iron Ore mines at Tom Price and Paraburdoo in the north-west of Western Australia. Total shipments have exceeded 40 Million tonnes per annum. At Tom Price, the orebody has a strike length of nearly 7km, and the pit slopes will be formed predominantly in a weak, highly fractured shale which will attain total heights of up to 400m. Structurally, the orebody consists of a major east-west synclinal fold, on which north-south cross folding and sub-parallel en echelon folding have been superimposed. Faulting and surface erosion has led to a complex surface representation, as shown in Fig. 1. Figures 2, 3 show typical sections through the main orebody.

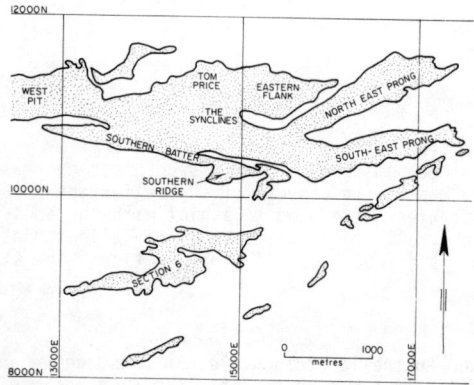

Fig. 1 Main Orebody Locations, Tom Price

In order to design the final pit slopes of this mine it was necessary to

(i)   identify the location and scale of the folding within the wall rocks;

(ii) measure the orientations and nature of the discontinuities influencing slope stability;

(iii) measure the shear strength parameters of these discontinuities;

(iv) measure the groundwater situation;

and model this information in a manner which would economically maximise the design slope angles.

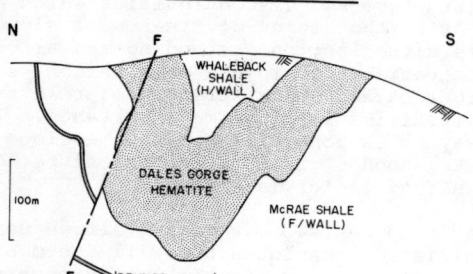

Fig. 2 Section through 16000mE

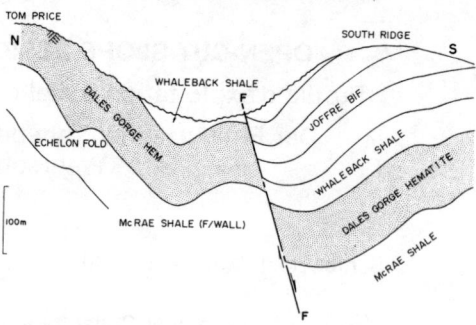

Fig. 3 Section through 14500mE

The orebody shape at Tom Price dictated that in many locations final batters were to be cut without 'pushbacks'. Consequently there was limited opportunity to map structure and discontinuities prior to the final batters being formed. Analysis of available drillcore and discontinuity data did however indicate considerable spatial variation in discontinuity orientation and nature.

A comprehensive data collection program was therefore initiated, which ultimately included:

(i)   more than 3000m of oriented diamond-drillcore in the wall rocks;

(ii) detailed discontinuity mapping of more than 3000m of available faces in the pit;

(iii) a program of 240 shear tests on discontinuities;

(iv) back analysis of several existing slides.

Figure 4 shows a flowchart of rational slope design adopted at Hamersley. Obviously, for a smaller pit or simpler structure, less data would be required.

Because of the volume of data which required processing, a computer was used to store details of diamond drillcore logging and face mapping. Computer programs, adapted from Ph.D. thesis by Rosengren (7) produced stereonets of selected discontinuity data. This system of data files and programs formed the basis of a comprehensive computer assisted slope stability system dealing with:

. Data storage and retrieval

. Comprehensive stereographic analysis

. Trend analysis

. Statistical analysis of data

. Slope Design Modelling

. Sensitivity analysis

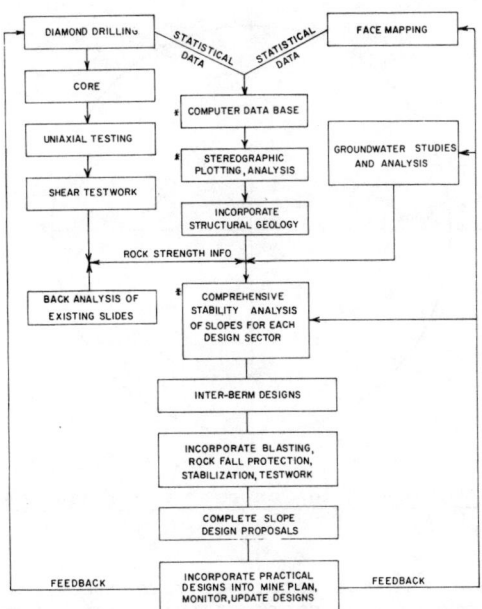

Fig. 4   Flowchart of Rational Slope Design

3.   DESIGN APPROACH

From stereonet analysis of the oriented drill-core, typical structures which would intersect the final walls were identified. These are indicated in Fig. 5, together with the appropriate models used in the slope analysis. This preliminary analysis of discontinuity orientation data showed the need for a probabilistic analysis, so as to consider the effect of measured variabilities on the stability of any design.

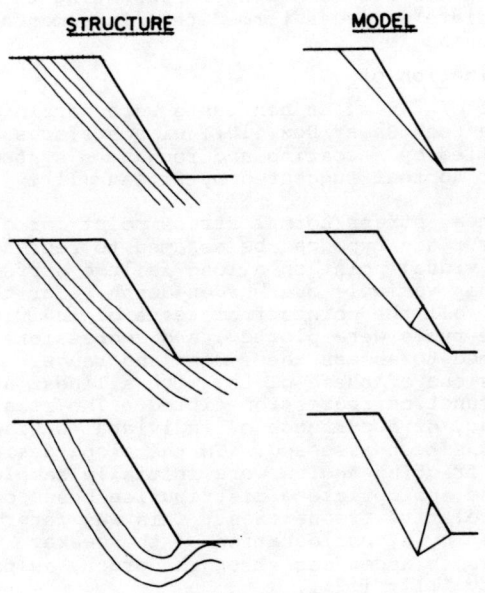

Fig. 5   Failure Modes and Analytical Models

The following sequence of events was therefore followed in the design of the slopes on each section at Tom Price:

(i)   compute the mean dip and variation of discontinuities in individual domains;

(ii)   regenerate this data appropriately in an analytical model;

(iii)   measure the discontinuity length and spacing data if relevant;

(iv)   measure and model the appropriate shear strength data;

(v)   derive the optimum interberm slope angles and incorporate design berms;

(vi)   carry out sensitivity analyses, allowing for critical parameters;

(vii)   combine optimised sections to form practical final wall designs.

The following section discusses four aspects of the study, which have applicability to other situations. Space does not permit a full description of the system developed at Hamersley, however.

4.   SOME ASPECTS OF DATA COLLECTION, ANALYSIS AND MODELLING

4.1   Core Orientation Technique

A novel adaptation of two stereographic projections was developed by an author to assist with orienting drillcore. Using the technique described by Morriss (8), it was possible to orient drillcore very simply whilst at the drill rig, reducing the chance of logging 'bad' data. The modified equal area projection is shown in Figure 6, and consists of a 'polar' net with an 'equatorial' net rotated through 90° to give an east-west rotation axis. Poles to planes on the core can simply be rotated on this net *without an overlay* into their correct location.

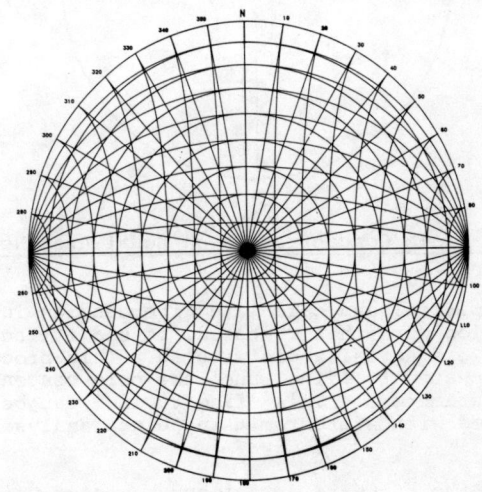

Fig. 6   Equal Area Net for Orienting Drillcore

4.2 Analysis and Modelling of Discontinuity Data

Discontinuity orientation data is commonly plotted as poles on a stereographic projection. In any domain, poles to planes which could affect the slope stability commonly plot as a binormally distributed set or group. It was therefore necessary to determine parameters describing each group to enable it to be reproduced by a Monte Carlo Sampling procedure. These are:

(i)   the mean direction and dip;

(ii) the plane of symmetry through the concentration;

(iii) the standard deviations along two perpendicular axes of symmetry.

A practical technique was developed to achieve these requirements, and is briefly described below.

Fig. 7 shows a typical equal area projection of a pole concentration representing planes within a single domain. The concentration was rotated on the sphere such that the major axis of symmetry lay in the north-south axis, whilst the specified mean orientation was moved to the downward vertical (centre of the projection). Fig. 8 shows the rotated concentration. The standard deviations were then calculated on the north-south and east-west axes using poles from two 'corridors'.

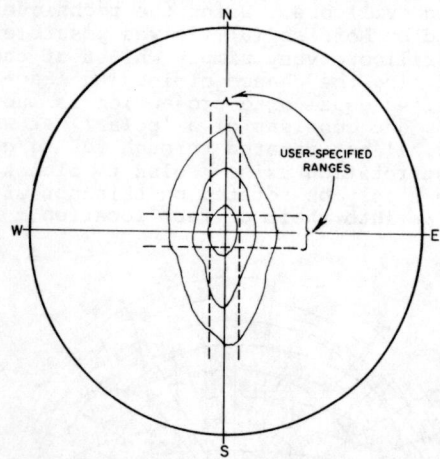

Fig. 7   Pole Concentration on Equal Area Net

The empirical and calculated frequencies in each direction were then compared, to see if the poles were normally distributed. This process was repeated to 'fine tune' the pole concentration parameters. This 'fine tuning' can be assisted with a program using eigen analysis (9).

Whilst the standard deviations computed using this method are not the actual angular deviations, poles were regenerated for the failure analysis models using the inverse process. The technique is therefore self-consistent.

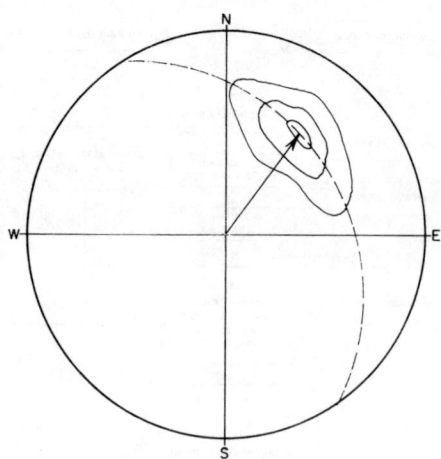

Fig. 8   Rotated Pole Concentration, Equal Area Net

Adjustments to the basic programs allow various interpretations of stereographic data. Some of these include production of stereonet diagrams with variable contouring, cylindrical folding correction for unoriented diamond drillcore and examination of jointing with respect to bedding planes.

4.3 Modelling of Representative Shear Strength Data

The shear strength available on a large scale failure surface in a rock mass is a function of the surface (discontinuity) coating, its waviness, continuity, and the groundwater conditions. For a continuous surface, the shear strength at equilibrium can be expressed by:

$$\zeta = c + (\sigma_n - u)\tan(\phi - i)$$

Where ϕ is the 'fundamental' friction angle, and i is a geometric component representing the appropriate waviness termed the 'incremental friction angle'.

Determination of ϕ

A total of 240 shear box tests were carried out using a Hoek Shear Box (10), with surfaces classified by a coating and roughness system similar to that suggested by Piteau (11).

Each shear stress/normal stress point for a given surface type can be assumed to represent an individual point on a long failure surface which has variable overburden depth. For this reason, all the points from tests of individual surface types were plotted, and regressions performed to assess the ϕ mean and range. Fig. 9 shows one of these plots, with a linear and power function regression fitted. The relative frequency of occurrence of individual surface types was then assessed. In the slope design model, friction angles were initially sampled from the appropriate ϕ distribution based on these relative frequencies. This was later replaced by a consideration of the weaker surfaces in ascending strength order, as proposed by Lilly (12).

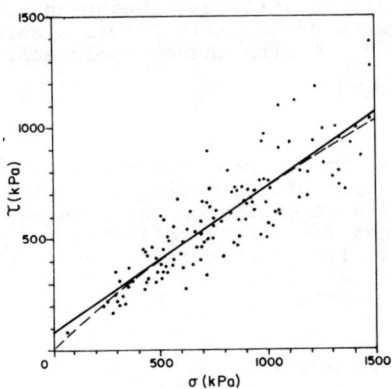

Fig. 9 $\zeta - \sigma$ Diagram for all Shear Tests on
One Surface Type. (After Lilly)

Determination of i

The incremental friction angle i is in this
situation the 'step angle' which a sliding block
must overcome for failure to occur. This angle
is a function of the scale of the failure, as
indicated in Fig. 10. McMahon (13) estimates
that the waviness wavelength pertinent to an
analysis is between 1 and 10% of failure plane
length as at less than 1% a slide requires
little dilation to overcome asperities, whilst
at greater than 10% it constitutes a structure
requiring analysis as a separate domain. At Tom
Price, i values were determined using several
techniques:

(i)   analysis of drillcore and face mapping
      data using the Cumulative Sums Technique
      (Lewis (14), Piteau (15);

(ii) analysis of stereographic pole data
      which exhibited double-peaked maxima;

(iii) detailed appraisal of constructed
      sections from oriented drilling.

There is insufficient space to expand on these
techniques here. The i values were found to be
log-normally distributed, with means of up to 6°
and standard deviations of around 3°

4.4   The Slope Modelling Technique

The slopes were modelled on the computer on a
section-by-section basis, using the collected
orientation, shear strength density and ground
water data. The sections were then joined to
produce a smooth design with allowances made for
sensitivity analyses.

Monte Carlo Simulation techniques were used to
sample the failure surface orientations and
failure plane strengths as follows:

(2) For a possible failure geometry occurring
    over some portion of the design slope height
    on an individual section;

    (i)   one plane direction and dip was
          selected from an 'ideal' binormal dis-
          tribution scattered on the equal area
          projection;

    (ii) the strength value ϕ could be sampled
          from distributions of particular
          failure plane types specified by the
          user;

(iii) the additional strength value i was
      calculated as described in Section 4.3;

(iv) a safety factor was calculated for a
      specified slope direction height and
      angle;

(v)   steps (i) to (iv) were repeated up to
      1000 times, and safety factor distri-
      butions were compiled;

(vi) the procedure was repeated for each
      slope height that the geometry was
      applicable to, and for a range of
      inter-berm angles typically from 35 to
      65 degrees

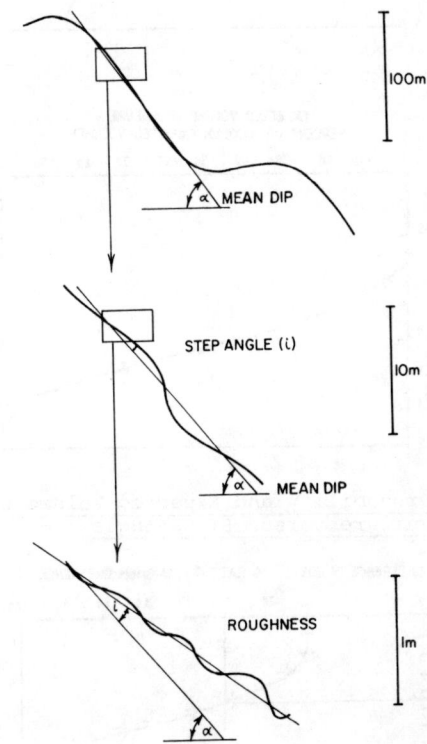

Fig. 10   Geometric Component of Shear Strength i

For each slope height and slope angle, the
volumes and number of all analyses with
safety factor less than 1.0 were calculated.
Graphs of probability of failure and
expected volume of failure as described by
McMahon (2) were therefore computed.
Examples are given in Fig. 11. The economic
optimisation of the slopes was then carried
out by considering;

(i)   the initial costs of excavation;

(ii) the cost of removal of failed material;

(iii) the fixed cost associated with the
      loss of a haul road, for example.

Optimisation must then always occur at some
finite value of probability of failure, as
decreasing costs of critial excavation must
be balanced by increasing costs of removing
material etc for increasing slope angles.
Fig. 12 shows an example of economic
optimisation.

TOM PRICE PLANAR FAILURE SH=135m

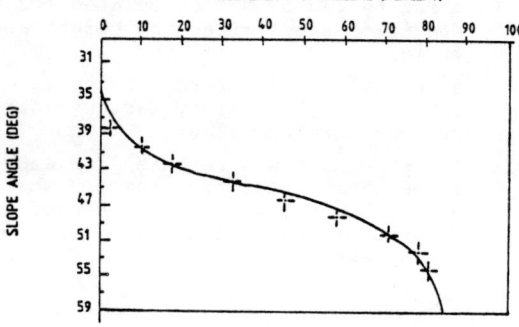

PROBABILITY OF FAILURE (PERCENT)

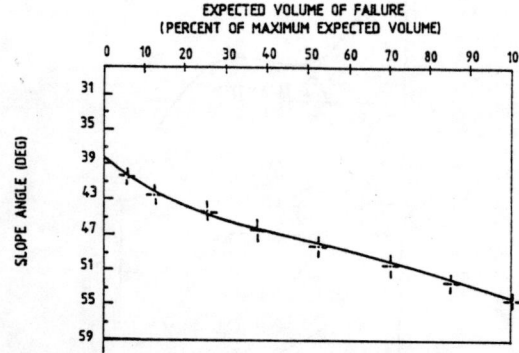

EXPECTED VOLUME OF FAILURE
(PERCENT OF MAXIMUM EXPECTED VOLUME)

Fig. 11 Probability and Expected Volume of
Failure versus Slope Angle

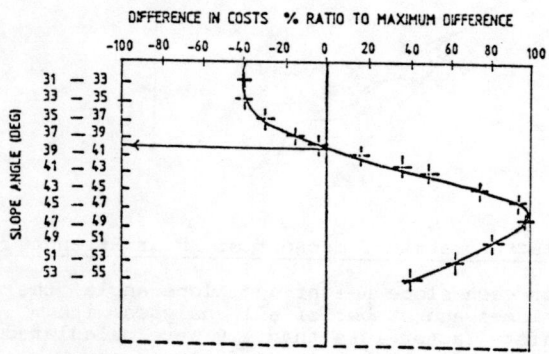

DIFFERENCE IN COSTS % RATIO TO MAXIMUM DIFFERENCE

MAXIMUM COST DIFFERENCE: 33.44   OPTIMUM SLOPE ANGLE: 39.0

Fig. 12 Economic Slope Optimisation

There are some practical limitations related to
failure in a berm not being independent of failure in a geometry behind the berm. The strength
in this approach is a detailed analysis which
incorporates the whole spread of observed or
implied conditions.

(2) For passive-active failure geometry, the
    'normal' and 'reverse' bedding directions
    were sampled from two binormal distributions.
    The procedure was then the same as for the
    planar failure mode.

(3) Other failure analyses - Wedge and Janbu
    type - with Monte Carlo sampling were avail-
    able, but not with the economic analysis.

5. SUMMARY

Probabilistic design procedures are not applicable to all open pit situations. However, given
the complexity of geological structures at the
Hamersley Iron mines, the collection of sufficient data has allowed successful application of
such statistical procedures to the design of
their open pit slopes.

6. ACKNOWLEDGEMENTS

The Authors wish to extend their thanks to the
Management of Hamersley Iron Pty. Limited and
CRA Consultants Pty. Limited for permission to
publish this paper. Thanks are also due to
B.K. McMahon, D. Piteau and K.J. Rosengren for
their valuable assistance given during the
development of the slope design program.

7. REFERENCES

(1) McMahon, B.K., A Statistical Method for the
    Design of Rock Slopes, Proc. 1st Aust-New
    Zealand conf. Geomechanics, Melbourne 1971.

(2) McMahon, B.K., Probability of Failure and
    Expected Volumes of Failure in High Rock
    Slopes, Proc, 2nd Aust-New Zealand conf.
    Geomechanics, Brisbane, 1975

(3) Moss, A.S.E. and Steffen, O.K.H., Geotech-
    nology and Probability in Open-Pit Mine
    Planning, 11th Comm. Mine Metall. Cong.,
    Hong Kong, May 1978.

(4) Sessano, A.A. and Castillo, E., A New
    Concept about the Stability of Rock Masses,
    ISRM 3rd Congress, 1974. Vol. 2-B,
    pp82-86.

(5) Beacher, G.B. and Einstein, H.H., Slope
    Reliability Models in Pit Optimization,
    16 APCOM Symposium

(6) Vanmarcke, E.H., Probabilistic Analysis of
    Earth Slopes, Engineering Geology, 16
    (1980), pp 29-50.

(7) Rosengren, K.J., Rock Mechanics of the
    Black Star Open Cut, Mt. Isa. Ph.D. Thesis
    ANU Can., 1968

(8) Morriss P., Core Orientation in Soft Ground,
    AMF Course 187/82, Rock Mechanics in Open
    Pit Mines, Adelaide, May 1982.

(9) Markland, J., The Analysis of Principal
    Components of Orientation, ISRM, vol 11
    pp 157-163, 1974.

(10) Hoek, E. and Bray, J.W., Rock Slope
     Engineering, 3rd ed. IMM, 1981

(11) Piteau, D.R., Geological Factors Signifi-
     cant to The Stability of Slopes Cut in
     Rock, SYmp. on Planning Open Pit Mines,
     (ed. P.W.J. Van Rensburg, SAIMM.
     Johannesburg, 33-53, 1970).

(12) Lilly, P.A., The Shear Behaviour of Bedding Planes in an Australian Shale with Implications for Rock Slope Design, Int. J. of Rock Mechanics 1982, No. 4.

(13) McMahon, B.K., Personal communication, April 1981.

(14) Lewis, I.H., Measurement of Bedding Waviness from Drill Hole Data with Examples from Section 6, Hamersley Iron Internal Note, July, 1981.

(15) Piteau, D.R. and Russell, L., Cumulative Sums Technique: A New Approach to Analysing Joints in Rock, Proc. 13th Int. Symp. on Rock Slopes, 1971.

J

SITE SPECIFIC APPROACH FOR THE DESIGN OF OPEN-PIT COAL MINES IN CANADIAN FOOTHILLS

Evaluation spécific de sites pour la conception de mines de charbon à ciel ouvert dans les contreforts des Montagnes Rocheuses au Canada

Eine spezifische Geländeuntersuchungsmethode zum Entwurf eines Kohle-Tagebaus im kanadischen Vorgebirge

F. D. McCosh and T. Vladut
Group Leader, Engineering Geology and Senior Geotechnical Engineer, respectively,
Techman Engineering Ltd., Calgary, Canada

SYNOPSIS

Open pit coal operations in the Alberta foothills will encounter difficult geomechanical conditions including steeply dipping coal bearing strata and inner weak coal and bentonite seams. The variation of the rock mass rating can be considered as a model of the variation of the geomechanical conditions of the mine, and was achieved by using a site specific classification. A stress-strain approach to shear strength evaluation of the rock behaviour allows the utilization of the full strength along the discontinuities in the design of slopes for the mines where important stress relief is expected.

RESUME

Les mines de charbon à ciel ouvert des contreforts des Rocheuses de l'Alberta vont rencontrer des conditions géoméca-niques difficiles, y compris des interlits de bentonite, des niveaux charbonneux à pendage raide, ainsi qu'un charbon à faible résistance. La variation de la valeur totale pour la masse rocheuse (RMR) peut être considérée comme un modèle de la variation des conditions géomécaniques dans la mine. Ce modèle est réalisé en utilisant une classification spécifique, développée pour le site. Une approche par contrainte-déformation pour évaluer la résistance au cisaillement permet l'uti-lisation de la pleine résistance le long des discontinuités lors de la conception des pentes pour les mines où un impor-tant relâchement des contraintes est envisagé.

ZUSAMMENFASSUNG

Der Übertagekohlenabbau im Vorgebirge Albertas stößt auf schwierige geomechanische Bedingungen, einschließlich steil eintauchender Kohleschichten mit Weichkohle- und Bentonitsäumen. Die jeweiligen Gebirgsgüten stellen die geomechanische Grundbedingung eines Kohlebergwerks dar, was durch eine spezielle Gebirgsklassifizierung erreicht wird. Spannungs-Ver-formungsmessungen zeigen, daß der Gesamtwiderstand entlang Trennflächen in Rechnung gestellt werden kann. Beim Abböschen sind bedeutende Spannungsentlastungen zu erwarten.

Future large open pit coal mines in the Canadian Rocky Mountain foothills will extend for several kilometers (15 to 25 km) along the seam sub-crops. Prior to development, studies must be carried out to evaluate the technical and economic feasibility of the project. The results of these studies are submitted to the Alberta Energy Resources Conservation Board to obtain approval of a mine development permit and subsequent license to mine under the Coal Conservation Act. The licensing process ensures that the proposed mine operation will maximize resource extraction while minimizing environment and social-economic impact. Geotechnical evaluation of the operation is required during this process to allow proper evaluation of the concerns. Assessment of the viability of the coal exploitation by open pit methods requires definition of the geomechanical condition in the field and design of slopes, mine dewatering systems, and dumps. The geomechanical parameters have a significant impact on both the technical and economic feasibility of the proposed development.

Proposed open pit coal operations in the Alberta foothills will affect large areas. During development of these coal deposits mine operations will encounter difficult geological conditions including steeply dipping coal bearing strata of low strength which will control the design of both the highwall and footwall slopes. In evaluating the stability of the pit slopes the geological structure and its influence on kinematics and rock mass strength should be assessed. This requires the determination of both the shear strength of discontinuities and the strength of the intact rock. Geomechanical evaluation of such large areas makes any assessment of rock structures and extrapolation of rock properties very difficult. Any field observation or laboratory test data could be challenged as being non-representative. Certainly a variation in rock strength is as important as the variation in structural factors. As the mining method employed tends to parallel the geological structure, the possible modes of failure tend to be similar along the entire pit wall. However, the most critical failure mechanism in any one section of the pit could vary with the local geological conditions. The stratigraphy of the Alberta foothills consist of a complex sequence of interbedded sandstones, siltstones, and mudstones, with local bentonite seams associated with the coal. The intense fracturing and related system of discontinuities dictates that evaluation of the strength along joints be fully understood and assessed prior to final mine design.

Because both the geologic conditions and the expected rock behavior will vary within the stratigraphic sequence, pit slope design may vary within individual operations. Ideally, under such conditions, more than one mining method would be recommended. However, it is generally not practical to utilize more than one mining method (ie. dragline vs. truck and shovel) in an actual operation.

Design of highwall and footwall pit slopes involves the identification of the principal kinematically possible failure modes including buckling, translation failure and non-circular slip surfaces. This is based on the evaluation of the distribution of discontinuities in areas of similar geological conditions. After idenfification of the principal mode of failure, an estimate of the stability of the slope is made through the factor of safety as a ratio of the maximum shearing resistance that can be mobilized in the direction of movement to the shearing forces active in the slope.

For stability assessment both the geological pattern in which different failure modes are defined; and the available strength pattern must be considered. Stable slopes for mining purposes are defined by using the available shearing resistance. Definition of geological patterns rely on systematic collection, presentation and evaluation of the geological structure. Typically this data is presented on stereoplots illustrating the distribution in relation to kinematically possible failure modes for individual slopes.

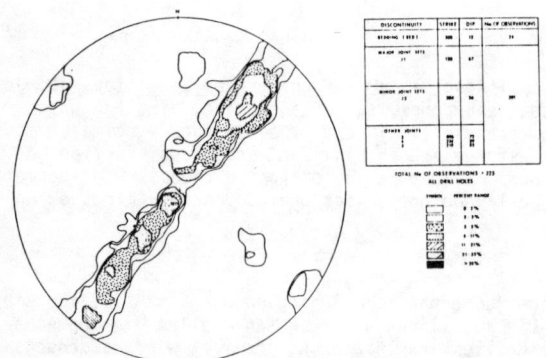

FIGURE 1. CONTOURED STEREOGRAM OF POLES OF DISCONTINUITIES MINE A.

Definition of reliable slopes requires at least knowledge of the structural geology and the relationship to mechanical strength. Evaluation of slope stability requires the incorporation of available strength data. This is done by considering the whole range of strength in the determination of conceptual slopes which could be approached by the selected or preferred mining scheme. This includes ensuring the stability during operations over the entire length of time the slope may be exposed. The importance of time dependent deterioration of the slope is a function of the particular mining method proposed and rock conditions.

The assessment of typical strength data requires a more general evaluation of the geomechanical conditions considering the geological conditions. One way to develop an assessment of a very large amount of geological data is to use a more general approach in the evaluation of the geomechanical properties of the rock mass. Empirical systems of evaluating geomechanical data have been developed to assist in the assessment of rock mass behavior in underground civil engineering structures. These rating systems allow adaption of engineering procedures to changing field conditions. It should be recognized in rock slope engineering much of the solutions are control-

led by the structural implications. The major consideration in slope assessment for mining of open pits is focused on the real operational mining procedure in which choices between mining operations are much fewer.

The geomechanical conditions of the proposed open pit mines was assessed by applying an empirical rock mass rating developed for the specific site. In selecting an existing rock mass classification system the applicability and suitability of the system to open pit slope design had to be assessed. This involved evaluation of three conditions: (1) the range of each parameter employed in the system compared with the actual range of properties for the site; (2) the form of the data required for the classification compared to the study information collected in the field; (3) the suitability of the system in assessing the engineering problems which control the slope design.

In evaluating the use of the Q system /1/ the range of the different parameters, i.e. block size, inter-block shear strength and the active stresses used in the Q system were assessed. The range of each factor is illustrated in three diagrams: the block size which is the ratio between fragmentation (RQD) and joint set number (Jn) in Figure 2; the interblock shear strength which is the ratio between joint roughness (Jr) and alteration (Ja) in Figure 3; and the active stress, which is the ratio between the joint water reduction factor and stress reduction factor (SRF) in Figure 4. The estimation of the ranges of variation of each factor was defined for coal related rocks such as sandstones, siltstones, mudstones. The data used in defining the class of coal related rocks should not be considered exhaustive. The references are related to open pit coal mines in Alberta and U.K. /2/. The data from the minesite relating interblock shear-strength (b) and active stresses (c), fall within a narrower range than that for coal related rocks, in general (Figure 5).

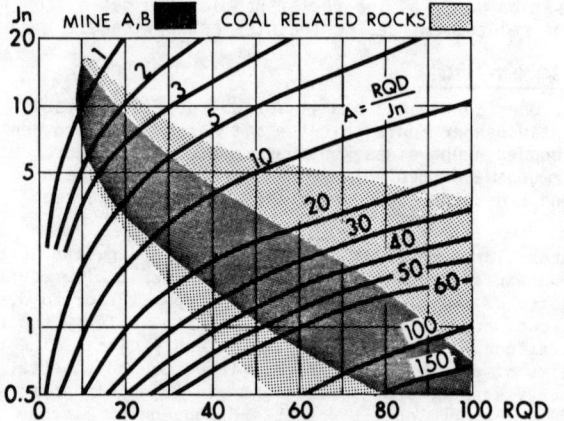

FIGURE 2: ESTIMATION OF THE VARIATION OF THE BLOCK SIZE FACTOR IN THE Q SYSTEM FOR COAL BEARING ROCKS.

The large range of the Q parameter is primarily a result of the large-range in possible block sizes (Figure 6.), but for the particular site of mines a narrower area is defined. This refers to the site of the mines as possible to use a simpler classification. Consideration was given to a classification of rocks as referred by Bieniawski's Geomechanical Classification /3/. However, the rating for joint condition similar to that employed in the Q system were incorporated into the final system employed in the evaluation of the stability of the pit slope. Joint roughness was excluded in the rating but was incorporated in the actual stability analysis.

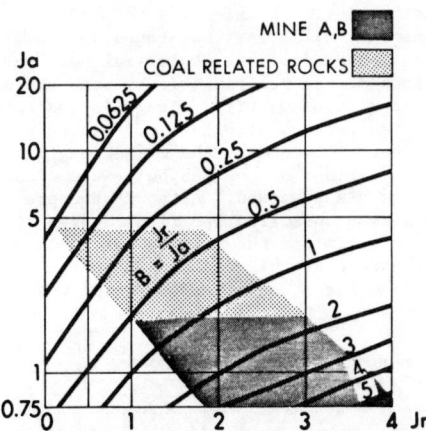

FIGURE 3: ESTIMATION OF THE VARIATION OF THE INTER-BLOCK SHEAR STRENGTH FACTOR IN THE Q SYSTEM FOR COAL BEARING ROCKS.

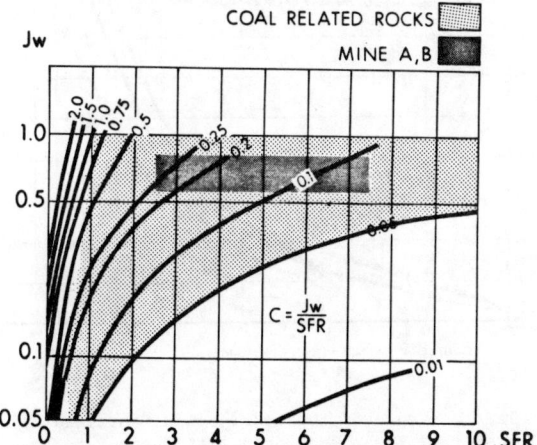

FIGURE 4: ESTIMATION OF THE VARIATION OF ACTIVE STRESS FACTOR IN THE Q SYSTEM FOR COAL BEARING ROCKS.

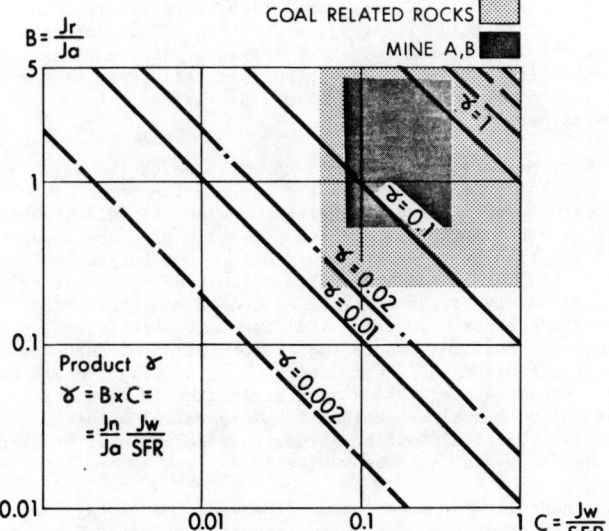

FIGURE 5: EVALUATION OF THE SHEAR STRENGTH AND ACTIVE STRESS FACTORS IN THE Q SYSTEM FOR COAL BEARING ROCKS.

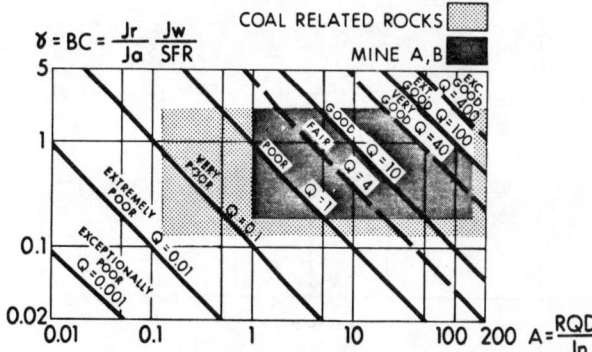

FIGURE 6: EVALUATION OF THE EFFECT OF BLOCK SIZE AND STRENGTH STRESSES IN THE Q SYSTEM FOR COAL BEARING ROCKS.

The classification finally employed in the analysis also used a modified rating for intact rock strength because of the consistently low strength of the rocks at the site.

The detailed evaluation of conditions of joints considered continuity, alteration by weathering, joint separation and type of filling between the joints (Table 1).

Table 1

Consideration of condition of discontinuity in the site specific classification

Condition of joints	Not continuous		Continuous		
			no gouge	with gouge	
Rating	5		3	0	
Weathering	unweathered	slightly	moderately	highly	completely
Rating	10	7	5	3	1
Separation of joints	< 0.1 mm		0.1-1 mm	1-5 mm	> 5 mm
Rating	5		4	3	0
Surface conditions	planar rough	planar moderately rough	planar moderately smooth	planar smooth	slickenside
Rating	5	4	3	2	1

At the proposed mine site the intact rock strength was determined from a large set of point-load strength indexes. A total of 966 rock core samples were tested in a portable point-load testing machine during core logging in the field. The test results indicate (Figure 7) a broad range of point-load indices related to the unconfined compressive strength. Different conversion factors, K, were defined for each class of rock strength and sample orientation in testing device relative to bedding. Several conversion factors have been developed.

For example, Frankline /4/ indicates a value of 25 is representative for all rocks while studies on sedimentary rocks in the U.K. /5/ indicate a value of 29. Evaluation of the data from the foothills site indicates that the conversion factor for tests parallel to bedding was about 35. For rocks with higher strength in the range of 0.5 to 8Mpa the conversation factor became closer to the reference value of 25. Results of tests perpendicular to bedding indicated significantly lower conversion factors of about 15 to 17. The results of the conversion factor are summarized in Figure 8. The figure shows curves based on conversion factors for data from tests conducted perpendicular to bedding, a combination of the results from tests conducted both perpendicular and parallel to bedding and the conversion factor of 25 employed by Franklin.

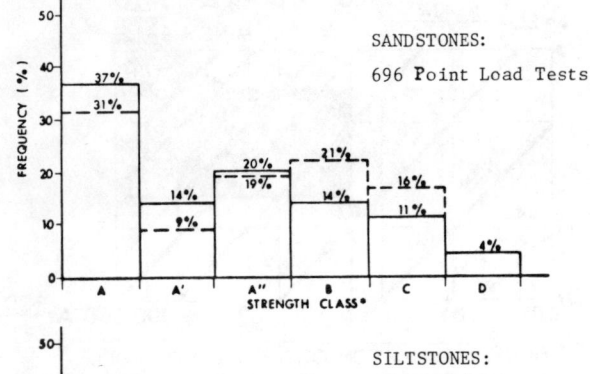

SANDSTONES:

696 Point Load Tests

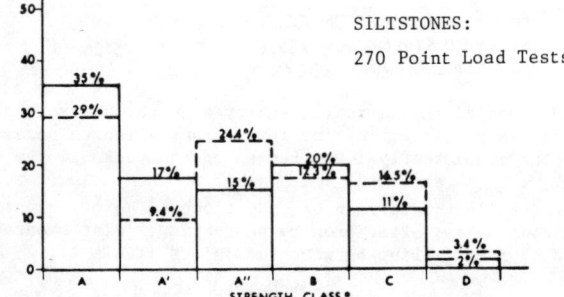

SILTSTONES:

270 Point Load Tests

TOTAL 966 POINT LOAD TESTS

— MINE B
--- MINE A

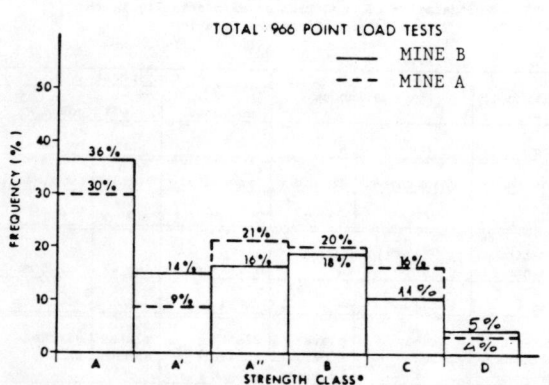

Definition of strength classes for point-load tests;
A 0.01-0.3 MPa A = 0.3-0.5 MPa = 0.5-1 MPa
B = 1-2 MPa C = 2-4 MPa D = 4-8 MPa

FIGURE 7. THE POINT-LOAD TEST DATA FOR THE COAL
MINES A AND B.

The evaluation of the field data by a site specific rating system allowed the construction of a geomechanical model of the properties of rock within the proposed pit area. An example of the computer output for rock mass classification analysis is presented in appendix A. Before the acceptance of the geomechanical model a detailed assessment of its confidence limits was made.

The assessment was based on two types of evaluations.
(a) An evaluation of the model output with respect to the geological information obtained during field investigations was made. Because of the limited exposure of rock in the area, the evaluation was based primarily on the geological data base available from the exploration drilling. This encompassed information from a total of 300 boreholes including core and drilling logs and a complete suite of geophysical logs generally consisting of single point resistance, focused beam resistivity, neutron density and caliper logs.

Initially the general framework indicated by the geological and geophysical data was checked against the trend

of rock mass rating. The highly interbedded character of the sandstone, siltstone, mudstones and coal seams indicated by geophysical and geological data was reflected by the lack of any relationship between the depth and evaluation of rock properties. Analysis indicated very low correlation factors of 0.2 to 0.3 between rock mass rating and depth. Secondly, the effect on the value of the rock mass ratings of the bentonite associated with the coal seams was assessed. Although the presence of bentonitic seams controlled the final design of the highwall slopes and limited the maximum height of the footwalls the presence of this geological feature did not significantly affect the rock mass rating results.

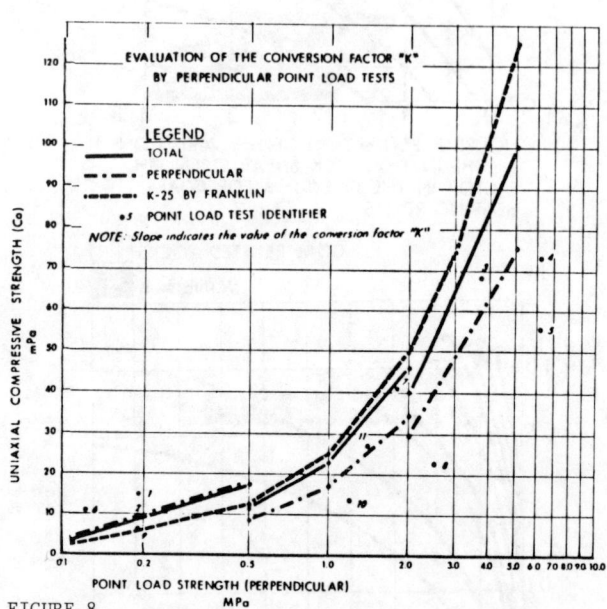

FIGURE 8.

(b) The relevance of the model output obtained using rock mass rating was evaluated with respect to assessment of pit wall stability. A statistical analyses correlating different types of field observations with the rating values was made. Good correlations were found between the rating (RMR) and the rock quality designation (RQD). Better correlations were referred by joint spacing (JS-Table 2), and rating. This correlation for the larger mine (A) which develops along a 25 km length is exemplified in appendix B. All types of correlations referred to in Table 2 are linear type of correlations.

The computer output presents the results of the statistical analysis. Comparing the correlation between the rock mass rating and joint spacing with the relationship between the joint spacing and RQD which presents lower correlation factors, some suggestions for future approaches were considered. The definition of RQD is related mainly to the criteria of recovery of cores in length greater than twice its diameter. For special geological conditions where the fragmentation of seams is more intense maybe the criteria of 10 cm length could be too conservative. Using smaller lengths in definition of the rock quality designation is parallel with the classical definition /6/ better consideration of smaller rock fragments could be achieved.

Low correlations were obtained between the rating and assessment of the groundwater conditions (GW). The combination of groundwater rating with joint conditions, (j-defined in Table 1), indicated an improved correlation with the total rating than that indicated only by the groundwater conditions.

Table 2.

Statistical evaluation of correlation between different
components of rock mass rating*.

x	y	Mine	No. of Points	Correlation factor	Slope (b)	Intercept (a)
RQD	RMR	A	590	0.68	1.72	21.8
		B	320	0.89	1.83	24.5
J.S.	RMR	A	590	0.88	1.87	29.5
		B	320	0.90	2.49	25.4
J.S.	RQD	A	590	0.43	0.37	11.4
		B	320	0.73	0.98	3.79
J	RMR	A	590	0.88	2.2	14.7
		B	320	0.61	4.06	16.7
GW	RMR	A	590	0.69	3.84	23.5
		B	320	0.54	4.2	20
J+GW	RMR	A	590	0.68	2.0	3.9
		B	320	0.64	15.2	2.57
PLR	RMR	A	590	0.48	2.89	43.7
		B	320	0.56	4.00	39.7

* Description of symbols are referred in the text. Y = a + bx.

Generally, the assessment of groundwater conditions in
rock mass evaluation is weak. Therefore more detailed
consideration of groundwater conditions was required for
the design of the slopes. Using a large set of data from
piezometer observations (57 piezometer in mine A and 29
in mine B), the regional underground flow system was i-
dentified (Figure 9) and flow properties were evaluated
by pumping tests.

was also employed in assessing the shear strength variations using Hook and Brown /7/ approaches. Unfortunately
the highly interbedded character of the stratigraphic sequence yield the available laboratory data insufficient
for such an evaluation.

Stress-strain consideration on mine slope stability.

The highly interbedded nature of the stratigraphic sequence associated with coal deposits is a general characteristic of coal deposits. The interaction of the
sandstone, mudstones, and siltstones make the evaluation
of stability a difficult task. These sandstones and
siltstones   are softening type rocks while benton-
ites, mudstones and other materials with high clay contents tend to typically harden. The contrasting behavior
of rock softening and hardening during shear introduce a
number of supplementary uncertainties when considering
the rock strength as one of the basic parameters in determining the factor of safety of slopes. Such consideration was referred to during the International Symposium
on Weak Rocks in 1981 by ISRM /8, 9, 10, 11/.

The difficulty in evaluating the contrasting rock behavior is further complicated if high horizontal stresses
are also present. Field observations including the presence of core discing indicates that this additional factor is present in the area of the Alberta foothills /12/.

The effect of the horizontal stress field was evaluated
with regard to   slope stability for the
deepest areas of the mines which are around 100 m in
depth. This stress relief behavior during mining requires that slope stability be evaluated for different
stages of slope deformations in order to asses the time
related stability.

The stress-strain consideration of rock properties allows
development of safety evaluation of slopes in which both long term and short term stability are considered using the same shear test data. The character of rock hardening or softening can be determined from laboratory test data by considering the shear strength at different levels of strain development. It is difficult to define

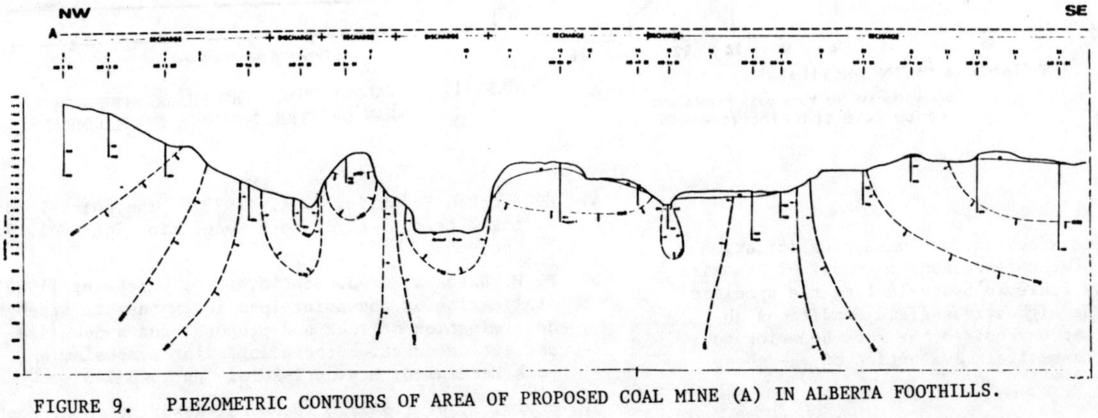

FIGURE 9. PIEZOMETRIC CONTOURS OF AREA OF PROPOSED COAL MINE (A) IN ALBERTA FOOTHILLS.

The references between different components of the rock
mass   rating assessed with the site specific
ratings,were found close to the weight referred by the
last format of the geomechanics classification of joint-:
ed rock masses developed by Bienawski, /3/.

The general correspondence between the geomechanical model obtained by rock mass rating and the geological conditions of the mine site facilitated the evaluation of the
pattern of variation of the rock properties along a very
long mine (25 km mine and 15 km Mine B). The final rock
mass model was employed in extrapolating geomechanical
data for the stability assessment of the slopes in the
mine. The variation of the rock properties with rock
type is presented in Figure 10. Reference was also made
to the lower and upper bounds of the ranges, parallel
with the reference to the variation of structural data,
the dip of the seams. The site specific rock mass rating

and assess the strain required to initiate slope failure,
for geological structures. The persistency of the basic
components of shear strength, i.e., cohesion and friction
is important. Because of the general inability to predict an allowable strain which would not affect safety
considerations, the slopes should be designed using residual strengths. The utilization of residual shear
strength in evaluation of short term stability may result
in generation of conservative slopes.

Such consideration was given in the assessment of mine
slope stability by investigating the whole variation of
strengths between the peak and residual strength. For
the rock hardening type of materials, i.e., coal and mudstones, such evaluations allowed consideration of cohesion was determined and further used in safety consideration of slopes. Also, for larger allowable strains in

the slope such consideration allowed use of some cohesion
developed by rock softening materials (sandstone) during
shearing. However, relatively low differences in the
peak and residual friction angle were observed during
shear testing (Figure 11).

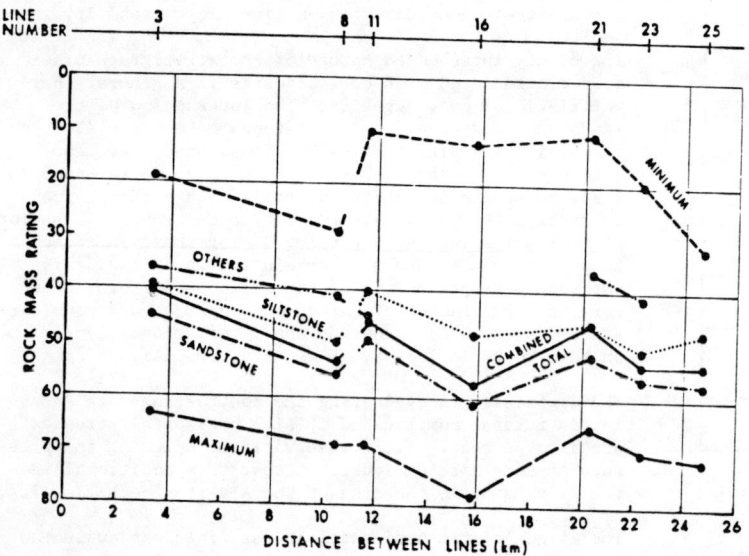

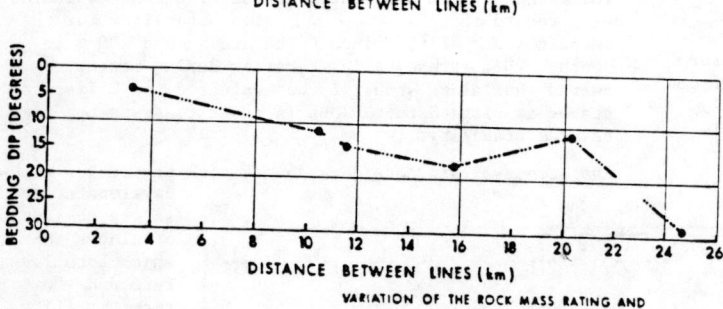

VARIATION OF THE ROCK MASS RATING AND
THE DIP ANGLE OF THE ROCK FORMATION

FIGURE 10.

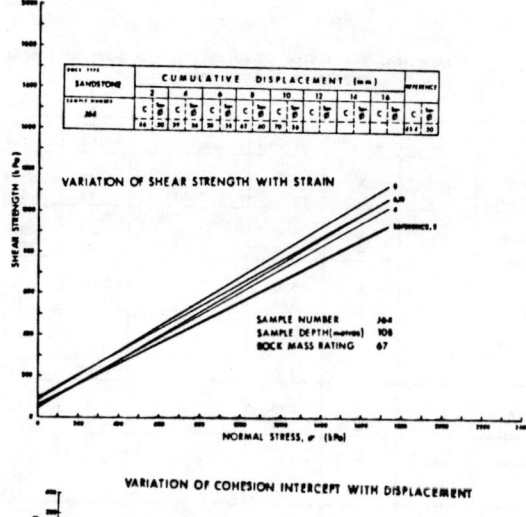

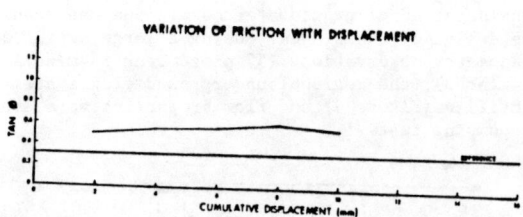

FIGURE 11. EVALUATION OF THE SHEAR STRENGTH BY
MEAN OF STRESS-STRAIN DEVELOPMENT

CONCLUSION

Application of a site specific rock mass classification
for the assessment of the geomechanical model of the site
of the mine could be achieved by evaluating the specific
rock conditions. The high stress field conditions observed at the mine sites required the rock behavior to
be assessed by a stress-strain evaluation of the shear
strength.

Acknowledgement

The authors are grateful for the support of Techman
Engineering Ltd., Calgary, Canada.

References

1. N. Barton. (1975). Estimation of support requirements for un-erground excavations. Sixteenth Symposium on Rock Mechanics, University of Minnesota,
   U.S.A. pp. 163-177.

2. M. J. Scoble and W.J.P. Leigh (1982). Factors governing the stability of rock slopes in British surface
   coal mines, National Coal Board, Newcastle, England.
   pp. 1-8.

3. Z. T. Bieniawski (1979). The Geomechanics Classification is rock engineering applications. 4th International Congress on Rock Mechanics, Montreux.
   Proceedings Volume 2, pp. 91-98.

4. E. Brock & Franklin, J. A. (1972). The point-load
   strength test, Int. J. Rock Mech. Min. Sci. Vol. 9,
   pp. 669-697.

5. F. P. Hassani, J. J. Scoble, B. N. Whittaker (1980).
   Application of the point-load index test to strength
   determination of rock and proposal for a new size-
   correction chart. Proceedings 21st Symposium on
   Rock Mechanics, Missouri-Rola. pp. 543-548.

6. Deere, D. U. (1963). Technical description of rock
   cores for engineering purposes. Int. J. Rock Mechanics, Min. Sci. Vol. 1, pp. 18-22.

7. E. Hoek & E. T. Brown (1980). Empirical strength
   criterion for rock masses. Journal of the Geotechnical Engineering Division, A.S.C.E., Proc.
   Paper 15715.

8. F. H. Tinoco ¢ D. A. Salcedo (1981). Analysis of
   slope failure in weathered phyllite. Proceedings
   of the International Symposium on Weak Rock, Tokyo.
   Vol. 1, pp. 50-57.

9. T. Vladut (1981). Engineering consideration associated with the full strength along the joints, Weak
   Rock Volume 3, edited by A. A. Balkema. Theme 1,
   pp. 1-6.

10. A. A. Loiselle, T. Vladut, Y. Lavoie (1979). Design
    of steep rock slopes using the full strength along
    the joints. 4th Pan American Conference on Soil

Mechanics and Foundation Engineering, Lima, Peru.
Vol. 3, pp. 379-396.

11. D. I. Gough and J. S. Bell (1981). Stress orientations from oil well fractures in Alberta and Texas, Canadian Journal of Earth Sciences, Vol. 18, No. 3, pp. 638-645.

APPENDIX A.

Computer output for the site specific rock mass classification (Mine A).

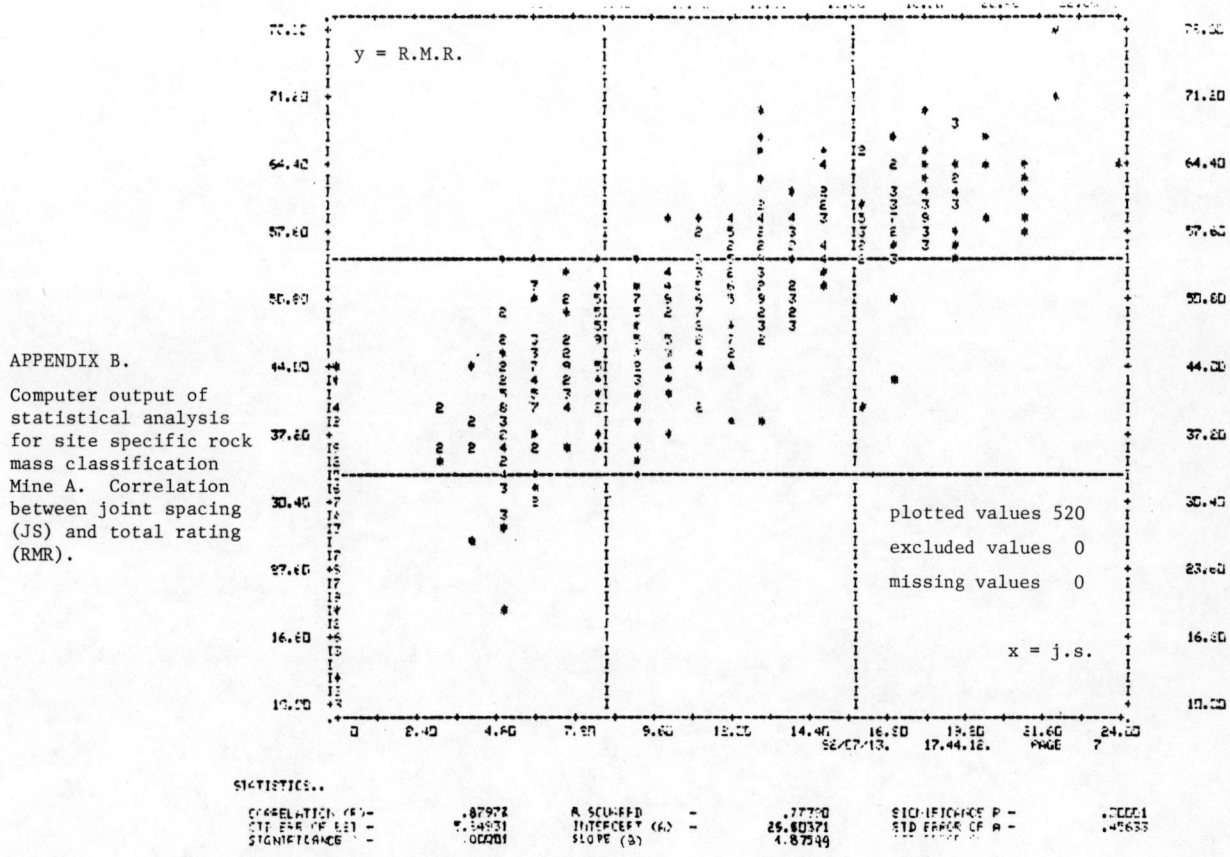

APPENDIX B.

Computer output of statistical analysis for site specific rock mass classification Mine A. Correlation between joint spacing (JS) and total rating (RMR).

BLASTING AND SLOPE STABILITY

Tirs de mines et stabilité des talus

Sprengen und Standsicherheit von Felsböschungen

Z. V. Solymar

Asst. Chief Geotechnical Engineer, Monenco Ltd., St. Catharines, Ontario, Canada

SYNOPSIS

This paper describes case examples where blasting damage to excavated rock faces was observed. One of the examples describes a major slide which commenced immediately following a blast.

RESUME

Cet article décrit des exemples de dégâts causés par des explosions aux fronts de taille en galeries. Un des exemples concerne en glissement majeur qui a commencé immédiatement après un tir.

ZUSAMMENFASSUNG

Dieser Bericht gibt einige typische Fälle der Wirkung von Sprengen auf Felsböschungen. Eines der Beispiele beschreibt eine Felsrutschung, welche sich als Folge einer Sprengung entwickelt hat.

When blasting work has to be performed in the neighborhood of man-made structures the ground vibration is often the factor which finally decides how the blasting is to be performed. Comprehensive data has been collected and published during the last two decades on seismic monitoring, however, there is still some difficulty in establishing limit values for varying degrees of damage. There have been relatively few cases where damage could be proven to be associated to any man made structure.

Less attention is directed during excavation to blast damage of unexcavated rock mass and excavated slopes and rock faces. In practice, if the damage is small, scaling will usually rectify the problem. However, if the damage is extensive, remedial measures such as secondary blasting or rock bolting in combination with shotcrete treatment is often considered.

Damage to rock slopes can be caused by blasts set off next to or near an excavated face. This paper presents case histories of the damage to rock slopes which is considered caused by blasting.

2. ROCK FRACTURING

Knowledge of the mechanisms of rock fracturing is a necessary part of all blasting operations to avoid potential damage to a nearby rock surface that is to remain in place.

Three major zones, namely crushed, fractured and seismic zones develop around an explosion. Concerning rock damage and slope stability, the fractured and seismic zones are the more important. Fracturing is a function of the energy release and the rock

the elastic limit and no fragmentation occurs, except where compressive stresses are transformed on reflection into tensile stresses which may cause the rock to fracture or spall.

2.1 Stress Waves

Approximately 5 to 20% of the total energy released in detonation of explosives is transmitted directly into the surrounding rock mass in the form of stress waves. The amount and the percentage of total energy which goes into stress waves depends on (a) type of explosive (impedance), (b) weight of explosives, (c) burden and (d) length of delay interval between explosive groups.

It is possible to obtain some idea of the comparative effects of different explosives since the impedance of the explosives is equal to its mass density multiplied by the detonation velocity. As a comparison, a 90% strength gelatin dynamite has an impedance ratio of around 1000 gs/cm^3 and an ammonium nitrate (ANFO) blasting agent only about half of this. Selecting the proper explosive is therefore an important factor in reducing vibration levels.

The stress wave is distributed all around a charge, a part of its energy will be distributed within the angle of breakage and the remainder of its energy will travel through the rock at a velocity of 2000 - 5000 m/s. If it is possible for the burden in front of a blast hole to move forward freely and the ignition of the next hole in the row occurs with an adequate delay, a smaller part of the energy will go into the rock. Consequently, if the burden approaches infinity, as for example around a presplit hole, a larger part of the energy will be transferred to the rock. It is, therefore, quite possible that at sites where presplitting is used the greatest vibration problem may be associated with the presplit shots as suggested by Devine et al., 1965.

The stress waves associated with ground motions observed at a given point are dependent upon (a) the energy transmitted by the stress waves, (b) the distance between the detonation and observation point, and (c) the transmission characteristics of the rock mass.

From seismic works it is known that the two material properties that are of significance in wave transmission are the modulus of deformation and the density. Changes in both of these properties for wave transmission in the rock means that a part of the wave will be reflected. The extreme example is when a free face is encountered, the compression wave is reflected as a tension wave and may cause scabbing of a part of the rock near the surface. From a bedding or joint plane reflected waves can cause fracturing far behind the blast. This is due to the tensile strength of the rocks being much less than the compressive strength.

Transmission characteristics of the rock mass can vary from site to site. From three different sites particle velocity observation data were plotted against scaled distance on log-log co-ordinates, as shown in Figure 1. The effect of differences, in charge weight, was eliminated from the data by dividing the scaled distance by the cubic root of the total or per delay charge weight, as suggested by Hendron (1970). The charge weight W used in Figures 1a and 1b is the maximum weight per delay if the delay interval was more than 1/4 of the transit time. The level of vibration was significantly larger for Site A than for Sites B and C, thus indicating different transmission characteristics of the rock.

2.2 Expansion of Gases

In the last stage of rock breakage from an explosion, under the influence of the pressure of the gases from the explosive, the primary cracks expand and the free rock surface yields and is moved forward as described by Langefors et al., 1963.

The majority of the kinetic energy (approximately 50% of the total) remaining in these gases from the explosion is likely to be the more important source of rock breakage. The maximum effect per drill hole and quantity of change is attained if it is possible for the burden in front of a hole to move forward freely.

3. EFFECTS OF BLASTING ON SLOPE STABILITY

The most common visually observed damage caused by blasting are:

3.1 Crushing of Rock

This type of damage is sometimes associated with pre-shear holes as shown in Figure 2 and the overall stability of the slope is not significantly affected. The crushed zone typically extends to about twice the charge radius.

3.2 Backbreak

The most commonly observed damage to a rock face caused by presplitting and by bench blasts is backbreak. This occurs in or behind the fracture zone.

Presplit blasting affects the unexcavated rock by generating the highest particle velocities of any type of blast as shown in Figure 1b for Site B. Fracturing of the rock near the top of a slope can result in instability and extra excavation as shown in Figure 2.

The crushed zone can be eliminated and fracturing minimized by using different strengths of igniter cords instead of dynamite in presplit holes. Figure 3 shows clear cutting of a channel face in poorly cemented, weak sandstone by using 100 g/m igniter cord.

The stability of the slope behind the presplit line can be further affected by inadequately delayed bench blasting, since the presplit failure plane between the blast area and the rock face is not very effective in reducing vibration levels.

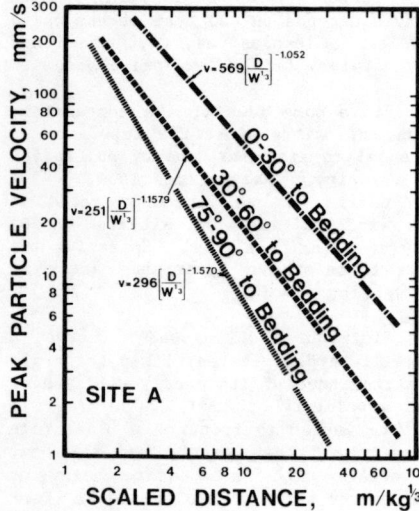
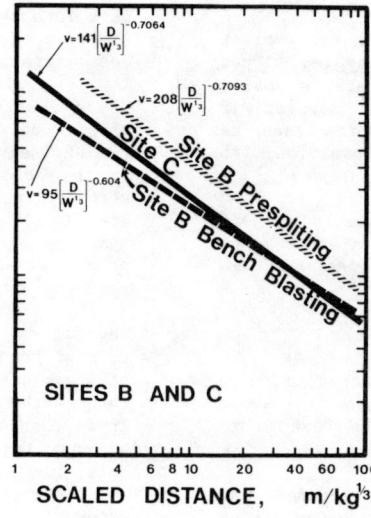
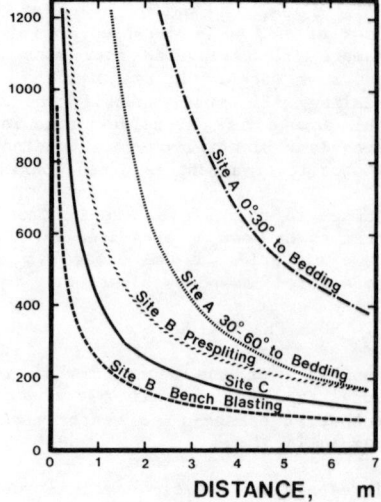

Fig. 1 Particle velocity versus scaled distance for three sites

Fig. 2 Presplitted rock face with some crushed rock and radial fractures around drill holes. Large overbreak at top

Fig. 3 General view of a channel slope achieved by using igniter in presplit holes

Figure 4 shows a typical fracture created by presplitting at the surface. The crack was not continuous, but probably occurs many times. Even assuming that the crack is continuous, tests by Devine et al., (1965) indicate that the presence of a vertical presplit fracture plane between the blast point and observation points has no or very little effect on the slope of the particle velocity distance data. Similar

interpretation can be made from the peak particle velocity recordings made at Site A and presented in Figure 1a. At this site the stratification (bedding) was the most important structural element. The average thickness of the subvertical greywacke beds was 0.4 m and 0.2 m for the slate layers. Analysis of particle velocity measurement data obtained from more than 140 blasts show that the vibration in a direction parallel or near parallel to the bedding plane (0-30°) is more severe than in a direction perpendicular to the bedding (75-90°) as indicated in Figure 1a. Therefore, it is concluded that more than one separating plane is required between the point of detonation and the blast before considerable reduction in vibration levels can occur.

Fig. 4 Closeup view of discontinuous presplit line between 70 mm diameter holes at 300 mm spacing

Figure 5 shows an open joint associated with backbreak extending up to 6 m behind a presplit face in a gnessic rock. Similar backbreak or opening of an existing joint is shown approximately 5 m behind a large hole diameter inadequately delayed bench blast in greywacke and slate in Figure 6.

A further but not so obvious damage from ground vibration occurs below the blasted level. This is where disputes and controversies between the contractor and the owner are common and may lead to claims for additional cost due to changed conditions. Figure 7 shows a well executed portion of a presplit face just below the original ground surface. In the same area, presplitting for the top part of the second bench was successful due to major damage to the rock from the first bench excavation as shown in Figure 8

Fig. 5   Open joint behind presplit face in gneiss

Fig. 6   Opened joints behind an excavated slope

The level of ground vibration necessary to cause damage to rock can be estimated from case history studies where the ground vibrations were measured and the resulting damage correlated with the level of ground vibrations.

Holmberg et al., (1981) reported that damage to rock in an open pit mine was introduced at a vibration level of 700-1000 mm/s. However, it is expected that opening of joints or creation of new joints could occur at much lower vibration levels in different types of rocks. At Site A opening of joints was observed by the author at much lower vibration level i.e. as low as 200 mm/s. As an example for the extent of possible fracture zone, in Figure 1c the calculated peak particle velocities for 100 kg charges are plotted against the horizontal distance from the centre of the blast. If damage is limited to 750 mm/s, a small diameter hole bench blast 3 m from the face could cause damage to the rock behind the final face at Site A. If, however, the threshold

limit is lower, say 200 mm/s, the size of a fractured zone for this particular site would be much higher.

Fig. 7   Presplit blasting for the first bench

Fig. 8   Presplit blasting for the second bench in granite

Another significant backbreak phenonema in horizontally or near horizontally stratified or jointed rock is lifting of large blocks and opening of joints behind the face by expanding gases. This can cause unstable and unsafe rock conditions. Figure 9 shows a large block moved by expanding gases.

Fig. 9   Large block moved by expanding gases

3.3 Minor or Major Slope Failures

A minor slope failure (slide) is shown in Figure 10. The blast set off under the partially excavated slope released the rock from the V-notch.

Fig. 10   Rock bolted slope in greywacke and slate (Site A)

Immediately following another blast upstream of this slope a rotatory rock movement commenced in the right half of the 1h:12v slope shown in Figure 10 and continued for a full two days. Location of the blast in relation to the area of rock movement is presented in Figure 11. The blast contained between 2523 and 3236 kg of explosives, fired for all practical purposes, instantaneously, and was located only 20 m from the area where the rotational slide occurred. The maximum charge per 5 ms delay interval ranged between 450 and 650 kg.

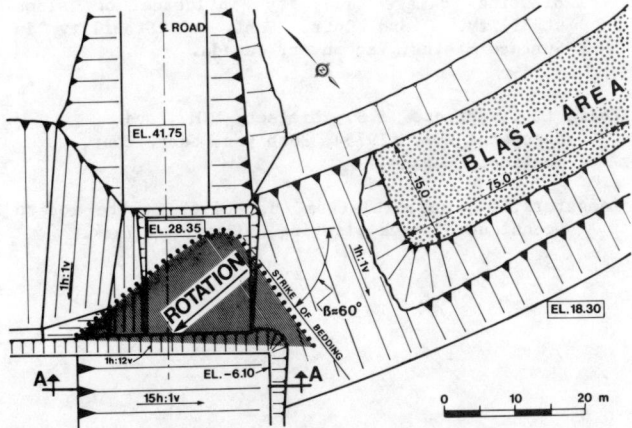

Fig. 11   Plan showing status of excavation and location of blasted area and failed slope

The 62 mm diameter holes were between 1 and 12 m deep on 1.5 x 1.5 m centers and were loaded with 40%

dynamite. The excavation method applied produced heavy burden in relation to the bench height. Large back-break and flying rock indicated that the blast was "tight", that is it did not break to a free face, which means that a large percentage of the explosive energy was absorbed by the surrounding rock. The 5 ms delays used between rows were insufficient to move the rock and reduce back pressure. For the 4 m burden the delay time between rows should have been at least 17 ms.

The slide movement comprised a rotational movement towards the southeast with a maximum displacement of some 4.5 metres at the top, followed by a sliding action along the bedding.

The rotation of the rock layers resulted in detachment of rock slabs on the front of the moving mass as shown in Figure 12. At the toe of the rotation the displacement equalled zero. Immediately after the blast some of the loosened rock was pushed down to the 28 m bench by bulldozer. The blast provided the external force and the same time loosened the rock structure.

Fig. 12   Failed slope

The second external load, from bulldozing broken rock on the bench, escalated the process. The detachment of rock started at once after the blast and reached a peak two days later. Halting the rock removal from the blast area resulted in stabilization of the rotating mass.

The rotational failure was a slow process and after the blast loosened some of the rock and rockbolts by spalling, the near vertical slabs started to deform which lead to a new state of equilibrium. However, by loading the top of the bench, the balance between acting and reacting forces could not be reached until the loading of the bench stopped. It was a very unique situation. By adding a few more tonnes of rock, the rotating mass tipped a little more and more loose rock fell from the face. A considerable volume of material was disturbed and it has been removed mechanically by controlled blasting to achieve acceptable foundations.

In assessing the stability of the rock mass, there are basically two partially related analyses which must be completed, an analysis of the static stability of the rock slope, which can be established by conventional and well documented methods of analysis, and an analysis of the stability of the rock mass considered in conjunction with outside forces.

The rotation of the entire mass could have taken place only through differential movement of single layers. The rotating mass indicates that the frictional resistance along the bedding planes must have been overcome. The movement was favoured by the vertical sets of joints and by the smaller number of horizontal or near horizontal joints.

Forces acting on a block which is resting on an inclined surface and on several other blocks is shown in Figure 13.

The calculated factor of safety against overturning prior to the blast with ß = 74° and ∅ = 33° was around 7.2. The blast adjacent to the slide area gave extrapolated vibration levels of:

$$v_L = 105\text{--}125 \text{ mm/s}$$
$$v_V = 65\text{--} 75 \text{ mm/s}$$
$$v_T = 65\text{--} 76 \text{ mm/s}$$

and displacement values of:

$$\delta_L = 1.315\text{--}1.813 \text{ mm}$$
$$\delta_V = 0.757\text{--}1.000 \text{ mm}$$
$$\delta_T = 0.511\text{--}0.654 \text{ mm}$$

for a D = 24 m and W = 450 and 650 kg respectively. The calculated frequency and acceleration values are:

$$f_L = 12.7\text{--}11 \text{ c/s} \qquad a_L = 8387\text{--}8641 \text{ mm/s}^2$$
$$f_V = 13.8\text{--}12 \text{ c/s} \qquad a_V = 5730\text{--}5652 \text{ mm/s}^2$$
$$f_T = 20.4\text{--}19 \text{ c/s} \qquad a_T = 8427\text{--}9106 \text{ mm/s}^2$$

where L stands for longitudinal, V for vertical and T for transversal component.

By using the longitudinal acceleration, the factor of safety is between 1.19 and 1.16 during the blast and by using the vector sum of each individual quantity the factor of safety against overturning is 0.83 and 0.80 respectively. The actual value is probably between the above values.

4. CONCLUSIONS

The detonation of an explosive confined in a hole generates a large volume of gas at high pressure. This pressure generates a compressive stress pulse in the rock, which constitutes the source of the ground vibrations. The primary causes of excessive vibration levels which could cause weakening of the rock structure and slope failure, are high speed explosives, large burden (underloading), insufficient delay intervals and too large a maximum charge per delay interval.

The risk of causing damage to rock slopes that lie beneath an area of blasting and are to be excavated later can be minimized if peak particle velocities are limited to between 300 and 600 mm/s as suggested by Keil et al., 1975. However, in certain highly jointed rock and unsupported slopes with a particular geometry and joint orientation the limit must be set as low as 200 mm/s.

Opening of joints by expanding gases along the crest of slopes is a major source of overbreak and creates unsafe slope conditions. Orientation of major joint systems should be included in the design of the blast to minimize overbreak and to increase slope stability.

NOTE: Opinions expressed by the writer have been distilled from personal observations and field experience.

REFERENCES

Devine, J.F., Beck, R.H., Meyer, A.V.C., and Duvall, W., (1965). Vibration levels transmitted across a presplit fracture plane. Bureau of Mines, Report of Investigation 6695.

Hendron, A.J., (1970). Ground vibrations and damage caused by blasting in rock. Acres 1970 Geotechnical Seminar, Proceedings.

Holmberg, R., and Mäki, K., (1981). Case examples of blasting damage and its influence on slope stability. 3rd Intr. Conf. on Stability in Surface Mining, Vancouver, Canada.

Keil, L.D., Burgess, A.S., Nielsen, N.M., and Koropatnick A. (1975), 28th Can. Geot. Conf., Montreal, Canada.

Langefors, U., and Kihlström, B., (1963). The modern technique of rock blasting, Wiley, New York.

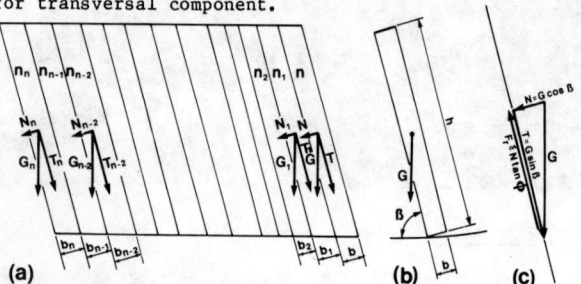

Fig. 13 Forces acting on a single layer and on n number of layers

A TOPPING FAILURE COMPUTATION METHOD USING OPTIONALLY NON-LINEAR INTRINSIC RESISTANCE

Une méthode pour le calcul de stabilité des blocs incorporant facultativement l'emploi d'un modèle de résistance intrinsèque non-linéaire

Eine Methode zur Berechnung der Kippstandfestigkeit mit wahlweiser Benutzung einer nicht-linearen Beziehung für den Reibungswiderstand

J. M. Del Corral
Civil Engineer, Hidroeléctrica de Cataluña, Barcelona, Spain

Mi. F. Bollo
Civil Engineer, Computer Scientist, Madrid, Spain

F. Hacar
Tech. Engineer, Madrid, Spain

J. M. Hacar
System Manager, Madrid, Spain

SYNOPSIS

We present a simple computational model useful for studying the influence of the different parameters that determine toppling stability. The analytical equations are derived in detail, which makes it possible to incorporate a non-linear intrinsic resistance model. The computational algorithm is described, making probabilistic studies easy to perform via simulation. The program may be used also as a building block in a more complex geological-geotechnical integrated model. It is being applied to the Airoto Pumping Storage Project in the Spanish Pyrenees.

RESUME

On décrit une méthode pour le calcul de stabilité des blocs rocheux utilisant facultativement un modèle de résistance intrinsèque non-linéaire. Le procédé de calcul est détaillé, ce qui permet de réaliser des études de probabilité par simulation. Le programme peut être intégré dans un modèle géologique-géotechnique intégré. On l'applique actuellement au projet de pompage-emmagasinage d'Airoto dans les Pyrénées espagnoles.

ZUSAMMENFASSUNG

Der Beitrag beschreibt ein einfaches rechnerisches Modell zur Berechnung des Einflusses der verschiedenen Parameter, die die Schichtstandfestigkeit bestimmen. Die analytischen Gleichungen werden im einzelnen abgeleitet, wodurch ein nicht-lineares Modell des wirklichen Widerstandes eingegliedert werden kann. Der Rechenvorgang wird beschrieben, welches Wahrscheinlichkeitsstudien mittels Simulation möglich macht. Das Programm, welches beim Airoto Pumpstauwerk in den spanischen Pyrenäen angewandt wird, kann als Baustein in einem komplizierteren geologisch-geotechnischen integrierten Modell benutzt werden.

1. INTRODUCTION

Bidimensional toppling failure models for jointed rock slopes are popular since 1971, when Ashby and Cundall published their first papers on the subject. Since then, many authors have worked on improvements to the model. Goodman and Bray (1976) have described the analysis of the different toppling failure modes. The main hypothesis are bidimensionality, existence of plane parallel joints with angle Ψ and homogeneity in every block.

These conditions are often aproximatively met by slopes in layered rock with parallel strata diving in the opposite direction of the cut. (See Fig. 1)

Failure occurs when the lower blocks, more affected by wheathering, slide or topple, and is transmited, usually slowly, until occasionally very important volumes become involved.

We have developed a simple computational model to derive the influence of the different parameters, and study the possible refinements to represent reality more accurately.

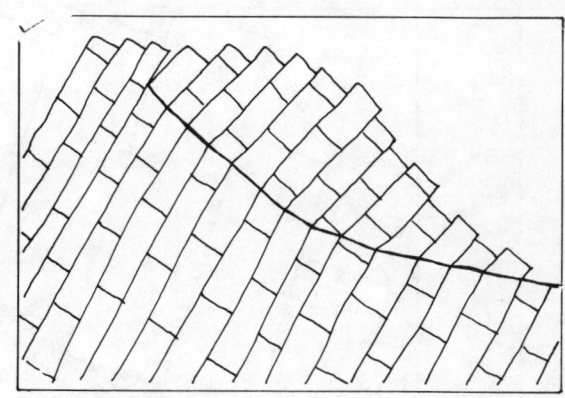

Fig. 1 Tipical case of toppling failure

2. SIMPLE COMPUTATIONAL MODEL.

For a preliminar analysis of the influence the different parameters exert, we have further simplified the calculus introducing the following additional hypothesis:

- homogeneous intrinsic resistance in all block borders, characterized by Mohr-Coulomb's model with no cohesion and a basic friction angle $\emptyset_r$.

- regular spacing of joints producing blocks with uniform thickness b.

- rock slope cut formed by three straight lines: horizontal foot, slope with angle θ, and top with angle μ.

- sliding or toppling occurs along a straight bottom line with slope β.

With these conditions, the stability of the block model may be computed if we define the following additional geometric data:

- height of cut, H. If we impose a natural number of blocks the measured height H will be substituted by its nearest value satisfying that condition, named H_c.

- lenght of first block, h_1.

In order to program the calculus, the following intermediate variables may be defined: (See Fig. 2).

- exposed lenght of a block in the cut slope:

$$a_1 = b \cdot \tan(\theta - \alpha) \qquad (1)$$

where $\alpha = 90 - \psi$

- exposed lenght of a block in the top slope:

$$a_2 = b \cdot \tan(\alpha - \mu) \qquad (2)$$

- exposed lenght between consecutive feet of blocks along the failure line:

$$a_3 = b \cdot \tan(\beta - \alpha) \qquad (3)$$

- Position i of apex block (limit between slope blocks and top blocks):

$$i = INT((H + (a_1 - h_1)\cos\alpha)/(b\sin\alpha + a_1\cos\alpha) + .5) \qquad (4)$$

- computed height H_c:

$$H_c = i \cdot b \cdot \sin\alpha + (h_1 + (i-1) \cdot a_1) \cdot \cos\alpha \qquad (5)$$

- coordinate x_i of apex block:

$$x_i = i \cdot b \cdot \cos\alpha - ((i-1) a_1 + h_1) \sin\alpha \qquad (6)$$

- total height H_m:

$$H_m = (\tan\beta(x_i\tan\mu - H_c) + b\tan\mu(\sin\alpha - \cos\alpha\tan\beta)/D \qquad (7)$$

where $D = \tan\mu - \tan\beta$

- total number of blocks:

$$n = INT((H_m - b\sin\alpha)/(SQR(b^2 + a_3^2)\sin\beta) + 1) \qquad (8)$$

3. STABILITY ANALYSIS

Using Mohr-Coulomb's linear model, we may express the trhee equilibrium equations for a general block. We denote by N_j the normal force acting on block j, T_j the shear force on the same block, and W_j the weight of the block.

Once we are able to express N_{j-1}, T_{j-1} as a function of N_j, T_j, W_j we may consider the problem solved, as N_j, T_j are known and equal to cero on the last block.

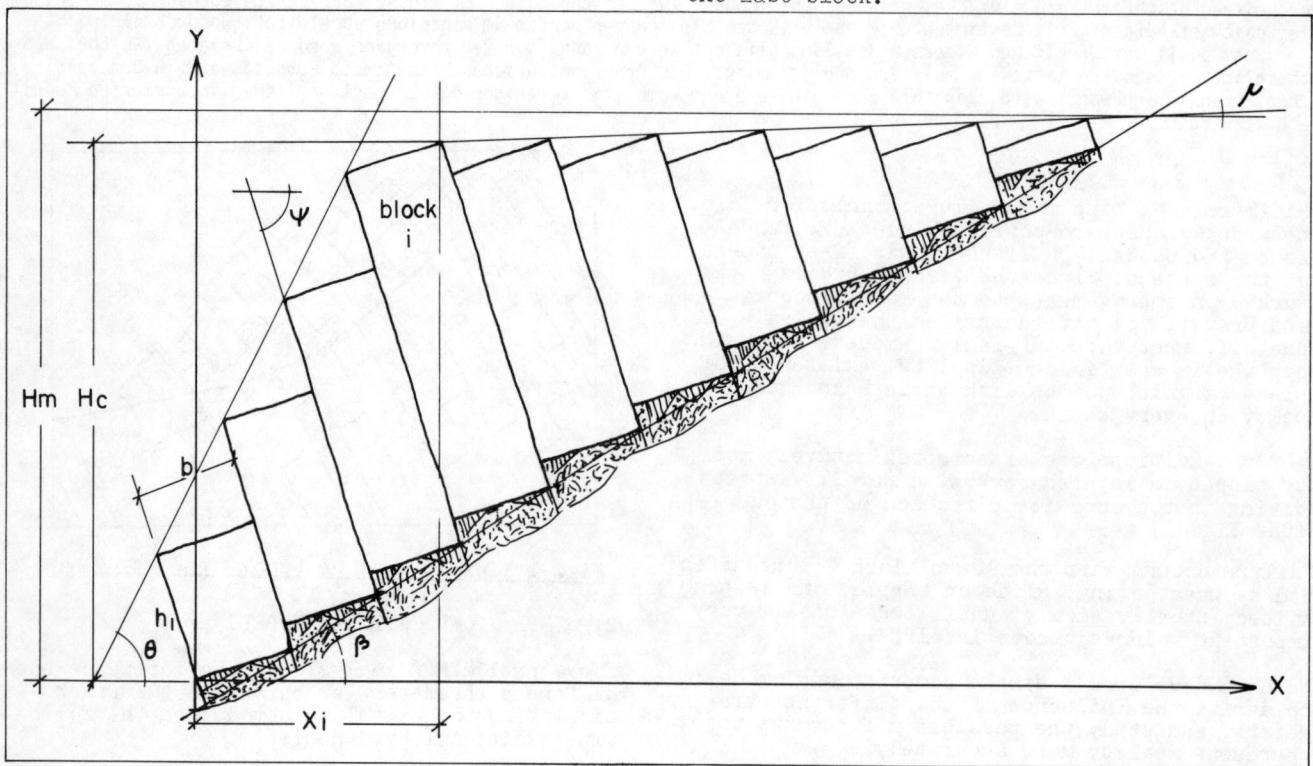

Fig. 2 Geometrical outlay.

Assuming the block j topples, the reactions of the ground are situated on its inferior apex. - We name them R_j (normal to the block base) and S_j along it.

If we assume that N_j, T_j act on the apex of - - block j, we may express the torque equilibrium equation and derive $N_{j-1,t}$, that is N_j in case of toppling.

$$N_{j-1,t} = (N_j(b\tan\phi_r - h_j) + W_j(b\tan\phi_r \cos\alpha - h_j \sin\alpha)/2)/(a_1 - h_j) \qquad (9)$$

where h_j is the block lenght, computable from h_1^j and formulas (1) through (3)

The equilibrium of forces dictates: (See Fig.3)

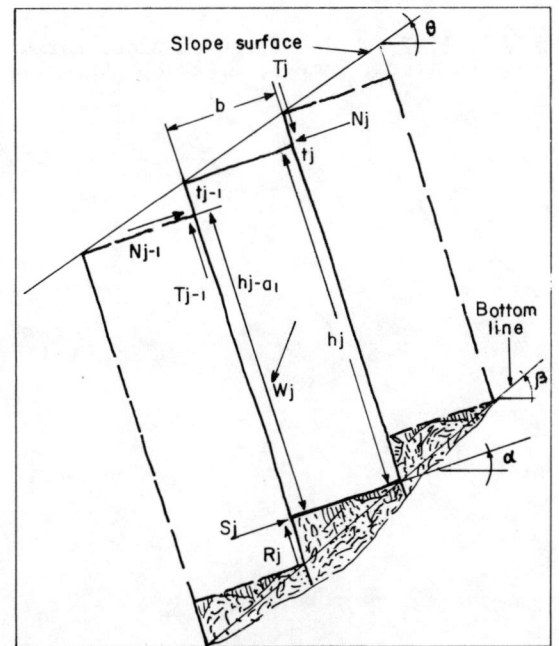

Fig. 3 Forces acting on a block.

$$R_j = \tan\phi_r (N_j - N_{j-1,t}) + W_j \cos\alpha \qquad (10)$$

$$S_j = (N_j - N_{j-1,t}) + W_j \sin\alpha \qquad (11)$$

Assuming no tensile strenght exists on the bo--ttom line of the blocks, R_j should be positive or cero. Also $S_j/R_j \leq \tan\phi_r$. As block j slides over j-1, we may assume $T_{j-1,t} = N_{j-1,t} \tan\phi_r$

If block j slides, we cannot assume that R_j passes by the lower apex or N_{j-1} acts on the - - upper apex of block j-1. Denoting N_{j-1} in the case of sliding by $N_{j-1,d}$, and assuming $T_{j-1,d} = N_{j-1,d} \tan\phi_r$ and $S_j = R_j \tan\phi_r$, we derive:

$$N_{j-1,d} = N_j - W_j(\cos\alpha \tan\phi_r - \sin\alpha)/(1 - \tan^2\phi_r) \qquad (12)$$

The computation algorithm is presented in fig.4

4. USE OF NON-LINEAR INTRINSIC RESISTANCE

Many authors have indicated the convenience of taking into account the non linear component of intrinsic resistance in jointed rock. For a dis cussion on the various formulas see Bollo (1983)

In order to use a non-linear intrinsic resistance model for block joints such as:

$$\tau = \sigma\tan\phi_r + \sigma(\tan\phi_i - \tan\phi_r)\Delta\sigma_c^2/(\sigma^2 + \Delta\sigma_c^2) \qquad (13)$$

where:

τ = shear pressure
σ = normal pressure
ϕ = basic friction angle
ϕ_i^r = initial friction angle
$\Delta\sigma_c$ = critical stress deviator

the effective friction angle ϕ_e is computed as ϕ for the lower apex of the toppling blocks - and:

$$\phi_e = ATN(\tan\phi_r + (\tan\phi_i - \tan\phi_r)/(1 + (\sigma/\Delta\sigma_c)^2)) \qquad (14)$$

in all other cases.

The average σ is computed from N_j, S_j and the geometrical data.

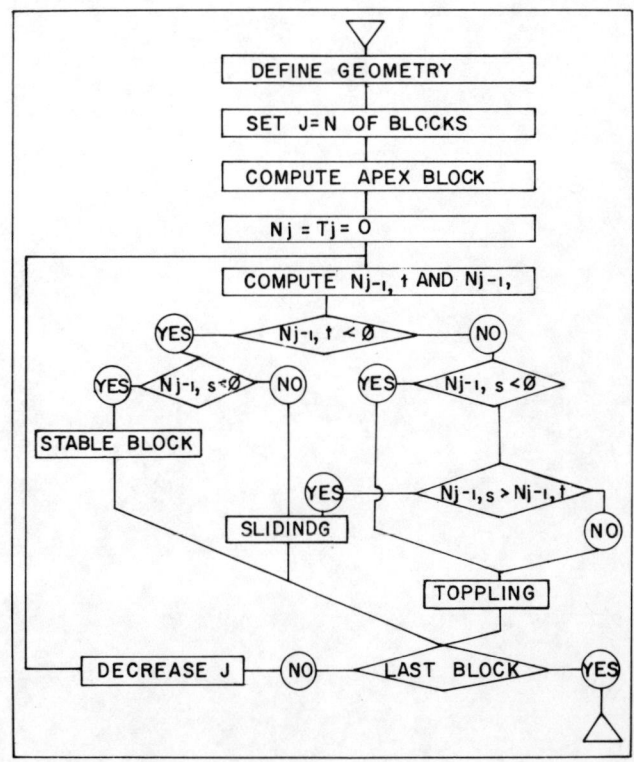

Fig. 4 Computational algorithm.

5. USE OF THE MODEL

The simple computational model we have descri-bed is useful as a tool for:

- sensitivity analysis on the influence of the various parameters. It is easy to compute -- the distribution of safety factors by simula tion, when the input data are defined by pro bability distributions.

- incorporating a toppling analysis building - block into an integrated geological-geotech-nical model. (See Del Corral, 1982).

We are using it for the Pumping storage Project of Airoto, in the Pyrenees (Lerida, Spain).

6. REFERENCES

ASHBY, J. (1971). Sliding and toppling modes - of failure in model and jointed rock slopes: Imperial College, Royal School of Mines, London

BOLLO, M.F., BOLLO, Mi.F. (1974). Etude d'une relation de resistance intrinséque non lineaire dans le project des talus: IIIrd Congress ISRM Denver.

DEL CORRAL, J.M., FERNANDEZ-BOLLO, Mi. (1982). A systematic approach to the geotechnical consideration of lithoclases in dam design: 14th ICOLD, Rio de Janeiro.

BOLLO, Mi.F., HERRERO, E., BUIL, J.M. (1983). A real time interpretation methodology for large scale (2m^3) in-situ rock shear test: 5th ISRM - Congress Melbourne.

CUNDALL, P.A. (1971). A computer model for simulating progressive large scale movements in blocky rock systems: Symposium on Rock Fracture, - Nancy, France.

CUNDALL, P.A., VOGELE, M.D., FAIRHURST, C. (1975) Computerized design of rock slopes using interactive graphics for the input and output of -- geometrical data: 16th Symposium on Rock Mechanics, Minessota.

GOODMAN, R.E., BRAY, J.W. (1976). Toppling of - rock slopes: Rock Engin. for foundations and - slopes: Special Publication ASCE, Colorado University.

HOEK, E., BRAY, J. (1977). Rock slope engineering: The Institution of Mining and Metallurgy, London.

ZAMBAK, C. (1979). Discussion on slope failures: 4th ISRM Congress, Tome 3, p.422, Montreux.

GEOTECHNICAL ASPECTS OF THE KENYIR DAM PROJECT, TRENGGANU, PENINSULAR MALAYSIA

Aspects géotechniques du projet de barrage Kenyir au Trengganu, Malaisie

Geotechnische Gesichtspunkte des Kenyir Damm Projektes, Trengganu, Halbinsel Malaysia

Boon Kong Tan
Lecturer, National University of Malaysia, Bangi, Malaysia

SYNOPSIS

The Kenyir Dam is one of the largest hydroelectric schemes in Peninsular Malaysia. It is an earth and rockfill dam sited in granitic terrain just downstream from the confluence of the Trengganu and Kenyir rivers in northern Trengganu. The project involves excavation and foundation preparations for the main dam and associated saddle dams, excavation of two diversion tunnels, four pressure tunnels, spillway and power station, as well as quarry works. Though simple in site geology (granitic rocks only), various construction problems have been encountered due to fracturing or jointing of the granite, weathering, and the occurrence of dolerite dykes and veins. The problems discussed include rock slope stability, tunnel and spillway excavation problems, grouting and quarry works.

RESUME

Le barrage du Kenyir est l'un des projets hydroélectriques les plus importants de la péninsule Malaise. C'est un barrage en terre et en enrochements situé sur terrain granitique juste en aval du confluent entre les rivières Trengganu et Kenyir au nord de l'état de Trengganu. Ce projet comporte des travaux préliminaires d'excavation et de fondations du barrage principal et de petits barrages d'ensellement, l'excavation de deux galeries de dérivation, quatre galeries de pression, un déversoir et une centrale éléctrique ainsi que des carrières d'alimentation en matériaux. Bien que situé dans une unité géologique unique (roche granitique seulement) divers problèmes de construction se sont posés dus aux fractures et aux joints du granite ainsi qu'à l'altération et à la présence de dykes et de filons de dolérite. On examine les problèmes de la stabilité des pentes des talus de même que ceux de l'excavation des galeries et du déversoir, des injections et des travaux en carrières.

ZUSAMMENFASSUNG

Der Kenyir Damm is Teil eines der größten Wasserkraftprojekte der Malayischen Halbinsel. Er is ein erd- und felsgeschütteter Damm, der auf granitischem Fels unmittelbar unterhalb des Zusammenflusses des Trengganu- und des Kenyirflusses im Norden Trangganus gelegen ist. Das Projekt umfaßt Aushub und andere Vorbereitungen zur Gründung des Hauptdammes und von Satteldämmen, die Konstruktion von zwei Umleitungsstollen, vier Druckstollen, der Hochwasserentlastungsanlage und des Kraftwerkes sowie Steinbrucharbeiten. Obwohl die geologischen Bedingungen einfach sind (gravitischer Fels), wurden verschiedene Probleme angetroffen, die durch die Zerklüftung des Granites, seine Verwitterung und das Auftreten doleritischer Gänge und Adern verursacht wurden. Die hier diskutierten Probleme umfassen Untersuchungen zur Böschungsstabilität, zum Aushub der Tunnel und der Hochwasserentlastungsanlage, zum Verpressen und zu den Steinbrucharbeiten.

1. INTRODUCTION

The Kenyir Dam, currently under construction, represents one of the largest hydroelectric schemes in Peninsular Malaysia. It is sited across the Trengganu River at about 200 m downstream from the confluence of the Trengganu and Kenyir rivers in northern Trengganu. The main dam consists of an earth core with rockfill embankments, measuring 150m in height and 800 m long. The storage volume of the resulting reservoir is 1.3×10^{10} m^3, and covers a surface area of 36,900 hectares. In addition, the scheme also includes eight saddle dams ranging in heights from 8 m to 45 m, which consist solely of earthfill. Other structures associated with the main dam include two diversion tunnels, four pressure tunnels, spillway and power station. The general lay-out of the Kenyir Dam scheme is as shown in Figure 1.

2. GEOLOGIC SETTING

The regional geology of the project area covering the entire catchment area of the Trengganu River, consists of a series of sedimentary(mainly sandstone and shale) and metamorphic rocks(mainly quartzite and phyllite) intruded by granitic masses. The fold axes of the sedimentary rocks strike in the NW to NNW directions. A major syncline striking NNW occurs within the catchment area. Various dolerite dykes intrude the granite, having strike directions ranging from NE to E. Only minor faulting have been encountered in the area.

However, the Kenyir Dam and its eight saddle dams are all sited in granitic terrains. Locally at the main dam site, the underlying bedrock consists of a coarse-grained biotite granite with numerous dolerite dykes. These dykes have widths varying mostly from 5 cm to 2 m, striking NE to E

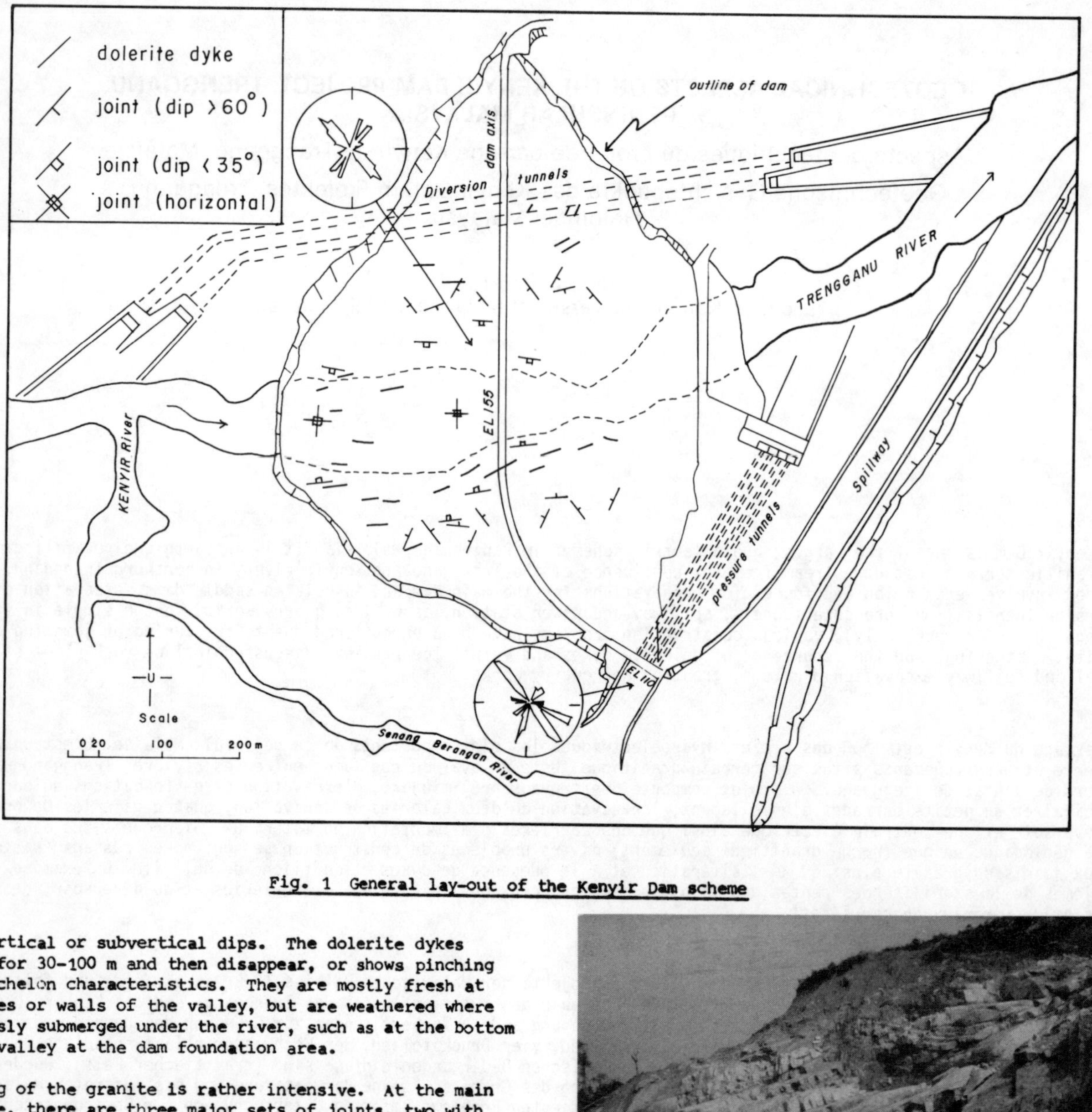

Fig. 1 General lay-out of the Kenyir Dam scheme

with vertical or subvertical dips. The dolerite dykes extend for 30-100 m and then disappear, or shows pinching or en-echelon characteristics. They are mostly fresh at the sides or walls of the valley, but are weathered where previously submerged under the river, such as at the bottom of the valley at the dam foundation area.

Jointing of the granite is rather intensive. At the main dam site, there are three major sets of joints, two with vertical or subvertical dips striking NNE and SSE, and one set with horizontal or subhorizontal dips, striking parallel to the valley walls. The joints mostly dip towards or into the valley, as shown in Figure 2. The joint spacings vary from 0.6m-1.6m. They are weathered or open at the surface with occasional clay fillings. At greater depths, however, the joints are cleaner and tighter. Some joints at the saddle dam areas are commonly filled with quartz. In addition to the three major sets of joints are several minor joints in between the major ones, which in certain localities render the granite highly fractured.

The granite is also highly weathered, as is typical of all granites in tropical Malaysia which are subjected to intensive chemical as well as physical weathering. Depths of weathering vary from place to place, but are generally of the order of tens of metres. As such, thick layers of residual soils riginally cover the granite bedrock, and a complete weathering profile showing grades of weathering from Grade I (fresh, unweathered rock) to Grade VI (residual soil) is observed throughout the dam sites.

Fig. 2 Joints dipping into the valley at main dam site

Though the lithology in the dam site areas is rather simple, namely only granite with minor dolerite dykes, much of the construction and excavation works are dictated by the degree and nature of jointing and weathering grades of the rock, as seen from the following sections.

3. ROCK EXCAVATIONS

Excavation of soils and rocks were necessary in the construction of the main dam and saddle dams, as well as the associated structures such as the diversion tunnels, pressure tunnels, spillway and the power station.

3.1 Main dam

Most of the excavation works occur at the main dam foundation area, where excavation has to be carried down to at least Grade II materials. Removal of the thick overburden soil and weathered rocks(grades IV to VI) was done by ripping and stripping, while the harder materials(grades I to III) required blasting. The sporadic occurrence of granite boulders or corestones in the residual soils, especially abundant in the saddle dam areas as well as in the main dam site, hinder excavation works and at times require additional secondary blasting. Special cleaning and washing of the river alluvium at the main dam site was also necessary to prepare the foundation for further works such as grouting, etc. Since the main dam spans the entire width of the valley and reaches considerable heights, much of the valley slopes have to be terraced or benched for ease of operation as well as safety reasons. This often involves the trimming of the rock slopes by pre-split blasting or smooth-wall blasting to obtain a smooth rock face which enhances stability. An example of a smooth rock face obtained by pre-split blasting is shown in Figure 3, located at the entrance of the pressure tunnels area. Depending on the intensity of jointing or fracturing of the granite, the result obtained from pre-split blasting may not always be successful.

Fig. 3 Smooth rock face obtained by pre-split blasting

3.2 Diversion tunnels

Prior to the actual foundation excavation for the main dam, two diversion tunnels have to be excavated first and these were located on the left bank of the river as shown in Fig. 1. The diversion tunnels, each of 14 m in diameter, were excavated by the drill and blast method since they pass through fresh, sound granite (Grade I material). Much of

the tunnels are at depths of greater than 30 m below the general ground surface. Throughout the diversion tunnels the rock is rather massive and stable, except for some sections where the subhorizontal sheet joints occuring at the roof portions of the tunnels might introduce some instability. In these sections, the subhorizontal sheet joints have dissected the granite into layers of rock reminiscent of sedimentary rocks. Several rock bolts were installed in these sections to tie the slabs of granite to the interior, more massive granite. Several dolerite dykes were also intersected in these tunnels, but they are fresh and in tight contact with the granite and as such pose no instability problems. The major portions of instability of the diversion tunnels are at the portal areas(as is generally the case for tunnels) which is in more weathered rock materials. These portal sections were supported by concrete lining or arches with some steel sets, as shown in Figure 4, taken during the construction stage. Part of a vertical dolerite dyke running between the two tunnels can also be seen in the same figure.

3.3 Pressure tunnels

Four pressure tunnels are required in the project and these are located on the right bank of the river. To-date only two of these four pressure tunnels are in the process of excavation. The pressure tunnels are also excavated by the drill and blast method since they are sited in fresh, sound granite. Jointing of the granite traversed by the tunnels are mostly subvertical to vertical and tight, and thus pose

Fig. 4 Diversion tunnels

little stability problems within the tunnels. At the entry portal area, however, the subvertical to vertical jointing has resulted in numerous vertical slabs of granite which are prone to sliding and thus require numerous, closely spaced, grouted rock anchors for stabilization, as shown in Figure 5. The tunnel portal sections are further supported with concrete arches coupled with some steel sets. On completion, the pressure tunnels will be lined with steel lining.

3.4 Spillway

The spillway is located just next to the pressure tunnels and is excavated from the surface down to fresh rock. Its excavation thus involves much ripping of the overburden soils as well as the blasting of harder materials. One of the key problems in the excavation for the spillway is its alignment which is often dictated by the major joints

Fig. 5 Pressure tunnels: note rock anchors for stabili-
sing granite slabs above the tunnels

in the rock. An example of this is shown in Figure 6.
The dominant influence of these major joints makes it
necessary to change slightly the original alignment of the

Fig. 6 Major joints at Spillway

spillway. On completion, the spillway will be lined with
concrete.

4. GROUTING

The presence of various joints in the granite at the surface and at greater depths, as observed during excavation
works and also from drilling records, necessitates grouting of the main dam and saddle dams foundations to reduce
possible seepage losses under the dams. Borehole water
pressure tests at various depths also indicate possible
water losses ranging in values from 2-10 lugeons, with
some sections having much higher values (upto 180 lugeons)
corresponding to core loss sections or severe weathering
along joints or higher fracture intensity.

At the main dam site, five rows of grout holes, spaced 3m
apart, have been proposed. The middle row, grouting to
depths of 40m - 50m, would form the main curtain grout;

while the other four rows reaching to depths of 5m only
would form the blanket grout. Previous water pressure
tests have indicated that at depths of greater than 50 m,
little or no water loss occurs.

At the saddle dams, the grouting is done in two stages.
Primary grout holes are spaced 6 m apart and grouted to
10 m depths. In between the primary holes are then inserted the secondary grout holes upto depths of 5 m only.
The cement:water ratio used was 1:3, with grouting pressures of 1 kg/m^2 (for 5 m depths), and 1-2 kg/m^2 (for 10m
depths). Since the saddle dams are long and narrow
(Figure 7), a single row of grout holes located along its
axis is deemed sufficient. The common occurrence of dolerite dykes in the granite was revealed during excavation
of the saddle dams, but they are similarly grouted.

In general, the saddle dams granite is less fractured or
jointed, when compared to that in the main dam area. As
such, grouting in the saddle dams areas will be to shall-
ower depths (upto 10 m only) as compared to the 50 m
depths of grouting at the main dam. The lack of horizontal
or subhorizontal sheet joints also excludes the necessity
of blanket grouting in the saddle dams.

Prior to grouting, the joints have to be cleaned or flushed
out of debris such as clay fillings or other weathered
materials. This is especially important at the main dam
which was previously under the river.

Fig. 7 Saddle dams

5. CONSTRUCTION MATERIALS

Abundant suitable construction materials are available at
or in the immediate vicinity of the dam sites. These include the abundant granite used for the rockfill embankments
of the main dam and as concrete aggregates, the soil borrow
pits consisting of residual granite soils(sandy silt or
sandy clay: SM or SM-SC), and the river sand (which also
occur in the alluvial terraces) for the filter zones.

Much of the rockfill/aggregates are obtained from a quarry

Fig. 8 Coffer dam consisting mainly of rockfill completed at the end of 1981

sited just upstream from the entrance into the diversion tunnels, which incidentally also yield considerable quantities of dolerite besides the granite. However, the dolerite rocks are also suitable for the purpose. In addition, granite blocks are also obtained from the excavations for the various other structures such as the tunnels and the spillway.

The thick layers of residual soils (tens of metres thick) that have developed over the granite yield plentiful supply of suitable borrow materials for the earth core of the main dam as well as for the earthfill of the saddle dams. These are obtainable at or near the various dam sites, but would require some selection(especially the removal of top soil) and proper compaction to yield the density and impermeability required.

At the main dam site, suitable filter sand are obtainable from the river and also from alluvial terraces that occur along the Trengganu River. These consist of clean sands, with sizes ranging from fine to coarse. Since the amount needed for the filter zones is relatively small, there is sufficient sand available though the sizes might require grading and selection.

Figure 8 shows a view of the coffer dam which has been completed at the end of 1981. The coffer dam consists mainly of rockfills, and will be incorporated into the main dam proper.

6. CONCLUSIONS

Though simple in site geology, namely granite only with some minor dolerite dykes, the Kenyir Dam project illustrates the significant influence of jointing and weathering grades on engineering constructions and excavations. Much of the excavation and remedial works carried out in the Kenyir Dam project were dictated by these two factors, namely jointing and weathering of the rock, the two factors that often influence the engineeringperformance of a rock mass.

REFERENCES

Chow,W.S.(1979). Engineering geology of the Kenyir Dam site, Trengganu, Malaysia. M.Sc. thesis, Univ.Malaya.

Tan,Boon Kong (1981). An overview of engineering geologic problems in Peninsular Malaysia: Proc. IV GEOSEA, Manila.

Zukeri,M.(1982). Geologi kejuruteraan tapak empangan Kenyir, Trengganu. B.Sc.(Hons.) thesis, Univ. Kebangsaan Malaysia.

COST ESTIMATES FOR FOUNDATION INVESTIGATIONS OF DAMS
Estimation des coûts pour les études de fondation de barrages
Kostenanschlag für Fundament-Untersuchung von Talsperren

A. Van Schalkwyk
Professor of Engineering Geology, University of Pretoria, Pretoria, South Africa

SYNOPSIS

Cost estimates for foundation investigations of dams are often required during early planning stages when very little information on the type and cost of structure or foundation conditions is available. Based on available geological and topographical site characteristics and an "investigation ratio", which is related to the reliability required, the volume of material to be sampled during different investigation stages can be determined. By relating the amount of professional time required to the volume of sample obtained and applying known unit costs to time and sampling work, a total investigation cost is derived.

RESUME

L'estimation des frais d'investigation préalable à la construction d'un barrage est souvent nécessaire aux stades préliminaires de la planification, alors que les données sur le type et les coûts de structures ou sur les conditions de fondations sont très peu nombreuses. En se basant sur les caractéristiques topographiques et géologiques du site et un "coefficient d'investigation" qui sera fonction de la fiabilité voulue, la quantité du matériau à échantilloner lors des divers stades d'études peut être définie. En corrélant le temps de travail affecté au volume d'échantillons et en utilisant les unités de coût, pour le temps d'étude et d'échantillonnage, on peut estimer le coût total des investigations.

ZUSAMMENFASSUNG

Kostenanschläge für die Fundamentsuntersuchungen von Stauwerken werden oft im Anfangsstadium der Planung benötigt, wenn nur sehr wenig Information über die Art und die Kosten des Baus, oder über die Fundamentsbedingungen, zur Verfügung stehen. Auf der Basis von anstehenden geologischen und topographischen Bauplatzcharakteristiken und einem "Untersuchungsverhältnis", welches mit der erwünschen Zuverlässigkeit in Beziehung steht, kann die Menge des während der verschiedenen Untersuchungsphasen zu erprobenden Materials ermittelt werden. Indem man ein Verhältnis zwischen dem benötigten fachmännischen Zeitaufwand und der Größe der erhaltenen Probe aufstellt, und indem man bekannte Kosten per Einheit auf die Zeit und die Arbeit der Probenentnahme bezieht, können die Kosten der Gesamtuntersuchung ermittelt werden.

1. INTRODUCTION

Dam foundations are subjected not only to loading by the dam wall, but also to the effects of water in the form of hydrostatic pressure, seepage forces and erosion or piping. For the safe design of a dam, an adequate knowledge of the often complex geological conditions within the large volume of foundation material is essential and therefore, comprehensive site investigation programmes have become accepted prerequisites for these structures.

During the early planning stages of a dam construction project, the client or owner often requires a fairly reliable estimate of the cost of such an investigation programme for budgeting purposes, while later on, the designer or contractor may wish to compare the cost of the investigation and the reliability thereof with acceptable norms established under similar conditions elsewhere.

The extent of a foundation investigation is often measured in terms of it's cost and expressed as a percentage of the capital cost of the project involved. Figures of 0,25 to 1 per cent are quoted as indications of the requirements for a fairly uncomplicated site whereas more complex sites may require site investigations costing 5 per cent or more (Price and Knill, 1974). De Beer (1981) presented the following relationship between cost of investigation (I) as percentage of project cost and the project cost (P), in millions for sites in Holland

$$I = \frac{1,9 + 0,5 \, P}{0,6 + P} \qquad (1)$$

Price and Knill (1974) have expressed the magnitude of a foundation investigation as the ratio of the volume of material sampled by drilling, to the volume of foundation material loaded by the structure.

The limitations of the above methods can be summarized as follows:

(i) The project cost is mostly not related to the area or intensity of foundation loading and is seldom available during the

early planning stages.
(ii) For calculating the volume of loaded
    foundation material, the type of dam must
    be known and furthermore, no provision is
    made for sampling methods other than
    drilling.
(iii) The complexity of geological conditions
    and the availability of geological
    information in the form of rock outcrops
    are not taken into account.

In this paper, an attempt has been made to
improve the above methods by introducing a few
refinements and general guide lines, based on
data obtained from a number of foundation
investigations.

2. FACTORS AFFECTING FOUNDATION INVESTIGATION COSTS

Foundation investigation costs can be subdivid=
ed into (i) fees for professional geological
and geotechnical services, usually on a time-
and cost basis and (ii) cost for work done by
contractors such as geophysical surveys, dril=
ling, testing, etc. for which quotations are
normally obtained.

The factors affecting the amount of work to
be done in respect of items (i) and (ii) above,
are (a) the number of alternative sites to be
considered, (b) the stage of the investigation,
(c) the size and type of structure, (d) the
complexity of geological conditions (e) the
cost-effectiveness of site investigation
methods and (f) the reliability required of the
investigation results.

2.1 Stages of a foundation investigation

The various investigation methods are usually
not applied simultaneously but the investiga=
tion programme is subdivided into well-defined
stages of reconnaissance, feasibility, design
and construction (Fookes, 1967 and Oliviera,
1979). The total investigation effort during
the first three stages is often referred to as
the "pre-construction" investigation. For the
purpose of this paper, investigations for
construction materials and investigations
during the construction stage are not taken
into consideration.

2.1.1 Reconnaissance investigation

A reconnaissance investigation comprises two
parts namely (i) a desk study to collect and
study all available information and (ii) a
site visit to check the correctness and inter=
pretation of the above and to obtain additional
data. Drilling work, geophysical surveys or
laboratory tests are not normally carried out.

The purpose of a reconnaissance investigation
is to collect all available information, iden=
tify important geological problems, determine
general foundation conditions, make a provi=
sional choice between alternative sites and
make recommendations regarding the
feasibility investigations.

2.1.2 Feasibility investigation

The purpose of a feasibility investigation is
to provide geological data of sufficient detail
to make the final choice amongst alternative
sites, prepare a preliminary design of the
structure and obtain a fairly accurate cost
estimate for the project.

During this stage, use is made of a number of
site investigation methods such as engineering
geological mapping, geophysical surveys, dril=
ling, test holes, soil profiling, joint surveys,
etc. The object is to describe the various
foundation materials and to classify them in
terms of their engineering properties. The
intensity of investigations at a particular
location is limited and it is attempted to
determine the general foundation conditions at
a number of alternative locations rather than
specific foundation properties along a
particular centre-line.

2.1.3 Design investigation

The purpose of this investigation is to obtain
detailed quantitative information on foundation
properties in specific areas in order to
prepare the design drawings and specifications
for construction.

Most of the investigation methods used during
the feasibility investigation are used again
but in greater detail and, in addition, more
sophisticated laboratory and in situ tests are
carried out in order to determine the required
design parameters.

2.2 Size and type of structure

The size and type of structure can be expressed
in terms of the loaded foundation area or
volume (Price and Knill, 1974). For the
purposes of a preliminary cost estimate, how=
ever, it is seldom possible to determine these
parameters accurately since they depend on the
type of dam and arbitrary boundaries of the
loaded volume ("pressure bulb"). An equivalent
loaded foundation volume (V) in cubic metres,
which is related to the actual volume of loaded
material for all types of dams and compares well
with the loaded volume for concrete dams as
proposed by Price and Knill (1974), can be
obtained from the following empirical relation=
ship:

$$V = 2.L_c.H_a^2 \qquad (2)$$

where L_c = length of centre-line in metres
and H_a = average height of dam wall in metres.

Similarly an equivalent loaded foundation area
(A) in square metres can be obtained as follows:

$$A = L_c.H_a \qquad (3)$$

2.3 Geological conditions

The variability of distribution and properties
of geological materials can seldom be fully
described even after construction has been
completed. For the purpose of these estimates,
the following simple classification system with
an adjustment factor (F_g) for each class, based
on information which is normally available on
geological maps, has been proposed:

Class A: Flat-lying or slightly tilted
         sedimentary formations (F_g = 0,2)
Class B: Moderately tilted or folded
         sedimentary formations as well as
         igneous and metamorphic rocks (F_g = 0,5)
Class C: All rock types disturbed by intense
         folding, faulting or intrusives
         (F_g = 1,0)

The presence or absence of solid outcrops of <u>in situ</u> rock on site affects the number of test holes or boreholes required and is taken into account by estimating the percentage of foundation area occupied by rock outcrops and expressing this as an adjustment factor (F_r)

$$F_r = \frac{\text{Percentage of rock outcrop area}}{100} \qquad (4)$$

2.4 Cost effectiveness of foundation investigation methods

Since the extent of a pre-construction investigation is largely measured by the volume of material sampled, the effectiveness of the investigation methods must be related to the cost of recovering or exposing a unit volume of foundation material.

2.4.1 Volume of sample

For diamond drilling and soil sampling, the volume of sample depends upon the size of drill bit or sampling tube but since N-size equipment is normally used, the volume of sample (V_d) in cubic metres can be approximated as follows:

$$V_d = 0,002 \, L_d \qquad (5)$$

where L_d is length drilled in metres

Large diameter augering, test trenches and test adits yield much larger volumes of sample but most of it is completely disturbed and information is normally obtained by means of a line survey along the length of the exposure and from samples taken at certain intervals. The following general formula to relate the size of the opening to usable sample volume (V_t) in cubic metres has been proposed:

$$V_t = 0,002 \, L_t + 0,0005 \, L_t C_t \qquad (6)$$

where L_t is the length of hole/adit in metres and C_t is the circumference in metres.

By means of small diameter augering only disturbed samples are obtained and due to their limited value, the volume of sample (V_a) in cubic metres is considered equal to $0,0005 \, L_a$, where L_a is the length of hole drilled.

2.4.2 Cost of sampling

Based on estimates of 1982 unit costs, the cost-effectiveness of various sampling methods is compared in Table I.

Table I : Cost-effectiveness of various sampling methods

Method	Length (m)	Circumference (m)	Sample volume (m^3)	Cost R*	Unit cost (R/m^3)
Diamond drilling	30 m	0,236	0,06	4500	75000
Large auger	30 m	2,36	0,095	600	6316
Adit	30 m	7,85	0,178	15000	84270
Trench	30 m	6	0,15	3000	20000
Small auger	30 m	0,345	0,015	450	30000

(R* = S.A. Rand ≈ U.S. $ 0,85 approx.)

2.5 Reliability of foundation investigations

The reliability of professional work in connection with reconnaissance investigations and the interpretation of site investigation results during the other stages, are difficult to evaluate since they depend not only upon the amount of information or volume of sample available but also on the knowledge and experience of the interpreter.

Time and cost analyses on a number of recently completed foundation investigations have yielded some useful average figures for the amount of professional time spent on the various aspects of a foundation investigation programme. It must, however, be pointed out that these figures depend on a large number of variables, including availability of information, complexity of geological conditions, type of structure, etc. and can only be used as general guide lines.

For determining the volume of sample (borehole core) required during a pre-construction investigation, Price and Knill (1974) have recommended a volume ratio (ratio of volume sampled to volume loaded) of $1:10^5$ (see Figure 1)

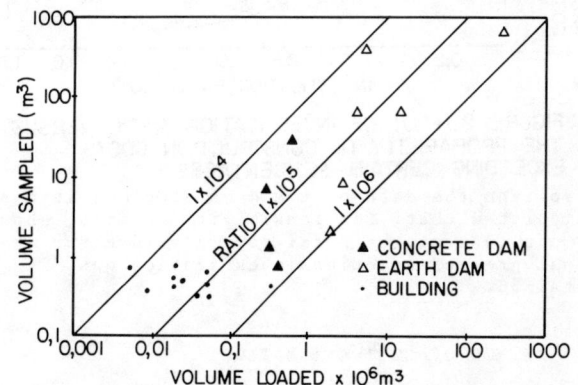

FIGURE 1. RELATIONSHIP BETWEEN VOLUME SAMPLED BY DRILLING AND VOLUME LOADED BY STRUCTURE.
(AFTER PRICE & KNILL, 1974)

The reliability of an investigation can also be defined as the probability that construction costs will not exceed a given percentage above the design cost estimate or tender price as a result of unexpected foundation conditions (Van Schalkwyk, 1980). Instead of using the term volume ratio, it was found more convenient to use the term "investigation ratio" (R)

where $R = \dfrac{\text{volume of sample required or obtained}}{\text{volume of loaded foundation}}$

$$\qquad (7)$$

In order to determine the investigation ratios normally achieved during foundation investigations for dams, a detailed study of 22 large dams in South Africa was undertaken. For each dam, the equivalent loaded foundation volume, the adjustment factors for geological conditions and the volume of samples actually obtained during all pre-construction investigations were calculated and the investigation ratios compared with records of construction cost increases as a result of unexpected foundation conditions (Van Schalkwyk, 1980).

The small number of dams studied and the lack of accurate information on construction cost increases, prevented a thorough statistical analysis of the above data but plots of the investigation ratio versus the probability of exceeding certain percentages of the construction cost (see Figure 2) clearly indicate the validity of Price and Knill's recommended ratio.

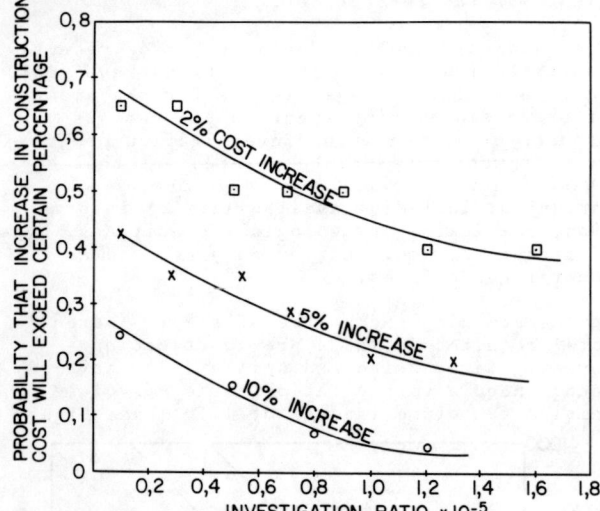

FIGURE 2. PLOT OF INVESTIGATION RATIO VERSUS THE PROBABILITY OF CONSTRUCTION COSTS EXCEEDING CERTAIN PERCENTAGES

Also from the data on the dams studied, it was calculated that, for feasibility studies, about 60 per cent of the total sample volume for the final pre-construction investigation was obtained.

3. METHOD OF COST ESTIMATION

Cost estimates for the various stages of investigation will be considered separately.

3.1 Reconnaissance investigation

Since subsurface investigation methods, other than hand augering, are seldom employed, contractual costs are normally not incurred.

Professional costs are due mainly to transport cost and cost for time spent on travelling, data retrieval, site inspection and reporting. Based on information from a number of site investigations (see Figure 3) the following simplified formula can be applied for the cost (C_R) of a reconnaissance investigation

$$C_R = C_{Tr} + C_M \qquad (8)$$

where C_{Tr} is the cost in Rand for Time T_r in hours,

$$T_r = A^{0,45}, \qquad (9)$$

and C_M is the cost for transport and travelling time. (A site visit for one person is allowed for every 50 hours of working time (T_r) or part thereof)

Where more than one site has to be investigated for the same project, the following rules can be applied:

(i) For sites within a radius of 5 km and on similar geological formations, the foundation areas are added together before T_r is calculated:

$$T_r = (A_1 + A_2 + A_3 + \ldots A_n)^{0,45} \qquad (10)$$

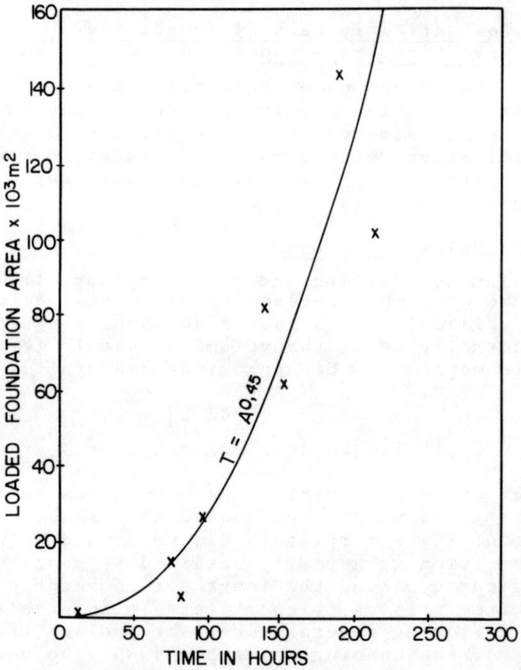

FIGURE 3. PLOT OF LOADED FOUNDATION AREA VERSUS TIME SPENT ON RECONNAISSANCE INVESTIGATIONS FOR A NUMBER OF DAMS IN SOUTH AFRICA.

(ii) Where sites are further apart or on dif= ferent geological formations the times are calculated separately and added together.

$$T_r = T_r 1 + T_r 2 + T_r 3 + \ldots T_r n \qquad (11)$$

3.2 Feasibility investigation

The volume of sample required (V_f) is calculated as follows:

$$V_f = 0,6.V.F_g(1-F_r) \times 10^{-5} \qquad (12)$$

where V is from equation (2) and F_g and F_r are the geological factors defined in paragraph 2.3.

The most economical way to obtain a representa= tive sample of volume (V_f) from the loaded foundation volume can be determined by consult= ing the cost-effectiveness of various methods on Table I and the actual cost of sampling can then be determined by applying known unit costs or costs obtained from quotations.

A rough approximation of professional time (T_f) required for the organization of the investigation programme, the evaluation of results and the compilation of a report may be obtained from the following equation:

$$T_f = 330 \, V_f \qquad (13)$$

where T_f is the time in hours
and V_f is the volume sampled in cubic metres

The professional costs C_F can then be
calculated as follows:

$$C_F = C_{Tf} + C_o \qquad (14)$$

where C_{Tf} the cost of time T_f
and C_o the cost for other items
which can be taken as 10 per cent of C_{Tf}

3.3 Design investigation

The volume of sample required (V_d) is
calculated as follows:

$$V_d = V.F_g(1-F_r) \times 10^{-5} - V_{fs} \qquad (15)$$

where V_{fs} is the volume within the loaded
foundation volume sampled during feasibility
investigations.

The cost of obtaining the sample is determined
in the same way as for feasibility
investigations.

Professional time (T_d) is calculated as
follows:

$$T_d = 500 \ V_d \qquad (16)$$

Professional costs are calculated in the same
way as for feasibility investigations.

4. EXAMPLE OF CALCULATIONS

Assume a dam site for which the following site
characteristics have been determined:

Length of centre-line $\qquad (L_c) = 400$ m

Average height of dam wall $\quad (H_a) = 18,25$ m

Class B geology : $\qquad\qquad F_g = 0,5$

Percentage of rock outcrop=20 : $F_r = 0,2$

With the above input data, the following
values can be calculated:

From equation (3) : $A = 400 \times 18,25 = 7300 \ m^2$
From equation (2) : $V = 2 \times 400 \times (18,25)^2 = 262800 m^3$
From equation (9) : $T_r = 7300^{0,45} = 54$ hours
From equation (12) : $V_f = 0,6 \times 262800 \times 0,5 \times 0,8 \times 10^{-5}$
$\qquad\qquad\qquad\qquad = 0,63 \ m^3$

Assuming that $0,44 \ m^3$ of this was obtained from
within the actual loaded volume, then from
equation (15):

$$V_d = 262800 \times 0,5 \times 0,8 \times 10^{-5} - 0,44$$
$$= 0,60 \ m^3$$

The cost for the reconnaissance investigation
is then from equation (8):

C_R =54 hours at say R40 per hour $\quad$ =R2160
+ transport costs for 2 site visits $\quad$ =R 280
$\qquad\qquad\qquad\qquad\qquad$ Total $\quad$ =R2440

Assuming that for the feasibility study, only
diamond drilling is undertaken, then from

equation (5) the length of drilling is

$$0,63 \div 0,002 = 310 \text{ m}$$

Cost for sampling is then R150x310 = R46500
From equation (14), the professional cost is:

$$C_F = C_{Tf} + C_o$$
$$= 330 \times 0,63(1+0,1) \times R40,00 = R\ 9147$$
$$\underline{\text{Total} \qquad = R55647}$$

Assuming that for the design investigation, a
trench 80 m long and 2 m by 2 m is required.

From equation (6):

$$V_t = 0,002 \times 80 + 0,0005 \times 80 \times 6$$
$$= 0,4 \ m^3 \text{ at a cost of say R 8000}$$

With the remaining $0,2 \ m^3$, 100 m of
drilling can be done at a cost of $\qquad$ R15000

From equations (14) and (16), profes=
sional cost $C_D = C_{Td} + C_o$
$$= 500 \times 0,6(1+0,1).40 = \underline{R13200}$$

$$\underline{\text{Total} \qquad = R36200}$$

5. CONCLUSIONS

By obtaining a few site characteristics from
existing information on maps and employing a
number of empirical relationships developed
through back-analyses of completed projects,
more cost effective foundation investigation
programmes can be planned and fairly accurate
cost estimates obtained for the various stages
of investigation.

It must be emphasised, however, that the above
relationships are based on average results ob=
tained from a limited number of actual site
investigations and for some cost estimates,
further allowances may have to be made for
special conditions regarding the accessibility
of sites, local geological conditions, require=
ments of the structure, availability of staff
and equipment, and time limitations imposed.

REFERENCES

DE BEER, E.E. (1981). Verbal presentation
during 10th Int. Conf. on Soil Mechanics
and Foundation Engineering, Stockholm.

FOOKES, P.G. (1967). Planning and stages of
site investigations. Engineering
Geology, 2, 63-134.

OLIVIERA, R. (1979). Engineering geological
problems related to the study, design and
construction of dam foundations. Bull.
Int. Ass. Engng. Geol., 20, 4-7.

PRICE, D.G. and KNILL, J.L. (1974). Scale in
the planning of site investigations.
Proc. 2nd Cong. Int. Ass. Engng Geol.,
San Paulo, Vol. II, III(3.1) - III(3.8).

VAN SCHALKWYK, A. (1980). Die invloed van
geologie op die ontwerp en konstruksie
van groter damme in Suid-Afrika. DSc
thesis, Univ. of Pretoria.

ANCHORING OF THE AGUAYO PENSTOCKS
Ancrage de la conduite forcée de Aguayo
Verankerung der Aguayo Druckwasserleitung

F. Muzas
Dr. Civil Engineer, Cimentaciones Especiales, S.A. (RODIO) — Madrid, Spain

J. J. Elorza
Dr. Civil Engineer, Electra de Viesgo-Santander, Spain

SYNOPSIS

The Aguayo hydroelectric system has a head of 320 m between the Aguayo reservoir and the reversible plant at Alsa. As penstocks, two parallel pipes of 3,800/3,400 mm diameter were designed without expansion joints. Each pipe is almost 1,300 m long with six fixed points which have to withstand the forces due to its own weight, water load and thermal variations. In the foundations of the four intermediate fixed points, and in order to do away with enormous volumes of concrete, great prestressed forces had to be applied by means of permanent anchors grouted into the ground.

RESUME

Le système hydroélectrique de Aguayo a une chute de 320 m entre le barrage de Aguayo et la centrale reversible de Alsa. Pour la conduite forcée on a prévu deux tubes parallèles de 3.800/3.400 mm de diamètre sans joints de dilatation. Cette conduite de presque 1.300 m de longueur a 6 points d'appui fixes où il faut absorber les charges dues aux poids propre, charge d'eau et variations thermiques. Pour les fondations des quatre points fixes intermédiaires, et afin d'éliminer d'énormes volumes de béton, il a été nécessaire d'appliquer de grands efforts de précontrainte, au moyen de tirants définitifs ancrés au terrain.

ZUSAMMENFASSUNG

Das Wasserkraftprojekt von Aguayo verfügt über ein Gefälle von 320 m zwischen dem Aguayo Stausee und der Alsa Umkehrzentrale. Die Druckleitung wurde als zwei parallel liegende Röhren ohne Dehnungsfugen projektiert, die einen Durchmesser von 3.800/3.400 mm haben. Die Leitung hat eine Länge von ca. 1.300 m und ist an 6 Punkten fixiert. Diese Punkte müssen das Eigengewicht der Röhren, die Wasserlast und die Lasten infolge thermischer Unterschiede aufnehmen. Für die Fundierung von 4 Fixierpunkten wurden allzu große Volumen von Beton dadurch vermieden, daß man große Vorspannkräfte mittels Daueranker aufbrachte.

1. INTRODUCTION

The Aguayo Project, in Spain, is a reversible-hydroelectric development, promoted by ELECTRA DE VIESGO, S.A. in the municipal districts of Bárcena de Pie de Concha and San Miguel de Aguayo, in the Cantabria region, north-west of Spain.

The Alsa reservoir, in existence since 1920, forms the lower basin the capacity of which has been raised to 22 Hm3 by increasing the height of the dam 7 m aprox.

The upper basin with 10 Hm3 capacity is an artificial reservoir located on a plateau to the south of Ano Peak, formed by a rock-fill dam abutted against the Peak, and constructed with the material excavated from the floor of the basin itself. The dam, whose length at crest is 3,000 m, has a maximum height of 32 m. The dam upstream face has an asphalt membrane, whereas the slopes cut in the rock and the basin bottom have no lining whatsoever, relying for watertightness on the nature of the rock, Triassic sandstones and lutites (Bundsandstein).

The power-plant is located next to the old Alsa dam and is connected to the upper basin by two telescopic high pressure penstocks 3,800/3,400 mm in diameter and 1,300 m long, and to the Alsa reservoir by means of four low pressure pipes, one for each group, 3,500 mm in diameter and approximately 70 m long. The gross mean head is 320 m an the total power capacity of the plant is 424 MVA.

The path of the high pressure pipelines crosses the above mentioned Triassic sandstones, except for a short stretch, approximately 300 m long, which passes through a Keuper zone, with outcrops of ophites, severely affected by spheroidal weathering.

2. PENSTOCK PROJECT

The two 1,300 m long penstock pipelines were designed parallel along the centre axis and without expansion joints. Their foundations have six
fixed points to take up the forces due to their
own weight, water load and thermal variations du
ring assembly and the later operating period.
Fig. 1 shows the penstocks in construction phase.

Fig. 1 - Aguayo penstocks in construction phase

In order to determine the stresses in the pipelines and the forces at each fixed point, the
following hypothesis were considered:

1) Penstock: empty.
   Temperature variation: 20ºC

2) Penstock: filled of water.
   Temperature variation: -50ºC

3) Penstock: filled of water.
   Temperature variation: 0ºC

4) Penstock: filled of water.
   Temperature variation: -20ºC

5) Lower part of penstock: filled of water.
   Temperature variation: 0ºC

   Upper part of penstock: empty.
   Temperature variation: 15ºC

Based on these hypothesis, the normal force N,
the moment M and the tangential force T (see
Fig. 2) acting on the basis of the fixed points
were calculated.

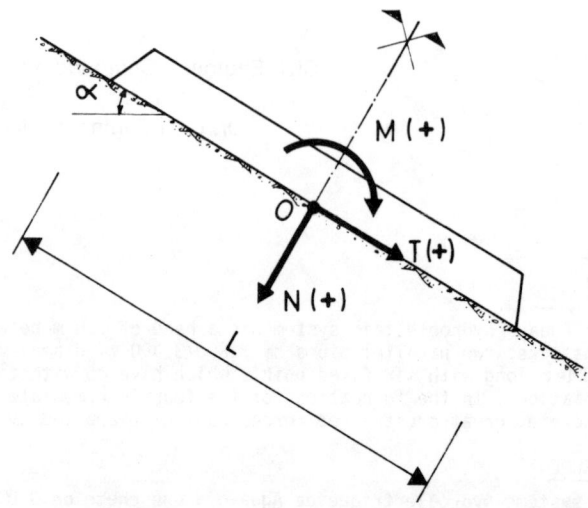

Fig. 2 - Forces acting on the basis of a fixed
point.

Table I resumes the most unfavorable forces to
be taken up at the four intermediate fixed points,
with the notation corresponding to that of Fig.2.

Table I - Summary of the most unfavorable forces
acting on the basis of each pipeline at the four
intermediate fixed points (see Fig. 2).

Fixed point	α (deg)	L (m)	Hypo thesis	N (MN)	M (MN × m)	T (MN)
A	32.0	6.75	1	-9.65	-4.93	-1.54
			5	-3.42	45.97	14.37
B	12.7	7.30	1	0.43	-5.71	-2.36
			2	-0.82	30.19	12.22
C	9.9	7.50	1	-1.33	-2.02	-0.54
			5	0.09	32.45	11.18
D	12.7	8.20	1	0.90	-6.16	-2.16
			2	-2.28	21.17	5.99

As a result of these forces, if classic gravity
blocks had been used for the foundations of the
se fixed points, enormous volumes of concrete
would have been required. The result would have
been an uneconomical and virtually unfeasible so
lution.

3. FOUNDATION SOLUTION

The solution adopted for the foundation of the
four intermediate fixed points consisted of ins
talling permanent prestessed anchorages in such
a way that, for the two most unfavorable hypothe

sis, the resultant forces R_I and R_{II} (see Fig.3) passed through the ends of the central third part of the supporting base, and with symmetrical dips in respect of it.

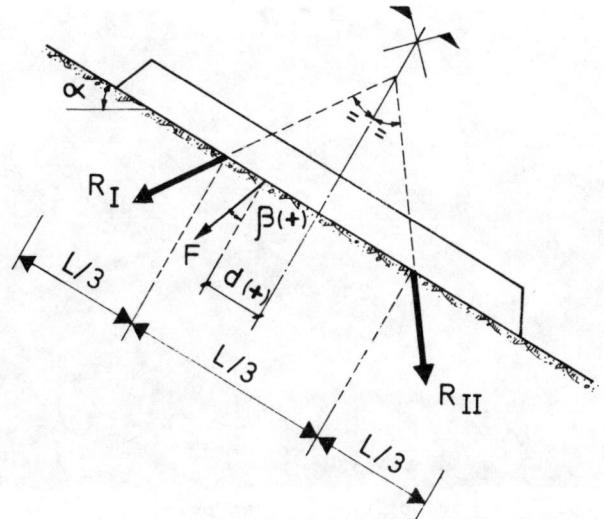

Fig. 3 – Schematic definition of anchorage force "F" and resultant forces R_I and R_{II} for the two most unfavorable hypothesis.

The total anchorage force, F, to be applied as well as its position and orientation, defined by eccentricity "d" and angle "β", as shown in Fig. 3 was calculated. The summary of these values is presented in Table II.

Table II – Anchorage forces per penstock, position and orientation (see Fig. 3).

Fixed point	F (MN)	d (m)	β (Deg)
A	29.64	0.584	10.34
B	15.84	0.870	19.31
C	15.26	0.994	19.21
D	11.22	0.900	17.39

4. ANCHORS CHARACTERISTICS

A permanent-type anchor, commercially known as "Rofix", was chosen. Its installation is as follows:

1. The drilling process in carried out, according the characteristics of the ground.

2. A "Rofix" steel pipe, fitted with non-return grout valves in the bonding length is installed and sealed to the ground with cement mixture.

3. The bonding length is created by grouting cement mixture at high pressure through the valves of the "Rofix" pipe.

4. The anchorage reinforcement consisting of various high tensile steel ropes, protected by

a P.V.C. coating only in the free anchor length, is installed.

5. The reinforcement is sealed to the "Rofix" pipe in the bonding length with cement grout.

6. Prestressing is carried out using a Freyssinet-type anchor head.

7. The space between the PVC coating and the Rofix pipe is permanently filled with an appropriate mixture to protect the cable.

8. The anchor head is duly protected.

5. FOUNDATION CHARACTERISTICS

The Figs. 4 and 5 show a foundation design with the orientation of anchors disposed symetrically respect to the total anchorage force, F, obtained by calculation.

Table III resumes the principal characteristics of the four fixed point foundations.

Table III – Principal characteristics of the foundations (see Figs. 4 and 5).

Characteristic		Fixed point			
		A	B	C	D
α_1	(deg)	15.9	15.2	4.6	14.7
α_2	(deg)	48.0	10.3	15.2	10.8
α_3	(deg)	32.0	12.7	9.9	12.7
β_1	(deg)	55	42	40	42
β_2	(deg)	50	37	35	37
β_3	(deg)	45	32	30	32
β_4	(deg)	40	27	25	27
β_5	(deg)	35	22	20	22
β_6	(deg)	30	–	–	–
No. anchors		24	20	20	20
Ø holes	(mm)	165	153	153	140
Ø Rofix pipe	(mm)	133 125	121 113	121 113	108 100
No. steel ropes		24	12	12	12
Ø rope	(in.)	0.5	0.6	0.6	0.5
Tens.strength	(MPa)	1,900	1,730	1,730	1,900
Elas.limit	(MPa)	1,720	1,580	1,580	1,720
Steel section	(cm2)	23.69	17.57	17.57	11.84
Stress load	(MN)	2.94	1.86	1.89	1.39
Anchor length	(m)	27.5	25.0	20.0	20.0
Bonded length	(m)	17.5	13.0	10.0	9.0

The foundations of the fixed points "A", "C" and "D" were carried out in Triassic sandstones, while point "B" was located in the mentioned Keuper zone consisting of severely weathered ophites, which necessitated a longer bonded length. Moreover, with the aim of enabling the stresses of the foundation slab –of 13x11 m and

2 m high- to be taken up, the ground was reinforced by fitting 59 "Ropress" micropiles, each 12 m long and an admissible load of O.55 MN.

Fig. 6 and 7 show different views of the foundation construction works.

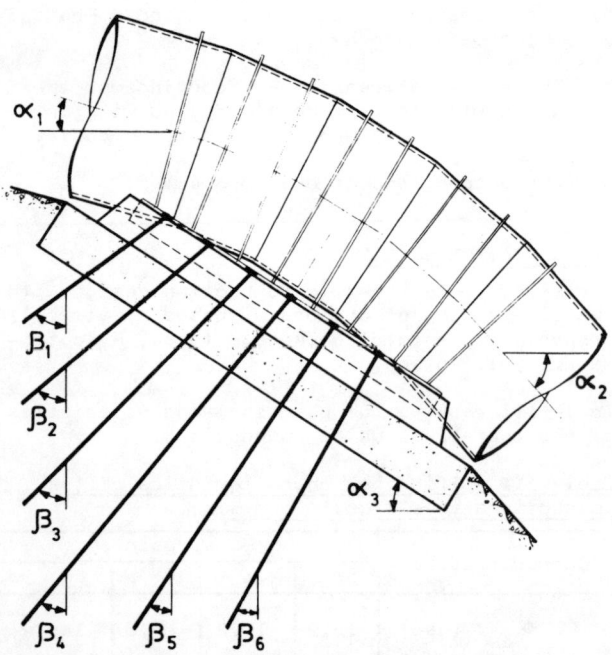

Fig. 4 - Longitudinal section of a fixed point foundation

Fig. 6 - Foundation of fixed point "A"

Fig. 7 - Instalation of penstocks in fixed point "D"

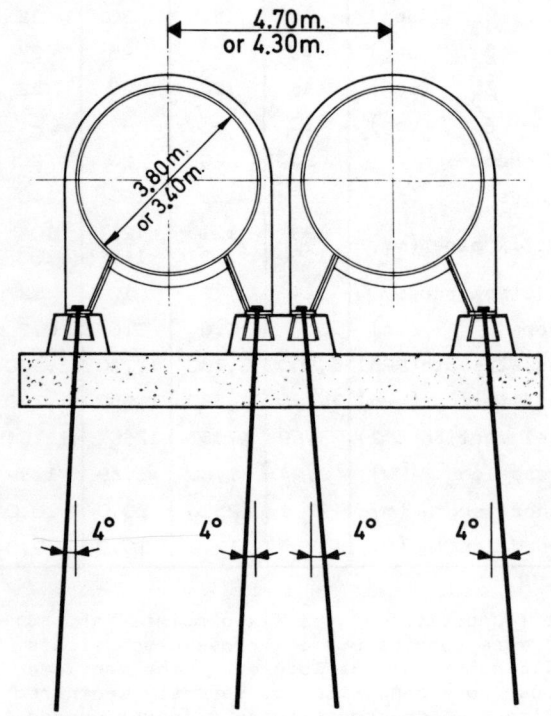

Fig. 5 - Cross section of a fixed point foundation

6. CONCLUSIONS

The use of permanent anchors of high capacity in the case reported in the present paper enabled an exceptional foundation problem to be solved. With this solution it was possible to avoid the construction of enormous blocks of concrete, the execution of which, moreover, was virtually unfeasible.

FOUNDATION ROCK BEHAVIOUR OF THE PASSANTE DAM (ITALY)

Le comportement du rocher de fondation du barrage du Passante (Italie)

Das Verhalten des Gründungsfelsens der Passante Staumauer (Italien)

P. Bonaldi
ISMES (Istituto Sperimentale Modelli e Strutture), Bergamo

G. Manfredini
S. Martinetti
ENEL (Ente Nationale Energia Elettrica) Construction Division. Rome

R. Ribacchi
Professor of Rock Mechanics, Faculty of Engineering. University of Rome

T. Silvestri
ENEL Design and Construction Centre. Naples

SYNOPSIS

The Passante dam is a 70 m high concrete gravity-dam founded on gneiss and crystalline schists having poor mechanical characteristics. The behaviour of the foundation is controlled by means of inverted plumb-lines. long-base extensometers (rock meters), electro-acoustic piezometers, and by measuring the rotations of the dam base. In this paper data collected during the dam construction and the first fillings of the reservoir are analysed by means of an FEM model. The rock deformability was identified and compared with the values provided by the in situ tests performed during the design stage. The presence of a superficial low-module layer was evidenced; a considerable rheological behaviour was also observed.

RESUME

Le barrage du Passante est un barrage poids en béton de 70 m de hauteur. Sa fondation est constituée de gneiss et de schistes avec médiocres propriétés mécaniques. Le comportement du rocher de fondation est contrôlé par des pendules inversés, des extensimètres à longue base, des piézomètres acoustiques et par la mesure des rotations de la base du barrage. Les mesures effectuées pendant la construction du barrage et pendant la mise en eau ont été analysées en utilisant les méthodes d'éléments finis. Cette analyse a permis d'établir la déformabilité du rocher et de la comparer avec les valeurs obtenus par des essais in situ réalisés au stade préparatoire. On a pu mettre en évidence la présence d'une couche superficielle plus déformable ainsi que de nettes déformations rhéologiques pendant la construction et la mise en eau.

ZUSAMMENFASSUNG

Die Passante Staumauer ist eine 70 m hohe Gewichtsstaumauer, die auf Gneis und kristallinen Schiefern mit schlechten mechanischen Eigenschaften gegründet ist. Das Verhalten des Baugrundes wird durch inverse Pendel, Dehnungsmesser mit langer Basis, elektro-akustische Piezometer und durch die Messung der Rotationen der Dammbasis erfaßt. Im vorliegenden Beitrag werden die Daten, die während des Dammbaues und der ersten Auffüllungen gesammelt wurden, mittels der Finite Element Methode analysiert. Durch diese Untersuchung ist es möglich gewesen, die Formänderung des Gesteins zu identifizieren und mit den Werten von in situ Versuchen während des Bauentwurfes zu vergleichen. Das Vorhandensein einer Schicht stark verformbaren Gesteins und ein bemerkenswertes rheologisches Verhalten während des Baues und der ersten Auffüllungen werden hervorgehoben.

1. INTRODUCTION

The Passante reservoir is part of the hydroelectric scheme which exploits the waters flowing down the eastern slopes of the Sila massif in Calabria (Fig.1). The Passante dam was built across the upper course of the river bearing the same name at about 1100 metres elevation; the reservoir thus formed has a maximum capacity of 35 .$10^6 m^3$, being fed by a catchment area of about 70 km^2. It will supply water to two hydroelectric plants in series for an overall head of 900

Fig. 1 Location of the Sila hydroelectric plants

m and a capacity of 75 MW.

The dam is a gravity dam with triangular section
and a central spillway; its features are summarized in Table 1 and illustrated in Fig. 2.

Its construction was completed in 1977; however,
due to delays in the completion of other structures and in particular of the nine-kilometre-
long diversion tunnel, the reservoir has so far
never been completely filled.

Monitoring devices were installed in the dam and
in the foundation rock so as to follow the behaviour of the structure right from the initial
stages of construction.

Table 1   Characteristics of the Dam

Max. height (above foundations)	71 m
Max. width	48 m
Number of blocks	27
Length of crown	450 m
Volume	310,000 m^3
Upstream slope	88.3°
Downstream slope	54°
Depth of grout curtain	35 m
Depth of drainage curtain	8 m

2.   GEOLOGICAL SITUATION

The reservoir is located in metamorphic rocks of
the Calabride complex.

In the gorge area two main units, known as "biotitic gneiss" and "white schists" are present
(Fig.2). The latter one outcrops in the lower
part of the valley slopes and is therefore the
foundation rock of the higher blocks. Its petrographic characteristics vary considerably but
the most common lithotype is a weakly schistous
rock, composed of quartz and feldspar with a low
percentage of phyllosilicates; its texture is
characterized by the presence of large highly
fissured crystals and of microgranular catacla-
stic layers (Fig. 3). Other less frequent litho-
types present a higher percentage of muscovite
and marked schistosity.

Fig. 3 Thin section of a gneissic rock in the
"white schists" formation. The photo shows a mi-
crofissured quartz grain, distorted mica plates
and microgranulation layers

3.   MECHANICAL PROPERTIES

3.1   Rock Material

The mechanical characteristics of the rock material are summarized in Table II and in the histo-
grams of Fig.4. The tests relate to the most
common, weakly-schistous, lithotype.

The high scatter of the strength and deformability characteristics may be attributed to the variable conditions of microfissuring and, in some
cases, to the presence of macroscopic weakness
surfaces.

The intensity of microfissuring is evidenced by
the difference between seismic velocity in dry
and saturated conditions, by the low values of
the Young modulus and by its dependence on the
applied stresses.
According to Deere's diagram (Fig.5) the rock
can be classified as having medium-low strength
and mean deformability. The triaxial strength
values can be represented by means of a Coulomb
law, even though a power law gives a better approximation; the relevant parameters are shown in
Fig. 6.

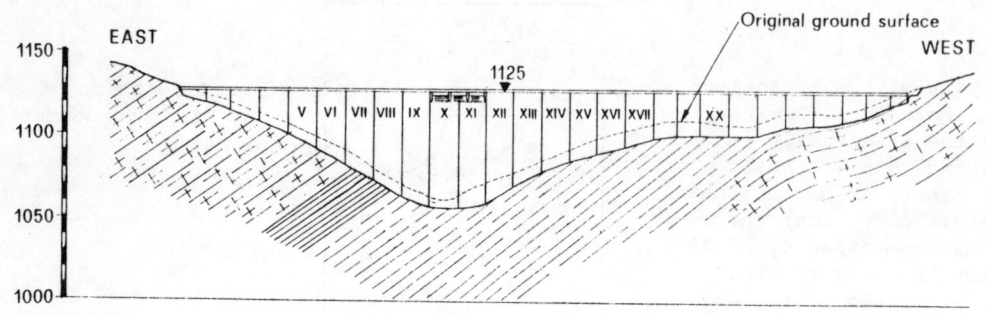

Fig. 2 Geological situation at the dam site

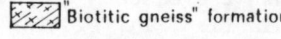
"Biotitic gneiss" formation

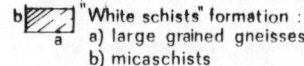
"White schists" formation :
a) large grained gneisses
b) micaschists

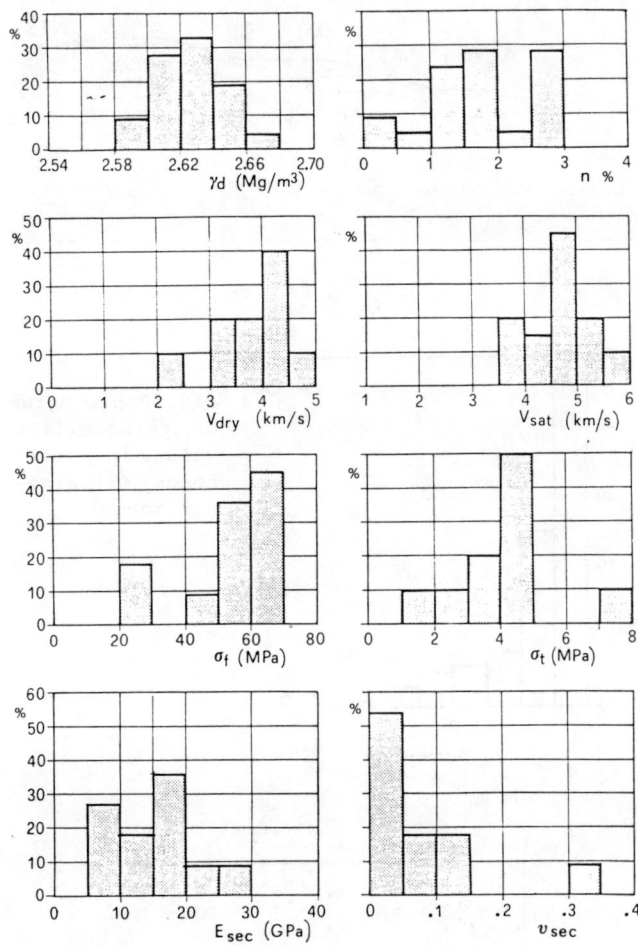

Fig. 4 Histogram of the mechanical properties
of the foundation rock

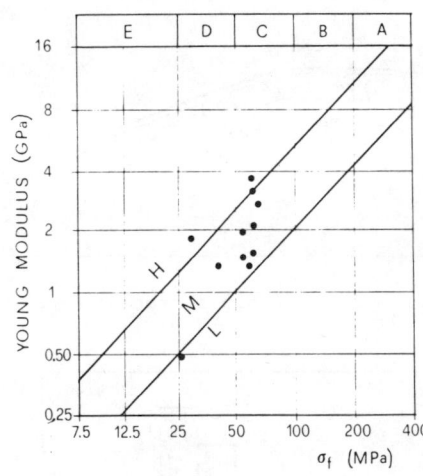

Fig.5 Position of the rock in Deere's diagram

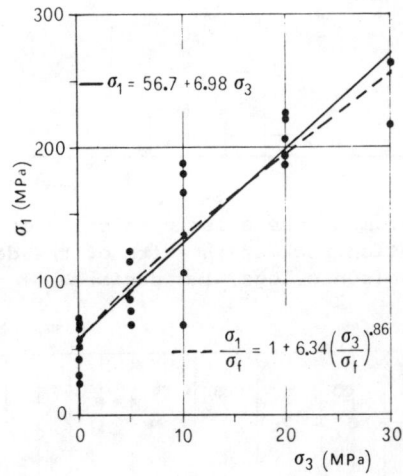

$$\sigma_1 = 56.7 + 6.98\,\sigma_3$$

$$\frac{\sigma_1}{\sigma_f} = 1 + 6.34\left(\frac{\sigma_3}{\sigma_f}\right)^{.86}$$

Fig. 6 Results of triaxial compression tests

3.2 Rock mass

A seismic refraction survey (Fig.7) indicated the presence of a superficial layer of low quality rock characterized by a seismic velocity of 1.7-2.5 km/s; this layer is virtually absent at the bottom of the valley, whereas its thickness increases towards the higher elevation along the slopes. The underlying rock is characterized by a velocity of 3.5-4 km/s.

Sonic logs on a 1 m base were carried out in boreholes underneath the foundation of the dam; Fig. 8 shows the histogram of the velocity for 10 boreholes located underneath the foundation of the deeper blocks (IX and XII). The mean velocity is 3.92 km/s, that is markedly lower than the values determined on saturated laboratory samples (4.97 km/s).

The longitudinal velocity was also determined by means of the crosshole technique on 20 m bases, aligned at perpendicular directions with the dam. Fig. 9 shows a general increase in seismic velocity versus depth below the surface and the influence of the stress variations caused by the

TABLE II Mechanical properties of the rock

	N	x̄	(%)
Dry density (Mg/m³)	21	2.625	0.04
Matrix density (Mg/m³)	8	2.670	0.06
Porosity (%)	21	1.71	31
Seismic velocity - dry samples (km/s)	20	3.68*	21
Seismic velocity - saturated samples (km/s)	20	4.97*	10
Tensile strength (MPa)	10	4.0	38
Uniaxial compressive strength (MPa)	11	52.2	110
Secant Young modulus ($\sigma = 10$ MPa)	11	13.1*	40
Secant Poisson coefficient	11	0.07	150
Cohesion (MPa)	-	10.7	-
Friction angle	-	49°	-

N = number of samples   = variation coefficient
x̄ = mean value   * = harmonic mean

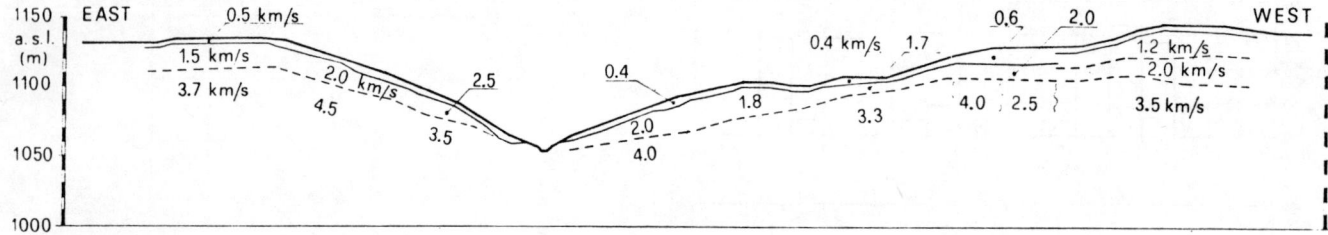

Fig.7 Results of a seismic refraction survey at the dam site

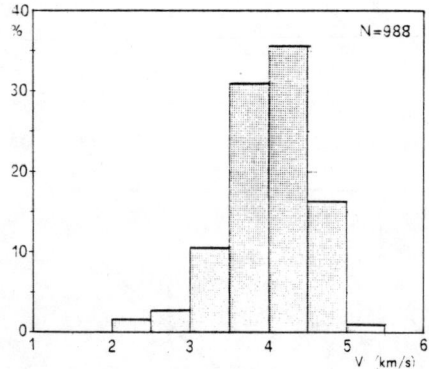

Fig. 8 Histogram of the seismic velocity in the foundation rock at the site of the deeper blocks (sonic logs, 1 m basis)

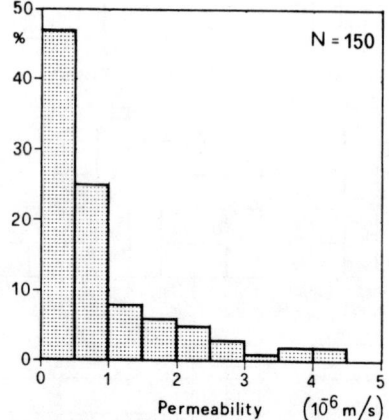

Fig.10 Histogram of permeability values in boreholes (5-metre stretches)

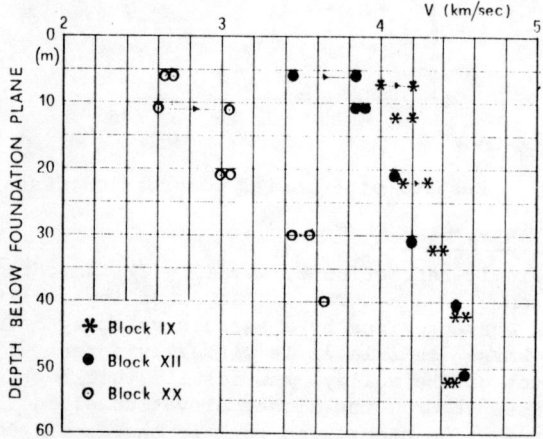

Fig. 9 Average seismic velocities at various depths below the foundation plane. The velocity increments measured after the construction of the dam are indicated by the arrows. The values measured along the valley side (block XX) are markedly lower than those on the valley bottom (blocks IX and XII)

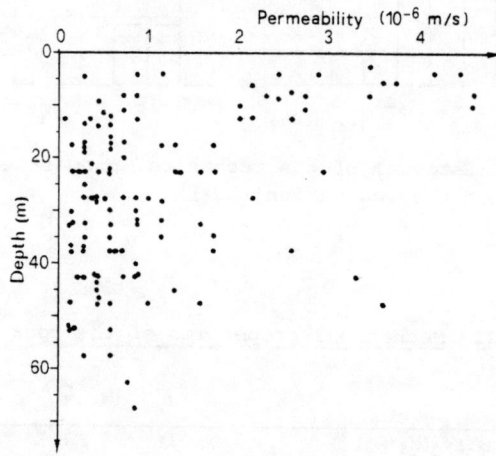

Fig. 11 Permeability versus depth below the ground surface

construction of the dam. The velocity values are markedly lower below block XX, which is located at a higher elevation on the valley slopes.

Average fracture spacing in boreholes below the central block was 0.09 m and the RQD value about 51%. Some open and oxidized fractures were found also at great depth (beyond 50 m).

Water adsorption tests at a pressure of 0.35 MPa on 5 metre stretches were carried out in 21 boreholes along the dam foundation (Fig.10); permeability values are more scattered and generally higher near the ground surface (Fig.11). The influence of the position of the boreholes along the valley slopes proved on the contrary to be less important.

Four plate-loading tests were performed by ISMES in an exploratory tunnel excavated in the right abutment. An accurate technique,which was described by BORSETTO et al.(1981), was adopted;

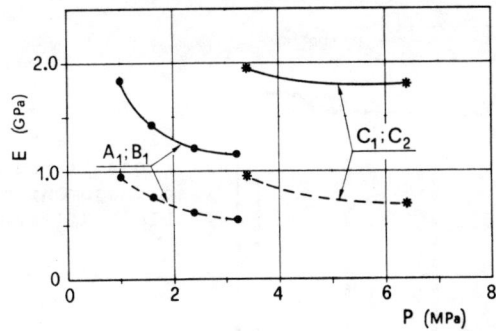

Fig.12 Values of elasticity (E_e) and deformation (E_d) moduli as a function of the applied pressure on the plate. In the C_1-C_2 tests the load was applied in a direction parallel to the schistosity plane, in the A_1-B_1 tests the load was applied in a direction forming a 60° angle to the schistosity plane

the load was applied by means of an annular flat jack and the displacements were measured not only at the surface but also at various depths underneath the centre of the loaded area. For interpreting the results, the measurements taken in the central hole were used, because they allow a more accurate consideration of the influence of the superficial loosening of the rock mass.

Fig. 12 shows the trend of the deformation (E_d) and elasticity (E_e) moduli as a function of the load; E_d was calculated on the basis of total deformation measured at a given load and E_e was calculated on the basis of the recoverable deformation during unloading. These values were determined disregarding the contribution of a loosened superficial layer 25 cm thick, just below the loading surface. The values of E_d and E_e appear to be very low even for a low quality rock mass; the fact that the tunnel had been excavated some years before the performance of the test may have influenced the results.

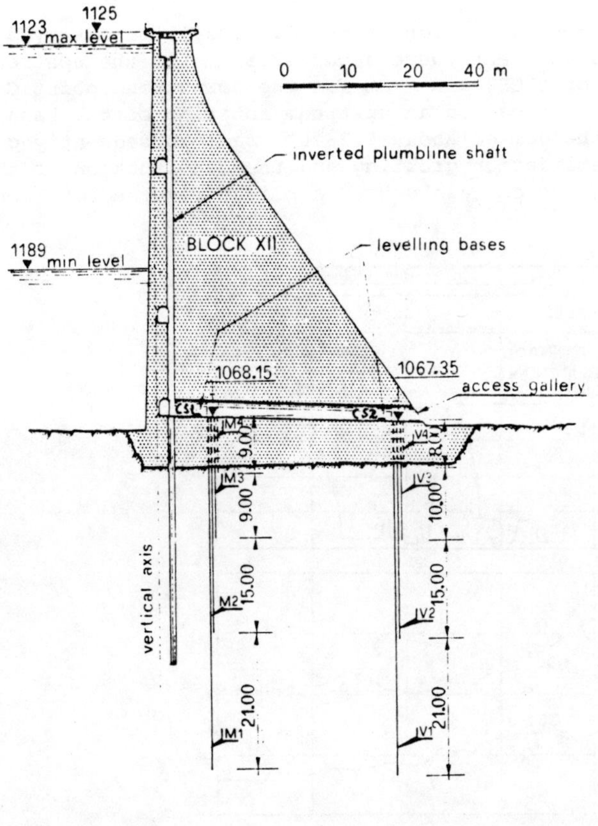

Fig.13 Monitoring system in the foundation rock of block XII
   - inverted plumbline anchored 30 m below the foundation plane
   - 2 rod extensometers with 4 sensing points
   - 2 levelling bases in the access drift

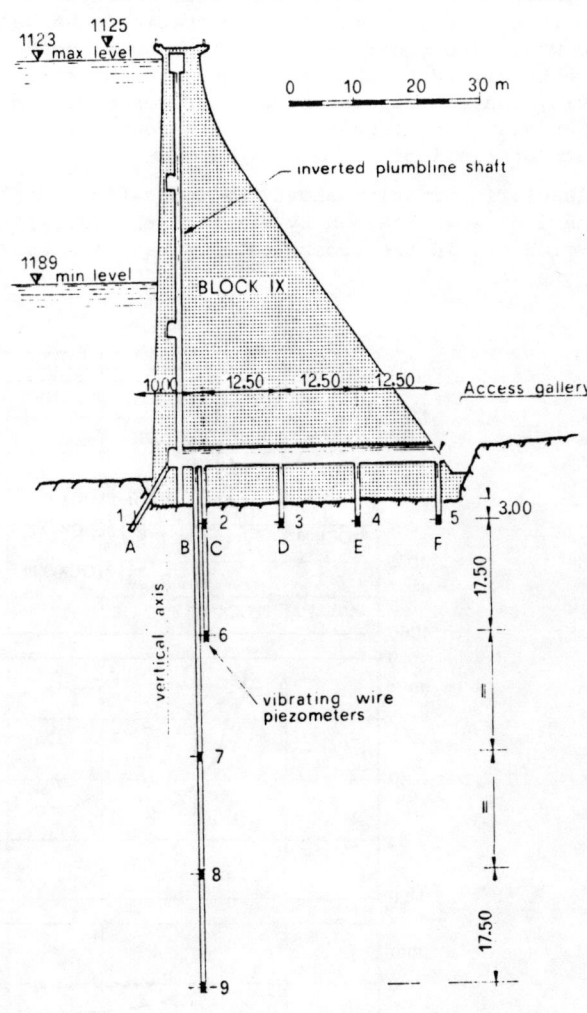

Fig.14 Piezometric system below block IX

4. INSTRUMENTATION

The dam was equipped with the usual monitoring instruments placed inside the dam body: thermometers located inside the concrete, plumb lines, collimation bases for controlling the horizontal displacements of the crown, geometrical levellings at various heights.

The instrumentation in the foundation rock includes:
- an inverted plumb-line anchored at 30 m below the foundation plane (Fig.13);
- multipoint rod extensometers underneath the upstream and downstream faces of the two blocks, XII and XVII (Fig.13);
- vibrating wire Mahiak piezometers (Fig.14).

Furthermore, underneath three blocks of the dam (IX, XII, XVII) a permanent network for measuring the seismic velocity was installed, according to the scheme illustrated in Fig. 15. In each cross-section the receivers (hydrophones characterized by great stability over time) were cemented to the rock in four holes at 20 m spacing; the seismic waves are generated in a fifth hole located downstream. With this technique, the seismic velocity can be measured at various depths in three stretches that globally cover the foundation width of the dam.

Valuable information about the foundation rock behaviour was obtained by means of the levellings carried out in the access gallery near the base of the dam.

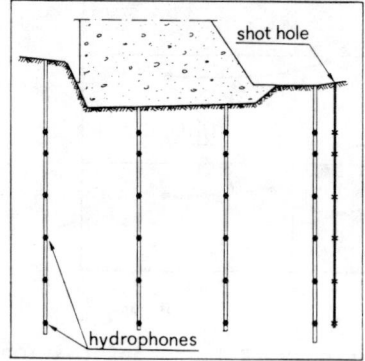

Fig.15 Permanent crosshole seismic arrangement below block XII

5. BEHAVIOUR OF THE STRUCTURE

The vertical displacements of the foundations of block XII, which is representative of the higher central blocks, were measured by the levelling of points CS1 and CS2, the positions of which are shown in Fig. 13 Measurements started already during construction, when the structure had reached half its final height. Fig.16 illustrates the settlements during the final construction stage and during the subsequent partial fillings.

Upon completion of the dam body, the observed settlements were about 12.5 mm for the upstream point CS1 and 6 mm for the downstream point CS2, which caused an upstream rotation of the base of the dam of about $0.2 \cdot 10^{-3}$ rad; subsequently consolidation grouting and the construction of the grout curtain caused a slight reverse rotation.

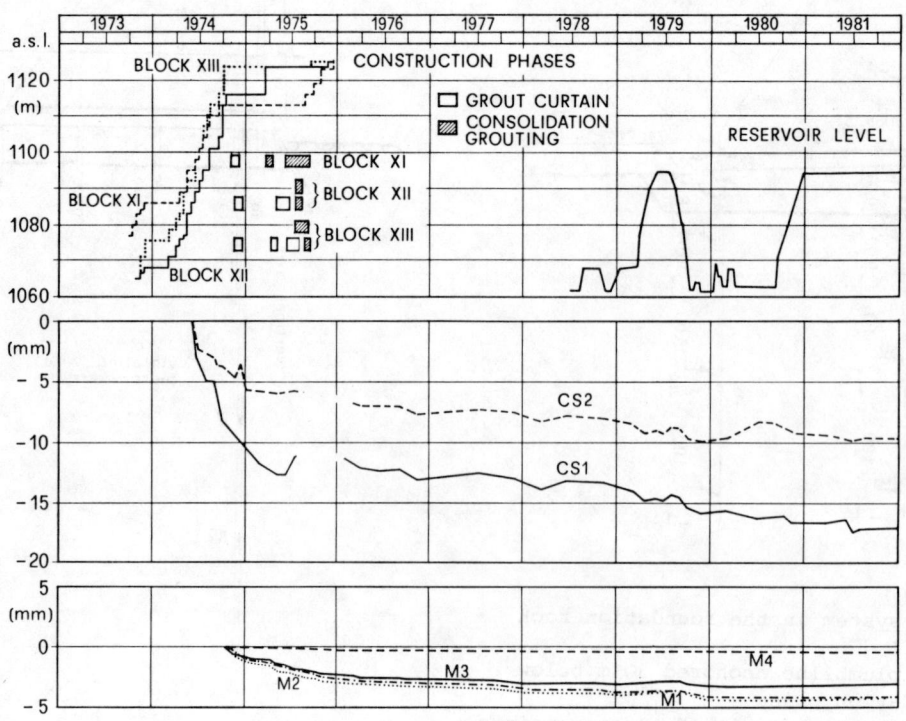

Fig.16 Settlements at points CS1 and CS2 and deformation recorded by the rock metres

The foundation rock showed a marked rheological behaviour which was clearly noted during a pause in construction activities at elevation 1116. Upon completion of the dam body the rheological settlements continued, even if at a reduced rate, being, over the last 6 years, on average about 0.6 mm per year for the downstream point CS2 and 1 mm for the upstream point CS1. These rates are almost proportional to the respective settlements at the end of construction; the rheological effect therefore, caused a further increase in the upstream rotation of the dam.

A close look at Fig.16 shows that seasonal changes have some influence on the settlement trend. Partial fillings already carried out did not cause downstream rotations of the structure.

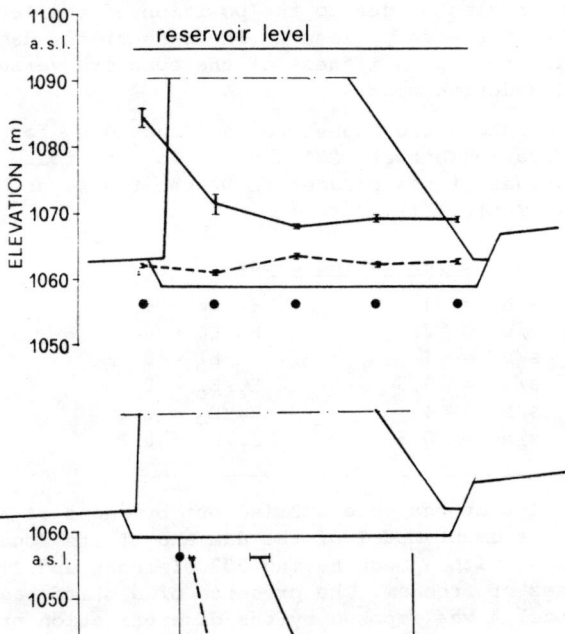

Fig. 17 Measured piezometric heights below block IX before filling (dashed lines) and during partial impounding (solid lines)

The extensometers in the foundation rock were placed when the dam had almost reached its final height, and so most of the deformations detected at the end (about 3 mm) are to be ascribed to rheological effects that are located mainly in the more superficial layer (about 10 m thick). In the following years there was a further increase in the rheological settlements, at a rate of about 0.3 mm per year, which is markedly lower than the values obtained with levelling.

Indications provided by the Mahiak piezometers clearly showed the relationship between water level in the reservoir and pore pressures in the foundation rock (Fig.17). Measurements taken when the reservoir was empty indicate that in depth the piezometer levels are higher than the foundation plane of the lower blocks; this fact had already been detected during the preliminary investigation stages. The filling of the reservoir causes an increase in uplift pressures, downstream of the grout and drainage curtain, corresponding to about 20% of the total head increment.

6. BACK ANALYSIS OF THE BEHAVIOUR OF THE FOUNDATION ROCK

Over the last years the application of simulation techniques by means of numerical models have afforded a considerable contribution to the interpretation of the behaviour of structures such as dams and their foundations, both during construction and during operation (BONALDI et al.,1981; BONALDI et al.,1982; FANELLI et al.,1979).

As to the Passante dam, numerical simulation has been applied so far mainly to the construction stage, in order to identify the deformability characteristics of the foundation rock.

This process of identification is complicated by the fact that the foundation rock cannot be considered homogeneous, its more superficial layer having higher deformability. This difficulty could have been overcome by measuring rock displacements at various depths by means of rockmeters; these instruments, however, were not operating at the time of the most significant stages of construction.

The simplest models for keeping account roughly of the rock foundation dishomogeneities are the

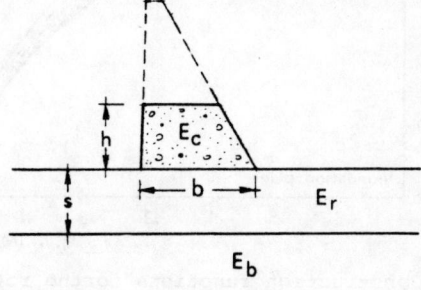

Fig. 18 Two-layer model of the rock foundation of the dam

"two layers" models (Fig.18) which are characterized by the following three parameters:
- E_r modulus of the superficial layer of the rock;
- s/d ratio between thickness of the superficial layer and the width d of the base of the dam;
- E_r/E_b ratio between the modulus of the superficial layer and that of the basement.

An even more simple model (rigid base model) can be obtained by assuming $E_r/E_b=0$.

The settlements of the two datum points allow one to identify two parameters of these models (i.e. E_r and s/b) by assuming the third parameter a priori.

A formally similar but more satisfactory procedure in practice is the following:
- evaluation of the modulus of the superficial layer on the basis of the values of the rotation of the dam (difference in settlement of the two levelling points);
- evaluation of the thickness of this deformable layer on the basis of the mean settlement of the two datum points.

Actually, the rotation of the dam base (proportional to the moment of the applied loads) depends essentially on the modulus of the superficial layer (provided that it is not too thin) and it is only slightly influenced both by the position of the rigid base and by the bidimensional schematization adopted in the model.

Analytically the generic displacement δ or rotation of a point of the dam under the applied loads due to the weight of the structure itself, may be expressed by the sum of the two terms, of which one is proportional to the deformability of the concrete, the other to that of the foundation rock.

$$\delta = \frac{1}{E_c} f_c(h_c) + \frac{1}{E_r} f_r(h_r)$$

where E_c and E_r are the moduli of the concrete and of the foundation rock, h_c is the height reached by the construction above the foundation plane. The function f_r depends not only on h_c but also on the mechanical and geometric parameters of the model adopted for the foundation rock. In the "two-layers" model the following can be assumed:

$$f_r = f_r(h_c, E_r/E_c, E_r/E_b, s/b)$$

The function f_c depends on the very same parameters, but in the specific case their influence can be neglected due to the position of the levelling points - very close to the base of the dam - and to the high stiffness of the concrete versus the foundation rock.

Fig. 19 shows the "construction functions" for the rotation between CS1 and CS2 for the following values of the parameters of the two layer model (Table III).

TABLE III   Investigated Cases

A	s/b	=	12	E_r/E_b	=	0
B	s/b	=	28	E_r/E_b	=	0
C	s/b	=	1	E_r/E_b	=	0
D	s/b	=	0.5	E_r/E_b	=	0
C'	s/b	=	1	E_r/E_b	=	0.2
D'	s/b	=	0.5	E_r/E_b	=	0.2

The calculations were carried out by means of a finite element model of the dam and of its foundations, with a mesh having 502 elements and 2122 degrees of freedom. The presence of a stiff base in model A was imposed by the discretization process, but the results are virtually equivalent to those which could be obtained for a homogeneous and indefinite foundation; a rigid base at the same depth is present also in models C' and D'.

An examination of the figure confirms that the values of function f_r are only slightly influenced by the thickness of the superficial loosened layer and by the basement modulus.

By comparing the measured values of the rotation with the theoretical ones it was possible to identify the characteristics of the foundation rock layer; the modulus of the concrete was taken as being 20 MPa.

This assessment may or may not take into account the rheological deformations of the rock mass that will very likely be less important in the operation cycles following the first ones. Therefore, Table IV shows the results of the two back analyses; in the first one only the displacements measured during a period of rapid progress in construction (from el. 1086 to el. 1116) were taken into account; in the other, the overall displacements measured from the beginning of the observa-

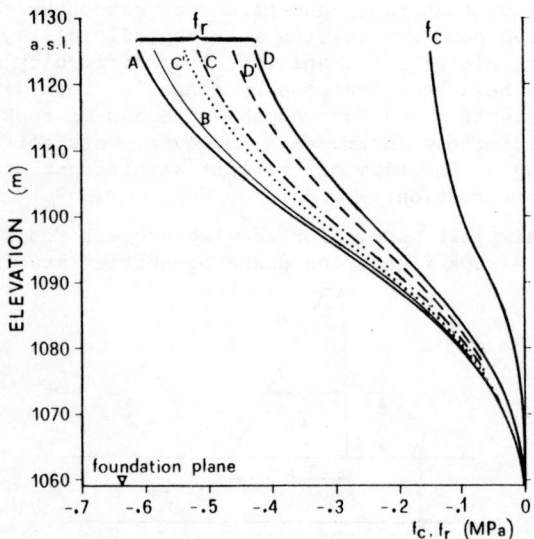

Fig. 19 "Construction functions" for the rotation of the dam between points CS1 and CS2; the different geomechanical models are indicated in Table III

TABLE IV   Deformation moduli E_r (GPa) of the foundation rock on the basis of the observed tilt between CS1 and CS2 during the construction of the dam

	SHORT TERM	LONG TERM
A	2.40	1.93
B	2.22	1.80
C	1.97	1.60
D	1.62	1.33
C'	2.03	1.64
D'	1.78	1.42

Table V   Identification of the elastic and geometrical parameters of the model

Model	Short-term behaviour	Long-term behaviour
$E_r/E_b = 0$ (rigid base)	$E_r = 2.07$ GPa $s/b = 1.5$	$E_r = 1.80$ GPa $s/b = 2.0$
$E_r/E_b = 0.2$	$E_r = 1.7$ GPa $s/b = 0.5$	$E_r = 1.60$ GPa $s/b = 0.9$

tion to the stage immediately preceding the grouting operations were considered.

The values thus obtained, which are only slightly different for the various models, point to a high deformability of the foundation rock, which is however lower than the value determined by means of plate-loading tests.

The functions f_r and f_c relating to the average settlement of the two datum points were subsequently calculated. Fig. 20 points out that in this case, unlike what occurs for the rotations, both the depth of the basement and the moduli of the deeper layers bear considerable influence. It must be pointed out that by adopting a simplified bidimensional model, the average settlements indefinitely increase as the depth of the rigid base increases; for very large depths the model is therefore not realistic because the actual tridimensional geometry of the dam becomes a critical factor.

By using at the same time the observed rotation values and the settlements, the parameters of the model in Fig. 18 can be identified (Table V).

In Fig. 21 the settlement values observed are compared with the theoretical ones evaluated on the basis of the moduli corresponding to the short-term behaviour; at the scale of the drawing there is no significant difference between the results given by the two models, which means that no discrimination can be made between the two models on the basis of the available experimental data.

7.   STORAGE PHASE

The behaviour of the structure during operation can be appropriately described by storage functions f'_r, f'_c which are similar to those described in para. 6 but which, in this case, are related to the reservoir level h_w. Both f'_r and f'_c can be broken down to two terms, the first related to the pressure of the water on the upstream face of the dam, and the other depending on the seepage gradients inside the rock mass and on the uplift pressure on the base of the dam (MANFREDINI et al.,1976).

In the case being examined here, the functions relating to the rotation of the base of the dam must be considered. The water pressure on the upstream face causes a downstream rotation which increases more than linearly with the water level in the reservoir. The values of the function f'_r essentially depend on the characteristics of the rock material directly underneath the dam and therefore in the rigid base model they are only slightly influenced by the depth of the rigid base.

The effect of the seepage varies widely depending

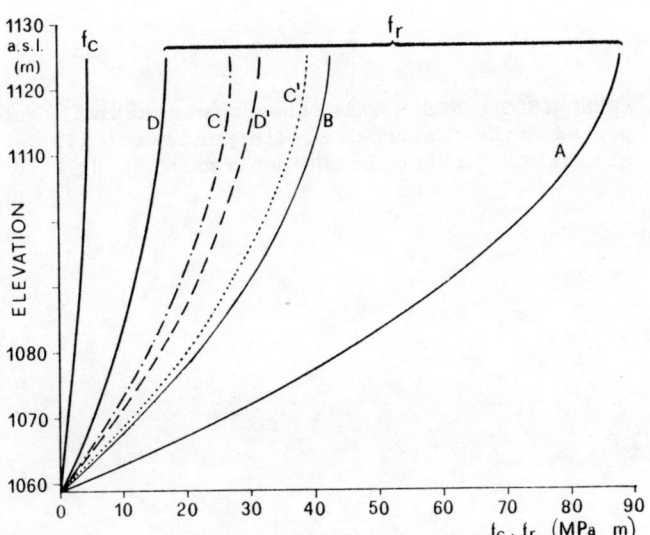

Fig. 20 "Construction functions" for the average settlement of the points CS1 and CS2

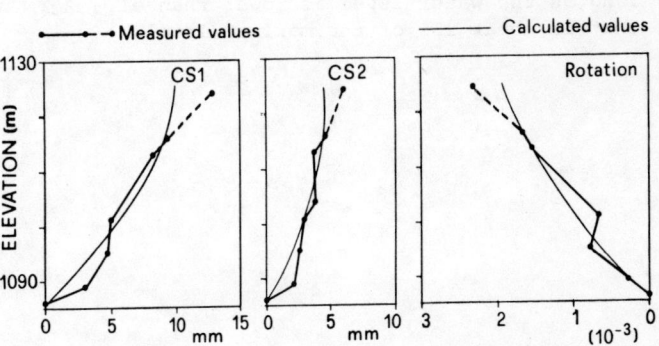

Fig. 21 Comparison between the measured and the theoretical values of the settlements of the points CS1 and CS2 and of the rotation of the dam base

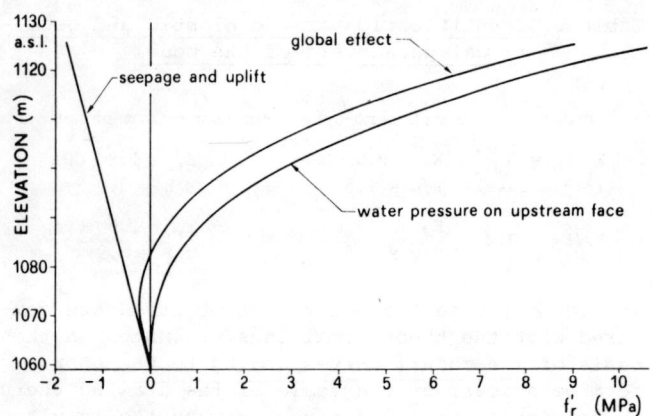

Fig. 22 "Storage function" f'$_r$ and f'$_c$ for the rotation of the dam between points CS1 and CS2

on the elasticity characteristics of the foundation and of the pore pressure distribution which is influenced also by the effectiveness of the grout curtain and of the drainage network. In many cases, however, seepage causes an upstream rotation of the structure, which increases almost linearly with the water level in the reservoir.

The global effect of increases in the water level therefore entails an upstream rotation at first, and subsequently a downstream rotation of the dam. This type of behaviour was observed in many Italian gravity dams (CAPOZZA et al.,1966).

As an example, Fig. 22 shows the storage functions for the rigid-base model, of which the geometric and mechanical parameters were identified on the basis of the behaviour during construction (s/b = 2; E$_r$ = 1.80 GPa); for determining the seepage conditions, the permeability of the foundation rock was taken as being uniform, and the grout curtain was considered to have an effective width of 3 m and a permeability 10 times lower than that of the rock.

The figure points out that the filling of the reservoir produces a slight upstream rotation so long as the water level is lower than el. 1082 that is, about 35% of the maximum level.

It can be also observed that the effects of the reservoir filling up to the level reached so far (el. 1095 m) are quite small and can be therefore easily masked by the converse rheological effects due to the dead weight of the dam.

REFERENCES

Bonaldi, P., Giuseppetti,G., Ribacchi,R., Selleri,G. (1981). Identification of rock foundation deformability of a large dam in operation: ISMES Publication n. 162

Bonaldi, P., Giuseppetti,G., Guccione,R., Ribacchi,R., Selleri,G.(1982). Evaluation of rock foundation behaviour for two dams in operation. XIV Congrès des Grands Barrages, (1) 927-942, Rio de Janeiro.

Borsetto,M., Ribacchi,R., Rossi,P.(1981). Long term and cyclic plate-loading tests in weak rocks. Int. Symp. on Weak Rock, (1) 137-142, Tokyo.

Capozza,F., Marazio,A., Penta,P. (1966). Déformabilité de la roche de fondation dans le cas de quelques barrages italiens. Proc. 1st Congr. Int. Soc. Rock Mech. (2) 603-616, Lisboa.

Fanelli,M., Giuseppetti,G., Riccioni,R. (1979) Experience gained during control of static behaviour os some large Italian dams. 13th Congress of Large Dams. (2) New Delhi.

Manfredini,G., Martinetti,S., Ribacchi,R. (1975). Mutual influence of water flow and state of stress in the analysis of dam foundations. Int. Symp. Criteria and assumptions for numerical analysis of dams. 881-897. Swansea.

ACKNOWLEDGEMENTS - This research was partially supported by the "Centro di Studio per la Geologia Tecnica del CNR" and by the CNR contract n. 80.02707.07.

MAJOR DAM FOUNDATIONS IN MADHYA PRADESH (INDIA)

Fondations pour des barrages importants au Madhya Pradesh (Inde)
Gründungen grösserer Staudämme in Madhya Pradesh (Indien)

V. M. Chitale

Engineer in Chief, Irrigation, Bhopal, Madhya Pradesh, India

SYNOPSIS

The understanding of foundation conditions of major dams in a realistic manner before embarking on actual construction is a must to avoid risks and failures. The statistics have shown that most of the failures of structures are caused by the failure of foundations. This necessitates studies of geological discontinuity of foundation rocks. In the Sixth Five Year Plan (1980-85) a major irrigation development programme has been launched in Madhya Pradesh, the heart of India. A few examples of projects where difficult foundation conditions were encountered are described below.

RESUME

Il est absolument nécessaire de saisir d'une manière réaliste les conditions de fondations des barrages importants avant de se lancer à la construction proprement dite si l'on veut éviter les risques et les ruptures. Les statistiques montrent que la plupart des ruptures des ouvrages sont dues aux ruptures des fondations. Cela exige des études sur la discontinuité géologique des roches des fondations. Dans le cadre du sixième plan quinquennal (1980-85), un vaste programme de développement pour l'irrigation a été lancé au Madhya Pradesh, au coeur de l'Inde. Quelques exemples d'aménagements où l'on a rencontré des conditions difficiles de fondations sont décrits ci-après.

ZUSAMMENFASSUNG

Das Verständnis der Fundamentbedingungen grösserer Staudämme in einer realistischen Weise vor Beginn der aktuellen Bautätigkeit ist unbedingt erforderlich, um Risiken und ein eventuelles Scheitern zu vermeiden. Statistiken haben gezeigt, daß die meisten Staudammbrüche auf Fundamentschwächen zurückzuführen sind. Deshalb ist ein Studium der geologischen Trennflächen der Fundamentfelsen erforderlich. Der 6. Fünfjahresplan (1980-85) enthält ein größeres Bewässerungsentwicklungsprogramm in Madhya Pradesh in Mittel-Indien. Einige Projektbeispiele, bei denen schwierige Fundamentbedingungen angetroffen wurden, werden beschrieben.

TAWA DAM

Forty metre high Tawa dam with drainage area of 5789 Sq.km. is located on the tributary of the Narmada river. The foundation rock of the dam consisted of fine to coarse grained gritty sand stone intercolated with shale and carbonaceous shale. The foundations dipped at 15° to 25° in the upstream direction. The sand stones were thick bedded and the bedding joints were most prominent. Percolation tests conducted in the right transition indicated that these joints were tight. The shale layers were one cm. to 1.5 m thick and extended from 3 to 60 metres. The results of unconfined compression tests were generally low and 50 per cent of the tested samples showed crushing strength values less than 90 kg/cm². A few samples crumbled after soaking in water. The weak spots were distributed in the whole foundation area and they were not concentrated in any specific zone. The crushing strength values varied erratically and there was no improvement with depth. The results of insitu shear tests on shales indicated that though the predominant value of $\emptyset$ was about 30°, there was possibility

of occurrence of $\emptyset$ of about 15° and cohesion of 0.54 kg/cm². Thus the stability of Tawa dam was likely to be governed by sliding along shale bands. Triaxial shear tests with Mohr's envelope gave minimum value $p = 45^\circ$ and $c = 6$ kg/cm² for sand stones. The values of modulus of elasticity of masonry was 0.135 to 0.225 x 10⁶ kg/cm². This necessitated non-linear analysis of foundation stress.

Because of the low crushing strength of the foundation sand stones the conventional factors of safety of 4 to 5 against crushing were not obtained. Analysis of the tests conducted on the saturated rock samples from the spillway foundation showed that 50 per cent of them were greater than 86 kg/cm². Corresponding values with 75 and 90 per cent reliability were 37 and 11 kg/cm² respectively. These crushing strength values compared to the applied stress of about 18 kg/cm² produced quite low factors of safety. The condition would be still more adverse with non-linear stress analysis which would give stresses about 2 to 3 times the linear stresses. Due to the low shear strength of the shales the factor of safety against sliding along the assumed shale bands were also rather low. With the solitary minimum value of 15°, the shear friction factors were just higher than unity. The usual recommended values of 4 with average values of 'c' and $\emptyset$ were not obtained.

To improve the factors of safety, the following arrangements were adopted:-

- An upstream batter of 1 in 2 below elevation 1065 feet was provided in the section to increase the base width and to bring down the maximum stress and reduce the eccentricity of the resultant forces.

- The foundations were taken about 3 metre deeper to provide some entrenchment of the dam into the rock.

- Dental treatment by removing shale/ shear seams and soft sand stone pockets to achieve rock cover of 3 m back filled with concrete to become part of the foundation.

- The monolith sizes were increased from 18 to 36 metres. A 300 metre radius upstream curvature was introduced in the spillway axis to make the blocks act together.

- A second drainage gallery was provided near the downstream toe of the dam in addition to the conventional upstream foundation gallery in order to minimise the uplift pressures acting under the dam.

The details are shown in fig. 1.

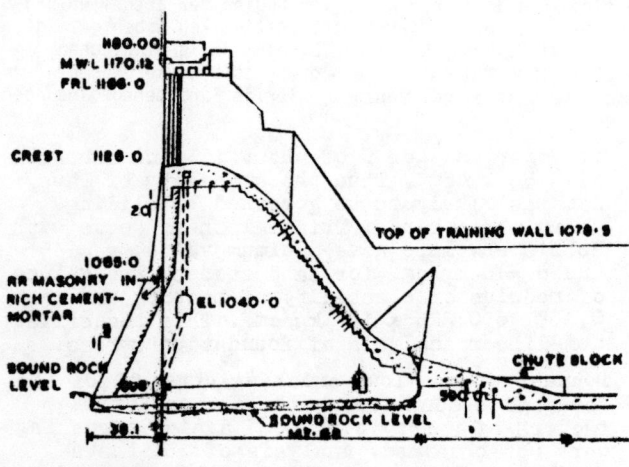

FIG.1 TAWA DAM — REMEDIAL MEASURES.

BARNA DAM

In 48 metre high Barna dam with drainage area of 1176 Sq.km. located on the tributary of the Narmada river, the presence of clay seams in the foundation necessitated a review of design against sliding along the weak bedding planes below the foundation. The preliminary geological investigations did not reveal the presence of clay seams which were observed during subsequent

geological investigations and drilling. The majority of the clay seams occurred at 1 to 3 metre vertical intervals in the foundation with their thickness varying from about 2 to 4 cm. and dipping about 15° towards the downstream at the bedding planes with sand stone. The aerial extent of the clay seams was varying from block to block and estimated as 40 per cent in the blocks 8 to 11, 50 per cent in block 12 and 70 per cent in the blocks 13 to 16 of the foundation area of each block. The clay seams had continuity in the strike direction and they were supposed to be discontinuous in the dip direction, the dip being 15° towards upstream. The shear friction factors were as low as 1.35 in blocks 13 to 16 with 70 per cent clay seams, 1.89 in block 12 with 50 per cent clay seam and 1.15 in blocks 8 to 11 with 40 per cent clay seams at the critical foundation level EL 298. Further studies indicated that the blocks 8 to 12 required nominal treatment while blocks 13 to 16 needed elaborate arrangements. Due to the presence of clay seams a review of design of dam against sliding along the weak bedding planes below the foundation was made and following remedial measures to obtain acceptable shear friction factor were adopted (Refer fig.2).

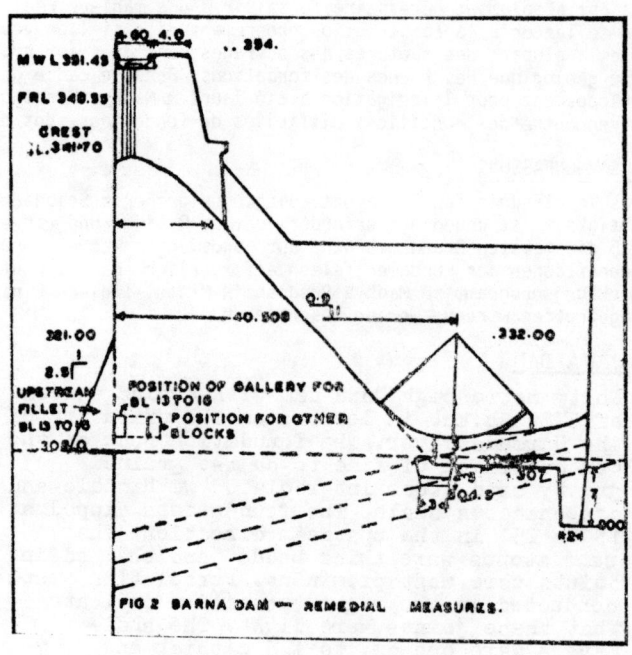

FIG 2 BARNA DAM — REMEDIAL MEASURES.

- A shear key of 7 metre depth at the toe of the roller bucket was provided in block 13 to 16 (river portion) in order to intercept the clay seams existing at about elevation 300 where shear friction factor of 2.5 would only be realised without any treatment.

- A shear key of 2.5 metre depth was also provided at the toe of the remaining dam portion.

- The thickness of the concrete in the slotted bucket was increased from 1.5 to 3 metre to mobilise the passive resistance of the rock downstream of the toe and also to bring into play the concrete slab of the roller bucket against the force of sliding.

- The consolidation grouting in the entire roller bucket area and the provision of the drainage holes was made to consolidate the foundation rock, improve its shear parameters and to reduce uplift pressure.

- An upstream fillet for dam was provided in the river portion with consequent shifting of the drainage gallery.

BANSAGAR DAM

Bansagar masonry dam 69 metres high above the lowest foundation level with drainage area of 18648 Sq.km. is located on the river Sone. In this there was problem due to the occurrence of fault zone consisting of breciated rock mass with chlorite schist seams under the heel of block 5 to 8. The fault zone turns towards the downstream of block 5 and after traversing the dam seat emerges at the toe of block 3. Fault zone has also been met with in block 4 which traverses the dam from upstream to downstream. In addition fractured zones with or without gauge material have been met with in the main fault zone. The above geological features are depicted in fig. 3.

The deformations traversing through major portion of foundation such as faults, fractures and shear zones have been classified as below for the purpose of treatment :-

- Faults, fractures, weak zones of width less than 0.5 metre, between 0.5 to 5 metre with deformations filled with weak material/chlorite schist.

- Fault zone filled with gauge material which could not be washed for grouting and was not groutable.

- Fissured and fractured zones capable of being grouted.

Treatment of the deformations which preceded the consolidation and curtain grouting so that voids incompletely filled by grout was adopted as under:-

- <u>Deformations of width less than 0.5 metre:</u> The weak zones were excavated to a depth of twice the width in 'V' shape and then back filled with M 150 concrete.

- <u>Deformations of width more than 0.5 metre but less than 5 metre:</u> The objectionable weak materials from the entire width of the weak zone were excavated and back filled with concrete M 200 reinforced. The depth was decided as per formula developed for foundations of Shasta dam as under (See fig. 4)

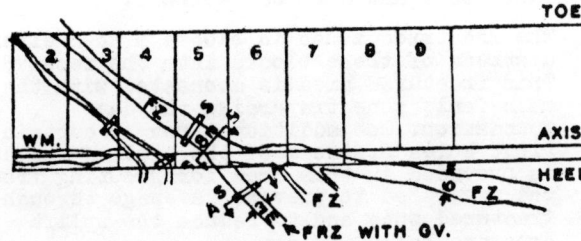

FIG. 3 PLAN SHOWING GEOLOGICAL FEATURES.

I N D E X

FZ = FAULT ZONE
FRZ = FRACTURED ZONE
FRZ FRACTURED ZONE.
WITH GV WITH GAUGE VEINS.
WM = WEAK MATERIALS
S = SHAFT.

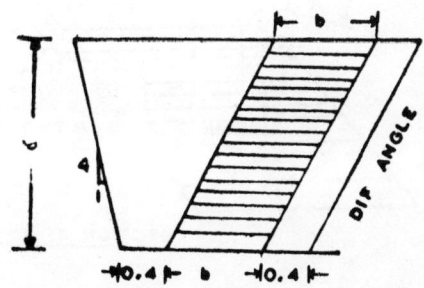

FIG. 4 SIZE OF V-TRENCH.

d = 0.3 b 1.5 (for h the height of dam equal or more than 46 m)
Where b = Width of weak zone in metres.
d = depth of excavation of weak zone below surface of adjoining sound rock in metres.

Typical reinforcement details are given in fig. 5.

M

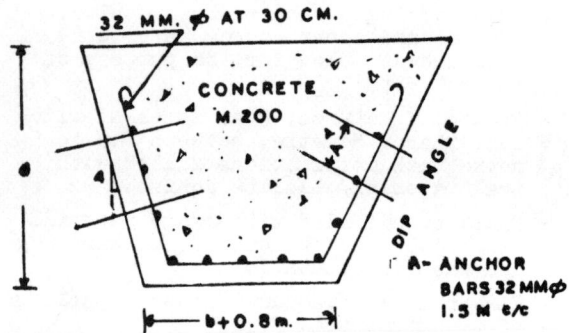

FIG.5 TREATMENT OF DEFORMTION.

Deformations of width more than 5 metres but less than 8 metres and filled with incompetent material: These zones were excavated to depth as per formula described above and filled with M 200 reinforced cement concrete. Typical reinforcement details are given in fig. 6.

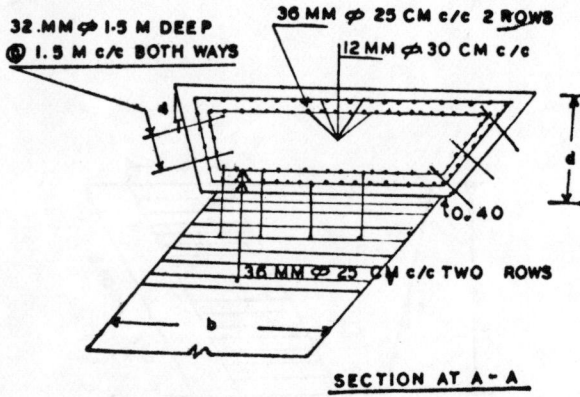

SECTION AT A-A

FIG.6 TREATMENT OF DEFORMATION HAVING WIDTH >5 M & <8 M.

- Fault zone filled with gouge material traversing across the dam from upstream to downstream: This material cannot be washed for grouting. The seepage is expected through such zones from reservoir and, therefore, to keep down the exit gradient below the safe limit, the creep length required is taken as 2.5 H, where H is the difference between MWL and the foundation level.

The treatment of the fault zone filled with incompetent material is the same as for deformations described above. In addition to the length so treated within the foundation zone from upstream end to downstream, additional length required for safe exit gradient is obtained by providing shaft. The details of shaft are given in fig. 7.

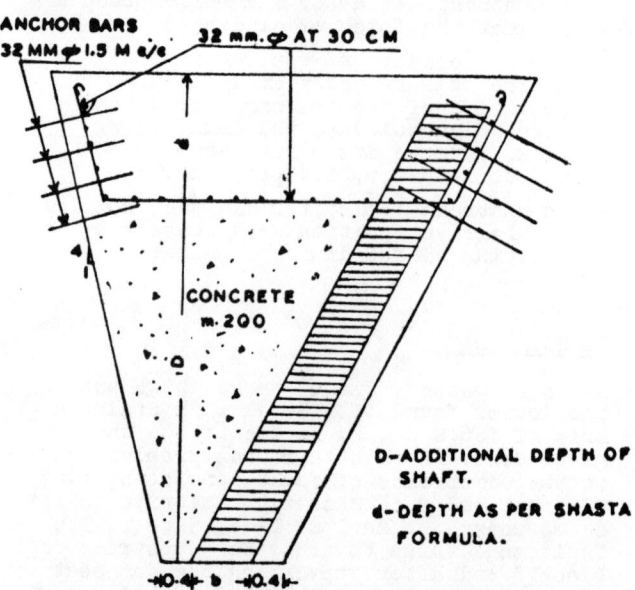

D-ADDITIONAL DEPTH OF SHAFT.

d-DEPTH AS PER SHASTA FORMULA.

FIG.7 FAULT ZONE FILLED WITH GOUSE MATERIAL.

Where the width of the weak zone exceeds 5 metres, a double row reinforcement as shown in fig. 6 has been provided for the plug instead of single row reinforcement shown in fig.5. Contact grouting will be done to seal the gap due to shrinkage of concrete in the fault zone and the rock surface.

The fractured zones in blocks 5 to 8 extend upstream of these blocks into the reservoir. This fractured area is connected with the main fault zone traversing the dam foundation. One additional row of curtain grout holes upstream of the heel of the dam is provided in this area for grouting from the river bed to prevent seepage through the fractured zone and to reduce the uplift pressure below the dam.

BARGI DAM

Bargi dam on the Narmada river with a drainage area of 14556 Sq.km. is 69 metre high. The dam site is underlain by five horizons of massive basalt lava flows. The top most layer is a dense and dark basalt (4-5m) underlain by amygdaloidal basalt (5-10m) and then by dense basalt (5-10m). At the base of this there is an impersistent amygdaloidal basalt (1-2m) underlain by 'red bole' consisting of

purple clayey material. Further below, the
sequence is repeated in the decending order.
The red bole is in the top of the bottom
flow and is a soft and weak zone whose thickness varies from 2.5 to 5 metres. The dense
basalt is fresh and sound and is an excellent
material for founding the dam. The
amygdaloidal basalt is comparatively less
strong depending on the amount of amygdales
present. In this flow there are lenses of
weathered pockets containing red bole and
amygdales. The red bole is nowhere day-
lighted in the basin and is in a confined
state 15-25 metre below dense basalt. The
red bole is the top of the bottom flow and
is a soft and weak zone. For the stability
of the dam it is necessary to establish
whether this soft zone could be grouted.
There is no water loss in the red bole and
the core recovery is 100 per cent. In
curtain grouting with primary holes at 3
metre centre to centre and grouted with
1.75 times the hydro-static pressure, no
attempt was made to deliberately wash and
remove the impervious red bole material.
The entire foundation has proved to be
sufficiently watertight, as the total
consumption of cement in curtain grouting
in 13 out of 37 blocks is only 76 tonnes.
In drainage holes which have the same depth
as the grout curtain, the red bole is
encountered. It is proposed to insert a
slotted pipe in the hole to cover this
horizon with a gravel pack around it (See
fig. 8).

In addition to the red bole problem on the
right bank close to the present course of
the river, a burried channel of 510 metre
width was located where the river was
earlier flowing. This was highly pervious
zone of sand and silty material. The
foundation treatment of this portion
consisted of a R.C.C. diaphragm 0.6 metre
thick and extending 0.6 metre inside hard
rock in panels of 6 metre length. The
process of construction of R.C.C. diaphragm
consisted of preparing pre-trench and slurry
tank, excavation through bentonite slurry,
lowering of form tubes, steel cage and
tremie pipes, concreting, grouting panel
joints and foundation rock, and construction
of cap and plastic bulb.

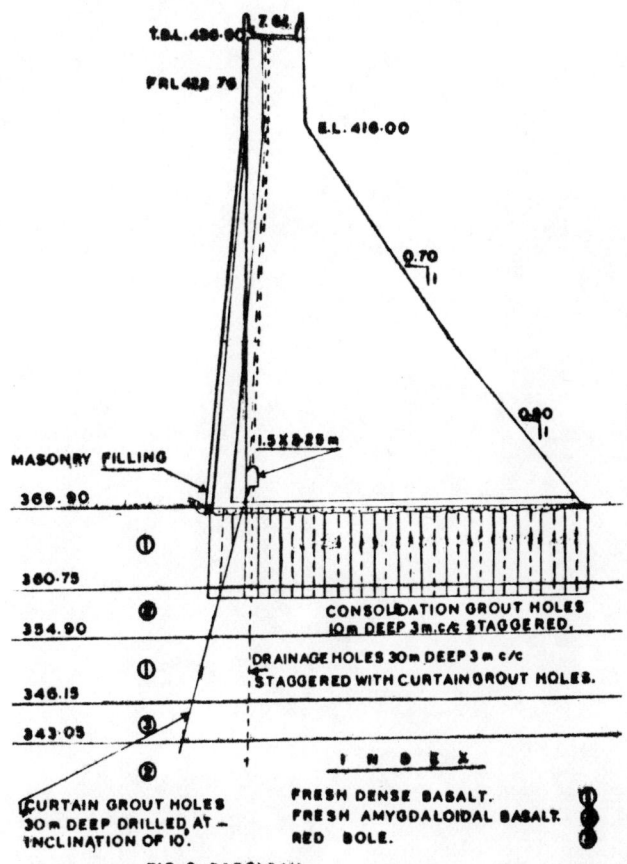

FIG. 8 BARGI DAM.

CONCLUSION

The design of the dam continues till its
completion. During the course of construction
when more data is available, the design of
the dam undergoes a change. This is
necessary as construction of a dam involves
a very serious responsibility in the public
sector. For fulfilling this task a wider
employment of experimental methods in the
analysis of foundation problem is necessary.

PRE-FAILURE DILATANCY AND THE STRESS DISTRIBUTION IN A CLOSELY JOINTED ROCK MASS

Dilatation de pré-rupture et distribution de l'effort dans une masse rocheuse à fissures rapprochées

Dilatation und Spannungsverteilung vor dem Bruch in einem eng geklüfteten Fels

M. J. Pender
Civil Engineering Dept., University of Auckland, New Zealand

C. J. Graham
W. J. Gray
Formerly Graduate Students, Civil Engineering Dept., University of Auckland

SYNOPSIS

The behaviour of dense sand, a rough joint, and models representing closely jointed media lead to the observation that, well before failure is reached, these materials increase in volume, i.e. the onset of dilatancy occurs before the peak strength is reached. The purpose of this paper is to investigate, in a rather simple manner, the effect of this pre-failure dilatancy on the stress distribution in a rock mass beneath a surface loading and adjacent to a pile.

RESUME

Le comportement d'un sable dense, d'une fissure grossière et de modèles représentant des milieux fissurés de façon dense conduit à observer que le volume de ces matériaux croît bien avant que le point de rupture soit atteint, c.a.d. que le début de la dilatation se produit avant que la force maximum soit atteinte. Le but de cet article est d'étudier, d'une façon assez simple, l'effet de cette dilatation d'avant rupture sur la distribution de l'effort dans une masse rocheuse placée sous une charge de surface et adjacente à un pilotis.

ZUSAMMENFASSUNG

Das Verhalten von dichtem Sand, rauhen Trennflächen und Modellen, die eng geklüftete Felsen darstellen, führt zu der Beobachtung, daß sich das Volumen beträchtlich vor dem Eintreten des Bruchs vergrößert, d.h. Dilatation tritt vor Erreichen der Höchstfestigkeit ein. In dem Aufsatz wird der Einfluß der Dilatation auf die Spannungsverteilung im Gebirge, infolge einer Oberflächenlast und in der Umgebung eines Pfahls in einer vereinfachten Form untersucht.

INTRODUCTION

The stress-strain behaviour of dense granular soils, models representing closely jointed media, and the shear load–displacement behaviour of a single rough joint surface, all indicate a net increase in volume before the peak shear resistance is mobilised. This phenomenon, volume change on the application of shearing stress, is commonly referred to as dilatancy. The effect of dilatancy on the behaviour of a rock mass is of some interest. Although the effect in the post-failure region is well known the effect of prefailure dilatancy seems to have received little attention. This paper presents the results of some preliminary studies of the topic.

Unfortunately no field evidence giving a direct indication of prefailure dilatant behaviour is known to the authors. This is not surprising in view of the difficulty of field measurement in a jointed rock mass. However there is good evidence from carefully controlled laboratory studies on model materials representing a closely jointed rock mass. For example Brown (1976) presented volumetric strain curves for an assembly of accurately fitting 'bricks' of various shapes. Some results are reproduced in Fig. 1. It is clear that the volume change behaviour is dependent on the arrangement of the separate blocks relative to the direction of loading. However in at least one of the cases (H30) there is a very pronounced increase in volume right from the start of loading. As another example Gerogiannopolous and Brown (1978) reported triaxial test results on granulated marble (Rosengren and Jaeger (1968)). These results are reproduced in Fig. 2. At the start of the loading process the material decreases in volume, but before half the peak strength has been mobilised the incremental volumetric strain gives a volume increase. As is well known this volume increase continues well past the point at which the peak strength is mobilised. Thirdly Barton (1972) performed an interesting series of tests on slabs of model rock material which were intersected by a series of approximately orthogonal rough joints. The shear and normal displacement results for some of these tests are reproduced in Fig. 3. Once again it is evident that well before the peak shear strength is mobilised there is an opening of the joints.

In the three examples quoted above there is no constraint on the volume change or normal displacement. The boundary conditions are such that the confining stress remains constant. However if there is some deformational restraint the dilatancy is suppressed by an increase in the confining pressure. This point is made by

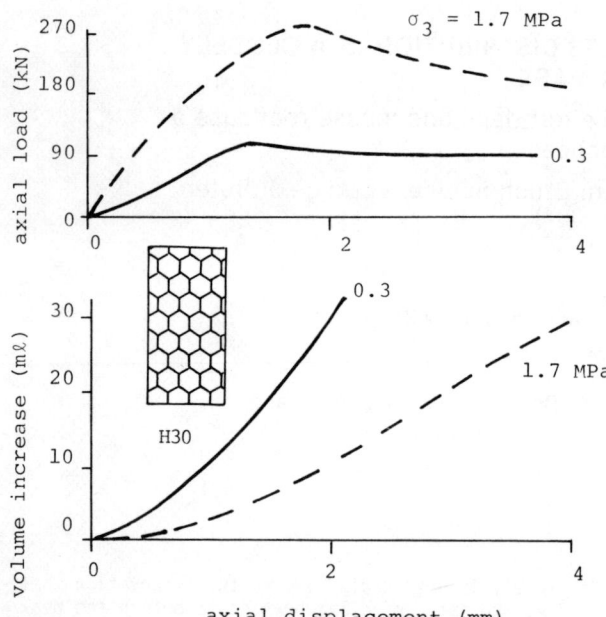

Fig. 1 : Stress-strain behaviour of the H30 configuration of plaster blocks, Brown (1976)

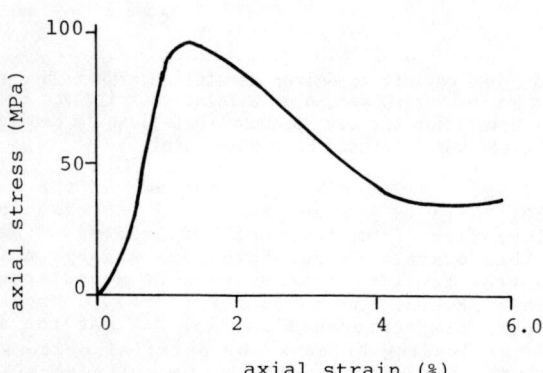

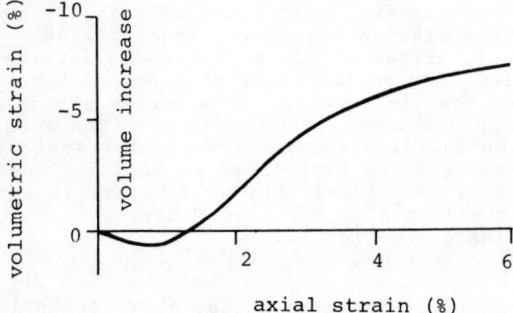

Fig. 2 : Stress-strain behaviour of granulated Carrara marble, σ_3 = 5.17 MPa. (Gerogiannopoulos and Brown (1978))

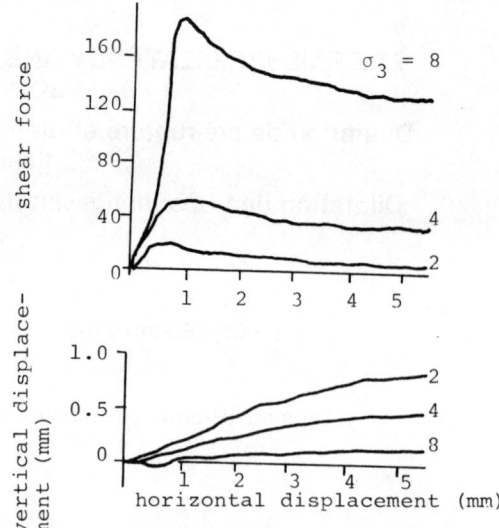

Fig. 3 : Shear force - displacement characteristics of a tension joint in a model material, Barton (1972)

Goodman (1974). Thus it might be expected that in a dilatant medium subject to deformational restraint, such as the rock mass surrounding the shaft of a socketed pile, the confining pressures would be higher than in an equivalent non-dilatant medium.

A limiting case in the spectrum of possible jointed rock masses is a randomly jointed medium. If the joint spacing is small in relation to the scale of the structure of interest then it is possible, at least as a first approximation, to idealise the rock mass as a homogeneous isotropic continuum. This is the viewpoint adopted in this paper.

The purpose of this paper is to illustrate the effect of prefailure dilatancy in a rock mass beneath a footing and adjacent to the shaft of a socketed pile. The calculations are done by means of a finite element analysis.

MATHEMATICAL MODELLING OF DILATANT BEHAVIOUR

The dilatant volume change is calculated herein by relating it to the change in octahedral shear stress. Octahedral stresses and strains are defined as:

$$\sigma_{oct} = \frac{1}{3}(\sigma_x + \sigma_y + \sigma_z) \tag{1}$$

$$\tau_{oct} = \frac{1}{3}\{(\sigma_x - \sigma_y)^2 + (\sigma_y - \sigma_z)^2 + (\sigma_z - \sigma_x)^2 + 6(\tau_{xy}^2 + \tau_{yz}^2 + \tau_{zx}^2)\}^{\frac{1}{2}} \tag{2}$$

$$v_{oct} = (\varepsilon_x + \varepsilon_y + \varepsilon_z) \tag{3}$$

$$\gamma_{oct} = \frac{2}{3}\{(\varepsilon_x - \varepsilon_y)^2 + (\varepsilon_y - \varepsilon_z)^2 + (\varepsilon_z - \varepsilon_x)^2 + 6(\gamma_{xy}^2 + \gamma_{yz}^2 + \gamma_{zx}^2)\}^{\frac{1}{2}} \tag{4}$$

It is assumed that the rock mass behaves as a

conventional linear elastic material with the addition of a dilatant contribution. As has been explained in the introduction the objective of this paper is to explore in a simple manner the consequences of prefailure dilatancy. Thus the method of modelling the dilatancy has been kept very simple. This is the reason for idealising the rock mass as a homogeneous isotropic continuum. It is then possible to use octahedral stresses and strains so that the dilatancy gives coupling between the octahedral shear stress and the volumetric strain. Expressed in incremental form the constitutive relationship used is:

$$dv_{oct} = \frac{d\sigma_{oct}}{K} - \frac{d\tau_{oct}}{D} \qquad (5)$$

$$d\gamma_{oct} = \frac{d\tau_{oct}}{G} \qquad (6)$$

where dv_{oct}, $d\gamma_{oct}$ are the octahedral volumetric and shear strain increments respectively

$d\sigma_{oct}$, $d\tau_{oct}$ are the octahedral normal and shear stress increments respectively

K is the bulk modulus

G is the shear modulus

D is the modulus of dilatancy

The negative sign in equation 5 is because the increase in volume, in a system where compressive strains are positive, is a negative strain.

The volumetric strain curve in Fig. 2 could be modelled accurately by equation 5 if D (and presumbaly K and G) was a function of stress.

The initial part of the volumetric strain curve can be modelled as a conventional isotropic elastic material with increasing Poisson's ratio. However once the minimum volume is passed a Poisson's ratio greater than 0.50 would be required to model the volume change behaviour. To get over this difficulty the dilatant term has been introduced above. However in the paper the calculations are restricted to linear dilatancy, i.e. once the dilatancy starts D is held constant.

As the rock mass is deformed positive work is done in shearing, whilst the dilatant component of deformation does work against the confining pressure and hence does negative work. The values of D used herein were checked at all stages during the calculations to ensure that the net work done in deforming the rock mass was positive.

FINITE ELEMENT PROCEDURE

The finite element calculations were done by adapting a program developed by Graham (1982). The program uses a parabolic isoparametric element. The incorporation of prefailure dilatancy set out below is appropriate both to plane strain and axisymmetric conditions.

From equations (5) and (6) the strain increment has elastic and dilatant components:

$$\{d\varepsilon\} = \{d\varepsilon^{e}\} + \{d\varepsilon^{d}\} \qquad (7)$$

where $\{d\varepsilon^{e}\}$ is the elastic strain increment vector
$\{d\varepsilon^{d}\}$ is the dilatant strain increment vector

The elastic response is

$$\{d\sigma\} = [D]\{d\varepsilon^{e}\} \qquad (8)$$

where $[D]$ is the matrix of elastic stress-strain coefficients

Substituting from (7) gives:

$$\{d\sigma\} = [D](\{d\varepsilon\} - \{d\varepsilon^{d}\}) \qquad (9)$$

Considering the incremental change in dilatant volumetric strain:

$$dv^{d} = - d\,\tau_{oct}/D$$

Expanding $d\,\tau_{oct}$ in terms of $\partial\tau_{oct}/\partial\sigma_{x}$ and $d\sigma_{x}$ etc., gives:

$$d\tau_{oct} = \frac{\partial\tau_{oct}}{\partial\sigma_{x}} d\sigma_{x} + \frac{\partial\tau_{oct}}{\partial\sigma_{y}} d\sigma_{y}$$
$$+ \frac{\partial\tau_{oct}}{\partial\sigma_{z}} d\sigma_{z} + \frac{\partial\tau_{oct}}{\partial\tau_{xy}} d\tau_{xy}$$

and from the definition of τ_{oct}, equation (2), the following are obtained:

$$\frac{\partial\tau_{oct}}{\partial\sigma_{x}} = \frac{2\sigma_{x} - \sigma_{y} - \sigma_{z}}{9\tau_{oct}},$$

$$\frac{\partial\tau_{oct}}{\partial\sigma_{y}} = \frac{2\sigma_{y} - \sigma_{x} - \sigma_{z}}{9\tau_{oct}}$$

$$\frac{\partial\tau_{oct}}{\partial\sigma_{z}} = \frac{2\sigma_{z} - \sigma_{x} - \sigma_{y}}{9\tau_{oct}},$$

$$\frac{\partial\tau_{oct}}{\partial\tau_{xy}} = \frac{6\tau_{xy}}{9\tau_{oct}}$$

The dilatant part of the volumetric strain increment can thus be expressed as:

$$\begin{Bmatrix} d\varepsilon_{x}^{d} \\ d\varepsilon_{y}^{d} \\ d\varepsilon_{z}^{d} \\ d\gamma_{xy}^{d} \end{Bmatrix} = -\frac{1}{27\,\tau_{oct}D} \begin{bmatrix} 2\sigma_{x}-\sigma_{y}-\sigma_{z}, & 2\sigma_{y}-\sigma_{x}-\sigma_{z}, & 2\sigma_{z}-\sigma_{x}-\sigma_{y}, & 6\tau_{xy} \\ '' & , & '' & , & '' \\ '' & , & '' & , & '' \\ 0 & , & 0 & , & 0 \end{bmatrix} \begin{Bmatrix} d\sigma_{x} \\ d\sigma_{y} \\ d\sigma_{z} \\ d\tau_{xy} \end{Bmatrix}$$

or $\{d\varepsilon^{d}\} = [H]\{d\sigma\} \qquad (10)$

where [H] is the matrix of stress components derived above.

Equation (10) shows that the incremental dilatant strain components are linearly related to the stress increments when D is constant.

Considering incremental changes in equation (9)

$$\{d\sigma\} = [D](\{d\varepsilon\} - \{d\varepsilon^{d}\})$$

substituting equation (10) gives,

$$\{d\sigma\} = [D](\{d\varepsilon\} - [H]\{d\sigma\})$$

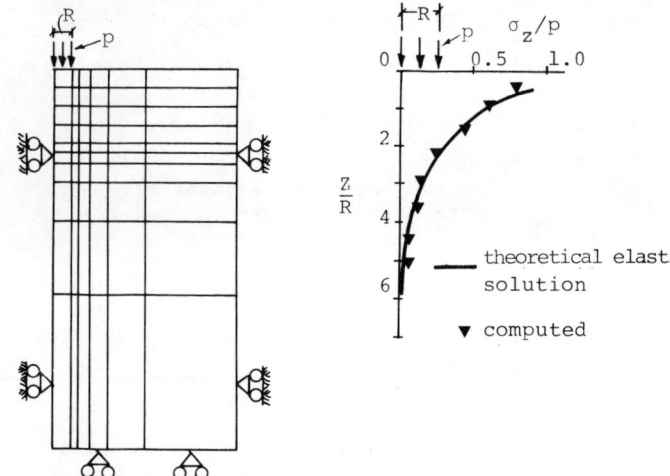

therefore

$$d\sigma = \left[\{[I] + [D][H]\}^{-1}[D]\right]\{d\varepsilon\}$$

where   [I] is the identity matrix

This is written as:

$$\{d\sigma\} = [D^*]\{d\varepsilon\} \qquad (11)$$

where $[D^*]$ is the matrix of stress strain coefficients including both elastic and dilatant contributions

Equation (11) provides a general relationship between increments of stress and increments of strain for the finite element analysis. It can be developed further for the specific cases of axial symmetry and plane strain.

In the calculations reported in this paper the loading was applied in four equal increments. For the first of these increments no dilatant effects were included. This was done in an attempt to model the volume change behaviour for granulated marble given in Fig. 2 in which there is a volume decrease before the onset of dilatancy.

The finite element results discussed give the changes that are caused in the rock mass by the

Fig. 4 : Finite element mesh and vertical stress distribution beneath the centre of a circular pressure loading

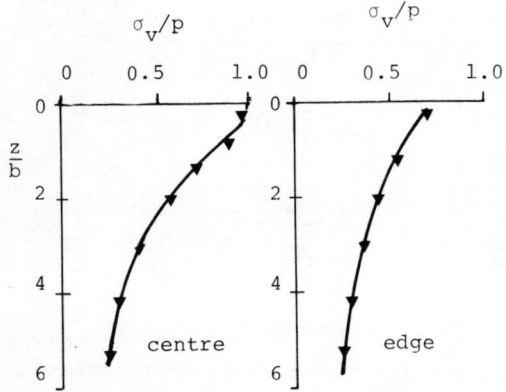

a vertical stress distribution

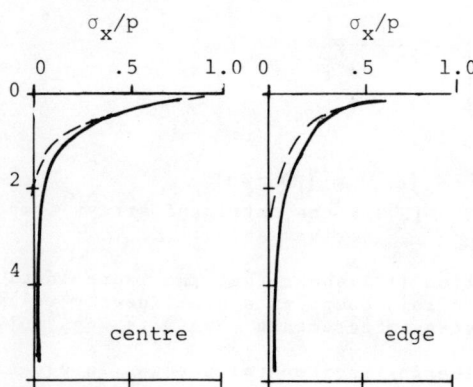

b lateral stress distribution

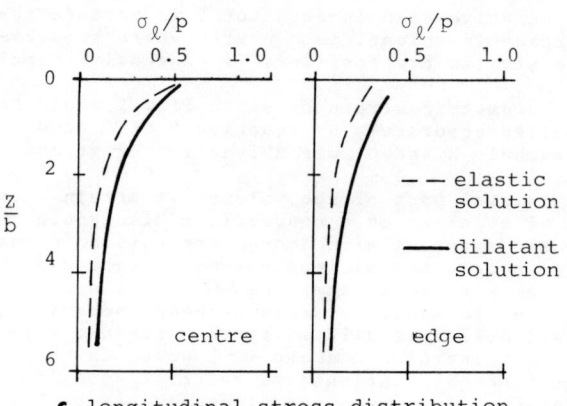

c longitudinal stress distribution

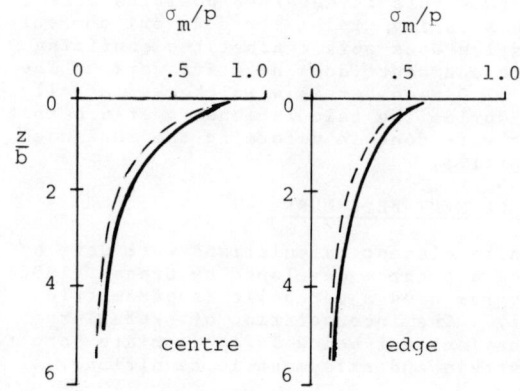

d Octahedral normal stress distribution

Fig. 5 : Stress distribution beneath the centre and edge of a strip load

applied loading. These would normally be added to the in situ stresses in the rock mass to give the final state of stress. This step is not taken here so there is no need to introduce the additional consideration of in situ stresses.

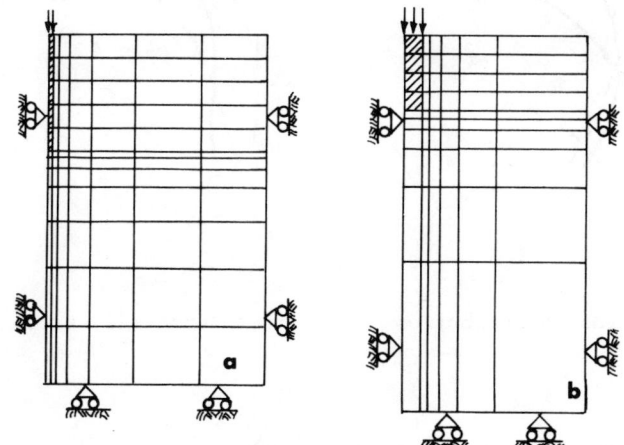

Fig. 6 : Finite element meshes for the piles
(a) Piles 1 and 2   (b) piles 3 and 4

SURFACE LOADINGS

Two cases of surface pressure loading (i.e. flexible loadings) were investigated - a strip load and a circular load. The finite element mesh is given in Fig. 4. Also in this figure the finite element vertical stress distribution is compared with the theoretical elastic vertical stress distribution beneath the centre of a circular load. This confirms that the element subdivision is reasonable. The use of isoparametric quadrilateral elements means that good modelling is possible with a relatively small number of elements.

In Fig. 5 the dilatant and elastic stress distributions for the strip loading are compared. The Poisson's ratio, ν, value used for the rock mass was 0.25. The values used for D are related to the Young's modulus by the parameter β (= $E/3D$). The value used was 0.63, this value is representative of the behaviour of granulated marble.

Of greatest interest in Fig. 5 is the fact that the vertical stress distribution for the dilatant case is nearly identical with that for the elastic case. On the other hand the remaining normal stress components are affected by the dilatancy. For example the longitudinal

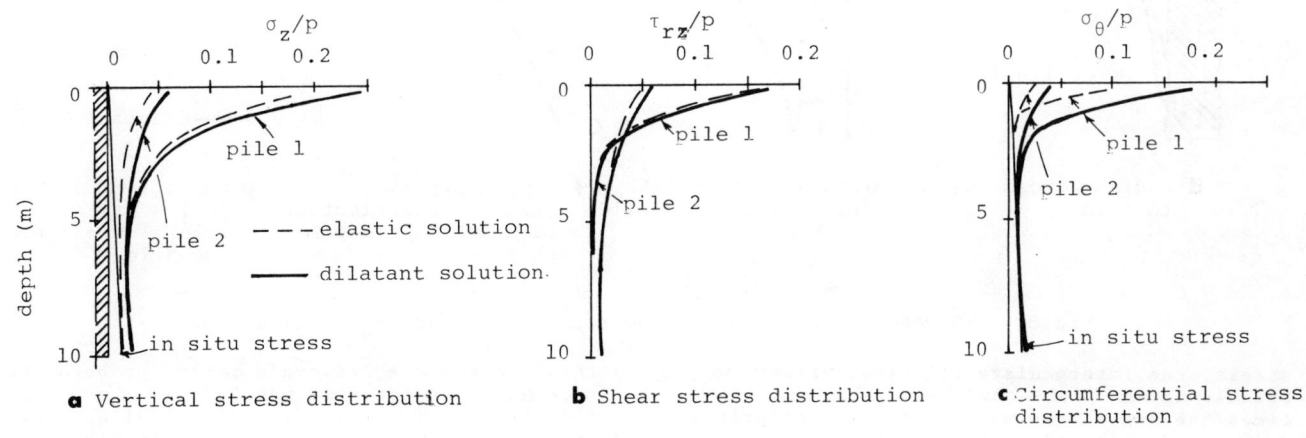

a Vertical stress distribution   b Shear stress distribution   c Circumferential stress distribution

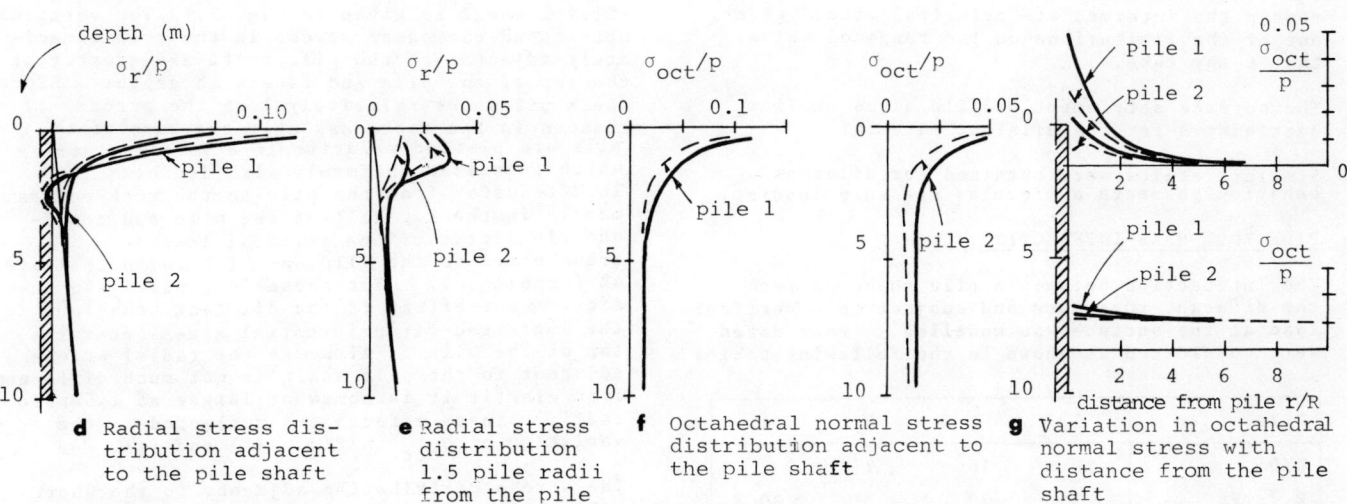

d Radial stress distribution adjacent to the pile shaft
e Radial stress distribution 1.5 pile radii from the pile
f Octahedral normal stress distribution adjacent to the pile shaft
g Variation in octahedral normal stress with distance from the pile shaft

Fig. 7   Stress distribution in the rock mass adjacent to piles 1 and 2

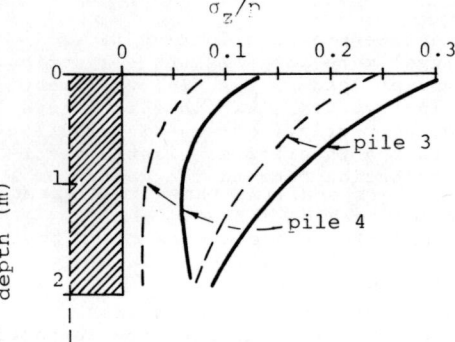 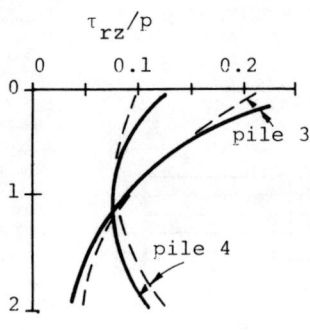 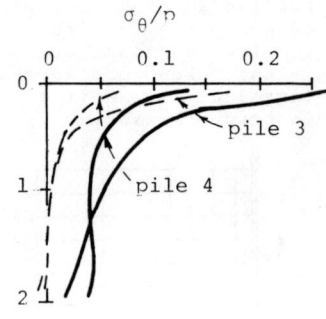

a Vertical stress distribution   **b** Shear stress distribution   **c** Circumferential stress distribution

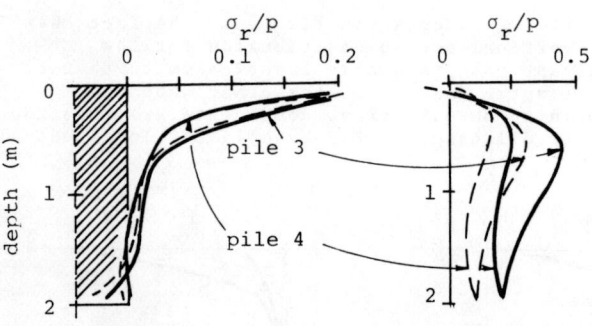 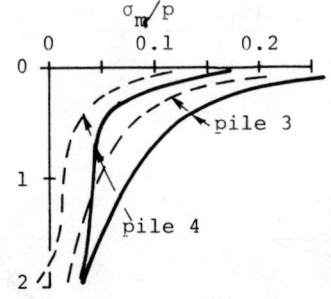 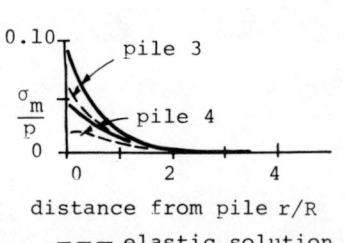

--- elastic solution

—— dilatant solution

d Radial stress adjacent to pile   **e** Radial stress 1.5 pile radii from pile   **f** Octahedral normal stress distribution adjacent to the pile shaft   **g** Variation in octahedral normal stress with distance from the pile

Fig.8 : Stress distribution in the rock mass adjacent to piles 3 and 4

stress, the intermediate principal stress because of the plane condition, which is ν times the sum of the major and minor principal stresses for elastic behaviour, is seen to be increased because of the dilatancy. In fact the condition that the longitudinal stress must remain the intermediate principal stress gives one of the limitations on the range of values that β may take.

The surface settlement profile gives smaller settlements for the dilatant material.

Similar results were obtained for dilatant behaviour beneath a circular pressure loading.

PILE ROCK MASS INTERACTION

The interaction between a pile socketed into the dilatant rock mass and subject to a vertical load at the surface was modelled. Four cases were considered as shown in the following table:

Pile:	1	2	3	4
L/D:	10	10	2	2
E_{pile}/E_{rock}	3	30	3	30

In all cases the Poisson's ratio for both the rock mass and the pile was 0.25. The value of β (= E/3D) was 0.63. The finite element meshes used for the two cases are given in Fig. 6.

The stress distribution in the rock adjacent to piles 1 and 2 is given in Fig. 7. The vertical stress and the shear stress in the rock immediately adjacent to the pile shaft are greater at the top of the pile and fade with depth. Since these piles are relatively long the stress changes in the rock mass near the base of the pile are plotted relative to a vertical stress which increases uniformly with depth. The load transfer from the pile to the rock occurs mostly in the top half of the pile and consequently little of the vertical load is transferred to the pile base (2% for pile 1 and 4% for pile 2). For these long piles the most significant effect of the dilatant behaviour is the increased circumferential stress near the top of the pile. Although the radial stress adjacent to the pile shaft is not much different from elastic it is somewhat larger at 1.5 pile radii. These effects are reflected in the variation of σ_{oct} in Fig. 7(f) and (g).

The stress distribution adjacent to the short piles, piles 3 and 4, is shown in Fig. 8. It is once again clear that the major effect of dilatancy is on the circumferential stress adjacent to the pile shaft. Although in this case the vertical stress in the rock is also increased. The shear stress distributions although slightly different at the top and base of the pile are consistent with the overall vertical equilibrium of the pile. It is clear from Fig. 8(d) and (e) that the increase in radial stress occurs away from the pile rather than adjacent to it. As with piles 1 and 2 there is an increase in σ_{oct} for a volume of rock with a diameter about four pile diameters.

DISCUSSION AND CONCLUSIONS

To appreciate the main significance of the stress distributions in Figs. 7 and 8, failure of the rock mass needs to be considered. This has not been done in the paper but it is clear enough from Figs. 7(f) and (g) and 8(f) and (g) that, assuming the shear strength of the jointed rock mass is a function of the octahedral normal stress, the available shear strength of the material surrounding the pile socket will be greater in a medium which has dilatant behaviour prior to failure. Furthermore if the stiffness of the rock mass increases with increase in the octahedral normal stress then prefailure dilatancy will reduce the settlement of the pile under working load. Thus prefailure dilatancy can be expected to give a net improvement in the behaviour of a socketed pile. Similar comments apply to a rock bolt and a ground anchor fixed into closely jointed material.

The effect of the dilatancy could be made more significant by increasing the value of β, i.e. reducing the value of D relative to E. At some value greater than the 0.63 used herein, which is representative of granulated marble, the negative work done by the dilatant component of the volumetric strain would be greater than the positive work done by the other strain components. The value of β at which this occurs has not been investigated. However it is not likely to be much greater than 0.63 because of related work, discussed by Gray (1981) and Pender et al (1982), on the excavation of a tunnel in a closely jointed rock mass. In this it was found that the condition which had to be satisfied so that the longitudinal

stress in the rock mass remained the intermediate principal stress, and the work done in deforming the rock mass remained positive, was $1 - \beta - \nu > 0$. For a Poisson's ratio of 0.25 this gives a limiting value of 0.75 for β.

ACKNOWLEDGEMENTS

The work of C.J. Graham was supported by a study grant from the N.Z. Ministry of Works and Development, whilst W.J. Gray was supported by the Structures Committee of the N.Z. Road Research Unit.

REFERENCES

Barton, N. (1972) A model study of rock joint deformation. *Int. J. Rock Mech. Min. Sci.* 9, 579-602

Brown, E.T. (1976) Volume changes in models of jointed rock. *J. Geotech. Eng. Div. Am. Soc. Civ. Engs.* 102, GT3, 273-276

Gerogiannoupoulos, N.G. and Brown, E.T. (1978) The critical state concept applied to rock. *Int. J. Rock Mech. Min. Sci.* 15, 1-10

Goodman, R.E. (1974) The mechanical properties of joints. *Proc. 3rd Congress ISRM,* Denver, Vol. 1A pp. 127-140

Graham, C.J. (1982) Non Linear Soil Modelling. *University of Auckland, Civil Engineering Department,* Report No. 294

Gray, W.J. (1981) Dilatancy and the behaviour of a closely and irregularly jointed rock mass. *University of Auckland, Civil Engineering Department,* Report No. 256

Pender, M.J., Gray, W.J. and Graham, C.J. (1982) Dilatancy effects about a circular opening in an idealised rock mass. *In preparation*

Rosengren, K.J. and Jaeger, J.C. (1968) The mechanical properties of an interlocked low-porosity aggregate. *Geotechnique* 18, No. 3, 317-326

GEOMECHANICS INVESTIGATION AT THE PROPOSED RAUPUNGA DAMSITE, MOHAKA RIVER, NEW ZEALAND

Les investigations en mécanique des roches sur le site du barrage proposé de Raupunga, Rivière Mohaka, Nouvelle Zélande

Geotechnische Untersuchungen zum geplanten Dammprojekt am Mohaka Fluss, Raupunga, Neuseeland

B. D. Hegan
S. A. L. Read
Engineering Geologists, New Zealand Geological Survey, D.S.I.R., Lower Hutt, New Zealand
P. J. Millar
Geomechanics Engineer, Ministry of Works and Development, Wellington, New Zealand

SYNOPSIS

The preliminary results of the geomechanics investigations at the Raupunga damsite on the proposed Mohaka River hydro-development are presented. The geology of the site comprises a marine sedimentary sequence of soft, low strength silt-stone and very low strength sandstone of Tertiary age. In situ behaviour of the rocks, determined from plate bearing tests, and deformation measurements of a trial enlargement in a test adit have shown a good agreement with strength and deformation characteristics obtained from laboratory testing.

RESUME

On présente ici les résultats préliminaires des investigations géomécaniques du site de Raupunga sur la rivière Mohaka, proposée pour un développement hydro-électrique. La géologie du site comprend une succession sédimentaire marine de roches limoneuses tendres et peu résistantes et de grès Tertiaire. Les données sur le comportement in situ des roches, déterminées à partir des essais de chargement sur plaque et des mesures de la déformation d'un test d'élargissement dans la galerie d'essai sont en accord avec les caractéristiques de résistance et de déformation établies en laboratoire.

ZUSAMMENFASSUNG

Die vorläufigen Resultate des Untersuchungsprogramms der geplanten Hydroentwicklung am Mohaka Fluß bei Raupunga liegen vor. Die Geologie der Baustelle ist durch sedimentäre Schichten von weichem, schwachem Schlammstein und Sandstein aus dem Tertiäralter gekennzeichnet. In situ Verhalten der Felsen, festgestellt durch Belastungsversuche, und Deformierungs-messungen, zeigen gute Übereinstimmung mit Versuchen im Laboratorium.

1. INTRODUCTION

The Mohaka River rises near the central volcanic plateau of the North Island, New Zealand, and flows eastwards discharging into Hawke Bay (Fig. 1). It has a catchment area of 2400 km^2, average mean flow of 77 m^3/sec and 1000 year design flood of 4700 m^3/sec. Studies for the development of the hydro-electric potential of the Mohaka River to date have focused on three sites in the lower 60 km of the river. Raupunga is the lowest station and has a potential generating capacity of 85 MW.

At the Raupunga site the Mohaka River has cut a narrow valley about 110 m deep and 400 m wide in marine siltstone and sandstone of late Pliocene to early Pleistocene age. The earth dam design at present under consideration has a maximum height of 80 m. Investigations commenced in November 1980 and have included field mapping, drilling, a test adit and both in situ and laboratory testing.

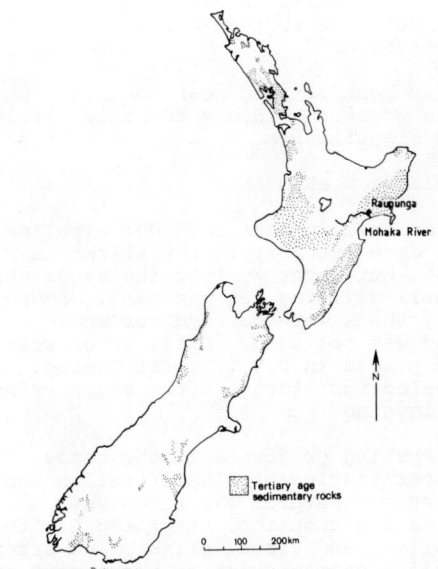

Fig. 1 Location, New Zealand

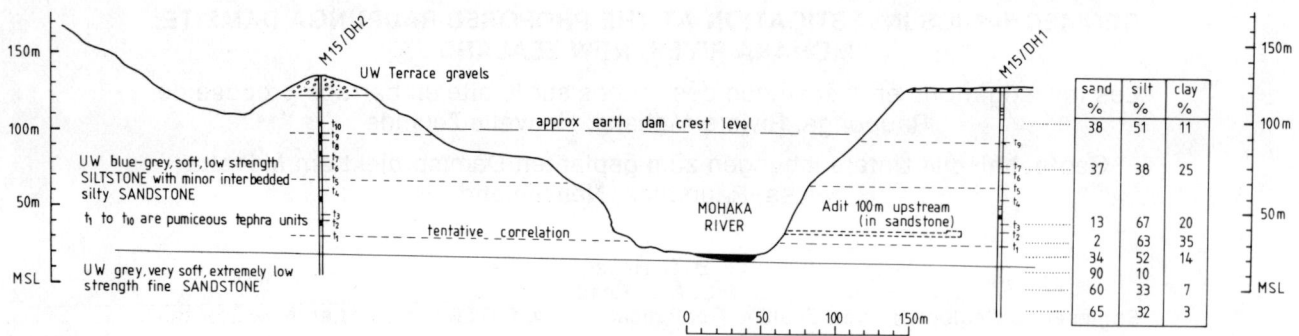

Fig. 2   Geological Section at Damsite

2.   GEOLOGY

The geology at the Raupunga site consists of a marine sedimentary sequence of light blue-grey homogeneous, very low to extremely low strength fine sandstone overlain by grey homogeneous low strength siltstone (Fig. 2). Numerous thin beds of volcanic tephra occur throughout the sequence which dips downstream at about 5°. Within the siltstone thin layers of silty sandstone or sandy siltstone form a transition zone immediately above each tephra bed (Hegan, 1981).

Alluvial gravels of late Quaternary age have been deposited on the eroded marine sequence. Subsequently, the present river has cut a 400 m wide valley to a depth of 110 m. On the left bank shallow meander channels, now truncated by the valley, are preserved. Both the terraces and the channel floors are veneered with gravel, loess and air-fall tephras.

Two sets of tight vertical joints are observed in the siltstone outcropping at river level while the sandstone generally appears free of defects. Stress relief joints in the siltstone, that strike parallel to the valley walls, are being investigated.

One fault has been located near the site, but is judged on present evidence to be of little engineering significance.

3.   INVESTIGATION PROGRAMME

Investigation drilling has used HQ3 wireline equipment. Core recovery in the siltstone exceeded 95%, but recovery from the sandstone was poor until drilling mud was used. Downhole permeability tests were carried out where drilling mud was not used. Drill cores were immediately placed in P.V.C. split tubing. The core was sealed for storage after engineering geological logging.

Laboratory testing performed on the cores included determination of classification and index properties (natural water content, particle size distribution, Atterberg Limits, slake durability and visual slake), and strength and deformation characteristics (uniaxial and triaxial compression). All testing was performed

in accordance with the appropriate New Zealand, International Society for Rock Mechanics (I.S.R.M.) or Canadian Centre for Energy and Mineral Technology (CANMET) standards or recommended procedures.

A 100 metre long (1.8 m wide by 2.6 m high) test adit, with a 20 metre long enlargement (3.8 m by 4.1 m), was excavated in the sandstone on the right bank by drill and blast methods (Fig. 2). The adit was logged in detail and the in situ mechanical behaviour of the sandstone determined by plate bearing tests, borehole extensometer, undercoring and convergence measurements.

Seismic refraction surveys were undertaken along the dam alignment, and down-hole geophysical logging of selected drill holes was carried out (neutron, gamma density, sonic velocity).

4.   CLASSIFICATION AND INDEX PROPERTIES

Typical classification properties of the three materials at the site, siltstone, fine sandstone and transition, are given in Table 1. The properties of the siltstone and fine sandstone are relatively uniform, although the transition materials from above the tephra layers, are variable. The siltstone has a low plasticity, and its plastic limit is markedly higher than the natural water content. The sandstone is non plastic. All materials exist at or near saturation, regardless of their position with respect to the groundwater table. The mineralogy of the siltstone and sandstone is similar to that of many other Tertiary age sedimentary rocks of New Zealand (Table 1 in Read et al., 1981).

Limited slake durability and jar slake index testing indicates the siltstone has a moderate to high resistance to slaking. Also, it did not exhibit significant swelling stress (< 100 kPa) or swelling strain (< 1%) indices when saturated from the oven dry condition. A pre-consolidation pressure of 12 MPa was obtained from consolidation testing. The sandstone is uncemented and disintegrates readily during the slake durability test, its behaviour being similar to that of a very dense highly over-consolidated sand.

C 174

Table I. Classification Properties

mean (std dev)	Particle Size Distribution			Atterberg Limits			Dry Density ρ_d (t/m³)	Natural Water Content w_n (%)	Solid Density of Particles ρ_s (t/m³)	Saturation Ratio Sr (%)	Void Ratio e
	Clay (%)	Silt (%)	Sand (%)	LL (%)	PL (%)	PI (%)					
Siltstone -12 samples	18 (6)	61 (11)	21 (12)	45 (6)	28 (3)	18 (5)	1.80 (0.03)	18.7 (1.2)	2.72 (0.03)	98 (2)	0.51 (0.03)
				1 non plastic							
Transition -5 samples	15 (2)	55 (8)	30 (7)	42 (3)	22 (1)	20 (3)	1.85 (0.16)	16.8 (4.8)	2.70 (0.04)	97 (3)	0.46 (0.12)
				1 non plastic							
Fine Sandstone -13 samples	3 (3)	21 (11)	75 (13)	Non Plastic			1.79 (0.05)	18.1 (1.5)	2.71 (0.01)	95 (5)	0.52 (0.04)

Table II. Uniaxial Compression Test Results

mean (std dev)	Uniaxial Compressive Strength qu (MPa)	Tangent Modulus of Deformation Et (MPa)	Failure Axial Strain εf (%)
Siltstone -22 specimens	7.5 (1.4)	1480 (320)	0.85 (0.14)
Transition -5 specimens	3.0 (1.1)	495 (360)	1.23 (0.49)
Fine Sandstone -32 specimens	1.0 (0.6)	160 (95)	1.03 (0.33)

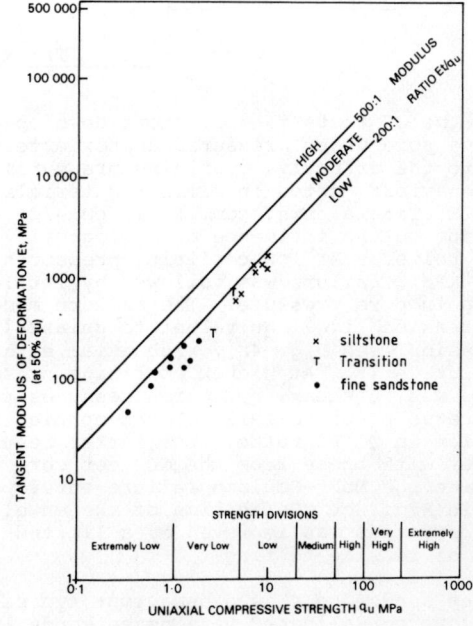

Fig. 3 Uniaxial Strength Rock Classification

5. LABORATORY STRENGTH AND DEFORMATION CHARACTERISTICS

5.1 Uniaxial Compression

Uniaxial compression tests were performed at natural water content, with axial strain measured over the whole specimen length. Significant differences in strength, modulus and axial strain at failure were obtained for the siltstone, fine sandstone and transition materials (Table II). In Fig. 3 the rocks are shown to have a low modulus ratio according to the classification of Deere and Miller (1966). The lowest points for the siltstone on Fig. 3 are representative of samples taken from the valley walls and indicate the effects of stress relief due to valley formation.

5.2 Triaxial Compression

Consolidated drained (CD) and undrained (CU) triaxial compression tests were carried out on back pressure saturated specimens. Effective confining pressures varied from 50 kPa to 1200 kPa. The Q'_f v P'_f plots at failure are given on Fig. 4a, and the derived Mohr-Coulomb failure envelopes on Fig. 4b.

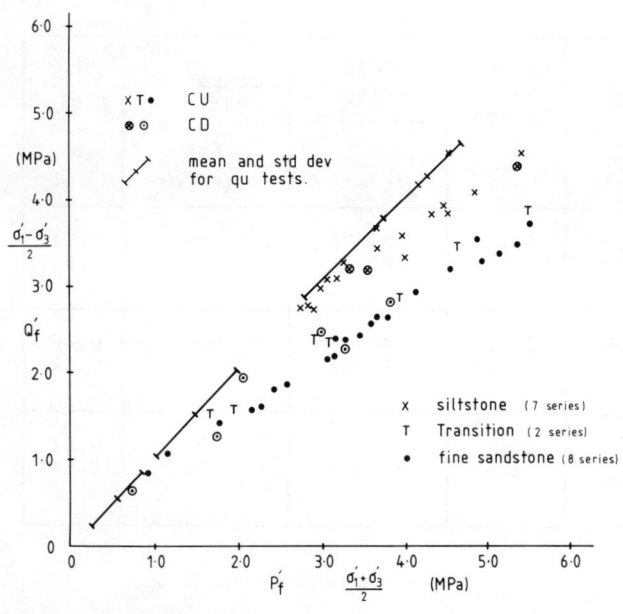

Fig. 4a Q'_f v P'_f

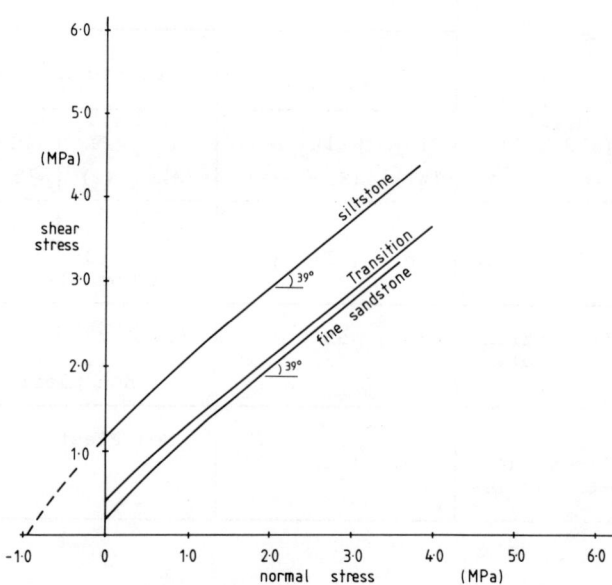

Fig. 4b Mohr-Coulomb Failure Envelopes

Fig. 4 Triaxial Compression Test Results

During the CU tests the siltstone developed positive pore water pressures approximately equal to the effective confining pressures. This behaviour, noted in similar materials in Japan (S. Tanaka pers. comm.), is considered to be due to the influence of geological stress relief. At low confining pressures (< 400 kPa), failure was followed by little or no drop in pore pressure. The failure mode was extension (i.e. equivalent to uniaxial compression, see Fig. 4a), with axial strains at failure < 1%. At higher confining pressures failure was in shear, and pore pressures tended to decrease prior to failure, accompanied by a reduction in Q'/P' ratio. The latter results together with those from the CD test were used to define the Mohr-Coulomb failure envelope given in Fig. 4b. Definition of the envelope at low stresses was improved by a limited amount of tensile testing.

The fine sandstone showed behaviour typical of highly overconsolidated very dense sands in both the CU and CD tests. Large negative pore pressures developed during CU tests and axial strains at failure were higher than for the siltstone. The transition materials showed a similar behaviour to the siltstone, although the pore pressures at failure were markedly lower and generally negative. Both materials allowed the confident development of the Mohr-Coulomb failure envelopes in Fig. 4b.

The failure envelopes for the three materials all had an angle of friction of 39°, but the cohesion varied widely.

6. PERMEABILITY AND GROUNDWATER CONDITIONS

Permeability tests were carried out in those sections of the drill holes where mud was not used. A single mechanical packer was used, but problems in achieving a good seal means some of the results may be high. Subsequently, an HQ size gas inflated double packer has been developed. In the siltstone and fine sandstone water losses were in the order of 1 Lugeon. Higher water losses (> 10 Lugeons) were recorded in some tephra layers, notably layer t_1 adjacent to the siltstone sandstone contact (Fig. 2) and in the valley walls where stress-relief joints are present in the siltstone.

Judged on measurements made during drilling and later in standpipe piezometers, the recorded water table is controlled by tephra layer t_1.

7. IN SITU TESTING IN ADIT

7.1 Plate Bearing Tests

Vertical and horizontal plate bearing tests have been carried out at 3 stations along the test adit. A 200 tonne hydraulic loading system was used with plate sizes ranging from 375 mm to 600 mm in diameter.

Bearing surfaces were prepared by removing a minimum of 200 mm of sandstone, followed by the placing of a thin layer of quick setting mortar for facing. This was placed after final assembly of the system so that a small load of 0.5 to 1.0 tonnes could be applied to extrude excess mortar and ensure good contact and alignment.

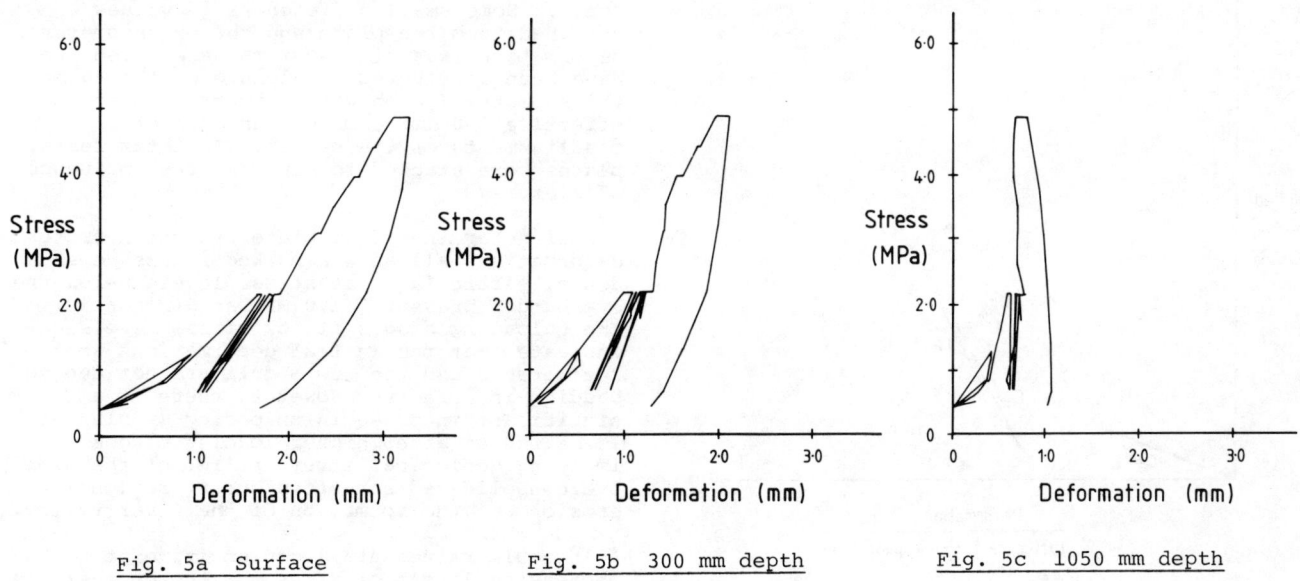

Fig. 5a Surface   Fig. 5b 300 mm depth   Fig. 5c 1050 mm depth

Fig. 5 Horizontal Plate Bearing Test Stress v Deformation Curves (Station 99m, 0.2m² plate)

Table III. Horizontal Plate Bearing Test Summary (Station 99)

APPLIED STRESS MPa	PLATE AREA (m²) DEPTH (mm)	Initial Secant Modulus E_{is} (MPa)				Final Secant Modulus E_{fs} (MPa)				Unloading Modulus E_{eu} (MPa)				Reloading Modulus E_{el} (MPa)			
		0.3	0.2	0.15	0.1	0.3	0.2	0.15	0.1	0.3	0.2	0.15	0.1	0.3	0.2	0.15	0.1
0.6	0	430	320			340					190				330		
	300		520								150						
	675	370				280				260							
	1050		100													160	
0.8	0		390												400		
	300		360												370		
	675																
	1050		160														
1.0	0	660	690			580				710							
	300		540														
	675	500				490				490							
	1050																
1.7	0	800	750			630	680			1580	1230			1550	1250		
	300		800				600				1600				1530		
	675	700				640				700	1210			600	1110		
	1050																
3.0	0			970		770	790	980									
	300						770										
	675			850		660	990	810									
	1050						740										
4.9	0				1180	890				1970							
	300				1500	820				2110							
	675					860				1020							
	1050					750											
6.9	0			1070				980			1480						
	300			1680													
	675			840				810									
	1050																
9.9	0				1230				1100				1720				
	300				1230				1080				1730				
	675				1030				1000								
	1050								1300								

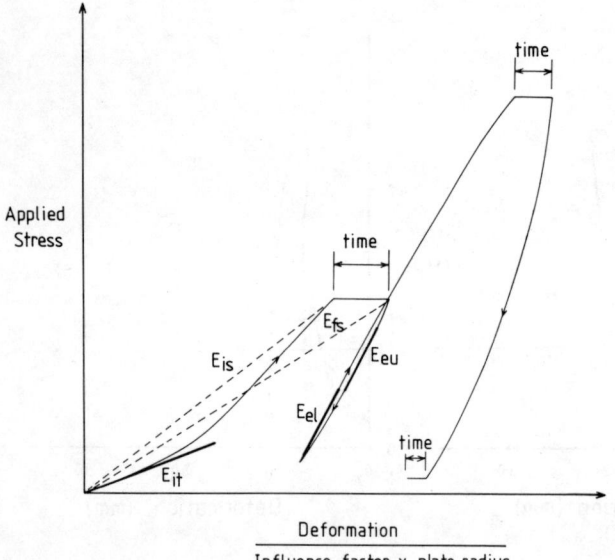

Influence factor x plate radius

E_{it}: initial tangent modulus

E_{is}: initial secant modulus - depending on loading rate

E_{fs}: final secant modulus - after creep rate <0.025mm/hr

E_{eu}: unloading modulus (elastic modulus)

E_{el}: loading modulus

Fig. 6 Definition of Modulus Terms for Plate Bearing Tests

Loads were applied at a rate of 25 kN/min and measured using a load cell while plate deformations were monitored by dial gauges mounted on an independently supported frame. In more recent tests the gauges have been supplemented by displacement transducers connected to a field data logging system. A multi-point borehole extensometer, using calibrated potentiometers for monitoring, was installed behind each plate to measure deformations at depths up to 1.05 metres.

The largest plate (600 mm diameter) was used first, followed by the smaller plates which allowed higher stress levels to be achieved. This procedure also provided an indication of the influence of plate size. A typical loading sequence, including cyclic loading, unloading stages and creep tests, is shown on Fig. 5.

Published plate bearing test results indicate a wide variation in the definition of modulus terms, for example, Lama and Vutukuri, 1978; and Oehadijono, 1979. These differences appear to have arisen because of variations in creep, cyclic loading and unloading sequences, and the resulting uncertainty in defining reference points on subsequent loading stages. As no commonly accepted system is available, the terms used in this paper are those defined in Fig. 6.

The moduli given on Table III have been calculated assuming elastic theory and a rigid plate. Variations of modulus with depth have been determined from the extensometer measurements. Some small differences in values obtained from the plate and the extensometer deformation measuring systems were noted, and have been attributed to flexure of the 40 mm thick plates (which were supported on an effective 300 mm diameter) and slight misalignments of the system. In later tests, plates were stacked to minimise the influence of flexure.

Moduli determined from plate and extensometer deformations all show a marked stress dependence, particularly at stress levels below the overburden pressure (1.0 mPa at Station 99). The unloading stages at low stress levels indicate that the initial deformations are recoverable and the low moduli are not due to bedding-in effects. However, there is a significant increase in unloading modulus at low stresses after higher loading stages implying geological stress relief of the highly overconsolidated sandstone during regional erosion and the formation of the river valley.

All moduli values are less sensitive at increasing levels of stress, although there is still a trend for them to increase with both decreasing size of plate and depth behind the plate.

7.2 Extensometer Measurements in the Enlargement

The 20 m long enlargement at the end of the test adit was instrumented with 5 three-point borehole extensometers to provide information on both the in situ stress field and the deformation characteristics of the fine sandstone. The enlargement was excavated in stages shown in Fig. 7a.

The three-point extensometers consisted of concentric tubes and rods with shell anchors. Dial gauges were used to monitor deformations relative to the measuring head. These instruments were simple, inexpensive and generally performed successfully, although some difficulties were encountered in achieving a satisfactory anchorage of the shells in the soft rock.

The deformations recorded as the excavation progressed are shown in Fig. 7b. The magnitude of the deformations measured are influenced by both the horizontal to vertical in situ stress ratio and the unloading modulus (E_{eu}) of the rock. However, the distribution of the deformations with depth in both directions provides sufficient additional information for the calculation of the in situ stresses and deformation characteristics using an iterative finite element method of analysis (see Millar, 1977).

Initial results indicate the maximum principal stress direction to be vertical with a horizontal stress ratio of about 0.7. At the enlargement the overburden pressure is 1.0 MPa giving an average unloading modulus (E_{eu}) of 240 MPa from the extensometer measurements.

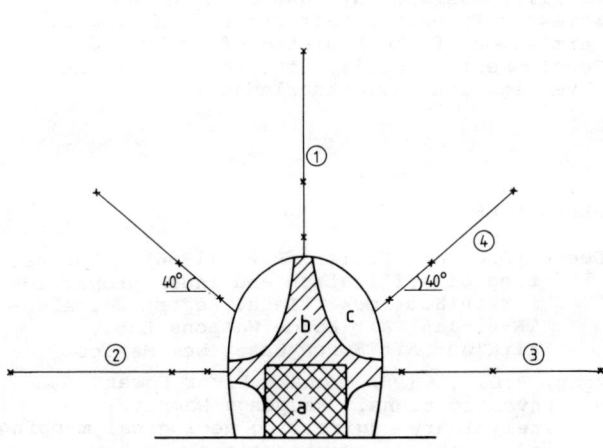

Excavation stages:

(a) 10 m long pilot drive to instrument station.
(b) 'Rat' holes (750 mm) for installation of 12 m long extensometers.
(c) Enlargement of pilot drive followed by installation of 10 m long inclined extensometers.
(d) Excavation to 10 m beyond test station by pilot drive, then enlargement.

Fig. 7a   Excavation Sequence and Location of Extensometers

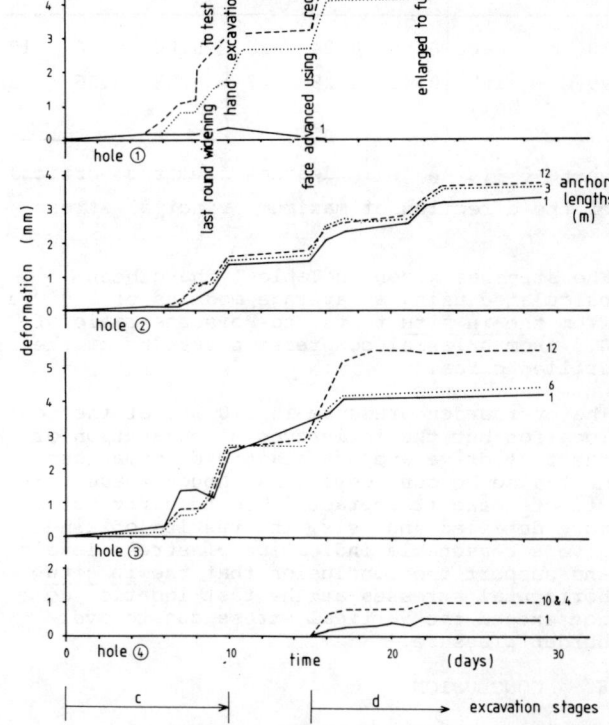

Fig. 7b   Extensometer Movements

Fig. 7   Extensometers in Enlargement

Relaxation of the sandstone prior to the installation of extensometers from the 'rat' holes is estimated from an axisymmetric elastic model to be, on average, 12 per cent of the total movement; with 50 to 70 per cent of the movement occurring prior to installation of the inclined extensometers. These values are in reasonable agreement with the deformations observed in the enlargement and shown in Fig. 7b. This demonstrates the advantage gained by placing the extensometers from 'rat' holes in the fine sandstone where deformations occurred rapidly after excavation.

7.3   Surface Stress Measurements

An indication of the stresses in the face and walls of the test adit was obtained using the undercoring stress relief technique described in Hooker et al. (1974). After positioning two rings of six equally-spaced measuring points on radii of 100 mm and 125 mm, a 250 mm deep hole was drilled into the face using a 150 mm diameter thin walled diamond core barrel. Stresses in the face were calculated from the convergence of diametrically opposite pins using equations based on isotropic linear elastic theory for the radial displacement of a circular hole at a free boundary.

The principal stresses σ_y, σ_x are calculated from:

$$\sigma_y, \sigma_x = \frac{E}{12a} \frac{u_1 + u_2 + u_3}{M}$$
$$\pm \frac{\sqrt{2}}{N} \{ (u_1 - u_2)^2 + (u_2 - u_3)^2 + (u_3 - u_1)^2 \}^{\frac{1}{2}} \quad (1)$$

$$\theta_1 = \frac{1}{2} \tan^{-1} \frac{\sqrt{3}(u_2 - u_3)}{2u_1 - u_2 - u_3} \quad (2)$$

where u_1, u_2, u_3 are measurements of convergence of the hole and

θ_1 is measured from u_1, r = radius of pins
a = radius of hole,   k = a/r
$M = \frac{(1+\nu)k}{2}$ $N = M \{4(1-\nu) - k^2\}$

E, ν = elastic constants

Table IV. Undercoring Test Results

LOCATION	125 mm RADIUS PINS			100 mm RADIUS PINS		
	σ_y (MPa)	σ_x (MPa)	$^1\theta_y$	σ_y (MPa)	σ_x (MPa)	θ_y
106 m - Face	0.55	0.24	-18	0.86	0.36	-17
99 m - Right Wall	0.43	0.29	3	0.53	0.35	2

[1] where θ_y is angle in degrees from the vertical of the direction of maximum principal stress

The stresses given in Table IV have been calculated using an average modulus of 240 MPa from the in situ tests and Poissons ratio of 0.3 from uniaxial compression testing of the drilled cores.

The overburden pressure is 1.0 MPa at the test location but the influences of excavation of the test drive and non-elastic deformations must also be considered. Although these effects make it impracticable to carry out a more detailed analysis, the results obtained give a reasonable indication of stress levels, and support the conclusion that the in situ horizontal stresses at the test location do not exceed the vertical stress due to overburden pressure.

8. CONCLUSION

The initial results of the geomechanics investigations at the Raupunga damsite show that the properties of the homogeneous soft Tertiary age siltstone and fine sandstone are similar in performance to uncemented, highly overconsolidated silt and sand. There has been good agreement between laboratory and field measurements of strength and deformation characteristics and these show the marked influence of stress relief during regional erosion and formation of the Mohaka river valley.

ACKNOWLEDGEMENTS

The authors wish to express their appreciation to all those who have assisted in the investigations, in particular Mr N. Logan. Permission of the Ministry of Works and Development to publish the results of the investigations is acknowledged.

REFERENCES

Deere, D.U. and Miller, R.P. (1966). Engineering classification and index properties for intact rock. Tech. Report No. AFWL-TR-65-116, Air Force Weapons Lab., Kirkland Air Force Base, New Mexico.

Hegan, B.D. (1981). Mohaka River Power Investigations. Raupunga Damsite. Preliminary engineering geological mapping and proposed investigations. N.Z. Geological Survey Engineering Geology Report EG 353.

Hooker, V.E., Aggson, J.R., Bickel, D.L., and Duvall, W. (1974). Improvement in the three component borehole deformation gauge and overcoring technique. U.S.B.M. Rep. Inv 7894; with Appendix by Duvall on the undercoring method.

Lama, R.D. and Vutukuri, V.S. (1978). Handbook on mechanical properties of rocks. Volume III. 1st Ed. 406 p Trans Tech Publications, Clawsthal, Germany.

Millar, P.J. (1977). The design of Rangipo underground powerhouse. NZIE Proc. Tech Groups (3) Issue 3(6) 6.46-6.59. Tunnelling in New Zealand.

Oehadijono, I. (1979). Java's Sempor dam and irrigation scheme. Water power and dam construction, Feb. 1979, P31-38.

Read, S.A.L., Millar, P.J., White, T., and Riddolls, B.W. (1981). Geomechanical properties of New Zealand soft sedimentary rocks. Proc Int Symp on Weak Rock, (1), 33-38, Tokyo.

HORIZONTAL BEHAVIOUR OF PIER FOUNDATION ON A SOFT ROCK SLOPE

Comportement horizontal de la fondation de piles sur une pente en roche tendre

Horizontale Verformungskomponenten an Pfählen in einer Böschung in Halbfestgesteinen

Hiroshi Maeda
The Tokyo Electric Power Company Inc.

SYNOPSIS

Through in-situ horizontal loading tests on full-scale pier foundations constructed on a soft-rock slope, it was found that the behaviour of soil-pier systems could be explained by the beam theory on elasto-plastic foundations using the ground reaction, i.e. the displacement relationship obtained from the test at each point in the ground. Further, it was established that the upper limits of ground reaction could be determined from a bearing capacity formula employing Coulomb's law of shearing resistance.

RESUME

En effectuant des essais de charge horizontale in situ sur les fondations de piles profondes et construites à échelle rëelle sur une pente de roches tendres, il s'est avéré que le comportement des systèmes sol-piles pourrait s'expliquer par la théorie de la poutre sur les fondations élastoplastiques en utilisant le rapport entre la réaction du sol et le déplacement obtenu par les essais à divers points du sol. En outre, nous avons constaté que les limites supérieures de réaction des sols pourraient être déterminées par une formule de capacité portante incorporant la loi de la résistance au cisaillement de Coulomb.

ZUSAMMENFASSUNG

Durch horizontale Ortbeton-Belastungsversuche auf Pfeilerfundamenten im Maßstab 1:1, die auf einem Hang aus faulem Gestein errichtet worden waren, wurde erkannt, daß das Verhalten von Bodenpfeilern mit der Balkentheorie für elasto-plastische Fundamente unter Benutzung der Bodenreaktion, d.h. dem Verschiebungsverhätnis, das durch die Versuche an den verschiedenen Bodenstellen festgestellt wurde, erklärt werden kann. Ferner wurde erkannt, daß die oberen Grenzen der Bodenreaktion mit der Belastbarkeitsformel unter Anwendung der Coulombschen Gleichung für Schubfestigkeit bestimmt werden können.

1. PREFACE

Approximately 70% of Japan is covered by mountainous terrain. As a result, construction site of transmission towers for the central electric power system is often limited on soft-rock slopes, thus fairly deep foundations are common. However, the relationship between ground reaction and displacement of soil-pier systems located on soft-rock slopes is not clearly understood. Specific attention has been focused on the horizontal resistance mechanism by way of carrying out horizontal loading tests using two pier foundations constructed on a ridge.

The following is a report on test results for ultimate bearing capacity obtained through observation of the failure mechanism, study of ground reaction characteristics and analytical simulation by employing the beam theory.

2. OUTLINE OF THE TEST

2.1 Test site

The major part of the geology at the test area consists of volcanic-clastic rock from the Miocene epoch and a covering layer of loamy soil of pleistocene origin.

Table 1 Site geology and soil properties

DEPTH (m)	GEOLOGY	STANDARD PENETRATION TEST N—Value	TRIAXIAL COMPRESSION TEST COHESION Cu (MPa)	TRIAXIAL COMPRESSION TEST INTERNAL FRICTION ANGLE ϕu (Degree)	TRIAXIAL COMPRESSION TEST MODULUS OF DEFORMATION E_{50} (MPa)	PLATE BEARING TEST MODULUS OF DEFORMATION D (MPa)	UNIT WEIGHT γt (kN/m³)
0—3	(a) LOAM	2	0.013	24	8	–	12
3—5	(b) TUFFY CLAY	20	0.078	17	19	93	20
5—10	(c) TUFFY BRECCIA (WETHERING)	–	0.26	51	324	422	21
10—	(d) TUFFY BRECCIA	–	4.90	77	–	2680	26

NOTE: 1MPa=10.197kgf/cm²   1kN/m³=0.10197gf/cm³

The volcanic-clastic rock consists of a rhyolitic tuffy breccia. This layer is rather hard at 10 m or more in depth, but due to progressive weathering effect, there are regions where the rock is extremely brittle at a depth of 10 m or less. Soil properties obtained at the test site are shown in Table 1.

2.2 Testing setup

Slopes with angles of 20° and 30° were shaped from a mountain ridge and a RC pier foundation ("Pier A") of 10 m in depth by 3.5 m in diameter and another one ("Pier B") with dimensions of 10 m in depth by 3.0 m in diameter were constructed as shown in Fig.1.

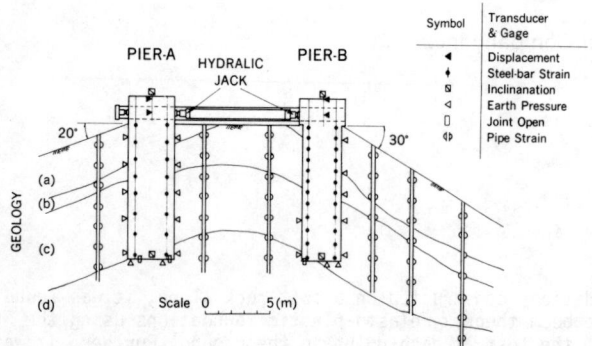

Fig. 1 Testing set and instrument arrangement

Between the two piers, reaction beams, hydraulic jacks and PC steel bars were installed so as to function as an alternating loading apparatus. Also shown in the figure is a various instrumentions to measure strain, pressure, deflection and displacement of the piers and the surrounding soil.
Loading conditions performed in the test are:

1) Alternating loading (0 ∿ 2.45 MN: design load level)
2) Constant loading (4.17 MN: Single direction loading)
3) Large displacement loading (0 ∿ 9.81 MN: Single direction loading)

3. TEST RESULTS

The following section describes the observed behavior of the soil-pier system at Pier B during the large displacement loading test.

3.1 Characteristic of pier displacement

Figure 2 shows the load versus pierhead displacement curve up to the maximum loading level.

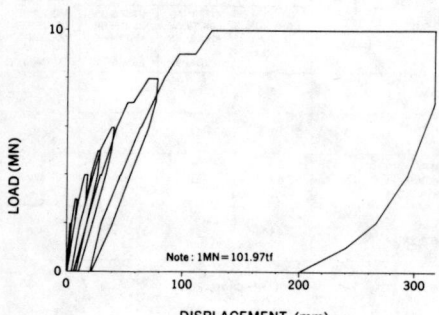

Fig. 2 Load vs. pierhead displacement curve

When the load reached 9.81 MN, the pierhead displacement increased suddenly. At the same loading level, there was an indication in the strain data of reinforced bars that the pier as a whole had not reached the ultimate resistance. Therefore, most of the ground around the pier seemed to have reached a plastic condition.

3.2 Characteristics of ground reaction

Figure 3 shows the distribution of ground reaction measured by the earth pressure gauge buried to the front of Pier B.

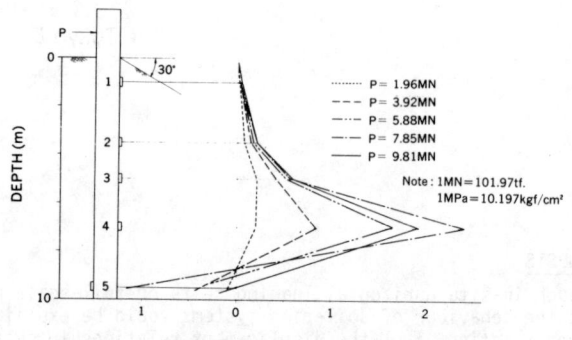

Fig. 3 Distribution of ground reaction

The ground reaction at each point initially increases linearly with an increase in load. However, as can be seen at measurement points 1 and 2 for the load over 3.92 MN, and at point 3 for the load over 5.88 MN, the rate of increase became smaller. At measurement point 4, the ground reaction declined at 9.81 MN. These data suggests that the plastic region of the soil extended gradually from the upper to the lower layers. At measurement point 5 located at the tip of the pier, the reaction is in opposite direction from the upper side of the pier. Thus, rotation of the pier must have taken place with the rotational center being located at a depth of 8.0 m to 9.0 m.

3.3. Failure process of ground

Figure 4(a) shows the final condition of the cracks occurring in the area of Pier B, and the difference in the crack pattern are distinguished by symbols ①, ② and ③ in the figure. The development process of these cracks is described below.

a. When 3.43 MN was loaded, the first tensile Cracks ① appeared and extended in a 45° direction towards the load from the side face of the pier.
b. The openings of the Cracks ① enlarged as the load increased. Then radial Cracks ② appeared in the foreground of the pier as the load increased from 4.41 MN to 7.85 MN.
c. When the load exceeded 8.83 MN, the foreground was apparently thrust upwards, and simultaneously Cracks ③ occurred in a direction perpendicular to the radial Cracks ② and subsequently formed a circular shape. Cracks ③ finally connected with the Cracks ①

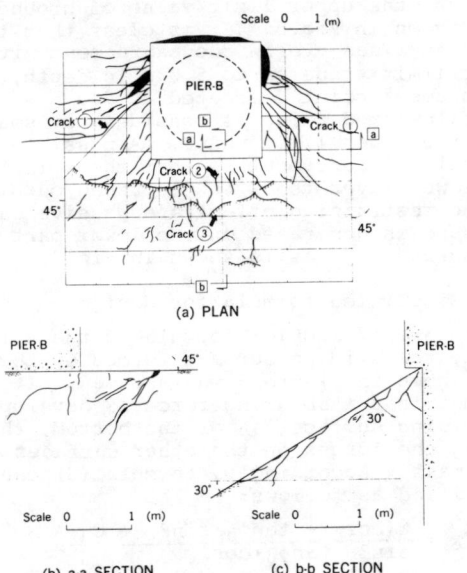

(a) PLAN

(b) a-a SECTION   (c) b-b SECTION

Fig. 4   Condition of crack occurrence in ground

Through visual inspections of excavated test area, it was found that Cracks ① were extended at an angle of about 45° to the horizontal surface, as is shown in Fig.4(b). On the other hand, Cracks ③ cut vertically though the slope at a 30° as is shown in Fig. 4(c). This angle nearly corresponds to (45°- φ/2) of classical sliding surface, assuming φ = the angle of internal friction of the ground.

4. SIMULATION BY BEAM THEORY

The observed behavior of the soil-pier system was simulated by a two dimentional beam theory. In this analysis, distribution pattern of the ground reaction was considerd to be equal to that is shown in Fig 3, and the intensity of the ground reaction was normalized with respect to the applied horizontal load.

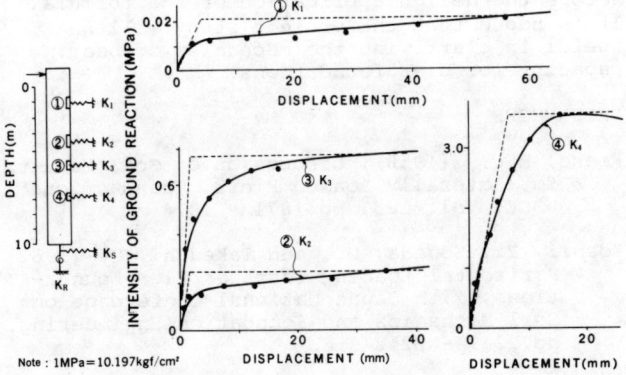

Note : 1MPa=10.197kgf/cm²

Fig. 5   Ground reaction vs. displacement curve

Figure 5 shows the ground reaction versus displacement curve assumed in the analysis for each measurement point. A nonlinear moment-flexural rigidity relationship was assumed for the pier.

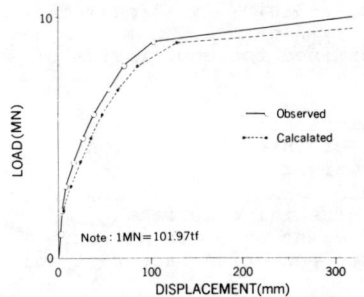

Fig. 6   Results of simulation for load vs. pierhead displacement

Figure 6 shows the calculated results for the load versus pierhead displacement. As can be seen in the figure, the analytical results show a good agreement with the test results. Good correlation was also obtained in a simpler analysis assuming a bilinear curve for the ground reaction versus displacement relationship as shown by the broken line in Fig. 5. Furthermore, the upper limit of the ground reaction is considered to be equivalent to the intensity of ultimate bearing capacity at each measurement point.

5. CALCULATION FOR ULTIMATE BEARING CAPACITY OF GROUND

5.1 Sliding model for ground

Observations of the cracking mechanism of the soil-pier system in the test led to the assumption that the mass of soil as shown in Fig.7(a) thrust upwards.

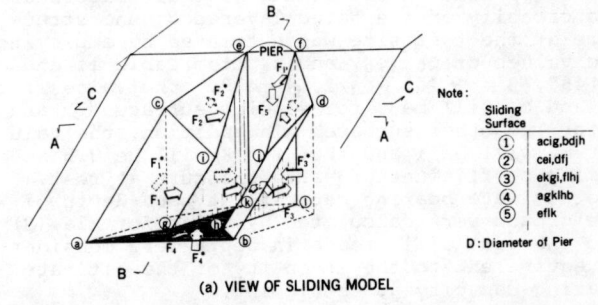

Note:
Sliding Surface
① acig,bdjh
② cei,dfj
③ ekgi,flhj
④ agklhb
⑤ eflk
D : Diameter of Pier

(a) VIEW OF SLIDING MODEL

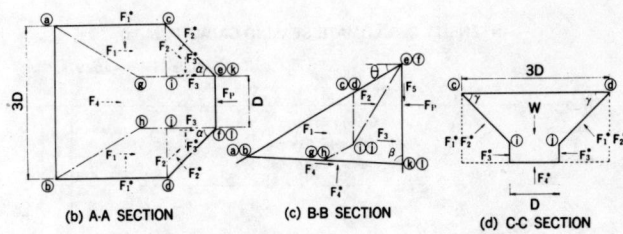

(b) A-A SECTION   (c) B-B SECTION   (d) C-C SECTION

Fig. 7   Sliding model for ground

The extent of the lateral spread of the thrust of the soil mass is limited to 3 times the diameter of the pier. It was further assumed that the force, F_i, acts on each sliding surface ① of the soil mass in accordance with Coulomb's law of shearing resistance,

$$F_i = M_o \cdot (A_i \cdot a_o \cdot C_u + F_i^* \cdot b_o \cdot \tan\phi_u) \quad i = 1 \sim 4 \quad (1)$$

where F_i^* are the normal forces on each face, and assumed in the following form:

$$F_i{}^* = \int_A (K \cdot \gamma_t \cdot Z \cdot \sin\gamma + \gamma_t \cdot Z \cdot \cos\gamma)\, dA_i \quad i = 1,2 \quad (2)$$
$$F_3{}^* = \int_A K \cdot \gamma_t \cdot Z \, dA_3$$

$F_4{}^*$: Determined for equilibrium of soil mass

where:

C_u: Cohesion
ϕ_u: Angle of internal friction
γ_t: Unit weight
z: Depth
A_i: Area of sliding surface ①
K: Coefficient of earth pressure
a_0: Stength reduction factor of soil by failure
$0 \le a_0 \le 1$
b_0: Shear transfer factor across crack
$0 \le b_0 \le 1$
m_0: Degree of mobilization
$0 \le m_0 \le 1$

At surface ⑤ , considering only ultimate bearing capacity F_p and ignoring shear resistance F_5, F_p is derived considering the equilibrium between formula (1) and the weight of soil mass W.

$$F_p = \left[W - 2 \sum_{i=1}^{2} F_i{}^* \cdot \cos\gamma + \left(2 \sum^{3} F_i \cdot + m_0 \cdot A_4 \cdot C_u \right) \cos\beta \right]$$

$$\times \frac{m_0 \cdot \tan\phi_u \cdot \sin\beta + \cos\beta}{\sin\beta - m_0 \cdot \tan\phi_u \cdot \cos\beta} - 2 F_2{}^* \sin\gamma \cdot \sin\alpha$$

$$+ \left(2 \sum_{i=1}^{3} F_i + m_0 \cdot A_4 \cdot C_u \right) \sin\beta \quad (3)$$

where α, β and γ express angles of a sliding surface as shown in Fig.7.

5.2 Application of formula for ultimate bearing capacity

As an application of formula (3), ultimate bearing capacity of the three layered ground structure at the test site was estimated by employing the values of C_u, ϕ_u, and γ_t from Table 1, and $\alpha = 45°$, $\beta = \theta + 45° + \phi_u/2$, $\gamma = 45°$. The values for a_0 and b_0 will be 0 for sliding surface ② and 1 for the other surface. In addition, the value for m_0 will be 1 and that for K will be 0.5 assuming coefficient of earth pressure at rest. The ultimate bearing capacity at each depth of the ground were calculated by using formula (3) and the rate of increase in depth were considered equivalent to the intensity of the ultimate bearing capacity at each point.

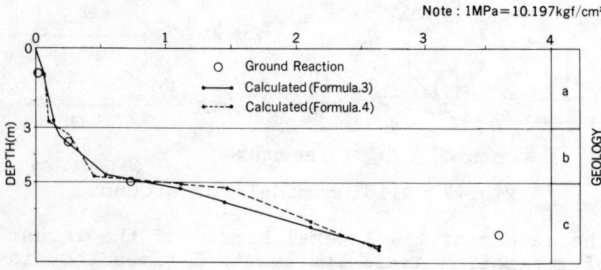

INTENSITY OF ULTIMATE BEARING CAPACITY(MPa)

Note : 1MPa=10.197kgf/cm²

Fig. 8 Result of calculations for intensity of ultimate bearing capacity

The calculated results for the intensity of the ultimate bearing capacity are shown by the solid line in Fig.8. The intensity of ground reaction designated by symbol "O" in this figure corres-

ponds to the upper limit value of ground reaction shown in Fig.5. It is clear that the calculated values are in good agreement with the upper limit value up to 5.0 m in depth, but at measurement point 4 located at 7.0 m in depth, the calculated value is considerably smaller than the upper limit value. A possible reason for this disagreement is that the soil mass in the lower layer tends to be pushed out laterally by the restraining effect and passive earth pressure is generated on the lower part of sliding surface ③ as is shown in Fig.7.

5.3 Simplified formula for design

To propose of simpler formula of the sliding model, the sliding surface formed by the broken line shown in Figure 8 was assumed. It was also assumed that a shearing force is developed only at sliding surface ④ of the bottom, thus ignoring the forces on the other surfaces of the soil mass. Accordingly, formula (3) can be simplified as follows:

$$F_p = \frac{W(\cos\beta + \tan\phi_u \cdot \sin\gamma) + C_u \cdot A}{\sin\beta - \tan\phi_u \cdot \cos\beta_u} \quad (4)$$

where $\beta = \theta + 45° + \phi_u/2$ and A is the area of the sliding surface at the bottom.
The results of calculations using formula (4) are shown in Figure 8 by the broken line and agree well with the results using formula (3).

6. CONCLUSIONS

It was found that the actual behavior of a pier foundation on a soft-rock slope is well explained by a simulation employing the beam theory on an elasto-plastic foundation and through determining the relationship between a ground reaction and displacement. Therefore, it becomes most important to obtain a relationship between a ground reaction and a displacement of the surrounding soil by means of soil exploration. Furthermore, the present study indicated that the ultimate bearing capacity of the test site was well predicted by a proposed formula employing Coulomb's law of shearing resistance, however further investigation should be required before the design application of the formula. It is hoped that the present study will be useful in clarifying the mechanism of bearing capacity for pier foundations.

REFERENCES

Reese, L.C., (1958), Discussion on Soil Modulus for Laterally Loaded Piers, Transactions, ASCE. Vol. 123, pp.1071 - 1074

Yoshii, Y., Yoneda, O., and Takeuchi, T. (1982), Horizontal Loading Tests of Pier Foundations, 17th Japan National Conference on Soil Mechanics and Foundation Engineering, pp.2249 - 2252

IMPROVEMENTS OF KARSTIC GROUND FOUNDATION FOR A LARGE HYDRAULIC STRUCTURE

Amélioration des fondations en terrain karstique pour une installation hydraulique importante

Verbesserung eines karstigen Untergrundes für die Gründung eines grossen hydraulischen Bauwerkes

A. J. da Costa Nunes
Eng? Professor, President of Tecnosolo, S.A.

Ediberto M. Vasconcelos
Geol. Professor, Consulting of CODEVASF

Waldir B. Santiago
Geologist, Engineer of Tecnosolo, S.A.

José Luiz Couto de Souza
Eng?, Engineer of Technosolo, S.A.

SYNOPSIS

The irrigation complex known as the Jaiba Project, administered by the state-owned company CODEVASF, in the north of the state of Minas Gerais, Brazil, includes the construction of a large pumping station EB-1, with a water discharge of 80m³/sec. As such a structure is founded over a karstic formation, a special treatment is required. Several technical solutions for foundations were studied, but after a large in situ test it was concluded that the best solution would be the improvement of the foundation ground by cement-clay grouts with a high friction angle.

RESUME

Le complexe d'irrigation connu sous le nom de Jaiba Project, administré par la Société d'Etat CODEVASF, et situé au nord de l'Etat de Minas Gerais, Brésil, comporte une grande station de pompage EB-1, avec un débit de 80 m³/sec. Cette station est située sur un terrain de formation karstique et a nécessité un traitement spécial pour ses fondations. Plusieurs solutions ont été étudiées et après un essai in situ, on a décidé que la meilleure solution serait d'améliorer le soubassement par des injections de coulis de ciment-argile à angle de frottement élevé.

ZUSAMMENFASSUNG

Der als Jaiba Projekt bekannte und von der staatlichen Gesellschaft CODEVASF verwaltete Bewässerungskomplex im Norden des Staates Minas Gerais in Brasilien umfaßt den Bau einer großen Pumpstation, EB-1, mit einer Pumpleistung von 80 m³/s. Da dieses Bauwerk auf einer Karstformation gegründet ist, wurde dafür eine Spezialbehandlung benötigt. Verschiedene Gründungsarten wurden untersucht. Nach Ausführung eines Großversuches kam man jedoch zu der Erkenntnis, daß die Verbesserung des Untergrundes für die Gründung durch die Injektion von Zement-Ton-Gemischen mit großem Reibungswinkel die beste Lösung sein würde.

1. GENERAL INFORMATION

The main pumping station, named EB-1, of the Jaiba irrigation project, which is owned by Companhia de Desenvolvimento do Vale do São Francisco - CODEVASF, a branch of the Interior Secretary of Brazil, will provide a water flow of 80m³/sec from the São Francisco river, thus allowing the agricultural use of a 100.000 ha area.

The Station occupies an area of 20 x 80m² and is located on the right bank of the São Francisco river, very near to a village called Mocambinho, in the state of Minas Gerais, about 250km north of Montes Claros, an important center in the northern Minas Gerais.

The elevation of the local terrain is about +444 and the foundation grade at elevation +427m, in need, therefore, of 17m excavation while the normal water level of the São Francisco river is around +443m, that is at about 16m above the foundation grade elevation.

The type of foundation for the major hydraulic structure

initially planned, as a slab, imparting a 350 kN/m² (3.5kgf/cm²) pressure on the soil.

The special geotechnical problem to be overcome is tied to the nature of the foundation, made up essentially by karstic limestone, whose surface sinkholes (Figure 1) comprised the evidence for its karsticity, in the course of the prospection studies that were performed.

2. GEOLOGICAL SETTING

From the many borings, drilled in part by the Instituto de Pesquisas Tecnológicas do Estado de São Paulo - IPT, and later by the Company which elaborated the studies and the soil improvement, Tecnosolo S.A., it became well known the geological configuration of the foundation sub-grade (Figure 2).

Betwenn elevations +412m and +443m, which represent the better investigated interval, the sub-grade comprises

the following sequence, from bottom upwards:

- Compacted limestone, gray, of the Bambuí Group up to elevations +417m to +420m. This limestone still preserves karstic caves, not been assured that such voids aren't present below elevation +412m.

- Highly karstified limestone, whitish, up to an elevation close to EB-1 foundation level, that is, +427m, of the same Bambuí Group. Cavities are filled-in by clays, plastic, whitish-yellow, with variable sand fractions, and also by yellowish sands.

- Sandy-clay sediments, grading into sands with low clay contents at the surface, covering a huge area in the region.

The geotechnical study of this formation, besides its own value for the foundation of an exceptional project, has also a high general interest due to the great extension of similar terrains, assumed to be present in the São Francisco river basin.

Fig. 1 Karstic sinks at construction site for pumping station EB-1 of Jaiba complex

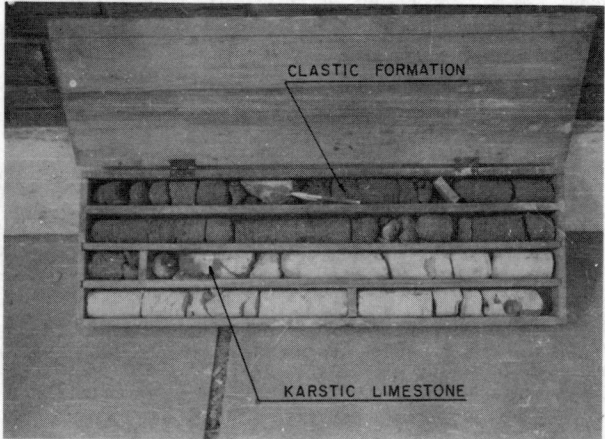

Fig. 2 Geological configuration of the foundation sub-grade EB-1 according core box

3. FOUNDATION TYPE CONCEPT

Three concepts have been proposed for EB-1 foundation type, taking in account the karstic structure:

a) Large diameter piles, to be bored mechanically with metallic casing which would confine the piles at the voided portions.

b) Small diameter piles, again with lost casing, associated by consolidation grouting of the sub-soil formations.

c) Shallow foundation, set on a 15m thick improved terrain, by means of clay-cement mixture grouting, yielding a compacted slab of ground, strong enough to assure foundation stability, independently of cavity distribution in the limestone skeleton below.

Initially, the profession's specialists that were involved preferred solution a), large-diameter piles, based mainly in the assumption that the sub-grade treatment would mean a difficult process to evaluate, comprising very high, and unforeseen, cement consumptions.

Solution b), small diameter piles, was deemed more expansive than solution a).

Other professionals held the opinion that solution a) was more expensive and, besides, that the thickness of the karstified limestone would be necessarily overpassed; it would be preferable to adopt:

1) foundation on improved ground, for EB-1, with an alternative to small-diameter piles (as solution b).

2) foundation on grouted mini-piles for the auxiliary works, like access bridge, tube suport and discharge structure.

Due to the difference in oppinions, CODEVASF decided to allow an "in situ" test, within the EB-1 general area, to really evaluate the solution of ground treatment by grouting.

4. TEST AREA RESULTS AND CHARACTERISTICS

The test area had dimensions of 10 x 10m^2, located near the central portion of EB-1, in the zone where the boring work indicated the most unfavorable ground conditions by its intense karstification. Test modeling scale was, then, in the order of 1:16.

Drilling for the grout work were by rock-drill, with the following procedure aiming exclusively the treatment of the ground between elevations +427m and 412m:

1) Insertion of a 5" casing down to 6m of depth, simultaneous with the drilling. This operation avoided ground slumps from the surface soil, which was highly sandy.

2) Insertion of a rigid PVC tube, 3" in diameter, within a 4" hole down to elevation +427m, sealing the tube at this elevation to prevent against grout losses into the upper section, which would be later scrapped out for foundation grading.

3) Finally, drilling holes into the limestone of 2 1/2" down to elevation +412m.

A special cement-clay grout mixture was evaluated, using local materials, which had more than 1,5 MN/m^2 (15 kgf/cm^2) strength at 7 days, and also had a high friction angle and viscosity to prevent against losses outside the treatment area.

The most appropriate mixture had the following composition:

- water 27%

- cement 36,5%
- soil 36,5%

The injections were performed from the boundary area towards the center, to minimize losses, and comprised primary, secondary and tertiary holes.

Maximum grouting pressures in the tertiary holes reached 1,2 MN/m^2 (12kgf/cm^2) pratically without absorptions.

Careful check-borings proved that the grout had hardened, filling in three types of pre-existing discontinuities:

a) voids in the compacted limestone.

b) zones in the voids originally filled with by plastic clay, which became strongly compressed.

c) voids in the weathered limestone.

Due to the grouting treatment, the consistency of the ground, specially where the filling material was present, in no case was less than 15/30, in the SPT test.

A satisfactory treatment was obtained with less than 12% of the ground volume.

The success of this treatment was strongly helped by the previous experince accumulated in the karstic grounds of Votorantim and Capão Bonito cement factories (Costa Nunes and alli 1976).

5. EB-1 FOUNDATION TREATMENT

5.1 General Aspects

Based upon the results obtained in the test area, it was designed and applied a treatment for the pumping station sub-soil.

The specifications were prepared by CODEVASF, through their consulting geologist.

The treatment was performed from the area's boundary towards the center, by 1st., 2nd. and 3rd. order holes, applying also additional holes in the zones which still contained voids detected by inspection borings.

Figure 3 indicates holes location and the interfaces between fully treated zones.

Grouting pressures increased from 0,5 MN/m^2(5kgf/cm^2), at the beginning, up to the final pressure of 1,2 MN/m^2 (12 kgf/cm^2).

The mostly used cement-soil mixtures, of high friction angle, had a proportion mix of 1 cement: 1 soil: 0,9 water. (Figure 4).

Fig. 4   Test showing cone produced by cement-clay grout

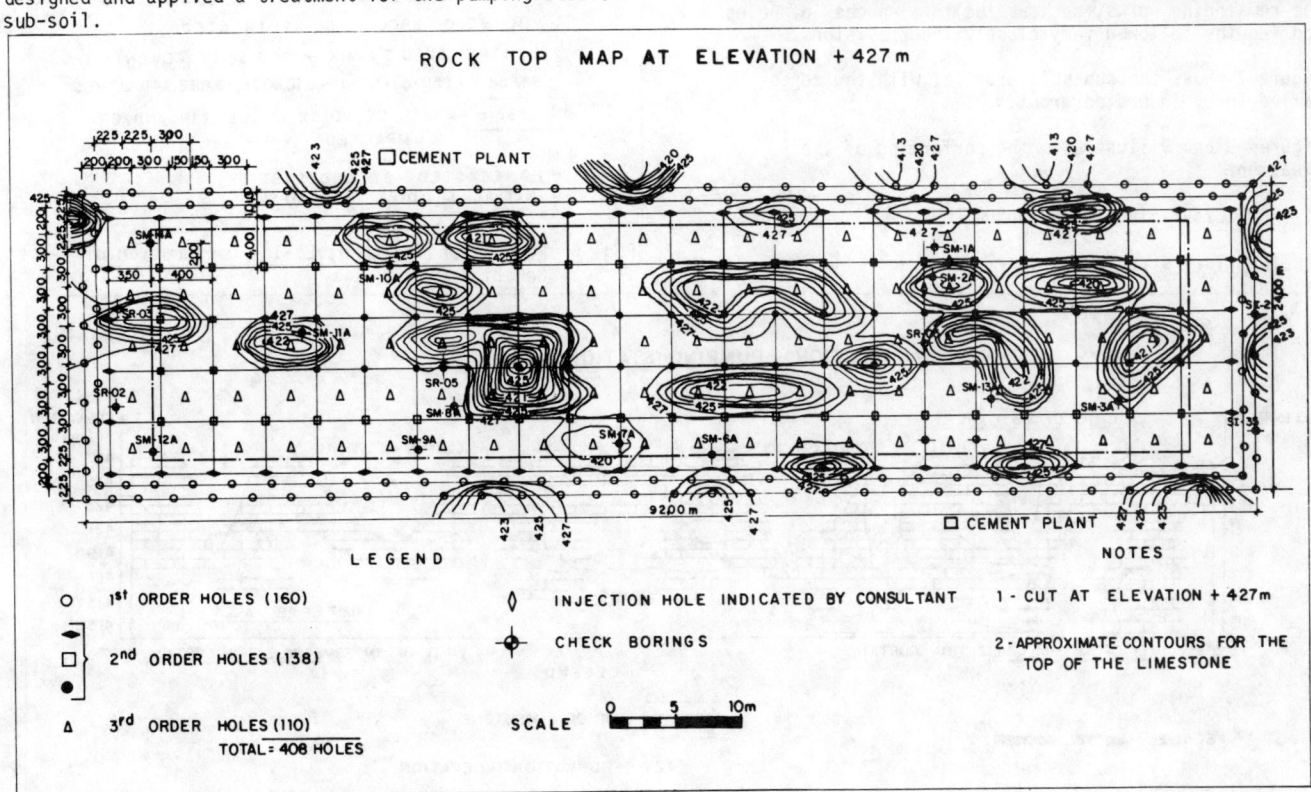

ROCK TOP MAP AT ELEVATION + 427 m

□ CEMENT PLANT

□ CEMENT PLANT

LEGEND

o   1st ORDER HOLES (160)

□   2nd ORDER HOLES (138)

●

Δ   3rd ORDER HOLES (110)
TOTAL = 408 HOLES

◊   INJECTION HOLE INDICATED BY CONSULTANT

⊕   CHECK BORINGS

SCALE   0   5   10m

NOTES

1 - CUT AT ELEVATION + 427m

2 - APPROXIMATE CONTOURS FOR THE TOP OF THE LIMESTONE

Fig. 3 Contour lines for limestone - Jaiba

Also used were cement-water grouts in the proportion mix of 1 cement: 0,5 water, in the 1st. and 3rd. order injections.

5.2   Check Borings

Check borings, with SPT in the materials of lesser consistency and rotary-drill with coring in the limestone and cement zones, showed that the grouting filled-in the clay bands of small consistency and also the fractures joints and caves in the limestone.

The soil within the grouted area became highly compacted by the injections, yielding high values of SPT.

In a few zones where the treated ground was less compact, additional injections were performed, finishing up the treatment.

5.3   Final Evaluation. Conclusions

The soil profiles from the boring campaigns allowed to evaluate the nature of the foundation ground voids. (Figure 5).

These voids, after the treatment, were able to withstand without any absorption, grout pressures of 700 kN/m^2 at least, while maximum working loads will be 350 kN/m^2 .

Cement and soil consumptions are presented graphically in Figure 6.

It can be seen that the real consumptions have been smaller than forecasted, which was an aim to keep within the project budget and resulted from the fact that the test area picked was located in the most karstified zone.

The remainning works, such as the total number of holes and lengths followed very closely the previsions.

Figure 7 shows the sub-soil profile, with the zones filled-in by indurated grout.

Figures 8 and 9 ilustrates the performing of the treatment.

6.   AUXILIARY STRUCTURES FOUNDATIONS

The auxiliary structures comprised an access bridge, a

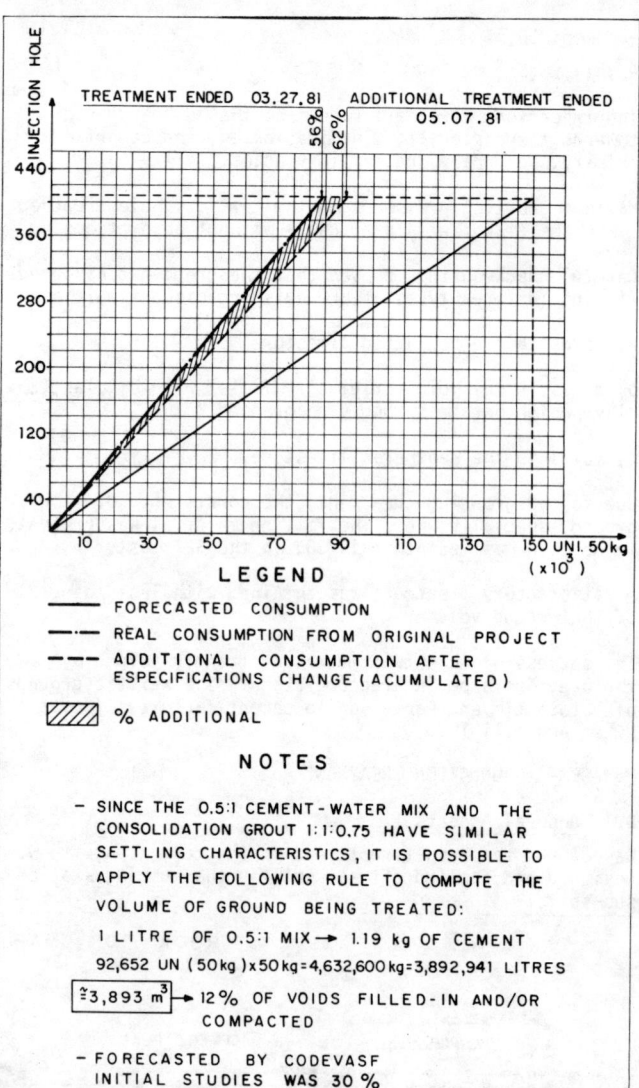

LEGEND

⎯⎯⎯⎯ FORECASTED CONSUMPTION

⎯·⎯·⎯ REAL CONSUMPTION FROM ORIGINAL PROJECT

⎯ ⎯ ⎯ ADDITIONAL CONSUMPTION AFTER ESPECIFICATIONS CHANGE (ACUMULATED)

▨▨▨ % ADDITIONAL

NOTES

- SINCE THE 0.5:1 CEMENT-WATER MIX AND THE CONSOLIDATION GROUT 1:1:0.75 HAVE SIMILAR SETTLING CHARACTERISTICS, IT IS POSSIBLE TO APPLY THE FOLLOWING RULE TO COMPUTE THE VOLUME OF GROUND BEING TREATED:

1 LITRE OF 0.5:1 MIX ⟶ 1.19 kg OF CEMENT

92,652 UN (50 kg) x 50 kg = 4,632,600 kg = 3,892,941 LITRES

$\boxed{\cong 3,893 \; m^3}$ ⟶ 12 % OF VOIDS FILLED-IN AND/OR COMPACTED

- FORECASTED BY CODEVASF INITIAL STUDIES WAS 30 %

Fig. 6   Foundation Treatment EB-1 - Consumption of Cement + Clay

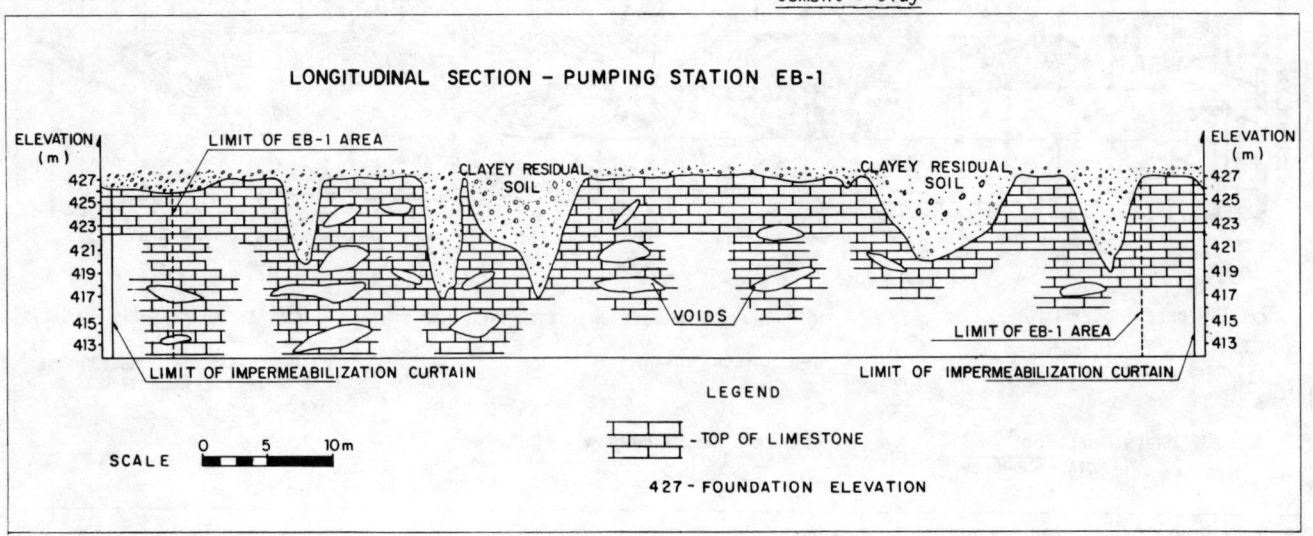

Fig. 5   Profile of Caves at Pumping Station EB-1 - JAIBA

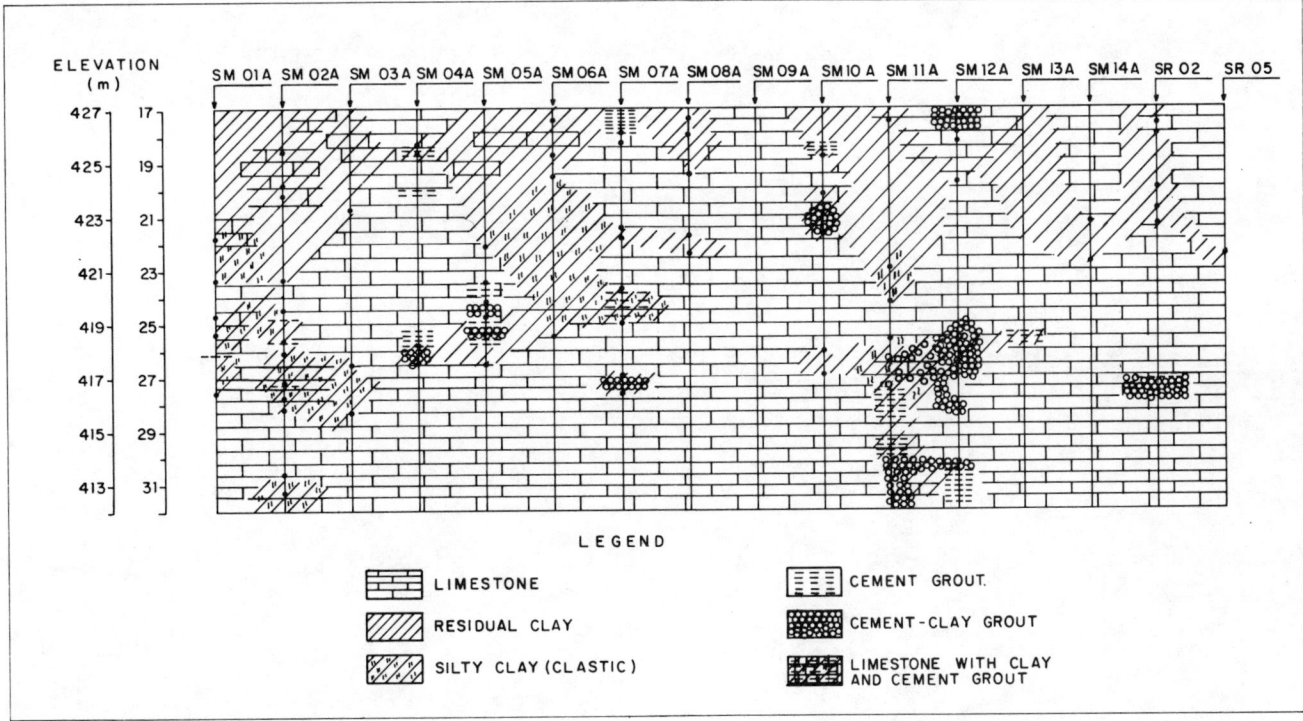

LEGEND

▤ LIMESTONE		▦ CEMENT GROUT.
▨ RESIDUAL CLAY		▦ CEMENT-CLAY GROUT
▩ SILTY CLAY (CLASTIC)		▨ LIMESTONE WITH CLAY AND CEMENT GROUT

Fig. 7 Geotechnical Profiles - Check Borings-EB-1

Fig. 8 Panoramic view of EB-1 site

discharge structure and tube support blocks.

These structures, due to their small areas, did not require a foundation treatment of the same kind as done for the pumping station.

For these auxiliary structures, foundations with small-diameter (0,10m) piles were applied, built as grouted tie-backs, good for loads up to 60t(~ 600 kN), which we call press-anchors.

The foundations for the access bridge have already been performed, with success, confirmed by tests which apply the same testing technique for tiebacks, resulting that all piles are then load tested.

The conception of this type of foundation is outside the scope of this paper.

Fig. 9 ⌀ 4 1/2" Drilling for Protective casing installation

7. BIBLIOGRAPHY

COSTA NUNES, A.J.; SILVA FILHO, J.G. and VASCONCELOS, E.M. (1976) - Problems of foundations over metamorphic karstics formations (in portuguese). Proceedings First Brasilian Congress on Engineering Geology. Rio of Janeiro - Vol. 2 Associação Brasileira de Geologia de Engenharia.

ITAIPÚ MAIN DAM FOUNDATIONS: DESIGN AND PERFORMANCE DURING CONSTRUCTION AND PRELIMINARY FILLING OF THE RESERVOIR

Barrage principal d'Itaipú: Conception et comportement pendant la construction et le remplissage préliminaire du réservoir

Gründung der Haupt-Talsperre Itaipú: Entwurf und Verhalten während des Baues und des ersten Teilstaues

Ricardo Antonio Abrahão
Geologist, Promon Engenharia, S.A.

João Francisco Alves Silveira
Civil Engineer, Promon Engenharia, S.A.

Fernão Paes de Barros
Geologist, Itaipu Binacional, Professor of Engineering Geology, Catholic University of Campinas, Brazil

SYNOPSIS

The dam for the Itaipú hydroelectric development has as its central portion a concrete hollow gravity structure founded on basalt rock with sub-horizontal discontinuities. These discontinuities make structural stability dependent on the sliding resistance of the foundations. Since the construction of the dam in its final phase and partial filling of the reservoir has been concluded, preliminary conclusions, based upon the installed instrumentation, suggest that a strengthening of the foundation has occurred due to the load increase and the introduction of concrete shear keys in the foundation. The design estimates related to foundation stiffness have not always been confirmed but reflect the average behaviour of the foundations. Analyses of piezometric levels have indicated that the grout curtain has great efficiency, as shown by hydrostatic uplifts and little water flows, both of which are smaller than previously foreseen for the project in this phase.

RESUME

Le barrage du projet hydroélectrique d'Itaipú se compose d'une partie centrale à poids élevé, construit en roche basaltique avec des joints sub-horizontaux, qui font que la stabilité structurale dépend de la résistance au glissement des fondations. Vu que la construction du barrage est dans sa phase finale, et qu'une partie du réservoir est déjà remplie, les conclusions préliminaires, basées sur les instruments installés, indiquent qu'il y a eu une plus forte déformation dans les fondations, due au surcroît de charge et à la construction de clavettes de cisaillement en béton. L'appréciation de la rigidité des fondations n'a pas toujours été confirmée et les données obtenues ne donnent qu'une idée du comportement moyen des fondations. L'analyse des niveaux piézométriques a montré l'efficacité des injections d'imperméabilisation et elle a permis de constater que l'élévation du niveau hydrostatique ainsi que les écoulements d'eau sont inférieurs aux prévisions.

ZUSAMMENFASSUNG

Der im Flußbett erbaute Teil der Talsperre der Wasserkraftanlage Itaipú ist eine hohle Schwergewichtsmauer, fundiert auf Basaltfelsen mit subhorizontalen Diskontinuitäten. Die Standsicherheit des Bauwerks ist vom Gleitwiderstand dieser Diskontinuitäten abhängig. Da der Bau der Talsperre fast beendigt und der Speicher schon teilweise gefüllt ist, liegen erste Meßergebnisse vor. Es zeigt sich, daß die Gründung wegen der Gewichtserhöhung und wegen der in die Gründung eingebauten Betonscherverfüllung standsicher ist. Die Festigkeit des Gründungsfelsens entsprach nicht immer der Entwurfsschätzung, gibt aber ein mittleres Verhalten der Gründung an. Die Analysen der Standrohr-Wasserstände zeigen die Effektivität des Injektionsschleiers an, da der Auftrieb und die Sickwassermengen kleiner sind als im Entwurf vorgesehen war.

1. INTRODUCTION

The Itaipu Hydroelectric Project is a bi-national enterprise located in the central reach of the Paraná River. The main dam is a concrete hollow gravity structure 185 m high located immediately upstream of the powerhouse and flanked by buttress type concrete dams, a diversion control structure, a spillway and rockfill and earthfill dams.

A complete description of the project can be found in Cotrim et al (1), and geotechnical-geological conditionings in the design of the main dam in Caric et al. (2).

All structures are founded on a sequence of basaltic flows of the Paraná River Sedimentary Basin. These rocks consist of approximately sub-horizontal layers with typical rock species of basaltic flows such as dense basalt, contact breccia, vesicular-amygdaloidal basalt, etc., and structures which vary from columnar jointing in the center of the flows to sub-horizontal discontinuities at the contacts, or shear zones in the river bed area. The discontinuities which follow the contacts have regional characteristics and are associated with the origin of the basaltic flows showing, at times, evidence of displacements such as slicken sided walls, which are not too prominent.

Other discontinuities, also with sub-horizontal dipping, are apparently more recent. Their origin is due to vertical unload caused by the erosion of the valley and the eventual occurrence of tectonic stresses, as discussed in Paes de Barros and Guidicini (3).

II. FOUNDATION DESIGN

The subsurface exploration of the dam foundation include a large number of drill holes and the opening of shafts and exploration galleries, which subsequently formed part of the foundation drainage system. Geologic maps made during excavation improved the information on the rock mass and made possible an estimate of the geotechnical parameters based on values obtained through "in situ" deformation and shear tests performed in shafts and galleries close to the area under discussion.

At that time it was estimated that three main levels of discontinuties occurred from El 30,0 m to El. 10,0 m approximately. The average level of the dam foundations is at about El. 40 m.

The discontinuity at El. 20,0 m is of the contact type and the others of the shear type. The contact type discontinuity imposed sliding stability problems, and the others, due to their greater deformability, are more associated with the settlement aspect of the structure-foundation complex.

The aspects related to construction schedule, deformability and shear strength of the foundation, determined which final solution of foundation treatment would be adopted.

As discussed in detail in Souza Lima et al. (4), the foundation design, can be described in three main phases. The first one is related to the structural dimensioning by means of limit equilibrium method and analyses of stress-strain fields by FEM. The loads were considered as conventionally stated. The following geomechanical parameters were adopted: rock mass considered elastic, isotropic, non-resistant to tension with deformation moduli of 20 and 15 GPa for the sound basalt and vesicular basalt with breccia, respectively. The discontinuity at El. 20,0 m was idealized as a joint with elastic-plastic behavior with the following

parameters: $\emptyset = 25^o$, $C = 0,0$ MPa, $k_t = 5$ MPa/cm and $kn = 15$ MPa/cm as the friction angle, cohesion and transversal and normal unit stiffnesses, respectively. The results of this phase showed that the safety factor required by the Design Criteria could not be achieved, unless some modifications were introduced in the design.

In order not to modify the geometry of the concrete structures it was decided to improve the drainage system, so as to reduce to a minimum the uplift forces. This decision was the starting point of the second phase of the foundation design. To accomplish this, a system of drainage was developed by introducing a drainage gallery in the discontinuity around the perimeter of the area under the dam blocks in question. This system allowed the adoption of an uplift pressure diagram considerably reduced in comparison to that previously adopted. However, safety factors for this solution were marginal; therefore, it was decided to adopt a more effective solution, which triggered the third phase.

Thus a system of shear keys was incorporated to the design, consisting of excavated tunnels in the rock mass, intercepting discontinuities and later filled with concrete originating an 8 x 8 orthogonal net of shear keys within the area surrounded by the drainage system. (Fig. 1 and 2). The minimum cross section of each shear key is 2,5 x 3,5 m, reaching up to 7,0 m of height when various levels of discontinuties are superposed. This system was developed so as to increase the shear strength and the unit stiffnesses of the discontinuities.

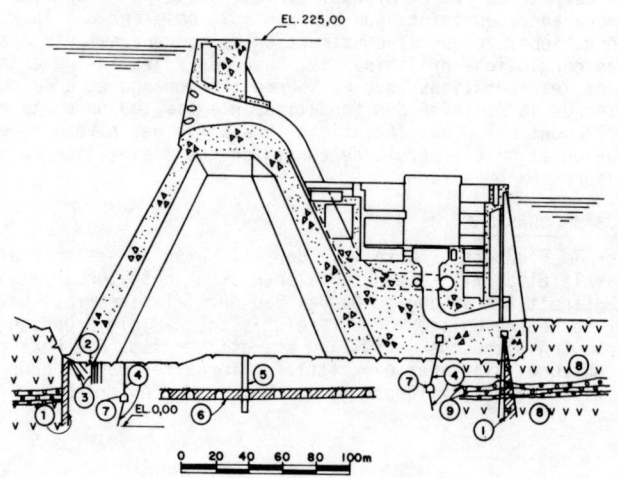

1 - Main Grout Curtain
2 - Contact Grout
3 - Consolidation Grout
4 - Drainage Curtain
5 - Access and Prospection Shaft
6 - Concrete Shear Keys
7 - Drainage Tunnel
8 - Dense Basalt
9 - Breccia and Vesicular Basalt

FIG.1
MAIN DAM AND POWER HOUSE CROSS SECTION

This treatment was performed using very careful excavation methods, due to preservation of the remaining rock and the concrete that was poured simultaneously to the excavation, what required particle velocities from blasting to be limited in the vicinity of the concrete according to the various times of setting.

After the excavation, the rock surfaces were washed with air and water jets to assure concrete bond. The concrete used, had a characteristic strength of fck = 280 kgf/cm^2 at one year and a maximum aggregate diameter of 3.8 cm, and was placed into the tunnels through 6" diameter holes drilled with down-the-hole equipment from the surface and spaced from 4 to 8 m apart. This concrete was normally vibrated in place up to a certain lift height and subsequently by means of vibrators introduced into the concrete placement holes. Where the tunnels were very high the concrete was placed in various lifts. In the last lift, near the tunnel roof, the concrete was placed with an expansive admixture in order to enhance the contact at irregular zones whenever possible. The vertical concrete construction joints were, wherever possible, positioned halfway between the intersection of the shear keys.

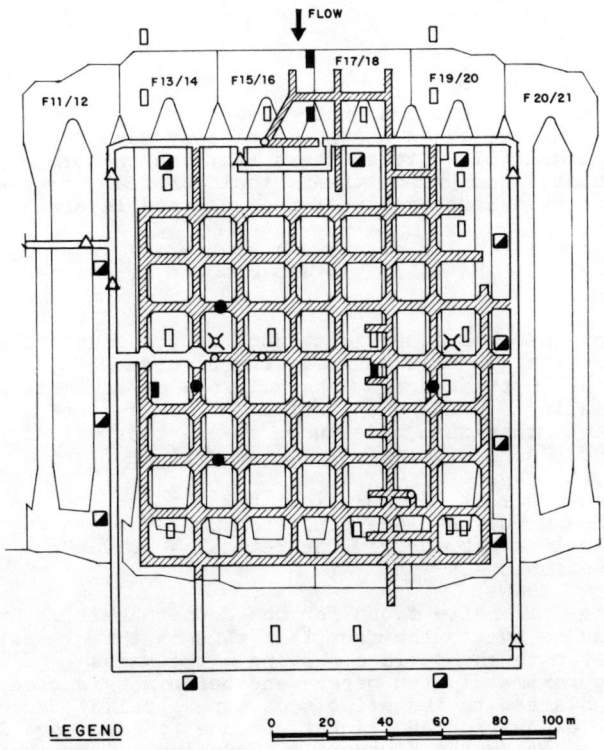

FIG.2
MAIN DAM AND POWER HOUSE
INSTRUMENTATION AT SHEAR - KEYS ELEVATION

The top of the shear keys were finally grouted from the surface to assure the concrete-rock bond and the consolidation of the rock affected by blasting in this area.

III.   FOUNDATION INSTRUMENTATION

In order to monitor the behavior of the basaltic rock mass in the foundation of the Itaipu structures located in the river bed, three blocks were selected and provided with a instrumentation system as complete as possible (Fig. 2 and 3). In addition to these blocks, an attempt was made to provide one of each two consecutive blocks with piezometers in the foundation and, in all contraction joints between blocks, with joint meters in the vicinity of the foundation, to enable observation of the differential displacement between blocks.

Instruments were installed to perform the following monitoring observations in the key blocks:

3.1   Settlement

Multiple rod extensometers were selected for monitoring the settlement of the concrete structures. These extensometers were generally installed with three rods, with the deepest one fixed well below the foundation discontinuities so as to assure that this rod could be used as a reference.

As much as possible, these extensometers were installed in the initial construction stage so that the monitoring of the settlement from the beginning was assured.

3.2   Horizontal Displacement

For monitoring the horizontal displacements of the key blocks respect to deep points in the foundation multiple rod extensometers and inverted pendula were installed. Some extensometers were installed sub-horizontally upstream, while the inverted pendula were installed so as to take advantage of the inspection shafts drilled in the foundation.

3.3   Displacements

The differential displacements between blocks are being monitored by means of 2 detachable joint meter basis installed in the vicinity of the foundation, one in a vertical plane and the other in a horizontal plane. Joint meter basis were also installed in the side walls of the buttresses, in order to monitor the differential displacements between monoliths.

The drainage tunnels excavated in the foundation along the discontinuities were used for installing the triorthogonal joint meters on the walls, for monitoring opening-closure movement and shearing displacements along these discontinuities.

3.4   Uplifts

Uplifts are monitored by standpipe piezometers installed upstream, in the rock-concrete contact and in the more pervious discontinuities of the foundation. A few piezometers were installed in the key blocks, immediately upstream and downstream of the grout curtain, with the basic objective of evaluating its performance. These piezometers were installed shortly after the grout curtain was completed, so as to enable the observation of the

hydrogeotechnical behavior of the basaltic rock mass prior to the preliminary filling of the reservoir.

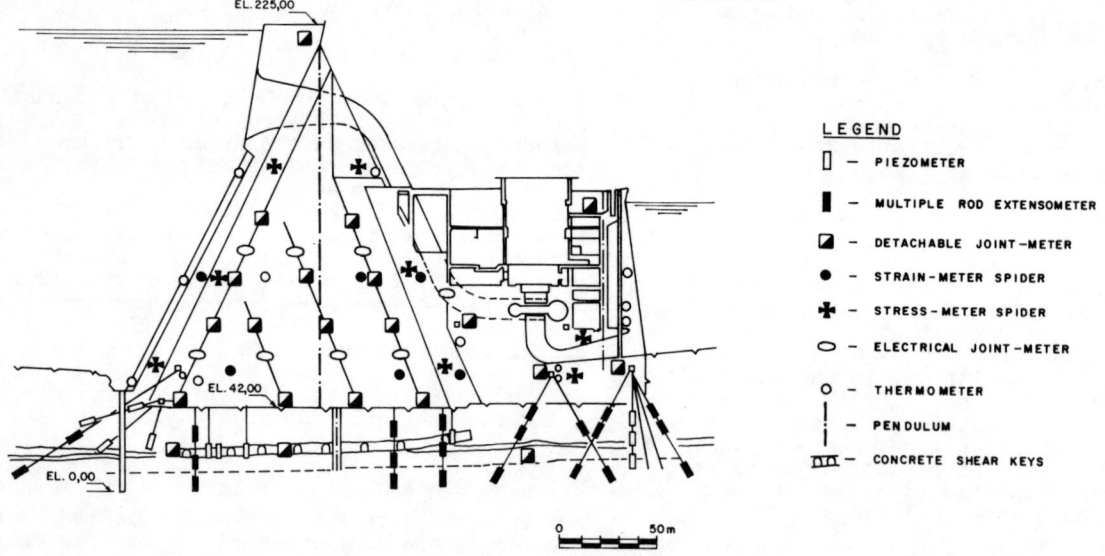

FIG. 3
MAIN DAM AND POWER-HOUSE
LAY-OUT INSTRUMENTATION ON A KEY BLOCK

3.5 Drainage Flows

Drainage flows are being monitored by triangular or trapezoidal flowmeters installed in the perimetral drainage tunnel.

3.6 Stresses in the Shear Keys

In order to monitor the shear keys behavior during the construction phase and reservoir filling, four strainmeter rosettes were installed, two of them in the transversal shear-keys and two in the longitudinal ones.

IV. PERFORMANCE DURING CONSTRUCTION

The multiple extensometers installed vertically in the foundation of the river bed blocks were of special importance in evaluating the influence and behavior of the treatment, considering that the shear keys were excavated, concreted and grouted while the concrete dam monoliths were under construction.

As seen in Fig. 4, the excavation of the shear-keys underneath the dam buttress caused a sudden increase in the settlement, thus leading to a recommendation to concrete and grout the tunnels immediately after excavation. In some locations where the discontinuities exhibited ramifications the tunnels had to be excavated up to 7,0 m of height such as downstream of blocks F18 and F19. The settlement rate indicated by the multiple extensometers indicated that the concreting of the dam blocks should be temporarily stopped until the shear-keys were completed. As the concrete placement of the shear-keys proceeded, the settlement rates decreased sharply and returned to the initially recorded rates or even less.

To analyse the deformability variation of the foundation during construction and evaluted the effect of the treatment concerning deformability,

various layers had their deformability moduli calculated based on the settlement measured by the multiple extensometers (between anchorage points) and on the theoretical vertical stresses of the central point of the analyzed layers.

Typical results of these analyses are shown in Fig. 5. The following aspects can be pointed out:

For the more superficial zone of the foundation (depths up to 30 meters), the deformability moduli obtained from the multiple extensometers readings during the construction phase were lower than the values used in the theoretical studies, as follows:

Dense basalt: E theoretical = 20,0 GPa; E measured = 5,5 GPa (superficial layer) Vesicular basalt and breccia: E theoretical = 10,0 GPa; E measured 7,0 GPa.

The difference found for the dense basalt layer may be due to the fact that this is the closest layer to the surface, having a thickness of approximately ten meters and being more directly subjected to the effects of stress relief caused by excavation and vibrations due to blasting. At greater depths however, a significant increase in the deformability moduli of the rock mass occurs as indicated by data obtained during the preliminary reservoir filling.

The maximum differential settlement between blocks, as indicated by the joint meters installed near the foundation, occurred between blocks F18 and F19, and reached a maximum of 2,5 mm during the construction period. Fig. 6 shows the differential settlements between the central blocks in the river bed.

Considering the low deformability moduli of the rock foundation, the constant settlement rates observed in the area, and the more pronounced

C 194

differential settlements between blocks F18 and F19, it was decided to treat them by consolidation grouting under upstream part of the foundation of Blocks F14 to F19.

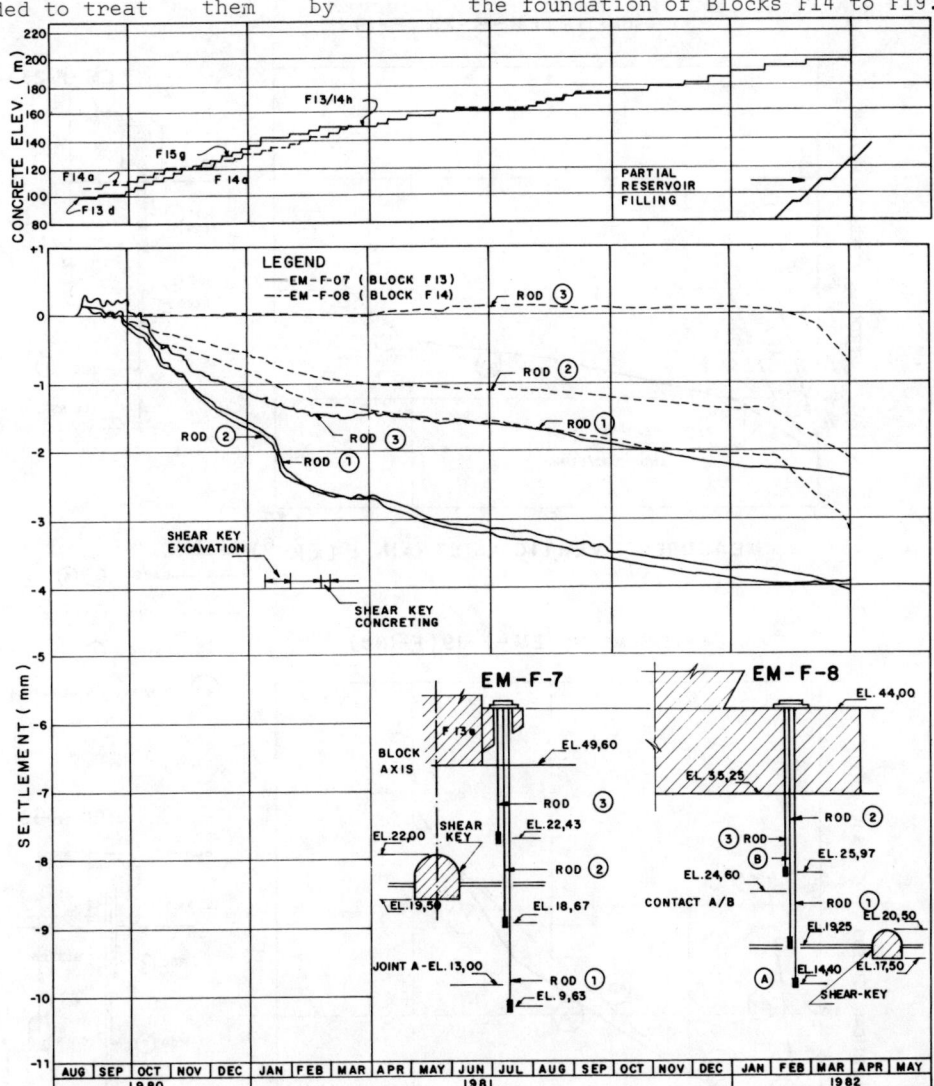

FIG. 4 MULTIPLE EXTENSOMETERS
MEASURED SETTLEMENT IN THE FOUNDATION OF A CENTRAL BLOCK

V. BEHAVIOR DURING THE PRELIMINARY FILLING OF THE RESERVOIR

The preliminary filling of the Itaipu reservoir consisted in filling the area between the upstream cofferdam and the concrete structure of the main dam, with the purpose to observe the behavior of these structures and particularly the foundations with a partial water head. This filling was up to El. 139 m, which resulted in a water column of 100 meters in the highest blocks of the dam, which correspond to a hidrostatic load of approximately 30% of the total one, to which these blocks will be subjected after final reservoir filling.

The preliminary filling was performed in five stages. At the end of each stage the filling was stopped for 5 days to allow the stabilization of the instrumentation readings and the interpretation analyses. The velocity of the water level increases was controlled, utilizing the following values:

TABLE I

Stage El.(m.)	Water level increasing rate (m./day)
40- 80	5,0
80- 95	2,5
95-110	1,0
110-125	1,0
125-139	1,0

The following is an analysis of these data, compared to the values furnished by the design criteria or calculated by the finite element method.

- Settlement and differential movements between blocks

During the pre-filling of the reservoir the settlements of the central blocks of the dam occured essentially in the upstream part. Table II shows the settlements observed during this phase as measured by the multiple extensometers located upstream, together with the theoretical settlement range. These values were calculated by adopting the deformability parameters for the various layer and the rock discontinuities at the limits of a probable range.

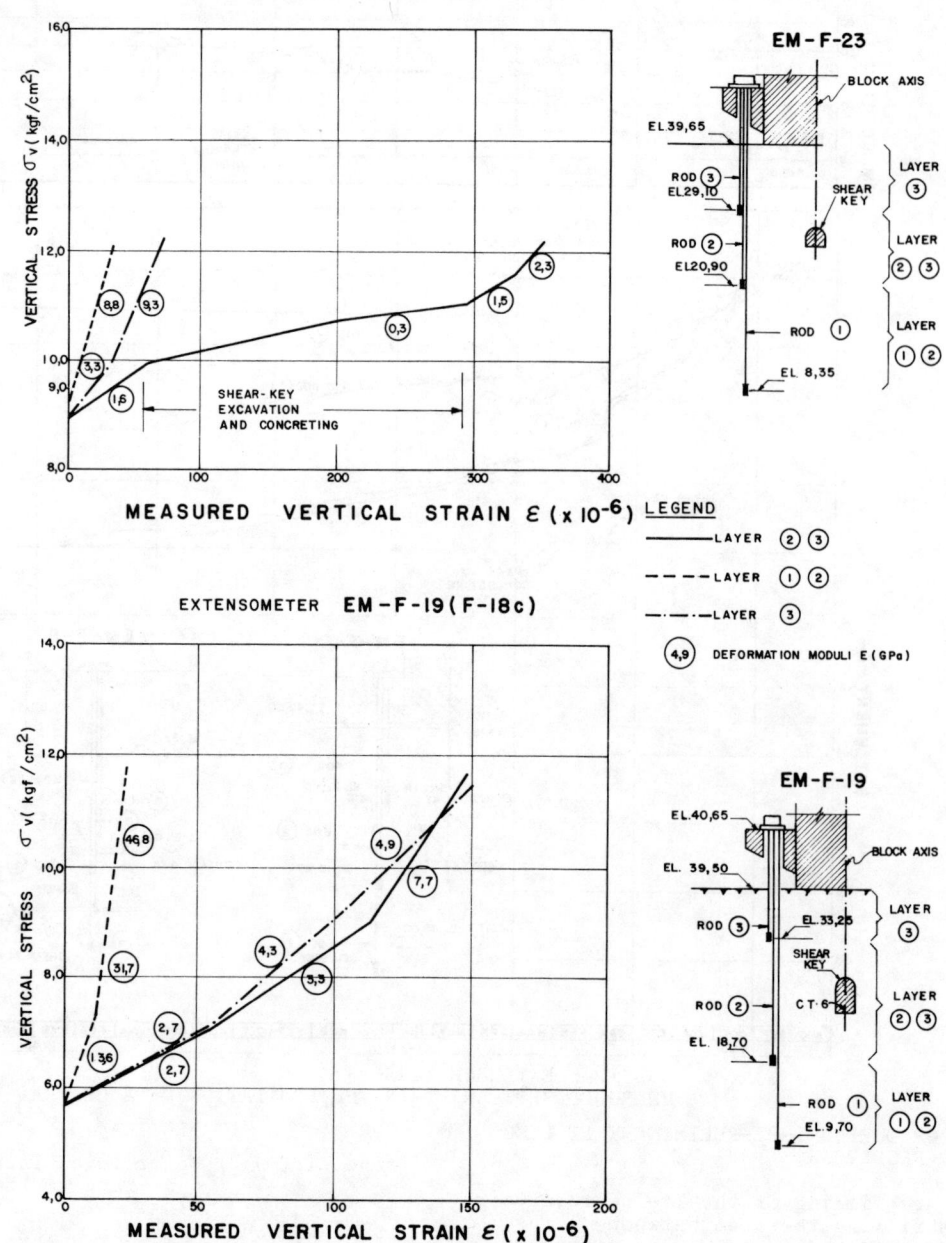

FIG. 5 - EVOLUTION OF THE ROCK DEFORMATION MODULI DURING THE CONSTRUCTION PERIOD

TABLE II

MEASURED AND THEORETICAL SETTLEMENTS DURING RESERVOIR PRELIMINARY FILLING

Multiple Extensometer	Dam Block	Settlements (mm)		
		Measured	Theoretical Min.	Theoretical Max.
EM-F-8	F14	1,3	0,9	2,6
EM-F-11	F16	1,5	1,3	3,4
EM-F-20	F18	1,6	1,3	3,3
EM-F-22	F19	1,6	1,3	3,2

It is worth mentioning that:

a) The settlements indicated by these four multiple extensometers have indicated a good rock mass uniformity regarding deformability. The fact the settlements of blocks F18 and F19 are identical, can be attributed to the efficiency of the consolidation grouting, provided that during the construction period the maximum differential settlement between blocks, occurred exactly between these two blocks, with

F18 having settled 2,5 mm more than F19. The settlements indicated in Table II refer to the total deformation measured by the lowest rod.

b) The measured settlements were closer to the minimum theoretical values, which indicate a foundation on the stiffer side as related to the established theoretical values.

- Horizontal displacements

The inverted pendula installed in the foundation of the highest blocks of the Main Dam, approximately 30 meters deep, do not indicate any measurable displacement of the dam in the horizontal direction.

Theoretically, the tensile stresses in the upstream foundation area should have reached, during this phase of the preliminary filling, values up to 2 MPa; however, there was no evidence of cracks in the upstream toe of the structures, which indicates that the range of natural stress should be different from the estimated values.

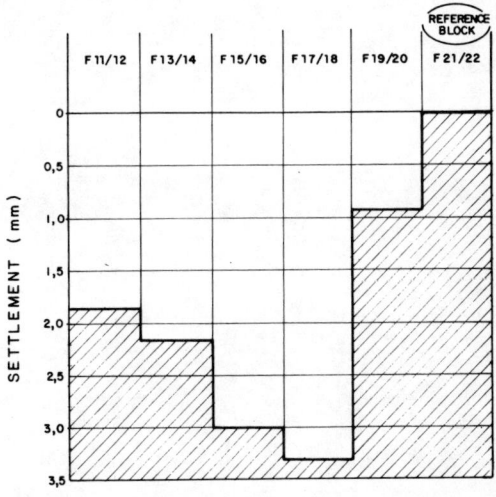

FIG.6
DIFFERENTIAL SETTLEMENTS OF THE CENTRAL BLOCKS DURING THE CONSTRUCTION PERIOD

TABLE III

MEASURED AND THEORETICAL DRAINAGE FLOWS DURING RESERVOIR PRELIMINARY FILLING

TUNNEL	GAGE	DRAINAGE FLOWS (l/min.)		
		BEFORE FILLING	AFTER FILLING	THEORETICAL
El. 60	MV-F-2	1,5	2,5	17
El. 55	MV-F-3	2,2	6,0	17
El. 20	MV-F-6	-	-	10
	MV-F-7	0,7	0,7	10
	MV-F-8	13,0 (*)	11,9	46
	MV-F-9	9,8 (*)	9,0	13
	MV-F-10	13,0 (*)	12,7	71

(*) Readings affected by water coming from construction

- Uplift Pressures

The piezometers installed in the discontinuities of El. 20 m, and in breccia "B" layer (between El. 70 and 80 m), upstream and downstream of the grout curtain are indicating a good performance of this curtain, particularly in discontinuity at El. 20 m.

The observed uplift pressures in the foundations are at all times lower than the ones established by the design criteria.

- Drainage Flows

Table III shows the flows observed in the drainage tunnels before and after the preliminary filling of the reservoir, together with the theoretical values.

The following conclusions are of particular interest:

a) The measured flows are 1/3 to 1/6 of the theoretical values, which may be due to the efficiency of the grout curtain and to the absence of tension cracks upstream, in the rock foundation.

b) The increase in drainage flows in tunnel at El. 20 m during the preliminary filling of the reservoir phase was of little significance.

VI. REFERENCES

1. COTRIM, J.R.; KRAUCH, W.H.; ROCHA, J.G.; GALLICO, A. (Oct.Nov.1977) and SARKARIA; G.S. "The bi-national Itaipu Hydropower project", Water Power and Dam Construction.

2. CARIC, D.M.; URIARTE, J.A.; NIMIR, W.A.; EIGENHEER, L.P. and NITTA, T.(1982) "Itaipu main-dam: geological and geotechnical features affecting the design", Proc. 14th ICOLD Congress, Rio de Janeiro, Brazil.

3. PAES DE BARROS, F. E GUIDICINI, G. (1981). "Um Processo de Alívio de Tensões e o Projeto de Drenagem das Fundações da Barragem de Itaipu", Proc. 14th National Seminar on Large Dams, Recife, Brazil.

4. SOUZA LIMA,V.M.; ABRAHÃO,R.A.; PINHEIRO, R.; DEGASPARE,J.C.(1982);"Rock Foundations with marked descontinuities. Criteria and assumptions for stability analyses" (1981). Proc. 14th ICOLD, Rio de Janeiro, Brazil.

5. SOUZA LIMA, V.M.; ABRAHÃO, R.A.(1981) "Two pratical examples of numerical approaches for solving discontinuity problems in dam design". Simposium of Implementation of Computer Procedures and Stress-strain laws in Geot. Eng., Chicago, EUA.

6. MORAES, J.: VILLALBA, J.R.; BARBI, A.L.; PIASENTIN, C.(1982) "Subsurface Treatment of Seams and Fractures in Foundation of Itaipu Dam", 14th ICOLD, Rio de Janeiro, Brazil.

7. SILVEIRA, J.F.A.; MIYA, S. e YENDO, M.(1978) "Geomechanical Parameters Computed from Instrumentation Measurements at Água Vermelha Dam Foundations", Proc. Int. Symp. on Rock Mechanics Related do Dam Foundations - ISRM-Rio de Janeiro, Brazil.

ANALYTICAL EXPERIENCE ON THE STABILITY OF THE HIGH DAM FOUNDATION OF THE LIU-JIA-XIA HYDRO-POWER STATION

Expérience analytique sur la stabilité des fondations du barrage poids de la centrale hydraulique de Liu-Jia-Xia

Erfahrungen bei der Stabilitätsanalyse des Fundamentes des Hochdammes des Liu-Jia-Xia Wasserkraftwerkes

Fu Bin-Jun
Research Institute of water Conservancy and Hydro-electric Power

Zhu Zhi-Jie
Institute of Geophysics, Academia Sinica

Li Guang-Zong
Chengdu Design Institute, Ministry of water-conservancy and Hydro-electric Power

SYNOPSIS

This paper describes the stability analysis of the main concrete gravity dam of the Liu-Jia-Xia Hydro-Power Station on the Yellow River in China. It contains the geological analysis of the regional tectonics and structure of the dam site, test results of the rock mass, foundation treatment, determination of the design parameters, long term monitoring data etc. Using the monitoring data, a back analysis of the dam stability is carried out.

RESUME

Cette communication rend compte des résultats d'analyses de la stabilité du barrage poids principal de la centrale hydraulique de Liu-Jia-Xia. Elles portent sur les structures géologiques de la région et du site du barrage, sur des tests de mécanique des roches, sur le traitement des fondations, sur le choix des paramètres du projet et sur les données de contrôle à long terme. En se basant sur les données de contrôle on a effectué une rétro-analyse de la stabilité du barrage.

ZUSAMMENFASSUNG

Die vorliegende Arbeit behandelt Stabilitätsanalysen des Hauptstaudammes des Liu-Jia-Xia Wasserkraftwerkes. Sie analysiert die Tektonik des Gebietes und die Struktur des Dammplatzes, prüft die Felsmechanik, die Beschaffenheit des Fundaments, die Wahl der Parameter und die Langzeitbeobachtung. Die Ergebnisse der Beobachtung des Dammfundaments benutzend, wird eine Rückanalyse der Stabilität des Bauwerkes durchgeführt.

1. INTRODUCTION

The Liu-Jia-Xia Hydropower station is a huge key project situated on the upper reaches of Yellow River. The main dam is of concrete gravity type with a maximum height of 147 M, and a total generating capacity of 1.225 million kw. The main dam is built on the Pre-Sinian micaceous quartz schist with high compressive strength and low permeability. For justifying the stability of the large dam, according to the principle put forward by Prof. Tan Tjong-Kie, i.e. the study on the stability of rock mass should be carried out on the basis of understanding its previous history, present situation and future behavior, the following works have been performed.

2. GEOLOGY

2.1 Brief description on regional geology

On the basis of existing regional literatures we know that the dam site is situated in the southeast part of the central uplift belt of the Qilian Geosyncline. From Palaeozoic to Mesozoic, this region was a broad maritime area, where sediments of corresponding ages were deposited. The Qilian mountain range was a Caledonian fold zone. It stretches northwest with the same direction of the subduction belts and enormous crustal fractures. Since Palaeozoic up to now, the regional north west tectonic line possesses the characteristics of tectonic inheritance. New fractures have always traced the old ones.

Since late Tertiary the China Plate has been subjected to the pushing forces of Pacific Plate from the east and India Plate from the southwest; simultaneously constrained by the Siberian Plate in the north and Philippine Plate in the southeast. The stress trajectories in China Plate under this boundary condition run as the dash lines shown in Fig.1. This trend of stress distribution has also been proved by the analysis of earthquake data. The result of fault plane solution for 173 shallow earthquakes occurred during the past 40 years (1937-1977) in China show that the horizontal maximum principal compressive stress in our region stretches NE(indicated by black arrows in Fig.1), which is in accordance with the above mentioned trajectories. Under the effect of this stress field, most of the big faults in this region repeated again along the old fractures with a left lateral dislocation during the tectonic movements.

Using the method of remote sensing for interpreting the linear structures around the reservoir area, it shows that the tectonic feature is in consistent with the geocynclines (Fig.2). The faults F_2 and F_9 which control the tectonic system around the reservoir area sheared with a left lateral feature, consequently caused a series of compressive-shear faults (f_3, f_5, f_{17}, f_{18} etc) and tensile faults (f_{10}, f_{11} etc) existing in the earth block between F_2 and F_9. The forming mechanism is illustrated in Fig.2-a.

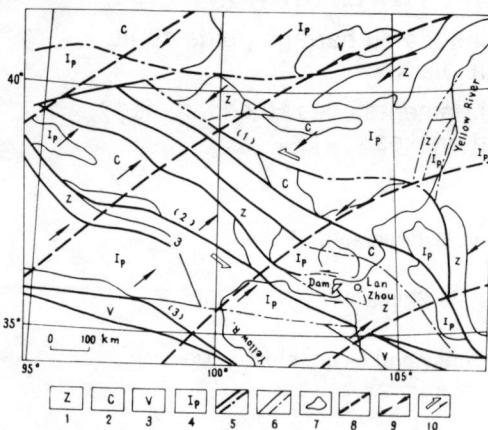

Fig.1 Neotectonic stress field and activity of fractures in region of Qilian Geosyncline 1-Pre-Cambrian 2-Caledonian 3-Varican 4-Post-Indosinian 5-Crustal fractures 6-Secondary fault 7-boundary of strata 8-max. principal stress trajectories 9-max. principal stress obtained by focal mechanism solution 10-sliding direction of fault wall.
(1), (2)-Palaeozoic subduction zone (3)-Mesozoic subduction zone.

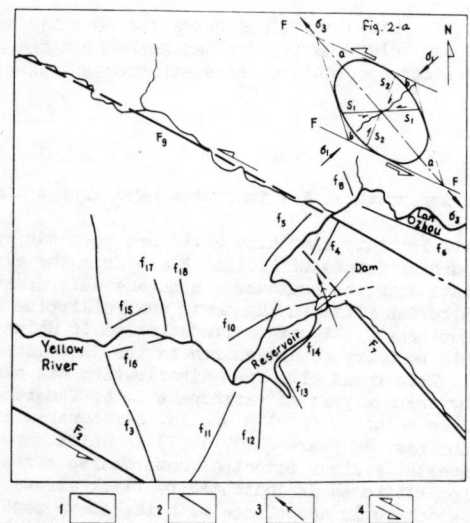

Fig.2 Map of linear structures around the reservoir area (interpretation of remote sensing) 1-primary fault 2-secondary fault 3-Hong-Liu-Gou anticline 4-direction of shearing

2.2 Geological condition of the dam site

The main dam is situated in a narrow v-shaped gorge with steep bank slopes. The bedrock is predominately composed of micaceous schist and hornblende schist of Pre-Sinian which were partly intruded by a small amount of granite veins and lamprophyre dykes.

The characteristic figures for the schist are as follows: sound and slightly weathered schist $E=370 \times 10^3 kg/cm^2$ (in situ), compressive strength $\sigma=1300 kg/cm^2$; moderate weathered schist $E=18 \times 10^3 kg/cm^2$, $\sigma=1000 kg/cm^2$.

The geological structure of the dam site is governed by a slightly plunged anticline-Hong-Liu-Go Anticline with its axial plane striking NNW which is a secondary structure of fault F_9, occurred during Tertiary. Detail geological mapping is shown on Fig.3.

Note that there are five sets of discontinuities distributed in the dam site.

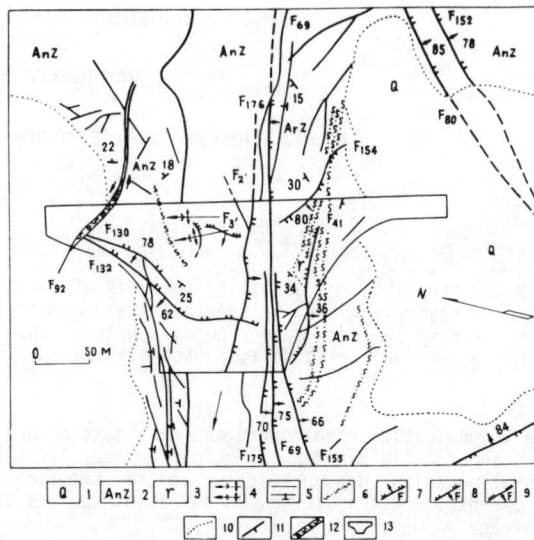

Fig.3 Geological map of the dam area
1-loess 2-micaceous quartz schist 3-granite vein 4-fold axis and dips 5-tectonic and bedding fissure 6-tectonic crushing belt 7-normal fault 8-reverse fault 9-thrust 10-boundary of strata 11-dip and strike of beds 12-fault zone 13-outline of the dam.

Figure 4 shows the statistical results of the discontinuities mapped at the dam site. It is obvious that all of them are of high dip angles(more than 60°).

The forming period of all these fractures distributed in the dam area can be traced to Palaeozoic or even earlier. However the last activity of these fractures according to the field observation of F_{69} (tensile fault) which cut through the Red Beds of Tertiary, but is overlain disconformably by undisturbed loess of Q_3(150 thousand years from recent), demonstrates that the fore-mentioned fractures were moved again by the combined force of Pacific Plate and Indian Plate during Tertiary. This force became weaker since late Pleistocene. So it might be inferred that under the effect of further earthquake, the occurence of new fault movement could be out of consideration.

For the purpose to study the stability of the dam, a detail analysis of the possibility of deep sliding failure is carefully discussed. The conclusion is that the sliding could only occur at the contact zone between the concrete dam and rock foundation.

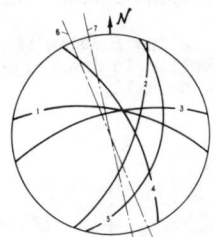

1-shearing fractures
2-shearing fractures
3-tensile fractures
4-compresive shearing fractures
5-bedding compressive shear zone
6-direction of the axial plane of Hong-Liu-Gou anticline
7-direction of the dam axis

Fig.4 Equatorial plane project of the discontinuities in the dam site (by means of the projection of lower hemisphere)

3.ROCK MECHANICS FIELD TESTS

For justifying the slide resistance of rock masses, 13 groups of in-situ direct shear tests with concrete blocks cast on rock surface were performed.
Only for the tectono-clastic rocks there happened the deep shear failure within the rock mass. All of the others were characterized by plane shear failure along the contact zone of concrete blocks and rocks with the behavior of brittle fracture and rather high shear resistance. Taking example of the test on slightly weathered micaceous quartz schists, the coefficient of friction $f=1.27$ and cohesion $c=10.59 kg/cm^2$ for the initial tests and $f=1.09$, $c=2.47 kg/cm^2$ for the repeatative tests. The average shear deformation, roughly corresponding to proportional limit τ_1 is found to be 0.035 mm apporoximately whereas the average deformation correspsnding to peak value τ_2, being 0.303 mm. The ratio of τ_1 to τ_2 is estimated at about 80%.

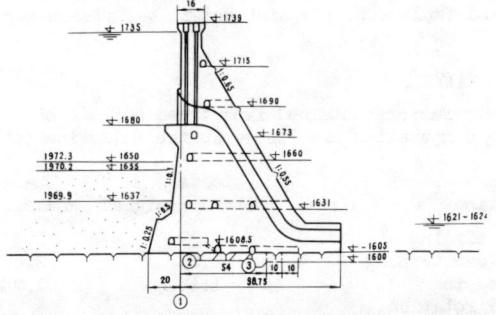

Fig.5 The cross section of the main dam and sediment deposit 1)Dam axis 2)Drainage galleries 3)Water sump

The design parameters against sliding was selected by considering primarily the test results of the slightly weathered micaceous quartz schist.Besides,other factors such as geological conditions of the excavated foundation, the foundation treatment etc. were also taken into consideration.
Finally, design parameters $f=0.80$, $c=0$ for the dam blocks I-VI and $f=0.75$, $c=0$ for the blocks VII-X were adopted. The cross section of the typical dam blocks is shown on Fig.5.

4.FOUNDATION TREATMENT (Fig.6)

4.1 Excavation

The foundation of all dam blocks was excavated to sound rock masses. At the left bank, the bedrock under the dam blocks IX and X was cut back 20 to 25 m in avoiding

of the fractured zone.

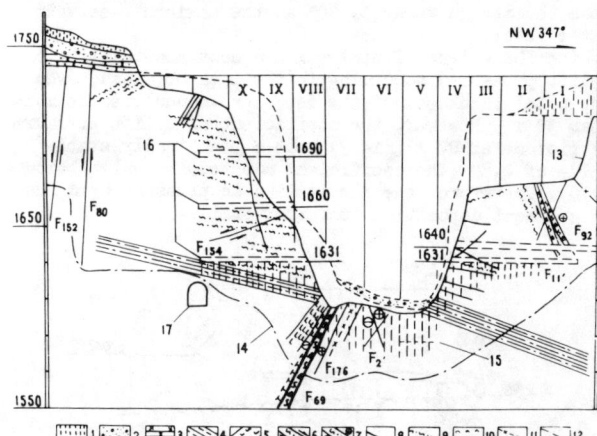

Fig.6 Longitudinal section showing the treatment of foundation for the main dam. 1-loess; 2-sand gravel; 3-Reddish sand stone; 4-Micaceous quartz schist; 5-Lamprophyre vein; 6-Normal fault and its crushed zone; 7-Thrust fault and its crushed zone; 8-Tectonic fissure; 9-Tectonically crushed zone; 10-Tectonically influenced zone; 11-Line showing the boundary line of weakly weathered and moderately weathered rocks; 12-Original groudsurface; 13-Excavation Line; 14-High pressure grout curtain; 15-Drainage well curtain; 16-Drainage gallery; 17-Diversion tunnel.

4.2 Grouting and drainage

One row of deep grout curtain of cement was provided for dam blocks II to X with the depth of 50 m. On the left bank, another row of cement grouting curtain with depth ranging from 80 to 120 m was arranged. In those parts of the foundation where fractures developed and rocks moderately weathered, consolidation grouting was performed. Down stream of the grout curtain, 3 rows of longitudinal drainage galleries and 2 rows of transverse ones were provided.
In this manner, comprehensive measures of grouting and drainage, with drainage playing the primary role were performed.

4.3 Treatment of fault

Stress was put on the improvement of F_{69}. A vertical shaft 3m×4m in cross-section and 15 m deep was excavated and back-filled with concrete to serve as a cut-off wall, under which a deep grout curtain 36m deep was set up. A trench was excavated in the fault zone, also back-filled with concrete to form a plug. Cement grouting for consolidation to depth of 15m was also carried out on both sides of the fault.
Further more, after reservoir impounding the measure of grouting for the transverse joints between the dam blocks was accomplished; thus the unfavourable condition of blocks on left abutment could be improved and the aseismatic ability to earthquake of the dam was raised as the dam could work as a whole.

5. MONITORING DATA

Long term observed data was obtained by means of monitoring instruments. Concerning the dam stability, following conclusions might be drawn:

5.1 Uplift pressure in the dam foundation

As shown on Fig.7 the total uplift pressure is less than the design value by 40% at the maximum reservoir water level.

As for the effect of upstream sediment accumulation on uplift pressure, according to the data obtained, even the total thickness of the deposits accumulated is more than 50 m , however, the coefficient of uplift pressure at piezometer U7-1 (Fig.7) kept a practically stable value of 0.84 (the coefficient adopted in design being 0.70). Therefore, the dissipation in pressure head due to sediment deposition is rather small.

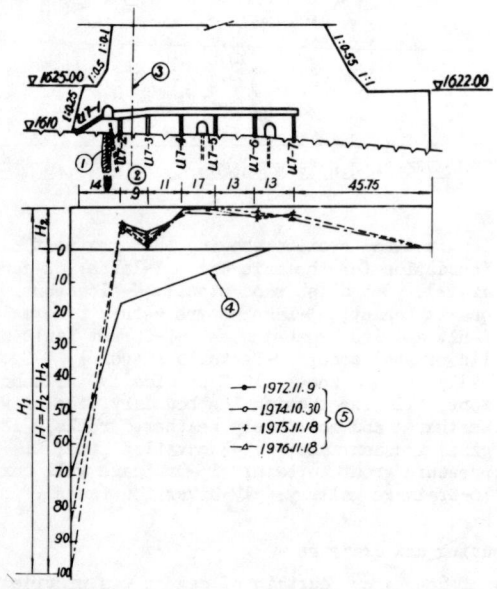

Fig.7 Distribution of uplift pressure in block VII of the main dam. U_{7-1} to U_{7-7} Number of piezometers 1)Grout curtain 2)Drainage curtain 3)Dam axis 4)Designed pressure 5)Measured pressure.

5.2 Working condition of the dam with the rock masses of the abutments and the fault F_{69}.

From Fig.8, it can be seen that the dam blocks V,IX on the abutments and the block VII on the fault zone of F_{69} are tightly combined with the bed rock.

5.3 The stresses in the heel zone of the dam

During operation the compressive stresses in the heel zone of the dam vary within the range of 15-20kg/cm² .

5.4 The deformation observation

The deformations concerning horizontal displacement, settlement, deflection and inclination were found to be rather small. The measured maximum settlement of the dam during the period from 1967-1978 varied in the range of 4-5 mm, which was characterized by time-effect and irreversibility. The horizontal displacement on the top of the typical dam blocks V, VII, IX varied periodically by year. The maximum values were measured to be 6-8 mm towards downstream in winter seasons during reservoir filling and to be 0.2 mm reversed in summer

seasons during reservoir emptying. They have been varing steadily without any sudden changes.

Other monitoring informations, such as the results of stress state and temperature distribution in dam proper also show that it has been operating normally.

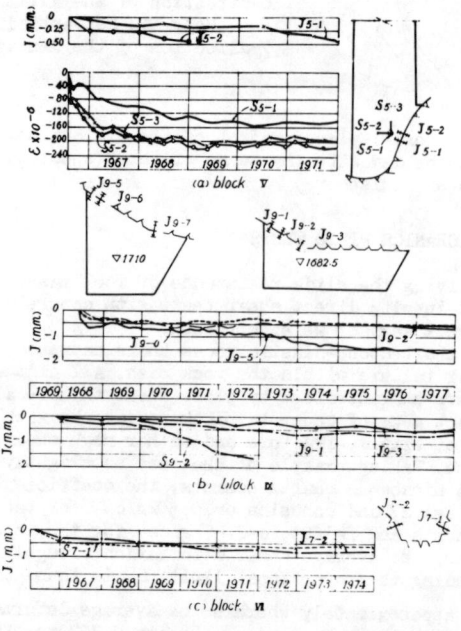

Fig.8 Combination of the dam with rock abutments and fault zone J-joint meter S-stress meter

6. BACK ANALYSIS

The safety factors obtained from three methods of checking computation are listed in the following table.

Ckecking computation	Loads normal designed	loads measured
Against sliding	1.37	1.53
Against overturning	2.23	2.63
According to critical rotation	1.64	1.75

Obviously, under the actually measured loads, the safety factor against sliding is 11.6% higher than that of designed. The finite element method has also been used to analyse the stress state of the dam and foundation.

A brief summary:

1. Based on above mentioned data, the main dam of Liu-Jia-Xia Hydropower Station could be considered being in good work condition.

2. It is of great importance to carry out the stability analysis with the method of geomechanics in combination with rock mechanics.

3. The stability against sliding for the dam has been much increased due to the decresing of uplift pressure primarily,So it is of vital important to ensure the normal working condition of the seepage control measures.

4. In reducing the uplift pressure , the effect of upstream sedimentation is not obvious for weak permeable rock formation.

ACKNOWLEDGMENTS:

Sincere thanks are due to Miss Liu Tse Fen, Wen Xiao Lei amd Mr. Wu Shang Yang for their help in drawing the figures. We are in debt to Mr. Jiang Guo-Cheng, Huang Ren-fu, especially Mr. Mei Jian-Yuen for their valuable suggestions.

REFERENCES:

Hsu Tseng-yen, Ho Sui-Hsin and Fu Ping-chun, 1976, Control of seepage through the dam foundation at the Liu-chia-hsia Hydropower Station, Scientia Sinica, Vol. XIX No. 5

Li Chunyu, 1980, A Preliminary study of Plate Tectonics of China, Bulletin Chinese Aced. Geol. Sci, Series I, Vol.2 pp. 11-12

Tan Tjong Kie, 1982. Crucial is the Correct Conception, Engineering Geology and Hydrogeology Vol.2 pp.5-10. 1982. On the Geodynamics of China Plate, Scientia Sinica, in Press.

Yan Jia Quan et. al., 1979.Some Features of the Recent Tectonic Stress Field of China and Environs, Acta Seismologica Sinica, Vol.1, No.1, pp.9-24.

SLIDING STABILITY OF FOUNDATION ROCK WITH SHEAR ZONES

Résistance au glissement des roches de fondation comprenant des zones broyées

Die Gleitsicherheit von Gründungsfelsen mit Scherzonen

Xu Lin Xiang
Engineer, Project Design Division

Gong Zhao Xiang
Senior Engineer and Head, Structure Research and Experimental Laboratory

Lin Wei Ping
Enginerer and Deputy Head, Rock Mechanics Service, Yangtze Valley Planning Office
(YVPO)

SUMMARY

At the beginning of this paper the authors give a brief description of the main geological conditions of the foundation rock under the second-channel spillway of the Gezhouba Project built across the Yangtze. They subsequently describe a comparative study of measures taken to improve the stability against sliding along the shear zones which lie several metres below the structure of the spillway gate. Finally, the degree of safety against sliding is evaluated by means of mathematical and physical models.

RESUME

Les auteurs commencent par présenter brièvement l'environnement géologique général des roches de fondation des vannes de décharge de la seconde voie d'eau du complexe hydraulique de Gezhouba construit sur le Yangtse. On décrit, ensuite, les études comparatives sur diverses mesures de stabilisation destinées à prévenir le glissement des intercalations argileuses. Enfin, on évalue la limite de sécurité de résistance au glissement au moyen de modèles mathématiques et physiques.

ZUSAMMENFASSUNG

Die Autoren geben am Anfang des Beitrages eine kurzgefaßte Darstellung der hauptsächlichen geologischen Verhältnisse im Gründungsfelsen des zweiten Hochwasserentlastungskanals im Projekt Gezhouba, das quer über den Yangtze gebaut wird. Dann beschreiben sie die vergleichenden Studien der getroffenen Maßnahmen zur Erhöhung der Stabilität gegen die Abgleitung entlang der Scherzonen, die einige Meter tief unter dem Überlauf liegen. Schließlich wird der Sicherheitsgrad gegen Abgleiten mit Hilfe von mathematischen und physikalischen Modellen beurteilt.

The Gezhouba Project is the first multiple purpose project built on the main stem of the Yangtze. Its first-stage construction, comprising a 27-bay Spillway with a maximum flood discharging capacity of 84,000 cu.m. per sec, a Second-channel Powerplant housing 7-turbogenerator units with an aggregate capacity of 965 MW, a 27m-lift large ship lock with chamber dimension of 34m wide, 280m long, 27m-lift medium sized ship lock with chamber dimension of 18m wide, 120m long, and a 6-bay silt-scouring sluice, has been in operation since Jan.1981; and its second-stage construction is progressing successfully.

The foundation of the project site is a series of cyclothem, terrestrial and clastic rock of the lower Cretaceous period. The lower part of the formation is conglomerate, and interbeds of sandstone, siltstone and clayey siltstone dominate the middle and upper parts with intercalated layers of claystone or silty claystone. The thickness of the clay beds varies greatly, from a few to scores of centimeters. Some of the clay beds spread out extensively, and others exist with varying sedimentary facies, sometimes as lenses in thicker strata. With the exception of conglomerate, the bedding planes are poorly cemented, the strength of the rocks is low and deformability high.

After the deposition of the red series in the Cretaceous period, the region of the project had experienced for several times the influences of tectonic movements, with large scale faults, striking NW and NNW, flanking the east and west side of the site, about 20 to 40 kilometers away. But only slight tectonic features can be traced at the site and its vicinities, such as small syclines and anticlines. The beds of the foundation rock strike N 20° to 40° E, dip 4° to 8° SE. The main structural fracture is characteristically signified by interstratal dislocation, and consequent action of groundwater, of a few clay beds. As revealed by careful exploration, there is such a shear zone, designated as No. 202, distributed extensively under the Spillway gate structure. For the purpose of building safe, economic, lasting hydraulic structures on such a rock mass, a series of comprehensive and in-depth studies have been initiated. The focus of the study has been placed on the following two aspects:
(1) Study of major geotechnical problems, both

at field and in laboratory, and (2) Study of alternative measures for enhancing the stability against sliding along the shear zone several meters underneath the spillway gate structure, by physical and mathematical models.

Study of Major Geotechnical Problems

as aforementioned, shear zones are the main stuctural feature of the foundation rock mass. The low shear strength of th e shear zone makes it a key to the settlement of the question of stability against sliding. Upon the following points the studies are thus centred.
(1) The trend of change of the shear zones du ring long-term operation of the project after impounding, (2) The proper shear strength of the shear zone, and (3) The failure mechanism of the rock mass, comprising thin and weak beds and shear zones, in the immediate vicinity of the downstream toe, acting as resistant block, of the Spillway gate structure.

The conclusions drawn from these studies were the contents of a written discussion contributed to the 4th Congress, ISRM, 1979, and published in volum 3 of the Proceedings of that Congress.

Alternative Measures for Enhancing Stability Against Sliding

It was found out by preliminary analysis, with conventional limit equilibrium method as stated below, of stability against sliding of the spill way gate structure along the shear zone directly underneath, that the factor of safety of the left most five bays of the gate structure did not meet the criterion, 1,30 for normal operation condition and 1,10 for extreme cases, set by the Design Code. The sketch in Fig 1 shows the loads and their distributions, and also the possible sliding plane which followed first the main ship plane of the shear zone and then cut through the resistant block along the assumed first plane of rupture.

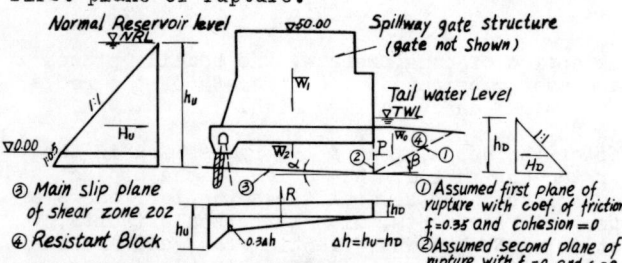

Fig.1 — Sketch showing Loads acting on the Spillway gate structure and assumed planes of rupture within the Resistant Block.
(for stability . analysis by Conventional Limit Equilibrium Method)

The factor of safety against sliding, Fs, can be estimated by the following formula:

$$Fs = \frac{f[(\Sigma W - R)\cos\alpha - \Sigma H \sin\alpha + P\sin\alpha] + P\cos\alpha}{\Sigma H \cos\alpha + (\Sigma W - R)\sin\alpha} \ldots (1)$$

in which, ΣH, ΣW, R, f, α, P denotes respectively the total hydrostatic thrust, total weight above the main slip plane including weight of rock mass above it, uplift acting on the main slip plane, friction factor of the shear zone (i.e. of the main slip plane), dip angle of the main slip plane, and resisting force offered by the resistant block.

As it is known that the stabilizing effect of the resistant block plays an important role in stability against sliding, and that it is probable that the resistant block composed of thin and weak layers would deform greatly under hydrostatic thrust, so that estimation of the resisting force, P, must be made on the safe side. The following assumptions are, therefore, proposed for defining the planes of rupture within resistant block (or the boundary surfaces of the resisting wedge): (1) the first plane of rupture tracks the most unfavourable sets of fissures and joints, say, with as the inclination angle of the plane, and takes only the probable lowest friction factor, f, or tg as the shear strength to be mobilized during the resisting force, P, coming into action; (2) there is no shear resistance on the second plane of rupture. The resisting force, P, can therefore be obtained as follows:

$$P = Wo \ Tg(\beta + \phi) \ldots (2)$$

Where Wo is the weight of the resisting wedge.

Following alternative measures were proposed for comparative studies:
(1) Deep cutoff: The main feature of the option is to deepen the existing cutoff, formed by the foundation gallery at the upstream heel of the gate structure, to a depth about 4 meters below the shear zone considered most vulnerable so far as stability against sliding is concerned. Its effectiveness lies in its capacity of direct utilization of the resistance against sliding offered by the huge rock mass underneath the gate structure. Its disadvantages overtake its merits. Not only deep trench construction makes difficulties unsurmountable, but also the harmful rebound resulting from deep excavation seems detrimental. Therefore, the deep cutoff option is not satisfactory. (2) Bored pile foundation: The idea was to build the gate structure on latge diameter, say, 4m or so, bored piles. Both onsite trial boring test and statical model studies in laboratory had proved it a promising scheme. But the option had to be abandoned since there were no such a large number of large diameter boring rigs available then to meet the number required to exsecute the scheme in tight construction schedule. (3) Strengthening of resistant block: As aforementioned that the downstream part of the left-most five bays of the gate structure is situated in a region with relative worse geological conditions, it deemed necessary to reinforce the resistant block in order to tie the thin and weak layers down to the harder sandstone stratum with bored reinforced concrete piles for accomplishing dual purposes of preventing buckling in the first place and of improving shear resistance in the next. Bored piles of two different diameters, one of 800mm and the other of 219 mm, were cast in place in an alternate pattern. The scheme is taken as a supplementary measure. And (4) Concrete impervious blanket: An impervious concrete blanket is placed in front of the gate structure with a foundation gallery construc ted at its upstream end. The blanket is effective to reduce theuplift acting on the shear zone and thus increase the effective weight above it, and, in turn, the stability against sliding is raised. The length of the blanket required to conform to the Design Code Standard if 30m. The advantages of this option are mainly two: first, it is

much easier to place a concrete blanket on grade: second, it incorporates the function of scour protection, so it is much cheaper as compared with other alternatives. The fact that the blanket is more than 30m under the normal reservoir level brings out significant deficiencies in the lime light: it is difficult to repair if there occurs some unexpected defects such as excessive seepage detrimental to normal functioning of the blanket. Preventive cures should be taken in advance to kill defects, and can probably be averted through careful supervised quality control of the blanket during construction. It is evident that the advantages out weight the disadvantages. Concrete impervious blanket is therefore selected after prudent considerations as the key measure.

Evaluation of Degree of safety against sliding with Mathematical and Physical Models

Studies with the aid of both finite element method and geomechanical models can bring out a more clearer view about the degree of safety against sliding along a deep-seated shear zone.

Comparative Studies of the Alternative Measures by Geomechanical Models:

The importance of in-depth understanding of the sailient features of the foundation rock and its behaviou under external loads can not be overemphasized for the success to tackle the stability problems. Simulating the major geological condition characterizes the geomechanical model as one of the most suitable tools for studying stability problems. For constructing geomechanical models, special materials of different compositions for simulating various characteristics of the foundation rock, complying in full with similitude, had been used. The models were fabricated according to the simplified geological profile as shown in Fig 1 (with and without the assumed planes of rupture depicted). In the following Table, the test results from comparative studies are condensed.

Alternative strengthening measures			Mode of failure		Relative Merits%
Serail NO.	Description	Condition of Resistant Block	Description		(Referred to ultimate Failure)
1	No Strengthening Measure	Without Assumed Plane of Rupture	Sliding along Shear zone No.202		100
2	Deep Cutoff		ditto		120
3	Resistant Block strengthened with Bored piles		Sliding along Shear zone No.202 until bored piles fractured by shearing		150
4	Impervious Blankt		Sliding along Shear zone NO.202		108
1	No Strengthening Measures	With Assumed Plane of Rupture	Sliding first along Shear zone No.202 and then following the assumed plane of ruptuye		100
2	Deep cutoff		ditto		129
3	Resistant Block strengthened with Bored Piles		Sliding along a weak layer first and then following the assumed plone of rupture		129
4	Impervious Blanket		Same trend as those of alternatives NO.5 1 and 2		115

Conclusions can be reached careful perusal of the Table, that the concrete impervious blanket option coupled with strengthened resistant block is reliable.
Fig 2

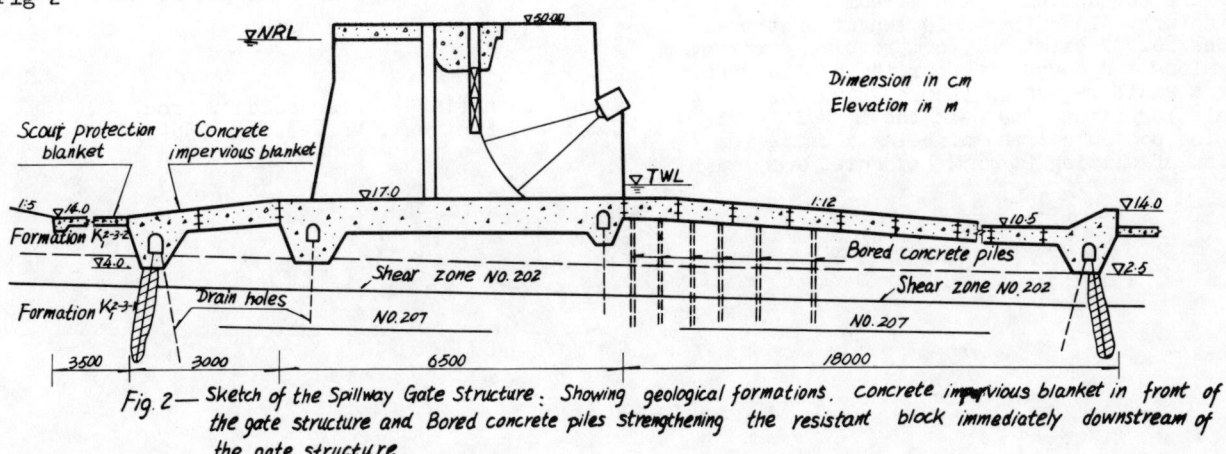

Fig. 2— Sketch of the Spillway Gate Structure. Showing geological formations. concrete impervious blanket in front of the gate structure and Bored concrete piles strengthening the resistant block immediately downstream of the gate structure.

Study of Degree of Safety against Sliding of the Adopted Scheme by Finite Element Method.

Conditions set out for analysis: The problem is treated as a 2-D one, with uplift taken from 3-D electric analogy model. The mesh for the foundation composes of 350 triangular elements and 60 Goodman jiont elements for the shear zone. The simplified geological profile depicted in Fig 1

is followed. The beneficial effects of bored piles in the resistant block have been purposely left as a safety margin. Essential results have been compiled in following Tables and Paragraphs.

Fig 3

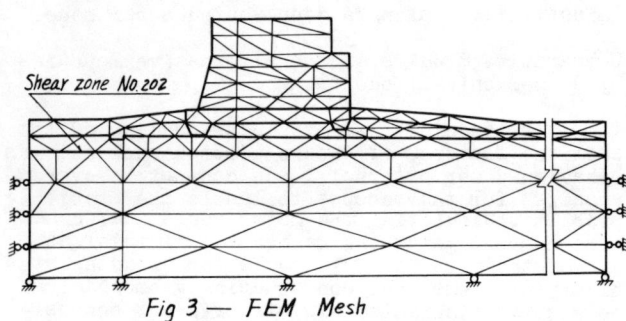

Shear zone No.202

Fig 3 — FEM Mesh

(1) Percentages of hydrostatic thrust shared by various parts of the foundation under normal operation conditions. (in %).

Location	Shear Zone underneath Impervious Blanket	Underneath Gate Structure	Resistant Block	Total
% shared	29.3	36.4	34.3	100

(2) Factors of Safety against Sliding, Fs. Where assumed plane of rupture is not in existence in the resistant block, Fs=2.46, with friction factor, f=0.20, for the shear zone No.202 is considered only; and in case there exists in the resistant block an assumed plane of rupture, fs'= 2.19, with friction factor, f=0.20 and f_1 =0.35 considered for the shear zone No.202 and the assumed plane of rupture respectively.

(3) Stress Distributions and Effects: The stresses in the foundation under the gate structure are all compressive ones except there is a tensile-region with small tensile stress occurred in the part of the foundation just upstream of the gate structure. Along the whole lenght of the shear zone No.202 exist only compressive stresses, acting, along the downstream portion of the shear zone, on a small region in the resistant block is relatively large than those in the immediate vicinity. The possible trend of being a cause for initiation of sliding is worthy of note, even though

their absolute values are not very significant if the peak values of shear strength of the shear zone obtained by conventional test method are taken as criterion for evaluating the probability of sliding. Further studies show that this very region extends to surrounding areas gradually and encompasses very limited area (which is still far away from the grout curtain) even under an assumed extreme condition with the friction factor halved.
It can be said with certainty that the scheme is safe.

Conclusions
1. Sliding stability of the spillway gate structure hangs on the sliding resistance offered by the deep-lying, gently dipped downstream shear zones intercalated in the foundation rock. The advantage of stabilizing effect of the resistant rock block lying immediately downstream of the gate structure should be taken of in sliding stability analysis. Safety can be raised by strengthening the resistant block composed of thin and weak rock formations with bored concrete piles.

2. Degree of safety against sliding should properly be evaluated either with non-linear FEM or geomechanical model testing.

3. The Spillway gate structure behaves, as evidenced by measured control variables such as horizontal displacements and vertical settlements, so far, much better than expected in operation since Jan.1981.

The authors are grateful to Prof. Dr. Tan Tjong Kie and the late Prof. Gu De Zhen, of Institute of Geophysics and Institute of Geology, Academia Sinica, respectively, for their constructive suggestions in the couse of implementing the field tests and numerical analysis. Appreciation is extended to Tsao Lo An of YVPO for reviewing and proposals for editing the paper.

Reference:
Tsao, Lo an,(1979), Discussions, Proc. 4th ISRM Congress, Montreux, Vol. 3, pp. 256-257

FOUNDATION OF A RADIO-TELESCOPE AT AN ALTITUDE OF 3000 m

Fondation d'un radio-télescope situé à une altitude d'environ 3000 m

Gründung eines Radio-Teleskopes in fast 3000 m über dem Meeresspiegel

H.-U. Werner
Director, Dorsch Consult Institute for Soilmechanics and Foundation Engineering,
München, Germany

SYNOPSIS:

The safe erection of the polygonal reinforced concrete tower for a 30 m radio-telescope in the Sierra Nevada, Spain called for a subsoil improvement of the weathered and highly fissured rock mass (metamorphic schist). Adoption of cement grouting with steady intensive controls proved to be an adequate measure. A concept was developed to check, by indirect means, that the dynamic modulus of elasticity of the rock in the foundation area, required by the structural analysis, could be achieved. The paper describes the geotechnical investigations, the grouting works, the parameter studies, the control tests and the essential results obtained.

RESUME

Pour la fondation d'un radio-téléscope de 30 m, établi sur une substructure polygonale en béton armé, il a fallu améliorer la roche. Il s'agit d'un schiste métamorphique altéré et très fissuré. La stabilisation du sous-sol a été réalisée avec des injections successivement contrôlées de ciment. On a développé une conception qui a permis la détermination indirecte du modulus dynamique d'élasticité de la roche améliorée, requise par l'analyse statique pour l'ouvrage. L'auteur décrit les investigations géotechniques, les travaux d'injection, les études paramétriques, les essais de contrôle et les principaux résultats obtenus.

ZUSAMMENFASSUNG

Um den polygonalen Unterbau für ein 30 m Radio-Teleskop in der Sierra Nevada, Spanien, sicher gründen zu können, mußte der anstehende verwitterte und stark zerklüftete Fels (Glimmerschiefer) im Gründungsbereich verfestigt werden. Zur Untergrundverbesserung wurden Zementinjektionen mit laufender Erfolgskontrolle durchgeführt. Mit Hilfe eines besonderen Konzeptes konnte indirekt überprüft werden, ob das behandelte Gebirge über den gemäß Statik geforderten dynamischen Elastizitätsmodul verfügte. Der Bericht zeigt die geotechnischen Untersuchungen auf, beschreibt die Injektionsmaßnahmen, Parameterstudien und Kontrollversuche und gibt zusammenfassend die wesentlichen Ergebnisse aller Arbeiten wieder.

1. INTRODUCTION

The construction of an institute for radio astronomy in the millimetre range (IRAM), with a telescope of 30 m in diameter (Fig. 1), is the objective of a joint German/French project in the Sierra Nevada, Spain. Both the institute and the radio-telescope are located on the Loma de Dilar at an altitude of almost 3000 m.

A reinforced concrete tower serves as substructure for the steel construction of the telescope. The lower part of this substructure, the foundation of which is described here, is 10 m high and has a polygonal floor area with an outer diameter of 17 m; its upper part is designed like a frustum of a pyramid. The adjoining technic-tract is underground so as to meet the demands of not interfering with an existing ski slope.

The original location on the Pico Veleta, at an altitude of 3,400 m, had to be abandoned. On the one hand, the expected intensive build-up of ice on the telescope, along with extreme wind velocities and the fact that snow conditions would permit free access for only three months of the year, would have led to high construction and maintenance costs. On the other hand, fissures

running parallel to the mountain ridge and indications of creeping of talus were observed in the originally proposed foundation area.

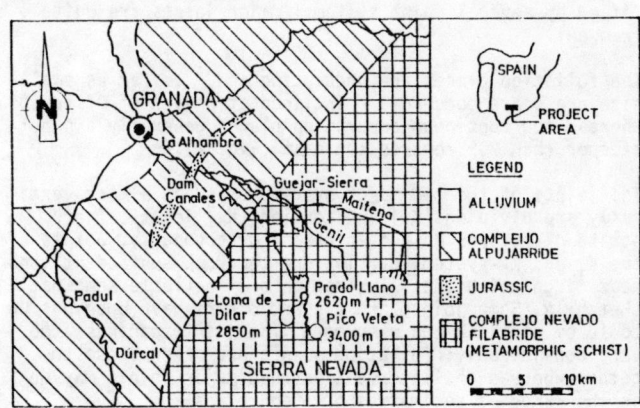

Fig. 1   Location Map

Because of the very purpose of the radio-telescope, the technical demands for the sub-structure were extremely strict. Thus it was stipulated, for example, that the foundation remains practically firm even under the impact of strong winds with gusts of up to 200 km/h. The given theoretically permissible tilting of the foundation was one second of arc resulting in 0,07 mm diametric differential settlement, which is far below the standard construction tolerances. Besides, the chosen site is located in a seismic zone; however, it could not be determined how many of the 208 earthquakes with a strength of I to VIII on the Mercalli scale, registered in the Province of Granada during this century, also affected the Loma de Dilar.

2. THE SUBSOIL

2.1 Investigations

The geotechnical investigations of the chosen site included geological surface mapping. Eleven test pits were dug to determine the characteristics and thickness of the weathered overburden. The rock was explored by means of 13 core borings (diameters 145 mm to 76 mm) drilled to depths of 10 m to 40 m. The heavy losses of drilling water during the operations, indicated the continuation of the open joints detected at the surface within the interior of the rock mass. In view of the partly high degree of weathering, some drilling holes had to be cased to much greater depths than originally anticipated. Despite a core recovery of mostly more than 95% it was hardly difficult to obtain RQD-values, partly so to great depths.

No optical monitoring of the drill holes were made since the expected supplementary information did not justify the cost and, in particular, the time required for such additional surveys.

In compliance with the envisaged technical problems, the modulus of compressibility of the weathering product as well as the compressive strength and modulus of elasticity of the metamorphic shist were determined at the laboratory.

2.2 Characteristics of the rock

In terms of geology, the mountains in the wider neighbourhood of the project area belong according to Melendez and Fuster, 1965, to the metamorphic formations of the so-called "Compleyo Nevado-Filábride" of the "Zona Betica". Petrographically speaking, this is a very monotonous series of quartzitic but also clayey metamorphic shists. These are partly highly weathered and generally interstratified by several joint systems; major joints are quite frequent.

The foliation planes are undulating and - as far as small-size areas are concerned - inclined at angles of 0° to 40°, whereas the continous separation planes generally dip not steeper than 20° towards the east (Fig. 2).

The joints of the two major systems standing almost vertically are dividing (K_1) and tension (K_2) joints. Their strike directions do not intersect orthogonally. Joints of the K_1 and K_2-systems partly form the rock surface at the near ridge which the illustration only reflects somewhat sketchily. Some joints are closed, others have openings up to 10 cm of width and even more. The joint spacings, too, vary considerably: foliation joints are often close together whereas the intervals between major joints may be in the range of decimetres or even metres.

The thickness of the weathered overburden in the wider project area was verified to a depth of 14 m. A shifting in location made it possible to limit its average thick-

ness in the foundation area to 3 m. Because of its origin, the weathered overburden consists mainly of flaky rock fragments whose size varies substantially (mm to dm). But it also contains a high percentage of fine and finest particles, i.e. fine sand and clay.

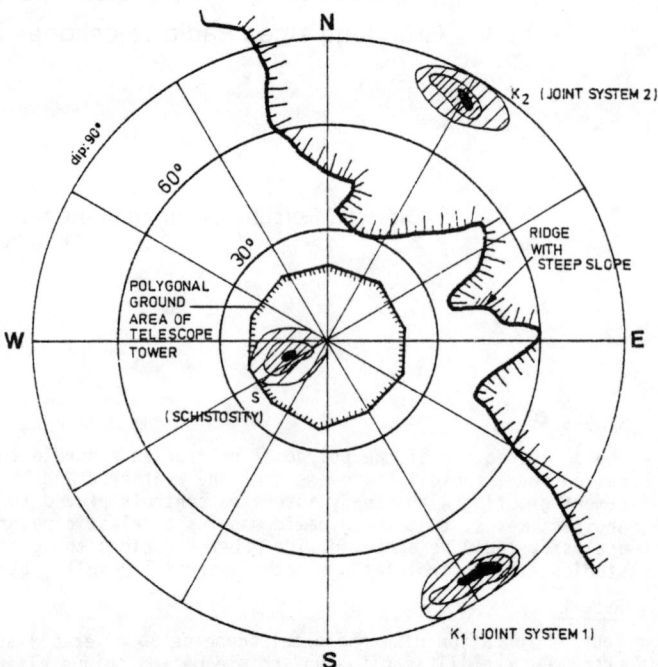

Fig. 2   Schematic layout with main joint systems

Tests on individual samples of the weathered material yielded moduli of compressibility of only 4 MN/m². Depending on the degree of weathering, the static moduli of the rock were found to be between 2,700 and 17,000 MN/m², with a weighted average of about 10,000 MN/m². The tests were conducted with reference to DIN 1048, part 3 (modulus of elasticity) and DIN 52 105 (compressive strength). Following a number of loading cycles, the load was increased until failure occurred. Stepped shear planes could be observed, i.e. the planes of failure were not identical with the foliation planes or joints.

3   SPECIAL DEMANDS REGARDING SUBSOIL

Vibration tests on the structure/subsoil-system yielded a required dynamic modulus of elasticity of the rock mass of E_{dyn} = 4,000 MN/m² in the load propagation area under the ring foundation. According to the results of the laboratory tests, this demands the presence of unfissured rock, which shows only a low degree of weathering. Since the encountered rock masses do not meet these requirements at all, its open joints had to be filled at first by grouting. A complete exchange of the subsoil under the foundations against concrete had to be ruled out because of the sudden change from high to low moduli of elasticity in the contact area soft rock/reinforced concrete. Besides, it had to be checked whether and to what extent the subsoil improvement measures were in fact expected to yield the required characteristics as regards the dynamic elasticity. For this purpose, the following points were considered:

a) The highly weathered shist of the upper rock stratum, with rock values of $E_{dyn} < 8,000$ MN/m² shall be removed and replaced by lean concrete, permitting a tolerance of 20% relative to the mean value.

b) For the grouted area, it is expected that (especially under dynamic loads) there may be:

- no exceeding of rock strength in highly loaded singular junction surfaces;

- no abrupt deformation (e.g. due to earthquake); and

- no "squeezing-out" of plastic joint fillings.

c) The influence of joint-fillings on the characteristics of the improved rock mass was investigated by way of variations of their possible dynamic moduli of elasticity. Such parameter studies showed, depending upon the dynamic modulus of elasticity of the joint fillings and applying, for instance, the average modulus of elasticity of the rock mass of 10,000 MN/m², that the accumulative thickness of the individual joints of the nearly horizontal schistosity per vertical metre of rock mass must not exceed 6 mm, 11 mm or 25 mm, respectively (Fig. 3).

d) Upon completion of the injections, the values "modulus of elasticity/relative thickness of joint-filling" had to be determined by very precise control-borings and on the basis of the shaft, to be excavated at the transition to the technic-tract, so as to prove their compatibility with the approaches of the parameter studies.

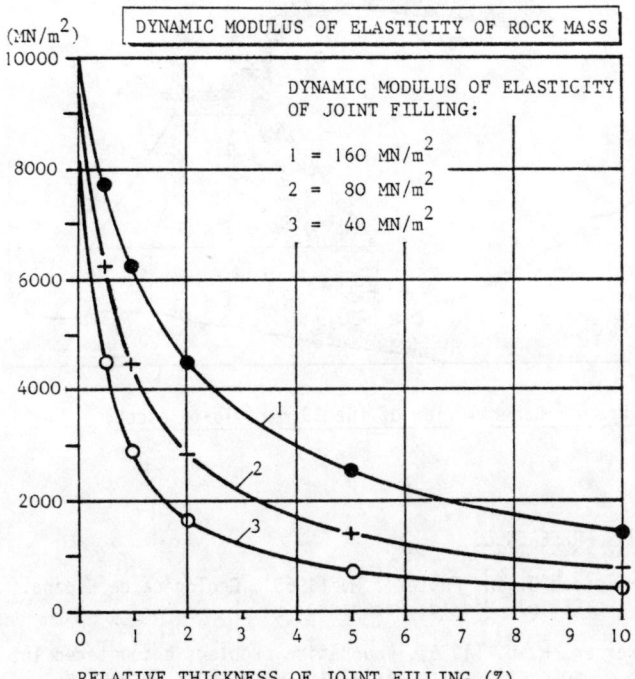

Fig. 3 Influence of soft joint fillings on the modulus of elasticity of rock mass

4. SUBSOIL IMPROVEMENT BY WAY OF GROUTING

The grouting work (injections) had to fill all voids around the ring foundation and to flush or at least enclose the existing joint fillings. The rock mass to be grouted consisted of a 5 m high ring with trapezoidal cross-section and a diameter of 16 m. Its design volume served as basis for the billing per cubic metre of grouted

rock (Fig. 4). Special positions covered: the closing of joints at the nearby steep slope and the so-called "form-work" grouting. These items provided for the special risk of loss of grouting material which under such topographical conditions is often impossible to prevent (Werner, 1976).

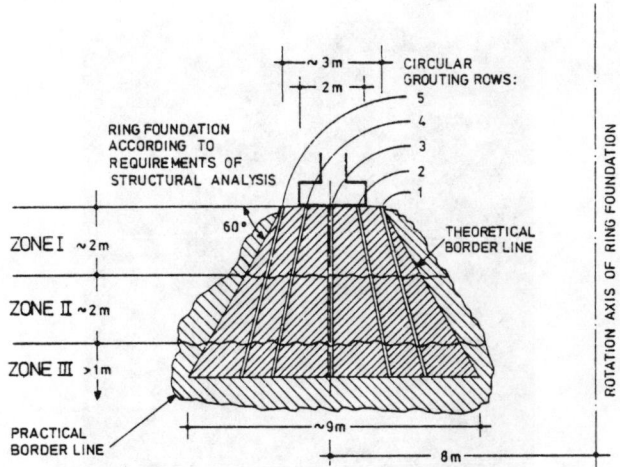

Fig. 4 Standard cross-section of grouted body

The grouting techniques were adapted to the difficult local conditions and their successful application was verified by field tests. The grouting pattern called for five concentric rows of grouting holes, the outer of which were grouted first. The ring was divided in 32 sectors, which where treated step by step in a binary raster. Fig. 4 also shows three zones which made it possible that, in general, no grouting section was longer than 2 m. Almost all grouting holes were treated twice, i.e. an injection from top to bottom in a first phase, and in opposite direction in a second phase.

Using hardly any pressure, the initial mixes were extremely thick (water/cement value 0.5 with addition of max. 10% sand). For the final grouting work, the applied mixes were as thin as possible in order to just achieve the saturation pressure. However, water/cement value was limited to 4 because of the danger of segregation or loss of strength. In view of the possibility of deflections of the surface and the escape of grouting material at the nearby steep slope, the saturation pressures were limited as follows:

- Zone I   0.5 MN/m²   (5 bar)
- Zone II   0.8 MN/m²   (8 bar)
- Zone III 1.0 MN/m²   (10 bar)

The average need was about 250 kg cement per m of borehole, with maximum values reaching about 600 kg/m. The relatively low overall cement consumption indicates that the voids within the joints only accounted for 70% of the volume defined on the basis of the preliminary investigations. More details about grouting for this project may be obtained from a special paper by Werner et al., 1981.

5. PERFORMANCE CONTROL

Apart from controls to ensure that the specified shape of the grouted body could be obtained, it had to be checked whether the concept described in Section 3 above could be successfully applied in practice. Control work was done on the basis of 14 core drillings, more than 400 water pressure tests, individual compression tests on grouted cores out of the grouted body (Fig. 5) and a precise survey of the rock surface laid open when excavating the area for the connection to the technic-tract.

Fig. 5   Core sample (Ø:85 mm) from control boring P 3 after grouting

It could be observed that:

- narrow as well as wide joints had properly been injected;
- loose rock fragments been well enclosed by the grout;
- closed separation planes had been bust open ("claquages") and filled up with grouting material.

With a few exceptions, where additional post-grouting was required, the remaining joint-fillings complied with the specified criteria. Indirectly, this proved that the rock area under the telescope tower has the characteristics required for the safe and quasi-immovable support of the ring-foundation for the substructure of the telescope.

6   CONCLUSIONS

In view of the extreme climatic conditions at the site, the altitude of about 3,000 m and the very short annual construction period it was imperative to plan and prepare the foundation work with utmost care. It proved to be benefitial that even more unfavourable site conditions had been considered in the planning concept than investigated so as to offset any time losses during construction.

The concept of verifying the required improved rock characteristics, as basis for the foundation, indirectly and with the aid of parameter studies proved to be a most successful method in the case at issue.

The presence on site of a grouting specialist, supplied by the Consultant, throughout the rock improvement operations paid high dividends since the estimated foundation cost could be reduced and the substructure was ready for the assembly of the telescope in due time. This was the precondition that the astronomic research with the 30 m radio-telescope at the Loma de Dilar (Fig. 6) can commence as scheduled.

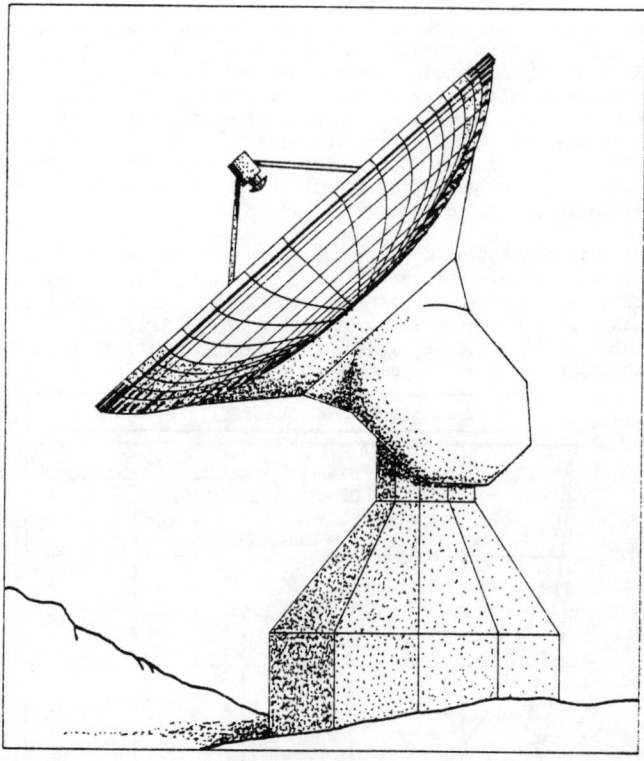

Fig. 6   General view of the 30 m radio-telescope

7   REFERENCES

Melendez B. and Fuster J.M. (1965). Geologica de Espana. 688 pp. Madrid: Paraninfo.

Werner, H.-U. (1976). Foundation problems encountered in connection with aerial tramways: Rock Mechanics, Suppl. 5, 81-100, Wien: Springer.

Werner, H.-U. et al. (1981). Improvement of subsoil conditions of a radio-telescope on the Loma de Dilar, Sierra Nevada, Spain: Proc. 3rd National Conference on Engineering Geology, 169-177, Ansbach.

CONCEPTUAL GEOMECHANICAL MODELS: THEIR EVOLUTION DURING THE DESIGN AND CONSTRUCTION OF DAMS

Modèles géomécaniques: Leur évolution aux stades de la conception et de la construction de barrage

Entwicklung geomechanischer Modelle für Talsperrenfundierungen vom Entwurf bis zur Bauausführung

C. M. Nieble
Engineer, MsC, ENGEVIX, S.A., Rio de Janeiro, Brazil

S. Bertin Neto
Senior Geol., ENGEVIX, S.A., Brasília, Brazil

SYNOPSIS

This paper deals with the conceptual geomechanical models for the foundation of the Right Wing Dam of Itaipú and the Spillway and Intake/Power House of Tucuruí. It shows the evolution of the models from the earlier studies up to the final design and construction. In each phase the compatibility of the conceived model with the available data is emphasised.

RESUME

Cette communication concerne les modèles géomécaniques de la fondation du Barrage d'Aile Droite d'Itaipú ainsi que du déversoir et de l'Admission/Centrale de Tucuruí. Elle montre l'évolution des modèles à partir des études initiales jusqu'à la conception définitive et la réalisation du projet. A chaque étape on actualise le total des données dans le modèle à l'aide des éléments déjà obtenus.

ZUSAMMENFASSUNG

In dieser Arbeit wird die Entwicklung geomechanischer Modelle für die Fundierungen auf dem rechten Flügel der Itaipú-Talsperre sowie für die Bauwerke der Hochwasserentlastung, des Einlaufes und des Krafthauses der Tucuruí-Talsperre von den ersten Studien bis zum entgültigen Entwurf und zur Bauausführung dargelegt.

I. INTRODUCTION

Conceptual Geomechanical models of dam foundations are of great importance for the analysis of stability, seepage and stress-deformation studies by the finite element method (Camargo et al., 1978).

Depending upon the geological condition of the site in some cases it can be a difficult task to conceive a two or three dimensional model from a physical reality.

This paper shows two examples of idealizing two dimensional geomechanical models for the foundation of three types of concrete structures of two dams in the different phases of design and construction.

In one of them the geological, geometric and geomechanical conditions are quite definitly assessed in the preliminary design, whereas in the other example the complexity of geology only allows an adequately conceived model during the excavation of the foundations.

II. GEOLOGY

II.1 - ITAIPU

Itaipu dam site (Fig. 1) is underlain by a sequence of basalt flows of the Paraná Sedimentary Basin: five flows have been identified, named A, B, C, D and E, in a rising order. These flows dip 1° to 2° toward Northeast, and its mean characteristics are summed up as follows (Moraes et al., 1982):

- thickness varying from 20 to 60 meters
- near-horizontal discontinuities usually located at the base of the transition zones of the different basalt flow types and at the contact between flows.

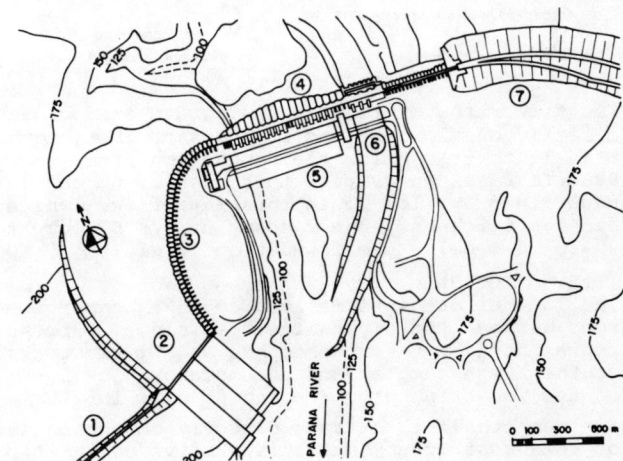

Fig. 1 - Itaipu Project - General Lay-Out

(1) Right Earthfill Dam
(2) Spillway
(3) Right Wing Dam
(4) (5) Main Dam & Power House
(6) Diversion Structure & Channel
(7) Rockfill Dam

- horizontal/vertical permeability ratio greater than one.

Each lava flow shows the following types of basalts (Ubiarte et al., 1982).

- dense basalt, microcrystaline, high density and high static deformation modulus, exibiting a columnar jointing.
- amygdaloidal vesicular basalt, less dense and much less jointed than the dense basalt. Shows lower values of deformability modulus.

Basaltic breccia occurs between lava flows, showing a lightly vesicular lava surrounding angular blocks of basalt and sedimentary materials (sandstone and siltstone), and exhibiting irregular cavities partially filled with carbonate, zeolite and amorphous and crystalline quartz.

Some average values obtained from laboratory and "in situ" tests in these materials are shown in table 1.

At the right wing Dam the three superior flows (E,D and C), breccia D and C and the horizontal discontinuity D are of special interest.

TABLE 1

TYPE OF MATERIAL	"IN SITU" TESTS			LABORATORY TESTS		
	E	$\emptyset$	C	γ	μ	σ_c
Dense Basalt	> 20	—	—	2,95	—	95
Vesicular Basalt	10-15	—	—	2,60	0,12	50
Breccia	9	35	3,6	2,20	0,25	25
Subhorizontal Discontinuities	—	35	0	—	—	—

E (GPa) - static modulus of deformability.

$\emptyset$ (degree) - friction angle.

C (MPa) - cohesion.

γ ($10^3 kg/m^3$) - density.

μ - Poisson ratio.

σ_c (MPa) - unconfined compressive strength.

The structures of the right wing Dam are founded directly on flow "E", and its contact with breccia D is 15 to 20 m below the foundation line. Breccia D has an average thickness of 6m and is underlined by flow D. In this lava flow, which has 18m thickness, the discontinuity "D" occurs 20 to 30 m below the foundation level.

II.2. - TUCURUÍ

The Tucuruí dam site is located on a transition zone between the crystaline shield rocks represented by the metamorphic and igneous rocks of the Xingu complex and the metamorphic rocks of the Tocantins group (Caúman et al., 1982).

The dam itself is sitting directly on the rocks of this last group, which is subdivided in two units:

- the first one, older, occurs on the right bank and in part of the river bed. Phyllites, clorite schists, quartzites and metabasic rocks are the predominant lythological types.

- the second, younger, is constituted by graywacke metasediments and metabasalts, and occurs on the left bank and part of the river bed.

These rocks have been affected by some tectonic movements whose expression in terms of intense faulting is evident in the area. The bedding planes show attitudes mainly northeast, with values ranging from 10° to 35°.

The Spillway and the Intake/Power House Structures lie directly on the graywacke metasediments. The principal structural features observed in these rocks are the additional major fault systems, besides the thrust fault, which does not affect the concrete structures (Fig. 2).

There are at least four additional major fault systems at the site whose expression is best observed in the metasediments:

- the first system occurs as a serie of subhorizontal thrust faults formed, possibly, at the time of the major regional thrust fault. This extremely low angle thrust fault system has been named the Principal Fault System, since it presents the most important stability problem for the Intake/Power House structures, although below the Spillway these faults are not important;

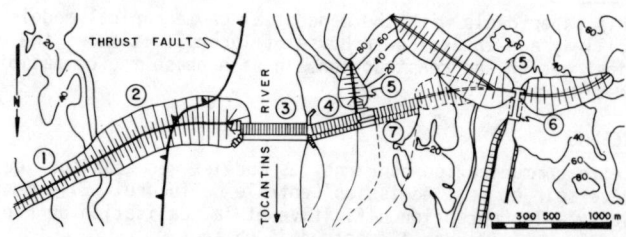

Fig. 2 - Tucuruí Project - General Lay-Out

① Right Bank Earthfill Dam
② River Bed Rockfill Dam
③ Spillway
④ Intake & Power House
⑤ Left Bank Earthfill Dam
⑥ Upper Navigation Lock
⑦ Intake & Power House (Second Stage)

- the next oldest system is developed as many secondary east dipping thrust faults which are approximately parallel to the beeding planes. This fault system was named the F1 System;

- a third system of intermediate age, named the F2 System, is represented by steeply inclined normal faults which cut and offset the older F1 thrust fault system;

- a younger system of subvertical or stick-slip or wrench faults cut the other three systems. These faults have the thickest and most decomposed fault gouges.

The F1 faults are truncated by F2 and F3 systems and therefore have not enough continuity to be considered potencial sliding planes (see fig. 7).

The principal faults are cut only by the F3 faults with little offsets and are, therefore, potencial sliding planes.

Table 2 presents the geotechnical and geomechanical characteristics of the metasediments and faults, as obtained by laboratory and in situ tests.

TABLE 2

TYPE OF MATERIAL	"IN SITU" TESTS			LABORATORY TESTS		
	E	$\varnothing$	C	γ	μ	σ_c
Metasediment	> 15	—	—	2,75	0,25	130
F1 and F2 Faults	—	38*	0*	—	—	—
F3 Faults (decomposed zones)	3,5	27*	0,02*	—	0,30*	—
Principal Faults	—	34	0	—	—	—

* - estimated. $E, \varnothing, C, \gamma, \mu, \sigma_c$ as on Table 1

III. GEOMECHANICAL MODELS

III.1 Introduction

A conceptual geomechanical model can be understood as "a two or three dimensional representation which adequately portrays the geological features and the physical and mechanical properties of the rock mass".

It is a basic element for the analysis of behavior of the concrete structure-foundation, either for stability, stress-strain and seepage, etc, and should be developed through sucessive lay-outs, adding to each previous studies all new data obtainable.

It should include the lithological types, discontinuities, contacts, altered zones, etc, of the rock mass and impermeabilization, grouting, drainage, bolting, from the design in a simple and clear way.

The process is repetitive and the model should be reappraised with all new information available such as observations and investigations in each phase of design and construction.

III.2 Evolution

III.2.1 Preliminary Design

Geological sections are the initial basic elements for idealizing the conceptual geomechanical models. In this stage, the available data are obtained from geological surveys and from rotary drill logs, mapping of pits and galleries.

Geological, geometric and geomechanical parameters for the basaltic lava flows of the foundations of Itaipu dam are quite well defined at this phase, due to its geological origin, without important tectonical perturbations, and due to preceedent experiences (Nieble et al., 1974). For the Tucuruí dam site, however, due to the geo-structural complexity of foundations, the geomechanical models conceptualized in basic design represents only a rough idea of reality. The position, attitude, inter-relations and persistence of the different fault systems are extremely difficult to define at this stage. Due to these conditions, it is prudent to be conservative in extrapolating data and defining geomechanical parameters. Consequentely, the results obtained from the analysis and studies are generally on the conservative side.

But, although not definitive, the geomechanical models conceived in this phase of design are very important to help defining the following investigations, for a preliminary assessment of

monitoring and to antecipate eventual stability and deformation problems.

III.2.2 Design and Construction

Direct observation of the principal weak zones of the rock masses during excavation are extremelly usefull to conceive the final geomechanical models. Some additonal "in situ" tests may be performed to check adopted values of geomechanical parameters in "critical" geological features.

III.3 Geomechanical models for the Itaipu Right Wing Dam

The preliminary design model was only a little modified, mainly the geometric and geomechanical parameters (see fig.3), but the original conception was maintained.

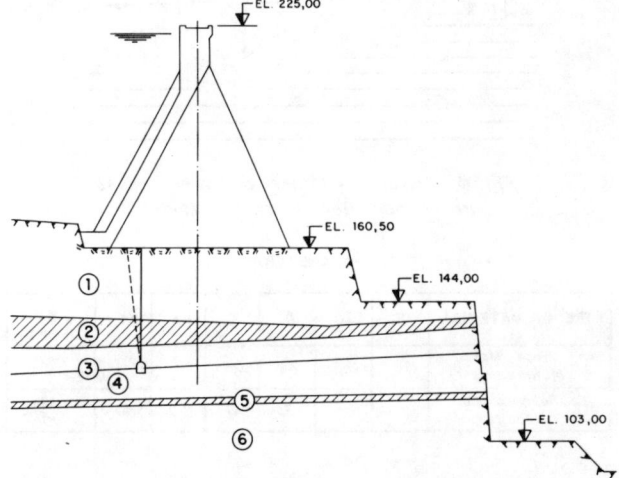

Fig.3 - Itaipu Right Wing Dam - Geomechanical Model.

PARAMETERS

MATERIAL / PARAMETERS		① FLOW E	② BRECCIA D	③ JOINT D	④ FLOW D	⑤ BRECCIA C	⑥ FLOW C
Preliminar	E	20	8	—	20	8	20
	μ	0,18	0,22	—	0,18	0,22	0,18
	$\varnothing$	45	30	38	45	30	45
	C	1,5	0,8	0	1,5	0,8	1,5
	Kn	—	—	200	—	—	—
	Kt	—	—	70	—	—	—
	K	10^{-5}	10^{-4}	10^{-3}	10^{-5}	10^{-4}	10^{-5}
Final	E	20	10	—	20	7	20
	μ	0,20	0,20	—	0,20	0,20	0,20
	$\varnothing$	55	40	40	55	45	55
	C	2,5	1,5	0,5	2,5	1,5	2,5
	Kn	—	—	500	—	—	—
	Kt	—	—	200	—	—	—
	K	10^{-5}	10^{-4}	10^{-3}	10^{-5}	10^{-4}	10^{-5}

Kn ($10^6 kg/m^3$) - unit normal stiffness. K ($10^2 m/s$) - permeability coefficient.

Kt ($10^6 kg/m^3$) - unit shear stiffness. $E, \mu, \varnothing, C$ as on Table 1.

III.4 Geomechanical models for the Tucuruí Intake/ Power house

In the preliminary design a serie of horizontal faults at regular intervals was conceived for the foundation (see fig.4). No information was available for the F2 and F3 fault systems. The final model showed only two principal (horizontal) faults - upper and lower -, and the position and thickness of the F3 faults were defined (see fig.5).

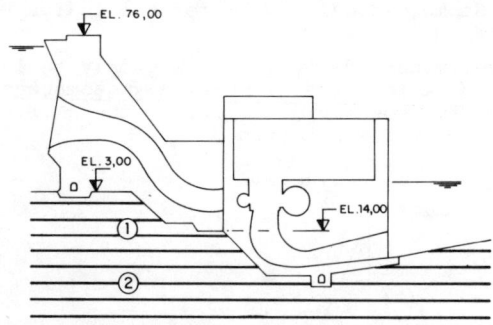

Fig. 4 – Tucurui – Intake & Power House Preliminar Geomechanical Model

PARAMETERS

TYPE OF MATERIAL	E	μ	$\emptyset$	C	Kn	Kt	$\overset{K}{\underset{2}{\uparrow}}$
① Rock Mass (Metasediment)	18	0,2	42	0,8	–	–	1 = 10⁻⁵ 2 = 10⁻⁵
② Subhorizontal Faults	–	–	33	0	150	50	1 = 10⁻⁵ 2 = 10⁻¹

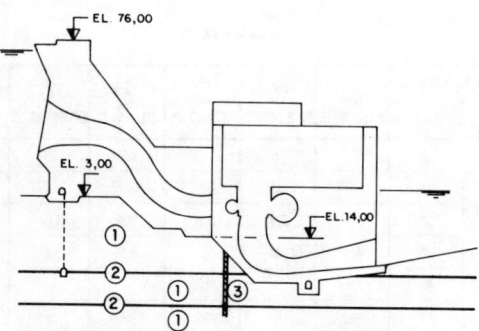

Fig. 5 – Tucurui – Intake & Power House Final Geomechanical Model

PARAMETERS

TYPE OF MATERIAL	E	μ	$\emptyset$	C	Kn	Kt	$\overset{K}{\underset{2}{\uparrow}}$
① Rock Mass (Metasediment)	20	0,2	45	1,0	–	–	1 = 10⁻⁵ 2 = 2 x 10⁻⁴
② Principal Faults (upper and lower)	–	–	34	0	100 to 250	50 to 300	1 = 10⁻⁵ 2 = 5 x 10⁻³
③ F3 Faults (decomposed zones)	35	0,3	27	0,02	–	–	1 = 10⁻⁵ 2 = 10⁻⁵

III.5 Geomechanical Model for Tucuruí Spillway

Geomechanical models for Tucuruí Spillway were idealized mainly for stability analysis. The preliminar model (see fig.6) assumed that the potencial sliding plane coincides with the F1 faults, parallel to the stratification. The average apparent dip of these faults in the upstream-downstream direction was about 12° and the shear strength parameters were those of the F1 faults, with a coesion intercept due to the offsets improved by the F2 and F3 systems.

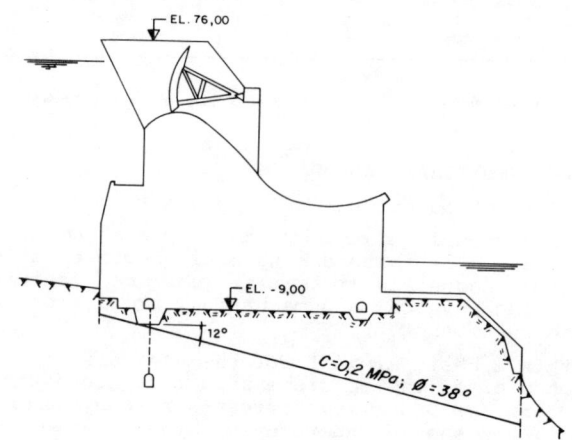

Fig. 6 – Tucurui – Spillway Preliminar Geomechanical Model for Stability Analysis

The excavation for the foundations allowed some detailed mapping and investigations, and a three dimensional aspect was found that had been neglected in the preliminary design. The attitude of the stratification (therefore F1 faults) was variable for each block of the spillway.

Besides that, a mechanism of rupture was studied so as to be compatible with the fact that the direction of these planes was inclined in relation to that of a possible movement of the block (upstream-downstream) (see fig. 7).

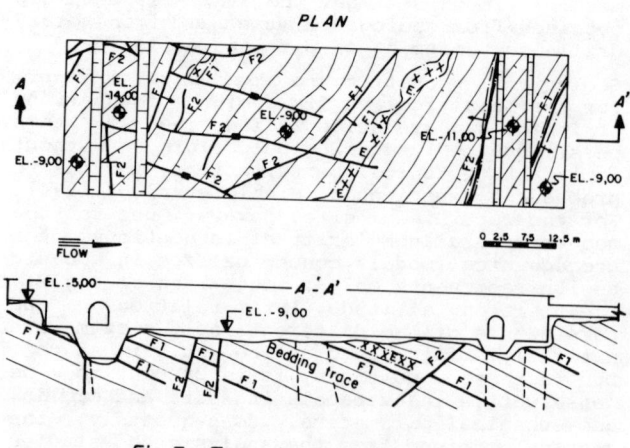

Fig. 7 – Tucurui – Spillway – Block 11 Geological Map

A hypothetic "wedge" of rupture were considered, formed by the F1 faults and by an idealized symetric plane, so that the intersection coincided with the possible direction of movement of the block.

This three-dimensional problem was reduced to a bidimensional one by the Kovary & Fritz formulation (1976) in order to obtain the "equivalent friction angle" (see fig. 8). The coesion intersept was established for each block based on the offset of the F1 faults.

A model was idealized for each block of the spillway, in order to consider local variations for three dimensional geometry.

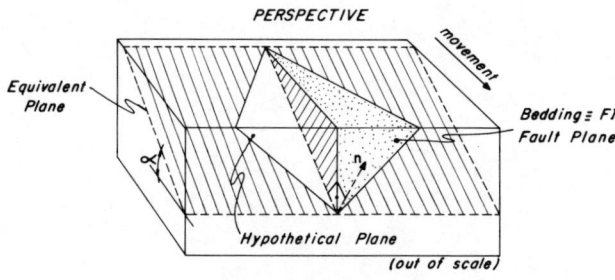

PERSPECTIVE

Equivalent Plane

movement

Bedding ≡ F1 Fault Plane

Hypothetical Plane

(out of scale)

VERTICAL PLANE THROUGH THE INTERSECTION

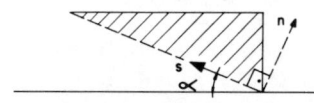

PLANE NORMAL TO THE INTERSECTION

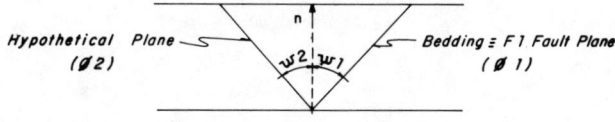

Hypothetical Plane (Ø2)   Bedding ≡ F1 Fault Plane (Ø1)

Fig. 8 - "Wedge Effect"

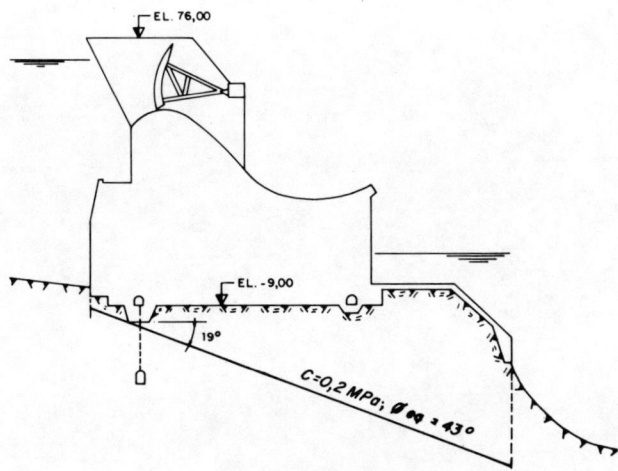

EL. 76,00

EL. -9,00

19°

C=0,2 MPa; Øeq = 43°

Fig. 9 - Tucurui - Spillway - Block 11
Final Geomechanical Model
for Stability Analysis

IV. CONCLUSIONS

Conceptual geomechanical models represent an abstraction from a physical reality. Generally, a three dimensional problem is converted in a two dimensional one, and geologic-geometric simplifications are made, where contacts are linearized, some features are blown up and some supressed. Even so, the model should represent, as close as possible, the relevant geomechanical properties and geometry of the rock mass.

In the early phases of design it is possible to idealize a model which, in some cases, will suffer only minor modifications to be final. In other situations, where the complexity of geological conditions are only adequately understood during excavations, models elaborated for preliminary design are generaly on the conservative side and should be improved or even substantially modified in the final design stages.

ACKNOWLEDGEMENTS

The authors wish to thank ITAIPU BINACIONAL and ELETRONORTE - Centrais Elétricas do Norte do Brasil, owners of Itaipu Dam and Tucuruí Dam, respectively, for permission to publish this paper.

REFERENCES

(1) Cadman, J. D.: Buosi, M.D; Bertin Neto, S.; Bastos, A.C. (1982) - "The Complex Metamorphic Rock Foundation Conditions of the Tucuruí Hydroelectrical Project" - XIV ICOLD, Vol. II, Q.53 - R. 51 - Rio de Janeiro.

(2) Camargo, F.P.; Leite, C.A.G.; Bertin Neto,S; Maldonado, F.; Cruz, P.T. (1978) - "Development of Conceptual Geomechanical Models for Foundations of Concrete Dams - Approach Applied to Three Projects" - Int. Symp. on Rock Mech. Related to Dam Foundation. Vol. 1 - Rio de Janeiro.

(3) Kovari, K.; Fritz, P, (1976) - "Stabilitatsberechnung Ebener und Felsboschungern" - Rock.Mech. Journal, Vol.8 nº 2.

(4) Moraes, J., Rodrigues Villalba, J., Barbi A. L. e Piasentin, C. (1982) - "Subsurface Treatments of Seams and Fractures in Foundation of Itaipu Dam". 14th Congress on Large Dams, vol. 2, Q.53 - Rio de Janeiro.

(5) Nieble, C.M. Midea, N.F., Fujimura, F., Bertin Neto, S. (1974) - "Shear Strenght of Typical Features of Basaltic Rock Mass, Paraná River, Brasil, 3rd International Congress on Rock Mechanics, ISRM, Denver.

(6) Ubiarte, J.A, Caric, D.M., Nimir, W.A., Eigenheer, L.P. and Nitta, T, 1982 - "Itaipu Main Dam - Geological and Geotechnical Features Affecting the Design" - 14th Congress on Large Dam, vol. 2, 53 -R-12, Rio de Janeiro.

STABILITY ANALYSIS OF THE EXCAVATION FOR THE KARAKAYA ARCH DAM AND POWER PLANT

Calcul de la stabilité de la fondation du barrage et de la centrale hydro-électrique de Karakaya

Stabilitätsberechnung des Aushubes für die Staumauer und das Krafthaus von Karakaya

M. Gavard
Dr. B. Gilg
Engineers in charge of design and supervision for the Karakaya Power Scheme,
Electrowatt Engineering Services Ltd., Zurich, Switzerland

SYNOPSIS

After a short description of the Karakaya Power Plant, the results of the geological investigations are summarized in order to permit a good understanding of the extensive stabilization work which was necessary for the protection of both valley flanks and especially of the left one. The selection of the rock properties to be introduced in the analysis is explained on the basis of site observations and laboratory tests. Furthermore, the applied method of computation for the dimensioning of the rock anchors is mentioned and the system implemented is indicated in a general view.

RESUME

Après une brève description de l'aménagement hydro-électrique de Karakaya, les résultats de l'étude géologique sont résumés de façon à ce que le lecteur puisse comprendre les travaux importants qui se sont avérés nécessaires pour la stabilisation des deux flancs rocheux, spécialement en ce qui concerne celui sur la rive gauche. Le choix des paramètres est expliqué sur la base des observations effectuées sur le site et des essais en laboratoire. Mention est faite de la théorie appliquée pour le dimensionnement des ancrages et une vue d'ensemble de leur positionnement est donnée.

ZUSAMMENFASSUNG

Nach einer kurzen Beschreibung der Kraftanlage Karakaya werden die Ergebnisse der geologischen Untersuchungen zusammengefaßt, damit der Leser versteht, warum die Stabilisierung der beiden Felsufer, vor allem des linken, so umfangreiche Arbeiten erforderte. Aufgrund der Beobachtungen an Ort und Stelle und der Laboratoriumsversuche wird die Wahl der Berechnungsparameter erläutert. Ebenso wird die angewandte Berechnungsmethode dargelegt, welche der Dimensionierung der Felsanker zu Grunde liegt. Schlußendlich wird ein Überblick über die verlegten Spannanker gegeben.

1. INTRODUCTION

The engineer dealing with the design of large dams is facing today in an increasing manner major geological problems. The reasons are evident : on one hand the water demand becomes more and more important, on the other hand the favourable dam sites are already used for water storage. Thus there is a real worldwide run on every location where a new storage lake might be created either for irrigation, water supply, power or simply for the flood control and discharge regulation.

In several parts of the world, water management is not yet well advanced and many regions do not yet dispose of the necessary water regulation and distribution. Considering the overall growing population numerous countries having used until now their water resources without big concern, begin to be aware of the major difficulties they will incur in the coming decades with respect to the water supply problems.

For every dam engieer, it is, therefore, a big challenge not to give-up too easily a dam site even when the geotechnical conditions appear very complex. The modern technics of foundation improvement by means of guniting, grouting and anchoring are a big help for all kinds of stabilization works and it is not seldom that a dam site can be rescued with additional expenditures, perhaps rather high in absolute figures but quite justified in view of the large benefits generated by the development scheme.

2. DESCRIPTION OF KARAKAYA DAM AND POWER PLANT

The Karakaya hydro-power scheme is located in Eastern Turkey on the Firat (Euphrates) river, some 170 km downstream of the Keban dam and 75 km East of Malatya. The storage lake is considerably smaller than that of Keban, which achieves also the river regulation for Karakaya.

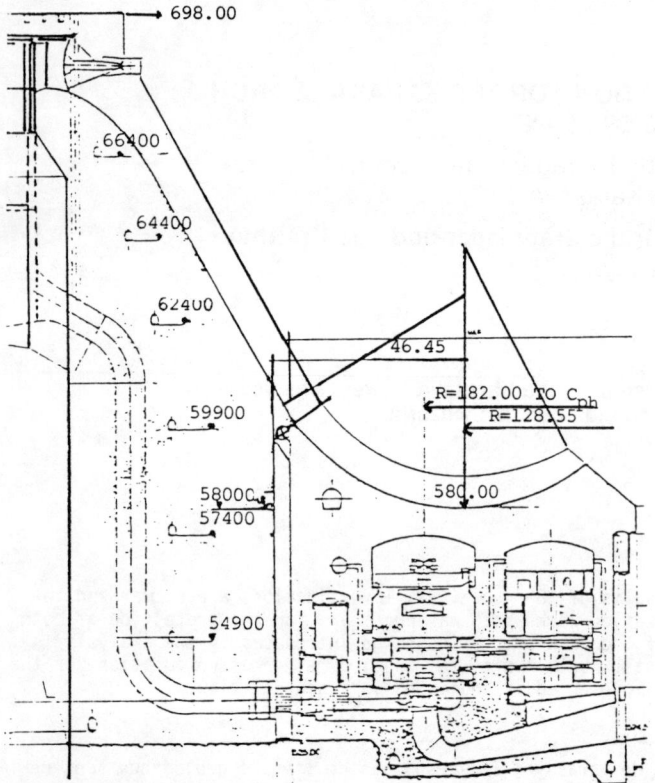

Fig. 1 - Cross Section of Karakaya Dam and Power House

The main features of the hydro-electric plant are the following :

Hydrology
- Drainage area   80'538 km2
- Average discharge   725 m3/s
- 10'000 years flood   19'300 m3/s
- Annual run off   22'880 Mio m3

Storage Lake
- Total storage   9'580 Mio m3
- Useful storage   5'400 Mio m3
- Surface area   298 km2

Dam (Arch type)
- Crest elevation   698 m.a.s.l
- Total height   173 m
- Crest length   462 m
- Crest width   10 m
- Crest radius   220 m
- Foot width   50 m
- Total concrete volume   2.0 Mio m3

Spillway (on the dam crest)
- Number of gates   10
- Type of gates   Segment / hydraulic
                                    operation
- Size of gates   14 m x 13 m
- Max. total discharge   17'000 m3/s

Spillway Chute (on the downstream face of the dam)
- Length   } on the dam   95 m
           } on the power house   85 m
- Max. velocity   40 m3/s
- Height of ski jump above
  tailwater   25 - 50 m

Power House (located at the dam foot beneath the ski
                                jump)
- Rated gross head   146.7 m
- Number of units   6
- Type of turbine   Francis
- Turbine discharge   233 m3/s
- Installed capacity   1800 MW

River Diversion (=Bottom Outlet)
- Number of tunnels : 2 (1 only for bottom outlet)
- Tunnel length   550 m resp. 680 m
- Tunnel diameter   11.50 m
- Tunnel lining thickness   80 - 100 cm
- Gate type for bottom outlet 2 x 2 sliding gates

3. GEOLOGICAL DATA

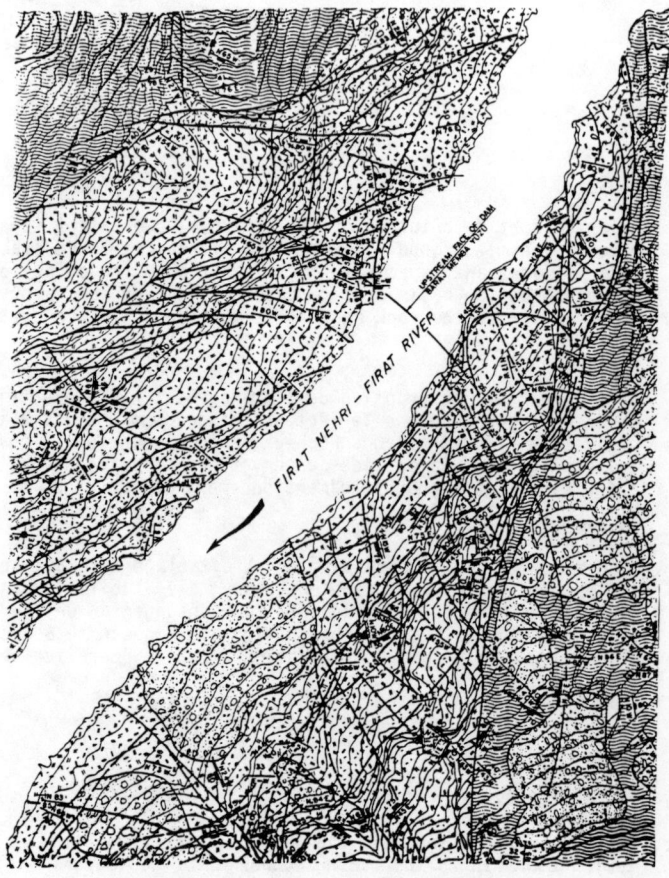

Fig. 2 Geological map of the dam site

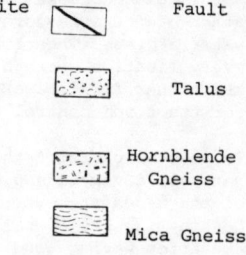

Fault

Talus

Hornblende
Gneiss

Mica Gneiss

The Firat course between Keban and Karakaya is divided
in a general manner in three sections :

- a narrow valley, about 40 km long, downstream of the
  Keban dam site
- the large basin of Malatya
- the gorge like valley through the Taurus chain

For this paper only the last section is of interest
and shall, therefore, be described in more detail.
At the Karakaya dam site the narrow river bed is flanked
on both sides by generally steep rock walls ; alluvial
fans or terraces are not available. Tributaries have locally created relatively limited cones of rockfall debris. The superficially weathered metamorphic rocks of
the slopes also produce rockfall and talus.

The highly metamorphosed formation strikes dominantly
West to East and dips northwards at 20º to 40º. Local
undulation of the strike direction is due to the lenticular shape of rock bodies which intercalate the isocli-
nal structure. Several mylonites, generally only some
centimeters thick, strike more or less parallel to the
general trend of the formation, but the dip is more
inclined and, therefore, cuts the strata. At the dam
site a complex of hornblende-gneisses and amphibolites
having minor intercalations of hornblende-schists is
prevailing. The upper portion of the section is built of
mica-gneisses and mica-schists which alternate with
phyllites. Intercalations of lenticular shaped marble
have been observed about 2 km upstream of the site
on the eastern bank. Along its boundary with the overlying mica-gneisses, the hornblende-complex shows
slight serpentinization.

Different possible dam axes have been investigated
but the geological most favourable site is in the section within the hornblende rich complex near the Karakaya village. The overlying mica-gneisses are hard
and relatively little tectonized, and the contact
zone between hornblende-gneisses and mica-gneisses shows
only a very thin coat of mylonitic facies. Few thin
mylonites cutting the hard hornblende-gneisses follow
the general trend of the local structure and correlate
with the lenticular rockbody.

The gentle undulation of these coarse-grained and
therefore groutable mylonitic discontinuities trends
locally towards the river, but since the mylonitic
zones are only a few centimeters thick, there is a minimum of rock disintegration. Clay-filling has been observed only at few places and grained fill is dominant.
On the left bank the mylonites have an outcropped level
parallel to the valley and dip at 12º to 42º NE.
Ancient displacements along the mylonites vary between
45 and 80 cm.

Although the left abutment is not more jointed and
fractured than the right one, it is of somewhat inferior
appearance. This is due to locally less favourable
intersections and especially to the different inclination of the schistosity and bedding planes. At the right
abutment these planes are dipping towards the interior
of the rock masses and therefore do not affect their
stability. At the left abutment, the planes dip
20º to 40º from the abutment towards the river and
consequently can cause locally limited creep phenomena
along superficial joints and some rockfalls, the debris
of which can be found at the bottom of several small ra-
vines.

However these joints and their associated features must
not be interpreted as stress-relief phenomena which
would have very deeply affected the rock structure of
the left abutment. On the contrary the various adits
and investigation galleries clearly show that the number of the joints as well as their width decrease with
increasing distance from the surface, and in deeper zones
they are even partially recrystallised and closed.
This is confirmed by numerous water pressure tests in
drill holes, which showed a high permeability near the
surface, decreasing rapidly within short distance
from the surface.

As far as ancient displacements can be observed, they
appear to have occurred along small faults rather than
along joints (these faults being an almost classic
feature of ancient rock structures at this site). Indeed,
there is no indication of any recent displacement
due to stress relief or a subsidiary thrust towards
the river valley.

The existing systems of joints and fractures are all of
a rather similar type, and no joint of large extent
has been discovered so far which cuts the rock in a
plane which would be unfavourable or dangerous for
the transmission of the dam thrust.

The geological investigations lead to the following
conclusions :

- an arch dam with a height of about 200 m is
  feasible
- the excavation depth in the rock will be of the
  order of 20 m
- the overburden excavation is of minor significancy
- the excavation slopes can be stabilized by
  different means going from rock bolts to prestressed
  anchors
- the rock can be watertightened by usual grouting
  work.

4. DESIGN OF THE EXCAVATION WORKS

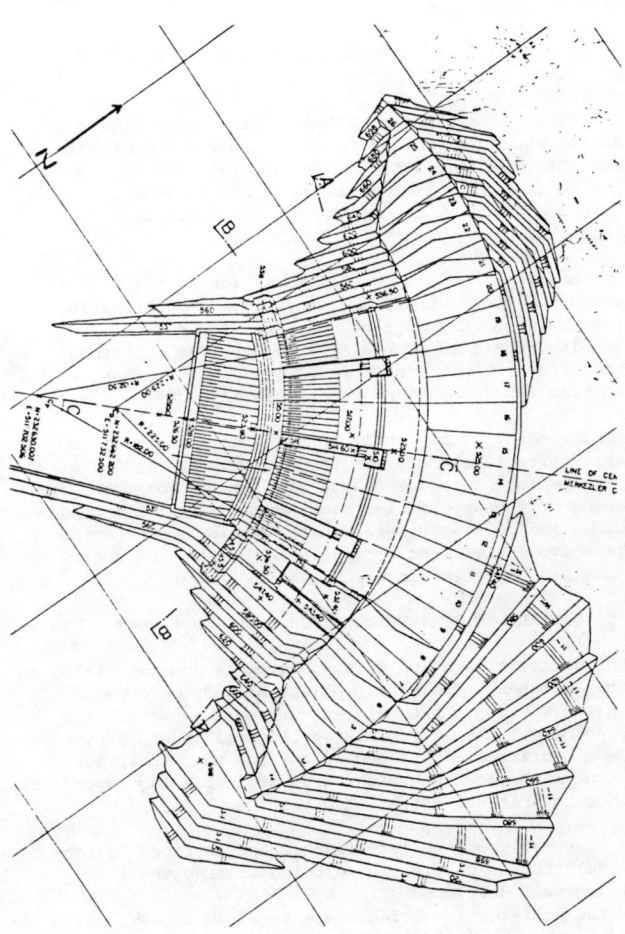

Fig. 3 Excavation and Foundation Layout

The excavation design was based from the very beginning on the assumption that the stabilization of the slopes would be a major concern during all the construction period. Therefore, the vertical distance of the corresponding berms has been reduced to 20 m and their width fixed at 5 m. This measure was intented to give a sufficient guarantee for the overall safety of the valley flanks and was planned for both sides of the gorge in a similar manner. The excavations cover an area of nearby 100'000 m2, not only for the dam but also for the rock flanks above the powerhouse at the downstream dam foot.

The slope between the berms was mainly fixed at 5:1, partly 3:1 only; on the left side up-stream of the dam abutment, the presence of mica schists has required the design of flatter slopes, decreasing from 5:1 continuously until approx. 1:1.

Being aware of the necessity of various protection measures the following possibilities
- rock bolting
- wire and steel meshes
- guniting and shotcreting
- normal anchoring
- prestressed anchors
- grouting
- drilling drainholes

have been taken into account in the tender documents. It was, however, obvious that the final extension would have to be defined during construction.

5. EXPERIENCE MADE DURING THE CONSTRUCTION OF THE RIVER DIVERSION SYSTEM

The first excavation work was for the construction of the river diversion system on the right abutment. For the inlet and outlet open cuts rather high slopes had to be realized. Several small rockfalls occurred and have demonstrated the necessity of anchorage, and of relatively long prestressed anchors.

On the other hand, a rough computation of the required amount of such anchors and the relevant prestressing force proved once more how sensible the result is with regard to the various assumptions concerning the rock mechanic properties, inside-water pressures, continuity of fractures and other weak zones.

Furthermore it was confirmed that for anchors, being fully effective, they had to be installed rapidly after the opening of the excavated zone because of the loosening of the superficial rock parts by climatic conditions and stress relief. Gunite and shotcrete proved to be unavoidable, when optimal behaviour of the rock mass had to be obtained.

6. DETAILED SITE INVESTIGATIONS AND LABORATORY TESTS

During the excavation in the abutments and especially on the left bank, the same problems already observed in the diversion tunnel open cuts were incurred. The findings of the preliminary geological investigation were confirmed and it became obvious that their influence on the rock behaviour were of increasing importance. In detail it was established that

- especially on the left flank steep inclined "diagonal" faults or fissures dipping towards the river are well developed
- the schistosity surfaces are inclined at 40°, which creates on the left side potential slip planes

- steeper and even some subvertical fractures which might be a consequence of decompression could form disconnection plans in combination with slip surfaces
- the lack of systematic clay filling allows for a good self-drainage which prevents the building up of water pressure along faults.

The detailed studies performed from the very beginning of the main excavation work allowed to recognize different natural sliding phenomena and to assess the limit equilibrium conditions of some slides, generally of the wedge shape.

As a general rule it was admitted that the shear strength on continuous failure planes merely consists of friction. This mechanical phenomenon can be decomposed in a "classical" friction, which can be expressed as angle of internal friction, and in the so-called surface roughness i. Assuming that the friction is equal on both planes forming the wedge and that the weight is the only acting force regardless of any other external force, the (φ + i) can be determined from the geometric readings of the strike and dip of both planes :

$$\tan \; (\varphi + i) = \tan \beta \; \frac{\sin(\theta_1 + \theta_2)}{\sin \theta_1 + \sin \theta_2} \qquad \text{(safety s = 1.00)}$$

where β : dip of the intersection line between the two wedge planes

θ_1, θ_2 : apparent dip of the two planes within the plane normal to the intersection line of the wedge

Under these conditions and by means of a back calculation of slide zones where the geologist was able to survey the surfaces position, the following friction properties have been determined :

Table 1 : Friction Properties

Location left bank	Elev.	Slide Shape	Rock nature	φ + i
Main access road	785	Wedge failure	Mica-gneiss	41°
Berm	740	"Spoon"	Mica-gneiss	34°
Berm X-180	720	Wedge failure	Mica-gneiss	52°
Berm X-205	720	Plane failure	Mica-gneiss	45°
D/S Berm	680	Wedge failure	Hornblende gneiss	52°

It is of course necessary to point out that the wedge failures did not always present a well defined shape and that the measurements (generally 6 readings at each place) cannot provide valuable statistical distribution. It results, therefore, a certain inaccuracy and a dispersion of the values but the order of magnitude of these is nevertheless significant.

Furthermore it has to be mentioned that some simple tilt-tests have also been performed on the spot on rock samples presenting a discontinuity in order to assess the friction coefficient at small scale and under very low normal stress. The results are listed in Table 2 herebelow but should only be regarded as comparative values with the laboratory test results :

Table 2 : Tilt tests results

	$\varphi + i$ (°)	Berm
- Hornblende gneiss, rough surface	550	698
- ditto, smooth slity coated surface	38	698
- mica-gneiss, undulating schistosity surface	44	740

Finally some rock samples have been taken from the boreholes and also directly on the spot in order to evaluate the corresponding shear strength along well perceptible discontinuities by means of laboratory tests.

It was then determined :

- the rock unit weight γ
- the shear strength versus normal stress, following the BARTON's type rule [1] :

$$\tau = \sigma \tan \left[JRC \log_{10} \frac{JCS}{\sigma} + \varphi \right]$$

where
τ : peak shear strength
σ : effective normal stress
JRC: joint roughness coefficient
JCS: joint wall compressive strength
φ : friction angle of planar joint surface

The first term of the expression between breakets can also be considered as being the average deviation angle i resulting from the roughness with regard to the shear stress direction : $\tau = \sigma \tan (\varphi + i)$, i being function of σ.

The tests have been carried out in the rock mechanics Laboratory of the Swiss Federal Institute of Technology at Lausanne [2].

The following method were applied :

- with constant normal load and displacement in both directions x and y (y = dilation)
- with variable normal stress and controlled vertical displacement

and the results are summarized in Table 3

Table 3 : Laboratory Tests

Sample / Depth (m)		Y (kN/m3)	σ (MN/m2)	φ (°)	$(\varphi +i)$ =1MN/m2	Remarks
A 8	5.50	30.0	204	33	48°	(a)
A 8	8.50			35	51°	(b)
A 8	9.50	30.2	89*	33	50°	
A 8	9.50			35	-	(c)
A 8	29.50			34	42°	(d)
Block Berm 698				33	37°	(e)
Block Mica-gneiss 740				31	38°	(f)

* = rupture along a plane of schistosity
(a) = average peak, 2 directions
(b) = JRC = 11.3 ; JCS = 23.5 MN/m2
(c) = sawn surface
(d) = less rough surface
(e) = flat silty coated joint
(f) = $(\varphi + i)$ = 49° pour c = 0.4 MN/m2

(1) = N. Barton and V. Choubey: "The Shear Strength of Rock Joints in Theory and Practice" Rock Mechanics 10 (1 - 54) 1977
(2) = Directed by Prof. F. Descoeudres, who was the special consultant for Karakaya rock mechanic problems

The following comments have to be made concerning the above results :

1. The hornblende gneiss has a unit weight in the order of 30 kN/m3 that corresponds to the density of the amphiboles. The in-situ density of the rock mass should not be much lower, taking into account that loosening phenomena as well as displacements along joints are kept to a minimum to preserve the peak shear strength.

2. The angles of internal friction of flat surfaces corresponding to the residual shear strength, lie within a very narrow range :

   $34° \pm 1°$ for the hornblende gneiss

   $31°$ (only one figure, but quite plausible) for the mica-gneiss along schistosity surfaces

3. As it was to be feared while testing small size samples showing local variations in roughness, there is a relatively marked dispersion of $(\varphi + i)$ values at peak shear strength. The value in the order of 50° obtained from three hornblende gneiss samples presenting fairly rough surfaces corresponds to a normal stress of approx. 1 MN/m2. This value can be compared with those given before for the wedge failure below berm 680 (52°) and with the tilt test made under almost nil normal stress (55°).

On the basis of the various results from site and laboratory, it was decided to use the following parameters for the stability computation :

- in the hornblende gneiss

$$\gamma = 29 \text{ kN/m3} \quad (\varphi + i) = 50°$$

This friction value may appear rather high with reference to some test results but takes in fact into account that there is a large scale surface undulation that participates in the overall imbrication, compensating for local reduced roughness. This of course remains true as long as bank deformations are avoided by implementing prestressed anchors immediately after the excavation works.

- in the mica-gneiss

$$\gamma = 29 \text{ kN/m3} \quad (\varphi + i) = 40°$$

7. FINAL COMPUTATION AND DESIGN OF SLOPE STABILIZATION MEASURES (Fig. 4)

The stability was calculated under the assumption that the sliding surfaces coincide with the fracture or schistosity planes and that there exists a vertical tension crack along this plane up to the surface. The acting forces are then :

W = weight of sliding block (including load if any)

U = uplift on the sliding plan

V = water pressure in the vertical tension crack (if any)

T = prestressing load

When the vertical tension crack is under water pressure, this pressure increases hydrostatically from the water table level down to the corner point between the vertical crack and the sliding plane ; from there the uplift decreases linearly to the point where this plane reaches the rock surface.

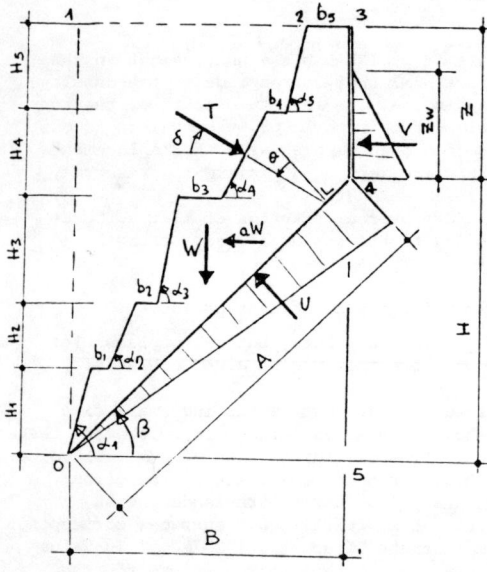

Fig. 4   Principle Shape of Sliding Wedge

The safety S against instability is thus given
by the formula

$$S = \frac{c\,A + (W\cos\beta - U - V\sin\beta - aW\sin\beta + T\cos\theta)\ tg\,\varphi}{W\sin\beta + V\cos\beta + aW\cos\beta - T\sin\theta}$$

$$(S = \frac{total\ force\ resisting\ sliding}{total\ force\ inducing\ sliding})$$

where :

a = earthquake acceleration
φ = friction angle
c = cohesion
A = sliding surface area
α = slope face dip
β = sliding surface dip
θ = angle of T with the perpendicular to sliding
    surface

For a given safety factor S the necessary prestressing
effort T is given by the expression

$$T = \frac{W\sin\beta + V\cos\beta + aW\cos\beta - \frac{cA}{S} - (W\cos\beta - U - V\sin\beta - aW\sin\beta)\frac{tg\,\varphi}{S}}{\sin\theta + \cos\theta\ \frac{tg\,\varphi}{S}}$$

Analoguous formulae can be obtained for more complicated rock masses sliding on surfaces with a polygonal
shape. With a simple computer programm all sliding
cases even with complex cross sections may be easily
controlled. It is especially possible to vary the water pressure and uplift conditions in the fractures
and sliding planes which are always a major reason
for the development of instability.

In stability analysis, one of the most important
decisions is the selection of a reasonable safety
factor. The respective choice depends on one hand
on the reliability of the basic data and the mechanical
properties. On the other hand it depends also on the
life time of the structure. When a slope protection
has only to be effective during the construction time,
for instance during 5 years, then lower factors can be
accepted especially in view of the earthquake safety.
However, for a protection measure, which is planned
for the life time of the dam, i.e. 100 - 200 years,
the safety factor should of course be higher.

The following safety factors have been considered
for the Karakaya project :

Load Case	Permanent Protection	"Temporary" Protection
Without earthquake	1.50	1.30
With earthquake	1.20	1.05

For the design of the anchors the following principles
were applied :

Direction of cables	:	in general perpendicular to the bench face
Steel section	:	strands of 25 t loading capacity prestressed at 60%, i.e. 15 t
Anchoring length	:	$L = \dfrac{T}{\pi \cdot \emptyset \cdot 10\ kg/cm2}$

L = length
T = nominal load
$\emptyset$ = diameter of borehole
10 kg/cm2 = bond stress

On the left bank the design has been performed basing
on a theoretical discontinuity parallel to the main
fault system (42°) and day lighting at elevation
579.50 (power house roof). The sliding mass boundary
is fixed by the subvertical fault beginning at elevation 698.00 m (dam crest).

After having determined the prestressing anchors as
above, the stability has been checked for all other
potential slip faces inside on the left abutment.
For highly extended sliding planes going up to
elevation 720 m or even higher, a reasonable
lowering of the safety factors was assumed for taking
into account that such big extensions without any
interruption are practically not possible.

On the berms 698.0 m (crest of dam) and 720.0 m
a vertical load of 5 t/m2 has been assumed for
taking into account the condition prevailing during
construction.

The overall stabilization of the left abutment needs
a total anchor force of about 1200 t per meter of excavation. Since the length of excavation from upstream
to downstream is about 200 m, the total anchorage force
reaches about 200'000 t or 1200 rock anchors with an
average tension of 170 t.

On the right abutment the number of the required anchors
is much less because of the more favourable angle of
the potential sliding planes.

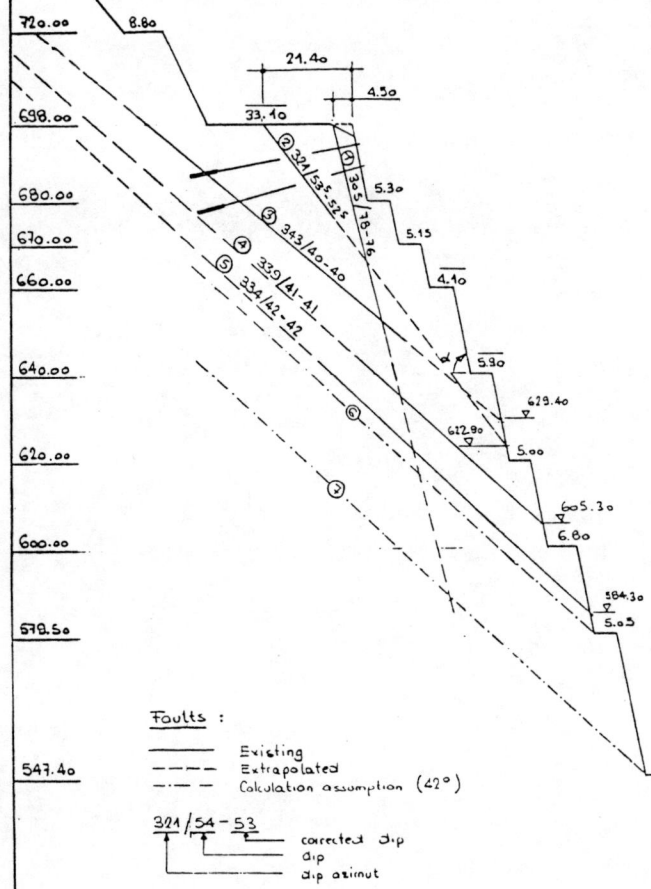

Fig. 5 Principle of computation of the left
bank, (1) to (7) : potential sliding planes

8. CONCLUSIONS

The design of the slope protection in Karakaya has confirmed that it is difficult to forecast in detail the
required means as long as the excavation works have
not yet started. Therefore in the case of voluminuous
and delicate dam foundations the design has to provide
for all possible means in view of various kinds of behaviour of the excavation.

These means are :

- simple wire meshes
- wire meshes with gunite or shotcrete protection
- rock bolts
- medium size anchors with limited stress action
- long anchors (bars or cables) with
  prestressing load
- protection wall with medium size anchors or prestressed cables.

For all these measures tentative quantities have to
be introduced in the bill of quantities, in order to
avoid difficulties during construction. These quantities should not be underestimated.

In the case of Karakaya the stabilization work will
exceed the forecast sign. On the left flank, there are
about 1200 anchors covering a surface of more or less
40'000 m2, i.e. an average density of 1 anchor per
30 m2, varying between 1 per 20 m2 until 1 per 40 m2.
But even if the corresponding cost amounts to
about US$ 10 Mio, this sum is rather modest in compa-
rison to the overall construction expenditure. It is,
therefore, unquestionable that several Million $
are justified for increasing the safety and giving
the guarantee that no unfavourable or even catastrophic event may occur during the construction period and afterwards the operation life of the plant.

Q

DYNAMIC STABILITY OF CONCRETE DAM FOUNDATIONS
Stabilité dynamique des fondations de barrages en béton
Dynamische Standfestigkeit der Gründungen von Betonstaumauern

G. A. Scott
K. J. Dreher
Geotechnical Engineer, Bureau of Reclamation, Denver, Colorado. Formerly with the
Bureau of Reclamation, Denver, Colorado

SYNOPSIS

Dynamic foundation loading from a concrete dam can be determined from finite element analyses utilizing initial forces, nodal displacements, and stiffness matrices. Dynamic three-dimensional analyses of potentially unstable foundation rock masses can then be performed by including static and time-varying dam and inertia forces. Permanent cumulative displacements can be estimated by double integration of the relative acceleration when the factor of safety against sliding for a rock mass drops below 1.0. Recent research indicates that the dynamic behavior of rock discontinuities may also be an important consideration.

RESUME

La charge dynamique que représentent les fondations d'un barrage en béton peut se déterminer à partir d'analyses d'éléments finis, utilisant les forces initiales, les déplacements nodaux, et les matrices de raideur. En incluant les forces — à la fois statiques et variables dans le temps — du barrage et de l'inertie, il serait alors possible de réaliser des analyses dynamiques à trois dimensions des roches de fondation potentiellement instables. Les déplacements cumulatifs permanents peuvent être calculés au cas où le facteur de sécurité du massif baisserait au moins de 1,0. De récentes recherches indiquent que le comportement dynamique des discontinuités des roches pourrait aussi figurer comme une considération importante.

ZUSAMMENFASSUNG

Die dynamische Belastung der Gründung einer Betonstaumauer kann durch Erfassung der Initialkräfte, Nodalverschiebungen und Steifigkeitsmatrizen mit Hilfe eines Finite-Elemente-Verfahrens ermittelt werden. Unter Berücksichtigung der statischen und zeitabhängigen Staudamm- und Trägheitswerte ist es sodann möglich, dynamische dreidimensionale Analysen von potentiell labilen Gebirgsmassivgründungen durchzuführen. Bleibende kumulative Verschiebungen können berechnet werden, wenn der Sicherheitsfaktor der Gesteinsmassen unter 1,0 abfällt. Neuere Forschungsergebnisse lassen darauf schließen, daß das dynamische Verhalten von Gesteinsdiskontinuitäten auch ein wichtiger Faktor sein kann.

1. INTRODUCTION

The application of three-dimensional finite element analyses to determine the response of major concrete dams to various static and dynamic loads is relatively common. Response history finite element analyses of concrete dams are generally accepted as being necessary to adequately assess safety for significant earthquakes. However, comparable analyses to estimate foundation behavior during earthquakes are usually not performed. Foundation stability is often determined by pseudo-static assessments whereby the inertial effect of peak or effective ground acceleration is simply added to the maximum loading from the dam. This approach is deficient because the maximum loads resulting from the dynamic response of the dam rarely occur at the same instant in time as the peak or effective ground acceleration. Although the simple superposition of maximum loads results in conservative estimates of the magnitude of foundation loading, more realistic assessments could result in savings of foundation treatment. Additionally, a more critical condition may occur when all loads are not at their maximum.

This paper describes techniques for computation of three-dimensional, time-varying foundation loads from dynamic analyses and for performance of stability analyses of potentially unstable rock masses. The techniques account for time-varying ground accelerations as well as the time-varying magnitude and direction of foundation loads from the dam.

2. DAM LOADS FROM FINITE ELEMENT ANALYSES

A common means of analytically modeling the foundation of a concrete dam is by assuming a two-dimensional elastic half-space under each abutment section where loading from the dam is applied. Unfortunately, this assumes that foundation deformations are independent of the shape of the foundation surface and that movements at a particular abutment section are due only to loads applied directly at that section. A related approach is to place spring elements at abutment sections. The stiffness of the elements is usually based on elastic half-space considerations, and the coupling effects between the abutment sections are again ignored. Occasionally, the abutments are assumed to be fixed and the foundation loads are obtained from integration of stresses over some portion of the abutment. Integration of stresses may not result in an adequate estimate of the directions of resultant loads. In addition, not providing for abutment deformations does not allow for appropriate load distributions, particulary in the case of arch dams.

A more realistic approach is to model the foundation with the same continuum finite elements used to model the dam. For each element of the dam connected to a foundation element where the rock mass is potentially unstable, the following matrix equation is solved:

$$\{F\} = [S] \{D\} + \{I\} \tag{1}$$

in which $[S]$ is the element stiffness matrix, $\{D\}$ is the vector of computed displacements for nodal points defining the element, $\{I\}$ is the vector of any initial forces calculated for nodal points in the element, and $\{F\}$ is the vector of resultant element nodal forces. Concentrated nodal forces applied at the dam-foundation juncture are not accounted for by equation (1) and must be added separately. However, this type of loading is seldom encountered in concrete dam analysis. The vector of initial element forces is required for temperature, gravity, and pressure loading. For example, zero displacement associated with temperature loading results in large stresses and therefore large loads. If for this case the initial element forces are neglected in equation (1), the resulting forces would be incorrectly calculated as zero since the displacements are zero. The accuracy of the calculated loads may be checked by setting the stiffness and loading of the dam structure to zero and applying the calculated loads directly to the foundation. The resulting foundation deformation should be identical to that from the original analysis.

Total load components to be applied to the foundation are calculated by the summation of components for each nodal point connect-ing a dam element with the potentially unstable portion of the foundation. The percentage of element load to be applied to a potentially unstable rock mass can be determined by utilizing the ratio of the element surface area contacting the potentially unstable block to the total foundation contact area of the element.

3. DYNAMIC RIGID-BLOCK STABILITY ANALYSIS

Consider the potential localized instability created by a powerplant excavation at the base of an arch dam, as shown in Figure 1.

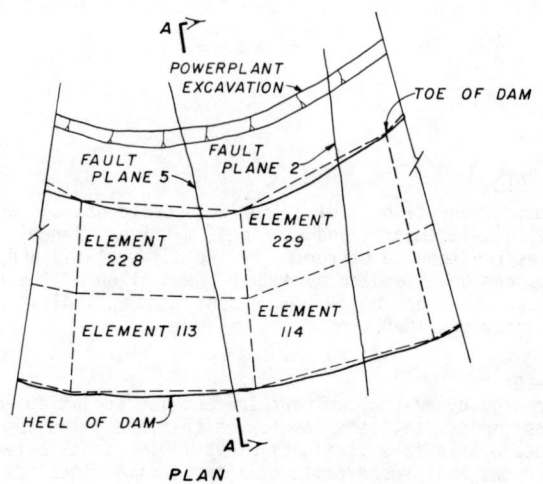

PLAN

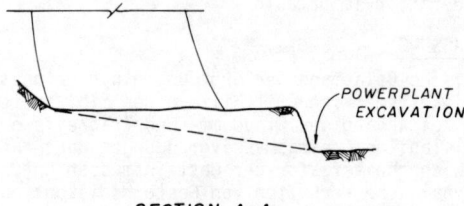

SECTION A-A

Figure 1. Example of a potentially unstable rock mass.

Two diverging vertical fault planes cross the foundation from upstream to downstream. A prominent and continuous joint set dips slightly downstream and daylights in the powerplant excavation between the fault planes. A "worst case" potentially unstable rock wedge includes a plane contacting the heel of the dam and daylighting at the base of the powerplant excavation. An isometric view of the resulting potentially unstable rock mass is shown in Figure 2. The orientations of the planes forming the rock mass are summarized in Table 1, along with shear strengths and water forces. The shear strength of the fault planes is approximated by the nonlinear failure envelope shown in Figure 3. The water forces act normal to each plane and are determined from a seepage analysis.

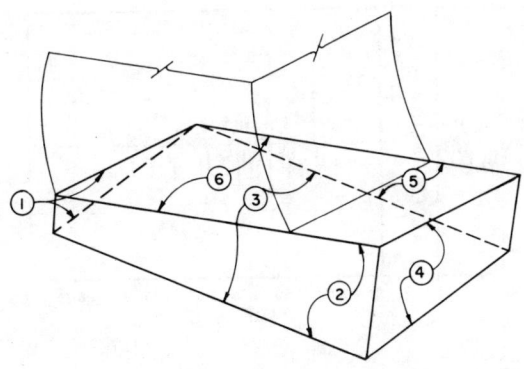

Figure 2. Isometric view of example wedge.

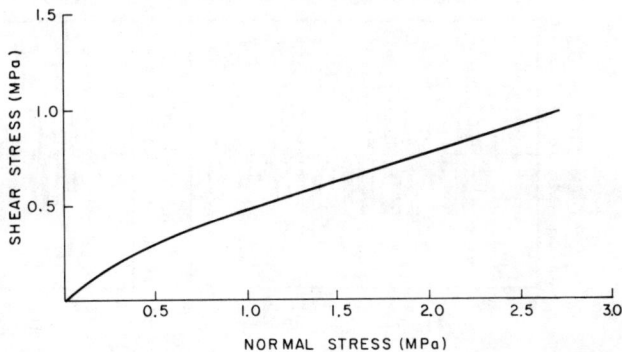

Figure 3. Nonlinear shear strength envelope.

Table 1. Summary of planes for example problem (See Figure 2)

Plane	Strike	Dip	Water Force (MN)	Area (m²)	Shear strength
1	N57°E	90°	225	111	phi=45°
2	N30°W	90°	202	408	Fig. 3
3	N57°E	8°SE	1,606	4,617	phi=55°
4	N58°E	90°	0	-	0
5	N35°W	90°	94	552	Fig. 3
6	Horizontal		0	-	0

In addition to the static forces, the dam is subjected to a Richter Magnitude 6.5 earthquake represented by three synthetic accelerograms shown in Figure 4 (Tarbox, et. al.,1979). Three-dimensional forces acting on the potentially unstable rock mass are computed at each time step during the earthquake from a finite element analysis of the dam, utilizing the method described in the previous section. The resulting histories of forces acting on the rock wedge are shown in Figure 5. Inertia forces acting on the rock wedge opposite to the directions of ground acceleration are also computed for each time step by multiplying the accelerations at each instant in time by the mass of the rock block and changing the sign of the resulting force.

A computer program is utilized to perform rigid-block limit-equilibrium analyses of the potentially unstable foundation block at each time step during the earthquake. The program described by Von Thun (1976) has been modified for this purpose. The static dam forces, weight of the foundation block, and water forces multiplied by plane normal direction cosines are added to the time-varying forces. The resultant forces acting on the block at each time step are resolved into components normal and parallel to each plane according to:

$$\{T\} = -[C]\{F_r\} \qquad (2)$$

in which [C] is a matrix containing direction cosines of the strike unit vector, dip unit vector, and normal unit vector, $\{F_r\}$ contains the resultant global forces acting on the block, and $\{T\}$ contains force components parallel to the plane strike, dip, and normal unit vectors. The normal unit vector is calculated from the cross product of the strike and dip unit vectors and is directed into the foundation block. The negative sign in equation (2) results in a positive force being directed out of the block. When the normal force on a plane is compressive, potential sliding along that plane is considered, provided that the direction of the shear force does not indicate movement into another plane. This is checked by solving the following equations:

$$\{F_s\} = \{S_c\}[A] \qquad (3)$$

$$Z = \{F_s\}\{C_n\} \qquad (4)$$

in which $\{F_s\}$ is the vector of global shear force components, $\{S_c\}$ contains the components of shear in the strike and dip directions of the potential sliding plane, [A] contains the direction cosines of strike and dip unit vectors for the potential sliding plane, and $\{C_n\}$ contains the direction cosines for the normal of a plane potentially blocking movement. If Z is greater than zero, movement is blocked. However, movement along the intersection of the two planes is possible. This is checked by calculating a unit vector in the direction of the intersection, determined by normalizing the cross product of the normals to the two planes. The components of force perpendicular to each plane and along their intersection, Q , are calculated as:

$$\{Q\} = -[N^T]^{-1}\{F_r\} \qquad (5)$$

in which [N] is a matrix containing the direction cosines of the potential sliding plane normal, the potential blocking plane normal, and their intersection. Blockage of movement is again checked as before, except:

$$\{F_s\} = Q3\ \{I\} \qquad (6)$$

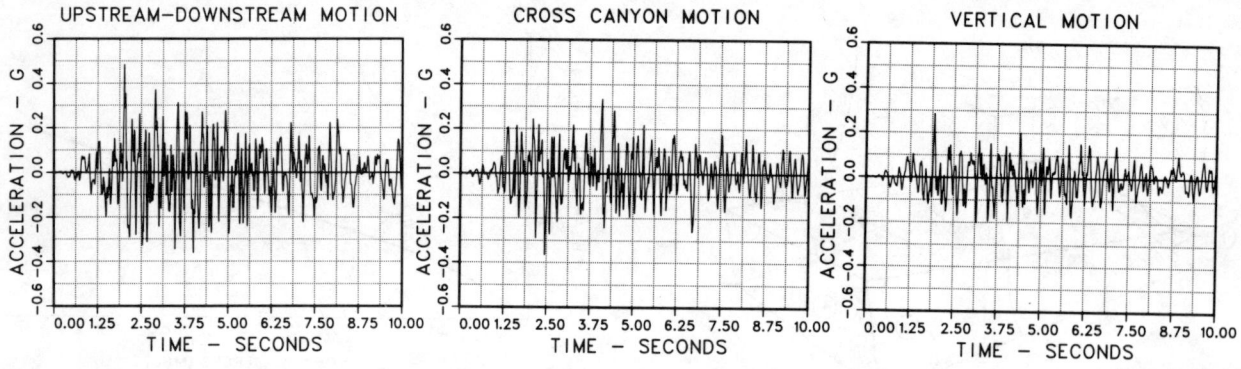

Figure 4. Synthetic ground accelerations for Richter Magnitude 6.5 earthquake (Tarbox, et. al., 1979)

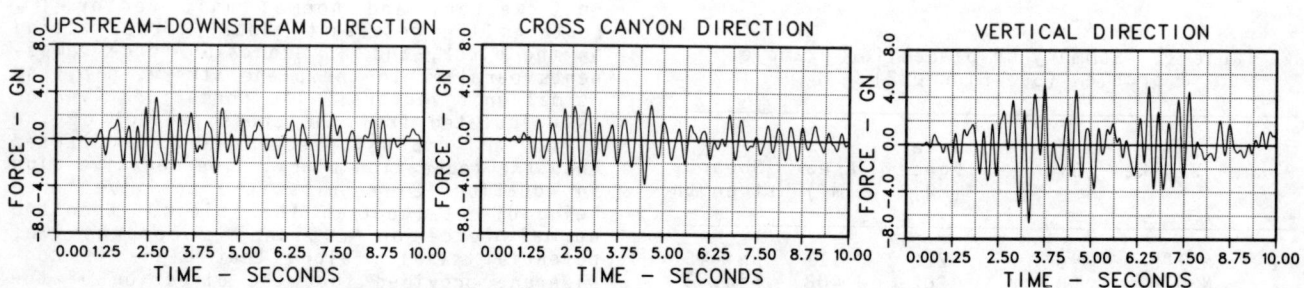

Figure 5. Loads from dam during Richter Magnitude 6.5 earthquake.

in which Q3 is the force along the intersection and $\{I\}$ contains the direction cosines of the intersection.

intersection is determined to be possible, the factor of safety against sliding is calculated by summing the shear resistances for planes with compressive normal forces and dividing the sum by the driving shear force. The potential mode of instability is checked at each time step during the dynamic analysis, and factors of safety are computed as appropriate.

The potential mode of sliding for the example problem described earlier is shown plotted against time in Figure 6. The potential mode of instability changes during the earthquake due to the time-varying forces. The factor of safety against sliding of the example rock mass is plotted against time in Figure 7. The safety factor drops below 1.0 twice during the earthquake. However, the total time the safety factor is below 1.0 is less than 0.05 seconds, and it is difficult to conclude that the rock mass is unstable. However, if only the maximum dam loads and peak inertia forces were considered, the pseudo-static factor of safety for the wedge would be less than 1.0, and the conclusion might be made that the mass is unstable.

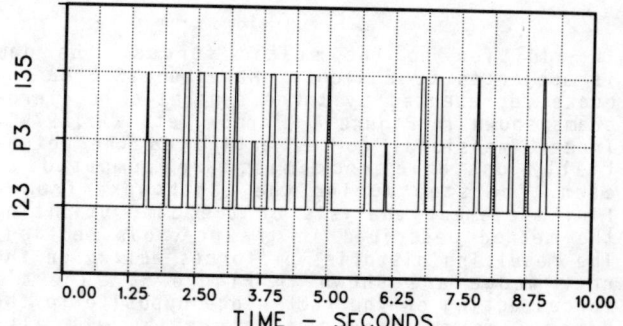

Figure 6. Modes of potential instability versus time. I35 indicates sliding on intersection 3/5, P3 indicates sliding on plane 3, and I23 indicates sliding on intersection 2/3 (see Figure 2).

4. DISPLACEMENT CONSIDERATIONS

If the cumulative time that the factor of safety against sliding is less than 1.0 for a rock mass is considered to be significant, the permanent displacement of that rock mass can be estimated. Von Thun and Harris (1981) describe a method for estimation of cumulative permanent displacements of slopes subjected to time-varying forces in two directions. The material shear strength is assumed to follow rigid-perfect plastic behavior, ignoring

C 230

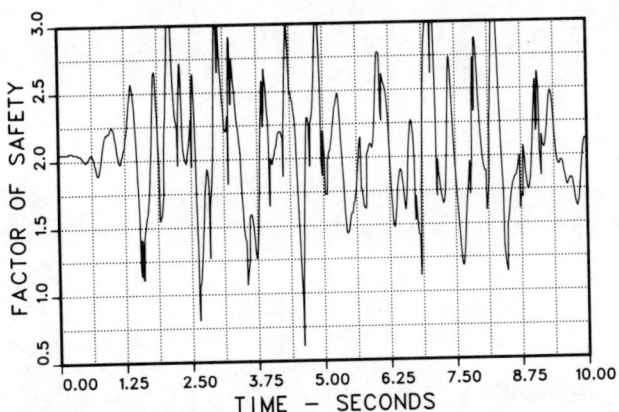

Figure 7. Factor of safety against
sliding versus time.

elastic shear deformation, and movement is
considered to occur only when the shear
strength of the resisting plane is exceeded.

This method can be extended to three dimensions
to estimate permanent deformations of rock
masses. Displacement occurs only when the
factor of safety drops below 1.0. Linear
interpolation between input time steps may be
utilized to find the exact time at which
movement initiates, if it does not occur at an
even time step. When the safety factor drops
below 1.0, the unbalanced force, F_u, acting
on the unstable rock mass at any time step is
given by:

$$F_u = D - R \qquad (7)$$

in which D is the driving force and R is
the resisting force, both of which vary
with time. Movement occurs in the direction
of the driving force. For the case of
sliding on a single plane, the resisting
force is a function of the normal stress
and shear strength of the single plane, and
the driving force is represented by the
magnitude and direction of the shear force in
the plane. In the case of sliding on an
intersection, the resisting force is a
function of the normal stress and shear
strength of both planes forming the intersection, and the driving force is the component of the resultant force in the direction
of the intersection. In case of lifting, the
resisting force is zero and the driving force
is the total resultant acting on the wedge.

When sliding occurs, the effects of displacement softening behavior must be considered. In addition, if sliding occurs
along the intersection of two planes, the
shear strengths utilized in the analysis
must be developed at compatible displacements.

The relative acceleration between the
unstable mass and the underlying rock at a
particular instant in time is given by:

$$A = \frac{F_u}{m} \qquad (8)$$

where m is the mass of the unstable rock.
Assuming the acceleration varies linearly
between input time steps:

$$A(t) = A(T_{n-1}) + \frac{A(T_n) - A(T_{n-1})}{\Delta T} t \qquad (9)$$

where ΔT is the time step between T_{n-1}
and T_n, and t varies from 0 to ΔT. The
relative velocity for each time step,
calculated by integration of the relative
acceleration, is:

$$V(T_n) = V(T_{n-1}) + \int_0^{\Delta T} A(t)\, dt$$

$$= V(T_{n-1}) + \frac{\Delta T}{2} [A(T_{n-1}) + A(T_n)] \qquad (10)$$

Another integration yields the relative
displacement for the time step.

$$U(T_n) = U(T_{n-1}) + \int_0^{\Delta T} \int_0^{\Delta T} A(t)\, dt$$

$$= U(T_{n-1}) + V(T_{n-1})\, \Delta T$$

$$+ [2A(T_{n-1}) + A(T_n)] \frac{\Delta T^2}{6} \qquad (11)$$

Movement stops when the relative velocity
becomes zero. The exact time that movement
stops may be determined by quadratic interpolation between input time steps, since velocity
is a quadratic function of time. The total
displacement magnitude and direction is determined by vector addition of the displacements
at each time step. The displacement calculated
in this manner is approximate and only appropriate for small displacements, since the mode
of instability is assumed to change instantaneously without repositioning the block.

For the example considered in previous sections, the cumulative displacement of the rock
mass estimated in this manner is 7.9 mm directed along plane 3 at a bearing of 147.3 degrees
and a plunge of 7.6 degrees. The effect of this
displacement on the dam structure must then be
evaluated.

5. DYNAMIC MATERIAL RESPONSE

Dynamic material and pore pressure response,
when different from static behavior, could
be incorporated into the response history
analysis previously described.

Crawford and Curran (1981) have studied the
effects of relative velocity on the shear

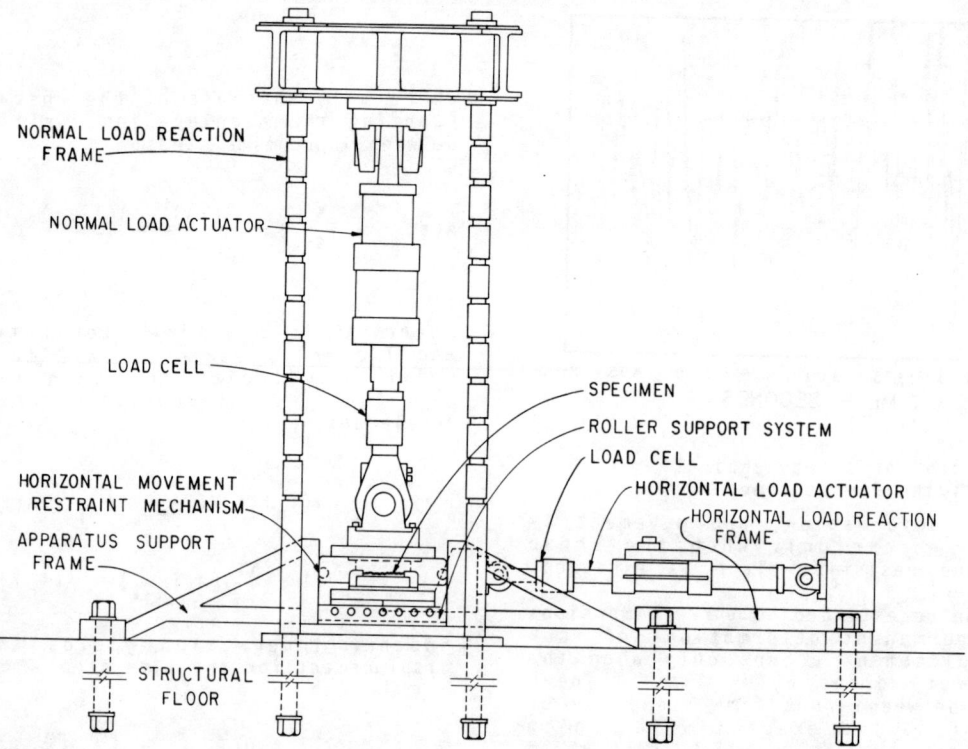

Figure 8. Schematic of dynamic direct shear apparatus.

resistance of rock discontinuities. Artificial saw-cut joints in five rock types were lapped with silicon carbide grit and tested in direct shear. A dynamic direct shear apparatus was used to displace the samples at various velocities under various constant normal loads. It was concluded that, in general, the shear resistance of harder rocks decreases with increasing velocity greater than a variable critical velocity, and the shear resistance of softer rocks increases with increasing shear velocity up to a critical velocity.

Similar research is being supported by the Bureau of Reclamation at the University of Colorado in Boulder, Colorado. Artificial rough joints with a shear area larger than 325 cm2 were made by tensile splitting specimens of Loveland sandstone. The rock has a uniaxial compressive strength of about 150 MPa. The samples were tested in a servo-controlled, dynamic direct shear apparatus developed for the research, shown schematically in Figure 8. A sinusoidal displacement function was used to displace the samples under various frequencies, displacement amplitudes, and constant normal stresses. Shear displacements were measured with LVDTs (linearly varying differential transducers) attached directly to the sample. The maximum relative test velocity is given by:

$$V_{max} = 2\pi f d_{max} \qquad (12)$$

in which d_{max} is the displacement amplitude and f is the frequency of the test.

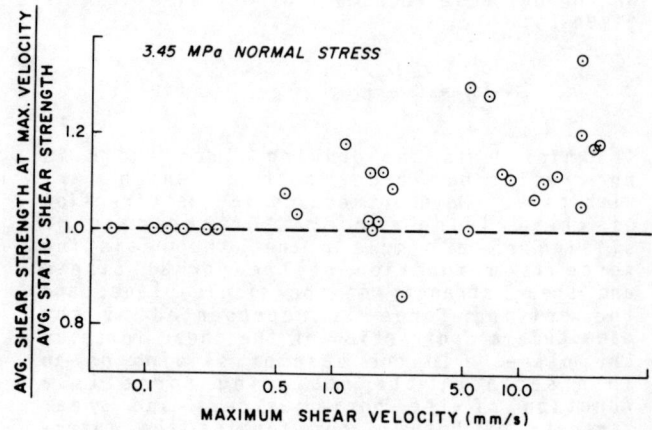

Figure 9. Preliminary results from dynamic direct shear tests.

Figure 9 shows preliminary results from tests on five samples. The shear strength at maximum velocity, normalized with respect to the "static" shear strength, is plotted against shear velocity. The results show generally increasing strength with increasing shear velocity. This is not inconsistent with the results reported by Crawford and Curran, since based on uniaxial compressive strength, the Loveland sandstone is in the middle range of the rock tested by Crawford and Curran. However, the roughness of the joints may impact on the velocity effects. The

C 232

relatively large amount of scatter indicated in Figure 9 is likely due to the rough character of the tested joints. More research is required to develop comprehensive models for dynamic behavior of rock joints including strength at higher velocities, changes in interstitial joint water pressure during dynamic loading, and effects due to fatigue and dilation.

6. CONCLUSIONS

It is considered practical to perform response history, dynamic rigid-block analyses for some concrete dam foundations when evaluating the effects of earthquakes is important. Time-varying forces determined from a dynamic finite element analysis of the dam are combined with static and inertia forces to determine time-varying resultant forces acting on potentially unstable rock masses. The potential mode of instability and factor of safety are determined at each time step during the earthquake. If the factor of safety falls below 1.0, cumulative permanent displacements can be estimated by double integration of the relative acceleration. Recent research has indicated that the dynamic material behavior may be an important consideration. Further research is required to develop adequate material models for the behavior of rock joints subjected to time-varying forces.

ACKNOWLEDGEMENTS

The example described is based on work performed for Auburn Dam. However, details were modified for illustration purposes. The authors wish to thank T. M. Tharp for his initial development work and M. Gould, S. Sture, and H. Y. Ko for the results from their research efforts at the University of Colorado.

REFERENCES

Crawford, A. M., and J. H. Curran, (1981). "The Influence of Shear Velocity on the Frictional Resistance of Rock Discontinuities," International Journal of Rock Mechanics, Mining Sciences and Geomechanics Abstracts, Vol 18., pp. 505-515.

Tarbox, G. S., K. J. Dreher, and L. Carpenter, (1979). "Seismic Analysis of Concrete Dams," Proceedings, 13th ICOLD Congress, New Delhi, pp. 963-994.

Von Thun, J. L. (1976). "Stability Analysis of Cut Slopes at Auburn dam," Proceedings, ASCE Specialty Conference on Rock Engineering for Foundations and Slopes, Boulder, Colorado, Vol. 1, pp. 349-360.

Von Thun, J. L., and C. W. Harris. (1981). "Estimation of Displacements of Rockfill Dams Due to Seismic Shakings," Proceedings, International Conference on Recent Advances in Geotechnical Earthquake Engineering and Soil Dynamics, St. Louis, Vol. 1, pp. 417-423.

PREDICTION OF THE PERFORMANCE OF SIDE RESISTANCE PILES SOCKETED IN MELBOURNE MUDSTONE

Prédiction de la performance des pieux de soutien latéral emboîtés dans de l'argilite de Melbourne

Vorhersage des Lastverformungsverhaltens von durch seitliche Schubspannungen belasteten und im Melbourne Lehmstein gegründeten Pfählen

H. K. Chiu
Geotechnical Engineer, Pilecon Engineering Sdn. Bhd., Kuala Lumpur, Malaysia. Formerly Postgraduate Scholar, Monash University

I. B. Donald
Associate Professor in Civil Engineering, Monash University, Clayton, Melbourne, Australia

SYNOPSIS

Piles socketed into rock are frequently required to carry their load entirely in side-shear. For soft rocks, with non-linear behaviour occurring at low stress levels, no simple method exists for predicting the load-deformation behaviour. This paper describes elasto-plastic and non-linear elastic finite element analyses for sockets in Melbourne Mudstone, using thee models of rock behaviour. Comparison with field tests shows that the best agreement is obtained with a non-linear elastic model which allows for work softening in the rock. Theoretical and measured stresses are also compared.

RESUME

Des pieux emboîtés dans de la roche sont souvent destinés à porter leur charge entièrement en cisaillement latéral. Dans le cas des roches tendres à comportement non-linéaire à bas niveau d'effort il n'existe aucune méthode simple de prédiction de comportement de déformation sous charge. En se référant à trois modèles de comportement de roches, cette communication décrit les analyses d'éléments finis élastoplastiques non-linéaires pour les emboîtements dans de l'argilite de Melbourne. La comparaison avec des essais in situ montre que la meilleure concordance s'obtient avec un modèle élastique non-linéaire qui tient compte d'un ameublissement de la roche dû au travaux. Les résistances théoriques et mesurées sont également mises en comparaison.

ZUSAMMENFASSUNG

Felsgegründete Pfähle müssen ihre Last oft durch seitliche Schubspannungen tragen. Für weichen Fels, wo nicht-lineares Verhalten bei kleinen Spannungen auftritt, existiert keine einfache Methode, um das Lastverformungsverhalten vorherzusagen. In diesem Beitrag wird eine elasto-plastische und nicht-linear elastische Finite Elemente Berechnung für Gründungen in Melbourner Lehmstein unter Verwendung von drei Modellen für das Felsverhalten beschrieben. Vergleiche mit Feldbeobachtungen zeigen, daß beste Übereinstimmung für ein nicht-linear elastisches Modell, welches das Nachgeben des Felses berücksichtigt, erzielt wird. Theoretische und gemessene Spannungen werden verglichen.

1. NOTATION

B	half width of rock mass model
c, c'	cohesion, drained or effective cohesion
c'_r	drained residual cohesion
D	pile diameter
E_m, E_{sec}	mass modulus; secant modulus at half peak deviator stress
F	factor giving degree of mobilization of shear strength
f_s	mobilised side resistance
G	shear modulus
G_1, G_2	constants relating to shear modulus
H	height of rock mass model
K	bulk modulus
K_1, K_2	constants relating to bulk modulus
L	pile length
m	exponent relating to shear modulus
n	exponent relating to bulk modulus
p	exponent relating to degree of shear strength mobilization
q	applied stress on pile
s_{ij}	deviatoric stress vector
w	saturated water content
α	side resistance reduction factor
γ	rock unit weight

δ_{ij}	Kronecker delta
ε_{ij}	strain vector
ν, ν_m	Poisson's ratio; Poisson's ratio of rock mass
ρ	pile settlement
σ_m	mean stress, $(\sigma_1 + \sigma_2 + \sigma_3)/3$
$\sigma_1, \sigma_2, \sigma_3$	principle stresses
τ, τ_f	shear stress; peak shear stress
ϕ, ϕ'	friction angle; drained or effective friction angle
ϕ'_r	drained residual friction angle

2. INTRODUCTION

In the Melbourne area large diameter piles carrying high loads are frequently constructed by socketing into the Melbourne Mudstone. Although in general a socket can carry appreciable proportions of its load in both end bearing and side shear there are cases when only the side resistance is significant. Piles may be constructed with a deliberate void at the base, particularly for test loading purposes, or it may be considered that sufficient compressible debris, which cannot be removed economically, will accumulate at the base of a bored pile hole to cause the pile to function effectively as a side-resistance-only pile. Piles resisting uplift forces will also rely entirely on side shear. For materials such as Melbourne Mudstone, which show non-linear behaviour at relatively small strains, no simple method of analysis exists for predicting the pile load-deformation behaviour. Socketed piles in Melbourne Mudstone are currently designed on a basis of permissible settlement, with a check on ultimate bearing capacity, and a satisfactory theory must be able to predict both types of behaviour sufficiently accurately. This paper describes computer studies for predictions well into the non-linear range, and validation checks which have been made against field measurement.

3. THEORETICAL CONSIDERATIONS

3.1 Review

Many theoretical solutions are now available for the load-settlement behaviour of socketed piles in the linear elastic range for a wide range of geometrical conditions and material properties, e.g. Osterberg & Gill (1973), Pells & Turner (1979), Donald, Sloan & Chiu (1980). However, it was shown by Williams (1980) that non-linear behaviour for piles in Melbourne Mudstone commences at relatively low settlements, well within the working range of the pile, and more sophisticated methods of analysis are required. This paper describes finite element studies using three models of rock behaviour and compares displacements and stresses with field measurements obtained by Williams (1980).

3.2 Rock models

Two variable moduli models, VM1 and VM2 were developed (Chiu and Donald (1982a) for use in a non-linear elastic finite element program VMOD4 Chiu (1982).

The variable moduli models used in this work are based on models developed by Nelson (1970) and Richards (1978, 1980). The model can be described by the following equations

$$\dot{\varepsilon}_{ij} = \frac{1}{2G} \dot{s}_{ij} + \frac{1}{3K} \delta_{ij} \dot{\sigma}_m \qquad (1)$$

$$K = K_1 \sigma_m^n + K_2; \quad \sigma_m > 0 \ \& \ \sigma_m \geqslant \sigma_{m_{max}} \qquad (2)$$

$$G = G_1 \sigma_n^m F + G_2; \quad \dot{J}'_2 > 0 \ \& \ \tau > \tau_{max} \qquad (3)$$

$$F = 1 - \left[\frac{\tau}{\tau_f}\right]^p \qquad (4)$$

Assuming a Mohr-Coulomb failure criterion,

$$\frac{\tau}{\tau_f} = \frac{\sigma_1 - \sigma_3}{(\sigma_1 + \sigma_3) \sin\phi + 2c \cos\phi} \qquad (5)$$

Once $\frac{\tau}{\tau_f} \to 1$, i.e. $F \to 0$, the shear modulus, G,

tends to zero and plastic yielding occurs without any loss of strength (VM1 model). However, if strain softening occurs (VM2 model), the shear strength drops abruptly to the residual value before plastic yielding follows.

It should be noted that equations 1-4, can also be used for unloading and reloading situations, however, only the virgin loading case is of interest in this work. Chiu (1981) and Chiu and Donald (1982a) have presented more details on both the VM1 and VM2 models together with their implementation into a finite element non-linear elastic program VMOD4. The program uses 4-noded isoparametric quadrilateral elements and incorporates a facility to redistribute excess shear and tensile stresses. A comprehensive user guide for program VMOD4 has been presented by Chiu (1982).

Two of the piles were also analysed by another finite element program, ELASP4, developed by Chiu (1981), using an elastic, ideally plastic rock model (EIP). This program uses 4 or 8 noded isoparametric quadrilateral elements and either a fully associated or non-associated flow rule. It can also handle limited tension analyses, as detailed by Chiu and Donald (1981).

3.3 Derivation of model parameters

This has been covered in greater detail by Chiu (1981) and Chiu and Donald (1982a) and only the main points are stated. The VM models are totally based on laboratory results of drained triaxial tests on the mudstone. From more than

100 triaxial tests on the moderately to highly weathered mudstone (Chiu 1981) it was found that the following water content correlations describe the basic deformation and strength properties of the mudstone.

$$\log E_{sec} \text{ (MPa)} = 3.511 - 0.078 \text{ w (\%)} \quad (6)$$

$$\nu = 0.097 + 0.009 \text{ w (\%)} \quad (7)$$

$$\log c' \text{ (kPa)} = 3.459 - 0.0603 \text{ w (\%)} \quad (8)$$

$$\tan \phi' = 0.986 - 0.0208 \text{ w (\%)} \quad (9)$$

$$c'_r = 0 \quad (10)$$

$$\phi'_r = 26^\circ \quad (11)$$

Using the above basic properties and by assuming different values for the variables of equations 1-4 in a finite element analysis of a triaxial test it was found that for the mudstone equations 2-4 become

$$K = K_1 = \frac{E_{sec}}{3(1-2\nu)} \quad (12)$$

$$G = G_1 = \frac{E_{sec}}{2(1+\nu)} \cdot F \quad (13)$$

$$F = 1 - \left[\frac{\tau}{\tau_f}\right]^p \quad (14)$$

where

$$p = 6 \quad (15)$$

$$K_2 = G_2 = m = n = 0$$

and E_{sec} and ν are given by equations 6 and 7.

For the VMI model, $c'_r = c'$ and $\phi'_r = \phi'$ but for the strain softening VM2 model equations 10 and 11 are used. Parameters for the EIP model were derived from quations 6 to 9.

4. FIELD TESTS

The five field tests analysed in this paper were selected from results published by Williams (1980). Construction details of the piles are summarised in Table 1 and geometries and test loadings are shown in Fig. 1.

The profiles of all socket walls were traced immediately prior to concrete casting and the average water content of the rock in the walls was determined for parameter prediction, as described in Section 2.3

Piles S1, S3, S5 and S15 were loaded at their upper surfaces and the vertical displacements measured. In addition, attempts were made to measure the radial movements of the socket walls for piles S3 and S5 (Williams, 1980). Pile M1 was loaded from below by flat jacks reacting against the base of the socket. Piles S1 and S3 were fitted with strain gauged briquettes placed at 4 levels within each pile for the determination of vertical and lateral stresses. Details of instrumentation have been given by Williams, Donald & Chiu (1980). The measured load-settlement curves are presented in Figs. 2 to 6 and the verticaland radial stress distributions in Figs. 7 and 8.

5. FINITE ELEMENT ANALYSIS

The general geometry and boundary conditions for the F.E. analyses are shown in Fig. 1 (c). The mudstone parameters derived from the water content measurements are given in Table 11, together with dimensions and details of the meshes used. All meshes were generated automatically.

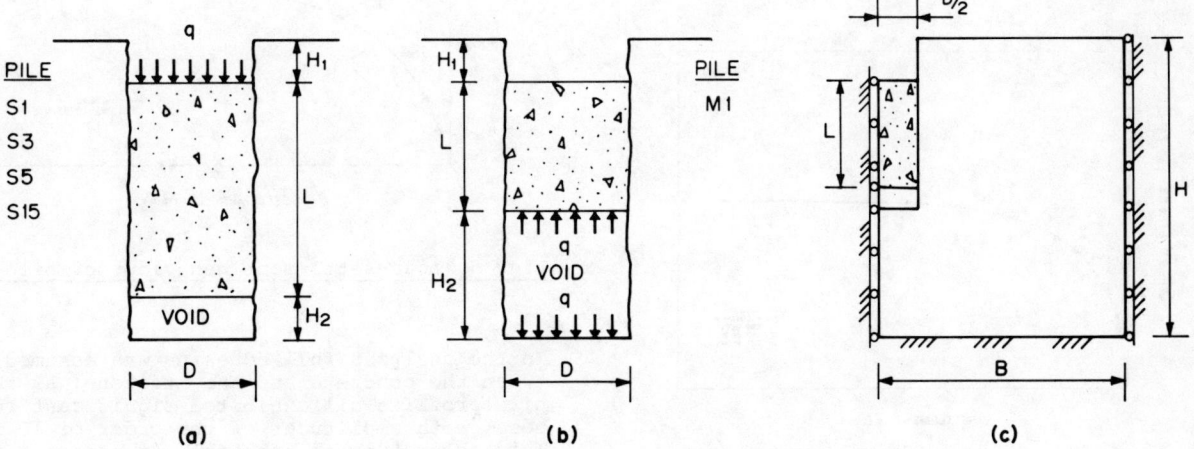

Fig. 1 Geometry and boundary conditions for side-resistance only piles

Table 1. Construction details and geometry of piles

PILE	L (m)	L (m)	L/D	H1 (m)	H2 (m)	Comments
S1	1.52	0.55	2.3	1.0	0.25	Drilled with 3 flight auger, soaked 12 hours, remoulded cake removed, water pumped out, concrete poured.
S3	2.59	1.12	2.3	1.0	0.25	As for S1, but socket wall roughened with a grooving tool
S5	2.51	1.17	2.1	1.0	0.25	As for S1
S15	0.87	0.395	2.2	0.1	0.15	Drilled with core barrel, asperities cut with 12° slope, 10 mm height, 100 mm pitch.
M1	2.0	1.22	1.6	10.5	3.0	Drilled with bucket auger, cast under bentonite - approx. 5 mm filter cake.

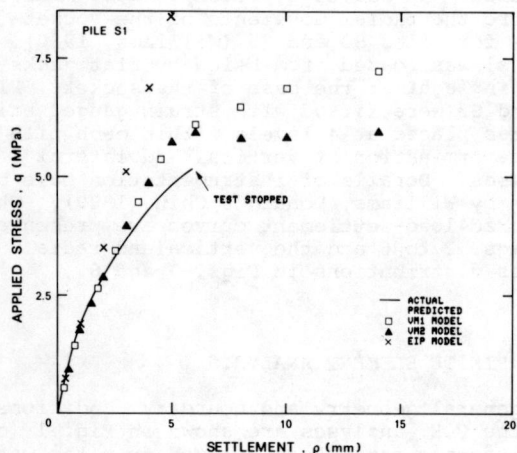

Fig. 2   Load-settlement behaviour of pile S1

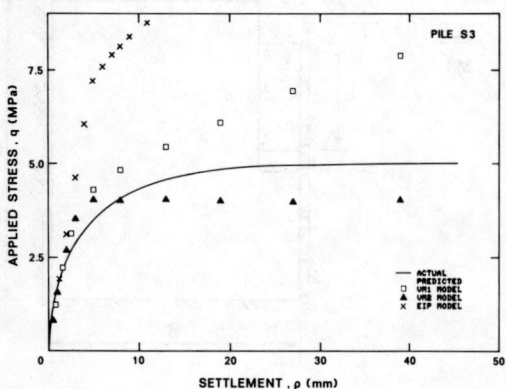

Fig. 3   Load-settlement behaviour of pile S3

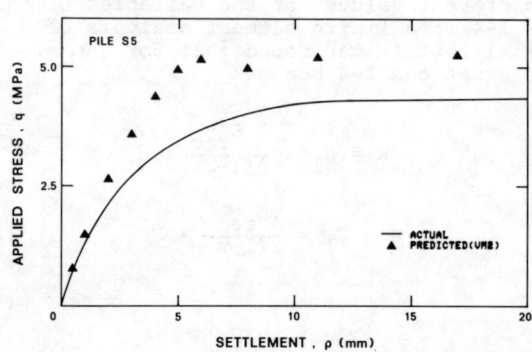

Fig. 4   Load-settlement behaviour of pile S5

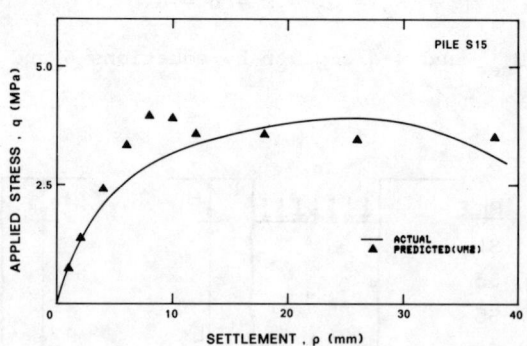

Fig. 5   Load-settlement behaviour of pile S15

In the analyses full adhesion was assumed between the concrete and the mudstone, as the pile profiles all exhibited significant roughness, with amplitudes of the order of 10 mm. This assumption is debatable in general, and undoubtedly invalid for the pile cast under bentonite and it will be discussed further in Section 7.

Table 11. F.E. Analyses - rock parameters and mesh details

| | | | | ROCK PARAMETERS | | | | | | | MESH DETAILS | | |
PILE	WATER CONTENT (%)	E_m (MPa)	ν_m	K_1 (kPa) $\times 10^3$	G_1 (kPa) $\times 10^3$	P	c' (kPa)	ϕ' (deg.)	ϕ'_r (deg.)	γ (kN/m^3)	B (m)	H (m)	Number of Elements
S1	16.3	260	0.24	167	105	6	300	32.9	26	23.5	12	25	338
S3	18.5	410	0.26	285	163	6	220	31.0	26	23.5	20	30	338
S5	18.1	420	0.26	292	167	6	233	31.4	26	23.5	20	30	338
S15	18.2	76	0.26	521	298	6	230	31.3	26	23.5	8	12	343
M1	10.4	275	0.19	148	116	6	585	36.8	26	23.5	20	40	330

*for the piles, E_c = 35 GPa and ν_c = 0.2

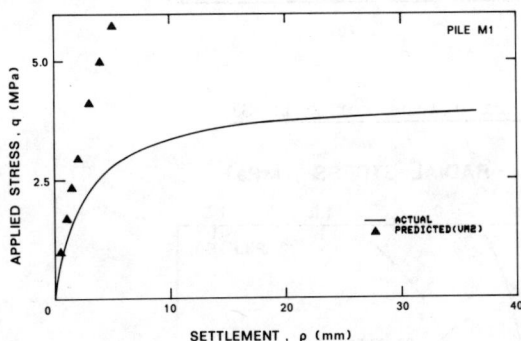

Fig. 6 Load-settlement behaviour of pile M1

The other extreme was treated by Rowe & Pells (1980) who assumed that all failure was localised at the interface and used a simple dilating joint model in an otherwise elastic medium. The differences in these assumptions reflect mainly the differences in much observed field behaviour for piles in Sydney Sandstone (Rowe & Pells) and Melbourne Mudstone, brittle failure predominating in the former and work hardening in the latter. The complete story for either rock type is not really so clearcut and a combination of approaches is under development.

All five piles were analysed using the strain-softening model VM2, and piles S1 and S2 were analysed additionally using the VM1 and EIP models. The results are plotted in Figs. 2 to 8 for comparison with field measurements.

6. COMPARISON OF RESULTS

6.1 Load-settlement behaviour

Reference to Figs. 2 to 6 shows that the linear

elastic range is small, both for theory and measurement, thus justifying the need for more advanced analytical techniques. All experimental curves, except for S15, either approach a constant load or exhibit mild work hardening behaviour and even S15 does not show work softening until the settlement has reached 27 mm - well beyond the normal design limit.

The VM2 (strain-softening) model gives reasonable predictions of peak strengths and of the general trend of the applied stress-settlement curves, except for pile M1 where the presence of a 5 mm cake of bentonite has obviously reduced both the strength and stiffness of the socket. The field test for pile S1 was stopped before failure because of experimental difficulties, but extrapolation of the measured curve on the basis of the other four results indicates a likely peak strength only slightly in excess of the predicted value.

For piles S1 and S3 the VM1 and EIP models both predict a stiffer than measured response after the initial linear elastic section, with the EIP results grossly overestimating the point of yielding. The VM1 model predicts the behaviour reasonably well until significant yielding has occurred, after which it indicates a steady work hardening effect which was not found in the experiments. This shows the importance of allowing for brittle failure of individual elements, as is done with the VM2 model, even though the overall field response generally exhibits no loss in load capacity after reaching the peak value. The EIP Model performed poorly and this can be attributed in part to the type of element used. Many investigators have experienced difficulty with elasto-plastic analyses in predicting collapse loads of foundations accurately and Chiu & Donald (1982b) have presented a comprehensive summary of the relative performances of a range of elements and integration techniques.

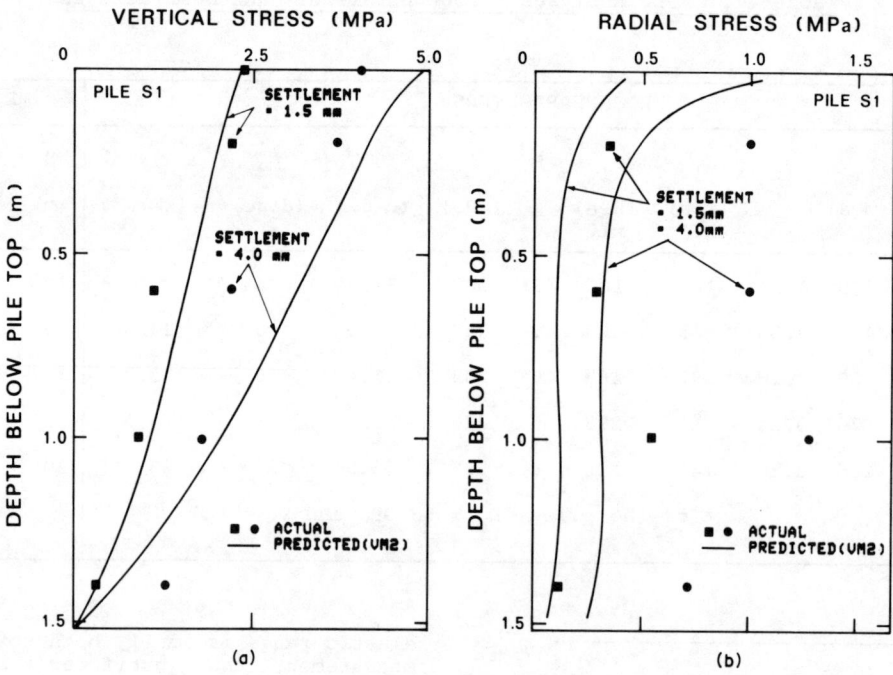

Fig. 7 Vertical and radial stress distributions for pile S1

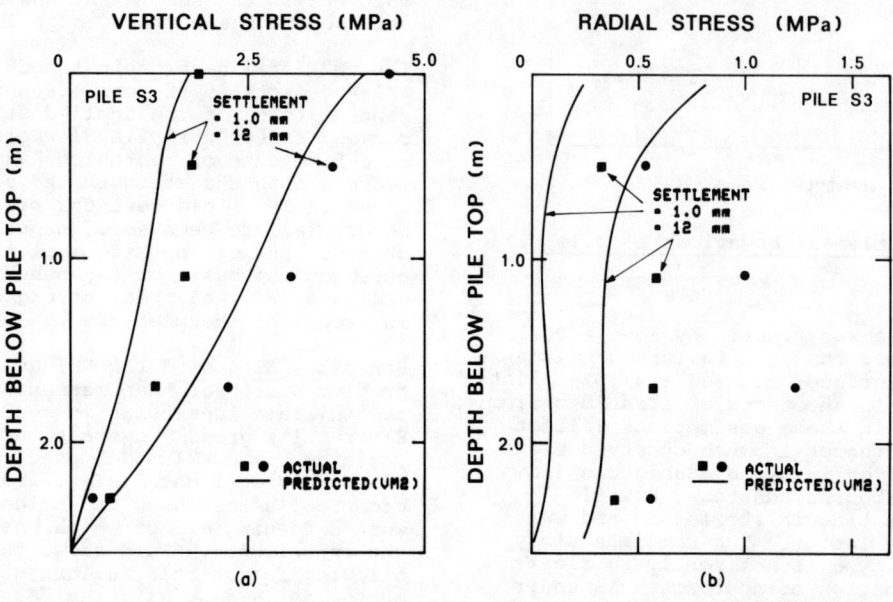

Fig. 8 Vertical and radial stress distributions for pile S3

6.2 Stress distributions

A method of analysis cannot be considered valid unless it is capable of predicting correct stress distributions as well as accurate settlement behaviour. The vertical and radial stress distributions for piles S1 and S2, calculated using the VM2 model, are compared with measurements by Williams (1981) in Figs. 7 and

8. Comparisons are given at settlements of 1.5 and 4.0 mm for S1 and 1.0 and 12 mm for S3, representing points in the elastic and significantly yielded regions respectively.

The vertical stress distribution is reasonably well predicted for both piles, allowing for the unavoidable scatter in the experimental results.

The curves could also be approximated accept-
ably by a linear relationship, implying a uniform shear stress, f_s, down the concrete-rock
interface. This is not surprising, as the
piles were all rather short, and the effects of
pile compressibility would be minor.

It is interesting to note that the theory predicts an average value of side reduction
factor

$$\alpha = \frac{f_{su}}{q_a}$$ (as defined by Williams, Johnston &

Donald, 1980) of 0.6, which agrees well with
the correlation between α and "unconfined"
compression strength, q_a, published by Williams,
Johnston & Donald.

The radial stress distributions are not well
predicted, theoretical values consistently
under-estimating the field stresses. Any tendency towards slip and dilation at the interface would increase lateral stresses significantly, as indicated by the profile analysis of
Dight & Chiu (1981), although accurate prediction and measurement of radial stresses remains
difficult. In spite of these problems it can
be said that both theory and experiment indicate a significant increase in lateral stress
with loading, with consequently an increase in
shearing resistance at the interface which
masks any strength loss due to progressive
brittle failure in the rock.

7. DISCUSSION

The VM2 model gives generally excellent results
for the four piles cast without bentonite in
the socket and may be considered to have been
validated as a design tool by the field experiments. The VM1 model is less satisfactory at
large strains as it ignores the experimental
fact that the drained residual shear parameters
for the mudstone, c_r' and ϕ_r', are much smaller
than the peak values, c' and ϕ'. This assumption also contributes to the poor predictive
performance of the EIP model, exaggerating the
effects of the numerical errors inherent in
elasto-plastic analyses.

All analyses were performed assuming full adhesion between concrete and rock, which although
a valid assumption for a very rough interface
might not be entirely correct for a surface
with relatively low asperity heights and angles.
Williams (1980) excavated his side resistance
piles after test and observed that failure, in
most instances, appeared to have occurred by
shear through the roots of the rock asperities,
with a shear zone extending up to 100 mm into
the rock mass. Separation between the rock and
the concrete was seldom evident and these observations lend considerable credence to the
assumption made for the analyses. However, it is
not certain that absolutely zero slip occurred
at the interfaces and future consideration
should be given to the addition of some form of
sliding, dilating joint element to the program.
This will be essential for analysing piles with
bentonite or other inclusion at the interface.

Recently a promising analytical method based on
interface profiles was published by Dight &
Chiu (1981). Their method uses the Ladanyi &
Archambault (1970) equation for dilatant joint
behaviour and a relatively simple calculation
which can be done on a pocket calculator. Chiu
(1981) has compared results obtained from the
profile method with the VM2 calculations described earlier, and his graphs are reproduced
in Figs. 9 and 10 for piles S1 and S3. The
load-settlement behaviour predicted from profiles underestimates the failure loads by a
greater amount than the VM2 calculations, but
the curve shapes are more realistic, lacking
the sudden failure of the VM2 model. The profile curves exhibit greater initial stiffness,
probably as a result of ignoring pile compressibility in the calculations. From the limited
evidence presented there is little to choose
between the two methods as reasonable predictors
of field settlement behaviour.

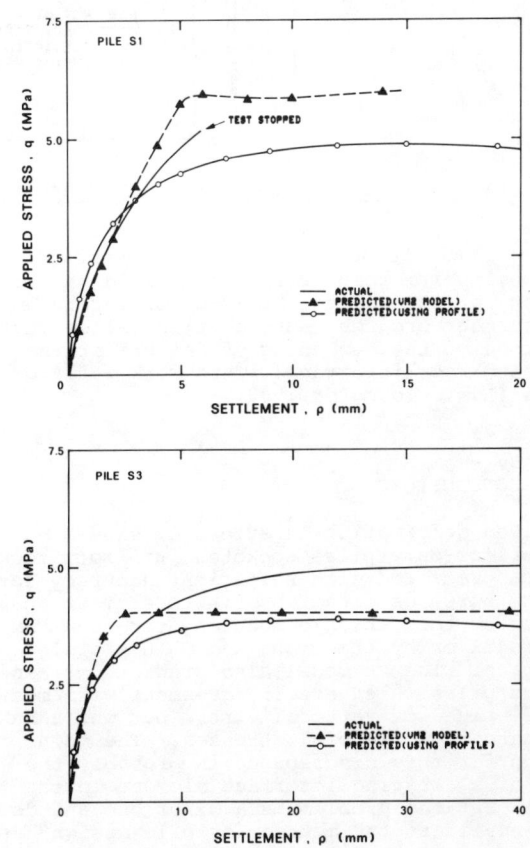

Fig. 9 Comparison of predicted and actual load-
settlement behaviour of Piles S1 and S3

Higher radial stresses are predicted by the
dilating interface profile model, although they
still appear less than the field measurements.
The profile method assumes a constant radial
stiffness over the length of the pile, which
leads to underestimates of radial stress at the
top and base levels. Williams (1980) reported
the possibility of large errors in the measured
radial stresses at low loads, but considered
that near failure the error would have been

R

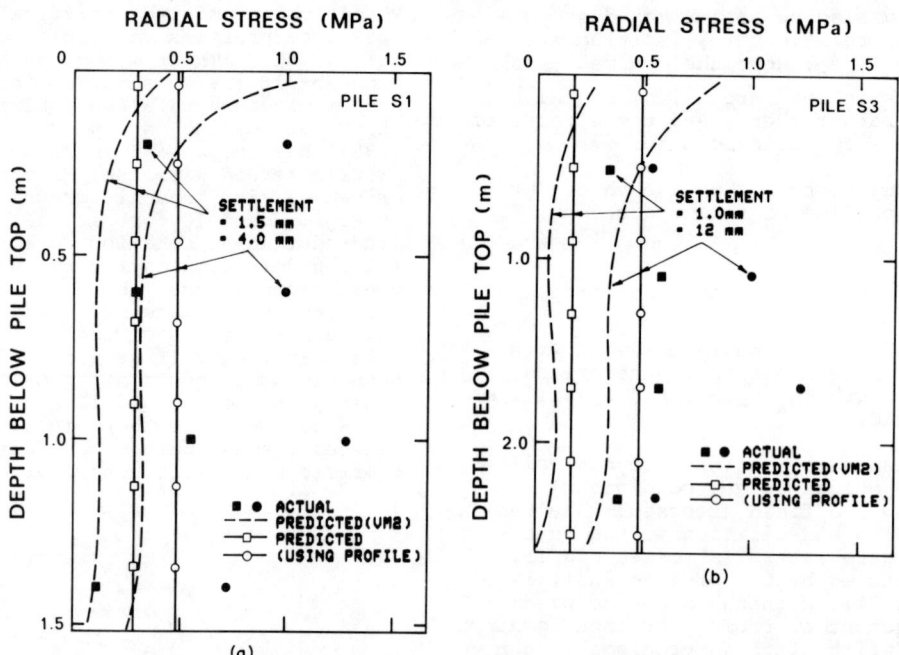

Fig. 10   Comparison of predicted and actual load-settlement behaviour of piles S1 and S3

only ± 20%. It must therefore be concluded that neither method of analysis yields a satisfactory picture of the build-up of lateral stress and further progress must await detailed investigation of the mechanics of failure at the concrete/rock interface. This work is in progress (Lam & Johnston, 1982).

8   CONCLUSIONS

The load-deformation behaviour of side-resistance-only piles socketed into soft rock may be predicted with sufficient accuracy for design purposes either by finite element analyses using the variable modulus, strain softening VM2 model or by the Dight and Chiu profile approach. The VM model also predicts stresses in the piles which are in agreement with measured valued of vertical stress but which underestimate field lateral stresses. The main reason for this discrepancy is probably the lack of a dilating interface element in the model, but the problem is a minor one and does not invalidate the method for pile design based on a settlement criterion. The analyses assume intact rock properties adjacent to the interface and will overestimate pile/rock stiffness if the interface contains an appreciable thickness of bentonite or remoulded rock. The methods were developed for mudstone with a small number of clean, tight joints, modelling conditions at the test sites.

9   ACKNOWLEDGEMENTS

The work described in this paper forms part of a continuing programme of research at Monash University, under the direction of Associate Professor I.B. Donald and Dr. I.W. Johnston, into the performance of foundations on Melbourne Mudstone, with special reference to rock socketed piles. Grateful acknowledgement is made of permission to use the field data gathered by Dr. A.F. Williams and of the contribution of P.M. Dight to the profile method.

10   REFERENCES

Chiu, H.K. (1981). Geotechnical properties and numerical analyses for socketed pile design in weak rock: Ph.D. Thesis, Monash University, Melbourne.

Chiu, H.K. (1982). Program VMOD4 - For Finite Element Non-Linear Elastic Analysis of Foundations: Research Report 82/01/G, Dept. Civil Eng., Monash University, Melbourne.

Chiu, H.K. and Donald, I.B. (1981). Limited tension analysis of socketed piles: Proc. 10th Int. Conf. Soil Mech. and Fndn. Eng., (2), 659-662, Stockholm.

Chiu, H.K. and Donald, I.B.(1982a). Prediction of the Behaviour of Foundations on Melbourne mudstone using a Variable Moduli Model: Research Report 82/02/G, Dept. Civil Eng., Monash University, Melbourne.

Chiu, H.K. and Donald I.B. (1982b). Finite element techniques to predict collapse loads of foundations: Proc. Int. Conf. F.E.M., (1), 341-347, Shanghai.

Dight, P.M. and Chiu, H.K. (1981). Prediction of shear behaviour of joints using profiles: Int. J. Rock Mech. Min. Sci. &

Geomech. Abstr., (18), 369-386.

Donald, I.B., Chiu, H.K. and Sloan, S.W. (1980). Theoretical analyses of rock socketed piles. Proc. Int. Conf. Structural Foundations on Rock (1), 303-316, Sydney: Balkema.

Ladanyi, B. and Archambault, G. (1970). Simulation of shear behaviour of a jointed rock mass: Proc. 11th Rock Mech. Symp. Amer. Inst. Min. Met., 105-125.

Lam, S.K.T. and Johnston, I.W. (1982). A constant stiffness direct shear machine: Proc. 7th S.E. Asian Geotech. Conf. Hong Kong.

Nelson, I. (1970). Investigation of Ground Shock Effects in Non-Linear Hysteretic Media: Report S-68-1, Modelling the Behaviour of a Real Soil, U.S. Army Eng. Waterways Experiment Station, Vicksburg, Mississippi.

Osterberg, J.O. and Gill, S.A. (1973). Load transfer mechanism for piers socketed in hard soils or rock: Proc. 9th Canadian Rock. Mech. Symp., 235-261, Montreal.

Pells, P.J.N. and Turner, R.M. (1979). Elastic solutions for the design and analysis of rock socketed piles: Canadian Geotech. J. (16), 481-487.

Richards, B.G. (1978). Application of an experimentally based non-linear constitutive model of soils in laboratory and field tests: Aust. Geomech. J., G8, 20-30.

Richards, B.G. (1980). Automatic joint elements generation to simulate strain softening yeild behaviour in earthen materials: Proc. 3rd A.N.Z. Conf. Geomech. (2), 233-246, Wellington.

Rowe, R.K. and Pells, P.J.N. (1980). A theoretical study of pile-rock socket behaviour: Proc. Int. Conf. Structural Foundations on Rock, (1), 253-264, Sydney: Balkema.

Williams, A.F. (1980). The Design and Performance of Piles Socketed into Weak Rock: Ph.D. Thesis, Monash University, Melbourne.

Williams, A.F., Donald, I.B. and Chiu, H.K. (1980). Stress distributions in rock socketed piles: Proc. Int. Conf. Structural Foundations on Rock (1), 317-325, Sydney: Balkema.

Williams, A.F., Johnston, I.W. and Donald, I.B. (1980). The design of socketed piles in weak rock: Proc. Int. Conf. Structural Foundations on Rock, (1), 327-347, Sydney: Balkema.

INFLUENCE OF COMPLEX GEOTECHNICAL CONDITIONS ON THE STRESS-STRAIN FIELD NEAR LARGE-SPAN BRIDGE FOUNDATIONS. STATIC AND DYNAMIC ANALYSIS

L'influence des conditions géotechniques complexes sur le champ de contraintes-déformations dans le voisinage des fondations d'un pont à grande travée. Analyse statique et dynamique

Einfluss der komplizierten geotechnischen Bedingungen auf das Spannungs- und Verformungsfeld an der Gründung einer Brücke grosser Spannweite. Statische und dynamische Analyse

Ibrahim Jasarević, Civ.Eng., Ph.D.Sen., Lect.
Nenad Bićanić, Civ.Eng., Ph.D.Sen., Lect
Marko Vrkljan, Civ.Eng.
Branimir Saban, Geol.Eng.
Verica Grubisić, Civ.Eng.
Miroslav Andrić, Geoph.Eng.
Faculty of Civil Engineering, University of Zagreb, Yugoslavia

SYNOPSIS

Complex geological and geotechnical conditions (faults and discontinuities) at the site for the large-span bridge (over the 100 m deep canyon near the town of Rijeka) have indicated the necessity for a detailed analysis of the stress-strain field near foundation blocks. Extensive geological, geophysical and geotechnical investigation, together with microseismology studies have been conducted to obtain parameters for geotechnical analysis. Both, 2D and 3D, static and dynamic analysis have been performed using the displacement-based finite element method. The soil-structure model is composed from 3D linear soil/rock elements (constant strain) whereas the bridge structure is modelled using 3D beam elements. Dominant faults and discontinuities have been analysed for three extreme states - full bond, free sliding and completely independent movement of neighbouring sides. Dynamic sensitivity analyses indicated significant influence of boundary condition and discontinuity idealisations on fundamental frequency. Superposition of stresses due to selfweight and seismic forces showed predominant influence of discontinuities on the stress-strain field and indicated the need for prestressing anchors. Surveying observations are constantly performed to follow the rock mass improvement and to measure the displacements of control points.

RESUME

Cet article présente les études nécessaires à la réalisation des fondations d'un pont à grande travée dans les conditions géologiques et géotechniques complexes. On y a souligné les problèmes qui apparaissent lors de la fondation de telles constructions, ainsi que l'amélioration des masses rocheuses. L'influence essentielle des discontinuités dans la masse rocheuse sur le champ de contraintes-déformations dans le voisinage des fondations a été constatée à l'aide de calculs géostatiques pour les modèles tridimensionels et bidimensionels en utilisant la méthode des éléments finis. Plusieurs calculs dynamiques pour les différentes suppositions sur les failles dominantes ont été effectués afin de pouvoir estimer la susceptibilité du système construction-sol lors des activités séismiques. On a constaté que les valeurs des fréquences propres varient considérablement dans le cas où l'influence du sol et des failles est négligée. Après avoir analysé le champ de contraintes-déformations autour du bloc de la fondation du pilier, près d'une faille, on a trouvé qu'il est indispensable d'améliorer la masse rocheuse. Cette amélioration est effectuée à l'aide de tirants d'ancrage précontraints. L'effet de l'amélioration est contrôlé par la méthode de calcul citée ci-dessus.

ZUSAMMENFASSUNG

Die Darstellung der Gründung einer Brücke größerer Spannweite mit verwickelten ingenieur-geologischen und geotechnischen Bedingungen weist auf die Probleme hin, die beim Grundbau größerer und schwerer Bauten auftreten, wobei auch die Verbesserung der Gesteinsmassen eingeschlossen ist. Durch geostatische Berechnungen für ein ebenes und ein räumliches Modell wurde, mittels der Methode der Finiten Elemente, ein wesentlicher Einfluß der Diskontinuität in der Gesteinsmasse auf das Spannungs- und Verformungsfeld in der Nähe des Fundaments festgestellt. Um die Empfindlichkeits des Systems "Konstruktion-Boden" bei seismischen Einflüssen zu begutachten, wurden mehrere dynamische Berechnungen unter verschiedenen Voraussetzungen in den maßgebenden Verwerfungen durchgeführt. Man stellte fest, daß die Werte der eigenen Frequenzen bedeutend variieren, wenn man den Einfluß des Bodens bzw. der Verwerfung in Betracht nimmt. Durch die Analyse des Spannungs- und Verformungsfeldes um den Fluß des Fundamentsblockes neben der Verwerfung wurde festgestellt, daß der Einbau von vorgespannten Ankern berechtigt ist.

1. INTRODUCTION

On the location where bypass road crosses loo m depp canyon, two parallel large span bridges (45m+98m+45m) are planned, with beam heights varying from 2,9o to 4,75 m. For both bridges cantilever type construction process is adopted, and therefore anchorage blocks have been fixed on the rock mass with Prestressed snchors. Side spans of the bridgs are cast on the scaffolding, and the middle span is formed, using cantilever construction, in sections of 5 m. Three dimensional structural and geotechnical model is given on Fig. 1, whereas Fig. 2 illustrates the

canyon and site conditions in May 1981.

2. ENGINEERING - GEOLOGICAL INVESTIGATIONS

In wider geological sense the bridge area is a part of carst of the Dinarides, being of a complex tectonic structure and NW-SE strike (as the Dinarides). The bridge is located in tectonic unit defined by compression structure and deformations of tangential type (reverse faults, overturned, overlap and fan folds, Fig. 3).
To be more precise, the bridge is located on the southwest flank of the anticline stretching out NW-SE, direc-

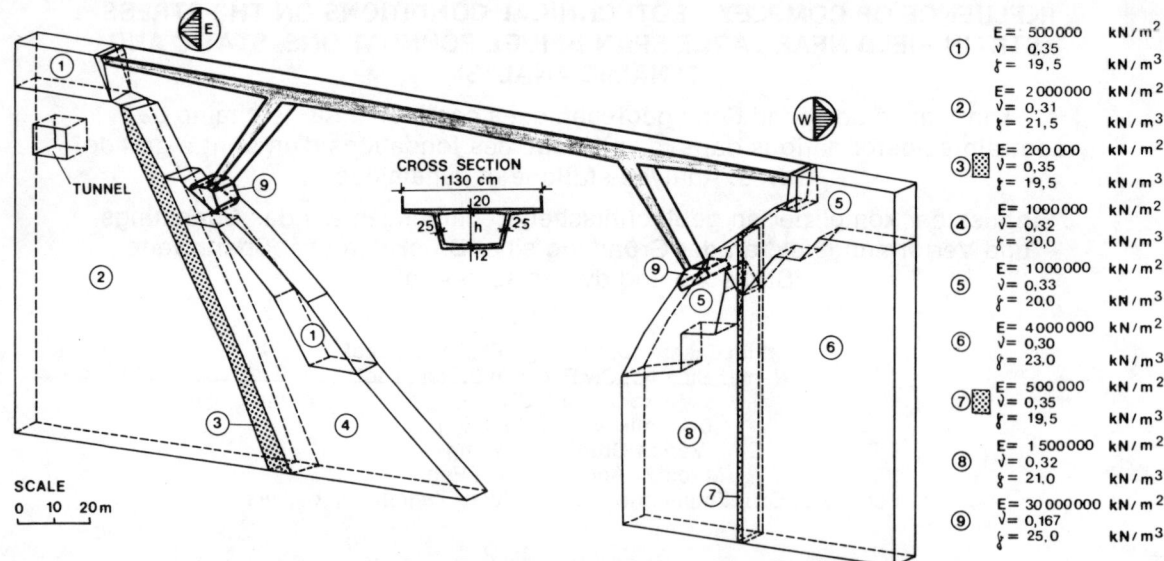

	E=	500 000	kN/m²
①	v=	0,35	
	f=	19,5	kN/m³
	E=	2 000 000	kN/m²
②	v=	0,31	
	f=	21,5	kN/m³
	E=	200 000	kN/m²
③	v=	0,35	
	f=	19,5	kN/m³
	E=	1 000 000	kN/m²
④	v=	0,32	
	f=	20,0	kN/m³
	E=	1 000 000	kN/m²
⑤	v=	0,33	
	f=	20,0	kN/m³
	E=	4 000 000	kN/m²
⑥	v=	0,30	
	f=	23,0	kN/m³
	E=	500 000	kN/m²
⑦	v=	0,35	
	f=	19,5	kN/m³
	E=	1 500 000	kN/m²
⑧	v=	0,32	
	f=	21,0	kN/m³
	E=	30 000 000	kN/m²
⑨	v=	0,167	
	f=	25,0	kN/m³

CROSS SECTION
1130 cm

Fig. 1 Schematic soil structure model

Fig. 2 View of the Rječina canyon and the bridge under construction (1981)

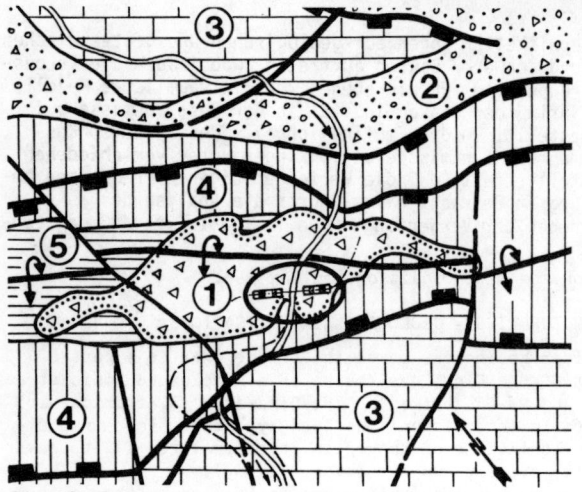

Fig. 3 Tectonic situation of the wider area around the bridge

tly along its crest. The north-eastern flank of the anticline is overturned while the southwest one slopes steeply. Dolomites and limestones of the Upper Cretaceous in the southwest flank of the anticline lean, along the reverse fault, against the younger Upper Cretaceous limestones.

In addition to these rocks the location also contains Tertiary breccias and conglomerates, the so called Tertiary clastic sediments. Contacts of the Cretaceous and Tertiary deposits are tectonic-erocional. By stepwise subsidence of Tertiary clastic sediments during consolidation and later on, a series of cracks and diaclases was formed. As opposed to compression deformations within the Cretaceous and Eocene deposits these Eocene - Oligocene deformations are "tensile" - open cracks.

In the area of the bridge and on its location detailed engineering - geological mapping and profiling operations were performed in 1:5oo and 1:loo scale, with special emphasis being given to shooting the situation and profiles in excavations for the bridge foundation in 1:5o scale - anchoring blocks, abutment piers and diagonal bracing. The hillslope on the eastern bank of the bridge consists of compact consolidated clastic sediments and

C 246

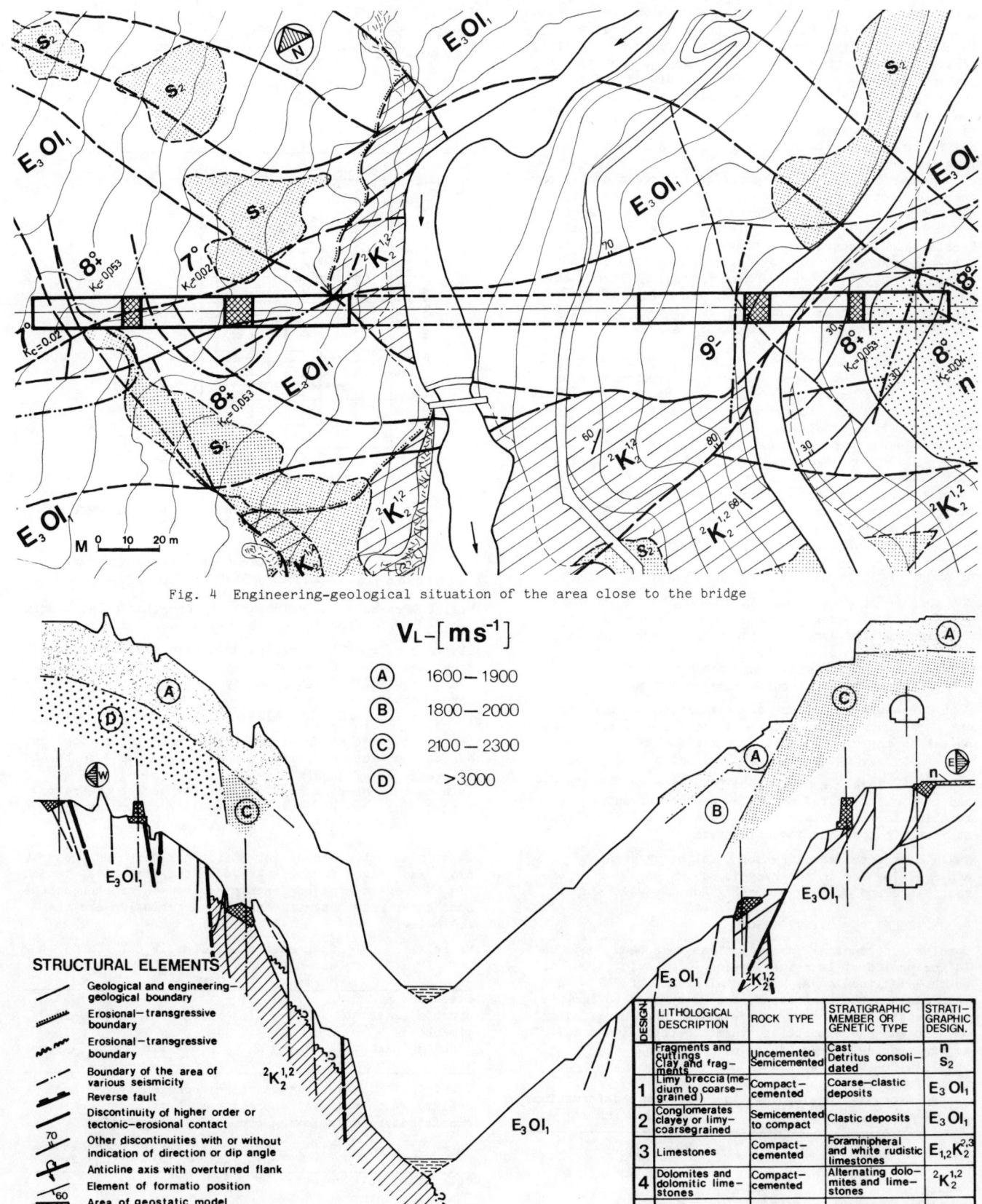

Fig. 4 Engineering-geological situation of the area close to the bridge

$V_L - [ms^{-1}]$

(A) 1600—1900

(B) 1800—2000

(C) 2100—2300

(D) >3000

$E_3\,OI_1$

$E_3\,OI_1$

$E_3\,Ol_1$

$E_3\,OI_1$

$^2K_2^{1,2}$

$E_3\,OI_1$

$E_3\,OI_1$

$^2K_2^{1,2}$

n (E)

STRUCTURAL ELEMENTS

———	Geological and engineering-geological boundary
∿∿∿	Erosional—transgressive boundary
∼∼∼	Erosional—transgressive boundary
—·—·—	Boundary of the area of various seismicity
▬▬▬	Reverse fault
———	Discontinuity of higher order or tectonic-erosional contact
⤢70	Other discontinuities with or without indication of direction or dip angle
⟊	Anticline axis with overturned flank
⤲60	Element of formatio position
▭▭▭	Area of geostatic model
—··—··	Axis of built structures, tunels

DESIGN.	LITHOLOGICAL DESCRIPTION	ROCK TYPE	STRATIGRAPHIC MEMBER OR GENETIC TYPE	STRATI-GRAPHIC DESIGN.
	Fragments and cuttings	Uncemented	Cast	n
	Clay and fragments	Semicemented	Detritus consolidated	s_2
1	Limy breccia (medium to coarse-grained)	Compact-cemented	Coarse-clastic deposits	$E_3\,OI_1$
2	Conglomerates clayey or limy-coarsegrained	Semicemented to compact	Clastic deposits	$E_3\,OI_1$
3	Limestones	Compact-cemented	Foraminipheral and white rudistic limestones	$E_{1,2}K_2^{2,3}$
4	Dolomites and dolomitic limestones	Compact-cemented	Alternating dolomites and limestones	$^2K_2^{1,2}$
5	Dolomitic breccias	Compact-cemented	Transition clastic deposits	$K_{1,2}$

Fig. 5 Engineering-geological and geophysical profile along the bridge axis

C 247

carbonate rocks (calcitic breccia and dolomitic limestones, Fig. 4). Calcitic breccias are unstratified to poore-
ly stratified and dolomitic limestones are well to platy
stratified. Contacts between sediments are tectonic or
tectonic-erosional. Recorded discontinuities and cracks
are perpendicular to the bridge axis, while the dip angles
down the slope are larger than the slope dip itself. The
majority of cracks are opened in upper sections without
any filling, while in the lower sections it contains filling (red soil).
The western hillslope of the bridge is predominantly composed of unstratified to poorly stratified Tertiary clastic sediments-calcitic breccias. The Cretaceous carbonate
rocks (dolomitic limestones) form the base of clastic
Tertiary sediments and the contact between them is erosi-
onal-transgressive, not easily recognized and without
usually frequent accompanying morphological forms (caverns, caves and zones of red soil). The strike of discontinuities and cracks is parallel or almost parallel to
the strike of the bridge axis, while slopes are very
steep or vertical.
Discontinuities were formed within the homogeneous rocks
(unstratified limy breccia) during the tectonic movements
and consolidation of sediments. Later on they were wide-
ned and filled with clay (red soil) in their lower (deeper) parts. The condition of the rock along the discontinuities is defined as highly weathered (RQD < 5o%) while
the rock sections between the discontinuities are less
weatherad (RQD > 5o%) (Fig. 8).
The performed exploration jobs have determined the position and length of anchoring sections as well as the
elements required for calculating the bearing capacity
and stability of slopes on both sides of the bridge.

3. GEOPHYSICAL INVESTIGATIONS AND SEISMIC MICROZONING

The purpose of geophysical investigations was twofold
- indication of quasi homogenous zones, where the deformation moduli, dynamic properties and the degree of
  fracturing are similar,
- determination of seismic parameters and maximum
  expected gound accelerations on bridge supports.

Both down-hole and cross-hole methods for compression
and shear wave speeds have been used, and microtremors
have been employed to determine maximum accelerations -
different wave speed zones are indicated on Fig. 5.

Seismicity level is determined following Medvedev's method [4], with amplification at particular support points
to allow for poor state of the rock and infill material
at local faults and discontinuities.

Maximum expected accelerations, following Kanai [2], [3]
and Seed [5] are in the range o,089 g - o,095 g on the
east slope and o,089 g - o,098 g for the west slope.

4. GEOTECHNICAL INVESTIGATIONS

Complex geotechnical investigations have been conducted
in the period of last three years - 18 borings (out of
which 9 inclined ones) with total length of 352 m have
been made on the east slope, and 22 borings (6 inclined)
with total length 38o m on the west slope. Longitudinal
wave speeds nad respective dynamic deformation moduli for
typical borings are given below.
VERTICAL BORING B_V -1
(pier foundation block, west side, L = 15 m, β = 9o$^{\circ}$)

depth interval (m)	longitudinal wave speed (m/sec)	dynamic deformation modulus (MN/m^2)
0 - 3	1395	1.71o
3 - 6	3756	23.21o
6 - 9	2856	15.35o
9 -12	2727	13.2oo
12 -15	4615	39.49o

INCLINED BORING SK-1
(pier foundation block, east side, L = 25 m, β = 7o$^{\circ}$)

dept interval (m)	longitudinal wave speed (m/sec)	dynamic deformation moduluc (MN/m^2)
0 - 5	1678	1.95o
5 -10	3o49	12.54o
10 -15	2283	6.97o
15 -20	1946	6.8oo

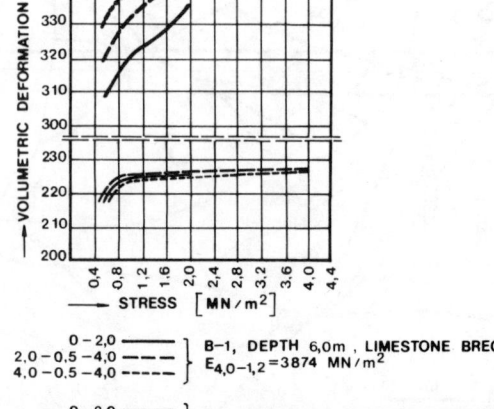

0 -2,0 ⎯⎯⎯
2,0 -0.5 -4,0 ⎯ ⎯ ⎯ } B-1, DEPTH 6,0m, LIMESTONE BRECCIA
4,0 -0.5 -4,0 ⎯⎯⎯⎯ $E_{4,0-1,2}$ =3874 MN/m^2

0 -2,0 ⎯⎯⎯
2,0 -0.5 -3,8 ⎯ ⎯ ⎯ } B-2, DEPTH 6,2m, WEATHERED CLAY
4,0 -0.5 -4,0 ⎯⎯⎯⎯ $E_{3,2-1,7}$ =456 MN/m^2

Fig. 6 Pressure vs. volumetric deformation relationship
for different depths (pressiometer diagram)

Static moduli of deformation have been evaluated using
pressiometer and this indicated an approximate correlations between the static and dynamic deformation moduli
for limestone breccia

$$K = E_{din}/E_{stat} = 4,2\mskip - 5,4o$$

The fault gouge material (clay type material in dominant
discontinuities) deformation moduli were found to be in
the range 163-263 MN/m^2 .
As a measure of rock mass damage, the degree of fracturing defined as the ratio of wave speeds through rock
mass and rock monolith

$$\xi = v_s^s / v_s^m$$

This is almost constant (o,43) for any depth on the east
side, and ranges from o,4o-o,7o on the west side, indicating a large degree of fracturing even on greater depths.
Some other relevant rock mass characteristics are the
following.

Material	Compressive strength MN/m^2	Tensile strength MN/m^2	CaCO$_3$ content %
limestone breccia (east side)	39,6-73,8	4,23-6,81	96,o-99,7
limestone breccia (west side)	45,6-158,2	3,02-9,1o	56,5-99,9

The infill material within the discontinuity has been
analysed with triaxial shear stress test under drained
conditions, and following properties are obtained.

Material	Angle of internal friction ($^{\circ}$)	Cohesion (kN/m^2)
Dolomite (fault breccia)	27°42'	46
Clay like material (fault gauge)	22°	28

5. STATIC AND DYNAMIC SOIL STRUCTURE INTERACTION ANALYSIS

On the basis of geophisical and geological investigations quasi homogenous zones have been identified. Very detailed 2D finite element model (triangles and quadrilaterals) has been adopted in the static analysis for different stages of bridge construction. As linear isotropic material model has been used, the results can be regarded only as an approximation for the anisotropic and nonhomogenous rock mass. Model boundaries include fixed vertical and bottom boundary conditions, whereas all other nodes in the finite element mesh are assumed to have three degrees of freedom. The obtained FEM results for stresses on the bridge foundation/rock contact surface show significant difference compared to the results for the isotropic, homogenous half space (Fig. 7). The stresses in the discontinuity layer (indicated with ③) are very evenly distributed, but in some of the analysed load cases, the Mohr envelope for fault gauge material is reached. Foundation block edge points (nodes 161 and 169) displace cca -3 mm and -1o mm in Y and Z direction respectively.

To analyse the effect of possible prestressing anchors in the discontinuity zone, two extra load cases have been conducted - firstly, evenly distributed prestressing force on 8 m length had been applied, and secondly the bridge weight has been added (phases I and II). It has been concluded that the state of the rock mass in the areas around the bridge pier supports should be improved by means of anchors. Slope stability factor of safety for the improved rock has shifted from original $F_s = 0{,}912$ to $F_s = 1{,}19$.

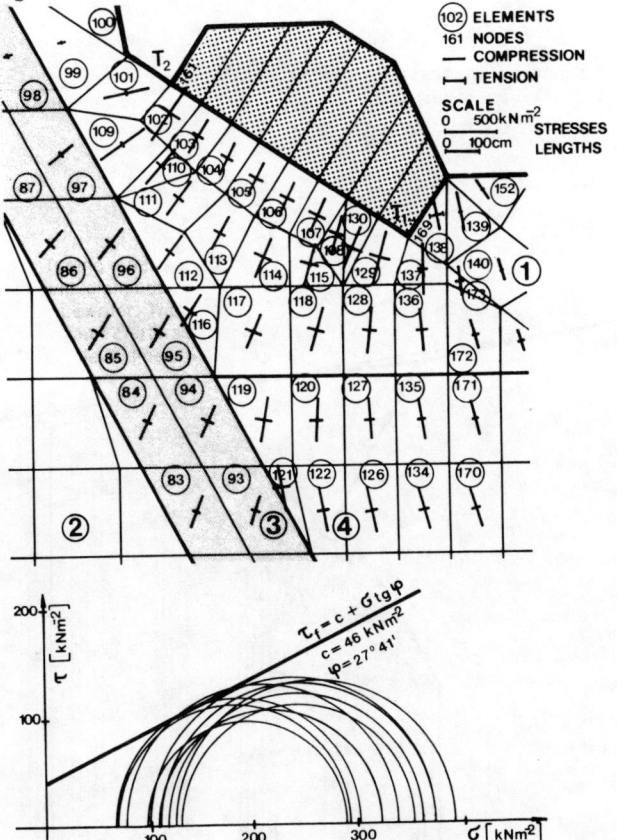

Fig. 7 Detail of the 2D finite element mesh arround bridge pier foundation block and Mohr stress circles (elements 83-97) for discontinuity layer ③ compared with Mohr envelope (BC_2, D_1)

Static and dynamic soil structure interaction analyses have been conducted for three different model boundary conditions and three different assumptions for dominant discontinutities. Three dimensional model is adopted and outside model boundaries follow roughly the dominant vertical faults. The bridge itself has been modelled using 3D beam elements, and the rock mass was modelled using four layers of 8 noded brick elements.

The first, extreme, boundary condition idealisation (BC_1) assumes full fixity only on the bottom edge of the model, whereas all other sides are free.

The second, more realistic, idealisation (BC_2) assumes full fixity on the bottom and on the vertical side model boundaries. The third boundary conditions idealisation (BC_3) adopts the other extreme, and assumes the full fixity of the bridge structure on the contact with rock mass. Existing large faults and dicontinuities are discretized using double row of nodes in the finite element mesh.

First discontinuity idealisation (D_1) assumes full displacement continuity, modelled via rigid connections of neighbouring nodes. In the second discontinuity idealisation (D_2) only forces normal to the discontinuity plane can be bransmitted. Third, extreme, case of discontinuity idealisation (D_3) assumes that each of the neighbouring nodes moves independently. Every nodal point representing rock mass (if not fixed) has three degrees of freedom, whereas nodal points on the bridge model have additional rotations, and therefore all six degrees of freedom. The effect of different boundary conditions and discontinuity idealisation of fundamental frequency and mode of vibration are given in Table 1 and Fig. 11.

6. CONTROL MEASUREMENTS DURING BRIDGE CONSTRUCTION

The deformation state of improved rock is continuosly monitored on several control points on rock surface - additional control points are located on anchorage block, bridge abutment, bridge pier foundation block etc. The convergence is measured in the tunnel, near the anchorage block on the east slope (Fig. 2). The convergence of torcrete tunnel lining ranges from -2,78 mm to 1,9o mm, approximately 3o days after the bridge abutment and the anchorage block have been completed.

Three test anchors 32 Ø 7, L_s = 8,o m, L_u = 17,2 m in anchor block (Fig. 13) exibit significant ratio of the inelastic to elastic deformation $0{,}42 < \Psi < 0{,}5o$ and $0{,}28 < \Psi < 0{,}69$ corresponding to prestressing forces of 12oo kN and 15oo kN respectively - indicating heterogeneity and anisotropy of the rock mass.

To evaluate the use of linear elastic material model in finite element analysis of the global soil structure interaction model, laboratory measurements on BBRV test anchors, where rock mass was replaced by concrete have been compared with linear elastic analysis using finite element method. As expected, the correlation is good only for the very low stress levels - for higher stress levels linear elastic model for concrete (or rock mass accordingly) can only present a very crude representation, as the model does not account for any stress redistribution along the anchor length.

To obtain some information on realistic stress transfer distribution along the anchor length, one of the anchors is instrumented with several extensometers on the injection paste/rock mass contact surface. Long term change of the prestressing force with time is also monitored. Finally, extensive surveying of control point displacements (Fig. 8) (precision ±0,5 mm) is also carried out after each of the principal construction phases is completed. The displacements measured so far are much lower (cca 3o%) than expected values from the numerical prediction analyses. As most important construction phases, and corresponding significant changes of the stress state, are still come, future comparisms may be more interesting.

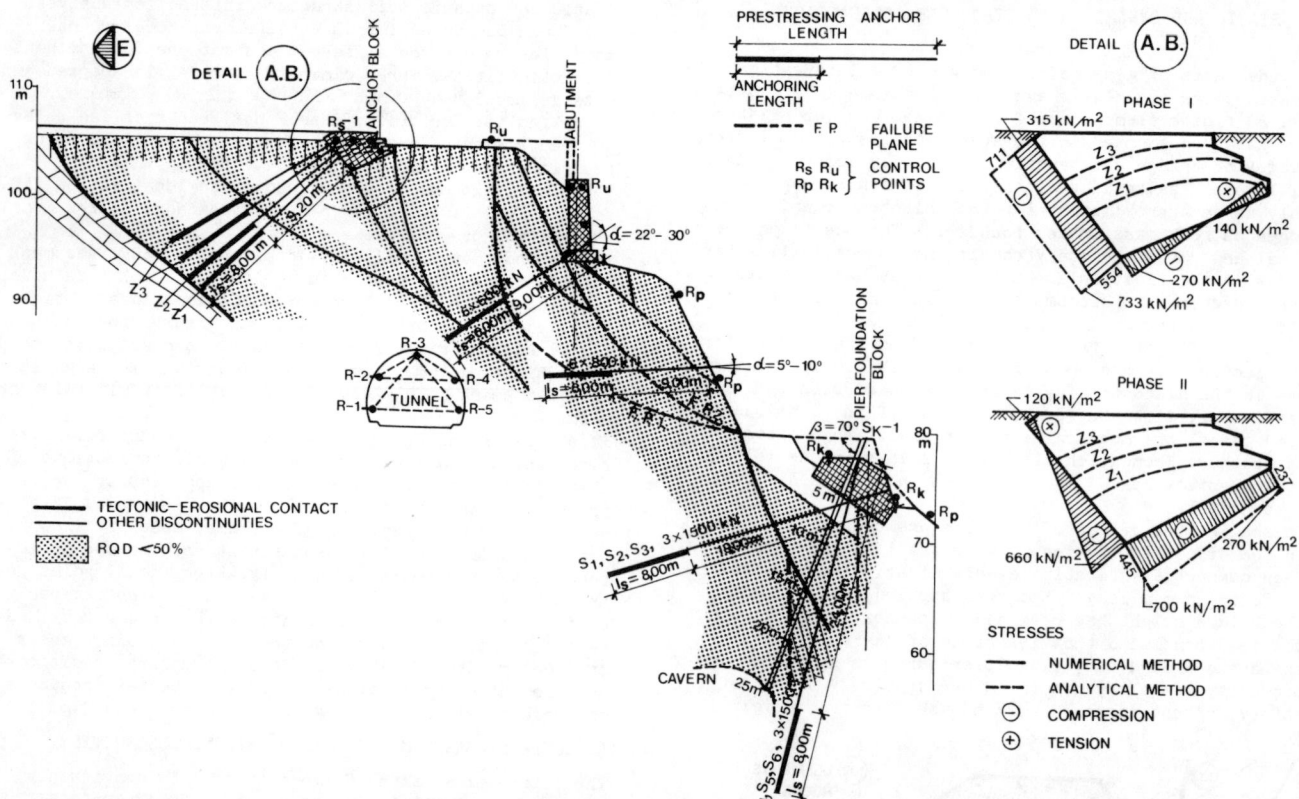

PRESTRESSING ANCHOR LENGTH

ANCHORING LENGTH

F.P.   FAILURE PLANE

Rs Ru } CONTROL
Rp Rk } POINTS

DETAIL (A.B.)

PHASE I

315 kN/m²
711
Z3
Z2
Z1
⊗
140 kN/m²
554
270 kN/m²
733 kN/m²

PHASE II

120 kN/m²
⊗
Z3
Z2
Z1
231
660 kN/m²
445
270 kN/m²
700 kN/m²

STRESSES

——— NUMERICAL METHOD
– – – ANALYTICAL METHOD
⊖ COMPRESSION
⊕ TENSION

DETAIL (A.B.)

ANCHOR BLOCK

Rs-1

l=8,00 m
ls=800 m

Z3 Z2 Z1

ABUTMENT

Ru
Ru

α=22°-30°

Rp
Rp

α=5°-10°

R-3
R-2   R-4
R-1 TUNNEL R-5

PIER FOUNDATION BLOCK

β=70° SK-1
5 m
Rk
Rk
Rp

S1, S2, S3, 3×1500 kN
ls=800 m

CAVERN   25m

S4, S5, S6, 3×1500 kN
ls=800 m

——— TECTONIC-EROSIONAL CONTACT
—— OTHER DISCONTINUITIES
▨ RQD <50%

Fig. 8 Positions of prestressing anchors for rock mass improvement and contact stress states at anchorage block prior (phase I) and after (phase II) prestressing

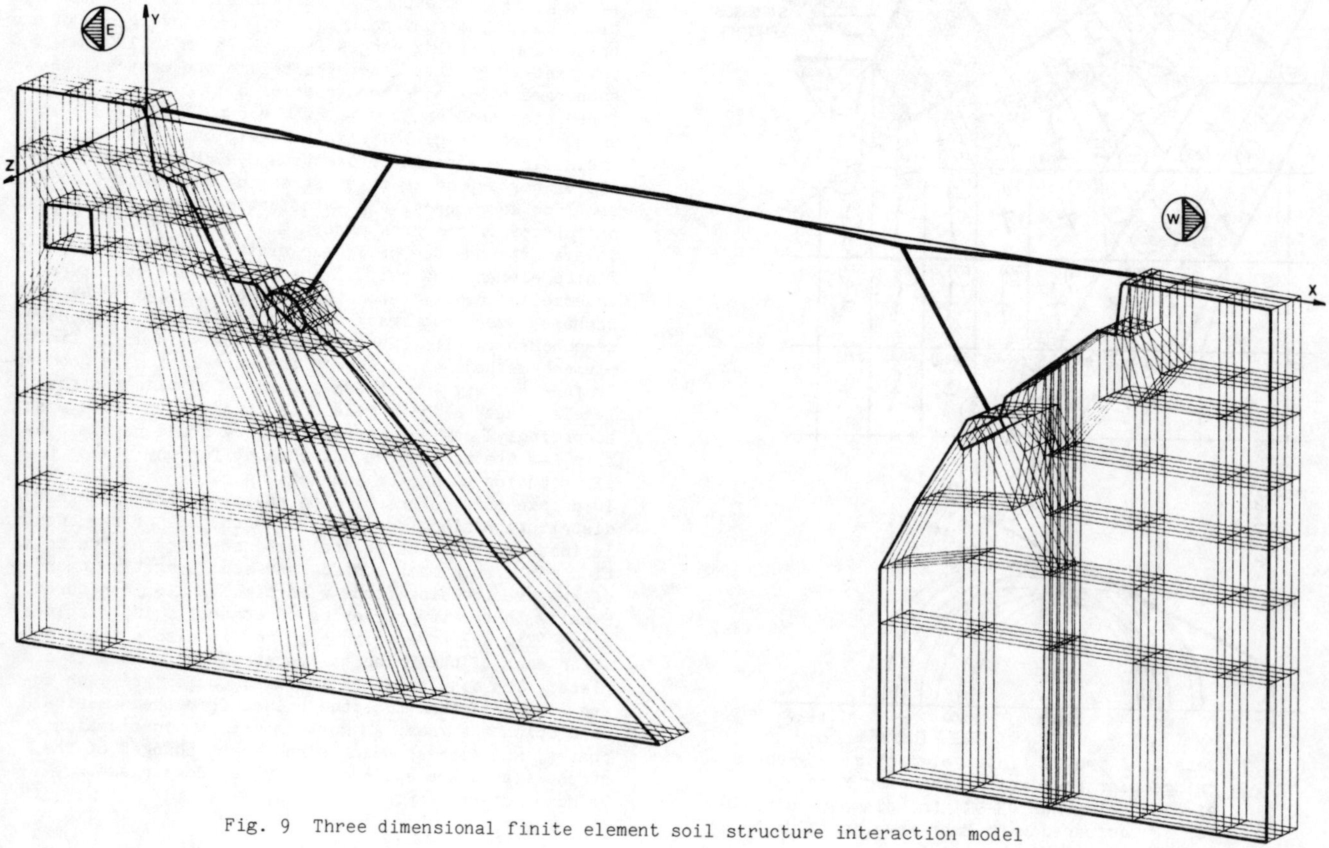

Fig. 9 Three dimensional finite element soil structure interaction model

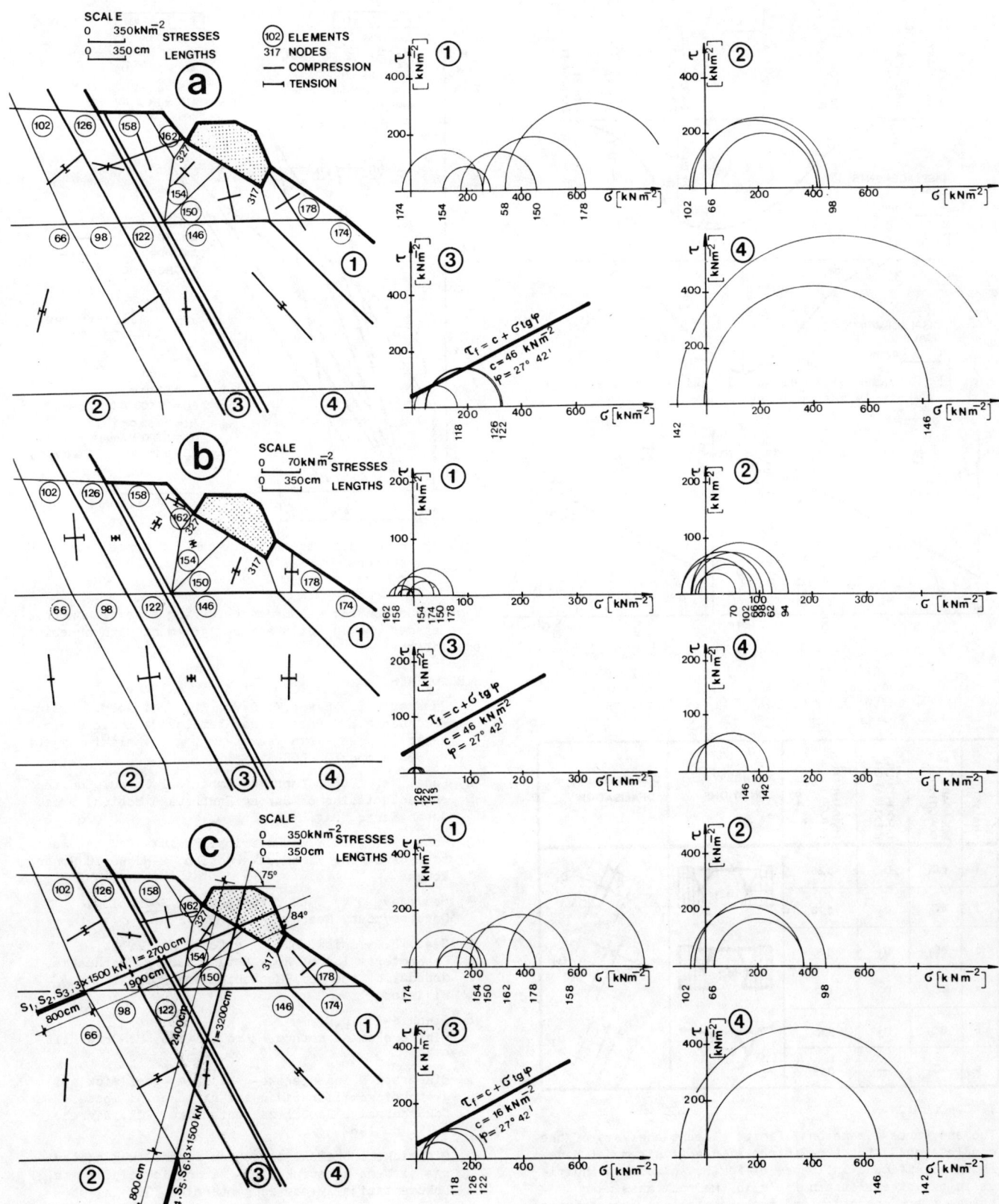

Fig. 10 Principal stresses at bridge pier foundation block (BC₂ . D₂) a)Selfweight, b) Earthquake loading
k_c = 0,053, c) Selfweight + Prestressing anchors S₁ to S₆

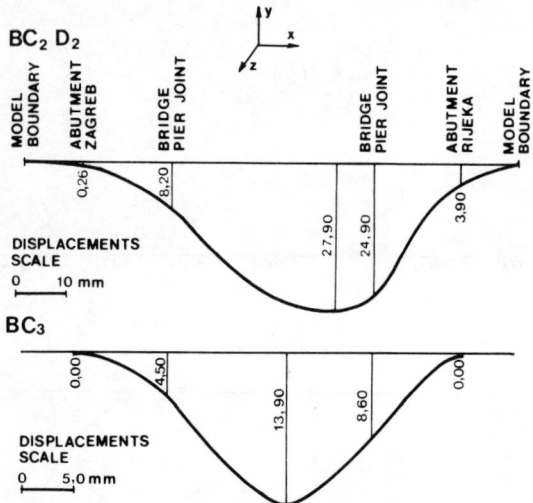

BC$_2$ D$_2$

DISPLACEMENTS
SCALE

0   10 mm

BC$_3$

DISPLACEMENTS
SCALE

0   5,0 mm

Fig. 11   Fundamental models of vibration for several
analysis assumptions

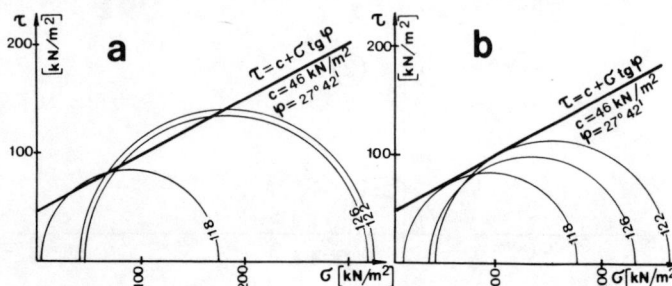

Fig. 12   Principal stress states (Mohr circles) for
discontinuity layer ③ near bridge pier
foundation block (BC$_2$, D$_2$) ; a) Selfweight +
+ Earthquake, b) Selfweight + Earthquake +
+ Prestressing

ANALYSIS	BOUNDARY CONDITIONS	DISCONTINUITY IDEALISATION	FUNDAMETAL PERIOD [sec]	BOUNDARY CONDITIONS	DISCONTINUITY IDEALISATION
1	BC$_1$	D$_1$	3,68	BC$_1$	D$_1$
2	BC$_1$	D$_2$	3,78		D$_2$
3	BC$_1$	D$_3$	3,80		D$_3$
4	BC$_2$	D$_1$	1,32	BC$_2$	
5	BC$_2$	D$_2$	2,10		
6	BC$_3$	—	0,94	BC$_3$	

7. CONCLUSION

Two and three dimensional finite element analyses of the
complete soil structure interaction analysis model have
shown significant influence of faults and discontinuities in the stress/strain field in the rock mass near
bridge pier foundation and anchorage block. The analysis
assumptions for boundary conditions and discontinuity
idealisations (BC$_2$, D$_2$) seem to be most realistic,
and the use of prestressing anchors to improve the
overall stress state in the discontinuity region, is

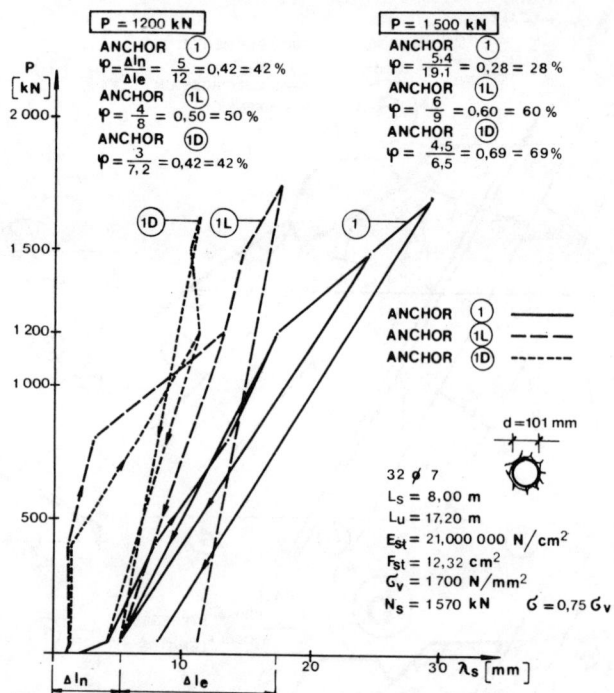

Fig. 13   Force displacement diagram for three test
anchors

indicated.
Continuous control measurements during several construction stages are in satisfactory agreement with numerical
predictions.

BIBLIOGRAPHY

1   Jašarević,I., Šavor,K.,Šavor,Z., Vrkljan,M., Vulić,
Ž., Andrić,M. 198o: Prikaz temeljenja mosta u slo-
ženim i inženjersko-geološkim i geotehničkim uvjeti-
ma, 5. simpozij JDMSPR, Split.

2   Kanai, K. 1966.: Improved Empirical Formula on the
characteristics of Strong Earthquake Motion, Proc.
Japan Eartq. Eng. Sym. Tokyo.

3   Kanai, K. 196o.: An Empirical Formula for the Spectrum of Strong Earthquake Motion, 2-Nd World Congr.
Tokyo.

4   Medvedev, S.V.,1962.: Engineering Seizmology,
Gosstroyzdat, Moscow.

5   Seed, B., Idriss, M.I., Keifer,F.M., 1969.:
Characteristics of Rock Motion During Earthuakes,
Journal of the Soil Mechanics and Foundation
Division

6   Zienkiewicz,D.C.
Finite element method, 3 rd edition, Mc Graw Hill,
1977.

7   Elaborati o inženjersko-geološkim, geofizičkim i
geotehničkim istraživanjima na lokaciji mosta,
Dokumentacija Zavoda za geotehniku, FGZ, Zagreb,
1977. do 1982.

8   Dinamički proračuni za ocjenu zajedničkog djelova-
nja sistema konstrukcija - tlo na lokaciji mosta,
Dokumentacija Zavoda za geotehniku FGZ, Zagreb,
1982.

CONTRIBUTION A L'INVESTIGATION ET AUX CONNAISSANCES DES PROPRIETES MECANIQUES DES ROCHES POUR LES FONDATIONS DE BARRAGES

Contribution to the investigation and the knowledge of mechanical properties of the rock masses for dam foundations

Ein Beitrag zur Kenntnis der mechanischen Eigenschaften von Fels für Staumauerfundamente

Dr Dusan Milovanaović

Professeur à la Faculté de Génie civil de l'Université de Belgrade, Yougoslavie

RESUME

On présente des investigations des propriétés mécaniques de la masse rocheuse pour fondations de barrages. En conclusion on donne des résultats qui démontrent la nécessité de modifier des opinions tenues pour valables jusqu'à présent.

SYNOPSIS

The author presents investigations of mechanical properties of the rock masses for dam foundations. In conclusion he gives the results which confirm the necessity to change some past opinions on these problems.

ZUSAMMENFASSUNG

Es wird über Untersuchungen über die mechanischen Eigenschaften von Fels im Hinblick auf Staumauergründungen berichtet. Die Ergebnisse bestätigen, daß einige ältere diesbezügliche Vorstellungen aufgegeben werden müssen.

1. INTRODUCTION

Dans le cadre de l'investigation de la roche pour les appuis d'un barrage-voûte on a fait de très amples essais en vue de mieux connaître et d'éclaircire quelques questions fondamentales dans le domaine des relations entre la théorie et la pratiques du comportement de la roche de fondation des barrages en béton. Le nombre des essais, leur ampleur et leur compléxité, la compétance de la roche du point de vue de la généralisation, ainsi que plusieurs essais précédants et quelques essais postérieurs ont permis la généralisation des attitudes se rapportant à la masse rocheuse ayant des modules de déformations à la compression de plus de 3000 MPa. C'est de la masse rocheuse sur la quelle on peut appuyer une barrage-voûte de dimensions moyennes ou grandes.

2. QUESTIONS PRINCIPALES D'INVESTIGATION, CONSEPTION ET DISPOSITIF D'ESSAIS

Dans le domaine de l'interaction du barrage et de la roche les questions principales sont:

- quelles sont les propriétés mécaniques de la masse rocheuse nécessaires à l'analyse d'un barrage-voûte (d'un barrage à voûtes multiples ou tout autre type de barrage en béton)

- comment trouver les valeurs numériques de ces propriétés,

- comment les introduire dans l'analyse statique et dynamique du barrage.

D'après les problèmes qui se posent il est évident que pour leur solution l'investigation de la roche "in situ" est indispensable.

L'auteur a fait ces investigations et recherches dans le cadre des essais géotechniques pour chacun des barrages-voûtes ou voûtes multiples dont il a fait les projets. Ces essais sont amples et couteux et pratiquement on peut les accomplir presque uniquement dans le cadre des travaux de recherche pour les barrages, particulièrement à cause des travaux de préparation qu'ils exigent. Avec le temps le nombre des résultats est devenu considérable. De telle façon on a obtenu les résultats d'un grand nombre d'essais dans diverses roches et de diverses envergures. Dans tous ces essais on retrouve la même conception, les mêmes idées et le même but - la résolution du problème de l'interaction barrage-roche, et entre autre: "grande échelle", épreuve comme modèle géotechnique, même ou semblable régime d'application des charges. En même temps on a toujours en vue l'effet d'échelle, cooperation de la roche hors de la surface chargée etc.

A chaque nouvel essai on a tenu compte de l'expérience obtenue au cours des essais précédents en éliminant les défauts remarqués. Les contraintes au contact béton-rocher effectuées par charges appliquées se tenaient dans les intervales du travail réel du contact béton-rocher dans le barrage, jusqu'à une augmentation de 50%. Dans d'autres cas ces contraintes arrivaient dans le domaine de l'equilibre limite et de la rupture, les essais étant traité comme modèles réologiques.

Sur un ouvrage, une fois toute autre condition remplie, on a conçu une investigation complete, au moyen de laquelle on a obtenu des réponses à un certain nombre de questions, jusqu'à présant non résolues, dans le domaine du travail mécanique de la masse rocheuse en interaction avec le barrage.

Ceci a rendu possible la généralisation d'un certain nombre de solutions dans un cadre déterminé de propriétés mécaniques de la masse rocheuse, ou tout au moins le rapprochement de cette généralisation. Elle n'est pas possible, du moins pour le moment, dans un vaste spectre de masse rocheuse, ce qui, de l'opinion de l'auteur, n'est même pas indispensable.

Pour compléter les résultats des recherches expérimentales dans la roche, pour faire des analyses parametriques, mieux voir les fonctions et pour arriver à la forme d'applica-tion dans l'analyse d'un barrage-voûte, on a fait un grand nombre de modeles mathematiques au moyen des éléments finis. Enfin on a formulé une méthode pour introduire les résultats des essais dans l'analyse statique du barrage-voûte. Dans ce rapport on passe en revue les résultats des recherches expérimentales.

Le dispositif des "essais integraux" consiste en ce qui suit.

1. Deux groupes d'investigations identi-ques ont été réalisés dans l'élargissement de chacune des deux galeries dans les flancs de l'emplacement du barrage. Le site entier du bar-

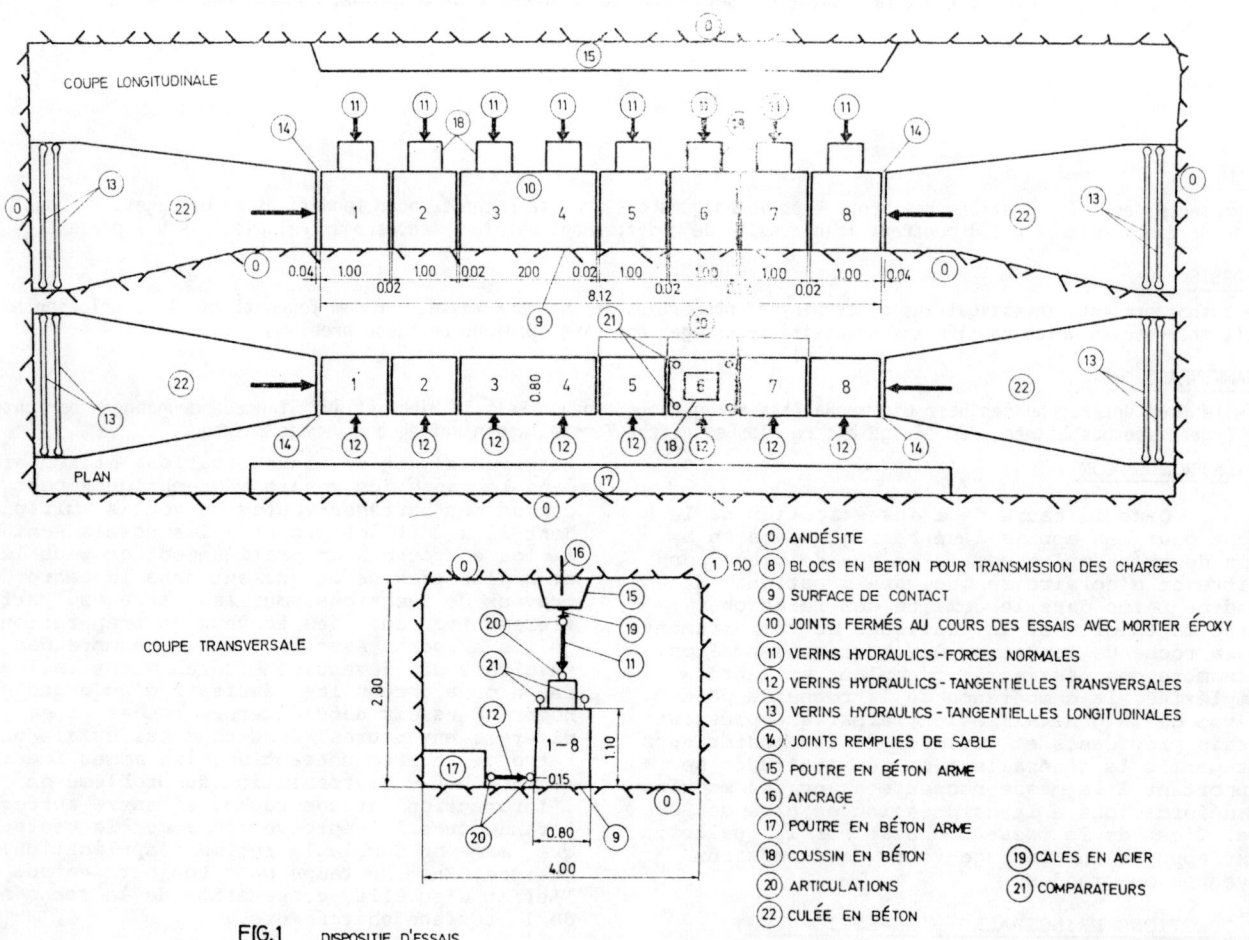

FIG.1   DISPOSITIF D'ESSAIS

rage est formé de dacitoandésite qui dans les deux galerie présente une légère différence en qualité mécanique du matrix de la roche ainsi qu'en structure microtectonique. Ceci d'une part augmente le nombre des essais et d'autre part celui des résultats obtenus dans des conditions très semblables. La masse rocheuse dans les deux galeries est quasihomogène.

2. La surface de contact béton-rocher dans les deux galeries se compose de huit recta-ngles 1,00x0,80 côte à côte (avec espace de 2 cm) en tout 8,14x0,80 m². Apres des essais sur les blocs de surface de contact 1,0x0,80 on les a

réliés au moyen de mortier d'époxy d'abord par 2, puis par 4, ensuite 6 et enfin tous les 8. De cette manière on a essayé les surfaces de contact 1,00x0,80, 2,02x0,80, 4,06x0,80, 6,10x x0,80 et 8,14x0,80 m². (Fig. 1).

3. L'excavation de la galerie a été réa-lisée très soigneusement par de faibles explos-ions, les avant-derniers 0,25 m avec marteau pneumatique et les derniers 0,25 m à la main au moyen de ciselet. On a fait des relevés de structure de la roche et des mesures microsi-smiques au dessous de la surface de contact de la roche.

C'est ainsi qu'on a confirmé que la masse rocheuse est représentative, compétente et convient pour les essais et la généralisation.

3. CONCLUSIONS

Les résultats des investigations et recherches sont présentés dans ce rapport sous forme de conclusions par suite du nombre limité de page. En premier lieu on a utilisé les "essais integraux" susmentionnés pour un barrage, complétés par les résultats des essais sur d'autres barrages et sur d'autres roches comme il a été mentionné au chapitre 2.

1. Tous les essais, a partir de 1965, et particulièrement ces derniers, ont confirmé l'opinion de l'auteur sur la nécessité de l'introduction du module de déformation au cisaillement dans le calcul de l'intéraction barrage-rocher, (outre le module de déformation à la compression).(1).

Le module de déformation au cisaillement de la roche est important pour les résultats de l'analyse statique du barrage; on l'obtient par des essais adéquates "in situ", mentionnés dans ce rapport. Cette propriété mécanique est une des valeurs numeriques montrant l'anisotropie de la masse rocheuse, considérée par l'auteur comme la plus importante, et qui rentre dans le calcul statique et dynamique (sismique) du barrage. Les relations analytiques entre les coefficients de compression et de cisaillement d'apres Vogt appliquées jusqu'à présent sont loin des relations réelles.

2. Dans les masses rocheuses dans lesquelles on a fait des investigations et qui

FIG. 2a

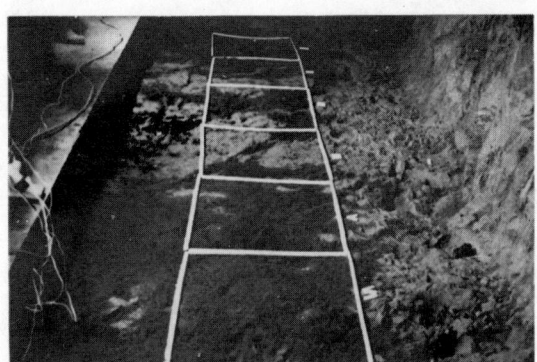

FIG. 2b

FIG. 2c

FIG. 2d

constituent la plus grande partie des appuis du barrage-voûte, et en même temps la partie représentative et compétente, la relation des contraintes en compression et des déplacements est pratiquement linéaire. Ceci dans l'intervalle des contraintes sur la surface des appuis dans lequel travaille le barrage-voûte. Cette relation linéaire peut etre introduite dans le calcul avec une exactitude suffisante. Les divergences sont petites et les differences de résultats négligeables. La relation linéaire a été prouvée de façon très convaincante par les "essais integraux". Cette propriété doit etre prouvée par des essais pour chacune des roches servant d'appuis de barrage. (Fig. 3 à 8).

3. Au moyen d'essais de cisaillement faisant partie de "l'essai intégral" on a trouvé

l'intervalle des contraintes de cisaillement pour lequel la fonction des déplacements de cisaillement est linéaire pour des contraintes de compression déterminé. On remarque qu'il n'était pas question d'une masse rocheuse particulierement bonne, mais au contraire très médiocre, avec des valeurs de modules de déformation en compression relativement faibles (environ 3000 MPa).

Dans les cas des roches où, le plus souvent un barrage-voûte est appuyé, d'après les résultats obtenues, on a le droit de supposer qu'il existe un intervalle de contraintes τ et σ pour lesquelles la relation des contraintes de cisaillement et des déplacements de cisaillement est linéaire.

Par suite d'une série d'analyses et de

calculs on conclut que le plus souvent l'intervalle des contraintes τ et σ (cisaillement et compression) dans lequel travaille un barrage-voûte à la surface d'appuis, s'insère dans l'intervalle des relations linéaires trouvées dans les essais, tant pour les compressions que pour les cisaillements. (Fig. 9). Cette conclussion rend possible l'application d'une série d'algoritmes basés sur la théorie de l'élasticité avec certaines corrections. Par exemple, l'emploie, dans un "sémiespace limité", des modèles mathématiques linéaires "step by step", des modèles à modules bilinéaires, ou bien orthotropes.

4. Le changement des déplacements de cisaillement (déplacement d'un point de la surface de contact en direciton de la force de cisaillement) en fonction du changement des contraintes de compression et de cisaillement est une particularité de la masse rocheuse d'une grande importance pour l'intéraction du barrage et de la roche dans le domaine du travail normal du barrage (et non seulement dans le domaine de l'équilibre limite). Cette propriété de la masse rocheuse a été étudié dans le cadre de "l'essai integral" et elle doit être régulièrement essayée.

5. On a étudié le comportement de la surface de contact béton-masse rocheuse chargée par moment de torsion (il semble que cela ait été fait pour la premiere fois "in situ"). On a cherché la relation entre ces résultats et les résultats obtenus par cisaillement. (Fig. 10).

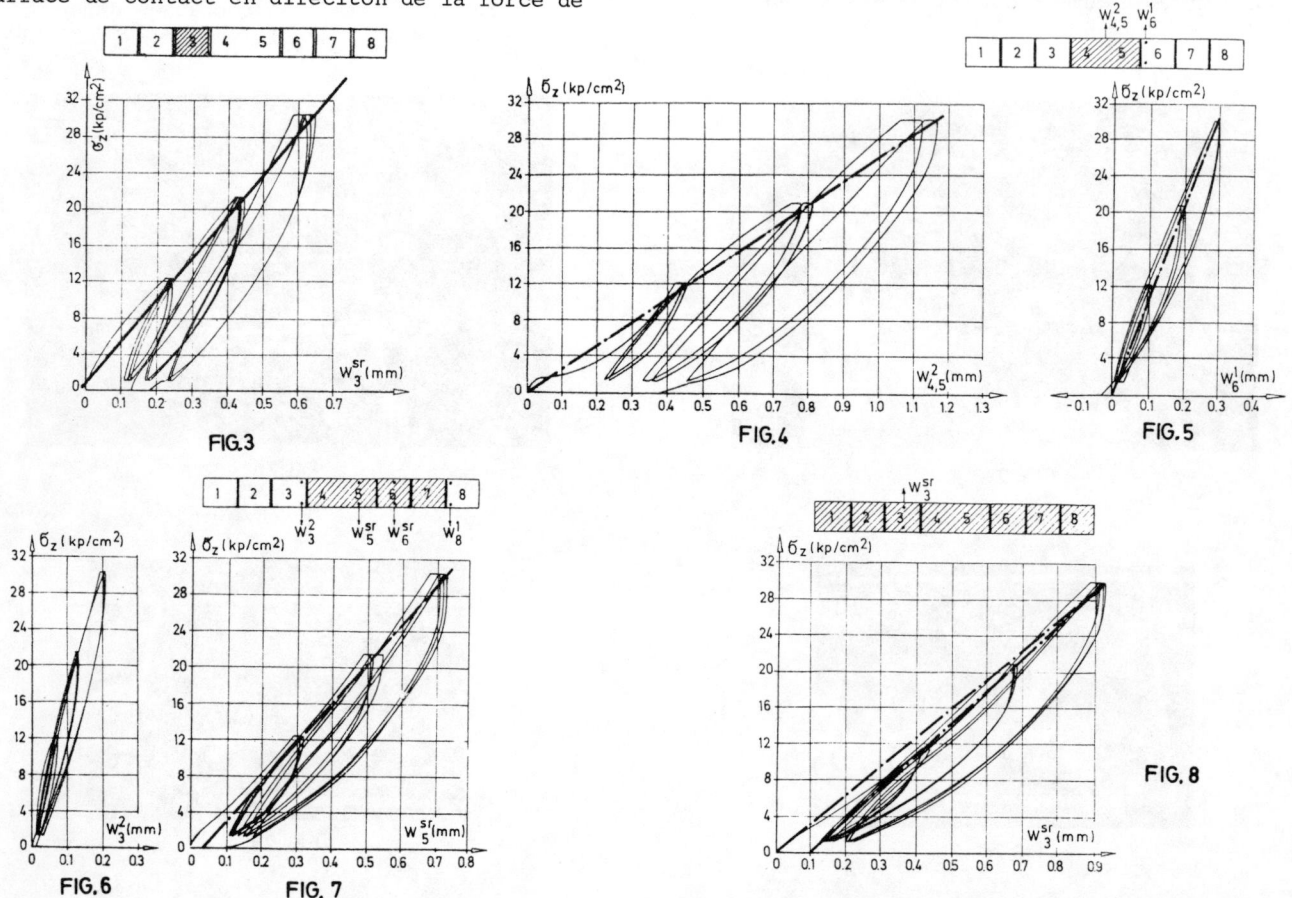

FIG.3

FIG.4

FIG.5

FIG.6

FIG. 7

FIG. 8

6. On a obtenu des résultats de charges alternatives de cisaillement dans les deux sens d'une même direction.

7. Les recherches dans le cadre de ce travail ont montré une très petite intéraction (coopération) de la partie de la roche se trouvant directement sous la surface chargée et des parties latérales hors de celle-ci.

On a constaté au moyen d'essais que les rapports entre les déplacements de la surface directement chargée et de la surface latérale nonchargée n'étaient pas les rapports obtenus par les calcul pour un milieu élastique et que les résultats dans les cas recherchés par essais sont très différents des résultats précédents. Le degré de coopération de la roche dans le voisinage de la surface directement chargée doit être déterminée par les mesures dans chacun des cas de fondation de barrage. Les essais des propriétés mécaniques de la roche "in situ" faits jusqu'à présent doivent être augmentés dans ce sens autant pour la compression que pour le cisaillement. Les recherches faites dans le cadre de ce travail sont les premières dans ce sens (en dehors de quelques tentatives faites auparavant pour un autre barrage).

8. Les mesures des déplacement en profodeur au dessous de la surface d'appui faites dans des essais ultérieurs ont donné d'excelants résultats et ont ouvert de nouvelles possibilités d'interprétation des résultats d'essais de la roche. Il est utile de faire toujours ses mesures quand c'est possible.

9. De ce qui a été mentionné jusqu'à

présent on peut conclure que les équations et les coefficients de Vogt pour l'introduction des déformations de la roche dans le calcul statique des barrages-voûtes ne sont pas réels - les conditions aux appuis introduites comme cela peuvent différer beaucoup des conditions réelles. Les coefficients de Vogt qui sont fonction du rapport des cotés de rectangle ne sont pas confirmés par les essais, tout au contraire. Si on calcule le module de déformation au moyen de quelque équation basée sur la solution de Boussinecq avec des données obtenues par mesures et si on rentre avec cette valeur de module dans les équations de Vogt on obtient facilement des

déformations deux ou trois fois plus grandes que les déformations réelles.

L'intéraction (la coopération) de la roche hors de la surface chargée obtenue par la méthode des éléments finis (ainsi qui par Vogt) est inexacte et loin de la réalité en ce qui concerne les roches sur lesquelles on a fait les essais. Les déplacements de cisaillement, d'après Vogt, dépendent du module d'élasticité en compression, tandis que ces déplacements doivent réellement dépendre du module de déformation au cisaillement. Ces déplacements de cisaillement, d'après Vogt, ne dépendent pas des contraintes normales (des forces normales)

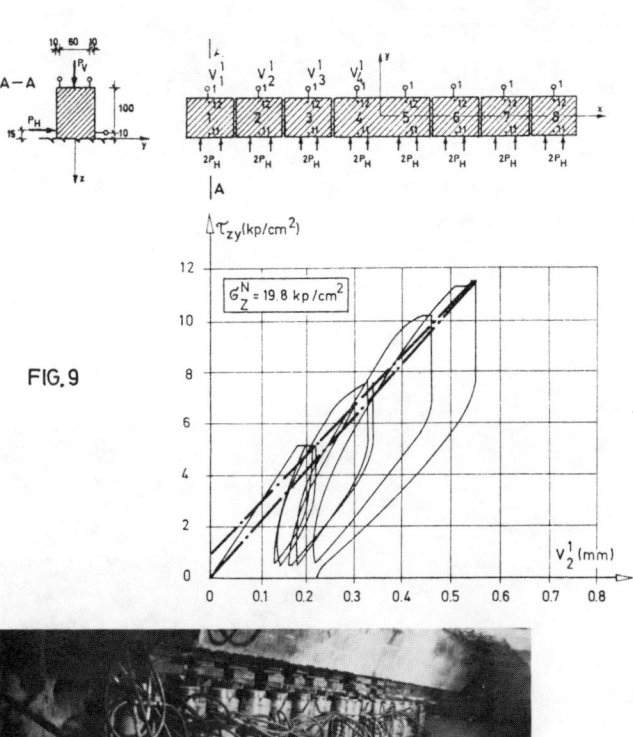

FIG. 9

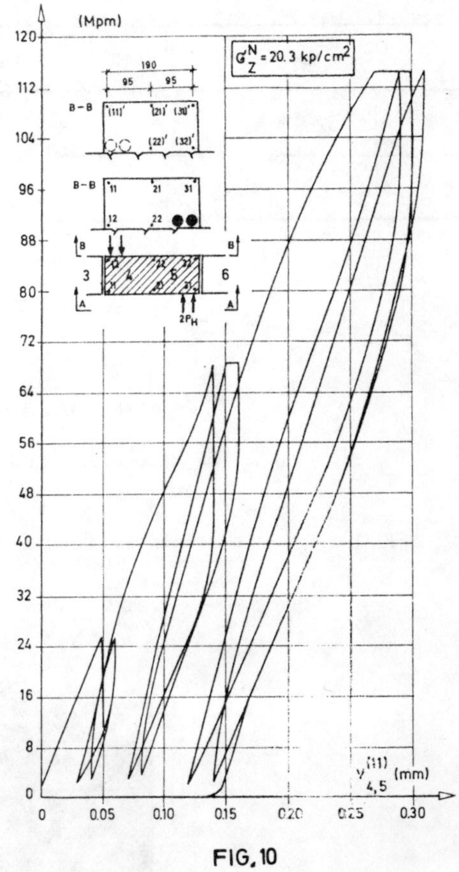

FIG. 12

FIG. 10

tandis que réellement elles doivent en dépendre.

10. L'effet d'echelle dépend surtout des discontinuités, de l'hétérogénéité et du degré de l'anisotropie, ainsi que de la position des discontinuités ou bien de la surface d'anisotropie dans l'espace et par rapport à la surface de contact.

L'effet d'echelle est plus ou moins manifesté dans chaque roche et il doit être examiné dans chacun des cas, en même temps que les autres propriétés mécaniques. Ceci a été fait dans le cadre des essais exposés dans ce travail, ce qui a donné de bons résultats. (Fig. 11).

11. On a démontré par les essais que les fissures apparaissaient au contact béton-rocher comme conséquence des contraintes de tractions.

12. Dans tous les essais ont été appliqués des programmes se composant de groupes de cycles en 3 ou 4 niveaux avec stabilisation. Avec de tels programmes on a obtenu: des relations réelles des contraintes normales et des déplacements, puis des contraintes normales, de cisaillement et de déplacements au cisaillement, ensuite des intervales dans lesquels ces relations sont linéaires, ainsi que des rapports entre les déformations élastiques et plastiques,

la vitesse de convergence des déformations plastiques etc.

Avec un tel programme on reproduit le comportement de la roche dans le temps sous l'influance des charges alternatives des barrages. Par l'introduction dans le calcul statique des déplacements totaux on prend en compte l'effet le plus défavorable de l'intéraction du barrage et de la roche.

13. Dans "l'essai integral" une même sur-

face est chargée à la compression et au cisaillement en deux directions et en deux sens d'une même direction. Ceci permet une interprétation plus réelle et plus exacte.

14. "Les essais intégraux" des propriétés mécaniques de la masse rocheuse, exposés dans ce travail, sont des plus grands réalisé jusqu' à présent quant à l'ampleur, aux espèces d'essais, au temps de la durée et aux nouveautés.

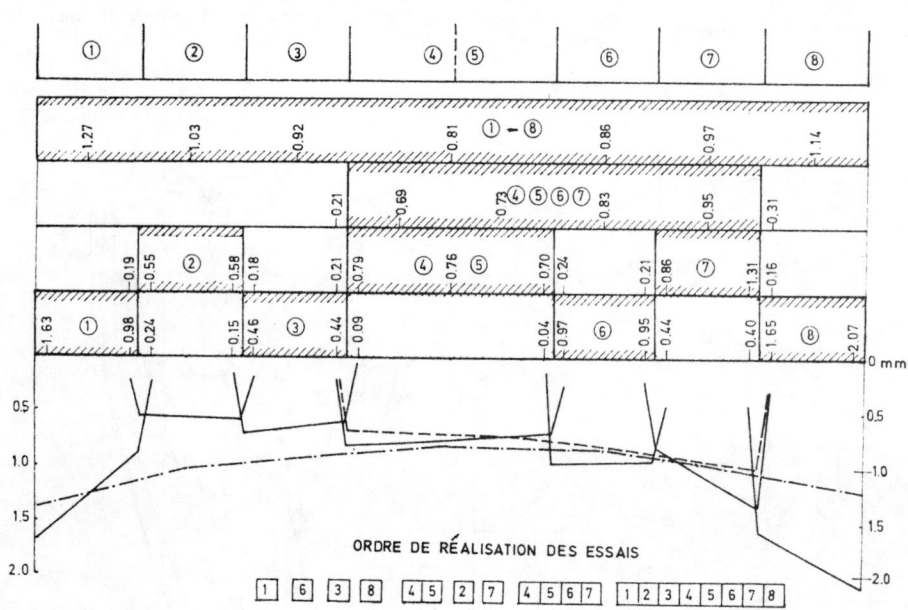

FIG. 11   DIAGRAMME DES DÉPLACEMENTS NORMAUX
——— UN ET DEUX BLOCS
——— 4 BLOCS
—·—·— 8 BLOCS

FIG. 13

Essais sur une surface 2x6,oox1,oo avec mesures en profondeur

FIG. 14

Instruments hors de la surface chargée

(1) Milovanović, D. De certaines caractéristiques mécaniques et de quelques propriétés de la masse rocheuse soumise à de grandes contraintes de cisaillement dans les fondations de la construction. Proceedings of the geotechn. conference Oslo (1967).

(2) Les essais "in situ" ont été réalisés par la Division géotechnique de l'Institut "Jaroslav Černi" de Belgrade, d'apres les conceptions, les programmes et les projets de l'auteur de ce rapport.

FELSBAU UND FELSMECHANISCHE UNTERSUCHUNGEN FÜR EINE FUNKÜBERTRAGUNGSSTELLE IM HOCHGEBIRGE

Construction in rock and rock-mecanical investigation for a telecommunication station in a high mountain region

La construction en roche et recherches de la méchanique de roche pour un relais de télécommunication en haute montagne

H. O. Hellerer
Dipl.-Geol., Lehrstuhl und Prüfamt für Grundbau, Bodenmechanik und Felsmechanik
Technische Universität München

H. Ostermayer
Dr.-Ing., Lehrstuhl und Prüfamt für Grundbau, Bodenmechanik und Felsmechanik
Technische Universität München

ZUSAMMENFASSUNG:
Auf Deutschlands höchstem Berg, der Zugspitze (2965 m) wurde eine Funkübertragungsstelle gebaut. Unter erschwerten Bedingungen - ständiger Touristenverkehr mit 3 Seilbahnen, Permafrost, hohe Windgeschwindigkeiten - wurde unter bestehenden Gebäuden im entspannten Fels ein Felseinschnitt ausgebrochen und ein Tunnel aufgefahren. Eine Antennenbrücke wurde vom Tunnel aus über eine Felsschlucht vorgeschoben und auf einer schlanken Betonscheibe nahe an der Landesgrenze aufgelagert. Es wurden neue Verfahren der Felsankerung angewendet, der Felsausbruch erfolgte durch hydraulisches Keilen.

SYNOPSIS:
On Germany's highest mountain, the Zugspitze (2965 m) a wireless telecommunication station was established. Below existing buildings a rock cut and a tunnel were constructed under extreme conditions, e.g. tourist traffic by 3 cable cars, permafrost, high wind velocities, highly disintegrated and stress relieved rock. One of two antenna bridges was pressed forward from the tunnel to a reinforced concrete disc near the Austrian border. New techniques for rock anchoring were applied. The rock cut was done with hydraulic wedges.

RESUME
On a installé un relais de télécommunication sur la Zugspitze (2965 m). Une tranchée ainsi qu'une galerie ont été creusées dans la roche dans des conditions extrèmement difficiles — p.ex. circulation touristique au moyen de trois téléphériques, sol gelé en permanence, rafales de vent, roche fortement désagrégée. On a avancé un pont à antennes à partir de la galerie jusqu'à un point près de la frontière autrichienne où on l'a posé sur un disque en béton armé. De nouvelles techniques de tirant d'ancrage ont été utilisées et la tranchée a été réalisée hydrauliquement.

1. EINLEITUNG

Auf dem Zugspitzgipfel (2965 m) hat die Deutsche Bundespost im August 1982 die Funkübertragungsstelle Garmisch 2 in Betrieb genommen. Über diese Station der Richtfunkbrücke München bzw. Brauneck (Deutschland) - Cima Gallina (Italien) bzw. Patscherkofel (Österreich) werden Telefonverkehr und Fernsehprogrammaustausch abgewickelt.

Da auf dem fast vollständig bebauten Gipfelgrat keine Antenne errichtet werden durfte, wurde unterhalb des Gipfels eine 36 m lange Antennenbrücke auf der Nordseite (Richtung Deutschland) mit einer 30 m langen Antennenbrücke auf der Südseite (Richtung Italien und Österreich) durch einen ca. 10 m unter dem Münchner Haus (Alpenvereinshütte, errichtet 1897) verlaufenden Tunnel verbunden (Bild 1). Das 3-geschossige, 30 m lange Technikgebäude liegt südlich unterhalb der Terrasse des Münchner Hauses in einem Felseinschnitt und wird von einer tonnenartigen Hüllenkonstruktion aus 4 Bogenträgern und Fachwerkstäben überspannt (EGGER, 1982). Die Antennenbrücken sind als Dreigurt-Rohrfachwerkträger ausgebildet. Die südliche Brücke liegt vor dem Technikgebäude auf zwei Vorsprüngen der Bogenträgerfundamente. Die nördliche Brücke wurde im Tunnel montiert und auf das nördliche Auflager vorgeschoben. Dieses besteht aus einer schlanken, 17 m hohen Stahlbetonscheibe unmittelbar an der Staatsgrenze.

Die Lage der Station ist durch den unter 30 bis 40° nach Südost zum Schneeferner einfallenden Gipfelhang gekennzeichnet. Die Nordwestseite des Gipfels wird durch eine annähernd vertikale, teilweise auch überhängende Felswand gebildet, in derem steilstem Teil der Tunnel mit der Brücke über das Schneekar austritt. Der westliche Zugspitzgipfel mit dem österreichischen Zugspitzhaus fällt mit 60 bis 80° ebenfalls steil zum Schneekar ab. In dieser Steilwand mußte auf deutschem Gebiet ein Auflager für die nördliche Antennenbrücke geschaffen werden.

Neben den beengten räumlichen Verhältnissen auf dem pultartigen Felsgrat und der hohen Lage der Baustelle spielten die extremen klimatischen Bedingungen eine Rolle: Windgeschwindigkeiten bis ca. 240 km/h, Schneehöhen von ca. 10 bis 12 m, eine geschlossene Schneedecke während ca. 8 bis 9 Monaten, Tagestemperaturen bis -40° und im Fels auch während der Sommermonate Permafrost.

Die geologischen Voruntersuchungen und Gefügeaufnahmen mußten ebenfalls unter extremen Bedingungen durchgeführt werden. Der Geologe wurde an den Felswänden unterhalb des Münchner Hauses und des Österreichischen Zugspitzhauses abgeseilt. Während der Felsarbeiten wurden laufend weitere felsmechanische Untersuchungen und Messungen durchgeführt, um die Annahmen zu überprüfen bzw. zu ergänzen. Mit dem ingenieurgeologischen Gutachten und der laufenden Beratung zur Durchführung der Felsarbeiten war Prof. Dr.-Ing. R. Jelinek der TU München betraut.

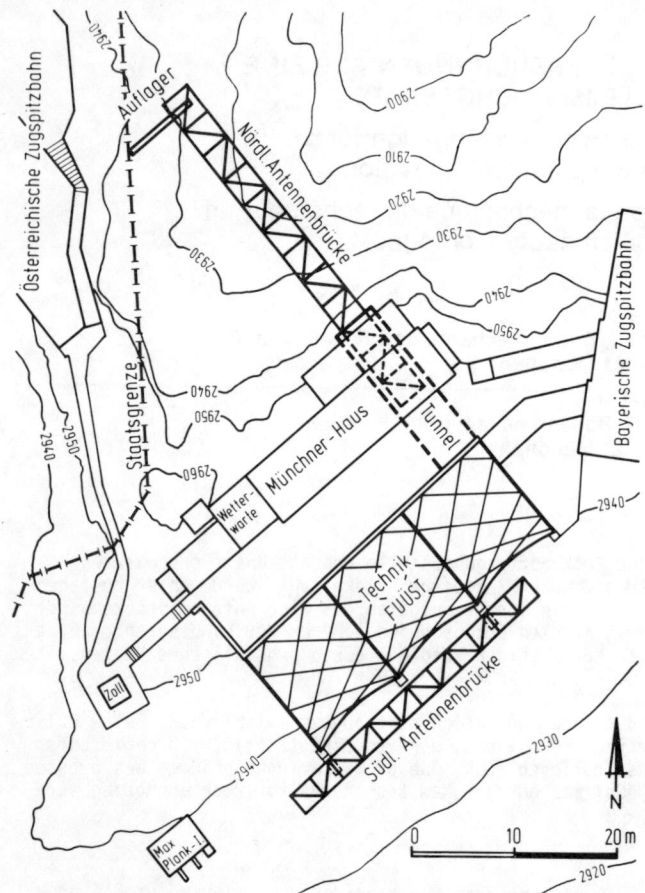

Bild 1 Lageplan

2. GEOLOGISCHE UND FELSMECHANISCHE VERHÄLTNISSE

Die Zugspitze liegt am nordwestlichen Rand des Wettersteingebirges, am Nordrand einer großtektonischen Mulde, die nach ENE abtaucht. Im Muldeninnern liegt der Schneeferner, der einzige deutsche Gletscher, zu dem die Schichtflächen vom Gipfel her einfallen.

Der hier anstehende Wettersteinkalk der alpinen Trias tritt teils plattig, teils grob gebankt auf. Feinschichtige Bereiche, wie am Gipfel und in der Südwand, zeigen eine intensive Kleinklüftung und neben "ripple marks" weit gewellte Schichtklüfte. Grobbankige Felsbereiche zeigen überwiegend mit Lehm und Eis verfüllte Vertikalklüfte.

An der Südseite des Grates ist die Felswand mit etwa 40° steiler als die mit etwa 30° einfallende Schichtung (Bild 2). Hier wurden 2-3 cm weit klaffende Klüfte, die im Sommer eisfrei sind, angetroffen. Dieser Teil des Berges liegt im Bereich des Frost-Tau-Wechsels.

Auf der auch im Sommer vereisten Nordseite herrscht Permafrost. Die Felswand ist hier steiler als 85° (Kluftrichtung K_1, Bild 2 und 3), weist Felsnasen und Überhänge auf und wird durch eine dichte Tapete aus Fels und Eis gegen weitere Erosion und Auflockerung geschützt.

Im Gebirgsstock wurden vertikale, mit Kluftlehm, Mylonit und Eis gefüllte Klüfte gefunden, die während der Ausbrucharbeiten besonders gesichert werden mußten.

Bereits bei den Voruntersuchungen wurden, nach Teilbereichen getrennt, Kluftaufnahmen ausgewertet. Entspre-

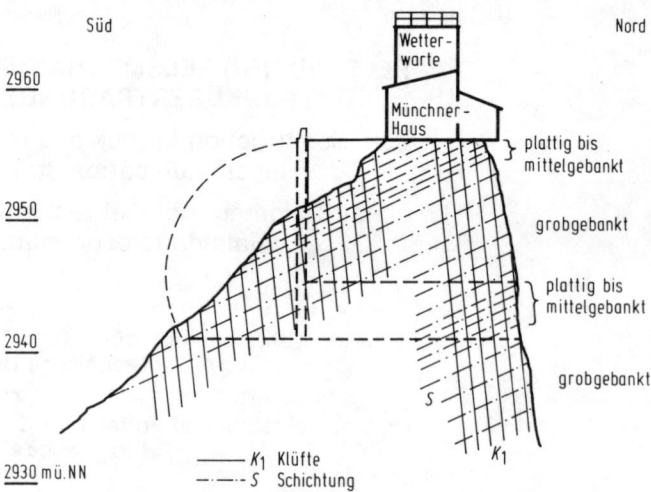

Bild 2 Schnitt durch den Zugspitzgipfel mit zwei Kluftsystemen

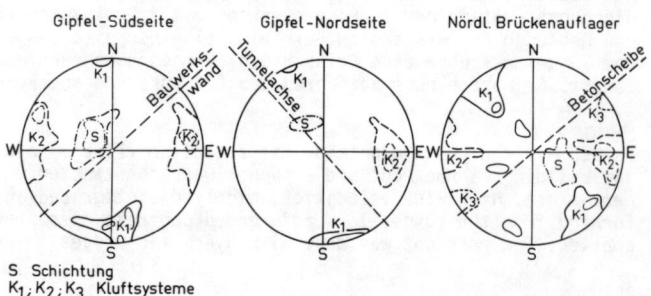

S Schichtung
$K_1; K_2; K_3$ Kluftsysteme

Bild 3 Poldiagramme der Kluftflächen in Lagekugeldarstellung von 3 Untersuchungsbereichen

chend der tektonischen Geschichte dieses Gebirgsstocks waren die Kluftsysteme allgemein ähnlich, sie zeigten aber je nach Standort auch größere Unterschiede. Geringe Unterschiede ergaben sich zwischen Nord- und Südseite des Felsgrates, größere Unterschiede zwischen diesen Bereichen und dem nördlichen Brückenauflager (Bild 3). Große, den Gebirgsstock durchziehende Scherklüfte bildeten hier Trennlinien zwischen kleintektonischen Bereichen.

Besonders in der steilen Wand beim nördlichen Brückenauflager wurden Großklüfte festgestellt, die nicht den 3 Hauptkluftrichtungen in den Diagrammen der Voruntersuchungen entsprachen, und die meist als hangparallele Entspannungsklüfte zu erkennen waren. Die Kluftaufnahmen wurden auch während der Felsarbeiten ständig ergänzt, um hierdurch rutschgefährdete Bereiche zu erkennen und um günstige Richtungen für Felsnägel und Anker festzulegen.

Aufgrund der Voruntersuchungen an der Oberfläche war anzunehmen, daß das Gebirge stark aufgelockert und ohne die Eisverkittung nach der Klassifizierung von PACHER überwiegend der Klasse II (stark nachbrüchiges Gebirge) zuzuordnen ist. Beim Felsabtrag wurde jedoch unter den oberen Schichten überwiegend Klasse I (standfestes bis gering nachbrüchiges Gebirge) angetroffen.

3. FELSARBEITEN

3.1 Besondere Probleme der Baustelle

Wegen der extremen klimatischen Bedingungen konnten verschiedene klassische Sicherungsmethoden, wie z.B. Spritzbetonarbeiten, nicht durchgeführt werden. Der touristische

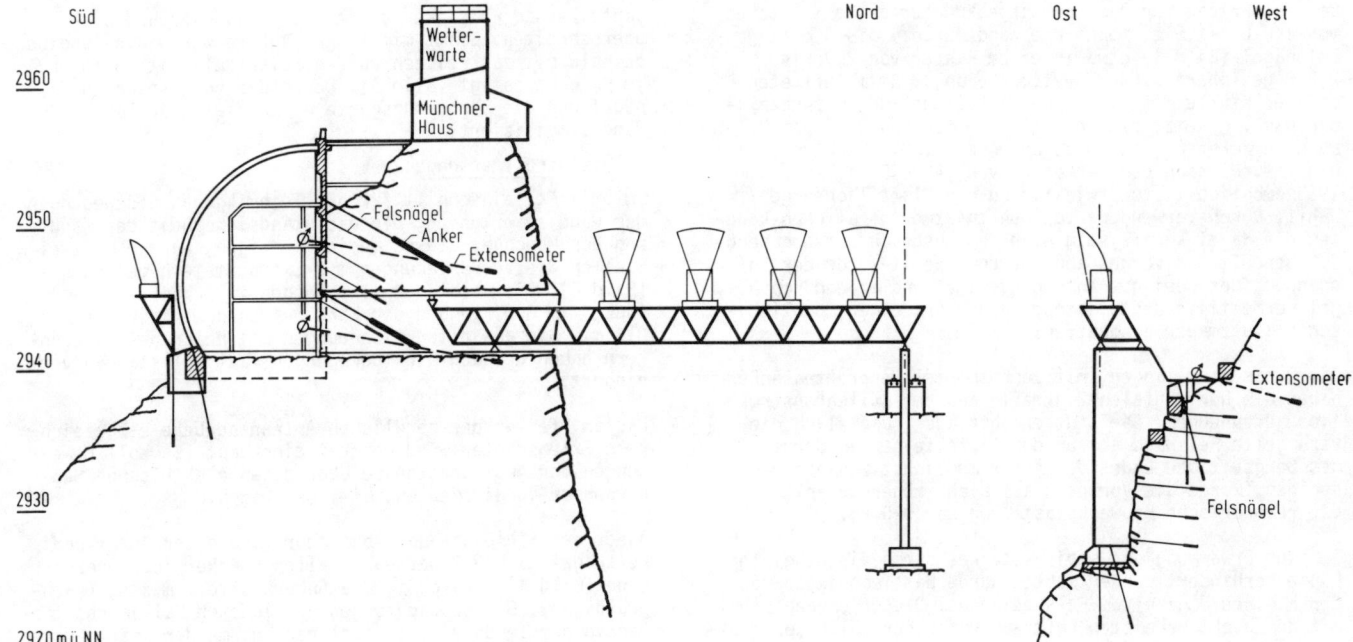

Süd   Nord   Ost   West

Wetter-warte

Münchner-Haus

Felsnägel
Anker
Extensometer

Extensometer

Felsnägel

2960
2950
2940
2930
2920 mü.NN

Bild 4 Schnitte durch die Funkübertragungsstelle

Verkehr durfte nicht gestört und die Gebäude oder technischen und wissenschaftlichen Einrichtungen (3 Seilbahnen, meteorologische Stationen) nicht gefährdet werden (z.B. durch Sprengerschütterungen).

Im Einzelnen waren die Hauptaufgaben (Bild 4):

a) Am Gipfel mußte ein Felseinschnitt ausgeführt und durch den verbleibenden, ca. 17 m breiten Gebirgsstock ein Tunnel ausgebrochen werden. Der Gipfel war als entspannt mit stark entfestigten Zonen anzunehmen. Die Verkittung des Gebirgsstockes durch eisgefüllte Klüfte mußte dabei erhalten, d.h. eine Erwärmung verhindert werden. Der Felsabtrag und die Einleitung der Zugkräfte aus den Bauwerken durfte den Gleichgewichtszustand möglichst wenig ändern. Erschütterungen waren weitgehend zu vermeiden.

b) Das Auflager der nördlichen Antennenbrücke mußte in 17 m Tiefe auf deutschem Grund in der sehr steilen und brüchigen Wand unterhalb des Österreichischen Zugspitzhauses geschaffen werden, da eine höher gelegene, auf österreichischem Gebiet verankerte Kragkonstruktion nicht ausgeführt werden durfte.

c) Es war zu klären, ob zur Aufnahme von Zugkräften aus den Bauwerken und zur Sicherung des Felsausbruchs und des gesamten Gipfelstocks Felsanker und Felsnägel im Permafrost einwandfrei hergestellt werden können.

3.2 Felsanker

Für die Wahl der Felsanker war entscheidend, daß ein Kriechen in den eisgefüllten Klüften verhindert werden mußte, d.h. daß das Gebirge weder entspannt noch konzentriert belastet werden durfte. Die Anker mußten eine ausreichende Federwirkung aufweisen und vorspannbar sein. Die Kräfte sollten über eine möglichst große Verankerungsstrecke in das Gebirge eingeleitet werden. Außerdem mußte sichergestellt sein, daß auch im Permafrost eine einwandfreie Herstellung und Kraftübertragung möglich ist.

Gewählt wurden Freispielanker, d.h. für Felssicherungen und geringe Bauwerkslasten bis 150 kN 4 bis 6 m lange "Felsnägel" Ø 16 mm aus Gewindestahl (St 1325/1470) und

für größere Bauwerkslasten bis 380 kN Bündelanker System Polensky und Zöllner, 12 Ø 7 mm (St 1470/1670).

In der Kältekammer des Prüfamts für Grundbau, Bodenmechanik und Felsmechanik wurden beide Ankertypen in verkürzter Ausführung bei -15 °C hergestellt und belastet. Es zeigte sich, daß bei den Felsnägeln der Bohrdurchmesser, die Kunstharzmörtelpatrone und die Spitze am Ankerstahl genau aufeinander abgestimmt sein müssen, damit eine einwandfreie Durchmischung von Harz, Härter und Füllstoff bei einer gewünschten Haftstrecke von ca. 1,2 m erreicht wird. Nach etwa 50 Stunden Abbindezeit konnten die Anker voll belastet werden.

Bei den Felsankern 12 Ø 7 mm handelt es sich um klassische, auf ganze Länge korrosionsgeschützte Daueranker, die jedoch zusätzlich im Bereich der Verankerungsstrecke mit einer Heizspirale und einem Temperaturfühler versehen waren. Damit konnte die Temperatur der Zementsuspension während der Abbindezeit auf +10° bis +20° eingestellt werden. Selbst bei verkürzter Verankerungslänge von 2,2 m traten bei der maximalen Prüflast von 490 kN nur geringe Verschiebungen auf.

Nach dem erfolgreichen Abschluß der Laborversuche konnte auf die ursprünglich vorgesehenen Anker verzichtet werden, die durch den ganzen Gipfel geführt und auf der Gegenseite an der Nordwand befestigt werden sollten.

Auf der Baustelle wurden die Anker wie in der Kältekammer hergestellt. Die Temperatur im Fels betrug jedoch nur -3° bis -5 °C. Die 130 Felsnägel wurden in der Regel mit 200 kN geprüft und mit 150 kN vorgespannt. Bei den 8 Felsankern betrug nach DIN 4125, Teil 2, die Prüflast 450 kN und die Festlegelast 300 kN.

3.3 Felsabtrag und Sicherung des Gipfels

Wegen der unter 3.1 beschriebenen besonderen Gefahren bei Erschütterungen hat die ausführende Firma auf Sprengungen verzichtet und den Fels durch Vorbohren und hydraulisches Keilen gelöst. Durch Anwendung dieses felsschonenden Abbauverfahrens konnte in vielen Bereichen auf die ursprünglich vorgesehene Sicherung mit Betonvorsatzplatten und Felsnägeln verzichtet werden.

Der Felsabtrag für das Technikgebäude erfolgte in Stufen von 1,0 bis 1,5 m, wobei die Wand durch 4 bis 6 m lange Felsnägel (Bild 4) etwa in einem Raster von 2,0 bis 2,5 m gesichert wurde (Systemankerung). Beim Auftreten offener Klüfte mit oder ohne Eisfüllung wurden zusätzliche Nägel gesetzt bzw. bei Kluftweiten von 0,05 bis 0,1 m auch verankerte Stahlgurte angeordnet. Die Richtung der Nägel wurde nach dem Verfahren von TALOBRE (s. MÜLLER, 1963) möglichst stumpfwinklig zu den Kluftflächen gewählt. Durch Verwendung von Kunstharzmörtelpatronen konnten die Nägel kurzfristig nach dem Ausbruch entsprechend 3.2 geprüft und vorgespannt werden. Somit wurde der Entspannung des Gebirges entgegengewirkt und eine durch Nägel verfestigte Gesteinszone in der Art einer monolithischen Stützmauer geschaffen.

Die 9 bis 11 m langen, mit Zementmörtel verpreßten Anker haben die horizontalen Zugkräfte aus der Hüllenkonstruktion aufzunehmen. Sie stützen aber auch zusätzlich die vernagelte Felswand ab. Um den Gipfelfelsen im Bereich des bergseitigen Endes der Anker möglichst wenig zu entspannen, wurde die Vorspannlast nicht höher gewählt als die rechnerische Bauwerkslast von ca. 300 kN.

Da eine Erwärmung des Gebirgsstockes durch die beheizten Räume verhindert werden mußte, wurde das Technikgebäude durch einen Umgang vom Fels getrennt. Dieser Umgang steht mit dem nicht beheizten Luftraum unter der Hüllenkonstruktion in Verbindung.

3.4 Ausbruch und Sicherung des Tunnels

Für den Ausbruch des etwa 10 m unter dem Münchner Haus verlaufenden Tunnels waren ähnliche Überlegungen anzustellen wie für den Felsabtrag für das Technikgebäude. Besonders kritisch war eine wenige Meter östlich an der Außenwand liegende Felsnase sowie die an der Nordwand beobachteten gelockerten Felsplatten. Es wurde deshalb auch hier mit hydraulischem Keilen gearbeitet.

Das Kreisprofil mit 4,3 m Durchmesser konnte im standfesten Gebirge aufgefahren werden. Alle offenen Klüfte waren mit Eis verkittet. Auch beim Durchbruch zu der hier überhängenden Nordwand war der Fels fester als nach den Vorerkundungen zu erwarten war. Trotzdem wurde z.T. mit Felsnägeln gesichert und nach dem Ausbruch eine Auskleidung mit verschraubten Wellblechtafeln (liner plates) eingebracht. Der Zwischenraum zwischen Wellblechtafeln und Gebirge wurde mit Zementmörtel verpreßt. Dieser schwere Ausbau ist damit in der Lage, den gesamten Felsstock von etwa 17 m Breite auszusteifen und die Bebauung zu sichern. Außerdem dient er als Auflager für die nördliche Antennenbrücke, die von hier aus zum nördlichen Auflager vorgeschoben wurde (Bild 4).

3.5 Nördlicher Auflagerpunkt der Antennenbrücke

Der Bau des Pfeilers für die nördliche Antennenbrücke war in der Ausführung die schwierigste und gefährlichste Aufgabe der gesamten Baumaßnahme. Die etwa 17 m hohe und 0,6 m breite Stahlbetonscheibe mußte in der im allgemeinen 70° steilen, durch Erosion im Frost-Tau-Wechsel-Bereich tiefgründig aufgelockerten Felswand gegründet werden. Der Wettersteinkalk ist hier von verschiedenen Mylonit- und Ruschelzonen durchzogen und weist sehr unterschiedliche Kluftrichtungen auf (Bild 3). Der Abgang von Felsstücken und im Winter von Eiswächten war ständig zu erwarten. Vor dem Ausbruch und den Bauarbeiten mußten umfangreiche Felssicherungen vorgenommen werden. Es wurden in 3 bis 10 m Abstand horizontale Stahlbetongurte den Fels entlang zur Festigung und als Erosionsschutz eingebaut, die mit Felsnägeln gesichert wurden. Der Fels wurde schonend mit Brechstangen und durch hydraulisches Keilen abgetragen.

Bauteile und Felsanker durften die Landesgrenze nicht überschreiten. Statt mit langen Ankern wurde die Scheibe deshalb mit zahlreichen kurzen Felsnägeln mit je 150 kN im Fels befestigt (Bild 4). Durch die vorgespannten Felsnägel und die Sicherungsgurte wurde die zuvor labile Wand stabilisiert.

3.6 Kontrollmessungen

Um beim Felseinschnitt für das Technikgebäude Bewegungen der Wand zu erkennen, die die Standsicherheit der Wand und des Münchner Hauses beeinträchtigen, wurden an zwei Stellen die Verschiebungen mit Extensometern gemessen (Bild 4). Die Relativverschiebungen zwischen dem "Fixpunkt" im Gebirge und der Wand betrugen weniger als 0,1 mm. Die gewählten Sicherungen mit Felsnägeln und Ankern haben demnach Gleitbewegungen fast vollständig verhindert.

Das am Pfeiler der nördlichen Antennenbrücke eingezeichnete Extensometer wird derzeit eingebaut. Es soll Bewegungen der Auflagerscheibe über einen elektrischen Wegaufnehmer in die Meßzentrale übertragen.

Außer den Eignungs- und Abnahmeprüfungen der Anker und Felsnägel nach 3.2 werden an allen 5 Ankern der oberen Lage (Bild 4) regelmäßig die Ankerkräfte gemessen (Nachprüfungen). Die Beobachtungen des letzten Halbjahres ließen in der Tendenz eine mittlere Abnahme der Kräfte um etwa 4 kN erkennen. Aufgrund der Ergebnisse der Eignungsprüfungen ist zu erwarten, daß sich trotz des bisher unbekannten Einflusses des Permafrostes auf das Kriechen nach wenigen Jahren ein Gleichgewichtszustand einstellen wird. Alle Ankerkräfte liegen derzeit noch mit Werten zwischen 305 und 340 kN über der rechnerischen Kraft von 300 kN. Falls erforderlich, können die Anker später nachgespannt werden.

4. SCHLUßBEMERKUNG

Die vom Bauherrn, der Oberpostdirektion München, veranlaßten ständigen Kontrollen und Messungen ermöglichten es, die Baumethoden den angetroffenen Verhältnissen ohne Verringerung der Sicherheit anzupassen und in vielen Fällen zu vereinfachen.

5. LITERATUR

EGGER, H. (1982): Die Tragwerke der Funkübertragungsstation Garmisch 2 auf der Zugspitze. Bauingenieur (57), S. 215-223

HELLERER, H.O. (1979): Felsmechanische und geologische Untersuchungen für eine neue Funkübertragungsstelle auf der Zugspitze. Festschrift zum 65. Geburtstag von Prof. Dr.-Ing. R. Jelinek, S. 74-89

JELINEK, R. (1977): Neubau der Funkübertragungsstelle Garmisch 2 auf der Zugspitze. Unveröffentlichtes Gutachten 7748, München

KNAUER, J. (1933): Die geologischen Erkenntnisse beim Bau der Bayerischen Zugspitzbahn, München. Abhandlung der Geologischen Landesuntersuchung am Bayer. Oberbergamt, Heft 10

KÖRNER, H., ULRICH, R. (1965): Geologische und felsmechanische Untersuchungen für die Gipfelstation der Seilbahn Eibsee-Zugspitze. Geologica Bavarica Nr. 55, S. 404-421, München

MÜLLER, L. (1963): Der Felsbau. Bd. 1, F. Enke Verlag, Stuttgart

CRITICAL GEOMETRY AND MATERIAL PARAMETERS OF SHALLOW TUNNELS

Paramètres géométriques et paramètres du matériau critiques pour des tunnels à faible profondeur

Kritische Geometrie- und Materialparameter bei oberflächennahen Tunneln

H.-B. Mühlhaus
Lehrstuhl für Felsmechanik Universität Karlsruhe Bundesrepublik Deutschland

SYNOPSIS

In the article a method developed by the author is applied to the problem of the evaluation of critical geometry- and material-parameters of shallow tunnels. The limit load problem of the shallow tunnel is formulated as a linear eigenvalue problem. As the author has shown previously [2] this is possible although a nonlinear material law is underlaid. The eigenvalue problem is solved by the Finite-Element Method. The eigensolutions of the eigenvalue problem may be interpreted as the velocity field at the collapse of the system (collapse mechanism). Results are presented and discussed in detail for some ratios of the tunnel diameter to the overburden at different values of some dimensionless material parameters.

RESUME

Cette communication décrit un nouveau procédé pour la détermination des paramètres inhérents à la géométrie critique dans le cas des tunnels à faible profondeur. Le problème de charge limite des tunnels à faible profondeur y est formulé comme si les valeurs propres étaient linéaires. Comme l'auteur l'a déjà montré [2], cela est possible bien que le comportement du matériau ne soit pas linéaire. Les données relatives aux valeurs propres sont étudiées par la méthode des éléments finis. Les solutions spécifiques des valeurs propres étaient celles des champs de vitesses lors de la rupture (mécanisme de rupture). On présente les détails de résultats obtenus pour quelques valeurs du rapport - rayon de tunnel: épaisseur de couverture, et cela pour différentes valeurs de paramètres de matériaux.

ZUSAMMENFASSUNG

In dem Beitrag wird ein von dem Verfasser entwickeltes Verfahren auf die Problematik der Ermittlung kritischer Geometrie- und Materialparameter bei oberflächennahen Tunneln angewendet. Dabei wird das Traglastproblem des oberflächennahen Tunnels als lineares Eigenwertproblem (EWP) formuliert. Wie in [2] gezeigt wurde, ist dies möglich, obwohl das Materialverhalten nicht linear ist. Das EWP wird nach de FE-Methode gelöst. Die Eigenlösungen des EWP sind die Geschwindigkeitsfelder beim Kollaps (Kollapsmechanismus). Im einzelnen werden Ergebnisse für verschiedene Verhältnisse von Tunnelradius zur Oberdeckung über der Firste und verschiedene Verhältnisse der Materialparameter vorgestellt und erläutert.

1. INTRODUCTION

An infinite half plane with a circular hole (Fig. 1a) is taken as the geometrical model of a shallow, unlined tunnel. The deformations are assumed to be plane and it is furthermore assumed that undrained conditions prevail. The latter implies that the material can be regarded as incompressible. As explained in detail in chapter 4 of this paper, the material behaviour is described by a power law which, in the case of pure shearing (Fig. 1b), reduces to:

$$\tau = \tau_0 \left(\frac{\varepsilon}{\varepsilon_0}\right)^N . \tag{1}$$

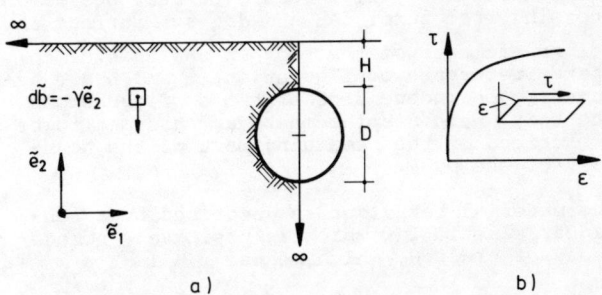

Fig 1 (a) Geometry of the problem. γ is the specific weight.
(b) Stress-strain relation in pure shearing.

Hence the dimensionless parameters of the problem sketched in Fig. 1 are H/D, τ_O, N and $\gamma (H+D/2)\cdot\frac{N}{\tau_O}$. In the following investigation a function f of the above parameters is evaluated (a relevant part of it, resp.) such that $f < 0$ if the stability of the tunnel is (mathematically) assured. At $f = 0$ the equilibrium is said to be critical.

If the material is assumed to be ideal plastic and if furthermore the displacement gradients are assumed to be infinitesimal, the evaluation of the function f is a typical problem of classical limit analysis (Koiter [1]). According to the practical importance of the above problem, upper and lower bound solutions, based on the classical theory, are existing (e. g. Davis et al., [2]). In contrast to calculations based on the classical theory, in the present calculation the effect of finite values of the displacement gradients is taken into consideration. If a power law is underlaid this in fact is necessary in order to render critical equilibrium states possible. If the effect of finite values of the displacement gradients is neglected a critical equilibrium state can exist only if $N = 0$.

In [3] and [4] the author has shown that, if a power law is used, the critical equilibrium problem leads to a linear eigenvalue problem. The basic assumptions which lead to this result are given in chapter 2 of this paper. Further results of [3], [4] which are used in the present calculation are summarized in chapter 3. To gain perspective the presentation of the assumptions and the results are given in more general terms than actually necessary. In chapter 4 the critical equilibrium problem is defined for the shallow tunnel and finally the results of the calculation are discussed.

2. ASSUMPTIONS AND DEFINITIONS

The loads prescribed on a part of the boundary of the body under consideration and the body forces acting on the particles of the body are given as

$$d\tilde{P} = \alpha\tilde{t}\ dA_R \qquad (2.1)$$

and

$$d\tilde{f} = \alpha\tilde{b}\ dV_R \qquad (2.2)$$

respectively. In (2.1) α is a positiv scalar increasing from zero. dA_R and dV_R are surface and volume elements of the body in the undeformed state (reference configuration). $\tilde{t}$ and $\tilde{b}$ are assumed to be independent of the configurations of the body. Zero displacements are assumed to be prescribed on the remaining part of the boundary of the body.

The material behaviour is described by a non-linear elastic law which is positively homogeneous of order N, which means:

$$\sigma_{ij} = f_{ij}(\varepsilon_{mn})\ ;\ i,j,m,n = 1,2,3 \qquad (2.3)$$

and

$$f_{ij}(\alpha\ \varepsilon_{mn}) = \alpha^N f_{ij}(\varepsilon_{mn}) \qquad (2.4)$$

A special form of such a material law will be presented in chapter 4 of this paper. σ, ε is any stress-strain pair conjugate in energy (Hill [5]). Note that if infinitesimal displacement gradients are assumed, all stress-strain pairs are equivalent and ε reduces to the symmetric part of the displacement gradient.

Assume now that there is a range of values of α (including $\alpha = 0$) where the boundary value problem is uniquely solvable. The range of unique solvability is bounded by a critical value of α which will be denoted as α_c in the following. For $\alpha = \alpha_c$ two possibilities arise:

1.) There is a neighbourhood of α_c where more than one solution of the boundary value problem (b.v.p.) under consideration is existing. Then α_c is denoted as bifurcation point.

2.) There is a neighbourhood of α_c where no solution of the b.v.p. is existing. Then α_c is denoted as limit point.

Classical examples for both of the two situations (buckling of perfect and imperfect beams, plates and shells) are discussed by Hutchinson [6]. Some examples from geomechanics are represented in the book of Biot [7]. Rudnicki and Rice [8] treated the spontaneous formation of a narrow shear band in an initially homogeneously deformed rock specimen as a bifurcation problem. Vardoulakis [9] solved the problem of the barreling of an ideal cylindrical sand specimen as a bifurcation problem. (Besides those mentioned above there are more authors working in this field.)

Before proceeding a final assumption has to be made:

It is assumed that displacement gradients are infinitesimal for $\alpha \leq \alpha_c$.

This assumption implies that all stress-strain pairs are equivalent for $\alpha \leq \alpha_c$. Note that the rates of the stress strain pairs are equivalent only for $\alpha < \alpha_c$ (strict inequality).

The linear eigenvalue problem (3.1) mentioned in the introduction, from which α_c is evaluated now follows from the homogeneity property (2.4) and the last assumption ([3], [4]).

3. CALCULATION OF CRITICAL VALUES OF α

The evaluation of the critical values of α is conducted in two steps. The numerical procedure underlaing both steps is the finite-element method (in its variant as displacement method).

1. step:

Evaluate the nodal point displacement vector $\bar{u}^1$ corresponding to the nodal point vector $\bar{F}^1$ of the external forces. The material behaviour be described by a law of type (2.3) with the property (2.4). The force vector $\bar{F}^1$ results from (2.1) and (2.2) at $\alpha = 1$ and the shape functions.

2. step:

Evaluate the smallest eigenvalue of the eigenvalue problem

$$\{\underline{K}_T(\bar{u}^1) - \lambda\underline{K}_G(\bar{u}^1)\}\bar{v} = \bar{o} \qquad (3.1)$$

In (3.1) $\underline{K}_T(\bar{u}^1)$ and $\underline{K}_G(\bar{u}^1)$ are the tangent- and the geometric stiffness matrix at $\bar{u}^1$. As shown in [3] and [4] the N'th power (2.4) of the smallest (positive) eigenvalue λ_c is the critical value α_c of α. If α_c is a limit point then the components of the corresponding eigenvector $\bar{v}$ are nodal point velocities at the collapse (collapse mechanism).

The condition for λ_c to be a limit point is, that the angle enclosed by $\bar{F}^1$ and the eigenvector $\bar{v}$ corresponding to λ_c is not equal to 90°, which means:

$$\bar{v}^T \bar{F}^1 \neq O \qquad (3.2)$$

The above statement may be interpreted more physically: The left (hand) side of (3.2) is the virtual work of the applied forces in the eigenvector corresponding to λ_c. If the virtual work does not vanish, then λ_c^N is (according to the definition given here) a limit point.

If more than one linearly independent eigenvectors correspond to λ_c, then it is sufficient that the virtual work does not vanish in one of them for λ_c^N to be a limit point.

4. CRITICAL PARAMETERS OF A SHALLOW TUNNEL

4.1 Constitutive relation

The geometry of the tunnel and a typical finite-element mesh are shown in Fig. 2a and 2b respectively. The collaps of the tunnel will usually

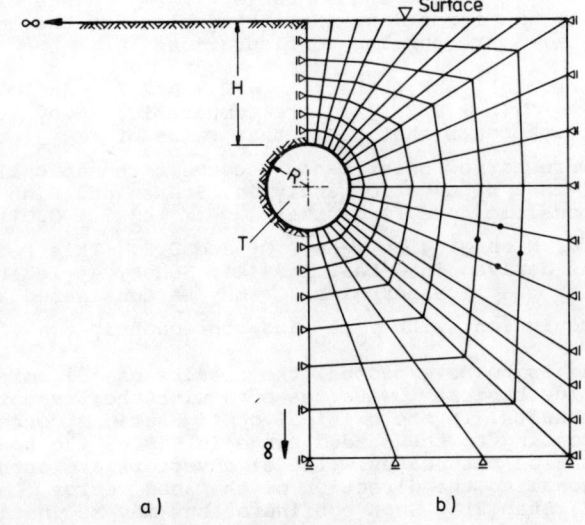

Fig. 2 (a) Geometry of the tunnel
(b) Finite element mesh at H/D = 1.5, 133 eight node isoparametric elements, 2 x 2 Gauss integration.

be a sudden event (e. g. due to a sudden loss of the tunnel pressure [2]). Hence it will be as-

sumed that undrained conditions prevail which implies that the material may be considered as incompressible. It remains to relate the deviatoric part of the stress tensor $s_{ij} = \sigma_{ij} - p\delta_{ij}$, $p = \frac{1}{3}\sigma_{kk}$ to the strain tensor ε_{ij}. (Because of the assumed incompressibility the pressure p will first remain undeterminate.) For this purpose a relationship of the deformation theory of plasticity (Hill 10) is used. The simplest constitutive relation of this theory is:

$$s_{ij} = h \, \varepsilon_{ij} \quad , \quad \varepsilon_{kk} = O \quad , \qquad (4.1)$$

where h is a scalar function of $\tau = (\frac{1}{2}s_{ij}s_{ij})^{\frac{1}{2}}$.

Formally (4.1) is a nonlinear elastic law. However there is an interesting possibility to interpret (4.1) within the scope of the vertex theories of plasticity (e. g. Budianski [11]).

Assuming the material to be rigid hardening the function h is defined as:

$$h = 2 \frac{\tau_o}{\varepsilon_o} (\frac{\tau}{\tau_o})^{\frac{N-1}{N}}, \qquad (4.2)$$

which implies (Fig. 3):

$$\tau = \tau_o(\frac{\varepsilon}{\varepsilon_o})^N; \quad \varepsilon = (2 \, \varepsilon_{ij}\varepsilon_{ij})^{\frac{1}{2}}, \qquad (4.3)$$

In (4.2), (4.3) τ_o, ε_o and N are material parameters. In connection with (4.2) and (4.3) eq. (4.1) is homogeneous of order N in ε_{ij}. Hence the method described in chapter 2 is applicable.

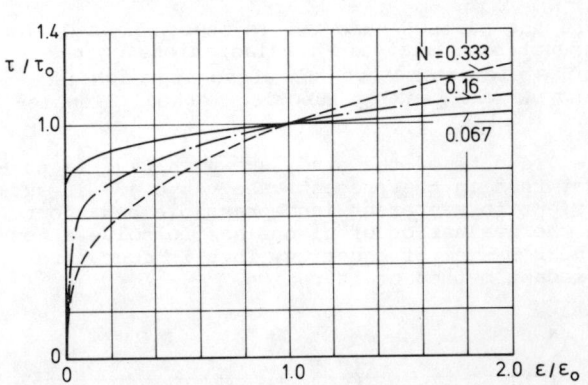

Fig. 3 Stress-strain diagram, showing (4.3) at different values of the hardening index N

Under the conditions of a biaxial test ($\varepsilon_1/\varepsilon_2 = -1$), ε reduces to $\varepsilon = 2\varepsilon_1$, $\varepsilon_1 > O$ and τ reduces to $\tau = \frac{1}{2}(\sigma_1 - \sigma_2)$, $\sigma_1 > \sigma_2$. The quantities ε_1, ε_2 and σ_1, σ_2 are the principal values of the stress- and strain-tensor respectively. Nahrgang [12] performed biaxial tests on slightly overconsolidated clay specimen. In these tests, depending on the watercontend, N was found to be ranging between 0.14 and 0.2.

In the following calculation it will be assumed that τ_o/ε_o^N is constant with depth, although in general, this will not be the case. However, there are many situations where the above assumption is reasonable. It should be mentioned, that the assumed constancy of τ_o/ε_o^N is not necessary for the method applied here to be applicable.

4.2 Calculation of critical parameters

In the evaluation of $\bar{u}^1$ (Chapt. 2, Step. 1), at fixed ratio H/D, the amount of the dimensionless body force $\gamma^* = \gamma(H + D/2) \varepsilon_o^N/\tau_o$ is set equal to unity. Once $\bar{u}^1$ is calculated, the matrixes $\underline{K}_T$ and $\underline{K}_G$ can be established. The solution of the eigenvalue problem (3.1) yields the smallest eigenvalue λ_c, and the critical value of γ^* is obtained as (Chapt. 2, Step 2):

$$\gamma(H + D/2)\varepsilon_o^N/\tau_o\Big|_c = \lambda_c^N \qquad (4.4)$$

This procedure is repeated at different ratios of H/D. Accordingly, γ_c^* is obtained as a function of H/D. Then the function f, introduced in the introduction, may be defined as:

$$f = \gamma(H+D/2)\varepsilon_o^N/\tau_o - \lambda_c^N(H/D). \qquad (4.5)$$

In the numerical solution of the problem the solid was treated as <u>nearly</u> incompressible rather than incompressible. The near incompressibility was achieved by introducing a bulk modulus

$$K = \frac{h(1 + \nu)}{3(1 - 2\nu)} \qquad (4.6)$$

were the poisson number ν is chosen near 0.5. Infunctional analytic considerations of the finite element method the above approach is known as the penalty function method (Zienkiewicz [13]).

The results of the study shown in Fig. 4a suggest that in the present case $\nu = 0.499$ is sufficient to represent <u>incompressible</u> behaviour. In the evaluation of u^1 one has to solve a nonlinear system of equation. This is done here by a secant method of iteration.

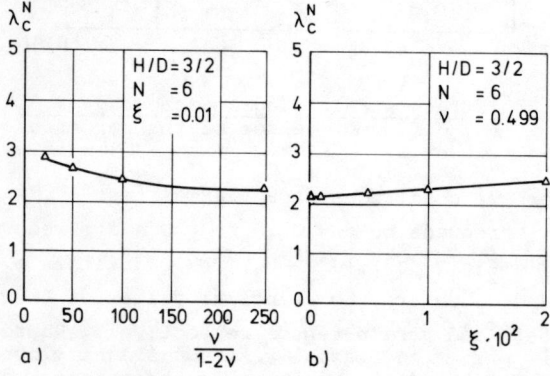

a)   b)

Fig. 4(a)   Dependence of λ_c^N on the Poisson ratio ν, (b) Dependence of λ_c^N on the truncation number. The smallest value of ξ at which an eigenvalue was calculated was ξ = 0.0001

Convergence of the iteration is declared if

$$\| \bar{u}^1_{r+1} - \bar{u}^1_r \|_E \leq \xi \| \bar{u}^1_{r+1} \|_E \qquad (4.7)$$

In (4.7) $\| \cdot \|_E$ denotes the Euclidean norm, r is the numer of the iterational step and ξ is a suitable truncation number. Based on the results of the study shown in Fig 4b, ξ was chosen equal to 0.001.

In the following, at fixed ν and ξ, critical values of γ^* were evaluated for different values of the hardening index N and ratios H/D. The results are shown in Fig. 5.

Assuming rigid-plastic behaviour of the solid in [2] lower bounds on the critical values of $\bar{\gamma}=\gamma(H + D/2)/C_u$ were calculated by the method of characteristics. (C_u denotes the shear strength measured under undrained conditions in a biaxial test- the so-called undrained cohesion). For comparison some of the results of [2] are also shown in Fig. 5.

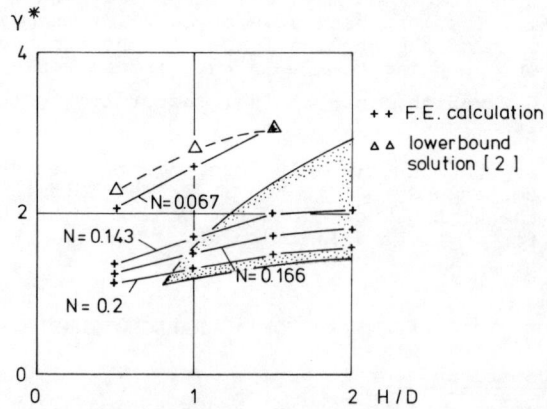

Fig. 5   Critical values of $\gamma^* = \gamma(H + D/2)\varepsilon_o^N/\tau_o$. In the shaded range $\bar{v}^T F^1 = 0$ hence bifurcations are taking place in the corresponding configurations

One might ask wether $\bar{\gamma} = \gamma(H + D/2)/C_u$ and $\gamma^* = \gamma(H + D/2)\varepsilon_o^N/\tau_o$ are comparable. However, if one accepts that C_u is that value of $\tau = \frac{1}{2}(\sigma_1-\sigma_2)$ were narrow shear bands become mathematically (-in a bifurcation analysis, Stören and Rice [14]) possible, one finds that $C_u = \tau_o/\varepsilon_o^N(1 \pm 0.01)$ for N ranging between 0.01 and 0.2. (This result is derived in detail in [4]). Hence, at least for practical purposes, C_u may be considered as equivalent with τ_o/ε_o^N and consequently $\bar{\gamma} \approx \gamma^*$.

As is to be expected, the results of [2] coincide best at low values of N with the present results. In the critical configurations corresponding to the shaded range in Fig. 5 the solution bifurcates. (The eigenvectors are orthogonal to the direction of the load vector (Cap. 2, Step 2). Such configurations may be unstable or not, whereas configurations corresponding to a limit load are always unstable.

In order to find out wether the system is stable after a bifurcation has taken place, one has to investigate the behaviour of the solution in the neighbourhood of the bifurcation point. However this be reserved for a subsequent paper.

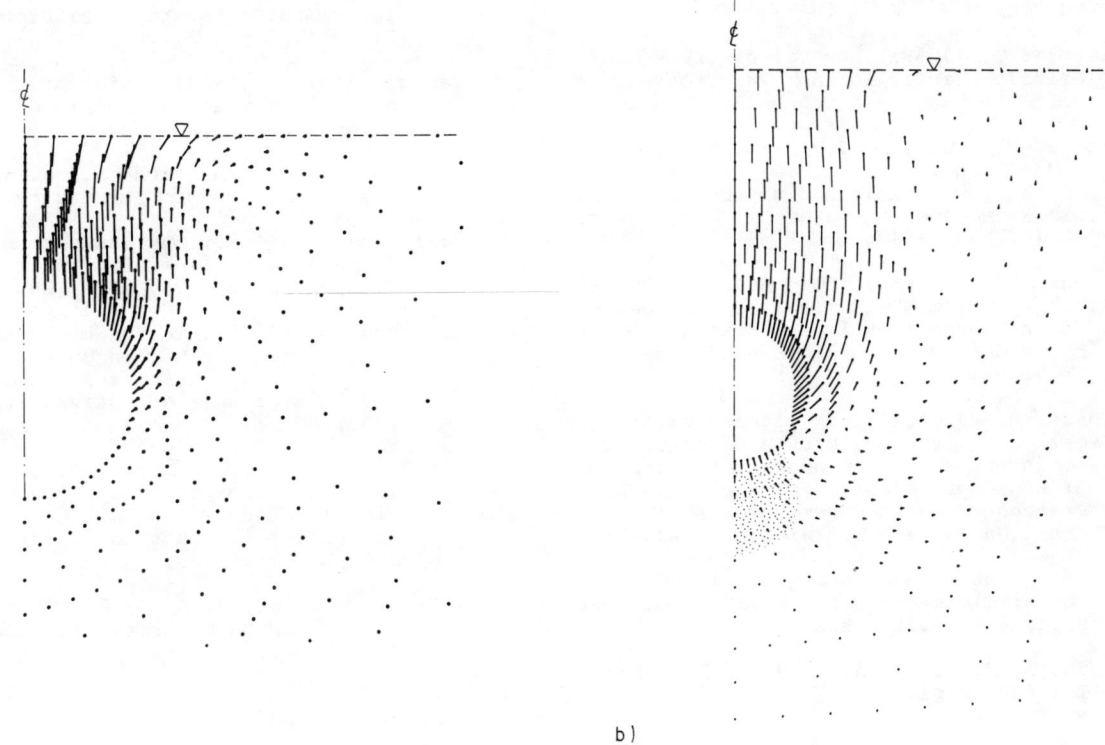

a)   b)

Fig. 6   Eigenvector fields (a) H/D = O.5, N = O,2, (b) H/D = 3/2, N = 15

In Fig. 6 the nodal point components of two eigenvectors are shown.The nodal point vectors are
nowhere orthogonal to the direction of the gravity load. Hence, according to the explanations
in chapter 2, both eigenvectors are collapse
mechanisms.

In the calculation the material behaviour was
described by the nonlinear elastic law (4.1) -
(4.3). Accordingly the stiffness of the material
is indifferent with respect to a reversal of the
loading direction. However in a real material
the stiffness is significantly higher in unloading than in loading. In this sense a virtual
solid (the so-called 'comparison solid' [6] whose
behaviour is identical with the behaviour of the
'real' solid in the loading regime, but differs
in the unloading regime) is described by (4.1) -
(4.3). In [6] it is shown that the critical
points of the comparison solid are lower bounds
on the critical points of the 'real' solid. As is
shown furthermore in [6] the critical points of
the comparison solid are mostly identical with
the critical points of the 'real' solid, provided the critical points are bifurcation points.
However in the case of limit points there may
be differences. Considering in this context the
marked region below the tunnel in Fig. 6b one
notes, that the real material would be in its
unloading regime there. In a subsequent paper
the author will show, that the eigenvector corresponding to a critical point of the comparison
solid can be used to evaluate an upper bound on
the critical point of the 'real' solid.

Final remark:
Due to the incompressibility of the material, the
resultsshown in Fig. 5 are also valid if uniform

pressures of equal amounts are acting on the
surfaces S and T (Fig. 2a).

5. CONCLUSIONS

In conventional finite element solutions of
limit load problems it is assumed that the material behaves elastic-plastic and that the displacement gradients are infinitesimal. The external loads are increased stepwise. The limit
load is found if the determinant of the tangential stiffness matrix vanishes at a certain
load level. However, mostly such a load is not
found because the calculation becomes highly
sensitive with respect to round off error in the
fully plastic range. In the method applied here,
the stiffness matrix does not need to become singular for a limit load to exist. Therefore this
method is (numerically) more robust than the
conventional method. The destabilizing effect of
finite values of the dispacement gradients (which
plays an important role for nonvanishing hardening index N) is taken into consideration. Furthermore one has the possibility to distinguish
between configurations which are critical in the
sense that the loads have reached their limit
values and critical configurations where the
solution bifurcates. It turned out that at fixed
ratio H/D the hardening index corresponding to
a bifurcation point is always higher than the
hardening index corresponding to limit point
(Fig. 5).

6.   ACKNOWLEDGEMENT

This work was supported by Deutsche Forschungsgemeinschaft, Grant NA 69/11 and Na 69/14-2.
The author gratefully acknowledges the admitted
funds.

6.   REFERENCES

1 Koiter, W. T. (1958). General theorems for elastic-plastic solids. In: Progress in solid mechanics; North Holland Publ.

2 Davis, E. H., Gunn, M. J., Mair, R. J. and Seneviratne, H. N. (1980). The stability of shallow tunnels and underground openings in cohesive material. Géotechnique 30 (4), 397 - 416.

3 Mühlhaus, H.-B. (1982). Lösung von Verzweigungsproblemen und Traglastproblemen bei einer Klasse nichtlinearer Randwertprobleme der Kontinuumsmechanik. Ingenieur-Archiv, 52.

4 Mühlhaus, H.-B. (1982). An eigensolution method for the evaluation of critical loads of certain boundary value problems of nonlinear elasticity. Report to the Deutsche Forschungsgemeinschaft (DFG) Grant Na 69/14-2. To be published 1983.

5 Hill, R. (1968). On constitutive inequalities for simple materials - I. J. Mech. Phys. Solids, 16, 229 - 242.

6 Hutchinson, J. W. (1974). Plastic buckling. In: Adv. appl. Mech., 14, Acad. Press, N. Y., S. F., London.

7 Biot, M. A. (1965). Mechanics of incremental deformation. Wiley, N. Y.

8 Rudnicki, J. W., Rice, J. R. (1975). Conditions for the localization of deformation in pressure-sensitive, dilatant materials. J. Mech. Phys. Solids, 23, 371 - 394.

9 Vardoulakis, I. (1979). Bifurcation analysis of the triaxial test on sand samples. Acta Mechanica, 32, 35 - 54.

10 Hill, R. (1950). The mathematical theory of plasticity. Oxford University Press.

11 Budianski, B. (1959). A reassessment of deformation theories of plasticity. J. appl. mech., 26, 259 - 264.

12 Nahrgang, E. (1974). Verformungsverhalten eines weichen bindigen Untergrunds. Veröffentl. des Instituts für Bodenmechanik und Felsmechanik der Universität Karlsruhe, 60.

13 Zienkiewicz, O. C., Godbole, P. V. (1975). Viscous, incompressible flow with special reference to non In: Finite Elements in Fluids, Vol. 1, John Wiley & Sons.

14 Stören, S., Rice, J. R. (1975). Localized necking in thin sheets. J. Mech. Phys. Solids, 23, 421 - 441.

Address of the author:

Mühlhaus, H.-B., Dr.-Ing.
Lehrstuhl für Felsmechanik
Universität Karlsruhe
7500 Karlsruhe 1
Federal Republic of Germany

STANDISCHERHEIT EINER TIEFEN BAUGRUBE IM DIAGENETISCH VERFESTIGTEN TONGESTEIN

Stability of a deep excavation in diagenetically consolidated mudstone

Stabilité d'une tranchée profonde dans une argile compacte consolidée diagénétiquement

R. Grüter
Dipl.-Ing., Bundesbahndirektor Bundesbahndirektion Stuttgart Deutsche Bundesbahn
W. Wittke
o. Prof. Dr.-Ing., Institut für Grundbau, Bodenmechanik, Felsmechanik und
Verkehrswasserbau Rheinisch-Westfälische Technische Hochschule Aachen

ZUSAMMENFASSUNG

Im Zuge des Neubaus des 5,5 km langen Hasenbergtunnels der S-Bahn Stuttgart wurde eine ca. 25 m tiefe, 220 m lange und 20-30 m breite Baugrube in diagenetisch verfestigten Tonsteinen des Lias α und des Knollenmergels hergestellt. Die beim Aushub dieser Baugrube gemessenen horizontalen Wandverformungen von max. 65 mm und die Sohlhebungen von etwa 45 mm waren nach den umfangreichen Voruntersuchungen nicht erwartet worden. Zur Interpretation dieser Verformungen wurden bereits während des Baus zahlreiche FE-Berechnungen durchgeführt. Es wurde deutlich, daß sich die gemessenen Verformungen nur durch zusätzliche Horizontalspannungen von 1-2 MN/m^2, die sich aus der Entstehungsgeschichte der anstehenden Gesteine ableiten lassen, erklären lassen.

SYNOPSIS

During the construction of the 5.5 km long Hasenberg Tunnel for the S-Bahn Stuttgart an approximately 25 m deep, 220 m long and 20-30 m wide open pit was excavated in the diagenetically consolidated mudstone of the Lias α and Knollenmergel formation. The horizontal displacements of the side walls of up to 65 mm and the heaving of the invert of 45 mm measured during construction were not expected though extensive explorations were carried out before construction started. Parallel to the excavation extensive F.E. analyses were carried out to interpret these displacements. It could be concluded, that the observations can only be explained by the presence of horizontal in situ stresses of 1-2 MN/m^2, which are in accordance with the history of origin of the rock.

RESUME

Au cours de la construction du tunnel de Hasenberg du métro de Stuttgart d'une longueur de 5,5 km, une tranchée d'environ 25 m de profondeur, 220 m de long et 20 à 30 m de large a été creusée dans des argiles compactes diagénétiquement consolidées du Lias α et du Knollenmergel. Les déformations horizontales de la paroi d'une valeur maximale de 65 mm et sur les soulèvements du fond d'environ 45 mm qui ont été mesurés lors du déblaiement n'avaient pas été prévus malgré les investigations préliminaires importantes. Lors de la construction, de nombreux calculs par la méthode des éléments finis ont été effectués pour interpréter ces déformations. Il est devenu clair, que les déformations mesurées ne peuvent être expliquées que par des contraintes horizontales supplémentaires de 1 à 2 MN/m^2, ce qui est en conformité avec la genèse de la roche.

PROJEKTÜBERSICHT

Bevor über die Erfahrungen und Erkenntnisse berichtet wird, die beim Bau des Hasenbergtunnels und der Baugrube für die S-Bahn-Station "Universität" in den diagenetisch verfestigten Tongesteinen des Knollenmergels und des Lias α gemacht wurden, soll zum besseren Verständnis das gesamte Tunnelprojekt kurz vorgestellt werden (Grüter, 1979).

Das Netz der S-Bahn Stuttgart besteht aus 6 Strecken, die am Hauptbahnhof Stuttgart einmünden und die durch eine rund 8,5 km lange unterirdische Verbindungsbahn mit 2 Strecken aus dem Süden Stuttgarts vom Flughafen und von Böblingen verbunden werden (Bild 1). Von diesem Netz sind

die 6 Strecken auf der Seite des Hauptbahnhofs und 3 km der unterirdischen Verbindungsbahn mit den 4 unterirdischen Haltestellen im Stuttgarter Stadtgebiet bereits in Betrieb. Die S-Bahn-Züge wenden zur Zeit alle in der 1,6 km langen Wendeschleife im Anschluß an die derzeitige Endstation "Schwabstraße" (Bild 1, 2).

Mit einem Kostenvolumen von ca. 700 Millionen DM werden jetzt die restlichen 5,5 km dieser unterirdischen Verbindungsbahn und die Strecken nach Böblingen und zum Flughafen für den S-Bahn-Verkehr ausgebaut. Das Kernstück dieser Baumaßnahmen ist dabei mit einem Kostenaufwand von allein 250 Millionen DM der Weiterbau der unterirdischen

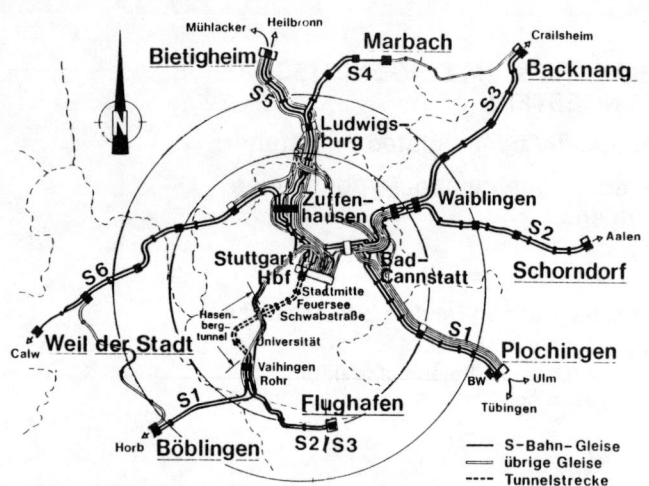

Bild 1   Netz der S-Bahn Stuttgart

Verbindungsbahn über den derzeitigen Endpunkt, die Station "Schwabstraße", hinaus unter dem Hasenberg und die neue Universität hindurch bis zum Anschluß an die vorhandene Eisenbahnstrecke Stuttgart - Horb - Zürich, die sogenannte Gäu-bahn (Bild 2).

Dieser neu zu bauende Teil der Verbindungs-strecke, der sogenannte Hasenbergtunnel, hat einschließlich einer Station im Bereich der

neuen Universität eine Länge von 5,5 km. Der Tunnel steigt nach der Station "Schwabstraße" mit ca. 35 ‰ an und erreicht nach 4,5 km unter Überwindung eines Höhenunterschieds von 154 m seinen höchsten Punkt bei der Station "Univer-sität". Er fällt dann mit 38 ‰ wieder zur Ein-mündung in die bestehende Eisenbahnstrecke ab. Die Überdeckung beträgt maximal 125 m unter dem Hasenberg und minimal 12 m im Bereich der Uni-versität und im anschließenden Wohngebiet.

Neben dem eigentlichen Tunnelbauwerk und der Station "Universität" waren noch drei Fenster-stollen und drei Schächte zu errichten. Sie die-nen zur Zeit teilweise zur Auffahrung des Tun-nels und später als Notausstiege und zur Lüftung des Tunnels.

Vor und während der Planungsphase wurde der Un-tergrund eingehend untersucht. Neben 60 Boh-rungen längs der Trasse wurden eine Reihe von felsmechanischen Versuchen in den Fensterstollen und Schächten, die wir vorab gebaut haben, in praktisch allen zu erwartenden Gebirgsschichten durchgeführt.

Der geologische Aufschluß (Bild 3) zeigt, daß in-folge des zu überwindenden Höhenunterschieds von 154 m eine Vielzahl von Gebirgsschichten mit dem Tunnel durchfahren werden, angefangen vom Gips-keuper über die Bunten Mergel, den Stubensand-stein, den Knollenmergel bis zum Lias. Insgesamt handelt es sich um 10 verschiedene Gebirgsarten mit teilweise extrem verschiedenen felsmechani-schen Verhalten.

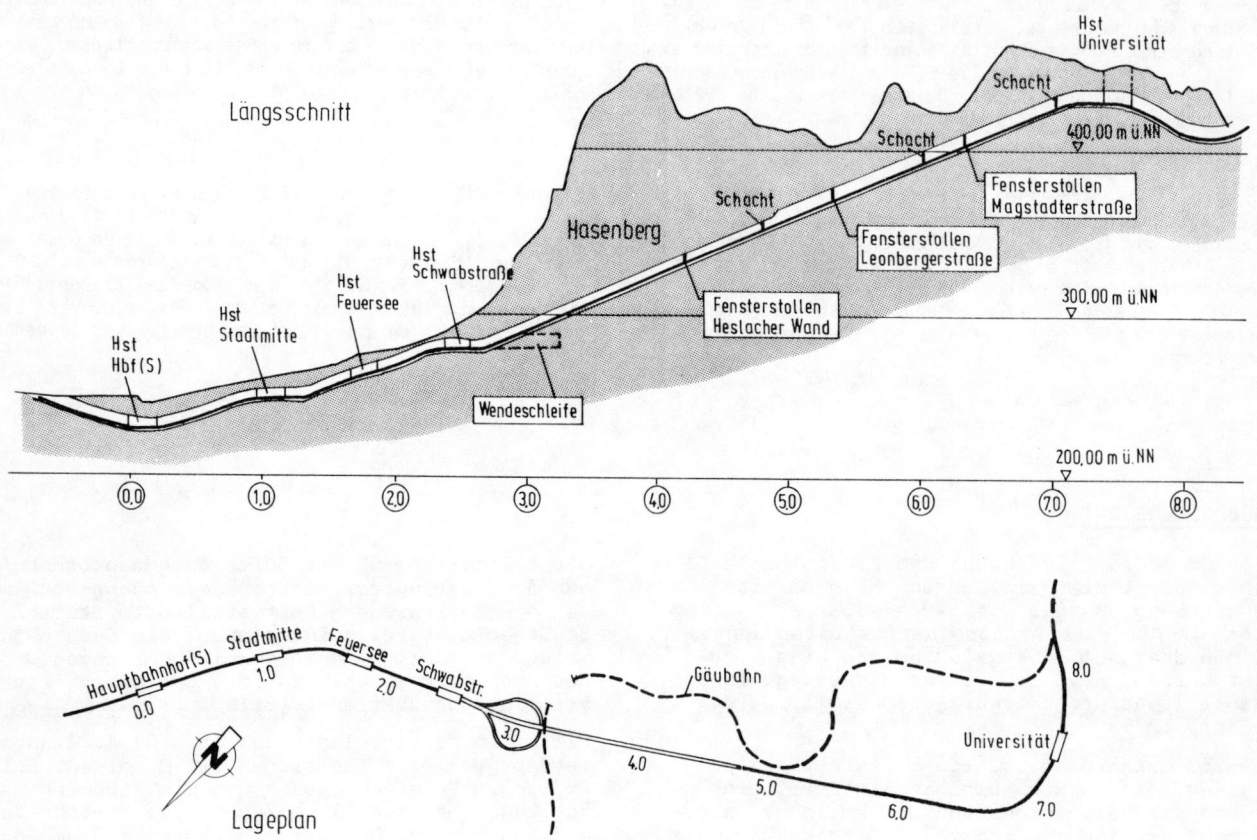

Bild 2   Grundriß und Längsschnitt der Tunneltrasse

C 270

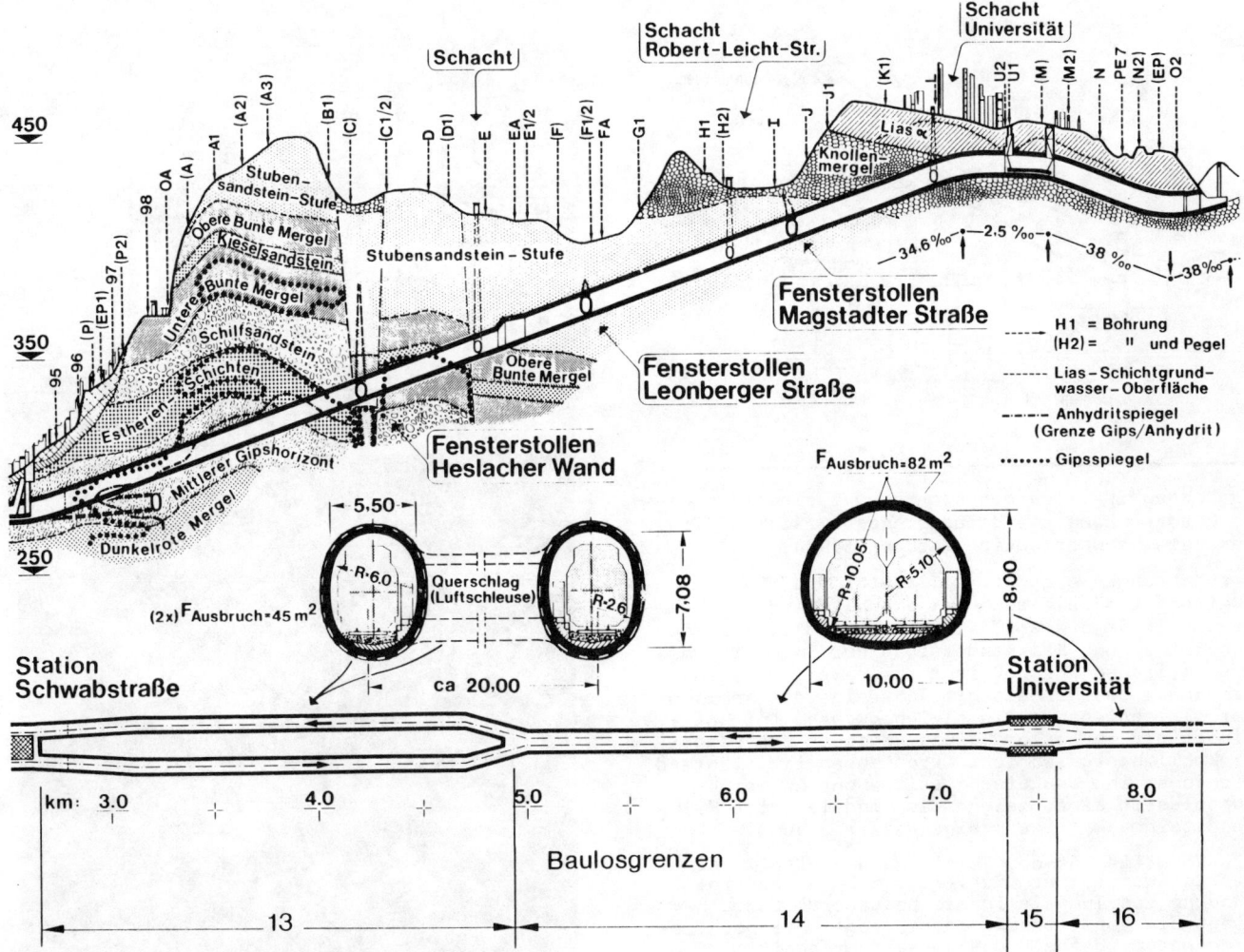

Bild 3   Geologischer Längsschnitt

Entsprechend vielseitig waren und sind die Probleme, die bei der Planung, Ausschreibung und jetzt bei der Ausführung zu berücksichtigen und zu lösen sind.

Der Tunnel wird wegen der großen Überdeckungen von bis zu 125 m ausschließlich in bergmännischer Bauweise gebaut. Nur die Station "Universität" wird in offener Baugrube gebaut. Das Gestein wird überwiegend durch gebirgsschonendes Sprengen gelöst. Nur da, wo besonders dicht unter der Bebauung gearbeitet werden muß, werden auch Teilschnittmaschinen eingesetzt. Gesichert wird mit Spritzbeton, Ankern und soweit erforderlich mit Streckenbögen.

Es sind zwei unterschiedliche Tunnelprofile vorgesehen. Im unteren Bereich, wo der Tunnel im sulfathaltigen Gestein (Gipskeuper und Bunte Mergel) aufgefahren werden muß, wurden 2 eingleisige Tunnel mit einem Ei- oder Ellipsenprofil gewählt, das besonders geeignet ist, die möglichen hohen Quelldrücke in diesem Gebirge aufzunehmen. In den übrigen Bereichen ist es wirtschaftlicher, ein zweigleisiges Maulprofil zu bauen.

Der Tunnel erhält eine Innenschale aus bewehrtem Beton, und zwar im Bereich des zweigleisigen Profils aus wasserundurchlässigem Beton. Im Bereich der 2 eingleisigen Tunnel muß eine Abdichtung vorgesehen werden, um den Beton vor dem dort vorhandenen betonangreifenden Wasser zu schützen.

Um eine möglichst kurze Bauzeit zu erreichen - vorgesehen sind für den Rohbau ca. 3 Jahre - wurde die Tunnelstrecke in 4 Baulose (13 - 16) unterteilt. In allen 4 Baulosen wird zur Zeit noch gebaut.

Gegenstand des vorliegenden Beitrags ist die in den diagenetisch verfestigten Tonsteinen des Knollenmergels und des Lias α liegende Haltestelle "Universität", das Baulos 15. Diese Haltestelle wird in einer 25 m tiefen, 220 m langen und zwischen 20 und 30 m breiten offenen Baugrube erstellt (Bild 4).

VORUNTERSUCHUNGEN

Da im Lias α und im Knollenmergel noch keine Tunnel und auch keine Baugruben der hier vorgesehenen Größenordnung gebaut wurden und daher auf keine Erfahrungen bei ähnlichen Bauvorhaben

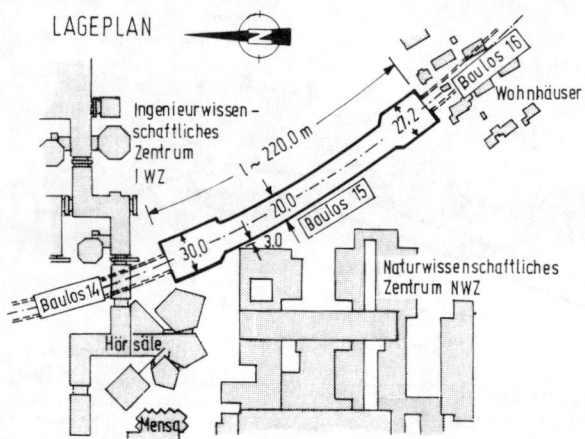

Bild 4   Grundriß der Station "Universität"

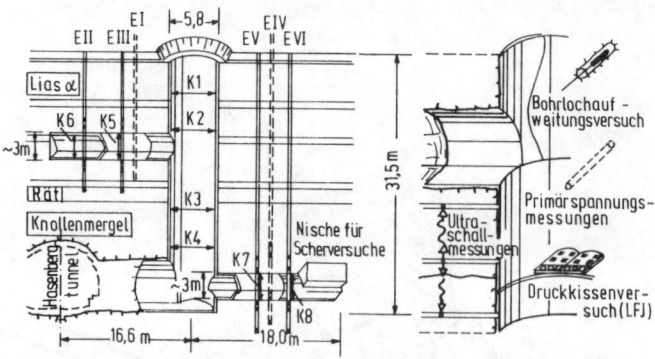

Bild 5   Felsmechanisches Untersuchungsprogramm
im Schacht "Universität"

zurückgegriffen werden konnte, war eine intensive Untersuchung des Untergrundes gerade in diesem Bereich unerläßlich (Grüter, 1980).

Kernbohrungen ergaben, daß sowohl die Station "Universität" als auch das anschließende Tunnelbaulos 16 im oberen Teil des Querschnitts in den Tonsteinen und Kalksandsteinen der Formation des Lias α liegen und daß im unteren Teil bzw. an der Sohle des Tunnels die Tonsteine des Knollenmergels anstehen. Die gleichen Verhältnisse finden wir im Baulos 14 vor, nachdem die mächtige Knollenmergelschicht durchfahren ist. Während die Tonsteine des Lias α weitgehend wasserundurchlässig sind, zeigen die eingelagerten Kalksandsteinbänke eine mäßige Wasserführung.

Die Verwitterung der Tonsteine des Lias α reicht bereichsweise sehr tief. In der Nähe der Einmündung des Tunnels in die bestehende Gäubahn liegt die Verwitterungsgrenze sogar nur noch wenige Meter oberhalb des Tunnelquerschnitts.

Die unter dem Lias α anstehenden Ton- und Schluffsteine des Knollenmergels sind nahezu wasserundurchlässig.

Bekanntlich ist die Aussagekraft der Ergebnisse von Kernbohrungen und darin durchgeführter Feldversuche hinsichtlich der für den Tunnelbau maßgebenden geologischen und felsmechanischen Eigenschaften des Gebirges nur begrenzt. So kann man aus diesen Aufschlüssen beispielsweise keine vollständigen Angaben über die Geometrie und die Festigkeit des Trennflächengefüges sowie die großmaßstäbliche Verformbarkeit des Gebirges gewinnen. In einer vorgezogenen Baumaßnahme wurden deshalb zwei ohnehin erforderliche Fensterstollen sowie zwei ebenfalls erforderliche Schächte aufgefahren und in ihnen umfangreiche felsmechanische Untersuchungen durchgeführt. Die bei der Ausführung dieser Stollen und Schächte gesammelten Erfahrungen und Erkenntnisse fanden Eingang in die Ausschreibungsunterlagen.

Für die Planung und den Bau der Haltestelle "Universität" sind die beim Bau des Schachts "Universität" gewonnenen Erfahrungen von besonderer Bedeutung (Bild 3, 5). Dieser Schacht mit einem Durchmesser von 5,80 m durchörtert die Schichten des Lias α und den oberen Bereich des Knollenmergels.

Sowohl im Bereich des Lias α als auch im Knollen-

Bild 6   Lias α

mergel wurde je ein ca. 3 m hoher Querschlag aufgefahren. Mit Hilfe von Konvergenzmeßquerschnitten, vor Baubeginn eingebauten Extensometern sowie Oberflächennivellements wurden die durch den Aushub des Schachts und den Vortrieb der Stollen bedingten Verformungen gemessen. Darüberhinaus wurde durch Großscherversuche die Scherfestigkeit der in den Tonsteinen des Knollenmergels angetroffenen Harnischflächen ge- messen. Die Verformbarkeit der Tonsteine des Lias α und des Knollenmergels wurde in Bohrloch- aufweitungsversuchen und in großmaßstäblichen Druckkissenversuchen mit Belastungsflächen von ca. 2 m^2 ermittelt. Darüberhinaus wurden zur Er- mittlung der Verformbarkeit der Gesteine noch Ultraschallmessungen durchgeführt und mit Hilfe einschlägiger Meßverfahren, in unserem Fall von Triaxialzellen, die Primärspannungen im Knollen- mergel gemessen.

Die Schacht- und Stollenwände wurden eingehend geologisch kartiert und fotografiert.

Bild 6 läßt die horizontale Schichtung des Ge- steins und auch die annähernd senkrecht stehen- den Klüfte erkennen. In die Tonsteine des Lias α sind bis zu 80 cm dicke Kalksandsteinbänke ein- gelagert, die ebenfalls horizontal liegen, aber eine deutlich höhere Festigkeit und eine ge- ringere Verformbarkeit besitzen als die Tonstei- ne. Eine solche Kalksandsteinbank ist in Bild 6 erkennbar.

Bei der geologischen Kartierung wurden alle er- kennbaren Klüfte hinsichtlich ihrer Raumstellung und Erstreckung eingemessen (Bild 7). Die sta- tistische Auswertung der gemessenen Streich- und Fallwinkel ergab, daß im Tonstein und in den

eingelagerten Kalksandsteinbänken zwei annähernd senkrecht einfallende Kluftscharen K 1 und K 3 ausgebildet sind (Bild 8).

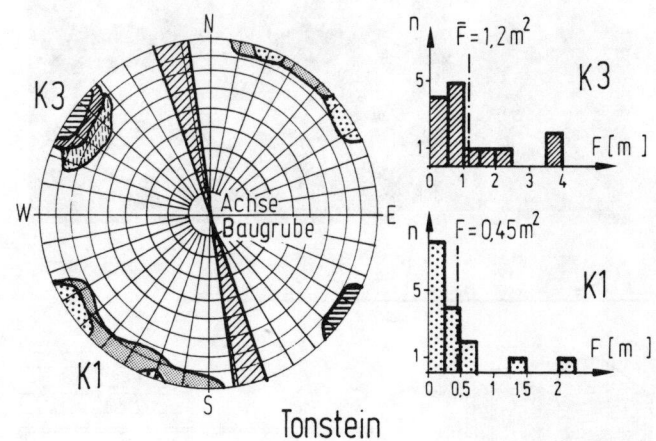

Tonstein

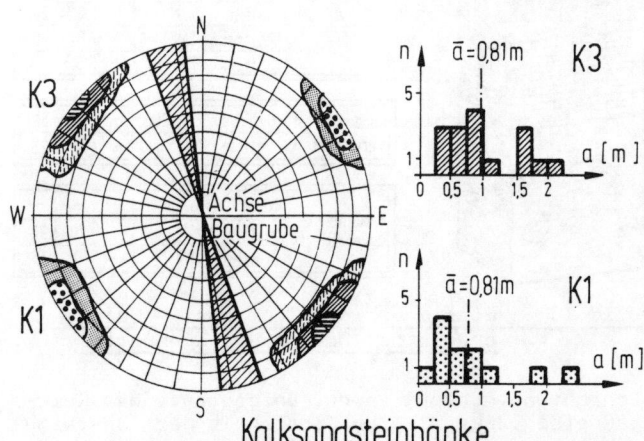

Kalksandsteinbänke

Bild 8   Trennflächen im Lias α - Lotpunktdia- gramme, Abstände a und Größe F

Die Klüfte der Schar K 3 schneiden die Baugru- benachse unter ca. 60°, während die Klüfte der Schar K 1 teilweise spitzwinklig zur Baugruben- achse streichen.

Eine Auswertung der Kartierung hinsichtlich der Flächengrößen und Abstände ergab, daß die Er- streckung der im Schacht "Universität" einge- messenen Klüfte mit Flächengrößen von ca. 0,5 - 1,2 m^2 und auch ihre Häufigkeit in den Tonstei- nen verhältnismäßig gering war. Dagegen ergaben sich in den Kalksandsteinbänken Kluftabstände von nur 80 cm, wobei die Klüfte die jeweiligen Bänke sowohl in horizontaler als auch in verti- kaler Richtung nahezu vollständig durchtrennten. Im Knollenmergel haben vor allem die im Mittel unter 30° einfallenden glatten, gewellten Har- nischflächen eine besondere Bedeutung für die Festigkeit (Bild 9).

Einen deutlichen Einfluß auf die Festigkeit und Verformbarkeit der Tonsteine des Lias α hat auch die annähernd horizontal verlaufende

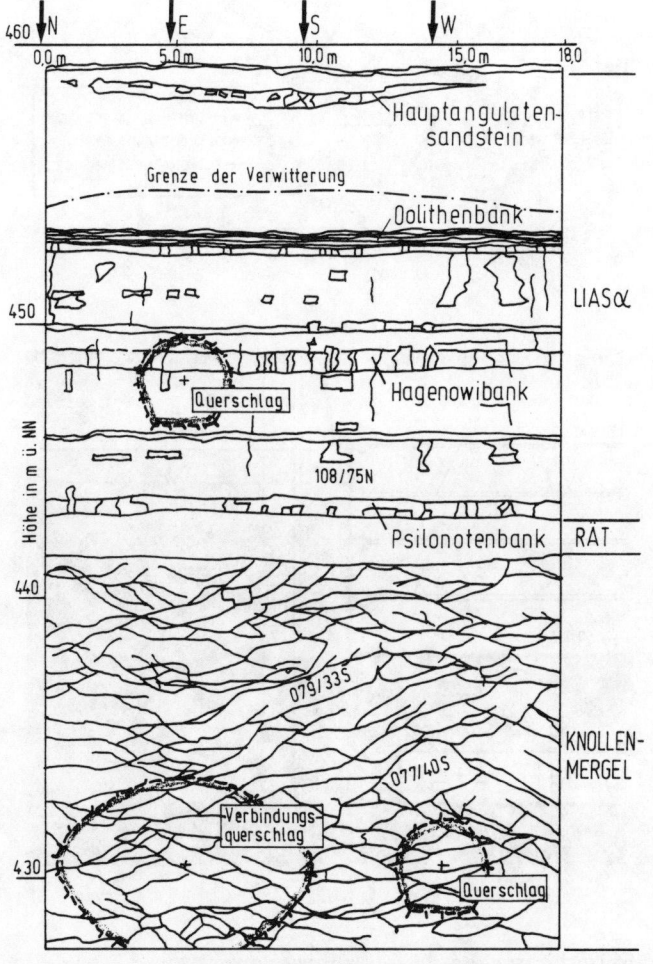

Bild 7   Schacht Universität - Geologische Kartierung der Schachtwand

Bild 9   Harnische im Knollenmergel

Bild 11   Los 15 - Blick in die teilweise ausgehobene Baugrube

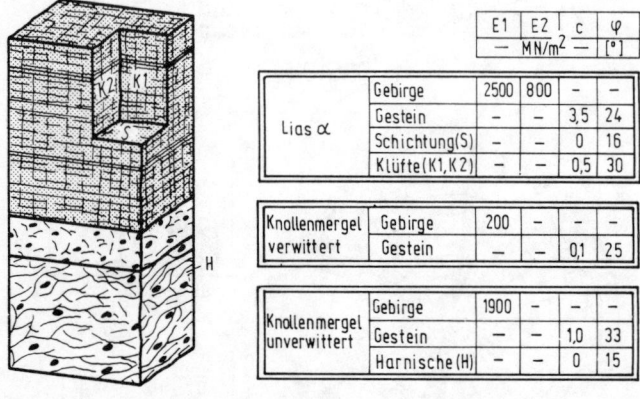

		E1	E2	c	φ
		— MN/m² —		—	[°]
Lias α	Gebirge	2500	800	—	—
	Gestein	—	—	3,5	24
	Schichtung(S)	—	—	0	16
	Klüfte(K1,K2)	—	—	0,5	30
Knollenmergel verwittert	Gebirge	200	—	—	—
	Gestein	—	—	0,1	25
Knollenmergel unverwittert	Gebirge	1900	—	—	—
	Gestein	—	—	1,0	33
	Harnische(H)	—	—	0	15

Bild 10   Lias α und Knollenmergel - Felsmechanisches Modell und Kennwerte

Schichtung. Sowohl an der Untergrenze der Kalksandsteinbänke als auch innerhalb der Tonsteinschichten selbst wurden weit durchgehende, zum Teil glatte und ebene mit Ton gefüllte Schichtfugen angetroffen. Aber auch die latent ausgebildete Schichtung hat einen festigkeitsmindernden Einfluß.

Die Ergebnisse der Kartierung, der felsmechanischen Versuche und der Interpretation der bei der Schachtabteufung gemessenen Verformungen führten zu dem in Bild 10 dargestellten felsmechanischen Modell und den angegebenen mittleren felsmechanischen Kennwerten. Danach wurde für die Tonsteine des Lias α mit deutlich unterschiedlichen E-Moduln (senkrecht zur Schichtung ~ 800 MN/m² und parallel zur Schichtung ~ 2500 MN/m²) gerechnet. Ebenso wurde parallel zur Schichtung eine sehr geringe Scherfestigkeit angenommen, während die Scherfestigkeit parallel zur Klüftung aufgrund der geringen im Schacht "Universität" gemessenen Erstreckung dieser Trennflächen verhältnismäßig hoch angesetzt wurde. Primärspannungsmessungen konnten aus versuchstechnischen Gründen nur in den Gesteinen des Knollenmergels durchgeführt werden. Aus diesen Versuchen und der Interpretation der aushubbedingten Verformungen wurde geschlossen, daß im Knollenmergel mit einem verhältnismäßig hohen Seitendruckbeiwert von ca. $K_0 = 2$ gerechnet werden muß. Für den Lias α ergaben die Vor-

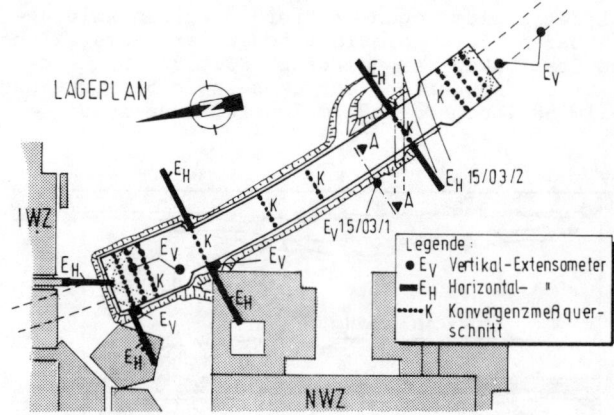

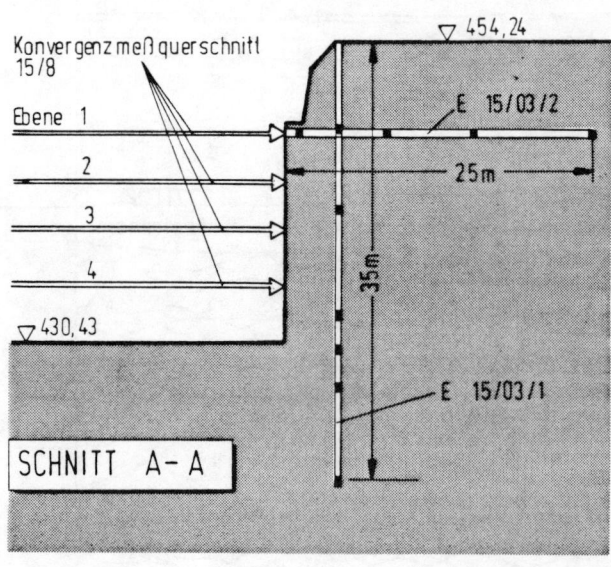

Bild 12   Los 15 - Meßprogramm

untersuchungen dagegen keine Hinweise auf erhöhte horizontale Primärspannungen. Mit diesen Annahmen wurden Standsicherheitsuntersuchungen für die Baugrube durchgeführt.

Die Baugrube sollte danach im unverwitterten Bereich des Lias α mit einer 10 cm starken bewehrten Spritzbetonschale und einer systematischen Ankerung aus 2,40 m langen im Raster von 2 x 2,3 m angeordneten SN- bzw. Perfo-Ankern gesichert werden. In einzelnen Bereichen zum Beispiel unter Gebäudefundamenten waren zusätzliche Verpreßanker vorgesehen (Bild 11).

AUSHUB DER BAUGRUBE IM BAULOS 15

Mit Beginn der Arbeiten im Baulos 15 zeigten sich einige Phänomene, die mit den Erkenntnissen aus den Voruntersuchungen und den daraus abgeleiteten Schlußfolgerungen nicht in Einklang zu bringen waren.

Zur Überwachung der Standsicherheit der Baugrube und der angrenzenden Gebäude wurde bereits vor Baubeginn ein Programm zur Messung der aushubbedingten Verformungen konzipiert (Bild 12). Das Meßprogramm besteht aus je drei Konvergenzmeßquerschnitten am Nordkopf und Südkopf und vier Konvergenzmeßquerschnitten im mittleren, schmaleren Teil der Baugrube, in denen die aushubbedingten, horizontalen Relativverschiebungen der Baugrubenwände in drei bzw. vier übereinander liegenden Meßebenen gemessen wurden. Außerdem wurden die Horizontalverschiebungen des Baugrundes in den an die Baugrubenwände anschließenden Bereichen mit sechs bis zu 30 m langen 5fach Extensometern gemessen. Seitlich der Baugrubenwände angeordnete, bis unter die Sohle reichende Vertikalextensometer ermöglichten die Messung von Vertikalkomponenten der Baugrundverformungen. Mit den innerhalb der Baugrube am Nordkopf liegenden 2 Vertikalextensometern konnten auch die aushubbedingten Sohlhebungen gemessen werden. Darüberhinaus wurden Nivellements der Geländeoberfläche und der Gebäude neben der Baugrube durchgeführt.

Auf größere Länge und bis zur endgültigen Tiefe wurde die Baugrube zuerst im Bereich des Nordkopfs ausgehoben (Bild 13).

Bild 13   Los 15 - Nordkopf

Im folgenden sollen deshalb die in diesem Bereich gemessenen Verformungen erläutert werden. In Bild 14 sind die im Bereich des naturwissenschaftlichen Zentrums (NWZ) gemessenen Verformungen über der Zeit aufgetragen. Teilt man die in den Konvergenzmeßquerschnitten gemessenen horizontalen Relativverschiebungen zu gleichen Teilen auf die beiden Wände auf, so ergibt sich für die erste und zweite Ebene nach Beendigung des Aushubs in diesem Bereich eine Wandverschiebung von 58 mm. Dagegen wurde in dem unter das NWZ-Gebäude reichenden Extensometer E 15/02/2 nur eine Relativverschiebung zwischen dem in der Wand liegenden Festpunkt und dem in einem Abstand von 30 m seitlich der Baugrubenwand gelegenen Meßpunkt von ca. 27 mm gemessen. Das sind nur ca. 40 % der gemessenen Gesamtverschiebung von 58 mm. Daraus ergibt sich, daß ca. 60 % der Gesamtverformung auf einen Bereich verteilt sind, der weiter als 30 m von der 25 m tiefen Baugrube entfernt ist.

Interessant ist ferner das Ergebnis der Messung am Vertikalextensometer E 15/02/1, der an der Geländeoberfläche nur Hebungen in der Größenordnung von ca. 1 - 2 mm zeigt. Ähnliche Ergebnisse zeigen auch die erwähnten Nivellements der Geländeoberfläche. Das bedeutet, daß die Horizontalverformungen seitlich der Baugrube das 30 - 60fache der Vertikalverformungen betragen.

Größere Hebungen wurden dagegen mit ca. 45 mm für den in der Baugrubensohle liegenden Extensometer E 15/6 gemessen. Diese Hebungen erstrecken sich auf einen Bereich von ca. 5 - 10 m unterhalb der Baugrubensohle und klingen darunter rasch ab.

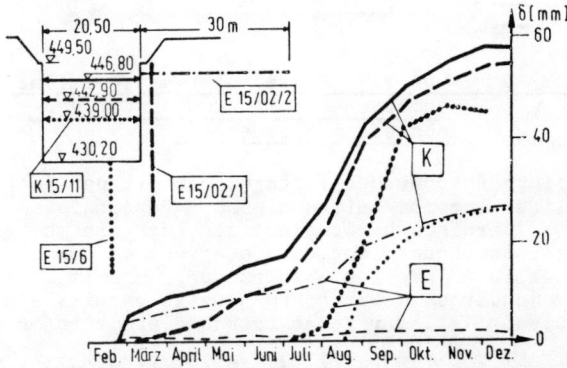

Bild 14   Los 15 - Verformungen der Baugrube am NWZ

Im Bild 15 sind die in der Ebene 1 des vorher erwähnten Konvergenzmeßquerschnittes am NWZ-Gebäude und mit dem Extensometer gemessenen Horizontalverformungen in Abhängigkeit von der Aushubtiefe dargestellt. Insbesondere für die Konvergenzmessungen sieht man hier sehr deutlich, daß die Verformungen mit dem Aushub der Kalksandsteinbänke wesentlich stärker zunehmen als beim Aushub der dazwischenliegenden Tonsteinschichten. Dies ist besonders gut im Bereich der Oolithen- und Hagenowibank zu erkennen. Die am NWZ gemessenen Horizontalverformungen der Baugrubenwand für verschiedene Bauzustände sind in Bild 16 in einem Vertikalschnitt durch die Baugrube dargestellt. Verbindet man die für die verschiedenen Meßebenen erhaltenen Wandverschiebungen miteinander, so ergibt sich nur eine

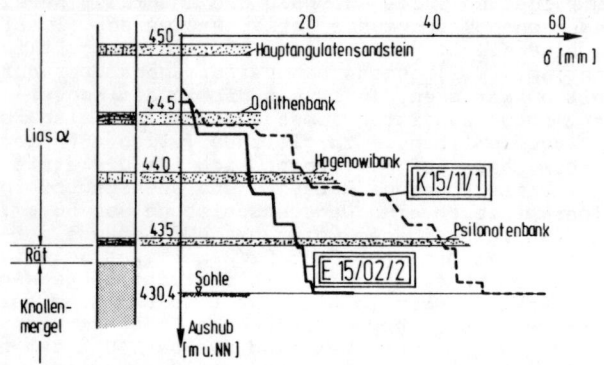

Bild 15   Los 15 – Verformungen der Baugrube
in Abhängigkeit von der Aushubtiefe

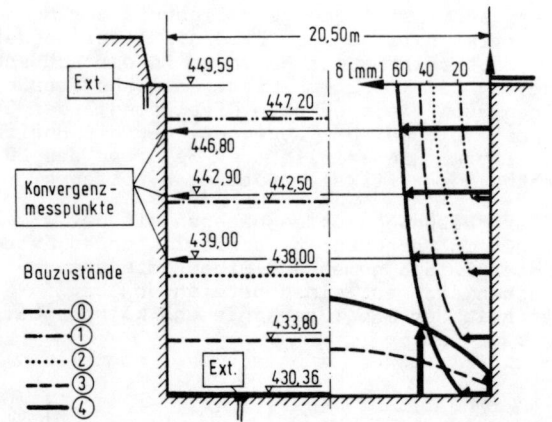

Bild 16   Los 15 – Querschnitt der Baugrube mit
gemessenen Verformungen für verschiedene Bauzustände

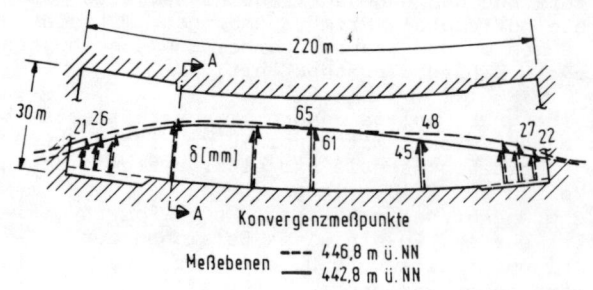

Bild 17   Los 15 – Grundriß der Baugrube mit
gemessenen Horizontalverformungen

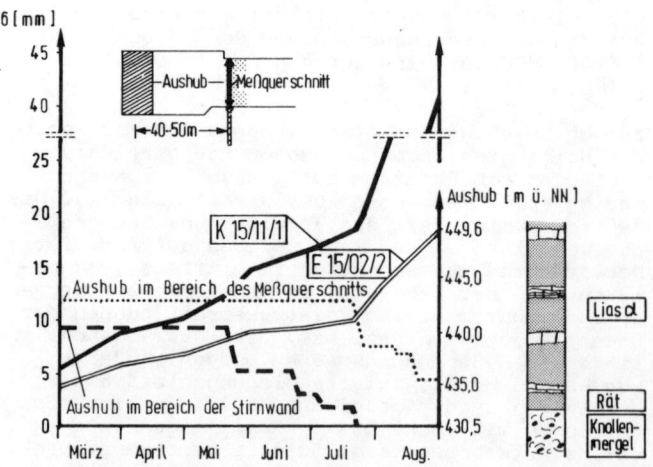

Bild 18   Los 15 – Wandverschiebungen als
Folge des Aushubs in 40 – 50 m
Abstand

geringe Abnahme der Verformungen mit der Tiefe.
Bei der Interpretation dieses Meßergebnisses muß
man allerdings berücksichtigen, daß die meßbaren
Verschiebungen nicht der Gesamtverschiebung der
jeweiligen Meßebene entsprechen, da beim Einbau
der Konvergenzmeßstrecken jeweils bereits ein
gewisser Teil der Verschiebungen eingetreten ist.

Interessant ist ferner die Darstellung der in
den verschiedenen Konvergenzmeßquerschnitten
für den Endzustand gemessenen Horizontalverformungen im Grundriß (Bild 17). Man erkennt
hier, daß die Horizontalverformungen in der Nähe
der Stirnwände deutlich geringer sind als im
mittleren Bereich der Baugrube.

In dieser Darstellung wird eine außerordentlich
weitreichende, aussteifende Wirkung der Stirn-
wand der Baugrube deutlich. Die weitreichende
Wirkung des Aushubs auf die Verformungen zeigte
sich auch in der Tatsache, daß eine Vertiefung
des Aushubniveaus in einem Teil der Baugrube
einen Anstieg der Verformungen in über 50 m entfernten Konvergenzquerschnitten bewirkte (Bild
18). So nahmen im Konvergenzmeßquerschnitt
K 15/11/1 die Verformungen zu, als in dem 50 m
entfernten Bereich ausgehoben wurde. Mit Beginn
des Aushubs im unmittelbaren Bereich des Meß-
querschnitts stieg dann die Verformungskurve
entsprechend stärker an.

Interessant ist auch die Beobachtung, daß die
Verformungen nach einem Stillstand der Aushub-
arbeiten sehr rasch, innerhalb nur weniger Tage
wieder abklangen, d.h., daß das Gebirge ein ver-
gleichsweise geringes zeitabhängiges Verformungsverhalten aufweist.

Neben den unverhältnismäßig großen horizontalen
Wandverschiebungen wurden beim Aushub der Baugrube große Erstreckungen der Klüfte in Tonsteinschichten des Lias α festgestellt, wie sie
aufgrund der Ergebnisse der geologischen Kartierung im Schacht Universität nicht vermutet werden konnten. Während, wie vorher erwähnt, im
Schacht "Universität" die Klüfte der Scharen K 1
und K 3 Größen von ca. 0,5 – 1,2 m^2 aufwiesen,
ergaben sich in freigelegten Abschnitten der
Baugrubensohle in horizontaler Richtung Erstreckungen von mehreren Dekametern. Das Foto
eines Sohlabschnittes der Baugrube am Südkopf
und die entsprechende Kartierung zeigen die
große Erstreckung der Klüfte sehr deutlich
(Bild 19).

INTERPRETATION DER GEMESSENEN VERFORMUNGEN

Zur Interpretation der unerwartet großen Wandverschiebungen wurden während und nach dem Bau-
grubenaushub FE-Berechnungen mit einem am

C 276

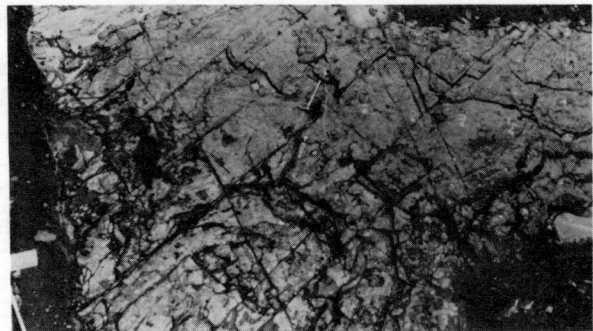

F O T O   5

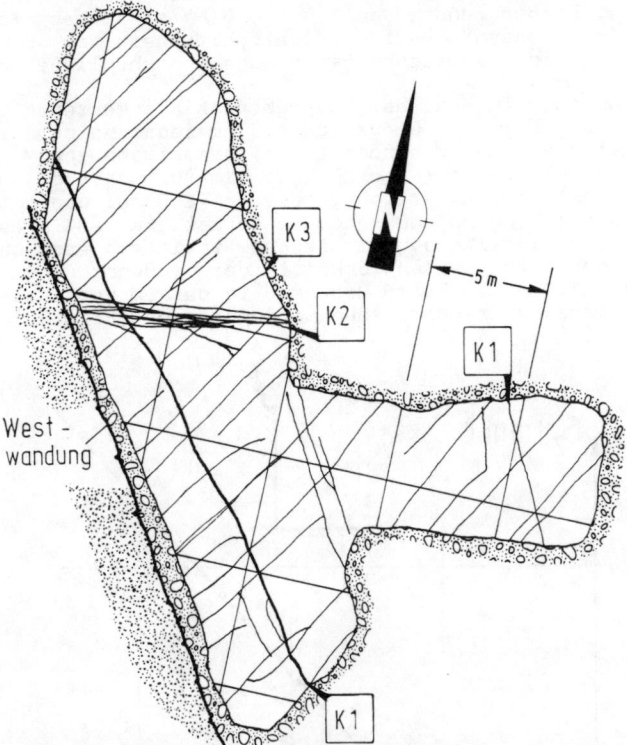

Bild 19   Los 15 - Hauptklüfte im Tonstein

Entstehungsgeschichte des Gesteins bedingter
Spannungen nicht erklärbar schienen. Es kann
hier nur auf wenige der zahlreichen durchgeführten Berechnungen eingegangen werden.

In Bild 20 ist das Ergebnis einer FE-Berechnung
dargestellt, in der der Lias α elastisch angenommen und die Kalksandsteinbänke nicht besonders nachgebildet wurden. Bei einem E-Modul des
Lias α in horizontaler Richtung von E_1 = 2500 MN/m^2
wurde eine zusätzlich zu den aus dem Eigengewicht und der Querdehnung wirkende horizontale
Primärspannung von $\Delta\sigma_{hor}$ = 1,8 MN/m^2 angenommen.
Auf dem Plot der Hauptnormalspannungen für den
Endaushubzustand geben die Striche die Größe und
Richtung der Spannungen an. Man erkennt, daß die
horizontalen Primärspannungen um die Baugrubensohle umgeleitet werden und daß ein sehr großer
Untergrundbereich seitlich der Baugrubenwände
horizontal nahezu vollständig entlastet wird.
Dementsprechend erstrecken sich die daraus resultierenden Horizontalverformungen des Baugrundes, die im unteren Teil des Bildes dargestellt
sind, über einen weiten Bereich seitlich der Baugrubenwände, was mit den Beobachtungen überein-
stimmt.

Für den Endaushubzustand sind die aus dieser Berechnung resultierenden horizontalen Verschiebungen der Baugrubenwand in Bild 21 dargestellt
und den Meßergebnissen gegenübergestellt. Geht
man von der Annahme aus, daß die in der Natur
nicht meßbaren Verformungsanteile in etwa den in
der Berechnung des Falls E 7 nicht erfaßten
plastischen Anteilen der Verformungen entsprechen, so ergibt sich mit dem angenommenen E-Modul

Verlauf der Hauptspannungen

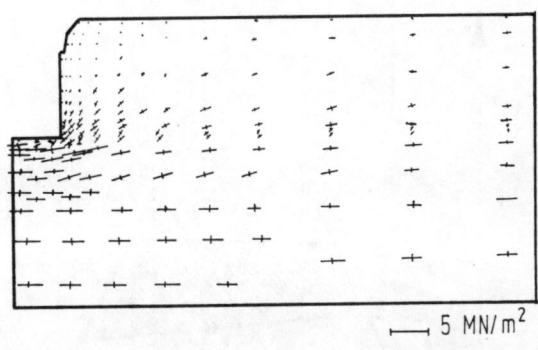

$\longmapsto$ 5 MN/m^2

Verformungen

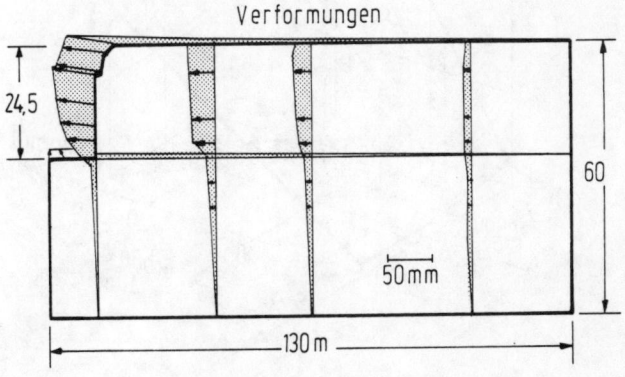

Bild 20   FE-Berechnungen für die Baugrube
($\Delta\sigma_{hor}$ = 1,8 MN/m^2)

Institut für Grundbau, Bodenmechanik, Felsmechanik und Verkehrswasserbau entwickelten Programm
durchgeführt (Semprich, 1980). In den Berechnungen wurden die elastischen Konstanten des
Lias α, die Festigkeit in den Klüften und Schichten sowie die Eigenschaften des Knollenmergels
variiert. Außerdem wurde der Einfluß der Kalksandsteinbänke in 2 verschiedenen Modellen berücksichtigt.

Im ersten Fall wurden für das gesamte Schicht-
paket des Lias α, einschließlich der Kalksandsteinbänke, einheitliche Kennwerte angenommen und
damit die Schichtenfolge des Lias α als quasi-
homogen angenommen. Im zweiten Fall wurden die
Kalksandsteinbänke dagegen mit ihren spezifischen Kennwerten im Rechenmodell nachgebildet.
Vor allem aber wurden in den Berechnungen unter-
schiedlich große horizontale Primärspannungen
angenommen, da die großen gemessenen Horizontalverformungen ohne die Annahme solcher durch die

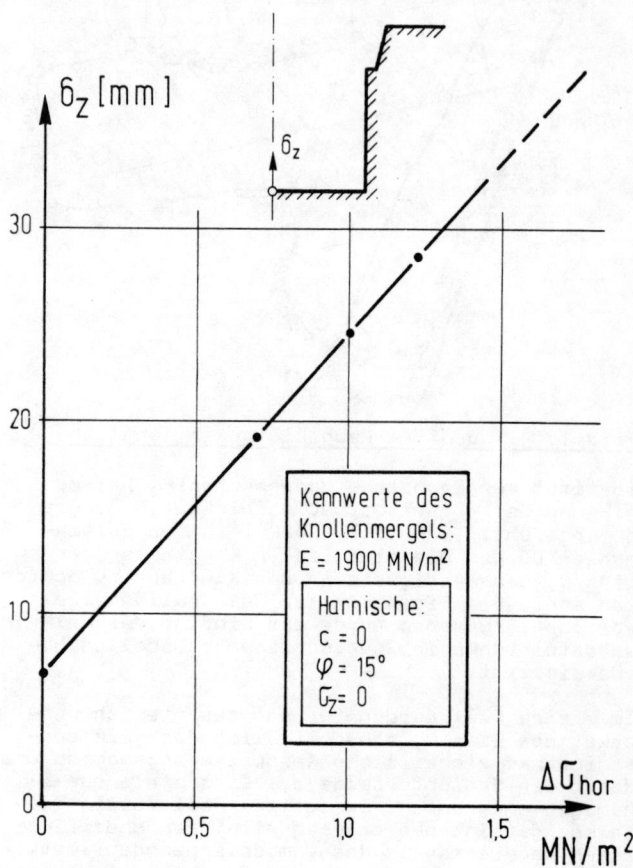

	E	ν
	[MN/m²]	—
Lias α	2500/800	0,25/0,10
Knollen- mergel	1900	0,45

$$\Delta\bar\sigma_{hor} = 1{,}8 \; [MN/m^2]$$

Konvergenzmessung K 15/11

FE-Rechnung Fall E7

Extensometermessung E 15/02/2

455,1 449,6 446,8 442,9 439,0 430,5

27   58   54   51

55 52 51 50 47 44

448,5 438,5

δ [mm]

Bild 21 Vergleich der Verformungen
– Messungen / FE-Berechnungen –

δ_{hor} [mm]

100 80 60 40 20

gemessene Verformungen

1250 2000 2500 3500 5000

Größe von $\bar E_h$

$\bar E_h$ [MN/m²]

$\Delta\bar\sigma_{hor}$ [MN/m²]

2,5 2,0 1,0

Größe von $\Delta\bar\sigma_{hor}$

Bild 22 Rechnerisch ermittelte Wandverformungen in Abhängigkeit von E-Modul und Seitendruck

von 2500 MN/m² bei einer horizontalen Primärspannung von $\Delta\sigma_{hor}$ = 1,8 MN/m² eine gute Übereinstimmung zwischen errechneten und gemessenen Verformungen.

In Vergleichsberechnungen wurde der Einfluß des in horizontaler Richtung wirksamen E-Moduls des Lias α und der horizontalen Primärspannungen untersucht. Die aus der Rechnung resultierenden Horizontalverformungen der Baugrubenwand in Höhe der Berme sind in Bild 22 dargestellt. Man erkennt, daß sich die im mittleren Bereich der Baugrube gemessenen Horizontalverformungen von 45 - 60 mm (Bild 17) mit den aus den Druckkissenversuchen erhaltenen E-Moduln in der Größenordnung von 2000 - 3500 MN/m² ohne die Annahme zusätzlicher Horizontalspannungen nicht annähernd erklären lassen. Erst mit Horizontalspannungen in der Größenordnung von 1,0 - 2,0 MN/m² lassen sich die gemessenen Horizontalverschiebungen von 45 - 60 mm in elastischen Berechnungen nachvollziehen.

Auch die in Höhe der Baugrubensohle gemessenen Hebungen in der Größe von 4,5 cm lassen sich nur mit der Annahme erhöhter Horizontalspannungen sinnvoll interpretieren. Mit dem für den Knollenmergel im Großversuch gemessenen E-Modul und mit sehr gering angenommenen Scherfestigkeiten in den flach einfallenden Harnischen erhält man erst bei Annahme von Seitendrücken in der Größenordnung von 1,0 - 2,0 MN/m² Hebungen in der im Vertikalextensometer gemessenen Größenordnung (Bild 23).

δ_z [mm]

30 20 10 0

δ_z

Kennwerte des Knollenmergels:
E = 1900 MN/m²

Harnische:
c = 0
φ = 15°
$\bar\sigma_z$ = 0

0   0,5   1,0   1,5

$\Delta\bar\sigma_{hor}$

MN/m²

Bild 23 Rechnerisch ermittelte Sohlhebungen in Abhängigkeit vom Seitendruck

Es stellt sich natürlich die Frage, ob es überhaupt möglich ist, daß im Lias α derart hohe horizontale Primärspannungen, wie sie sich aufgrund der Überlegungen ergeben, auftreten können. Anhand einer einfachen Modellvorstellung soll deshalb versucht werden, eine Erklärung für die Entstehung dieser Spannungen zu geben (Bild 24).

Ausgangspunkt hierfür ist die Erkenntnis der Geologen, daß die Tonsteinschichten in diesem Bereich früher eine Mächtigkeit von ungefähr 600 m hatten. Sie entstanden durch allmähliche Sedimentation von Tonteilchen in dem damals vorhandenen Meer.

Die Horizontalspannungen σ_H in der heute noch anstehenden, ca. 20 m dicken Tonsteinschicht des Lias α sind das Ergebnis dieses Sedimentationsprozesses, der Konsolidierung und Verfestigung der Tonschichten sowie der nachfolgenden Abtragung der Überlagerung durch Erosion. Betrachtet man zunächst die Horizontalspannungen für eine Überlagerung T = 600 m über der derzeitigen Geländeoberfläche (Bild 24, Punkt 1), so muß man bei der Berechnung der Horizontalspannung mit Hilfe des Seitendruckbeiwertes $\nu/(1-\nu)$ berücksichtigen, daß parallel zum Anwachsen der Überlagerung auch bereits eine Verfestigung der Tonablagerungen erfolgt. Dies führt zu einer Abnahme der Poissonzahl ν in Abhängigkeit von der Überlagerungshöhe T, wie das in Bild 24 unter Punkt 2 qualitativ dargestellt ist.

Die Horizontalspannung σ_H ergibt sich dann aus dem in Bild 24 unter Punkt 3 dargestellten Integral.

An die Sedimentation schloß sich eine weitgehende Abtragung der Tonablagerungen durch Erosion an. Schließlich verblieb nur die ca. 20 m mächtige Tonsteinbank des Lias α, die wir auch heute noch antreffen. Durch die Wegnahme der Auflast von ca. 600 m verringerte sich auch die Horizontalspannung (Bild 24, Punkt 4). Die jetzt vorhandene Spannung σ_H ergibt sich aus der Differenz der Spannung bei einer Überlagerungshöhe von 600 m und der Abnahme der Spannung $\Delta\sigma_H$ infolge der Abtragung dieser Überlagerung. Für diese Spannungsabnahme ist aber die Poissonzahl ν_E nach Abschluß des Verfestigungsvorgangs entsprechend Punkt 2 in Bild 24 maßgebend. Da ν_E

kleiner ist als die Poissonzahlen während des Sedimentationsvorgangs, verbleibt im Tonstein auch nach Abtragung der Überlagerung eine Restspannung, die größer ist als sie sich allein aus der jetzigen Überlagerungshöhe ergibt.

In Bild 24 ist unter Punkt 5 an einem Beispiel mit den dort angegebenen Annahmen die mögliche Horizontalspannung bei der heutigen Schichthöhe ausgerechnet worden.

Mit den gewählten Annahmen würden im Tonstein trotz der geringen heutigen Überlagerung noch horizontale Restspannungen in der Größenordnung von ca. 7 MN/m² vorhanden sein. Dies ist natürlich ein sehr hoher Wert, wie er heute sicher nicht mehr anzutreffen ist. Man muß dabei aber berücksichtigen, daß bei diesen hohen Spannungen Festigkeitsüberschreitungen in den Gebirgsschichten aufgetreten sind, die zu Brüchen und Aufschiebungen bzw. Gleitbewegungen einzelner Schichtpakete geführt haben und daß dadurch die hohen Horizontalspannungen teilweise abgebaut wurden. Solche Bruchflächen sind im Lias deutlich sichtbar. In Bild 25 wurde dieser Vorgang vereinfacht dargestellt. Eine weitere Entspannung des Gebirges trat an Geländeeinschnitten auf, wodurch sich die verbleibenden Restspannungen weiter verminderten.

Es ist also durchaus möglich, daß im Lias α horizontale Primärspannungen in der Größe von 1 - 2 MN/m² vorhanden sind, wie sie aus der Rückrechnung der in Los 15 gemessenen Verformungen ermittelt wurden.

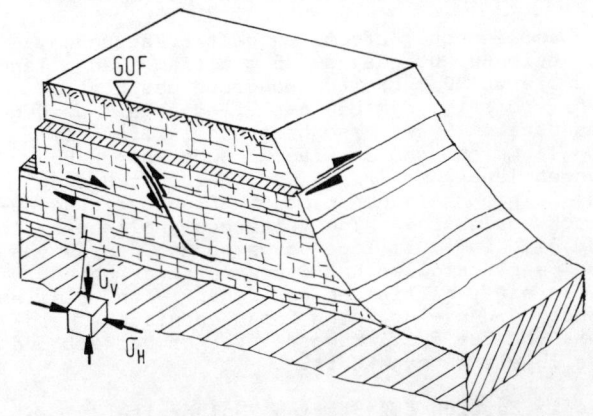

Bild 25   Lias α - Teilweise Entspannung durch Bruchvorgänge und Geländeeinschnitte

SCHLUSSFOLGERUNGEN

Aufgrund der neuen Erkenntnisse über Horizontalspannungen und Klüftigkeit im Lias α wurde das Konzept der Sicherungsmaßnahmen für die Baugrube und in diesem Gebirge liegenden, an die Baugrube angrenzenden Tunnelabschnitte entsprechend angepaßt.

So wurde in der Baugrube "Universität" eine wesentlich stärkere Vernagelung der Baugrubenwände mit bis zu 5,4 m langen SN-Ankern und eine systematische Ankerung mit horizontalen Verpreßankern hauptsächlich in den Kalksandsteinbänken vorgenommen. Dadurch sollte in erster Linie ein Abgleiten von Felskeilen in die Baugrube verhindert

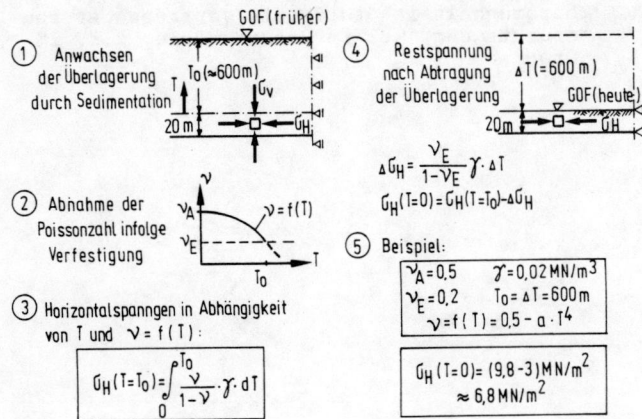

Bild 24   Entstehung der Horizontalspannungen im Lias α

werden. Die Möglichkeit eines solchen Abgleitens war durch die beim Aushub festgestellte größere Erstreckung der Großklüfte gegeben. Die Verformungen konnten durch diese Maßnahmen jedoch nicht wesentlich beeinflußt werden, was aufgrund der großen räumlichen Ausdehnung dieser Verformungen auch nicht zu erwarten war. Die an die Baugrube angrenzenden Baulose 14 und 16 wurden im Kalottenvortrieb mit anschließendem Strossenabbau aufgefahren. Sie wurden mit einer mit Baustahlgewebe armierten, relativ starken Spritzbetonschale von 20 - 30 cm Dicke und einer intensiven Ankerung mit 3,9 m langen SN-Ankern gesichert. Außerdem wurden kurze Abschlagslängen und eine durchgehende Sicherung der Kalottensohle ausgeführt. Die Verformungen und die Spannungen wurden laufend durch ein Meßprogramm, bestehend aus Konvergenzmeßquerschnitten, Extensometern, Nivellements und Spannungsmeßdosen in der Spritzbetonschale überwacht (Wittke, Grüter 1982).

ZUSAMMENFASSUNG

Im Netz der S-Bahn Stuttgart wird der 5,5 km lange Hasenbergtunnel die Verbindung zwischen der derzeitigen Endstation "Schwabstraße" und der vorhandenen Eisenbahnstrecke Stuttgart - Horb - Zürich schaffen. Wegen der großen Überdeckung zwischen 12 und 125 m wird der Tunnel mit Ausnahme der in offener Baugrube hergestellten Station "Universität" bergmännisch aufgefahren. Der Tunnel durchörtert eine Vielzahl von Gebirgsschichten mit stark unterschiedlichen felsmechanischen Eigenschaften.

Der vorliegende Aufsatz enthält Erfahrungen, die beim Aushub der ca. 25 m tiefen, 220 m langen und bis zu 30 m breiten Baugrube der Station "Universität", dem Baulos 15, gesammelt wurden. Das Baulos 15 des Hasenbergtunnels liegt im Knollenmergel und im Lias α. Außer durch Bohrungen längs der Trasse wurden zur Erkundung der felsmechanischen Eigenschaften der diagenetisch verfestigten Tonsteine des Knollenmergels und des Lias α im Zuge vorgezogener Baumaßnahmen Fensterstollen und Schächte aufgefahren. Neben einer Kartierung der Schacht- und Stollenwände wurden eine Reihe felsmechanischer Feldversuche zur Bestimmung der Verformbarkeit und Festigkeit durchgeführt.

Die im Bereich der Station "Universität" installierten Meßeinrichtungen zeigten beim Aus-

hub der Baugrube große horizontale Wandverformungen (bis 65 mm), die das 30 - 60fache der gemessenen Hebungen am Rand der Baugrube betragen. Größere Hebungen wurden mit ca. 45 mm in der Baugrubensohle gemessen. Besonders bemerkenswert ist, daß die großen Horizontalverformungen mit zunehmendem Abstand von der Baugrube nur langsam abnehmen. Weiterhin wurde beim Aushub der Baugrube für die Klüfte in den Tonsteinschichten des Lias α eine wesentlich größere Erstreckung festgestellt, als aufgrund der Vorerkundungen vermutet wurde.

Zur Interpretation der unerwartet großen Wandverschiebungen beim Aushub der Baugrube wurde eine große Zahl von Finite-Element-Berechnungen durchgeführt. Dabei wurde deutlich, daß sich die gemessenen Verformungen nur erklären lassen, wenn man Horizontalspannungen in der Größenordnung von 1 - 2 MN/m^2 annimmt. Derartig hohe Horizontalspannungen lassen sich bei Berücksichtigung einer hier in vorgeschichtlichen Zeiten vorhandenen Überlagerung von ca. 600 m aus der Entstehungsgeschichte der Tonsteine des Lias α erklären. Aufgrund der neuen Erkenntnisse über Horizontalspannungen und Klüftigkeit im Lias α wurde die Sicherung der Baugrubenwände verstärkt. Auch in den benachbarten Tunnelabschnitten wurden die Sicherungsmaßnahmen entsprechend angepaßt.

LITERATURVERZEICHNIS

Grüter, R.: Planung des Hasenberg-Tunnels in Stuttgart; Berichte 2. Nationale Tagung für Ingenieurgeologie, Fellbach 1979, S. 45 - 51

Grüter, R.: S-Bahn Stuttgart, Feldmessungen als Grundlage für Entwurf und Ausschreibung des Hasenbergtunnels; 4. Nationale Tagung über Felsmechanik, Aachen 1980

Semprich, S.: Berechnung der Spannungen und Verformungen im Bereich der Ortsbrust von Tunnelbauwerken im Fels; Veröffentlichungen des Institutes für Grundbau, Bodenmechanik, Felsmechanik und Verkehrswasserbau der RWTH Aachen; Heft 8, Aachen 1980

Wittke, W., R. Grüter: Erfahrungen beim Bau der im Lias α und im Knollenmergel liegenden Abschnitte der Baulose 14 - 16 des Hasenbergtunnels in Stuttgart; Vorträge der Baugrundtagung 1982 in Braunschweig, z.Z. im Druck

AUSCULTATION PENDANT LE DOUBLEMENT D'UN TUNNEL AUTOROUTIER DANS DES MARNES

Instrumentation during the doubling of a road tunnel in marls

Messungen im Zuge des zweiröhrigen Ausbaus eines in Mergeln ausgebohrten Strassentunnels

J. C. Bailly
Agence du Sud-Est — SCETAUROUTE — France

A. Bouvard
Département Géotechnique — Coyne et Bellier — France

G. Colombet
Département Géotechnique — Coyne et Bellier — France

M. Laboure
Agence du Sud-Est — SCETAUROUTE — France

M. Panet
Directeur Technique — Laboratoire Central des Ponts et Chaussées — France

RESUME

Le doublement du tunnel autoroutier de Las Planas dans des marnes a nécéssité une étude de prévision de comportement et de contrôle de l'ouvrage existant comprenant:
- l'interprétation des mesures de déformations pour déterminer l'état de sollicitation du revêtement
- un modèle numérique aux éléments finis
- l'auscultation du massif pendant les travaux

SYNOPSIS

During the construction in the marls of the second tube of the LAS PLANAS motorway tunnel, the analysis and control of the behaviour of the existing tube has been performed. The study included:
- interpretation of measurements of displacements to determine the stress distribution in the concrete lining
- computer analysis by Finite-element model
- measurements of displacements during the works.

ZUSAMMENFASSUNG

Der zweiröhrige Ausbau des Las Planas Straßentunnels in Nice weist gewisse Gefahren für das existierende Tunnelrohr auf. Die eingehende Untersuchung dieser Frage besteht aus:
- der Analyse der Beanspruchungen auf die Auskleidung vor dem zweiröhrigen Ausbau
- der Aufstellung eines FEM Modells
- der zerstörungsfreien Prüfung des Komplexes im Laufe der Tunnelausbohrung.

1. INTRODUCTION

L'autoroute A.8., reliant AIX-EN-PROVENCE et l'ITALIE, a été concédée à la Société de l'AUTO-ROUTE ESTEREL COTE D'AZUR (ESCOTA) qui a confié à SCETAUROUTE les études et la direction des travaux, auxquelles collabore COYNE et BELLIER.

A proximité de NICE, l'autoroute A.8. contourne l'agglomération dans un site très urbanisé et constitué d'une succession de vallées profondes et de massifs franchis transversalement.

Actuellement, une seule chaussée est en service. Devant l'augmentation rapide du trafic, la construction de la seconde chaussée a été engagée ; elle consiste à doubler tous les ouvrages d'art importants et en particulier les tunnels.

Parmi eux, le tunnel de LAS PLANAS comprend un premier tube à 3 voies de 1100 m de long réalisé en 1975. Le doublement de cet ouvrage par un tube à 2 voies implanté à 1,4 fois le diamètre seulement du premier est donc prévu.

Ces ouvrages traversent des marnes plastiques pliocènes sur 400 m de longueur et sous 100 m de couverture. Dans ces formations, de sérieuses difficultés, dont un éboulement majeur, avaient été rencontrées pendant le creusement du premier tube effectué avec la méthode dite "Nouvelle Méthode Autrichienne".

Devant les risques non négligeables que présente pour le premier la construction du second tube, un programme d'études a été élaboré sur les bases suivantes :

(i) bilan aussi exact que possible de l'état de sollicitations du tube existant,

(ii) étude théorique par éléments finis de l'action du creusement du 2ème tube sur le 1er,

(iii) auscultation du massif pendant les travaux afin de contrôler l'évolution réelle des sollicitations dans le premier tube, et donc de pouvoir agir sur les méthodes d'excavation du second, et également de confronter les résultats du calcul aux mesures.

2. GEOLOGIE PARTIELLE DU TUBE DE LAS PLANAS

La partie du tunnel de Las Planas intéressée par l'étude traverse une épaisse formation pliocène comprenant des poudingues et des marnes gris-bleu.

Du fait de leur origine deltaïque, ces faciès s'interpénètrent et se répartissent de façon très aléatoire. Elles sont d'autre part extrêmement hétérogènes et ces hétérogénéités ont été la cause principale des incidents qui se sont produits lors du creusement du premier tube.

Les essais sur les marnes ont donné les caractéristiques suivantes :
. Cohésion : Cuu de 0,4 à 0,8 MPa
. Frottement limite : ϕ de l'ordre de 19°
. Vitesse sismique : 1 000 à 2 000 m/s

3. ETAT DU PREMIER TUBE AVANT DOUBLEMENT

3.1 Auscultation du tunnel

Les déformations du premier tube sont régulièrement mesurées depuis sa construction au moyen d'extensomètres TELEMAC noyés dans le béton (sections A et B - Fig.1a) et de mesures de convergence relative au distomètre ISETH à fil invar du CETE de Lyon (sections C et D - Fig.2a).

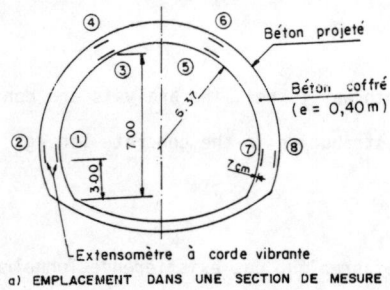

a) EMPLACEMENT DANS UNE SECTION DE MESURE

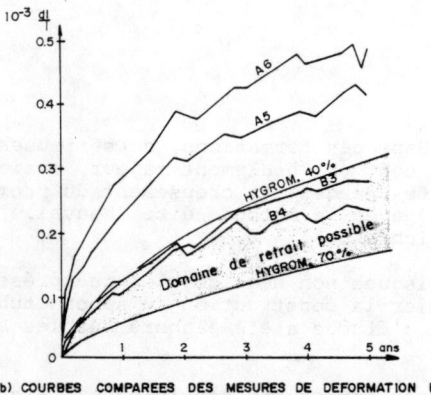

b) COURBES COMPAREES DES MESURES DE DEFORMATION ET DES VALEURS EMPIRIQUES DU RETRAIT

Fig.1 : Extensomètres à cordes vibrantes dans le revêtement du premier tube.

Les mesures effectuées avec ces deux types d'instruments sont concordantes. Elles montrent sur cinq années une déformation tangentielle dl/l comprise entre 0,195.10-3 et 0,490.10-3 (Fig.1b) et une convergence du revêtement dR comprise entre 1,7 et 2,6 mm (Fig.2b).

Si l'on admet que le module différé du béton est de l'ordre de 10 000 MPa, ces mesures laissent a

priori supposer que les contraintes dans le revêtement sont de l'ordre de 2 à 5 MPa.

Une campagne de mesures directes de contraintes dans la peau du revêtement par vérins plats a été lancée en 1981. Nous avons alors été surpris des valeurs faibles constatées : 1,5 à 3,5 MPa en clé, 0 à 2 MPa ou même parfois des tractions en reins et en piédroits.

Cette différence apparente dans les mesures montre que, dans l'interprétation des déformations, d'autres facteurs que la mise en charge et le fluage doivent être pris en compte. Les déformations thermiques et le retrait sont également à considérer.

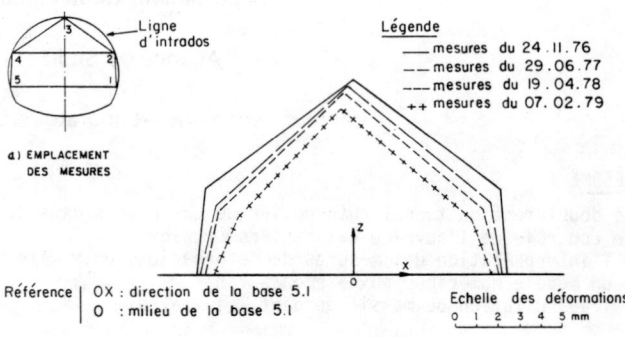

a) EMPLACEMENT DES MESURES

b) MESURES DE LA SECTION C

Fig.2 : Mesures de convergence relative dans le premier tube (d'après le CETE de Lyon).

3.2. Analyse des mesures de déformation

3.2.1. Déformations thermiques

Les variations thermiques saisonnières entraînent des fluctuations de déformation visibles sur les courbes (Fig.1b) qui sont faibles vis-à-vis des déformations totales enregistrées (généralement moins de 10 %). La variation est de l'ordre de 0,01 à 0,03.10-3 entre l'hiver et l'été.

3.2.2. Déformations de retrait

Le raccourcissement du béton non chargé, provoqué par la variation d'hygrométrie du ciment au cours de sa prise et de son durcissement, n'a pu être évalué directement, en l'absence d'échantillons témoins.

Nous nous sommes donc basés sur les recommandations du code modèle CEB-FIP (1978) qui ont permis une interprétation satisfaisante des mesures de déformations en adoptant les hypothèses qui suivent. Le retrait ou raccourcissement $\varepsilon(t)$ dans le temps s'exprime par la relation :

$$\varepsilon(t) = \varepsilon_r \cdot f(t)$$

où : - f(t) est une fonction du temps comprise entre 0 et 1
- ε_r est l'amplitude du retrait, produit de deux facteurs, ε_c fonction de l'hygrométrie ambiante et k_3 fonction de l'épaisseur du revêtement.

a)   Influence de l'hygrométrie (εc)

L'hygrométrie moyenne normale de l'atmosphère est de 70 %. La ventilation et les courants d'air qui se produisent dans le tunnel conduisent à une dessication du béton beaucoup plus importante qu'en atmosphère normale. Ceci nous amène à considérer un milieu ambiant dans le tunnel plus sec, mais dont l'hygrométrie est difficile à estimer. Le calcul du retrait a donc été réalisé avec deux valeurs extrêmes : 70 % (atmosphère normale) et 40 % (atmosphère très sèche)

avec hygrométrie 70 % : $\varepsilon_c = 0,32.10^{-6}$
avec hygrométrie 40 % : $\varepsilon_c = 0,52.10^{-6}$

b)   Influence de l'épaisseur du revêtement (k_3)

Le revêtement de 0,40 m d'épaisseur n'est en contact avec l'air qu'à l'intrados. L'épaisseur fictive ε_m à prendre en compte est le double, soit 0,80 m. De plus, cette épaisseur est affectée d'un coefficient correcteur selon l'hygrométrie. ε_m est donc égal à 1,2 m ou 0,80 m selon que l'hygrométrie est de 70 % ou de 40 %.
Le code modèle CEB-FIP indique que pour ces deux valeurs d'hygrométrie et d'épaisseur fictive, le coefficient k3 vaut sensiblement 0,75.

Dans ces conditions, le coefficient ε_r vaut 0,24 ou $0,39.10^{-3}$ selon l'hygrométrie.

c)   Evolution du retrait dans le temps (f(t))

La fonction f(t) est donnée par le CEB-FIP. Elle dépend de l'épaisseur fictive, de la température et de la nature du ciment.

Dans les deux cas d'hygrométrie considérés, pour une température moyenne de 15° et l'utilisation d'un ciment à prise rapide, la courbe d'évolution du retrait est indiquée sur la figure 1b.

3.2.3.   Influence du retrait sur l'état de contrainte dans le revêtement

L'estimation faite jusqu'ici suppose que les déformations dues au retrait sont homogènes et se produisent librement. En fait :

- le retrait s'effectue différemment à l'intrados et à l'extrados du revêtement, de façon d'autant plus importante que l'humidité est maintenue à l'extrados ;

- il se produit un retrait superficiel sur les 25 premiers centimètres qui, comme le montre l'Hermite (1979), peut atteindre en parement le double de celui atteint au coeur du béton.

Ces deux phénomènes provoquent des contraintes internes pouvant mettre le parement en traction et créer les fissures de retrait ;

- la forme du revêtement (anneau pratiquement fermé, variations de courbure), les conditions de contact entre le terrain et le revêtement (frottement en piedroit, décollement de la voûte) empêchent les déformations de retrait de s'effectuer librement.

Ces différentes circonstances peuvent expliquer la différence des contraintes en parement entre les piedroits et la voûte ainsi que les déformations généralement mesurées plus fortes à l'intrados qu'à l'extrados.

3.2.4.   Conclusion sur l'état de sollicitation du premier tube

La figure 1b indique les déformations mesurées par le couple de cordes vibrantes donnant les valeurs les plus faibles (3 et 4 de la section B) et par celui donnant les valeurs les plus fortes (5 et 6 de la section A). On constate que ces courbes sont homothétiques aux courbes théoriques de retrait.

La déformation due à la mise en contrainte du revêtement est égale à la déformation totale mesurée, diminuée du retrait. Pour le profil le moins chargé (section B), la déformation mécanique est faible et peut-être nulle. Pour le profil le plus chargé (section A), la déformation mécanique dl/l après 5 ans est comprise entre 0,1 et $0,3.10^{-3}$.

Le module mesuré sur carottes à 28 jours est de 23 000 MPa. Il subsiste trop d'inconnues sur la mise en charge du revêtement pour appliquer au fluage la même méthode que celle utilisée pour le retrait. En considérant cependant un module de béton après 5 ans de chargement progressif égal à 10 000 MPa, on estime que la contrainte en voûte ne dépasse pas 3 MPa.

L'ensemble de ces considérations explique donc les valeurs mesurées au vérin plat.

4.   CALCUL PAR ELEMENTS FINIS

Un calcul par éléments finis a permis d'étudier l'influence du creusement du deuxième tube sur le premier et d'analyser la nature du risque encouru pour l'ouvrage en exploitation.

Les caractéristiques des marnes prises en compte avaient été définies à partir des essais de laboratoire et in situ réalisées pour le projet du premier tunnel. La validité de ces hypothèses avait été vérifiée pendant l'excavation de celui-ci.

Le calcul a été mené en élastoplasticité avec c = 0,7 MPa, ϕ = 19°, E = 500 MPa et ν = 0,42.

La figure 3 montre le maillage utilisé. Le calcul a été fait en trois phases successives :

- Phase 1 : Creusement du premier tube et mise en place du soutènement provisoire (simulé par une pression de confinement permettant de caler le modèle en phase 2).

- Phase 2 : Mise en place du revêtement définitif dans le premier tube. L'état de contrainte et de déformation dans le béton correspond, en fin de cette phase de calcul, à l'état décrit dans le paragraphe 3.2.4.(calage du modèle).

- Phase 3 : Excavation du deuxième tube. Etude de l'évolution des contraintes et des déformations dans le premier tube suivant les valeurs de confinement dans le deuxième tube (c'est-à-dire suivant le mode de creusement et de soutènement adopté).

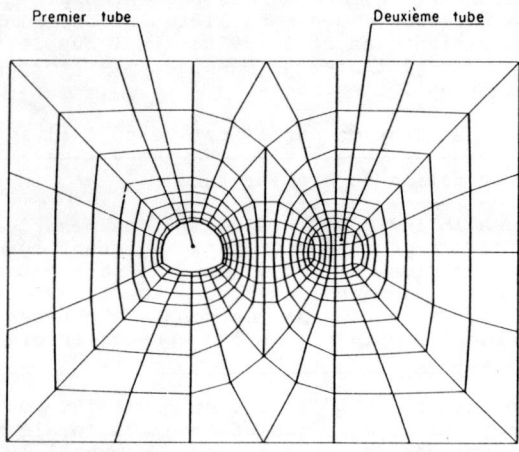

Premier tube   Deuxième tube

Fig.3 : Maillage

La figure 4 montre la déformée maximale du tube
ancien due au creusement du tube voisin. On constate un aplatissement de sa voûte et un déplacement de son piédroit en direction du tube excavé.
Ceci se traduit pour les travaux par un risque
d'apparition de traction à l'intrados de la voûte.

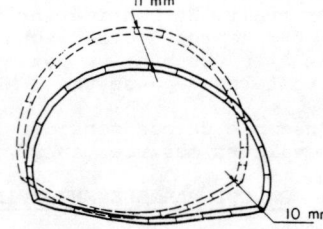

Fig.4 : Déformée du
       premier tube
       à l'excavation
       du deuxième.

Suite à ces calculs, il a été décidé de creuser
le deuxième tube avec un machine à attaque ponctuelle, de placer un revêtement provisoire en
voûte constitué de 15 cm de béton projeté armé
de cintres métalliques et de fermer la section
en radier par une voûte inversée en béton chaque
fin de semaine.

5.   NATURE DE L'INSTRUMENTATION MISE EN PLACE
     POUR SUIVRE LES TRAVAUX DE DOUBLEMENT

Pour contrôler l'influence des travaux de doublement sur le premier tube et confronter le
calcul à la réalité, l'auscultation décrite ci-
après a été prévue.

a) Auscultation du premier tube

L'auscultation déjà en place (convergence relative et extensomètres dans le revêtement) permet
de suivre les déformations du premier tube pendant les travaux.

b) Auscultation du deuxième tube

Au moment de la rédaction de cet article, le
deuxième tube est ouvert en demi-section supérieure seulement. Les instruments de mesure suivants ont été placés au front de taille (Fig.5):

- trois sections de mesure primaire équipées
chacune de trois extensomètres en forage (Distofor de TELEMAC) de 7,5 m de longueur avec trois
capteurs distants de 2,5 m,

- une section de mesure secondaire tous les
20 mètres équipée de mesures de convergence relative (distomètre du CETE de Lyon).

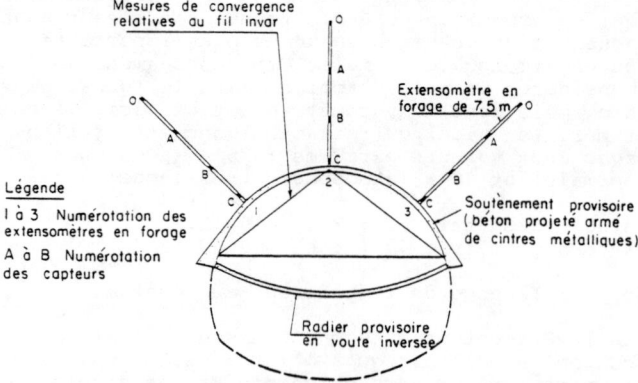

Fig.5 : Appareils d'auscultation du deuxième tube.

c) Auscultation du massif entre les deux tubes

Un extensomètre en forage (Distofor de TELEMAC)
de 18 m de longueur a été placé à partir du premier tube en deux sections différentes pour mesurer les déformations du massif au passage du
front de taille (Fig.6).

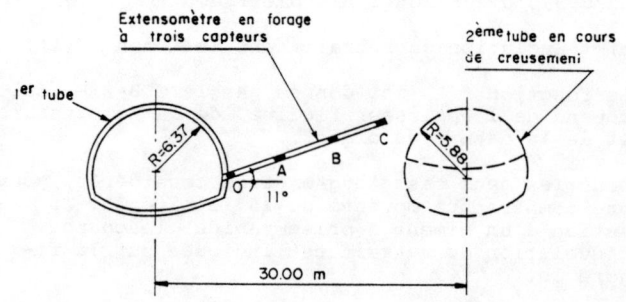

Fig.6 : Appareils d'auscultation
        entre les deux tubes.

6.   RESULTAT DES MESURES A L'EXCAVATION DE LA
     DEMI-SECTION SUPERIEURE DU DEUXIEME TUBE

Le creusement du deuxième tube en demi-section
supérieure a commencé au début de l'année 1982.
La section excavée n'étant alors que partielle,
son creusement a eu un effet réduit sur le tube
existant. Néanmoins, les mesures relevées à cette
occasion montrent que les déformations sont dans
l'ensemble de l'ordre de grandeur de celles calculées. Elles confirment également la réalité
des risques prévus par le calcul : le chargement
de la voûte du premier tube risque de faire appa-
raître des tractions à l'intrados de la voûte.

Les mesures les plus caractéristiques sont présentées ci-après.

La figure 7 indique les résultats des mesures
de déformations obtenus dans l'une des sections
primaires présentées en figure 5.

La figure 8 indique les déformations mesurées
par un extensomètre présenté en figure 6.

Dans les deux cas, on constate un ralentissement
très net des mouvements ce qui témoigne de la
stabilité de l'excavation.

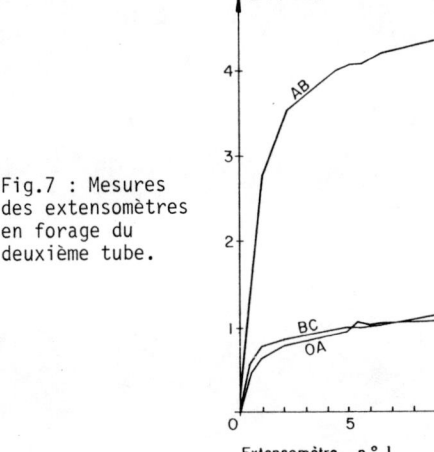
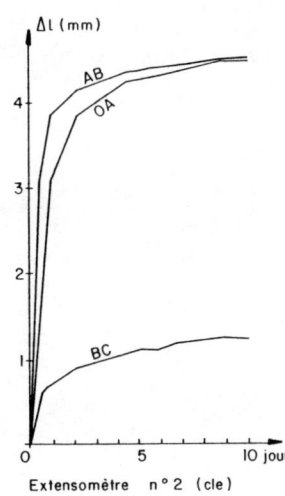
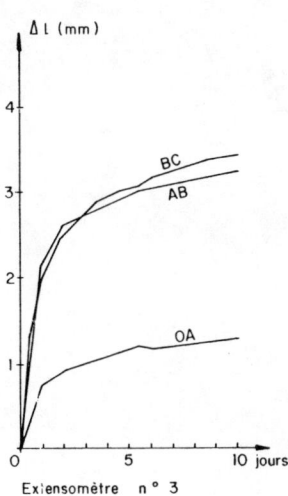

Fig.7 : Mesures des extensomètres en forage du deuxième tube.

Extensomètre n° 1   Extensomètre n° 2 (cle)   Extensomètre n° 3

On constate également sur la dernière figure que les déplacements apparaissent 20 mètres (soit 1,5 diamètre) devant le front de taille. Les déplacements sont encore faibles (10 % des déplacements mesurés) lorsque le front atteint la section de mesure. Ceci montre que la zone plastique apparaît seulement après le passage du front de taille et la mise en place du soutènement provisoire, ce qui est sécurisant.

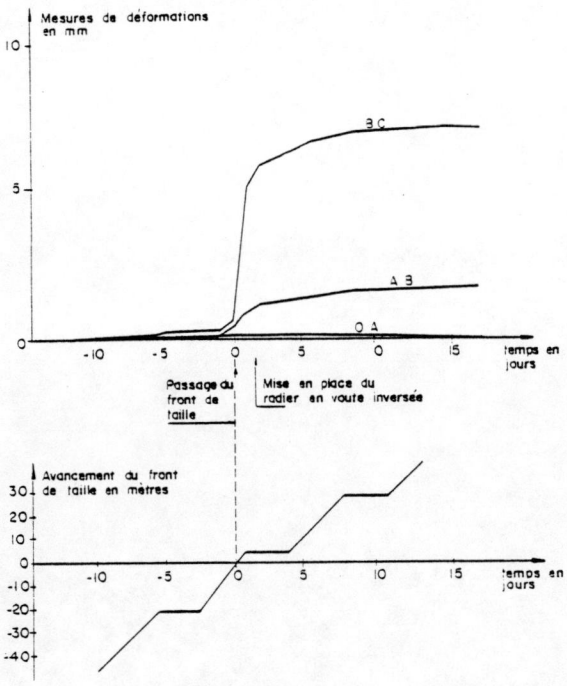

Fig.8 : Mesures des extensomètres en forage entre les deux tubes.

7. CONCLUSIONS

Ce cas concret présente les difficultés rencontrées dans l'interprétation des mesures et montre la nécessité d'utiliser plusieurs types d'appareils pour recouper les valeurs obtenues et tenir compte des nombreux facteurs mal connus et trop souvent négligés (tel ici le retrait).

Les modèles numériques ont indiqué la nature des risques encourus par le premier tube et les précautions à prendre pendant le creusement du deuxième tube. L'analyse ainsi faite doit être complétée par une auscultation pendant les travaux, pour voir au moins qualitativement si les phénomènes vont bien dans le sens prévu.

REFERENCES

BOUVARD A. (1979) Calcul du tunnel de Las Planas - Tunnels n° 32 - Mars Avril 1979 - p. 122 à 124.

Code-modèle CEB-FIP pour les structures en béton (1978) - 3ème édition des Recommandations Internationales CEB-FIB - Annexe e - Comportement dans le temps du béton - p. 319 à 331.

L'HERMITE R. (1979) Expériences et théories sur la technologie du béton - Annales de l'ITBTP n° 375 - Septembre 1979 - Série béton n° 190.

LONDE P. (1977) Field measurements in tunnels - Proceedings of the International Symposium Zurich on Field measurements in rock mechanics - Volume 2 - p. 619 à 638.

PANET M. (1976) Stabilité et soutènement des tunnels - La mécanique des roches appliquée aux ouvrages de génie civil - p. 145 à 166.

ROCHET L. (1976) Auscultation des ouvrages et des massifs rocheux encaissants - La mécanique des roches appliquée aux ouvrages de génie civil - p. 183 à 214.

NEAR-SURFACE UNDERGROUND CONSTRUCTIONS IN MIXED ABRUPTLY CHANGING GROUND

Constructions à faible profondeur dans terrain mixte et brusquement variable

Oberflächennahe Konstruktionen bei gemischten, plötzlich abwechselnden Untergrundverhältnissen

C. Valore

Researcher in Geotechnical Engineering Palermo Univ.; Lecturer in Earth Retaining Structures — Calabria Univ. — Italy

SYNOPSIS

Constitution and geotechnical characteristics of the subsoil of Palermo, groundwater circulation and associated processes are outlined. Design criteria, based on observational approach, for near-surface underground structures built by tunnelling or trenching in mixed, abruptly changing ground, are suggested. Effectiveness of recommended approach is appreciated with reference to a long sewer trunk, about 5 m diam., presently under construction.

RESUME

On illustre la constitution et les caractéristiques géotechniques du sous-sol urbain de Palerme, dont la composition et les propriétés physico-mécaniques sont brusquement variables, ainsi que les caractéristiques de la circulation d'eau et les phénomènes d'érosion interne liés à l'écoulement des eaux souterraines. Par référence à cette situation on propose des critères de conception pour ouvrages souterrains à faible profondeur réalisés en galerie ou par excavation à ciel ouvert. Les critères exposés sont fondés sur la méthode d'observation; leur validité est évaluée relativement à la construction d'un long égout collecteur de 5 m de diamètre.

ZUSAMMENFASSUNG

Die Zusammensetzung und geotechnischen Daten des Untergrundes der Stadt Palermo, des Grundwasserverlaufes und die dazugehörigen Entwicklungen werden kurz erwähnt. Entwurfskriterien, die sich auf Beobachtungsmethoden stützen, werden für oberflächennahe Konstruktionen, in Tunnel oder offener Bauweise, bei gemischten plötzlich wechselnden Untergrundverhältnissen empfohlen. Die Wirksamkeit dieser Empfehlung wurde schon während der Durchführung einer Strecke eines Abwasserkanals mit 5 m Durchmesser bestätigt.

1. INTRODUCTION

The subsoil of the city of Palermo is composed, up to depths relevant to civil engineering activities, of rocks and soils of widely different physico-mechanical properties, so intricately combined and associated to form up, almost always, "structurally complex formations" (A.G.I., 1977; 1979). The latter term is intended to denote a portion of the subsoil characterized by rapid variation, in the space, of composition and geotechnical properties and for which, however, the assumption of random distribution of composition, structural pattern and mechanical properties, is not warranted. These formations are not marked by disorder; they, on the contrary, present an intricately ordered arrangement, difficult to discover and to decipher. Precise knowledge of the anatomy of such masses, is an essential unavoidable step in modelling their behaviour; when the structural arrangement of the mass can not be reliably ascertained, recourse must be made to an explicit observational approach (Valore, 1978).

Due to sharp variation, vertically and laterally, of macrostructural features, deformability, shear strength, permeability and erodibility of rocks and soil present in the subsoil of Pa-

lermo, the prediction of settlements and of perturbations to ground water circulation, associated with construction of large or long underground structures may prove to be very difficult if not impossible. Difficulties are increased by possible modifications of natural processes still taking place in the ground. Modelling of ground conditions is furthermore

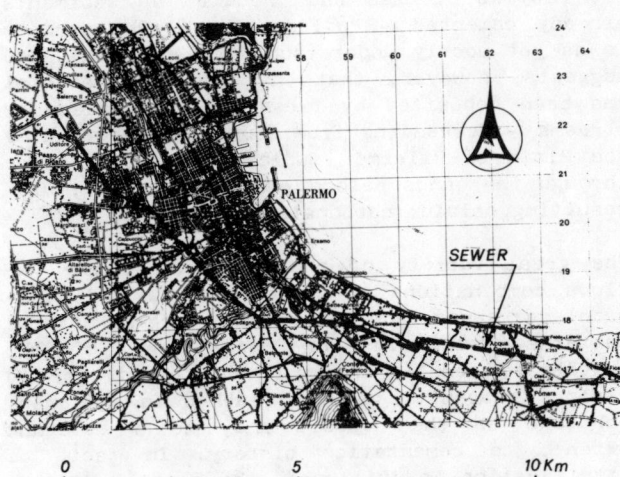

Fig. 1 — Map of Palermo and sewer layout

complicated by the existence in the subsoil of artificial and natural cavities, the last beeing frequently of small size and of complex configuration.

Utilization of underground space of the town is taking place at increasing rate; it requires, therefore - at least - a clear position of geotechnical problems associated with its use. The realization currently underway of a sewer about 13 Km long, that underpasses the urban area, fig. 1, offers the opportunity to point out the peculiar constitutive and geotechnical characters of the subsoil, to discuss problems and design criteria and to report on the experience as yet collected during the construction of the first trunk of the sewer.

2. CONSTITUTION AND GEOTECHNICAL CHARACTERISTICS OF THE SUBSOIL

Rocks and soils belong to a marly clay formation,oligo-miocenic in age, and to a quaternary complex including an upper calcarenitic deposit and a lower grey sands and clayey silts deposit. Quaternary soils and rocks overly, everywhere in the urban area, the marly clay formation. The top of this latter, located at depths from few to 70 m, is irregulary shaped due to "ups and downs" of tectonic origin that confer to the contact quaternary complex-clay basement a step--like configuration. Quaternary sediments - clayey silt and sands - have been transported and deposited under changing environmental conditions largely controlled by the irregularities of the top of clay formation, prevailing marine currents and by the location of paleo-rivers (Ruggieri - Sprovieri, 1975; Jappelli et al., 1981).Subsequently, sediments of upper part of the deposit underwent a lithification process, mainly by calcium carbonate cementation of sands. This process, still in progress although at a reduced rate, is coupled with internal erosion of sand included among or in contact with layers, lenses and "noduli" of sediments already cemented - "calcarenites". The process is as yet poorly understood. Available evidence suggests, however, that the cementing material has been deposited by fresh waters supplied by streams - descending from mountains surrounding the Plain of Palermo - which flowed by seepage through the sands before reaching the sea, with resulting calcium carbonate deposition.

The great variety of characteristics, the various combinations of cemented rocks and cohesionless sands and macro and megastructural patterns found in the upper part of the quaternary complex may well be traced back to various diagenitic stages reached in different portions of the deposit and, to a relevant extent, to cementation history. In fact, as lithification is initiated, the spatial distribution of permeability coefficients is modified, the conditions of water circulation are altered

and additional changes take place in the cementation and in the internal erosion process of uncemented sands.Internal erosion, transport and redistribution of sands by action of circulating waters, can still be observed in the calcarenitic complex; they are responsible for the formation of cavities, later enlarged or partly filled with loose sands, sometimes well stratified, fig. 2. The cementation, and probably the dissolution processes, are evidently still taking place at shallow depth in the vicinity of the free surface of groundwater table, as documented by figures. 3 and 4, taken during the construction of the sewer, 20 days after placement of shotcrete on tunnel walls.

The multiform pattern of the cementation process gives rise to strangely shaped concretions, fig. 5, 5 ÷ 30 cm mean diam. - subsequentely referred to as "noduli" - variously mixed with sands, to poorly cemented porous or to strongly cemented calcarenites. These latter rocks take the form of lenses (0,5 ÷ 2 m thick) or of irregular shaped inclusions surrounded by "noduli" and sand (fig. 6); this type of rock type frequently results from welding, by carbonatic cement, of previously formed concretions, fig. 7; sometimes strongly cemented layers include lenses and pockets of fine gravel and sand, fig. 8, or, when the sand has been eroded, cavities with small stalactites hanging on their walls (fig. 9). When the shortage of cement or water circulation conditions prevent the completion of the welding process, mixed interconnected noduli and loose sand aggregates result; sands and noduli are sometimes disposed in ordered sequences, fig. 10. In both cases, subsequent removal of sands by underground flowing waters, produces small cavities giving rise to aggregates of noduli, stable or collapsible depending on interlocking degree.

Main rock types found in the calcarenitic complex are the following:
- Evenly cemented yellow calcarenites; very weak to soft, depending on degree of cementation; porosity from 0,02 to 0,25; frequently stratified (CL);
- Strongly cemented yellow calcarenites; low porosity; found as irregular lenses, less than 2 m thick, or as inclusions of various sizes (CFC);
- Vacuolar calcarenites, well - but very irregularly cemented; contain minute cavities and frequent shells (CV);
- Irregularly shaped concretions - noduli -, strongly cemented; mean diam. from 3 to 30 cm; internodular space empty or filled with yellowish sands; structural arrangement may be stable or collapsible depending on relative proportion in comparison to sands and on interlocking degree, (CN);
- Yellowish medium sands; from very loose to dense (SC);

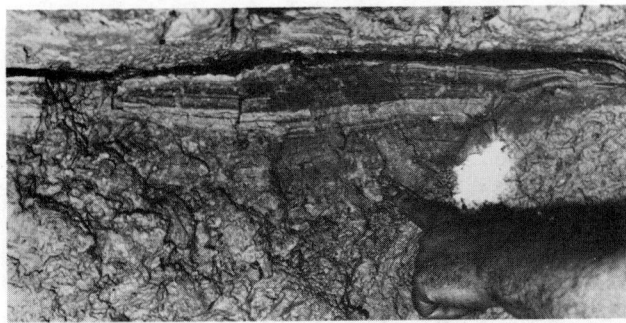

Fig. 2 - Laminated loose sands, filling a previously existing cavity, bounded by strongly cemented calca
renites

Fig. 3 - Calcium carbonate concretion on tunnel wall, 1 cm
thick, deposited shortly after temporary lining
placement (tunnel in calcarenitic rocks)

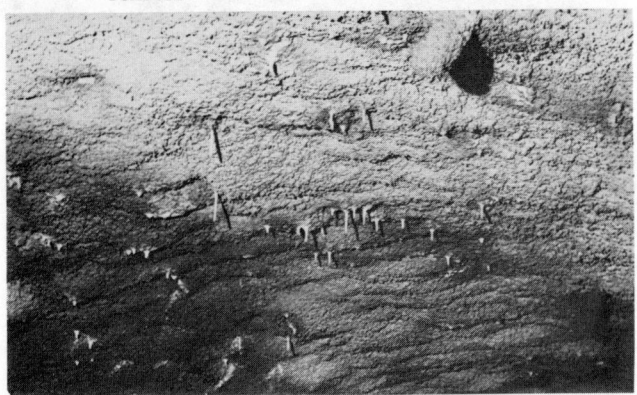

Fig. 4 - Minute calcitic stalactites hanging on tunnel
roof, formed 20 days after spritz-beton placement
(tunnel in calcarenitic rocks)

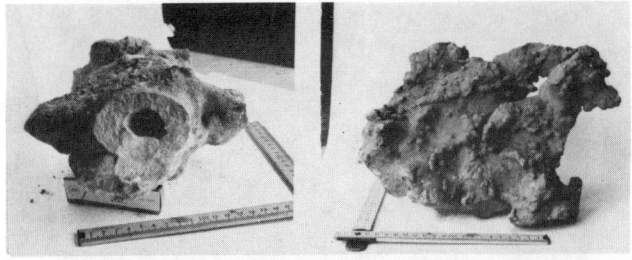

Fig. 5 - Irregularly shaped calcarenitic "noduli"; note
central hole on left "piece"

Fig. 6 Strongly cemented rock (CFC calcarenite) inclusion
surrounded by "noduli"; internodular spaces were
previously filled with yellowish sand

Fig. 7 - Vacuolar calcarenite, overlain by "noduli". The
rock results from welding, not yet completed, of
"noduli", still recognisable (Photo width: 0.7 m)

Fig. 8 - Sandy fine gravel lense between vacuolar calcarenite layers

Fig. 9 - Irregular-shaped cavity; stalactites have partly closed the cavity

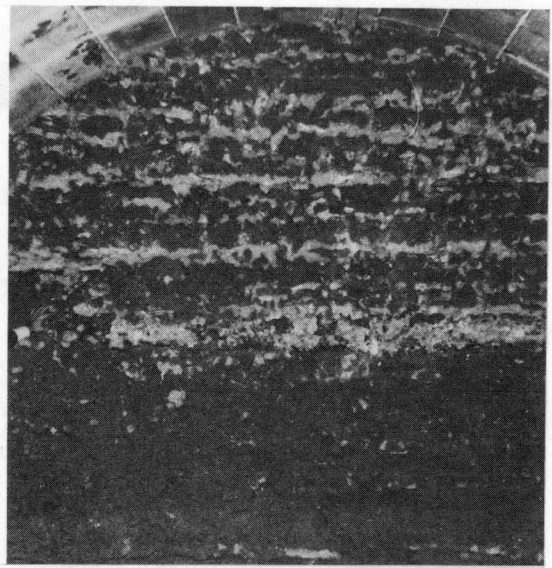

Fig. 10 - Alternating calcarenitic concretions and yellow sand layers (Photo width: 3 m)

- Fine grey sandy gravels; subangular to rounded, medium dense, (SG).

These "basic" types are found variously combined; it is not rare that as many as ten different associations be present within a volume of few cubic meters.

Sampling is a most difficult task, sample recover never exceeding 50%; RQD values are less than 10 %. In this situation, reliable reconnaissance of the rock sequence may be accomplished by means of exploratory pits and adits, excavated whenever possible in advance to underground work construction. Comparison of information from direct observation of pit's wall and that from adjacent boreholes is used and suggested as

the best way to calibrate results from the latter.

Mechanical behaviour ranges from that typical of loose sand to that of strong rock. As an example unconfined compressive strength of CFC calcarenites may reach 10 kN/cm^2; N number of SPT tests in sands mixed with noduli may result as low as 5 or very high up to refuse, being anyway misleading; cohesion intercept of CL calcarenites specimens ranges from a few to more than 500 N/cm^2, being strikingly influenced by porosity.Tensile strength of CFC, CV and CL calcarenites is low as may be deduced from open tension cracks observed in horizontal slabs of these rock spanning over flat cavities, 3 ÷ 6 m long, during tunnel construction. Permeability coefficients of CL and CV calcarenites specimens range in the wide interval $10^{-2} ÷ 10^{-8}$ cm/sec.

The lower part of quaternary deposit is formed by grey sands (S) rich in fossil shell fragments, grey sandy silts (LS) and clayey silts (A); these soils are soft, normally consolidates or slightly overconsolidated (OCR = 2); shear strength parameters ranges: c' = 0 ÷ 1 N/cm^2; φ' = 20° ÷ 30°; constrained modulus values, at usual working stress level typically range between 500 and 1000 N/cm^2; permeability coefficient of grey sands ranges between 10^{-8} and 10^{-6} cm/sec.

Marly clay formation includes weathered soft clays (ABG), tectonized, brecciated clays (AB), stiff marly clays (MA) and hard quartzarenitic lenses (QZ). MA clays present closely spaced persistent subvertical striated joints. Except ABG clays, the other terms of the formation are heavily overconsolidated. Shear strength parameters of AB clays are: c' = 1 ÷ 3 N/cm^2; φ' = 20° ÷ 25°; intact specimens of MA clays are much more resistant, mechanical behaviour is, however controlled by structural pattern of joints and by strength along discontinuity surfaces. When joints are tightly closed the clays present very low permeability.

A single engineering work may involve, at a given location, rocks and soils of the first, the first and second or all three deposits or formations.

On account of constitutive characters of the calcarenitic complex, above sinthetically summarized, it is not surprising that modelling the rock mass for geotechnical engineering prediction purposes escapes our present capabilities and that observational approach must be invoked as the most effective way to handle engineering problems.

3. GROUNDWATER CIRCULATION

Free water surface is found from some to about 40 m below ground level. Water circulation takes place prevailingly at the interior of calcarenitic complex and is directed toward the sea: it is controlled by complicate boundary hydraulic conditions and by "internal" factors such as the irregular, variable geometric configuration of contacts between calcarenitic complex and the lower, much less permeable, grey sands and silts or marly clays, the existence in the subsoil of ancient streams and valleys, today filled with heterogeneous materials, the presence of natural and artificial cavities.

Prediction of perturbations of ground water regime, and of their possible propagation, associated with water injection or withdrawal or with underground structures that obstacle water movement, is hardly feasible at the present state of knowledge of the subsoil. Apart from possible variations that may be induced in effective stress state of fine or medium grained soils, and their associated settlemens, long--term perturbations are relevant to internal erosion, dissolution and cementation processes, which might be appreciably altered, provoking detrimental effects on existing buildings. On this respect, it must be noted that water free surface seasonal oscillations of some meters do not appear to have been of major concern in the past.

4. GEOTECHNICAL DESIGN CRITERIA FOR NEAR-SURFACE UNDERGROUND STRUCTURES

Selection of design criteria and choice of construction sequences are dictated by constitution and geotechnical characters of subsoil, ground water circulation, physico-mechanical processes taking place in the ground, existence of buildings, in the heavily built-up urban area, and by the presence, at shallow depths, of networks of underground facilities. Underground space development perspectives are greatly conditioned by the above factors.

Although the suggested design philosophy is thought to be of general validity, it will be discussed with reference to the sewer in fig. 1, that may well be considered a good example of long underground work. Problems are best identified, and design principles presented, by taking successive "overlies" of the ground, each focussing on a single aspect of structure-ground interaction.

The first overly refers to groundwater circulation and its possible modifications consequent to construction of the sewer. This must be, of course, watertight in order to avoid flow of sewage fluids in the subsoil and, reciprocally, to prevent the sewer functioning as a drain. It is also of paramount importance that the contact soil or rock - outer surface of the structure be continous, and where necessary treated in order to discourage preferential ground water flow parallel to sewer axis, that might provoke diversion in flow lines and the connection of presently distinct water flow domains.Due to sewer length and taking account of the flow direction - almost everywhere normal to its axis - it is evident that the structure must be located so to avoid the formation of an underground barrage. Control of possible perturbations of water flow regime appears to be essential for preservation of present equilibrium of calcium carbonate distribution and internal erosion processes. Besides, possible detrimental impact on these processes of fresh water injection, currently under consideration for ground table rechargement purposes, are to be stressed. It is evident that first choice solutions must consider the possibility of locating the structure above free water surface, when feasible; this proved to be the case for some trunks of the sewer under discussion. It is not possible, because of space restrictions, to discuss all types of relationships between structure, soil and rock sequence and groundwater. Some typical situations are shown in fig. 11. Long-term interference will be almost negligible when the structure is wholly embedded in soils of very low permeability; limited when structure lies well below water surface at the interior of highly permeable rocks of the calcarenitic complex if this is more than about 3 sewer diameters thick; interaction is relevant when the structure restricts the flow section when passing through thin calcarenitic layers or when it is partly keyed into low-permeability soils that are overlain by permeable water-bearing rock strata: in this case bypasses must be provided to permit flow around lower part of the structure.

Stability and deformability problems - both during construction and in the long-term - are readily apparent from information given before

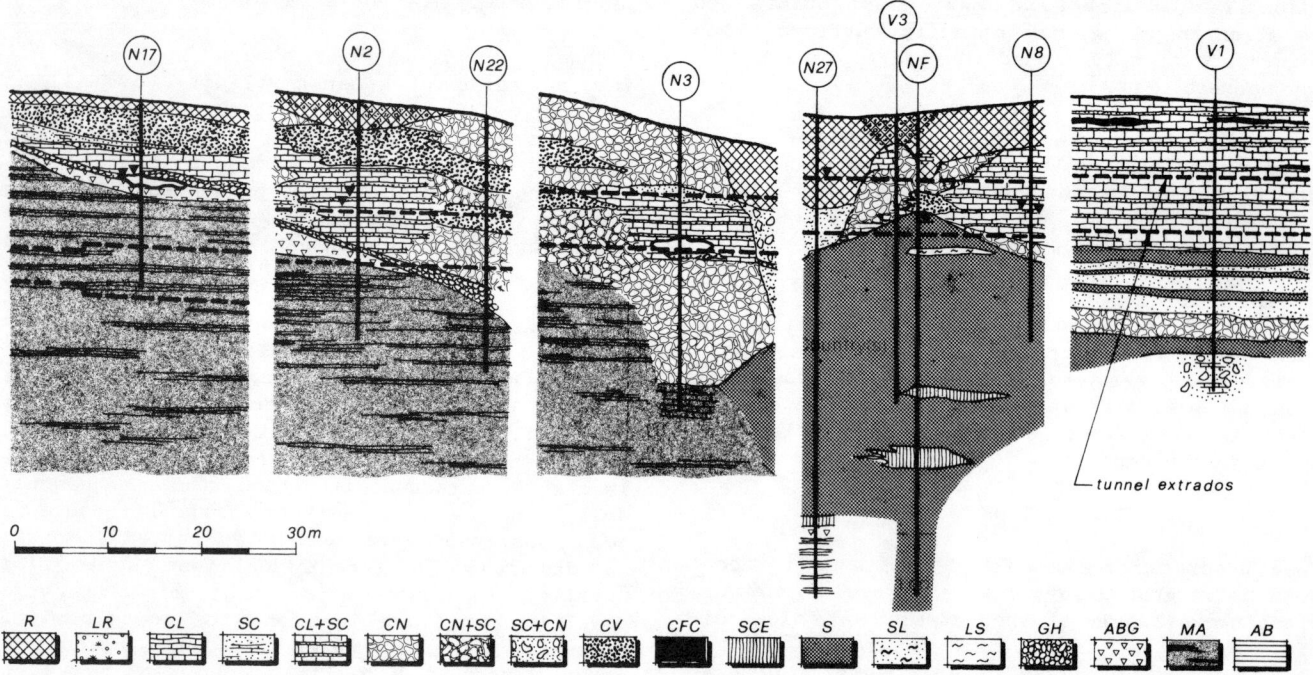

Fig. 11 - Typical subsoil profiles, showing relationships among rocks and soils, groundwater table, and tunnel. R, LR recent organic silty soils; GH gravel; SCE grey cemented sands; other symbols defined in text.

and from examination of typical situations such as those shown in fig. 11. In this regard, it is emphasized strict control of settlements and lateral movement of rock in every stage of construction, in order not to endanger existing underground facilities and buildings. To this end, two basic construction modalities have been envisaged, namely by tunnelling and by open excavation. The last method is reserved to sewer trunks at shallow depths-less than 6 m - when passing through zones not yet crossed by underground services and when the structure directly bears on grey sands and silts. Support to excavation walls is provided by concrete dia-phragm panels, poured prior to excavation. A typical working sequence is schematically shown in fig. 12.

The most serious difficulty associated with this method is the piping tendency of grey sands or of yellow calcarenitic sands mixed with "nodu-li". The problem may be overcome by performing excavation and pouring of sewer base-slab underwater. Diaphragms are strutted to prevent lateral movements.

Many more problems arise when tunnelling, due to mixed and abruptly changing ground; the most relevant ones concern face and roof stability and water inflow control during construction. When the tunnel passes entirely in calcarenitic rocks, full face excavation is possible by tunneller equipped with picks; when it passes through the contact calcarenites-grey silty sands, partial section excavation is necessary. A successful construction sequence for the latter case is shown in fig. 13. In advance dewatering by pumping in wells, spaced 40 ÷ 70 m, not more than 3 m deep below invert elevation, has been satisfactorily carried out. Face and roof stability problems may be handled by cement-sand-water low fluidity admixtures injection, prior to excavation. Caution, how-ever, must be exerted because the grout, if sufficiently fluid, travels in calcarenitic rocks to very great distances of the order of 100 m.

Placement of temporary lining must be carried

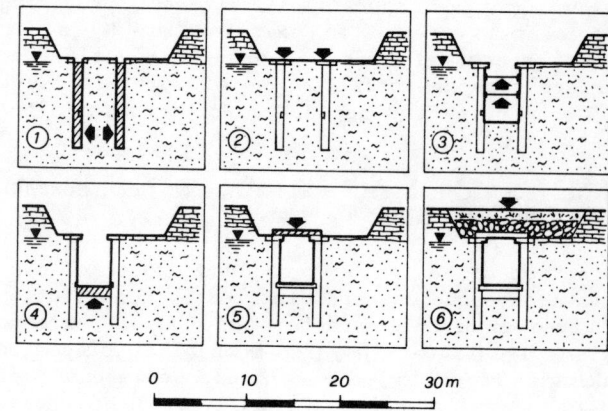

Fig. 12 - Construction phases of sewer in open excavation supported by diaphragm walls

out immediately after excavation. Stability problems, however, are not deemed troublesome on the basis of available experience, as an ample range of measures that can be rapidly carried out, if necessary are at disposal: namely, steel ribs, rock bolts, grout injections. The construction can be, therefore, accomplished safely on the basis of observational approach, guided by frequent convergence, lateral and vertical ground displacements surveys. This same learn-as-you-go methodology is indicated for evaluation of the effectiveness of bypasses and on detection of flow perturbations. On this regard, timely interpretation of piezometric surveys - along a strip about 300 m wide on both sides of the sewer - presents some uncertainty since reference long-period past data are lacking.

Response of the structure to seismic actions is hardly predictable in the situation at hand; full cross section watertight joints, spaced 20 ÷ 30 m, are suggested to reduce bending moments.

5. EXPERIENCE IN NEAR-SURFACE TUNNELLING AND TRENCHING

Construction of the first trunk of sewer, totalling more than 4 Km, of wich 1 Km by tunnelling in mixed, abruptly changing ground (figures 13 and 14), was carried out according to above principles. Selected construction sequences are shown in figures 12, 13 and 15. Along the trunk built by open excavation, lateral movements were effectively controlled to desired values, established not to exceed one centimeter; a single case of piping initiaton in yellow calcarenitic sands mixed with "noduli" was incurred on, because of deviation from specifications; it was promptly stopped by ceasing incautious pumping of water at excavation bottom; subsequently construction proceeded underwater.

Unforeseen difficulties were due to the presence in the ground of heavy quantities of toxic pollulants (hydrocarbons) along a stretch of the sewer 1.5 Km long.

Construction by tunnelling, at depths of cover from 5 to 8 m, according to the sequence of fig. 13, proved satisfactory in controlling movements of tunnel face and walls. A 2 meter thick lense of loose material - yellow sand mixed with noduli - was present above tunnel crown along a 200 m stretch. It was deemed profitable enlarging the tunnel section upwards rather than grouting the loose material. During the work the necessity of immediate support and protection against erosion of grey fine sands present in the lower half tunnel section was definitely proved, the rate of erosion being very rapid (fig. 16). When excavating the last 30 m long tunnel stretch, involving alternating yellow sands and concretions layers in which innumerable 5 ÷ 15 cm diam. cavities were present, abrupt collapse of these latter induced lateral and vertical movements amounting to about 5 cm. Movements were accompanied by acoustic emissions and stopped soon after. Here again the trouble was due to deviation from specification. In fact, a tunnel length of about 30 m, was left unsupported. Work was resumed according to

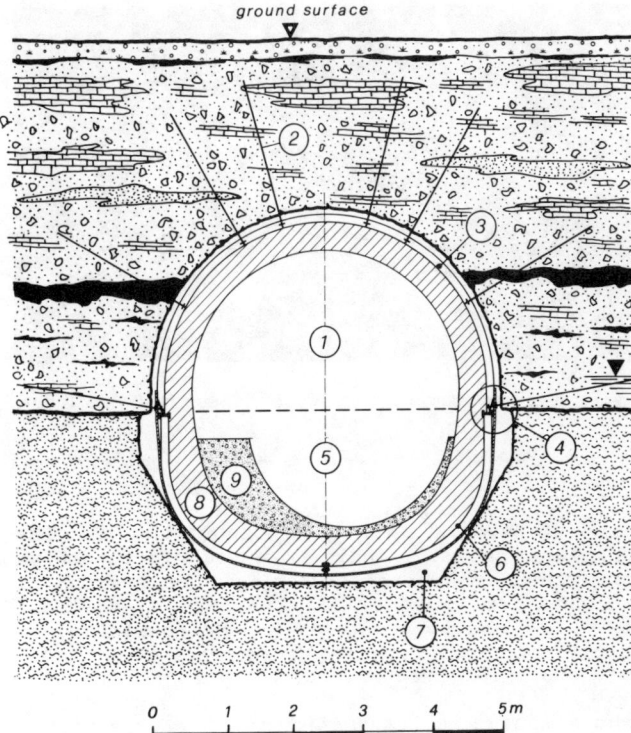

Fig. 13 - Mixed face tunnelling trough calcarenitic rocks-silty sand contact. First phase: 1 rock excavation by mole; 2 bolt installation; 3 wire-mesh and shotcrete; 4 horizontal H steel beam. Second phase: 5 silty sand excavation; steel ribs and precast concrete segments; 7 concrete injection and bypasses placement; 8 final lining; 9 sewer section completion. (See fig. 11 for graph. symb.)

Fig. 14 - Abrupt passage from calcarenitic rocks to silty sands: note erosion of the latter. Tunnel diam. 5 m

design schemes, temporary lining supplemented with radial bolts regularly spaced at 1 m intervals. Subsequently, daily inspection revealed after three months the appearance of frequent capillary cracks that were taken as

Fig. 15 - Tunnelling work in progress: second phase (compare fig. 13)

Fig. 16 - Sand erosion by flowing water, when protective measures lack behind

signs of anomalous behaviour, requiring immediate placement of final lining. Unfortunately, convergence measuring device was not available, and decisions outght to be assumed on the base of crude measurements. During construction, observations on cementation process were made, previously referred to.

Excavation of calcarenitic rocks was performed by a Paurat boom type mole equipped with picks; picks comsumption mean value was about 1 pick/m^3 of excavated rock; peak value up to 3 picks/m^3 were recorded in vacuolar strongly cemented calcarenites; in that case, pick frequently ruptured because of impacts against rock walls of small cavities.

6. CONCLUSIONS

The subsoil of Palermo city is composed of rocks and soil of widely different macro and megastructural features and physico-mechanical properties, associated and interconnected according to complicate threedimensional sequences, changing, moreover, even over very short distances - as compared to engineering structure size. Ground water circulation regime, associated with internal erosion, solution and cementation processes, is controlled by permeability coefficient distribution, characterized, in turn, by abrupt variations.

Reliable geotechnical characterization and modelling of subsoil for engineering behaviour prediction purposes is still beyond our capabilities. Notwithstanding this conclusion, the knowledge, for some aspects hardly qualitative, of physical and geotechnical processes taking place in the ground or that might be initiated by engineering activities, coupled with learn-as-you-go procedures, enables to carry out safely the construction of near-surface underground works, as confirmed by experience as yet collected.

REFERENCES

A.G.I. (Ital. Geotech. Assoc.) (1977) - Proc. Int. Symp. on Geotechnics of Structurally Complex Formations - voll. I and II.

A.G.I. (1979) - Some italian experiences on the mechanical characterization of structurally complex formations. Int. Congr. Rock Mech. - Montreux.

JAPPELLI R., CUSIMANO G., LIGUORI V., VALORE C. (1981) - Contributo alla conoscenza geotecnica del sottosuolo di Palermo - Atti Riunione Gruppo Ingegneria Geotecnica del C.N.R.- Roma.

RUGGIERI G., SPROVIERI R. (1975) - Ricerche sul Siciliano di Palermo: le argille del Fiume Oreto - Boll. Soc. Geol. It. 94, pp 889 - 917.

VALORE C. (1978) - Previsioni dei cedimenti di dighe su formazioni complesse - Proc. XIII It. Geot. Conf. vol. II, pp 161 - 168.

ACKNOWLEDGMENTS

The study referred to in the paper was suggested by the construction of a long sewer presently being carried out by Comune di Palermo and financed by Cassa per il Mezzogiorno. A. wishes to thank Prof. R. Jappelli for his valuable suggestions. Research work on the subsoil of Palermo is carried out with partial financial support of National Research Council of Italy, CNR.

ROOF STABILITY OF SHALLOW TUNNELS IN ISOTROPIC AND JOINTED ROCK

Stabilité des tunnels à faible profondeur creusés en milieu isotrope et fissuré

Standsicherheit von seichten Tunneln in isotropem und geklüftetem Gebirge

P. Egger

Head of Section Laboratory of Rock Mechanics, Swiss Federal Institute of Technology, Lausanne

SYNOPSIS

The paper proposes a simple approach to the problem of roof stability of shallow tunnels which is based upon the nature of frequently observed failures. Using the principal stress trajectories and the failure criteria respectively for the rock mass and discontinuities, the ground pressures acting upon a tunnel are evaluated for three typical geological situations: isotropic, stratified and cross-jointed rock. The paramount importance of the value of joint friction is clearly demonstrated. The optimal shape of the tunnel roof — defined by the major principal stress trajectories — and criteria for the tunnel roof being self-supporting are given for the three mentioned cases.

RESUME

La communication propose une approche simple du problème de la stabilité des tunnels à faible profondeur, basée sur la nature des ruptures fréquemment observées. A l'aide des trajectoires des contraintes principales et des critères de rupture pour la roche et ses discontinuités, les pressions des terrains s'exerçant sur le tunnel sont évaluées pour trois situations géologiques typiques: massif isotrope, stratifié et traversé de deux familles de fissures. L'importance du frottement sur les joints est clairement démontrée. Pour les trois cas mentionnés, la forme optimale du toit du tunnel — définie par les trajectoires des contraintes principales majeures — ainsi que des critères pour l'autostabilité du toit sont indiqués.

ZUSAMMENFASSUNG

Für das Problem der Standsicherheit der Kalotte von seichten Tunneln wird eine einfache Näherungslösung vorgeschlagen, die auf der Natur häufig beobachteter Verbrüche beruht. Mit Hilfe der Hauptspannungstrajektorien und der Bruchkriterien für den Fels und die Diskontinuitäten werden die auf den Tunnel wirkenden Gebirgsdrücke für drei typische geologische Situationen ermittelt: isotropes, geschichtetes und normal aufeinander geklüftetes Gebirge. Die Wichtigkeit der Größe der Kluftreibung tritt deutlich hervor. Die optimale Form der Tunnelkalotte — gegeben durch die Trajektorie der größten Hauptspannung — sowie Kriterien für die Standsicherheit des ungestützten Tunnels werden für die drei erwähnten Fälle mitgeteilt.

1. INTRODUCTION

Despite of the comparably simple geometry of a shallow circular tunnel, no general solution for the stress and strain fields above its roof is yet available. For the static conditions depend intricately on the initial stresses and the constitutive law of the rock mass as well as on the adopted construction method, and can only be quantified for few particular cases.

For these reasons the consideration of statically possible states of limit equilibrium was soon proposed by various authors, e.g. TERZAGHI (1943), d'ESCATHA-MANDEL (1974), ATKINSON (1975), in order to find out the order of magnitude of the ground pressures acting upon the tunnel. Though depending on the assumed failure mechanism, these methods appear to be valuable approaches of the problem. More recently, the theory of stress trajectories was applied to particular aspects of the tunnel problem by GUDEHUS (1974) and KOLYMBAS (1982).

All approaches known to the writer deal with the case of isotropic ground. In order to cope with frequently encountered geological situations, the present paper proposes an extension of the theory of stress trajectories which furnishes the complete stress field for isotropic, stratified and cross-jointed rock masses.

2. ISOTROPIC ROCK MASS

The observation of collapsed shallow tunnels in the nature and in physical models shows that frequently the failure surfaces are nearly vertical and tangent to the tunnel, yielding chimney like breakdowns. Therefore we consider the limit equilibrium of a parabolic arch situated above the tunnel [Fig. 1] :

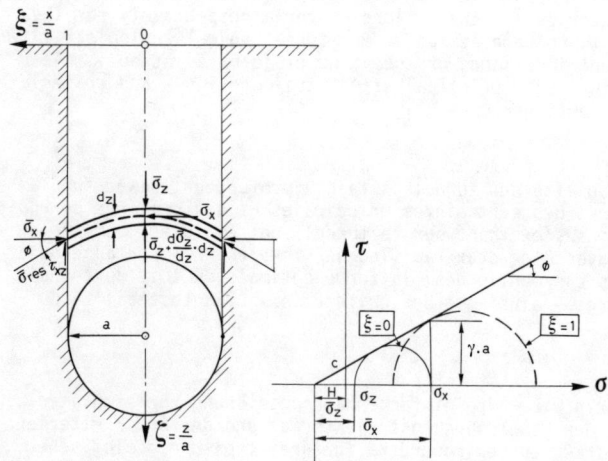

Fig. 1 : Assumed failure mechanism for a shallow tunnel

tunnel axis $\quad \overline{\sigma}_x = \lambda_p \cdot \overline{\sigma}_z \qquad (1)$

abutment $\quad \tau_{xz} = \left(\gamma - \dfrac{d\overline{\sigma}_z}{dz}\right) \cdot a = \overline{\sigma}_x \cdot tg\,\phi \qquad (2)$

and find by integration, similarly to TERZAGHI, the solution for $\overline{\sigma}_z$ in the tunnel axis :

$$\overline{\sigma}_z = \frac{\gamma \cdot a}{\lambda_p \cdot tg\,\phi} \cdot \left(1 - e^{-\frac{\lambda_p \cdot tg\,\phi \cdot z}{a}}\right) \qquad (3)$$

with $\gamma \qquad$ density of the rock mass

$\phi \qquad$ angle of internal friction

$\lambda_p = \dfrac{1+\sin\phi}{1-\sin\phi} \qquad$ coefficient of passive ground pressure

$$\left.\begin{array}{l} \overline{\sigma}_x = \sigma_x + H \\[2mm] \overline{\sigma}_z = \sigma_z + H \end{array}\right\} \begin{array}{l} \text{equivalent stresses where} \\ H = c \cdot cot\,\phi \end{array}$$

$c \qquad\qquad\qquad$ cohesion

For z increasing the expression (3) tends rapidly to an upper limit :

$$\overline{\sigma}_{z,lim} = \frac{\gamma \cdot a}{\lambda_p \cdot tg\,\phi} \qquad (4)$$

hence $\qquad \overline{\sigma}_{x,lim} = \dfrac{\gamma \cdot a}{tg\,\phi} \qquad (5)$

In the following calculations, these upper limit values will be considered in order to remain on the safe side. For the evaluation of the complete stress field we start from the conditions [Fig. 2] :

$$\overline{\sigma}_x = \frac{\gamma a}{tg\,\phi} = const. \quad \text{and} \quad \tau_{xz} = \gamma \cdot x = \gamma \cdot a \cdot \xi \qquad (6)$$

with $\xi = \dfrac{x}{a}$

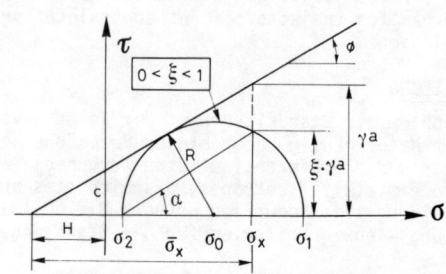

Fig. 2 : Stress conditions for isotropic rock

By introducing the failure criterion of the rock, we obtain :

$$\frac{\overline{\sigma}_0}{\gamma a} = \frac{1}{\sin\phi \cdot \cos\phi} \cdot \left[1 - \sin\phi \cdot \sqrt{1-\xi^2}\right] \qquad (7)$$

$$\frac{\overline{\sigma}_z}{\gamma a} = \frac{2}{\sin\phi \cdot \cos\phi} \cdot \left[1 - \sin\phi \cdot \sqrt{1-\xi^2}\right] - cot\,\phi \qquad (8)$$

$$\frac{\overline{\sigma}_2}{\gamma a} = \frac{1}{tg\,\phi \cdot (1+\sin\phi)} \cdot \left[1 - \sin\phi \cdot \sqrt{1-\xi^2}\right] \qquad (9)$$

The expression for the major principal stress trajectory writes as follows :

$$\operatorname{tg}\alpha = \frac{d\zeta}{d\xi} = \frac{\tau_{xz}}{\bar{\sigma}_x - \bar{\sigma}_2} = \frac{\xi}{1 + \sqrt{1-\xi^2}} \cdot \operatorname{tg}\left(\frac{\pi}{4} + \frac{\phi}{2}\right) \quad (10)$$

and after integration :

$$\zeta = \operatorname{tg}\left(\frac{\pi}{4} + \frac{\phi}{2}\right) \cdot \left[\ell n \frac{1 + \sqrt{1-\xi^2}}{2} + 1 - \sqrt{1-\xi^2}\right] \quad (11)$$

with $\zeta = \dfrac{z}{a}$

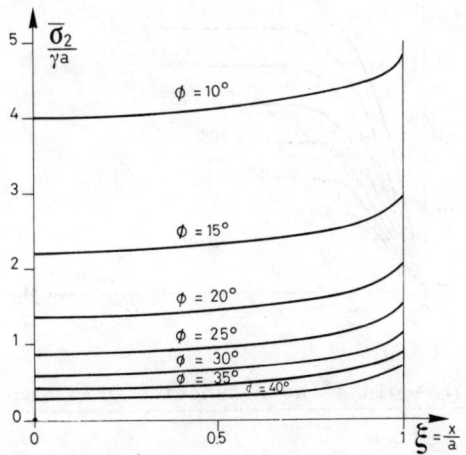

Fig. 3 : Curves of the minor principal stress for isotropic rock

Using eq. (9), the curves of the minor principal stress $\bar{\sigma}_2$ are plotted in Fig. 3 for various values of ϕ. As an example, Fig. 4 shows the net of the principal stress trajectories and the equivalent contact stresses along the wall of a circular tunnel, for $\phi = 30°$. It should be kept in mind that the required stresses to be applied by the tunnel support are $\sigma_r = \bar{\sigma}_r - H$ and $\tau_{supp} = \tau_\beta - H \cdot \operatorname{tg}\phi = \tau_\beta - c$.

The rock pressures are seen to increase from the tunnel axis towards the sidewall, and we obtain from eq. (9) :

$$\frac{\bar{\sigma}_{2,\xi=1}}{\bar{\sigma}_{2,\xi=0}} = \frac{1}{1-\sin\phi} \quad (12)$$

For the case that the tunnel roof follows the shape of the $\bar{\sigma}_1$-trajectory, the condition for the roof being self-supporting writes as follows :

$$\frac{\sigma_{2,max}}{\gamma a} = \frac{\bar{\sigma}_{2,max} - H}{\gamma a} < 0 \quad (13)$$

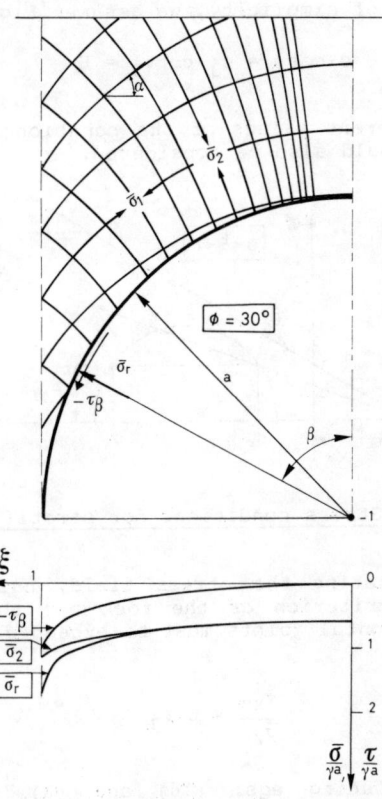

Fig. 4 : Stress trajectories and equivalent contact stresses for a circular tunnel in isotropic rock ($\phi = 30°$)

Thus for $\bar{\sigma}_{2,max} = \bar{\sigma}_{2,\xi=1}$ and $c = H \cdot \operatorname{tg}\phi$ we obtain the minimal possible value for the cohesion :

$$\frac{c}{\gamma a} > \frac{1}{1 + \sin\phi} \quad (14)$$

For a circular tunnel the conditions $\sigma_x = \tau_{supp} = 0$ at $\xi = 1$ yield :

$$\frac{c}{\gamma a} > 1 \quad (14a)$$

The thus required cohesion is proportional to the tunnel radius but does not depend on the depth of the tunnel. It should be noted, however, that the stability of the sidewall depends upon both the radius and the depth.

3. STRATIFIED ROCK MASS

We consider a thinly bedded or horizontally laminated rock mass, characterized by the strength parameters (ϕ, c) for the rock and (ϕ_j, c_j) for the horizontal weakness planes.

For sake of simplicity, we assume [Fig. 5]

$$c \cdot \cot \phi = c_j \cdot \cot \phi_j = H \qquad (15)$$

but different values of the cohesion along the joints could also be considered.

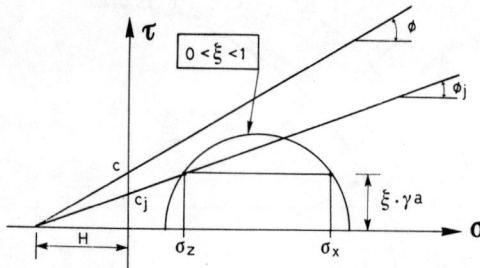

Fig. 5 : Stress conditions for stratified rock

For evaluating the stress field, neither the failure criterion of the rock nor that along the horizontal joints must be exceeded :

$$\frac{\tau_{xz}}{\sigma_z} < \operatorname{tg} \phi_j \qquad (16)$$

By introducing eqs. (16) and (6) into the relations derived from MOHR's circle we find

$$\frac{\overline{\sigma}_0}{\gamma a} = \frac{1}{2 \cdot \operatorname{tg} \phi} \left(1 + \xi \cdot \frac{\operatorname{tg} \phi}{\operatorname{tg} \phi_j} \right) \qquad (17)$$

$$\frac{\overline{\sigma}_{12}}{\gamma a} = \cot \phi \cdot \left[\frac{1}{2} \left(1 + \xi \cdot \frac{\operatorname{tg} \phi}{\operatorname{tg} \phi_j} \right) \pm \sqrt{ \xi^2 \cdot \operatorname{tg}^2 \phi + \frac{1}{4} \left(1 - \xi \cdot \frac{\operatorname{tg} \phi}{\operatorname{tg} \phi_j} \right)^2 } \right] \qquad (18)$$

Comparing eq. (7) which is derived from the failure criterion of the rock, with eq. (17) which is based on the failure along the joints, we find that both criteria can only be satisfied for values ξ near to 1, if

$$\operatorname{tg} \phi_j > \operatorname{tg} \phi_{j,cr} = \frac{\sin 2\phi}{3 - \cos 2\phi} \qquad (19)$$

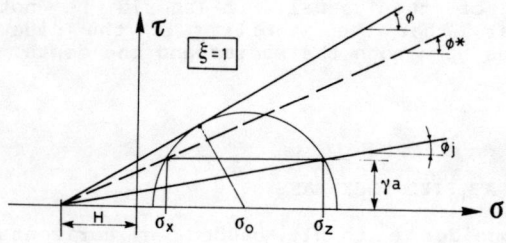

Fig. 6 : <u>Definition of angle ϕ^* for stratified rock</u>

For smaller values of ϕ_j, a study by means of MOHR's circle [Fig. 6] yields the expression :

$$\frac{\overline{\sigma}_x}{\gamma a} = \frac{2}{\operatorname{tg} \phi_j \cdot \cos^2 \phi} \cdot \left(1 - \sqrt{ 1 - \frac{\cos^2 \phi}{\cos^2 \phi_j} } - \frac{\cos^2 \phi}{2} \right) = \frac{1}{\operatorname{tg} \phi^*} \qquad (20)$$

This relation between ϕ, ϕ_j and ϕ^* is illustrated on the Fig. 7.

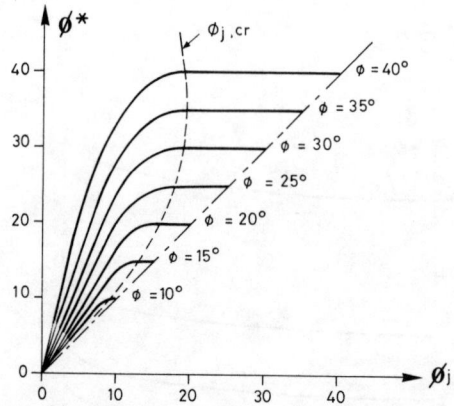

Fig. 7 : Angle ϕ^* as a function of ϕ and ϕ_j

Consequently, the eqs. (7) and (9) become

$$\frac{\overline{\sigma}_0}{\gamma \cdot a} = \frac{1}{\operatorname{tg} \phi^* \cdot \cos^2 \phi} \cdot \left[1 \pm \sqrt{ 1 - \cos^2 \phi \cdot \left(1 + \xi^2 \cdot \operatorname{tg}^2 \phi^* \right) } \right] \qquad (21)$$

$$\frac{\overline{\sigma}_2}{\gamma a} = \frac{1}{\operatorname{tg} \phi^* \cdot (1 + \sin \phi)} \cdot \left[1 \pm \sqrt{ 1 - \cos^2 \phi \cdot \left(1 + \xi^2 \cdot \operatorname{tg}^2 \phi^* \right) } \right] \qquad (22)$$

the minus sign to be taken for ξ near zero, the plus sign for ξ near 1.

Whereas eqs. (17) and (18) write now

$$\frac{\overline{\sigma}_0}{\gamma a} = \frac{1}{2} \left[\frac{1}{\operatorname{tg} \phi^*} + \frac{\xi}{\operatorname{tg} \phi_j} \right] \qquad (23)$$

$$\frac{\overline{\sigma}_{12}}{\gamma a} = \frac{1}{2} \left[\frac{1}{\operatorname{tg} \phi^*} + \frac{\xi}{\operatorname{tg} \phi_j} \right] \pm \sqrt{ \xi^2 + \frac{1}{4} \left(\frac{\xi}{\operatorname{tg} \phi_j} - \frac{1}{\operatorname{tg} \phi^*} \right)^2 } \qquad (24)$$

Using eqs. (21) and (23), it can be found that for values of ξ inferior to

$$\xi_{jm} = \frac{tg^2\phi_j \cdot \cos^2\phi}{tg^2\phi^* \cdot (\cos^2\phi + 4\ tg^2\phi_j)} \qquad (25)$$

failure occurs in the rock, for higher values along the joints. In the case $\phi_j > \phi_{j,cr}$, a second zone appears near $\xi = 1$ where the failure occurs in the rock.

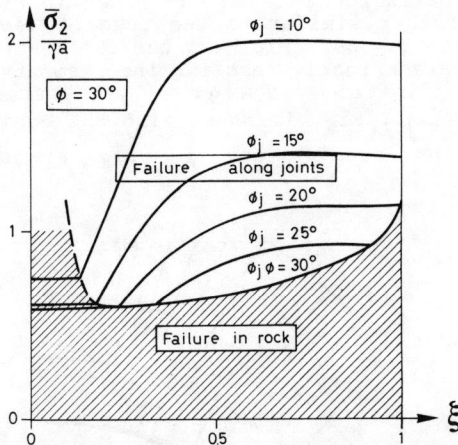

Fig. 8 : Curves of the minor principal stress for stratified rock ($\phi = 30°$, ϕ_j variable)

As an example Fig. 8 shows the curves of the minor principal stress $\overline{\sigma}_2$ for $\phi = 30°$ and various values of ϕ_j. The high increase of the rock pressure with decreasing friction ϕ_j along the horizontal joints can clearly be seen : e.g. for $\phi_j = 10°$, $\overline{\sigma}_2 \doteq 2,0 \cdot \gamma a$ in the outer half of the tunnel whereas $\overline{\sigma}_2 = (0,66$ to $1,16) \cdot \gamma a$ for isotropic rock.

The inclination of the trajectory of the major principal stress is given by the expression

$$tg\,\alpha = \frac{\tau_{xz}}{\overline{\sigma}_x - \overline{\sigma}_2} = \frac{d\zeta}{d\xi} = \frac{\xi}{a + \sqrt{\xi^2 + a^2}}$$

$$(26)$$

$$\text{with } a = \frac{1}{2}\left(\frac{1}{tg\,\phi^*} - \frac{\xi}{tg\,\phi_j}\right)$$

For the case $\phi = 30°$, $\phi_j = 10°$, the Fig. 9 shows the net of the stress trajectories, $\overline{\sigma}_2$ and the equivalent contact stresses σ_r and τ_β along the wall of a circular tunnel. The radial pressure σ_r is seen to pass through a maximum at $\xi = 0,65$ ($\beta \doteq 40°$) what is frequently confirmed by monitoring results, e.g. EGGER (1975).

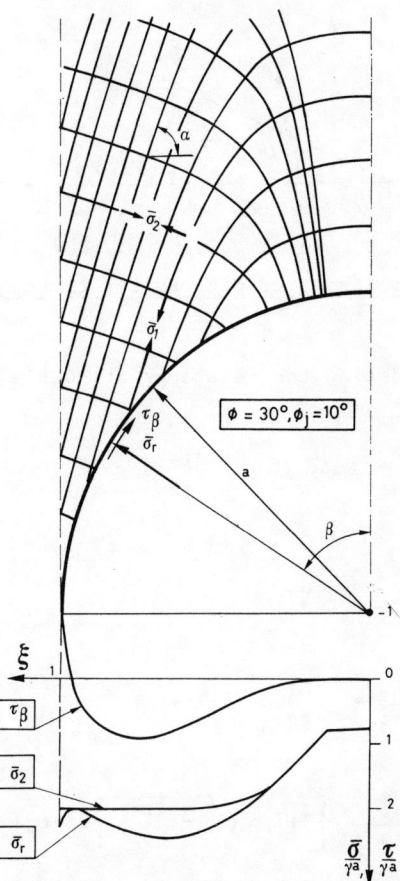

Fig. 9 : Stress trajectories and equivalent contact stresses for a circular tunnel in stratified rock ($\phi = 30°$, $\phi_j = 10°$)

Due to the steep inclination of the $\overline{\sigma}_1$-trajectories, the shear stresses τ_β along the tunnel wall are directed towards the axis; this means that rock bolts as a tunnel support should preferably be set steeper than radially in order to ascertain limit equilibrium in an optimal way.

Like in the case of isotropic rock, a tunnel roof following the $\overline{\sigma}_1$-trajectory is self supporting if $\sigma_{2,max} = \overline{\sigma}_{2,max} - H_{tr} < 0$. For a circular tunnel $\sigma_{r,max} = \overline{\sigma}_{r,max} - H_c < 0$ must be satisfied. In our example ($\phi = 30°$, $\phi_j = 10°$) we find a difference due to the shape of the roof of $H_c - H_{tr} = (2,42 - 2,00) \cdot \gamma a = 0,42 \cdot \gamma a$.

4. CROSS-JOINTED SOUND ROCK

The case of a horizontally bedded sound rock with vertical joints can be treated in a similar manner as before. No failure is assumed to happen in the rock itself, both sets of joints are characterized by the same values of (ϕ_j, H_j) for sake of simplicity [Fig. 10]. But different parameters can be introduced for the horizontal and vertical joints at the price of somewhat longer calculations.

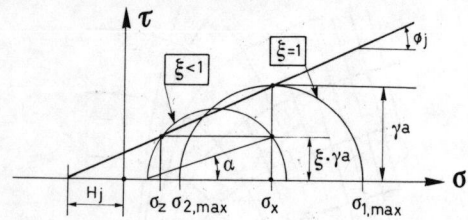

Fig. 10 : Stress conditions for cross-jointed
          rock

By considering the relations at MOHR's circle
we find :

$$\frac{\overline{\sigma}_x}{\gamma a} = \cot\phi_j \ ; \qquad \frac{\tau_{xz}}{\gamma a} = \xi \qquad (27)$$

$$\frac{\overline{\sigma}_z}{\gamma a} = \frac{\xi}{\tan\phi_j} \qquad (28)$$

$$\frac{\overline{\sigma}_0}{\gamma a} = \frac{1}{2 \cdot \tan\phi_j} \cdot (1 + \xi) \qquad (29)$$

$$\frac{\overline{\sigma}_{12}}{\gamma a} = \frac{1}{2 \cdot \tan\phi_j} \cdot \left[1+\xi \pm \sqrt{(1-\xi)^2 + (2\xi \cdot \tan\phi_j)^2} \right] \quad (30)$$

$$\tan\alpha = \frac{2\xi \cdot \tan\phi_j}{1-\xi + \sqrt{(1-\xi)^2 + (2\xi \cdot \tan\phi_j)^2}} \qquad (31)$$

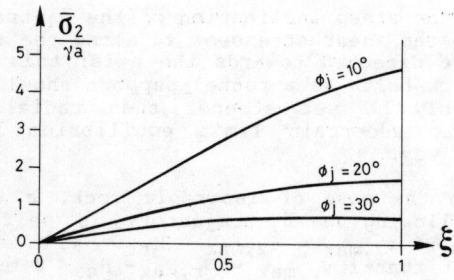

Fig. 11 : Curves of the minor principal stress
          for cross-jointed rock

Fig. 11 shows the curves of the minor principal
stress $\overline{\sigma}_2$ for various values of ϕ_j. As mentioned above, they are valid as long as no failure
occurs in the rock itself, the condition

$$\frac{\sigma_c}{\gamma \cdot a} > \frac{\sigma_{1,max}}{\gamma \cdot a} = \frac{\overline{\sigma}_{1, \xi=1} - H_j}{\gamma a} = \frac{1+\tan\phi_j}{\tan\phi_j} - \frac{H_j}{\gamma a} \qquad (32)$$

being on the safe side.

In this case, $\overline{\sigma}_2$ is zero at the tunnel axis and
increases monotonously towards the spring
line. For a given ϕ_j, the $\overline{\sigma}_2$-curve remains
below the corresponding curve for the isotropic
rock mass, except at the point

$$\xi_m = \frac{\cos^2\phi}{2 - \cos^2\phi} \qquad (33)$$

where $\alpha = \frac{\pi}{4} - \frac{\phi}{2}$; there both curves are tangent
to each other. The net of the principal stress
trajectories for $\phi_j = 30°$ is given in Fig. 12;
it is rather similar to the case of isotropic
rock with $\phi = 30°$ [Fig. 4], but the $\overline{\sigma}_1$-trajec-
tories are slightly less inclined, so at $\xi = 1$
the inclination is $\tan\alpha = 1$ instead of
$\tan\left(\frac{\pi}{4} + \frac{\phi}{2}\right)$. Fig. 12 shows also the equivalent
contact stresses $\overline{\sigma}_r$ and τ_β for a circular
tunnel.

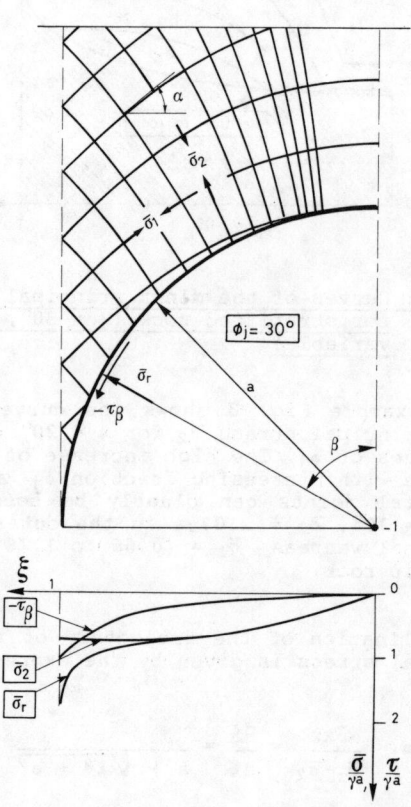

Fig. 12 : Stress trajectories and equivalent
          contact stresses for a circular
          tunnel in cross-jointed rock
          ($\phi_j = 30°$)

5. DISCUSSION

The proposed method assumes a ground body
situated above the tunnel and limited by two

C 300

vertical planes to be in a state of limit equilibrium. Observations show that the assumption of vertical failure planes tangent to the tunnel holds generally quite well as long as the overburden does not exceed two to three diameters.

Taking into account the conditions at the borders and the respective failure criteria for the rock and for the joints, the expressions for a statically possible complete stress field can be derived. But it is clear that this method does not allow any indications about the displacements.

The calculations have been performed for the three cases of isotropic, stratified and cross-jointed rock. Their results allow to formulate the following remarks :

- The trajectory of the major principal stress which defines the optimal shape of the tunnel roof is flatter than a circle for isotropic and cross-jointed rock, but considerably steeper for stratified rock with a low angle of friction along the horizontal discontinuities.

- The ground pressures increase monotonously from the tunnel axis towards the sidewall for isotropic and cross-jointed rock, but they show a maximum at the shoulders of the tunnel for stratified rock.

- The required strength respectively of the rock or along the joints for the tunnel roof being self-supporting depends on the tunnel diameter and the shape of the roof, but not on the depth of the tunnel.

REFERENCES

Atkinson, J.H. et al. (1975). Collapse of shallow unlined tunnels in dense sand : Tunnels & Tunnelling (7), 3, 81-87.

D'Escatha, Y. and Mandel J. (1974). Stabilité d'une galerie peu profonde en terrain meuble : J. Ind. Minérale, 45-53.

Egger, P. (1975). Erfahrungen beim Bau eines seichtliegenden Tunels in tertiären Mergeln : Rock Mech. Suppl. 4, 41-54.

Gudehus, G. (1974). Konstruktion statisch möglicher Spannungsfelder in Erdkörpern : Straße Brücke Tunnel, 6, 157-161.

Kolymbas, D. (1982). Vereinfachte statische Berechnung der Firste eines Tunnels in massigem Fels : Rock Mech. (14), 201-207.

Terzaghi, K. (1943). Theoretical Soil Mechanics. New York : Wiley.

COMPARISON OF FIVE EMPIRICAL TUNNEL CLASSIFICATION METHODS — ACCURACY, EFFECT OF SUBJECTIVITY AND AVAILABLE INFORMATION

Comparaison de cinq méthodes empiriques de classification des roches dans les tunnels — précision des prédictions, influence de la subjectivité et de l'information disponible

Vergleich von fünf empirischen Gebirgsklassifizierungsmethoden im Tunnelbau — Vorhersage, Einfluss der Subjektivität und der zur Verfügung stehenden Information

H. H. Einstein
Prof. of Civil Engineering, Mass. Inst. of Tech., Cambridge, U.S.A.

D. E. Thompson
Senior Vice-President, Haley & Aldrich, Inc., Cambridge, U.S.A.

A. S. Azzouz
Assist. Prof. of Civil Engineering, Mass. Inst. of Tech., Cambridge, U.S.A.

K. P. O'Reilly
M. S. Schultz
S. Ordun
Former Research Assistants, Mass. Inst. of Tech., Cambridge, U.S.A.

SYNOPSIS

The Rock Load (Terzaghi) Approach, the RQD-, RSR-, and "Geomechanics"-methods and the Q-system were applied at Porter Square Station, a 21 m span subway station cavern in Cambridge Argillite. Rock classification and support predictions were made in four phases (boring records, exploration shaft, pilot tunnel, cavern survey) by two independent investigators. The predictions were compared to actual pilot tunnel and cavern support. Results of the study show that rock classification and support predictions are affected to some extent by the increasing information as one goes from phase to phase; subjectivity only affects prediction of spatial variations. For most probable (average) rock conditions the results of all empirical methods are comparable, while most methods are not suited to handle unfavorable conditions in such a large cavern.

RESUME

La classification des roches dans la station souterraine de Porter Square a été effectuée en utilisant les méthodes de Terzaghi, les méthodes RQD, RSR et "Geomechanics" et le système Q. Il s'agit d'une galerie souterraine d'une largeur de 21 m dans de l'argilite de Cambridge. La classification des roches et la prédiction des dimensions des supports ont été effectuées séparément par deux personnes en quatre phases (sondages, puits de reconnaissance, galerie pilote et observations dans la galerie finale). Les prédictions ont été comparées aux observations dans la galerie pilote et dans la caverne. Les résultats indiquent que les prédictions sont légèrement influencées par l'accumulation d'information d'une phase à l'autre; par contre la subjectivité joue un rôle seulement dans l'estimation de la variabilité spatiale. Pour les conditions géologiques moyennes les prédictions obtenues par toutes les méthodes sont comparables; par contre pour une galerie d'une telle largeur, la plupart des méthodes se sont révélées peu adaptées à des conditions géologiques défavorables.

ZUSAMMENFASSUNG

Die Terzaghi-, RQD-, RSR- und "Geomechanics"-Methoden und das Q-System wurden für die Gebirgsklassifizierung der Porter Square Untergrundbahnstation angewendet. Es handelt sich um eine Kaverne von 21 m Spannweite in Cambridge Argillit. Gebirgsklassifizierung und Ausbaudimensionen wurden unabhängig von zwei Personen und in vier Phasen bestimmt, nämlich aufgrund von Bohrungen eines Erkundungsschachtes, eines Pilotstollens und aufgrund der Aufnahme in der ausgebrochenen Kaverne. Es zeigt sich, daß die von Phase zu Phase zunehmende Information einen gewissen Einfluß auf Gebirgsklassifizierung und Ausbauvorhersagen hat. Subjektivität hingegen beeinflußt nur die vorhergesagte Variation längs der Kavernenachse. Die Resultate alle Methoden sind vergleichbar, solange die geologischen Bedingungen durchschnittlich sind. Für ungünstige Extrembedingungen in einer so großen Kaverne sind die meisten Methoden dagegen nicht geeignet.

INTRODUCTION

Formal empirical rock classification methods for tunneling have played a role for over 100 years and new methods have developed at regular intervals during the last 40 years; particularly since the late 60's the creation of new methods has accelerated: Terzaghi (1946), Stini (1950), Lauffer (1958), NATM (1965, Rabcewicz (1965), Muller (1978), RQD-Deere (1969), RSR (1974), Q (1974), Geomechanics-RMR (1973-79), Louis (1974), Franklin (1975,1976). Users and creators of the methods are naturally interested in the accuracy of prediction and correlation between methods. A significant number of comparative studies have thus been conducted either for specific cases or as general comparisons (e.g., Barton, 1977; Blackey, 1979; Bieniawski, 1980, 1979b; Steiner et al., 1980; McCusker, 1980; Rose et al., 1981) without reaching any definite conclusions as to superiority of a particular method or generally valid correlations. The investigation on which this paper is based does not pretend to provide this answer either. However, the study's context and several of the issues it addresses are different and should therefore provide additional insight. These differences lie in the size of the opening (a 21m (70ft), span subway station), in the fact that not only empirical, but also analytical and numerical methods are compared, and in the investigation of effects of subjectivity and of available information.

The comparison of five empirical methods (Rock Load -Terzaghi, RQD, RSR, Geomechanics-RMR, Q) with regard to the influence of subjectivity and of available information is the topic of this paper. Information increased through 3 phases of exploration to the 4th phase, a survey in the final cavern. In most of these phases the methods were applied by two independent investigators. Predictions can be compared with the actually placed support and performance of the pilot tunnel and of the final cavern.

After introducing the project and exploration program (Section 2) the comparative procedure will be described (Section 3) followed by the presentation of results (Section 4), their discussion (Section 5) and conclusions (Section 6).

2. PORTER SQUARE STATION CAVERN -PROJECT AND EXPLORATION PROGRAM

2.1 Project

Porter Square Station is an underground rock chamber of 21 m (70 ft) span, 14 m (45 ft) height and 150 m (500 ft) length, with a transverse crossover tunnel (span 12 m, height 14 m (Figures 1,2). The rock is Cambridge Argillite. Figure 2 shows the staged excavation procedure that started from an access shaft, which together with the pilot tunnel had been previously excavated (see 2.2). Regular cavern supports consist of bolts (dowels), W8 x 40 steel sets at 1.5 m (5 ft) spacing and continuously blocked and embedded in shotcrete with welded wire fabric (Figures 2a,b). Support installation followed immediately behind the excavation of 1 round (5 ft). An extensive monitoring program with extensometers, inclinometers, strain gages and convergence

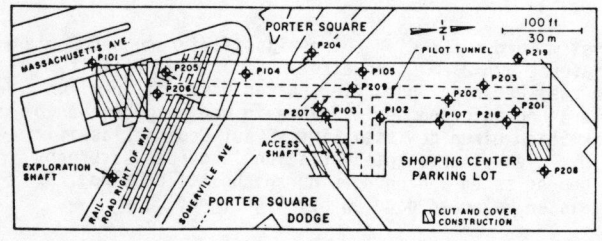

Figure 1 SUBSURFACE EXPLORATION LOCATION PLAN.

measurements in nine test sections served for construction control and was also used to compare predictions with performance.

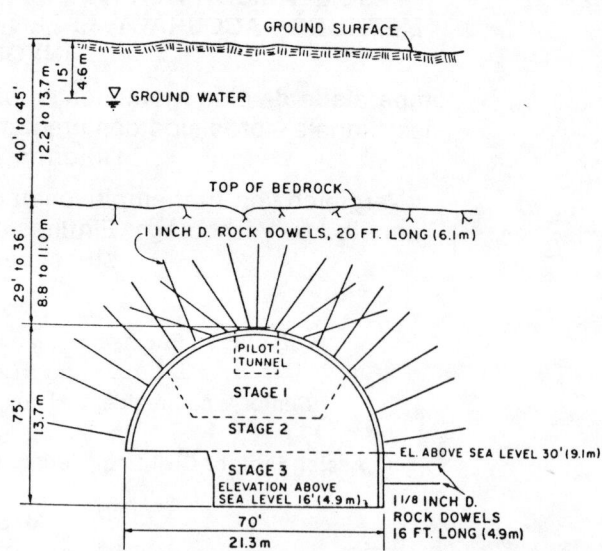

Figure 2a TYPICAL TRAINROOM CROSS SECTION.

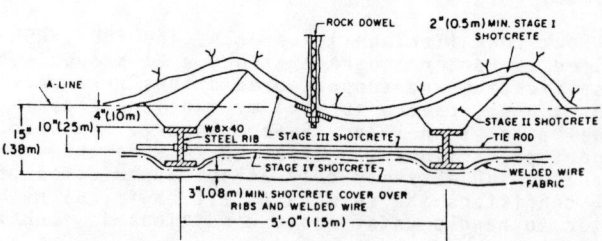

Figure 2b TYPICAL TRAINROOM VAULT LINING.

2.2 Exploration Program and Geologic -Geotechnical Site Conditions

The exploration program consisted of three phases:

I: 26 vertical and inclined borings with logging of RQD, fracture spacing, lithology and water conditions (Figure 1 shows some of these borings).

II: 1 m (3 ft) diameter exploration shaft of 33.5 m (110 ft) depth (Figure 1) with visual inspection and logging as above. Phases I and II were part of the predesign exploration program; for details see Haley and Aldrich (1978).

III: Pilot tunnel of ≈3.6 x 3.6 m (12 ft) cross section in the crown of the future cavern (Fig. 2), and associated access shaft. The pilot tunnel surfaces were mapped in detail. Use of this information in the final design and inspection by the bidding contractors was the main purpose of the pilot tunnel.

At this point additional details on geological-geotechnical site conditions have to be given. Cambridge Argillite is a slightly metamorphosed mudstone with an average unconfined compressive strength of 27×10^3 psi (~140 MPa). Three joint sets were mapped: (1) relatively flat parallel to bedding (N55 to 85W, 5-15 S); (2) East-West and steeply dipping (N75W ($\mp$25), 45 to 90N); and (3) North-South and steeply dipping (N20E($\mp$10), 60 to 90 E or W). Set (1) has spacings of ~0.1 to 1.2 m (0.5 to 4 ft), Set (2) of 0.3 to 1.2 m (1 to 4 ft) with generally non-persistent joints. Set (3) has spacings of 0.3 to 3 m (1 to 10 ft) with many joints being persistent over at least 6 m (20 ft) Shears, i.e., discontinuities with some shear displacements, are parallel to sets (1) and (3). Two faults intersect the site, one .05-.15 m wide (N60E, 50NW) near the south end and the other ~0.5 m (1.5 ft) wide (N25E, 60NW) with a parallel andesite dike in the center. To conclude this brief description, it has to be understood that the conditions vary significantly along the axis as will be seen shortly.

3. CLASSIFICATION AND COMPARISON PROCEDURE

Some details on the classification and on the subsequent comparisons need to be given for a better understanding of the results.

3.1 Classification

Phase I: Use of boring records to determine parameters and rock classes (best, worst, most probable - see comments below).
(2 investigators)

Use of rock classes (parameters) to estimate support requirements representative of the entire site (best, worst, and most probable).
(2 investigators)

Phase II: Use of exploration shaft records to determine classes or parameters (best, worst, most probable).
(1 investigator)

Phase III: Determination of parameters and rock classes in the pilot tunnel -variability of classes along the tunnel and best, worst, most probable for entire tunnel.
(2 investigators)

Use of rock classes to estimate pilot tunnel supports - variability along tunnel and best, worst, most probable for entire tunnel.
(2 investigators)

Use of rock classes to estimate main cavern support requirements best, worst, and most probable.
(2 investigators)

Phase IV: Determination of parameters and rock classes in the cavern -variability along cavern and best, worst, most probable for entire cavern.
(1 investigator)

Use of rock classes to estimate cavern support requirements -- var-iability along cavern and best, Use of rock classes to estimate main cavern support requirements best, worst, and most probable.
(2 investigators)

Phase IV: Determination of parameters and rock classes in the cavern -variability along cavern and best, worst, most probable for entire cavern.
(1 investigator)

Use of rock classes to estimate cavern support requirements -- var-iability along cavern and best, worst, most probable for entire cavern.

To have a basis for comparison, the best, worst, and most probable rock classes (parameters) for the entire site and corresponding support requirements were determined in each phase. In addition, variability along tunnel and cavern axes was also predicted when such information was obtainable (Phases III, IV).

3.2 Comparisons

There are thus several investigators using five methods to provide estimates in each of four phases; these estimates can be compared among each other and with the actually used support in the pilot tunnel and main cavern. The many possible comparisons have to be systematically structured:

Comparison 1: Change in classification and support prediction (support in pilot tunnel and cavern) due to increasing information as one goes from borings to exploration shaft.

Comparison 2a: Change in classification and support prediction as one goes from boring (exploration shaft), i.e., Phases I(II), to pilot tunnel observation (Phase III).

Comparison 2b: Predicted and actual support in pilot tunnel

Comparison 3:   (a) Differences in Phases I(II) due to investigator
            (b) Differences in Phase III due to investigator

Comparison 4: Spatial variability in pilot tunnel classification (Phase III)

Comparison 5a: Change in classification and support prediction as one goes from Phases I(II) to Phase III to Phase IV (cavern observation)

Comparison 5b: Predicted and actual support in cavern and comment on performance.

Parameters (classes) and supports are compared separately to demonstrate that differences in classes or parameters do not necessarily mean different supports.

4. DETAILS AND RESULTS OF THE CLASSIFICATION

4.1 Classification Details

Complete information on exploration results and on the individual parameter measurements is given in "Porter Square, 1982". Some details are mentioned here to provide the basis for the results and especially because they illustrate the use of information available in a particular phase.

Phase I (Borings)

Boring logs of 19 of the 26 borings were used over their full length for lithologic and water inflow estimation, while RQD was only determined for the tunnel influence zone (pilot tunnel: tunnel plus 3.6 m (12 ft) above crown; cavern - cavern plus 10.5 m (35 ft) above crown). It is interesting to observe that RQD values for the two zones do not vary greatly (average pilot tunnel 90, cavern 85; worst pilot tunnel 70, cavern 64). From the borings only joint sets (1) (shallow dip) and (2) (steep dip, perpendicular to tunnel axis) were determined. With regard to the specific methods, several problematic points need to be mentioned. The Rock Load (Terzaghi) classification is not suited for direct interpretation from boring logs unless one has significant experience. This was tried by Investigator 2. Investigator 1 used the relations between RQD and Rock Load (Monsees, 1970).

In the RSR method, parameter B has to reflect joint orientation, which was only possible to a limited extent; parameter C reflects water conditions, which had to be estimated from general conditions and pressure tests.

The Geomechanics classification was also affected by the joint-orientation problem. There is also some uncertainty in associating joint spacing to specific sets and regarding the representativeness of small cores to describe joint conditions.

The Q system is also affected by the latter two problems, which concern Joint Set Number and Joint Roughness Number in this method. The Stress Reduction Factor was estimated using the ratio "unconfined compressive strength / largest in situ stress" as recommended by Barton et al. (1974) for competent rock.

Phase II (Exploration Shaft)

The information lacking so far on number of joint sets and orientation becomes now available, and especially the "missing" joint set (3) is now visible. Also, water inflow can be observed in tunnel-like conditions (confirming, however, the low inflow interpreted from the borings).

Phase III (Pilot Tunnel)

All five methods were applied by classifying the rock conditions every 3 m (10 ft) along the tunnel axis. Given the three-dimensional exposure and the large surfaces, no problems were encountered in direct classification (Rock Load method) or parameter determination, possibly with the exception of RQD. RQD is defined as modified core recovery and not as surface exposure. Approximate estimates were obtained in considering by eye 3-foot sections on the tunnel surface (see also Phase IV for another estimation).

The classes were related to pilot tunnel and cavern support. It should be recalled that the classes do not reflect different size openings but that some of the relations between class and support do. The rock load includes opening width, but it is doubtful that the method was developed to represent also large caverns (Steiner et al., 1980). Similar reservations have to be made for the RSR method, where the transformation into rock loads was used to evaluate the cavern support. The RQD method as modified by Cording et al. (1971) covers such openings, while the Geomechanics classification does not differentiate according to opening size. The Q-system not only reflects span but also the difference between a temporary pilot tunnel and a station cavern.

Phase IV (Station Cavern)

In principle, the survey in the cavern was conducted analogously to the one in the pilot tunnel. The survey concentrated on the vertical face of the permanent bench at elevation 23 (see Figure 2). Direct classification and parameter measurements took place every 10 ft unless prevented by muck piles (see Figure 3 for exact locations). In addition, line surveys for RQD

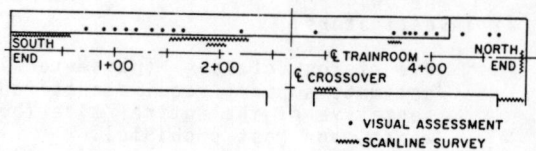

Figure 3   PLAIN VIEW OF PORTER SQUARE STATION SHOWING PARAMETER ASSESSMENT STATIONS FOR PHASE IV.

determination were made using the formal method by Priest and Hudson (1976) to transform fracture spacing into RQD. It would naturally have been desirable to survey the entire interior rock surface of the cavern. This was not possible due to the construction procedure which required application of shotcrete immediately after blasting. However, mapping the face of the cavern after each round provided information on how the results of the survey at Elevation 23 corresponds to the rest of the rock mass.

4.2 Results

The following tables and figures summarize the results of the respective phases.

Phase I

TABLE I: PHASE I ROCK CLASSIFICATION AND SUPPORT PREDICTIONS

NOTE: In ROCKLOAD, RQD, RSR - either bolts OR steel sets;
in GEOMECHANICS, Q - combined supports

INVEST METHOD	CLASS OR PARAMETER BEST 1	BEST 2	WORST 1	WORST 2	MOST PROBABLE 1	MOST PROBABLE 2	SUPPORT / PILOT TUNNEL CAVERN — BEST 1	BEST 2	WORST 1	WORST 2	MOST PROBABLE 1	MOST PROBABLE 2	COMMENT
	1	2	1	2	1	2	1	2	1	2	1	2	
ROCKLOAD (TERZAGHI)	CLASS 2	CLS 1	CLS 5	CLS 4	CLS 3	CLS 3	B > 2.2	NONE	B 0.8	B 1.2	B 2.2	B 2.2	(1) BOLTS:125kN
							B > 1.0 S > 1.0	NONE NONE	B0.5÷0.6 S0.2÷0.4	B0.6÷0.9 S0.4÷0.9	B > 1.0 S ≥ 1.0	B > 1.0 S ≥ 1.0	(2) STEEL SETS: (3) W8 X 40
RQD	95	100	65	70 (64)	87	89	B Occ.	B Occ	B1.2÷1.8	B1.2÷1.8	B Occ.	B Occ.	BOLTS:125kN
							B 1.5(ℓ 5.5)		B 0.3(ℓ 7.5) (3) S 0.9÷2.3		B 1.1(ℓ 7.5)		STEEL SETS: W8 X 40
RSR	78	88	52	49	65	69	B SPOT	NONE	B 1.5	B 1.4	B 2.3	B 2.7	BOLTS:125kN
							B 2.8 S > 5.	B NONE S > 5.	B 0.6 S 0.4	B 0.6 S 0.4	B 0.9 S 0.9	B 1.1 S 1.3	(4)STEEL SETS: W8 X 40
GEOMECHANICS*	80	85	47	55	64	74	II	I	III	III	II	II	
							II	I	III	III	II	II	
Q*	25.2	53.3	2.9	0.8	8.7	7.9	NONE	NONE	NONE	25	NONE	NONE	
							15	11	24	28	19 or 20	19 or 20	

~ same support

(1) Terzaghi rockload transformed into bolt spacings according to Cording et al. (1971)
(2) Rockload method is questionable for large cavern; support requirements are impractical.
(3) Bolts for these conditions questionable according to Merritt (1972)
(4) Method probably not applicable to span > 10m (30')
* See Table Ia for support description and explanation of abbreviations

TABLE Ia:

EXPLANATIONS:

B = Bolts - spacing in m
TB = Tensioned Bolts - spacing in m
S = Steel sets - spacing in m
Sh = Shotcrete - thickness in m
W = Wiremesh
ℓ = Bolt length in m
c = Crown
w = Wall
÷ = to

SUPPORT DESCRIPTION FOR GEOMECHANICS ROCK CLASS

I   Spot bolting
II B 2.5(ℓ3)c + W + Sh 0.05c
III B 1.5÷2.0(ℓ4) + W + Sh 0.05÷0.1c, 0.03w
IV B 1.0÷1.5(ℓ4÷5) + W + Sh 0.1÷0.15c, 0.1w
    + S(light)1.5

SUPPORT DESCRIPTION FOR Q-SYSTEM SUPPORT
CATEGORY

11 TB 1.5÷2.0 + chain link mesh
15 TB 1.5÷2.0 + chain link mesh
19 TB 1.÷2. + Sh 0.1÷0.2 + W
20 TB 1.0÷2.0 + Sh 0.1÷0.2 + W
24 TB 1.0÷1.5 Sh 0.1÷0.15 + W
25 TB 1.0 + Sh 0.05 + W
28 TB 1.0 + Sh 0.2÷0.3 + W
32 TB 1.0 + Sh 0.4÷0.6 + W

Phase II

TABLE II: CHANGES DUE TO EXPLORATION SHAFT

METHOD	RANGE PRIOR / AFTER	MOST PROB. PRIOR / AFTER	CHANGE IN MOST PROBABLE SUPPORT FOR PILOT TUNNEL
RQD	65-95 / SAME	87 / SAME	NONE
RSR	52-78 / 50-73	65 / 61	B2.3 / B2.0
GEOMECHANICS RMR	47-80 / 48-73	64 / 62	NONE
Q	2.9-25.2 / 2.9-5.6	8.7 / 3	NONE

C 307

Phase III

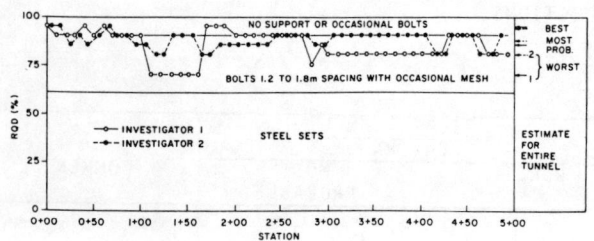

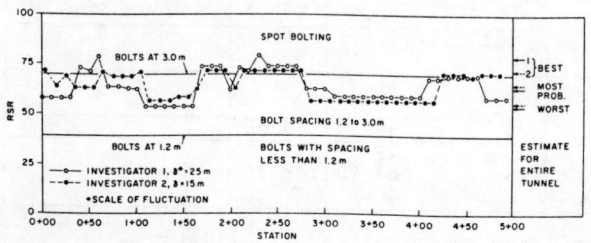

INVESTI-GATOR	ROCK LOAD CLASSES			SUPPORTS		
	MOST PROBABLE	BEST	WORST	MOST PROBABLE	BEST	WORST
1	3	2	5	B 2.2	NONE	B 0.5
2	3	3	3	B 2.2	B 2.2	B 2.2

Figure 4a ROCK LOADS AND RQD AND CORRESPONDING SUPPORT IN PILOT TUNNEL.

Figure 4b RSR AND CORRESPONDING SUPPORT IN PILOT TUNNEL.

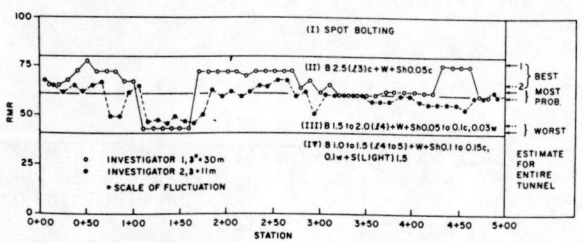

Figure 4c GEOMECHANICS (RMR) AND CORRESPONDING SUPPORT IN PILOT TUNNEL.

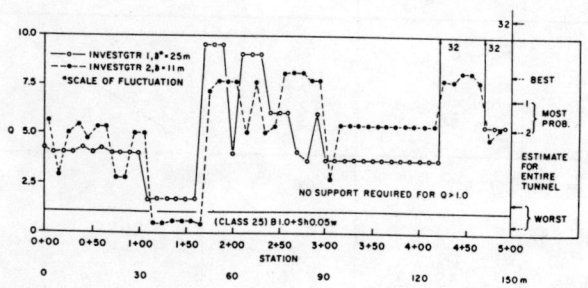

Figure 4d Q AND CORRESPONDING SUPPORT IN PILOT TUNNEL.

Phase IV

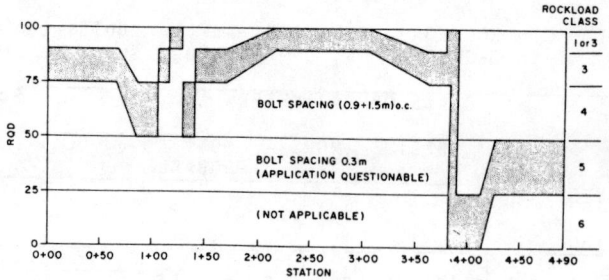

Figure 5a ROCKLOAD, RQD AND CORRESPONDING SUPPORT (ACCORDING TO CORDING ET AL. 1971) IN CAVERN.

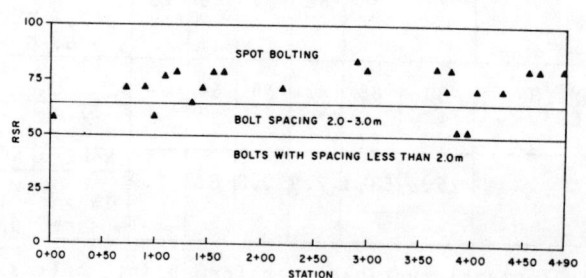

Figure 5b RSR CLASSIFICATION AND CORRESPONDING SUPPORT IN CAVERN.

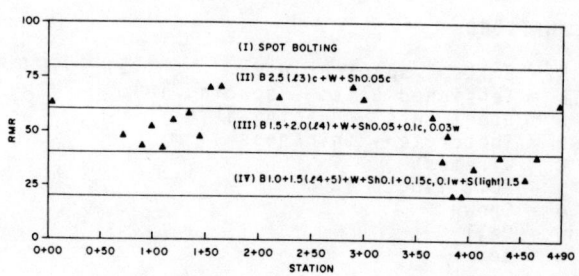

Figure 5c GEOMECHANICS RMR CLASSIFICATION AND CORRESPONDING SUPPORT IN CAVERN.

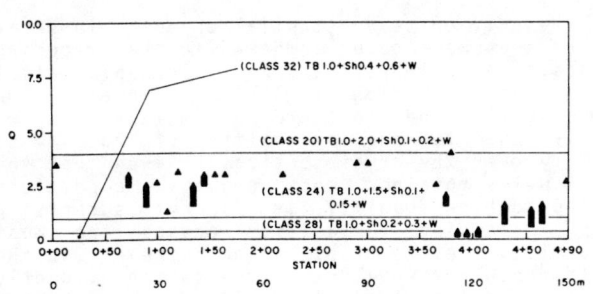

Figure 5d Q-CLASSIFICATION AND CORRESPONDING
SUPPORT IN CAVERN.

Table III shows that parameter estimation
changes only slightly in going from Phase I to
III, mostly (but not throughout) leading to a
lowering of most probable values as well as of
the extremes. As with the exploration shaft,
this does reflect the unfavorably oriented joint
set (3) -- but not very strongly, since the
flatly inclined set (1) (which was recognizable
in the borings) causes most of the parameter
reductions due to joint orientation. The decrease of Q values is also caused by a lowering
of J_r and supports the previously made statement
that estimation from small cores may be unrepresentative.

These slight differences in parameters become
even less significant when transformed into
support requirements; the most probable support
requirements of both phases are largely identical. Thus the value of Phase III lies not so
much in this most probable prediction but rather
in relating locations and support estimations
(Figure 4).

The pilot tunnel was supported in the crown by
2.4 m (8 ft) long resin grouted bolts spaced at
1.5 m (5 ft) along the axis and 0.9 to 1.2 m (3
to 4 ft) laterally; also chain link mesh was
installed in the crown. Based on the borings,
all classification methods predict for most
probable conditions somewhat less support than
was actually placed (Tables I,III). Given the
consistency in predictions, one could thus have
used less support in the pilot tunnel. It is
important to note that the pilot tunnel support
also provided some "ahead of the face support"
for the cavern excavation, thus probably justifying the placed support quantities.

5. DISCUSSION OF RESULTS AND COMPARISON

This section follows the sequence outlined in
Section 3.2.

5.1 Effect of Exploration Shaft

Table II shows that, except for RQD, the ranges
between extremes and the most probable values
decreased, reflecting mainly the unfavorably
directed joint set (3) that was discovered in
the shaft. However, only in the RSR method does
this shift change the support requirements. (As
will be seen many times in the following,
parameter fluctuation is often "dampened" in
predicting support requirements.)

5.2 Classification and Support Prediction from Phases I (Boring) and III (Pilot Tunnel) and Actual Support in Pilot Tunnel

TABLE III: COMPARISON OF CLASSIFICATION AND SUPPORT PREDICTIONS FOR PILOT TUNNEL

METHOD	Investigator	PHASE I (BORINGS) PARAMETERS OR CLASSES			PHASE I (BORINGS) SUPPORT	PHASE III (PILOT TUNNEL) PARAMETERS OR CLASSES			PHASE III (PILOT TUNNEL) SUPPORT	ACTUAL SUPPORT
		WORST	BEST	MOST PRO.	MOST PROBABLE	WORST	BEST	MOST PRO.	MOST PROBABLE	
ROCK LOAD	1	5	2	3	B 2.2	5	2	3	B 2.2	Bolts 1.5 x ~1.0 chain link
	2	4	1	3	B 2.2	3	3	3	B2.2	
RQD	1	65	95	87	B Occ.	70	95	85	B Occ.	
	2	70	100	89	B OCC.	80	95	87	B Occ.	
RSR	1	52	78	65	B 2.3	54	79	63	B 2.1	
	2	49	88	69	B 2.7	57	72	64	B 2.2	
GEOMECH-ANICS*	1	47	80	64	II	43	77	64	II	
	2	55	85	74	II	46	67	60	III (or II)	
Q*	1	2.9	25.2	8.7	NONE	1.6	32	6.7	NONE	
	2	0.8	53.3	7.9	NONE	0.4	8.0	5.2	NONE	

see Table Ia for description of supports.

C 309

5.3 Effect of Subjectivity in Phases I and III

5.3.1 Subjectivity in using the boring records (Phase I)

Table I can serve as a basis for this comparison. Considering first the parameters, one notices differences in the order of 5%, except for Q where the differences are greater -- a somewhat deceiving impression, as will be discussed in Section 5.4. Probably the most interesting result is that the investigators are not consistently conservative or unconservative. Looking at the reasons for the differences (not evident from the tables, but from the underlying data) one notices that Investigator 1 is more conservative with regard to joint spacing (RQD), while Investigator 2 consistently assigns lower values to joint surface conditions and water inflow. The results will be affected differently, depending on the weight that the particular method assigns to these factors.

Whatever the differences between parameter estimates, their effect is even less significant in the support predictions (see Table I and Figure 4). The fact that ranges of parameters are associated with support requirements serves as an equalizer of subjectivity effects. (A similar equalizing effect occurs with the Rock Load method where rock loads of neighboring classes, and thus the support requirements, overlap.)

5.3.2 Subjectivity in pilot tunnel classification (Phase III)

Figure 4 and Table III show that there are again slight differences in parameter assessments (see also Section 5.4). The aforementioned equalizing effect in support assignments occurs again. Except for the Geomechanics classification, where Investigator 2's conservatism regarding water and joint conditions causes a difference (just at the limit, however), the investigators predicted identical supports.

5.4 Spatial Variability in Pilot Tunnel Classification

Figure 4 expresses spatial variability of parameters (classes) and supports. As mentioned several times, Q seems to be more variable than the other parameters. However, the spatial variability of support predictions is the same as for the other methods, since support requirements in the Q system are related to greater parameter ranges. More interesting are the differences in spatial variation as predicted by the two investigators. Scale of fluctuation, which is defined as the average distance over which a parameter is above or below its mean, varies considerably between the two investigators.

5.5 Classification and Support Predictions for Cavern

Table IV summarizes parameter (class) assessments and support predictions for the cavern as one proceeds from boring records (Phase I) to pilot tunnel (Phase III) to final cavern (Phase IV). Table IV also includes the listing of the actual support (see also Figure 5 for cavern survey).

In Phase IV quite substantial changes in parameter assessment occur compared to the preceding phases. Both worst and most probable values (with the exception of RSR) are lower. The decrease in "the most probable values" may be mostly due to the variability of rock mass quality over the cross section. Given that the previouslly obtained most probable conditions reflected high quality, it is not surprising that increasing variability causes the most probable values to drop. (The increase in the most probable RSR values is due to an evidently overly pessimistic assessment of parameters B (jointing) and C (water) in the borings and pilot tunnel.)

In contrast to the reasonably comparable most probable parameters, there is a substantial decrease in the parameters describing the worst conditions. When going from Phases I, II, III to Phase IV one recognizes in Figure 5 that these lower values come from two zones, one at station 1 + 40 and one at station 4 + 00 and beyond. Comparison with Figure 3 and the discussion of the exploration results in Section 2.2 lead one to realize that the zone at 1 + 40 corresponds to the fault there and has been well recognized in preceding phases. The more extreme condition at station 4 + 00 did not enter previous phases. It is a disturbed zone existing in the lower west wall of the cavern and dipping eastward, i.e., it is inclined away from the upper parts of the cavern. This means that the most sensitive parts in the crown and upper part of the cavern walls are not affected. The fact that the cavern survey concentrated on the bench wall clearly overemphasizes the significance of this zone; one may however speculate about a slightly different dip directing this zone just outside the pilot tunnel but inside the cavern! As was the case in previous phases, parameter changes are dampened in the support predictions. Nevertheless there are changes in support requirement predictions by all methods compared to the preceding phases. This is in contrast to Phases I, II, and III where practically no changes occurred. Although not surprising to the practitioner, the geology related increase of support requirements when going from borings or small surface surveys (shaft, pilot tunnel) to the large cavern is a reason for concern in exploration interpretation and design.

This leads to the comparison of predicted support requirements among themselves and with the actual support. The actually placed support exceeds all predictions, even those for worst conditions (the Q-system category 32 comes close, however). Observations of support deformation and stresses (steel set loads) confirm this; the steel sets are used to 15-25% of their capacity. In comparing the predictions of the various methods for the most probable conditions, one notices that the Q system is somewhat more conservative than the others. This may be due to the stress reduction factor (SRF); rather than selecting SRF = 2.5 one could have interpolated between SRF = 2.5 and 1.0. With regard to the Rock Load and RSR methods and possibly also the Geomechanics method, one has to be aware that these methods may not be applicable

TABLE IV: CLASSIFICATION AND SUPPORT PREDICTIONS FOR CAVERN FROM 3 PHASES AND ACTUAL CAVERN SUPPORT

		BORINGS (Phase I)				PILOT TUNNEL (Phase III)				CAVERN (Phase IV)			
		PARAMETER OR CLASS		SUPPORTS		PARAMETER OR CLASS		SUPPORTS		PARAMETER OR CLASS		SUPPORTS	
METHOD	INVEST-IGATOR	WORST	MOST PROB.	WORST	MOST PROB.	WORST	MOST PROB.	WORST	MOST PROB.	WORST	MOST PROB.	WORST	MOST PROB.
ROCK LOAD (2)	1	5	3	B 0.5÷0.6 S 0.2÷0.4	B,S>1	5	3	B 0.5÷0.6 S 0.2÷0.4	B,S>1				
	2	4	3	B 0.6÷0.9 S 0.4÷0.9	B,S>1	3	3	B,S 1	B,S>1				
	3									6 (1)	4	NA	B 0.6÷0.9 S 0.4÷0.9
RQD	1	65	87	B 0.3 S 0.9÷2.3	B 1.0	70	85	B 1.0 S 1.0	B 1.1 S 1.3				
	2	64	89	SAME	SAME	80	87	SAME	SAME				
	3									10	78	NA	B 0.9÷1.5
RSR (2)	1	52	65	B 0.6 S 0.4	B 0.9 S 0.9	54	63	B 0.7 S 0.5	B 0.9 S 0.7				
	2	49	69	B 0.6 S 0.4	B 1.1 S 1.3	57	64	B 0.7 S 0.6	B 0.9 S 0.9				
	3									51	72	B 0.6 S 0.3	B 2.0 S 3.3
GEOMECH-ANICS* (2)	1	47	64	III	II	43	64	III	II				
	2	55	74	III	II	46	60	III	II				
	3									21	49	IV	II or III
Q*	1	2.9	8.7	24	19 or 20	1.6	6.7	24	20				
	2	0.8	7.9	28	19 or 20	0.4	5.2	28	20				
	3									0.13	2.26	32	24

NOTE: ACTUAL CAVERN SUPPORT -- Steel Sets W8 x 40 spaced at 1.5m + Bolts (ℓ6 to 7.2m) 0.6 to 2m + Shotcrete 0.42m = Wiremesh

(1) Derived from RQD
(2) Applicability to 70' (21m) cavern questionable
*   See Table Ia for description of supports

to a 21 m span cavern, given that the base cases of the methods do not include such spans (or if so, in different geologic conditions). The RQD method requires the least support, although not by a large margin. For the most probable conditions one can thus say that the predictions by all methods fall into a relatively small range, even including methods whose applicability may be questionable. The range of most probable support predictions is less than the difference between actual support and the average of the predictions.

For the worst case conditions the conclusions are different, particularly because only one method, the Q-system, is applicable on the basis of the underlying cases. Category 32 of the Q system comes close to the actually placed support.

6. Conclusions

Purposely first overstating the results, one can say:

1) The predicted most probable and extreme support requirements were already obtained from the (extensive) boring program. The exploration shaft and pilot tunnel did not change this.

2) Increased support requirements that resulted from the cavern survey were not predicted in any of the exploration phases.

3) The examined methods are not affected by subjectivity of the user.

Although correct, these statements have to be viewed together with more subtle differences that are very important in tunneling.

The borings did not reveal the third joint set, while exploration shaft and pilot tunnel did. Only because other conditions had led already to lower classes did this unfavorably oriented set not affect the predictions. This may be different in other applications and the three-dimensional exposure (in shaft or tunnel) will have a greater effect.

The spatial variability, i.e., knowing where a particular condition occurs, is often as important in tunneling as most probable and extreme conditions. A pilot tunnel or analogous exploratory procedure is necessary for this purpose.

The fact that the encountered conditions in the cavern deviate from those predicted in a very extensive exploration program causes concern. The conclusion to be drawn is that either a conservative design has to be chosen, even with such an exploration program, or that an adaptable design construction approach is required. In applying such an adaptable approach, the empirical methods can be used to decide on the support needed, but only if the methods cover the entire range of conditions.

In the Porter Square case a conservative design was chosen. An adaptable approach would have used less support in the better quality sections.

Although subjectivity seems to be largely excluded from support predictions, one has to be aware that this is mostly due to the dampening of parameter differences, when relating parameters (rock classes) to supports. Where the effect of subjectivity did not disappear is in the spatial fluctuations of rock class predictions. These fluctuations, which indicate the length over which a construction procedure would be used, are important in longer tunnels.

In comparing methods with each other, this case clearly shows that the most probable predicted support requirements fall into a relatively narrow range below the actually placed support. The methods are thus roughly equivalent for this purpose. However, predictions for the worst conditions in the cavern seem to be possible only with the Q-system, the only method which covers the particular combination of large span and low quality conditions.

The study and this paper have thus shown that a well planned multiphase exploration program does influence empirical classification and support predictions, but on a more subtle level than originally thought. Subjectivity caused only differences when assessing spatial variability. Obviously, none of the methods can predict support requirements for extreme conditions that are not discovered in exploration. Design conservativism or adaptable design construction methods (the latter combined with on site use of applicable empirical methods) are thus necessary.

Acknowledgements

The research for this paper and the underlying study is sponsored by Urban Mass Transportation Administration DOT under no. MA-06-0127; Mr. H Evoy is project monitor. The work is conducted in cooperation with the Massachusetts Bay Transportation Authority, Mr. J. McGowan. The authors are grateful for the support by these agencies and their representatives. They would also like to acknowledge the work by Mr. G. Butler who initiated this study and by the contractor, Slattery, McLean and Grove, whose interest and help was central in conducting the work.

REFERENCES

BARTON, C. M. (1977), "A Geotechnical Analysis of Rock Structure and Fabric in the C.S.A. mine, Cobar, New South Walls," C.S.I.R.O. Australia Division of Applied Geomechanics Technical Paper No. 24, 30 pp.

Barton, N. R.; Lien, R.; Lunde, J. (1974), "Engineering Classification of Rock Masses for the Design of Tunnel Supports," Rock Mechanics, Vol. 6, No. 4, Springer-Vering Vienna, pp. 189-236.

Bieniawski, Z. T. (1974), "Geomechanics Classification of rock masses and its application in tunneling," Proc. 3rd Int. Cong. on Rock Mechanics, ISRM, Denver, USA, Vol. II A, pp. 27-32.

Bieniawski, Z. T. (1979a), "The Geomechanics Classification in Rock Engineering Applications," Proceedings 4th International Congress on Rock Mechanics, Montreux, Vol. 2, pp. 41-48.

Bieniawski, A. T. (1979b), " Tunnel Design by Rock Mass Classificationls," Technical Report GL-79-19, Report to Office, Chief of Engineers, U.S. Army, Washington, D.C., 131 pp.

Bieniawski, Z. T. (1980), "Rock Classifications: State of the Art and Prospects for Standardization," Transportation Research Record, No. 783, pp. 2-9.

Blackey, A. E. (1979), Presentation on Park River Auziliary Tunnel Given to ASCE Convention, Boston, April 2-6, 1979.

Cameron-Clarke I. S. Budarari, S. (1981), "Correlation of rock mass classification from bore core and in-stiu observations," Eng. Geology, Vol. 17.

Cording, E. J.; Hendrom, A. J.; and Deere, D. U. 91971), "Rock Engineering for Underground Caverns," Symposium on Underground Rock Chambers, ASCE, Phoenix, pp. 567-600.

Deere, D. U.; Peck, R. B.; Monsees, J. B.; and Schmidt, B., (1969), "Design of Tunnel Liners and Support Systems," Report of University of Illinois to OHSGT-U.S. DOT, NTIS No. PB-183799

Einstein, H. H., Baecher, G. B., Steiner, W. (1979), "Assessment of Empirical Design Methods for Tunnels in Rock," Proc., RETC, Atlanta.

Franklin, J. A. (1976), "An Observational Approach to the Selection and Control of Rock Tunnel Liner," Proceedings Engineering Foundation Conference on Shotcrete for Ground Support, Easton Md, ASCE, New York and ACI, SP-54, Detroit pp. 556-596.

Haley & Aldrich, Inc., "Geotechnical Data Report, Massachusetts Bay Transportation Authority, Red Line Extension NW - Harvard to Davis, Porter Square Station Pilot Tunnel," Volumes I and II, submitted to Cambridge Seven Associates, Inc., Cambridge, 1979.

Haley & Aldrich, Inc., "Summary Geotechnical Data Report, Massachusetts Bay Transportation Authority, Red Line Extension NW - Harvard to Davis, Porter Square Station," submitted to Bechtel, Inc., West Somerville, MA, 1978.

Lauffer, H. (1958), "Gebirgsklassifizierung fur den Stollenbau," (in german), Geologie und Bauwesen, Springer, Vienna, Austria, Vol. 24, No. 1, pp. 46-51.

Louis, C. (1974), "Reconnaissance des Massifs Rocheux par Sondages et Classification Geotechnique des Roches," annales I.T.B.T.P., No. 319, Juillet-Aout, pp. 97-122.

McCusker, T. G. (1980), "A Review of Empirical Approaches to Tunnel Support Design," Tunneling Technology newsletter, No. 30, June 1980, pp. 1-8.

Merritt, A. H. (1972), "Geologic Predictions for Underground Excavation," Proceedings, 1st RETC, AIME, New York, pp. 115-123.

Monsees, J. E. (1970), "Design of Support Systems for Tunnels in Rock," Ph.D. Thesis, University of Illnois, 253 pp.

Muller, L. (1978), Der Felsbau, 3rd Volume, Tunnelbau, Enke Verlag.

Ordun, S., "Empirical and Numerical analysis of the Porter Square Subway Station," a Thesis presented to the Mass. Inst. of Technology, in Cambridge, Mass., in partial fulfillment of the degree of Master of Science.

"Porter Square (1982)" Series of reprots on Porter Square Station Design and Construction, in preparation.

Proctor, R. V.; and White, T. L. (1946, 1968), "Rock tunneling with Steel Supports," Commercial Shearing and Stamping Company, Youngstown, Ohio.

Priest, S. S., Hudson, J. (1976), "Discontinuity Spacing in Rock," Int. J. Rock Mech. and Mi. Sci., Vol. 13.

Rabcewicz, L. V. (1965), "The Stability of Tunnels under Rock Load," Water Power, Part 1, June 1969, pp. 225-229; Part 2, July 1969, pp. 266-273; Part 3, July 1969, pp. 297-302.

Rose, D., Kuboli, P., Mayes, R. (1981), "Influence of Geologic Logs and Descriptions on Tunnel Design and Cost," Proc., 22nd U.S. Symposium on Rock Mechanics.

Rutledge, J. C., and Preston, R. L. (1978), "New Zealand Experience with Engineering Classifications of Rock for the Prediction of Tunnel Support," International Tunnel Symposium, Tokyo, Japan, pp. A-1,3-7.

Steiner, W., Einstein, H. H. (1980), "Improved Design of Tunnel Supports," Vol. 5, Emp. Methods in Rock Tunneling, Report UMTA-MA-06-0100-80-8.

Stini, T. (1950), "Tunnelbaugeologie," Wien, Springer-Verlag.

Terzaghi, K. (1946), "An Introduction to Tunnel Geology," in Rock Tunneling with Steel Supports, by Proctor, R. V. and White, T. L., The Commercial Shearing and Stamping Co., Youngstown, Ohio, U.S.A.